W9-CMM-477

YOUR ACCESS TO SUCCESS

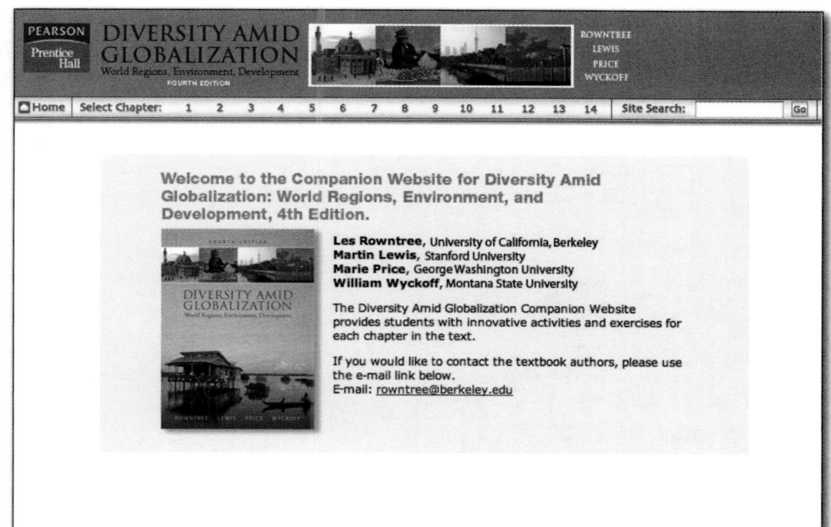

Students with new copies of Rowntree/Lewis/Price/ Wyckoff, *Diversity Amid Globalization,* Fourth Edition have full access to the book's Companion Website—a 24/7 study tool with review exercises, Web destinations, and other features designed to help you make the most of your study time.

Just follow the easy Website registration steps listed below...

Registration Instructions for *www.prenhall.com/rowntree*

1. Go to *www.prenhall.com/rowntree* and follow the link for Companion Websites.
2. Click the cover for Rowntree/Lewis/Price/Wyckoff, *Diversity Amid Globalization,* Fourth Edition.
3. Click the "Companion Website" link, then the "Register" button.
4. Using a coin (not a knife) scratch off the metallic coating below to reveal your Access Code.
5. Complete the online registration form, choosing your own personal Login Name and Password.
6. Enter your preassigned Access Code exactly as it appears below.
7. Complete the online registration form by entering your School Location information.
8. After your personal Login Name and Password are confirmed by e-mail, go back to *www.prenhall.com/rowntree*, click your book's cover, enter your new Login Name and Password, and click "Log In."

Your Access Code is:

[illegible]A-YZKZL-ALL [illegible]

If there is no metallic coating covering the Access Code above, the code may no longer be valid and you will need to purchase online access using a major credit card to use the Website. To do so, go to *www.prenhall.com/rowntree*, click the cover for Rowntree/Lewis/Price/ Wyckoff, *Diversity Amid Globalization*, Fourth Edition, then click the "Companion Website" link and the "Get Access" button, and follow the instructions under "Students."

Important: Please read the Subscription and End-User License Agreement located on the "Log In" screen before using the Rowntree/Lewis/Price/ Wyckoff, *Diversity Amid Globalization*, Fourth Edition, Companion Website. By using the Website, you indicate that you have read, understood, and accepted the terms of the agreement.

Minimum system requirements

PC Operating Systems:

Windows 2000, XP

Pentium II 233 MHz processor. 64 MB RAM In addition to the minimum memory required by your OS.

Internet Explorer 6.0, Netscape Navigator 7.2, or Firefox 1.0.x

Macintosh Operating Systems:

Macintosh Power PC with OS X (10.2, 10.3)

In addition to the RAM required by your OS, this product requires 64 MB RAM, with 40MB free RAM, with Virtual Memory enabled

Netscape Navigator 7.2, Safari 1.0, 1.3, or Firefox 1.0.x

Macromedia Shockwave(TM)

8.50 release 326 plugin

Macromedia Flash Player 6.0.79 & 7.0

Acrobat Reader 6.0.1

800 x 600 pixel screen resolution

Technical Support Visit our support site at *http://247pearsoned.custhelp.com*
E-mail support is available 24/7.

Diversity Amid Globalization

WORLD REGIONS, ENVIRONMENT, DEVELOPMENT

FOURTH EDITION

Les Rowntree
University of California, Berkeley

Martin Lewis
Stanford University

Marie Price
George Washington University

William Wyckoff
Montana State University

Upper Saddle River, NJ 07458

Library of Congress Cataloging-in-Publication Data

Diversity amid globalization: world regions, environment, development / Les Rowntree. –
4th ed.
p. cm.
Includes bibliographical references and index
ISBN 0-13-600554-3
1. Geography. I. Rowntree, Lester,
G128 .D58 2009
910–dc22
2007047878

Publisher, Geosciences: Daniel Kaveney
Editor in Chief, Science: Nicole Folchetti
Project Manager: Tim Flem
Development Editor: Ginger Birkeland
Vice President, Executive Director of Development: Carol Trueheart
Media Editor: Andrew Sobel
Associate Editor: Amanda Brown
Marketing Manager: Amy Porubsky
Assistant Managing Editor, Science: Gina M.Cheselka
Director of Operations: Barbara Kittle
Operations Supervisor: Alan Fischer
Art Director: Suzanne Behnke
Interior Designer: Suzanne Behnke
Cover Designer: Bruce Kenselaar
Page Composition: GGS Book Services
Director of Production Services, GGS Book Services: Cindy Miller
Art Project Manager: Connie Long
Cartographer/Art Studio: Spatial Graphics
Production Manager, Spatial Graphics: Kevin Lear
Director, Image Resource Center: Melinda Patelli
Manager, Rights and Permissions: Zina Arabia
Interior Image Specialist: Beth Boyd-Brenzel
Cover Specialist: Karen Sanatar
Photo Researcher: Barbara Salz

Cover Photographs:
Iraqis walk past the damaged Askariya shrine following an explosion in Samarra. Credit: HAMEED RASHEED / AP Wide World Photos; *Traditional residential houses along the Huangpu River.* Credit: Keren Su/China Span / Aurora Photos; *Like giant pinwheels, turbines spin at the Middelgrunden Wind Park.* Credit: SARAH LEEN/National Geographic Image Collection; *Peruvian street vendor.* Credit: Photograph by Rob Crandall; *A house on stilts in southern Cambodia on the Mekong River.* Credit: MICHAEL S. YAMASHITA / National Geographic Image Collection.

Pearson Prentice Hall
Pearson Education, Inc.
Upper Saddle River, New Jersey 07458

Printed in the United States of America

10 9 8 7 6 5 4 3 2 1

ISBN–10: 0–13–600554–3
ISBN–13: 978–0–13–600554–4

Pearson Education LTD., *London*
Pearson Education Australia PTY, Limited, *Sydney*
Pearson Education Singapore, Pte. Ltd
Pearson Education North Asia Ltd, *Hong Kong*
Pearson Education Canada, Ltd., *Toronto*
Pearson Educación de Mexico, S.A. de C.V.
Pearson Education — Japan, *Tokyo*
Pearson Education Malaysia, Pte. Ltd

Brief Contents

Environmental Statement

This book is carefully crafted to minimize environmental impact. Pearson Prentice Hall is proud to report that the materials used to manufacture this book originated from sources committed to sustainable forestry practices, tree harvesting, and associated land management. The binding, cover, and paper come from facilities that minimize waste, energy usage, and the use of harmful chemicals.

Equally important, Pearson Prentice Hall closes the loop by recycling every out-of-date text returned to our warehouse. We pulp the books, and the pulp is used to produce other items such as paper coffee cups or shopping bags.

The future holds great promise for reducing our impact on Earth's environment, and Pearson Prentice Hall is proud to be leading the way in this initiative. From production of the book to putting a copy in your hands, we strive to publish the best books with the most up-to-date and accurate content, and to do so in ways that minimize our impact on Earth.

Contents

5 The Caribbean 178

6 Sub-Saharan Africa 222

9 The Russian Domain 380

10 Central Asia 432

11 East Asia 472

12 South Asia 526

13 Southeast Asia 576

14 Australia and Oceania 622

Preface

Diversity Amid Globalization is an issues-oriented textbook for college and university world regional geography classes that explicitly recognizes the geographic changes accompanying globalization. With this focus, we join the many who argue that globalization is the most fundamental reorganization of the planet's socioeconomic, cultural, and geopolitical structure since the Industrial Revolution. The explicit recognition of this premise provides the point of departure for this book. As geographers, we think it essential for students to understand two interactive tensions. First, they need to appreciate and critically ponder the consequences of converging environmental, cultural, political, and economic systems through globalization. Second, they need to deepen their understanding of the creation and persistence of geographic diversity and difference. The interaction and tension between these opposing forces of homogenization and diversification form a current running throughout the following chapters and, as well, is reflected in our title, *Diversity Amid Globalization*.

Chapter Organization

As is true of all other regional geography textbooks, *Diversity Amid Globalization* is structured to explain and describe the major world regions of Asia, Africa, the Americas, and so on. Our 12 regional chapters, however, depart somewhat from traditional world regional textbooks. Instead of filling these regional chapters with descriptions of individual countries, we place most of that important material on the textbook Website. This leaves us free to develop five important themes as the structure for each regional chapter. We begin with Environmental Geography, which discusses the physical geography of each region as well as current environmental issues. Next, we assess the Population and Settlement geography, in which demography, land use, and settlement (including cities) are discussed. Third is a section titled Cultural Coherence and Diversity that examines the geography of language and religion and also explores current cultural tensions resulting from the interplay of globalization and diversity. Then a section on each region's Geopolitical Framework treats the dynamic political geography of the region, including microregionalism, separatism, ethnic conflicts, global terrorism, and supranational organizations. We then conclude each regional treatment with a section titled Economic and Social Development, in which we analyze each region's economic framework as well as its social geography, including gender issues.

This regional treatment follows two substantive introductory chapters that provide the conceptual and theoretical framework of human and physical geography necessary to understand our dynamic world. In the first chapter, students are introduced to the notion of globalization and are asked to ponder the costs and benefits of the globalization process, a critical perspective that is becoming increasingly common and important to understand. Following this, the geographical foundation for each of the five thematic sections is examined. This discussion draws heavily on the major concepts fundamental to an introductory university geography course. The second chapter, The Changing Global Environment, presents the themes and concepts of global physical geography, including landforms and geology, climate, global warming, hydrology, and biogeography.

Chapter Features

Within each regional chapter, several unique features complement the thematic pedagogy of our approach:

- *Comparable Maps.* Of the many maps in each regional chapter, 7 are constructed on the same theme and with similar data so that readers can easily draw comparisons between different regions. Thus, in every regional chapter readers will find maps of physical geography and place-names, climate, environmental issues, population density, language geography, religion, and a map showing the geopolitical issues of the region.
- *Other Maps.* In addition, each regional chapter also has 7 or 8 other maps illustrating such major themes as urban growth, ethnic tensions, social development, and linkages to the global economy. As well, the fourth edition presents 30 new maps of specific subregions. These new maps are linked to a specific theme (or themes) within each chapter and are explicitly discussed in the text. These new subregion maps help the text provide a stronger emphasis on the local scale and an emphasis on how global processes affect places.
- *Comparable Regional Data Sets.* Again, to facilitate comparison between regions, as well as to provide important insight into the characteristics of each region, each chapter contains two tables. The first provides population data, including population density, level of urbanization, total fertility rates, proportion of the population under 15 and over 65 years of age, and net migration rates for each country within the region. The second presents economic and social development data for each country including GNI per capita, GDP growth, life expectancy, percentage of the population living on less than $2 per day, infant mortality rates, and the UN gender equity index.
- *Sidebars.* Within each chapter, a standard set of sidebars complement text material. These sidebars complement the text in a variety of ways.

- "Setting the Boundaries" sidebars give an overview of the region; discuss its role in the world today, raise questions about its geographic coherence by accentuating its diversity; and look at the vexatious issue of regional boundaries.
- "Cityscapes" sidebars combine text, maps, and photos to convey a sense of place for a major city within each regional chapter.
- "People on the Move" sidebars capture the human geography behind contemporary migration as people move—legally and not so legally—between familiar and new places, responding and reacting to the varied currents of globalization.
- "Global to Local" sidebars highlight how global trends affect local areas and local people. The sidebars foreground the fact that people often respond to homogenizing global forces very differently, and this variation in responses gives unique character to local people and places.
- "Geography in the Making" sidebars highlight late-breaking newsworthy content of geographic importance in each region.
- "Geographic Tools" sidebars discuss the various tools—such as GIS, aerial photography, field surveys, repeat photography, and interviewing—geographers use to understand the world. These materials include case studies to illustrate how these tools are used in various contexts.

■ *Review and Research Questions.* There are two sets of review and research questions at the end of each regional chapter. The first set helps readers review basic terminology and concepts. The second set, "Thinking Geographically," asks readers to draw together more abstract themes and problems by doing further research on an issue. Resources for answering these critical thinking exercises are found in the chapter bibliographies, in the instructor's manual, and on the textbook's Website.

New to the Fourth Edition

■ *Expanded Treatment of Globalization.* Globalization has become increasingly important during the history of this book and, as a result, globalization is more central to the study of world regional geography than ever before. Furthermore, many experts would say that globalization itself has become more complex in the last decade as instantaneous global communications have become more sophisticated; new, less hierarchical international relationships have developed; and a powerful array of new players have emerged on the national and international scene. Thus we have once again expanded our initial discussion in Chapter 1 to present varying perspectives and viewpoints on globalization. Each regional chapter also has new information on globalization and its many-faceted implications.

■ *Increased Emphasis on Global Warming and Climate Change.* There is little question that global warming is one of the major environmental, political, and economic problems facing the world and, because of the importance of this complicated issue, we have expanded our coverage of the projected effects of climate change on different regions of the world, along with the different policies and measures taken to mitigate the human and economic costs of this problem.

■ *Updated Maps/Graphics and New Design.* The maps and illustrations for the fourth edition have been updated to an open, visually appealing approach while retaining the accuracy and pedagogical focus of previous edition maps and illustrations. In addition, the text design has been refreshed to capture students' attention with inviting visual content.

■ *Current Data and Information.* All text, tables, and maps have been updated with the most current information.

Supplements: The Teaching and Learning Package

We have been pleased to work with Prentice Hall to produce an effective supplements package that will enhance teaching and learning. Not only does this package contain the traditional supplements that students and professors have come to expect from authors and publishers, but it also folds together new digital technologies that complement and enhance the learning experience.

For the Instructor

■ **Instructor Resource Center on DVD (0-13-601173-X):** Everything instructors need all in one place. The Prentice Hall Instructor Resource Center helps make instructors more effective by saving them time and effort. All digital resources are in one well-organized, easy-to-access place. The IRC on DVD includes:
- *Photographs:* Every photograph from the text is included on the IRC.
- *Maps and Illustrations:* JPEGs of every map and illustration from the text are included.
- *PowerPoint™:* Pre-authored slides outline the concepts of each chapter with embedded art and can be used as is for lecture or can be customized to fit instructors' lecture presentation needs. Also included are Classroom Response System Questions for Personal Response System (PRS) or H-ITT systems.
- *TestGen:* The TestGen software, questions, and answers are included.
- *Instructor's Manual and Test Item File*

■ **Transparencies (0-13-601168-3):** Include more than 150 full-color images from the text, all enlarged for excellent classroom visibility, along with every map in the book.

■ **Prentice Hall World Regional Geography Video Series (0-13-159348-X):** This two-DVD set is designed to enhance any world regional geography course. It contains 10 full-length video programs covering a wide array of issues affecting people and places in the contemporary world, including international immigration, the HIV/AIDS epidemic, urbanization, homelessness, poverty, and environmental destruction.

These DVDs are designed to function in computer-based DVD drives in addition to traditional component DVD players. The videos included on these DVDs are offered at the highest quality to allow for full-screen viewing on your computer and projection in large lecture classrooms. Video programs include:

DVD 1
- The Millennium Goals—Dream or Reality?
- Cash Flow Fever
- Sowing Seeds of Hunger
- Yemeni Futures
- Blue Danube?

DVD 2
- Cheated of Childhood
- The Real Leap Forward
- Slum Futures
- Helping Ourselves
- My Hanoi

Average length: 26 minutes

The Prentice Hall World Regional Geography video series is available either as a stand-alone or at a very substantial discount when packaged with *Globalization Amid Diversity*. Please see your local Prentice Hall representative for details.

- **Instructor's Manual (0-13-601172-1)**: Includes answers to the end-of-chapter questions in addition to key terms lists, learning objectives, thought and discussion boxes, and activity and exercises boxes.
- **Test Item File (0-13-601167-5)**: Includes more than 150 multiple-choice, true/false, and essay questions per chapter. The test item file that accompanies the fourth edition has been extensively reviewed, revised, and updated for clarity and question effectiveness.
- **TestGen (0–13–605416–1)**: TestGen is a computerized test generator that lets you view and edit testbank questions, transfer questions to tests, and print tests in a variety of customized formats.
- **On-line Course Management Systems:** Prentice Hall offers content specific to *Diversity Amid Globalization*, Fourth Edition, ready for easy importation into the BlackBoard and WebCT course management system platforms. Each of these platforms lets you easily post your syllabus, communicate with students online or offline, administer quizzes, and record student results and track their progress. Please call your local Prentice Hall representative for details.

For the Student

- **Companion Website (http://www.prenhall.com/rowntree)**: Prentice Hall provides an online study guide and resource center for students using the text. The online study guide provides both mapping exercises and immediate feedback to practice quizzes. In addition, the guide provides Web links for students to use in exploring the region at hand, along with exercises and discussion questions about key global issues in different world regions.
- **Study Guide (0-13-601169-1)**: Designed to function as a test preparation guide, the study guide includes additional learning objectives, a complete chapter outline, a regional summary for students to fill out, practice quizzes, and critical thinking exercises and problems.
- **Mapping Workbook (0-13-601174-8)**: This workbook, which features the base maps from the book printed in black and white, can be used in conjunction with the textbook or an atlas. The workbook contains a political and a physical base map for every region in the book, along with a list of key map items for each region.
- ***Goode's World Atlas:*** Prentice Hall and Rand McNally are pleased to announce that Prentice Hall is now distributing the *Goode's World Atlas*—the number-one atlas used by college professors—to colleges and universities worldwide. The *Goode's World Atlas* is the world's premiere educational atlas, and for good reason. It features nearly 250 pages of maps, from definitive physical and political maps to important thematic maps that illustrate the spatial aspects of many important topics. The current 21st edition of the atlas features fully updated content, and has been vetted by an academic board comprising some of the most trusted names in geography today. Prentice Hall offers the atlas at a dramatically reduced price with *Diversity Amid Globalization*. See your local Prentice Hall representative for details.

Acknowledgments

We have many people to thank for their help in the conceptualization, writing, rewriting, and production of *Diversity Amid Globalization*. First, we'd like to thank the thousands of students in our world regional geography classes who have inspired us with their energy, engagement, and curiosity; challenged us with their critical insights; and demanded a textbook that better meets their need to understand the diverse people and places of our dynamic world.

Next, we are deeply indebted to many professional geographers and educators for their assistance, advice, inspiration, encouragement, and constructive criticism as we labored through the different stages of this book. Among the many who provided invaluable comments on various drafts and editions of *Diversity Amid Globalization* are:

Dan Arreola, *Arizona State University*
Bernard BakamaNume, *Texas A&M University*
Brad Baltensperger, *Michigan Technological University*
Laurence Becker, *Oregon State University*
Max Beavers, *Samford University*
James Bell, *University of Colorado*
William H. Berentsen, *University of Connecticut*
Kevin Blake, *Kansas State University*

Michelle Calvarese, *California State University, Fresno*
Craig Campbell, *Youngstown State University*
Elizabeth Chacko, *George Washington University*
Philip Chaney, *Auburn University*
David B. Cole, *University of Northern Colorado*
Malcolm Comeaux, *Arizona State University*
Jonathan C. Comer, *Oklahoma State University*
Catherine Cooper, *George Washington University*
Jeremy Crampton, *George Mason University*
Kevin Curtin, *University of Texas at Dallas*
James Curtis, *California State University, Long Beach*
Dydia DeLyser, *Louisiana State University*
Francis H. Dillon, *George Mason University*
Jason Dittmer, *Georgia Southern University*
Jerome Dobson, *University of Kansas*
Caroline Doherty, *Northern Arizona University*
Vernon Domingo, *Bridgewater State College*
Roy Doyon, *Ball State University*
Jane Ehemann, *Shippensburg University*
Doug Fuller, *George Washington University*
Gary Gaile, *University of Colorado*
Sherry Goddicksen, *California State University, Fullerton*
Reuel Hanks, *Oklahoma State University*
Steven Hoelscher, *University of Texas, Austin*
Peter J. Hugil, *Texas A&M University*
Eva Humbeck, *Arizona State University*
Ryan S. Kelly, *University of Kentucky*
Richard H. Kesel, *Louisiana State University*
Rob Kremer, *Front Range Community College*
Robert C. Larson, *Indiana State University*
Alan A. Lew, *Northern Arizona University*
Catherine Lockwood, *Chadron State College*
Max Lu, *Kansas State University*
Luke Marzen, *Auburn University*
Kent Matthewson, *Louisiana State University*
James Miller, *Clemson University*
Bob Mings, *Arizona State University*
Sherry D. Morea-Oakes, *University of Colorado, Denver*
Anne E. Mosher, *Syracuse University*
Tim Oakes, *University of Colorado*
Nancy Obermeyer, *Indiana State University*
Karl Offen, *Oklahoma University*
Jean Palmer-Moloney, *Hartwick College*
Bimal K. Paul, *Kansas State University*
Michael P. Peterson, *University of Nebraska–Omaha*
Richard Pillsbury, *Georgia State University*
Brandon Plewe, *Brigham Young University*
Patricia Price, *Florida International University*
Erik Prout, *Texas A&M University*
David Rain, *United States Census Bureau*
Rhonda Reagan, *Blinn College*
Craig S. Revels, *Portland State University*
Scott M. Robeson, *Indiana State University*
Paul A. Rollinson, *Southwest Missouri State University*
Yda Schreuder, *University of Delaware*
Kay L. Scott, *University of Central Florida*
Duncan Shaeffer, *Arizona State University*
Dimitrii Sidorov, *California State University, Long Beach*
Susan C. Slowey, *Blinn College*
Andrew Sluyter, *Louisiana State University*
Joseph Spinelli, *Bowling Green State University*
William Strong, *University of Northern Alabama*
Philip W. Suckling, *University of Northern Iowa*
Curtis Thomson, *University of Idaho*
Suzanne Traub-Metlay, *Front Range Community College*
Nina Veregge, *University of Colorado*
Gerald R. Webster, *University of Alabama*
Keith Yearman, *College of DuPage*
Emily Young, *University of Arizona*
Bin Zhon, *Southern Illinois University at Edwardsville*
Henry J. Zintambila, *Illinois State University*

In addition, we wish to thank the many publishing professionals who have been involved with this project. It has been a privilege to work with you. We thank Paul F. Corey, President of Pearson's Arts and Sciences division, for his early—and continued—support for this book project; Geosciences Publisher and good friend Dan Kaveney, for his daily engagement, professional guidance, enduring patience, unyielding discipline, high standards, and warm companionship; Project Manager Tim Flem, for his daily miracles and steady hand on the tiller; Media Editor Andrew Sobel, for his creative and stimulating additions to our digital program; Associate Editor Amanda Brown, for gracefully taking care of the thousands of tasks necessary to this project; Marketing Manager Amy Porubsky, for her effective sales and promotion work; Editorial Assistant Jessica Neumann, for gracefully meeting the incessant needs of four demanding authors who rarely remembered that she was also simultaneously doing the same for many other Geoscience authors; Developmental Editor Ginger Birkeland, for her clear articulation of what this project needed to be successful, her guidance on how that could be attained, and her firm insistence that we meet those standards; production editors Holly Henjum and Cindy Miller, for their work to somehow turn thousands of pages of manuscript into a finished book; copyeditor Sharon O'Donnell, for her eagle eyes and graceful hands; and photo researcher Barbara Salz, for her creative solutions to indulging four geographers in our quest for outstanding pictures from every part of the world. To all of you, your professionalism is truly inspirational and very, very much appreciated.

Finally, the authors want to thank that special group of friends and family who were there when we needed you most—early in the morning and late at night; in foreign countries and familiar places; when we were on the verge of crying, yet needed to laugh; for your love, patience, companionship, inspiration, solace, understanding, and enthusiasm: Eugene Adogia; Elizabeth Chacko; Meg Conkey; Rob Crandall; Karen Wigen; and Linda, Tom, and Katie Wyckoff. Words cannot thank you enough.

Les Rowntree
Martin Lewis
Marie Price
William Wyckoff

About the Authors

Les Rowntree is a Visiting Scholar at the University of California, Berkeley, where he researches and writes about environmental issues. This career change comes after three decades of teaching both Geography and Environmental Studies at San Jose State University in California. As an environmental geographer, Dr. Rowntree's interests focus on international environmental issues, biodiversity conservation, and human-caused global change. He sees world regional geography as a way to engage and inform students by giving them the conceptual tools needed to critically assess global issues. Dr. Rowntree has done research in Iceland, Alaska, Morocco, Mexico, Australia, and Europe, as well as in his native California. Current writing projects include a book on the natural history of California's coast, as well as textbooks in geography and environmental science.

Martin Lewis is a Senior Lecturer in History at Stanford University. He has conducted extensive research on environmental geography in the Philippines and on the intellectual history of global geography. His publications include *Wagering the Land: Ritual, Capital, and Environmental Degradation in the Cordillera of Northern Luzon, 1900–1986* (1992), and, with Karen Wigen, *The Myth of Continents: A Critique of Metageography (1997).* Dr. Lewis has traveled extensively in East, South, and Southeast Asia. His current research focuses on the geographical dimensions of globalization.

Marie Price is an Associate Professor of Geography and International Affairs at George Washington University. A Latin American specialist, Marie has conducted research in Belize, Mexico, Venezuela, Cuba, and Bolivia. She has also traveled widely throughout Latin America and Sub-Saharan Africa. Her studies have explored human migration, natural resource use, environmental conservation, and regional development. In 2006 she was a Visiting Scholar at the Migration Policy Institute in Washington, D.C., conducting research on immigration to the world's major cities. Dr. Price brings to *Globalization and Diversity* a special interest in regions as dynamic spatial constructs that are shaped over time through both global and local forces. Her publications include articles in the *Annals of the Association of American Geographers. Geographical Review, Journal of Historical Geography, CLAG Yearbook, Studies in Comparative International Development, the Brookings Institution Survey Series, International Journal of Urban and Regional Research*, and *Focus*.

William Wyckoff is a geographer in the Department of Earth Sciences at Montana State University specializing in the cultural and historical geography of North America. He has written and co-edited several books on North American settlement geography, including *The Developer's Frontier: The Making of the Western New York Landscape* (1988), *The Mountainous West: Explorations in Historical Geography* (1995) (with Lary M. Dilsaver), *Creating Colorado: The Making of a Western American Landscape 1860–1940* (1999), and *On the Road Again: Montana's Changing Landscape* (2006). In 2003 he received Montana State's Cox Family Fund for Excellence Faculty Award for Teaching and Scholarship. A World Regional Geography instructor for 26 years, Dr. Wyckoff emphasizes in the classroom the connections between the everyday lives of his students and the larger global geographies that surround them and increasingly shape their future.

1
Diversity Amid Globalization

The modern skyline of Shanghai, showcase of China's recent economic boom, stands in stark contrast to traditional housing along the muddy Huangpu River. (*Keren Su/Aurora & Quanta Productions*)

The most important challenges facing the world in the twenty-first century are associated with **globalization**, the increasing interconnectedness of people and places through converging processes of economic, political, and cultural change. Formerly distant regions and cultures are now increasingly linked through commerce, communications, and travel. Although earlier forms of globalization existed, especially during Europe's colonial period, the degree of planetary integration is now stronger than ever. In fact, many observers argue that contemporary globalization is the most fundamental reorganization of the planet's socioeconomic structure since the Industrial Revolution. While few dispute the widespread changes brought about by globalization, not everyone agrees on whether the benefits outweigh the costs.

Although economic activities may be the main force behind globalization, the consequences affect all aspects of land and life in this new millennium. Cultural patterns, political arrangements, and social development are all undergoing profound change. Because natural resources are now global commodities, the planet's physical environment is also affected. Financial decisions made thousands of miles away alter local ecosystems, and the cumulative effect of these far-ranging activities often has negative consequences for the world's climates, oceans, waterways, and forests.

These immense, widespread global changes make understanding our world a challenging, yet necessary task. Our future depends on understanding globalization in its varied expressions because our lives are now deeply intertwined with this worldwide phenomenon. Although examining globalization cuts across many academic disciplines, world regional geography is an effective starting point because of its focus on regions, environment, geopolitics, culture, and economic and social development. This book seeks to impart such knowledge by carefully outlining the basic patterns of world geography and showing how they are being constantly reorganized by global interconnections (Figure 1.1).

DIVERSITY AMID GLOBALIZATION: A Geography for the Twenty-First Century

This chapter introduces a framework for studying world regional geography by examining the varied aspects of globalization in contemporary life. While the economic aspects of globalization dominate most discussions, one must also consider the cultural, environmental, and geopolitical expressions. Following this introduction to globalization is an overview of the major concepts of global geography. The last section introduces the five organizational themes found in each regional chapter of this textbook.

Converging Currents of Globalization

Most scholars agree that the major component of globalization is the economic reorganization of the world. Although different forms of a world economy have existed for centuries, a well-integrated and truly global economy is primarily the product of the last several decades. The attributes of this system, although familiar, bear repeating:

- Global communication systems that link all regions on the planet instantaneously (Figure 1.2);
- Transportation systems capable of moving goods quickly by air, sea, and land;
- Transnational corporate strategies that have created global corporations more powerful than many sovereign nations;
- New and more flexible forms of capital accumulation and international financial institutions that make 24-hour trading possible;
- Global agreements that promote free trade;
- Market economies that have replaced state-controlled economies and privatized firms and services formerly operated by governments;
- An abundance of planetary goods and services that have arisen to fulfill consumer demand (real or imaginary);

Figure 1.1 Global Communications The impacts of globalization, often through global TV, are everywhere, even in remote villages in developing countries. Here, in a small village in southwestern India, a rural family earns a few dollars a week by renting out viewing time on its globally linked television set. *(Rob Crandall/www.robcrandall.com)*

Figure 1.2 Global to Local Connections
A controversial aspect of globalization is the outsourcing of jobs from developed to developing countries. One example is the relocation of customer call centers away from Europe and North America to India where educated English-speaking employees work for relatively low wages. This photo shows a call center in Bangalore, India. *(© Sherwin Crasto/Reuters/Corbis)*

- Economic disparities between rich and poor regions and countries that drive people to migrate, both legally and illegally, in search of a better life;
- And an army of international workers, managers, and executives who give this powerful economic force a human dimension.

As a result of this global reorganization, economic growth in some areas of the world has been unprecedented over recent decades. International corporations, along with their managers and executives, have amassed vast amounts of wealth, profiting from the new opportunities unleashed by globalization. However, not everyone has profited from economic globalization, nor have all world regions shared equally in the benefits. While globalization is often touted as benefiting everyone through trickle-down economics, there is mounting evidence that this trickling process is not happening in all places nor for all peoples. As critics like to say, if globalization is promoted as lifting all boats as the tide of economic development floods the world, it is increasingly clear that this tide has lifted the yachts of the rich more than the rowboats of the poor.

Globalization and Cultural Change Economic changes also trigger cultural change. The spread of a global consumer culture that threatens to reduce local diversity often accompanies globalization, frequently setting up deep and serious social tensions between traditional cultures and new, external globalizing currents. Global TV, movies, and videos promote images of Western style and culture that are imitated by millions throughout the world. NBA T-shirts, sneakers, and caps now are found in small villages and large world cities alike.

Fast-food franchises are changing—some would say corrupting—traditional diets with the explosive growth of McDonald's, Burger King, and Kentucky Fried Chicken outlets in the world's cities (Figure 1.3). While these changes may seem harmless to North Americans because of their familiarity, they are expressions of the deeper cultural changes the world is experiencing through globalization.

Although the media give much attention to the rapid spread of Western consumer culture, nonmaterial culture also is becoming more dispersed and homogenized through globalization. Language is an obvious example. Many a Western tourist in Russia or Thailand has been startled by locals speaking an English made up largely of Hollywood or pop-song phrases. But far more than speech is involved, as social values also are dispersed globally. Changing expectations about human rights, the role of women in society, and the intervention of nongovernmental organizations also are expressions of globalization that may have far-reaching effects on cultural change.

It would be a mistake, however, to view cultural globalization as a one-way flow that spreads from the United States and Europe into the corners of the world. In actuality, when forms of American popular culture spread abroad, they are typically melded with local cultural traditions in a process known as *hybridization*. The resulting cultural "hybridities," such as world beat music or Asian food, can themselves echo across the planet, adding yet another layer to globalization.

Additionally, the ideas and forms from the rest of the world are also having a great impact on U.S. culture. The

Figure 1.3 Hybrid World Culture
Globalization is creating unique cultural expressions that often merge, or hybridize, the old with the new. Here, in Bangkok, a Thai woman eats lunch at a well-known hamburger chain under a poster for a uniquely Thai traditional dish, a tart papaya salad, which the hamburger chain is marketing as "Thai Spicy McSalad Shaker." *(AP/Wide World Photos)*

growing internationalization of American food, the multiple languages spoken in the United States, and even the spread of Japanese comic book culture among American children are all examples of globalization's effects within the United States (Figure 1.4).

Globalization and Geopolitics Globalization also has important geopolitical components. To many, an essential dimension of globalization is that it is not restricted by territorial or national boundaries. For example, the creation of the United Nations following World War II was a step toward creating an international governmental structure in which all nations could find representation. The simultaneous emergence of the Soviet Union as a military and political superpower led to a rigid division into Cold War blocs that slowed further geopolitical integration. With the peaceful end of the Cold War in the late 1980s and early 1990s, the former communist countries of eastern Europe and the Soviet Union were opened almost immediately to global trade and cultural exchange. These political developments coincided with the economic and technological changes we now see as the early waves of globalization.

Some observers argue that economic imperialism has replaced the ideological divisions of the Cold War; thus, today, economic activity and politics are more intertwined than ever. For example, when governments interact with the World Trade Organization (WTO) and other trading blocs, expanded global economic activity is often considered as complementary to national interests.

Environmental Concerns The expansion of a globalized economy also is creating and intensifying environmental problems throughout the world. **Transnational firms**, which do global business through international subsidiaries, disrupt local ecosystems in their incessant search for natural resources and manufacturing sites. Landscapes and resources previously used by only small groups of local peoples are now thought of as global commodities to be exploited and traded on the world marketplace. As a result, native peoples are often deprived of their traditional resource base and displaced into marginal environments. On a larger scale, economic globalization is aggravating worldwide environmental problems like climate change, air pollution, water pollution, and deforestation. And yet it is only through global cooperation, such as the UN treaties on biodiversity protection or the Kyoto Protocol on global warming, that these problems can be addressed.

Social Dimensions Globalization has a clear demographic dimension as well. Although international migration is nothing new, increasing numbers of people from all parts of the world are crossing national boundaries, often permanently (Figure 1.5). Migration from Latin America and Asia has drastically changed the demographic configuration of the United States, just as migration from Africa and Asia has transformed western Europe. Countries such as Japan and South Korea that

Figure 1.4 Global Culture in the United States
The multilingual welcome offered by a public library in Montgomery county, Maryland, speaks not only to the many different languages spoken by people in the suburbs of Washington, DC, but also reminds us that expressions of globalization are found throughout North America. *(Rob Crandall/www.robcrandall.com)*

have long been perceived as ethnically homogeneous now have substantial immigrant populations. Even a number of relatively poor countries, such as Nigeria and the Ivory Coast, encounter large numbers of immigrants coming from even poorer countries, such as Burkina Faso. Although international immigration is still curtailed by the laws of every country—much more so, in fact, than the movement of goods or capital—it is still rapidly mounting, propelled by the uneven economic development associated with globalization (see "People on the Move: The Geography of Global Migration").

Finally, there also is a significant criminal element to contemporary globalization, including terrorism (discussed later in this chapter), drugs, pornography, slavery, and prostitution. Illegal narcotics, for example, are definitely a global commodity (Figure 1.6). Some of the most remote parts of the world, such as the mountains of northern Burma, are thoroughly integrated into the circuits of global exchange through the production of opium and, therefore, to the world heroin trade. Even many areas that do not directly produce drugs are involved in their global sale and

Figure 1.5 International Migration Workers from southern India dig a hole to install a street sign in Dubai, UAE. This Persian Gulf emirate is experiencing a massive construction boom as it shifts from an oil-based economy to one based upon real estate, tourism, and international finance. As a result, temporary migrant workers from India and Pakistan constitute much of the labor force. *(Stephanie Kuykendal/ Corbis)*

transshipment. Nigerians often occupy prominent positions in the international drug trade, as do members of the Russian mafia. Many Caribbean countries have seen their economies become reoriented to drug transshipments and the laundering of drug money. Prostitution, pornography, and gambling have also emerged as highly profitable global businesses. Over the past decades, for example, parts of eastern Europe have become major sources of both pornography and prostitution, finding a lucrative but morally questionable niche in the new global economy.

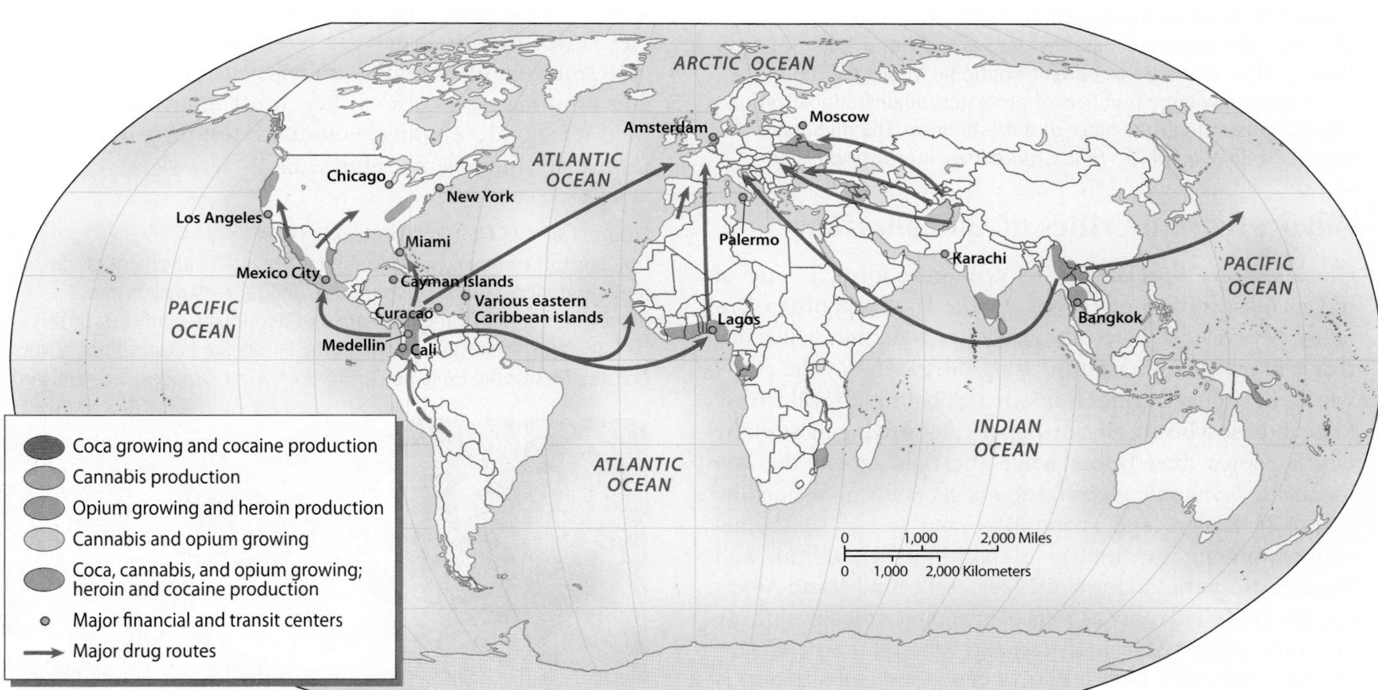

Figure 1.6 The Global Drug Trade
The cultivation, processing, and transshipment of coca (cocaine), opium (heroin), and cannabis are global issues. The most important cultivation centers are Colombia, Mexico, Afghanistan, and northern Southeast Asia, whereas the major drug financing centers are located mostly in the Caribbean, the United States, and Europe. Additionally, Nigeria and Russia also have significant roles in the global transshipment of illegal drugs.

Figure 1.7 Protests Against Globalization
Meetings of international groups such as the World Trade Organization, World Bank, and International Monetary Fund commonly draw large numbers of protesters against globalization. This demonstration took place at a Washington, DC, meeting of the World Bank and IMF. *(Rob Crandall/www.robcrandall.com)*

Advocates and Critics of Globalization

Globalization, especially in its economic form, is one of today's most contentious issues (Figure 1.7). Supporters generally believe that it results in greater economic efficiency that will eventually result in rising prosperity for the entire world. In contrast, critics think that globalization will largely benefit those who are already prosperous, leaving most of the world poorer than before while the rich exploit the less fortunate. Economic globalization is generally applauded by corporate leaders and economists, and it has substantial support among the leaders of both the Republican and Democratic parties in the United States. Beyond North America, free trade and other aspects of economic globalization are generally supported by moderate and conservative politicians in most countries. Opposition to economic globalization is widespread in the labor and environmental movements, and among many student groups worldwide. Hostility toward globalization is sometimes deeply felt, as massive protests at World Bank and WTO meetings have made obvious.

The Pro-globalization Stance Advocates argue that globalization is a logical and inevitable expression of contemporary international capitalism that will benefit all nations and all peoples. These new riches, they say, will eventually trickle down to enrich even the poorest of peoples in all the world's different regions. Economic globalization can work such wonders, they contend, by enhancing competition, allowing the flow of capital to poor areas, and encouraging the spread of beneficial new technologies and ideas. As countries reduce their barriers to trade, inefficient local industries will be forced to become more efficient in order to compete with the new flood of imports, thereby enhancing overall national productivity. Those that cannot adjust will most likely go out of business.

Every country and region of the world, moreover, ought to be able to concentrate on those activities for which it is best suited in the global economy. Enhancing such geographical specialization, the pro-globalizers argue, creates a more efficient world economy. Such economic restructuring is made increasingly possible by the free flow of capital, which will flow smoothly to those areas that have the greatest opportunities. By making access to capital more readily available throughout the world, economists contend, globalization should eventually result in a certain global **economic convergence**, meaning that the world's poorer countries will gradually catch up with the more advanced economies.

Thomas Friedman, one of the most influential advocates of economic globalization, argues the world has not only shrunk but it has also become flat in a sense so that financial capital, goods, and services flow freely from place to place. For example, the need to attract capital from abroad forces countries to adopt new economic policies. Friedman describes the great power of the global "electronic herd" of bond traders, currency speculators, and fund managers who either direct money to, or withhold it from, developing economies, resulting in economic winners and losers (Figure 1.8).

Figure 1.8 The Global "Electronic Herd"
One facet of economic globalization is the rapid movement of capital as bond traders, currency speculators, and fund managers direct money into or out of the economies of developing countries. In this picture, traders and clerks work on the Eurodollar Futures floor of the Chicago Mercantile Exchange. *(Tim Boyle/Getty Images, Inc.—Liaison)*

PEOPLE ON THE MOVE The Geography of Global Migration

Globalization has led to one of the largest migrations in human history as people leave their homes to take advantage of booming economies and better living conditions in other countries. Either desperate conditions at home (what geographers call "push forces") or the lure of promising pastures elsewhere (the "pull forces") can lead to migration. Often, though, it's a combination of the two that make people move.

While accurate data on migration are notoriously difficult to gather because people often migrate without documentation, the United Nations says that currently there are at least 190 million people legally living in a country different than where they were born. Most of those people have moved to cities in the developed Western world, but there is also considerable migration to magnet cities in the developing world, places such as Delhi, Dubai, Rio, and Mexico City. Furthermore, that UN statistic does not include those people who stay within their own country yet leave home, like the highland peasants of Bolivia moving to La Paz or a farmer from Iowa cashing in to take a job in Atlanta.

Nor does that statistic count the estimated 35 million people who are classified as refugees, those who have fled violence or natural disasters. Or the untold number who are not in established refugee camps but, instead, making do somehow on their own. Although the push and pull forces for refugees may not have been caused directly by globalization, often, if one digs deep enough, a connection can be found.

Because the human geography of global migration wears many different faces and tells many different stories—of desperate refugees fleeing violence and chaos in Iraq, Iran, or Palestine, or of comfortable middle-class workers leaving Los Angeles for Seattle—each of our chapters sheds a bit of light on this complicated process of migration through our "People on the Move" sidebars. Several examples show the kinds of migration stories told: in Chapter 3, North America, "California Dreamin' and California Leavin'" documents out-migration from the Golden State; in Chapter 8, Europe, "Smuggling Illegal Immigrants into Europe" discusses the worrisome practice of smuggling people into the European Union; in Chapter 10, Central Asia, "Afghan Refugees Return Home" recounts the recent influxes of Afghans returning from refugee camps in Pakistan; and in Chapter 6, Sub-Saharan Africa, "Cape Verde's Culture of Migration" describes how out-migration has helped define a national culture.

To the committed pro-globalizer, even the global spread of **sweatshops**—crude factories in which workers sew clothing, assemble sneakers, or perform similar labor-intensive tasks for extremely low wages—is to be applauded (Figure 1.9). People go to work in sweatshops because the alternatives in the local economy are even worse. Once countries achieve full employment through sweatshops, the argument goes, they can gradually improve conditions and move into more sophisticated, better-paying industries. Proponents of economic globalization also commonly argue that multinational firms based in North America or Europe often offer better pay and safer working conditions than do local firms and thus contribute to worker well-being. Extreme pro-globalizers go so far as to assert that poor countries ought to take advantage of their generally lax environmental laws in order to attract highly polluting industries from the wealthy countries since acquiring such industries would enhance their overall economic positions.

The pro-globalizers generally strongly support the large multinational organizations that facilitate the flow of goods and capital across international boundaries. Three such organizations are particularly important: the World Bank, the International Monetary Fund (IMF), and the World Trade Organization (WTO). The primary function of the World Bank is to make loans to poor countries so that they can invest in infrastructure and build more modern economic foundations. The IMF is concerned with making short-term loans to countries that are in financial difficulty—those having trouble, for example, making interest payments on the loans that they had previously taken. The WTO, a much smaller organization than the other two, works to reduce trade barriers between countries to enhance economic globalization. It also tries to mediate between countries and trading blocks that are engaged in trade disputes (Figure 1.10).

Figure 1.9 Global Sweatshops
One of the most debated aspects of economic globalization is the crude factories, or sweatshops, in which workers sew clothing, assemble sneakers, stitch together soccer balls, or perform similar labor-intensive work for low wages. *(Macduff Everton/The Image Works)*

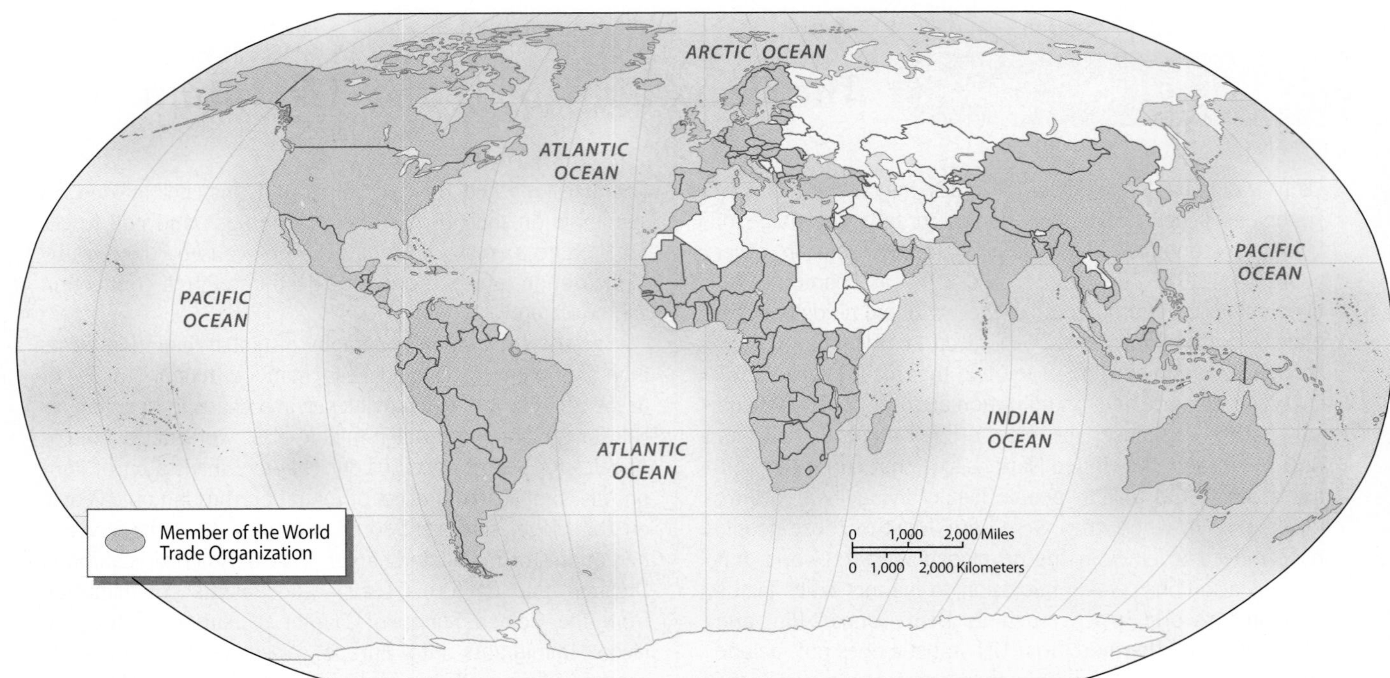

Figure 1.10 World Trade Organization
One of the most powerful institutions of economic globalization is the World Trade Organization (WTO), which was created in 1995 to oversee trade agreements, encourage open markets, enforce trade rules, and settle disputes. The WTO currently consists of 151 member countries. In addition to these member countries, more than 30 states have "observer status" including Russia (which is expected to become a full member in late 2008), Iran, and Iraq.

To support their claims, pro-globalizers argue that countries that have been highly open to the global economy have generally had much more economic success than those that have isolated themselves by seeking self-sufficiency. The world's most isolated countries, such as Burma (Myanmar) and North Korea, have become economic disasters, with little growth and rampant poverty, whereas those that have opened themselves to global forces, like Singapore and Thailand, in the same period have seen rapid growth and a substantial reduction of poverty.

Critics of Globalization Virtually all of the claims of the pro-globalizers are strongly contradicted by the anti-globalizers. Opponents often begin by arguing that globalization is not a "natural" process. Instead, it is the product of an explicit economic policy promoted by free-trade advocates, capitalist countries (mainly the United States, but also Japan and those of western Europe), financial interests, international investors, and multinational firms. Opponents also point out that the processes of globalization today are more pervasive than those of the past, even during the period of European colonialism. Thus, while global economic and political linkages have been around for centuries, their modern expression is unprecedented.

Because the globalization of the world economy appears to be creating greater inequity between rich and poor, the "trickle-down" model of developmental benefits for all people in all regions has yet to be validated. This inequity is demonstrated by the increase in the percentage of poor people in many world regions (Figure 1.11). A recent United Nations report notes that the social conditions of 60 countries are worse off now than 30 years ago and, further, that there is a clear trajectory in the direction of increasing inequity within the world. On a global scale, the richest 20 percent of the world's people consume 86 percent of the world's resources, whereas the poorest 80 percent use only 14 percent. The growing inequality of this age of globalization is apparent on both global and national scales. Globally, the wealthiest countries have grown much richer over the past two decades, while the poorest have become more impoverished. Nationally, even in developed countries such as the United States, the wealthiest 10 percent of the population have reaped almost all of the gains that globalization has offered; the poorest 10 percent have seen either their income decline in recent decades as wages have remained static or actually have become unemployed as jobs are lost to outsourcing.

Opponents also contend that globalization promotes free-market, export-oriented economies at the expense of localized, sustainable activities. World forests, for example, are increasingly cut for export timber rather than serving local needs. As part of their economic structural adjustment package, the World Bank and the International Monetary Fund encourage developing countries to expand their resource exports so they will have more hard currency to make payments on their foreign debts. This strategy, however, usually leads to overexploitation of local resources. Opponents also note that the IMF often requires developing countries to adopt programs of fiscal austerity that often entails a substantial reduction in public spending for education, health, and food subsidies. By adopting such policies, critics warn, poor countries will end up with even more impoverished populations than before.

Anti-globalizers also dispute the empirical evidence on national development offered by the pro-globalizers. Highly successful developing countries such as South Korea, Taiwan, and Malaysia, they argue, have indeed been engaged with the world market but they have generally done so on their own terms, rather than those advocated by the IMF and other advocates of full-fledged economic globalization. These countries have actually protected many of their domestic industries from foreign competition and have, at various times, controlled the flow of capital.

Furthermore, anti-globalizers contend that the "free-market" economic model commonly promoted for developing countries is not even the one that Western industrial countries used for their own economic development. In Germany, France, and even to some extent the United States, governments historically have played a strong role in directing investment, managing trade, and subsidizing chosen sectors of the economy.

Those who challenge globalization also worry that the entire system—with its instantaneous transfers of vast sums of money over nearly the entire world on a daily basis—is inherently unstable. The noted critic John Gray, for example, thinks that the same "electronic herd" that Thomas Friedman applauds is a dangerous force because it is susceptible to "stampedes." These international managers of capital tend to panic when they think their funds are at risk; when they do so, the entire intricately linked global financial system can quickly become destabilized, leading to a crisis of global proportions. Even when the "herd" spots opportunity, trouble may still ensue. As vast sums of money flow into a developing country, they may create a speculatively inflated **bubble economy** that cannot be sustained.

Such a bubble economy emerged in Thailand and many other parts of Southeast Asia in the mid-1990s. So much money flooded Thailand that property values in the capital city of Bangkok came to exceed those of most U.S. cities—clearly an economically inefficient arrangement when one considers how much less productive the Thai economy is than that of the United States. When the bubble burst in 1997, a severe depression crippled most of Southeast Asia (Figure 1.12). The depression has never really ended in Indonesia, the biggest country of the region. While this Asian crisis did not undermine the entire world economy, as many anti-globalizers feared it would, some evidence suggests that it very nearly did so.

A Middle Position? A number of experts, not surprisingly, argue that both the anti-globalization and the pro-globalization stances are exaggerated. Those in the middle ground tend to argue that economic globalization is indeed unavoidable; even the anti-globalization movement, they point out, is made possible by the globalizing power of the

Figure 1.11 Global Economic Inequity
The GINI Index measures the equity and inequity of income within a country. That is, a country where there are vast differences between rich and poor will score high on the GINI Index, with low scores going to countries where income is more evenly distributed. As you can see, the United States falls into the middle category.

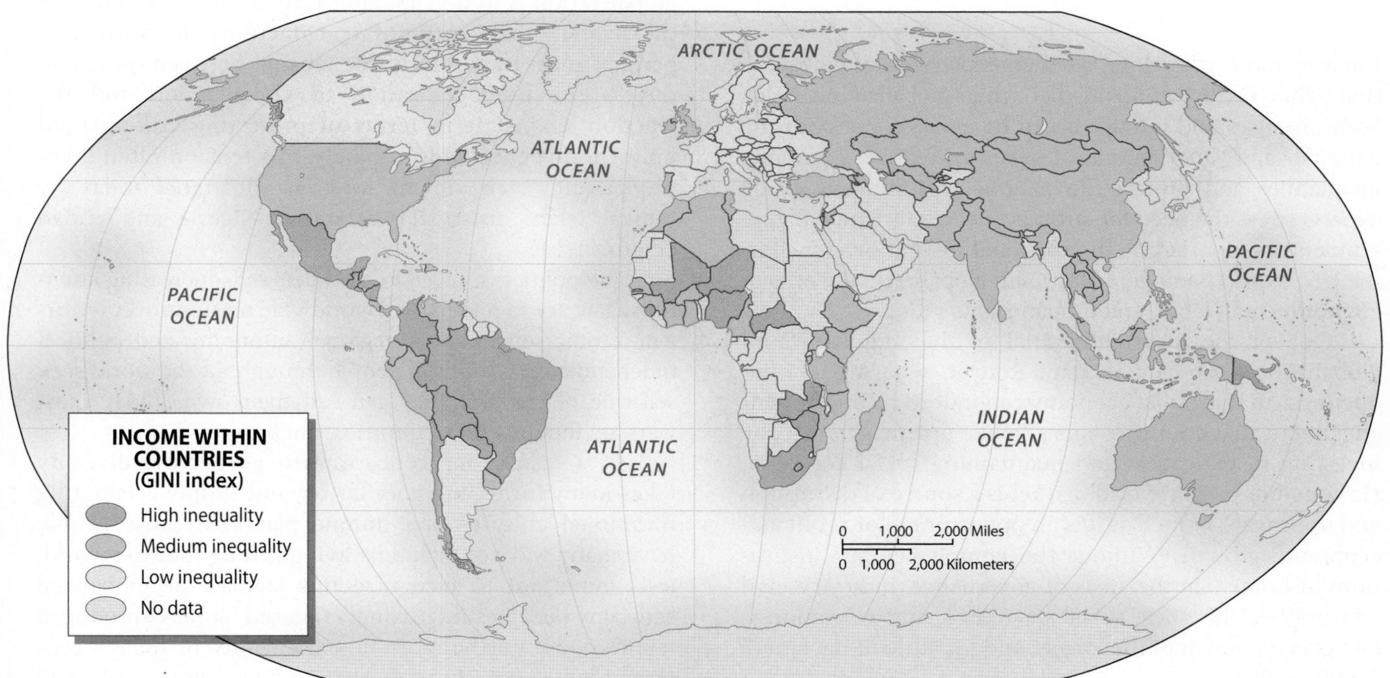

Figure 1.12 Economic Recession in Thailand
One of the unfortunate symbols of Thailand's 1997 economic financial crisis are more than 300 unfinished high-rise buildings that now litter Bangkok's downtown skyline. Before the economic collapse, Thailand's "bubble economy" led to widespread speculation in downtown property values and new buildings. When the bubble burst, bankrupt developers walked away from their high-rise office construction, leaving these abandoned steel skeletons. *(Agence France Presse/Getty Images)*

Internet and is, therefore, itself an expression of globalization. They further contend that while globalization holds both promises and pitfalls, it can be managed, at both the national and international levels, to reduce economic inequalities and protect the natural environment. These experts stress the need for strong yet efficient national governments, supported by international institutions (such as the UN, World Bank, and IMF) and globalized networks of environmental, labor, and human rights groups.

One of the most influential of the middle-ground globalization scholars is Dani Rodrik, who argues that openness to the global economy can indeed be highly beneficial, but that countries must "make openness work" by investing in education and maintaining social cohesion. He concludes, "The world market is a source of disruption and upheaval as much as it is an opportunity for profit and economic growth. Without the complementary institutions at home—in the areas of governance, judiciary, civil and political liberties, social insurance, and education—one gets too much of the former and too little of the latter" (1999, p. 96).

Globalization is one of the most important issues of the day, and one of the most complicated. While this book does not pretend to resolve the controversy, it does encourage readers to reflect on these critical points as they apply to different world regions.

Diversity in a Globalizing World

As globalization increases, many observers foresee a world far more uniform and homogeneous than today's. The optimists among them imagine a universal global culture uniting all humankind into a single community untroubled by war, ethnic strife, or resource shortage—a global utopia of sorts.

A more common view, however, is that the world is becoming blandly homogeneous as different places, peoples, and environments lose their distinctive character and become indistinguishable from their neighbors. While diversity may be the hardest thing for a society to live with, it also may be the most dangerous thing to live without. Nationality, ethnicity, cultural distinctiveness—all are the legitimate legacy of humanity. If this diversity is blurred, denied, or repressed through global homogenization, humanity loses one of its defining traits.

But even if globalization is generating a certain degree of homogenization, the world is still a highly diverse place (Figure 1.13). One still finds marked differences in culture (including language, religion, architecture, foods, and many other attributes of daily life), economy, and politics—as well as in the natural environment. Such diversity is so vast that it cannot readily be extinguished even by the most powerful forces of globalization. In fact, globalization often provokes a strong reaction on the part of local people, making them all the more determined to maintain what is distinctive about their way of life. Thus, globalization is understandable only if one also examines the diversity that continues to characterize the world and, perhaps most importantly, the tension between these two forces—the homogenization of globalization and the reaction against it in terms of protecting cultural and political diversity. Unfortunately, this tension often takes unpleasant, even violent form as illustrated today by radical Islam or tribal warfare in Nigeria and Sudan (Figure 1.14).

The politics of diversity also demands increasing attention as we try to understand worldwide tensions over terrorism, ethnic separateness, regional autonomy, and political independence. Groups of people throughout the world seek self-rule of territory they can call their own. Today most wars are fought *within* countries, not *between* them.

As a result, our concern with geographic diversity takes many forms and goes far beyond simply celebrating traditional cultures and unique places. Because people have many ways of making a living throughout the world, it is important to recognize this fact as the globalized economy becomes increasingly focused on mass-produced retail goods. Furthermore, a stark reality of today's economic landscape isunevenness—while some people and places prosper, others suffer from unrelenting poverty. Unfortunately, this also is a form of diversity amid globalization.

In summary, globalization can be defined as the increasing interconnectedness of people and places through converging processes of economic, political, and cultural change. Globalization is pervasive and unrelenting, and is a dominant feature of the contemporary world. Furthermore, its benefits are uneven. Although economic restructuring is a prime cause, globalization is not simply the growth and expansion of international trade. It also is represented by converging and homogenizing forces that many people resist. Thus, an equally important theme is how geographic diversity comes in conflict with globalization—how globalization in different places and at different times is suppressed, renegotiated, hybridized, protected, preserved, or extinguished. Because of the importance of this theme, diversity and globalization should be examined as inseparable—often in conflict, yet at other times complementary.

Figure 1.14 Tribal Warfare
Rebels from the Sudan Liberation Movement (SLM) declared war against the Sudanese government in the province of Darfur. The two groups that originated the rebellion have now morphed into dozens of warring factions, complicating the search for peace amid a profound humanitarian crisis. *(Ramzi Haidar/AFP/Getty Images)*

Figure 1.13 Local Cultures
This family in Ghana reminds us that although few places are beyond the reach of globalization, many unique local landscapes, economies, and cultures still exist. For example, most of the family's belongings in this photograph are local in origin and come from long-standing cultural tradition. *(Caroline Penn/Panos Pictures)*

GEOGRAPHY MATTERS: Environments, Regions, Landscapes

Geography is one of the most fundamental sciences, a discipline awakened and informed by a long-standing human curiosity about our surroundings and the larger world environment. The term *geography* has its roots in the Greek words for "describing the Earth," and this discipline has been advanced since classical times by all cultures and civilizations. Along with the inherent satisfaction of knowing about different environments also come practical benefits through exploration, resource use, world commerce, and travel. In some ways, geography can be compared to history: while historians describe and explain what has happened over time, geographers explain Earth's spatial dimensions, describing how the world differs from place to place.

Given the broad scope of geography, it is no surprise that geographers have many different conceptual approaches to examining the world. At the most basic level,

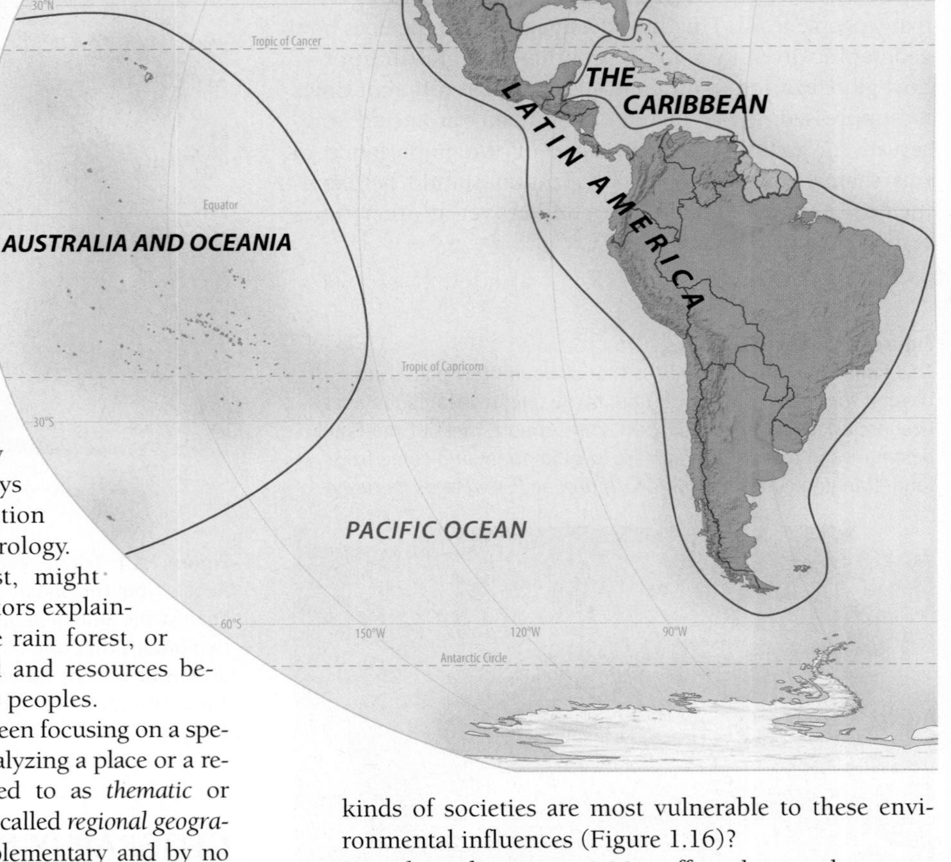

Figure 1.15 World Regions
These regions are the basis for the 12 regional chapters in this book. Countries or areas within countries that are treated in more than one chapter are designated on the map with a striped pattern. For example, western China is discussed in both Chapter 10, Central Asia, and Chapter 11, East Asia. Also, three countries on the South American continent are discussed as part of the Caribbean region because of their close cultural similarities to the island region.

geography can be broken into two complementary pursuits, physical and human geography. Physical geography examines climate, landforms, soils, vegetation, and hydrology, whereas human geography concentrates on the spatial analysis of economic, social, and cultural systems (see "Geographic Tools: From Satellites to Shovels").

A physical geographer, for example, studying the Amazon Basin of Brazil might be interested primarily in the ecological diversity of the tropical rain forest or the ways the destruction of that dense vegetation changes the local climate and hydrology. The human geographer, in contrast, might focus on the social and economic factors explaining the migration of settlers into the rain forest, or the tensions and conflicts over land and resources between these migrants and indigenous peoples.

Another basic division is that between focusing on a specific topic or theme contrasted with analyzing a place or a region. The former approach is referred to as *thematic* or *systematic geography*, while the latter is called *regional geography*. These two perspectives are complementary and by no means mutually exclusive. This textbook, for example, draws upon a regional scheme for its overall architecture, dividing the globe into 12 separate world regions, yet it organizes each chapter thematically, examining the topics of environment; population and settlement; cultural coherence, and diversity; geopolitics; and economic social development. In doing so, each chapter synthesizes physical and human geography, as well as thematic and regional geography (Figure 1.15).

Human–Environment Interaction

A fundamental component of geographic inquiry is the examination of the basic yet highly complex relationship between humans and their environment. Though this analysis may take many different forms, three overarching questions form the foundation:

- Where and under what conditions does nature control, constrain, or disrupt human activities, and what kinds of societies are most vulnerable to these environmental influences (Figure 1.16)?
- How have human activities affected natural systems such as the world's climate, vegetation, rivers, and oceans, and what are the consequences of these actions?
- Is there any evidence that humans can live in balance with nature so that human use of natural resources might be sustainable over a longer period of time? Or is environmental damage an inevitable consequence of human settlement?

While there are no simple answers to these three questions, the study of world regional geography offers valuable insights from different cultural and regional perspectives.

Areal Differentiation and Integration

Geography is considered the spatial science, charged with the study of Earth's space or surface area, the uniqueness of places as well as the similarities between them. One component of that responsibility is describing and explaining the differences that distinguish one piece of the world from another. The geographical term for this is **areal differentiation** (*areal* means "pertaining to area"). Why is one part of Earth humid and lush while another just a few hundred kilometers away is an arid desert (Figure 1.17)?

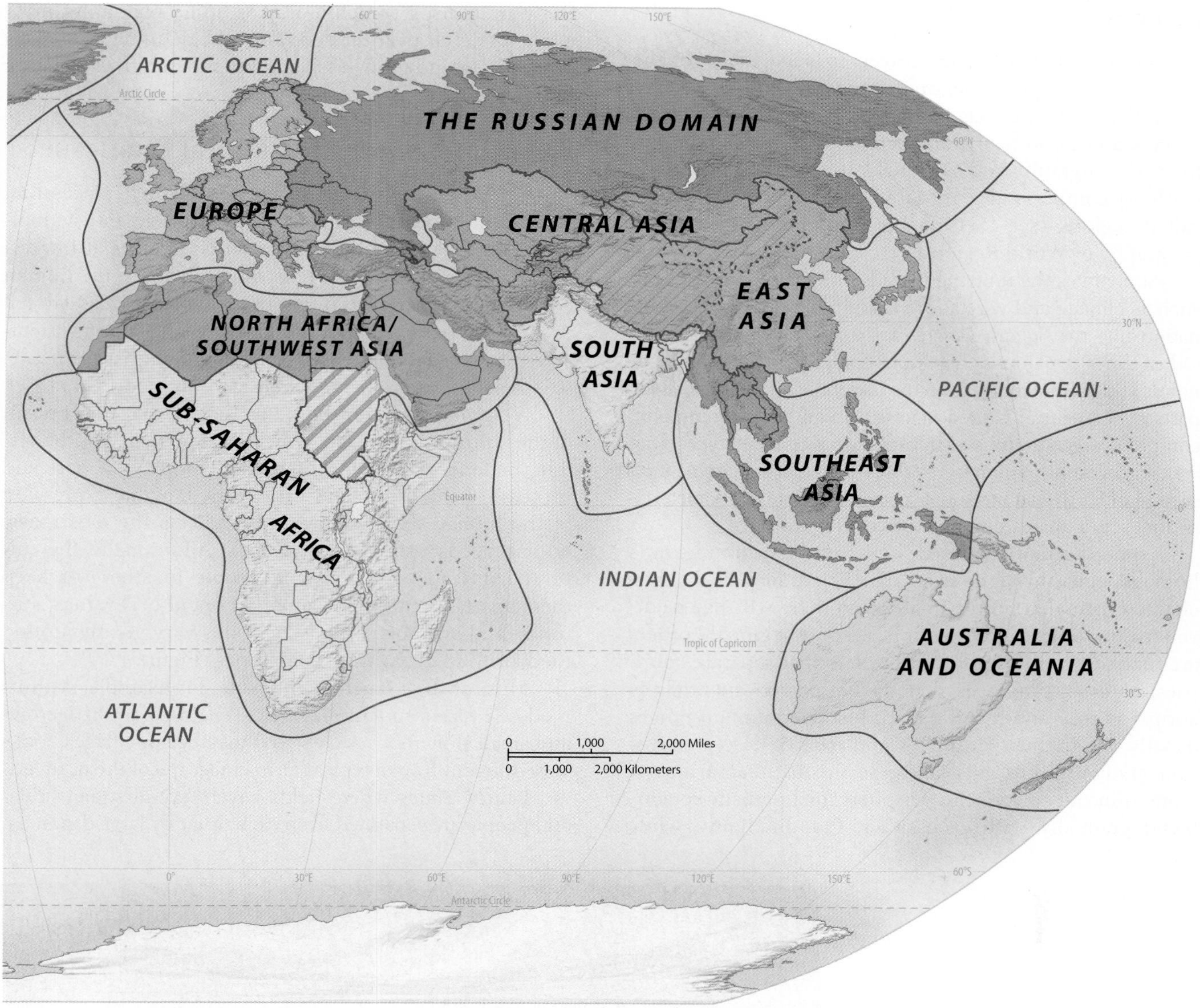

Geographers also are interested in the connections between different places and how they are linked. This concern is one of **areal integration**, which is the study of how places interact with each other. How and why are the economies of Singapore and the United States closely intertwined even though the two countries are situated in entirely different physical, cultural, and political environments? The questions of areal integration, obviously, are becoming increasingly important because of the global linkages that come with globalization.

Figure 1.16 Natural Hazards
One pressing question in geographic inquiry is how people and societies are affected by natural hazards. Often, as a population increases in an area, more and more people are threatened by natural hazards, such as hurricanes, floods, and drought. Shown here is a flood in Bangladesh. *(James Blair/NGS Image Collection)*

Regions

The human intellect seems driven to make sense of the universe by lumping phenomena together into categories of similarity. Biology has its taxa of living objects, history its eras and periods of time, geology its epochs of Earth history. Geography, too, organizes information about the world by compressing it into units of similarity, which are called **regions** (see "Setting the Boundaries: The Metageography of World Regions").

Sometimes the unifying threads of a region are physical, such as climate and vegetation, resulting in a regional designation like the *Sahara Desert* or the *Amazonian rain forest*. Other times the threads are more complex, combining economic and cultural traits, as in the use of the term *Corn Belt* for parts of the central United States. Human beings commonly compress large amounts of information into stereotypes; in a way, a geographic region is just that—a spatial stereotype for a section of Earth that has some special signature or characteristic that sets it apart from other places (Figure 1.18).

Some caution is advised, however. First, no region is homogeneous throughout its area. There may be a single characteristic that unites an area, but there will also be diversity and differences within it. Second, regional borders are fuzzy. Except for political borders that separate countries, states, or provinces, regional boundaries are rarely as abrupt as they appear on a map. Just as mountain ranges usually rise gradually from lowland areas or deserts transition gradually into grasslands, so do manufacturing regions thin out toward the periphery, or linguistic regions blend gradually with each other. Put differently, while those traits used to define a geographic region are most explicit and clear in the core area, the defining traits usually weaken and grow less apparent moving from the core to the boundary.

Space into Place: The Cultural Landscape

Human beings transform space into distinct places that are unique and heavily loaded with meaning and symbolism. This diverse fabric of *placefulness* is of great interest to geographers because it tells us much about the human condition as it varies throughout the world. Places can tell us how humans interact with nature and among themselves; where there are tensions and where there is peace; where people are rich and where they are poor.

A common tool for the analysis of place is the concept of the **cultural landscape**, which is, simply stated, the visible, material expression of human settlement, past and present. The cultural landscape is the tangible expression of the human habitat. It visually reflects the most basic human needs—shelter, food, work. Additionally, the cultural landscape acts to bring people together (or keep them apart), because it is a marker of cultural values, attitudes, and symbols. Because cultures vary greatly around the world, so do cultural landscapes (Figure 1.19).

Most of the French countryside, for example, with its modestly sized agricultural fields enclosed by hedgerows and small stone houses clustered together in villages, looks very different from a typical farm landscape of the midwestern United States where fields cover several square miles and people are separated from each other by large distances.

Figure 1.17 Areal Differentiation The California–Mexico border, with different agricultural landscapes on opposite sides, reminds us of the many variables that shape cultural landscapes. The larger red pattern to the upper left is irrigated agricultural fields of the United States, while to the bottom right, the much smaller fields of Mexico show a different landscape signature. Explaining these differences (and the similarities) is a central focus of geography. *(Earth Satellite Corporation/SPL/Photo Researchers, Inc.)*

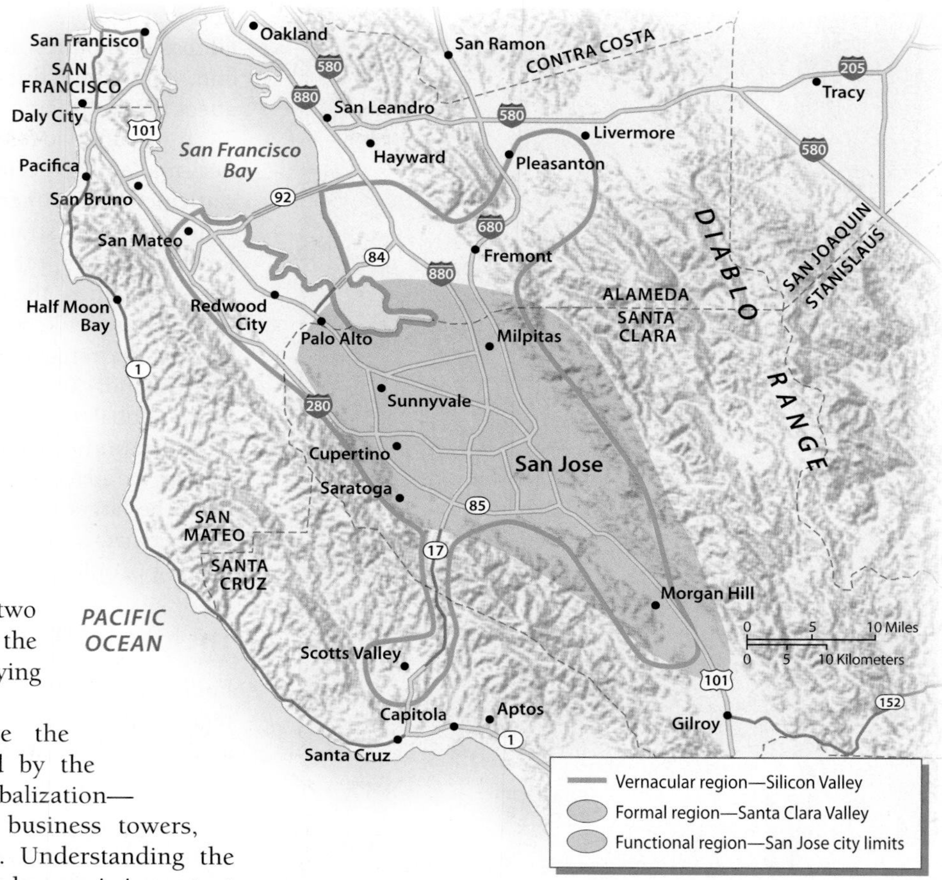

Figure 1.18 Geographic Regions This map illustrates three different kinds of regions. A vernacular region has vague, cognitive borders, and is used by the public to refer to a general area. Examples of vernacular regions are "the Midwest" or "Silicon Valley," which is shown here. Usually, vernacular regions are a combination of many traits, real or imagined. In contrast, formal regions have distinct boundaries that are constructed by using one specific trait. In the case of the Santa Clara Valley, that trait is the topographic boundary of the valley. Last, a functional region is based on a certain activity or organization, such as the civic government of San Jose, which is then delimited by the city limits.

From the contrasting look of these two landscapes, we can learn much about the social and economic systems underlying the two different societies.

Increasingly, however, we see the uniqueness of places being eroded by the homogeneous landscapes of globalization—shopping malls, fast-food outlets, business towers, theme parks, industrial complexes. Understanding the forces behind the spread of these landscapes is important because they tell us much about the expansion of global economies and cultures. While a modern shopping mall in Hanoi, Vietnam, may seem familiar to someone from North America, this new landscape represents yet another component of globalized world culture that has been implanted into a once remote and distinctive city (Figure 1.20).

Scales: Global to Local

There is a sense of scale to all systematic inquiry, whatever the discipline. In biology, for example, some scientists study the smaller units of cells, genes, or molecules, while others take a larger view, analyzing plants, animals, or ecosystems.

Figure 1.19 The Cultural Landscape Despite globalization, the world's landscapes still show great diversity. Geographers use the concept of the cultural landscape to better understand how people shape their environment to create unique places. This landscape is of a village and its rice terraces on the island of Luzon, the Philippines. *(Maurice Joseph/Robert Harding World Imagery)*

Figure 1.20 **Global Shopping Malls**
Once a fixture only of suburban North America, the shopping mall is now found throughout the world, as illustrated by this mall in downtown Kuming, the capital city of Yunam province, China. *(Sean Sprague/The Image Works)*

Similarly, some historians may focus on a specific individual at a specific point in time, whereas others take a broader view of international events over many centuries.

Geographers also work at different scales. Some may concentrate on analysis of a local landscape—perhaps a single village in southern India; others will focus on a broader regional picture, examining all of southern India. Others do research on a still larger global scale, perhaps studying emerging trade networks between southern India's center of information technology in Bangalore and North America's Silicon Valley, or investigating how India's monsoon might be connected to and affected by the Pacific Ocean's El Niño. But even though geographers may be working at different scales, they never lose sight of the interactivity and connectivity between local, regional, and global scales, of the ways that the village in southern India might be linked to world trade patterns, or how the late arrival of the monsoon (delayed, perhaps, by El Niño) could affect agriculture and food supplies in different parts of India.

This ability to move between different scales—global, regional, local—is critical for understanding contemporary world regional geography because of the way globalization links all peoples and places. Few villages today, however remote, are unconnected to the modern world. Global economies draw upon local crops and resources, and, conversely, fluctuations in world commodity prices affect the well-being of local people (Figure 1.21). Global TV, films, and videos introduce foreign styles, ideas, mannerisms, and expectations into small towns and remote settlements in formerly isolated parts of the world. Global tourism brings strangers (some welcome, some not) to far-flung places, often corrupting and compromising the very uniqueness the tourists had sought.

In this book, we use the phrase "global to local" to capture the range of scales necessary for placing regions, places, landscapes, and people in their globalized context (see "Global to Local: Is This Nebraska or India? The Globalization of Back-Office Jobs"). Because our premise is that no place in today's world is completely isolated from the larger currents of a globalizing society, we need this perspective to examine the complicated linkages connecting people and places with the larger, dynamic world.

Themes in *Diversity Amid Globalization*

Following two introductory chapters, this book adopts a regional perspective, grouping all of Earth's countries into a framework of world regions. We begin with a region

Figure 1.21 **Local to Global**
Given the daily fluctuations of crop prices, farmers throughout the world are plugging into internet sites that provide them with update information so they can get the best prices for their crops. In this photo a farmer in Guilin, China checks grain prices. *(Bill Bachmann/PhotoEdit Inc.)*

GEOGRAPHIC TOOLS From Satellites to Shovels

Geographers use a wide array of tools—from satellites to shovels—to explain the world and solve environmental problems (Figure 1.2.1). Some geographers and environmental scientists get their information from aerial photos and images from airplanes and satellites; others work at ground level, digging holes to sample soils, mapping trees and plants, or even interviewing people in their homes and at the mall. Although geographers' toolboxes and methods may differ, the goals are the same: a better understanding of global environments and the people who inhabit them.

Figure 1.2.1 Satellites to Shovels
A graduate student gathers soil samples for her master of science thesis on the ecology of rare native plants in the hills of coastal California. *(Les Rowntree)*

Many of our tools are commonplace, like maps, compasses, handheld GPS (Global Positioning System) devices, or even computers. Other tools, though, are highly specialized, such as multispectral satellite images of vegetation, computer models of refugee migration, or soil-boring tools to capture prehistoric pollen samples. As readers of a book about the world, you might be interested in how geographers collect and analyze information about different places and different issues. Toward that goal, each chapter includes a sidebar called "Geographic Tools" that illustrates how specific tools and methods are used to expand our understanding of the world.

In Chapter 2 (The Changing Global Environment), for example, we briefly explore the tools used to study climate change, while in Chapter 4 (Latin America), we explain how "participatory mapping" has helped native peoples establish better legal claims to their Central American lands. Further, in Chapter 10 (Central Asia), we assess how satellite images have been used to monitor the disappearing Aral Sea; and in Chapter 14 (Australia and Oceania), we describe the tools used by fire ecologists in Australia to understand—and prepare for—wildfires.

Many colleges and universities now offer courses in the use of these tools, particularly GIS (geographic information system), remote sensing (interpreting aerial photographs and satellite imagery), and computer-aided cartography (mapmaking). Such tools can provide students with technical skills that are much sought after in today's job market. If you are interested, ask your world regional geography professor about those opportunities.

familiar to most of our readers, North America, and then move to Latin America, the Caribbean, Africa, the Middle East, Europe, Russia, and the different regions of Asia before concluding with Australia and Oceania. Each of the 12 regional chapters employs the same five-part thematic structure—environmental geography; population and settlement; cultural coherence and diversity; geopolitical framework; and economic and social development. The concepts and data central to each theme are discussed in the following section.

Background on global environmental geography is found in Chapter 2, The Changing Global Environment, which outlines the global environmental elements fundamental to human settlement—climate, topography, vegetation, and hydrology—and discusses the linkages between environmental issues and globalization.

POPULATION AND SETTLEMENT: People on the Land

The human population is far larger than it has ever been, and there is considerable debate about whether continued growth will benefit or harm the human condition. This concern is not only about the total population of the world, but also about the geographic patterns of human settlement. As the world map shows, while some parts of the world are densely populated, other places remain almost empty (Figure 1.22).

With more than 6.6 billion humans on Earth, we currently add 129 million people each year at a rate of 14,000 births per hour. About 90 percent of this population growth takes place in the developing countries of Africa, South and East Asia, and Latin America. Because of rapid growth in developing countries, perplexing questions dominate discussions of

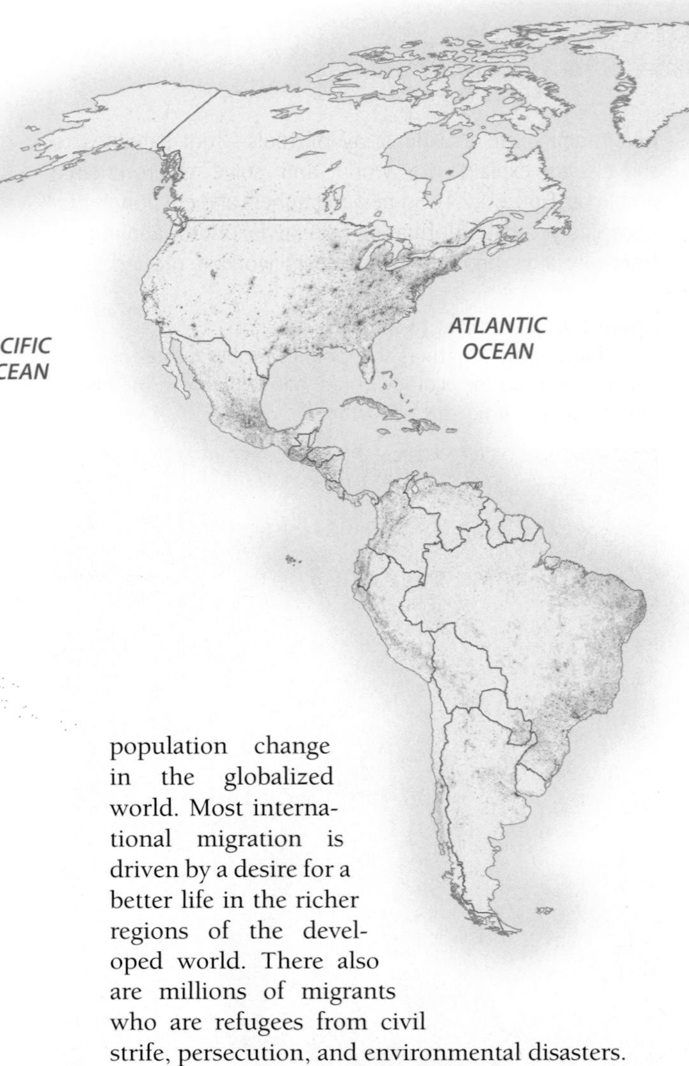

Figure 1.22 World Population
This world population map shows the differing densities of population in the regions of the world. East Asia stands out as the most populated region, with high densities in Japan, Korea, and eastern China. The second most populated region is South Asia, dominated by India, which is second only to China in population. In North Africa and Southwest Asia, population clusters are often linked to the availability of water for irrigated agriculture, as is apparent with the population cluster along the Nile River. Higher population densities in Europe, North America, and other countries are usually associated with large cities, their extensive suburbs, and nearby economic activities.

many global issues. Can these countries absorb the rapidly increasing population and still achieve the necessary economic and social development to ensure some level of well-being and stability for their population? What role, if any, should the developed countries of North America, Europe, and East Asia play in helping developing countries with their population problems? While population is a complex and contentious topic, several points help focus the issues:

- Very different rates of population growth are found in different regions of the world. While some countries are growing rapidly, others have no natural growth at all. Instead, any population growth comes from in-migration. India is an example of the former; Italy of the latter.
- Population planning takes many forms, from the fairly rigid one-child policies of China to the "please have children" programs of western Europe. Further, while some programs attempt to achieve their goals through coercion and force, others rely on incentives and social cooperation (Figure 1.23).
- Not all attention should be focused on natural growth because migration is increasingly the root cause of population change in the globalized world. Most international migration is driven by a desire for a better life in the richer regions of the developed world. There also are millions of migrants who are refugees from civil strife, persecution, and environmental disasters.
- The greatest migration in human history is now going on as people move from rural to urban environments. Soon, within the next year or two, more than half the world's population will live in cities. As with natural population growth, the developing countries of Africa, Latin America, and Asia are experiencing the most rapid changes in urbanization.

Figure 1.23 Family Planning Policies
Many countries in the developing world have concluded that unrestrained population growth may keep them from realizing their development goals, and therefore have put family planning policies in place. This poster in Vietnam urges smaller families. *(Chris Stowers/Panos Pictures)*

Population Growth and Change

Because of the centrality of population growth, each regional chapter in this book includes a table with population data for the countries within that region (see Table 1.1). Although the statistics in these tables might seem daunting at first glance, this information is crucial to understanding the population geography of the regions.

Natural Population Increase A common starting point for measuring demographic change is the **rate of natural increase** (abbreviated as RNI), which depicts the annual growth rate for a country or region as a percentage. This statistic is produced by subtracting the number of deaths in a given year from the births. Important to remember is that gains or losses through migration are not considered in the RNI.

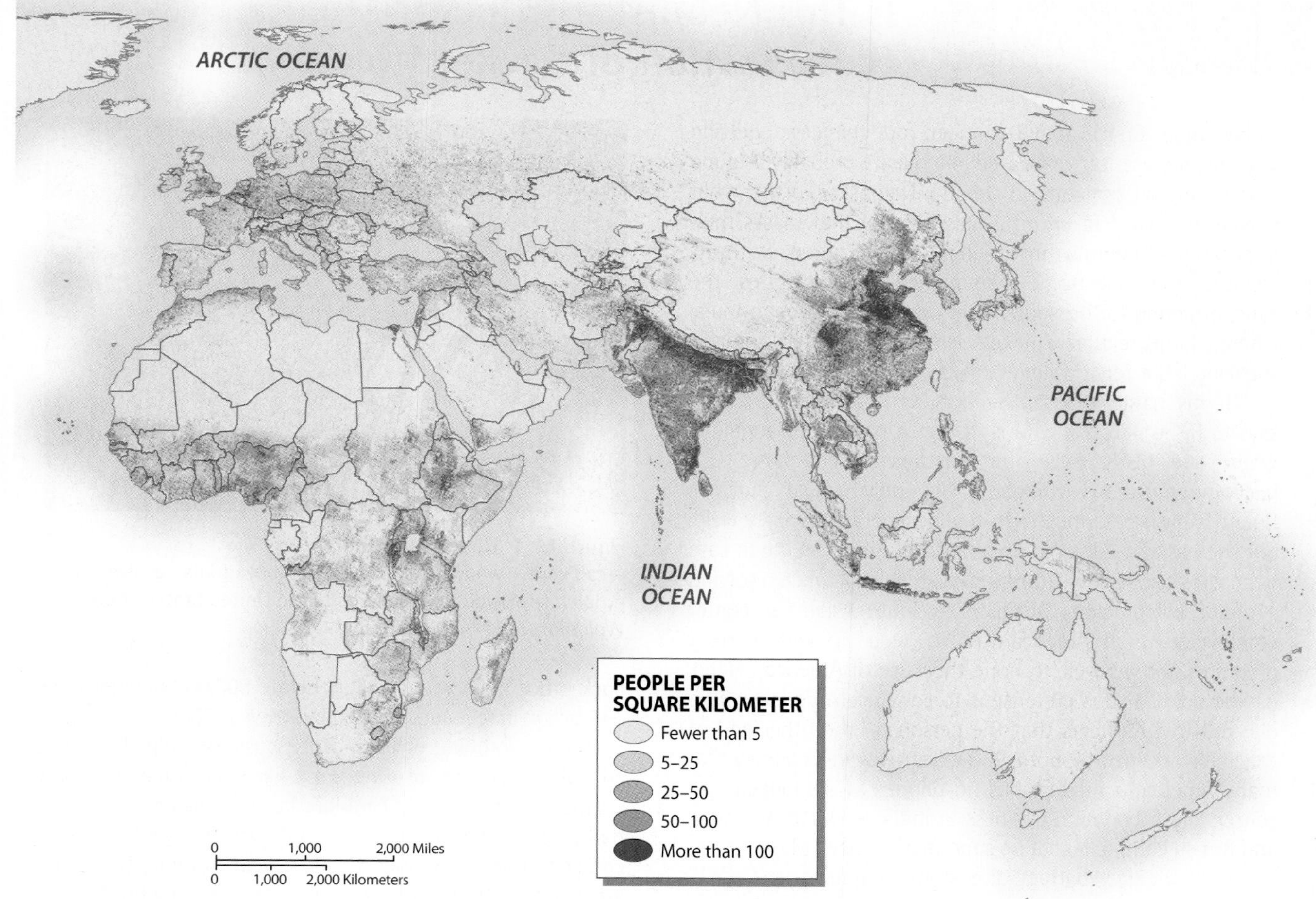

Further, instead of using raw numbers for a large population, demographers divide the gross numbers of births or deaths by the total population, thereby producing a figure per 1,000 of the population. This is referred to as the *crude birthrate* or the *crude death rate*. For example, current data for the world as a whole show a crude birthrate of 21 per 1,000 and a crude death rate of 9 per 1,000. Thus, the natural growth rate is 12 per 1,000. This is expressed as the RNI by converting that figure to a percentage; as a result, the current rate of natural increase for the world is 1.2 percent per year.

Because birthrates vary greatly between countries (and even regions of the world), rates of natural increase also vary greatly. In Africa, for example, many countries have crude rates of more than 40 births per 1,000 people.

TABLE 1.1 Population Indicators of the 10 Largest Countries

Country	Population (Millions, 2007)	Population Density (per Square Kilometer)	Total Fertility Rate	Percent Urban	Percent <15[a]	Percent >65[b]	Net Migration (per 1,000, 2000–05)
China	1,318	138	1.6	44	20	8	–0.3
India	1,132	344	2.9	28	33	5	–0.3
United States	302	31	2.1	79	20	12	4.0
Indonesia	232	122	2.4	42	28	6	–0.9
Brazil	189	22	2.3	81	28	6	–0.1
Pakistan	169	213	4.1	34	40	4	–2.4
Bangladesh	149	1,035	3.0	23	33	4	–0.5
Nigeria	144	156	5.9	44	45	3	–0.3
Russia	142	8	1.3	73	15	14	0.6
Japan	128	338	1.3	79	14	21	0.4

[a]*Percentage of population younger than 15.*
[b]*Percentage of population older than 65.*
Sources: Population Reference Bureau, World Population Data Sheet, 2007; *net migration rate data from* UN International Migration, 2006.

GLOBAL TO LOCAL

Is This Nebraska or India? The Globalization of Back-Office Jobs

Today, when you call an 800 number, your chances of getting a service representative located in India are probably as good as speaking to someone in Omaha, Nebraska. Further, that person may have taken on a new identity that masks their international identity. Increasingly, globalization, with its instantaneous electronic communication, facilitates the subcontracting (or outsourcing) of back-office jobs to India, where a large, relatively inexpensive, well-educated, English-speaking labor force willingly takes on these tasks.

"Hi, my name is Susan Sanders, and I'm from Chicago," says C. R. Suman, 22, who is in fact a native of Bangalore, where she fields calls from customers of a large U.S. telecommunications company. Not only would Suman's fluent English pass muster in the bleachers of Wrigley Field, but she has also conjured up a fictional American life in case her callers ask personal questions as they chat while resolving a phone bill problem. Often these South Asian call center employees watch American TV to keep up with current programs and, as well, to hone their use of American slang.

The point of this pretense is to convince Americans who dial toll-free numbers that the person on the other end of the line works nearby, not 8,300 miles away in a country that many Americans assume has no phones at all. Indeed, call centers are a booming business in India as North American and British companies set up supermarket-sized phone banks to handle the daily barrage of customer inquiries.

Call centers, though, are only the low end of a much larger industry of Indian software designers, accountants, Website designers, and animation artists who work on projects for foreign companies from their offices in the booming cities of Bangalore, Mumbai (Bombay), and Hyderabad (Figure 1.4.1). In fact, some experts predict that India is on its way to becoming the back office of the world. One consultant predicts that back-office assignments will generate 800,000 new jobs and $17 billion in revenue for India by the year 2008.

Figure 1.3.1 Nebraska or India?
A call center worker in Mumbai (Bombay), India, resolves problems for welfare recipients in the United States. *(Amy Waldman/The New York Times)*

The outsourcing of back-office business helps U.S. firms cut labor costs for routine work like processing insurance claims or settling credit card bills because wages in India are half those in the United States. But even though salaries for call center jobs pay only $2,000 a year in India, these jobs are coveted. "In the U.S., these jobs are taken by housewives or kids who haven't decided what they want to do with their lives," said K. Ghanesh, the founder of Customer Asset. "Here they are career jobs for college graduates."

Sources: Adapted from Mark Landler, "Hi, I'm in Bangalore (But I Dare Not Tell)," *New York Times*, March 21, 2001, p. 1; and Katherine Boo, "Letter from India: The Best Job in Town—The Americanization of Chennai," *The New Yorker*, July 5, 2004, pp. 54–69.

Because their death rates are generally less than 20 per 1,000, this results in rates of natural increase greater than 3 percent per year; these are among the highest population growth found anywhere in the world.

Total Fertility Rate Though the crude birthrate gives some insight into current conditions in a country, demographers place more emphasis on the **total fertility rate**, or **TFR** to predict future growth. The TFR is a synthetic hypothetical number that measures the fertility of a statistically fictitious, yet average group of women moving through their child-bearing years. If women marry early and have many children over a long span of years, the TFR will be a relatively high number. Conversely, if data show that women marry late and have few children, the number will be correspondingly small. Important to note is that any number less than 2.1 implies a population is not growing at all since it takes a minimum of two children to replace their parents. From population data collected in the last decade, the current TFR for the world is 2.7. While that is the average for the whole world, the variability between regions is striking. To illustrate, the current TFR for Africa is 5.0, whereas in slow- and no-growth Europe it is only 1.5.

Young and Old Populations One of the best indicators of the momentum (or lack of) for continued population growth is the youthfulness of a population because these data show the proportion of a population about to enter their prime reproductive years. The common statistic for this measure is the *percentage of a population under the age of 15.* As a global average, 28 percent of the population is younger than age 15, but in fast-growing Africa, that figure is 41 percent, with several African countries approaching 50 percent (Figure 1.24). This suggests strongly that rapid population growth in Africa will continue for at least another generation despite the tragedy of the AIDS epidemic. In contrast, Europe has only 16 percent of its population under 15, and North America 20 percent.

Figure 1.24 Fertility and Mortality Birthrates and death rates vary widely around the world. Fertility rates result from an array of variables, including state family-planning programs and the level of a woman's education. This family is in Bangladesh, a country that is working hard to reduce its birthrate through education programs. *(Trygve Bolstad/Panos Pictures)*

The other end of the age spectrum is also important, and it is measured by the *percentage of a population over the age of 65*. That figure is useful for inferring the needs of a society in providing social services for its senior citizens and pensioners. Japan and most European countries, for example, have a relatively high proportion of people over age 65. As a result, concerns are raised about whether there will be enough wage earners to support the social security needs of a large elderly population.

The structure of a population, which includes the percentage of young and old, is presented graphically as a **population pyramid**. This graph plots the percentage of all different age groups along a vertical axis that divides the population into male and female (Figure 1.25). The large percentage of young people in a fast-growing population provides a wide base and the small percentage of elderly a narrow tip, thereby giving the graph its pyramidal shape. Older populations, in contrast, with fewer young people and an aging population give the graph a very

Figure 1.25 Population Pyramids
The term *population pyramid* comes from the form assumed by a rapidly growing country such as Nigeria when data for age and sex are plotted graphically as a percentage of the total population. The broad base illustrates the high percentage of young people in the country's population, which indicates that rapid growth will probably continue for at least another generation. This pyramidal shape contrasts with the narrow bases of the slow and negative growth countries, the United States and Germany, which have fewer people in the childbearing years.

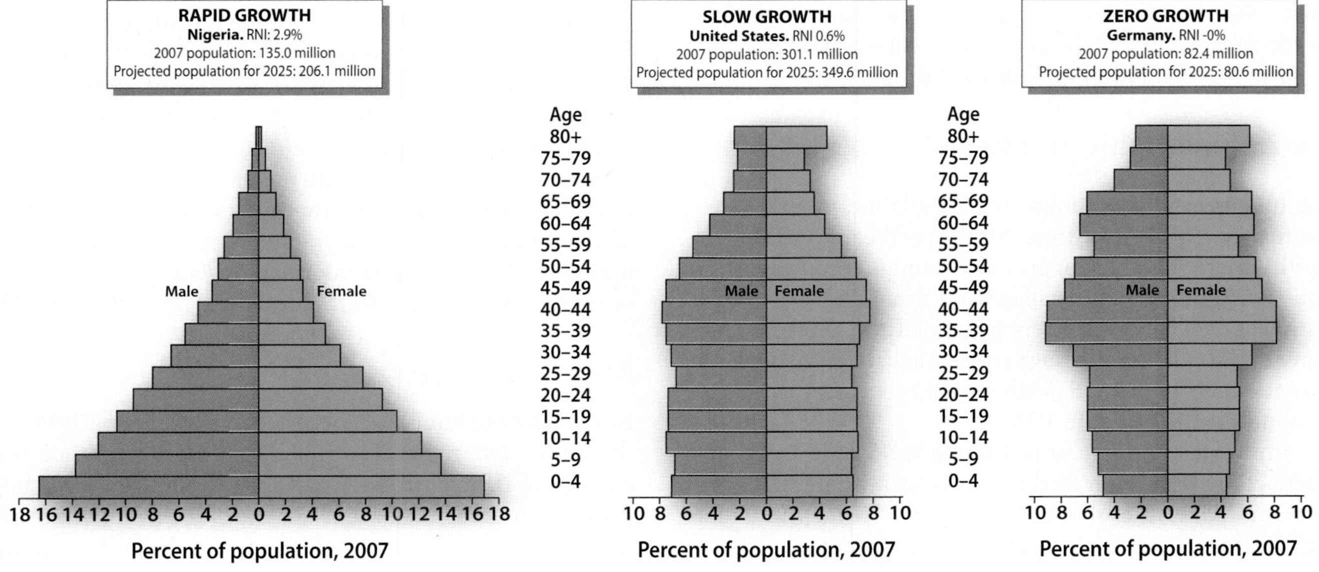

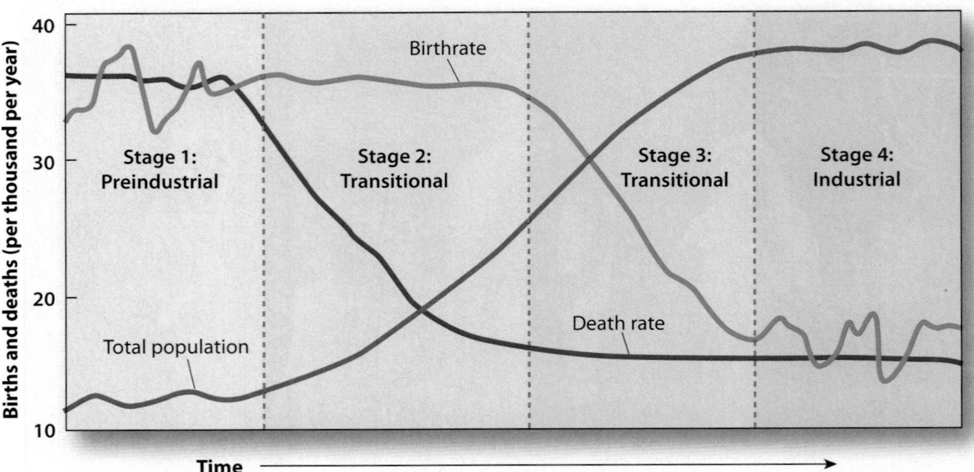

Figure 1.26 Demographic Transition As a country goes through industrialization, its population moves through the four stages in this diagram, referred to as the *demographic transition.* In Stage 1, population growth is low because high birthrates are offset by high death rates. Rapid growth takes place in Stage 2 as death rates decline. Stage 3 is characterized by a decline in birthrates. The transition ends with low growth once again in Stage 4, resulting from a relative balance between low birthrates and death rates.

different shape with a narrow base and broader upper segment.

Life Expectancy Another demographic indicator that contains information about health and well-being in a society is *life expectancy*, which is the average length of a life expected at the birth of a typical male or female in specific country. Because a large number of social factors such as health services, nutrition, and sanitation influence life expectancy, these data are often used as an indicator of the level of social development in a country. Since this book uses life expectancy as a social indicator, life expectancy data are found in the economic and social development tables, not with other population statistics (see Table 1.2, page 43).

Not surprisingly, because social conditions vary widely around the world, so do life expectancy figures. In general, though, life expectancy has been increasing over the decades, implying that the conditions supporting life and longevity are improving. To illustrate, in 1975, the average life expectancy figure for the world was 58 years, whereas today it is 66. In sub-Saharan Africa, however, life expectancy has changed very little over the last 30 years because of the HIV/AIDS epidemic. As a result, the life expectancy for the region is the same (46) as it was in 1975. In Russia, life expectancy has also fallen in the last two decades because of the deterioration of social services accompanying economic restructuring in the post-Soviet era.

The Demographic Transition

The historical record shows that population growth rates commonly change over time. More specifically, in Europe, North America, and Japan growth declined as countries became increasingly industrialized and urbanized. From this historical data, demographers generated the **demographic transition** model, a four-stage conceptualization that tracks changes in birthrates and death rates through time as a population urbanizes (Figure 1.26).

In the demographic transition model, Stage 1 is characterized by both high birthrates and death rates, leading to a very slow rate of natural increase. Historically, this stage is associated with Europe's preindustrial period, a period that also predated common public health measures such as sewage treatment, the scientific understanding of disease transmission, and the most fundamental aspects of modern medicine. Little wonder that death rates were high and life expectancy short. Unfortunately, these conditions are still found in some parts of the world.

In Stage 2, death rates fall dramatically while birthrates remain high, producing a rapid rise in the RNI. Again, in both historical and modern times, this decrease in death rates is usually associated with the development of public health measures and modern medicine. One of the assumptions of the demographic transition model is that these services become increasingly available after some degree of economic development and urbanization takes place.

However, even as death rates fall, it takes time for people to respond with lower birthrates, which begin in Stage 3. This, then, is the transitional stage in which people apparently become aware of the advantages of smaller families in an urban and industrial setting. In Stage 4, a very low RNI results from a combination of low birthrates and very low death rates. As a result, there is very little natural population increase. Today, the United States, Japan, and most European countries, along with other industrial nations, are clearly in Stage 4 of the demographic transition.

For many professionals, the demographic transition model is a reasonable predictor of where and when we can expect population growth rates to slow in the developing world. However, it also has detractors who see too many complicating variables and differences between countries for universal application in the twenty-first century. Regardless of this criticism, the demographic transition concept is a useful tool for organizing information about the vast differences in birthrates and death rates found around the world.

Migration Patterns

Never before in human history have so many people been on the move. Today over 190 million people live outside the country of their birth, thus are officially designated as migrants by international agencies. Much of this international migration is directly linked to the new globalized economy

Setting the Boundaries

The Metageography of World Regions

One of the perplexing issues of world regional geography is the definition and demarcation of the regions themselves. For example, historically Europe has been separated from Asia by the Straits of Bosporus, the narrow waterway that divides Turkey into European Turkey to the west and Asiatic Turkey to the east. But because Turkey is unified by language, religion, and culture, any traveler to that country can see that this traditional division is misleading.

The problem of dividing the world into a small number of regional units has been apparent since the earliest days of geography. The early Greeks, for example, conceptualized the world as divided into three continents. As seafaring explorers, they chose to divide the great land masses of Europe, Asia, and Africa based upon seas and waterways. As a result, the Bosporus and the Red Sea became the landmarks separating those continent-based world regions. Later, European explorers agreed with this threefold division even though a large country—Russia—had to be separated into European Russia west of the Ural Mountains and Asiatic Russia to the east.

In spite of this problem, European explorers continued to divide the world into continents: North and South America were referred to as separate continents although they were clearly joined by the narrow isthmus of Panama (or pan-America, as some call it); Australia was added as a continent even though many argued it was simply a large island. But if Australia was a continent, shouldn't Madagascar also be a continent? Not according to European explorers, who were quite comfortable with considering that large island nation part of the African continent. Finally, Antarctica was given continental status even though it was totally uninhabited.

This seven-part "myth of the continents" seemed primarily useful to allow Europeans to differentiate themselves from other cultures, specifically Asian and African cultures, for it was easier to draw cultural and ethnic boundaries if Asians and Africans were placed on distinct continents, separated from Europe by mountains, seas, and waterways. This scheme, however, makes little sense in today's world for there is no such thing as an Asian culture. Nor, for that matter, is there a single African culture.

World War II and the subsequent aid programs focused greater emphasis on cultural and geopolitical regions than on continents. Asia was subdivided into a handful of regions roughly comparable to historically defined civilizations; thus, East Asia, centered on China and Japan, was differentiated from South Asia, anchored by India. Though progress was made by moving away from the myth of continents, there is still a good deal of quibbling about where to draw the boundaries for different world regions. For readers interested in this topic, each of our regional chapters includes a sidebar titled "Setting the Boundaries" that examines in more detail this complicated issue of regional definition and boundary-drawing.

since half of the migrants live either in the developed world or in those developing countries with vibrant industrial, mining, or petroleum extraction economies. In the oil-rich countries of Kuwait and Saudi Arabia, for example, the labor force is composed primarily of foreign migrants. In total numbers, fully a third of the world's migrants live in seven industrial countries: Japan, Germany, France, Canada, the United States, Italy, and the United Kingdom. Moreover, most of these migrants have moved to cities; in fact, fully 20 percent of the migrants live in just 20 world cities. Further, because industrial countries usually have very low birthrates, immigration accounts for a large proportion of their population growth. For example, about one-third of the annual growth in the United States comes from in-migration.

But not all migrants move for economic reasons. War, persecution, famine, and environmental destruction cause people to flee to safe havens elsewhere. Accurate data on refugees are often difficult to obtain for a number of reasons (such as not legally crossing international boundaries or countries deliberately obscuring the number for political reasons), but UN officials estimate that some 35 million people should be considered refugees. More than half of these are in Africa and western Asia (Figure 1.27).

Push and Pull Forces While its causes may be complicated, the migration process can often be understood through three interactive concepts. First, *push* forces, such as civil strife, environmental degradation, or unemployment, drive people from their homelands. Second, *pull* forces, such as better economic opportunity or health services, attract migrants to certain locations, within or beyond their national boundaries. Connecting the two are the *informational networks* of families, friends, and

Figure 1.27 Refugees
About 35 million people are refugees from ethnic warfare and civil strife. Most of these refugees are in Africa and western Asia. These women wait in line for water at a refugee camp in southern Sudan. *(Liz Gilbert/Corbis/Sygma)*

sometimes labor contractors who provide information on the mechanics of migration, transportation details, and housing and job opportunities.

Net Migration Rates The amount of immigration (in-migration) and emigration (out-migration) is measured by the **net migration rate**, a statistic that depicts whether more people are entering or leaving a country. A positive figure means the population is growing because of in-migration, whereas a negative number means more people are leaving than arriving. As with other demographic indicators, the net migration rate is provided for the number of migrants per 1,000 of a base population. To illustrate, the net migration rate for the United States is 3.52 per 1,000 people, whereas Canada, which receives even more immigrants than the United States, has a net migration rate of 6.02. In contrast, Mexico, the source of many migrants to North America, has a net migration rate of −2.65.

Some of the highest net migration rates are found in countries that depend heavily on migrants for their labor force, such as the United Arab Emirates, with a net migration rate of 26, or Kuwait at 16. Countries with the highest negative migration rates are American Samoa, −21, Federated Micronesia, also −21, as well as a number of the Caribbean island states with net migration rates around −10.

Settlement Geography

Geography has long been interested in the different patterns of human settlement on the landscape. While in many world regions people commonly cluster together in villages and towns, in other regions—and the plains of North America are an example—people live in a more dispersed settlement pattern of widely spaced farms and homesteads. The contrast between these two settlement forms, clustered and dispersed, tells us much about different socioeconomic conditions and cultural preferences.

Settlement patterns also convey information about historical conditions, of how and when and under what conditions the countryside was settled. In many world regions—Africa, Latin America, and Asia serve as examples—European colonial powers commonly relocated indigenous people into new settlement patterns to suit their own purposes. Plantation labor needs, for example, were often satisfied by moving people from their native villages to centralized settlements. In Africa, the slave trade caused massive disruption to traditional settlement patterns as well.

Today settlement geography offers insight into the profound changes resulting from globalization (Figure 1.28). More specifically, contemporary settlement patterns give clues to many important issues such as whether people are leaving the countryside or staying; whether jobs from new plants and factories are decentralized into the countryside or agglomerated around large cities; whether mechanization is replacing farm workers who then must relocate other places; and whether civil and ethnic strife are ravaging the countryside, forcing people to flee as refugees.

Population Density Data on *population density*, which is the average number of people per area unit (square mile or square kilometer), often conveys important information about settlement in a specific country. In Table 1.1 one sees the striking difference between the high population density of India contrasted to the much lower figure for the United States. Flying over these two countries and looking down at the settlement patterns would explain this contrast. While much of the United States is covered by farms covering hundreds of acres, with houses and barns several miles from their neighbors, the landscape of India is made up of small villages, distanced from each other by only a mile or so. This results in a population density three times higher in India than in the United States. Japan has a similar settlement pattern as India even though it is primarily an urban, industrial country.

Figure 1.28 Settlement Landscapes
City and village landscapes differ widely because of the interplay between modern and historical forces that are both cultural and socioeconomic in nature. (a) A small Peruvian village in the Andes that is populated by indigenous people and expresses their local traditions. (b) A town in Venezuela, where the Spanish colonial presence is still found in the plaza, street pattern, and architecture. *(Rob Crandall/www.robcrandall.com)*

(a)

(b)

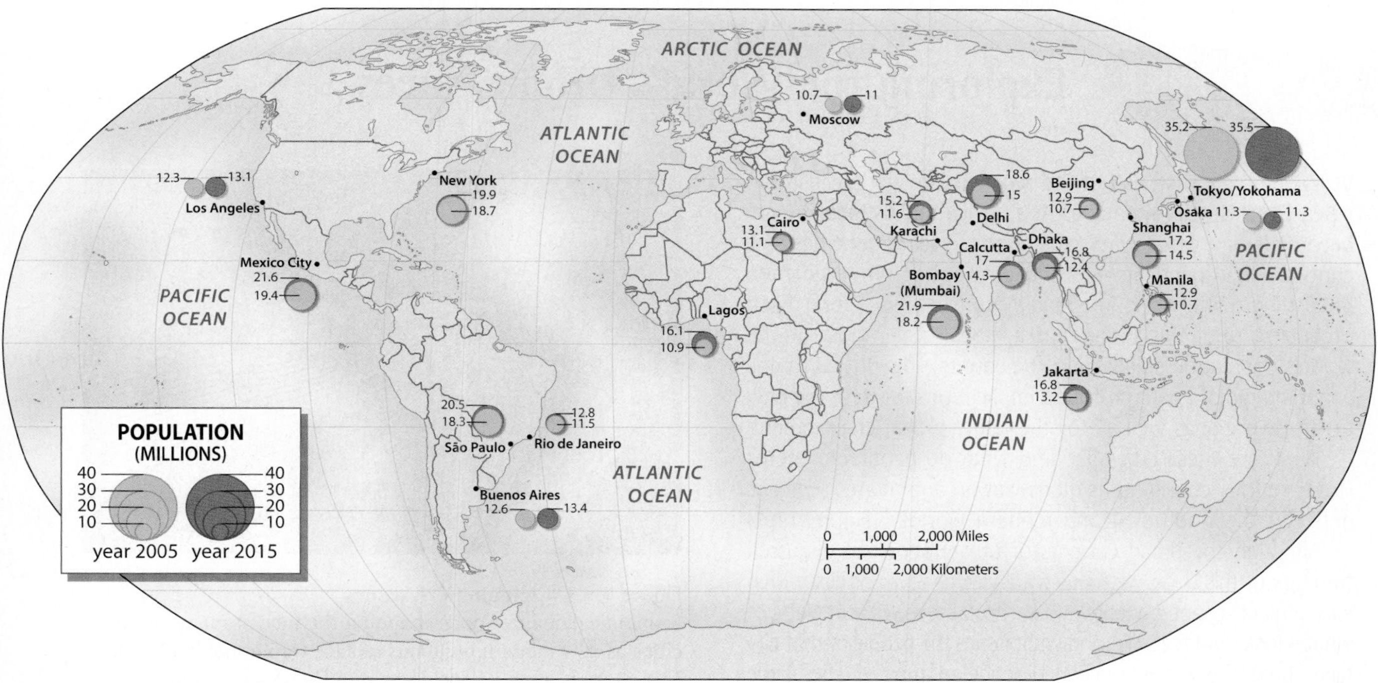

Figure 1.29 Growth of World Cities
This map shows the 20 largest cities in the world along with predicted growth by the year 2015. Note that the greatest population gains are expected in the large cities of the developing world, such as Lagos, Nigeria; Karachi, Pakistan; and Mumbai (Bombay), India. In contrast, large cities in the developed world are predicted to grow slowly over the next decades. Tokyo, for example, will add fewer than a million people.

Bangladesh, one of the most densely settled countries in the world, must squeeze its large and rapidly growing population into a limited amount of dry land built by the delta of two large rivers, the Ganges and the Brahmaputra.

While a population density figure is calculated for the total area of a country, that figure can vary greatly between rural and urban settlement. Many of the world's cities, for example, have densities of over 30,000 people per square mile (77,700 per square kilometer), with areas of Paris, Mumbai (Bombay), and Hong Kong, twice that amount. Most North American cities, in contrast, have densities of less than 10,000 people per square mile (25,900 per square kilometer) because of the cultural preference for single-family dwellings on individual lots instead of the high-rise apartment buildings that characterize most of the world's cities.

An Urban World Cities are the focal points of the contemporary, globalizing world, the fast-paced centers of deep and widespread economic, political, and cultural change (see "Cityscapes: Exploring the World's Urban Places"). Because of this vitality and the options cities offer to impoverished and uprooted rural peoples, they also are magnets for migration. The scale and rate of growth of some world cities is staggering: estimates are that both Mexico City and São Paulo (Brazil), cities of more than 20 million, are adding about 10,000 new people each week (Figure 1.29). Urban planners predict that these two cities may actually double in size within the next 15 years.

Assuming that predictions about the migration rate to cities are correct, the world has approached the point where more than half its people are urban dwellers. Evidence comes from data on the **urbanized population**, which is the percentage of a country's population living in cities. Further, many demographers predict that the world will be 60 percent urbanized by the year 2025.

Tables in this book's regional chapters include data on the urbanization rate for each country. To illustrate, more than 80 percent of the populations of Europe, Japan, Australia, and the United States live in cities. Generally speaking, most countries with such high rates of urbanization also are highly industrialized because manufacturing tends to cluster around urban centers (Figure 1.30). In

Figure 1.30 Industry and Cities
Historically, there is a close relationship between a country's level of industrial development and its level of urbanization. That relationship is linked to a number of factors, including transportation, labor, financing, and markets. This photo of a Seattle steel mill illustrates those connections. *(Dan Lamont/Corbis/Bettman)*

CITYSCAPES Exploring the World's Urban Places

Very soon more than half of the people in the world will live in cities. From North America to Asia, Africa to Australia, we are becoming an urban species, clustering in ever larger cities that continue to expand upward and outward at unprecedented rates of growth. Although many countries have long been urbanized, until this decade, at a global scale, most of the world's population still lived in the countryside. But all that is changing rapidly, with profound implications for both the new urban dwellers, as well as for the larger global environment.

What are these cities like and what do people do there? How are they changing as globalization aggravates tensions between the traditional and modern worlds? (Figure 1.5.1)

Although the broad currents of urbanization are discussed throughout this book, we focus on a specific town or city within each regional chapter to convey a particular sense of that place, of what it looks and feels like, and what issues the people of that city face. These "Cityscapes" sidebars describe an array of cities, large and small, rich and poor (but usually a combination of both), and offer a series of regional snapshots that illustrate how these special localities are growing and changing. Here's a preview:

In Chapter 3 (North America), we explore the dynamic global city of Miami, followed in Chapter 4 (Latin America) by the border city of Juarez, Mexico. In Chapter 5 (the Caribbean), Havana is described, a city closely linked to Miami in many different and complex ways. Then, in Chapter 6 (Sub-Saharan Africa) we examine the influences of globalization on the traditional city of Accra, Ghana. In Chapter 7 (North Africa), Dubai, United Arab Emirates, an old Islamic city undergoing rapid change, offers another "Cityscapes" that illustrates the profound social and cultural changes being experienced in that portion of the world.

Figure 1.5.1 Changing Cityscapes
Economic globalization is changing the face of many world cities as new modern buildings replace traditional landscapes. This photo is of new development in Quingdao, China. *(Howard Davies/Exile Images)*

Berlin, Germany, and St. Petersburg, Russia, two historic capital cities propelled into the twenty-first century by recent political changes, are highlighted in Chapters 8 (Europe) and 9 (Russia); and the Silk Road trade city of Samarkand, Uzbekistan, a UN World Heritage site, is featured in Chapter 10 (Central Asia). We visit Tokyo, one of the world's largest cities, in East Asia (Chapter 11), while Mumbai (Bombay), India, a rapidly growing world city, is described in Chapter 12 (South Asia). Manila, the dominant city of the Philippines, is the "Cityscapes" focus for Chapter 13 (Southeast Asia), while the small town of Apia, capital of Samoa, is the focus in Chapter 14 (Australia and Oceania).

contrast, the urbanized rate for developing countries is usually less than 50 percent, with figures closer to 40 percent not uncommon. Urbanization figures also show where there is high potential for urban migration. If the urbanized population is relatively small, as in Zimbabwe (Africa), where only some 34 percent of the population lives in cities, the probability of high rates of urban migration in the next decades is great.

Conceptualizing the City Given the important role of cities in world regional geography, it seems appropriate to offer a brief conceptual framework for examining urban settlements.

- Cities are interdependent with other cities and towns, both regionally and globally, and are linked in an urban system of interactive settlements. As some cities expand or come to specialize in a specific function, such as banking or manufacturing, the relationships and interactions within that urban system may change. The largest city within the city system may have **urban primacy**, a term used to describe a city that is disproportionately large and dominates economic, political, and cultural activities within the country. In many countries,

Figure 1.31 Urban Form
This photo of downtown Caracas, Venezuela, shows how city landscapes express both local and global influences. In the foreground are low-income "ranchitos" where local migrants live; in the background are the high-rise buildings of the financial center, linked to global trade. *(Rob Crandall/www.robcrandall.com)*

one large primate city usually stands out in that country's urban system. Examples include Bangkok, Mexico City, Cairo, Paris, and Buenos Aires.

- Although cities are highly complicated settlements, they often share traits of **urban structure**, a term that refers to the distribution and patterning of land use within the city. Land use is linked to specific functions, such as finance, governmental activities, housing, retailing, and industry. Almost all large cities, for example, contain a *central business district (CBD)* in which banks and business headquarters are located. Because these functions are not randomly distributed, but instead follow well-understood location principles, it is possible to generate models of urban structure that convey the commonalties between cities within a specific region. An example is the model of Latin American cities discussed in Chapter 4.
- Closely related to urban structure is the landscape of the city, the **urban form** or physical arrangement of buildings, streets, parks, and architecture that gives each city its unique sense of place. For example, Cairo, Egypt, clearly looks and feels very different from New Delhi, India, or Caracas, Venezuela (Figure 1.31). The reasons for the range of urban landscapes include differences in histories, economies, planning policies, and cultural values.
- Because urban migration has taken place so rapidly in developing countries, many cities have reached a state of **overurbanization**, in which the urban population grows more quickly than the provision of necessary support services such as housing, transportation, waste disposal, and water supply. Unfortunately, this is a fairly common situation in the developing world. The hundreds of thousands of "street people" of Kolkata (Calcutta), India are legendary, although the phenomenon of homelessness is by no means restricted to South Asia.
- Overurbanization often results in **squatter settlements**, illegal developments of makeshift housing on land neither owned nor rented by their inhabitants. Such settlements often are built on steep hillside slopes or on river floodplains that expose the occupants to the dangers of landslide and floods. Squatter settlements also are often found in the open space of public parks or along roadways, where they are regularly destroyed by government authorities, usually to be quickly rebuilt by migrants who have no other alternatives (Figure 1.32). Even for those who have found more permanent housing, water and sewer connections are often rudimentary at best. Water service can be sporadic and restricted to a few hours a day, and open sewers are a disturbingly common feature of many cities. The World Bank estimated recently that 60 percent of the people in less-developed cities lived in homes that are not connected to sewers.

Figure 1.32 Squatter Settlements
Because of the massive migration of people to world cities, adequate housing for the rapidly growing population becomes a daunting problem. Often, migrant housing needs are filled with illegal squatter settlements, such as this one in New Delhi, India. *(Porterfield/Chickering/Photo Researchers, Inc.)*

Figure 1.33 **Ethnic Tensions**
Unfortunately, much contemporary cultural change is characterized by violence between different ethnic groups. In this photo a commuter in the Indian state of Gujarat passes by a burning car resulting from rioting between Muslims and Hindus.
(AP Photo/Ajit Solanki)

CULTURAL COHERENCE AND DIVERSITY: The Geography of Tradition and Change

If culture is the weaving that keeps the world's diverse social fabric together, one glance at the daily news suggests that tapestry is unraveling because of the frequency and intensity of cultural conflict around the world. With the recent rise of global communication systems (satellite TV, films, videos, etc.), stereotypic Western culture is spreading at a rapid pace. But while some cultures accept these new cultural influences willingly, others resist what they consider to be a new form of cultural imperialism through protests, censorship, and even terrorism.

The geography of cultural cohesion and diversity, then, entails an examination of tradition and change, of language and religion, of group belonging and identity, and of those complex and varied currents that underlie and pervade twenty-first-century ethnic factionalism and separatism (Figure 1.33).

(a)

(b)

(c)

(d)

Figure 1.34 **Folk, Ethnic, Popular, and Global Culture**
A broad spectrum of cultural groups characterizes the world's contemporary cultural geography. Representing global culture (a) is a Korean woman in downtown Washington, DC, with cell phone and laptop computer. In Amsterdam, the Netherlands, two visitors (b) wear the universal clothes of popular culture, while at the opposite end of the spectrum are the ethnic cultures of (c) Garifuna women in Honduras and (d) men in Peru performing the Chonguinada dance. *(Rob Crandall/www.robcrandall.com)*

Culture in a Globalizing World

Given the diversity of cultures around the world, coupled with the dynamic changes associated with globalization, traditional definitions of culture must be stretched somewhat to provide a viable conceptual framework. **Culture** is learned, not innate, and is shared, not individual, behavior that is held in common by a group of people, empowering them with what could be called, for lack of a better term, a "way of life."

Additionally, culture has both abstract and material dimensions: speech, religion, ideology, livelihood, and value systems, but also technology, housing, foods, and music. These varied expressions of culture are relevant to the study of world regional geography because they tell us much about the way people interact with their environment, each other, and the larger world. Finally, especially given the widespread influences of globalization, it is best to think of culture as dynamic rather than static. That is, culture is a process, not a condition, and something that is constantly adapting to new circumstances. As a result, there are always tensions between the conservative, traditional elements of a culture and the newer, more progressive forces promoting change (Figure 1.34).

When Cultures Collide

Cultural change often takes place within the context of international tensions. Sometimes one cultural system will replace another; at other times resistance will stave off change. More commonly, however, a newer, hybrid form of culture results that is an amalgamation of two cultural traditions. Historically, colonialism was the most important perpetuator of these cultural collisions; today, globalization in its varied forms can be thought of as the major vehicle of cultural collision and change.

Cultural imperialism is the active promotion of one cultural system at the expense of another. Although there are still many expressions of cultural imperialism today, the most severe examples occurred in the colonial period when European cultures spread worldwide, sometimes overwhelming, eroding, and even replacing indigenous cultures. During this period, Spanish culture spread widely in Latin America; French culture diffused into parts of Africa; and British culture entered India. Of course, North America also was highly influenced by these three cultures. New languages were mandated, new education systems implanted, and new administrative institutions took the place of the old. Foreign dress styles, diets, gestures, and organizations were added to existing cultural systems. Many vestiges of colonial culture are still evident today. In India the makeover was so complete that many are fond of saying, with only slight exaggeration, that "the last true Englishman will be an Indian."

Today's cultural imperialism is seldom linked to an explicit colonizing force, but more often comes as a fellow traveler with globalization. Though many expressions of cultural imperialism carry a Western (even U.S.) tone—such as McDonald's, MTV, Marlboro cigarettes, and even the use of English as the dominant language of the Internet—these facets result more from a search for new consumer markets than from deliberate efforts to spread modern American culture throughout the world.

The reaction against cultural imperialism, **cultural nationalism**, is the process of protecting and defending a cultural system against diluting or offensive cultural expressions while at the same time actively promoting national and local cultural values. Often, cultural nationalism takes the form of explicit legislation or official censorship that simply outlaws the unwanted cultural traits. Examples of legislated cultural nationalism are common. France has long fought the Anglicization of its language by banning "Franglais," the use of English words such as "Le weekend," in official French. More recently, France has sought to protect its national music and film industries by legislating that radio DJs play a certain percentage (40 percent at this writing) of French songs and artists each broadcast day. France also is erecting tax and tariff barriers to Hollywood's products. Additionally, many Muslim countries censor Western cultural influences by restricting and censoring international TV, which they consider the source of many undesirable cultural influences. Most Asian countries, as well, are increasingly protective of their cultural values, and many are demanding changes to tone down the sexual content of MTV and other international TV networks.

As noted previously, the most common product of cultural collision is the blending of forces to form a new, synergistic form of culture. This is called **cultural syncretism or hybridization** (Figure 1.35). To characterize India's culture

Figure 1.35 Cultural Syncretism
The hybridization of culture is clearly visible in this photo of South Asians (India vs. Pakistan) playing cricket, which was, before the colonial period, a uniquely English sport. Note, too, how this traditional sport now carries the logos and advertisements of global culture. *(Arko Datta/Reuters/Corbis/Bettmann)*

as British, for example, would be to grossly oversimplify and exaggerate England's colonial influence. Instead, Indians have adapted many British traits to their own circumstances, infusing them with their own meanings. India's use of English, for example, has produced a unique form of "Indlish" that often befuddles visitors to South Asia. Nor should we forget how India has added many words to our own vocabulary—*khaki*,. *pajamas*, *veranda*, *bungalow*. In sum, both the Anglo and Indian cultures have been changed by the British colonial presence in South Asia.

Language and Culture in Global Context

Language and culture are so intertwined that, in the minds of many, language is the characteristic that best differentiates and defines cultural groups (Figure 1.36). Furthermore, because language is the primary means for communication, it obviously folds together many other aspects of cultural identity, such as politics, religion, commerce, and customs. Additionally, language is fundamental to cultural cohesiveness. It not only brings people together but also sets them apart; it can be an important component of national or ethnic identity and a way of creating and maintaining boundaries of regional identity.

Because most languages have common historical (and even prehistoric) roots, linguists have grouped the thousands of languages found throughout the world into a handful of language families. These are simply a first-order grouping of languages into large units based upon common ancestral speech. For example, about half of the world's people speak languages of the Indo-European family, which includes not only European languages such as English and Spanish, but also Hindi and Bengali, widespread languages of South Asia.

Within language families are smaller units that also give clues to the common history and geography of peoples and cultures. *Language branches and groups* (also called *subfamilies*) are closely related subsets within a language family in which there are usually similar sounds, words, and grammar. Well known are the similarities between German and English, or between French and Spanish. Because of those similarities, these languages are placed into the same linguistic grouping.

Even individual languages often have very distinctive forms associated with specific regions. Such different forms of speech are called *dialects*. Although dialects of the same language have their own unique pronunciation and grammar (think of the distinctive differences, for example, between British, North American, and Australian English), they are—sometimes with considerable effort—mutually intelligible.

Additionally, when people from different cultural groups cannot communicate directly in their native languages, they

Figure 1.36 World Languages
Most languages of the world belong to a handful of major language families. About 50 percent of the world's population speaks a language belonging to the Indo-European language family, which includes languages common to Europe, but also major languages in South Asia, such as Hindi. They are in the same family because of their linguistic similarities. The next largest family is the Sino-Tibetan family, which includes languages spoken in China, the world's most populous country. *(Adapted from Rubenstein, 2005,* The Cultural Landscape: An Introduction to Human Geography, *8th ed., Upper Saddle River, NJ: Prentice Hall)*

Figure 1.37 English as Global Language
A bilingual traffic sign in Dubai, United Arabic Emirates, is a reminder that English has become a universal form of communication for transportation, business, and science. *(Rob Crandall/ www.robcrandall.com)*

often agree upon a third language to serve as a common tongue, or **lingua franca**. Swahili has long served that purpose between the many tribal languages of eastern Africa, and French was historically the lingua franca of international politics and diplomacy. Today English is increasingly the common language of international communications (Figure 1.37).

A Geography of World Religions

Another extremely important defining trait of cultural groups is religion (Figure 1.38). Indeed, in this era of a totalizing global culture, religion is becoming increasingly important in defining cultural identity. Recent ethnic violence and unrest in far-flung places such as the Balkans, Afghanistan, and Indonesia illustrate the point.

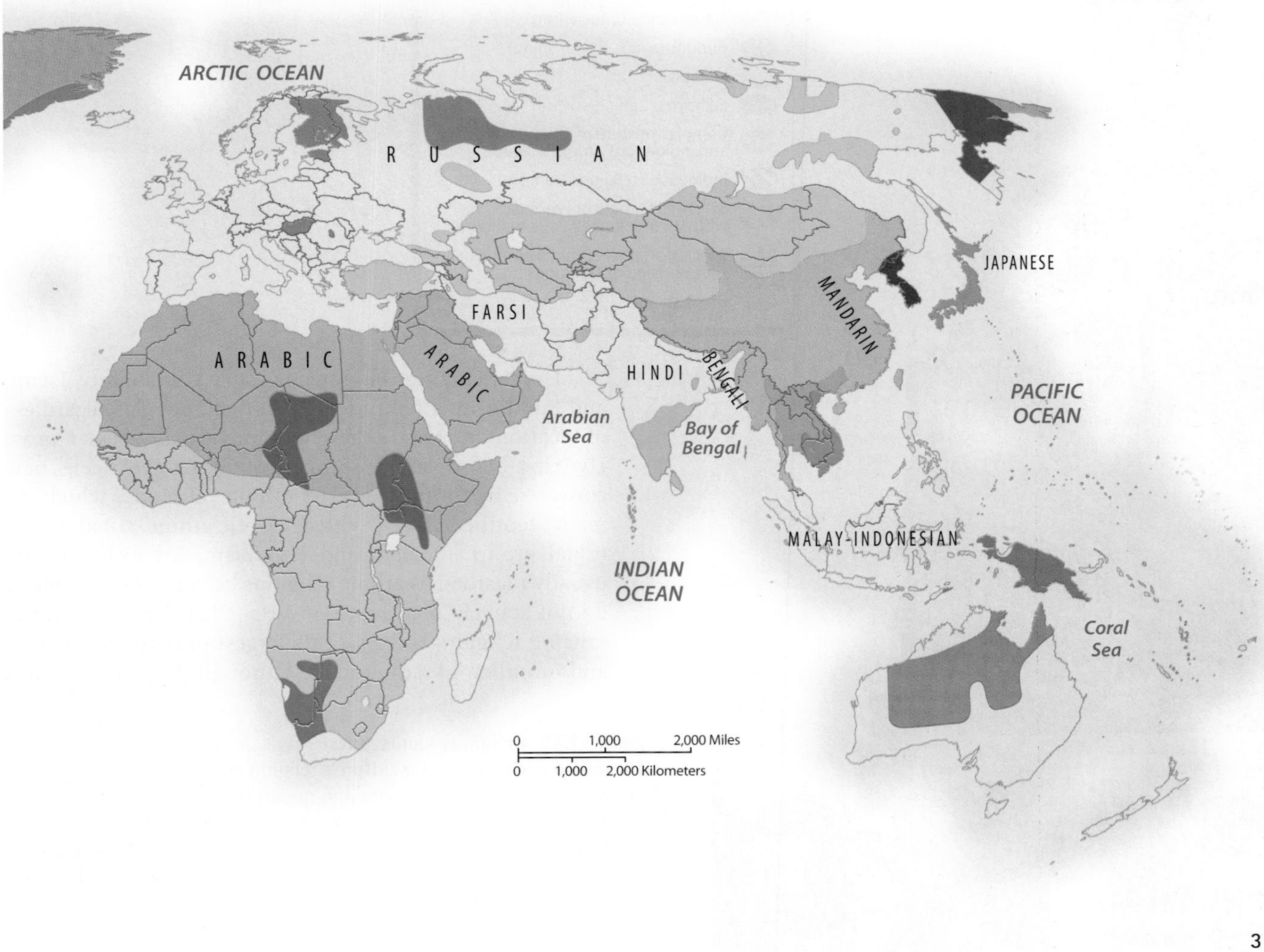

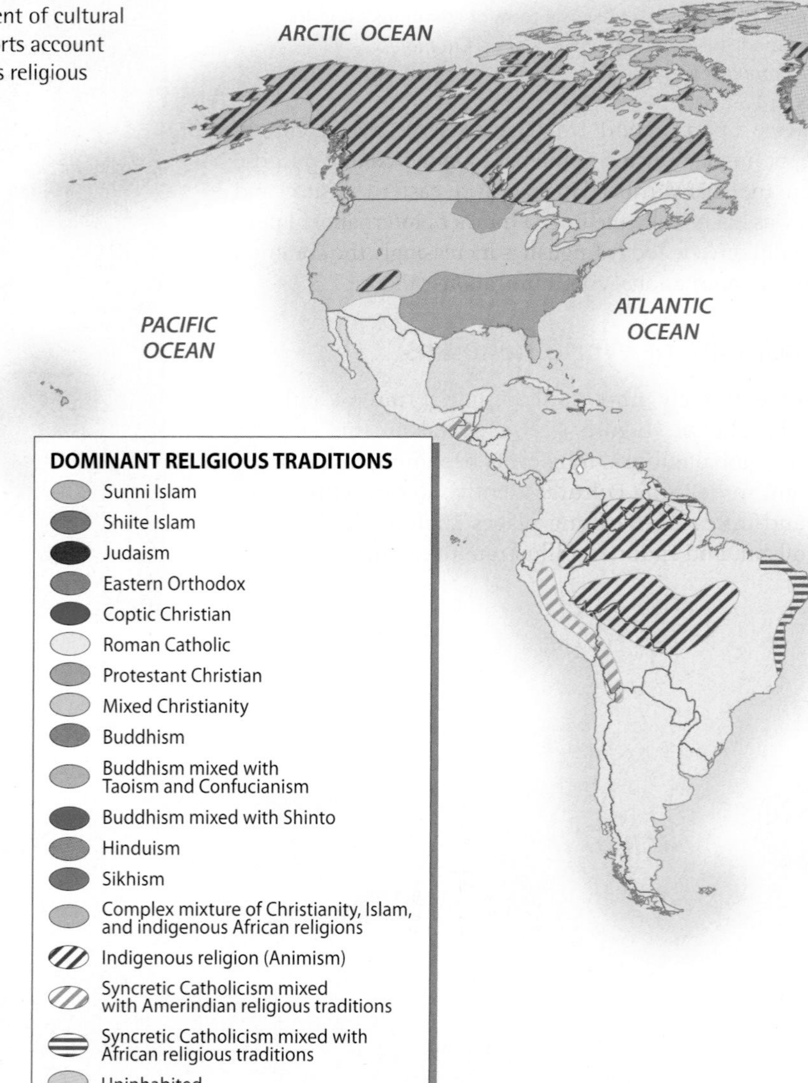

Figure 1.38 Major Religious Traditions
This map shows the major religions found throughout the world. For most people, religious tradition is a major component of cultural and ethnic identity. While Christians of different sorts account for about 34 percent of the world's population, this religious tradition is highly fragmented. Within Christianity, there are about twice as many Roman Catholics as Protestants. Islam accounts for about 20 percent of the world's population; Hindus make up about 14 percent.

Universalizing religions, such as Christianity, Islam, and Buddhism, attempt to appeal to all peoples regardless of location or culture; these religions usually have a proselytizing or missionary program that actively seeks new converts. In contrast are the **ethnic religions**, which remain identified closely with a specific ethnic, tribal, or national group. Judaism and Hinduism, for example, are usually regarded as ethnic religions because they normally do not actively seek new converts. Christianity, a universalizing religion, is the world's largest in both areal extent and number of adherents. Although fragmented into

Figure 1.39 Religious Landscapes
The varied expressions of a culture's religion commonly appear in the landscape, be they the mosques and minarets of Islam, the churches and cathedrals of Christianity, the synagogues of Judaism, the shrines of Hinduism, or the temples and statues of Buddhism. An example is the statue of the sea deity, *Kwun Yum*, at the Tin Hau Temple on Repulse Bay, Hong Kong. *(Rob Crandall/ www.robcrandall.com)*

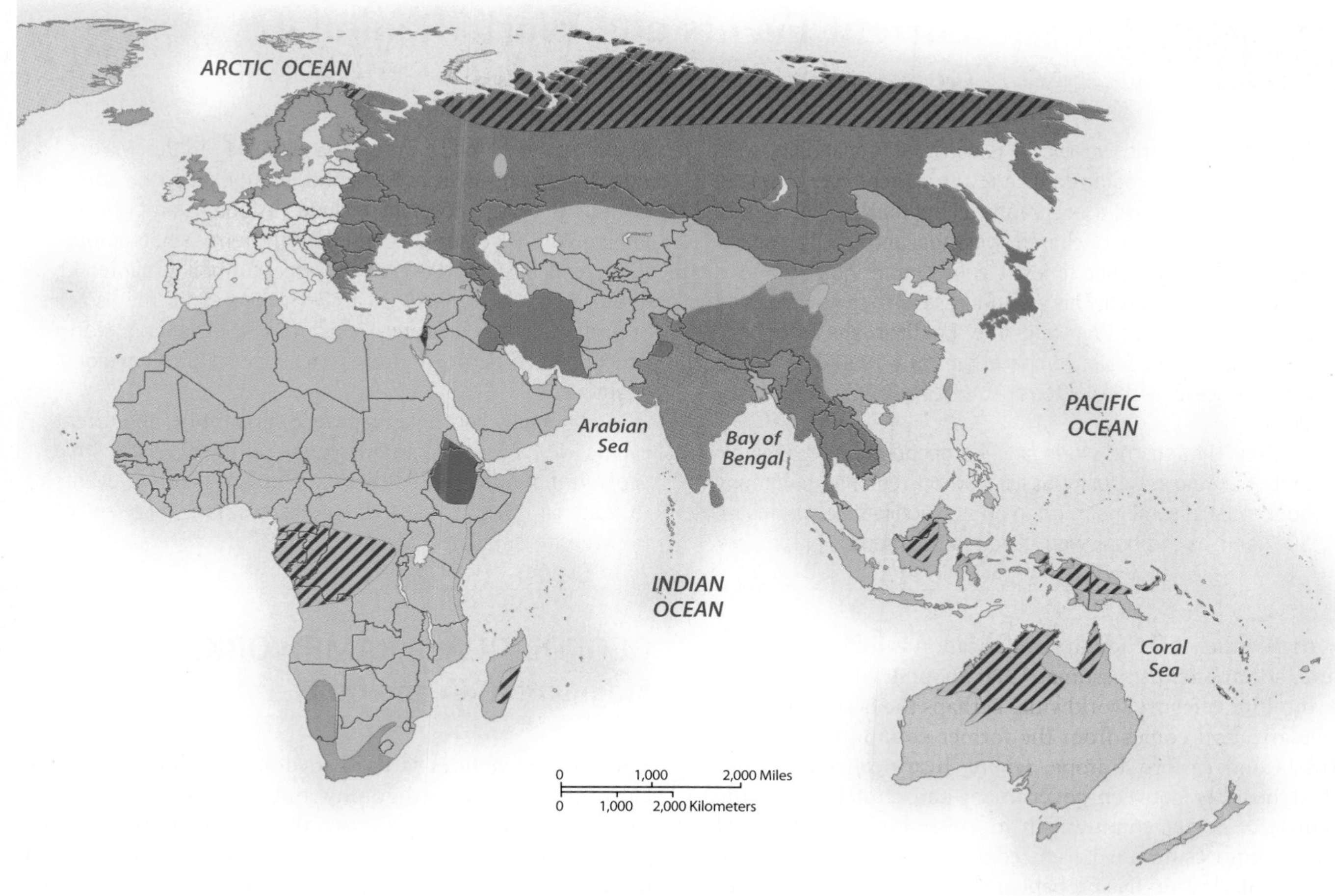

separate branches and churches, Christianity as a whole has 2.1 billion adherents, or about a third of the world's population. The largest number of Christians can be found in Europe, Africa, Latin America, and North America. Islam, which has spread from its origins on the Arabian Peninsula as far east as Indonesia and the Philippines, has about 1.3 billion members.

While not as severely fragmented as Christianity, Islam should not be thought of as a homogeneous religion because it also has split into separate groups. The major branches are Shiite Islam, which constitutes about 11 percent of the total Islamic population and represents a majority in Iran and southern Iraq, and the dominant Sunni Islam, which is found from the Arab-speaking lands of North Africa to Indonesia. Both of these forms of Islam are experiencing a fundamentalist revival in which proponents are interested in maintaining a purity of faith distanced from Western influences.

Judaism, the parent religion of Christianity, also is closely related to Islam. Although tensions are often high between Jews and Muslims because of the Israeli–Palestinian conflict, these two religions, along with Christianity, actually share historical and theological roots in the Hebrew prophets and leaders. Judaism now numbers about 14 million adherents, having lost perhaps a third of its total population from the systematic extermination of Jews by the Nazis during World War II.

Hinduism, which is closely linked to India, has about 900 million adherents. Outsiders often regard Hinduism as polytheistic because Hindus worship many deities. Most Hindus argue, however, that all of their faith's gods are merely representations of different aspects of a single divine, cosmic unity. Historically, Hinduism is linked to the caste system with its segregation of peoples based upon ancestry and occupation. Today, however, since India's democratic government is committed to reducing the social distinctions between castes, the connections between religion and caste are now much less explicit than in the past.

Buddhism, which originated as a reform movement within Hinduism 2,500 years ago, is widespread in Asia, extending from Sri Lanka to Japan and from Mongolia to Vietnam (Figure 1.39). In its spread, Buddhism came to coexist with other faiths in certain areas, making it difficult to accurately estimate the number of its adherents. Estimates of the total Buddhist population thus range from 350 million to 900 million people.

Finally, in some parts of the world, religious practice has declined for different reasons, giving way to **secularization**, in which people consider themselves either nonreligious or

GEOGRAPHY IN THE MAKING

Current Events and World Regional Geography

Because the world we live in is constantly changing, so is world regional geography. As a result, current events ensure that there is always a geography in the making. Sometimes the pace and nature of change is gradual and takes a familiar course, while at other times it comes more quickly and is largely unanticipated. This was the case with the unexpected end of the Cold War and the political and economic devolution of the Soviet Union in the early 1990s, and it was also the case with the terrorist attacks of September 11, 2001.

Even though the world and its geography are dynamic and changing on a daily basis, a world regional geography book must at some point go to press. Our deadline was in late 2007 and, as this book went through the production process, space was held open in the "Geography in the Making" sidebars for late-breaking current events. Some of the rapidly changing topics covered include "Afghanistan's Never-Ending War" (Chapter 10, Central Asia); "Terrorism in Europe" (Chapter 8, Europe); "Pakistan's Political Turmoil" (Chapter 12, South Asia), and "Walling in Bethlehem and the Israeli Security Barrier" (Chapter 7, Southwest Asia and North Africa). Other pertinent topics are covered in all regional chapters.

Additionally, knowing that we can't capture all current events, we have set up a new section on the *Diversity Amid Globalization* Website that links and updates current events to material in the book. This current events component of the Website is updated weekly.

outright atheistic. Though secularization is difficult to measure, social scientists estimate that about 1.1 billion people fit into this category worldwide. Perhaps the best example of secularization comes from the former communist lands of Russia and eastern Europe, where, historically, there was overt hostility between government and church. Since the demise of Soviet communism in 1991, however, many of these countries have experienced a modest religious revival.

Secularization has probably grown more pronounced in recent years in western Europe. Although historically and to some extent still culturally a Roman Catholic country, France now reportedly has more people (mostly migrants) attending Muslim mosques on Fridays than Christian churches on Sundays. Japan and the other countries of East Asia are also noted for their high degree of secularization.

GEOPOLITICAL FRAMEWORK: Fragmentation and Unity

The term *geopolitics* is used to describe and explain the close link between geography and politics. More specifically, geopolitics focuses on the interactivity between power, territory, and space at all scales, from the local to the global. There is little question that one of the dominant characteristics of the last several decades has been the speed, scope, and character of political change in various regions of the world (see "Geography in the Making: Current Events and World Regional Geography").

With the end of the Soviet Union in 1991 came opportunities for self-determination and independence in

Figure 1.40 End of the Cold War With the fall of Soviet communism, many nations rushed to remove the symbols of their former governments to forget the past and make room for a new future. In this photo taken in 1990, a monument to Lenin is toppled in Bucharest, Romania. *(Bi/Getty Images, Inc.—Liaison)*

Figure 1.41 Global Terrorism
Innocent people throughout the world have fallen victim to extremists attempting to advance their cause through acts of terrorism. England, for example, has suffered numerous attacks by groups advocating independence for Northern Ireland, as well as more recent attacks by terrorists allied with Islamic extremism. *(Peter Macdiarmid/GettyImages, Inc.—Liaison)*

eastern Europe and Central Asia, resulting in fundamental changes to economic, political, and even cultural alignments (Figure 1.40). Religious freedom helps drive national identities in some new Central Asian republics, but eastern Europe seems primarily concerned with economic and political links to western Europe. Russia itself wavers perilously between different geopolitical pathways. Phrases and terms common to an earlier generation may disappear; for example, the "Cold War" between the two superpowers, the United States and Soviet Union, has already become a historical artifact.

While wide-ranging international conflicts remain a nagging concern of diplomats and governments, perhaps more common are strife and tension within—rather than between—nation-states. Civil unrest, tribal tensions, terrorism, and religious factionalism have created a new fabric and scale of political tension.

Global Terrorism

In many ways, the September 2001 terrorist attacks on the United States underscore the need to expand our conceptualization of globalization and geopolitics. Previous terrorist acts were usually connected to nationalist or regional geopolitical aspirations to achieve independence or autonomy; the attacks on the World Trade Center and the Pentagon were different. Unlike Irish Republican Army (IRA) terrorist bombings in Great Britain or those of the extremist wing of Basque nationalists in Spain, the September 2001 terrorism went beyond conventional geopolitics as a small group of religious fanatics attacked the symbols of Western culture, power, and finance. Moreover, these attacks were a stark reminder of the increasingly close and often unpredictable interconnections between political activity, cultural identity, and the economic linkages that bind our contemporary world.

In geopolitical terms, global terrorism demands a new way of looking at the world to understand the attacks and assess their implications. More to the point, many experts argue that global terrorism is both a product of as well as a reaction to globalization (Figure 1.41). Unlike earlier geopolitical conflicts, the geography of global terrorism is not defined by a war between well-established political states. Instead, the Al Qaeda terrorists appear to belong to a web of small, well-organized cells located in many different countries. These cells are linked in a decentralized network that provides guidance, financing, and political autonomy. This network has used the tools and means of globalization to its advantage. Members communicate instantaneously via cell phones and the Internet. Transnational members travel between countries quickly and frequently. The network's activities are financed through a complicated array of holding companies and subsidiaries that traffic in a range of goods, such as honey, diamonds, and opium. To recruit members and political support, the network feeds upon the unrest and inequities (real and imagined) resulting from economic globalization. The network's terrorist acts then target symbols of those modern global values and activities it opposes.

Even though the 9/11 attacks focused terror on the United States, the casualties and resultant damage were truly international as citizens from more than 80 countries were killed in the World Trade Center tragedy. Additionally, the attacks pushed parts of the global economy from temporary economic slowdown to full-on recession. The devastating effects on U.S. and international airline companies as well as the loss of tourist revenue to moderate Muslim countries such as Morocco and Egypt serve as examples.

The military and political responses to global terrorism also demand an expanded conceptualization of geopolitics. In 2001, in Afghanistan, the military muscle of a nation-state superpower—the United States—was directed not at another nation-state but, rather, at a confederation of religious extremists, the Taliban and Al Qaeda. Military strategists refer to this kind of fighting as **asymmetrical warfare**, a term that aptly describes the differences between a superpower's military technology and strategy and the lower level technology and guerilla tactics used by Al Qaeda and the Taliban. Most superpower military strategists agree that the war on terrorism will be fought with this sort of political and battlefield asymmetry.

Nation-States

A traditional reference point for examining political geography is the concept of the nation-state. The hyphen links two concepts: the term *state*, which is a political entity with territorial boundaries recognized by other countries and internally governed by an organizational structure, and the term *nation*, referring to a large group of people who share numerous cultural elements such as language, religion, tradition, and simple cultural identity. The **nation-state**, then, is ideally a relatively homogeneous cultural group with its own fully independent political territory.

Nations, however, do not always fit neatly into the boundaries of actual states. In fact, perfect nation-state congruence is

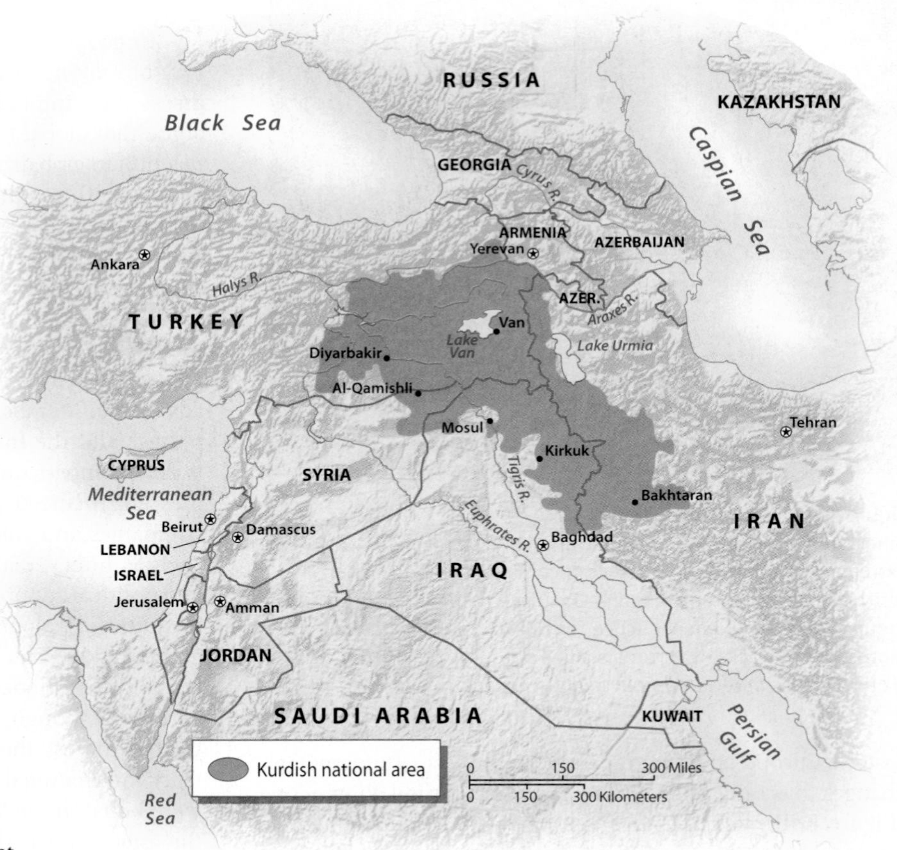

Figure 1.42 Nation Without a State Not all nations or large cultural groups control their own political territories, thus are without a state. As this map shows, the Kurdish people of Southwest Asia occupy a larger cultural territory that is in four different political states—Turkey, Iraq, Syria, and Iran. As a result of this political fragmentation, the Kurds are considered a minority in each of those four countries.

relatively rare. While some large cultural groups may consider themselves to be nations lacking recognized, self-governed territory (such as the Kurdish people of Southwest Asia), many, if not most independent states include cultural and ethnic groups within their established boundaries that seek autonomy and self-rule (Figure 1.42). In Spain, for example, both the Catalans and the Basques think of themselves as forming separate, non-Spanish nations, and thus seek political autonomy from the central government. On a world scale, out of the more than 200 different countries that now make up the global geopolitical fabric, only several dozen would qualify as true nation-states in the narrow sense of the word.

Centrifugal and Centripetal Forces

Cultural and political forces acting to weaken or divide an existing state are called **centrifugal forces** because they pull away from the center. Many have already been mentioned: a linguistic minority, ethnic separatism, the desire for territorial autonomy, along with disparities in income and well-being. Separatist tendencies in French-speaking Quebec (Canada) or the Basque region in Spain are good examples (Figure 1.43). Counteracting these dissipating forces are those that promote political unity and reinforce the state structure. These **centripetal forces** could include a shared sense of history, a need for military security, a coordinated economic structure, or simply the advantages of a larger political organization to maintain the infrastructure of highways, airports, and schools.

The overriding question in much of the world is whether centrifugal forces will increase in strength and furor so that they overcome the centripetal forces, thus leading to new, smaller independent units. Also important is whether this process takes place through violent struggle or peaceful means. Given the current widespread nature of these internal tensions, separatist struggles will continue to dominate regional geopolitics for decades to come.

Figure 1.43 Basque Separatism The demands for independence of the Spanish Basques in northeastern Spain are just one example of the ethnic separatism common in much of the world. *(AlbertoArzoz/Panos Pictures)*

Boundaries and Frontiers

A key component of the definition of a state is that its territory is demarcated by borders recognized by most other states. Although these boundaries may be one of the most recognizable features of a world map, the issue of agreed-upon borders is often complicated and contentious. Border wars were frequent in the twentieth century, and if the current trend toward an increasing number of independent states continues, boundary disputes may become an even more important part of the geopolitical landscape.

A cursory glance at the world political map reminds us that boundaries come in many forms. Some are drawn along physical features such as mountain ridges, rivers, or coastlines. Although historically it was thought that using clearly marked landscape features would lessen border tensions, this is not always the case. The Rio Grande between Texas and Mexico is notoriously porous to illegal boundary crossings. Similarly, the sinuous coastline of southern Spain offers North Africans ample opportunity for illegal entry into Europe. **Ethnographic boundaries**, which follow cultural traits such as language or religion, have long been a model for international politics because they complement the ideal of the nation-state. Many boundaries in Europe were adjusted ethnographically after World War I. Similarly, attempts were made to define the new states of Bosnia and Serbia with ethnographic boundaries after the dissolution of Yugoslavia in the early 1990s.

In contrast, **geometric boundaries** are perfectly straight lines, drawn without regard for physical or cultural features, that usually follow a parallel of latitude or a meridian of longitude. Historically, geometric boundaries often were drawn by colonial powers as a convenient way of quickly demarcating spheres of influence; the use of the 49th parallel to draw the western part of the boundary between Canada and the United States is an excellent example. Other illustrations can be found in the arid lands of North Africa.

Colonialism and Decolonialization

One of the overarching themes in world regional geography is the waxing and waning of European colonial power over much of the world. **Colonialism** refers to the formal establishment of rule over a foreign population. A colony has no independent standing in the world community, but instead is seen only as an appendage of the colonial power. Generally speaking, the main period of colonialization by European states was from 1500 through the mid-1900s, though even today a few colonies remain (Figure 1.44).

Decolonialization refers to the process of a colony's gaining (or regaining) control over its territory and establishing a separate, independent government. As was the

Figure 1.44 The Colonial World, 1914
This world map shows the extent of colonial power and territory just prior to World War I. At that time, most of Africa was under colonial control, as were Southwest Asia, South Asia, and Southeast Asia. Australia and Canada were very closely aligned with England. Also note that in Asia, Japan controlled colonial territory in Korea and northeastern China, which was known as Manchuria at that time.

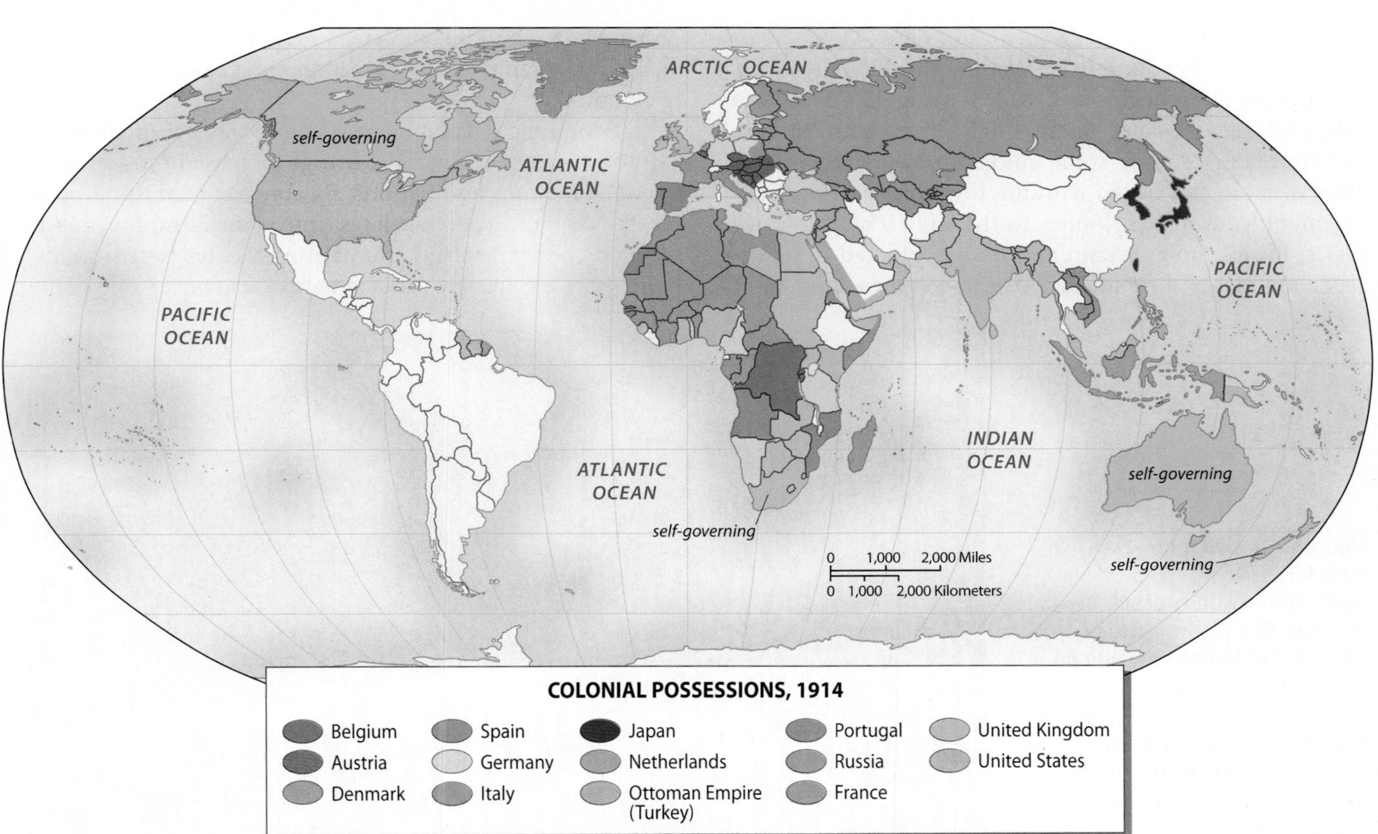

Figure 1.45 **Colonial Vestiges in Vietnam**
The red star flag of communist Vietnam flies in front of the Hotel de Ville in Ho Chi Minh City (formerly Saigon). This juxtaposition of the contemporary government's symbol and an artifact of the French colonial period captures the process of decolonialization and independence. *(Catherine Karnow/Woodfin Camp & Associates)*

case with the Revolutionary War in the United States, this process often begins as a violent struggle. As wars of independence became increasingly prevalent in the mid-twentieth century, some colonial powers recognized the inevitable and began working toward peaceful disengagement from their colonies. In the late 1950s and early 1960s, for example, Britain and France granted independence to most of their former African colonies, often after a period of warfare or civil unrest. This process was nearly completed in 1997 when Hong Kong was peacefully restored to China by the United Kingdom.

But decades and even centuries of colonial rule are not easily erased, thus the influence of colonialism is still commonly found in the new nations' governments, education, agriculture, and economies. While some countries may enjoy special status with, and receive continued aid from, their former colonial masters, others remain disadvantaged because of a reduced resource base. Some scholars regard the continuing economic ties between certain imperial powers and their former colonies as a form of exploitative neocolonialism. On the other hand, some remaining colonies, such as the Dutch Antilles in the Caribbean, have made it clear that they have no wish for independence, finding both economic and political advantages in continued dependency. Because the consequences of colonialism differ greatly from place to place, the final accounting of its effects is far from complete (Figure 1.45).

ECONOMIC AND SOCIAL DEVELOPMENT: The Geography of Wealth and Poverty

The pace of global economic change and development has accelerated dramatically in the last several decades. As a result, we now talk about global assembly lines, outsourcing, **transnational corporations**, commodity chains, international electronic offices, and so on. The overarching question, though, is whether the positive changes of economic globalization outweigh the negative. Answers vary and are often elusive and incomplete; a place to begin in world regional geography is to link economic change to social development.

Economic development is commonly accepted as desirable because it generally brings increased prosperity to people, regions, and nations. By conventional thinking at least, this usually translates into social improvements such as better health care, improved education systems, and more progressive labor practices. However, one of the

Figure 1.46 **Living on Less Than $2 a Day**
Over half the world's people live on less than $2 a day, which is the United Nation's definition for poverty. In South Asia, where these workers are making bricks, about three-quarters of the population is considered impoverished. *(Ajay Werma/Reuters/Corbis)*

more troubling expressions of recent economic growth has been the geographic unevenness of prosperity and social improvement. While some regions prosper, others languish and, in fact, fall farther behind. As a result, the gap between rich and poor regions has increased over the last several decades and this economic and social unevenness has, unfortunately, become a signature of globalization. According to the World Bank, about half the people in the world live on less than $2 a day, which is the commonly accepted definition of poverty (Figure 1.46).

These inequities are problems because of their inseparable interaction with political, environmental, and social issues. Political instability and civil strife within a nation, for example, are often driven by the economic disparity between a poor periphery and an affluent, industrial core. In the periphery, poverty, social tensions, and environmental degradation often cause civil unrest that ripples through the rest of the country. Because this social and economic unevenness is so prevalent, it is a major theme in the regional chapters of this book.

More- and Less-Developed Countries

Until the last century, economic development was centered in North America, Japan, and Europe, while most of the rest of the world remained gripped in poverty. This uneven distribution of economic power led scholars to devise a **core-periphery model** of the world. According to this scheme, the United States, Canada, western Europe, and Japan constituted the global economic core of the North, whereas most of the areas to the south made up a less-developed global periphery. Although oversimplified, this core-periphery dichotomy does contain some truth. All the G-8 countries—the exclusive club of the world's major industrial nations made up of the United States, Canada, France, England, Germany, Italy, Japan, and Russia—are located in the Northern Hemisphere. Additionally, many critics postulate that the developed countries achieved their wealth primarily by exploiting the poorer countries of the southern periphery, both historically through colonial relationships and still today with economic imperialism.

As a result, much is made today of "North–South tensions," a phrase implying that the rich and powerful countries of the Northern Hemisphere are still at odds with the poor and less powerful countries of the south. Over the past several decades, however, the global economy has grown much more complicated. A few former colonies of the south, most notably Singapore, have become very wealthy. Additionally, a few northern countries, namely Russia, have experienced economic decline in the last decades. Further, the developed countries of Australia and New Zealand never fit into the North–South division because of their Southern Hemisphere location. For these reasons, many global experts conclude that the term *North–South* is outdated and, thus, should be avoided. We agree.

The *Third World* is another term often used to refer to the developing world. This phrase suggests a low level of economic development, unstable political organizations, and a rudimentary social infrastructure. Historically, the term came from the Cold War vocabulary used to describe those countries that were independent and not allied with either the capitalist and democratic First World or the communist Second World superpowers of the Soviet Union and China. Today, however, because the Soviet Union no longer exists and China has changed its economic orientation, the term *Third World* has lost its original meaning. In this book, therefore, we also avoid the term and instead use relational terms that capture the complex spectrum of economic and social development—*more-developed country (MDC)*, and *less-developed country (LDC)*. The global pattern of more- and less-developed countries can be inferred from a map of gross national income (Figure 1.47), one of several indicators used to assess development and economic wealth.

Indicators of Economic Development

The terms *development* and *growth* often are used interchangeably when referring to international economic activities. There is, however, value in keeping them separate. *Development* has both qualitative and quantitative dimensions. Common dictionary definitions use phrases such as "expanding or realizing potential; bringing gradually to a fuller or better state." When we talk about economic development, then, we usually imply structural changes, such as a shift from agricultural to manufacturing activity that also involves changes in the allocation of labor, capital, and technology. Along with these changes are assumed improvements in standard of living, education, and even political organization. The structural changes experienced by Southeast Asian countries such as Thailand or Malaysia in the last several decades illustrate this process.

Growth, in contrast, is simply the increase in size of a system. The agricultural or industrial output of a country may grow, as it has for India in the last decade, and this growth may—or may not—have positive implications for development. Many growing economies, in fact, have actually experienced increased poverty with economic expansion. When something grows, it gets bigger; when it develops, it improves. Critics of the world economy are often heard to say that we need less growth and more development.

In this book, each of the regional chapters includes a table of economic and development indicators (Table 1.2). A few introductory comments are necessary to explain these data.

Gross Domestic Product and Income The traditional measure for the size of a country's economy is the value of all final goods and services produced within its borders, called its **gross domestic product** or **GDP**. When combined with net income from abroad, this domestic income constitutes a country's **gross national income**, or **GNI** (formerly referred to as gross national product, or GNP). Although commonly used, GNI is still an incomplete and sometimes misleading economic indicator because it ignores nonmarket economic activity such as bartering or household work, and also because it does not take into

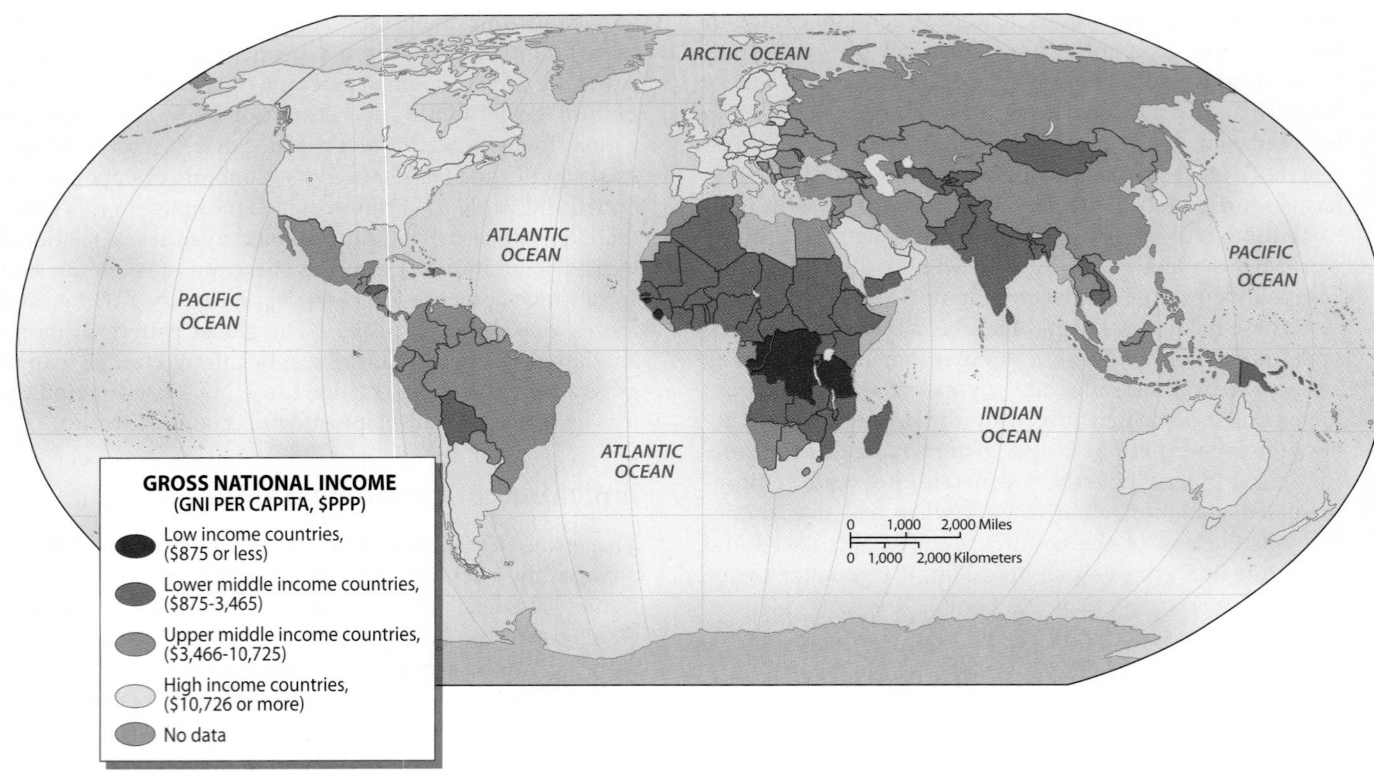

Figure 1.47 More- and Less-Developed Countries
Based upon GNI per capita, PPP adjusted, one sees the global pattern of more- and less-developed countries (MDCs and LDCs). Africa and the different parts of Asia stand out as the regions with the greatest number of developing countries.

account ecological degradation or depletion of natural resources. For example, if a country were to clear-cut its forest, an activity that could severely limit future growth if forest resources were in short supply, this cutting would nevertheless increase the GNI for that particular year. Further, diverting educational funds to purchase military weapons might increase a country's GNI in the short run, but the economy would likely suffer in the future because of its less well-educated population. In other words, the GNI is a snapshot of a country's economy for a specific period, not an infallible indicator of continued vitality or social well-being.

Because gross national income data vary widely between countries, and, additionally, are commonly expressed in mind-boggling numbers such as trillions and billions of

Figure 1.48 Village Food Market
The concept of purchasing power parity (PPP) is used to generate a sense of the true cost of living for different countries so that economic data can be adjusted for currency inflation. This outdoor produce market is in Chichicastenango, Guatemala. *(Rob Crandall/www.robcrandall.com)*

TABLE 1.2 Development Indicators of the Largest 10 Countries

Country	GNI[a] per Capita, PPP[b] (2005)	GDP[c] Average Annual % Growth (2000–05)	Life Expectancy (2007)	Percent of Population Living on Less Than $2 a Day	Under Age 5 Mortality Rate 1990	Under Age 5 Mortality Rate 2005	Gender Equity[d]
China	6,600	9.6	72	47	49	27	98
India	3,460	7.0	64	80	123	74	87
United States	41,950	2.6	78	—	11	7	109
Indonesia	3,720	4.7	69	52	91	36	97
Brazil	8,230	2.2	72	21	60	33	105
Pakistan	2,350	4.8	62	74	130	99	75
Bangladesh	2,090	5.4	62	83	149	73	101
Nigeria	1,040	5.9	47	92	230	194	82
Russia	10,640	6.2	65	12	27	18	110
Japan	31,410	1.1	82	—	6	4	98

[a]*Gross national income.*
[b]*Purchasing power parity.*
[c]*Gross domestic product.*
[d]*Ratio of female-to-male enrollments in primary and secondary school percentage. Numbers below 100 have more males in primary/secondary school; numbers above 100 have more females in primary/secondary schools.*
Source: World Bank, World Development Indicators, *2007; life expectancy 2007 and percentage of population living on $2 a day data from Population Bureau,* World Data Sheet, 2007.

dollars, a better comparison is to divide the GNI by the country's population, thereby generating a **gross national income (GNI) per capita** figure. This way one can compare large and small economies in terms of how they may (or may not) be benefiting the population. For example, the annual GNI for the United States is over $10 trillion. Dividing that figure by the population of 302 million results in a GNI per capita of about $42,000. Japan has a GNI about half the size of the United States; however, because that country also has a much smaller population of 128 million, its unadjusted GNI per capita is about $39,000, thus, one could conclude the economies are comparable.

An important qualification to these GNI per capita data is the concept of adjustment through **purchasing power parity (PPP)**, an adjustment that takes into account the strength or weakness of local currencies. When not adjusted by PPP, GNI data are based on the market exchange rate for a country's national currency as compared to the U.S. dollar. As a result, the GNI data might be inflated or undervalued depending on the strength or weakness of that currency. If the Japanese yen were to fall overnight against the dollar as a result of currency speculation, Japan's GNI would correspondingly drop despite the fact that Japan had experienced no real decline in economic output. Because of these possible distortions, the PPP adjustment provides a more accurate sense of the local cost of living (Figure 1.48). To illustrate, when Japan's GNI per capita is adjusted for PPP, which takes out the inflationary factor, the figure is $31,410, which is $8,000 lower than the unadjusted GNI per capita figure.

Economic Growth Rates The rate of economic growth of a country is measured by the average annual growth of the GDP over a five-year period, a statistic called *GDP average annual percent growth.* The average growth rate for developing countries such as China, India, and Nigeria is considerably higher than those of the developed countries of the United States and Japan (see Table 1.2). This difference in economic growth rates is expected given that developing countries are just that, developing; thus one would expect a higher annual growth rate than from a mature, developed economy like that of the United States. Nevertheless, these GDP growth data usually depict more than simply the difference between developed and developing economies. For example, the low growth rate for Japan worries many economists with concerns over economic stagnation. As well, the low growth rate for Brazil—which should be higher as a developing country—is also an issue. Russia's economy, which shows a healthy annual growth rate, indicates recovery from the meltdown of the 1990s following the end of the Soviet Union.

Indicators of Social Development

Although economic growth is a major component of development, equally important are the conditions and quality of human life. As noted earlier, the standard assumption is that economic development and growth will spill over into the social infrastructure, leading to improvements in public health, gender equity, and education. Unfortunately, even the briefest glance at the world reveals that poverty, disease, illiteracy, and gender inequity are still widespread despite a booming global economy. Even in China, which has experienced unprecedented economic growth in the last decades, half the population is still impoverished; in Pakistan, two-thirds live below the poverty line of $2 per day, and in some African countries that figure approaches 90 percent.

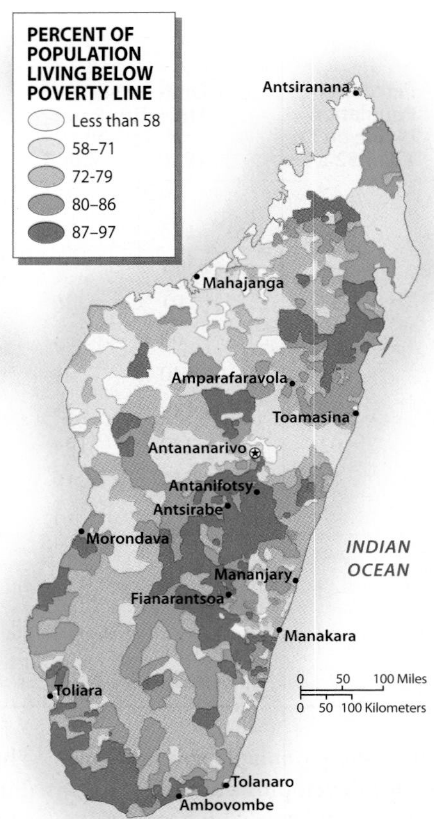

Figure 1.49 The Landscape of Poverty
As is true of most countries, the distribution of poverty within Madagascar is uneven, with clusters of abject poverty contrasting with regions that are less poor. This map shows the highest rates of poverty are in the central highlands where the country's population is concentrated and the density highest. In contrast, the northern and eastern lowlands are less poor. *(Map adapted from* Where the Poor Are: An Atlas of Poverty. *CIESIN. Center for International Earth Science Information Network 2006)*

However, if one looks closely, there are hints of improvement in some developing countries. For example, the percentage of those living in abject poverty, which the UN defines as living on less $1 per day, has fallen from 28 percent in 1990 to just under 20 percent today. Life expectancy, too, has increased in those countries, rising from 60 in 1990 to 65. Female literacy has also increased in developing countries, rising from 62 percent in 1990 to 70 percent in 2007.

Poverty and Infant Mortality The *percentage of the population living on less than $2 per day* represents the international definition of poverty. Deep poverty is defined as existing on less than $1 per day. Granted, the cost of living varies greatly around the world, yet the United Nations has found these definitions, when adjusted for local conditions, work well for measuring poverty and its associated social conditions. While poverty data is usually presented at the national level (see Table 1.2), the UN and other agencies are attempting to compile data at a local scale so as to better understand—and, hopefully, improve—the economic landscape within a country. The patterns of poverty in the African country of Madagascar provide an instructive example (Figure 1.49).

Under age 5 mortality depicts the number of children who die per 1,000 persons within that age bracket, and is another widely used indicator of social conditions. Aside from the tragedy of infant death, child mortality reflects the wider conditions of a society, namely the availability of food, health services, and public sanitation (Figure 1.50). If those factors are lacking, children under age five suffer most; thus their death rate is taken as a surrogate measure of whether a country has the necessary social infrastructure to sustain life. In the social development tables, child mortality data are given for two points in time, 1990 and 2005, to indicate whether the social structure has improved over

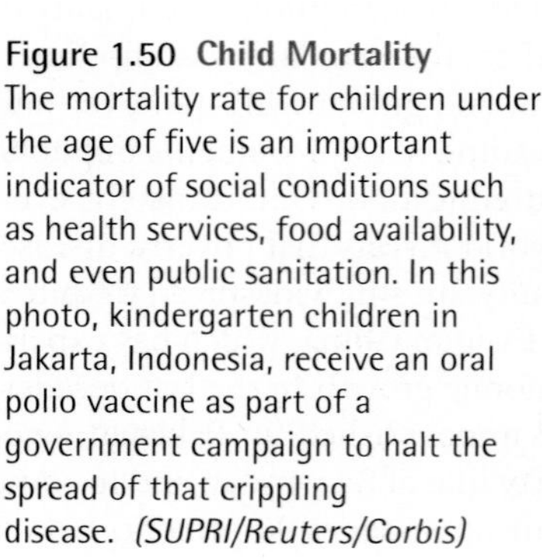

Figure 1.50 Child Mortality
The mortality rate for children under the age of five is an important indicator of social conditions such as health services, food availability, and even public sanitation. In this photo, kindergarten children in Jakarta, Indonesia, receive an oral polio vaccine as part of a government campaign to halt the spread of that crippling disease. *(SUPRI/Reuters/Corbis)*

Figure 1.51 Women and Literacy Gender inequities in education lead to higher rates of illiteracy for women. However, when there is gender equity in education, female literacy has a number of positive outcomes in a society. For example, educated women have a higher participation rate in family planning, which usually results in lower birth rates. *(Ron Giling/Panos Pictures)*

that period. Although most countries have seen improvement, disturbing differences in child mortality rates still exist on a global level.

Gender Equity The ratio of male to female students enrolled in primary and secondary schools is called *gender equity*. If the gender equity ratio is below 100, then more males are enrolled in schools than females. Conversely, a ratio over 100 means more females are enrolled than males in primary and secondary schools. Used by the UN as a measure of equality in education between the sexes, this statistic is linked closely to literacy rates, which in turn are linked to social development. The assumption is that if females are not enrolled in school, a high rate of female illiteracy results, which has negative consequences for social development. To illustrate, gender equity is high in the United States, Russia, and Brazil, and lower in countries such as India, Pakistan, and Nigeria (Table 1.2).

Although sex discrimination takes other forms, such as the preference for male children in India and China or the barriers erected against women in business and politics, the United Nations is currently focusing on correcting the gender inequity in education because illiteracy is much higher for women than men (Figure 1.51). More specifically, the UN has made neutralizing this educational disparity one of its Millennium Development Goals. Gender equity now has higher priority because increasing women's literacy has positive outcomes that complement social development. For example, some studies indicate a strong correlation between lower birth rates and female literacy, probably because literate women are more open to family planning measures.

Besides gender equity, there are many other linkages between economic and social development. This introductory chapter has discussed only a few and at a general scale; other issues, and in more detail, are examined in the regional chapters of this book.

Summary

- As the world becomes increasingly interdependent, globalization is driving a fundamental reorganization of economies and cultures through instantaneous global communication, the growth of transnational corporations, and the spread of Western consumer habits.
- The world seems to be converging and becoming more homogeneous because of globalization. Nonetheless, there is still great diversity. While some areas prosper, others stagnate; as global TV promotes a common world culture, small-group identity becomes increasingly important; as nation-states become world players through multinational organizations, separatist groups seek autonomy and independence. This tension between globalization and diversity gives world regional geography its primary focus.
- In most regions of the developing world, population and settlement issues revolve around four issues: rapid population growth, family planning (or its absence), migration to new centers of economic activity (both within and outside the region), and rapid urbanization.
- A major theme in cultural geography is the tension between the forces of global cultural homogenization and the countercurrents of local cultural and ethnic identity. Throughout the world, small groups are setting themselves apart from larger national cultures with renewed interest in ethnic traits, languages, religion, territory, and shared histories. Not merely a matter of colorful folklore and local customs, this cultural diversity is translated into geopolitics with outspoken calls for regional autonomy or separatism.

- Cultural tensions have caused the geopolitical issues of many world regions to be dominated by matters of global terrorism; ethnic strife and territorial disputes within nation-states; border tensions between neighbors of different cultural traditions; and new military approaches to deal with the ever-changing nature of national security.
- The theme of economic and social development is dominated by one issue—the increasing disparity between rich and poor, between countries and regions that already have wealth and are getting even richer through globalization, and those that remain impoverished. Often, blatant inequities in social development, schools, health care, and working conditions accompany these disparities in wealth.

Key Terms

areal differentiation *(page 13)*
areal integration *(page 13)*
asymmetrical warfare *(page 35)*
bubble economy *(page 9)*
centrifugal forces *(page 36)*
centripetal forces *(page 36)*
colonialism *(page 37)*
core-periphery model *(page 39)*
cultural imperialism *(page 29)*
cultural landscape *(page 14)*
cultural nationalism *(page 29)*
cultural syncretism or hybridization *(page 29)*
culture *(page 29)*
decolonialization *(page 37)*
demographic transition *(page 22)*
economic convergence *(page 6)*
ethnic religion *(page 32)*
ethnographic boundaries *(page 37)*
geometric boundaries *(page 37)*
globalization *(page 2)*
gross domestic product (GDP) *(page 39)*
gross national income (GNI) *(page 39)*
gross national income (GNI) per capita *(page 41)*
lingua franca *(page 31)*
nation-state *(page 35)*
net migration rate *(page 24)*
overurbanization *(page 27)*
population pyramid *(page 21)*
purchasing power parity (PPP) *(page 41)*
rate of natural increase (RNI) *(page 18)*
region *(page 14)*
secularization *(page 33)*
squatter settlements *(page 27)*
sweatshops *(page 7)*
total fertility rate (TFR) *(page 20)*
transnational firm *(page 4)*
transnational corporation *(page 38)*
universalizing religion *(page 32)*
urban form *(page 27)*
urban primacy *(page 26)*
urban structure *(page 27)*
urbanized population *(page 25)*

Questions for Review

1. Why is globalization a signature of the twenty-first century? What are the different components of globalization?
2. Summarize the major benefits and liabilities of globalization as discussed in this chapter.
3. Give examples of how changing scale from global to local might change an analysis of areal differentiation and integration for a specific area.
4. If the rate of natural increase (RNI) for the world remains consistent at 1.3 percent per year, what will the planet's population be in 10 years?
5. Explain the four stages of the demographic transition. How do birthrates and death rates differ in each stage? At what point is RNI highest?
6. What are the expressions and implications of overurbanization for cities in the developing world?
7. Why is it difficult to define *culture* in this globalized world?
8. Using a world map or atlas, come up with five examples each of ethnographic and geometric boundaries. Is there any clue about problems associated with these different borders?
9. What are the shortcomings of GNI? How are these rectified?
10. Evaluate the different indicators of social development as to their strengths and weaknesses.

Thinking Geographically

1. Select an economic, political, or cultural activity in your city, and discuss how it has been influenced by globalization.
2. Choose a specific country (or region) of the world to examine the benefits and liabilities of globalization. Remember to look at different facets of globalization, such as the environment and cultural cohesion and conflict, as well as the economic effects on different segments of the population.
3. Using the concepts of areal integration and city systems, discuss the global links between your city and other urban areas, both nationally and internationally.
4. How would you characterize your location in terms of regions? That is, in what physical, economic, and cultural regions are you located?
5. Compare the issues and problems with urban growth in your city to those of a city of similar size in a foreign country. Consider how urban structure and form might be similar and different in these two cities.
6. Discuss the cultural geography of your area, noting the distribution, landscapes, and interaction among different cultural groups.
7. Drawing upon information in current newspapers and magazines and on TV and the Internet, apply the concepts of cultural imperialism, nationalism, and syncretism to a region or place experiencing cultural tensions.
8. Choose a nation-state, and elucidate the different centrifugal and centripetal forces within that country. Based upon your findings, speculate on how that country might change in the next 10 years.
9. Apply the core-periphery concept to a country of your choice by delimiting what you think to be cores of economic development and contrasting them with peripheries of lower development. If possible, collect and analyze data that test your findings.
10. Using the tables of social indicators in the regional chapters of this book, identify traits shared by countries in which there is a high percentage of female illiteracy. What general conclusions result from your inquiry?

Bibliography

Anderson, Sarah, and Cavanagh, John. 1999. *Field Guide to the Global Economy*. New York: New Press.

Boo, Katherine. 2004. "Letter from India: The Best Job in Town—The Americanization of Chennai." *The New Yorker*, July 5, pp. 54–69.

Friedman, Thomas L. 2006. *The World Is Flat: A Brief History of the Twenty-first Century*. New York: Farrar, Straus and Giroux.

Gilpin, Robert. 2000. *The Challenge of Global Capitalism: The World Economy in the 21st Century*. Princeton, NJ: Princeton University Press.

Lewis, Martin W., and Wigen, Karen. 1997. *The Myth of Continents: A Critique of Metageography*. Berkeley: University of California Press.

Luttwak, Edward. 1999. *Turbo-Capitalism: Winners and Losers in the Global Economy*. New York: HarperCollins.

McFalls, Joseph, Jr. 2003. *Population: A Lively Introduction*. 4th ed. Washington, DC: Population Reference Bureau.

Rodrik, Dani. 1999. *The New Global Economy and Developing Counties: Making Openness Work*. Washington, DC: Overseas Development Council.

Schaeffer, Robert K. 2005. *Understanding Globalization: The Social Consequences of Political, Economic, and Environmental Change*. 3rd ed. Lanham, MD: Rowman and Littlefield.

The World Bank. 2007. *The World Bank Atlas*. Washington, DC: International Bank for Reconstruction and Development/The World Bank.

Additional bibliographic resources are at the *Diversity Amid Globalization* Website: http://www.prenhall.com/rowntree/.

2
The Changing Global Environment

Global warming is one of the most pressing environmental problems of the 21st century. While there are many implications to human-caused climate change, one of the most serious is the flooding of low lying islands and coastal areas by rising sea levels, which will result from the melting of polar ice caps and glaciers. This photo was taken in southwestern Greenland. (*Landov Photo*)

The human imprint is everywhere on Earth, from the highest mountains to the deepest ocean depths; from dry deserts to lush tropical forests; from frozen Arctic ice caps to the world's atmosphere. While hundreds of spent oxygen canisters clutter the heights of Mt. Everest, plastic litter and garbage soils coastlines and oceans everywhere (Figure 2.1). As deserts bloom in North America with irrigated cotton fields, tropical forests in Brazil are laid waste by logging. Although many changes to the global environment are intentional and have improved human life, other changes are inadvertent and have proven harmful to both human welfare and the environment. Because of the importance of these environmental issues, a study of the changing global environment is central to the study of world regional geography.

Environmental issues are also deeply intertwined with globalization and diversity. The destruction of tropical rain forests, for example, is a response to international demand for wood products and beef. Similarly, global warming through human-caused climate change is closely linked to patterns of world industry, commerce, and consumption. With about 6.7 billion people on Earth, the list is almost endless of the ways humans interact with and change the natural environment (see "Global to Local: The World of Bushmeat and Animal Poaching").

The purpose of this chapter is to provide an overview of Earth's environmental systems—geology, climate, hydrology (the study of water), vegetation, and agriculture—to set the scene for better understanding the environmental geography of the 12 world regions discussed in the following chapters.

Figure 2.1 The Human Imprint
Trash dumped at sea litters an isolated beach in the Outer Hebrides islands of the Atlantic, attesting to the fact that there's probably not an environment on Earth untouched and unaffected by human activities. *(Chris James/Peter Arnold)*

GEOLOGY AND HUMAN SETTLEMENT: A Restless Earth

Geology shapes the fundamental form of Earth's surface by giving a distinctive character to world landscapes through the physical fabric of mountains, hills, valleys, and plains. The geologic environment is also critical to a wide spectrum of human activities and concerns, such as the relationship between soil fertility and agriculture or the distribution of mineral resources like iron and coal. Additionally, the geologic environment presents humans with challenges and hazards in the form of devastating earthquakes, unstable landscapes, and explosive volcanoes. Clearly, a basic understanding of the physical processes that shape Earth's surface is crucial to comprehending human settlement in different parts of the world.

Plate Tectonics

The starting point for understanding the dynamic geology of Earth is **plate tectonics**, a geophysical theory that postulates that Earth is made up of a large number of geological plates that move slowly across its surface. This theory explains and describes both the inner workings of our planet as well as many surface landscape features. Further, plate tectonic theory gives clues about the world distribution of hazardous earthquakes and volcanoes.

Earth's interior is separated into three major zones with very different physical characteristics. Those three parts are the core, the mantle, and the outer crust. Tectonic plate theory is built on the assumption that a significant heat exchange takes place deep within Earth due to the cooling of the inner core. As a result, molten material is circulated through the mantle area (Figure 2.2).

This exchange of heat occurs in numerous **convection cells**, large areas of very slow-moving molten rock within Earth. Much like a stove provides heat for boiling water within a kettle, radioactive decay deep within Earth's core drives these convection cells. As the molten material reaches Earth's surface, it cools and becomes denser, causing the material to sink back into the mantle. Although geophysicists assume that convection cells move material at different rates, that movement is only a few inches or

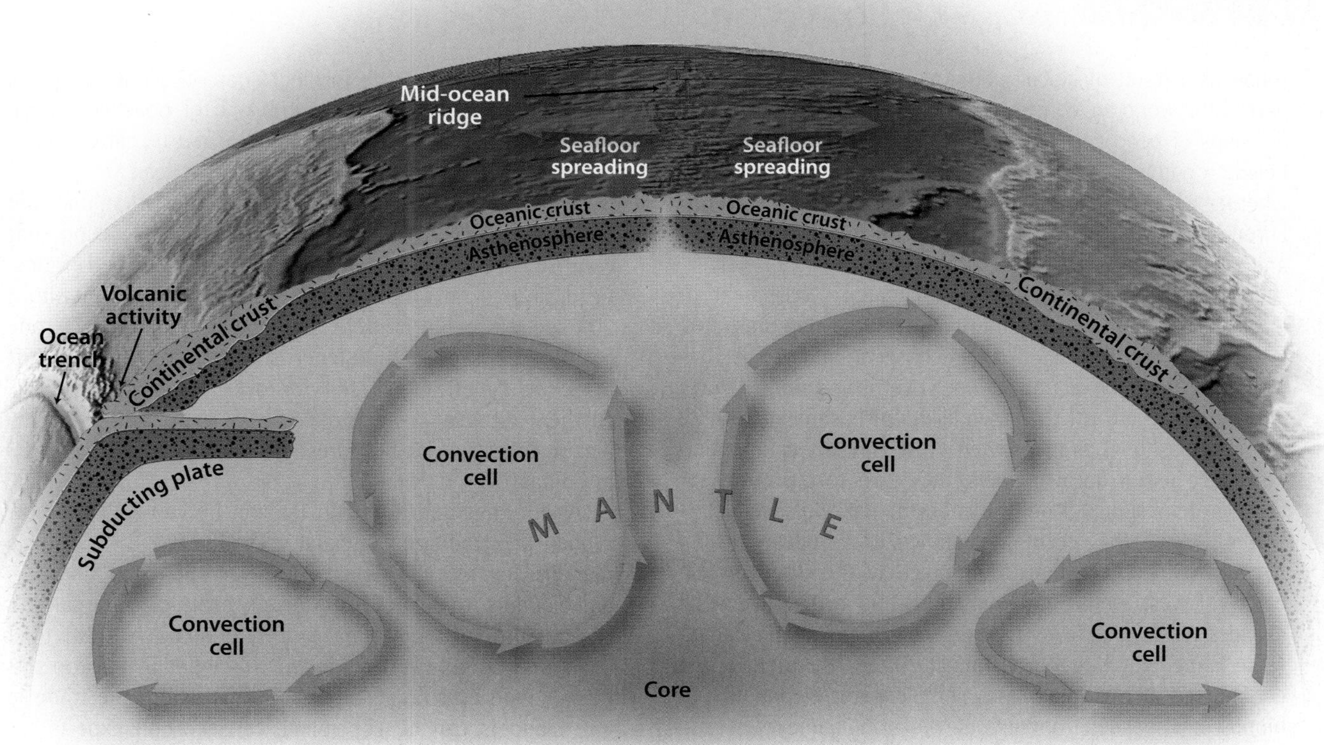

Figure 2.2 Plate Tectonics
According to plate tectonic theory, large convection cells circulate molten rock in different directions within Earth's mantle. Near the crust, this slow movement drags tectonic plates away from mid-ocean ridges, leading in other areas to the collision of tectonic plates where, eventually, one plate must go under the other in subduction zones.

centimeters per year, a rate roughly similar to the growth of human fingernails.

In turn, these convection cells drag **tectonic plates**, which are large blocks of rock that move slowly over the underlying semimolten material in different directions across Earth's surface. On top of the tectonic plates sit parts of the major continents and the major ocean basins. The map of world tectonic plates (Figure 2.3) shows that these plates vary significantly in size. It is important to note that the continents are not identical to the tectonic plates. In fact, rarely is there a close match between the two. More often continents straddle several tectonic plates, which, in turn, explains many of the world's diverse landscapes. Western North America, California specifically, lies atop two different tectonic plates, the Pacific Plate and the North American Plate, on what is called a **convergent plate boundary**. This means that the two plates are converging, or being forced together by convection cells deep within Earth. The notorious San Andreas Fault, which runs north and south through coastal California, is the actual plate boundary. Major mountain ranges have resulted from the tectonic convergence or collision of these two plates. The volcanic Cascade Range of Oregon and Washington and the massive Sierra Nevada of California are two examples. Similarly, in Latin America, the Andes Mountains are also products of two colliding plates, the eastward-moving Nazca Plate and the westward-moving South American Plate.

Often in these collision zones one tectonic plate dives below another, creating a **subduction zone** characterized by deep trenches where the ocean floor has been pulled downward by tectonic movement. Subduction zones exist off the western coast of South America and also near the Philippines, where the Mariana Trench forms the world's deepest ocean depths at 35,000 feet (10,700 meters).

In other parts of the world, tectonic plates move in opposite directions, forming a **divergent plate boundary**. As plates diverge, magma often flows from Earth's interior, creating mountain ranges with active volcanoes. In the North Atlantic, Iceland is formed on a divergent plate boundary that bisects the Atlantic Ocean. In other places, divergent boundaries form deep depressions, or **rift valleys**, such as that occupied by the Red Sea between northern Africa and Saudi Arabia. To the west, in Africa, a splinter of this boundary has created the extensive African rift, a part of which is the famous Olduvai Gorge, where tectonic activity has preserved the earliest traces of our human ancestors in volcanic soils (Figure 2.4).

Geologic evidence suggests that some 250 million years ago all the world's plates were tightly consolidated into a supercontinent centered on present-day Africa. Through time, this large area, now called *Pangaea*, was broken up as convection cells moved the tectonic plates

GLOBAL TO LOCAL

The World of Bushmeat and Animal Poaching

Today, as a result of globalization, plants and animals in one part of the world are commonly threatened by the desires and decisions of people thousands of miles away. A middle-aged Beijing gentleman's sex life, for example, may determine whether an endangered African rhino lives or dies; similarly, Europe's appetite for fish endangers African's lions, leopards, and great apes.

In Africa, the market for wild animal meat— "bushmeat" in local parlance—is a major conservation issue. Justin Brashares, a University of California–Berkeley conservation biologist, has revealed a close connection between overfishing along Africa's western coast and the increased killing of bushmeat on land. The relationship is simple: when West Africans cannot find affordable fish in local markets, they turn to bushmeat for their protein supply. And the problem has worsened considerably with the recent appearance of European fishing fleets off Africa's coastline. More precisely, the European fish catch off West Africa has increased 20-fold in the last several decades, resulting in fewer fish for local fishermen and leading to a higher kill rate of bushmeat. Often, threatened and endangered species—great apes, lions, leopards, elephants, and hippos—are poached from national parks and game reserves to fulfill the food needs of West Africa's growing population.

Also taking a heavy toll on the world's wildlife are global poaching networks that constitute a blackmarket activity comparable to the trade in illegal narcotics and weapons. Much of the activity in the trafficking of poached animal parts is thought to be run by international drug cartels and is estimated to be worth around $20 billion a year.

In Brazil, for example, more than 20 million animals are taken out of the country illegally each year, making that country one of the top sources for smuggled fauna. According to government officials, poaching has brought 208 species to the brink of extinction. A blue macaw, one of the Brazil's most threatened species with only 200 left in South America, will sell for over $50,000 in Europe. Further, a toucan is worth $7,000 in the United States, and a collection of rainforest butterflies sells for $3,000 in China.

As for that gentleman in Beijing, he can find a whole range of traditional sexual stimulants in his local marketplace, most of which come from poached animals—gall bladders from black bears, grizzly bears, or giant pandas; powder ground from rhino horns; or extracts from tiger paws and gonads. Although China is working to stop this illegal trade by subsidizing bear farms where the trade in gall bladders can be controlled, conservation experts say that while it's a step in the right direction, so far this program has had little effect on the huge blackmarket for poached animal parts.

Figure 2.3 Global Tectonic Plates
Akin to a fractured jigsaw puzzle, Earth's tectonic plates vary greatly in size and shape. Where plates converge and collide, active volcano and earthquake zones frequently appear, creating new mountains and significant environmental hazards. In other areas, one plate may dive beneath another, creating ocean trenches more than 30,000 feet (9,144 meters) deep, such as east of the Philippines. *(Adapted from McKnight and Hess, 2005,* Physical Geography: A Landscape Appreciation, *8th ed., Upper Saddle River, NJ: Prentice Hall)*

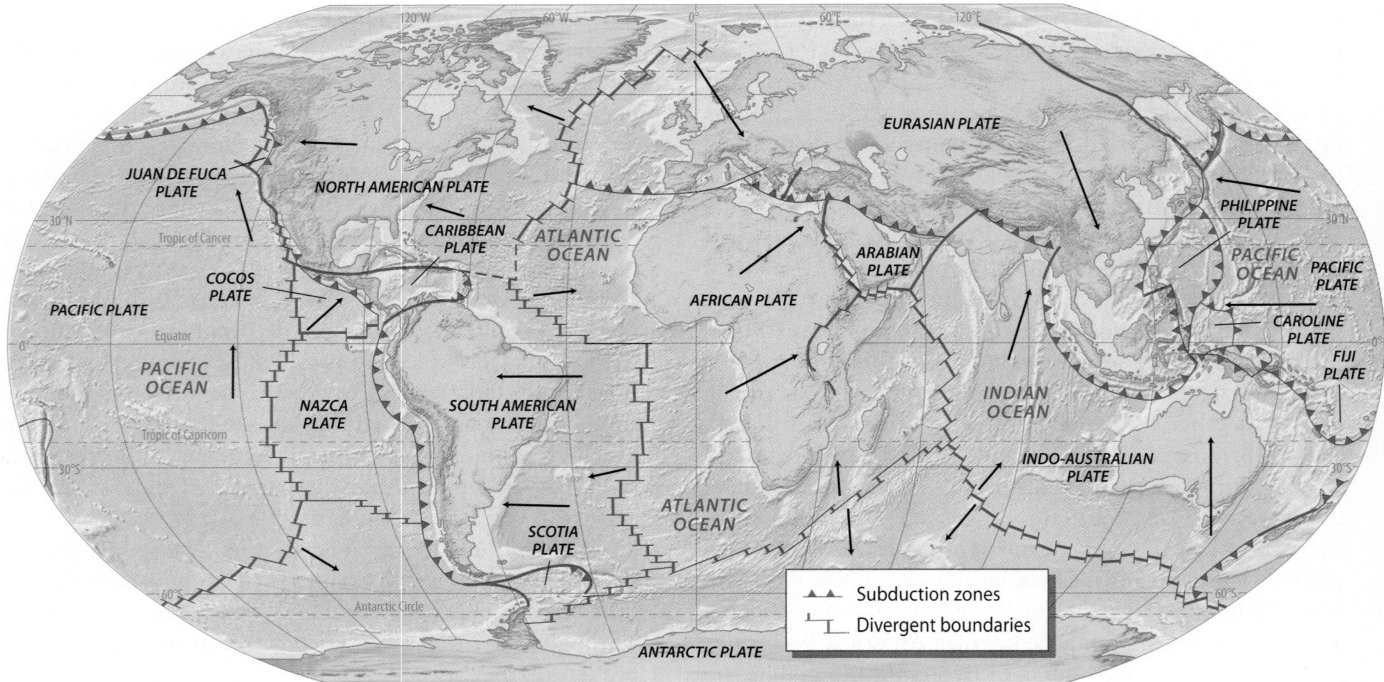

Figure 2.4 African Rift Valley
Where tectonic plates pull apart from one another, large, structural rift valleys create dramatic settings. This photo is of the African rift valley in southern Kenya, near Lake Magadi. *(Robert Caputo/Aurora & Quanta Productions, Inc.)*

apart. A hint of this former continent can be seen in the jigsaw puzzle fit of South America to Africa and of North America to Europe.

Although tectonic plate theory explains many of the world's large mountain ranges, it does not account for all highlands. For example, both the Himalaya and Alpine mountain ranges were created by the colliding forces of tectonic plates, as were the westernmost mountains of North America (Figure 2.5). However, there are also many mountain ranges far removed from tectonic boundaries. In North America, the Rocky Mountains serve as an illustration, as do the Ural Mountains in the center of Russia. This reminds us that local geologic forces also play an important role in shaping the landscape.

Geologic Hazards: Earthquakes and Volcanoes

Although the human toll from geologic hazards is not nearly as great as that from other natural disasters, namely floods and hurricanes, earthquakes and volcanoes can have a major impact on human settlement and activities. To illustrate, in October 2005, more than 75,000 people died when a large earthquake struck Pakistan (Figure 2.6). Similarly, in December 2003, southeastern Iran was devastated by a large quake that killed around 43,000 people. Generally, more than 100 earthquakes cause considerable damage to human settlements each year in different parts of the world.

Predicting earthquakes is much more difficult than warning people about hurricanes, tidal waves (tsunamis), or even volcanic explosions, and no scientific breakthrough is expected in the next decade that will make earthquake prediction more accurate. As a result, many cities in earthquake-prone regions emphasize building codes for stronger structures, land-use planning that discourages building in hazardous areas, or better preparation for postquake search and rescue. Unfortunately, these measures are expensive; therefore, only wealthier countries have been able to reduce the toll from seismic disasters, and damage from earthquakes remains high in poorer regions. For example, earthquakes in Iran or Mexico generally cause higher loss of life and more widespread damage than similar-magnitude earthquakes in Japan or California.

However, developed countries cannot dismiss the threat of major earthquakes. Professionals estimate that when the inevitable "big one" strikes California, the loss of life could be between 30,000 and 50,000 people, depending on the proximity of the earthquake to the major urban areas of Los Angeles or San Francisco. Despite planning and emergency preparedness, the threat of major destruction

Figure 2.5 The World's Mountains
Many—but not all—of the mountainous areas of the world are created by tectonic plate collisions. Good examples are the Alps of Europe and the Himalayas of Asia, both of which are found on tectonic borders. Some mountains, however, such as the Canadian Rockies shown in this photo are located far from plate boundaries. *(Paul A. Souders/Corbis)*

Figure 2.6 Earthquake Disasters Most earthquakes, such as this disastrous one in Pakistan that killed over 1000 people, are associated with stresses that build up along tectonic plate boundaries. While the magnitude of earthquakes varies greatly, building codes and construction are a major variable as to the toll taken on human life. *(Paul A. Souders/Corbis)*

and a disastrous loss of life from a catastrophic earthquake is a nagging reality along North America's West Coast from San Diego to Seattle, and even inland to Salt Lake City.

In addition to earthquakes, volcanic eruptions are also found along most tectonic plate boundaries and can cause major destruction (Figure 2.7). In 1985, for example, a volcanic eruption resulted in 23,000 deaths in Colombia, South America. In some cases eruptions can be predicted days in advance, which usually provides enough time for evacuation. During the 1991 eruption of Mt. Pinatubo in the Philippines, 60,000 people were evacuated, although 800 did die in the disaster (Figure 2.8). Because of this

Figure 2.7 Global Earthquakes and Volcanoes
The distribution of major earthquakes and volcanoes is strongly associated with tectonic plate boundaries around the world. The circum-Pacific zone of activity from the western Americas to East Asia is particularly active. Large populations found near zones of tectonic activity will continue to pay a price for living in these high-risk locations. *(Adapted from McKnight and Hess, 2005,* Physical Geography: A Landscape Appreciation, *8th ed., Upper Saddle River, NJ: Prentice Hall)*

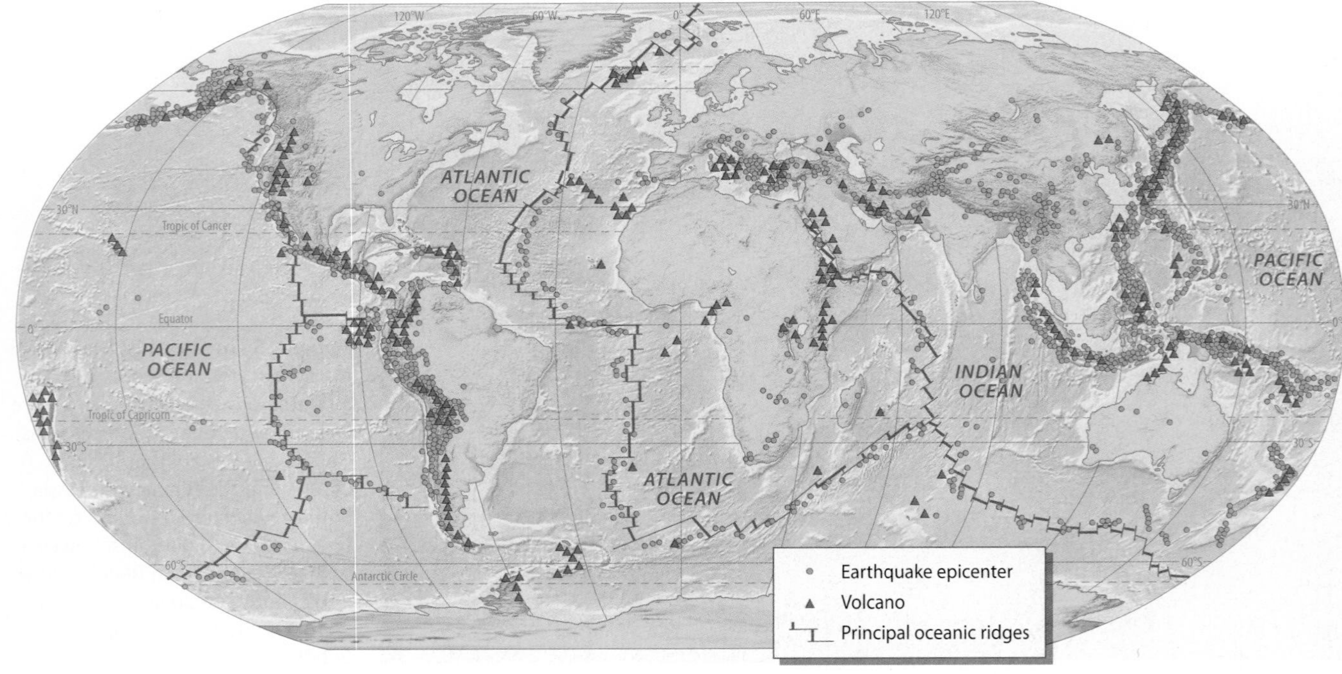

predictability, the loss of life from volcanoes is generally a fraction of that from earthquakes. In the twentieth century, an estimated 75,000 people were killed by volcanic eruptions while approximately 1.5 million died in earthquakes.

Unlike earthquakes, volcanoes also provide some benefits to people. In Iceland, New Zealand, and Italy, geothermal activity produces energy to heat houses and power factories. In other parts of the world, such as the islands of Indonesia, volcanic ash has enriched soil fertility for food crops. Additionally, local economies benefit from tourists attracted by scenic volcanoes in such places as Hawaii, Japan, and the Pacific Northwest.

GLOBAL CLIMATES: An Uncertain Forecast

Human settlement and food production are closely linked to patterns of local weather and climate. Where it is dry, such as in the arid parts of Southwest Asia, life and landscape differ considerably from the wet tropical areas of Southeast Asia. People in different parts of the world adapt to weather and climate in widely varying ways, depending on their culture, economy, and technology. Although some desert areas of California are covered with high-value irrigated agriculture that produces vegetables for the global marketplace year around, most of the world's arid regions support very little agriculture and barely participate in global commerce. Moreover, when drought hits one portion of the world, such as Russia's grain belt or Africa's Sahel, the socioeconomic repercussions are felt throughout the world. As a result, climate links us in our globalized economy, providing opportunities for some, hardships for others, yet challenges for all in the struggle to supply the world with food.

Understanding the complex meteorological processes that influence our global climates presents quite an intellectual challenge. Nevertheless, it is highly rewarding because it provides insight into one of the reasons the human condition varies so widely around the world.

Climatic Controls

Although weather and climate differ tremendously around the world, an accepted set of atmospheric processes control and influence meteorological conditions. More specifically, there are five main factors that should be understood—solar energy, latitude, interaction between land and water, world pressure systems, and global wind patterns.

Solar Energy Both the surface of Earth and the atmosphere immediately above it are heated by energy from the sun as our planet revolves around that large star. Solar energy is one of the most important variables that produce different world climates for it explains the great differences between the tropical climates near the equator and the cold climates closer to the poles. Most incoming solar energy, or **insolation**, is absorbed by Earth's land and water surfaces. These, in turn, heat the lower atmosphere through the process of re-radiation. This re-radiated energy is trapped by clouds and water moisture in the air adjacent to Earth, providing a warm envelope that makes life possible on our planet. Because there is some similarity between this process and the way a garden greenhouse traps sunlight to make that structure's interior warmer than the outside, this natural process of atmospheric heating is known as the **greenhouse effect** (Figure 2.9). Were it not for this process, Earth would be far too cold for human habitation. More specifically, without the greenhouse effect it is generally agreed our climate would be much like that on the planet Mars.

Figure 2.8 Volcanic Explosions Active volcanoes are also associated with plate tectonic borders throughout the world. Unlike earthquakes, there are many precursors to catastrophic volcanic explosions, thus allowing humans to evacuate the danger zone. This photo is from the 1991 eruption of Mt. Pinatubo in the Philippines. *(Alberto Garcia/Corbis)*

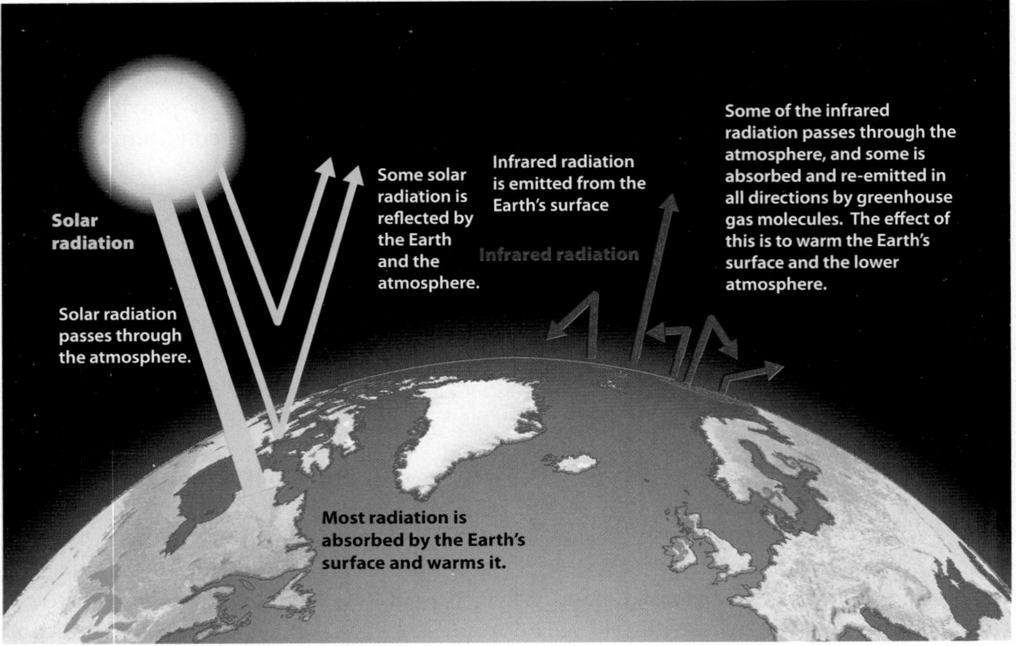

Figure 2.9 Solar Energy and the Greenhouse Effect
The greenhouse effect is a natural process that results in the trapping of solar radiation in the lower atmosphere, resulting in a warm envelope surrounding Earth.

Latitude Because of the curvature of Earth, the highest amounts of insolation are received in the equatorial region, which is the area between the equator and 23.5° north and south latitude. Poleward of this equatorial region, Earth receives far less insolation. As a result, not only are the tropics much warmer than the middle or high latitudes, but there is also a buildup of heat energy that must be redistributed through other processes, namely global wind systems, ocean currents, and even massive tropical storms such as typhoons and hurricanes (Figure 2.10).

Figure 2.10 Cumulus Clouds
A vivid signature of Earth's atmosphere, these massive cumulus clouds display the transfer of heat and moisture in the tropics. As heat energy builds up in low latitudes, ocean currents and wind and weather systems redistribute it globally. *(Jules Bucher/Photo Researchers, Inc.)*

Interaction Between Land and Water Because land and water differ in their abilities to absorb and re-radiate insolation, the global arrangement of oceans and continents is a major influence on world climates. Land areas heat and cool faster than bodies of water, which explains why temperature extremes such as hot summers and cold winters are always found away from coasts. Conversely, because water bodies retain solar heat longer than land areas, oceanic or maritime climates usually have moderate temperatures without the seasonal extremes found inland. The climatic characteristics of land and water differ so much that geographers use the term **continentality** to describe inland climates with hot summers and cold, snowy winters such as those found in interior North America and Russia. In contrast, **maritime climates**, or those close to the ocean, generally have cool, cloudy summers (Oregon, Washington, and British Columbia are good examples), with winters that are cold yet lack the extreme subfreezing temperatures of interior locations.

Global Pressure Systems The uneven heating of Earth due to latitudinal differences and the arrangement of oceans and continents produces a regular pattern of high- and low-pressure cells that, in turn, drive the world's wind and storm systems (Figure 2.11). For example, the interaction between high- and low-pressure systems in the North Pacific produces the storms that are driven onto the North American continent in both winter and summer. Those same processes in the North Atlantic produce winter and summer weather for Europe.

Farther south in the subtropical zones, large oceanic cells of high pressure cause different conditions. These high-pressure cells expand during the warm summer months because of the subsidence of warm air from the equatorial regions. As they enlarge, the cells produce the warm, rainless summers of the Mediterranean area of Europe and California. In the equatorial zone itself, summer weather spawns the strong tropical storms known as typhoons in Asia and hurricanes in North America and the Caribbean. The map of global pressure systems shows that this same alternating pattern of high- and low-pressure systems is also found in the Southern Hemisphere.

Global Wind Patterns The pressure systems also produce global wind systems. It is important to remember that air flows from high pressure to low (just as water flows from high elevations to low), thus, winds flow away from high-pressure and into low-pressure cells. This explains the monsoon in India, for example, which arrives in June as moisture-laden air masses flow from the warm Indian Ocean over land into the low-pressure area above northern India and Tibet. In the winter the opposite is true: as high pressure builds over these same areas, winds flow outward from cold Tibet and the snowy Himalayas toward the low pressure over the warm Indian Ocean.

World Climate Regions

The interaction of the meteorological processes just discussed produces the world's weather and climate. Before going further, though, it is important to note the difference between these two terms. *Weather* is the short-term day-to-day expression of atmospheric processes; our weather can be rainy, cloudy, sunny, hot, windy, calm, or stormy all within a short time period. This ever-changing weather is measured at regular intervals each day, often hourly. Data are then compiled on temperature, pressure, precipitation, humidity, and so on. Over a period of time, statistical averages from these daily observations provide a quantitative picture of common or usual conditions. From this long-term view, a sense of a regional *climate* is generated. Usually at least 30 years of daily weather data are required before climatologists and geographers construct a picture of an area's climate. In summary, weather is the short-term expression of highly variable meteorological processes, whereas climate is the long-term, average conditions (see "Geographic Tools: Reconstructing Climates of the Past").

Where similar conditions prevail over a larger area, boundaries are drawn cartographically around that area, which is called a **climate region**, so it can be folded into a larger classification of world climates. Knowing the climate type for a given part of the world not only conveys a clear sense of average rainfall and temperatures, but also allows larger inferences about human activities and settlement. If an area is categorized as desert, then we infer that rainfall is so limited that any agricultural activities require irrigation. Without supplemental watering, no crops can

Figure 2.11 Global Pressure Systems
Predictable patterns of low and high pressure result from the unequal heating of Earth's surface and the positioning of the continents. The intertropical convergence zone often causes unsettled weather near the equator, while fast-flowing westerlies move storm systems across the middle latitudes.

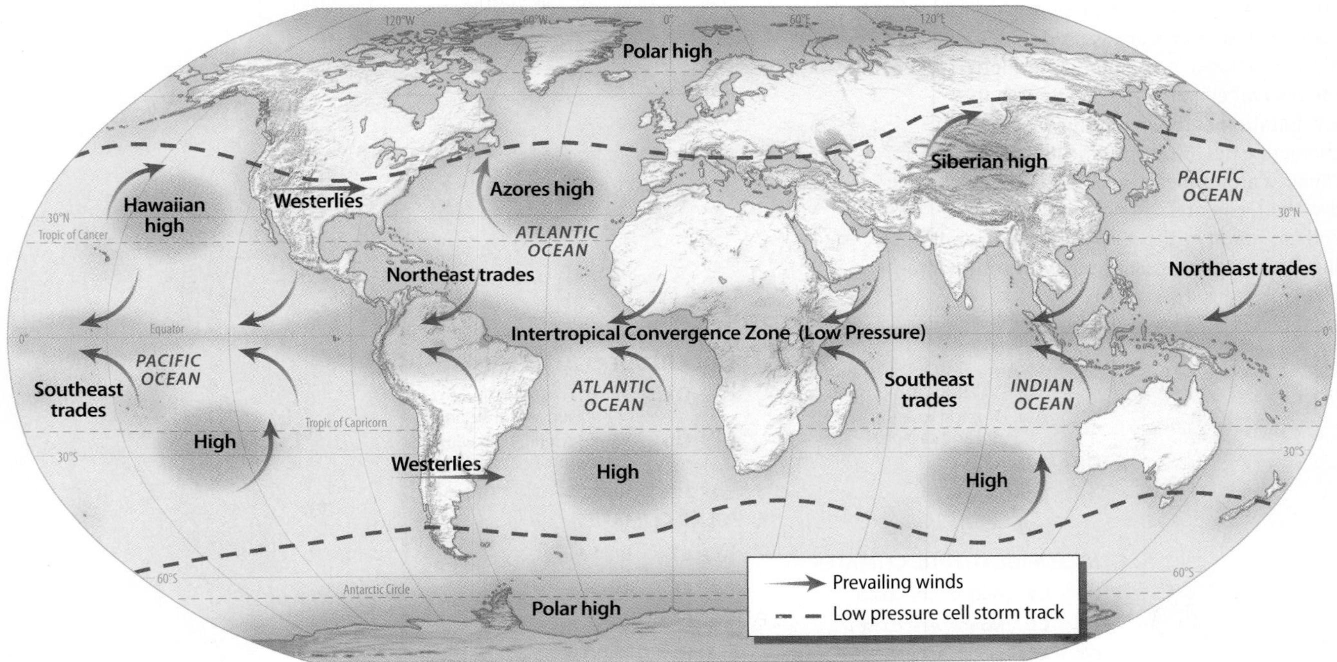

be grown, consequently people must turn to other strategies to obtain food. In contrast, if we see an area characterized by the climatic designation for tropical monsoon, we know there is enough rainfall for agriculture.

A standard scheme of climate types is used throughout this book, and each regional chapter contains a map showing the different climate regions in some detail. Figure 2.12 shows these climatic regions generalized to a world scale. The regional climate maps also contain **climographs**, graphs of average high and low temperatures and precipitation for an entire year. In Figure 2.12 the climate of Cape Town, South Africa, is presented as an example. Climographs for specific locations are used to illustrate different climatic regions of the world. Two lines for temperature data are presented on each climograph: the upper one plots average high temperatures for each month, while the lower shows average low temperatures. These monthly averages convey a good sense of what typical days might be like during different seasons.

Besides temperatures, climographs also contain bar graphs depicting average monthly precipitation. Not only is the total amount of rain and snowfall important, but the seasonality of this precipitation provides valuable information for making inferences about agriculture and food production (Figure 2.13). In many midlatitude regions, most moisture falls as rainfall during the warm summer months, which is good for agriculture because it comes during the growing season. However, in other parts of the world, precipitation falls primarily as snowfall during the cold winter months, when it cannot be used directly by crops but instead adds to soil moisture. In a few areas, such as in California and the Mediterranean region of Europe and northern Africa, the summers are dry, with rain falling during the winters. Only irrigated agriculture can thrive during the summer season.

Global Warming

Human activities connected with economic development and industrialization are changing the world's climate in ways that will have significant consequences for all living organisms, be they plants, animals, or humankind. More specifically, *anthropogenic*, or human-caused, pollution into the lower atmosphere is increasing the natural greenhouse effect so that worldwide **global warming**—an increase in the temperature of Earth's atmosphere—is taking place. This warming, in turn, will change rainfall patterns, increasing aridity in some areas; melt polar ice caps, causing a rise in sea levels; and possibly lead to a greater intensity in tropical storms. Moreover, because of climate change from global warming, the world may experience a dramatic change in food production regions. Whereas some prime agricultural areas, such as the lower Midwest of North America, may suffer because of increased dryness, other areas, namely the Russian steppes, may actually become better suited for agriculture. Although it is too early to tell when these changes may occur, there is little question these climate changes will

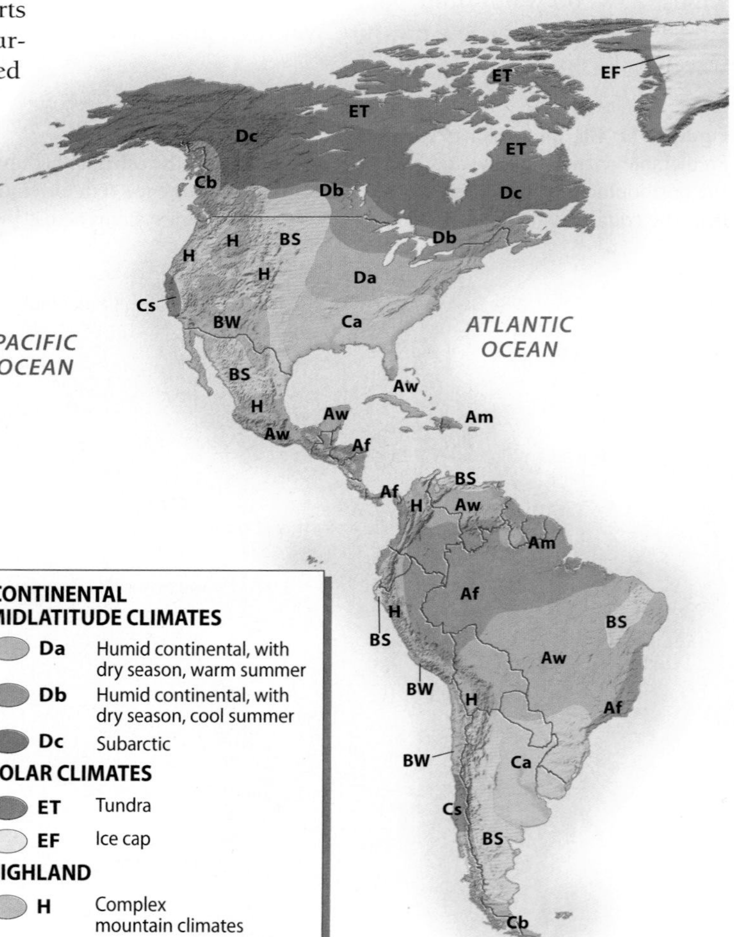

Figure 2.12 World Climate Regions
A standard scheme, called the *Köppen system* after the Austrian geographer who devised it in the early twentieth century, is used to describe the world's diverse climates. A combination of letters refers to the general climate type, along with its precipitation and temperature characteristics. More specifically, the *A* climates are tropical, the *B* climates are dry, the *C* climates are generally moderate and are found in the middle latitudes, and the *D* climates are associated with continental and high-latitude locations.

A WET CLIMATES
- **Af-Am** Tropical rainy and Tropical monsoon
- **Aw** Tropical wet and dry and savanna

B DRY CLIMATES
- **BS** Steppe
- **BW** Desert

C MILD MIDLATITUDE CLIMATES
- **Ca** Humid subtropical
- **Cb** Marine west coast
- **Cs** Mediterranean

D CONTINENTAL MIDLATITUDE CLIMATES
- **Da** Humid continental, with dry season, warm summer
- **Db** Humid continental, with dry season, cool summer
- **Dc** Subarctic

E POLAR CLIMATES
- **ET** Tundra
- **EF** Ice cap

F HIGHLAND
- **H** Complex mountain climates

have a dramatic effect on human settlement and world food supplies.

Causes of Global Warming As noted earlier, a natural greenhouse effect provides us with a warm atmospheric envelope that supports human life. This warmth comes from the trapping of incoming and outgoing (or re-radiated) solar radiation by an array of natural greenhouse gases in the atmospheric layer closest to Earth. Although natural greenhouse gases have varied somewhat over long periods of geologic time, they seem to have been relatively stable until recently. However, within the last 130 years, coincidental with the Industrial Revolution in Europe and North America, the composition and amount of those greenhouse gases have changed dramatically. This has resulted primarily from the burning of fossil fuels associated with industrialization and, as a result, the average temperature for Earth has increased markedly (Figure 2.14). Four major greenhouse gases account for the bulk of this change:

1. *Carbon dioxide* (CO_2) accounts for more than half of the human-generated greenhouse gases. The increase in atmospheric CO_2 is primarily a result of burning fossil fuels like coal and petroleum. To illustrate, in 1860 atmospheric CO_2 was measured at 280 parts per million (ppm). Today it is more than 380 ppm and is projected to exceed 450 ppm by 2050, assuming that fossil fuels remain a major source for the world's energy needs.
2. *Chlorofluorocarbons (CFCs)* make up nearly 25 percent of the human-generated greenhouse gases and come mainly from widespread use of aerosol sprays and refrigeration (including air conditioning). Although CFCs have been banned in North America and most of Europe, they are still increasing in the atmosphere at the rate of 4 percent each year. These gases are highly stable and reside in the atmosphere for a long time, perhaps as long as 100 years. As a result, their role in global warming is highly significant. Recent research has shown that a molecule of CFC absorbs a thousand times more infrared radiation from Earth than a molecule of CO_2.
3. *Methane* (CH_4) has increased 151 percent since 1750 as a result of vegetation burning associated with rainforest clearing, anaerobic activity in flooded rice fields, cattle and sheep effluent, and leakage of pipelines and refineries connected with natural gas production. Currently, CH_4 accounts for about 15 percent of anthropogenic greenhouse gases.
4. *Nitrous oxide* (N_2O) is responsible for just over 5 percent of human-caused greenhouse gases. The major source is from chemical fertilizers.

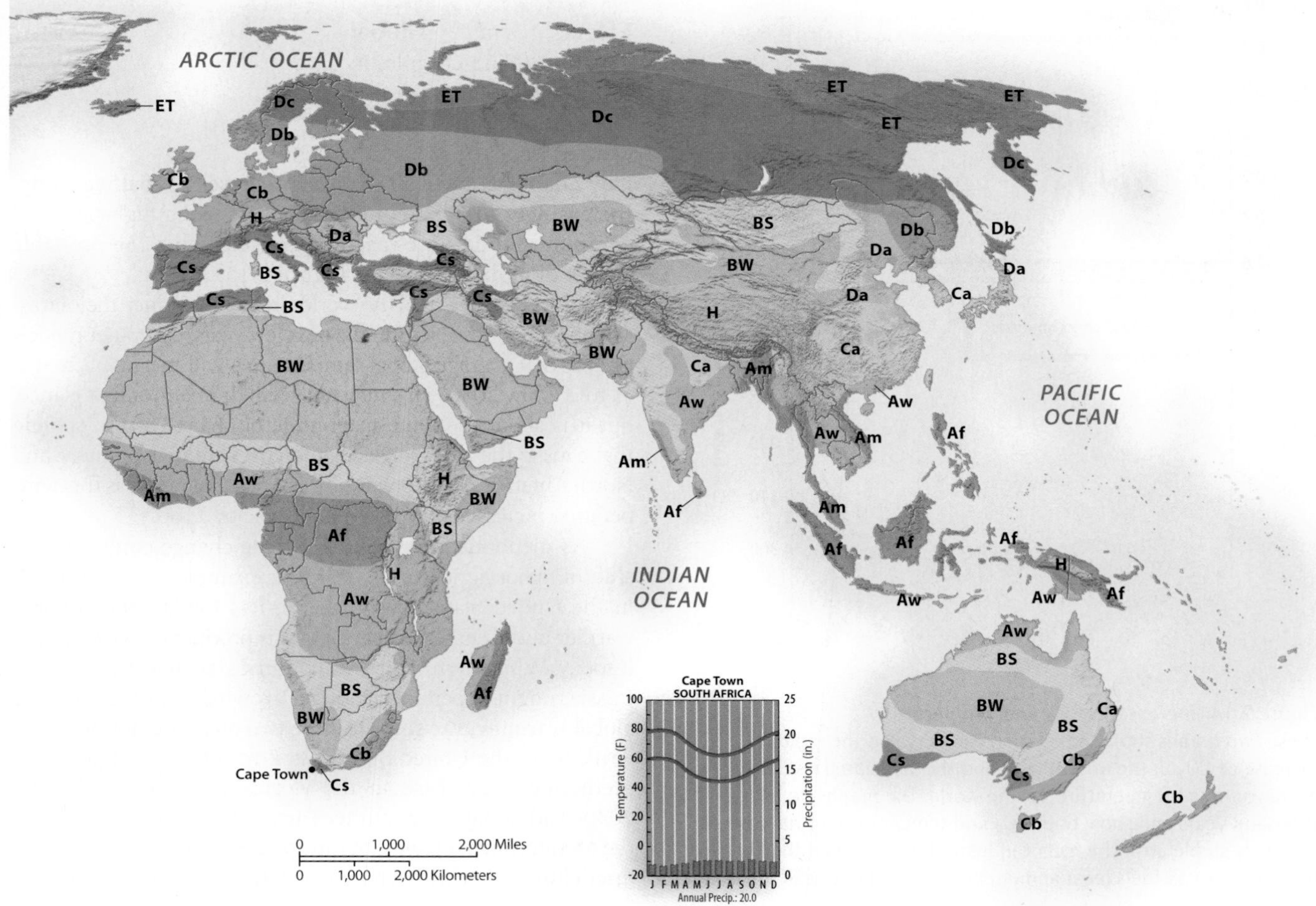

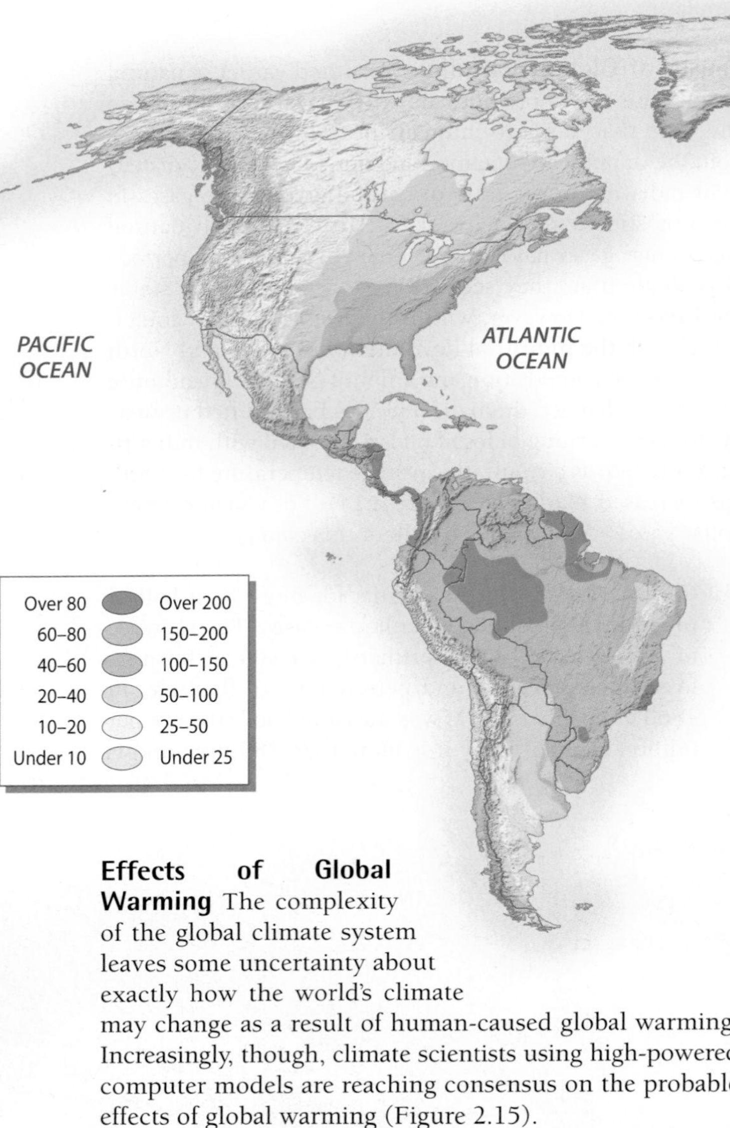

Figure 2.13 Global Precipitation
Note the vast range of annual precipitation totals around the world. As a general rule, the highest amounts are found in the tropics, particularly in places where mountains wring moisture from the tropical clouds. To illustrate, on the island of Kauai, Hawaii, the uplands experience annual totals of more than 400 inches (1,016 centimeters). Directly north and south of these wet tropics lie expansive deserts and steppes with low rainfall, caused by the subtropical high-pressure cells that dominate weather in these regions. Poleward of the arid lands lie the midlatitudes, where precipitation falls as both rain and snow and varies highly, depending on storm tracks, coastal location, and topography. *(Adapted from Clawson and Fisher, 2004,* World Regional Geography, *8th ed., Upper Saddle River, NJ: Prentice Hall)*

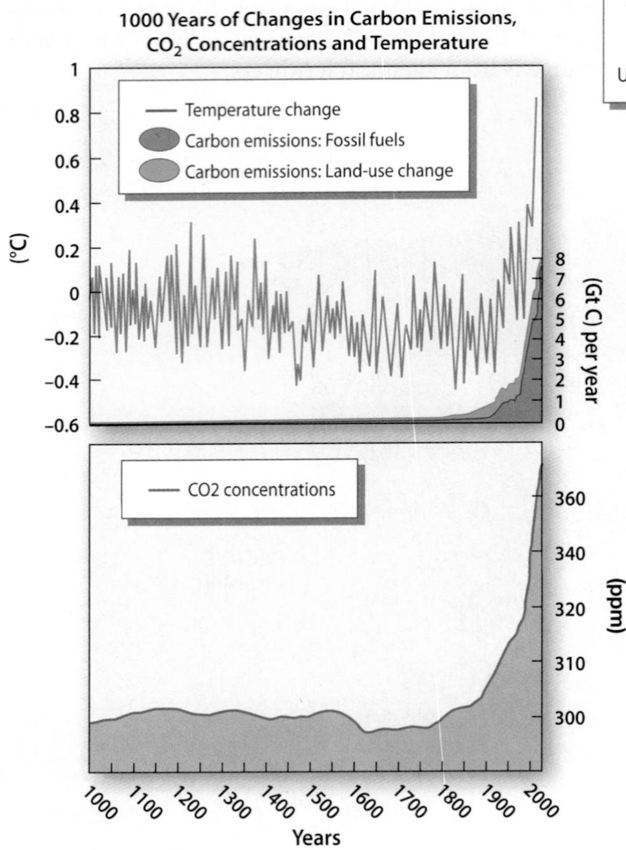

Figure 2.14 Increase in CO_2 and Temperature
These two graphs show the relationship between the rapid increase of CO_2 in the atmosphere and the associated rise in average annual temperature for the world. The graphs go back a thousand years and show both CO_2 and temperature to have been relatively stable until the recent industrial period when the burning of fossil fuels (coal and oil) began on a large scale.

Effects of Global Warming The complexity of the global climate system leaves some uncertainty about exactly how the world's climate may change as a result of human-caused global warming. Increasingly, though, climate scientists using high-powered computer models are reaching consensus on the probable effects of global warming (Figure 2.15).

Unless countries of the world drastically reduce their emission of greenhouse gases in the next few years, computer models predict that average global temperatures will increase 2 to 4°F (1 to 2°C) by 2030. This may not seem dramatic at first glance, but it is about the same magnitude of change as the cooling that caused the Ice Age glaciers to cover much of Europe and North America 30,000 years ago. More problematic is this temperature increase is projected to double by 2100.

As mentioned earlier, this climate change could cause a shift in major agricultural areas. For example, the Wheat Belt in the United States might receive less rainfall and become warmer and drier, endangering grain production as we know it today. While more northern countries, such as Canada and Russia, might experience a longer growing season because of global warming, the soils in these two areas are not nearly as fertile as in the United States. As a result, food experts are predicting a decrease in the world's grain production by 2030. Additionally, the southern areas of the United States and the Mediterranean region of Europe can expect a warmer and drier climate that will demand even more irrigation for crops.

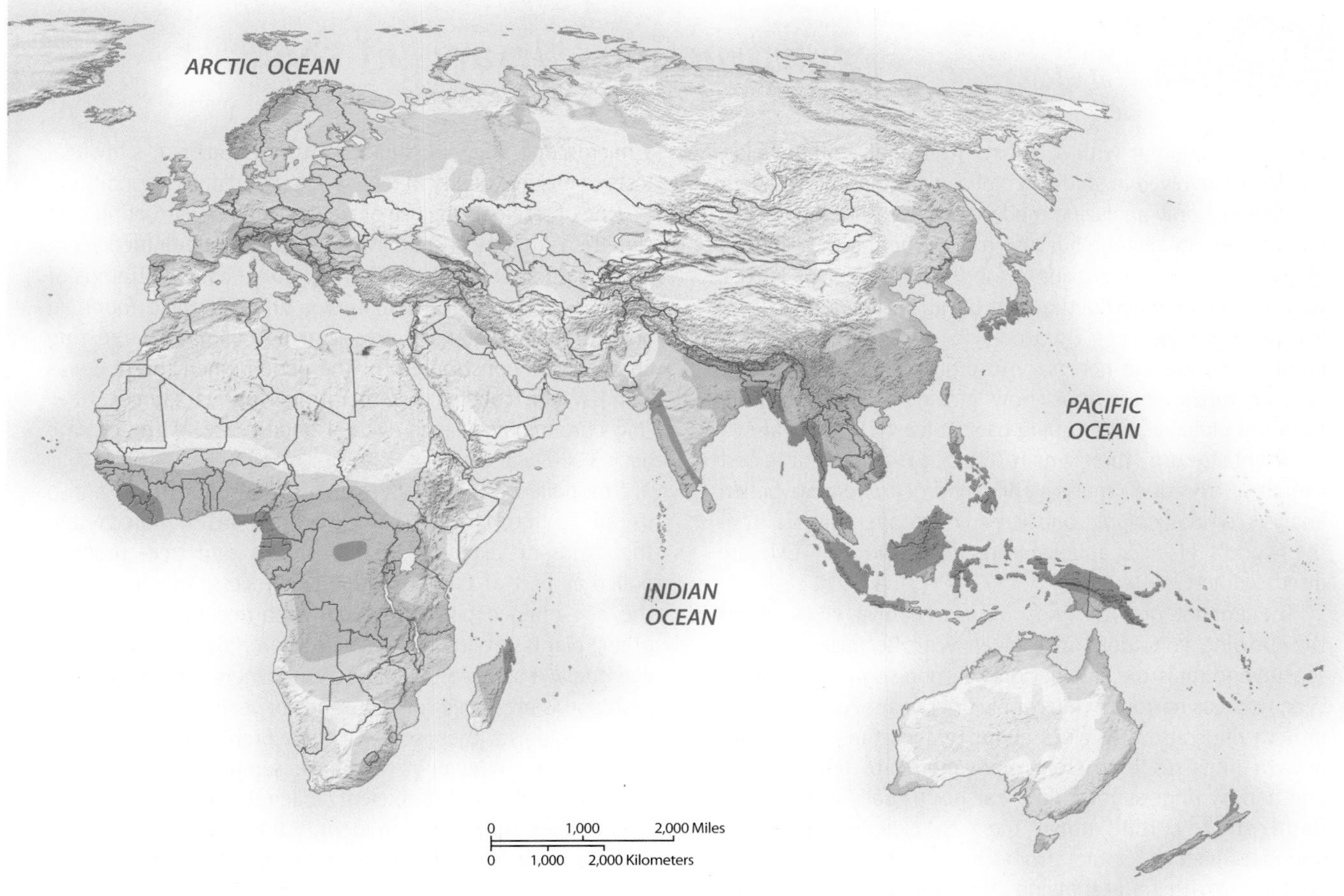

Figure 2.15 **Polar Bears Threatened by Global Warming** Ice floes are an important part of the polar bear's habitat. Because Arctic ice is melting more rapidly now due to global warming, the 20,000 bears living in the wild face an uncertain future. Recently, polar bears have been reported drowning trying to swim long distances between ice sheets. In this picture a mother and her cub with bloody snouts from eating a seal walk on thin ice. *(Bobbé Christopherson)*

Warmer global temperatures will also cause rising sea levels as oceans warm and polar ice sheets and mountain glaciers melt. Although projections differ on the predicted sea level rise—from several inches to almost 3 feet (1 meter)—even the smallest increase will endanger low-lying island nations throughout the world as well as coastal areas in Europe, Asia, and North America. Island nations in the Pacific and the Indian Ocean are particularly concerned, since they may be flooded out of existence.

Globalization and Climate Change: The International Debate on Limiting Greenhouse Gases By the early 1990s, the United Nations recognized that climate change was "a common concern of humankind" and that all nations have a "common but differentiated responsibility" for combating global warming. At the Rio de Janiero Earth summit in 1992 the first international agreement on global warming was presented. Signatories were bound by international law to reduce their greenhouse gas emissions by agreed-upon target dates. Within a year, 167 countries had signed. Unfortunately, those countries emitting the most greenhouse gases, namely, the United States, Japan, India, and China, did not come close to meeting the emission reductions agreed upon in Rio.

Consequently, another attempt to develop an international treaty was made at a December 1997 meeting in Kyoto, Japan. From these discussions emerged the current

GEOGRAPHIC TOOLS Reconstructing Climates of the Past

Geographers, along with scientists from many other fields, are deeply involved in studying global warming. This research involves not only analyzing modern-day weather records for signs of climate change but also reconstructing the climates of past centuries to provide a longer term view of Earth's weather and climate fluctuations. Because accurate weather instruments have been used only for the last 200 years, a number of tools are necessary to probe into the climate of past centuries; these tools now provide a fairly accurate record of global climates going back at least 15,000 years.

Scientists use three main tools to reconstruct past climates: tree-ring analysis, or *dendrochronology*; pollen analysis, or *palynology*; and ice-core isotope analysis from the world's glaciers and ice caps. Only the first two are discussed here.

The annual growth rings of a tree give clues about the growing conditions in a given year. Because trees are sensitive to moisture, a wider growth ring suggests a wetter year, whereas narrower rings imply drier years. Further, yearly growth rings also give clues about temperature, with wider growth rings resulting from more moderate temperatures and narrow rings suggesting a harsher (usually colder) year. Rather than cutting down trees to examine the rings, scientists instead extract from the trunk thin cores that contain growth rings showing the complete history of the tree's growth. After these rings are measured precisely under a microscope, they are subjected to a number of statistical tests to ensure validity in replicating past climatic conditions.

One obstacle to the use of tree rings for constructing past climates is that most trees live only for several hundred years. So how do scientists use tree rings to go back thousands of years? Logs found in old houses, or dead trees found in bogs or underwater in lakes, can be cored and tree ring chronologies constructed for the period when the tree was living. When cobbled together from various sources, tree-ring chronologies can provide a good sense of the climate back 3,000 years.

The pollen given off by plants, shrubs, and trees is also useful to reconstruct growing conditions hundreds and thousands of years ago. Plants, shrubs, and trees produce pollen as part of their reproductive process, and commonly the fine-grained pollen material is carried by the wind to other plants. Pollen also collects on lake surfaces, and eventually it sinks to the lake bottom where it collects in the mud and is preserved for thousands of years.

Palynologists, then, seek out these prehistoric lakes, many of which are now bogs, and take samples from the lake bottom to retrieve the ancient pollen. Using high-powered microscopes, they then can identify the pollen and link it to specific plant, shrub, and tree species, thus reconstructing the vegetation that covered the landscape thousands of years ago and, by inference, the climate of that period.

vehicle for international action on climate change—the Kyoto Protocol. In this agreement, 38 industrialized countries agreed to reduce their emissions of greenhouse gases below 1990 levels. For example, the United States, Japan, and Canada agreed to a reduction of about 7 percent below 1990 levels by the year 2012. However, the Protocol faced numerous obstacles between the Kyoto meeting in 1997 and early 2005 when the agreement finally became law. Within the United States, which until recently emitted the largest amount of greenhouse gas per year, the debate over global warming and the Kyoto Protocol was contentious, partisan, and divisive. Although many politicians saw the need for greenhouse gas controls, others emphatically resisted any action for fear that emission controls might constrain business, thereby slowing the economy, and increasing the cost of living for U.S. citizens. Early in 2001, while agreeing that global warming was a serious issue, President George W. Bush went on record opposing the Kyoto Protocol. Not only did he voice concern that restrictions on atmospheric emissions would harm the U.S. economy, but President Bush also objected to it on the grounds that large developing countries, namely China and India, were not yet bound to specific greenhouse gas reductions (Figure 2.16). As a result of his opposition, the United States was the only industrialized country to not ratify the Kyoto Protocol.

There is also tension between developed and less-developed countries over the Kyoto Protocol. Unrestricted emissions in the industrial world (North America, Europe, Japan) created the global warming problem, and these countries still today contribute about half of the total human-caused greenhouse gases (Table 2.1). Many international experts argue that these same developed countries should be required not only to take stringent steps to curb their own emissions, but also should subsidize and underwrite emission controls in developing countries. Understandably, the less-developed countries are reluctant to sign an emission control agreement that will constrain their economic future when, as they argue, they have contributed very little to the global warming problem up to now. Furthermore, these less-developed countries point the finger at the United States and say that it should take the lead in reducing global warming emissions by ratifying the Kyoto Protocol.

To become international law, the Kyoto Protocol had to be ratified by those countries emitting a total of 55 percent of the world's greenhouse gases. With the United States rejecting the agreement, the Protocol stalled for several

Figure 2.16 Auto Emissions in China
Reportedly, China has now surpassed the United States as the largest polluter of greenhouse gases. The main reason for China's huge emissions is the widespread use of coal to generate power. However, the rapid increase in private auto ownership is an increasing concern because car traffic has become a major cause of smoggy skies in China's urban areas. This traffic jam in Beijing and the leaden skies are expressions of that problem. *(Claro Cortes IV/Reuters/Corbis)*

years while Russia, the world's third-largest source of greenhouse gas emissions, pondered the costs and benefits of ratification. Finally, in late 2004, Russia signed the agreement, thus making Kyoto international law in early 2005. In total, over 100 countries have ratified the agreement. Although the 1997 Kyoto Protocol will not solve the global warming problem, it is an important first step toward international cooperation in mitigating climate change. A second step will be taken in 2012 when phase two of the Kyoto emission reduction process begins.

WATER ON EARTH: A Scarce and Polluted Resource

Freshwater is one of the most critical—and scarcest—resources on Earth for it is needed not only to sustain human life itself but also to support agricultural systems, industry, transportation, and even recreation. However, water is unevenly distributed about the world; some regions have abundant water resources, while others have serious shortages. Currently, about 40 percent of the world's population lives in areas of the world where water shortages are common.

The Global Water Budget

Any world map shows that more than 70 percent of the surface area of the world is covered by oceans. As a result, 97 percent of the total global water supply is salt water and only 3 percent is freshwater. Of the minuscule amount of freshwater on Earth, almost 70 percent is locked up in polar ice caps and mountain glaciers (see "Preserving Nature Through International Agreements: The Case of Antarctica"). Groundwater supplies account for almost 30 percent of the world's freshwater, which leaves less than 1 percent accessible from surface rivers and lakes.

Another way to conceptualize this limited amount of freshwater is to think of the total global water supply as 26 gallons (100 liters). Of that amount, 0.8 gallons (3 liters) would be freshwater; and of that small supply, only about half a teaspoon (0.003 liter) would be available to humans. Given this limited amount of available water coupled with growing demands for its use, political and economic tensions over efficient water usage are pronounced and problematic.

International planners use the concept of **water stress** to describe and predict where water resource problems will be greatest (Figure 2.17). Water stress data are generated from the amount of freshwater available on a per capita basis in different parts of the world. By dividing population growth rates into the amount of water available, one can develop a picture of future problem areas.

Africa stands out as a region of high water stress. More specifically, hydrologists predict that three-quarters of Africa's population will experience water shortages by the year 2025. Other problem areas will be northern China, India, much of Southwest Asia, Mexico, and even parts of Russia.

Agriculture accounts for 70 percent of current water usage. Many water and food experts predict that in com-

TABLE 2.1 The World's Major CO_2 Polluters

Country	Total CO_2 Emissions in MMT[a] (2005)	Percent Change (1995–2005)	Percent of World Total, CO_2 Emissions	Per Capita Emissions of CO_2, Metric Tons
United States	5,935	12	21	20.2
China[b]	5,323	87	19	3.6
Russia	1,696	4	6	11.7
Japan	1,230	14	4.4	9.9
India	1,166	35	4.1	1
Germany	844	–4	3	10.5
World Total	28,193		57.5	

[a]*Million metric tons.*
[b]*Some agencies maintain China's emissions surpassed the United States in 2007, but that data is not yet verified.*
Source: Energy Information Administration, International Energy Database

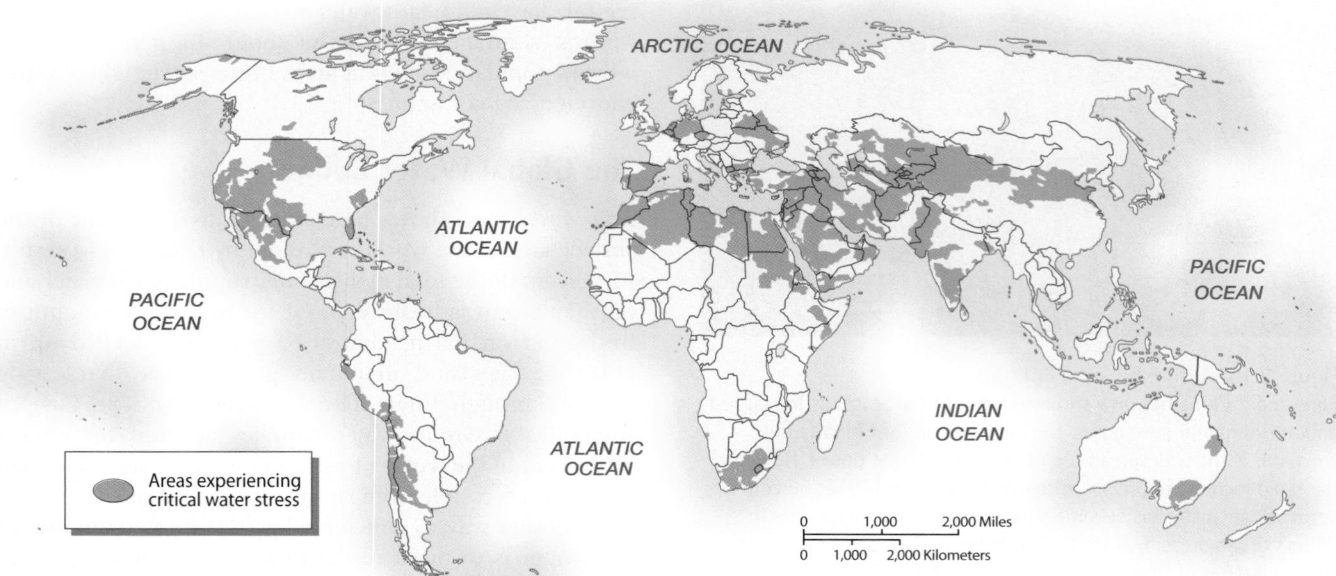

Figure 2.17 Water Shortages
This map shows those parts of the world forecast to have serious water shortages by the year 2025. Africa will suffer greatly; projections are that 75 percent of its population will face water shortages by the year 2025. Unfortunately, climate change from global warming will probably aggravate water issues even further.

ing years water shortages will limit global agricultural production. Additionally, the world's groundwater supplies, which currently provide water for about 2 billion people along with irrigation for vast acreages of agricultural land, are also showing signs of water stress. Another complicated water issue is the highly controversial issue of the privatization of domestic water supplies by international firms. This is happening in both developed and developing countries. In the latter case, often water that was traditionally provided by municipalities at low rates has become prohibitively expensive, forcing people to either reduce their water usage or turn to streams and rivers that are often polluted.

Figure 2.18 Flooding in the Philippines
Growing world populations create more opportunities for natural disasters. Increasingly, human settlements such as Makati City (near Manila) in the Philippines are being constructed in flood-prone areas, and this trend is likely to continue in the future. *(Bullit Marquez/AP/Wide World Photos)*

Flooding

Ironically, given the severe global water shortages, floods cause the most deaths of all natural disasters—far more than earthquakes, hurricanes, or volcanic eruptions. More specifically, they are responsible for almost 50 percent of deaths from natural disasters. Further, as the world's population increases, this figure seems to also be increasing (Figure 2.18). This trend can be attributed to several causes. As the population of a densely settled area increases from natural growth or from in-migration, people are often forced to settle in flood-prone areas, such as river floodplains and deltas. In Bangladesh, for example, an ever-expanding population has forced people to settle in low-lying delta lands subject to serious flooding during the monsoon season. This problem will undoubtedly become worse as sea level rise forces large numbers of people to move inland away from low-lying coastal areas.

HUMAN IMPACTS ON PLANTS AND ANIMALS:
The Globalization of Nature

One aspect of Earth's uniqueness compared to other planets in our solar system is the rich diversity of plants and animals that covers its continents. Geographers and biologists think

PRESERVING NATURE THROUGH INTERNATIONAL AGREEMENTS

The Case of Antarctica

Thanks to international cooperation, Earth's fifth largest continent—Antarctica—is dedicated to scientific research and environmental protection. From the first Antarctica Treaty in 1959 to the most current agreements on environmental protection, the global community is committed to preserving this huge world region from the ravages of economic exploitation.

Larger than Australia and almost one and a half times the size of the United States, Antarctica is the windiest, coldest, highest (based upon average elevation), and driest continent on Earth. Despite the fact that 98 percent of Antarctica is covered with ice, only about 2 inches (50 centimeters) of precipitation falls on the continent each year. While winter temperatures in the heart of Antarctica have reached −128°F (−89°C), balmy summer temperatures of 50°F (15°C) are not uncommon on the coast. In the interior, the Vinson Massif, with elevations of more than 15,000 feet (4,500 meters), rivals the world's highest peaks.

The first Antarctic Treaty, signed by a handful of counties in 1959, prohibited the establishment of military bases and the dumping of radioactive wastes. The treaty also banned all forms of economic exploitation—particularly mining. More than 40 countries are signatories to the most current version of the treaty, which focuses on environmental protection. These countries include superpowers, such as the United States, Great Britain, France, and Russia, neighboring nations (primarily Argentina, New Zealand, and Australia), and a number of developing countries, including China, India, Peru, and Bulgaria.

As a result of this treaty, there are 27 permanent research stations on the continent representing 16 national governments. During the Antarctic summer about 4,000 scientists are in residence; the number drops to 1,000 during the fierce winter. These scientists work on a wide range of topics relevant to contemporary environmental issues. To name just a few, studies are currently underway on the ozone hole's effects on plants and animals; the impact of global warming on the massive Antarctic ice shelves; reconstructing past climates through analysis of deep ice cores; and the effects of global warming on ocean life.

Despite this international protection, there are still serious problems in Antarctica. For example, the effects of tourism on the environment are a concern. Tourist visits (mainly on cruise ships) have increased to more than 25,000 people per year. More pressing still are problems associated with commercial fishing and oceanic warming as they affect Antarctic animals such as penguins and whales. Recent evidence suggests strongly that the penguin population is highly stressed because of reduced fish populations in nearby waters. Whether the scarcity results from overfishing or global warming is not clear.

There is no question, though, that illegal fishing is a serious issue in the vast Antarctic waters. To illustrate, 9,000 tons of Patagonian toothfish (marketed as Chilean sea bass) were poached recently by fishing boats from the distant island nation of Mauritius, which until recently was better known for sugarcane than for illegal fishing. Incidents such as this serve as warning that even Antarctica is not beyond the reach of economic globalization.

of the cloak of vegetation as the "green glue" that binds together life, land, and atmosphere. Humans are very much a part of this interaction. Not only are we evolutionary products of a specific **bioregion**, or assemblage of local plants and animals (most probably the tropical savanna of Africa), but also our human prehistory included the domestication of certain plants and animals. From this process has come agriculture. Further, humans have changed the natural pattern of plants and animals dramatically by plowing grasslands, burning woodlands, cutting forests, and hunting animals. The pace of such change has accelerated in recent decades, and these actions have led to a crisis in the biological world as ecosystems are devastated and entire species are exterminated. Many argue that the very vitality of our life-giving biosphere is threatened in many parts of the world.

Many of these problems can be explained by the globalization of nature and of local ecologies. Until the last half-century, tropical forests were primarily homes for small populations of indigenous peoples who made modest demands on local environments for their sustenance and subsistence. Today, however, these same tropical forests are capital for multinational corporations that clear-cut forests for international trade in wood products or search out plants and animals to meet the needs of far-removed populations. For example, Japanese lumber companies cut South American rain forests; German pharmaceutical corporations harvest medicinal plants in Africa; poachers kill North American bears and then sell their gallbladders on the Asian blackmarket as elixirs and sexual stimulants. Unfortunately, the list of human impacts on nature is a long one.

Biomes and Bioregions

Biome is the biogeographical term used to describe a grouping of the world's flora and fauna into a large ecological province or region. In this book we use the terms *biome* and *bioregion* interchangeably. Biomes are

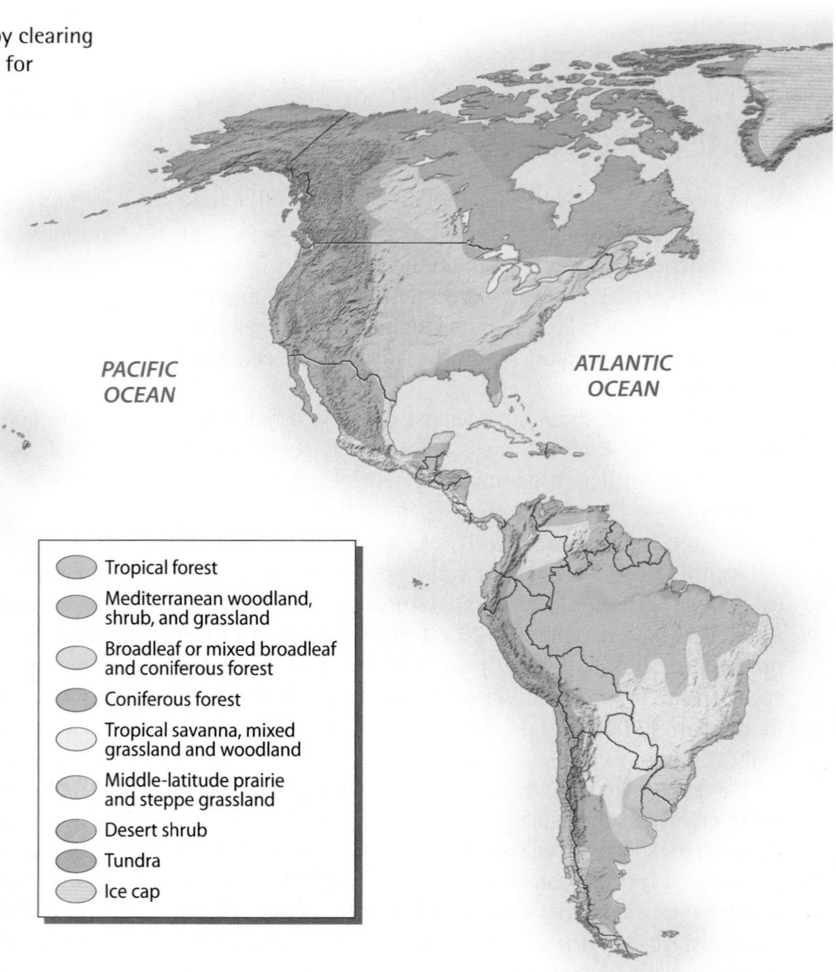

Figure 2.19 World Bioregions
Although global vegetation has been greatly modified by clearing land for agriculture and settlements and cutting forests for lumber and paper pulp, there is still a recognizable pattern to the world's bioregions, ranging from tropical forests to arctic tundra. Each bioregion has its own array of ecosystems containing plants, animals, and insects. These species constitute the biodiversity so necessary for robust gene pools. Think of biodiversity as the genetic library that informs life on Earth. *(Adapted from Clawson and Fisher, 2004, World Regional Geography, 8th ed., Upper Saddle River, NJ: Prentice Hall)*

closely connected with climate regions, since the major characteristics of a climate region—temperature, precipitation, and seasonality—are also the major factors influencing the distribution of natural vegetation and animals. A brief overview of the more important bioregions follows (Figure 2.19).

Tropical Forests and Savannas

Tropical forests are found in the equatorial climate zones of high average annual temperatures, long days of sunlight throughout the year, and heavy amounts of rainfall. This bioregion covers about 7 percent of the world's land area (roughly the size of the contiguous United States) in Central and South America, Sub-Saharan Africa, Southeast Asia, Australia, and on many tropical Pacific islands. More than half of the known plant and animal species live in the tropical forest bioregion, making it the most diverse of all biomes.

The dense tropical forest vegetation is usually arrayed in three distinct levels that are adapted to decreasing amounts of sunlight from the canopy to the forest floor (Figure 2.20). The tallest trees, around 200 feet (61 meters) high, receive open sunlight; the middle level (around 100 feet, or 31 meters) gets filtered sunlight; the third level is the forest floor, where plants can survive with very little direct sunlight. Even though much organic material accumulates on the forest floor in the form of falling leaves, tropical forest soils tend to be very low in stored nutrients. The nutrients are stored instead in the living plants. As a result, tropical forest soils are not well suited for intensive agriculture.

If the tropical rainfall is seasonal and there is a distinct dry season, the dense tropical rain forest is replaced by a more open forest where trees drop their leaves during the hot, dry season. At the poleward margins of the equatorial zone, where there are even longer dry seasons, the tropical forest is missing completely and is replaced with tropical grassland or savanna characterized by widely scattered trees and lush grassland during the wet season. The savanna is the environment of large mammals and, most agree, also the home of the earliest humans.

Deforestation in the Tropics

For a variety of reasons, tropical forests are now being devastated at an unprecedented rate, creating a crisis that tests our political, economic, and ethical systems as the world searches for solutions to this pressing problem.

Although deforestation rates differ from region to region, each year an area of tropical forest about the size of Wisconsin or Pennsylvania is denuded. Spatially, almost half of this activity is in the Amazon Basin of South America. However, actual deforestation appears to be occurring faster in Southeast Asia, where some estimates suggest that logging is occurring three times faster than in the Amazon. If those estimates are accurate, Southeast Asia could be completely stripped of forests within 15 years.

An important side effect of tropical deforestation is the release of CO_2 into the atmosphere. More specifically, current estimates are that fully 20 percent of all human-caused greenhouse gas emissions result from cutting and burning tropical forests.

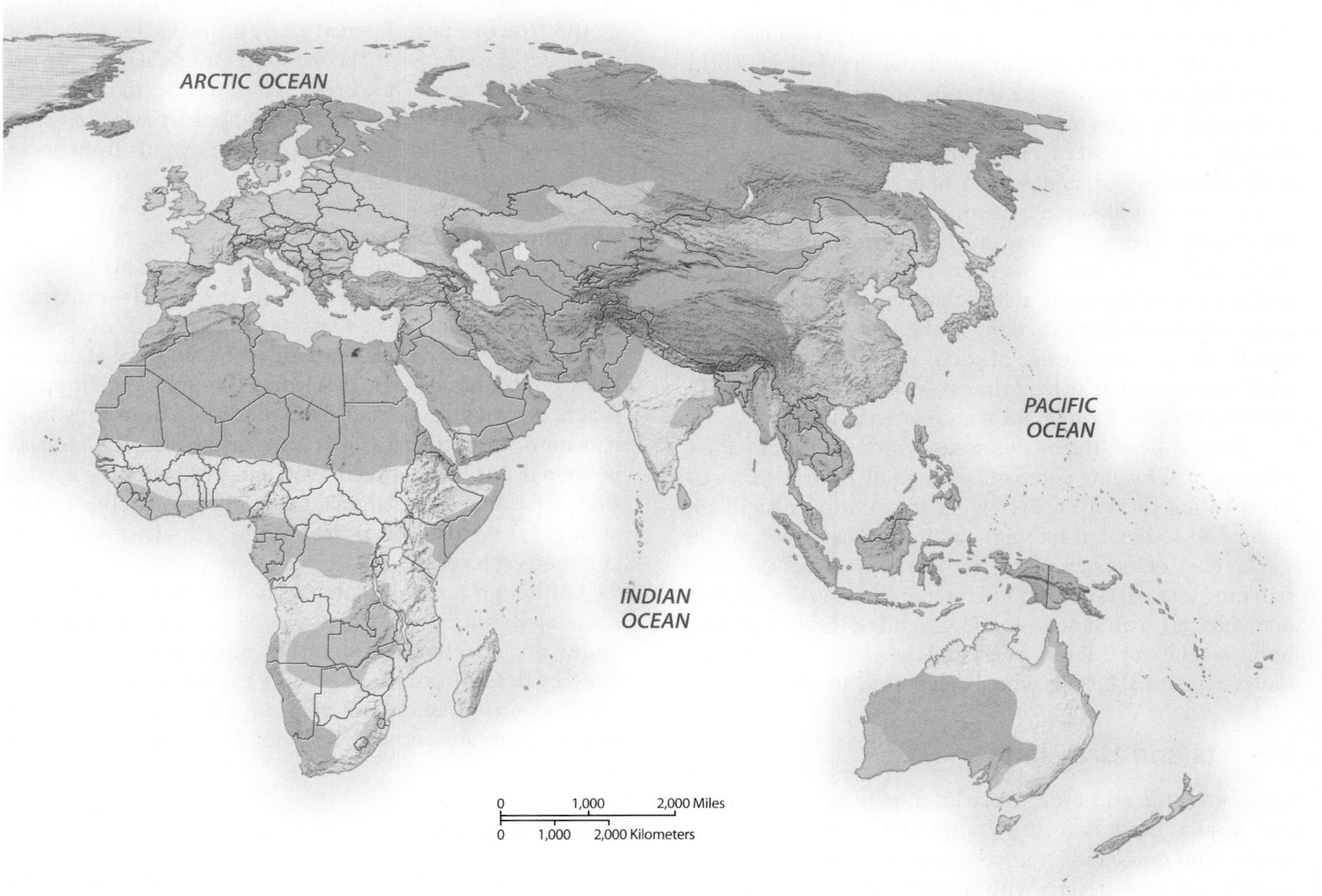

ARCTIC OCEAN
PACIFIC OCEAN
INDIAN OCEAN
0 1,000 2,000 Miles
0 1,000 2,000 Kilometers

Behind this widespread cutting of tropical forests lies the recent globalization of commerce in international wood products. By all accounts, Japan was the first to globalize its timber business by reaching far beyond its national boundaries to purchase timber in North and South America and Southeast Asia. This was a two-pronged strategy for meeting the increased demand for wood products in Japan while at the same time protecting its own rather limited forests for recreational use. Currently, about one-half of all tropical forest timber is destined for Japan. Unfortunately, much of it is used for throwaway items such as chopsticks and newspapers.

Another factor contributing to the rapid destruction of tropical rain forests is the world's seemingly insatiable appetite for beef. Cattle species originally bred to survive the hot weather of India are now raised on grassland pastures created by cutting tropical forests throughout the tropical world. Unfortunately, because tropical forest soils are poor in nutrients, cattle ranching is not a sustainable activity in these new grassland areas. After a few years, soil nutrients become exhausted, making it necessary to move ranching activities into newly cleared forestland. As a result, more forest is cut; more pasture created; more soil destroyed; and the process goes on at the expense of forestlands.

Figure 2.20 Tropical Rain Forest
As fragile as they are diverse, tropical rainforest environments feature a complex, multilayered canopy of vegetation. Plants on the forest floor are well adapted to receiving very little direct sunlight. *(Gary Braasch/Woodfin Camp & Associates)*

A third factor behind tropical forest destruction is that these forest areas are often the last settlement frontiers for the rapidly growing population of the developing world. Brazil has used much of its interior Amazon rain forest for settlement in order to ease population pressure along its densely settled eastern coast where rural lands are controlled by powerful landowners. Brazil had a choice: either address the troublesome issue of land reform and break up the coastal estates or open up the interior rain forests. It chose the path of least resistance, allowing settlers to clear and homestead the tropical forestlands of the interior. Many countries do the same, looking at the vast tracts of tropical rain forest as a safety valve of sorts, a landscape that can be used to temporarily deflect the pressures of land hunger by opening up land for migration and settlement.

Often these three processes work together. First, international logging companies are granted timber concessions to log the forest. Following this, cut-over lands are opened to settlement by migrants from other areas. At the same time, deforested lands are made available for cattle ranching by both domestic and foreign firms. When all three demands are combined—wood products, beef, and land hunger—the tropical rain forest and the plants, animals, and indigenous tribal groups who dwell within are the losers.

Deserts and Grasslands

Large areas of arid and semiarid climate lie poleward of the tropics, and here are found the world's extensive deserts and grasslands. Fully one-third of Earth's land area qualifies as true desert, with annual rainfall of less than 10 inches (25 centimeters). In areas receiving more rainfall, grassy plants appear, often forming a lush cover during the wet season. In North America, the midsection of both Canada and the United States is covered by grassland known as **prairie**, which is characterized by thick, long grasses. In other parts of the world, such as Central Asia, Russia, and Southwest Asia, shorter, less dense grasslands form the **steppe**.

The boundary between desert and grassland has always fluctuated naturally because of changes in climate. During wet periods, grasslands might expand, only to contract once again during drier decades. The transition zone between the two is a precarious environment for humans as the United States learned during the 1930s when the semiarid grasslands of the western prairie lands turned into the notorious "Dust Bowl." At that time, thousands of farmers watched their fields devastated by wind erosion and drought—a disaster that led to an exodus from these once-productive lands.

Farming marginal lands may actually worsen the situation, leading to **desertification**, or a spread of desertlike conditions into semiarid areas (Figure 2.21). This has happened on a large scale throughout the world—in Africa, Australia, and South Asia, to name just a few regions. In fact, in the last several decades an area estimated to be about the size of Brazil has become desertified through poor cropping practices, overgrazing, and the buildup of salt in soils from irrigation. In northern China an area the size of Denmark became desertified between 1950 and 1980 with the expansion of farming into marginal lands. Historically this region averaged 3 sandstorms a year; today 25 such storms are common each year.

Although some scientists say the case is somewhat overstated, the United Nations recently estimated that about 60 percent of the world's rangelands are threatened by desertification. According to this estimate, the desertification of such an amount of rangeland would threaten the agricultural livelihood of some 1.2 billion people.

Temperate Forests

The large tracts of forests found in middle and high latitudes are called *temperate forests*. Their vegetation is different from the low-latitude forests found in the equatorial regions. In temperate forests, two major tree types dominate. One type is softwood coniferous evergreen trees, such as pine, spruce, and firs, which are found in higher elevations and higher latitudes. The second category comprises deciduous trees, which drop their leaves during the winter. Examples are elm, maple, and beech. Because these trees are hardwood, hence harder to mill, they are

Figure 2.21 Desertification
Climatic fluctuations and human misuse combine to produce desertification in certain localities, where marginal lands are overcropped or grazed heavily, resulting in expansion of nearby deserts. Globally, many rangelands remain threatened by desertification. *(Alamy.com)*

generally less favored by the timber industry than softwood species.

In North America, conifers dominate the mountainous West, Alaska, and Canada's western mountains, while deciduous trees are found on the Eastern Seaboard of the United States north to New England. There the two tree types intermix before giving way to the softwood forests of Maine and the Maritime provinces of eastern Canada.

In the coniferous forests of western North America, the struggle between timber harvesting and environmental concerns remains controversial. Whereas timber interests argue increased cutting is the only way to meet high demand for lumber and other wood products, environmental concerns over the protection of habitat for endangered species (the spotted owl, for example) have caused the government to place large tracts of forest off-limits to commercial logging (Figure 2.22). Further complicating the future of western forests are global market forces. Many Japanese and Chinese timber firms pay premium prices for logs cut from U.S. and Canadian forests, thus outbidding domestic firms for these scarce resources. Since these trees are often cut from public lands, this facet of globalization raises an interesting question about the appropriate use of public forests that are maintained by tax money.

In Europe, hardwoods were the natural tree cover in western countries such as France and Great Britain before these once-extensive forests were cleared away for agriculture. In the higher latitudes of Norway and Sweden, coniferous species prevail in the remaining forests. These conifers also populate extensive forests across Germany and Eastern Europe, and through Russia into Siberia, creating an almost unbroken landscape of dense forests.

The Siberian forest remains a resource that could become a major source of income for financially strapped Russia. Some argue that if the Siberian forests are put on the market for global trade, it will reduce logging pressure on North America's western forests, which, in turn, could make it easier to enact and enforce comprehensive environmental protection in the United States. This example illustrates once again how the multifaceted forces of globalization are intertwined with the fate of local ecosystems.

FOOD RESOURCES: Environment, Diversity, and Globalization

If the human population continues to grow at expected rates, food production must double by 2025 just to provide each person in the world with a basic subsistence diet. Every minute of each day, 250 people are born who need food; during that same minute, about 10 acres of existing cropland are lost because of environmental problems such as soil erosion and desertification. Many experts argue that food scarcity will be the defining issue of the next several decades. Because of the close ties between world food problems, the environment, and globalization, we conclude this chapter by introducing some concepts underlying world food resource issues. Additional material on regional food problems is discussed in subsequent chapters.

Industrial and Traditional Agriculture

Food is produced in a great variety of ways around the world, and these different production strategies and techniques yield widely varying results. A useful, though necessarily simplified starting point for assessing these strategies is the differentiation between *industrial* and *traditional* food production systems. Industrial agriculture is practiced on about 25 percent of the world's croplands. It is characterized by the use of large amounts of fossil fuel, both to drive farm machinery and as the industrial basis for chemical fertilizers and pesticides (Figure 2.23).

Figure 2.22 Clear-Cut Forest
Commercial logging in Washington's Olympic Peninsula has dramatically reshaped this landscape. Throughout the Pacific Northwest, environmental lobbies have successfully restricted logging to protect habitat for endangered species and recreation. *(Calvin Larsen/Photo Researchers, Inc.)*

Figure 2.23 Industrial Farming in Iowa
The harvesting of corn in Iowa reveals the imprint of technology on modern agriculture. In such settings, large capital investments in machinery, fuel, and agrochemicals produce high yields while keeping labor costs low. *(Andy Levin/Photo Researchers, Inc.)*

Additionally, irrigation water is generally used to a high degree. Because of the high level of mechanization, labor requirements are low. In the United States, for example, less than 2 percent of the total workforce is employed in agriculture.

In contrast, traditional agriculture is practiced by almost half the people on Earth. Besides being labor-intensive, traditional agriculture is characterized by less mechanization, less fertilizer, infrequent applications of pesticides, and, as a result, generally lower yields. Until recently, traditional agriculture was synonymous with **subsistence agriculture**, farming that produces only enough crops or livestock for a farm family's survival. However, because of the increasing influence of export economies and globalization, many traditional farmers now also grow cash crops that can be traded on world markets to supplement their subsistence needs (Figure 2.24).

The amount of capital (including labor) necessary for an agricultural system defines the difference between *intensive* and *extensive* farming. To begin, intensive farming can be found in both industrial and traditional systems. In an industrial farming context, intensive agriculture requires high inputs of capital to support fossil fuels, machinery, and fertilizer. In return, this form of farming generally produces high yields. In contrast, traditional intensive farming requires high inputs of labor. Padi rice agriculture in Southeast Asia is an example. Extensive farming takes place when labor and capital inputs are considerably lower, such as in open-range livestock ranching in North America or nomadic herding in North Africa. In both cases, large amounts of grazing land are needed to support herd animals. Another example of extensive agriculture is traditional wheat farming in places such as Turkey and Iran. Here, animals rather than machinery are used for plowing and harvesting, thus, the capital input for fossil fuels or tractors is nonexistent.

The Green Revolution

During the last 40 years of the twentieth century, the world's population doubled. Even more remarkably, during the same period global food production also doubled to keep pace with this population explosion. This increase in food production came primarily from the expansion of intensive, industrial agriculture into areas that previously produced subsistence crops through extensive and traditional means.

More specifically, since 1950 the increases in global food production have come from the **Green Revolution**, which involved new agricultural techniques using genetically altered seeds, coupled with high inputs of chemical fertilizers and pesticides. The first stage of the Green Revolution combined three processes: first, the change from traditional mixed crops to monocrops, or single fields, of genetically altered, high-yield rice, wheat, and corn seeds; second, intensive applications of water, fertilizers, and pesticides; and, third, additional increases in the intensity of agriculture by reduction in the fallow, or field-resting time, between seasonal crops.

Since the 1970s, a second stage of the Green Revolution has evolved. This phase emphasizes new strains of fast-growing wheat and rice specifically bred for tropical and subtropical climates (Figure 2.25). When combined with irrigation, fertilizers, and pesticides, these new varieties allow farmers to grow two or even three crops a year on a parcel that previously supported only one. Using these methods, India actually doubled its food production between 1970 and 1992.

However, many argue that these agricultural revolutions also carry high environmental and social costs. Because these crops draw heavily on fossil fuels, there has been a 400 percent increase in the agricultural use of fossil fuels during the last several decades. As a result, Green Revolution agriculture now consumes almost 10 percent of the world's annual oil output. Earlier, cheap oil prices facilitated

Figure 2.24 Subsistence Farming in Burkina Faso
These West African women are harvesting sorghum. Such traditional agricultural operations remain pivotal in feeding people in the less-developed world, although traditional farmers are being increasingly lured into growing crops for the global market. *(Mark Edwards/Still Pictures/Peter Arnold, Inc.)*

much of this increased agricultural usage of fossil fuel, but today, rising oil prices raise the question of how this will affect food prices in the less-developed world.

The environmental costs of the Green Revolution include damage to habitat and wildlife from diversion of natural rivers and streams to agriculture; pollution of rivers and water sources by pesticides and chemical fertilizers applied in heavy amounts to fields; and increased regional air pollution from factories and chemical plants that produce these agricultural chemicals.

There is also evidence that the Green Revolution can bring high social costs and disruption to some areas. Because the financial costs to farmers participating in the Green Revolution are higher than with traditional farming, successful farmers must have access to capital or bank loans to purchase hybrid seeds, fertilizers, pesticides, and even new machinery. For those with high social standing, a family support system, or good credit, the rewards can be great, but for those without access to loans or family support, the toll can be heavy. Usually traditional farmers cannot compete against those raising Green Revolution crops in the regional marketplace. As a result, they often struggle at a poverty level while Green Revolution farmers prosper.

In some wheat-growing areas of India, the social and economic distance between the well-off Green Revolution farmers and poor, traditional farmers has become a major source of economic, social, and political tension. An area where once all shared a common plight has now become a highly stratified society of rich and poor. These social costs can be condoned only because of the pressing need for food in this rapidly growing country.

Figure 2.25 The Green Revolution in India
Because of its large and expanding population, India is a region where food supplies may be a problem in the near future. The country doubled its food production between 1970 and 1992 primarily due to the expansion of rice and wheat yields. Nonetheless, the future of food security in this large country is uncertain. In this photo, a woman is planting rice seedlings in the Rishi Valley of Andhra Pradesh state. *(Mark Edwards/Peter Arnold, Inc.)*

Problems and Projections

Even though agriculture has been able to keep up with population growth in the last decades, few are completely confident that pace will continue in the twenty-first century. Although the regional chapters of this book include fuller discussion of food and agriculture issues, four key points offer a starting point for understanding food security issues.

- While overall food production remains an important global issue, it is in fact local and regional problems that often keep people from obtaining food. To many experts, the issue is less global food production and more the widespread poverty and civil unrest at local levels that keep people from growing, buying, or receiving adequate food supplies.

  If the total global agricultural output were somehow shared equally among all people in the world, each person would have approximately four times his or her basic daily needs. Given this, some say that there is already enough food to feed the world's population—the real issue, they argue, is distribution and purchasing power.
- Political problems are usually more responsible for food shortages and famines than are natural events such as drought and flooding. Distribution of world food supplies is often highly politicized, starting at the global level and continuing down to the local. Food aid goes to political friends and allies, while enemies go without.
- Globalization is causing dietary preferences to change worldwide, and the implications of this change could be profound. Currently two-thirds of the world population is primarily vegetarian, eating only small portions of meat because it is so expensive. However, because of recent economic booms in some developing countries, an increasing number of people in developing countries are now eating meat, having moved up the food chain from a subsistence diet based on local resources to a global diet containing a greater proportion of meat (Figure 2.26). In many cases this change in diet comes not just from a new economic prowess, but also from changing cultural tastes and values as people are exposed to new products through economic and cultural globalization.

  There are, however, constraints on the world's ability to supply these new tastes because of the agricultural and environmental support system needed to produce and market meat. According to some food experts, the global food production system could sustain only half the current world population if everyone ate the meat-rich diet of North America, Europe, and Japan; this diet contains three or four times as much meat as the diet of people in developing countries.
- Most food supply experts agree that the two world regions of greatest concern are Africa and South Asia. Until 1970 Africa was self-sufficient in food, but since that time there have been serious disruptions and even breakdowns in the food supply system. The causes for

these problems in Africa are twofold—rapid population increase and civil disruption from tribal warfare. As a result, one of every four people faces food shortages in Sub-Saharan Africa. Chapter 6 contains more detail on these issues. South Asia's future is also problematic. The United Nations predicts that by the year 2010 almost 200 million people in South Asia will suffer from chronic undernourishment. Further discussion of these problems is found in Chapter 12.

There is some good news in this otherwise bleak picture. Because population growth rates are generally declining in the industrializing areas of East Asia and food production there is still increasing, the United Nations predicts that the percentage of undernourished people in that region will actually drop by almost 5 percent in the next 10 years. Similar gains are being made in Latin America because of lower population growth and higher agricultural production. World food experts predict that the proportion of chronically hungry in Latin America will fall to about 6 percent in 2010, which is half the rate of 1990.

Figure 2.26 Cattle Ranching in Brazil
Growing global demands for meat are reshaping the world's agricultural landscapes. Cattle ranching in western Brazil's rain forest reflects these changing market conditions, but the practice is also introducing new environmental problems. *(Martin Wendler/Peter Arnold, Inc.)*

Summary

- Globalization affects the environment in numerous ways, both good and bad. On the positive side, international agreements to preserve resources and protect the environment have become increasingly common. The Kyoto Protocol and treaties to protect Antarctica are examples.
- On the negative side (at least to those concerned with preserving the environment) is that economic globalization makes it easier for countries and corporations to fulfill their resource needs by reaching into remote corners of the world. Japan, for example, has long exploited forests around the globe in order to preserve its own woodlands, and, more recently, China's seemingly insatiable demand for wood products has become a major factor behind the continuing destruction of tropical rain forests.
- Climate change and global warming resulting from pollution of the atmosphere by greenhouse gases is a byproduct of industrialization, both past and present. A major facet of this problem is the burning of fossil fuels—petroleum and coal—that releases CO_2 into the atmosphere. Historically, the developed countries of Europe and North America were the major greenhouse gas producers; today, however, the developing economies of China and India have become major producers as well.
- Between the explosive population growth of the last century and the expansion of irrigated agriculture to feed hungry mouths, water has becoming an increasingly scarce resource. Global warming may aggravate this problem with higher temperatures and more frequent droughts.
- Plants and animals throughout the world face an extinction crisis because of habitat destruction from human activities, economic exploitation, and blatantly illegal poaching activities. Tropical forests are a focus of these problems since they contain the most plant and animal species of any bioregion, yet are threatened by numerous forces including the world's demands for wood products, cattle ranching for global meat-eaters, and resettlement of a country's own people.
- Even though the world's population is growing more slowly than in the past, and food supplies are increasing, there are still serious issues of food security for at least half the world's population. A raft of causes lie behind these problems, including governmental policies emphasizing export crops at the expense of crops fulfilling local food needs; international free trade policies and local protectionism that manipulates food markets; and political turmoil and favoritism that hinders effective food distribution within countries.

Key Terms

biome *(page 63)*
bioregion *(page 63)*
climate region *(page 55)*
climograph *(page 56)*
continentality *(page 54)*
convection cells *(page 48)*
convergent plate boundary *(page 49)*
desertification *(page 66)*
divergent plate boundary *(page 49)*
global warming *(page 56)*
greenhouse effect *(page 53)*
Green Revolution *(page 68)*
insolation *(page 53)*
maritime climate *(page 54)*
plate tectonics *(page 48)*

prairie *(page 66)*
rift valley *(page 49)*
steppe *(page 66)*
subduction zone *(page 49)*
subsistence agriculture *(page 68)*
tectonic plates *(page 49)*
water stress *(page 61)*

Questions for Review

1. Explain the role of convection cells in tectonic plate theory.
2. Describe several kinds of tectonic plate boundaries, along with the landforms and landscapes usually associated with these boundaries.
3. Describe the natural greenhouse effect.
4. How do continental and maritime climates differ? What causes this difference?
5. Which major human activities cause global climate change and warming? More specifically, how have greenhouse gases changed because of these activities?
6. Which natural disaster takes the highest human toll—earthquakes, volcanic eruptions, or flooding? Why?
7. Describe the ecological characteristics of a tropical rain forest.
8. What are the different causes of tropical forest deforestation? How are they linked to globalization?
9. What is desertification? Why, and where, is it a problem?
10. Describe the characteristics of the Green Revolution.

Thinking Geographically

1. What are the most threatening natural hazards in your region—earthquakes, tornados, hurricanes, floods, drought? Research local agencies or the public library to learn disaster preparedness plans for your community.
2. Which climate region do you live in? What are the major weather problems faced by people in your area? How do they adjust to these problems?
3. Visit some of the many Internet Websites that relate to global warming. Once you have an overview of the different positions, concentrate on one or two of the most contentious issues, such as the debate within the United States between environmentalists and business interests, or the difference in opinion between developed and developing countries. Discuss the different stakeholders and vested interests holding contrasting views.
4. How has the vegetation in your area been changed by human activities in the last 100 years? Has this change led to the extinction of any plants or animals or placed them on the endangered list? If so, what is being done to protect them or restore their habitat?
5. Study the globalization of wood products by acquainting yourself with the source areas for different items, such as building lumber, paper, furniture, or other items found in a local import store. More specifically, does the lumber used in construction in your area come from Canada or the United States? Which items in local stores come from tropical forests?
6. Conduct a detailed analysis of the food issues in a foreign country of your choice. Consider the following points. In general, has food supply kept up with population growth? Has the country suffered recently from food shortages or famine? If so, were these shortages the result of natural causes, such as drought or floods, or distribution problems resulting from civil disruption? Which segments of the population have food security and which segments have recurring problems obtaining adequate food? How is this inequity best explained? Finally, have food preferences or diets changed recently? If so, how and why?

Bibliography

American Association for the Advancement of Science. 2000. *AAAS Atlas of Population and Environment*. Berkeley: University of California Press.

Ecological Society of America. 2007. "Ecology in an Era of Globalization" [Special issue]. *Frontiers in Ecology and the Environment* 5(4).

Gleick, Peter, et al. 2002. *The World's Water: The Biennial Report on Freshwater Resources, 2002–2003*. Washington, DC: Island Press.

Hassol, Susan Joy. 2004. *Impacts of a Warming Arctic*. Cambridge, UK: Arctic Climate Impact Assessment, 2004, and Cambridge University Press.

IPCC. 2007. *Climate Change 2007: The Scientific Basis. A Report of Working Group 1 of the Intergovernmental Panel on Climate Change*. New York: United Nations Environmental Programme (UNEP).

IPCC. 2007. *Climate Change 2007: Synthesis Report. Summary for Policymakers. A Report of Working Group 1 of the Intergovernmental Panel on Climate Change*. New York: United Nations Environmental Programme (UNEP).

Lambin, Eric, and Geist, Helmut. 2003. "Regional Differences in Tropical Deforestation." *Environment* 45(6), 8ff.

Manning, Richard. 2000. *Food's Frontier: The Next Green Revolution*. New York: North Point Press.

Rosegrant, Mark, et al. 2003. "Will the World Run Dry? Global Water and Food Security." *Environment* 45(7), 24–36.

Shiva, Vandana. 2000. *Stolen Harvest: The Hijacking of the Global Food Supply*. Cambridge, MA: South End Press.

Additional bibliographic resources are at the *Diversity Amid Globalization* Website: http://www.prenhall.com/rowntree/.

3
North America

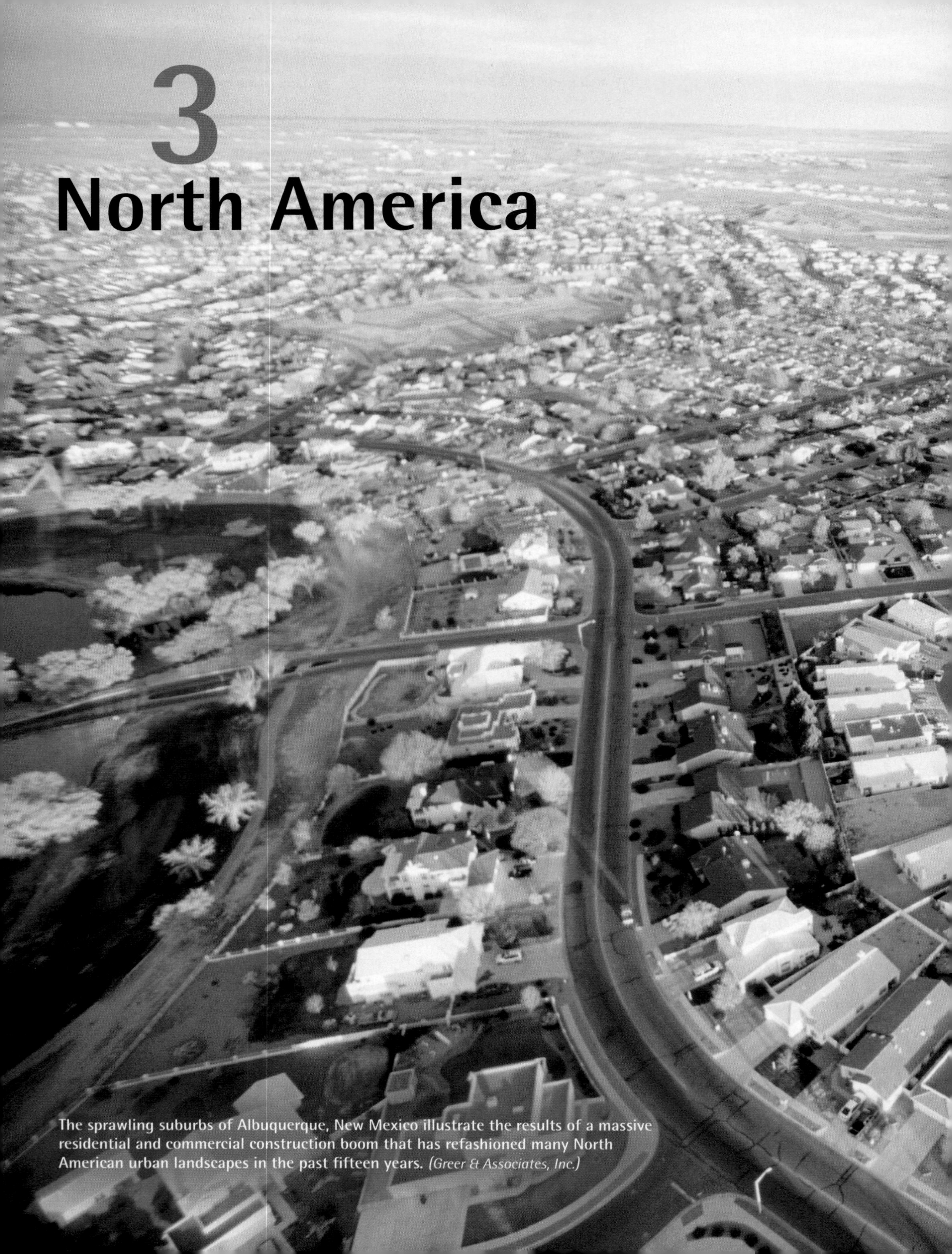

The sprawling suburbs of Albuquerque, New Mexico illustrate the results of a massive residential and commercial construction boom that has refashioned many North American urban landscapes in the past fifteen years. *(Greer & Associates, Inc.)*

ENVIRONMENTAL GEOGRAPHY

Fewer ski days in southern Quebec, widespread drought in the U.S. Southwest, and an increased frequency of hurricanes in the Gulf of Mexico may be only a few of the many consequences of global climate change within the North American region.

POPULATION AND SETTLEMENT

Sprawling suburbs characterize the expanding peripheries of hundreds of North American cities, creating a multitude of challenges including lengthier commutes for urban workers and the loss of prime agricultural lands along the edge of metropolitan areas.

CULTURAL COHERENCE AND DIVERSITY

Cultural pluralism remains strong in North America. Currently, there are more than 43 million immigrants living in the United States alone. The tremendous growth in Hispanic and Asian immigrants since 1970 has fundamentally reshaped the region's cultural geography.

GEOPOLITICAL FRAMEWORK

Cultural pluralism continues to shape political geographies within the region. American immigration policy remains hotly contested in the United States and persisting regional and native rights issues confront Canadians.

ECONOMIC AND SOCIAL DEVELOPMENT

The North American region plays a pivotal role in economic globalization, both as a leading driver of global change as well as a growing participant in an increasingly interconnected international economy.

Setting the Boundaries

The United States and Canada are commonly referred to as "North America," but that regional terminology can sometimes be confusing. In some geography textbooks the realm is called "Anglo America" because of its close and abiding connections with Britain and its Anglo-Saxon cultural traditions. The increasingly visible cultural diversity of the realm, however, has discouraged the widespread use of the term in more recent years. While more culturally neutral, the term *North America* also has its problems. As a physical feature, the North American continent commonly includes Mexico, Central America, and often the Caribbean. Culturally, however, the United States–Mexico border seems a better dividing line, although the growing Hispanic presence in the southwestern United States, as well as ever-closer economic links across the border, make problematic even that regional division. In some geography texts North America's boundaries extend northeastward to include Greenland, a colony of Denmark that is home to more than 55,000 indigenous people. To the west, North America reaches to Alaska's Aleutian Islands and to the state of Hawaii in the mid-Pacific Ocean. While the future may find Mexico even more intimately tied to its northern neighbors, our coverage of the "North American" realm concentrates on Canada and the United States, two of the world's largest and most affluent nation-states.

Stroll down any busy street in Toronto, Tucson, or Toledo, and count the ways in which international products, foods, culture, and economic connections shape the everyday scene. The simple visual lesson reveals that today North America stands at the center of the globalization process. From farmers to software engineers, most North Americans are employed in occupations that are either directly or indirectly linked to the global economy. Sizable foreign-born populations in each nation also provide direct links to every part of the world. Tourism brings in millions of additional foreign visitors and billions of dollars that are spent everywhere from Las Vegas to Disney World. In more subtle ways, North Americans embrace globalization in their everyday lives. They consume ethnic foods, tune in to televised international sporting events, enjoy the sounds of salsa and Senegalese music, surf the Internet from one continent to the next, and invest their pensions in global mutual funds.

Globalization also is a two-way street, and North American capital, culture, and power are ubiquitous. By any measure of multinational corporate investment and global trade, the region plays a dominant role that far outweighs its population of 335 million residents. North American consumer goods, information technology, and investment capital circle the globe. In addition, North American foods and popular culture are diffusing globally at a rapid pace. North American music, cinema, and fashion have also spread rapidly around the world. Soaring downtown skylines from Mumbai to Beijing increasingly resemble their North American counterparts.

The North American region encompasses the United States and Canada, a culturally diverse and resource-rich region that has seen unparalleled human modification and economic development over the past two centuries (see Figure 3.1 and "Setting the Boundaries"). The result is one of the world's most affluent regions, where two highly urbanized, mobile, and rapidly globalizing societies have the highest rates of resource consumption on Earth. Indeed, the region superbly exemplifies a **postindustrial economy** that is shaped by modern technology, by innovative financial and information services, and by a popular culture that dominates both North America and the world beyond.

Politically, North America is home to the United States, the last remaining global superpower. Such status brings the country onto center stage in times of global tensions, whether they are in the Middle East, South Asia, or West Africa. In addition, North America's largest metropolitan area of New York City (22 million people) is home to the United Nations and other global political and financial institutions.

North of the United States, Canada is the other political unit within the region. While slightly larger in area than the United States (3.83 million square miles [9.97 million square kilometers] versus 3.68 million square miles [9.36 million square kilometers]), Canada's population is only about 10 percent that of the United States.

Widespread abundance and affluence characterize North America. The region is extraordinarily rich in natural resources, such as navigable waterways, good farmland, fossil fuels, and industrial metals. The modern scene displays the results of combining that rich resource base with the acquisitive nature of its European colonizers and an accelerating pace of technological innovation. Indeed, contemporary North America displays both the bounty and the price of the development process. On one hand, the region shares the benefits of modern agriculture, globally competitive industries, excellent transport and communications infrastructures, and two of the most highly urbanized societies in the world. The cost of abundance, however, has been high: European settlers all but eliminated native populations, and later populations logged forests, converted grasslands into farms, eroded precious soils, threatened numerous species with extinction, diverted great rivers, and often wasted natural resources. Today, although home to only about 5 percent of the world's population, the region consumes 25 percent of the world's commercial energy budget and produces carbon

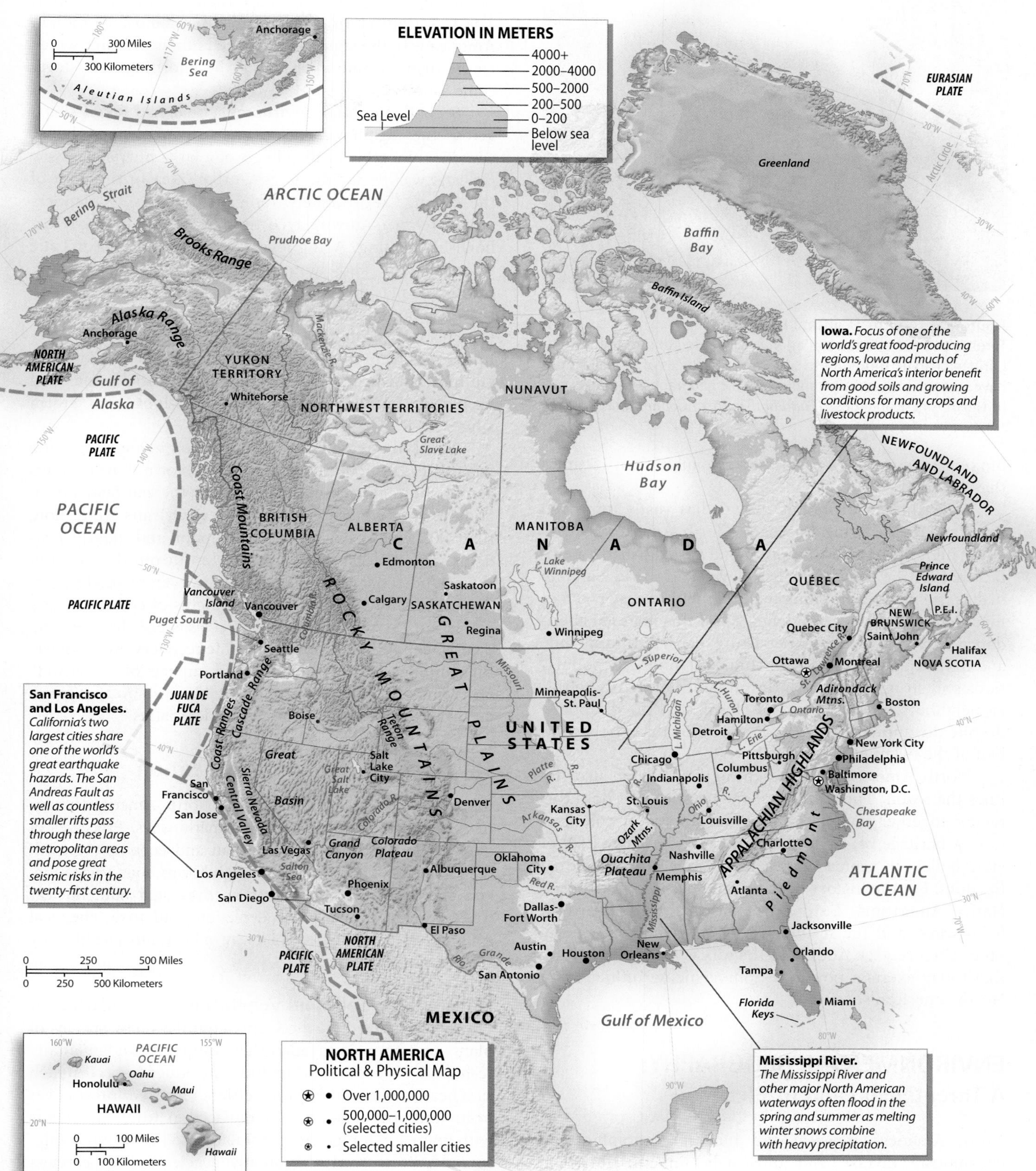

Figure 3.1 North America
North America plays a pivotal role in globalization. The region also contains one of the world's most highly urbanized and culturally diverse populations. With 335 million people and extensive economic development, North America is also one of the largest consumers of natural resources on the planet.

Figure 3.2 Toronto's Cultural Landscape
Toronto's varied ethnic population is celebrated during the Caribana Parade through the city. Powerful forces of globalization have reshaped the cultural and economic geographies of dozens of North American cities. *(Canadian Tourism Commission)*

dioxide emissions at a per person rate more than 15 times that of India.

Nevertheless, economic growth has vastly improved the standard of living for many North Americans, who enjoy high rates of consumption and varied urban amenities that are the envy of the less-developed world. Satellite dishes, sushi, and shopping malls are within easy reach of most North American residents. Amid this material abundance, however, there are persisting disparities in income and in the quality of life. Poor rural and inner-city populations still struggle to match the affluence of their wealthier neighbors, and poverty has been slow to disappear, despite the unprecedented economic growth of the second half of the twentieth century.

North America's unique cultural character also defines the realm. The cultural characteristics that hold this region together include a common process of colonization, a heritage of Anglo dominance, and a shared set of civic beliefs in representative democracy and individual freedom. But the history of the region has also juxtaposed Native Americans, Europeans, Africans, and Asians in fresh ways, and the results are two societies unlike any other (Figure 3.2). Adding to the mix is a popular culture that today exerts a powerful homogenizing influence on North American society.

ENVIRONMENTAL GEOGRAPHY: A Threatened Land of Plenty

North America's physical and human geographies are enormously diverse and intricately linked. Hurricane Katrina's tumultuous arrival along the Gulf coast in August 2005 exemplifies the complexities and significance of those interconnections. Three intertwined variables came into play. First, the storm itself was large and powerful, sweeping inland across southern Louisiana and coastal Mississippi with winds of more than 120 miles per hour. The U.S. Census Bureau estimates that almost 10 million Americans in the region experienced hurricane force winds (more than 75 miles per hour) the morning the storm came ashore.

Second, the built environment that had taken shape across the region was extensive, with many population centers vulnerable to the effects of a large hurricane. New Orleans was by far the largest urban area in the path of the storm. Unfortunately, for generations, city planners and developers downplayed the hazards of its low-lying, bowl-like setting. Aging levees were breached near Lake Pontchartrain and along several commercial canals, putting 80 percent of the city under water (Figure 3.3). To the east, populated coastal areas of southern Mississippi were also devastated by the storm: casinos built on shoreline barges were tossed off their foundations in Biloxi and Gulfport and nearby rural areas saw historic homes and farmlands inundated by Katrina's 30-foot tidal storm surge.

Third, the region's social geography made evacuating urban areas difficult in the face of the storm. Poverty rates across the region are almost double the national norm. Some of the worst-hit areas of New Orleans were more than 80 percent black. The city's poor and elderly residents were its least mobile, forced to ride out the storm in their flooded neighborhoods or in the overcrowded Superdome. Added to this were the problems of coordinating a viable emergency response to the disaster. The resulting deaths, damage, and slow economic recovery are a powerful reminder of how the costs and impacts of a "natural" environmental disaster are inevitably wed to a region's cultural, social, and economic characteristics.

The Costs of Human Modification

Katrina's story is a reminder that North Americans have modified their physical setting in many ways. Processes of globalization and accelerated urban and economic growth have transformed North America's landforms, soils, vegetation, and climate. Indeed, problems such as acid rain, nuclear waste storage, groundwater depletion, and toxic chemical spills are all manifestations of a way of life unimaginable only a century ago (Figure 3.4).

Transforming Soils and Vegetation Particularly after Europeans arrived on the scene, countless new species of plants and animals invaded the continent, fundamentally reshaping its ecology to include introduced species such as wheat, cattle, and horses. Settlers removed forest cover from millions of acres and replaced natural grasslands with grain and forage crops not native to the region. Widespread soil erosion was greatly accelerated by careless cropping and ranching practices, and many areas of the Great Plains and South suffered lasting damage. With twentieth-century urbanization, many earlier agricultural transformations were swept aside to make way for suburbs and shopping centers. From the nonnative lawns of

(a)

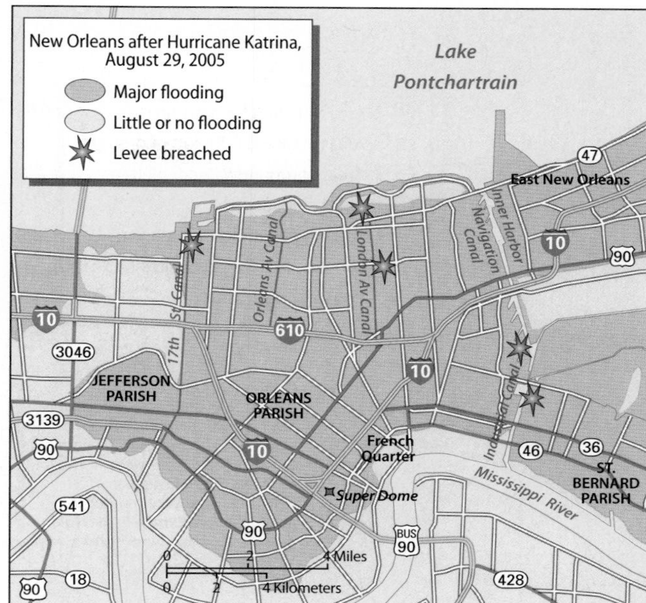

(b)

(c)

Figure 3.3 Hurricane Katrina
Approaching from the Gulf of Mexico (a), Hurricane Katrina put much of New Orleans under water (b), and inflicted particular devastation upon the city's poor, black population (c). *(a, NASA Jet Propulsion Laboratory; b, James M. Rubenstein,* The Cultural Landscape, 8e, *© 2008. Reprinted with permission of Pearson Prentice Hall; c, John McCusker/Landov LLC)*

Southern California to the wheat fields of Montana and Alberta, North Americans today are living in rural and urban environments that bear little resemblance to settings across the region two centuries ago.

Managing Water The United States is the third largest consumer of water in the world. Only population giants China and India consume more. Although conservation efforts and better technology have slightly reduced per capita rates of water use over the past 25 years, city dwellers still directly consume an average of more than 175 gallons daily. Indirectly, through other forms of food and industrial consumption, Americans may consume more than 1,400 gallons of water per day.

Many North American localities are threatened by water shortages. Metropolitan areas in eastern Canada and the United States struggle with outdated municipal water supply systems. Beneath the Great Plains, the waters of the Ogallala Aquifer are also being depleted. The largest in North America, this aquifer is a huge reserve of precious groundwater created during the last Ice Age, and today it irrigates about 20 percent of all U.S. cropland. Center-pivot irrigation systems are steadily tapping into the supply; water tables across much of the region have fallen more than 100 feet (30 meters) in the past 50 years; and the costs of pumping are rising steadily. Farther west, California's elaborate system of water management is a reminder of that state's ever-growing demands (Figure 3.5). Aqueducts move the precious resource from an already overutilized Colorado River system into Southern California's agricultural and metropolitan regions.

Issues of water quality are equally problematic. North Americans are exposed to water pollution in varied ways, and even mandates such as the U.S. Clean Water Act or Canada's Green Plan cannot stop ongoing abuses. Mining operations and industrial users such as chemical, paper, and steel plants generate toxic wastewater or metals that enter surface and groundwater supplies. Accidental chemical and petroleum spills are another ongoing threat. Metropolitan areas also produce vast amounts of raw sewage that annually costs billions to repurify.

Overall, the United States generates hazardous wastes at a much higher per capita level than either Canada or western Europe. America's most toxic localities include the petroleum-rich Texas and Louisiana Gulf coasts, the older industrial centers of the Northeast and Midwest, and in nuclear fuel and chemical warfare storage areas such as Hanford, Washington. In addition, the widespread use of fertilizers and pesticides in modern North American agriculture raises land productivity, but many of these chemicals are carried into natural runoff. In the U.S. Southwest, another critical water problem is salinization, a process in which irrigation water leaves behind a toxic buildup of salts in the soil.

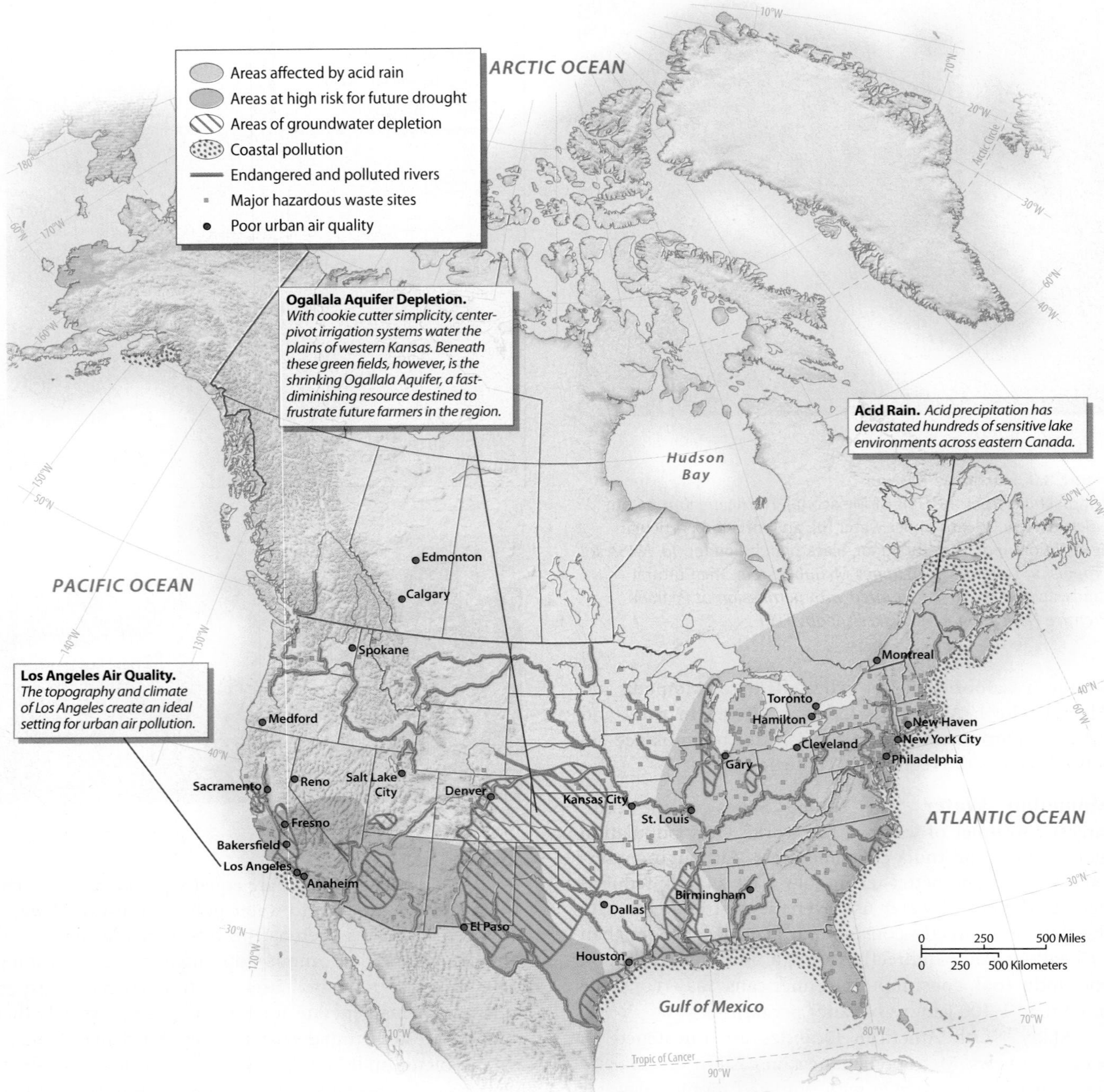

Figure 3.4 Environmental Issues in North America
Many environmental issues threaten North America. Acid rain damage is widespread in regions downwind from industrial source areas. Elsewhere, widespread water pollution, cities with high levels of air pollution, and zones of accelerating groundwater depletion pose health dangers and economic costs to residents of the region. Since 1970, however, both Americans and Canadians have become increasingly responsive to the dangers posed by these environmental challenges.

Altering the Atmosphere North Americans also modify the very air they breathe; in doing so, they change local and regional climates as well as the chemical composition of the atmosphere. Urban areas produce "islands" of heat, which on clear, calm nights generate nighttime temperatures some 9 to 14°F (5 to 8°C) warmer than nearby rural areas. Most significantly, air pollution affects large numbers of plants, animals, and people in many localities. At the local level, industries, utilities, and automobiles contribute carbon monoxide, sulfur, nitrogen oxides, hydrocarbons, and particulates to the urban atmosphere. While some of the region's worst offenders are U.S. cities such as Houston and Los Angeles, Canadian cities such as Toronto, Hamilton, and Edmonton also experience significant problems of air quality.

Figure 3.5 California Aqueduct
Thirsty Californians have extensively modified the geography of water in their state. Large aqueducts have dramatically reconfigured the distribution of this precious resource, promoting tremendous agricultural and metropolitan expansion within western North America. *(Alexander Lowry/Photo Researchers, Inc.)*

On a broader scale, North America is plagued by **acid rain**, industrially produced sulfur dioxide and nitrogen oxides in the atmosphere that damage forests, poison lakes, and kill fish. Many acid rain producers are located in the Midwest and southern Ontario, where industrial plants, power-generating facilities, and motor vehicles contribute emissions. Prevailing winds transport the pollutants and deposit damaging acidic precipitation across the Ohio Valley, Appalachia, the northeastern United States, and eastern Canada (Figure 3.4).

Air pollution is also going global, freely drifting into Canada and the United States on prevailing winds. One recent estimate suggests that about 30 percent of the region's ozone comes from beyond its borders. With China's rapid industrialization, airborne mercury from Asia is increasingly being deposited in North American rivers and lakes, putting fish populations at greater risk. Smog from Mexico's cities and power plants are also affecting settings such as California's Sequoia National Park or Texas's Big Bend National Park, reducing visibilities and air quality in these natural areas.

The Price of Affluence

Globalization has brought many benefits to North America, but with the accompanying urbanization, industrialization, and heightened consumption, the realm is also paying an environmental price for its affluence. Energy consumption within the region, for example, remains extremely high, imposing a growing list of environmental and economic costs both within North America and beyond. North Americans consume energy at a per capita rate almost double that of the Japanese and more than 16 times that of India's population. For all its economic growth and material wealth, the twentieth century in North America will also be remembered for its toxic waste dumps, frequently unbreathable air, and wildlands lost to development. Still, many environmental initiatives in the United States and Canada have addressed local and regional problems. For example, the improved water quality of the Great Lakes over the past 30 years is an achievement to which both nations contributed and from which both benefit. Tougher air quality standards have

Figure 3.6 Satellite Image of Chesapeake Bay
This view of the Mid-Atlantic coast reveals the intricate shoreline of Chesapeake Bay (lower center). The coast sector is characterized by drowned river valleys, barrier islands, and sandy beaches. The Piedmont zone and Appalachian Highlands appear to the northwest. *(Earth Satellite Corporation/Science Photo Library/ Photo Researchers, Inc.)*

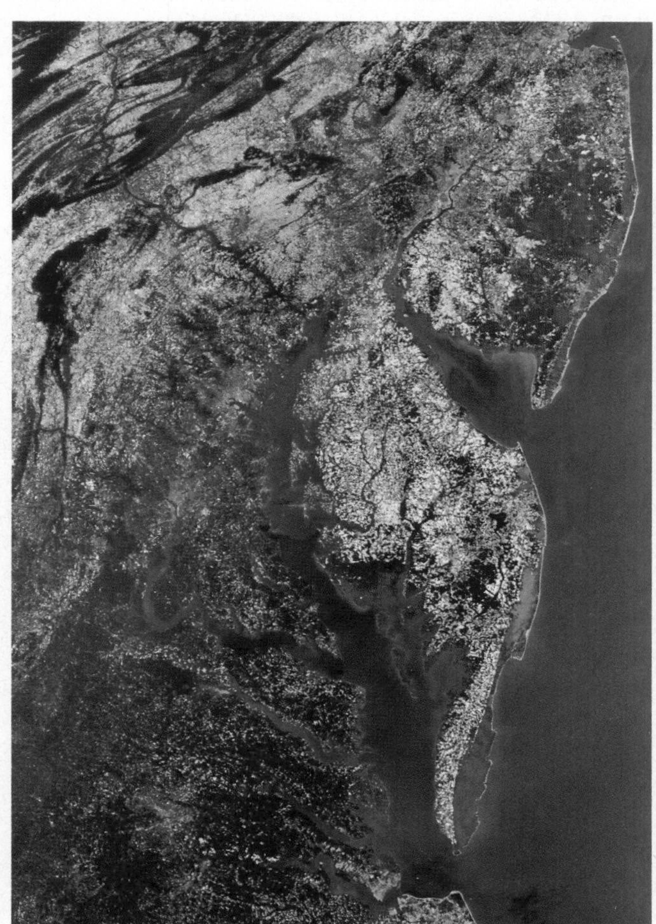

selectively reduced emissions in many North American cities. Perhaps most important, there is growing North American support for green industries and technologies. In the United States alone, Americans invested almost $30 billion in alternative energy sources in 2006 and recent court rulings and political initiatives (many at the state level) have encouraged environmentally friendly innovation.

A Diverse Physical Setting

North America's complex environmental challenges need to be understood in the context of its underlying physical geography. The regional landscape is dominated by vast interior lowlands bordered by more mountainous topography in the western portion of the region (Figure 3.1). Extensive coastal plains stretch from southern New York to Texas and include a sizable portion of the lower Mississippi Valley. These plains are flat, sometimes poorly drained, and prone to coastal and river flooding. The present coastline is an intricate landscape of drowned river valleys, bays, swamps, and low barrier islands (Figure 3.6).

The nearby Piedmont consists of rolling hills and low mountains that are much older and less easily eroded than the lowlands. West and north of the Piedmont are the Appalachian Highlands, an internally complex zone of higher and rougher country that typically crests between 3,000 and 6,000 feet (915 and 1,829 meters, respectively). From Pennsylvania south, folded layers of sedimentary rock, often rich in coal, create a diverse setting of ridges, valleys, and plateaus. North of Pennsylvania, the underlying rock is crystalline, and surface forms are more irregular. Here the highlands reach the Atlantic and include New England and the Canadian Maritimes. Far to the southwest, Missouri's Ozark Mountains and the Ouachita Plateau of northern Arkansas resemble portions of the southern Appalachians.

Much of the North American interior is a vast and easily accessible lowland (Figure 3.1). These mainly depositional plains have accumulated thick sediments, and they extend from west central Canada to the coastal lowlands near the Gulf of Mexico. Eastward, they include the southern Great Lakes and the lower Ohio River valley. To the west, the Great Plains are formed from younger sediments that have been eroded from the Rocky Mountains. Glacial forces, particularly north of the Ohio and Missouri rivers, also have actively carved and reshaped the landscapes of this lowland zone.

In the West, mountain-building (including large earthquakes and volcanic eruptions), alpine glaciation, and often spectacular processes of erosion produce a regional topography quite unlike that of eastern North America. The Rocky Mountains are a series of uplifts, many more than 10,000 feet (3,048 meters) in height, that stretch from Alaska's Brooks Range to northern New Mexico's Sangre de Cristo Mountains (Figure 3.7). West of the

Figure 3.7 Rocky Mountains
Montana's Glacier National Park reveals the characteristic signatures of alpine glaciation that are found in many portions of the Rocky Mountain region, both in the United States and in Canada. *(Joe Sohm/The Image Works)*

Rockies, the Colorado Plateau includes uplifted blocks of highly colorful sedimentary rock eroded into spectacular buttes and mesas. Nevada's vast and sparsely settled basin and range country features numerous north-south-trending mountain ranges interrupted by structural basins with no outlet to the sea. North America's western border is marked by the mountainous and rain-drenched coasts of southeastern Alaska and British Columbia; the Coast Ranges of Washington, Oregon, and California; the lowlands of the Puget Sound, Willamette Valley, and Central Valley; and the complex uplifts of the Cascade Range and Sierra Nevada.

Patterns of Climate and Vegetation

North America's climates and vegetation are as varied as its landforms. Not surprisingly, the region's size, latitudinal range, and varied terrain have contributed to a diversity of temperature and precipitation patterns (Figure 3.8). These, in turn, have shaped patterns of vegetation. Much of North America south of the Great Lakes is characterized by a relatively long growing season, 30 to 60 inches (76.2 to152.4 centimeters) of precipitation annually, and a deciduous broadleaf forest that has been extensively modified by commercial agriculture. From the Great Lakes north, the coniferous evergreen or **boreal forest** dominates the cooler continental interior, and these conditions extend to sparsely settled portions of eastern Quebec. Near Hudson Bay and across harsher northern tracts, trees give way to **tundra**, a mixture of low shrubs, grasses, and flowering herbs that grow briefly in the short growing seasons of the high latitudes.

The drier continental climates found from West Texas to Alberta feature tremendous seasonal ranges in temperature, frequent winds (including tornadoes), and unpredictable precipitation that averages between 10 and 30 inches (25.4 and 76.2 centimeters) annually. Indeed, the region straddles the transition zone between humid and arid portions of the North American interior, a boundary broadly defined as the 20-inch line of annual precipitation that runs near the 100th meridian. Much of this subhumid region is blessed with fertile soils and was originally clothed in **prairie** vegetation dominated by tall grasslands in the East and by short grasses and scrub vegetation in the West.

Western North American climates and vegetation are greatly complicated by the Pacific Ocean and the region's complex landforms. Marine west-coast climates dominate north of San Francisco, while a dry summer Mediterranean climate is present across central and southern California. Farther east, the intermontane interior and the Rocky Mountains experience the typical seasonal variations of the middle latitudes, but the patterns are modified by the effects of topography.

North America and Global Climate Change

Over the longer term, North America's present-day climates and vegetation are merely a snapshot in time, reflecting dynamic natural processes as well as the growing accumulation of human impacts. Accelerating rates of human-influenced climate change appear destined to complicate the situation further. Many quiet transformations are already underway. High latitude and alpine environments continue to be particularly vulnerable. Changes in arctic temperatures, sea ice, and sea levels have increased coastal erosion, impacted migrating whale and polar bear populations, and stressed traditional ways of life for Eskimo and

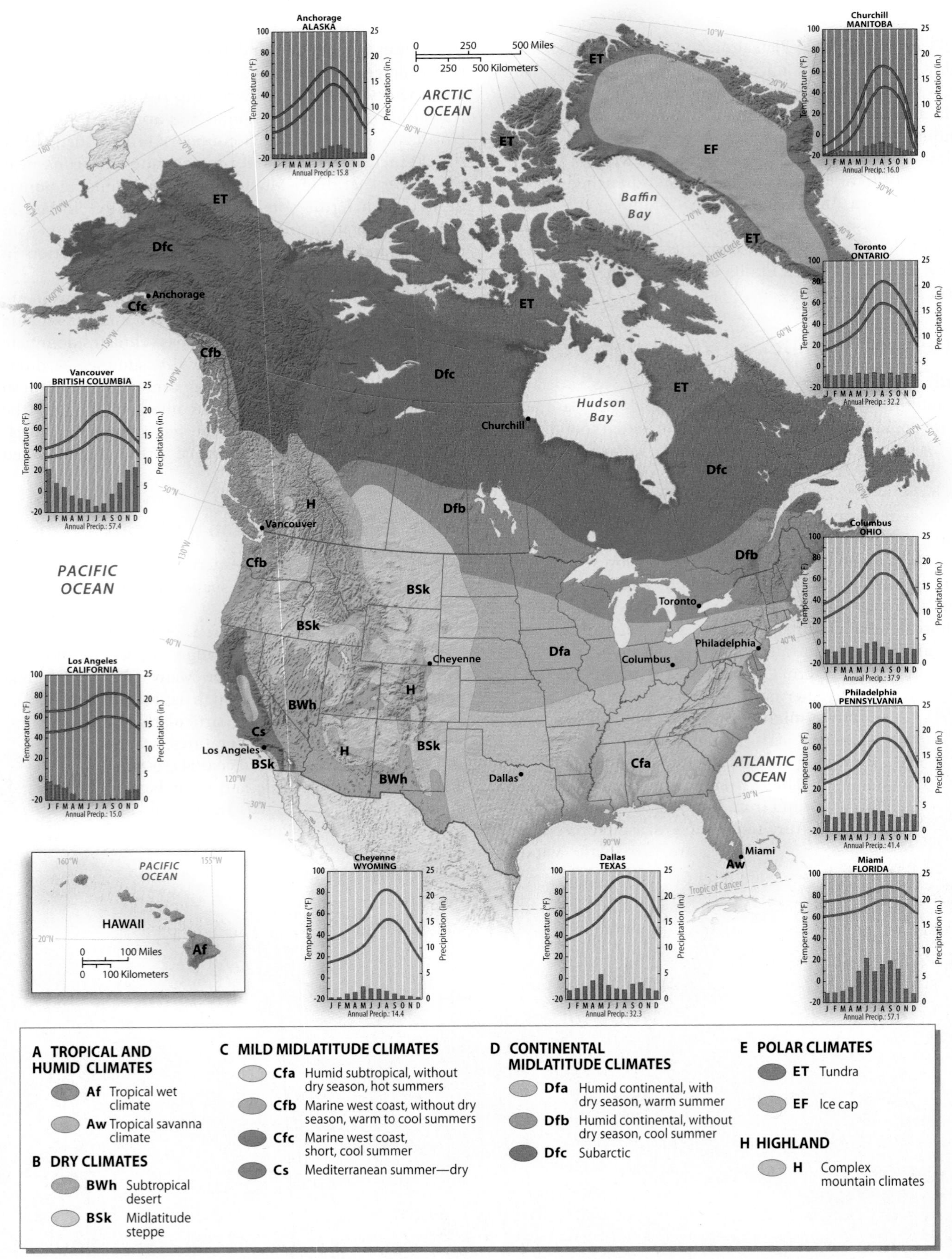

Figure 3.8 Climate Map of North America
North American climates include everything from tropical savanna (Aw) to tundra (ET) environments. Most of the region's best farmland and densest settlements lie in the mild (C) or continental (D) midlatitude climate zones.

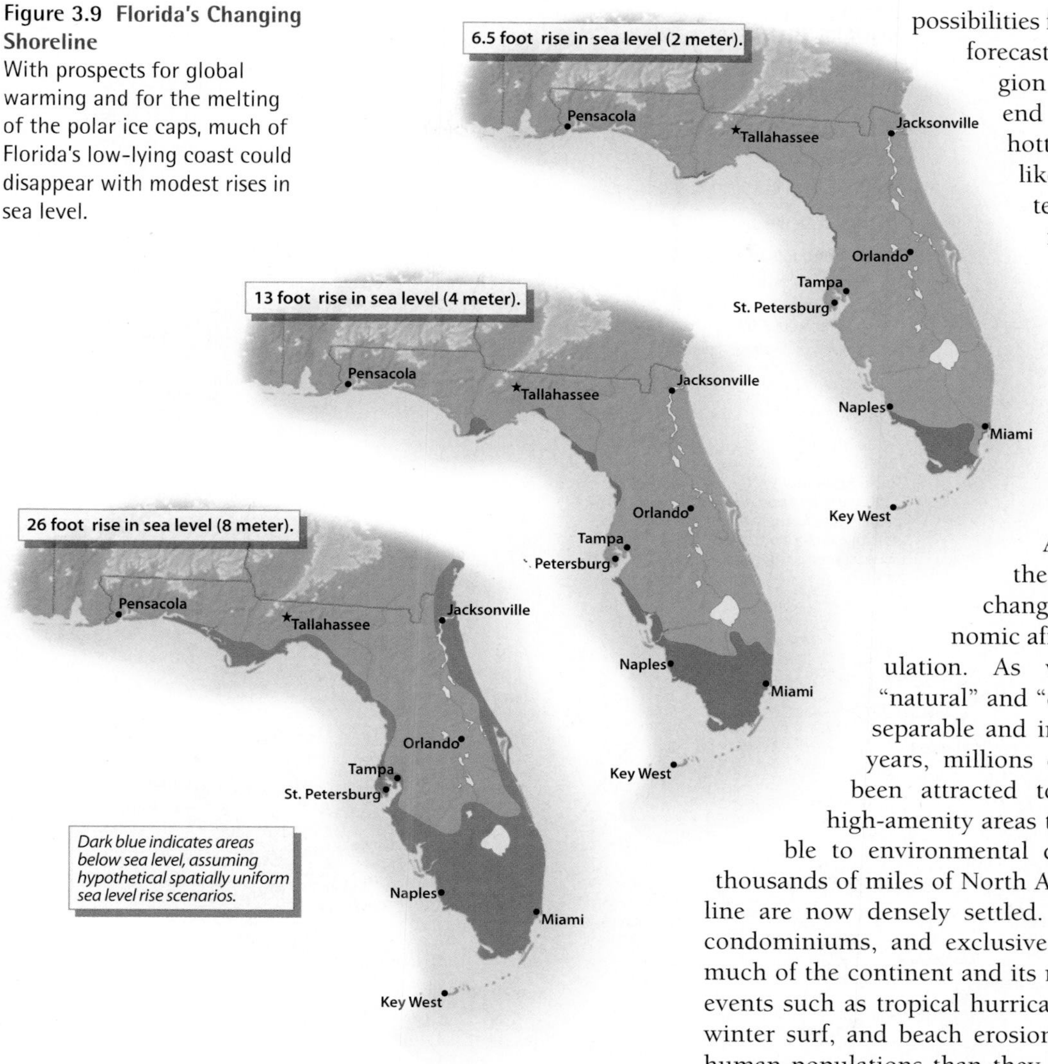

Figure 3.9 Florida's Changing Shoreline
With prospects for global warming and for the melting of the polar ice caps, much of Florida's low-lying coast could disappear with modest rises in sea level.

6.5 foot rise in sea level (2 meter).

13 foot rise in sea level (4 meter).

26 foot rise in sea level (8 meter).

Dark blue indicates areas below sea level, assuming hypothetical spatially uniform sea level rise scenarios.

Inuit populations. In western North American mountains, pikas are disappearing from their alpine habitats across the Great Basin and expanding mountain pine beetle populations are rapidly infesting lodgepole and whitebark pine forests in the northern Rocky Mountains. Elsewhere, Great Lakes water temperatures have warmed more than 4°F (2°C) since the 1970s, changing regional weather patterns. Mexican butterflies are breeding in Texas, armadillos (earlier natives of Texas) are marching into southern Illinois, and exotic fish species are migrating north into the Chesapeake Bay and offshore Southern California.

The cumulative long-term consequences of climate change for North Americans are enormous. Basic redistributions of native plants, animals, and crops are already underway and may accelerate. For example, high-latitude distributions of tundra and permafrost environments may shift more than 300 miles (480 km) poleward by 2050, dramatically impacting wildlife populations and human settlements in these regions. For northern U.S. and Canadian farmers, longer growing seasons may open up new possibilities for agriculture. Experts also forecast that the Great Lakes region will become wetter by the end of the century. But both hotter and drier conditions are likely in the Southwest, portending more western wildfires and growing resource conflicts in that rapidly growing region of the United States (Figure 3.4). Extreme weather events such as heat waves or tornadoes will also become more common.

Ironically, North America's vulnerability to the effects of global climate change is magnified by the economic affluence of the region's population. As with Hurricane Katrina, "natural" and "cultural" influences are inseparable and interwoven. In the past 50 years, millions of North Americans have been attracted to precisely those scenic, high-amenity areas that are especially vulnerable to environmental disruptions. For example, thousands of miles of North America's spectacular coastline are now densely settled. Resort hotels, retirement condominiums, and exclusive private estates now ring much of the continent and its nearby islands. As a result, events such as tropical hurricanes, coastal storms, heavy winter surf, and beach erosion present more hazards to human populations than they did a century or two ago. Future global warming will also raise global sea levels, increasing these hazards even more dramatically in coming centuries, particularly for low-lying zones along the East Coast, in the North American arctic, and in the Gulf of Mexico (Figure 3.9).

POPULATION AND SETTLEMENT: Reshaping a Continental Landscape

The North American landscape is the product of four centuries of extraordinary human change. During that period, Europeans, Africans, and Asians converged upon the region, displaced a continental expanse of Native American peoples, and created a new geography of human settlement. Today, more than 335 million people live in North America, and they are some of the world's most affluent and highly mobile populations. The contemporary North American scene dramatically displays how its population has changed the settlement landscape to meet the needs of a modern, postindustrial society.

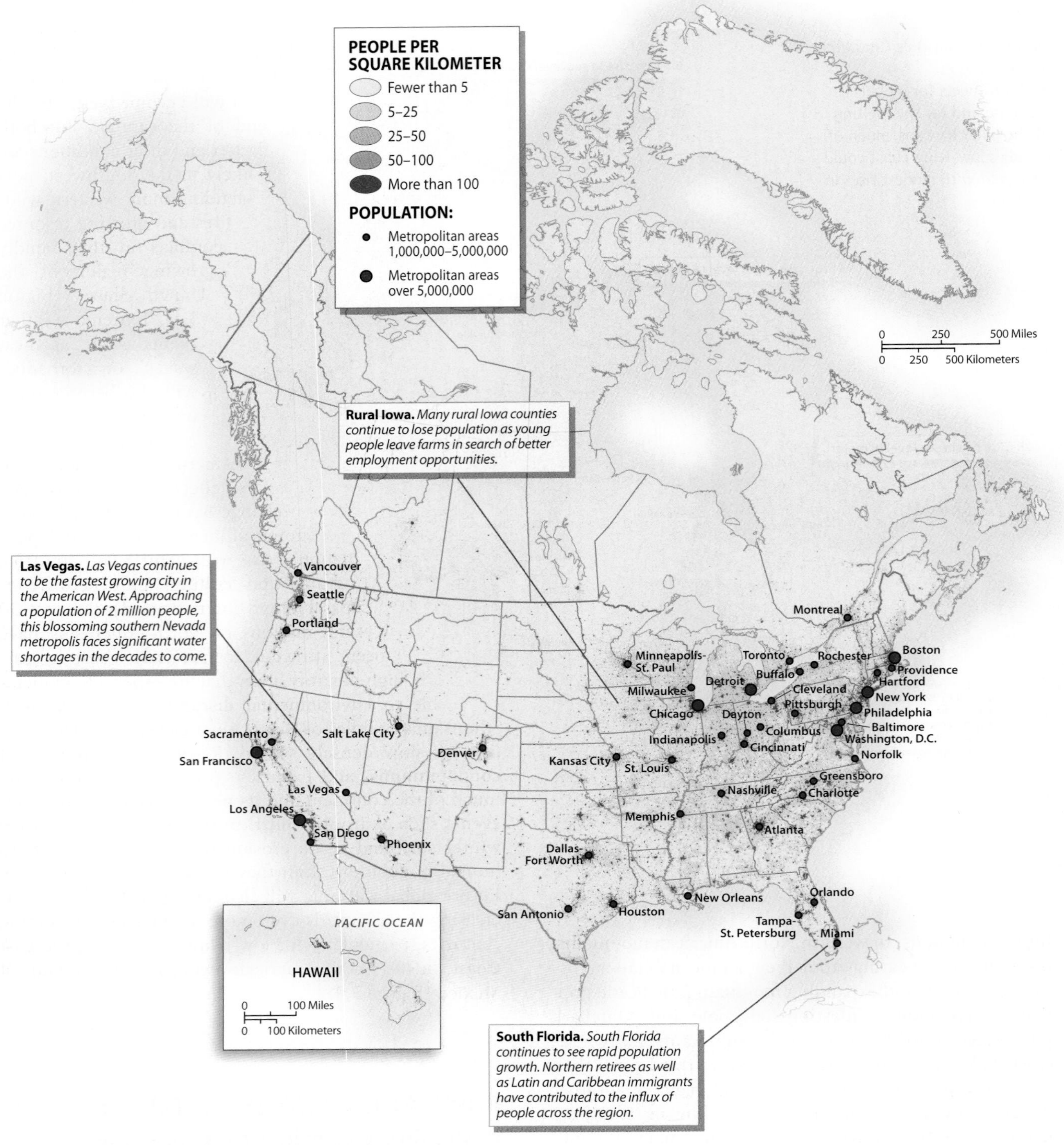

Figure 3.10 Population Map of North America
North America's geography of population reveals a strikingly clustered pattern of large cities interspersed with more sparsely settled zones. Notable concentrations are found on the Eastern Seaboard between Boston and Washington, D.C., along the shores of the Great Lakes, and across the Sun Belt from Florida to California.

Modern Spatial and Demographic Patterns

Metropolitan clusters dominate North America's population geography, producing strikingly uneven patterns of settlement across the region (Figure 3.10). In Canada, about 90 percent of the population is found within 100 miles (160 km) of the U.S. border. Within this broad region, Canada's "Main Street" corridor contains most of that nation's urban population, led by the cities of Toronto (5.4 million) and Montreal (3.7 million). The federal capital of Ottawa (1.2 million) and the industrial center of Hamilton (700,000) are also within the Canadian urban

TABLE 3.1 Population Indicators

Country	Population (Millions, 2007)	Population Density (per Square Kilometer)	Total Fertility Rate	Percent Urban	Percent <15 Years of Age	Percent >65 Years of Age	Net Migration (Rate per 1,000) 2000–2005
Canada	32.9	3	1.5	81	18	13	6.7
United States	302	31	2.1	79	20	12	4.0

Sources: Population Reference Bureau, World Population Data Sheet, *2007; net migration rate data from* UN International Migration, *2006.*

corridor. **Megalopolis**, the largest settlement agglomeration in the United States, includes Washington, DC (5.3 million), Baltimore (2.7 million), Philadelphia (6.4 million), New York City (22 million), and Boston (4.5 million). Beyond these two national core areas, other sprawling urban centers cluster around the southern Great Lakes, various parts of the South, and along the Pacific Coast.

North America's population has increased greatly since the beginning of European colonization. Before 1900, high rates of natural increase produced large families. In addition, waves of foreign immigration swelled settlement, a pattern that continues today. In Canada, a population of fewer than 300,000 Indians and Europeans in the 1760s grew to an impressive 3.2 million a century later. For the United States, a late colonial (1770) total of around 2.5 million increased more than 10-fold to more than 30 million by 1860. Both countries saw even higher rates of immigration in the late nineteenth and twentieth centuries, although birthrates gradually fell after 1900. After World War II, birthrates rose once again in both countries, resulting in the "baby boom" generation born between 1946 and 1965. Today, however, as in much of the developed world, rates of natural increase in North America are below 1 percent annually (Table 3.1), and the overall population is growing older. Still, the region continues to attract immigrants: more than 43 million foreign-born migrants now live in North America. These growing numbers, along with higher birthrates among immigrant populations, recently have led demographic experts to increase long-term population projections for the twenty-first century. Indeed, new predictions by the United Nations that by 2050 the region's population will reach 445 million (402 million in the United States and 43 million in Canada) may prove conservative.

Occupying the Land

When Europeans began occupying North America about 400 years ago, they were not peopling an empty land. North America was populated for at least 12,000 years by peoples as culturally diverse as those who came to conquer them. Native Americans were broadly distributed across the region and made diverse adaptations to its many natural environments. Their precontact numbers are impossible to reconstruct precisely, although cultural geographers estimate populations in 1500 CE at 3.2 million for the continental United States and another 1.2 million for Canada, Alaska, Hawaii, and Greenland. Agricultural societies became well established across many portions of the Northeast, South, and Southwest prior to contact with Europe. Mississippian culture, for example, thrived in the Midwest, supporting dense populations, well-tilled fields, and sizable villages. Elsewhere, hunting, gathering, and fishing supported native populations in regions such as the interior West and Northwest coast. European diseases and disruption decimated these Native American populations as contacts increased. Native societies could hardly have anticipated the magnitude of coming changes after 1600 as the European world expanded its reach. The continental sweep of settlement that followed took shape in three stages, and by the twenty-first century the results had fundamentally changed North America's human geography.

The first stage of this dramatic new settlement geography began with a series of European colonial footholds, mostly within the coastal regions of eastern North America (Figure 3.11). Established between 1600 and 1750, these regionally distinct societies were anchored on the north by the French settlement of the St. Lawrence Valley. English Puritans dominated nearby southern New England, imposing their own cultural traditions by the 1640s. Farther south, both the Dutch colony of New Netherlands (later English-controlled New York) and the English Quaker colony of Pennsylvania attracted a varied collection of farmers, merchants, and tradesmen. Bicultural European and African settlements concentrated in the plantation South, with the largest colonies in English-controlled Virginia and South Carolina. Additional French settlements concentrated along the Gulf of Mexico (New Orleans was founded in 1718), and there was an early Spanish presence in the Southwest (Santa Fe was founded in 1610) and Florida (St. Augustine was founded in 1565).

The second stage in the Europeanization of the North American landscape took place between 1750 and 1850, and it was highlighted by the infilling of much of the better agricultural land within the eastern half of the continent. Restrictive English colonial policies failed to deter frontier settlement in the upper Ohio and Tennessee valleys. Following the American Revolution (1776) and a series of Indian conflicts, pioneers surged across the Appalachians. They found much of the Interior Lowlands region almost

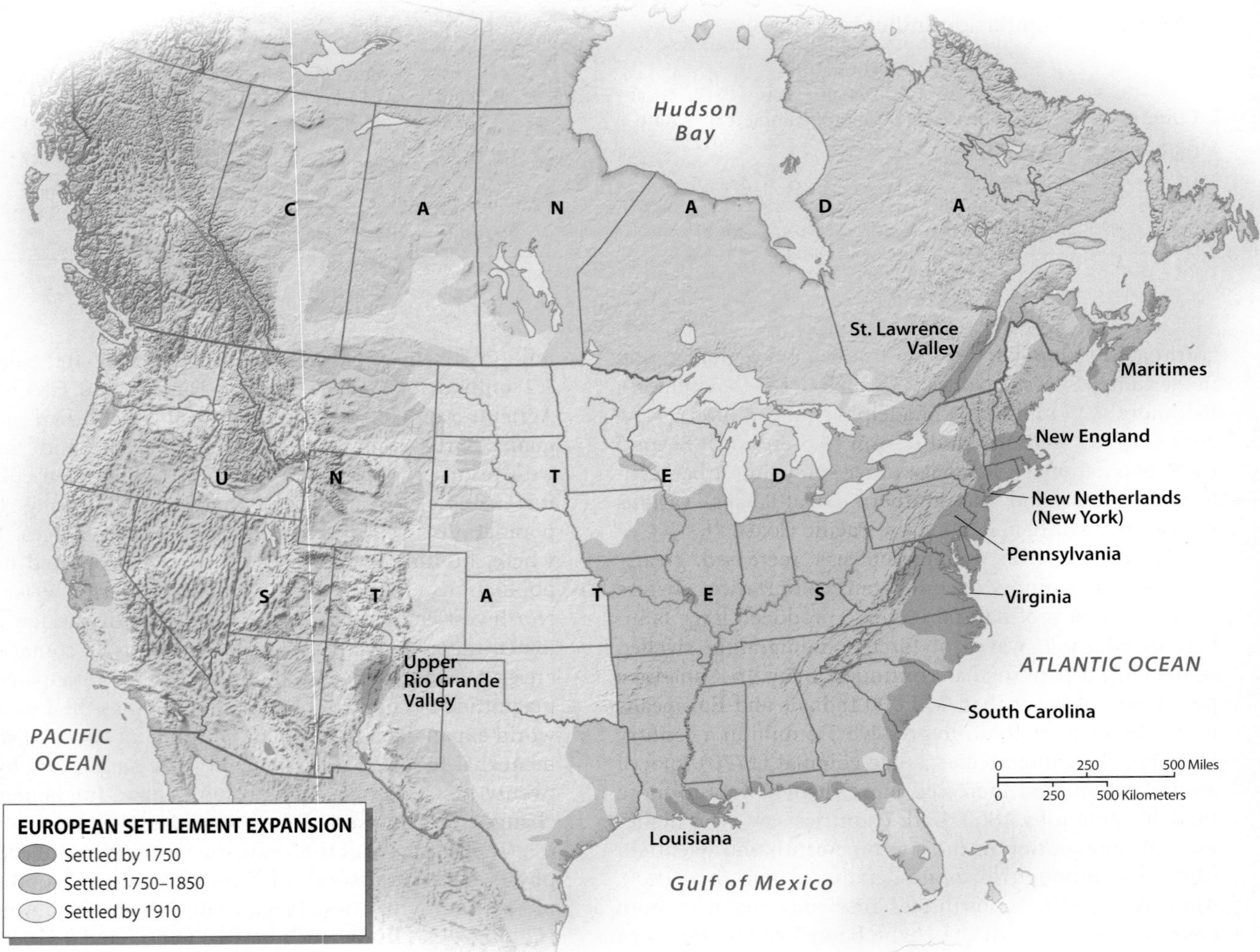

Figure 3.11 European Settlement Expansion
Sizable portions of North America's East Coast and the St. Lawrence Valley were occupied by Europeans before 1750. The most remarkable surge of settlement occurred during the next century as Europeans opened vast areas of land and dramatically disrupted Native American populations.

ideal for agricultural settlement. Expansion in early Canada was more modest. Few major changes in settlement came as the region shifted from French to English control in 1763. Much of southern Ontario, or Upper Canada, was opened to widespread development in 1791.

The third stage in North America's settlement expansion accelerated after 1850 and continued until just after 1910. During this period, most of the region's remaining agricultural lands were settled by a mix of native-born and immigrant farmers. Often, pioneers tested their luck and ingenuity in unfamiliar environmental surroundings. Farmers were challenged and sometimes defeated by aridity, mountainous terrain, and short northern growing seasons. In the American West, settlers were attracted by opportunities in California, the Oregon country, Mormon Utah, and the Great Plains. In Canada, few flocked to the far north, but thousands occupied southern portions of Manitoba, Saskatchewan, and Alberta. Mineral rushes led to initial development in areas such as Colorado, Montana, and British Columbia's Fraser Valley.

Incredibly, in a mere 160 years, much of the North American landscape was domesticated as expanding populations sought new land to occupy and as a globe-encircling capitalist economy demanded resources to fuel its growth. It was one of the largest and most rapid human transformations of the landscape in the history of the human population. This European-led advance forever reshaped North America in its own image, and in the process also changed the larger globe in lasting ways by creating a "New World" destined to reshape the Old.

North American Migration

From the mythic days of Davy Crockett and Daniel Boone to the twentieth-century sojourns of John Steinbeck and Jack Kerouac, North Americans have been on the move.

Indeed, almost one in every five Americans moves annually, suggesting that residents of the region are quite willing to change addresses in order to improve their income or their quality of life. Although interregional population flows are complex in both the United States and Canada, several trends dominate the picture.

Westward-Moving Populations The most persistent regional migration trend in North America has been the tendency for people to move west. Indeed, for two centuries people have followed the setting sun, and many North Americans continue that pattern to the present (see Figure 3.11). By 1990, more than half of the population of the United States lived west of the Mississippi River, a dramatic shift from colonial times. Since 1990, some of the fastest-growing states have been in the American West (including Texas, Arizona, and Nevada), as well as in the western Canadian provinces of Alberta and British Columbia.

Much of the extraordinary growth in the Mountain States is fueled by new job creation in high-technology industries and services, as well as by the region's scenic, recreational, and retirement amenities (Figure 3.12). In larger metropolitan areas such as Salt Lake City and Denver, the numbers continue to swell. Maricopa County, Arizona, which includes Phoenix, was the most rapidly growing county in the United States between 2000 and 2006, adding 700,000 new residents. Among smaller metropolitan centers, St. George, Utah, grew 40 percent between 2000 and 2006 and Bend, Oregon, saw its population increase by 29 percent. Many of these new migrants include out-bound Californians (see "People on

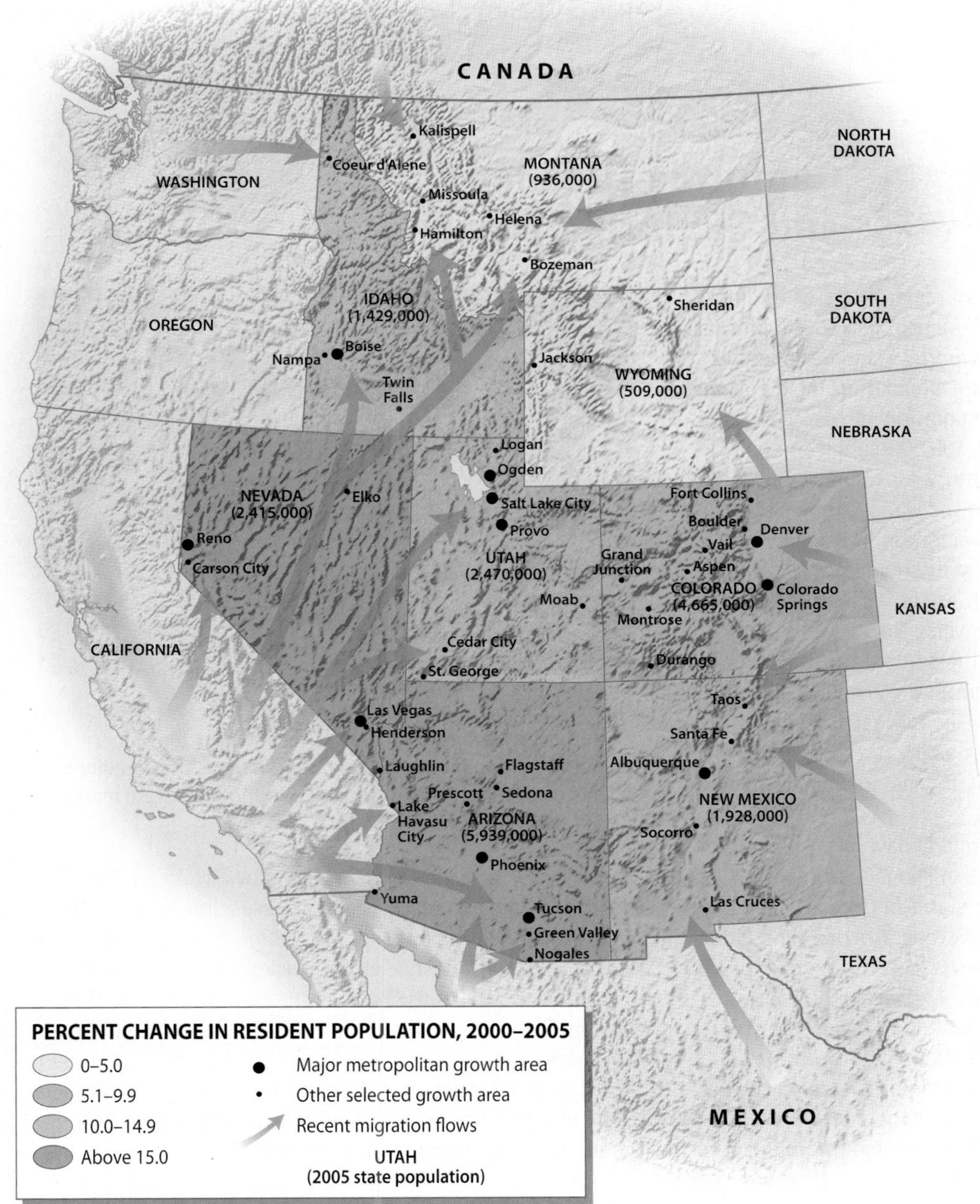

Figure 3.12 Intermountain West Growth, 2000–2005
Nevada (20.8 percent growth) and Arizona (15.8 percent growth) were the two fastest-growing states between 2000 and 2005. Many people are flocking to the region's larger cities to find employment, often in growing high-technology industries. Other amenity-bound migrants and retirees are attracted to the region's smaller towns and recreational areas. *(Modified from the U.S. Census, State and Metropolitan Area Data Book: 2006)*

the Move: California Dreamin' and California Leavin'"). The sustained move to the Interior West has many implications: the demand for water in the arid region continues to grow; the mix of natives and recent migrants creates cultural tensions; and the area's growing political and economic power is redefining its traditionally peripheral role in national affairs.

The spectacular growth of Las Vegas illustrates how these trends are playing out in particular settings within the West. As one of North America's fastest growing major metropolitan areas, Las Vegas represents a quintessential global niche city. It has emerged as a national and international destination resort, entertainment, and retirement center. It is home to more than 133,000 hotel rooms (the largest total in North America) and hosts more than 39 million visitors a year, many of them from Europe and Asia. The city's huge hotels, entertainment complexes, shopping areas, and golf courses are now globally recognized amenities.

What are the local consequences of this globally driven expansion? Most important, the region's population soared from 64,000 people in 1960 to more than 1.9 million in 2007. Predictions are for more than 3 million residents by 2020. The accompanying suburban expansion is a low-density assortment of single-family homes, retirement complexes, and decentralized shopping malls. The local environmental impacts of this space-extensive growth are dramatic. The metropolis has rapidly sprawled into the nearby desert environment both east and west of the central city (Figure 3.13). In fact, plans are already underway to create an Arizona suburb for Las Vegas, where 100,000 desert suburbanites will live, making the daily 50-mile commute into Nevada. The city has also made ever-increasing demands on rapidly dwindling regional water supplies, mostly for residential use. The city is now engaged in the nation's largest groundwater-pumping project, designed to tap aquifers in eastern Nevada's Basin and Range country. Meanwhile, the proliferation of housing

PEOPLE ON THE MOVE

California Dreamin' and California Leavin'

By any measure, the Golden State remains one of the most dynamic settings within North America. In 2005 California had almost 10 million foreign-born residents, about 25 percent of the nation's immigrant population. At the same time, in terms of domestic migration flows, the state experienced a *net loss* of almost 100,000 people annually in the first five years of the new century. Making sense of those numbers offers some insight into the complex role played by California in the larger story of migration in North America.

Thousands of immigrants are still attracted to the Golden State and, as a result, cities such as Los Angeles are among North America's most culturally diverse places. While many of these immigrants arrive legally, the state also attracts many illegal migrants, particularly from Latin America and Asia. Experts estimate that almost 1 million illegal immigrants live in Los Angeles County alone. For many, the state's job opportunities and the proximity of friends and family remain powerful draws.

But since 1990, many Californians have been pulling up stakes (Figure 3.2.1). By 2007, 5 million California natives were living outside their home state. Many have gone to nearby western states, such as Nevada, Arizona, and Utah, but others have ventured to lower cost housing and job opportunities in the South and Midwest. This net out-migration of Californians is not limited to wealthier whites. In fact recently, more Hispanics than Anglos were exiting the state, eager to find better opportunities in places such as Salt Lake City, Boise, or Las Vegas. Los Angeles County is even witnessing a decline in its illegal immigrant population.

Figure 3.2.1 Land of Tarnished Dreams
California's dynamic population is often on the move. This neighborhood in the Golden State reflects the combined effects of population turnover and a tough real estate market. *(David Young-Wolff/PhotoEdit)*

Some newly arriving migrants are now avoiding the city, instead heading to places such as Dallas and Atlanta. Other well-established migrants are on the move, as well. Between 2002 and 2004, Los Angeles County lost about 15,000 illegal immigrants, even as the national total grew by more than a million. Some are moving only a short distance, to suburban Southern California settings such as Riverside and San Bernardino counties, but others are watching the Golden State recede in their rearview mirrors as they head elsewhere, bound for greener economic pastures.

Figure 3.13 Sprawling Las Vegas
These satellite images from 1990 and 2000 reveal the extraordinary pace of change in Henderson, Nevada, the most rapidly growing suburb of Las Vegas. Henderson is located 8 miles southeast of the famous Las Vegas Strip. *(Landiscor)*

subdivisions threatens nearby wildlife species, such as the desert tortoise.

Social and cultural changes have also accelerated. The Clark County School District, which serves Las Vegas and its surrounding area, is now the nation's fifth largest, adding an average of 10 new schools per year. The city's cultural geography is becoming as diverse as nearby California's, with thousands of new Hispanic and Asian workers attracted to the area's strong job market. Chronic problems of crime, drug abuse, and gang violence have grown along with the population. What the Las Vegas Valley will look like in 50 years is anyone's guess.

Black Exodus from the South African Americans have shaped their own shifting migration patterns over time (Figure 3.14). Originally, black slave populations were concentrated in the plantation South. In fact, at the end of the American colonial period, African Americans constituted the majority of the population in southern states such as South Carolina. Even after the legal emancipation of blacks in the 1860s, most remained economically bound to the rural South, where they worked as sharecroppers, often for their former owners. Conditions changed in the twentieth century. Many African Americans migrated because of declining demands for labor in the agricultural

Figure 3.14 The Block II (1972)
Romare Bearden's collage of Harlem life was based on an actual neighborhood scene. Bearden's work as a black artist drew upon his own diverse American experiences in both the Northeast and South. *(Romare Bearden, "The Block, II." 1972. The Walter O. Evans Collection of African American Art.)*

South and growing industrial opportunities in the North and West. Two waves of migration, one between 1910 and 1920 and the other from 1940 to 1960, propelled blacks into many new American settings. Overwhelmingly, migrants ended up in cities where job opportunities beckoned. Boston, New York, Philadelphia, Detroit, and Chicago became key destinations for southern blacks. Los Angeles and the San Francisco–Oakland areas drew many blacks to the West. Since 1970, however, more blacks have moved from North to South. Indeed, Sun Belt jobs and federal civil rights guarantees now attract many northern urban blacks to growing southern cities. The net result is still a profound change from 1900: at the beginning of the century, more than 90 percent of African Americans lived in the South, while today only about half of the nation's 40 million blacks reside within the region.

Rural-to-Urban Migration Another persistent trend in North American migration has taken people from the country to the city. Two centuries ago, only 5 percent of North Americans lived in urban areas (cities of more than 2,500 people), whereas today about 80 percent of the North American population is urban. Shifting economic opportunities account for much of the transformation: as mechanization on the farm reduced the demand for labor, many young people left for new employment opportunities in the city.

Historically, rates of urbanization jumped first in the Northeast as industrialization spread through the region between 1800 and 1850. The scene of rapid urban expansion shifted westward to the Great Lakes region later in the nineteenth century as industrial growth transformed cities such as Toronto and Chicago. In the twentieth century, people were drawn by many urban employment opportunities in more peripheral zones, resulting in large migrations to cities in the South and West.

Growth of the Sun Belt South Twentieth-century moves to the American South are clearly related to other dominant trends in North American migration, yet the pattern deserves closer inspection. Particularly after 1970, southern states from the Carolinas to Texas grew much more rapidly than states in the Northeast and Midwest. Since 2000, the South has remained the primary regional destination for domestic migrants within the United States, adding people even more rapidly than the West. Florida, Texas, Georgia, and North Carolina have recently experienced sizable population gains, with migrants heading for job-rich suburbs as well as high-amenity retirement locations. Factors that have contributed to the South's growth are its buoyant economy, modest living costs, adoption of air conditioning, attractive recreational opportunities, and appeal to snow-weary retirees. Movements have been selective, however; many rural agricultural and mountain counties within the South have seen few new residents, while amenity-rich coastal settings and job-generating metropolitan areas have witnessed spectacular growth. Dallas–Fort Worth's bustling metropolitan area (6.0 million), for example, is now larger than that of Boston (4.5 million) (Figure 3.15).

The Counterurbanization Trend During the 1970s, certain nonmetropolitan areas in North America began to see significant population gains, including many rural settings that had previously lost population. Selectively, that pattern of **counterurbanization**, in which people leave large cities and move to smaller towns and rural areas, continues today. Some participants in counterurbanization are part of the growing retiree population in both Canada and the United States, but a substantial number are younger, so-called *lifestyle migrants*. They find or create employment in affordable smaller cities and rural settings that are rich in amenities and often removed from the perceived problems of urban America. In fact, recent nonmetropolitan population growth has exceeded metropolitan growth in most western states. Other smaller communities outside the West, such as Mason City, Iowa; Mankato, Minnesota; and Traverse City, Michigan, also are seen as desirable destinations for migrants interested in downsizing from their metropolitan roots.

Figure 3.15 Downtown Dallas, Texas
Rapid job creation has transformed Sun Belt cities such as Dallas. Healthy growth in office space, specialty retailing, and entertainment districts has fueled downtown Dallas's expansion and reshaped the look of the central-city skyline. *(Jeremy Woodhouse/Getty Images, Inc.–Photodisc)*

Settlement Geographies: The Decentralized Metropolis

North America's settlement landscape reflects the population movements, shifting regional economic fortunes, and technological innovations of the last century. The ways in which settlements are organized on the land—the actual appearance of cities, suburbs, and farms—as well as the very ways in which North Americans socially construct their communities have changed greatly in the past century. Today's cloverleaf interchanges, sprawling suburbs, outlet malls, and theme parks would have struck most 1900-era residents as utterly extraordinary.

Settlement landscapes of North American cities boldly display the consequences of **urban decentralization**, in which metropolitan areas sprawl in all directions and suburbs take on many of the characteristics of traditional downtowns. Although both Canadian and U.S. cities have experienced decentralization, the impact is particularly profound in the United States, where inner-city problems, poor public transportation, widespread automobile ownership, and fewer regional-scale planning initiatives have encouraged many middle-class urban residents to move beyond the central city. Even beyond North America, observers note a globalization of urban sprawl: many Asian, European, and Latin American cities are taking on attributes of their North American counterparts as they experience similar technological and economic shifts. Indeed, much as they have in Seattle and Albuquerque, suburban Wal-Marts, semiconductor industrial parks, and shopping malls may become increasingly familiar sights on the peripheries of Kuala Lumpur or Mexico City.

Historical Evolution of the City in the United States

Changing transportation technologies decisively shaped the evolution of the city in the United States (Figure 3.16). The pedestrian–horsecar city (pre-1888) was compact, essentially limiting urban growth to a 3- or 4-mile-diameter ring around downtown conveniently accessible by foot or horse-powered trolley cars. The invention of the electric trolley in 1888 propelled the urbanized landscape further into new "streetcar suburbs," often 5 or 10 miles from the city center. Indeed, the electric streetcar city (1888–1920) offered commuters affordable mass transit at the dizzying speed of 15 or 20 miles an hour. A star-shaped urban pattern resulted, with growth extending outward along and near the streetcar lines.

The biggest technological revolution came after 1920 with the widespread adoption of the automobile. The recreational automobile city (1920–1945) promoted the infilling of areas beyond the reach of the streetcar and added even more distant suburban rings in the surrounding countryside. Essentially, it allowed many middle-income residents, particularly whites, to leave the central city in favor of lower-density and less ethnically complex suburban settings. Following World War II, the outer city (1945 to the present) promoted even more dramatic decentralization along commuter routes as built-up areas appeared 40 to 60 miles from downtown.

Urban decentralization also reconfigured land-use patterns in the city, producing metropolitan areas today that are strikingly different from their counterparts of the early twentieth century. In the city of the early twentieth century, idealized in the **concentric zone model**, urban land uses were neatly organized in rings around a highly focused central business district (CBD) that contained much of the city's retailing and office functions. Residential districts beyond the CBD were added as the city expanded, with higher-income groups seeking more desirable, peripheral locations. Elements of the model are still relevant today, particularly in examining how eras of economic expansion add rings of residential construction around the periphery of a city.

Today's **urban realms model** highlights new suburban growth characterized by a mix of peripheral retailing, industrial parks, office complexes, and entertainment facilities. These areas of activity, often called "edge cities," have fewer functional connections with the central city than they have with other suburban centers. For most suburban residents of an edge city, jobs, friends, and

Figure 3.16 Growth of the American City
Many U.S. cities became increasingly decentralized as they moved through eras dominated by the pedestrian/horsecar, electric streetcar, automobile, and freeway. Each era left a distinctive mark on metropolitan America, including the recent growth of edge cities on the urban periphery.

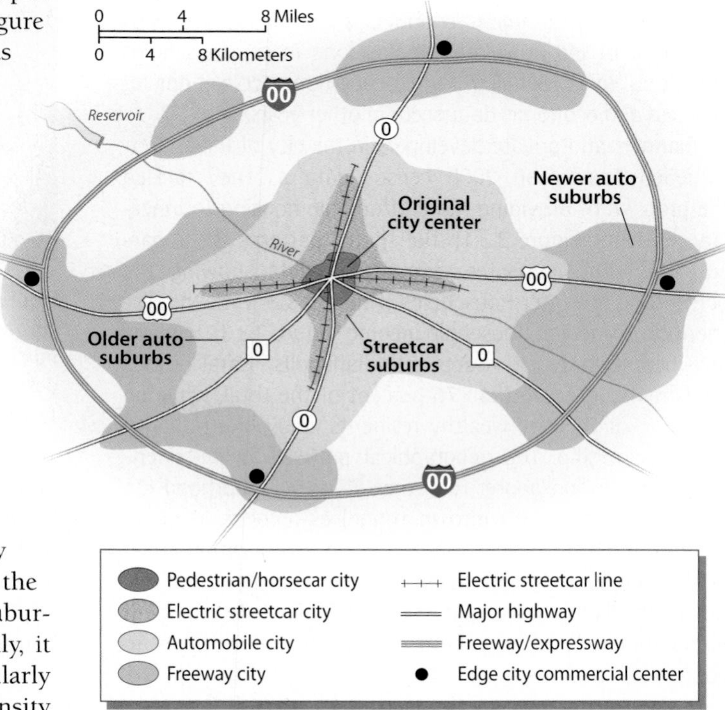

entertainment are located in other peripheral realms, rather than in the old downtown.

The Consequences of Sprawl The rapid evolution of the North American city continues to transform the urban landscape and those who live in it (see "Geographic Tools: Housing Affordability and the U.S. Census"). As suburbanization accelerated in the 1960s and 1970s, many inner cities, especially in the Northeast and Midwest, suffered losses in population, increased levels of crime and social disruption, and a shrinking tax base that often brought them to the brink of bankruptcy. Today, inner-city poverty rates average almost three times those of nearby suburbs. Unemployment rates remain above the national average. Central cities in the United States are also places of racial tension, the product of decades of discrimination, segregation, and poverty. Although the number of middle-class African Americans and Hispanics is growing, many exit the central city for the suburbs, further isolating the urban underclass that remains behind. In Detroit, for example, almost 90 percent of inner-city residents are black and the city's poverty rate is almost triple the national norm.

Amid these challenges, inner-city landscapes are also enjoying a selective renaissance. Referred to as **gentrification**, the process involves the displacement of lower-income residents of central-city neighborhoods with higher-income residents, the rehabilitation of deteriorated inner-city landscapes, and the construction of new shopping complexes, entertainment attractions, or convention centers in selected downtown locations. The older and more architecturally diverse housing of the central city is also a draw, serving as specialty shops and restaurants for a cosmopolitan urban clientele and offering residential opportunities for upscale singles who wish to live near downtown. Seattle's Pioneer Square, Toronto's Yorkville district, and Baltimore's Harborplace exemplify how such public and private investments shape the central city. Pittsburgh's recent urban renaissance

GEOGRAPHIC TOOLS Housing Affordability and the U.S. Census

The U.S. Census is one of the most widely used tools of spatial analysis by human geographers interested in solving a variety of demographic, social, and economic problems. Begun as a simple population count across the new nation in 1790, the census today collects data on a vast variety of subjects (including age, sex, race, income, housing, and occupation) and makes the information easily accessible in electronic and cartographic formats. Every 10 years (years ending in 0), the national government conducts a comprehensive census count of individual households (one in six households completes more detailed forms); additional collections of business and economic data occur in other years.

Planners and private developers in the city of Indianapolis relied heavily on 2000 census data as they tackled the problem of providing more affordable housing to inner-city residents (Figure 3.3.1). The spatial patterns are apparent when one compares census tract data showing the percentage of renter households (Figure 3.3.2) with the distribution of median household income (Figure 3.3.3). In many low-income inner-city tracts in Indianapolis, rental households make up more than 75 percent of the total, while on the edge of the city wealthy residents enjoy high rates of home ownership. The geographical patterns helped identify the older and poorer Fall Creek Place neighborhood (on the city's northern downtown fringe) as a focus of urban revitalization efforts. In 2003, a housing revitalization effort in the neighborhood garnered a national award from the American Planning Association. The 26-block development offers hundreds of new homeownership opportunities to poorer city residents.

Figure 3.3.1 Fall Creek Place, Indianapolis
This renovated inner-city neighborhood features a pleasing mix of century-old homes and new traditional-style houses designed to offer residents affordable housing. *(Mansur Real Estate Services)*

Anyone can explore the tools offered by the U.S. Census Bureau by visiting www.census.gov. A feature called "American FactFinder" is available on the site and offers professional planners, geographers, students, or interested individuals an easy way to obtain census information on any U.S. locality; to access current reports on population, housing, and business; and to create customized tables and thematic maps that quickly summarize statistical data.

offers an affordable housing market, an older highly skilled workforce, and mixed-use neighborhoods such as the SouthSide Works development, a 34-acre assortment of residences, offices, and stores created on the site of an old steel plant on the Monongahela River (Figure 3.17).

The suburbs are also changing. Construction of new corporate office centers, fashion malls, and industrial facilities has created true "suburban downtowns" in suburbs that are no longer simply bedroom communities for central-city workers. Indeed, such localities have been growing players in the continent's globalization process. Many of North America's key internationally connected corporate offices (IBM, Microsoft), industrial facilities (Boeing, Cisco, Oracle), and entertainment complexes (Disneyland, Walt Disney World, and the Las Vegas Strip) are now in such settings, and they are intimately tied to global information, technology, capital, and migration flows.

The edge-city lifestyle has also transformed the United States into a continent of suburban commuters in

Figure 3.17 **Pittsburgh's SouthSide Works Neighborhood** New investments have transformed Pittsburgh's SouthSide Works neighborhood into an upscale office, shopping, and entertainment district. (*Ralf Brown*)

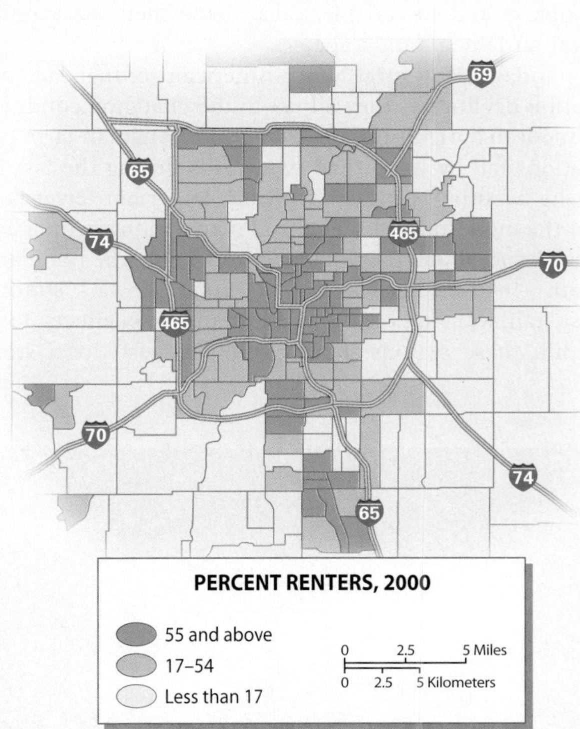

Figure 3.3.2 **Indianapolis, Percentage of Renter Households, 2000**
Older housing stock near the center of Indianapolis has often been converted into rental units. City planners and developers are trying to create more single-family housing opportunities in such localities. *(Reprinted from Rubenstein, 2005,* An Introduction to Human Geography, *8th ed., Upper Saddle River, NJ: Prentice Hall)*

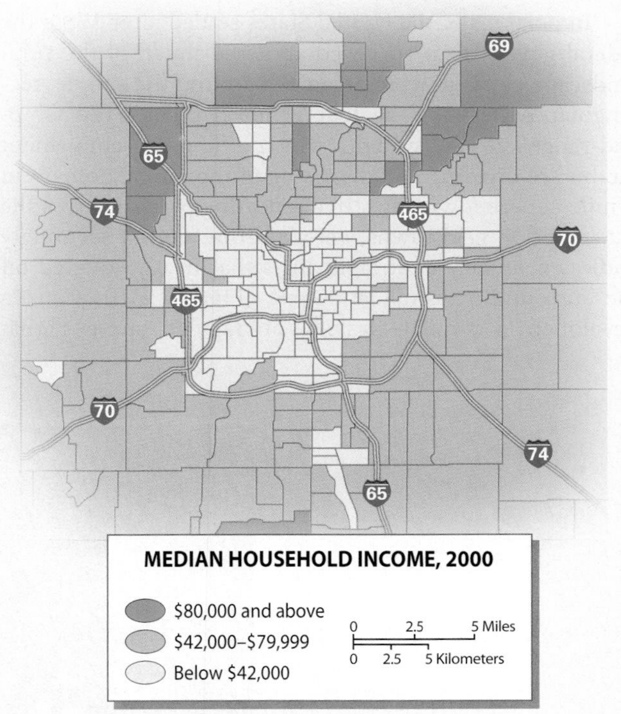

Figure 3.3.3 **Indianapolis, Median Household Income, 2000**
Higher-income residents tend to cluster toward the newer housing units, larger lot sizes, and expanded employment and entertainment opportunities of the urban periphery. In Indianapolis, the city's wealthy north and northeast sectors contrast dramatically with census precincts near the center of the city. *(Reprinted from Rubenstein, 2005,* An Introduction to Human Geography, *8th ed., Upper Saddle River, NJ: Prentice Hall)*

which people live in one suburb and work in another. The average daily commute now exceeds 30 miles (48 km) per person in cities such as Atlanta, Georgia, and Birmingham, Alabama. As urban settlements of the twenty-first century grow in area but decline in density, it will become more difficult to build roads, schools, and other public infrastructure in cost-effective ways. Particularly on the outer suburban fringe, the boundary between city and country is blurring; as urbanites reach toward the amenities and privacy of the rural landscape, modern technologies and metropolitan growth bring the countryside closer to the city.

Settlement Geographies: Rural North America

Rural North American landscapes trace their origins to initial European settlement. Although many New World communities replicated elements of the Old World, North Americans favored more dispersed rural settlement patterns. An abundance of available acreage, liberal land disposal laws, and perhaps a cultural preference for independence and privacy all contributed to a pattern of dispersed settlement in North America in which settlers lived on their own farmsteads.

In portions of the United States settled after 1785, the federal government surveyed and sold much of the rural landscape. Surveys were organized around the simple, rectangular pattern of the federal government's township-and-range survey system, which offered a convenient method of dividing and disposing of the public domain in 6-mile-square townships. In a repeating pattern across the U.S. interior, grid towns were laid out amid a seemingly endless geometry of straight lines that neatly squared off the surrounding rural landscape (Figure 3.18). Canada developed a similar system of regular surveys that stamped much of southern Ontario and the western provinces with a rectangular pattern. Exceptions to the predictable pattern were found in former American colonies such as Virginia, where irregular surveys dominated, and in Canada's St. Lawrence Valley, in which distinctive long lot lines lay perpendicular to the river to give more farmers access to the water.

Commercialized agriculture and technological changes further reorganized the settlement landscape. Railroads opened new corridors of development, provided access to markets for commercial crops, and facilitated the establishment of towns that served surrounding rural populations. By 1900, several transcontinental lines spanned North America, radically transforming the farm economy and the pace of rural life. After 1920, however, even greater changes accompanied the arrival of the automobile, farm mechanization, and better rural road networks. The need for farm labor declined with mechanization, and many smaller market centers became unnecessary as farmers equipped with automobiles and trucks could travel farther and faster to larger, more diverse towns. Other technological innovations, including center-pivot irrigation systems, better-yielding seeds, and new fertilizers and insecticides, also made their mark on the rural settlement landscape.

Today, many rural North American settings face population declines as they adjust to the changing conditions of modern agriculture. Both U.S. and Canadian farm populations fell by more than two-thirds during the last half of the twentieth century. Typically, larger but fewer farms dot the modern rural scene, and many young people leave the land to obtain employment elsewhere. The population drain in settings such as rural Iowa, southern Saskatchewan, and eastern Montana also affects towns within these regions. Fewer shoppers visit local stores;

Figure 3.18 Iowa Settlement Patterns
The regular rectangular look of this Iowa town and the nearby rural setting reveals the North American penchant for simplicity and efficiency. In the United States, the township-and-range survey system stamped such predictable patterns across vast portions of the North American interior. *(Craig Aurness/Corbis/Bettmann)*

doctors and dentists retire and move; and schools and churches are forced to close or consolidate. Visual reminders of abandonment, such as weed-choked driveways, empty farmhouses, roofless barns, and the empty marquees of small-town movie houses, tell the story more powerfully than any census or government report. Recently, a program in several small towns in Kansas has sparked interest as declining communities literally give away land in an effort to lure new residents.

Some rural settings show signs of economic health and population growth. No single explanation suffices for this expansion. Some localities are merely feeling the fringe effects of encroaching edge cities. Suddenly, almost overnight, such places are overrun with suburbanites who snap up farmhouses once beyond the reach of the city. Other growing rural settings lie beyond direct metropolitan influence but are seeing new populations who seek amenity-rich environments removed from city pressures. These impulses toward counterurbanization are shaping the settlement landscape from British Columbia's Vancouver Island to Michigan's Upper Peninsula. Newly subdivided land parcels, a collection of new real estate offices, the appearance of resort and golf complexes, and the telltale in-migration of espresso bars all signal the changes afoot in such surroundings.

CULTURAL COHERENCE AND DIVERSITY: Shifting Patterns of Pluralism

North America's cultural geography is both globally dominant and internally pluralistic. On one hand, history and technology have produced a contemporary North American cultural force that is second to none in the world. Many people outside the United States speak of cultural imperialism when they describe the increasing global dominance of American popular culture, which they often see as threatening the vitality of other cultural values. Yet North America is also a mosaic of different peoples who retain part of their traditional cultural identities and celebrate their pluralistic roots.

The Roots of a Cultural Identity

Historically, both Canada and the United States were strongly tied to Great Britain, and these links forged a common dominant culture within the region. With some exceptions, most early Europeans who played key cultural, economic, and political roles in the region were from the British Isles. Key Anglo legal institutions and social customs solidified the common set of core values that many North Americans shared with Britain and, eventually, with one another. Traditional beliefs, defined within the limits of the Anglo world view, emphasized representative government, separation of church and state, liberal individualism, privacy, pragmatism, and social mobility. The American Revolution (1776) and the Confederation of Canada (1867), while distancing the region from direct British authority, contributed to a cultural continuity that combined the Anglo world view with the emerging cultural identities of the two new countries.

As settlement and economic linkages spread across the continent in the late nineteenth century, so did these core values. From those traditional roots, particularly within the United States, consumer culture blossomed after 1920, producing a common set of shared experiences that included mass advertising, movies, national radio and television programming, and an increasingly secular society oriented toward convenience and consumption. Since World War II, these modern technologies and values, with the help of growing political and multinational corporate imperatives, have cumulatively shaped the world in North America's cultural image.

But North America's cultural coherence coexists with pluralism, the persistence of distinctive cultural identities. Closely related is the concept of **ethnicity**, in which a group of people with a common background and history identify with one another, often as a minority group within a larger society. The presence of numerous ethnic groups combined into a pluralistic society has created ongoing tensions within the two countries. For Canada, the early and enduring French colonization of Quebec complicates its contemporary cultural geography. Canadians face the challenge of creating a truly bicultural society where issues of language and political representation are central concerns. Within the United States, given its unique immigration history, a greater diversity of ethnic groups exists, and key ethnic areas are often defined both at local and at regional scales.

Peopling North America

North America is a region of immigrants. Quite literally, global-scale migrations made possible the North America we know today. Decisively displacing Native Americans in most portions of the realm, immigrant populations created a new cultural geography of ethnic groups, languages, and religions. Early migrants often had considerable cultural influence, despite minuscule numbers. Over time, varied immigrant groups and their changing destinations produced a culturally diverse landscape. Also varying between groups was the pace and degree of **cultural assimilation**, the process in which immigrants were absorbed by the larger host society.

Migration to the United States In the United States, variations in the number and source regions of migrants produced five distinctive chapters in the country's history (Figure 3.19). In Phase 1 (prior to 1820), absolute numbers of migrants were modest and English and African influences dominated. Other Europeans were important, particularly Irish, Dutch, French, and Germans, but it was the English who played the pivotal role. In the late eighteenth century, black slaves, mostly from West Africa, often outnumbered European arrivals. They

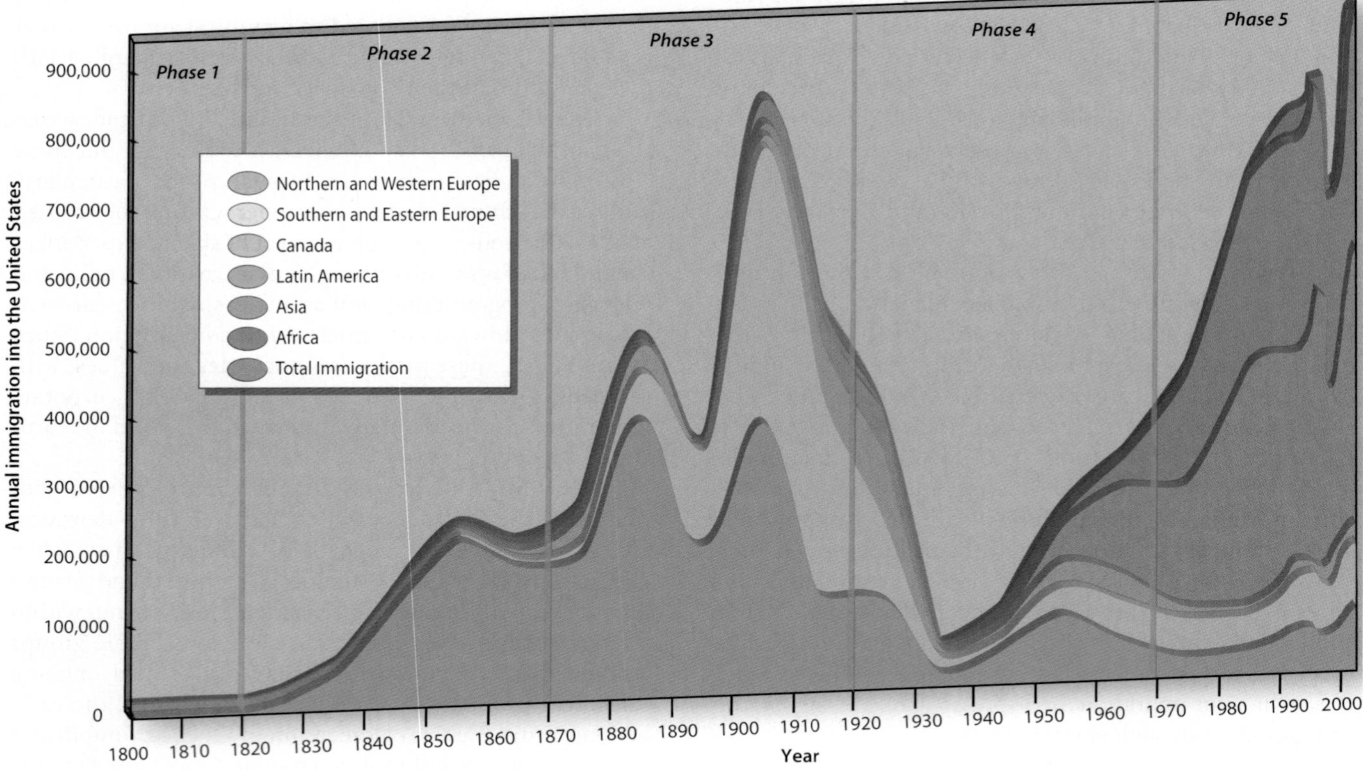

Figure 3.19 U.S. Immigration, by Year and Group
Annual immigration rates peaked around 1900, declined in the early twentieth century, and then surged again, particularly since 1970. The source areas of these migrants have also shifted. Note the decreased role of Europeans currently versus the growing importance of Asians and Latin Americans. *(Modified from Rubenstein, 2005,* An Introduction to Human Geography, *8th ed., Upper Saddle River, NJ: Prentice Hall)*

contributed additional cultural influences in the South and were part of a much larger forced migration to the Americas (see Chapters 4 and 5). Northwestern Europe served as the main source region of immigrants between 1820 and 1870 (Phase 2). But the emphasis shifted away from English dominance to include greater numbers of Irish and Germans. The pace of immigration also increased greatly: only 130,000 migrants arrived in the 1820s versus more than 2.8 million in the 1850s.

As Figure 3.19 shows, immigration reached a much higher peak around 1900, when almost 1 million foreigners entered the United States *annually*. During Phase 3 (1870–1920), the source regions of this great wave of migrants also shifted. Northwestern Europeans declined in importance while southern (particularly Italians) and eastern Europeans (Poles, Russians, Austro-Hungarians) dominated. Political strife and poor economies in Europe sharply contrasted with news of available land, expanding industrialization, and well-paying jobs in America. Between 1880 and 1920, Scandinavians also swept through the northern interior, where they settled on still-available farmland. By 1910, almost 14 percent of the nation was foreign-born. Very few of these immigrants targeted the job-poor American South, however, creating a cultural divergence that persists to the present.

Twentieth-century migrations also profoundly shaped the nation's cultural geography. Between 1920 and 1970 (Phase 4), a growing number of immigrants came from neighboring Canada and Latin America, but overall totals plunged, a function of more restrictive federal immigration policies (the Quota Act of 1921 and the National Origins Act of 1924), the Great Depression, and the disruption caused by World War II.

Since 1970 (Phase 5), the realm has witnessed a sharp reversal in numbers, and now annual arrivals surpass those of the early twentieth century (see Figure 3.19). Most legal migrants since 1970 originated in Latin America or Asia. The current surge was made possible by economic and political instability abroad, a growing postwar American economy, and a loosening of immigration laws (the Immigration Acts of 1965 and 1990, and the Immigration Reform and Control Act of 1986). Illegal immigration also rose after 1970, although several federal laws passed in 1996 greatly increased the number of U.S. border patrol agents and made it more difficult for immigrants, both legal and illegal, to receive federal welfare benefits. Today, about 12 million unauthorized immigrants live in the United States.

An estimated 12 million Mexican-born residents (or about 11 percent of Mexico's population) now live in

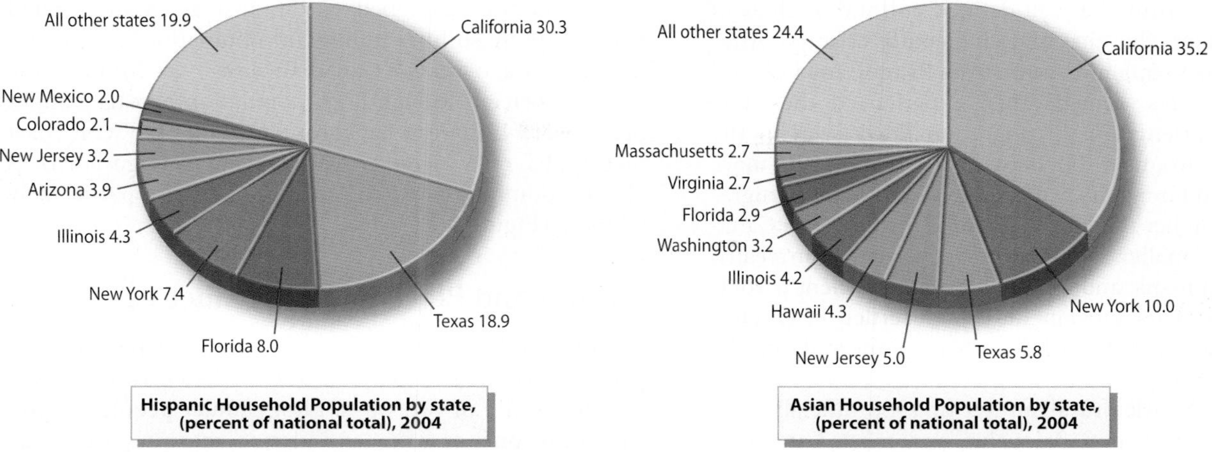

Figure 3.20 Distribution of U.S. Hispanic and Asian Household Populations, by State, 2004 California and Texas claim about half of the nation's Hispanic population, but a growing number are locating elsewhere. California alone is still home to more than one-third of the country's Asian population. *(U.S. Census, American Community Survey Reports, issued February 2007)*

the United States. Mexicans make up about 64 percent of the nation's Hispanic population, but those numbers are changing. In the next 25 years, most of the projected increase in the U.S. Hispanic population will be fueled by births within the country versus new immigrants. While half of U.S. Hispanics live in California or Texas, they are increasingly moving to other areas (Figure 3.20). Recently, states such as Wisconsin, Georgia, Kansas, and Arkansas have witnessed dramatic increases in their immigrant Hispanic populations, a trend likely to continue in the early twenty-first century. The cultural implications of these Hispanic migrations are profound: today the United States is the fifth largest Spanish-speaking nation on Earth.

In percentage terms, migrants from Asia constitute the fastest-growing immigrant group, and various Asian ethnicities, both native and foreign-born, account for 5 percent of the U.S. population. Chinese is now the third most spoken language in the United States (behind English and Spanish). California remains a key entry point for new migrants and is home to more than one-third of the nation's overall Asian population (Figure 3.20). Asian migrants often move to large cities such as Los Angeles, San Francisco, and New York City. Beyond these key gateway cities, Asians are also moving to growing communities in Washington, D.C.; Chicago; Seattle; and Houston. Diverse cultures make up the nation's Asian population. The largest groups represented include Chinese (23 percent), Asian Indian (19 percent), Filipino (18 percent), Vietnamese (11 percent), and Korean (10 percent).

The future cultural geography of the United States will be dramatically redefined by these recent immigration patterns (Figure 3.21). Indeed, the increasing ethnic diversity of the country is simply one more manifestation of the globalization process and the powerful pull of North America's economy and political stability. By 2070, Asians may total more than 10 percent of the U.S. population, and almost one American in three will be Hispanic. Indeed, it is likely that the U.S. non-Hispanic white population will achieve minority status by that date.

The Canadian Pattern The peopling of Canada broadly parallels the U.S. story, but with important differences. Early French arrivals concentrated in the St. Lawrence

Figure 3.21 Projected U.S. Ethnic Composition, 1999 to 2070
By late in the twenty-first century, almost one in three Americans will be Hispanic, and non-Hispanic whites will achieve minority status amid an increasingly diverse U.S. population. *(Modified from the U.S. Census, Census of Population, National Projections, 2020–2070)*

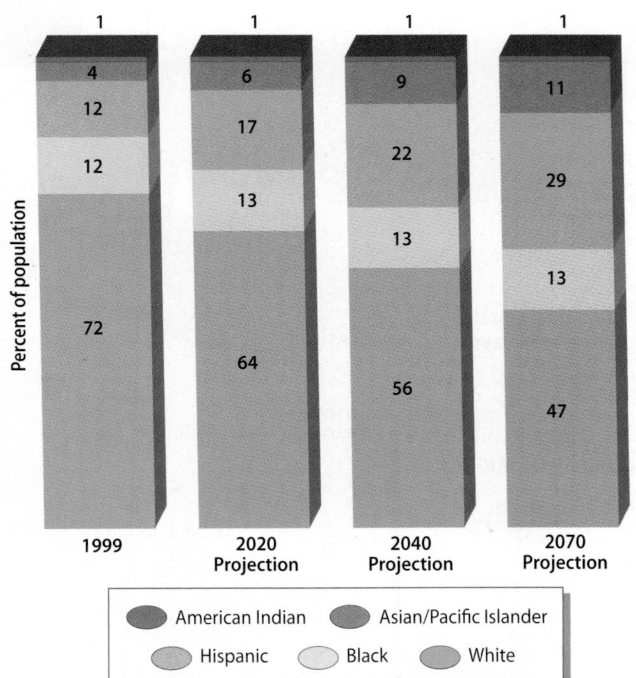

Valley and constituted a well-defined cultural nucleus of 60,000 residents by the mid-eighteenth century. After 1765, many new migrants came from Britain, Ireland, and the United States. Canada then experienced the same surge and reorientation in migration flows seen in the United States around 1900. Between 1900 and 1920, more than 3 million foreigners ventured to Canada, an immigration rate far higher than for the United States, given Canada's much smaller population. Eastern Europeans, Italians, Ukrainians, and Russians were prominent participants in these later movements. Some settled in Ontario cities such as Toronto, while others pioneered on the Canadian prairies.

Canada's Multiculturalism Act of 1988 officially recognized that basic national value and most Canadians support relatively open immigration policies. Today, about 60 percent of Canada's recent immigrants are Asians. As in the United States, more liberal immigration laws since the 1960s encouraged movement to Canada, and its 17 percent foreign-born population is among the highest in the developed world. Just as in the United States, rapidly shifting immigrant populations are changing the look of many Canadian cities. In Toronto, the city's 44 percent foreign-born population reveals a slight bias toward European backgrounds, while the west-coast metropolis of Vancouver (38 percent foreign-born) is clearly dominated by Asian populations, particularly Chinese (Figure 3.22).

Culture and Place in North America

Cultural and ethnic identity are often strongly tied to place. North America's cultural diversity is expressed geographically in two ways. First, similar people congregate near one another and derive meaning from the territories they occupy in common. Second, culture marks the visible scene: the everyday landscape is filled with the artifacts, habits, language, and values of different groups. Boston's Italian North End simply looks and smells different from nearby Chinatown, and rural

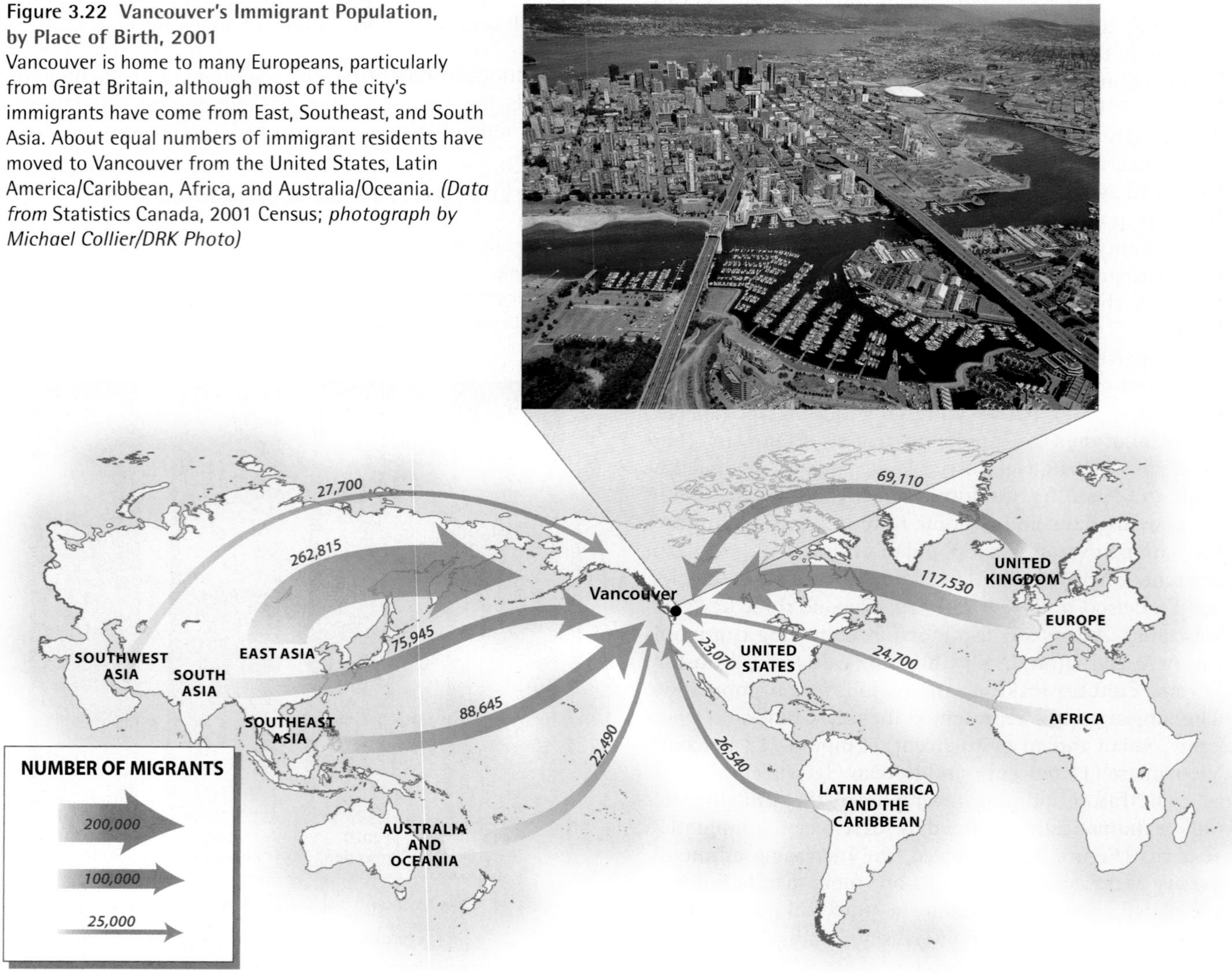

Figure 3.22 Vancouver's Immigrant Population, by Place of Birth, 2001
Vancouver is home to many Europeans, particularly from Great Britain, although most of the city's immigrants have come from East, Southeast, and South Asia. About equal numbers of immigrant residents have moved to Vancouver from the United States, Latin America/Caribbean, Africa, and Australia/Oceania. *(Data from* Statistics Canada, 2001 Census; *photograph by Michael Collier/DRK Photo)*

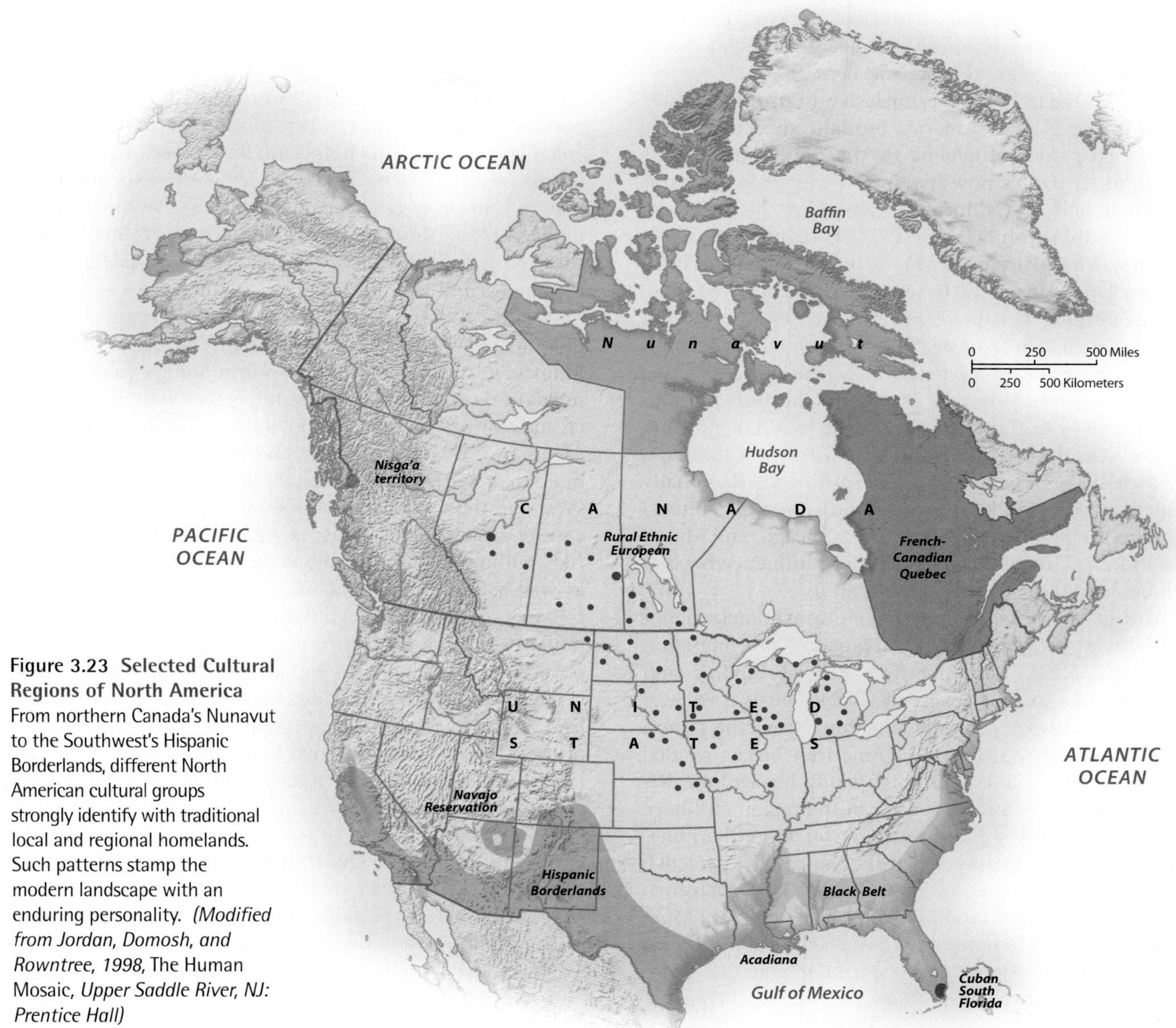

Figure 3.23 Selected Cultural Regions of North America From northern Canada's Nunavut to the Southwest's Hispanic Borderlands, different North American cultural groups strongly identify with traditional local and regional homelands. Such patterns stamp the modern landscape with an enduring personality. *(Modified from Jordan, Domosh, and Rowntree, 1998,* The Human Mosaic, *Upper Saddle River, NJ: Prentice Hall)*

French Quebec is a world away from a Hopi village in Arizona (Figure 3.23).

Persisting Cultural Homelands French Canadian Quebec superbly exemplifies the cultural homeland: it is a culturally distinctive nucleus of settlement in a well-defined geographical area, and its ethnicity has survived over time, stamping the cultural landscape with an enduring personality. Overall, about 23 percent of Canadians are French, but more than 80 percent of the population of Quebec speaks French, and language remains the cultural glue that unites the homeland. Indeed, policies adopted after 1976 strengthened the French language within the province by requiring French instruction in the schools and by mandating national bilingual programming by the Canadian Broadcasting Corporation (CBC). Ironically, many Quebecois feel that the greatest cultural threat may come not from Anglo-Canadians but rather from recent immigrants to the province. Southern Europeans or Asians in Montreal, for example, show little desire to learn French, preferring instead to put their children in English-speaking private schools.

Another well-defined cultural homeland is the Hispanic Borderlands (see Figure 3.23). It is similar in geographical magnitude to French Canadian Quebec, significantly larger in total population, but not specifically linked to a single political entity such as a state or province. Historical roots of the homeland are deep, extending back to the sixteenth century, when Spaniards opened the region to the European world. The homeland's core is in northern New Mexico, including Santa Fe and much of the surrounding rural hinterlands. A rich legacy of Spanish place-names, earth-toned Catholic churches, and traditional Hispanic settlements dot the rolling highlands of northern New Mexico and southern Colorado. From California to Texas, other historical sites, place-names, missions, and presidios also reflect the Hispanic heritage.

Unlike Quebec, however, massive twentieth-century migrations from Latin America brought an entirely new wave of Hispanic settlement to the Southwest. About 45 million Hispanics now live in the United States, with more than half in California, Texas, New Mexico, and Arizona combined. Indeed, by 2015, Hispanics will likely outnumber non-Hispanic whites in California. Within the homeland, Hispanics have created a distinctive Borderlands culture that mixes many elements of Latin and North America. These newer migrants augment the rural Hispanic presence in agricultural settings such as the lower Rio Grande Valley in Texas and the Imperial and Central Valleys in California. Cities such as San Antonio and Los Angeles also play leading roles in expressing the Hispanic presence within the Southwest. Regionally distinctive Latin foods and music add internal cultural variety to the region. New York City, Chicago, and Miami serve as key points of Hispanic cultural influence beyond the homeland (see "Cityscapes: Miami").

African Americans also retain a cultural homeland, but it has diminished in intensity because of out-migration (see Figure 3.23). Reaching from coastal Virginia and North Carolina to East Texas, the Black Belt is a zone of African American population remaining from the cotton South, when a vast majority of American blacks resided within the region. Today, although many blacks have left for cities, dozens of rural counties in the region still have large black majorities. Blacks account for more than one-quarter of the populations of Mississippi (37 percent), Louisiana (33 percent), South Carolina (29 percent), Georgia (29 percent), and Alabama (26 percent). More broadly, the South is home to many black folk traditions, including such music as black spirituals and the blues, which have now become popular far beyond their rural origins. Regrettably, even though the rural neighborhoods of the black homeland differ greatly in density and appearance from African American urban neighborhoods in the North, poverty plagues both types of communities.

A second rural homeland in the South is Acadiana, a zone of persisting Cajun culture in southwestern Louisiana (see Figure 3.23). This homeland was created in the eighteenth century when French settlers were expelled from eastern Canada and relocated to Louisiana. Nationally popularized today through their food and music, the Cajuns (derived from "Canadians") have a lasting attachment to the bayous and swamps of southern Louisiana.

Native American populations are also strongly tied to their homelands. Indeed, many native peoples maintain intimate relationships with their surroundings, weaving elements of the natural environment together with their material and spiritual lives. About 5.2 million Indians, Inuits, and Aleuts live in North America, and they claim allegiance to more than 1,100 tribal bands. Particularly in the American West and the Canadian and Alaskan North, native peoples control sizable reservations, including the 16-million-acre (6.5-million-hectare) Navajo Reservation in the Southwest, as well as self-governing Nunavut in the Canadian North (see Figure 3.23). Although these homelands preserve traditional ties to the land, they are also settings for pervasive poverty, health problems, and increasing cultural tensions (Figure 3.24). Within the United States, many Native American groups have taken advantage of the special legal status of their reservations and have built gambling casinos and tourist facilities that bring in much-needed capital, but also challenge traditional lifeways.

A Mosaic of Ethnic Neighborhoods North America's cultural mosaic is also enlivened by smaller-scale ethnic signatures that shape both rural and urban landscapes. For example, distinctive rural communities that range from Amish settlements in Pennsylvania to Ukrainian neighborhoods in southern Saskatchewan add cultural va-

Figure 3.24 Native American Poverty Navajo youngsters enjoy a game of basketball on the Navajo Reservation in Arizona. Poor housing, low incomes, and persistent unemployment plague many Native American settings across the rural West. *(Jim Noelker/The Image Works)*

CITYSCAPES Miami

Miami represents the quintessentially diverse North American city of the twenty-first century (Figure 3.4.1). With a blossoming regional metropolitan population of more than 5.5 million people (including the Fort Lauderdale area), Miami is one of the largest cities in the South. It features a unique, subtropical urban landscape, rich ethnic diversity, and an economy closely wed to nearby Latin America.

In some ways, Miami has always been defined by its exotic weather, by the warm Atlantic breezes that rustle through its palm trees, and by the ever-changing clouds that drift above its urban skyline, beach resorts, and nearby bays. In 1896 it was the warm weather (average January high temperature of 76°F [24°C]) and the region's prospects as a great winter resort that enticed business executive Henry Flagler to build a railroad into South Florida. Many northeastern urbanites, eager for winter warmth, came south and, particularly after World War II, a growing number retired there. Added to this mix was an expanding collection of Caribbean and Latin American migrants who trickled, then flooded into Miami as its regional economy grew. Much of the city's initial surge in Latin population came with Castro's takeover of Cuba in 1959. Between 1960 and 1980, hundreds of thousands of refugees left Cuba and went to Miami's Cuban community. Many brought capital and entrepreneurial skills, establishing hundreds of Cuban businesses in the city. By the late 1970s, Latin America's economic growth demanded closer ties with the United States. What better place than Miami to serve as a gateway?

Walk through Miami's bustling airport today (45 percent of all airport arrivals are international passengers) and you are immediately aware of the unique role this city plays in linking the economic and cultural worlds of Latin America with those of North America. Spanish-speaking travelers and international flight announcements dominate airport conversation as planes arrive from and depart to cities throughout Central and South America. As a result, modern Miami has become the leading North American corporate and trade center for Latin America. Indeed, some have termed it the "capital of the Americas." Today, Miami handles more than 44 percent of all U.S. trade with Latin America and the Caribbean. Airplane connections, money flows, and even the illegal drug trade demonstrate Miami's centrality. Newspapers such as the *Colombian Post*, *Venezuela Al Dia*, and the *Diario Las Americas* keep local residents and Latin visitors in touch with news from the home country.

Miami's neighborhoods reflect its unique history and ethnic richness. Southwest of bustling downtown, the culturally diverse neighborhood of Coral Gables contains

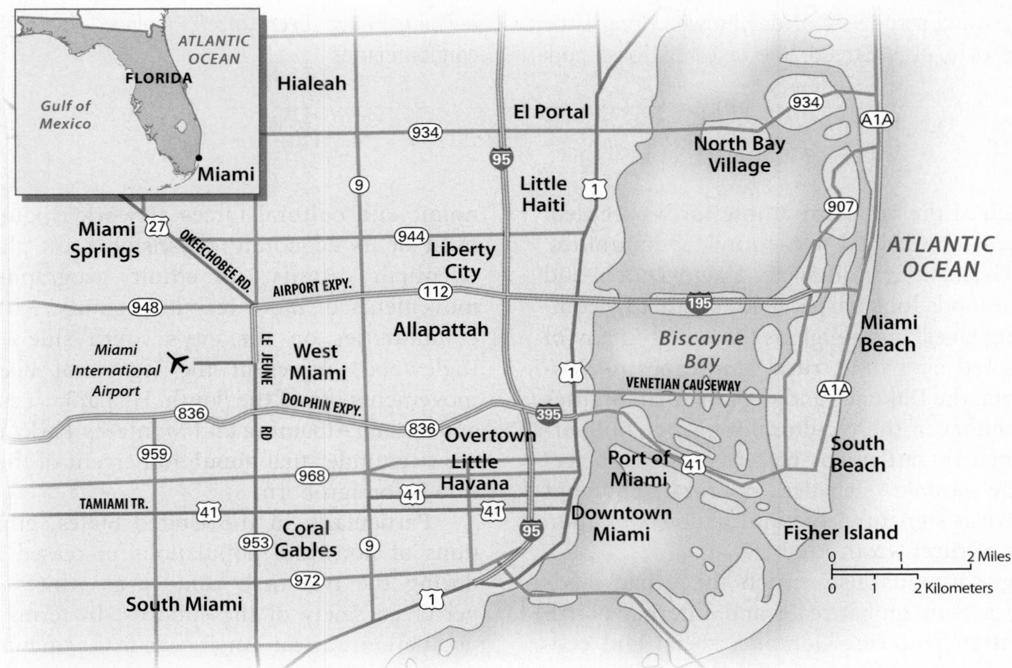

Figure 3.4.1 Miami
With a regional metropolitan population of more than 5.5 million, Miami displays an extraordinary diversity of cultural landscapes that reflect both its exotic character in the American South as well as its intimate links to Latin America.

(*continued on next page*)

Figure 3.4.2 Little Havana
Many of Miami's older Cuban residents cluster along the Calle Ocho on the city's west side and enjoy spending time socializing at Maximo Gomez Park. *(Rob Crandall/www.robcrandall.com)*

Figure 3.4.3 Miami Beach
Miami Beach's art deco district has achieved nationwide fame. Many of the resort town's older hotels, commercial businesses, and apartment houses have received a colorful facelift in the past 25 years. *(Rob Crandall/www.robcrandall.com)*

Spanish-style homes and affluent shopping and entertainment centers. But on the city's north side, the largely African American Liberty City and Overtown communities struggle with urban poverty and drug problems. Elsewhere, Caribbean influences shape the sights, sounds, and smells of Little Haiti and Allapattah (Little Dominican Republic). The city's largest Latin community is Cuban and traditionally is centered around Little Havana southwest of downtown. Along Little Havana's Calle Ocho (8th Street), the smell of roast pork mingles with cigar smoke and nearby Maximo Gomez Park is dotted with Cuban checkers and domino players (Figure 3.4.2). In nearby Miami Beach, a revitalized art deco–style commercial district features a mix of hotels and restaurants that cater to older Jewish retirees as well as to a younger generation of partygoers bound for the trendy South Beach nightclub scene (Figure 3.4.3). Elsewhere, many older buildings have been razed, replaced by high-rise luxury condominiums.

riety. When much of the agricultural interior was settled, immigrants often established close-knit communities. Among others, German, Scandinavian, Slavic, Dutch, and Finnish neighborhoods took shape, held together by common origins, languages, and religions. Although many of these ties weakened over time, rural landscapes of Wisconsin, Minnesota, the Dakotas, and the Canadian prairies still display a number of these cultural imprints. Folk architecture, distinctive settlement patterns, ethnic place-names, and the simple elegance of rural churches selectively survive as signatures of cultural diversity upon the visible scene of rural North America.

Ethnic neighborhoods also enrich the urban landscape and reflect both global-scale and internal North American migration patterns. Complex social and economic processes are clearly at work. Employment opportunities historically fueled population growth in North American cities, but the cultural makeup of the incoming labor force varied, depending on the timing of the economic expansion and the relative accessibility of an urban area to different cultural groups. The ethnic geography of Los Angeles exemplifies the interplay of the economic and cultural forces at work (Figure 3.25). Since most of its economic expansion took place during the twentieth century, its ethnic geography reflects the movements of more recent migrants. African American communities on the city's south side (Compton and Inglewood) represent the legacy of black population movements out of the South. Hispanic (East Los Angeles) and Asian (Alhambra and Monterey Park) neighborhoods are a reminder that about 40 percent of the city's population is foreign-born.

Particularly in the United States, ethnic concentrations of nonwhite populations increased in many cities during the twentieth century as whites exited for the perceived safety of the suburbs. In terms of central-city population, African Americans make up more than 60 percent of Atlanta, while Los Angeles is now more than 40 percent Hispanic. Ethnic concentrations are also growing in the suburbs of some U.S. cities: Southern California's Monterey Park has been called the "first suburban Chinatown," and growing numbers of middle-class African Americans and Hispanics shape suburban neighborhoods in metropolitan settings such as Atlanta and San Antonio.

Patterns of North American Religion

Many religious traditions also shape North America's human geography. Reflecting its colonial roots, Protestantism is dominant within the United States, accounting for about 60 percent of the population. In some settings, hybrid American religions sprang from broadly Protestant roots. By far the most successful are the Latter-day Saints (Mormons), regionally concentrated in Utah and Idaho, and claiming more than 5 million North American members. While many traditional Catholic neighborhoods have lost population in the urban Northeast, Catholic numbers are growing in the West and South, reflecting both domestic migration patterns and higher rates of Hispanic immigration and births. Within Canada, almost 40 percent of the population is Protestant, with the United Church of Canada and the Anglican Church claiming the largest numbers of followers. Roman Catholicism is important in regions that received large numbers of Catholic immigrants. French Canadian Quebec is a bastion of Catholic tradition and makes Canada's population (43 percent) distinctly more Catholic than that of the United States (24 percent).

Millions of other North Americans practice religions outside of the Protestant and Catholic traditions or are unaffiliated with traditional religions. Orthodox Christians congregate in the urban Northeast, where many Greek, Russian, and Serbian Orthodox communities were established between 1890 and 1920. The telltale domes of Ukrainian Orthodox churches still dot the Canadian prairies of Alberta, Saskatchewan, and Manitoba. More than 7 million Jews live in North America, concentrated in East and West Coast cities. In the United States, the rapidly growing organization known as the Nation of Islam also has a strong urban orientation, reflecting its appeal to many economically dispossessed African Americans. Many other Muslims (6 million), Buddhists (1 million), and Hindus (1 million) also live in the United States. While only about 8 percent of people in the United States classify themselves as nonbelievers, a recent survey showed that 30 percent of the population claimed to have a largely secular lifestyle in which religion was rarely practiced.

The Globalization of American Culture

North America's culture is becoming more global at the same time that global cultures are becoming more North American (influenced particularly by the United States). But the process of cultural globalization is becoming more complex. No longer can we think of simple flows of foreign influences into North America or the juggernaut of U.S. cultural dominance invading every traditional corner of the globe. In the twenty-first century, the story of cultural globalization will increasingly feature a mix of influences that flow in many directions at once, and that feature new hybrid cultural creations (see "Global to Local: Pentecostalism Redraws the Religious Map of the World").

North Americans: Living Globally More than ever before, North Americans in their everyday lives are exposed to people from beyond the region. With 43 million foreign-born migrants living across the region, diverse global influences are free to mingle in new ways. In addition, tens of thousands of foreigners arrive daily, mostly as tourists. In 2006, the United States recorded more than 50 million international visitors arriving in the country. At American colleges and universities, several hundred thousand international students add

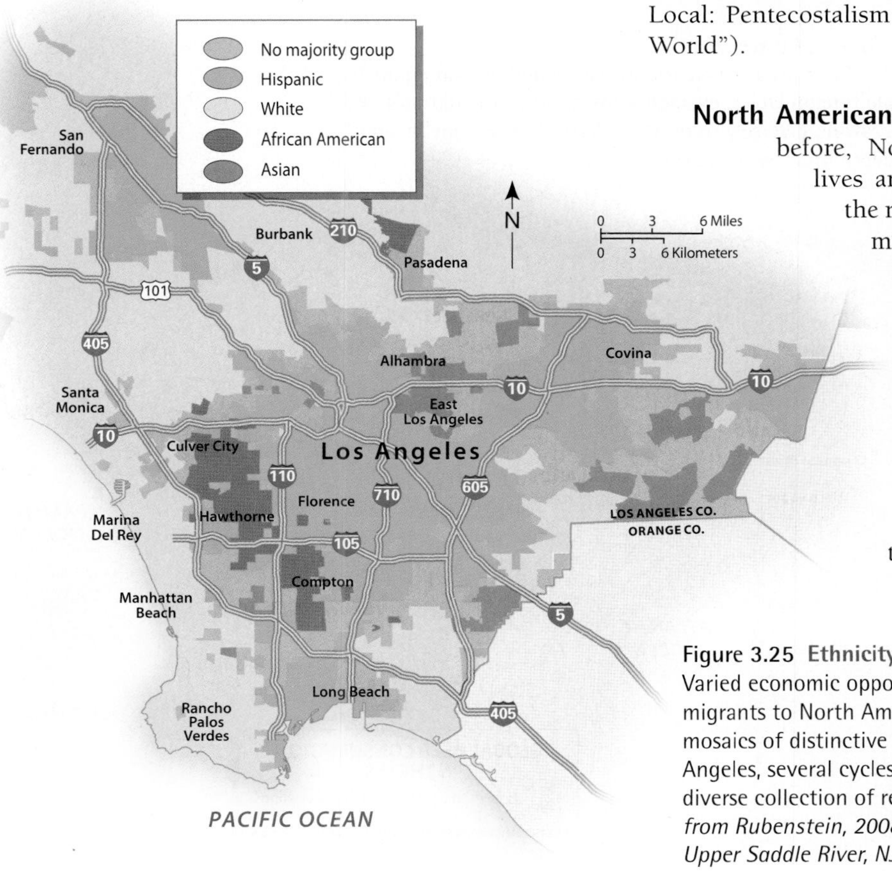

Figure 3.25 Ethnicity in Los Angeles
Varied economic opportunities have attracted a wide variety of migrants to North American cities, producing varying ethnic mosaics of distinctive neighborhoods and communities. In Los Angeles, several cycles of economic expansion have drawn a diverse collection of residents from around the globe. *(Reprinted from Rubenstein, 2008,* An Introduction to Human Geography, *Upper Saddle River, NJ: Prentice Hall)*

a global flavor to the ordinary curriculum. Canada experiences a similar saturation of such global influences, with the U.S. presence particularly dominant.

Globalization presents cultural challenges for North Americans. In the United States, one key issue revolves around the English language, which some have described as the "social glue" that holds the nation together. Since 1980, the continuing flood of non-English-speaking immigrants into the country has sharpened the debate over the role English should play in American culture. Many in the United States argue that English should be the country's only officially recognized language in order to compel immigrants to learn it and thus speed their assimilation into the host culture. Some immigrants suggest they need to maintain their traditional languages both to function within their ethnic communities and to preserve their cultural heritage.

At the same time, the evidence suggests that North America's immigrants are learning English more rapidly than ever before, seeing it as a powerful tool to accelerate their economic opportunities, both in the United States and Canada. A 2006 poll of Hispanic residents revealed that almost 60 percent of Hispanics agreed that "immigrants have to speak English to say that they are part of American society." The growing popularity of **Spanglish**, a hybrid combination of English and Spanish spoken by Hispanic Americans, also illustrates the complexities of North American globalization. Spanglish includes interesting hybrids such as *chatear*, which means to have an online conversation.

North Americans are going global in other ways. By 2008, the vast majority of Americans and Canadians had Internet access, opening the door for far-reaching journeys in cyberspace. North Americans also travel much more widely than ever before. Residents of the United States take more than 27 million overseas trips annually (75 percent for leisure travel, 25 percent for business travel). Within North America, the popularity of ethnic restaurants has peppered the realm with a bewildering variety of Cuban, Ethiopian, Basque, and Pakistani eateries. Chinese and Italian foods dominate the taste buds of all North Americans, with Mexican food a rapidly growing choice in the United States. The growing affinity for foreign beverages mirrors the pattern; imported beer sales in the United States are expanding much more rapidly than domestic sales. American consumers now drink more than 800 million gallons of imported beer annually (Figure 3.26). Americans also have increased their consumption of foreign red wines and have rapidly Europeanized their coffee-drinking habits. In fashion, *Gucci*, *Armani*, and *Benetton* are household words for millions who keep their eyes on European styles. Although British pop music has been an accepted part of North American culture for five decades, the beat of German techno bands, Gaelic instrumentals, and Latin rhythms also has become an increasingly seamless part of daily life within the realm. Indeed, from

Figure 3.26 Annual Beer Imports to the United States, 2002
Whether they are aware of it or not, North Americans are increasingly eating and drinking globally. Rising beer imports, including many upscale foreign labels, exemplify the pattern. The nation's beer drinkers know no bounds to their thirsts, drawing diversely from Asian, Australian, European, and Latin American producers. *(Data from* Modern Brewery Age, *2003)*

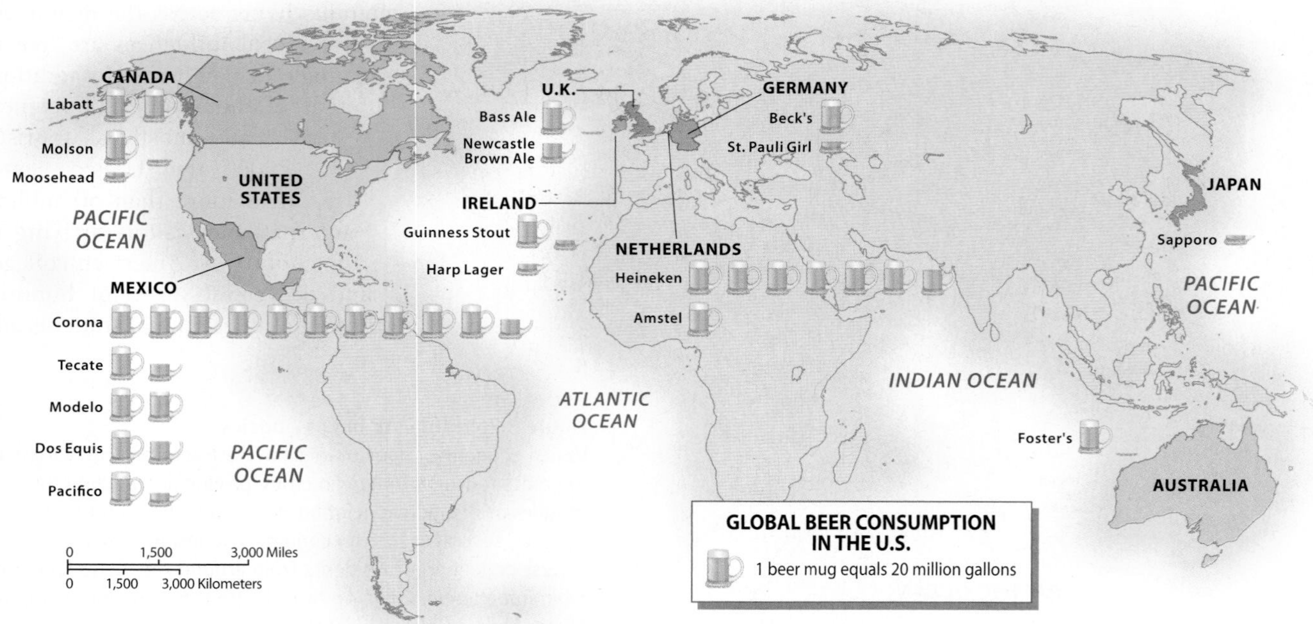

GLOBAL TO LOCAL

Pentecostalism Redraws the Religious Map of the World

Most people think about MTV, blue jeans, or Hollywood films when they ponder the ways in which North American culture has spread around the world. What about divine healing, speaking in tongues, and religious conversion? Evangelical Protestantism has been on the rise in the United States since the 1970s. What many people don't realize is its surging global popularity. Equally fascinating is the movement's ability to adapt its global message to local practices.

The Pentecostal movement is a branch of evangelical Christianity that emphasizes a personal relation to Jesus Christ, the literal interpretation of the Bible, and the power of religious conversion. Along with other conservative, charismatic Christian denominations, these Protestant revivalist movements have reshaped the religious geography of the United States since 1970. More recently, with the help of ambitious global missionary efforts, these American-style "renewalists" probably represent one of the world's fastest-growing religious movements (Figure 3.5.1).

Recent surveys suggest, for example, that this unique aspect of North American culture has redrawn the map of religion in large areas of Africa and Latin America. Renewalists now make up sizable minorities in countries such as Kenya and Brazil, particularly appealing to young, upwardly mobile populations. Often the local expressions of this global Pentecostal expansion reflect a blend of Western Christian and traditional religions. In Africa, for example, the movements can blend ancestor worship with evangelical Christianity and feature frenzied testimonials of conversion in churches such as Nigeria's Mountain of Fire and Miracles.

Guatemala has witnessed the explosive growth of Pentecostalism since 1990. Pentecostal services take place all over Guatemala City, from luxury hotels to abandoned buildings. The Fraternidad Cristiana de Guatemala is one of the city's largest Pentecostal organizations and it is constructing a 12,000-person church with a baptism pool for hundreds that may be the largest building in all of Central America. Converts respond to the energy and excitement of church services. Also blossoming in the country are dozens of charismatic Christian radio and television stations, all proclaiming loudly the arrival of a new religious fervor. The result is a popular hybrid cultural phenomenon that fuses conservative North American Protestantism with the changing lives of twenty-first-century Latin Americans. Commenting on the larger global movement, one sociologist noted it is "a potent mixture of the pre-modern and the post-modern, of the pre-literate and the post-literate, of the fiesta and the encounter group."

Figure 3.5.1 The Globalization of Pentecostal Christianity In a Pentecostal service in Kampala, Uganda, the faithful lift their hands during prayers for miracles led by "prophet" Pius Muwanguzi. *(James Akena/Reuters/Landov)*

Source: Adapted from "Christianity Reborn," *The Economist*, December 23, 2006.

acupuncture and massage therapy to soccer and New Age religions, North Americans are tirelessly borrowing, adapting, and absorbing the larger world around them.

The Global Diffusion of U.S. Culture In a parallel fashion, U.S. culture has forever changed the lives of billions of people beyond the region. Although the economic and military power of the United States was notable by 1900, it was not until after World War II that the country's popular culture reshaped global human geographies in fundamental ways. The Marshall Plan and Peace Corps initiatives exemplified the growing presence of the United States on the world stage even as European colonialism waned. Rapid improvements in global transportation and information technologies, much of them engineered in the United States, also brought the world more surely under the region's spell. Perhaps most critical was the marriage between growing global demands for consumer goods and the rise of the multinational corporation, which was superbly structured to meet and cultivate those needs.

The results of these connections are not a simple Americanization of traditional cultures or a single, synthesized global culture shaped in the U.S. image. Still, millions of people, particularly the young, are strongly attracted by the North American emphasis on individualism, consumption, youth, and mobility. The wide popularity of English-language teaching programs in places from China to Cuba is testimony to the cultural power of the United States. Indeed, English is already an official language in more than 75 countries around the world. Global information flows illustrate the country's cultural influence on the world. Companies such as Time Warner and Walt Disney increasingly dominate multiple entertainment media. U.S. book and magazine publishing is thriving, and there is an expanding international appetite for everything from technical journals to romance novels and science fiction. The United States also dominates television; as television sets, cable networks, and satellite dishes spread, so do U.S. sitcoms, CNN, and MTV. Movie screens tell a similar story. In western Europe, the United States claims two-thirds of the film market, and more than half of Japan's movie entertainment is dubbed fresh from Hollywood.

The United States shapes the popular cultural landscape of every corner of the globe. Global corporate advertising, distribution networks, and mass consumption bring Cokes and Big Macs to Moscow and Beijing, golf courses to Thai jungles, Mickey and Minnie Mouse to Tokyo and Paris, and Avon cosmetics to millions of beauty-conscious Chinese. Western-style business suits have become the professional uniform of choice, while T-shirts and jeans offer standardized global comfort on days away from work. In the built landscape, central-city skylines become indistinguishable from one another, suburban apartment blocks take on a global sameness, and one airport hotel looks the same as another eight time zones away.

But U.S. cultural control has not gone unchallenged, illustrating the varied consequences of globalization. As worldwide use of the Internet has grown, the online dominance of English-speaking users has dramatically declined globally from more than 71 percent in 1998 to only 22 percent in 2005. Not surprisingly, given the rapid diffusion of the Internet and China's growing influence, Mandarin may pass English as the leading global language of Internet users by 2010. Active resistance to U.S. cultural influence is also notable. For example, Canadian government agencies routinely chastise their radio, television, and film industries for letting in too much U.S. cultural influence. As an antidote to their overbearing southern neighbor, the government requires certain levels of Canadian content in much of the media programming. The French also criticize U.S. dominance in such media as the Internet. Public subsidies to France's Centre National de la Cinematographie are designed to foster filmmaking for a national audience deluged with English-language productions. Elsewhere, Iran banned satellite dishes and many U.S. films, although illegal copies of top box-office hits often find their way through national borders.

GEOPOLITICAL FRAMEWORK: Patterns of Dominance and Division

In disarmingly simple fashion, North America's political geography brings together two of the world's largest states. The creation of these states was the complex outcome of historical processes that might have created quite a different North American map. Once established, the states have coexisted in a close relationship of mutual economic and political interdependence. President John F. Kennedy summarized the links in a speech to the Canadian parliament in 1962: "Geography has made us neighbors, history has made us friends, economics has made us partners and necessity has made us allies." That cozy continental relationship has not been without its tensions, and some persist today. In addition, both nations have had to deal with fundamental internal political complexities that have not only tested the limits of their federal structures, but also challenged their very existence as states.

Creating Political Space

The United States and Canada sprang from very different political roots. The United States broke cleanly and violently from Great Britain. The American Revolution fostered a powerful sense of nationalism that sped the process of spatial expansion and contributed toward the creation of a continental, indeed global, power. By contrast, Canada was a country of convenience, born from a peaceful, incremental separation from Britain and then assembled as a collection of distinctive regional societies that only gradually acknowledged their common political destiny.

Uniting the States Turning back the clock to the early eighteenth century reveals a political geography very much in the making. Beyond scattered frontiers of European settlement lay vast domains of Native American–controlled political space. Although boundaries were not formally surveyed or mapped, native peoples carved up the continent in an elaborate geography of homelands, allied territories, and enemy terrain. The creation of the United States replaced this continental division of political space with another.

With amazing rapidity, Europe and then the United States imposed their own political boundaries across the realm. The 13 English colonies, sensing their common destiny after 1750, finally united two decades later and clashed violently with their colonial parent. By the 1790s, the young nation's political claims had reached the Mississippi River; the new republic was busily coercing land cessions from native peoples; and the Ordinance of 1787 had provided a template for western territory and state formation that served as the model of expansion for the next century. Soon the Louisiana Purchase (1803) nearly doubled the country's size, creating a new political domain that was as vast as it was unexplored. By mid-century, Texas had been annexed; treaties with Britain had secured the Pacific Northwest; and an aggressive war with Mexico had captured much of the Southwest. The territorial acquisition of Alaska (1867) and Hawaii (1898) eventually rounded out the present political domain of 50 states.

Assembling the Provinces Canada was created under quite different circumstances. The modern pattern of provinces assembled in a slow and uncertain fashion. After the American Revolution, England's remaining territorial claims in the region came under the control of colonial administrators in British North America. The Quebec Act of 1774 allowed for continued French settlement in the St. Lawrence Valley and provided the initial template for governing the region. Soon, however, Anglo settlers near Lakes Ontario and Erie pressed for more local colonial representation. The result was the Constitutional Act of 1791, which divided the colony into Upper Canada (Ontario) and Lower Canada (Quebec). Frustration with that system led to the Act of Union in 1840, thereby reuniting the two Canadas. In 1867 the British North America Act united the provinces of Ontario, Quebec, Nova Scotia, and New Brunswick in an independent Canadian Confederation. The peaceful separation from the mother country also guaranteed to Quebec special legal and cultural privileges that set the stage for some of the modern power struggles within the country.

Once created, the Canadian Confederation grew in piecemeal fashion, more out of geographical convenience than from any compelling nationalism at work to unite the northern portion of the continent. Within a decade, the Northwest Territories (1870), Manitoba (1870), British Columbia (1871), and Prince Edward Island (1873) joined Canada, and the continental dimensions of the country took shape. Soon, the Yukon Territory (1898) separated from the Northwest Territories; Alberta and Saskatchewan gained provincial status (1905); and Manitoba, Ontario, and Quebec were enlarged (1912) north to Hudson Bay. Newfoundland finally joined in 1949. The recent addition of Nunavut Territory (1999), carved from the Northwest Territories, represents the latest change in Canada's political geography.

Continental Neighbors

Geopolitical relationships between Canada and the United States have always been close: their common 5,525-mile (8,900-kilometer) boundary requires both nations to pay close attention to one another. During the twentieth century, the two countries lived largely in political harmony with one another. In 1909, the Boundary Waters Treaty created the International Joint Commission, an early step in the common regulation of cross-boundary issues involving water resources, transportation, and environmental quality. The St. Lawrence Seaway (1959) opened the Great Lakes region to better global trade connections. The two nations also joined in cleaning up Great Lakes pollution and in making plans to reduce acid rain in eastern North America.

It has been in the area of trade relations that the close political ties between these neighbors have mattered most. The United States receives almost 90 percent of Canada's exports and supplies more than two-thirds of its imports. Conversely, Canada is the United States' most important trading partner, accounting for roughly 20 percent of its exports and imports. One landmark agreement reached in 1989 was the signing of the bilateral (two-way) Free Trade Agreement (FTA). Five years later, the larger **North American Free Trade Agreement (NAFTA)** extended the alliance to Mexico. Paralleling the success of the European Union (EU), NAFTA has forged the world's largest trading bloc, including more than 400 million consumers and a huge free-trade zone that stretches from beyond the Arctic Circle to Latin America.

Political conflicts still divide the two countries (Figure 3.27). Environmental issues produce cross-border tensions, especially when environmental degradation in one nation affects the other. For example, Montana's North Flathead River flows out of British Columbia, where Canadian logging and coal mining operations periodically threaten fisheries and recreational lands south of the border. Agricultural and natural resource competition also causes occasional controversy between the two neighbors. The appearance of mad cow disease in Canadian livestock curtailed exports to the United States and elsewhere. Furthermore, when the disease appeared in U.S. cattle in 2003, Canadian sources were suspected, raising tensions between the two nations. Problems recently developed when Canadian wheat and potato growers were accused of dumping their products into U.S. markets, thus depressing prices and profits for U.S. farmers. Similar issues have arisen in the logging industry, although a 2006 agreement signed between the two countries has lessened tensions over that issue.

The Legacy of Federalism

The United States and Canada are **federal states** in that both nations allocate considerable political power to units of government beneath the national level. Other nations, such as France, have traditionally been **unitary states**, in which power is centralized at the national level. Federalism leaves many political decisions to local and regional governments and often allows distinctive cultural and political groups to be recognized as distinct entities within a country. Although both nations have federal constitutions, their origins and evolution are very different. The U.S. Constitution (1787), created out of a violent struggle with the powerful British nation, specifically limited centralized authority, giving all unspecified powers to the states or the people. In contrast, the Canadian Constitution (1867), which created a federal parliamentary state, was an act of the British Parliament. Originally, it reserved most powers to central authorities and maintained many political links between Canada and the British Crown. Ironically, the evolution of the United States as a federal republic produced an increasingly powerful central government, while Canada's geopolitical balance of power shifted toward more provincial autonomy and a relatively weak national government. For example, the federal government largely controls U.S. public lands, but in Canada provincial authorities retain power over public Crown lands.

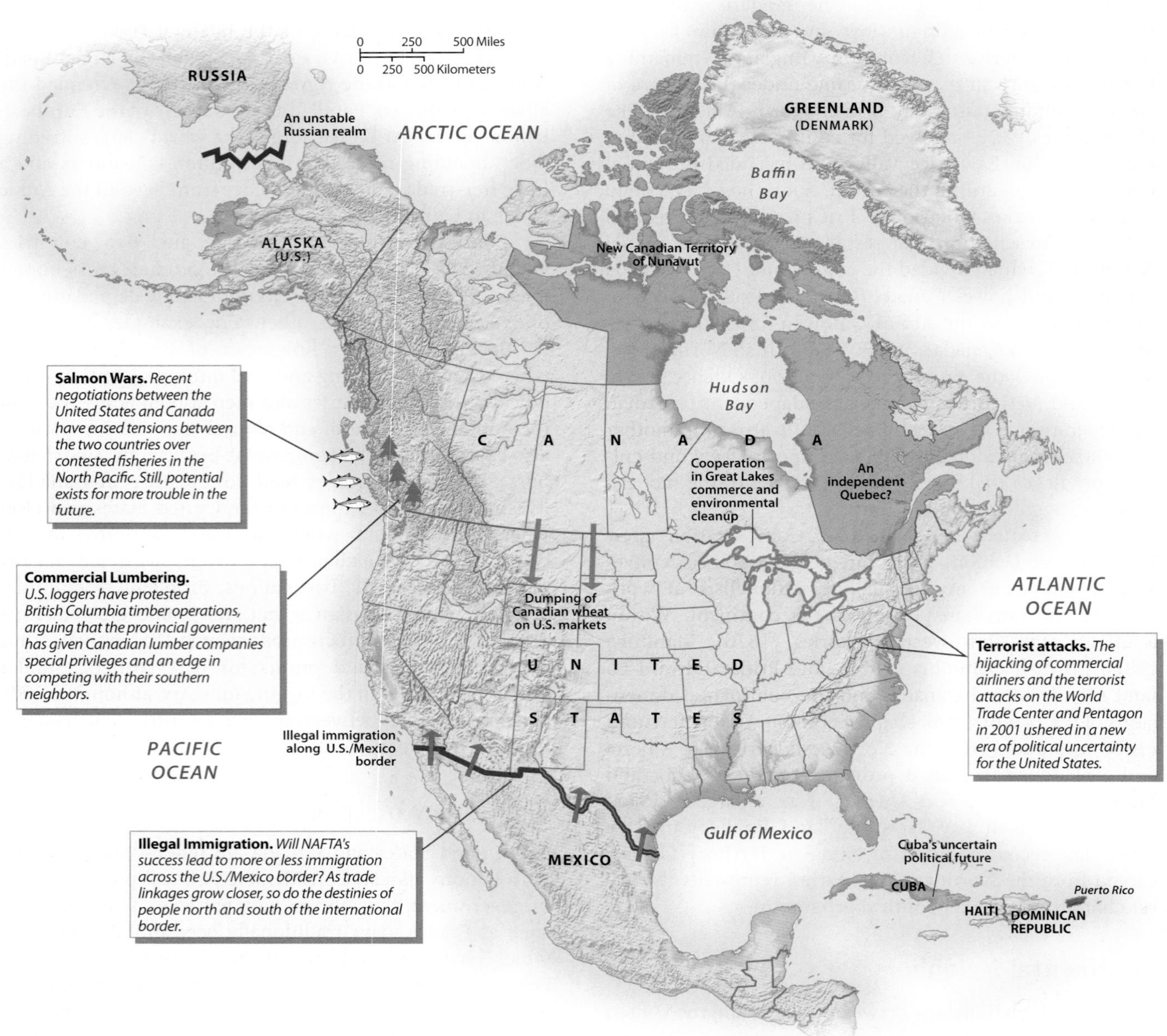

Figure 3.27 Geopolitical Issues in North America
Although Canada and the United States share a long and peaceful border, many political issues still divide the two countries. In addition, internal political conflicts cause tensions, particularly in bicultural Canada. Given its global prominence and the openness of its society, the United States also remains vulnerable to acts of global terrorism.

Quebec's Challenge The political status of Quebec remains a major issue within Canada (see Figure 3.27). Economic disparities between the Anglo and French populations have reinforced cultural differences between the two groups, with the French Canadians often suffering when compared with their wealthier neighbors in Ontario. Beginning in the 1960s, a separatist political party in Quebec (the Parti Quebecois) increasingly voiced French Canadian concerns. When the party won provincial elections in 1976, it declared French the official language of Quebec. Formal provincial votes over the question of remaining within Canada were held in 1980 and 1995. Both measures failed. Since then, support for separation has ebbed in favor of a more modest strategy of increased "autonomy" within Canada. In elections held in 2007, for example, the Parti Quebecois ran a poor third in Quebec, suggesting its separatist rhetoric is no longer as popular within the province.

Figure 3.28 Life in Nunavut
This scene on Baffin Island is from Iqaluit, the largest urban center in the new Canadian province of Nunavut. *(Alison Wright/Corbis)*

Native Peoples and National Politics Another challenge to federal political power has come from North American Indian and Inuit populations, in both Canada and the United States. Within the United States, the renewed assertion of Native American political power began in the 1960s and marked a decisive turn away from earlier policies of assimilation. Since passage of the Indian Self-Determination and Education Assistance Act of 1975, the trend has been toward increased Native American autonomy. The Indian Gaming Regulatory Act (1988) offered potential economic independence for many tribes. By 2006, Indian gaming operations nationally netted tribes almost $25 billion annually. In the western American interior, where Indians control roughly 20 percent of the land, tribes are solidifying their hold on resources, reacquiring former reservation acreage, and participating in political interest groups such as the Native American Fish and Wildlife Society and the Council of Energy Resource Tribes. In Alaska, native peoples acquired title to 44 million acres (18 million hectares) of land in 1971 under the Alaska Native Claims Settlement Act.

In Canada even more ambitious challenges to a weaker centralized government have yielded dramatic results. As natives pressed their claims for land and political power in the 1970s, Canada established the Native Claims Office (1975) and began negotiating settlements with various groups, particularly within the country's vast northern interior. Agreements with native peoples in Quebec, Yukon, and British Columbia turned over millions of acres of land to aboriginal control and increased native participation in managing remaining public lands. By far, the most ambitious agreement has been to create the territory of Nunavut out of the eastern portion of the Northwest Territories in 1999 (Figure 3.27). Nunavut is home to 30,000 people (85 percent Inuit) and is the largest territorial/provincial unit within Canada (Figure 3.28). Its creation represents a new level of native self-government in North America. Recently, agreements between the federal Parliament and British Columbia tribes have initiated a similar move toward more native self-government in that western province. The Nisga'a tribe, for example, now controls a 770-square-mile (1,992-square-kilometer) portion of the Nass River valley near the Alaska border (see Figure 3.23). Elsewhere, recent disputes over land claims in Ontario (near Toronto) and in oil-rich Alberta are reminders that Canadians, much like their neighbors to the south, are still struggling with defining the political status of their native populations.

The Politics of U.S. Immigration

Immigration policies are hotly contested within the United States. Recently, growing public protests on various sides of the dilemma have helped highlight the conflict (Figure 3.29). Three key issues remain at the center of the debate. First, there are ongoing disagreements concerning the overall numbers of legal immigrants that should be allowed into the country. Some groups, such as the Federation for American Immigration Reform (FAIR), suggest that sharply reduced numbers of immigrants would protect American jobs and allow for a more gradual assimilation of existing foreigners. But other groups such as the Cato Institute take the opposite position, proposing to loosen existing restrictions on immigrants

Figure 3.29 Immigration Protests
Growing political disagreements over U.S. immigration policy have recently sparked large public protests over the issue in many American cities. *(David McNew/Getty Images)*

in a move to spur economic growth and business expansion.

A second major issue, particularly along the border with Mexico, is how to tighten up on the daily illegal flow of immigrants. Since the terrorist attacks in 2001, many have argued that the country's wide-open southern border is really a national security issue. Recent federal legislation has allowed for more border patrols and for the construction of more fencing along the border, although the existing barrier is still far from the more than 700 miles (1,125 kilometers) of fence desired by some. Others argue such measures are pointless and merely sour relations with Mexico. Meanwhile, more than 1.2 million foreigners are apprehended along the border annually.

Finally, there is no political consensus on a fair policy to deal with existing undocumented workers within the United States. Some policymakers have advocated stricter felony-level penalties for illegal immigrants, while others have proposed loosening requirements for citizenship or some form of amnesty to enable illegal immigrants to more easily enter the mainstream of American society. Suggested political compromises have included expanding the country's guest worker program or requiring illegal immigrants to return to their home countries ("touching base") before they can reapply for legal status within the United States.

A Global Reach

The geopolitical reach of the United States, in particular, has taken its influence far beyond the bounds of North America. The Monroe Doctrine (1824) asserted that U.S. interests were hemispheric and transcended national boundaries, but it was not until after 1895 that the United States accelerated its global expansion. Two principal settings served as early laboratories for political imperatives in the United States. In the Pacific, the United States claimed the Philippines as a prize of the Spanish-American War (1898), and further annexations of Guam (1898) and the Hawaiian Islands (1898) began the country's twentieth-century dominance of the region. In Central America and the Caribbean, the growing role of the American military between 1898 and 1916 shaped politics in Cuba, Puerto Rico, Panama, Nicaragua, Haiti, Mexico, and elsewhere. Further, the country's role in World War I raised its stakes in European affairs.

The 1920s and 1930s briefly returned the United States to isolationist policies, but World War II and its aftermath forever redefined the country's role in world affairs. Victorious in both the Atlantic and Pacific theaters, postwar America emerged from the conflict as the world's dominant political power. Quickly, however, a resurgent Soviet Union challenged the United States, and the Cold War began in the late 1940s. In response, the Truman Doctrine promised aid to struggling postwar economies and actively challenged communist expansion in Europe and elsewhere. The United States also fashioned multinational political and military agreements, such as those establishing the North Atlantic Treaty Organization (NATO) and the Organization of American States (OAS), which were designed to cast a broad umbrella of U.S. protection across much of the noncommunist world. Violent conflicts in Korea (1950–1953) and Vietnam (1961–1973) pitted U.S. political interests against communist attempts to extend their Asian dominance beyond the Soviet Union and China. Tensions also ran high in Europe as the Berlin Wall crisis (1961) and nuclear weapons deployments by NATO- and Soviet-backed forces brought the world closer to another global war. The Cuban missile crisis (1962) reminded Americans that traditional political boundaries provided little defense in a world uneasily brought closer together by technologies of potential mass destruction.

Even as the Cold War gradually receded during the late 1980s, the global political reach of the United States continued to expand. A few examples suggest the pattern. Interventionist policies in Central America favored regimes friendly to the United States. President Carter's successful Middle East Peace Treaty between Israel and Egypt (1979) guaranteed a continuing diplomatic and military presence in the eastern Mediterranean. When Iraq's Saddam Hussein threatened Persian Gulf oil supplies in 1990, the United States led a United Nations coalition to contain the aggression. In the late 1990s, Serbian aggression within Kosovo prompted an American- and NATO-led intervention, which included major air attacks on the Serbian capital of Belgrade (1999) and a peacekeeping presence (with the UN) in the disputed area of Kosovo. The recent wars in Afghanistan (2001) and Iraq (2003), and their unsettled aftermath, offer other examples of the growing political presence of the United States beyond the bounds of North America.

Indeed, the current overhaul of the U.S. military is changing its worldwide distribution (Figure 3.30). Military planners argue that in the future less emphasis should be given to housing large numbers of troops in relatively friendly foreign "hub" settings such as Germany, South Korea, and Japan. A growing number of the U.S. military's 1.4 million active-duty personnel appear headed for more spartan assignments located near source regions for terrorists and in arenas of potential conflict. Military strategists identify a huge global "arc of instability" where U.S. interests need to be refocused. From the drug- and rebel-filled jungles of Colombia to the terrorist training camps of the Philippines, the region includes much of Africa, the Middle East, Central Asia, and Southeast Asia.

As recent military experiences in Afghanistan and Iraq demonstrate, however, there may be practical and strategic limits to the expanded U.S. role. Although some have compared the country's geopolitical aspirations to those of ancient Rome or the British Empire, other observers have cautioned that the economic expense and political costs involved with such operations may be very high. One thing is certain, however. The farflung global distribution of the U.S. military, while it may bring increased influence, will also bring greater risks to the world's last remaining superpower.

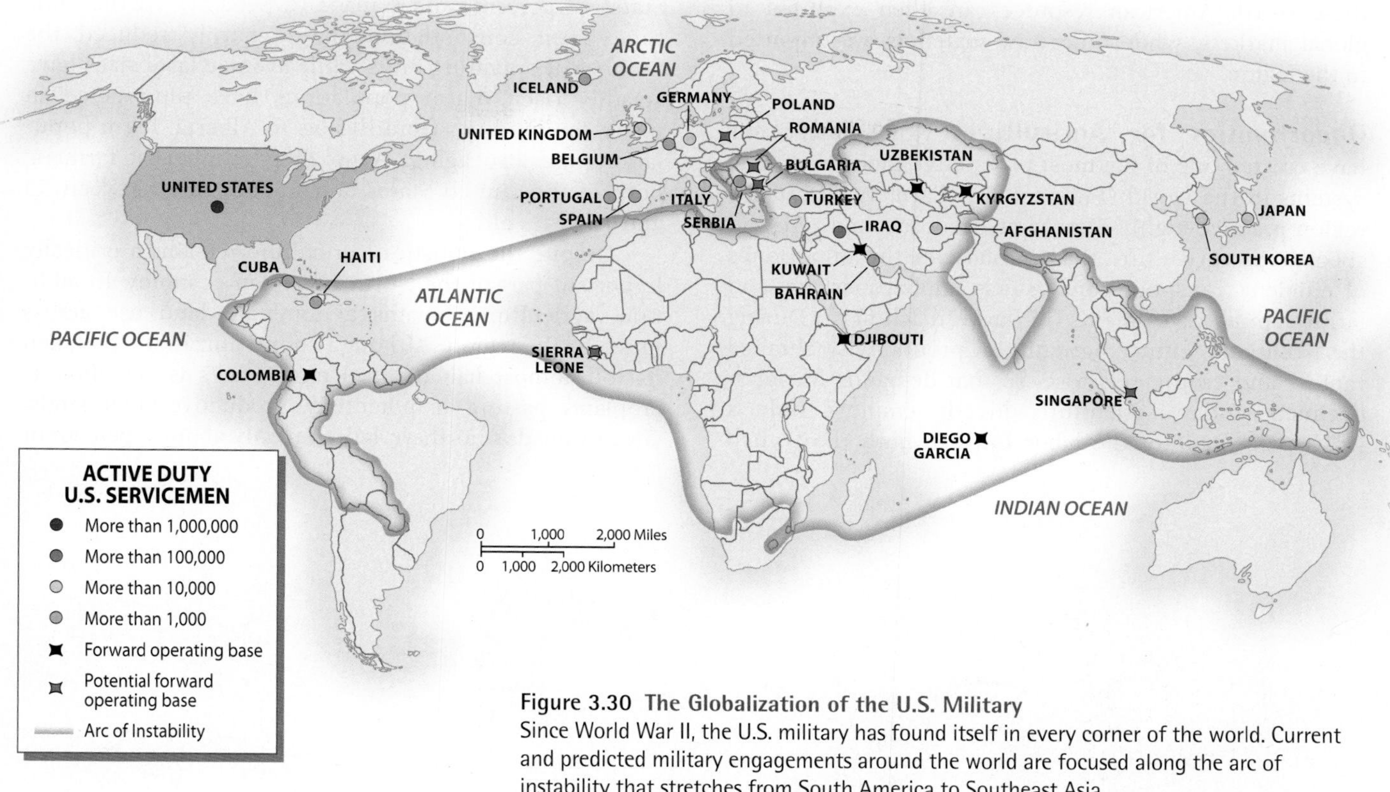

Figure 3.30 The Globalization of the U.S. Military
Since World War II, the U.S. military has found itself in every corner of the world. Current and predicted military engagements around the world are focused along the arc of instability that stretches from South America to Southeast Asia.

ECONOMIC AND SOCIAL DEVELOPMENT: Geographies of Abundance and Affluence

Along with its global political clout, North America possesses the world's most powerful economy and its most affluent population. Its 335 million people consume huge quantities of global resources but also produce some of the world's most sought-after manufactured goods and services. North America's size, geographic diversity, and resource abundance have all contributed to the region's global dominance in economic affairs. More than that, however, the region's human capital—the skills and diversity of its population—has enabled North Americans to achieve high levels of economic development (Table 3.2).

An Abundant Resource Base

North America is blessed with a varied storehouse of natural resources. The region's climatic and biological diversity, its soils and terrain, and its abundant energy, metals, and forest resources have provided a variety of raw materials for development. Indeed, the direct extraction of natural resources still makes up 3 percent of the U.S. economy and 6 percent of the Canadian economy. Some of

TABLE 3.2 Development Indicators

Country	GNI[a] per Capita, PPP[b] (2005)	GDP[c] Average Annual % Growth (2000–05)	Life Expectancy (2007)	Percent of Population Living on Less Than $2 a Day	Under Age 5 Mortality Rate, 1990	Under Age 5 Mortality Rate, 2005	Gender Equity[d]
Canada	32,220	2.5	80		8	6	106
United States	41,950	2.6	78		11	7	109

[a]*Gross national income.*
[b]*Purchasing power parity.*
[c]*Gross domestic product.*
[d]*Ratio of female-to-male enrollments in primary and secondary school percentage. Numbers below 100 have more males in primary/secondary school; numbers above 100 have more females in primary/secondary schools.*

Sources: World Bank, *World Development Indicators, 2007; life expectancy and percentage of population living on $2 a day data from Population Reference Bureau,* World Data Sheet, *2007.*

these North American resources are then exported to global markets, while other raw materials are imported to the region.

Opportunities for Agriculture North Americans have created one of the most productive food-producing systems in the world (Figure 3.31). Farmers within the region practice highly commercialized, mechanized, and specialized agriculture that emphasizes the importance of efficient transportation systems, global markets, and large capital investments in farm machinery. During the twentieth century, agricultural productivity climbed rapidly and crop yields rose, so that demands for labor declined. Today, agriculture directly employs only a small percentage of the labor force in both the United States (1 percent) and Canada (2 percent). At the same time, farm consolidations have sharply reduced the number of operating units while average farm sizes have steadily risen. Farm populations have plummeted in many communities from Illinois to Alberta. Farm populations are also aging. A growing number of farmers, both in the United States and Canada, are between 55 and 75 years old.

Despite the statistics, agriculture remains a critically important part of the North American economy. In addition, agriculture remains a dominant land use across much of the region. Although suburbanization has taken its toll, almost half of the United States is classified as cropland, pasture, or potentially productive range lands. Given Canada's northern latitude, only about 7 percent of

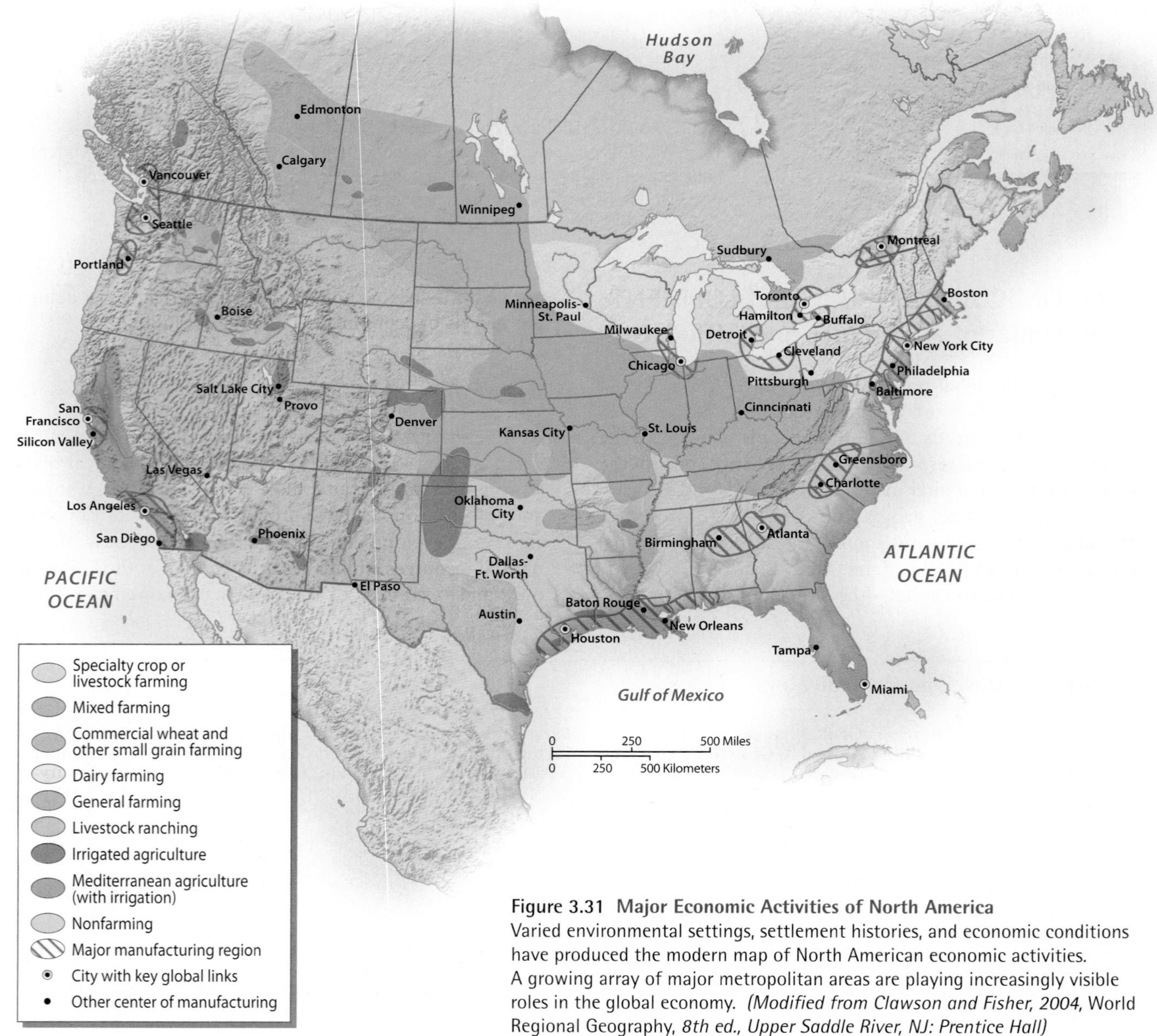

Figure 3.31 Major Economic Activities of North America
Varied environmental settings, settlement histories, and economic conditions have produced the modern map of North American economic activities. A growing array of major metropolitan areas are playing increasingly visible roles in the global economy. *(Modified from Clawson and Fisher, 2004,* World Regional Geography, *8th ed., Upper Saddle River, NJ: Prentice Hall)*

its land is potentially agricultural, but its extensive forests have made it one of the world's largest exporters of timber, pulp, and newsprint.

The geography of North American farming represents the marriage between diverse environments, varied continental and global markets for food, and historical patterns of settlement and agricultural evolution. In the Northeast, dairy operations and truck farms specializing in fresh produce capitalize on the proximity of the region to major cities in Megalopolis and southern Canada. Corn and soybeans dominate croplands across the U.S. Midwest and western Ontario, where a tradition of mixed farming combines feed grains with commercially raised livestock, particularly cattle and hogs. Growing interest in using corn as an alternative fuel in the form of ethanol has boosted that crop's prospects since 2005. Overall, more than 15 percent of the agricultural wealth of the United States sprouts from the three heartland states of Illinois, Iowa, and Nebraska. North of the Corn Belt, cooler, shorter growing seasons support more dairy operations across Wisconsin and central Michigan. To the south, only remnants of the old Cotton Belt remain, largely replaced by a diverse geography of subtropical specialty crops; poultry, catfish, and livestock production; and commercial forestry (Figure 3.32).

West of the 100th meridian, extensive, highly mechanized commercial grain-growing operations include portions of Kansas, the Dakotas, northern Montana, eastern Washington, Saskatchewan, and Alberta. Depending on surface and groundwater resources, irrigated agriculture across western North America also offers opportunities for more intensive operations. Indeed, California's agricultural output, nourished by agribusiness operations in the irrigated Central Valley, accounts for more than 10 percent of the overall farm economy of the United States. The state specializes in operations such as large-scale dairying, grape production, and other fruit crops.

Industrial Raw Materials North Americans produce and consume huge quantities of other natural resources. While the region is well endowed with a variety of energy and metals resources, the scale and diversity of the North American economy have led to the need to import additional raw materials. Petroleum use in the United States exemplifies the pattern. The country produces about 12 percent of the world's oil but consumes about 25 percent. As a result, the United States imports more than half of its oil, much of it coming from the Americas (Venezuela, Mexico, and Canada), the Middle East (Saudi Arabia), and Africa (Nigeria and Angola). Within the region, the major areas of oil and gas production are the Gulf coast of Texas and Louisiana; the Central Interior, including West Texas, Oklahoma, and eastern Kansas; Alaska's North Slope; and Central Canada (especially Alberta).

The most abundant fossil fuel in the United States is coal, but its relative importance in the overall energy economy declined in the twentieth century as industrial technologies changed and environmental concerns grew. Nonetheless, the country's 400-year supply of coal reserves (27 percent of the world's total) will be an important energy resource, both for domestic consumption as well as for export. The nation's leading coal-producing region is Appalachia, but demand has been increasing for the environmentally cleaner low-sulfur coals of the western interior (Great Plains, Intermountain West, and Alberta).

North America also remains a major producer of metals resources, although global competition, rising extraction costs, and environmental concerns pose challenges for this sector of the economy. Still, the realm is endowed with more than 20 percent of the world's copper, lead, and zinc reserves, and it accounts for more than 20 percent of global gold, silver, and nickel production.

Creating a Continental Economy

The timing of European settlement in North America proved pivotal in its rapid economic transformation. The region's abundant resources came under the control of varied European powers armed with new technologies that reshaped the landscape and reorganized its economic geography. By the nineteenth century, North Americans themselves were actively contributing to those technological changes, and as new resources were unlocked in the interior and new immigrant populations flooded the continent, the region took full advantage of its storehouse of raw materials as well as the benefits of the Industrial Revolution. In the twentieth century, although natural resources remained important, new industrial innovations and greatly expanded service employment diversified the economic base and extended its global reach.

Connectivity and Economic Growth Dramatic improvements in North America's transportation and communication systems laid the foundation for urbanization, industrialization, and the commercialization of

Figure 3.32 Specialty Agriculture in the American South Today, the agriculture of the American South focuses on specialty operations engaged in subtropical cropping, commercial forestry, and poultry, catfish, and livestock production. This Florida farm grows basil, a mint plant widely used as a cooking herb. *(Rob Crandall)*

agriculture. Indeed, the region's economic miracle was a function of its **connectivity**, or how well its different locations became linked with one another through an improved transportation and communication infrastructure. Those links greatly facilitated the potential for interaction between locations and dramatically reduced the cost of distance. Before 1830, North American connectivity gradually improved with the help of better roads, but it still took a week to travel from New York City to Ohio. For commercial traffic, canal construction also facilitated the movement of bulky goods. More than 1,000 miles (1,600 kilometers) of canals crisscrossed the eastern United States, including the Erie Canal (completed in 1825), which linked New York City with the North American interior. Other canals connected the Great Lakes and Ohio River systems, further integrating the Midwest into the Northeast economic core.

Tremendous technological breakthroughs revolutionized North America's economic geography between 1830 and 1920. Railroads created whole new economic relationships. By 1860 more than 30,000 miles (48,387 kilometers) of track had been laid in the United States, and the network grew to more than 250,000 miles (403,226 kilometers) by 1910. Farmers in the Midwest and Great Plains found ready markets for their products in cities hundreds of miles away. Industrialists collected raw materials from vast distances, processed them, and shipped manufactured goods to their final destinations. After 1869, travelers could board a train in Chicago and conveniently cross the continent in a few days. The telegraph brought similar changes to information: long-distance messages flowed across eastern North America by the late 1840s, and 20 years later undersea cables firmly linked the realm to Europe, another milestone in the larger process of globalization.

North America's integrated transportation and communications systems were redefined after 1920 as automobiles, mechanized farm equipment, paved highways, commercial air links, national radio broadcasts, and dependable transcontinental telephone service further reduced the cost of distance across the region. Cheap mass-produced automobiles reached many middle-class North Americans by the 1920s, and the subsequent construction of major North American highways by 1970 produced the world's largest integrated road system. Today, the United States has more than 46,000 miles of interstate highways. After World War II, continental connectivity also benefited from the St. Lawrence Seaway between the Atlantic Ocean and Great Lakes (1959), vast improvements in jet airline connections, and the increasing prevalence of television. Perhaps most important, the region has taken the lead in the global information age, integrating computer, satellite, and Internet technologies in a web of connections that facilitates the flow of knowledge both within the region and beyond (Figure 3.33).

The Sectoral Transformation Changes in employment structure signaled North America's economic modernization just as surely as its increasingly interconnected society. The **sectoral transformation** refers to the evolution of a nation's labor force from one highly dependent on the *primary* sector (natural resource extraction) to one with more employment in the *secondary* (manufacturing or industrial), *tertiary* (services), and *quaternary* (information processing) sectors. The sectoral transformation thus reflects the way innovation changes employment opportunities in a society. For example, with agricultural mechanization, lower demands for primary-sector workers are replaced by new opportunities in the expanding industrial sector. In the twentieth century, new services (trade, retailing) and information-based activities (education, data processing, research) created employment opportunities in these postindustrial sectors of the economy.

Both the United States and Canada have witnessed dramatic changes in employment that reveal the formative impacts of technological innovation and economic restructuring. In the late nineteenth century, primary-sector employment in agriculture and mining dominated the occupational structures of the two countries. Rapid

Figure 3.33 Cell Phone Nation
These college students stay connected with friends and family through picture-phone technology. The telecommunications and information services industries promise to continue reshaping North American lifestyles in the decades to come. *(Getty Images, Inc.–PhotoDisc)*

industrialization after 1870 contributed to steady growth in the secondary sector, and a century later it accounted for about 30 percent of the workforce. Modern technology has had the opposite effect on the primary sector, however, as machines replaced labor, and employment in the sector has fallen to less than 4 percent of the workforce.

In the later twentieth century, both nations also experienced relative declines in manufacturing employment compared with the rapid growth of the tertiary and quaternary sectors. Today these two latter sectors employ more than 70 percent of the labor force in both Canada and the United States. In addition, the service sector generates more than 75 percent of the region's annual economic output. These recent trends also reveal the tangible imprint of globalization on the North American labor force. Much of North America's sustained growth in the tertiary and quaternary sectors is directly tied to the ability of those industries to export innovations in services, financial management, and information processing to a worldwide clientele.

Eastern Canada's Maritime provinces illustrate the challenges of successfully engineering the sectoral transformation in more peripheral areas of North America. The region's long dependence on a small number of extractive activities, such as fishing, logging, and marginal agriculture, has encouraged many of its young people to leave and has boosted unemployment levels well above the national norm (Figure 3.34). Although new discoveries in the Hibernia offshore oil fields have given portions of the region a boost, the regional impacts of the energy industry are still limited. As its residents struggle to find reliable alternatives to its traditional economic base, their wages remain at only 80 or 90 percent of the national average. Canada's federal government has pledged to spend millions to support new businesses, foster the development of its offshore energy industry, and attract more high-wage jobs to the Maritimes, but the economy of the region is likely to lag as long as its reliance on the waning primary sector continues.

Regional Economic Patterns North America's industries reveal important regional patterns of concentration. **Location factors** are the varied influences that explain *why* an economic activity is located where it is, and North America's economic geography suggests that these factors change over time and that many influences, both within and beyond the realm, shape patterns of economic activity. Patterns of industrial location illustrate the concept (see Figure 3.31). The historical manufacturing core includes Megalopolis (Boston, New York, Philadelphia, and Baltimore), southern Ontario (Toronto and Hamilton), and much of the industrial Midwest. Several location factors account for the core's nineteenth-century development and the ongoing importance of its diverse industrial economy. Its proximity to *natural resources* (farmland, coal, and iron ore); increasing *connectivity* (canals and railroad networks, highways, air traffic hubs, and telecommunications centers); a ready supply of *productive labor*; and a growing national, then global, *market demand* for its industrial goods spurred sustained *capital investment* within the core. Traditionally, the core has dominated in the production of steel, automobiles, machine

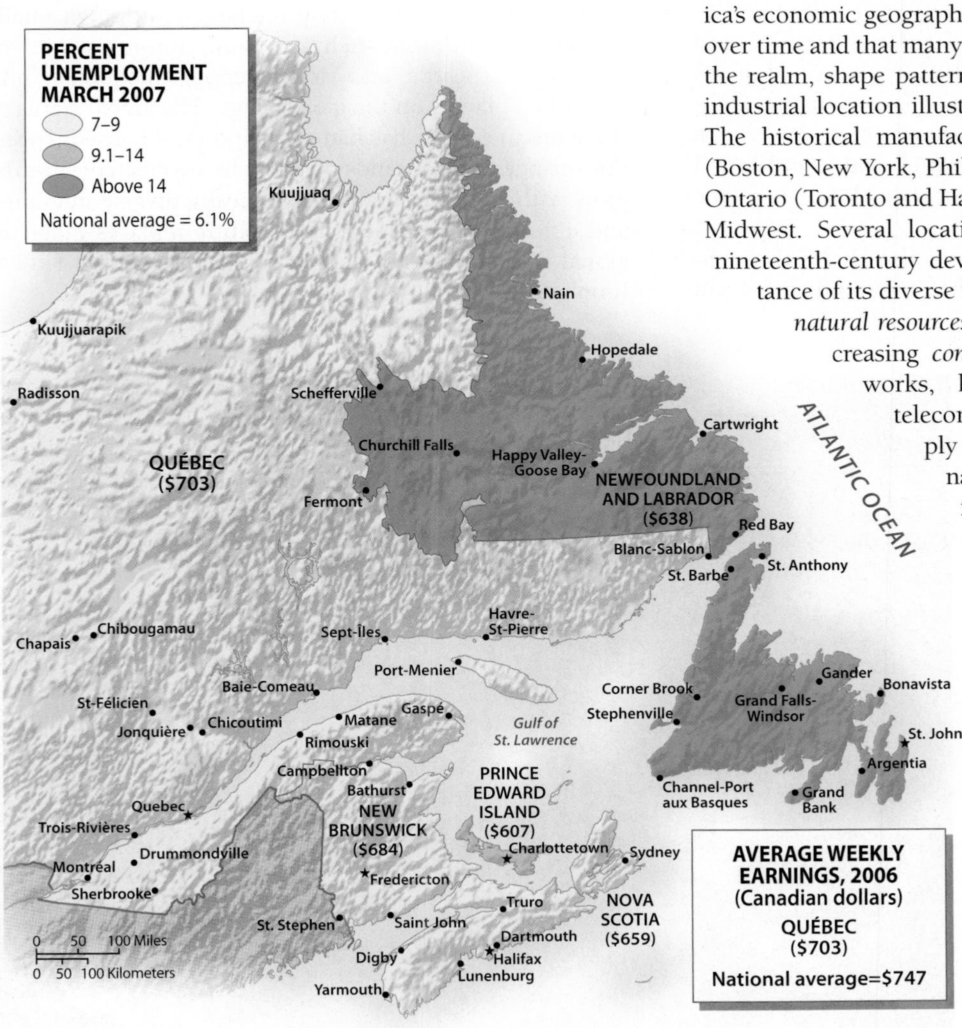

Figure 3.34 Eastern Canadian Unemployment
Eastern Canada's Maritime provinces continue to suffer from high unemployment as the regional economy struggles to shift away from dependence on primary-sector activities. Regional weekly earnings also remain well below the national average. *(Statistics Canada, 2007)*

tools, and agricultural equipment, as well as played a pivotal role in producer services such as banking and insurance.

In the last half of the twentieth century, however, industrial and service-sector growth has gravitated to peripheral centers in the South and West. Many older industrial centers in the core experienced drastic job losses in manufacturing, thus reducing its overall dominance within the region. Conversely, cities of the South's Piedmont manufacturing belt (Greensboro to Birmingham) have seen industrial expansion after 1960, partly because lower labor costs and Sun Belt amenities attracted new investment. By 2007, for example, the North Carolina "research triangle" area between Raleigh, Durham, and Chapel Hill emerged to become the nation's third largest biotech cluster behind California and Massachusetts. The Gulf coast industrial region remains strongly tied to nearby fossil fuels that provide raw materials for its many energy refining and petrochemical industries (Figure 3.35).

The varied West Coast industrial region stretches discontinuously from Vancouver, British Columbia, to San Diego, California (and beyond into northern Mexico), and it reveals the increasing importance of Pacific Basin trade. Large aerospace operations in the West also suggest the role of *government spending* as a location factor. Silicon Valley is now North America's leading region of manufacturing exports. Its proximity to Stanford, Berkeley, and other universities demonstrates the importance of *access to innovation and research* for many fast-changing high-technology industries and the advantages of *agglomeration economies*, in which many companies with similar and often integrated manufacturing operations locate near one another (Figure 3.36).

Many smaller industrial centers also thrive throughout North America, suggesting that some industries can be relatively footloose and succeed in many localities. Places such as Provo, Utah, and Austin, Texas, that specialize in high-technology industries demonstrate the growing role of *lifestyle amenities* in shaping industrial location decisions, both for entrepreneurs and for the skilled workers who need to be attracted to such opportunities.

North America and the Global Economy

Together with Europe and East Asia, North America plays a pivotal role in the global economy. In prosperous times, the region benefits from global economic growth, but in periods of international instability, globalization means that the region is more vulnerable to economic downturns. These links mean more to the region than abstract trade flow and foreign investment statistics. Increasingly, North American workers and localities find their futures directly tied to export markets in Latin America, the rise and fall of Asian imports, or the pattern of global investments in U.S. stock and bond markets.

The region is home to a growing number of truly "global cities" that serve as key connecting points and decision-making centers in the world economy (see Figure 3.31). New York City is the largest, although smaller metropolitan areas such as Miami, Toronto, Chicago, Los Angeles, Seattle, and Vancouver are emerging as pivotal urban players on the global stage. The global status of these urban centers has had profound local consequences. Life changes as thousands of new jobs are created; suburbs grow with a rush of new and ethnically diverse migrants; and a new, more cosmopolitan culture replaces older regional traditions (see "Geography in the Making: Skilled Immigrants Fuel Big City Growth").

Figure 3.35 Gulf Coast Petroleum Refining
Petroleum-related manufacturing has transformed many Gulf coast settings. Much of Houston's twentieth-century growth has been fueled by the dramatic expansion of oil-related industries. The port of Houston remains a major center of North America's refining and petrochemical operations. *(Walter Frerck/Odyssey Productions)*

Figure 3.36 Silicon Valley
The high-technology industrial landscape of California's Silicon Valley contrasts sharply with the look of traditional manufacturing centers. Here similar industries form complex links, benefiting from their proximity to one another and to nearby universities such as Stanford and Berkeley. *(George Hall/Woodfin Camp & Associates)*

The United States, with Canada's firm support, played a formative role in creating much of this new global economy and in shaping many of its key institutions. In 1944 allied nations met at Bretton Woods, New Hampshire, to discuss economic affairs. Under U.S. leadership, the group set up the International Monetary Fund (IMF) and the World Bank and gave these global organizations the responsibility for defending the world's monetary system and making key postwar investments in infrastructure. The United States also spurred the creation (1948) of the General Agreement on Tariffs and Trade (GATT). Renamed the **World Trade Organization (WTO)** in 1995, its 151 member states are dedicated to reducing global barriers to trade. In addition, the United States and Canada participate in the **Group of Eight (G-8)**—a collection of powerful countries (including Japan, Germany, Great Britain, France, Italy, and Russia)—that confers regularly on key global economic and political issues.

Patterns of Trade North America is prominent in both the sale and purchase of goods and services within the international economy. Both countries import diverse products from many global sources. Dominated overwhelmingly by the United States, Canada imports large quantities of manufactured parts, vehicles, computers, and foodstuffs. For the United States, imports continue to grow, creating a persistent global trade deficit for the country. Canada, Mexico, China, Japan, and world oil exporters supply the United States with a diversity of raw materials, low-cost consumer goods, and high-quality vehicles and electronics products.

Outgoing trade flows suggest what North Americans produce most cheaply and efficiently. Canada's exports include large quantities of raw materials (grain, energy, metals, and wood products), but manufactured goods are becoming increasingly important, particularly in its pivotal trade with the United States. Since 1994, trade initiatives with the Pacific Rim have offered Canadians new opportunities for export growth. The United States also enjoys many lucrative global economic ties, and its geography of exports reveals particularly strong links to other portions of the more developed world. Sales of automobiles, aircraft, computer and telecommunications equipment, entertainment, financial and tourism services, and food products contribute to the nation's flow of exports.

Patterns of Global Investment Patterns of capital investment and corporate power place the North American realm at the center of global money flows and economic influence. Given its relative stability, the region attracts huge inflows of foreign capital, both as investments in North American stocks and bonds and as foreign direct investment (FDI) by international companies. For Canada, U.S. wealth and proximity have meant that 80 percent of foreign-owned corporations in the country are based in the United States. In the United States, sustained economic growth and supportive government policies have encouraged large foreign investments, particularly since the late 1970s. Today, the United States is the largest destination of foreign investment in the world.

The impact of U.S. investments in foreign stock markets also suggests how flows of out-bound capital are transforming the way business is done throughout the world. Since 1980, aging U.S. baby boomers have poured billions of pension fund and investment dollars into Japanese, European, and "emerging" stock markets. U.S. investments in foreign countries also flow through direct investments made by multinational corporations based in the United States. In 2007, 11 of the world's 25 largest companies were based in the United States. Coca-Cola, Intel, and Procter & Gamble are household words from Bangladesh to Bolivia. The economic influence of these corporations exceeds that of many nation-states. ExxonMobil and General Electric are the two largest public companies on the planet. Valued at somewhere between \$350 billion and \$500 billion each, these two companies both enjoy huge global sales.

But the geography of twenty-first-century multinational corporations is changing, illustrating three recent shifts in broader patterns of globalization. These shifts have important consequences for North Americans. First, traditional American-based multinational corporations are adopting a new, more globally integrated model. For example, IBM now has more than 50,000 employees in India and in 2006 it held its annual investors' day in Bangalore, rather than in New York City (Figure 3.37). The company is now committed to putting people and jobs anywhere in the world "based on the right cost, the right skills and the right business environment . . . work flows to the places where it will be done best."

Second, a growing array of multinational corporations based elsewhere in the world, especially in places such as China, India, Russia, and Latin America, are buying up companies and assets once controlled by North American or European capital. In 2006, for example, Brazilians invested \$26 billion overseas (especially in North America), far outpacing the \$18 billion that foreign

Figure 3.37 **IBM in Bangalore**
The changing look of globally integrated multinational corporations such as IBM includes a greater investment in overseas plants and workers. *(Namas Bhojani/OnAsia)*

companies invested in their country. In Asia, China's Lenovo purchased IBM's huge Chinese personal computer business within that country. Indian multinationals are also buying dozens of foreign companies based in the more developed world.

Third, many of these same multinational companies are making huge investments of their own in other portions of the less developed world, from Africa to Southeast Asia, bypassing North American control altogether. Simply put, the late-twentieth-century top-down model of multinational corporate control and investment, traditionally based in North America, Europe, and Japan, is being replaced by a more globally distributed model of corporate control. This new model has many origins, many destinations, and new patterns of labor, capital, production, and consumption.

North Americans have certainly experienced direct consequences from these shifts in global capitalism. There has been a growing reaction in the United States, for example, to corporate **outsourcing**, a business practice that

GEOGRAPHY IN THE MAKING

Skilled Immigrants Fuel Big City Growth

While broader questions surrounding immigration continue to foster debate in both Canada and the United States, one economic trend is beyond dispute: the region's growing inflow of skilled immigrants is providing a powerful economic stimulus for many of North America's largest and most dynamic cities. Part of the story is simply the numerical impact of immigrant populations: the New York City, Los Angeles, San Francisco, and Boston metropolitan areas, for example, would all have actually lost population between 2000 and 2006 without substantial numbers of new immigrants. The economic consequences of those movements are also huge. William Frey, a demographer with the Brookings Institution, notes that "a lot of cities rely on immigration to prop up their housing market and prop up their economies."

Many immigrants become urban entrepreneurs, starting new businesses that cater both to their own communities as well as to larger metropolitan populations (Figure 3.6.1). Whether it is the Chinese in Vancouver or the Cubans in Miami, immigrants in many of North America's largest, most global cities have made huge capital and human investments in their adopted communities. Almost 30 percent of the Korean-born and 20 percent of the Iranian-born populations in the United States are self-employed, a strong indicator of business ownership. A 2007 report issued by the Center for an Urban Future argues that "immigrants have been the entrepreneurial spark plugs of cities from New York to Los Angeles." Is it any surprise that first-generation immigrants created 22 of the 100 fastest-growing companies in Los Angeles?

Figure 3.6.1 **Immigrant Entrepreneurs**
This Russian immigrant couple operates a deli in Los Angeles, California. *(Michael Newman/PhotoEdit)*

Additional statistics gathered by the U.S. Department of Homeland Security also point to the unique contributions of highly skilled immigrants. So-called H-1B visas are granted to special "temporary skilled workers" to encourage computer programmers, doctors, and other professionals to work in the United States. More than 400,000 such visas were issued in 2005, enabling these individuals to work within the United States, adding immeasurably to that country's creative human capital. The geography of the top 20 contributing countries

transfers portions of a company's production and service activities to lower-cost settings, often located overseas. In addition, millions of jobs in manufacturing, textiles, semiconductors, and electronics have effectively migrated to settings such as China, India, and Mexico as those localities offer low-cost, less regulated settings for production, both for local and for foreign firms. The results are complex: North American consumers benefit from buying cheap imports, but they may find their own jobs are threatened in the corporate restructurings that make such bargains possible.

Persisting Social Issues

Profound economic and social problems shape the human geography of North America. Even with its continental wealth, great differences persist and have increased between rich and poor. High median household incomes in the United States and Canada fail to reveal the differences in wealth within the two countries. Broader measures of social well-being suggest disparities in health care and education. In addition, both nations face problems associated with gender inequity and aging populations. One consequence of globalization is that many of these economic and social challenges are increasingly defined beyond the region. Poverty in the rural American South may be related to low Asian wage rates, for example, and a viral outbreak in Hong Kong might be only a plane flight away from suburban Vancouver.

Wealth and Poverty The North American landscape vividly displays contrasting signatures of wealth and poverty. Elite suburbs, gated and guarded neighborhoods, upscale shopping malls, and posh resorts are all geographical expressions of the spatial exclusivity that characterizes many wealthier North American communities, particularly within the United States (Figure 3.38). On the other hand, signatures of poverty are displayed in a great range of local and regional settings. Substandard housing, abandoned property, aging infrastructure, and unemployed workers

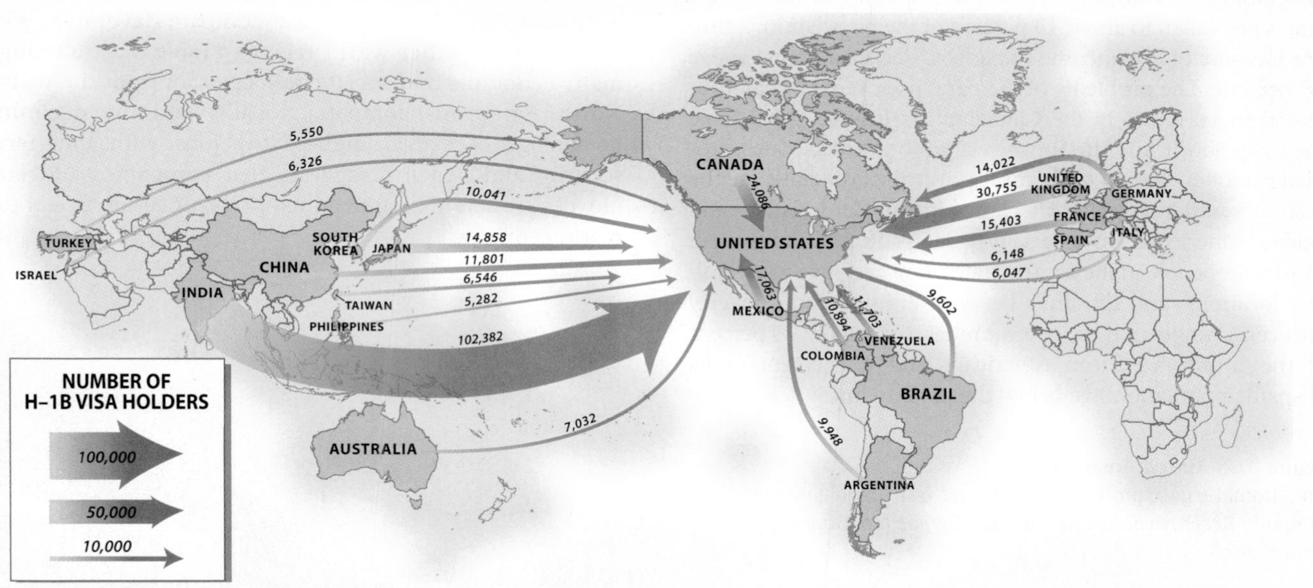

Figure 3.6.2 Origins of Temporary Skilled Workers (H-1B Visas) in the United States, 2005 (Top 20 Contributing Countries)
India's contribution to the skilled human capital of the United States is particularly notable. Both Mexico and Canada contribute many skilled workers, as do other nations of East Asia and western Europe. *(U.S. Department of Homeland Security, 2005)*

offers yet another snapshot of the economic impact of globalization and is a powerful reminder of another way in which the United States benefits in the process (Figure 3.6.2). While there are predictable ties to many European nations, the linkages with India (particularly through the computer, software, and electronics industries) are dramatic. Continental connections to nearby Mexico and Canada are also notable, along with growing links to Asian nations such as South Korea, Japan, the Philippines, and Taiwan. The patterns offer a powerful reminder that the economic evolution of both the United States and Canada remain intimately connected to skilled immigrant populations and that the region's large urban centers will continue to be a destination for these creative and talented individuals.

are visual reminders of the gap between rich and poor within the region (Figure 3.39). Specifically, within the United States, black household incomes remain only 67 percent of the national average, while Hispanic incomes fare slightly better at 80 percent of the national average.

The distribution of wealth and poverty varies widely across the United States and Canada. Many of America's wealthiest communities are suburbs on the edge of large metropolitan areas. Just outside New York City, Fairfield, Connecticut, is one of the nation's wealthiest counties; similar settings are found in suburban New Jersey, Maryland, and Virginia, as well as on the peripheries of dozens of southern and western cities. Resort and retirement communities are havens for the rich as well: Palm Beach, Florida, and Aspen, Colorado, have some of the nation's most desirable real estate and costliest housing. In terms of average income, the Northeast and West remain America's richest regions. In Canada, Ontario and British Columbia are the country's wealthiest provinces, with Vancouver's high housing prices vying with those of San Francisco.

Poverty levels have declined in both countries since 1980, but poor populations are still clustered in a variety of geographical settings. By 2007, poverty rates in the United States had fallen to about 13 percent of the population; a similar measure of low-income Canadians declined to just above 18 percent. The problems of the rural poor remain major regional social issues in the Canadian Maritimes, Appalachia, the Deep South, and the Southwest. In the United States, ethnicity is often linked with rural poverty, particularly in the case of southern blacks, Hispanics in the Borderlands, and Native Americans on reservation lands in the Southwest. Most poor people in the United States, however, live in central-city locations, and the links between ethnicity and poverty are even stronger in these communities. Nationally, 25 percent of the country's African Americans and 22 percent of its Hispanic population live below the poverty line.

Figure 3.39 Inner-City Neighborhood
Many poor inner-city neighborhoods, such as this community in New York City, suffer from substandard housing and aging infrastructure. Ethnic bonds often supply the social support necessary to ease the challenges of high unemployment and troubled family life. *(Erica Lananer/Black Star)*

Twenty-First-Century Challenges Measures of social well-being in North America compare favorably with those of most other world regions (Table 3.2), although many economic and social challenges confront the realm. North Americans compete globally, and the declining percentage of stable, high-paying jobs with long-term security and benefits suggests that companies and employees are scrambling to adjust to the uncertainties of the world economy.

Figure 3.38 Gated America
An automatic gate protects the entrance of a gated community in Apollo Beach, near Tampa, Florida. *(Dennis MacDonald/PhotoEdit, Inc.)*

Education is also a major public policy issue in Canada and the United States. Although political parties differ in their approach, most public officials agree that more investment in education can only improve North America's chances for competing successfully in the global marketplace. In the United States, dropout rates average about 13 percent for 18- to 24-year-olds, but rates are much higher in many poor urban neighborhoods and rural areas. In addition, American whites are two or three times more likely than blacks or Hispanics to hold a college degree. Another challenge, particularly in the United States, is debating an effective national education policy in the face of a very strong tradition of local control of education.

Gender remains a key social issue in both countries. Since World War II, both nations have seen great changes in the role that women play in society, but the "gender gap" is yet to be closed when it comes to salary issues, working conditions, and political power. Women widely participate in the workforce in both countries and are as educated as men (Table 3.2), but they still earn only about 75¢ for every $1 that men earn. Although the gap is shrinking, corporate America's "glass ceiling" still makes it extraordinarily difficult for women to advance to top managerial positions and pay scales. Even more daunting is the fact that women head the vast majority of poorer single-parent families in the United States and many of these women are unwed mothers.

Canadian women, particularly single mothers who work full-time, are also greatly disadvantaged, averaging only about two-thirds of the salaries of Canadian men. In fact, more than half the single women who head families in Canada are classified as low-income workers. In addition, political power remains largely in male hands, even though women make up the majority of the population and electorate. Although Canadian women have voted since 1918 and U.S. women since 1920, females in the early twenty-first century remain in the minority in the Canadian Parliament and the U.S. Congress.

Health care and aging are also key concerns within a region of graying baby boomers. A recent report on aging in the United States predicted that 20 percent of the nation's population will be older than 65 by 2050. Today, the most elderly senior citizenry (age 85+) constitute more than 12 percent of all seniors and they are the fastest-growing part of the population, particularly among women. Poverty rates are also higher for seniors. With fewer young people to support their parents and grandparents, officials debate the merits of reforming social security programs. Whatever the outcome of such debates, the geographical consequences of aging are already abundantly clear. Whole sections of the United States—from Florida to southern Arizona—have become increasingly oriented around retirement (Figure 3.40). Communities cater to seniors with special assisted-living arrangements, health-care facilities, and recreational opportunities. Florida and California have the largest number of senior citizens; Farm Belt states struggling to hold onto their younger residents also have higher percentages of the aged in their populations.

Although Canada and the United States have distinctive health-care delivery systems, they share many of the health-related benefits and costs of living within the realm. As long average lifespans and low childhood mortality rates suggest (see Table 3.2), the two countries reap many rewards from modern health-care systems that offer the latest in high technology. Still, North Americans voice concern over escalating health-care costs and uneven access to premium care. While U.S. residents spend more than 15 percent of gross domestic product (GDP) on health care (Canadians spend slightly less), there are 47 million Americans without any health-care insurance. Canada has long possessed a largely publicly financed health-care system, but it is facing a growing cost crunch and debates among the provinces over health-care policies.

The rising incidence of chronic diseases associated with aging (heart disease, cancer, and stroke are the three leading causes of death) will continue to pressure both health-care systems. In addition, hectic lives, often oriented around fast food and more meals eaten out (the average American consumed 603 more calories daily in 2000 than 20 years earlier), have contributed to rapidly growing rates of obesity. On a typical day, more than 30 percent of American children and adolescents report eating fast food. The long-term results are sobering: almost two-thirds of adult Americans are overweight and the trend is contributing to higher rates of heart disease and diabetes. Another cost has been the care and treatment of the region's 1 million HIV/AIDS victims. The price of the disease will be broadly borne in the twenty-first century, but particularly among poorer black (34 percent of AIDS cases in the United States) and Hispanic (18 percent) populations in the United States, in which rates of new infection are still on the rise.

Figure 3.40 Tomorrow's Baby Boom Landscape? This Arizona retirement community, boldly stamped on the desert landscape, suggests the twenty-first-century destination for many aging baby boomers. As North Americans grow older, entire subregions may be devoted to age-dependent demands for housing, recreation, and health services. *(Cont/Frank Fournier/Woodfin Camp & Associates)*

The region also displays striking variations in the geographies of disease. For example, a combination of poor diet, poor medical care, and genetic factors may contribute to the existence of the Stroke Belt in the eastern Carolinas and Georgia, where death rates from the disease are double the national norm. Although cancer is a leading cause of death in both countries, certain regions and populations appear prone to variations of the disease. For example, high rates of bladder and lung cancer in southern Louisiana appear strongly linked to the region's petrochemical industry and to problems of toxic waste and pollution.

Summary

- North Americans have reaped the natural abundance of their realm, and in the process they have transformed the environment, created a highly affluent society, and extended their global economic, cultural, and political reach. The environmental costs associated with these actions range from global warming and urban air pollution to water shortages to the extinction of native plants and animals.
- In a remarkably short time period, a unique mix of varied cultural groups contributed to the settlement of a huge and resource-rich continent that is now the world's most urbanized region, with super cities covering both coasts, from Boston to Washington, DC in the east and San Diego to San Francisco in the west.
- North American cities are among the world's most decentralized urban settings, with edge-city suburbs often employing far more metropolitan residents that older central cities. Examples are everywhere—Washington, DC, Los Angeles, Boston, San Jose, St. Louis, and Seattle.
- North Americans produced two societies that are closely intertwined, yet face distinctive national political and cultural issues. For both, the twenty-first century brings uncertainties. In Canada, the nation's identity remains problematic as it works through the persisting realities of its multicultural character and both the costs and benefits of its proximity to its ever-present, sometimes overbearing continental neighbor. For the United States, social problems linked to ethnic diversity and enduring poverty remain central concerns.
- The United States continues to play a pivotal and controversial role in the international economy, in the global balance of geopolitical power, and in the ongoing, but sometimes problematic expansion of popular culture, most notably through fast-food franchises, Hollywood films, and the varied forms of pop music.

Key Terms

acid rain *(page 79)*
boreal forest *(page 81)*
concentric zone model *(page 91)*
connectivity *(page 114)*
counterurbanization *(page 90)*
cultural assimilation *(page 95)*
ethnicity *(page 95)*
federal state *(page 107)*
gentrification *(page 92)*
Group of Eight (G-8) *(page 117)*
location factors *(page 115)*
Megalopolis *(page 85)*
North American Free Trade Agreement (NAFTA) *(page 107)*
outsourcing *(page 118)*
postindustrial economy *(page 74)*
prairie *(page 81)*
sectoral transformation *(page 114)*
Spanglish *(page 104)*
tundra *(page 81)*
unitary state *(page 107)*
urban decentralization *(page 91)*
urban realms model *(page 91)*
World Trade Organization (WTO) *(page 117)*

Questions for Review

1. Describe North America's major landform regions, and suggest ways in which their physical setting has shaped patterns of human settlement.
2. How are patterns of commercial agriculture related to underlying climatic patterns in North America? What key economic and technological trends have shaped agriculture?
3. Describe the dominant North American migration flows during the twentieth century.
4. Describe the principal patterns of land use within the modern U.S. metropolis. Include a discussion of (a) the central city and (b) the suburbs/edge city. How have forces of globalization shaped North American cities?
5. How have (a) railroads, (b) the township-and-range survey system, and (c) freeways shaped North America's settlement geography?
6. What are the distinctive eras of immigration in U.S. history, and how do they compare with those of Canada?
7. How do the political origins of the United States and Canada differ?
8. Cite examples of globalization that illustrate the impact of North American cultural, political, and economic influence elsewhere in the world.
9. What is the sectoral transformation, and how does it help explain economic change in North America?
10. Cite five types of location factors, and illustrate each with examples from your local economy.

Thinking Geographically

1. Compare and contrast the African American experience in the Black Belt and in urban America. Assess the unique challenges faced by each group.
2. Explain how "natural hazards" can be "culturally" defined. In other words, what role do humans play in shaping the distribution of hazards?
3. Summarize and map the ethnic background and migration history of your own family. Discuss how these patterns parallel or depart from larger North American trends.
4. Describe the strengths and weaknesses of federalism, and cite examples from both Canada and the United States.
5. The environmental price for North American development has often been steep. Suggest why it may or may not be worth the price, and defend your answer.
6. Immigration remains a key issue for North America. Argue the pros and cons of sharply curtailing immigration in the future. What about opening up the region to even larger flows of immigrants?
7. Who will America's leading trade partner be in 2050? Explain the reasons for your choice.
8. Assume you are a leader of a small African nation. Describe the chief benefits and costs of North American globalization in your part of the world.

Regional Novels and Films

Novels

Willa Cather, *My Antonia* (1918, Houghton Mifflin)

Ivan Doig, *This House of Sky* (1978, Harcourt Brace Jovanovich)

Frederick Philip Grove, *Settlers of the Marsh* (1925, Ryerson Press)

David Guterson, *Snow Falling on Cedars* (1995, Vintage Books)

William Kennedy, *Ironweed* (1983, Viking Press)

Annie Proulx, *The Shipping News* (1993, Scribner's)

Richard Russo, *Empire Falls* (2001, Knopf)

Wallace Stegner, *Angle of Repose* (1971, Doubleday)

John Steinbeck, *The Grapes of Wrath* (1939, Viking Press)

Nathaniel West, *The Day of the Locust* (1939, Random House)

Films

Avalon (1990, U.S.)

Black Robe (1991, Canada)

Chinatown (1974, U.S.)

Dances with Wolves (1990, U.S.)

The Deer Hunter (1978, U.S.)

Fargo (1996, U.S.)

Mystic River (2003, U.S.)

Thunderheart (1992, U.S.)

To Kill a Mockingbird (1962, U.S.)

Traffic (2000, U.S.)

Bibliography

Agnew, John A., and Smith, Jonathan M., eds. 2002. *American Space/American Place: Geographies of the Contemporary United States*. New York: Routledge.

Clay, Grady. 1994. *Real Places: An Unconventional Guide to America's Generic Landscape*. Chicago: University of Chicago Press.

Conzen, Michael P., ed. 1990. *The Making of the American Landscape*. Boston: Unwin Hyman.

Ford, Larry R. 1994. *Cities and Buildings. Skyscrapers, Skid Rows, and Suburbs*. Baltimore: Johns Hopkins University Press.

Harris, R. Cole, ed. 1987. *Historical Atlas of Canada. Volume 1: From the Beginning to 1800*. Toronto: University of Toronto Press.

Hudson, John C. 2002. *Across This Land: A Regional Geography of the United States and Canada*. Baltimore: Johns Hopkins University Press.

McIlwraith, Thomas F., and Muller, Edward K., eds. 2001. *North America: The Historical Geography of a Changing Continent*. Lanham, MD: Rowman and Littlefield.

Martin, Philip, and Midgeley, Elizabeth. 2006. *Immigration: Shaping and Reshaping America*. Population Reference Bureau, Volume 61, No. 4 (December).

Nostrand, Richard L., and Estaville, Lawrence E., eds. 2001. *Homelands: A Geography of Culture and Place Across America*. Baltimore: Johns Hopkins University Press.

Zelinsky, Wilbur. 1992. *The Cultural Geography of the United States: A Revised Edition*. Upper Saddle River, NJ: Prentice Hall.

Additional bibliographic resources are at the *Diversity Amid Globalization* Website: http://www. prenhall.com/rowntree/.

4
Latin America

Sprawling São Paulo is Brazil's largest city with nearly 20 million people. This Latin American megacity is part of an urban industrial core that spans from Rio de Janeiro to São Paulo. (*Alamy*)

ENVIRONMENTAL GEOGRAPHY
Tropical forests in Latin America, especially in the Amazon Basin, are one of the planet's greatest reserves of biological diversity. How this diversity will be managed is a critical question.

POPULATION AND SETTLEMENT
Latin America is the most urbanized region of the developing world with 75 percent of the population living in cities. Four megacities (10 million people or more) are found there.

CULTURAL COHERENCE AND DIVERSITY
Amerindian activism is on the rise in Latin America. Indigenous peoples from Central America to the Andes are finding their political voice by demanding territorial and cultural recognition.

GEOPOLITICAL FRAMEWORK
Recent elections in the region have seen liberal democrats and populists gaining power. These same politicians tend to be highly skeptical of the ability of neoliberal reforms and economic globalization to reduce poverty in the region.

ECONOMIC AND SOCIAL DEVELOPMENT
From NAFTA to Mercosur, regional economic integration is changing the way Latin Americans trade with each other and the world.

Beginning with Mexico and extending to the tip of South America, Latin America's regional coherence stems largely from its shared colonial history, rather than from the different levels of development seen today. More than 500 years ago, the Iberian countries of Spain and Portugal began their conquest of the Americas. Iberia's mark is still visible throughout the area: officially, two-thirds of the inhabitants speak Spanish, and the rest speak Portuguese. Iberian architecture and town design add homogeneity to the colonial landscape. The vast majority of the population is Catholic. These European traits blended with those of different Amerindian peoples. The Indian presence remains especially strong in Bolivia, Peru, Ecuador, Guatemala, and southern Mexico, where large and diverse indigenous populations maintain their native languages, dress, and traditions. After the initial conquest, other cultural groups were added to this mix of indigenous and Iberian peoples. The legacy of slavery imparted a strong African influence, primarily on the coasts of Colombia and Venezuela and throughout Brazil. In the nineteenth and twentieth centuries new waves of settlers came from Spain, Italy, Germany, Japan, and Lebanon. The result is one of the world's most racially mixed regions. The modern states of Latin America are multiethnic, with distinct indigenous and immigrant profiles and very different rates of social and economic development (Figure 4.1; see also "Setting the Boundaries").

Through colonialism, immigration, and trade, the forces of globalization have been embedded in the Latin American landscape. The early Spanish Empire concentrated on extracting precious metals, sending galleons laden with silver and gold across the Atlantic. The Portuguese became prominent producers of dyewoods, sugar products, gold, and, later, coffee. In the late nineteenth and early twentieth centuries, exports to North America and Europe fueled the region's economy. Most countries specialized in one or two products: bananas and coffee, meats and wool, wheat and corn, petroleum and copper. Such a primary export tradition, according to Latin American economists, led to an unhealthy economic dependence. They argued in the 1960s that Latin American economies were too specialized and faced unequal terms of trade that inhibited overall development.

Since then the countries of the region have industrialized and diversified their production, but they continue to be major producers of primary goods for North America, Europe, and East Asia. Today, neoliberal policies that encourage foreign investment, export production, and privatization have been adopted by many states. These policies exemplify the current impact of economic globalization on Latin America. The results are mixed, with some states experiencing impressive economic growth but increased disparity between rich and poor. Intraregional trade within Latin America stimulated by Mercosur (the Southern Common Market found in South America), as well as the impact of the North American Free Trade Agreement (NAFTA; see Chapter 3) and the newly formed **Central American Free Trade Association (CAFTA)**, are indicators of heightened economic integration in the hemisphere.

Roughly equal in area to North America, Latin America has a much larger and faster-growing population of 527 million people. Its most populous state, Brazil, has 190 million people, making it the fifth largest country in the world.

Setting the Boundaries

The concept of Latin America as a distinct region has been popularly accepted for nearly a century. The boundaries of this region are straightforward, beginning at the Rio Grande (called the *Rio Bravo* in Mexico) and ending at Tierra del Fuego. French geographers are credited with coining the term *Latin America* in the nineteenth century to distinguish the Spanish- and Portuguese-speaking republics of the Americas plus Haiti from the English-speaking territories. There is nothing particularly "Latin" about the area, other than the predominance of romance languages. The term stuck because it was vague enough to be inclusive of different colonial histories while also offering a clear cultural boundary from Anglo-America, the region referred to as North America in this text.

Latin America is one of the world regions that North Americans are most likely to visit. The trend, of course, is to visit the northern fringe of this region. Tourism is robust, especially along Mexico's northern border and in coastal resorts such as Cancun on the Yucatan peninsula. Unfortunately, visitors tend to visit one area in the region and generalize for all of it. Although it is historically sound to think of Latin America as a major world region, there are extreme variations in the physical environment, levels of social and economic development, and the influences of indigenous societies.

Because the region is so large, geographers often divide Latin America. The continent of South America is typically distinguished from Middle America (which includes Central America, Mexico, and the Caribbean). The term *Middle America* was created to identify an area culturally distinct from North America but physically part of it, as most of Mexico and all of Cuba rest on the North American Plate. Such a division has merit, but it also separates countries with very similar histories (such as Mexico and Peru) while joining countries that have very little in common (such as El Salvador and Jamaica). This book divides the Americas slightly differently. This chapter describes Latin America, consisting of the Spanish- and Portuguese-speaking countries of Central and South America, including Mexico. Chapter 5 examines the Caribbean, consisting of the islands of the Antilles, the Guianas, and Belize. This division emphasizes the important Indian and Iberian influences affecting mainland Latin America, and separates it from the unique colonial and demographic history of the Caribbean, discussed in Chapter 5.

Figure 4.1 Latin America
Roughly equal in size to North America, Latin America supports a larger population and far greater ecological diversity. The 17 countries included in this region share a history of Iberian colonization. Seventy-five percent of the region's 527 million people live in cities, making it the most urbanized region of the developing world. It is noted for its production of primary exports and manufactured goods, although rates of economic development vary greatly among states.

The next largest state, Mexico, has a population of 107 million. Collectively, many Latin American states fall into the middle-income category and support a significant middle class. But national debt, political scandal, currency devaluation, and triple-digit inflation have triggered grave economic hardships throughout the region, especially during the 1980s. Despite many social and economic gains, poverty remains a major concern for this region. It is estimated that one in four people in Latin America live on less than $2 per day. Yet unlike most areas of the developing world today,

Figure 4.2 **Tropical Wilderness** The biological diversity of Latin America—home to the world's largest rain forest—is increasingly seen as a genetic and economic asset. These forested mesas (called *tepuis*) in southern Venezuela are representative of the wild lands that many conservationists seek to protect. *(Rob Crandall/ www.robcrandall.com)*

Latin America is decidedly urban. Prior to World War II, most people lived in rural settings and worked as farmers. Today three-quarters of Latin Americans are city dwellers. Even more startling is the number of **megacities**. São Paulo, Mexico City, Buenos Aires, and Rio de Janeiro all have more than 10 million inhabitants. In addition, more than 40 cities have at least 1 million residents.

Despite the region's growing industrial capacity, extractive industries will continue to prevail, in part, because of the area's impressive natural resources. Latin America is home to Earth's largest rain forest, the greatest river by volume, and massive reserves of natural gas, oil, and copper. With its vast territory, its tropical location, and its relatively low population density (Latin America has half the population of India in nearly seven times the area), the region is also recognized as one of the world's great reserves of biological diversity. How this diversity will be managed in the face of global demand for natural resources is an increasingly important question for the countries of this region (Figure 4.2).

ENVIRONMENTAL GEOGRAPHY: Neotropical Diversity and Urban Degradation

Much of the region is characterized by its tropicality. Travel posters of Latin America showcase verdant forests and brightly colored parrots. Naturalists eager to understand the unique flora and fauna have long been attracted to the diversity and uniqueness of the neotropics, which are the tropical ecosystems of the Western Hemisphere. It is no accident that Charles Darwin's insights on evolution were inspired by his two-year journey in tropical America. Even today, scientists throughout the region work to understand complex ecosystems, discern new species, conserve genetic resources, and interpret the impact of human settlement, especially in neotropical forests. Not all of the region is tropical. Important population centers extend below the tropic of Capricorn, most notably Buenos Aires and Santiago. Much of northern Mexico, including the city of Monterrey, is north of the tropic of Cancer. Yet it is Latin America's tropical climate and vegetation that prevail in popular images of the region. Given the territory's large size and relatively low population density, Latin America has not experienced the same levels of environmental degradation witnessed in Europe and East Asia. The region's biggest environmental concerns are related to deforestation, the loss of biodiversity, and the livability of urban areas (Figure 4.3).

Huge areas of Latin America still remain relatively untouched, supporting an incredible diversity of plant and animal life. Throughout the region, national parks offer some protection to unique communities of plants and animals. A growing environmental movement in countries such as Costa Rica and Brazil has yielded both popular and political support for "green" initiatives. In short, Latin Americans enter the twenty-first century with a real opportunity to avoid many of the environmental mistakes seen in other regions of the world. At the same time, global market forces are driving governments to exploit minerals, fossil fuels, forests, and soils. The region's biggest natural resource challenge is to balance the economic benefits of extraction with the ecological soundness of conservation (see "Global to Local: Can Eco-Friendly Fair Trade Coffee Promote Biodiversity and Rural Development?"). Another major challenge is to improve the environmental quality of Latin American cities.

Rural Environmental Issues: Declining Forests and Degraded Farmlands

Perhaps the environmental issue most commonly associated with Latin America is deforestation. In rural areas, other pressing environmental issues occur on farmlands, and result from modern agricultural practices and soil erosion.

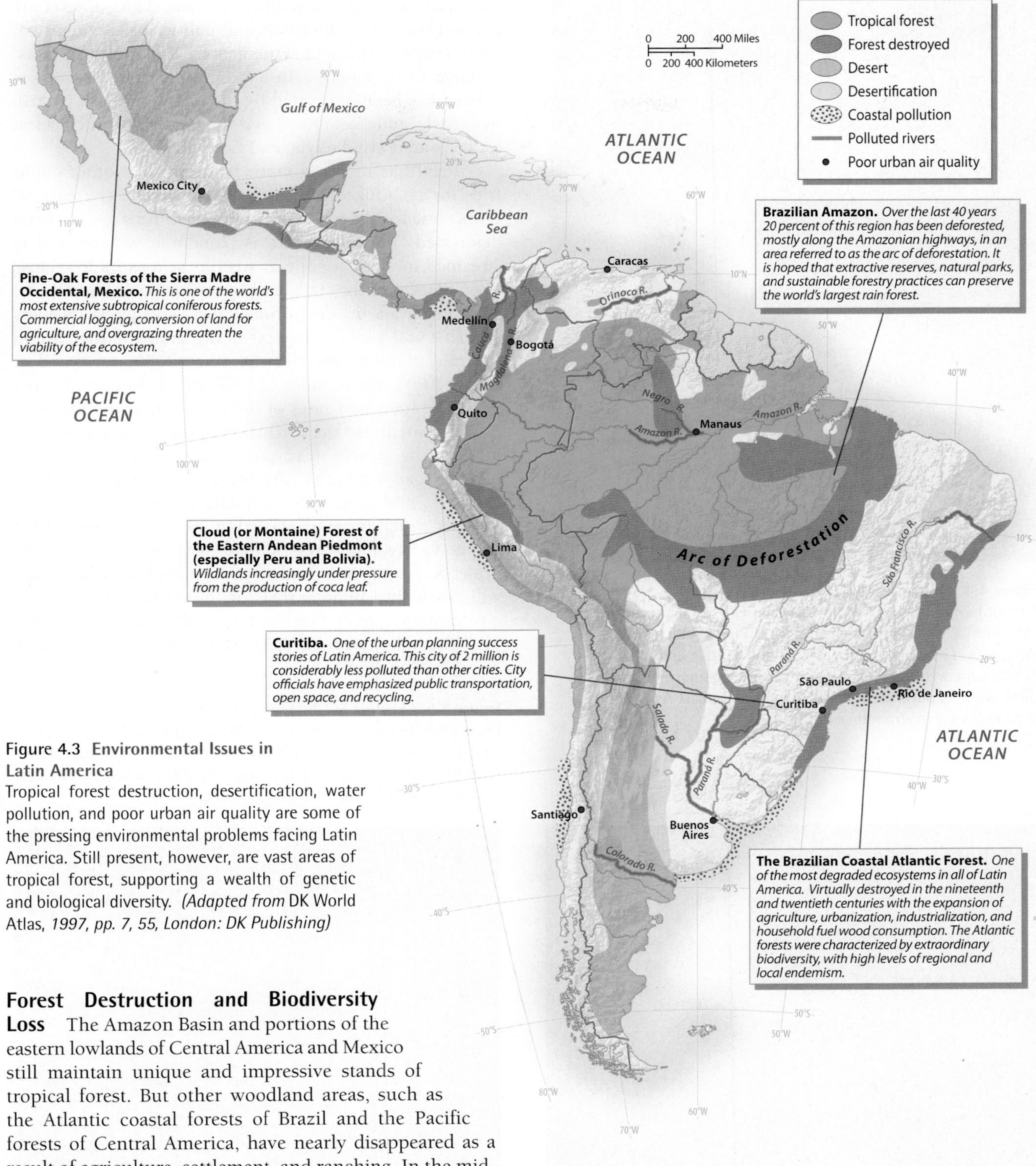

Figure 4.3 Environmental Issues in Latin America
Tropical forest destruction, desertification, water pollution, and poor urban air quality are some of the pressing environmental problems facing Latin America. Still present, however, are vast areas of tropical forest, supporting a wealth of genetic and biological diversity. *(Adapted from* DK World Atlas, *1997, pp. 7, 55, London: DK Publishing)*

Forest Destruction and Biodiversity Loss The Amazon Basin and portions of the eastern lowlands of Central America and Mexico still maintain unique and impressive stands of tropical forest. But other woodland areas, such as the Atlantic coastal forests of Brazil and the Pacific forests of Central America, have nearly disappeared as a result of agriculture, settlement, and ranching. In the midlatitudes, the ecologically unique evergreen rain forest of southern Chile (the Valdivian forest) is being cleared to export wood chips to Asian markets (Figure 4.4). The coniferous forests of northern Mexico are also falling, in part because of a bonanza for commercial logging stimulated by NAFTA.

The loss of tropical rain forests is most critical in terms of biological diversity. Tropical rainforests account for only 6 percent of Earth's land mass, but at least 50 percent of the world's species are found in this biome. Moreover, the Amazon Basin contains the largest undisturbed stretches of rain forest in the world. Unlike Southeast Asian forests, where hardwood extraction drives forest clearance, Latin American forests are usually seen as an agricultural frontier that state governments divide in an attempt to give land to

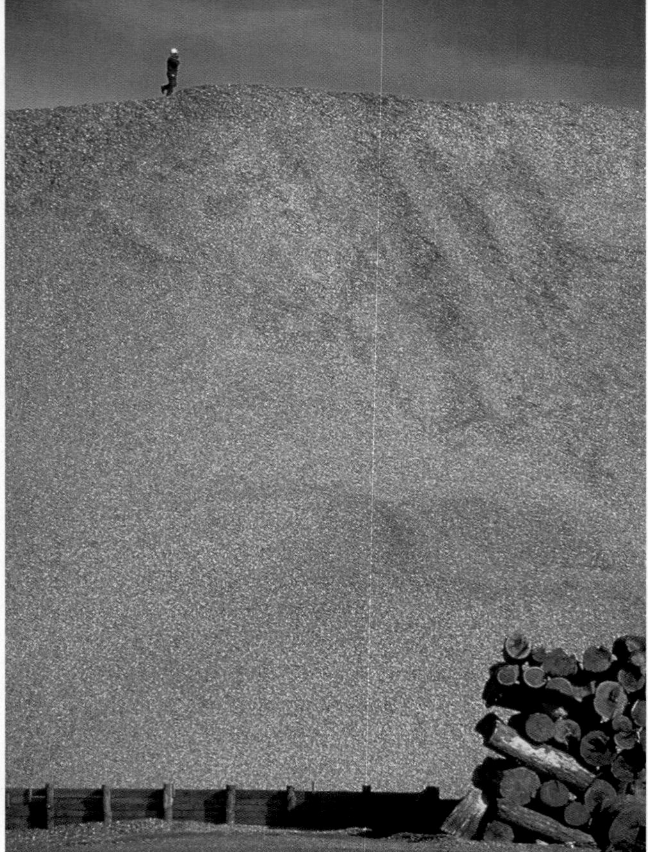

Figure 4.4 Chilean Wood Chips
A mountain of wood chips awaits shipment to Japanese paper mills from the southern Chilean port of Punta Arenas. The exploitation of wood products from both native and plantation forests supports Chile's booming export economy. Increasingly these wood exports are bound for markets in East Asia. *(Rob Crandall/www.robcrandall.com)*

the landless or reward political cronies. Thus, forests are cut and burned, with settlers and politicians carving them up to create permanent settlements, slash-and-burn plots, or large cattle ranches. In addition, some tropical forest cutting has been motivated by the search for gold (Brazil, Venezuela, and Costa Rica) and the production of coca leaf for cocaine (Peru, Bolivia, and Colombia).

Brazil has incurred more criticism than other countries for its Amazon forest policies. During the past 40 years close to 20 percent of the Brazilian Amazon has been deforested. In states such as Rondônia, where settlers streamed in along a popular road known as BR364, close to 60 percent of the state has been deforested (Figure 4.5). What most alarms environmentalists and forest dwellers (Indians and rubber tappers) is the dramatic increase in the rate of rainforest clearing since 2000, estimated at nearly 8,000 square miles (20,000 square kilometers) per year. The increased rates of deforestation in the Brazilian Amazon are due to the expansion of industrial mining and logging, the growth in corporate farms, the development of new road networks, the incidence of human-ignited wildfires, and continued population growth. Under the *Advance Brazil* program started in 2000, some $40 billion will go to new highways, railroads, gas lines, hydroelectric projects, power lines, and river canalization projects that will reach into remote areas of the basin. The Brazilian government, under President Lula da Silva, has created 59,000 square miles (150,000 square kilometers) of new

Figure 4.5 Frontier Settlement in Rondônia, Brazil
These satellite images of Rondônia, Brazil illustrate the dramatic change in forest cover. In 1975 (a) there was one major road and most of the region was tropical forest. By 2001, roads, pasture and farmland had replaced the natural land cover. *(UNEP)*

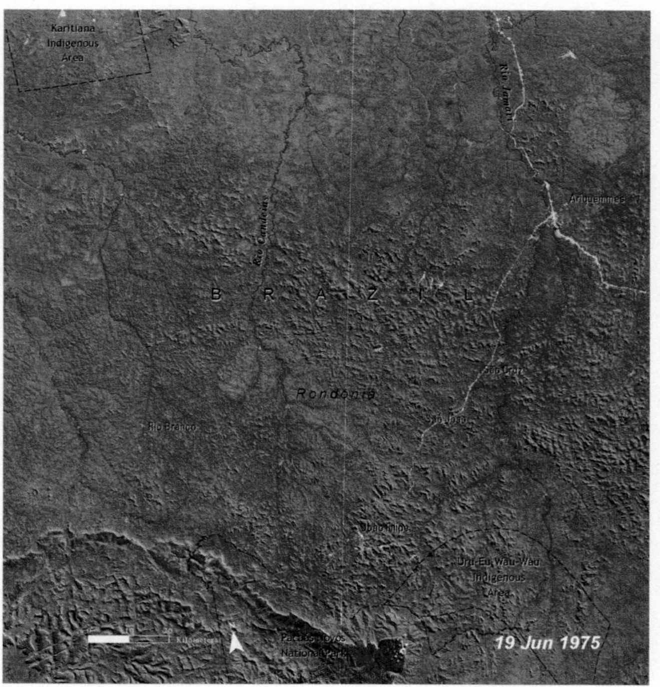

(a)

(b)

GLOBAL TO LOCAL

Can Eco-Friendly Fair Trade Coffee Promote Biodiversity and Rural Development?

Most of the coffee consumed in North America and Europe comes from Latin America (especially Central America, Mexico, Colombia, and Brazil). Environmental groups believe that if a small percentage of North American and European consumers demanded Fair Trade and organic shade-grown coffee (Figure 4.2.1) a vast area of habitat could be saved in northern Latin America and small farmers would earn more for their coffee harvest. But what is the link between organic shade-grown coffee and biodiversity?

As traditionally grown, under a canopy of shade trees and other vegetation, coffee is an environmentally friendly crop. The coffee bushes grow for years without the need for farmers to disturb the soil, and the surrounding vegetation offers valuable habitat to wildlife, especially birds. One study by the Smithsonian Migratory Bird Center found more than 150 species of birds on shade-coffee farms in Chiapas, Mexico. Many of these birds are migratory species such as the Baltimore oriole or the Tennessee warbler that summer in North America and winter in Latin America. In the last two decades, populations of these familiar migratory birds began a drastic decline, due in part to a loss of winter habitat in Latin America.

Figure 4.2.1 Shade-Grown Coffee
Shade-grown organic coffee in Latin America provides a biologically diverse habitat for many species and provides peasant farmers with an opportunity to earn more money from their harvest while being better stewards of their environment. *(Rob Crandall/www.robcrandall.com)*

From the 1970s until the early 1990s, thousands of acres of traditional shade-coffee plants were replaced by high-yielding varieties grown in full sun and requiring extensive chemical inputs to be productive. This monocrop production system resulted in very high yields but created a green desert that supported few other species. Millions of dollars were spent turning shade-grown plantations into so-called technified ones. With the loss of tropical forests and shade-coffee farms, scientists noted annual declines of migratory songbirds. Over time, farmers also found their profits dwindling because of the high price of chemical fertilizers and pesticides and the declining price of coffee. Coffee is an extremely important export crop for rural Latin America, second only to oil as the most valuable commodity in terms of world trade.

Recently, environmentalists and nongovernmental groups have promoted coffee certified as **Fair Trade**, which means the growers (often smallholders) receive a higher price for their crop (at least10 percent above market price). Coffee farmers from Central America, Mexico, and Haiti are finding that the use of organic methods lowers their costs and Fair Trade certification increases their earnings. In 2007, Fair Trade organic coffee earned farmers $1.51 per pound compared to the market price of about $1.08. An added benefit of shade-grown organic coffee is that it helps maintain an economically productive and biologically diverse habitat. Even Starbucks now markets shade-grown and organic coffee in some of its stores. Labels identifying coffee as Fair Trade, organic, or bird-friendly are making consumers aware of the connection between their daily cup of coffee and a more biologically diverse and economically sustainable environment.

conservation areas, many of them alongside the "arc of deforestation"—a swath of agricultural development along the southern edge of the Amazon Basin where the worst deforestation has occurred (see Figure 4.3). Government officials hope that the establishment of these protected areas, along with more aggressive enforcement by Brazil's environmental protection agency, will bring down current rates of deforestation.

Grassification The conversion of tropical forest into pasture, called **grassification**, is another practice that has contributed to deforestation. Particularly in southern Mexico, Central America, and the Brazilian Amazon, a hodgepodge of development policies from the 1960s through the 1980s encouraged deforestation to make room for cattle. The preference for ranching as a status-conferring occupation seems to be a carryover from Iberia. The image of the *vaquero* (cowboy) looms large in the region's history. Even poor farmers appreciate the value of having livestock. Like a savings account, cattle can be quickly sold for cash.

There are many natural grasslands suitable for grazing in Latin America, such as the Llanos in Colombia and Venezuela, and the Chaco and Pampas in Argentina. However, the rush to convert forest into pasture made ranching a scourge on the land. Ultimately, the result was a cattle bust in the 1980s when world prices collapsed. Even in cases in which domestic demand for beef increased, ranching in remote tropical frontiers was seldom economically self-sustaining.

Problems on Agricultural Lands The pressure to modernize agriculture has produced a series of environmental problems. As peasants were encouraged to adopt new hybrid varieties of corn, beans, and potatoes, an erosion of genetic diversity occurred. Efforts to preserve dozens of native domesticates are underway at agricultural research centers in the central Andes and Mexico. Nonetheless, many useful native plants may have been lost. Modern agriculture also depends on chemical fertilizers and pesticides that eventually run off into surface and groundwater supplies. Consequently, many rural areas suffer from contamination of local water supplies. Even more troublesome is the direct exposure of farm workers to toxic agricultural chemicals. Mishandling of pesticides and fertilizers can lead to exposure, resulting in rashes and burns. And in some areas, such as Sinaloa, Mexico, the widespread application of chemicals parallels a rise in serious birth defects.

Soil erosion and fertility decline occur in all agricultural areas. Certain soil types in Latin America are particularly vulnerable to erosion, most notably the volcanic soils and the reddish *oxisols* found in the humid tropical lowlands. The productivity of the Paraná basalt plateau in Brazil, for example, has declined over the decades due to the ease with which these volcanic soils erode and the failure to apply soil conservation methods. By contrast, the oxisols of the tropical lowlands can quickly degrade into a baked claypan surface when the natural cover is removed, making permanent agriculture nearly impossible. Ironically, the consolidation of the large-scale modern farms in the basins and valleys of the highlands tends to push peasant subsistence farmers into marginal areas. On these hillside farms, gullies and landslides reduce productivity (Figure 4.6). Lastly, the sprawl of Latin American cities consumes both arable land and water, eliminating some of the region's best farmland.

Figure 4.6 Degraded Bolivian Farmlands
The farmlands of Cochabamba have historically been Bolivia's breadbasket. In recent decades, increased aridity and severe soil erosion has led to a decline in agricultural productivity. Note here that soil erosion has produced deep gullies, which divide fields and remove valuable topsoil. *(Rob Crandall/www.robcrandall.com)*

Urban Environmental Challenges

Some of the most pressing environmental issues in Latin America are found in the region's urban areas. Mexico City, in particular, suffers from a host of environmental problems and is a good example of the kinds of environmental challenges facing modern Latin American cities. The severe problems facing this urban ecosystem underscore how environment, population, technology, and politics are interwoven in the effort to make Mexico City livable.

The Valley of Mexico's Environmental Problems At the southern end of the great Central Plateau of Mexico lies the Valley of Mexico, the cradle of Aztec civilization and the site of Mexico City, a metropolitan area of approximately 18 million people. This high-altitude basin with mild temperatures, fertile soils, and ample water surrounded by snow-capped volcanoes was an early center of Amerindian settlement and plant domestication, and the site of one of Spain's most important colonial cities. Given these features, it is no wonder that Mexico City has retained its primacy into the twenty-first century, even though the environmental setting that made it attractive centuries ago is now severely degraded. Some of the most pressing problems are air quality, adequate water, and subsidence (soil sinkage) caused by overdrawing the valley's aquifer (groundwater).

Air quality has been a major issue for Mexico City since the 1960s, driven in part by the city's phenomenal rate of growth. (Between 1950 and 1980, the city's annual rate of growth was 4.8 percent.) The city sits in a bowl 7,400 feet (2,250 meters) above sea level, where a layer of warm air (called a *thermal inversion*) regularly traps exhaust, industrial smoke, garbage, and fecal matter (Figure 4.7). During the worst pollution emergencies, when a thermal inversion layer traps pollutants, schoolchildren are required to stay indoors, and high-polluting vehicles are barred from the streets. Steps were finally taken in the late 1980s to reduce emissions from factories and cars. Unleaded gas is now widely available for the 3 million cars in the metropolitan area, and some of the worst polluting factories in the Valley of Mexico have closed. In fact, a study in 2001 suggested that the high levels of lead, carbon dioxide, and sulfur dioxide in the atmosphere were starting to decline. Nonetheless, the health costs of breathing such contaminated air are real, as elevated death rates due to heart disease, influenza, and pneumonia suggest.

One of Mexico City's most relentless environmental problems is water. Ironically, it was the abundance of water that made this site attractive for settlement. Large shallow lakes once filled the valley, but over the centuries most were drained to expand agricultural land. As surface water became scarce, wells were dug to tap the basin's massive freshwater aquifer. Today approximately 70 percent of the water used in the metropolitan area is drawn from the valley's aquifer. There is troubling evidence that the aquifer is being overdrawn and at risk of contamination, especially in areas where unlined drainage canals can leak pollutants into the surrounding soil, which then leach into the aquifer. To reduce reliance on groundwater, the city now pumps water nearly a mile uphill from more than 100 miles (160 kilometers) away.

A lesser-known but equally vexing problem is that Mexico City is sinking. As the metropolis grows and pumps more water from its aquifer, subsidence worsens. Mexico City sank 30 feet (9 meters) during the twentieth century. By comparison, Venice, an Italian city known for its subsidence problems, sank only 9 inches (23 centimeters) during the same period. Although the amount of subsidence varies across the metropolitan area, its impact is similar throughout the city. Building foundations are destroyed, and water and sewer lines rupture. Cherished landmarks, such as the national cathedral in the central plaza, have cracks, list to one side, and are supported by scaffolding. Because of subsidence, groundwater is no longer pumped from the city center, so that sinking has slowed to 1 inch (2.5 centimeters) per year. In the periphery, where the aquifer is actively tapped, some areas are sinking as much as 20 inches (50 centimeters) per year.

Ongoing Urban Challenges and Responses For most Latin Americans, air pollution, inadequate water, and garbage removal are the pressing environmental problems of everyday life. In this most urbanized region of the developing world, city dwellers do have better access to water, sewers, and electricity than their counterparts in Asia and Africa. Yet the inevitable environmental problems that come from dense urban settings ultimately require expensive infrastructural remedies. Since many urban dwellers tend to reside in unplanned squatter settlements, servicing these communities with utilities after they are built is difficult and costly.

Industrialization is a major cause of pollution. Factories, electricity generation, and transportation all contribute to urban pollution. A lax attitude about enforcing environmental laws tends to be the norm. In the worst cases, the consequences pose a serious threat to people and the environment. In the 1980s the Brazilian industrial center of Cubatão, near São Paulo, became synonymous with urban environmental nightmares. For years people complained of headaches and nausea from the belching factory smokestacks, but their complaints were ignored. In 1984 a leak developed in a gasoline pipeline that ran through one of the poorest squatter settlements. The smell of leaking gas went unnoticed because of the omnipresent stench of industrial pollutants in the valley. When the gas eventually ignited, the resulting explosion and fire incinerated as many as 200 people. A year later, a break in an ammonia pipeline forced the evacuation of 6,000 people and the hospitalization of 65. While industry downplayed these disasters as part of the risk of doing business, traumatized residents mobilized to address the worst abuses. The events in Cubatão, more than the destruction of rain forest, are credited with invigorating the environmental movement in Brazil.

Many environmental activists in Latin America focus their efforts on making urban environments cleaner by introducing "green" legislation and calling people to action. Not far from Cubatão in Brazil is Curitiba, the capital of Paraná state. Curitiba is the celebrated "green city" of Brazil because of some relatively simple yet innovative planning decisions. More than 2 million people inhabit this industrial and commercial center, yet it is significantly less polluted than other similar-sized cities. Because the location was vulnerable to flooding, city planners built drainage canals and set aside the remaining natural drainage areas as parks in the 1960s, well before explosive growth would have made such a policy difficult. Next, public transportation became a top priority. An innovative and clean bus system made Curitiba a model for transportation in the developing world. And last, a low-tech but effective recycling program

Figure 4.7 Air Pollution in Mexico City Pollution blankets Mexico City, which is notorious for its smog. Its high elevation and immense size make air quality management difficult. Although lead levels have declined with the introduction of unleaded gasoline, respiratory ailments are on the rise. *(Jorge Uzon/AFP/Getty Images)*

has greatly reduced solid waste. Cities such as Curitiba demonstrate that designing with nature makes sense both ecologically and economically.

Western Mountains and Eastern Shields

Latin America is a region of diverse landforms, including high mountains and extensive upland plateaus. The movement of tectonic plates explains much of the region's basic topography, including the formation of its geologically young western mountain ranges (again see Figure 4.1). In contrast, the Atlantic side of South America is characterized by humid lowlands interspersed with large upland plateaus called **shields**. Across these lowlands meander some of the great rivers of the world, including the Amazon.

Historically the most important areas of settlement in tropical Latin America were not along the region's major rivers, but across its shields, plateaus, and fertile intermontane basins. In these localities the combination of arable land, mild climate, and sufficient rainfall produced the region's most productive agricultural areas and its densest settlement. The Mexican Plateau, for example, is a massive upland area ringed by the Sierra Madre Mountains. The Valley of Mexico is located at the southern end of the plateau. Similarly, the elevated and well-watered basins of Brazil's southern mountains provide an ideal setting for agriculture. These especially fertile areas are able to support high population densities, so it is not surprising that the two largest cities, Mexico City and São Paulo, emerged in these settings.

The Andes Beginning in northwestern Venezuela and ending at Tierra del Fuego, the Andes are relatively young mountains that extend nearly 5,000 miles (8,000 kilometers). Created by the collision of oceanic and continental plates, the mountains are a series of folded and faulted sedimentary rocks with intrusions of crystalline and volcanic rock. The Andes are still forming, so active volcanism and regular earthquakes are common in this zone. The result is an ecologically and geologically diverse mountain chain with some 30 peaks higher than 20,000 feet (6,000 meters). Due to the violent and complex origins of this mountain chain, many rich veins of precious metals and minerals are found here. The initial economic wealth of many Andean countries came from mining silver, gold, tin, copper, and iron.

Figure 4.8 **Altiplano and Lake Titicaca**
The Altiplano is an elevated plateau straddling the Peruvian and Bolivian Andes. One of its striking features is beautiful Lake Titicaca at an elevation of 12,500 feet (3,810 meters). Amerindians inhabit this stark and windswept land, which is one of the poorer areas of the Andes. *(Hubert Stadler/Corbis/Bettmann)*

Figure 4.9 **Subsistence Farming in Guatemala**
A peasant farmer works his small plot of maize (corn) in the Guatemala highlands. The fertile volcanic soils, ample rainfall, and temperate climate of the Guatemala highlands have supported dense populations for centuries. *(David Alan Harvey/Woodfin Camp & Associates)*

Given the length of the Andes, the mountain chain is typically divided into northern, central, and southern components. In Colombia, the northern Andes actually split into three distinct mountain ranges before merging near the border with Ecuador. High-altitude plateaus and snow-covered peaks distinguish the central Andes of Ecuador, Peru, and Bolivia. The Andes reach their greatest width here. Of special interest is the treeless high plain of Peru and Bolivia, called the **Altiplano**. The floor of this elevated plateau ranges from 11,800 feet (3,600 meters) to 13,000 feet (4,000 meters) in altitude, and it has limited usefulness for grazing. Two high-altitude lakes, Titicaca on the Peruvian and Bolivian border and the smaller Poopó in Bolivia, are located in the Altiplano, as well as many mining sites (Figure 4.8). The southern Andes are shared by Chile and Argentina. The highest peaks of the Andes are found in the southern Andes, including the highest peak in the western hemisphere, Aconcagua, at almost 23,000 feet (6,958 meters). South of Santiago, Chile, the mountains are lower and the chain less compact. The impact of glaciation is most evident in the southernmost extension of the Andes.

The Uplands of Mexico and Central America In terms of settlement, the most important elevated lands are the Mexican Plateau and the Volcanic Axis of Central America. Most major cities of Mexico and Central America are located here. The Mexican Plateau is a large, tilted block that has its highest elevations in the south, about 8,000 feet (2,500 meters) around Mexico City, and its lowest, just 4,000 feet (1,200 meters), at Ciudad Juárez. The southern end of the plateau, the Mesa Central, contains a number of flat-bottomed basins interspersed with volcanic peaks. It also contains Mexico's megalopolis—a concentration of the largest population centers such as Mexico City, Guadalajara, and Puebla. (The Valley of Mexico, discussed earlier, is one of the basins of the Mesa Central.) Across the Mexican Plateau are rich seams of silver, copper, and zinc. The quest for silver drove much of the economic activity of colonial Mexico.

Along the Pacific coast of Central America lies a chain of volcanoes that stretches from Guatemala to Costa Rica. The Volcanic Axis of Central America is a handsome landscape of rolling green hills, elevated basins with sparkling lakes, and conical volcanic peaks. More than 40 volcanoes are found here, many of them still active. Their legacy is a rich volcanic soil that yields a wide variety of domestic and export crops. Most of Central America's population is also concentrated in this zone, in the capital cities or the surrounding rural villages. The bulk of the agricultural land is tied up in large holdings that yield export products including beef, cotton, and coffee. Yet most of the farms are small subsistence properties that produce corn, beans, squash, and assorted fruits (Figure 4.9). A major rift (a large crustal fracture) lies east of the Volcanic Axis in Nicaragua. In this low-lying valley are Lakes Managua and Nicaragua, the largest in Central America.

Figure 4.11 Patagonian Wildlife
Guanacos thrive on the thin steppe vegetation found throughout Patagonia. Native to South America, the numbers of guanacos fell dramatically due to hunting and competition with introduced livestock. *(Rob Crandall/www.robcrandall.com)*

Figure 4.10 Brazilian oranges
Most estate-grown oranges in Brazil are processed into frozen concentrate and exported. São Paulo and Paraná have some of the finest soils in Brazil. In addition to oranges, coffee and soybeans are widely cultivated. *(Stephanie Maze/NGS Image Collection)*

The Shields South America has three major shields—large upland areas of exposed crystalline rock that are similar to upland plateaus found in Africa and Australia. (The Guiana Shield will be discussed in Chapter 5.) The Brazilian Shield is the largest and more important in terms of natural resources and settlement. Far from a uniform land surface, the Brazilian Shield covers much of Brazil from the Amazon Basin in the north to the Plata Basin in the south. It is studded with isolated low ranges and flat-topped plateaus in the north. In southeastern Brazil a series of mountains (Serra da Mantiqueira and Serra do Mar) reach elevations of 9,000 feet (2,700 meters). In between these ranges are elevated basins that offer a mild climate and fertile soils. In one of these basins is the city of São Paulo, the largest urban conglomeration in South America.

The Paraná basalt plateau, located on the southern end of the Brazilian shield, is celebrated for its fertile red soils (*terra roxa*), which yield coffee, oranges, and soybeans. The basalt plateau, much like the Deccan plateau in India, is an ancient lava flow that resulted from the breakup of Gondwanaland. So fertile is this area that the economic rise of São Paulo is attributed to the expansion into this area of commercial agriculture (Figure 4.10).

The vast low-lying Patagonian shield lies in the southern tip of South America. Beginning south of Bahia Blanca and extending to Tierra del Fuego, the region to this day remains sparsely settled and hauntingly beautiful. It is treeless, covered by scrubby steppe vegetation, and home to wildlife such as the condor and guanaco (Figure 4.11). Sheep

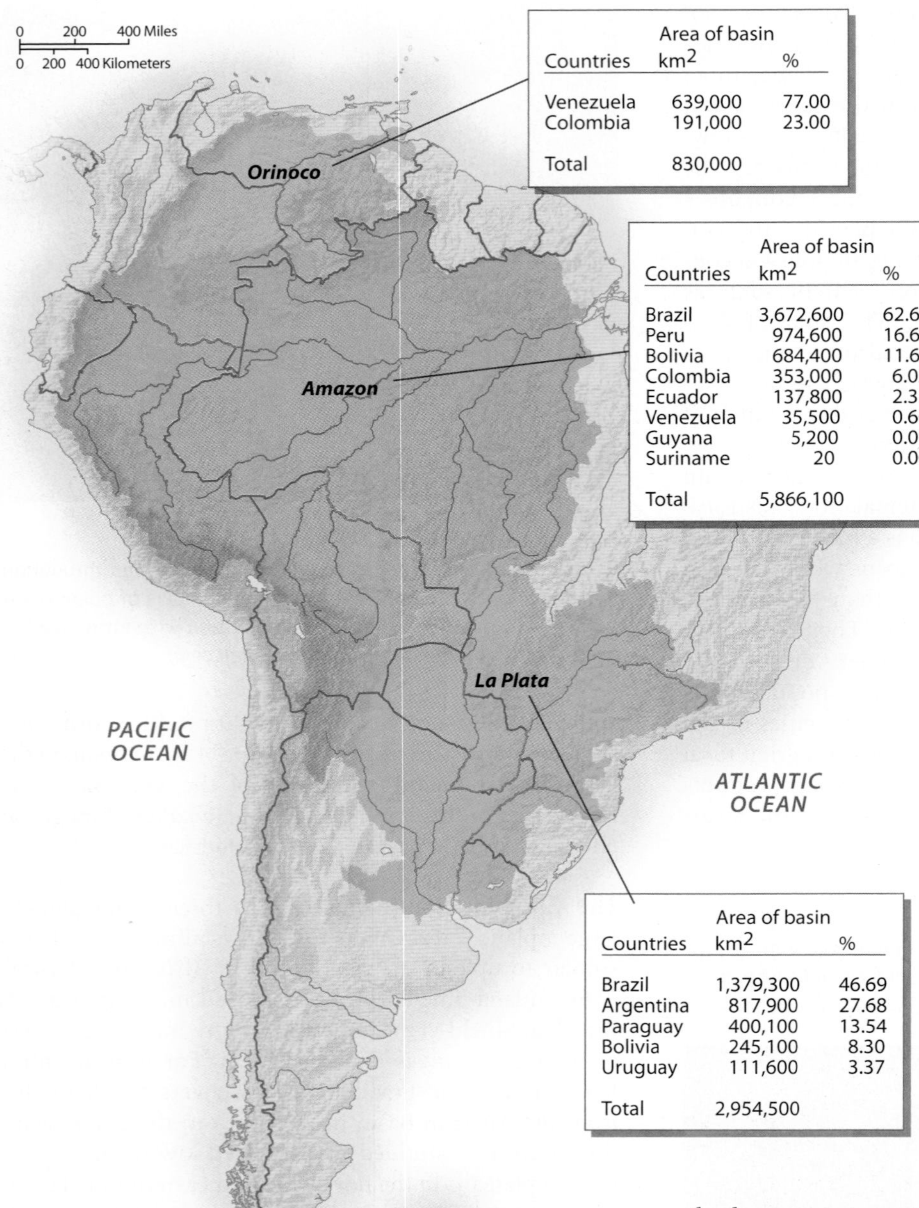

Countries	Area of basin km2	%
Venezuela	639,000	77.00
Colombia	191,000	23.00
Total	830,000	

Countries	Area of basin km2	%
Brazil	3,672,600	62.61
Peru	974,600	16.61
Bolivia	684,400	11.67
Colombia	353,000	6.02
Ecuador	137,800	2.35
Venezuela	35,500	0.66
Guyana	5,200	0.09
Suriname	20	0.00
Total	5,866,100	

Countries	Area of basin km2	%
Brazil	1,379,300	46.69
Argentina	817,900	27.68
Paraguay	400,100	13.54
Bolivia	245,100	8.30
Uruguay	111,600	3.37
Total	2,954,500	

Figure 4.12 South American River Basins
The three great river basins of the region are the Amazon, Plata, and Orinoco. The Amazon Basin covers 2.4 million square miles (6.6 million square kilometers), including portions of eight countries, but the majority of the basin is within Brazil. The Amazon is the largest river system in the world in terms of volume of water and area. The Plata Basin drains nearly 1.2 million square miles (3 million square kilometers) across five countries and is intensely farmed and used for hydroelectricity. The Orinoco Basin is shared by Venezuela and Colombia, covering nearly 390,000 square miles (1 million square kilometers).

were introduced to Patagonia in the late nineteenth century, spurring a wool boom. More recently, offshore oil production has renewed the economic importance of this area.

River Basins and Lowlands

Three great river basins drain the Atlantic lowlands of South America: the Amazon, Plata, and Orinoco (Figure 4.12). Within these basins are vast interior lowlands, less than 600 feet (200 meters) in elevation, that lie over young sedimentary rock. From north to south they are the Llanos, the Amazon lowlands, the Pantanal, the Chaco, and the Pampas. With the exception of the Pampas, most of these lowlands are sparsely settled and offer limited agricultural potential except for grazing livestock. Yet the pressure to open new areas for settlement and to exploit natural resources has created pockets of intense economic activity in the lowlands. Areas such as the Amazon and the Chaco have witnessed marked increases in resource extraction, soy cultivation, and settlement since the 1970s.

Amazon Basin The Amazon drains an area of roughly 2.4 million square miles (6.6 million square kilometers), making it the largest river system in the world by volume and area and the second longest by length. Everywhere in the basin, annual rainfall is more than 60 inches (150 centimeters), and in many places more than 80 inches (200 centimeters). The basin's largest city, Belém, averages close to 100 inches (250 centimeters) a year. Although there is no real dry season, there are definitely drier and wetter times of year, with August and September the driest months. The immensity of this watershed and its hydrologic cycle is underscored by the fact that 20 percent of all freshwater discharged into the oceans comes from the Amazon.

Since the Amazon Basin draws from eight countries, this watershed is an ideal network to integrate the northern half of South America. In fact, the river is navigable to Iquitos, Peru, some 2,100 miles (3,600 kilometers) upstream. Ironically, compared with the other great rivers of the world, settlement in the basin continues to be sparse, in large part due to the poor quality of the forest soils. The best soils are found on the floodplain, where natural levees reach heights of 20 feet (6 meters). Thousands of years of alluvium deposited on these levees have made them extremely fertile, as well as safe from normal flooding. Consequently, most of the older settlements, such as Manaus, are found on the levees. Active colonization of the Brazilian portion of the Amazon since the 1960s has boosted the population. According to the 2000 census, 12 million people live in the Brazilian Amazon, which is about 6 percent of the country's total population. Still, the population in the Amazonian states is increasing at nearly 4 percent a year. The development of the basin, most notably through settlements, roads, hydroelectric power, farms, and mines, is forever changing what was viewed as a vast tropical wilderness just a half century ago.

Plata Basin The region's second largest watershed begins in the tropics and discharges into the Atlantic in the midlatitudes. Three major rivers make up this system: the Paraná, the Paraguay, and the Uruguay. The Paraguay River and its tributaries drain the eastern Andes of Bolivia, the Brazilian Shield, and the Chaco. The Paraná primarily drains the Brazilian uplands before the Paraguay River joins it in northern Argentina. The Paraná and the considerably smaller Uruguay empty into the Rio de la Plata estuary, which begins north of Buenos Aires.

Unlike the Amazon, much of the Plata Basin is now economically productive through large-scale mechanized agriculture, especially soybean production. Arid areas such as the Chaco and inundated lowlands such as the Pantanal support livestock. The Plata Basin contains several major dams, including the region's largest hydroelectric plant, the Itaipú on the Paraná, which generates electricity for all of Paraguay and much of southern Brazil (Figure 4.13). As agricultural output in this watershed grows, sections of the Paraná River are being canalized and dredged to enhance the river's capacity for barge and boat traffic.

Orinoco Basin The third largest river basin by area is the Orinoco in northern South America. The Orinoco River meanders through much of southern Venezuela and part of eastern Colombia, giving character to a tropical grassland called the *Llanos*. Although it is only one-sixth the size of the Amazon watershed, its discharge roughly equals that of the Mississippi River. Like the Amazon, this basin is home to very few individuals; 90 percent of Venezuela's population lives north of the basin. With the exception of the industrial developments between Ciudad Guayana and Ciudad Bolívar, cities are rare. Much of the Orinoco drains the Llanos, which are inundated by several feet of water during the rainy season. Since the colonial era, these grasslands have supported large cattle ranches. Although cattle are still important, the Llanos have also become a dynamic area of petroleum production for both Colombia and Venezuela.

Climate Patterns

In tropical Latin America average monthly temperatures in localities such as Managua, Quito, or Manaus show little variation (Figure 4.14). Precipitation patterns do vary, however, and create distinct wet and dry seasons. In Managua, for example, January is typically a dry month, and June is a wet one. The tropical lowlands of Latin America, especially east of the Andes, are usually classified as tropical humid climates that support forest or savanna, depending on the amount of rainfall. The region's desert climates are found along the Pacific coasts of Peru and Chile, Patagonia, northern Mexico, and the Bahia of Brazil. Because of the extreme aridity of the Peruvian coast, a city such as Lima, Peru, which is clearly in the tropics, averages only 1.5 inches (4 centimeters) of annual rainfall.

Midlatitude climates, with hot summers and cold winters, prevail in Argentina, Uruguay, and parts of

Figure 4.13 Itaipú Dam
The largest dam in Latin America, the Itaipú blocks the flow of the Paraná River on the border between Paraguay and Brazil. The power station at Itaipú generates all of Paraguay's electricity needs and much of southern Brazil's power. *(Sue Cunningham/Worldwide Picture Library/Alamy.com)*

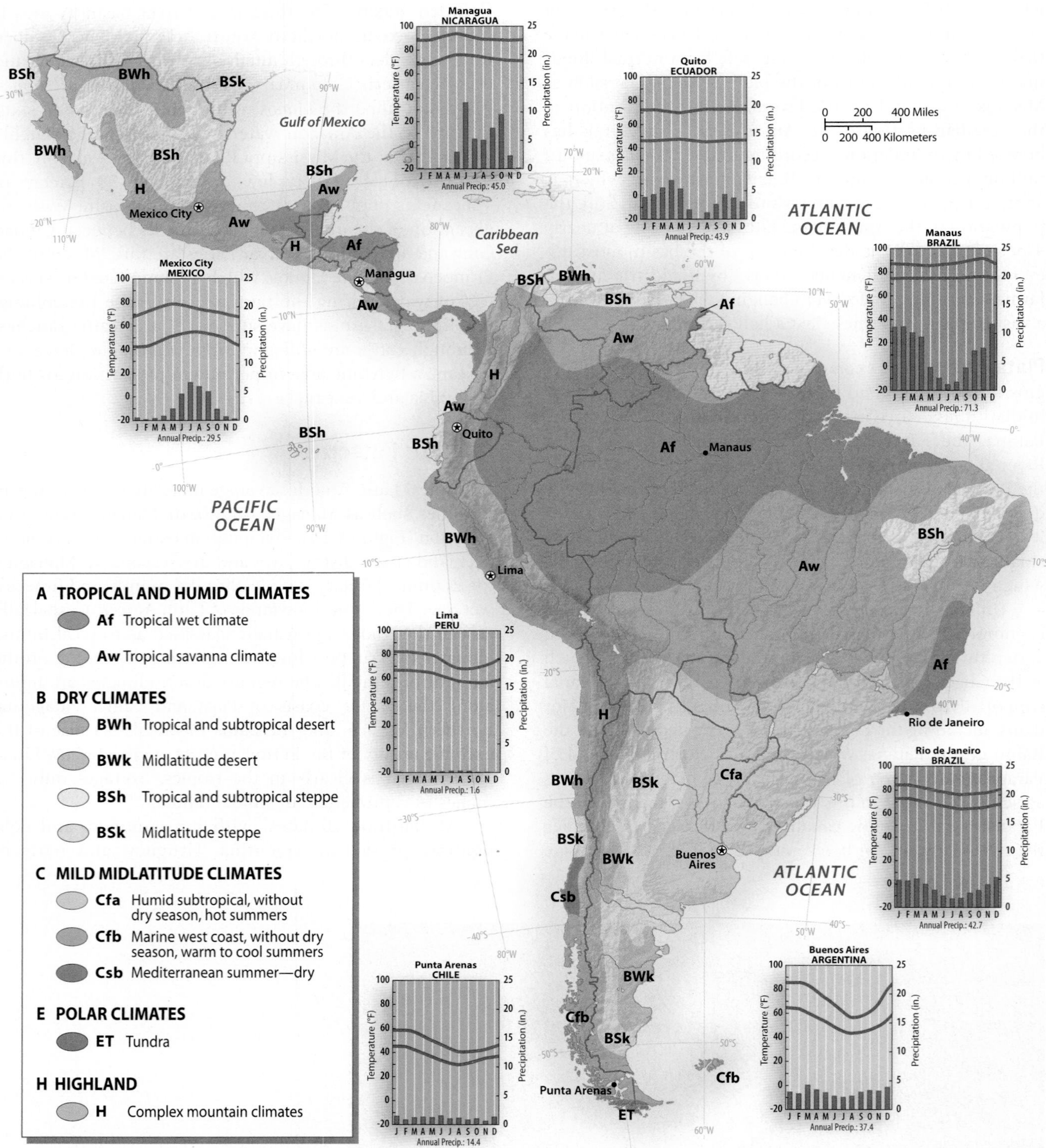

Figure 4.14 Climate Map of Latin America
Latin America includes the world's largest rain forest (Af) and driest desert (BWh), as well as nearly every other climate classification. Latitude, elevation, and rainfall play important roles in determining the region's climates. Note the contrast in rainfall patterns between humid Quito and arid Lima. *(Temperature and precipitation data from Pearce and Smith, 1984,* The World Weather Guide, *London: Hutchinson)*

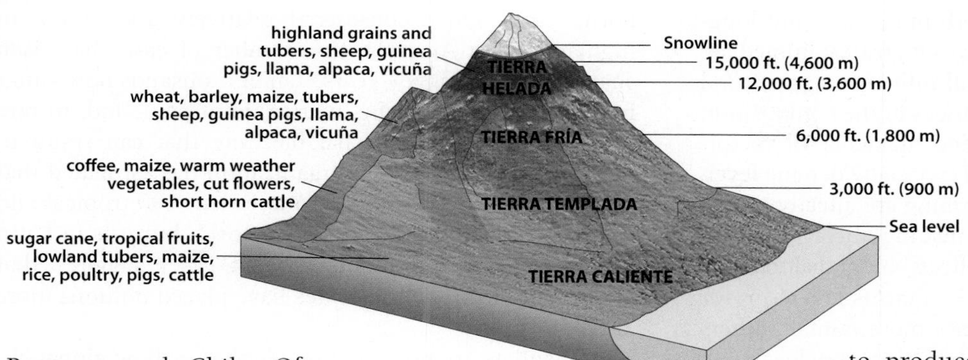

Figure 4.15 Altitudinal Zonation Tropical highland areas support a complex array of ecosystems. In the *tierra fría* zone (6,000 to 12,000 feet; 1,800 to 3,700 meters), for example, midlatitude crops such as wheat and barley can be grown. This diagram depicts the range of crops and animals found at different elevations in the Andes.

Paraguay and Chile. Of course, the midlatitude temperature shifts in the Southern Hemisphere are the inverse of those in the Northern Hemisphere, meaning cold Julys and warm Januarys. Chile's climate is a mirror image of the west coast of Mexico and the United States, with the Atacama Desert in the north (like Baja California), a Mediterranean dry summer around Santiago (similar to Los Angeles), and a marine west-coast climate with no dry season south of Concepción (like the coasts of Oregon and Washington). In the mountain ranges, complex climate patterns result from changes in elevation. To appreciate how humans adapt to tropical mountain ecosystems, one must understand the concept of **altitudinal zonation**, the relationship between cooler temperatures at higher elevations and changes in vegetation.

Altitudinal Zonation First described in the scientific literature by Prussian naturalist Alexander von Humboldt in the early 1800s, altitudinal zonation has practical applications that are intimately understood by all the region's native inhabitants. Humboldt systematically recorded declines in temperature as he ascended to higher elevations, a phenomenon known as the **environmental lapse rate**. According to Humboldt's description of the environmental lapse rate, temperature declines approximately 3.5°F for every 1,000 feet in elevation or 6.5°C for every 1,000 meters. Humboldt also noted changes in vegetation by elevation, demonstrating that plant communities common to the midlatitudes could thrive in the tropics at higher elevations. These different altitudinal zones are commonly referred to as the *tierra caliente* (hot land) from sea level to 3,000 feet (900 meters); the *tierra templada* (temperate land) at 3,000 to 6,000 feet (900 to 1,800 meters); the *tierra fría* (cold land) at 6,000 to 12,000 feet (1,800 to 3,600 meters); and the *tierra helada* (frozen land) above 12,000 feet (3,600 meters). Exploitation of these zones allows agriculturists, especially in the uplands, access to a great diversity of domesticated and wild plants (Figure 4.15).

The concept of altitudinal zonation is most relevant for the Andes, the highlands of Central America, and the Mexican Plateau. For example, traditional Andean farmers might use the high pastures of the Altiplano for grazing llamas and alpacas, the tierra fría for potato and quinoa production, and the lower temperate zone to produce corn. All the great precontact civilizations, especially the Incas and the Aztecs, systematically extracted resources from these zones, thus ensuring a diverse and abundant resource base. Yet these complex ecosystems are extremely fragile and have become important areas of research for the effects of climate change in the tropics.

El Niño Probably the most talked-about weather phenomenon in the world, **El Niño** (a reference to the Christ child) occurs when a warm Pacific current arrives along the normally cold coastal waters of Ecuador and Peru in December, around Christmastime. This change in ocean temperature happens every few years and produces torrential rains, signaling the arrival of an El Niño year. The 1997–98 El Niño was especially strong, resulting in devastating floods in Peru and Ecuador. Heavy May rains in Paraguay and Argentina caused the Paraná River to rise 26 feet (8 meters) above normal, while flooding in Peru drove some 350,000 people from their homes.

El Niño's impacts vary, however, across Latin America. While the Pacific coast of South and North America experienced record rainfall in the 1997–98 El Niño, Colombia, Venezuela, northern Brazil, Central America, and Mexico battled drought. Crop and livestock losses were estimated to be in the billions of dollars, and hundreds of brush and forest fires left their mark. In addition, extreme weather events unrelated to El Niño also affect Central America and Mexico, such as hurricanes and their associated heavy rain, flooding, and landslides.

Latin America and Climate Change

By World Bank estimates, Latin America represents about 8 percent of the world's population and produces about 6 percent of global greenhouse gas emissions. The rate of growth of greenhouse gas emissions in Latin America is dramatically lower than in all other regions of the developing world, except for Sub-Saharan Africa. The region's relatively low emissions can be explained by lower average energy consumption, higher reliance on renewable energy (especially hydropower and biofuels), and greater dependence upon public transportation. The burning of forest and brush, a common practice in the region, does produce spikes in carbon dioxide (CO_2) emissions, but the regrowth of vegetation also absorbs vast amounts of CO_2.

Global climate change has both immediate and long-term implications for Latin America. Of greatest immediate concern is how climate change will influence agricultural productivity, water availability, changes in the composition and productivity of ecosystems, and incidence of vector-borne diseases such as malaria and especially dengue fever. Changes attributable to global warming are already apparent in higher-elevation regions, making these concerns more pressing. The long-term effects of global climate change on lowland tropical forest systems is less clear; for example, some areas may experience more rainfall, others less. Other long-term impacts, such as rising sea level, will not cause the same levels of displacement in Latin America as predicted for the Caribbean or Oceania.

Climate change research indicates that highland areas are particularly vulnerable to global warming. Tropical mountain systems are projected to experience increased temperatures of 2° to 6°F (1° to 3°C), as well as lower rainfall. This will raise the altitudinal limits of various ecosystems, impacting the range of crops and arable land available to farmers and pastoralists. Research over the past 50 years has documented the dramatic retreat of Andean glaciers—some no longer exist and others will cease to exist in the next 10 to 15 years (Figure 4.16). Although this is a visible indicator of global warming, it also has pressing human repercussions. Many Andean villages, as well as metropolitan areas such as La Paz, Bolivia, get much of their water from glacial runoff. A major Bolivian glacier, Chacaltaya, has lost 80 percent of its area in the last 20 years. Thus as average temperatures increase in the highlands, and glaciers recede, there is widespread concern about future drinking water supplies.

Another immediate concern brought on by warmer temperatures is the sudden rise in dengue fever, a mosquito-borne virus. Once considered relatively uncommon in highland Latin America, the number of cases has risen sharply in the past few years. Ten of thousands now suffer from the fever, headache, nausea, joint pain, and, in rare cases, external and internal bleeding that can result in death. The World Health Organization has estimated that dengue is now widespread in more than 100 tropical and subtropical countries around the world, but it is in Latin America where the sudden rise in cases suggests that warmer highland temperatures have placed millions more at risk.

Scientists are not yet sure whether or how global climate change will impact the frequency and strength of El Niño cycles. If El Niño cycles intensify and occur more often as a result of global warming, increased flooding in western South America and a declining fishery (which is supported by nutrient upwelling in the normally cold currents) off the coast of Peru and Chile will result. Moreover, some evidence indicates that hurricane intensity will increase as ocean temperatures warm with climate change, heightening the impacts of these natural disasters. The effects of global warming on El Niño and extreme storm events are the subject of ongoing research, but are of particular concern to the Latin American region.

POPULATION AND SETTLEMENT: The Dominance of Cities

Latin America did not have great river basin civilizations like those in Asia. In fact, the great rivers of the region are surprisingly underutilized as areas of settlement or corridors for transportation. While the major population clusters of Central America and Mexico are in the interior plateaus and valleys, the interior lowlands of South America are relatively empty. Historically, the highlands supported most of the region's population during the pre-Hispanic and colonial eras. In the twentieth century, population growth and migration to the Atlantic lowlands of Argentina and Brazil, along with continued growth of Andean coastal cities such as Guayaquil, Barranquilla, and Maracaibo, have reduced the

Figure 4.16 Glacial Retreat
Research in areas such as the Cordillera Blanca of Peru over the past four decades indicate a dramatic decline in the size of Andean glaciers. An indicator of the pace of climate change in the tropics, highland glaciers are also sources of water for mountain villages and cities. Their demise has serious repercussions for the communities that depend upon them. *(Rob Crandall)*

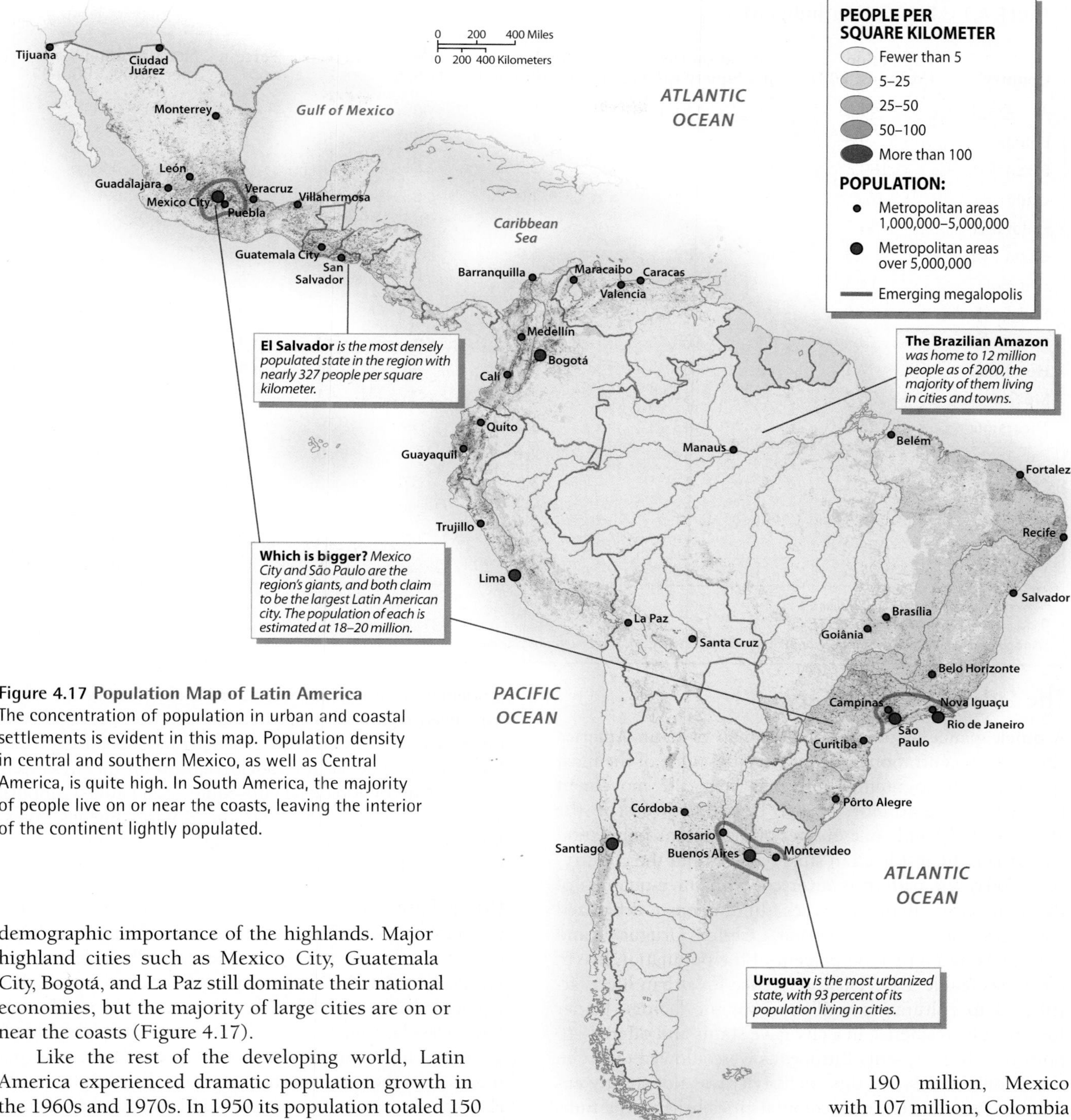

Figure 4.17 Population Map of Latin America The concentration of population in urban and coastal settlements is evident in this map. Population density in central and southern Mexico, as well as Central America, is quite high. In South America, the majority of people live on or near the coasts, leaving the interior of the continent lightly populated.

demographic importance of the highlands. Major highland cities such as Mexico City, Guatemala City, Bogotá, and La Paz still dominate their national economies, but the majority of large cities are on or near the coasts (Figure 4.17).

Like the rest of the developing world, Latin America experienced dramatic population growth in the 1960s and 1970s. In 1950 its population totaled 150 million people, which equaled the population of the United States at that time. By 1995 the population had tripled to 450 million; in comparison, the United States reached 300 million in 2006. Latin America outpaced the United States because its birthrate remained consistently higher as infant mortality rates dropped and life expectancy soared. In 1950 Brazilian life expectancy was only 43 years; by the 1980s it was 63 and by 2007 it was 72. In fact, between 1950 and 1980 most countries in the region experienced a 15- to 20-year improvement in life expectancy, which pushed up growth rates. Four countries account for over 70 percent of the region's population: Brazil with 190 million, Mexico with 107 million, Colombia with 46 million, and Argentina with 39 million (Table 4.1).

During the 1980s, population growth rates in Latin America suddenly began to slow, and by the 1990s most countries reported rates of less than 2 percent. By 2006, the regional rate of natural increase was 1.5 percent. This relatively sudden shift surprised demographers who had predicted in 1985 that the region's population would reach 750 million in 2025. Today's projection is for only 650 million by that date. One of the reasons for this fertility decline is the shift to urban living, which tends to reduce family size.

TABLE 4.1 Population Indicators

Country	Population (Millions, 2007)	Population Density (per Square Kilometer)	Total Fertility Rate	Percent Urban	Percent <15[a]	Percent >65[b]	Net Migration (per 1,000, 2000–05)
Argentina	39.4	14	2.5	89	27	10	−0.5
Bolivia	9.8	9	3.7	63	38	4	−2.3
Brazil	189.3	22	2.3	81	28	6	−0.1
Chile	16.6	22	2.0	88	25	8	0.4
Colombia	46.2	41	2.4	72	30	5	−0.9
Costa Rica	4.5	88	1.9	59	28	6	4.1
Ecuador	13.5	48	3.1	62	33	6	−3.9
El Salvador	6.9	327	2.9	59	35	5	−1.2
Guatemala	13.4	123	4.4	47	43	4	−5.0
Honduras	7.1	63	3.3	48	38	4	−0.9
Mexico	106.5	54	2.4	75	32	6	−3.9
Nicaragua	5.6	43	3.2	59	39	4	−3.8
Panama	3.3	44	2.4	64	30	6	0.5
Paraguay	6.1	15	3.5	57	36	5	−0.9
Peru	27.9	22	2.5	73	32	6	−2.2
Uruguay	3.3	19	2.1	93	24	13	−0.6
Venezuela	27.5	30	2.7	88	31	5	0.3

[a]*Percentage of population younger than 15.*
[b]*Percentage of population older than 65.*
Source: Population Reference Bureau, Word Population Data Sheet, *2007; net migration rate data from* UN International Migration, *2006.*

The Latin American City

A quick glance at the population map of Latin America shows a concentration of people in cities. One of the most significant demographic shifts has been the movement from rural areas to cities, which began in earnest in the 1950s. In 1950, only one-quarter of the region's population was urban; the rest lived in small villages and the countryside. Today the pattern is reversed, with three-quarters of the population living in cities. In the most urbanized countries, such as Argentina, Chile, Uruguay, and Venezuela, more than 85 percent of the population live in cities (see Table 4.1). This preference for urban life is attributed to cultural as well as economic factors. Under Iberian rule, residence in a city gave status and offered opportunity. Initially, only Europeans were allowed to live in the colonial cities, but this exclusivity was not strictly enforced. Over the centuries colonial cities became the hubs for transportation and communication, underscoring their primary role in structuring regional economies.

Latin American cities are noted for high levels of **urban primacy**, a condition in which a country has a primate city three to four times larger than any other city in the country. Examples of primate cities are Lima, Caracas, Guatemala City, Santiago, Buenos Aires, and Mexico City. Primacy is often viewed as a liability, as too many national resources are concentrated into one urban center. In an effort to decentralize, some governments have intentionally built new cities far from existing primate cities (for example, Ciudad Guayana in Venezuela and Brasília in Brazil). Despite these efforts, the tendency toward primacy remains. Moreover, the growth of urbanized regions has inspired the label of *megalopolis* for three areas in Latin America. Emerging megalopolises include Mexico City–Puebla–Toluca–Cuernavaca on the Mesa Central, the Niterói–Rio de Janeiro–Santos–São Paulo–Campinas axis in southern Brazil, and the Rosario–Buenos Aires–Montevideo–San Nicolás corridor in Argentina and Uruguay's lower Rio Plata Basin (see Figure 4.17).

Urban Form Latin American cities have a distinct urban morphology that reflects both their colonial origins and their contemporary growth (Figure 4.18). Usually a clear central business district (CBD) exists in the old colonial core. Radiating out from the central business district is older middle- and lower-class housing found in the zones of maturity and *in situ* accretion. In this model, residential quality declines as one moves from the center to the periphery. The exception is the elite spine, a newer commercial and business strip that extends from the colonial core to newer parts of the city. Along the spine one finds superior services, roads, and transportation. The city's best residential zones, as well as shopping malls, are usually on either side of the spine. Close to the elite residential sector, a limited area of middle-class housing is typically found. Most major urban centers also have a *periférico* (a ring road or beltway highway) that circumscribes the city. Industry is located in isolated areas of the inner city and in larger industrial parks outside the ring road.

Straddling the periférico is a zone of peripheral squatter settlements where many of the urban poor live in the worst housing. Services and infrastructure are extremely limited:

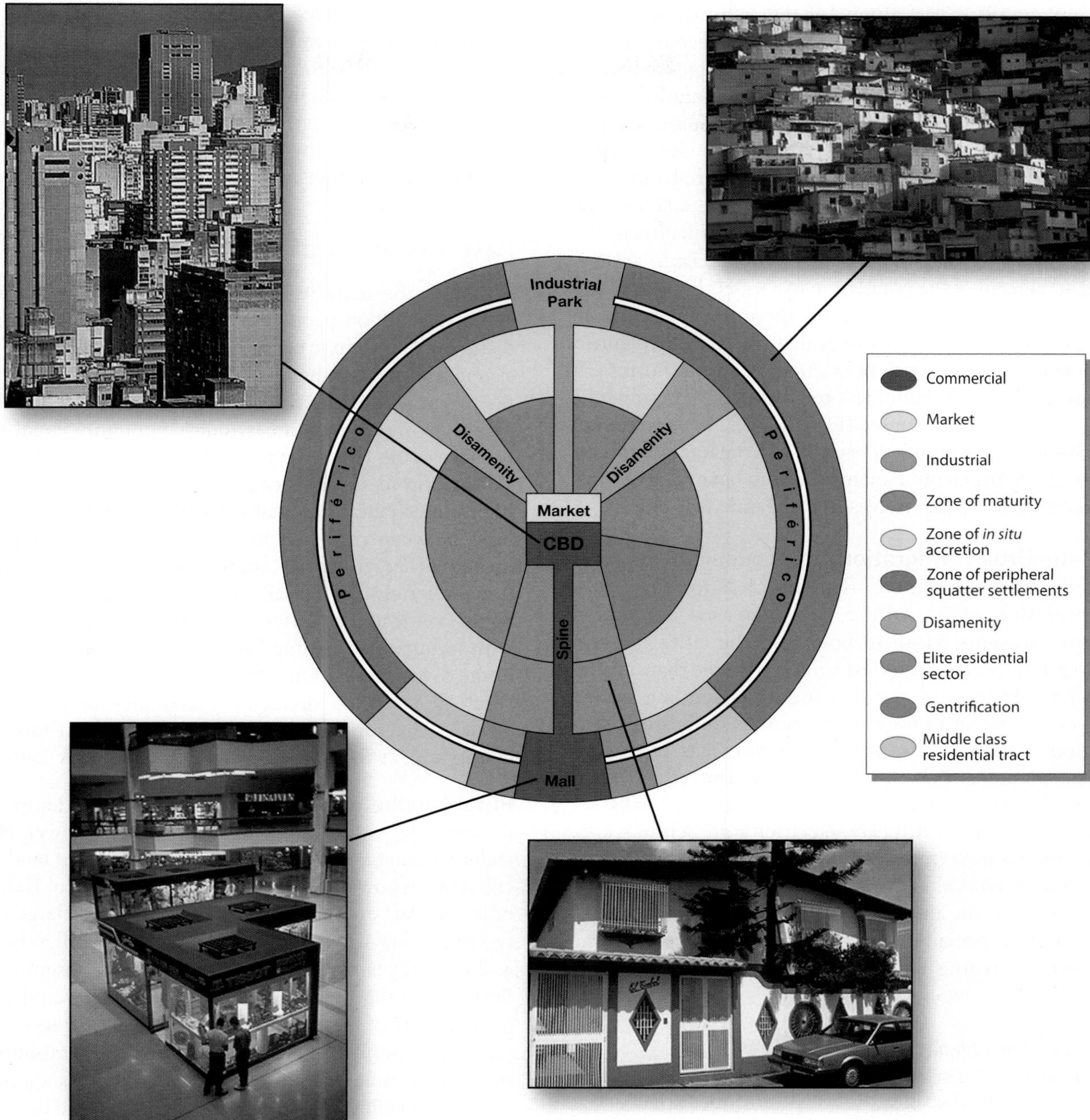

Figure 4.18 Latin American City Model
This urban model highlights the growth of Latin American cities and the class divisions within them. While the central business district (CDB), elite spine, and residential sectors may have excellent services and utilities, life in the zone of peripheral squatter settlements is much more difficult. In many Latin American cities, one-third of the population resides in squatter settlements. *(Model reprinted from Ford, 1996, "New and Improved Model of Latin American City Structure,"* Geographical Review 86*[3], 437–40. Photos by Rob Crandall/www.robcrandall.com)*

roads are unpaved; water is often trucked in; and sewer systems are nonexistent. The dense ring of squatter settlements (variously called *ranchos, favelas, barrios jovenes,* or *pueblos nuevos*) that encircle Latin American cities reflect the speed and intensity with which these zones were created. In some cities more than one-third of the population live in these self-built homes of marginal or poor quality. These kinds of dwellings are recognizable throughout the developing world, yet the practice of building one's home on the "urban frontier" has a longer history in Latin America than in most Asian and African cities. The combination of a rapid inflow of migrants, the inability of governments to meet pressing housing needs, and the eventual official recognition of many of these neighborhoods with land titles and utilities meant that this housing strategy was rarely discouraged. Each successful colonization encouraged more.

Among the inhabitants of these neighborhoods, the **informal sector** is a fundamental force that houses, services, and employs them. Definitions of the informal sector are much debated. The term usually refers to the economic sector that relies on self-employed, low-wage jobs (such as street vending, shoe shining, and artisan manufacturing) that are virtually unregulated and untaxed. Some scholars include as part of the informal sector illegal activities such as drug smuggling, prostitution, and sale of contraband items such as illegally copied movie or music CDs and tapes. One of the most interesting expressions of informality is the housing in the squatter settlements. In Lima, Peru, an estimated 40 percent of the population live in self-built housing, often of very poor quality. Typically these settlements begin as illegal invasions of open spaces that are carefully planned and timed to avoid the risk of eviction by city authorities. If the hastily built communities go unchallenged, squatters steadily improve their houses (Figure 4.19). The creation of these landscapes reflects a conscious and organized effort on the part of the urban poor, many of whom have rural origins, to make a place for themselves in Latin America's cities.

Rural-to-Urban Migration As conditions in rural areas deteriorated due to the consolidation of lands, mechanization of agriculture, and increased population pressure, peasants began to pour into the cities of Latin America in a process referred to as **rural-to-urban migration**. The strategy of rural households sending family members to the cities for employment as domestics, construction workers, artisans, and vendors has been well established since the 1950s. Once in the cities, rural migrants generally found conditions better, especially access to education, health care, electricity, and clean water.

It was not poverty alone that drove people out of rural areas, but individual choice and an urban preference. Migrants believed in, and often realized, greater opportunities in cities, especially the capital cities. Those who came were usually young (in their 20s) and better educated than those who stayed behind. Women slightly outnumbered men in this migrant stream. The move itself was made easier by extended kin networks formed by earlier migrants who settled in discrete areas of the city and aided new arrivals. The migrants maintained their links to their rural communities by periodically sending remittances and making return visits.

Figure 4.19 Lima Squatter Settlement
In arid Lima, squatters initially build their homes using straw mats. As settlements become established, residents invest in adobe and cinder block to improve their homes. Life on the urban frontier is harsh. Water is trucked in, electricity is irregular, and travel toward the city center is costly and slow. *(Rob Crandall/www.robcrandall.com)*

Patterns of Rural Settlement

Although the majority of Latin Americans live in cities, some 125 million people do not. Throughout the region a distinct rural lifestyle exists, especially among peasant subsistence farmers. In Brazil alone, more than 35 million people live in rural areas. Interestingly, the absolute number of people living in rural areas today is roughly equal to the number in the 1960s. Yet rural life has changed dramatically. In addition to subsistence agriculture, in most rural areas highly mechanized capital-intensive farming occurs. The links between rural and urban areas are much improved, meaning that rural folks are less isolated. Also, as international migration increases, many rural communities are directly connected to cities in North America and Europe, with immigrants sending back remittances and supporting home town associations. Much like the region's cities, the rural landscape is divided by extremes of poverty and wealth. The root of social and economic tension in the countryside is the uneven distribution of arable land.

Rural Landholdings The control of land in Latin America was the basis for political and economic power. Historically, colonial authorities granted large tracts of land to the colonists, who were also promised the service of Indian laborers as part of the *encomienda* system. These large estates typically took up the best lands along the valley bottoms and coastal plains. The owners were often absentee landlords, spending most of their time in the city and relying on a mixture of hired, tributary, and slave labor to run their rural operations. Passed down from one generation to the next, many estates can trace their ownership back a century or two. The allocation of large blocks of land to one owner also denied peasants access to land, so they were forced to labor on the estates. This entrenched practice of maintaining large estates is referred to as **latifundia**.

Although the pattern of estate ownership is well documented, peasants have always farmed small plots for their subsistence. This practice of **minifundia** can lead to permanent or **shifting cultivation**. Small farmers typically plant a mixture of crops for subsistence as well as for trade. Peasant farmers in Colombia or Costa Rica, for example, grow corn, fruits, and various vegetables alongside coffee bushes that produce beans for export. Strains on the minifundia system occur when demographic pressures create land scarcity or political elites reallocate land for their needs.

Much of the turmoil in twentieth-century Latin America surrounded the question of land, with peasants demanding its redistribution through the process of **agrarian reform** as a means to reduce poverty. Governments have

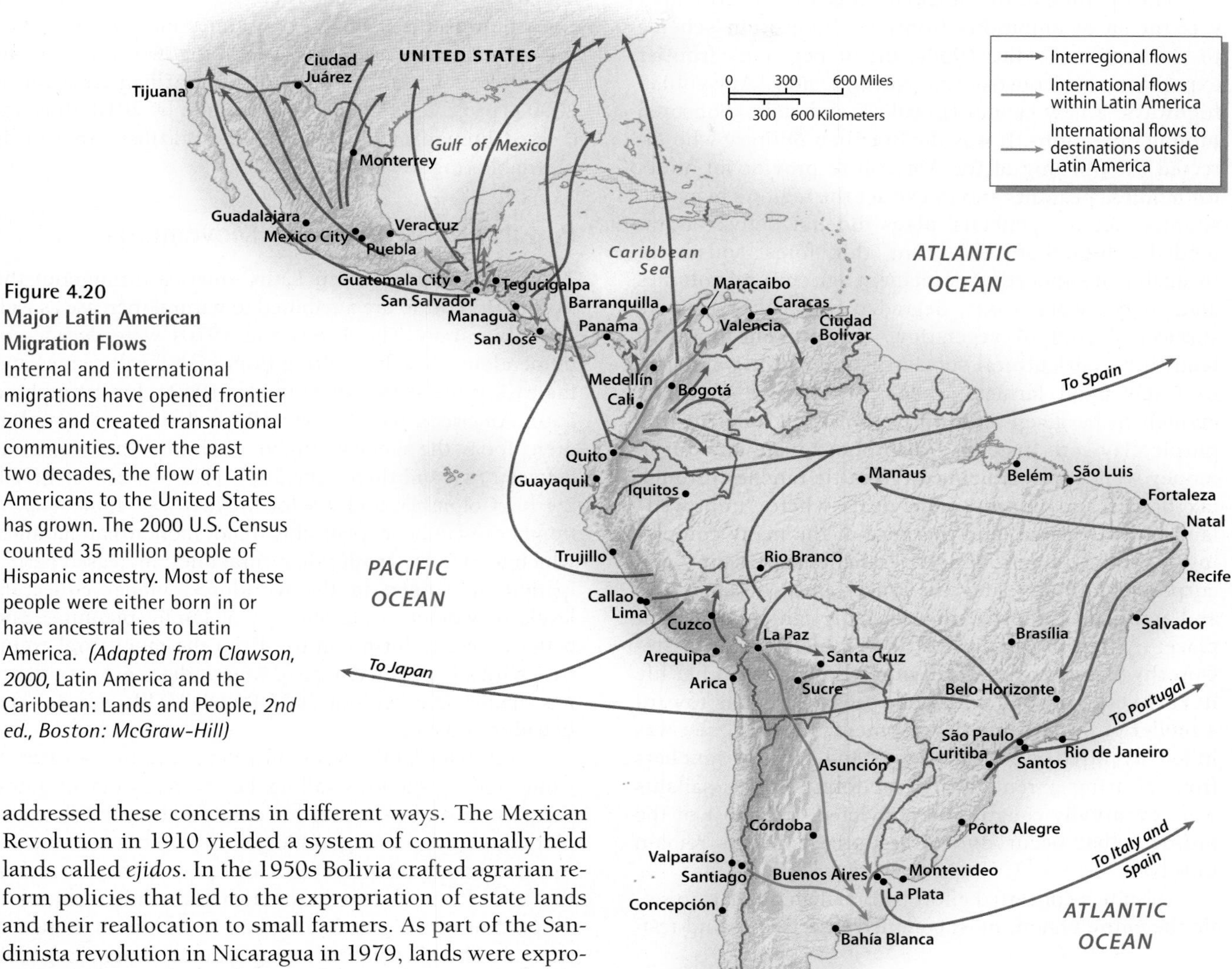

Figure 4.20
Major Latin American Migration Flows
Internal and international migrations have opened frontier zones and created transnational communities. Over the past two decades, the flow of Latin Americans to the United States has grown. The 2000 U.S. Census counted 35 million people of Hispanic ancestry. Most of these people were either born in or have ancestral ties to Latin America. *(Adapted from Clawson, 2000,* Latin America and the Caribbean: Lands and People, *2nd ed., Boston: McGraw-Hill)*

addressed these concerns in different ways. The Mexican Revolution in 1910 yielded a system of communally held lands called *ejidos*. In the 1950s Bolivia crafted agrarian reform policies that led to the expropriation of estate lands and their reallocation to small farmers. As part of the Sandinista revolution in Nicaragua in 1979, lands were expropriated from the political elite and converted into collective farms. In 2000, president Hugh Chavez ushered in a new era of agrarian reform in Venezuela and the Bolivian president, Evo Morales, introduced an agrarian reform program in 2006 aimed at giving land title to indigenous communities in the eastern lowlands. These programs have met with resistance and, at times, have proven to be costly politically. Eventually the path chosen by most governments was to make frontier lands available to land-hungry peasants. The opening of tropical frontiers, especially in South America, was a widely practiced strategy that changed national settlement patterns and began waves of rural-to-rural and even urban-to-rural migration.

Agricultural Frontiers The creation of agricultural frontiers serves several purposes: providing peasants with land, tapping unused resources, and shoring up political boundaries. A number of frontier colonization efforts in South America are noteworthy. In addition to settlement along Brazil's Trans-Amazon Highway, Peru developed its Carretera Marginal (Marginal Highway) in an effort to lure colonists into the cloud and rain forests of eastern Peru. In Bolivia, Colombia, and Venezuela, agricultural frontier schemes in the lowland tropical plains attracted peasant farmers and large-scale investors. Mexico sent colonists, some displaced by dam construction, into the forests of Tehuantepec. Guatemala developed its northern Petén region. El Salvador had no frontier left, but many desperately poor Salvadorans poured into the neighboring states of Honduras and Belize in search of land. In short, although the dominant demographic trend has been a rural-to-urban movement, an important rural-to-rural flow has turned previously virgin areas into agricultural zones (Figure 4.20).

The opening of the Brazilian Amazon for settlement was the most ambitious frontier colonization scheme in the region. In the 1960s, Brazil began its frontier expansion by constructing several new Amazonian highways, a new capital (Brasília), and state-sponsored mining operations. It was the Brazilian military who directed the opening of the Amazon to provide an outlet for landless peasants and to extract the region's many resources. Yet the generals' plans did not deliver as intended. Throughout the basin, thin forest soils were incapable of supporting permanent agricultural colonies and, in the worst cases, degraded into baked claylike surfaces devoid of vegetation. Government-promised land titles, agricultural subsidies, and credit were slow to reach small farmers, even if they were fortunate enough to be given land on *terra roxa*, a nutrient-rich purple clay soil. Instead, a disproportionate amount of money went to subsidizing large cattle ranches through tax breaks and improvement deals where "improved" land meant cleared land. Because of the many competing factions (miners, loggers, ranchers, peasants, and corporate farmers) and the uncertainty of land title and political authority, the Amazon can be a violent place. A recent victim of this violence was Sister Dorothy Stang, an American nun who had spent her life in Pará state working with poor peasant farmers toward a model of sustainable development. In 2005, she was killed by hired gunmen after trying to stop ranchers from clearing a remote area of forest. Her assailants were eventually caught and convicted, but most of the murders that occur due to resource conflicts remain unsolved.

Despite a concerted effort by Brazilian officials to settle the entire region, most commercial activities and residents are concentrated in a few sites and much resource extraction occurs illegally. Four times more people lived in the Amazon in the 1990s than in the 1960s. The population of 12 million people in the Brazilian Amazon in 2000 is projected to grow to 15 million by 2010; thus, increased human modification of the Brazilian Amazon is inevitable (Figure 4.21).

Figure 4.21 Settlers in the Amazon
In the Brazilian Amazon many settlers opted to raise cattle rather than grow crops because of the region's relatively poor soils and governmental incentives to clear forest for pasture. *(Alamy Images)*

Population Growth and Movements

The high growth rates in Latin America throughout the twentieth century are attributed to natural increase as well as immigration. The 1960s and 1970s were decades of tremendous growth resulting from high fertility rates and increasing life expectancy. In the 1960s, for example, a Latin American woman typically had six or seven children. Today the average number of children per woman is between two and three. The 2007 TFR was 2.3 for Brazil, 2.4 for Colombia, and 2.4 for Mexico (see Table 4.1). A number of factors explain this trend: more urban families, which tend to be smaller than rural ones; increased participation of women in the workforce; higher education levels of women; state support of family planning; and better access to birth control. Even the more rural countries with a high percentage of Amerindians are experiencing smaller families—in 2007 Bolivia's TFR was 3.7 and Ecuador's was 3.1.

Even with family sizes shrinking, and in the cases of Chile and Costa Rica falling below replacement rates, there is built-in potential for continued growth because of the relative demographic youth of these countries. Approximately 30 percent of the population are below the age of 15. In North America the similar cohort is 20 percent of the population, and in Europe it is just 16 percent. This means that a proportionally larger segment of the population has yet to enter into its childbearing years.

Waves of immigrants into Latin America and migrant streams within Latin America have influenced population size and patterns of settlement. Beginning in the late nineteenth century, new immigrants from Europe and Asia added to the region's size and ethnic complexity. Important population shifts within countries have also occurred in recent decades, as witnessed by the growth of Mexican border towns and the demographic expansion of the Bolivian plains. In an increasingly globalized economy, even more Latin Americans live and work outside the region, especially in the United States and Europe (see "People on the Move: Latin Americans Return to Iberia").

European Migration After gaining their independence from Iberia, Latin America's new leaders sought to develop their territories through immigration. Firmly believing in the dictum "to govern is to populate," many countries set up immigration offices in Europe to attract hardworking peasants to till the soils and "whiten" the **mestizo** population, those people of mixed European and Indian

PEOPLE ON THE MOVE Latin Americans Return to Iberia

A recent television newscast in Bolivia showed a new type of immigration service. Amerindian women, locally referred to as *cholas*, were being remade into European-looking women in order to ease their way into the Spanish labor market. Their long braids were cut into more stylish hairdos and their layered skirts were discarded for more contemporary pants and blouses. Most were planning to fly to Madrid as "tourists," but their real objective was to find jobs as domestics and earn Euros. Nearly all international migrations, eventually, lead to a counter migration. So it is surprising that the movement of Latin American immigrants to Iberia (Spain and Portugal) has taken so long to take hold. But taken hold it has, especially over the last decade.

Several factors have converged to make Spain and Portugal favored destinations for South American immigrants, especially immigrants from the Andean states. Until the late 1990s, most South Americans who emigrated chose the United States. Yet as both legal and illegal immigration to the United States became more costly and difficult, especially in the aftermath of September 11, 2001, Spain became an increasingly more attractive option. Until recently, no visa requirement existed for Latin Americans traveling to Spain. Thus potential immigrants would enter as tourists and stay on illegally as workers to serve the booming Spanish economy. The fact that they speak Spanish and understand the culture gives Latin Americans an added advantage over immigrants from Africa. Due to the onslaught of new arrivals, Spain responded by introducing new visa requirements along with a regularization plan that legalized many foreign workers. An estimated 1.8 million Latin Americans working in Spain (mostly Madrid, Barcelona, and Valencia) sent home $5 billion in remittances in 2006 (Figures 4.3.1 and 4.3.2). According to Spanish census data the main groups of Latin American immigrants are Ecuadorians (414,000), Colombians (269,000), Argentines (262,000), Bolivians (136,000), and Peruvians (126,000). The flow to Portugal is considerably smaller, but it is the European destination of choice for Brazilians.

Figure 4.3.1 **Latin American Immigrants in Spain**
An immigrant family from Ecuador poses for a portrait in the streets of a small town in central Spain. *(Larry Mangino/Image Works)*

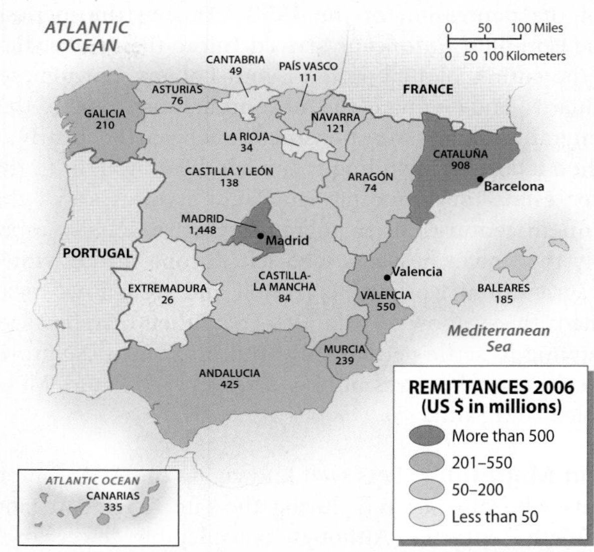

Figure 4.3.2 **Remittances from Spain to Latin America**
In 2006 immigrants from Latin America working in Spain sent $5 billion in remittances to their home countries. Most Latino immigrants reside in metropolitan Madrid and Barcelona.

The Iberian countries, with their long colonial history, were known until quite recently for producing emigrants, and were not noted as major immigrant destinations. With membership in the EU and growing prosperity, both states have seen their foreign-born populations soar with immigrants from North Africa, Sub-Saharan Africa, Latin America, and eastern Europe, creating an extremely diverse immigrant population. The speed of the Spanish transformation as an immigrant destination for Andean immigrants is the most dramatic. In 2000, the entire foreign-born population in Spain equaled just 4 percent of the total population; yet, by 2005, the foreign-born stock accounted for 11.1 percent of the total population—which is an unprecedented surge. In comparison, the immigrant stock in Portugal is 7.3 percent of the total population and it has increased more steadily over the years.

Source: Adapted from Inter-American Development Bank 2007, *Remittances from Spain to Latin America, 2006.*

ancestry. The Southern Cone countries of Argentina, Chile, Uruguay, and southern Brazil were the most successful in attracting European immigrants from the 1870s until the depression of the 1930s. During this period, some 8 million Europeans arrived (more than came during the entire colonial period), with Italians, Portuguese, Spaniards, and Germans the most numerous. Some of this immigration was state-sponsored, such as the nearly 1 million laborers (including entire families) brought to the coffee estates surrounding São Paulo at the start of the twentieth century. Other migrants came seasonally, especially the Italian peasants who left Europe in the winter for agricultural work in Argentina and were thus nicknamed "the swallows." Still others paid their own passage, intending to settle permanently and prosper in the growing commercial centers of Buenos Aires, São Paulo, Montevideo, and Santiago.

Asian Migration Less well known are the Asian immigrants who also arrived during the late nineteenth and twentieth centuries. Although considerably fewer, over time they established an important presence in the large cities of Brazil, Peru, Argentina, and Paraguay. Beginning in the mid-nineteenth century, the Chinese and Japanese who settled in Latin America were contracted to work on the coffee estates in southern Brazil and the sugar estates and guano mines of Peru. A son of Japanese immigrants, Alberto Fujimori was president of Peru from 1990 to 2000.

The Japanese in Brazil are the most studied Asian immigrant group. Between 1908 and 1978, a quarter-million Japanese immigrated to Brazil; today the country is home to more than 1 million people of Japanese descent. Initially, most Japanese were landless laborers, yet by the 1940s they had accumulated enough capital so that three-quarters of the migrants had their own land in the rural areas of São Paulo and Paraná states. As a group, the Japanese have been closely associated with the expansion of soybean and orange production. Today Brazil leads the world in exports of orange juice concentrate, with most of the oranges grown on Japanese-Brazilian farms. Increasingly, second- and third-generation Japanese have taken professional and commercial jobs in Brazilian cities. South America's economic turmoil in the 1990s encouraged many ethnic Japanese to emigrate to Japan in search of better opportunities. Nearly one-quarter of a million ethnic Japanese left South America in the 1990s (mostly from Brazil and Peru) and now work in Japan (Figure 4.22).

The latest Asian immigrants are from South Korea. Unlike their predecessors, most of the Korean immigrants came with enough capital to invest in small businesses, and they settled in cities, rather than in the countryside. According to official South Korean statistics, 120,000 Koreans emigrated to Paraguay between 1975 and 1990. While many have stayed in Paraguay, there seems to be a pattern of secondary immigration to Brazil and Argentina. Recent Korean immigrants in São Paulo have created more than 2,500 small businesses. Unofficial estimates of the number of Koreans living in Brazil range from 40,000 to 120,000. As a group they are decidedly commercial in orientation and urban in residence; their cities of choice are Asunción and Ciudad del Este in Paraguay, São Paulo in Brazil, and Buenos Aires in Argentina.

Figure 4.22 Japanese Brazilians
Retired Japanese Brazilians play the board game "Go" in a city plaza. Most Brazilians of Japanese ancestry live in the southern states of São Paulo, Paraná, and Santa Catarina. The majority are descended from Japanese who immigrated to Brazil in the first half of the twentieth century. *(Gary Payne/Getty Images, Inc.–Liaison)*

Latino Migration and Hemispheric Change Movement within Latin America and between Latin America and North America has had a significant impact on sending and receiving communities alike. Within Latin America, international migration is shaped by shifting economic and political realities. Thus, Venezuela's oil wealth during the 1960s and 1970s attracted between 1 and 2 million Colombian immigrants who tended to work as domestics or agricultural laborers. Argentina has long been a destination for Bolivian and Paraguayan laborers. And, of course, farmers in the United States have depended on Mexican laborers for most of the twentieth century (see Figure 4.20).

Political turmoil also sparked waves of international migrants. Chilean intellectuals fled to neighboring countries

Figure 4.23 Mexican–U.S. Border Crossing
Mexican day workers cross the border into El Paso, Texas, from Ciudad Juárez. Mexicans have long used these busy border crossings to enter the United States. Other Latino immigrants, especially from Central America, now join them. *(Rob Crandall/www.robcrandall.com)*

in the 1970s when General Pinochet wrested power from the socialist government led by Salvador Allende. Nicaraguans likewise fled when the socialist Sandanistas came to power in 1979. The bloody civil wars in El Salvador and Guatemala sent waves of refugees into neighboring countries, such as Mexico and the United States. With democratization on the rise in the region, many of today's immigrants are classified as economic migrants, not political asylum seekers.

Presently, Mexico is by far the largest country of origin of legal immigrants to the United States, followed by the Philippines, China, Korea, and Vietnam (Figure 4.23). Twenty-two million people claimed Mexican ancestry in the 2000 U.S. Census, of whom approximately 8 million were immigrants. Mexican labor migration to the United States dates back to the late 1800s when relatively unskilled labor was recruited to work in agriculture, mining, and railroads. This practice was formalized in the 1940s through the 1960s with the *Bracero* program, which granted temporary employment residence to 5 million Mexican laborers (much like the "guest workers" that West Germany recruited from southern Europe and Turkey). Today roughly 60 percent of the Hispanic population (both foreign-born and native-born) in the United States claims Mexican ancestry. Mexican immigrants are most concentrated in California and Texas, but increasingly they are found throughout the country. Although Mexicans continue to have the greatest presence among Latinos in the United States, the number of immigrants from El Salvador, Guatemala, Nicaragua, Colombia, Ecuador, and Brazil has steadily grown. The 2000 Census counted 35 million Hispanics in the United States (both foreign- and native-born). Most of this population has ancestral ties with peoples from Latin America and the Caribbean (see Chapter 5).

Today, Latin America is a region of emigration rather than one of immigration. The majority of states have negative rates of annual net migration, which means they are losing more people each year than they are gaining through immigration. Mexico and Ecuador have an annual rate of −3.9 per 1,000 and Guatemala's is –5.0. By comparison the average rate of net migration for the developing world as a whole is a much smaller figure of –0.5 (see Table 4.1). Both skilled and unskilled workers from Latin America are an important source of labor in North America, Europe, and Japan. Many of these immigrants send monthly **remittances** (monies sent back home) to sustain family members. In 2006, it was estimated that immigrants sent more than $60 billion to Latin America. Most of this money came from workers in the United States, but Latino immigrants in Spain, Portugal, Japan, Canada, and Italy also sent money back to the region.

Through remittances and technological advances that make communication faster and cheaper, immigrants maintain close contact with their home countries in ways that earlier generations could not. Scholars have labeled this ability to straddle livelihoods between two countries as **transnationalism**. A cultural and economic outcome of globalization, transnationalism highlights the social and economic links that form between home and host countries. Transnational migrants maintain a dual or hybrid identity, which is also seen as a cultural expression of globalization. Salvadorans working in Washington, DC, for example, maintain regular contract with their rural villages in El Salvador through sending goods and forming hometown associations while also developing vital immigrant social networks in their new home (Figure 4.24).

Figure 4.24 Salvadorans in the Suburbs
Day laborers wait for employment in a Maryland suburb outside Washington, DC. War and economic hardship drove many Salvadorans from their country in the 1980s and 1990s. Today, Salvadorans are the largest immigrant group in the Washington, DC, metropolitan area. *(Rob Crandall/www.robcrandall.com)*

PATTERNS OF CULTURAL COHERENCE AND DIVERSITY: Repopulating a Continent

The Iberian colonial experience (1492 to the 1800s) imposed a political and cultural coherence on Latin America that makes it distinguishable today as a world region. Yet this was not a simple transplanting of Iberia across the Atlantic. Often a syncretic process unfolded in which European and Indian traditions blended as indigenous groups were incorporated into either the Spanish or the Portuguese empires. In some areas such as southern Mexico, Guatemala, Bolivia, Ecuador, and Peru, Amerindian cultures have showed remarkable resilience, as evidenced by the survival of indigenous languages. Yet the prevailing pattern is one of forced assimilation in which European religion, languages, and political organization were imposed on the surviving fragments of native society. Later, other cultures, arriving as both forced and voluntary migrants, added to the region's cultural mix. Perhaps the single most important factor in the dominance of European culture in Latin America was the demographic collapse of native populations.

Figure 4.25 Tikal, Guatemala
This ancient Mayan city located in the lowland forests of the Petén was part of a complex network of cities located in the Yucatan and northern Guatemala. At its height Tikal supported over 100,000 residents, until it was abandoned in the late tenth century. Today, no modern Peten city has reached the size of Tikal. *(Rob Crandall/ www.robcrandall.com)*

Demographic Collapse

Beginning in 1492 and lasting until the nineteenth century, the cultural change and human loss resulting from the cataclysmic encounter between European and Indian worlds was enormous. Throughout the region archaeological sites are poignant reminders of the complexity of precontact civilizations. Dozens of stone temples found throughout Mexico and Central America, where the Mayan and Aztec civilizations flourished, attest to the ability of these societies to thrive in the area's tropical forests and upland plateaus (Figure 4.25). The Mayan city of Tikal flourished in the lowland forests of the Petén, supporting tens of thousands, before its mysterious collapse centuries before the arrival of Europeans. In the Andes, stone terraces built by the Incas are still being used by Andean farmers; earthen platforms for village sites and raised fields for agriculture are still being discovered and mapped. Evidence of the complexity of precontact civilizations are ceremonial centers such as Cuzco (the core of the great Incan empire that was nearly leveled by the Spanish) and the Incan site of Machu Picchu (unknown to most of the world until American archaeologist Hiram Bingham investigated the site in the early 1900s). The Spanish, too, were impressed by the sophistication and wealth they saw around them, especially in the incomparable Tenochtitlán, site of present-day Mexico City. Tenochtitlán was the political and ceremonial center of the Aztecs, supporting a complex metropolitan area with some 300,000 residents. The largest city in Spain at the time was considerably smaller.

The most telling figures of the impact of European expansion are demographic. Experts believe that the precontact Americas had 54 million inhabitants; by comparison, western Europe in 1500 had approximately 42 million. Of the 54 million, about 47 million were in what is now Latin America, and the rest were in North America and the Caribbean. The region had two major population centers: one in Central Mexico with 14 million people and the other in the Central Andes (highland Peru and Bolivia) with nearly 12 million. By 1650, after a century and a half of colonization, the indigenous population was one-tenth its precontact size. The human tragedy of this population loss is difficult to comprehend. The relentless elimination of 90 percent of the indigenous population was largely caused by epidemics of influenza and smallpox, but warfare, forced labor, and starvation due to a collapse of food production systems also contributed to the death rate.

The population low point for Amerindians was in 1650, but the tragedy continued throughout the colonial period and to a much lesser extent it continues today. After the indigenous population began its slow recovery in the Central Andes and Central Mexico, there were still tribal bands in southern Chile (the Mapuche) and Patagonia (Araucania) that experienced the ravages of disease three centuries after Columbus landed. Even now, the isolation of some Amazonian tribes has made them vulnerable to disease. Conflicts with outsiders who invade their territories in search of land or gold still occur. In an all-too-familiar story, a common cold can prove deadly to forest dwellers who lack the needed immunities.

The Columbian Exchange Historian Alfred Crosby likens the contact period between the Old World (Europe, Africa, and Asia) and the New World (the Americas) as an immense biological swap, which he terms the **Columbian exchange**. According to Crosby, Europeans benefited greatly from this exchange, and Amerindian peoples suffered the most from it. On both sides of the Atlantic, however, the introduction of new diseases, peoples, plants, and animals forever changed the human ecology. Consider, for example, the introduction of Old World crops. The Spanish brought their staples of wheat, olives, and grapes to plant in the Americas. Wheat did surprisingly well in the highland tropics and became a widely consumed grain over time. Grapes and olive trees did not fare as well, but eventually grapes were produced commercially in the temperate zones of South America. The Spanish grew to appreciate the domestication skills of Indian agriculturalists who had developed valuable starch crops such as corn, potatoes, and bitter manioc, as well as condiments such as hot peppers, tomatoes, pineapple, cacao, and avocados. Corn never became a popular food for Europeans, but many African peoples adopted it as a vital staple food. After initial reluctance, Europeans and Russians widely consumed the potato as a basic food. Domesticated in the highlands of Peru and Bolivia, the humble potato has an impressive ability to produce a tremendous volume of food in a very small area, even when climatic conditions are not ideal. This root crop is credited with driving Europe's rapid population increase in the eighteenth century when peasant farmers from Ireland to Russia became increasingly dependent on it as a basic food. This potato dependence also made them vulnerable to potato blight, a fungal disease that emerged in the nineteenth century and came close to unraveling Irish society.

Tropical crops transferred from Asia and Africa reconfigured the economic potential of the region. Sugarcane, an Asian transfer, became the dominant cash crop of the Caribbean and the Atlantic tropical lowlands of South America. With sugar production came the importation of millions of African slaves. Coffee, a later transfer from East Africa, emerged as one of the leading export crops throughout Central America, Colombia, Venezuela, and Brazil in the nineteenth century. Introduced African pasture grasses enhanced the forage available to livestock.

The movement of Old World animals across the Atlantic had a profound impact on the Americas. Initially, these animals hastened Indian decline by introducing animal-borne diseases and by producing feral offspring that consumed everything in their paths. The utility of swine, sheep, cows, and horses was eventually appreciated by native survivors. Draft animals were adopted, as was the plow, which facilitated the preparation of soil for planting. Wool became a very important fiber for indigenous communities in the uplands. And slowly, pork, chicken, and eggs added protein and diversity to the staple diets of corn, potatoes, and cassava. Ironically, the horse, which was a feared and formidable weapon of the Europeans, became a tool of resistance in the hands of skilled riders who inhabited the plains of the Chaco and Patagonia. Much like native peoples of North America, these tribal groups challenged European conquest by using their horsemanship in combat or for flight. With the major exception of disease, many transfers of plants and animals ultimately benefited both worlds. Still, it is clear that the ecological and material basis for life in Latin America was completely reworked through this exchange process initiated by Columbus.

Indian Survival Presently, Mexico, Guatemala, Ecuador, Peru, and Bolivia have the largest indigenous populations. Not surprisingly, these areas had the densest native populations at contact. Indigenous survival also occurs in isolated settings where the workings of national and global economies are slow to penetrate. The isolated Miskito Coast of Honduras is home to Miskito, Pech, and Garífuna. In the Brazilian state of Roraima in the northern Amazon, some 15,000 indigenous Pemong speakers organized in 2004 to create the Raposa/Serra do Sol Indian reservation. In these relatively isolated areas, small groups of people have managed to maintain a distinct way of life despite the pressures to assimilate.

In many cases Indian survival comes down to one key resource—land. Indigenous peoples who are able to maintain a territorial home, formally through land title or informally through long-term occupancy, are more likely to preserve a distinct ethnic identity. Because of this close association between identity and territory, native peoples are increasingly insisting on a recognized space within their countries. These efforts to define indigenous territory are seldom welcomed by the state but they are occurring throughout the region (see "Geographic Tools: Participatory Mapping").

From Amazonia to the highlands of Chiapas, many native groups are demanding formal political and territorial recognition as a means to redress centuries of injustice. Whether these efforts will actually reshape political and cultural space for Latin America's Amerindians is still uncertain. Yet there are hopeful signs of increased political participation by Amerindian peoples. In 2001, Peruvians elected President Alejandro Toledo, an Amerindian who rose from acute poverty to obtain a doctorate in economics from Stanford University. In 2005, Bolivians elected a former union organizer, Evo Morales, as their first Amerindian president. More so than Toledo, Morales is seen as a powerful symbol of the rise of indigenous political power in the Andes (Figure 4.26).

GEOGRAPHIC TOOLS Participatory Mapping

Latin America has been a critical region for the development of participatory mapping techniques. Indigenous groups, environmentalists, and human rights activists have come to appreciate the power of maps to serve society and the environment. Participatory mapping (PM) is a catchall label that refers to an array of community-based research approaches used to map places. These approaches rely upon the collaboration of local people with mapping technicians. Geographers Peter Herlihy and Greg Knapp explain that the "methodology rests on the philosophy that local populations have some of the best and most detailed knowledge of their surrounding lands and resources, and that knowledge can be collected and interpreted geographically." In this process, community representatives master collecting data with handheld global positioning systems, and learn to interpret air photos and satellite images. Working with a participatory researcher or a technical team, locals gather and interpret data through community meetings, administering questionnaires, drawing sketch maps, and recording place names and other spatial information directly on cartographic sheets.

An example of a participatory mapping product is Figure 4.4.1, showing the area around the Rio Plátano Biosphere Reserve in northeastern Honduras. This is the largest area of road-free rain forest in Central America. Using PM, indigenous peoples such as the Miskito, Pech, and Garífuna outlined their communal land-use zones (resource areas needed for subsistence). By mapping communal land-use zones, they could express their claims to this area in a powerful but less confrontational way. Potential problem areas between groups were made apparent through the PM exercise. For example, several ethnic groups rely on the Tinto-Ibans zone at the park's northern border. The existence of the map, moreover, records the Amerindian territorial presence in and around the park.

Participatory mapping is a simple tool that has real political importance for indigenous groups in the Americas and elsewhere. It has been used to document human land use and occupancy, design conservation plans, survey biodiversity, protect indigenous reserves, demarcate land claims, educate communities, and build consensus over natural resource use. It is one of many tools that geographers are using with indigenous groups to help them navigate their rapidly globalizing world.

Source: Adapted from Peter Herlihy and Gregory Knapp, 2003, "Maps of, by, and for the Peoples of Latin America," *Human Organization* 62(4), 303–14.

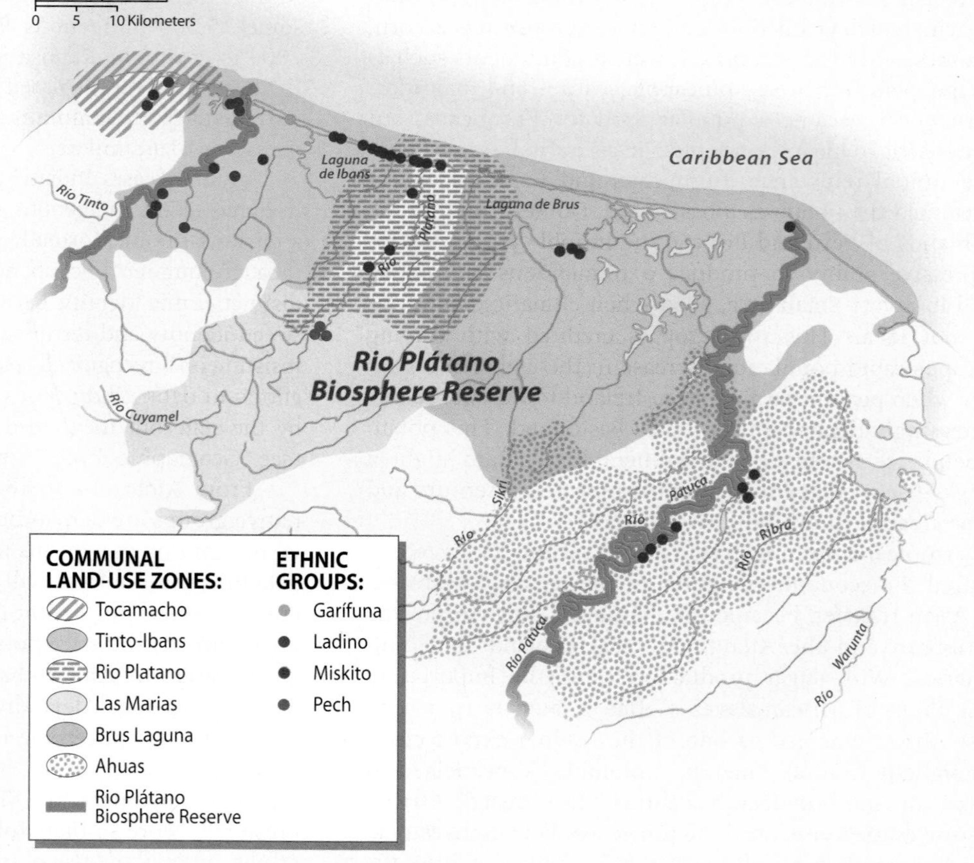

Figure 4.4.1 Communal Land-Use Zones and the Rio Plátano Biosphere Reserve Various ethnic communities, including the Garífuna, Miskito, Pech, and Ladinos, live within the Rio Plátano Biosphere Reserve. The first three groups are using participatory mapping to maintain access to parklands for fishing, hunting, and farming. *(Modified from Mopawi, Herlihy, and Leake, 1992,* Tierra Indigenas de la Mosquitia Hondureña-1992: Zonas de Subsistencia*)*

Figure 4.26 President Morales in Bolivia
Elected in 2005, Evo Morales is the first Amerindian president in Bolivia, a county where Amerindians are the majority. Formerly a union leader of coca growers, he is a powerful symbol of both the rise of the political left and indigenous activism. In this photo Morales is chatting with an Aymara woman in the city of El Alto. *(Getty Images, Inc./Agence France Presse)*

Patterns of Ethnicity and Culture

The Indian demographic collapse enabled Spain and Portugal to refashion Latin America into a European likeness. Yet instead of a neo-Europe rising in the tropics, a complex ethnic blend evolved. Beginning with the first years of contact, unions between European sailors and Indian women began the process of racial mixing that over time became a defining feature of the region. The courts of Spain and Portugal officially discouraged racial mixing but were unable to prevent it. Spain, which had a far larger native population to oversee than did the Portuguese in Brazil, became obsessed with the matter of race and maintaining racial purity among its colonists. An elaborate classification system was constructed to distinguish emerging racial castes. Thus, in Mexico in the eighteenth century a Spaniard and an Indian union resulted in a *mestizo* child. A child of a mestizo and a Spanish woman was a *castizo*. However, the children from a castizo woman and a Spanish man were considered Spanish in Mexico but a quarter mestizo in Peru. Likewise, *mulattoes* were the progeny of European and African unions, and *zambos* were the offspring of Africans and Indians.

After generations of intermarriage, such a classification system collapsed under the weight of its complexity, and four broad categories resulted: *blanco* (European ancestry), *mestizo* (mixed ancestry), *indio* (Indian ancestry), and *negro* (African ancestry). The blancos (or Europeans) continue to be well represented among the elites, yet the vast majority of people are of mixed racial ancestry. In Venezuela, for example, the phrase "cafe con leche" (coffee with milk) is used to describe the racial makeup of the majority of the population who share European, African, and Indian characteristics. Dia de la Raza, the region's observance of Columbus Day, recognizes the emergence of a new mestizo race as the legacy of European conquest. Throughout Latin America, more than other regions of the world, miscegenation—or racial mixing—is the norm, which makes the process of mapping racial or ethnic groups especially difficult.

Enduring Amerindian Languages Roughly two-thirds of Latin Americans are Spanish speakers, and one-third speak Portuguese. These colonial languages were so prevalent by the nineteenth century that they were the unquestioned languages of government and instruction for the newly independent Latin American republics. In fact, until recently many countries actively discouraged, and even repressed, Indian tongues. It took a constitutional amendment in Bolivia in the 1990s to legalize native-language instruction in primary schools and to recognize the country's multiethnic heritage (more than half the population is Indian; and Quechua, Aymara, and Guaraní are widely spoken) (Figure 4.27).

Because Spanish and Portuguese dominate, there is a tendency to neglect the influence of indigenous languages in the region. Mapping the use of indigenous languages, however, reveals important pockets of Indian resistance and survival. In the Central Andes of Peru, Bolivia, and southern Ecuador, more than 10 million people still speak Quechua and Aymara, along with Spanish. In Paraguay and lowland Bolivia there are 4 million Guaraní speakers, and in southern Mexico and Guatemala at least 6 to 8 million speak Mayan languages. Small groups of native-language speakers are found scattered throughout the sparsely settled interior of South America and the more isolated forests of Central America, but many of these languages have fewer than 10,000 speakers.

Blended Religions Like language, the Roman Catholic faith appears to have been imposed upon the region without challenge. Most countries report 90 percent or more of their population as Catholic. Every major city has dozens of churches, and even the smallest hamlet maintains a graceful church on its central square (Figure 4.28). In some countries, such as El Salvador and Uruguay, a sizable portion of the population attend Protestant evangelical churches, but the Catholic core of this region is still intact.

Exactly what native peoples absorbed of the Christian faith is unclear. Throughout Latin America **syncretic religions**, blends of different belief systems, enabled pre-Hispanic religious practices to be folded into Christian worship. These blends took hold and endured, in part because Christian saints were easy surrogates for pre-Christian gods and because the Catholic Church tolerated local variations in worship as long as the process

Figure 4.27 Language Map of Latin America
The dominant languages of Latin America are Spanish and Portuguese. Nevertheless, there are significant areas in which indigenous languages persist and, in some cases, are recognized as official languages. Smaller language groups exist in Central America, the Amazon Basin, and southern Chile. *(Adapted from* The Atlas of the World's Languages, *1994, New York: Routledge)*

of conversion was underway. The Mayan practice of paying tribute to spirits of the underworld seems to be replicated today in Mexico and Guatemala via the practice of building small cave shrines to favorite Catholic saints and leaving offerings of fresh flowers and fruits. One of the most celebrated religious icons in Mexico is the Virgin of Guadelupe, a dark-skinned virgin seen by an Indian shepherd boy in the sixteenth century who became the patron saint of Mexico.

Syncretic religious practices also evolved and endured among African slaves. By far the greatest concentration of slaves was in the Caribbean, where slaves were used to replace the indigenous population, which was wiped out by disease (see Chapter 5). Within Latin America, the Portuguese colony of Brazil received the most Africans—at least 4 million. In Brazil, where the volume and the duration of the slave trade were the greatest, the transfer of African-based religious and medical systems is most evident. West

Figure 4.28 The Catholic Church Churches, such as the Dolores Church in Tegucigalpa, Honduras, are important religious and social centers. The vast majority of people in Latin America define themselves as Catholic. Many churches built in the colonial era are valued as architectural treasures and are beautifully preserved. *(Rob Crandall/www.robcrandall.com)*

African-based religious systems such as Batuque, Umbanda, Candomblé, and Shango are often mixed with, or ancillary to, Catholicism, and are widely practiced in Brazil. So accurate were some of these religious transfers that it is common to have Nigerian priests journey to Brazil to learn forgotten traditions. In many parts of southern Brazil, Umbanda is as popular with people of European ancestry as with Afro-Brazilians. Typically a person becomes familiar with Umbanda after falling victim to a magician's spell by having some object of black magic buried outside his or her home. In order to regain control of his or her life, the victim needs the help of a priest or priestess.

The syncretic blend of Catholicism with African traditions is most obvious in the celebration of carnival, Brazil's most popular festival and one of the major components of Brazilian national identity. The three days of carnival, known as the Reign of Momo, combine Christian Lenten beliefs with pagan influences and feature African musical traditions epitomized by the rhythmic samba bands. Although the street festival was banned for part of the nineteenth century, Afro-Brazilians in Rio de Janeiro resurrected it in the 1880s with nightly parades, music, and dancing. Within 50 years the street festival had given rise to formalized samba schools and helped break down racial barriers. By the 1960s, carnival became an important symbol for Brazil's multiracial national identity. Today the festival—which is most associated with Rio—draws thousands of participants from all over world.

The Global Reach of Latino Culture

Latin American culture, vivid and diverse as it is, is widely recognized throughout the world. Whether it is the sultry pulse of the tango or the fanaticism with which Latinos embrace soccer as an art form, aspects of Latin American culture have been absorbed into a globalizing world culture. A dramatic example of the reach of Latino culture can be seen every Saturday on the Univision television network, based in Miami. There, a charismatic Chilean, Don Francisco, has hosted *Sabado Gigante* since 1986. This four-hour-long Spanish variety show is viewed in 28 countries and draws a weekly audience of 120 million viewers. In the arts, Latin American writers such as Gabriel García Marquez and Isabel Allende have obtained worldwide recognition. In terms of popular culture, new musical artists such as Colombia's Shakira and Brazil's hip-hop samba singer Max de Castro are gaining international audiences, while Latino superstars Jennifer Lopez and Ricky Martin are icons of world pop culture. Through music, literature, and even *telenovelas* (soap operas), Latino culture is being transmitted to an eager worldwide audience.

Telenovelas Popular nightly soap operas are a mainstay of Latin American television. These tightly plotted series are filled with intrigue and double deals. Unlike their counterparts in the United States, they end, usually after 100 episodes. Once standard fare for the working class, many telenovelas take hold and absorb an entire nation. During particularly popular episodes, the streets are noticeably calm as millions of people tune in to catch up on the lives of their favorite heroines. Brazil, Venezuela, and Mexico each produce scores of telenovelas, but the Mexican ones are international megahits.

Televisa, a Mexican production agency, has aggressively marketed its inventory of soap operas to an eager global public. Mexican telenovelas are avidly watched in countries as diverse as Croatia, Russia, China, South Korea, Iran, the United States, and France, as well as throughout Latin America. Predictably scripted as Mexican Cinderella stories, these sagas of poor underclass women (often domestics) falling in love with members of the elite, battling jealous rivals, and ultimately emerging triumphant seem to resonate with fans around the world. In addition to their broad appeal, telenovelas are big business, perhaps Mexico's largest international export. While Hollywood and Mumbai (formerly Bombay) grind out movies, much of Mexico's entertainment industry is geared toward producing this popular art form.

Soccer Perhaps the quintessential global sport, soccer has a fanatical following throughout much of the world. Yet it is Latin America, and especially in South America, where "*fútbol*" is considered a cultural necessity. Still largely a male game, although girls are beginning to play, young boys and men are constantly seen on fields, beaches, and blacktops playing soccer, especially in late afternoons and on the weekends. The great soccer stadiums of Buenos Aires (Bombonera) and Rio de Janeiro (Maracaña) are regarded as shrines to the game. Many individuals use the victories and losses of their national soccer teams as the important chronological markers of their lives.

It was Pelé, the Brazilian soccer phenomenon of the 1960s and 1970s, who introduced the free-flowing acrobatic style that became known as "the beautiful game." Today Latino soccer stars (especially Brazilians) play for corporate clubs all over Europe and earn millions. Latin Americans also fill up the few slots allotted to foreign players on the U.S. Major League Soccer teams. Yet the dream of many Latino soccer players is to be on the national team and bring home the World Cup. Visit any Latin American country when their team is playing a World Cup qualifying match and the streets are eerily quiet. During the 2006 World Cup, the business and financial centers of Argentina and Brazil virtually came to a standstill during the entire tournament. Somehow, as if by magic, it is believed that a World Cup victory will make things better, improve the economy, and even reduce crime. In short, soccer is regarded with near religious significance. And as Latin Americans emigrate (both as players and as laborers), they bring their enthusiasm for the sport with them.

National Identities Viewed from the outside, there is considerable homogeneity to this region; yet distinct national identities and cultures flourish in Latin America. Since the early days of the republics, countries celebrated particular elements from their pasts when creating their national histories. In the case of Brazil, the country's interracial characteristics were highlighted to proclaim a new society in which the color lines between Europeans and Africans ceased to matter. Mexico celebrated the architectural and cultural achievements of its Aztec predecessors while at the same time forging an assimilationist strategy that discouraged surviving indigenous culture and language.

Musical and dance traditions evolved and became emblematic of these new societies. The tango in Argentina, caporales in Bolivia, the vallenato and cumbia in Colombia, the mariachi in Mexico, and the samba in Brazil are easily distinguished styles that are representative of distinct national cultures (Figure 4.29). Literature also reflects the distinct identities found in Latin America. Writers such as Isabel Allende, Gabriel García Marquez, Mario Vargas Llosa, Carlos Fuentes, and Jorge Amado situate their stories in their native countries and, in so doing, celebrate the unique characteristics of Chileans, Colombians, Peruvians, Mexicans, and Brazilians. Distinct political cultures also evolved, which at times led to expansionist policies that brought neighbors into conflict.

Figure 4.29 Musical and Dance Traditions
a) Caporales is a popular Bolivian folkloric dance. During carnival and other holidays caporales dance groups take to the streets in friendly competition. b) Tango is the signature dance and music of Buenos Aires, Argentina. Today tango clubs are popular with tourists who want to see the dance and even learn how to do it. *(a, Rob Crandall; b, Christian Heeb/Aurora & Quanta Productions)*

(a)

(b)

GEOPOLITICAL FRAMEWORK: Redrawing the Map

Latin America's colonial history, more than its present condition, unifies this region. For the first 300 years after Columbus's arrival, Latin America was a territorial prize sought by various European countries but effectively settled by Spain and Portugal. By the nineteenth century, the independent states of Latin America had formed, but they continued to experience foreign influence and, at times, overt political pressure, especially from the United States. At other times a more neutral Pan-American vision of American relations and hemispheric cooperation has held sway, represented by the formation of the **Organization of American States** (OAS). The present organization was chartered in 1948, but its origins date back to 1889. Yet there is no doubt that U.S. policies toward trade, economic assistance, political development, and at times military intervention are often seen as compromising to the sovereignty of these states.

Within Latin America there have been cycles of intraregional cooperation and antagonism. Neighboring countries have fought over territory, closed borders, imposed high tariffs, and cut off diplomatic relations. Even today there are a dozen long-standing border disputes in Latin America that occasionally erupt into armed conflict. The 1990s witnessed a revival in the trade block concept with the formation of **Mercosur**, the Southern Common Market, which now includes Brazil, Uruguay, Argentina, Paraguay, and Venezuela as full members and Bolivia, Chile, Colombia, Ecuador, and Peru as associate members; and NAFTA, which includes Mexico, the United States, and Canada. As economic ties strengthen, these trade blocks could form the basis for a new alignment of political and economic interests in the region.

Iberian Conquest and Territorial Division

Because it was Christopher Columbus who claimed the Americas for Spain, the Spanish were the first active colonial agents in the Western Hemisphere. In contrast, the Portuguese presence in the Americas was the result of the **Treaty of Tordesillas**, brokered by the pope in 1493–94. By that time Portuguese navigators had charted much of the coast of Africa in an attempt to find a water route to the Spice Islands (Moluccas) in Southeast Asia. With the help of Christopher Columbus, Spain sought a western route to the Far East. When Columbus discovered the Americas, Spain and Portugal asked the pope to settle how these new territories should be divided. Without consulting other European powers, the pope divided the Atlantic world in half—the eastern half containing the African continent was awarded to Portugal, the western half with most of the Americas was given to Spain. The line of division established by the treaty actually cut through the eastern part of South America, placing it under Portugese rule. This treaty was never recognized by the French, English, or Dutch, who also asserted territorial claims in the Americas, but it did provide the legal apparatus for the creation of a Portuguese territory in America—Brazil—which would later become the largest and most populous state in Latin America (Figure 4.30).

Six years after the treaty was signed, Portuguese navigator Alvares Cabral inadvertently reached the coast of Brazil on a voyage to southern Africa. The Portuguese soon realized that this territory was on their side of the Tordesillas line. Initially they were unimpressed by what Brazil had to offer; there were no spices or major indigenous settlements. Quickly, however, they came to appreciate the utility of the coast as a provisioning site as well as a source for brazilwood, used to produce a valuable dye. Portuguese interest in the territory intensified in the late sixteenth century with the development of sugar estates and the expansion of the slave trade, and in the seventeenth century with the discovery of gold in the Brazilian interior.

Spain, in contrast, aggressively pursued the conquest and settlement of its new American territories from the very start. After discovering little gold in the Caribbean, by the mid-sixteenth century Spain's energy was directed toward developing the silver resources of Central Mexico and the Central Andes (most notably Potosí in Bolivia). Gradually the economy diversified to include some agricultural exports, such as cacao and sugar, as well as a variety of livestock. In terms of foodstuffs, the colonies were virtually self-sufficient. Some basic manufactured items, such as crude woolen cloth and agricultural tools, were also produced, but in general, manufacturing was forbidden in the Spanish American colonies in order to keep them dependent on Spain.

Revolution and Independence It was not until the 1800s, with a rise of revolutionary movements between 1810 and 1826, that Spanish authority on the mainland was challenged. Ultimately, elites born in the Americas gained control, displacing the representatives of the crown. In Brazil, the evolution from Portuguese colony to independent republic was a slower and less violent process that spanned eight decades (1808–89). In the nineteenth century Brazil was declared a separate kingdom from Portugal with its own monarch, and later it became a republic.

The territorial division of Spanish and Portuguese America into administrative units provided the legal basis for the modern states of Latin America (see Figure 4.30). The Spanish colonies were first divided into two viceroyalties (New Spain and Peru) and within these were various subdivisions that later became the basis for the modern states. (In the eighteenth century the Viceroyalty of Peru, which included all of Spanish South America, was divided to form three viceroyalties: La Plata, Peru, and New Granada.) Unlike Brazil, which evolved from a colony into a single republic, the former Spanish colonies experienced fragmentation in the nineteenth century. Prominent

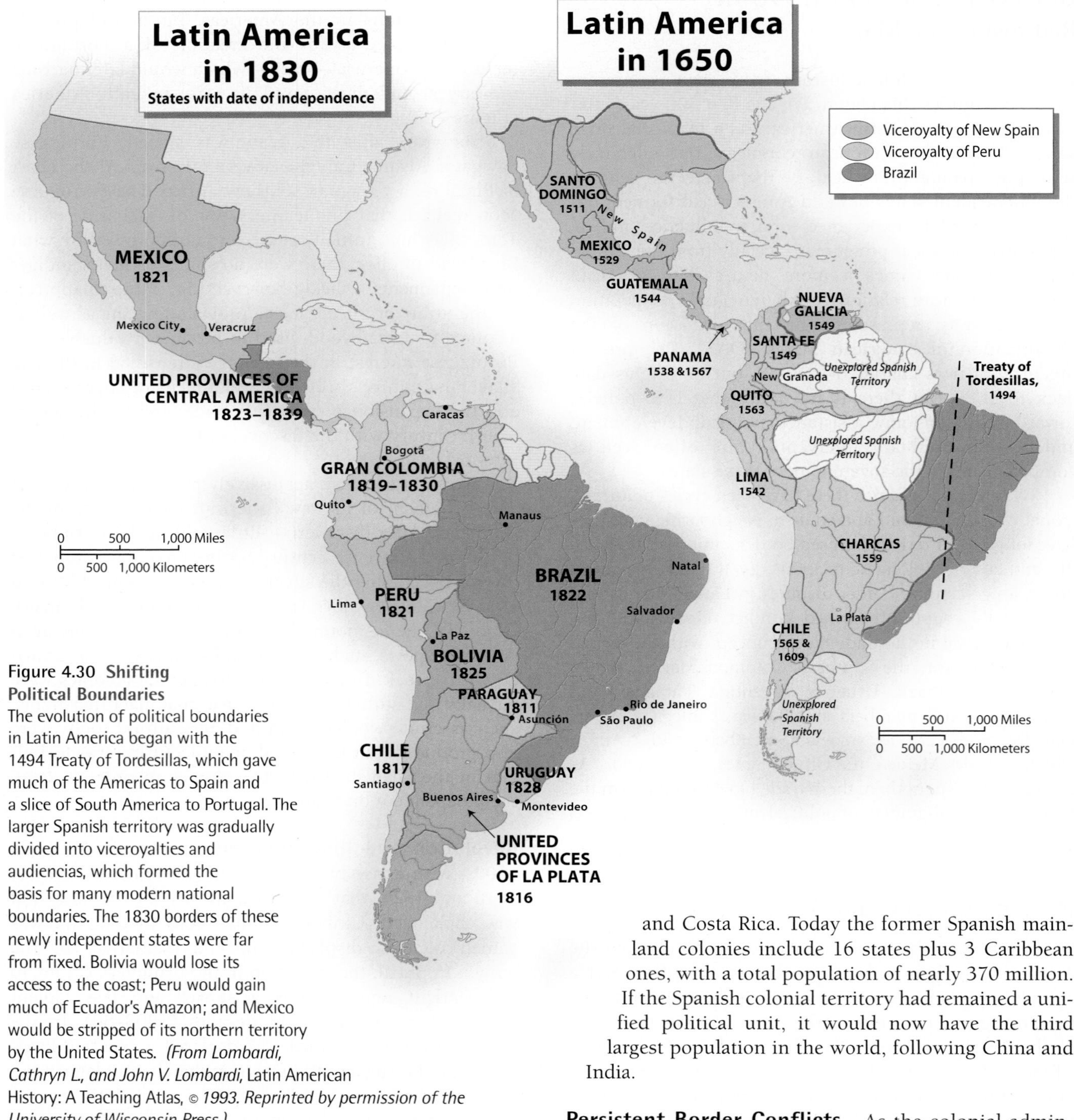

Figure 4.30 **Shifting Political Boundaries**
The evolution of political boundaries in Latin America began with the 1494 Treaty of Tordesillas, which gave much of the Americas to Spain and a slice of South America to Portugal. The larger Spanish territory was gradually divided into viceroyalties and audiencias, which formed the basis for many modern national boundaries. The 1830 borders of these newly independent states were far from fixed. Bolivia would lose its access to the coast; Peru would gain much of Ecuador's Amazon; and Mexico would be stripped of its northern territory by the United States. *(From Lombardi, Cathryn L., and John V. Lombardi,* Latin American History: A Teaching Atlas, *© 1993. Reprinted by permission of the University of Wisconsin Press.)*

among revolutionary leaders was Venezuelan-born Simon Bolívar (Figure 4.31), who advocated his vision for a new and independent state of Gran Colombia. For a short time (1822–30) Bolívar's vision was realized, as Colombia, Venezuela, Ecuador, and Panama were combined into one political unit. Similarly, in 1823 the United Provinces of Central America was formed to avoid annexation by Mexico. By the 1830s, this union also broke apart into the states of Guatemala, Honduras, El Salvador, Nicaragua, and Costa Rica. Today the former Spanish mainland colonies include 16 states plus 3 Caribbean ones, with a total population of nearly 370 million. If the Spanish colonial territory had remained a unified political unit, it would now have the third largest population in the world, following China and India.

Persistent Border Conflicts As the colonial administrative units turned into states, it became clear that the territories were not clearly delimited, especially the borders that stretched into the sparsely populated interior of South America. This would later become a source of conflict as new states struggled to demarcate their territorial boundaries. Numerous border wars erupted in the nineteenth and twentieth centuries, and the map of Latin America has been redrawn many times. Some of the more noted conflicts were the War of the Pacific (1879–82) in which Chile expanded to the north and Bolivia lost its access to the Pacific; warfare between

Figure 4.31 Simon Bolívar
Heroic likenesses of Simon Bolívar (the Liberator) are found throughout South America, especially in his native country of Venezuela. This statue stands in the central plaza of the Andean city of Mérida, Venezuela. *(Rob Crandall/www.robcrandall.com)*

Mexico and the United States in the 1840s, which resulted in the present border under the Treaty of Hidalgo (1848) and Mexico's loss of what became the southwestern United States; and the War of the Triple Alliance (1864–70), the bloodiest war of the postcolonial period, which occurred when Argentina, Brazil, and Uruguay allied themselves to defeat Paraguay in its claim to control the upper Paraná River Basin. It is estimated that this conflict resulted in the reduction of Paraguay's adult male population by nine-tenths. Sixty years later, the Chaco War (1932–35) resulted in a territorial loss for Bolivia in its eastern lowlands and a gain for Paraguay. In the 1980s Argentina lost a war with Great Britain over control of the Falkland, or Malvinas, Islands in the South Atlantic. And as recently as 1998, Peru and Ecuador skirmished over a disputed boundary in the Amazon Basin.

Outright war in the region is less common than ongoing and nagging disputes over international boundaries. Any of a dozen dormant claims erupt from time to time, given the political climate between neighbors. These include Venezuela and Guyana's border dispute, Ecuador's claims to the Peruvian eastern lowlands, and the maritime boundary dispute between Honduras and Nicaragua as well as one between Venezuela and Colombia. Many of these disputes are based on historical territorial claims dating back to the colonial period.

One of the more interesting geopolitical conflicts in the Southern Cone is based on territorial claims to Antarctica. Seven claimant states—Argentina, Chile, Norway, the United Kingdom, France, New Zealand, and Australia—plus five nonclaimant states (including the United States) seek resource rights to Antarctica. Even though an Antarctic Treaty has existed since 1959 stating that the land mass should be used for peaceful purposes, Chile and Argentina have incorporated their Antarctic territorial claims on national maps and even postage stamps. In the 1970s, Chilean President Augusto Pinochet spent a week touring the Chilean claim. Argentina held a national cabinet meeting on its portion of Antarctica in the 1970s and then sent a pregnant Argentine woman to give birth to the continent's first child. For the most part, these nationalist claims are symbolic. Recent treaties show an international inclination toward cooperation and the banning of commercial exploitation of the continent.

The Trend Toward Democracy Early in the twenty-first century, most of the 17 countries in this region will celebrate their bicentennials. Compared with most of the developing world, Latin Americans have been independent for a long time. Yet political stability is not a hallmark of the region. Among the countries in the region, some 250 constitutions have been written since independence, and military coups have been alarmingly frequent. Since the 1980s, however, the trend has been toward democratically elected governments, the opening of markets, and broader popular participation in the political process. Where dictators once outnumbered elected leaders, by the 1990s each country in the region had a democratically elected president. (Cuba, the one exception, will be discussed in Chapter 5.)

Democracy may not be enough for the millions frustrated by the slow pace of political and economic reform. In survey after survey, Latin Americans register their dissatisfaction with politicians and governments. Most of the newly elected democratic leaders are also free-market reformers who are quick to eliminate state-backed social safety nets, such as food subsidies, government jobs, and pensions. Many of the poor and middle class have grown skeptical about whether this brand of democracy could improve their lives. But the political left has not yet produced an alternative to privatization and market-driven policies, although there has definitely been a trend of leftist leaning politicians winning presidential elections in Brazil, Bolivia, Nicaragua, Ecuador, Peru, and Venezuela. In January 2007, 75,000 protesters converged on Mexico City's central plaza to protest the sudden jump in tortilla prices, the staple for the city's poor. U.S. demand for ethanol (produced from corn) drove up corn prices in 2006. And since the United States supplies about a quarter of Mexico's corn, higher prices in the United States impacted Mexican consumers. For opponents of Mexican president Felipe Carderón, who narrowly won a contested election in 2006, the protests were seen as another example of how free-market policies are hurting the poor. The president responded by freezing tortilla prices, albeit at the already higher price (Figure 4.32). For now, the status quo continues, although popular frustration with falling incomes, rising violence, corruption, and chronic underemployment are a recipe for political instability in Mexico, and throughout the region.

Regional Organizations

Even as democratically elected leaders struggle to address the pressing needs of their countries, political developments at the supranational and subnational levels pose new challenges to their authority. The most discussed **supranational organizations**—governing bodies that include several states—are the trade blocks. **Subnational organizations**, which are groups that represent areas or people within the state, often form along ethnic or ideological lines and can provoke serious internal divisions. Indigenous groups seeking territorial or political recognition and insurgent groups espousing Marxist ideas (such as the FARC in Colombia) have challenged the authority of the states. Finally, the financial and political force of drug cartels and gangs, especially for smaller countries, transcends state boundaries and undermines judicial systems.

Trade Blocks Beginning in the 1960s, regional trade alliances were attempted in an effort to foster internal markets and reduce trade barriers. The Latin American Free Trade Association (LAFTA), the Central American Common Market, and the Andean Community have existed for decades, but their ability to influence economic trade and growth is limited at best. In the 1990s, Mercosur and NAFTA emerged as supranational structures that could influence development (Figure 4.33). For Latin America, the lessons of Mercosur and NAFTA are causing politicians to rethink the value of regional trade.

Mercosur was formed in 1991 with Brazil and Argentina, the two largest economies in South America, and the smaller states of Uruguay and Paraguay as members. Since its formation, trade among these countries grew tremendously in the 1990s but faltered some in the last few years due to economic turmoil and political infighting. The size and productivity of this market have not gone unnoticed by other trade blocks. Between 1991 and 2001, trade between the European Union and Mercosur doubled, from $20 billion to $40 billion. Mercosur countries are the EU's largest trading partners in Latin America, and there have been renewed efforts to form a free-trade agreement between Mercosur and the EU. This development is significant in two ways: it reflects the growth of these economies and the willingness to put aside old rivalries (especially long-standing antagonisms between Argentina and Brazil) for the economic benefits of regional cooperation. (See "Geography in the Making: The Growth of Mercosur.")

Transportation and communication between South American countries will certainly improve with the expansion of Mercosur. Plans abound to fundamentally rework the flow of goods and communication in this area. As privatization of the telephone companies has benefited telecommunication, major joint engineering projects are being considered. Studies are underway to improve navigation along the Paraná and Paraguay rivers, which are already important arteries for the transport of grains. Other schemes include building a bridge across the Plata River estuary to form a more direct link between the capital cities of Uruguay and Argentina, a tunnel through the Andes between Chile and Argentina, and a regional network of natural gas pipelines that would link Bolivia, Argentina, Peru, and southern Brazil. In short, the elaboration of Mercosur could change the way individuals in the member countries relate to one another and think about themselves. Much like the European Union,

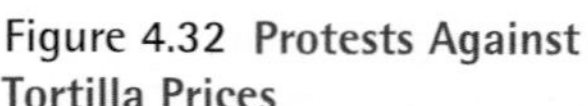

Figure 4.32 Protests Against Tortilla Prices
A peasant woman joined protestors in Mexico City in 2007 as Mexicans marched against the sudden increase in corn and tortilla prices, the staple foods of the country. People carried banners proclaiming, "without corn, there is no country." In the end the president of Mexico decided to freeze tortilla prices. *(Luis Acosta/ AFP/Getty Images)*

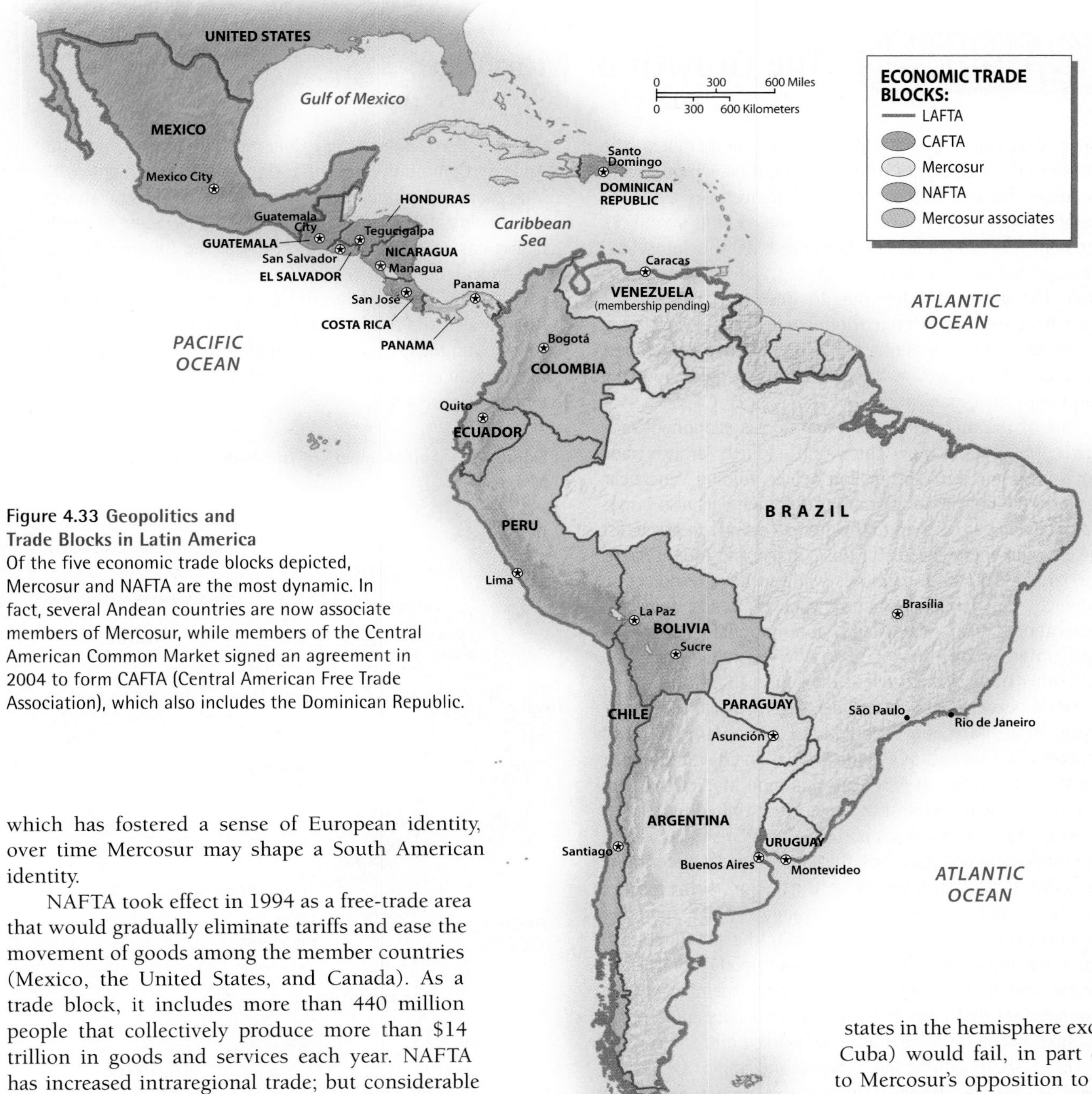

Figure 4.33 Geopolitics and Trade Blocks in Latin America
Of the five economic trade blocks depicted, Mercosur and NAFTA are the most dynamic. In fact, several Andean countries are now associate members of Mercosur, while members of the Central American Common Market signed an agreement in 2004 to form CAFTA (Central American Free Trade Association), which also includes the Dominican Republic.

which has fostered a sense of European identity, over time Mercosur may shape a South American identity.

NAFTA took effect in 1994 as a free-trade area that would gradually eliminate tariffs and ease the movement of goods among the member countries (Mexico, the United States, and Canada). As a trade block, it includes more than 440 million people that collectively produce more than $14 trillion in goods and services each year. NAFTA has increased intraregional trade; but considerable controversy rages about who are the beneficiaries and who are the losers in this trade block (see Chapter 3). NAFTA did stimulate trade and overall growth, but it also brought dislocation and change to the economic landscape of the countries involved. Although NAFTA did not detour immigration to the United States by creating jobs in Mexico, it did prove that a free-trade area combining industrialized and developing states was possible. By some measures the success of NAFTA inspired other free-trade agreements, the newest one being the Central American Free Trade Agreement (CAFTA) signed in 2004 to include the Dominican Republic along with Guatemala, El Salvador, Honduras, Nicaragua, and Costa Rica. The push for CAFTA emerged after it became clear that an overly ambitious Free Trade Area of the Americas (including all states in the hemisphere except Cuba) would fail, in part due to Mercosur's opposition to the plan. Like NAFTA, the CAFTA agreement requires opening markets for goods and services, dismantling protections of local industries, and enforcement of intellectual property rights. In return, the United Stated promises greater market access for Central American and Dominican products including textiles and agricultural commodities. CAFTA is likely to increase direct foreign investment and jobs in the block, although detractors worry about the rise of low-wage sweatshop employment and not more skilled labor. Interestingly, it is resistance from Costa Rica—the most prosperous member of the group—that has delayed its full implementation as of 2007.

GEOGRAPHY IN THE MAKING: The Growth of Mercosur

When Mercosur began in 1991 it had 4 full members—the Southern Cone countries of Brazil, Argentina, Uruguay, and Paraguay. Today Mercosur has 10 members: 5 associate members (all Andean countries), and 5 full members, pending the ratification of Venezuela as the latest full member. Bolivia may also be granted full member status soon. The dream of Mercosur, to integrate trade in the southern cone, has extended to include most of continent. Yet expansion has brought tensions among the members and associates, as well as an ideological struggle about the purpose of the organization.

Part of the motivation for expansion is economic trade power. Mercosur is now the world's fourth largest trade block, including over 260 million people. Adding Venezuela, and possibly Bolivia, to the club of full members also brings major oil and gas reserves to the group. This will greatly assist the growing energy needs for Chile, Brazil, and Argentina. As one large block, these countries will also have resources to enhance regional infrastructure such as roads, telecommunications, gas lines, and international waterways. Furthermore, economies of scale now exist in negotiation with other trade blocks, such as the European Union.

Yet Venezuela's populist leader, President Hugo Chavez, insists that Mercosur should be "decontaminated of neoliberalism" and act as a political block. In Chavez's mind, it is only a strong South American Common Market that can counter the neoimperialist actions of the United States. For Chavez, and some of his supporters, Mercosur should be first and foremost a political block that uses trade to foster more equitable social development in the region. However, the original intent of Mercosur, certainly by its founding members, was to build a trade block with the aim of regional economic integration in the European model. Brazil, the largest and most economically powerful member of the group, insists that economic integration is happening but it will take awhile for the details to be worked out.

Given these conflicting visions, and rapid expansion, there is intense upheaval within the group, making it unclear if a continental-sized Mercosur can be sustained. The difficulties lie in integrating states at such different scales (Uruguay has barely 3 million people and Brazil has nearly 190 million) and levels of development. At the same time, the expansion of Mercosur could make one of the region's oldest trade blocks, the Andean Community of Nations, unnecessary. Formed in 1969 with Colombia, Ecuador, Peru, Bolivia, and Chile as the Andean Pact, the Andean Community recently signed an agreement with Mercosur's full members allowing them to be associates of the organization. Thus a competing geography of trade blocks has emerged in South America. The full members of the Andean Community and Mercosur could peel away from each other, with Mercosur more oriented toward the Atlantic side and the Andean Community toward the Pacific. Or, through the gradual inclusion of Andean countries as full members of Mercosur, a single continental trade block could emerge (Figure 4.5.1).

Figure 4.5.1 Full Members of the Andean Community and Mercosur

Insurgencies and Drug Trafficking Guerrilla groups such as the Revolutionary Armed Forces of Colombia (FARC) have controlled large territories of their countries through the support of those loyal to their cause, along with theft, kidnapping, and violence. The FARC, along with the National Liberation Army (ELN), gained wealth and might in Colombia through the drug trade in the 1980s and 1990s. The level of violence in Colombia escalated further with the rise of paramilitary groups—well-armed vigilante groups that seek to "cleanse" areas of insurgency sympathizers. The paramilitary groups are blamed for the rise in politically motivated murders and there are clear indications that they received help from individuals in the government. As many as 2.5 million Colombians have been internally displaced by violence since the late 1980s, most fleeing rural areas for towns and cities. By 2002, the level of violence in Colombia had become intolerable, with an estimated 300,000 murders that year. In the major cities of Bogotá and Medellín, people did not go out at night for fear of crime and kidnapping, economic activity declined, and many people left the country. Colombian statistics suggest that over 3 million Colombians now live abroad. Fortunately, the situation in the last five years has improved considerably. Under President Uribe, the police presence increased throughout the state and negotiations with insurgencies stopped, ultimately reducing their power. The introduction of the Justice and Peace Law also began the process of disarming paramilitary groups. After years of violence, Colombians desperately want peace. For now, at least, the levels of violence have declined in Colombia. Yet drug cartels and gangs in states as diverse as Mexico, El Salvador, and Brazil have been blamed for increases in violence and lawlessness.

The drug trade is often seen as the root of many of the region's problems, but this illegal trade also generates billions of dollars for Latin Americans, from the small coca farmers in Bolivia to the cartel leaders in Colombia. The drug trade began in earnest in the 1970s, with Colombia at its center of distribution and Bolivia and Peru the zones of coca production. By the 1980s the Medellín Cartel was a powerful and wealthy crime syndicate that used narco-dollars to bribe or murder anyone who got in its way. The organization was weakened by the death of its leader, Pablo Escobar, in 1993 but by then the Cali Cartel was poised to take control. Its power was reduced by arrests of key leaders in 1995. No one cartel dominates the drug trade today; instead, it has decentralized into dozens of smaller productive syndicates.

Initially, most Latin governments cared little about controlling the drug trade as it brought in much-needed hard currency. Within the region, drug consumption was scarcely a problem. Some drug lords even became popular folk heroes, lavishly spending money on their communities for housing, parks, and schools. The social costs of the drug trade to Latin America became evident by the 1980s, when the region was crippled by a badly damaged judicial system. By paying off police, the military, judges, and politicians, the drug syndicates wielded incredible political power that threatened the civil fabric of the states in which they work. And eventually drug consumption and addiction became a problem, especially in Latin America's large cities. Years of counternarcotics work have done little to reduce the overall flow of drugs to North America and Europe, but the programs have changed the areas of production.

In the 1980s, the major production centers were in the cloud forests of Peru and Bolivia, and the producers were peasants working small plots. During the 1990s, the production of coca leaf (the main ingredient for cocaine) shifted to the rain forests of Colombia, where coca was grown on large, modern estates. Plan Colombia, introduced by President Bill Clinton in 2000, upped the stakes in the drug war with a commitment of $1.3 billion to Colombia alone. Since 1990, the United States has spent between $3 billion and $4 billion to reduce Andean coca and cocaine production. Proponents of this production-oriented strategy are claiming victory, as Bolivian and Peruvian production has tumbled (Figure 4.34). Yet the overall area of coca cultivation has not fallen significantly, from 516,450 acres (209,000 hectares) in 1996 to 457,000 acres (185,000 hectares) in 2000 and 431,000 acres (174,500 hectares) in 2005.

Figure 4.34 Coca-Growing Areas in South America
Although the oldest coca-growing regions are in Bolivia and Peru, Colombian traffickers turned the processing and distribution of cocaine into an international narcotics trade. By the late 1990s the bulk of coca production had shifted out of Bolivia and Peru and to Colombia. *(From U.S. Government, 2002, 2003, 2005* Latin American Narcotics Cultivation and Production Estimates, *Washington, DC: CIA Crime and Narcotics Center)*

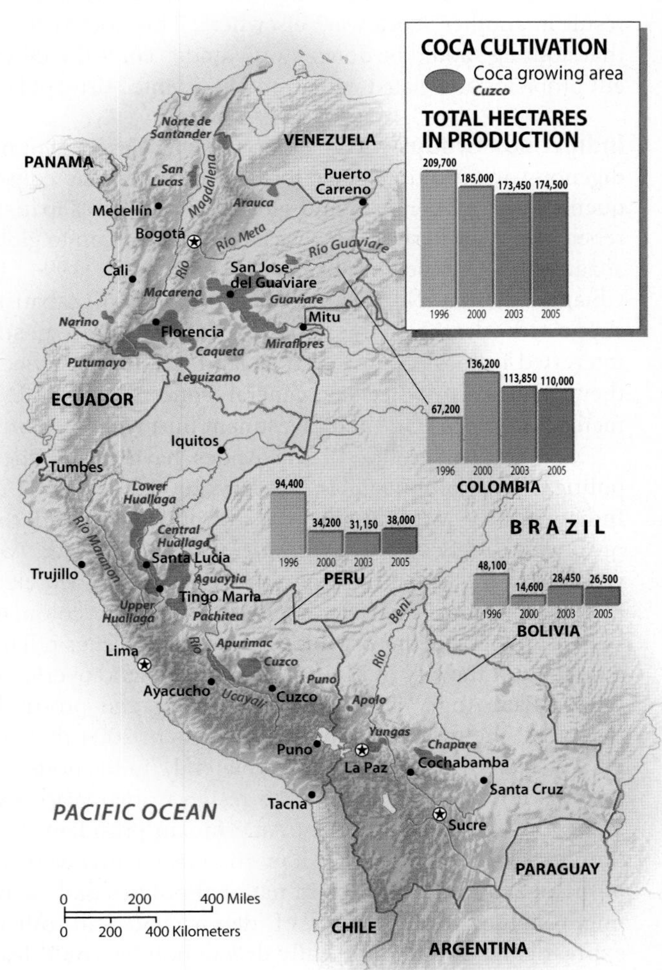

Less appreciated are the environmental consequences of coca production and coca eradication. Geographer Ken Young estimates that in Peru alone as many as 2.5 million acres (1 million hectares) of tropical forest were degraded in the 1980s by coca cultivation. Delicate tropical ecosystems and their waterways are being contaminated by the tons of chemicals used to produce cocaine. Unfortunately, serious ecological damage may also be occurring as a result of eradication efforts. The preferred method in southern Colombia is aerial spraying, but fumigation has not been that successful, and it causes serious health problems, destruction of legal crops, increased social tension, and damage to both terrestrial and aquatic ecosystems (Figure 4.35). The World Wildlife Fund has compared the potential environmental impact of fumigation in Colombia to that of Agent Orange in Vietnam.

It is profitability that drives the cocaine trade. As long as there is demand, the supply will be met. Although still confined to South America, coca can grow in other tropical areas, including Southeast Asia, the Pacific, and Africa. The poppy industry, the base for heroin, offers an instructive comparison. Once, poppies were almost an exclusive crop of Southeast Asia, but today Central Asia as well as Colombia and Mexico are major producers of poppies for the heroin trade. Given the global nature of the drug trade, it is likely that a victory in Colombia against coca cultivation will result in another "coca war" elsewhere. This doesn't mean that nothing should be done; it only underscores the inherent problems associated with a supply-oriented drug policy.

Indigenous Groups At the grassroots level, scores of indigenous organizations have formed to protest the consequences of neoliberalism and globalization. The Zapatista rebellion in southern Mexico was in part a reaction to globalization. The rebellion began on January 1, 1994, in Chiapas, the day NAFTA took effect. Although Zapatista supporters—largely Amerindian peasants—are mostly interested in local issues, such as land and basic services, their movement reflects a general concern about how increased foreign trade and investment hurt rural peasants.

Amerindian groups in the Andes have flexed their political muscle as well. In January 2000, thousands of Indians, allied with dissident army officers, forced the resignation of Ecuadorian president Jamil Mahuad. Two years later, the Indian movement in Ecuador helped elect Lucio Gutiérrez, one of the dissident army colonels. Bolivia witnessed the organized protests of Amerindians who inhabit the city of El Alto, on the Altiplano overlooking La Paz. Angry with their pro-mining and pro-trade president, the Indians set up road blocks in 2004 that cut off La Paz from supplies. In the end, indigenous-led protests forced two presidents to resign in two years and led to the election of the first Amerindian president, Evo Morales. Such experiences show the greater involvement of organized Indian groups in national politics and, some might argue, a deepening of democracy. Still others express concern that ethnically driven politics could lead to national fragmentation.

Figure 4.35 Coca Eradication
A member of an elite antidrug unit of Colombia's National Police watches as an aircraft dumps herbicide on an illicit coca crop in Colombia. Coca leaf, the raw material for cocaine, has become the leading cash crop in the area. With financial assistance from the United States, Colombians use aerial spraying in an effort to eradicate the crop. *(Eliana Aponte/Reuters/Corbis/Bettmann)*

ECONOMIC AND SOCIAL DEVELOPMENT: Dependent Economic Growth

Most Latin American economies fit into the broad middle-income category set by the World Bank. Clearly part of the developing world, their people are much better off than those in Sub-Saharan Africa, South Asia, and China. Still, the economic contrasts are sharp both between states and within them. Table 4.2 compares the purchasing power parity (PPP) per capita of the countries in the region. The highest figures are for Argentina and Chile ($13,920 and $11,470, respectively). Next is Mexico, with a PPP per capita figure of over $10,000, closely followed by Uruguay and Costa Rica. The poorest countries in the region are in the Andes and Central America. Bolivia and Honduras have the region's lowest PPP per capita figure of under $3,000. In terms of annual growth, all the countries have positive annual growth rates for the period 2000 to 2005. In fact, several countries, including Peru, Panama, Ecuador, Costa Rica, and Chile, had average annual growth rates of 4 percent or higher. Given the grave financial crises in Ecuador, Argentina, and Uruguay in 2000 and 2001, the positive growth figures for these countries are especially impressive. Even with its middle-income status, extreme poverty is evident throughout the region; nearly one-fourth of all Latin Americans live in extreme poverty. By this measure, Nicaragua is the region's poorest country given that 80 percent of its population lives on less than $2 a day (see Table 4.2).

This was not the future envisioned for the region in the mid-1960s when Brazil, Mexico, and Argentina all seemed poised to enter the ranks of the industrialized world.

TABLE 4.2 Development Indicators

Country	GNI[a] per Capita, PPP[b] (2005)	GDP[c] Average Annual % Growth (2000–05)	Life Expectancy (2007)	Percent of Population Living on Less Than $2 a Day	Under Age 5 Mortality Rate 1990	Under Age 5 Mortality Rate 2005	Gender Equity[d]
Argentina	13,920	2.2	75	23	29	18	111
Bolivia	2,740	3.0	65	42	125	65	93
Brazil	8,230	2.2	72	21	60	33	105
Chile	11,470	4.3	78	18	21	10	98
Colombia	7,420	3.5	72	37	35	21	104
Costa Rica	9,680	4.2	79	8	18	12	104
Ecuador	4,070	5.1	75	41	57	25	
El Salvador	5,120	2.2	71	41	60	27	100
Guatemala	4,410	2.5	69	32	82	43	91
Honduras	2,900	3.6	71	44	59	40	109
Mexico	10,030	1.9	75	20	46	27	101
Nicaragua	3,650	3.0	71	80	68	37	103
Panama	7,310	4.3	75	17	34	24	109
Paraguay	4,970	2.6	71	33	41	23	101
Peru	5,830	4.3	70	32	78	27	103
Uruguay	9,810	0.9	75	5	23	15	114
Venezuela	6,440	1.3	73	28	33	21	104

[a]*Gross national income.*
[b]*Purchasing power parity.*
[c]*Gross domestic product.*
[d]*Ratio of female-to-male enrollments in primary and secondary school percentage. Numbers below 100 have more males in primary/secondary school; numbers above 100 have more females in primary/secondary schools.*
Sources: World Bank, World Development Indicators, 2007; *percentage of population living on $2 a day data from* Population Reference Bureau, World Data Sheet, 2007; *gender equity data from* Millennium Development Goals.

Multilateral agencies such as the World Bank and the Inter-American Development Bank loaned money for big development projects: continental highways, dams, mechanized agriculture, and power plants. Yet the dream of the 1960s and 1970s became a nightmare by the 1980s. Argentina's economy, which was larger than Japan's in 1960, ranked 30th in 2003, while Japan's was ranked 2nd. The economic engines of the region are the two largest countries, Brazil and Mexico. As of 2004, Mexico was the 10th largest economy in the world. Higher oil prices have boosted the Mexican economy in recent years but the country's middle class was badly shaken by debt crises and currency devaluations in the early 1980s and the late 1990s. And Brazil, the 13th largest economy in the world in 2004, is the developing world's largest debtor. Brazil also maintains the region's worst income disparity: 10 percent of the country's richest people control nearly half the country's wealth, while the poorest 40 percent control less than one tenth.

Development Strategies

Development policies of the 1960s and 1970s accelerated industrialization and infrastructural development, but they also fostered debt and rural displacement. All sectors of the economy were radically transformed. Agricultural production increased with the application of Green Revolution technology and mechanization. State-run industries reduced the need for imported goods, and the service sector ballooned as a result of new government and private jobs. Protectionist tariffs ensured that domestically manufactured goods would be consumed over imports. Yet this rush to modernize produced victims. In rural areas, poverty, landlessness, and inadequate systems of credit hindered the productivity of small farmers. As rural people were pushed off the land, there were not enough jobs in industry to absorb them, so they created their own niche in the urban informal economy. In the end, most countries made the transition from predominantly rural and agrarian economies dependent on one or two commodities to more economically diversified and urbanized countries with mixed levels of industrialization.

In the 1990s, most Latin American governments radically changed their economic development strategies. National industries and tariffs were jettisoned for a set of free-market policy reforms that came to be known as the Washington Consensus. Through tough fiscal policy, increased trade, privatization, and reduced government spending, many countries in the 1990s saw their economies grow and poverty decline. Yet a series of economic downturns from 1999 to 2002 made these neoliberal policies highly unpopular with the masses, causing major political and economic turmoil. In particular, the value of increased trade and foreign investment was criticized as benefiting only a minority of the people in the region. In the past five years, political leaders such as Venezuela's Hugo Chavez, Brazil's Lula da Silva, and Bolivia's Morales have openly challenged the Washington Consensus as a path to development. However, an alternative Latin American approach, perhaps driven by more regional trade and greater state involvement, is still being articulated.

Industrialization Since the 1960s, most government development policies have emphasized manufacturing. Various strategies have been employed, from planned industrial centers and nationalized industries to import substitution. Since the 1990s, much of the growth in manufacturing in Mexico and Central America has been driven by investment by foreign companies, especially along the U.S.–Mexican border. The results have been mixed. Today at least 25 percent of the formally employed male labor force in Argentina, Brazil, Bolivia, Chile, Costa Rica, Ecuador, El Salvador, Mexico, Peru, Uruguay, and Venezuela is employed in industry (including mining, construction, and energy). Yet this is short of the hoped-for levels of industrial manufacturing, especially when the size of urban populations is taken into account. Moreover, the most industrialized areas tend to be around the capitals or planned industrial cities, such as Ciudad Guyana in Venezuela and the Mexican border cities of Ciudad Juárez and Tijuana.

In several noncapital cities, industry has thrived without direct state support. The cities of Monterrey, Mexico, Medellín, Colombia, and São Paulo, Brazil, all developed important industrial sectors initially from local investment. Long before Medellín (2 million people) was associated with cocaine, it was a major center of textile production, and was more industrialized than the larger capital of Bogotá. A popular image of Medellín's inhabitants—the South American Yankees—surfaced in the 1950s. Celebrated as hardworking and entrepreneurial to the core, residents developed a strong sense of regional pride so that being from the department of Antioquia (where Medellín is capital) meant more than being Colombian. Similar perceptions exist for Monterrey (3 million people), a city that is not well known outside Mexico but is perceived within Mexico as innovative, resourceful, and solidly middle class.

The industrial giant of Latin America is metropolitan São Paulo in Brazil. Rio de Janeiro has greater name recognition and was the capital before Brasília was built, but it does not have the economic muscle of São Paulo. This city of 18 million, which vies with Mexico City for the title of Latin America's largest, began to industrialize in the early 1900s when the city's coffee merchants started to diversify their investments. Since then, a combination of private and state-owned industries have agglomerated around São Paulo. Within a 60-mile radius of the city center, automobiles, aircraft, chemicals, processed foods, and construction materials are produced. There are also heavy industry and industrial parks. With the port of Santos nearby and the city of Rio de Janeiro a few hours away, São Paulo is the uncontested financial center of Brazil. A stunning sight to most first-time visitors to São Paulo is the forest of high-rises that greets them, a tropical version of Manhattan, only larger.

Maquiladoras and Foreign Investment The Mexican assembly plants that line the border with the United States, called **maquiladoras**, are characteristic of manufacturing systems in an increasingly globalized economy. Nearly 4,000 maquiladoras exist, employing 1.1 million people who assemble automobiles, consumer electronics, and apparel. Between 1994 and 2000, three of every 10 new jobs in Mexico were in the maquiladoras, which account for nearly half of Mexico's exports. Maquiladora employment peaked in 2001 with 1.3 million people employed. Since 2003, China has become a favorite destination for the labor-intensive assembly work that Mexico has specialized in for the last three decades. As Mexican wages have gone up, some companies have relocated factories to East Asia. Northern Mexico is still an attractive location but competition from China and even Central America may erode Mexico's various locational and structural advantages.

The policy of border industrialization began in the 1960s, long before offshore production was widely practiced. To accelerate industrialization, the Mexican government allowed the duty-free import of machinery, components, and supplies from the United States to be used for manufacturing goods for export back to the United States. Initially, all products had to be exported, but changes in the law in 1994 now allow up to half of the goods to be sold in Mexico. The program was slow to develop, but it took off in the 1980s as foreign companies realized tremendous profits from the inexpensive Mexican labor. An autoworker in a General Motors plant in Mexico earns in a day what the same worker in the United States earns in an hour and with far fewer benefits.

Considerable controversy surrounds this form of industrialization on both sides of the border. Organized labor in the United States complains that well-paying manufacturing jobs are being lost to low-cost competitors, while environmentalists decry serious industrial pollution resulting from lax government regulation. Mexicans worry that these plants are poorly integrated with the rest of the economy and that many of the workers are young unmarried women who are easily exploited. With NAFTA, foreign-owned manufacturing plants are no longer restricted to the border zone, which may result in slower rates of industrialization for the border towns. Foreign-owned plants are increasingly being constructed near the population centers of Monterrey, Puebla, and Veracruz. However, Mexican workers and foreign corporations continue to locate in the border zone because of the unique advantages of proximity to the U.S. border (see "Cityscapes: Life on the Streets of Ciudad Juárez").

Mexico's competitive advantage is twofold: its location along the U.S. border and its membership in NAFTA. However, other Latin American states are attracting foreign companies through tax incentives and low labor costs. Assembly plants in Honduras, Guatemala, and

CITYSCAPES Life in the Streets of Ciudad Juárez

Ciudad Juárez, with more than 1 million people, is much larger than El Paso, Texas, its sister city across the border. Maquiladora-based manufacturing has profoundly changed the look, size, and rhythm of the city. The factories are east of the old town, as are most of the middle- and upper-class houses. To the west, squatters have settled in shantytowns that take up three-fifths of the city's land area (Figure 4.6.1). Every morning buses grind their way through the west side of town, slowly picking up factory workers. The trip can take up to an hour each way. The workers—young men and women—are neatly dressed. Their day of labor will earn them anywhere from $10 to $20.

Founded in 1659 as Paso del Norte on the floodplain of the Rio Grande, the city was renamed in 1888 in honor of Mexican president Benito Juárez. Yet it is the sprawling, shabby newness of the place, and not its history, that impresses the visitor. Figures tell only part of the story. It is a city of young workers with a high birthrate, but no one really knows how many people live there: estimates for the year 2000 range from 1.3 to 2.4 million, although most experts favor the lower figures. Since the 1980s, several hundred acres of housing have been added to the city each year. In the west, the roads are mostly dirt, and the houses are built of cinder blocks, adobe, and plywood. Access to electricity is spotty, connection to sewer lines rare, and water is usually trucked in by men such as Salvador Durón. "The city keeps getting bigger and bigger" he said. "There is no way to get water to everyone who needs it." Many families get by on a couple of barrels of water delivered each week by Mr. Durón's truck (Figure 4.6.2).

The east side of town is another world. There one finds large paved roads filled with buses that efficiently move workers in and out of the factories. In between the assembly plants are the nicest suburbs, with sidewalks, garages, and neatly landscaped yards. Drive down the commercial strip and stop at McDonald's, Burger King, or the 7-Eleven.

Most geographers find border zones intriguing cultural landscapes that invite the question of how a line in the sand can create two distinct faces. At this particular bend of the Rio Grande, there are three faces: El Paso to the north and the dual faces of Ciudad Juárez's west and east sides.

(a)

(b)

Figure 4.6.2 Shanty and Elite Housing in Ciudad Juárez
(a) The majority of the city's newcomers reside in self-built homes like these on the urban periphery. (b) In contrast, small elite suburbs are interspersed with assembly plants on the east side of the city. *(Rob Crandall/www.robcrandall.com)*

Figure 4.6.1 Ciudad Juárez
The core of the city sits along the Rio Grande/Bravo, while the periphery radiates out for several miles. The industrial zone, filled with maquiladoras, is east of downtown. *(Modified from Arreola and Curtis, 1993,* The Mexican Border Cities, *Tucson: University of Arizona Press, p. 39)*

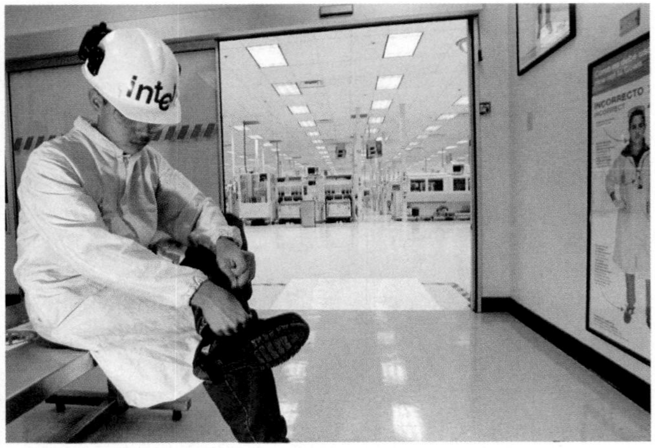

Figure 4.36 **High Tech in Costa Rica**
A Costa Rican worker straps on protective clothing before entering the manufacturing area of Intel's plant in Belen, not far from the capital city of San José. In 2000, Costa Rica earned more foreign exchange from exporting computer chips than it did from coffee, its traditional export. *(AP/Wide World Photos)*

El Salvador are drawing foreign investors, especially in the apparel industry. A recent report from El Salvador claims that not one of its apparel factories has a union. Making goods for American labels such as the Gap, Liz Claiborne, and Nike, many Salvadoran garment workers complain they do not make a living wage, work 80-hour weeks, and face mandatory pregnancy tests. With the signing of CAFTA, Central America states are hopeful that more foreign investment will flow into their countries and that wages and conditions will improve.

The situation in Costa Rica, which is now a major chip manufacturer for Intel, is quite different (Figure 4.36). With a well-educated population, low crime rate, and stable political scene, Costa Rica is now attracting other high-tech firms. Hopeful officials claim that Costa Rica is transitioning from a banana republic (bananas and coffee were the country's long-standing exports) to a high-tech manufacturing center. As a result, the Costa Rican economy averaged over 4 percent annual growth from 2000 to 2005. Uruguay is another example of a small country with a well-educated population that has recently emerged as the leader in Latin American **outsourcing** operations. Outsourcing, most commonly associated with India, is the practice of moving service jobs such as tech support, data entry, and programming to cheaper locations. Partnered with the Indian multinational company Tata, in the last few years TCS Iberoamerica in Uruguay has created the largest outsourcing operation in the region. Uruguay takes advantage of being in a similar time zone as the eastern United States. While India's top engineers sleep, Uruguayan engineers and programmers can serve their customers from Montevideo and no one is the wiser.

The Entrenched Informal Sector Even in prosperous Montevideo, Uruguay, a short drive to the urban periphery shows large neighborhoods of self-built housing filled with street traders and family-run workshops. Such activities make up the informal sector, the provision of goods and services without the benefit of government regulation, registration, or taxation. Most people in the informal economy are self-employed and receive no wages or benefits except the profits they clear. The most common informal activities are housing construction (in many cities as many as half of all residents live in self-built housing), manufacturing in small workshops, street vending, transportation services (messenger services, bicycle delivery, and collective taxis), garbage picking, street performing, and even line-waiting (Figure 4.37). These activities are legal. Illegal informal activities also exist: drug trafficking, prostitution, and money laundering, for example. The vast majority of people who rely on informal livelihoods produce legal goods and services.

No one is sure how big this economy is, in part because separating formal activities from informal ones is difficult. Visitors to Lima, Belém, Guatemala City, or Guayaquil could easily get the impression that the informal economy *is* the economy. From self-help housing that dominates the landscape to hundreds of street vendors that crowd the sidewalks, it is impossible to avoid. There are advantages in the informal sector—hours are flexible, children can work with their parents, and there are no bosses. Peruvian economist Hernando de Soto even argues that

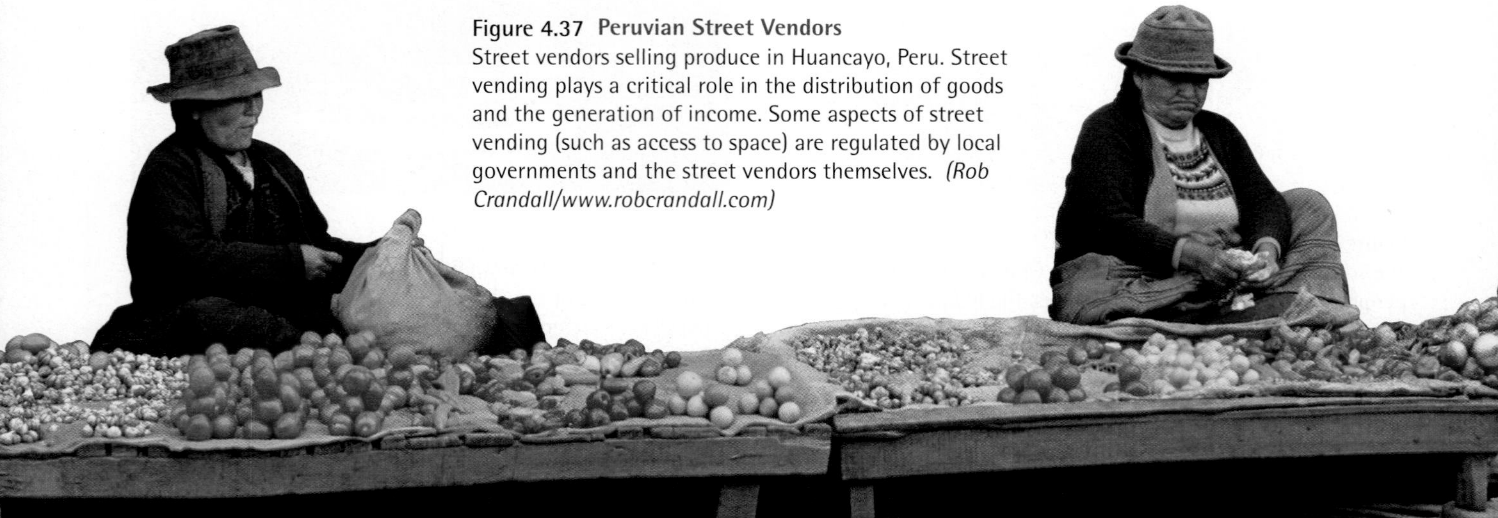

Figure 4.37 **Peruvian Street Vendors**
Street vendors selling produce in Huancayo, Peru. Street vending plays a critical role in the distribution of goods and the generation of income. Some aspects of street vending (such as access to space) are regulated by local governments and the street vendors themselves. *(Rob Crandall/www.robcrandall.com)*

this most dynamic sector of the economy should be encouraged and offered formal lines of credit. As important as this sector may be, however, widespread dependence on it signals Latin America's poverty, not its wealth. It reflects the inability of the formal economies of the region, especially in industry, to absorb labor. For millions of urban dwellers, formal employment that offers benefits, safety, and a living wage is still a dream. With nowhere else to go, the ranks of the informally employed continue to grow.

Primary Export Dependency

Historically, Latin America's abundant natural resources were its wealth. In the colonial period, silver, gold, and sugar generated tremendous riches for the colonists. With independence in the nineteenth century, a series of export booms introduced to an expanding world market commodities such as bananas, coffee, cacao, grains, tin, rubber, copper, wool, and petroleum. One of the legacies of this export-led development was a tendency to specialize in one or two major commodities, a pattern that continued into the 1950s. During that decade, Costa Rica earned 90 percent of its export earnings from bananas and coffee; Nicaragua earned 70 percent from coffee and cotton; 85 percent of Chilean export income came from copper; half of Uruguay's export income came from wood. Even Brazil generated 60 percent of its export earnings from coffee in 1955; by 2000, coffee accounted for less than 5 percent of the country's exports, yet Brazil remained the world leader in coffee production.

The economies of the region have industrialized and diversified. Since the 1990s, however, an increased demand from East Asia for primary exports (from fossil fuels to metals and soybeans) have driven up commodity prices, creating another boom in primary export commodities. While many Latin America states are riding this wave, from oil-rich Venezuela to copper-laden Chile, there is concern that Latin America may once again become too reliant on its bountiful natural resources.

Agricultural Production The trend in Latin America has been to diversify and to mechanize agriculture. Nowhere is this more evident than in the Plata Basin, which includes southern Brazil, Uruguay, northern Argentina, Paraguay, and eastern Bolivia. Soybeans, used for oil and animal feed, transformed these lowlands in the 1980s and 1990s. Brazil is now the second largest producer of soy in the world (following the United States), and Argentina is the third largest. Between 2000 and 2003, soy production nearly doubled in both countries. In addition, there are acres of rice, cotton, and orange groves, as well as the more traditional plantings of wheat and sugar. The speed with which the shields are being converted into soy fields alarms many but with soy prices high, the rush to plant continues (Figure 4.38).

Similar large-scale agricultural frontiers exist along the piedmont zone of the Venezuelan Llanos (mostly grains), the Pacific slope of Central America (cotton and some tropical fruits), and the Central Valley of Chile and the foothills of Argentina (wine and fruit production). In northern Mexico, water supplied from dams along the Sierra Madre Occidental has turned the valleys in Sinaloa into intensive producers of fruits and vegetables for consumers in the United States. The relatively mild winters in northern Mexico allow growers to produce strawberries and tomatoes during the winter months.

Figure 4.38 Soy Production
A worker sprays a field of soybeans with fertilizer in São Paulo State, Brazil. This field is an experimental field planted with genetically modified seeds. Brazil is now the second largest producer of soy in the world, following the United States. *(John Maier, Jr./The Image Works)*

In each of these cases, the agricultural sector is capital-intensive and dynamic. By using machinery, high-yielding hybrids, chemical fertilizers, and pesticides, many corporate farms are extremely productive and profitable. What these operations fail to do is employ many rural people, which is especially problematic in countries where a third or more of the population depends on agriculture for its livelihood. As industrialized agriculture becomes the norm in Latin America, subsistence peasant producers are further marginalized. The overall trend is for agricultural production to increase while employing proportionally fewer people. In Peru, 45 percent of the male labor force worked in agriculture in 1980; by 2002 the figure had plummeted to just 11 percent. In absolute terms, however, the number of people living in rural areas is about the same as it was in 1960 (roughly 125 million). Thus the modernization of agriculture has left behind many subsistence producers who make up the ranks of Latin America's most impoverished people.

Mining and Forestry The exploitation of silver, zinc, copper, iron ore, bauxite, and gold is an economic mainstay for many countries in the region. Like agriculture, mining has become more mechanized and less labor intensive. Even Bolivia, a country long dependent on tin production, cut 70 percent of its miners from the payrolls in the 1990s. The measure was part of a broad-based austerity program, yet it suggests that the majority of the miners were not needed. Similarly, the vast copper mines of northern Chile are producing record amounts of copper with few miners. In contrast, gold mining continues to be labor intensive, offering employment for thousands of prospectors. Gold rushes are occurring in remote tropical regions of Venezuela, Brazil, Colombia, and Costa Rica. The famous Serra Pelada mine in Brazil, an anthill of workers and ladders, shovels, and burlap sacks filled with gold-bearing soil, is the most dramatic example of labor-intensive mining.

Logging is another important, and controversial, extractive activity. Ironically, many of the forest areas cleared for cattle were not systematically harvested. More often than not, all but the most valuable trees were burned. Logging concessions are commonly awarded to domestic and foreign timber companies, which export boards and wood pulp. These one-time arrangements are seen as a quick means for foreign exchange, particularly if prized hardwoods like mahogany are found. Logging can mean a short-term infusion of cash into a local economy. Yet rarely do long-term conservation strategies exist, making this system of extraction unsustainable. There is growing interest in certification programs that designate wood products that have been produced sustainably. This is due to consumer demand for certified wood, mostly in Europe. Unfortunately, such programs are small and the lure of profit usually overwhelms the impulse to conserve for future generations.

Several countries rely on plantation forests of introduced species of pines, teak, and eucalyptus to supply domestic fuelwood, pulp, and board lumber. These plantation forests grow single species and fall far short of the complex ecosystems occurring in natural forests. Nonetheless, growing trees for paper or fuel reduces the pressure on other forested areas. Leaders in plantation forestry are Brazil, Venezuela, Chile, and Argentina. Considered Latin America's economic star in the 1990s, Chile relied on timber and wood chips to boost its export earnings. Thousands of hectares of nonnative trees (eucalyptus and pine) have been planted, systematically harvested, cut into boards, or chipped for wood pulp (see Figure 4.4). Japanese capital is heavily involved in this sector of the Chilean economy. The recent expansion of the wood chip business, however, has led to a dramatic increase in the logging of native forests.

The Energy Sector The oil-rich nations of Venezuela, Mexico, and Ecuador are able to meet their own fuel needs and to earn vital state revenues from oil exports. In fact, Venezuela and Ecuador were original members of OPEC, along with several Arab states. Venezuela is most dependent on revenues from oil, earning up to 90 percent of its foreign exchange from crude petroleum and petroleum products. In 2002, the United States imported as much oil from Venezuela and Mexico as it did from the Arab world (Figure 4.39).

Figure 4.39 Venezuelan Oil and Gas Production
Lake Maracaibo in western Venezuela is the oldest area of oil and gas production in the country. Below the surface of the shallow lake are some of Venezuela's largest reserves of fossil fuels. *(Rob Crandall/www.robcrandall.com)*

Venezuela and Bolivia have the largest reserves of natural gas in the region. A network of gas pipelines links Bolivia's lowlands to southern Brazil and Argentina. In 2006, Bolivia's President Morales nationalized the country's oil and gas industry in an effort to retain more of the industry's profits. Brazilian and Spanish firms had built much of the network, and in the end were forced to accept new contracts that were less favorable but still profitable. A new network of pipelines from Venezuela's gas fields in the Llanos through the Brazilian Amazon has also been proposed. Yet this idea has been deemed ecologically disruptive and too expensive. As demands for liquefied gas grow, especially from China, both Peru and Bolivia seek to fill this demand through pipelines and a dedicated port on the Pacific.

In the area of biofuels, Brazil offers a story of sweet success. In the 1970s when oil prices skyrocketed, oil-poor Brazil decided to convert its abundant sugarcane harvest into ethanol. It also induced its car industry to build cars that run on ethanol. Over the years, even when oil prices plummeted, Brazil continued to invest in ethanol, building mills and a distribution system that delivered ethanol to gas stations. One of Brazil's major technological successes was inventing flex-fuel cars that run on any combination of ethanol and gasoline. At the time, Brazil was motivated by its limited oil reserves but today, with interest in biofuels growing as a way to reduce CO_2 emissions, Brazil's support of its ethanol program looks visionary. In 2007, Brazil and the United States signed an agreement to spread the production and consumption of ethanol globally.

Latin America in the Global Economy

In order to conceptualize Latin America's place in the world economy, scholars from the region advanced **dependency theory** in the 1960s. The premise of the theory is that expansion of European capitalism created the region's underdevelopment. For the developed "cores" of the world to prosper, the "peripheries" became dependent and impoverished. Dependent economies, as those in Latin America, were export-oriented and vulnerable to fluctuations in the global market. Even when they experienced economic growth, it was subordinate to the economic demands of the core (North America and Europe).

Economists who embraced this interpretation of Latin America's history were convinced that economic development could occur only through self-sufficiency, growth of internal markets, agrarian reform, and greater income equality. In short, they argued for vigorous state intervention and an uncoupling from the economic cores. Policies such as import substitution, industrialization, and nationalization of key industries were partially influenced by this view. Dependency theory has its detractors. In its simplest form, it becomes a way to blame forces external to Latin America for the region's problems. Implicit in dependency theory is also the notion that the path to development taken by Europe and North America cannot be easily replicated. This was a radical idea for its time.

Latin America's century-long dependence on North America, especially the United States, as its major trading partner is still evident. However, growing trade with, as well as investment from, Europe and East Asia, suggests that a more complex and less U.S.-dependent pattern of trade is emerging. Latin America is linked to the world economy in ways other than trade. Figure 4.40 shows the changes in foreign direct investment (FDI) as a percentage of GDP from 1990 to 2004. For nearly every country in the region, the value of foreign investment in terms of a percentage of GDP went up in the 1990s. In terms of absolute amount of money invested, the two largest economies in the region, Brazil and Mexico, received the most foreign investment. In 1990, Brazil's FDI was less than $1 billion and Mexico's was $2.5 billion. By 2004, FDI in Brazil was valued at $18 billion and Mexico's was $17 billion.

Remittances are another important financial flow that reflects the integration of Latin America migrants into labor markets around the world, especially North America. In 2006, the value of remittances to Latin America was estimated at $60 billion. Mexico is the regional leader, receiving over $23 billion in remittance income (which is equal to over $200 per capita). Brazil was a distant second at $7 billion. The value of remittances began to soar in the late 1990s as many households saw their personal income decline and unemployment rise. To cope with these changes, family members abroad began to send more financial resources home. Reliance on remittances is an indicator of globalization from below. Rather than corporate-based foreign investment, millions of individual migrants are transferring money (often $200 to $300 a month) to support families and communities left behind. Scholars debate whether this flow of capital can actually lead to sustained development or if it is simply a survival strategy of last resort. The economic impact of remittances shown on a per capita basis is real (see Figure 4.40). El Salvador, a county of 7 million people, received over $3.3 billion in remittances in 2006, which is nearly $500 per capita. For many Latinos, remittances are the surest way to poverty alleviation, although they must depend upon an international migration system that is constantly changing and includes both legal and illegal channels of movement.

Neoliberalism as Globalization By the 1990s governments and the World Bank had become champions of neoliberalism as a sure path to economic development. **Neoliberal policies** stress privatization, export production, direct foreign investment, and few restrictions on imports. They epitomize the forces of globalization by turning away from policies that emphasize state intervention and self-sufficiency. Most Latin American political leaders are embracing neoliberalism and the benefits that come with it, such as increased trade and more favorable terms for debt repayment. Yet there are signs of discontent with neoliberalism throughout the region. Recent protests in Peru and Bolivia reflect the popular anger against trade policies that seem to benefit only the elite.

Chile is an outspoken champion of neoliberalism. Its average annual growth rate between 2000 and 2005 was 4.3 percent, one of the region's healthiest. In 1995 alone, however, the Chilean economy grew 10.4 percent, placing it in the same league as the Asian tigers. Consequently, it is the most studied and watched country in Latin America. By the numbers, Chile's 17 million people are doing well, but the country's accomplishments are not readily transferable. For example, the radical move to privatize state-owned business and open the economy occurred under an oppressive military dictatorship that did not tolerate opposition. Such dictatorships are rare in Latin America today, but in a democracy the pace of reform must be slower. Much of Chile's export-led growth has been based on primary products: fruits, seafood, copper, and wood. Although many of these products are renewable, Chile will need to develop more value-added goods before it will be labeled "developed." Furthermore, its relatively small and homogeneous population in a resource-rich land does not have the same ethnic divisions that hinder so many states in Latin America. Although neoliberalism has worked in the Chilean case, for many Latin American states the social and environmental disruptions associated with neoliberal policies have led to political upheaval, and an active search for alternatives is underway.

Dollarization As financial crises spread through Latin America in the late 1990s, governments began to consider the economic benefits of **dollarization**, a process by which a country adopts—in whole or in part—the U.S. dollar as its official currency. In a totally dollarized economy, the U.S. dollar becomes the only medium of exchange and the country's national currency ceases to exist. In 2000, Ecuador took this radical step to address the dual

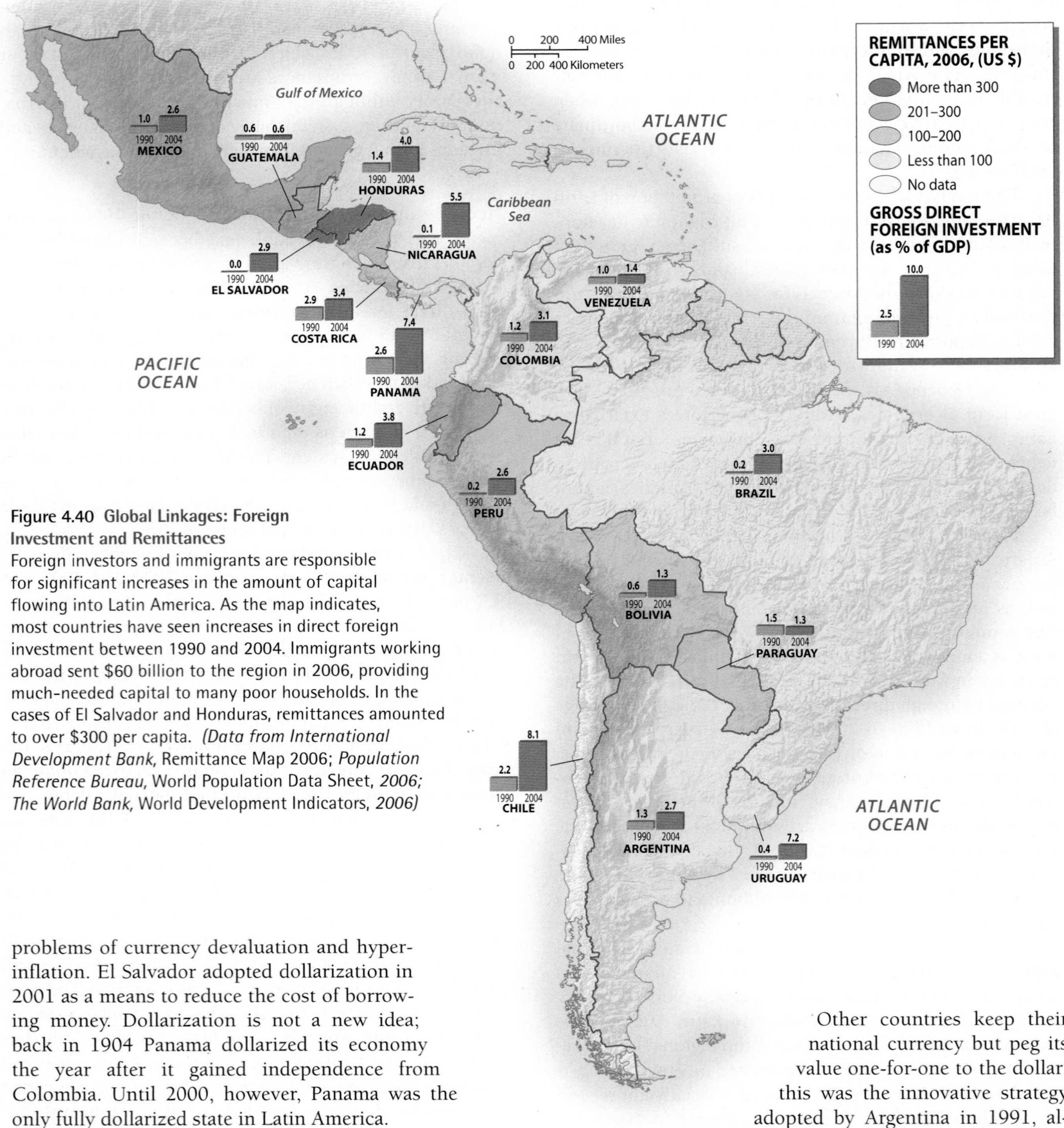

Figure 4.40 Global Linkages: Foreign Investment and Remittances
Foreign investors and immigrants are responsible for significant increases in the amount of capital flowing into Latin America. As the map indicates, most countries have seen increases in direct foreign investment between 1990 and 2004. Immigrants working abroad sent $60 billion to the region in 2006, providing much-needed capital to many poor households. In the cases of El Salvador and Honduras, remittances amounted to over $300 per capita. *(Data from International Development Bank,* Remittance Map 2006; *Population Reference Bureau,* World Population Data Sheet, *2006; The World Bank,* World Development Indicators, *2006)*

problems of currency devaluation and hyperinflation. El Salvador adopted dollarization in 2001 as a means to reduce the cost of borrowing money. Dollarization is not a new idea; back in 1904 Panama dollarized its economy the year after it gained independence from Colombia. Until 2000, however, Panama was the only fully dollarized state in Latin America.

A more common strategy in Latin America is limited dollarization, in which U.S. dollars circulate and are used alongside the country's national currency. Limited dollarization exists in many countries around the world, but most notably in Latin America. Since the economies of Latin America are prone to currency devaluation and hyperinflation, limited dollarization is a type of insurance. Many banks in Latin America, for example, allow customers to maintain accounts in dollars to avoid the problem of capital flight should a local currency be devalued.

Other countries keep their national currency but peg its value one-for-one to the dollar; this was the innovative strategy adopted by Argentina in 1991, although it led to a serious financial crisis in 2001 and was eventually stopped. Dollarization, partial or full, tends to reduce inflation, eliminate fears of currency devaluation, and reduce the cost of trade by eliminating currency conversion costs.

Dollarization has its drawbacks. The obvious one is that a country no longer has control of its monetary policy, making it reliant on the decisions of the U.S. Federal Reserve. Foreign governments do not have to ask permission to dollarize their economies. At the same time, the

United States insists that all its monetary policies be based exclusively on domestic considerations, regardless of the impact such decisions may have on foreign countries. The political impact of eliminating a national currency is serious. The case of Ecuador is instructive. In 1999, when President Jamil Mahuad announced his plan to dollarize the economy to head off hyperinflation, he was quickly forced out of office by a coalition of military and Indian activists. When Vice President Gustavo Naboa became president and the economic situation worsened, the country's political leadership went ahead with dollarization. In short, dollarization may help in a time of economic duress, but is not a popular policy.

Social Development

Over the past three decades Latin America has experienced marked improvements in life expectancy, child survival, and educational equality. One telling indicator is the steady decline in mortality rate for children below the age of five between 1990 and 2005 (Table 4.2). In 1990, Brazil had 60 deaths per 1,000 children under age five; in 2005, the number was 33. Nicaragua dropped from 66 deaths per 1,000 in 1990 to 37 in 2005. This indicator is important because when children younger than five years of age are surviving, it suggests that basic nutritional and health-care needs are being met. One can also infer that resources are being used to sustain women and their children. By comparison, the United States has a mortality rate of 7 per 1,000 children under age five, and Japan's rate is 4. Most other developing countries have much higher under-five death rates than Latin America; India's is 74 per 1,000 and Kenya's is 120. Despite economic downturns, the region's social networks have been able to mitigate some of the negative effects on children.

A combination of government policies and grassroots and nongovernmental organizations (NGOs) play a fundamental role in contributing to social well-being. In the past few years the Brazilian poverty reduction program, *Bolsa Família*, has grown to benefit over 46 million people. Families that qualify receive a monthly check from the state but are required to keep their children in school and take them to clinics for health check-ups. Such conditional cash transfer programs have the immediate impact of reducing extreme poverty and the long-term hope of improving educational attainment and health care. Mexico has adopted a similar program. For states with far fewer resources than Brazil and Mexico, international humanitarian organizations, church organizations, and community activists provide many services that state and local governments cannot. Catholic Relief Services and Caritas, for example, work with the rural poor throughout the region to improve their water supplies, health care, and education. Other groups lobby local governments to build schools or recognize squatters' claims. Grassroots organizations also develop cooperatives that market anything from sweaters to cheeses. Cooperative organizations are able to realize some economies of scale as well as improve access to credit.

Other important gauges for social development are life expectancy, gender educational equity, and access to improved water sources. In aggregate, 84 percent of the people in the region have access to an adequate amount of water from an improved source; slightly more girls receive education than boys; and life expectancy (men and women) is 72 years (see Table 4.2). Masked by this aggregate data are extreme variations between rural and urban areas, between regions, and along racial and gender lines. Despite the difficulty of city life, the poorest people in Latin America are in rural areas. The move to cities for many people is a development strategy.

A running joke in Latin America is that the economy is doing well but the people are not. Even today, when Chile or Costa Rica is heralded for making impressive economic gains, fear persists that a large segment of the population is excluded. The 1980s were dubbed the "lost decade" for Latin America due to the debt crisis, inflation, and slow-to-negative growth rates. During that period, several countries witnessed setbacks in both their economic and social indicators—most notably Peru, Bolivia, Nicaragua, Venezuela, Mexico, and Brazil. Many economic indicators improved in the 1990s and stabilized in the first years of the twenty-first century.

Within Mexico and Brazil tremendous internal differences exist in socioeconomic indicators (Figure 4.41). The northeastern part of Brazil lags behind the rest of the country in every social indicator. The country has a literacy rate of over 85 percent, but in the northeast it is only 60 percent. Moreover, within the northeast, literacy for city residents is 70 percent, but for rural residents is only 40 percent. In Mexico, the levels of poverty are highest in the more Indian south. In contrast, Mexico City, the states of Nuevo Leon (Monterrey is the capital), and Quintana Roo (home to Mexico's largest resort, Cancun) and Campeche have the highest GDP per capita. Noting this north–south economic divide, President Calderón explained, "there is one Mexico more like North America and another Mexico more like Central America. It is a very clear challenge for me to make them more alike." All countries have spatial inequities regarding income and availability of services, but the contrasts tend to be sharper in the developing world. In the cases of Mexico and Brazil, it is hard to ignore ethnicity and race when trying to explain these patterns.

Race and Inequality There is much to admire about race relations in Latin America. The complex racial and ethnic mix that was created in Latin America fostered tolerance for diversity. However, Indians and blacks are disproportionately represented among the poor of the region. More than ever, racial discrimination is a major political issue in Brazil. Headlines report organized killings of street children, most of them Afro-Brazilian. For decades, Brazil put forward its vision of a color-blind racial democracy. True, residential segregation by race is rare in Brazil and interracial marriage is common, but certain patterns of social and economic inequity seem best explained by race.

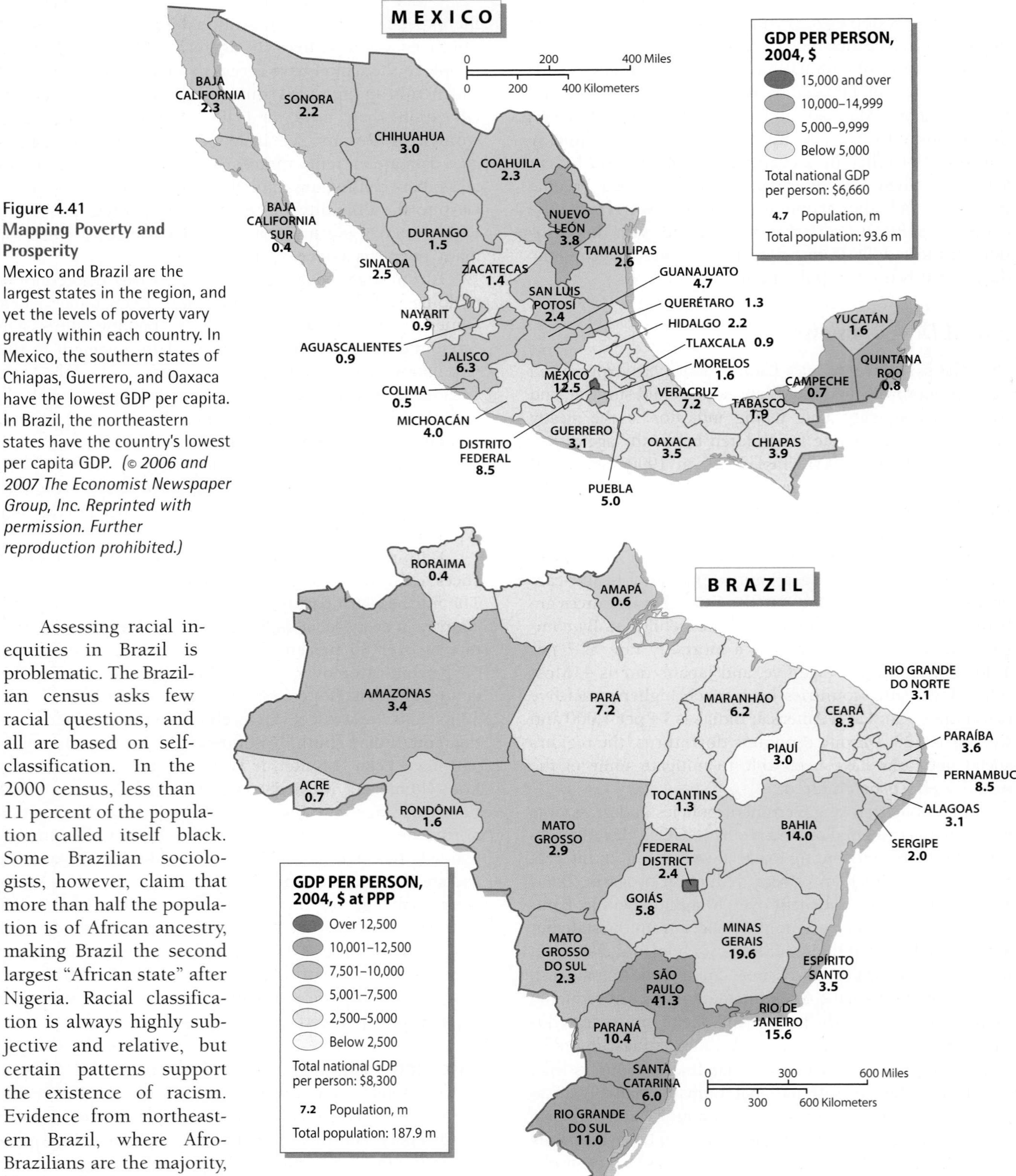

Figure 4.41
Mapping Poverty and Prosperity
Mexico and Brazil are the largest states in the region, and yet the levels of poverty vary greatly within each country. In Mexico, the southern states of Chiapas, Guerrero, and Oaxaca have the lowest GDP per capita. In Brazil, the northeastern states have the country's lowest per capita GDP. *(© 2006 and 2007 The Economist Newspaper Group, Inc. Reprinted with permission. Further reproduction prohibited.)*

Assessing racial inequities in Brazil is problematic. The Brazilian census asks few racial questions, and all are based on self-classification. In the 2000 census, less than 11 percent of the population called itself black. Some Brazilian sociologists, however, claim that more than half the population is of African ancestry, making Brazil the second largest "African state" after Nigeria. Racial classification is always highly subjective and relative, but certain patterns support the existence of racism. Evidence from northeastern Brazil, where Afro-Brazilians are the majority, shows death rates approaching those of some of the world's poorest countries. Throughout Brazil, blacks suffer higher rates of homelessness, landlessness, illiteracy, and unemployment. The last few years have seen revolution in Brazilian society as affirmative action measures have been implemented. From federal ministries to public universities, various quota systems are being tried to improve the condition of Afro-Brazilians. After years of denying that there was a race problem, it seems that Brazil is now trying to address it. In areas of Latin America where Indian cultures are strong, one also finds low socioeconomic indicators. In most countries, areas where native languages are widely spoken invariably correspond with areas of persistent poverty. In Mexico, the Indian south lags behind the booming north and Mexico City. Prejudice is embedded in the language. To call someone an *indio* (Indian) is an insult in Mexico. In Bolivia, women who dress in the Indian style of full, pleated skirts and bowler hats are called *cholas*, a descriptive term referring to the rural *mestizo* population that suggests backwardness and even cowardice. No one of high social standing, regardless of skin color, would ever be called a *chola* or *cholo*.

It is difficult to separate status divisions based on class from those based on race. From the days of conquest, being European meant an immediate elevation in status over the Indian, African, and mestizo populations. Class awareness is very strong. Race does not necessarily determine one's economic standing, but it certainly influences it. Most people recognize the power of the elite and envy their lifestyle.

An emerging middle class also exists that is formally employed, aspires to own a home and car, and strives to give its children a university education. The vast majority of people, however, are the working poor who struggle to meet basic food, shelter, clothing, and transport needs. These class differences are evident in the landscape. Go to any large Latin American city and find handsome suburbs, country clubs, and trendy shopping centers. High-rise luxury apartment buildings with beautiful terraces offer all the modern amenities, including maids' quarters. The elite and the middle class even show a preference for decentralized suburban living and dependence on automobiles, as do North Americans. Yet near these same residences are shantytowns where urban squatters build their own homes, create their own economy, and eke out a living.

The Status of Women The status of women in Latin America is a web of contradictions. The gender stereotypes of *machismo* and *marianismo* are breaking down in the younger generations. Many Latina women work outside the home. In most countries the formal figures hover between 30 and 40 percent of the workforce, not far off from many European countries, but lower than in the United States. Legally speaking, women can vote, own property, and sign for loans, although they are less likely than men to do so, which reflects the patriarchal tendencies in the society. Even though Latin America is predominantly Catholic, divorce is legal and family planning is promoted. In most countries, however, abortion remains illegal.

Overall, access to education in Latin America is good compared to other developing regions, and illiteracy rates tend to be low. In terms of gender parity, there are slightly more girls than boys getting primary and secondary educations. Throughout higher education in Latin America today, male and female students are equally represented. Consequently, in the fields of education, medicine, and law, women are regularly employed.

The biggest changes for women are the trends toward smaller families, urban living, and educational parity with men. These factors have greatly improved the participation of women in the labor force. And in nearly every country, women's participation in the labor force has climbed by at least 10 percent since 1980. In the countryside, however, serious inequities remain. Rural women are less likely to be educated and tend to have larger families. In addition, they are often left to care for their families alone as husbands leave seasonally in search of employment. In most cases, the conditions and prejudices facing rural women have been slow to improve.

Women play an increasingly active role in Latin American politics. A Mayan woman from Guatemala, Rigoberta Menchú, won the Nobel Peace Prize in 1992 for denouncing human rights abuses in her native country. In 1990 Nicaragua elected the first woman president in Latin America, Violeta Chamorro, the owner of an opposition newspaper. Nine years later, Panamanians voted Mireya Moscoso into power. And, in 2005 Chileans elected Dr. Michelle Bachelet. A pediatrician and single mother, one of President Bachelet's first acts in office was to extend the availability of state-subsidized child care (Figure 4.42). Women are also active organizers and participants in cooperatives, microenterprises, and unions. In a relatively short period, urban women have won a formal place in the economy and a political voice. Moreover, evidence suggests that this trend will continue, reflecting an improved status for women in the region.

Figure 4.42 **President Bachelet of Chile**
The first woman to be elected president in South America, Dr. Bachelet campaigned as a moderate socialist, emphasizing the need for expanded social services as well as pro-growth fiscal policies. *(Victor Rojas/AFP/Getty Images, Inc.—Agence France Presse)*

Summary

- Latin America and the Caribbean were the first world regions to be fully colonized by Europe. In the process, perhaps 90 percent of the native population died from disease, cruelty, and forced resettlement. The slow demographic recovery of native peoples and the continual arrival of Europeans and Africans resulted in an unprecedented level of racial and cultural mixing.
- Unlike other developing areas, most Latin Americans live in cities. This shift started early and reflects a cultural bias toward urban living with roots in the colonial past. The cities are large and combine aspects of the formal industrial economy with an informal one. The region is home to four megacities of over 10 million people; these are Mexico City, Buenos Aires, Rio de Janeiro and Sao Paulo.
- Compared to Europe and Asia, this region is still rich in natural resources and relatively lightly populated. Yet as populations continue to grow and trade in natural resources increases, there is a growing concern for the state of the environment. Of particular concern is the relentless cutting of tropical rainforests, unhealthy urban air pollution, and water shortages that plague both city dwellers and farmers alike.
- Uneven development and economic frustration have led many Latin Americans (both highly skilled and low skilled) to consider emigration as an economic strategy. Today, Latin America is a region of emigration, with migrants going to North America, Europe, and Japan to work. Collectively they send billions of dollars in remittances back to Latin America each year.
- Latin American governments were early adopters of neoliberal economic policies. While some states prospered, others faltered, sparking popular protests against the negative effects of globalization. Out of this frustration new political actors are emerging—from indigenous leaders to women—who are challenging old ways of doing things.

Key Terms

agrarian reform *(page 144)*
Altiplano *(page 134)*
altitudinal zonation *(page 139)*
Central American Free Trade Association (CAFTA) *(page 126)*
Columbian Exchange *(page 151)*
dependency theory *(page 170)*
dollarization *(page 171)*
El Niño *(page 139)*
environmental lapse rate *(page 139)*
Fair Trade *(page 131)*
grassification *(page 131)*
informal sector *(page 143)*
latifundia *(page 144)*
maquiladora *(page 166)*
megacity *(page 128)*
Mercosur *(page 157)*
mestizo *(page 146)*
minifundia *(page 144)*
neoliberal policies *(page 171)*
Organization of American States (OAS) *(page 157)*
outsourcing *(page 168)*
remittances *(page 149)*
rural-to-urban migration *(page 144)*
shields *(page 134)*
shifting cultivation *(page 144)*
subnational organizations *(page 160)*
supranational organizations *(page 160)*
syncretic religions *(page 153)*
transnationalism *(page 149)*
Treaty of Tordesillas *(page 157)*
urban primacy *(page 142)*

Questions for Review

1. Why is this region called *Latin America*?
2. Discuss the different environmental problems facing Mexico City. What special problems must megacities, such as Mexico City, have to contend with?
3. What is altitudinal zonation, and how is it relevant to agricultural practices and global climate change?
4. What was the Columbian Exchange? Discuss ways in which it still occurs today.
5. What are the origins of latifundia? Discuss the relationship between latifundia and minifundia.
6. What is the role of coca production in the region? Where is it produced and how has it impacted Latin American states?
7. Discuss the origins of the current political boundaries in Latin America. Where are boundary disputes occurring and why?
8. How are trade blocks reshaping international relations in Latin America?
9. What is dollarization, and how is it being used in Latin America?
10. What social and demographic shifts account for improvements in the lives of women in Latin America?

Thinking Geographically

1. Discuss the processes driving tropical deforestation in Latin America and how they compare with deforestation in other areas of the world.
2. Agrarian reform has been advocated throughout Latin America as a means to reduce social and economic inequalities. Has it worked in the region? How have land distribution programs fared? Where else in the world have agrarian reforms been attempted?

3. How is neoliberalism influencing the way Latin America interacts with the rest of the world? What are the social and environmental costs of neoliberalism? Is this model appropriate for understanding the impact of globalization in the developing world?
4. After examining the language map, what conclusions can you draw about the patterns of Indian survival in Latin America?
5. Given the dominance of cities in this region, describe the particular urban environmental problems facing cities in the developing world. How might Latin America's megacities use their size and density to reduce the environmental problems associated with urbanization?
6. Participatory mapping is being used by indigenous peoples and conservation groups throughout Latin America. Who gains from participatory mapping. How might participatory mapping be used in other developing regions of the world? Could it be useful in urban settings as well as rural ones?
7. Discuss the social, environmental, and economic consequences behind the modernization of agriculture. How does Latin America's experience with modern agricultural systems compare with North America's?
8. In this multiracial region, what is the importance of race in assessing social and economic development? How have Amerindian groups fared in comparison to Afro-Latino communities?
9. Latin America is a region of emigration. Discuss the impact mass emigration has had on this region. Why is it higher here than in other regions of the developing world? Which countries are producing the most immigrants and where are they going?
10. Is the dependency theory useful in understanding Latin America's position in the world economy? Can the theory be applied to other regions in the developing world? Explain.

Regional Novels and Films

Novels

Isabel Allende, *House of the Spirits* (1982, Barcelona Plaza & Janes)

Jorge Amado, *Dona Flor and Her Two Husbands* (1967, Losada)

Jorge Amado, *Gabriela, Clove and Cinnamon* (1958, Biblioteca Ayacucha)

Mario Bencastro, *Odyssey to the North* (1998, Arte Público)

Jorge Luis Borges, *Labyrinths* (1962, New Directions)

Carlos Fuentes, *The Old Gringo* (1985, Fondo de Cultura Economica)

Gabriel García Marquez, *Love in the Time of Cholera* (1988, Penguin)

Gabriel García Marquez, *One Hundred Years of Solitude* (1967, Biblioteca Ayacucha)

Edmundo Paz Soldán, *The Matter of Desire* (2004, Mariner Books)

Mario Vargas Llosa, *The Real Life of Alejandro Mayta* (1986, Farrar, Straus & Giroux)

Films

Black Orpheus (1958, Brazil)

City of God (2002, Brazil)

The Devil's Miner (2005, U.S./Germany)

El Norte (1983, U.S.)

Geraldo Off-Line (2000, U.S.)

Like Water for Chocolate (1993, U.S.)

Maria Full of Grace (2004, Colombia)

Traffic (2000, U.S.)

Transnational Fiesta (1992, U.S.)

Trinkets & Beads (1996, U.S.)

Bibliography

Clawson, David L. 2000. *Latin America and the Caribbean: Lands and Peoples*. 2nd ed. Boston: McGraw-Hill.

Crosby, Alfred. 1972. *The Columbian Exchange: Biological and Cultural Consequences of 1492*. Westport, CT: Greenwood Press.

Denevan, William M. 1992. *The Native Population of the Americas in 1492*. 2nd ed. Madison: University of Wisconsin Press.

Ezcurra, Exequiel, Mazari-Hiriart, Marisa, Pisanty, Irene, and Aguilar, Adrián Guillermo. 1999. *The Basin of Mexico: Critical Environmental Issues and Sustainability*. Tokyo: United Nations Press.

Gilbert, Alan. 1998. *The Latin American City*. 2nd ed. New York: Monthly Review Press.

Hecht, Susanna, and Cockburn, Alexander. 1989. *The Fate of the Forest: Developers, Destroyers and Defenders of the Amazon*. London: Verson Press.

Kent, Robert. 2006. *Latin America: Regions and People*. New York: Guilford Press.

Roberts, J. Timmons, and Thanos, Nikki Demetria. 2003. *Trouble in Paradise: Globalization and Environmental Crises in Latin America*. New York: Routledge.

Voeks, Robert. 1997. *Sacred Leaves of Candomblé: African Magic, Medicine and Religion in Brazil*. Austin: University of Texas Press.

Zimmerer, Karl. 1996. *Changing Fortunes: Biodiversity and Peasant Livelihood in the Peruvian Andes*. Berkeley: University of California Press.

Additional bibliographic resources are at the *Diversity Amid Globalization* Website: http://www.prenhall.com/rowntree.

5
The Caribbean

An aerial view of Marina Cay, British Virgin Islands in the Lesser Antilles. Many Caribbean island economies rely upon their scenic beauty and warm climate to attract foreign tourists. *(LAIF/Aurora & Quanta Productions)*

ENVIRONMENTAL GEOGRAPHY
Climate change threatens the Caribbean with the potential for stronger and more frequent hurricanes, loss of territory due to sea level rising, and destruction of coral reefs.

POPULATION AND SETTLEMENT
Large numbers of Caribbean peoples have emigrated from the region, leaving in search of economic opportunity and sending back millions of dollars in remittances.

CULTURAL COHERENCE AND DIVERSITY
Creolization, the blending of African, European, and even Amerindian elements, has resulted in many unique Caribbean expressions of culture, such as rara and steel drum bands.

GEOPOLITICAL FRAMEWORK
The first area of the Americas to be extensively explored and colonized by Europeans, the region has seen many rival European claims and, since the early twentieth century, has experienced strong U.S. influence.

ECONOMIC AND SOCIAL DEVELOPMENT
Environmental, locational, and economic factors make tourism a vital component of this region's economy, particularly in Puerto Rico, Cuba, the Dominican Republic, and the Bahamas.

The Caribbean was the first region of the Americas to be extensively explored and colonized by Europeans. Yet its modern regional identity is unclear, often merged with Latin America but also viewed as apart from it (see "Setting the Boundaries"). Today the region is home to over 41 million inhabitants scattered across 26 countries and dependent territories. They range from the small British dependency of the Turks and Caicos, with less than 12,000 people, to the island of Hispaniola, with 18 million. In addition to the Caribbean islands, Belize of Central America and the three Guianas—Guyana, Suriname, and French Guiana—of South America are included as part of the Caribbean. For historical and cultural reasons, the peoples of these mainland states identify with the island nations and are thus included in this chapter (Figure 5.1).

Historically, the Caribbean was a battleground between rival European powers competing for territorial control of these tropical lands. In the early 1900s, the United States took over as the dominant geopolitical force in the region, regularly sending troops and maintaining what some have called a neocolonial presence. As in many developing areas, external control of the Caribbean produced highly dependent and inequitable economies. The plantation was the dominant production system and sugar the leading commodity. Over time other products began to have economic importance, as did the international tourist industry. Increasingly, governments have sought to diversify their economies by expanding nontraditional exports (such as flowers, nuts, and processed fruits), banking services, and manufacturing to reduce the region's dependence on agriculture and tourism.

Generally, when one thinks of the Caribbean, images of white sandy beaches and turquoise tropical waters come to mind. Tucked between the Tropic of Cancer and the equator, with year-round temperatures averaging in the high 70s, the hundreds of islands and picturesque waters of the Caribbean often have inspired comparisons to paradise. Columbus began the tradition by describing the islands of the New World as the most marvelous, beautiful, and fertile lands he had ever known, filled with flocks of parrots, exotic plants, and friendly natives. Writers today are still lured by the sea, sands, and swaying palms of the Caribbean. Since the 1960s, the Caribbean has earned much of its international reputation as a playground for northern vacationers. There is, of course, another Caribbean that is far poorer and economically more dependent than the one portrayed on travel posters. Haiti, by many measures the poorest country in the Western Hemisphere, has the third largest population in the region with 9 million people. The two largest countries—Cuba (11.2 million) and the Dominican Republic (9.4 million)—also suffer from serious economic problems (see "Setting the Boundaries").

Setting the Boundaries

This culturally diverse area of the world does not fit neatly into current world regional schemes. In fact, in most textbooks it is scarcely mentioned, being folded into Latin America or appearing as part of Middle America (along with Mexico and Central America). By recognizing the Caribbean as a distinct world region, however, one can see clearly the long-term processes of colonization and globalization that have created it.

For much of the colonial period, the islands themselves were referred to as *the Indies* or *the Spanish Main*, and the sea surrounding them was called *Mar del Norte* (North Sea). It was not until the late eighteenth century that an English cartographer first applied the name *Caribbean* to the sea, a reference to the Carib Indians who once inhabited the islands. It took another 200 years for the term *Caribbean* to be generally applied to this region of islands and rimland.

The islands of the region surround the Caribbean Sea, extending from Cuba in the northwest along the island arc stretching to Trinidad and Tobago near the coast of South America. The rimland consists of the Caribbean coastal zone of the mainland, including Belize and the three Guianas. Other island groups are the Dutch islands of Aruba, Bonaire, and Curaçao (known as the ABC islands) off the coast of Venezuela; the British dependencies of the Cayman Islands and the Turks and Caicos on either side of Cuba; and the Bahamian archipelago northeast of Cuba. While technically the Bahamas and the Turks and Caicos are in the Atlantic Ocean and not the Caribbean Sea, convention places them as part of the Caribbean region. Finally, the British Overseas Territory of Bermuda is often linked with the Caribbean even though it is in the North Atlantic some 600 miles east of the South Carolina coast. Its dependence upon tourism and financial services, as well as its associate membership with the Caribbean Community and Common Market (CARICOM), justify its inclusion in the greater Caribbean.

The rationale for treating the Caribbean as a distinct area lies within its particular cultural and economic history. Culturally, this region can be distinguished from the largely Iberian-influenced mainland of Latin America because of its more diverse European colonial history and much stronger African imprint due to the elimination of Amerindian groups and their replacement by African slaves. In terms of economic production, the dominance of export-oriented plantation agriculture explains many of the social, economic, and environmental patterns in the region.

Today, the region is partially united by organizations such as CARICOM, which exist to foster the common interests of island and mainland states. Yet intraregional alliances often split along colonial lines, so that former Spanish, British, French, and Dutch colonies at times have more affinity for each other than for the region as a whole. As a crossroads of the Americas, the Caribbean has a distinctive, albeit peripheral, role in the global economy.

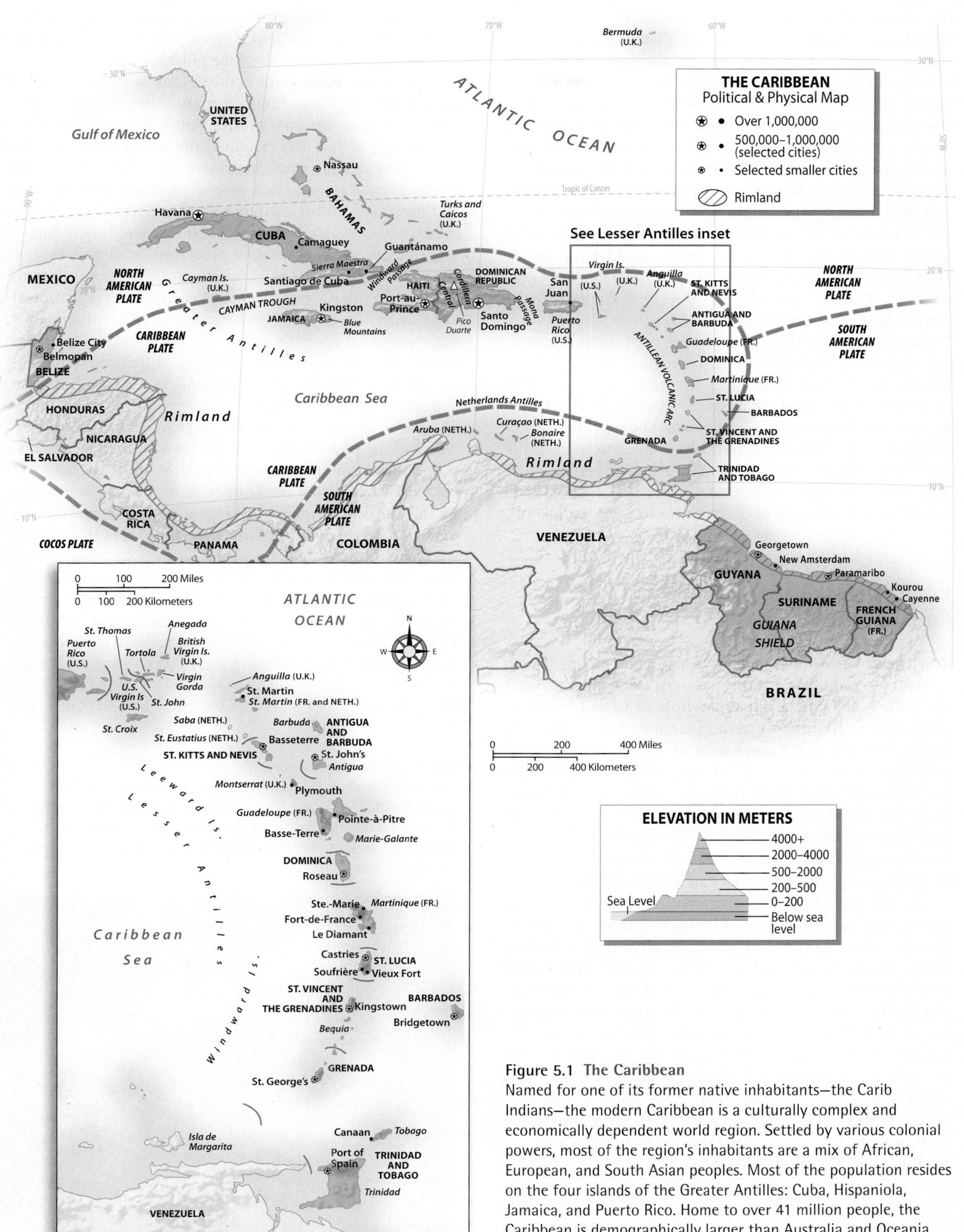

Figure 5.1 The Caribbean
Named for one of its former native inhabitants—the Carib Indians—the modern Caribbean is a culturally complex and economically dependent world region. Settled by various colonial powers, most of the region's inhabitants are a mix of African, European, and South Asian peoples. Most of the population resides on the four islands of the Greater Antilles: Cuba, Hispaniola, Jamaica, and Puerto Rico. Home to over 41 million people, the Caribbean is demographically larger than Australia and Oceania.

The majority of Caribbean people are poor, living in the shadow of North America's vast wealth. The concept of **isolated proximity** has been used to explain the region's unusual and contradictory position in the world. The *isolation* of the Caribbean sustains the area's cultural diversity (Figure 5.2) but also explains its limited economic opportunities. Caribbean writers note that this isolation fosters a strong sense of place and an inward orientation by the people. Yet the relative *proximity* of the Caribbean to North America (and, to a lesser extent, Europe) ensures its transnational connections and economic dependence. For example, each year Dominican workers in the United States send nearly $3 billion to family and friends in the Dominican Republic who rely on this money for their sustenance. History has shown the inhabitants of the Caribbean that their fate is not under their own control. Through the years, the Caribbean has evolved as a distinct but economically peripheral world region. This status expresses itself today as workers flee the region in search of employment, while foreign companies are attracted to the Caribbean for its cheap labor. The economic well-being of most Caribbean countries is precarious. Despite such uncertainty, an enduring cultural richness and attachment to place is witnessed here that may explain a growing countercurrent of returnees to their homelands.

ENVIRONMENTAL GEOGRAPHY: Paradise Undone

The levels of environmental degradation and poverty found in Haiti are unmatched in the Western Hemisphere. Haiti's story, although extreme, illustrates how social and economic inequities contribute to unsound natural resource choices, which in turn lead to more poverty. What was once considered France's richest colony—its tropical jewel—now has a per capita income of $480. Haiti's life expectancy (58 years) is the lowest in the Americas, while its levels of child malnutrition and infant mortality are the highest. The reasons for Haiti's misfortune are complex, but political economy, demography, and natural resources are critical factors.

In the late eighteenth century, Haiti's plantation economy yielded vast wealth for several thousand Europeans (mostly French) who exploited the labor of half a million African slaves. During the colonial period, the lowlands were cleared to make way for cane fields, but most of the mountains remained forested. The inequities of Haiti's plantation economy led to the world's first successful slave uprising, followed by Haitian independence in 1804. Even though Haiti was the second state in the Americas to achieve independence, social and economic inequities persisted. Most Haitians were subsistence farmers, planting food crops along with some coffee and cacao. In 1915, the United States sent in the Marines to quell political unrest and safeguard U.S. interests in the region. U.S. troops and advisors remained for nearly 20 years, during which time they rewrote the country's laws so that foreign companies could own land. The U.S. occupation resulted in improvements in Haitian infrastructure, but it also caused increased land pressure as foreign sugar companies purchased the best lands, forcing the majority of rural peasants onto the marginal soils of the hillsides.

By the mid-twentieth century, a destructive cycle of environmental and economic impoverishment was established that still confronts Haiti. Political corruption lasting from 1957 to 1986 benefited the wealthy but saw the conditions of the rural poor worsen. Sixty percent of Haiti's people are peasants who work small hillside plots and seasonally labor on large estates. As the population grew, people sought more land. They cleared the remaining hillsides, subdivided their plots into smaller units, and abandoned the practice of fallowing land in an effort to eke out an annual subsistence. When the heavy tropical rains came, the exposed and easily eroded mountain soils were washed away. As sediment collected in downstream irrigation ditches and behind dams, agriculture suffered, electricity production declined, and water supplies were degraded throughout the country.

Deforestation was further aggravated by the dependence of the population on wood for fuel. Because of their poverty and limited electricity supplies, most Haitians use charcoal (made from trees) to cook meals and heat water. By the late 1990s, only an estimated 3 percent of Haiti remained forested, and today the percentage is even less. Ongoing political turmoil continues to hamper reforestation programs, and the effects of forest removal have only worsened. In May 2004 torrential rains brought nearly 5 feet of water in seven days. Rain raced down the denuded hillsides; the silty rivers rose; and towns were washed away in the floods that followed. Nearly 3,000 people perished in Haiti and the bordering villages of the Dominican Republic. In the town of Mapou alone 1,600 Haitians died.

Figure 5.2 Carnival Dancer
A woman in elaborate costume celebrates carnival in Port of Spain, Trinidad. Linking a Christian pre-Lenten celebration with African musical rhythms, carnival has become an important symbol of Caribbean identity. *(Rob Crandall/www.robcrandall.com)*

Figure 5.3 Deforestation in Haiti Denuded slopes on the outskirts of Port-au-Prince are indicative of how much Haitians rely upon their island for fuelwood and agricultural land. Poverty and land scarcity force many Haitians to farm the hillsides. The search for wood fuels has led to widespread deforestation. With so many trees removed, soils are more vulnerable to erosion, and agricultural yields tend to decline. *(Biological Photo Service)*

In less than a lifetime, deforestation and soil erosion have become devastating environmental problems that disproportionately affect the rural poor in Haiti (Figure 5.3).

Environmental Issues

Ecologically speaking, it is difficult to imagine an area so completely reworked through colonization and global trade as the Caribbean. For nearly five centuries, the destruction of forests and the unrelenting cultivation of soils resulted in the extinction of many endemic Caribbean plants and animals, including various shrubs and trees, songbirds, large mammals, and monkeys. The beauty of the Caribbean belies its current economic and social instability, the result of severe depletion of biological resources (Figure 5.4.) Most of the environmental problems in the region are associated with agricultural practices, soil erosion, excessive reliance on wood and charcoal for fuel, and water and air pollution associated with sprawling and impoverished cities. Yet global climate change now threatens the Caribbean with problems originating outside the region, such as sea level rise, destruction of coral reefs, and variability in rainfall.

Deforestation on the Islands Much of the Caribbean was covered in tropical rain forests and deciduous forests prior to the arrival of Europeans. The great clearing of its forests began in earnest on the smaller islands of the eastern Caribbean in the seventeenth century and spread westward. The island forests fell to make room for sugarcane planting as well as to provide lumber for housing, fences, and ships. In contrast with soils in the midlatitudes, the newly exposed tropical soils easily eroded and ceased to be productive after several harvests, a situation that led to two distinct land-use strategies. On the larger islands of Cuba and Hispaniola and on the mainland, new lands were constantly cleared and older ones abandoned or fallowed (left untilled for several seasons) in an effort to keep up sugar production. On the smaller islands, such as Barbados and Antigua, where land was limited, labor-intensive efforts to conserve soil and maintain fertility were employed. In both cases, however, the island forests were replaced by a landscape devoted to crops for world markets.

While Haiti has lost most of its forest cover, Jamaica and the Dominican Republic still retain forests on nearly 30 percent of the land. On Puerto Rico one-quarter of the land has forest cover, whereas for Cuba the figure has dropped to 20 percent. Cuba has experienced a surge in charcoal production brought on by the economic and energy crises that began in 1990. Since much of the country's electricity is generated by imported fuel, and Cuba lacks hard currency for its purchase, locals have turned to the island's forests to make charcoal for domestic energy needs.

Managing Rimland Forests The Caribbean **rimland** is the coastal zone of the mainland, beginning with Belize and extending along the coast of Central America to northern South America. In general, the biological diversity and stability of the rimland states are less threatened than in the rest of the Caribbean. Thus, current conservation efforts stand a good chance of success. Even though much of Belize was selectively logged for mahogany in the nineteenth and twentieth centuries, healthy forest cover still supports a diversity of mammals, birds, reptiles, and plants. Today the public is aware of the negative consequences of deforestation. In the 1980s, local farmers outside Belmopan, the capital of Belize, created a community-managed wildlife sanctuary for the region's black howler monkeys (locally referred to as *baboons*). Once thought of as pests, foreign visitors and Belizeans come to see these monkeys, whose numbers have grown to more than 1,000. Also in the 1980s,

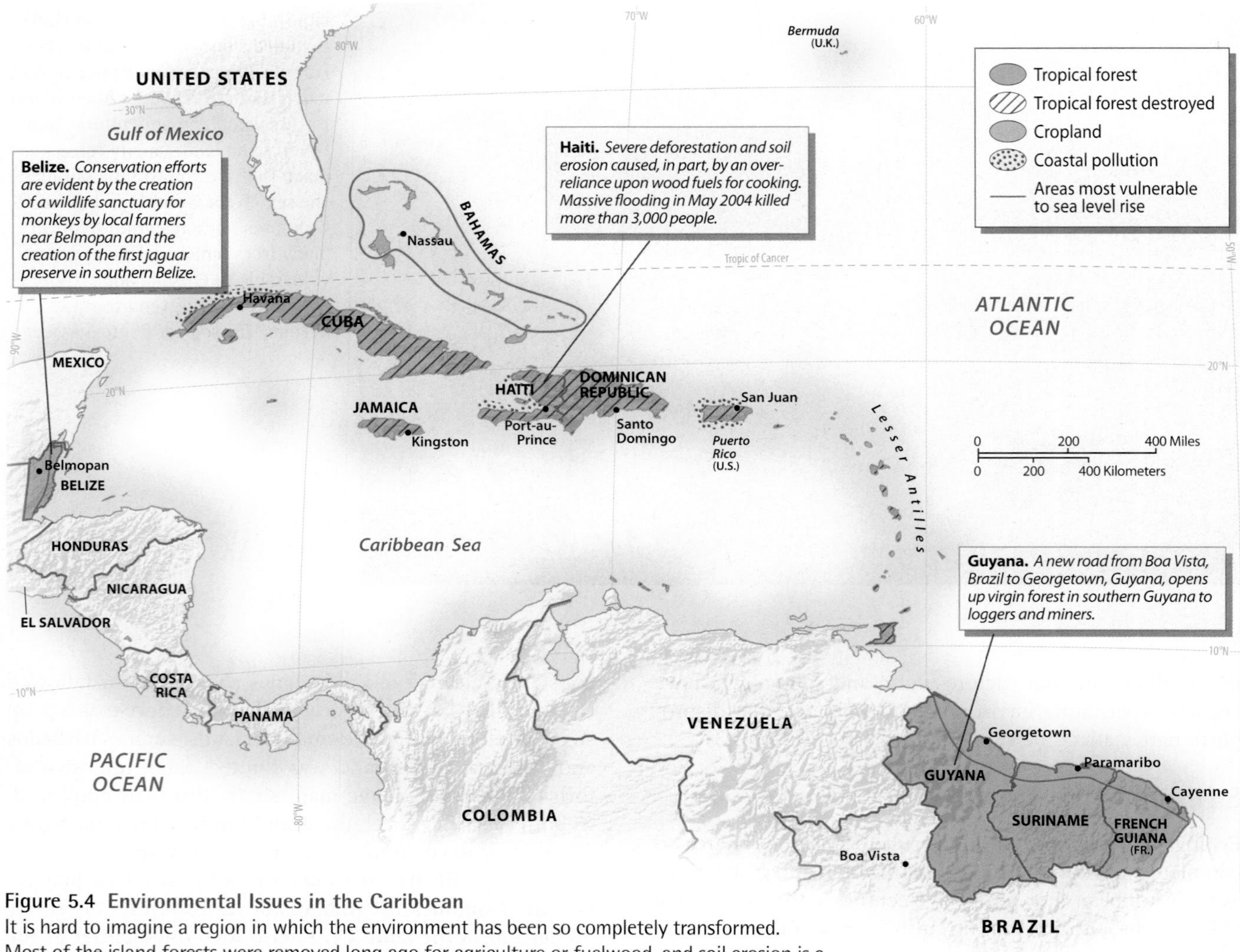

Figure 5.4 Environmental Issues in the Caribbean
It is hard to imagine a region in which the environment has been so completely transformed. Most of the island forests were removed long ago for agriculture or fuelwood, and soil erosion is a chronic problem. Coastal pollution is serious around the largest cities and industrial zones. The forest cover of the rimland states, however, is largely intact and is attracting the interest of environmentalists. *(Adapted from* DK World Atlas, *1997, pp. 7, 55, London: DK Publishing)*

the first jaguar preserve in the Americas was established in Belize's Cockscomb Mountains.

The relatively pristine interior forests of Guyana and Suriname are becoming a battleground among conservationists, indigenous peoples, and developers. For over a decade, the Guyanese and Surinamese governments have made the wood processing industry a priority by encouraging private investment and granting concessions to companies from Malaysia, Indonesia, Europe, and China for logging and sawmill operations. A new dry-season highway traverses the length of Guyana, connecting Boa Vista, Brazil, with Georgetown, Guyana. Not only does this road improve trade between Guyana and Brazil, giving Brazil access to the North Atlantic, it also provides access to the forest and mineral resources (chiefly gold) in southern Guyana. While governments in Guyana and Brazil are encouraged by the economic possibilities of this road, a coalition of local and foreign conservationists and indigenous peoples is attempting to establish a vast national park in part of this region where no commercial extraction could occur. Meanwhile, as the pace of logging and mining accelerates in Suriname, the government is meeting resistance from its maroon population, made up of escaped slaves who have lived in the forests for more than 200 years. They claim legal rights to the land based on old treaties with the Dutch.

Urban Environmental Problems The growth of Caribbean cities has led to environmental issues concerning water quality and proper waste disposal. The urban poor are the most vulnerable to health problems associated with overtaxed or nonexistent water and sewage services. According to 2003 World Bank estimates, less than half of Haiti's population has access to improved water sources, typically through shared neighborhood faucets. The figures are better for Jamaica and the Dominican Republic, where roughly 90 percent of city dwellers have access to improved water supplies. The expense of improving basic urban infrastructure far exceeds the capabilities of most island economies. Several dams exist on the larger islands to supply water, while some of the smaller islands rely on expensive desalination plants. Still, existing freshwater supplies fall far short of domestic needs. As tourism, offshore manufacturing, and a growing urban population demand more water and produce more waste, Caribbean countries will be forced to make hard decisions about their infrastructure.

In addition to being a public health concern, water contamination from improper waste disposal poses serious economic problems for states dependent on tourism. Governments are caught between a desire to fix the problem before tourists notice and a tendency not to discuss it at all. Doing nothing, however, may not be an option for countries that rely heavily on tourism. In the more industrialized Puerto Rico, it is estimated that half the country's coastline is unfit for swimming, mostly due to contamination from sewage.

The Caribbean and Climate Change

The effects of climate change on the Caribbean region include sea level rise, increased intensity of storms, variable rainfall leading to both floods and droughts, and loss of biodiversity (both in forests and coral reefs). The scientific consensus is that global warming could promote a sea level rise of 3 to 10 feet (1 to 3 meters) in this century. In terms of land loss to inundation, the low-lying Bahamas would be the most impacted country—losing nearly 30 percent of its land with a 10-foot (3-meter) sea level rise. Yet in terms of people affected by inundation, Suriname, French Guiana, Guyana, Belize, and the Bahamas would be the most severely impacted: just a 3-foot (1-meter) sea level rise would be devastating because most of the population lives near the coast. Should sea level rise reach 10 feet (3 meters), 30 percent of Suriname's population and 25 percent of Guyana's would be displaced (Figure 5.4).

The Caribbean has not been a major contributor of greenhouse gases, but this maritime region is extremely vulnerable to the negative impacts of climate change. In addition to land loss and population displacement due to sea level rise, other concerns focus on changes in rainfall patterns, leading to declines in agricultural yields and freshwater supplies, and increases in storm intensity—especially hurricanes—that cause destruction of infrastructure and other problems. All of these changes would negatively impact tourism and thus the gross domestic income of countries in the region. Some of the worst-case scenarios are catastrophic.

In terms of biodiversity, continued warming of ocean temperatures will further negatively impact the Caribbean's coral reefs, which are the most biologically diverse ecosystems of the marine world. These reefs, particularly those of the rimland, are already threatened by water pollution and subsistence fishing practices. Now there is mounting evidence of coral bleaching and die-off due to higher sea temperatures. Coral reefs are diverse and productive ecosystems that function as nurseries for many marine species. Healthy reefs also serve as barriers to protect populated coastal zones as well as mangroves and wetlands. As the reefs become more ecologically vulnerable, so too do the human populations that depend upon the many services that the reefs provide.

A division of the United Nations Environment Program (UNEP) that focuses on the world's coral reefs has supported a regional conservation effort in the Caribbean. The longest barrier reef in the Americas lies off the coast of Belize, where international researchers monitor the impact of warmer waters, disease, and pollution on the health of the coral. Examples of local conservation efforts are seen in Bonaire, where the National Parks Foundation actively manages the Bonaire Marine Park, recognized as one of the most effectively managed marine reserves in the Caribbean.

Throughout the Caribbean, protecting the environment and preparing for the effects of climate change are increasingly being recognized not as a luxury but a question of economic livelihood. In fact, CARICOM has monitored the threat of climate change for over a decade. Although many see increased environmental awareness as a hopeful sign, both urban and rural ecosystems suffer from serious degradation and are especially vulnerable to the long-term impacts of global warming.

Island and Rimland Landscapes

It is the Caribbean Sea itself, that body of water enclosed between the Antillean islands (the arc of islands that begins with Cuba and ends with Trinidad) and the mainland of Central and South America, that links the states of the Caribbean region (Again see Figure 5.2). Historically the sea connected people through its trade routes and sustained them with its marine resources of fish, green turtle, manatee, lobster, and crab. While the sea is noted for its clarity and biological diversity, the quantities of any one species are not great, so it has never supported large commercial fishing. The surface temperature of the sea ranges from 73° to 84°F (23° to 29°C), over which forms a warm tropical marine air mass that influences daily weather patterns. This warm water and tropical setting continues to be a key resource for the region, attracting millions of tourists to the Caribbean each year (Figure 5.5).

The arc of islands that stretches across the sea is the region's most distinguishing feature. The Antillean islands are divided into two groups: the Greater and Lesser Antilles. The rimland (the Caribbean coastal zone of the mainland) includes Belize and the Guianas, as well as the Caribbean shoreline of Central and South America.

Greater Antilles The four large islands of Cuba, Jamaica, Hispaniola (shared by Haiti and the Dominican Republic), and Puerto Rico make up the **Greater Antilles**. On these islands are found the bulk of the region's population, arable lands, and large mountain ranges. Given the popular interest in the Caribbean coasts, it still surprises many people that Pico Duarte in the Cordillera Central of the Dominican Republic is more than 10,000 feet (3,000 meters) tall, Jamaica's Blue Mountains top 7,000 feet (2,100 meters), and Cuba's Sierra Maestra is more than 6,000 feet (1,800 meters) tall. The mountains of the Greater Antilles were of little economic interest to plantation owners, who preferred the coastal plains

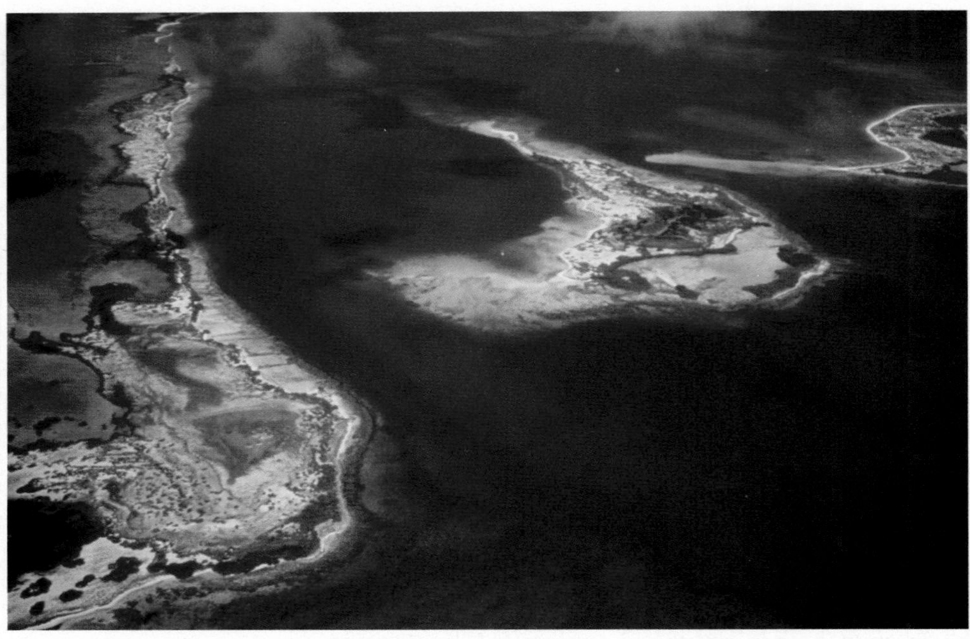

Figure 5.5 Caribbean Sea Noted for its calm, turquoise waters, steady breezes, and treacherous shallows, the Caribbean Sea has both sheltered and challenged sailors for centuries. This aerial photograph shows the southern Caribbean islands of Los Roques, off the Venezuelan coast. *(Rob Crandall/www.robcrandall.com)*

and valleys. Yet the mountains were an important refuge for runaway slaves and subsistence farmers, and thus figure prominently in the cultural history of the region.

The best farmlands are found in the central and western valleys of Cuba, where a limestone base contributes to the formation of a fertile red clay soil (locally called *matanzas*) and a gray or black soil type called *rendzinas* (also found in Antigua, Barbados, and lowland Jamaica). The rendzinas soils, which consist of a gravelly loam with a high organic content ideal for sugar production, are actively exploited for agriculture wherever found. Surprisingly, given the area's agricultural orientation, many of the soils are nutrient-poor, heavily leached, and acidic. These poor, *ferralitic* soils are found in the wetter areas where crystalline base rock exists (as in parts of Hispaniola, the Guianas, and Belize). They are characterized by heavy accumulations of red and yellow clays and offer little potential for permanent intensive agriculture.

Lesser Antilles The **Lesser Antilles** form a double arc of small islands stretching from the Virgin Islands to Trinidad. Smaller in size and population than the Greater Antilles, they were important early footholds for rival European colonial powers. The islands from St. Kitts to Grenada form the inner arc of the Lesser Antilles. These mountainous islands, with peaks ranging from 4,000 to 5,000 feet (1,200 to 1,500 meters), have volcanic origins. In this subduction zone, the heavier North and South American plates dive underneath the Caribbean Plate, producing volcanic activity. Erosion of the island peaks and the accumulation of ash from eruptions have created small pockets of arable soils, although the steepness of the terrain limits agricultural development. The latest round of volcanic activity began in July 1995 on Montserrat. A series of volcanic eruptions of ash and rock took several lives and forced most of the island's 10,000 inhabitants to nearby islands and even to London. Plymouth, the capital, was abandoned in 1997, although interim government buildings now exist on the northwest corner of the island and some residents have returned.

Just east of this volcanic arc are the low-lying islands of Barbados, Antigua, Barbuda, and the eastern half of Guadeloupe. Covered in limestone that overlays volcanic rock, these lands were inviting for agriculture, especially sugarcane. Trinidad and Tobago are on the South American Plate and consist of sedimentary rather than volcanic rock. These islands include alluvial soils and, more important, sedimentary basins that contain oil reserves.

The Rimland Much of low-lying Belize is limestone. Sugarcane dominates in the drier north, while citrus is produced in the wetter central portion of the state. The Guianas, however, are characterized by the rolling hills of the Guiana Shield. The shield's crystalline rock explains the area's overall poor soil quality. Most agriculture in the

Figure 5.6 Kourou, French Guiana The European Space Agency regularly launches rockets from its center in Kourou, French Guiana. This French territory, near the equator and on the coast, makes an ideal launching site. In this photo, the *Ariane 5* rocket is being moved to the launch pad. *(European Space Agency)*

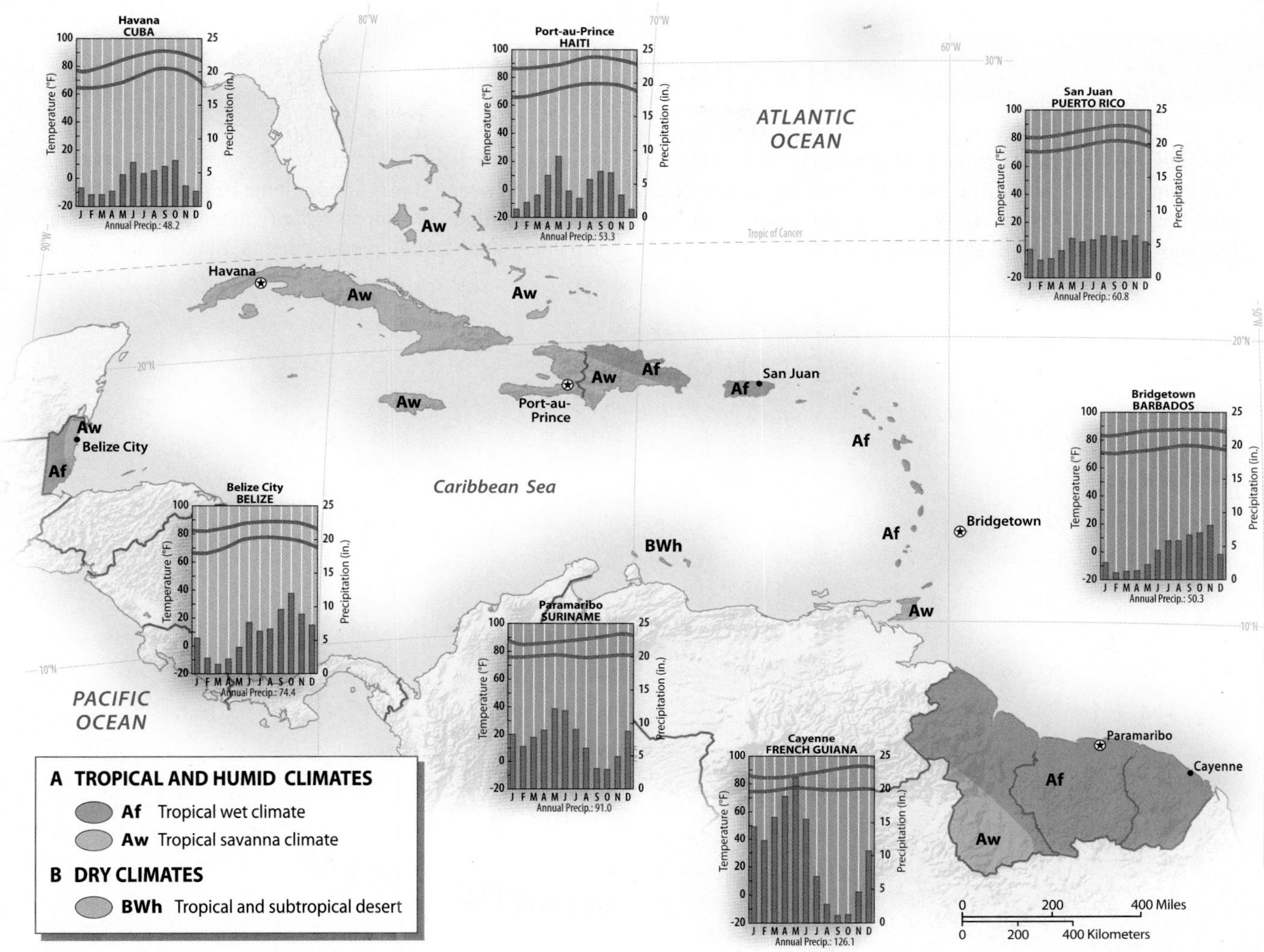

Figure 5.7 Climate Map of the Caribbean
Most of the region is classified as having either a tropical wet (Af) or tropical savanna (Aw) climate. Temperature varies little across the region, with highs slightly above 80 degrees and lows around 70 degrees. Important differences in total rainfall and the timing of the dry season distinguish different places. *(Temperature and precipitation data from Pearce and Smith, 1984,* The World Weather Guide, *London: Hutchinson)*

Guianas occurs on the narrow coastal plain, where sugar and rice are produced. Unlike the rest of the Caribbean, these rimland territories still contain significant amounts of forest cover. In Suriname alone, 90 percent of the territory is forest or woodland. Timber continues to be an important export for these rimland states. Metal extraction (bauxite and gold) also is vital to the economies of Guyana and Suriname. French Guiana, which is an overseas territory of France, relies mostly on French subsidies but exports shrimp and timber. It also is home to the European Space Center at Kourou (Figure 5.6).

Climate and Vegetation

Coconut palms framed by a blue sky evoke the postcard image of the Caribbean. This tropical region is warm year-round, and rainfall is abundant. Much of the Antillean islands and rimland receives more than 80 inches (200 centimeters) of rainfall annually and can support tropical forests. Amid the forests are pockets of naturally occurring grasslands in parts of Cuba, Hispaniola, and southern Guyana. Distinctly dry areas exist, such as the rain shadow basin in western Hispaniola. As explained earlier, much of the natural vegetation on the islands has been removed to accommodate agriculture and fuel needs, and only small forest fragments remain today.

As in many tropical lowlands, seasonality in the Caribbean is defined by changes in rainfall more than temperature. Although some rain falls throughout the year, the rainy season is from July to October. In Belize City and Havana, October is the wettest month. In Bridgetown and San Juan, the wettest month is November (Figure 5.7) when the Atlantic high-pressure cell is farthest north and easterly winds generate moisture-laden and unstable

GEOGRAPHIC TOOLS Tracking Hurricanes

The National Hurricane Center in Miami tracks and predicts hurricanes in the Atlantic, the Caribbean, the Gulf of Mexico, and the eastern Pacific. During hurricane season (July–October), surface reports and satellite data are constantly monitored to detect and track tropical depressions that may achieve hurricane status. The National Hurricane Center uses a combination of sophisticated models, geostationary satellites, and actual measurements of the approaching

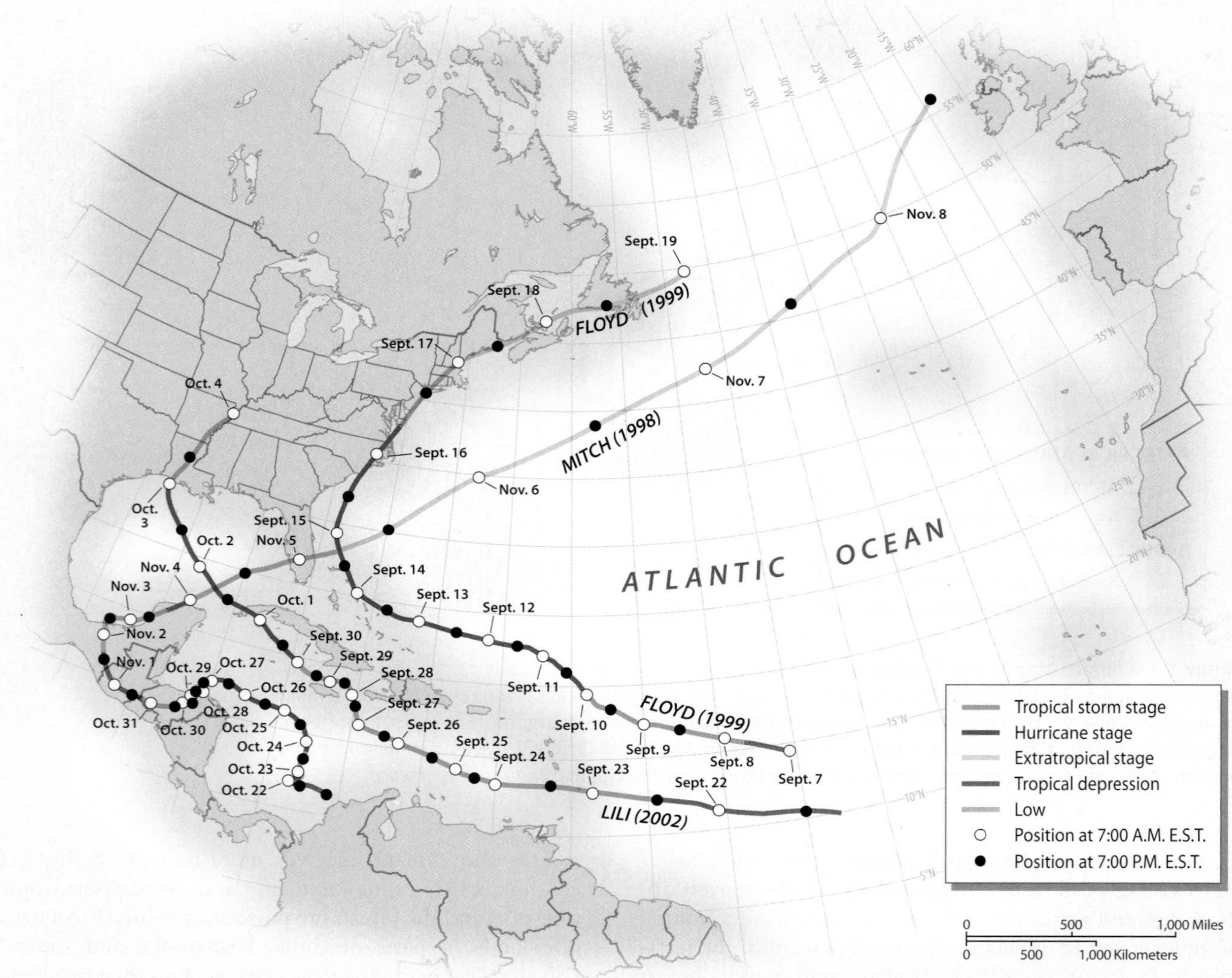

Figure 5.2.1 Tracking the Paths of Hurricanes Mitch, Floyd, and Lili
(From Aguado and Burt, 2004, Understanding Weather and Climate, *3rd ed., Upper Saddle River, NJ: Prentice Hall)*

atmospheric conditions that sometimes yield hurricanes. During the slightly cooler months of December through March, rainfall declines. This time of year corresponds with the peak tourist season.

The Guianas have a different rainfall cycle. These territories, on average, receive more rain than the Antillean islands. In Cayenne, French Guiana, an average of 126 inches (320 centimeters) falls each year. Unlike the Antilles, the Guianas experience a brief dry period in late summer (September to October). Also, January tends to be a wet period for the mainland, while it is a dry time for the islands. Climatically, the Guianas also are distinguishable from the rest of the region because they are not affected by hurricanes.

hurricanes through specially equipped aircraft that fly into hurricanes before they approach land. Using supercomputers that integrate dynamic data from a particular tropical depression along with laws of physics and historical patterns, meteorologists project the storm position, direction, wind speed, and air pressure every few hours. Figure 5.2.1. shows the actual path of three recent hurricanes over the course of two weeks in 12-hour intervals. A particular storm can take an irregular jog or have periods of slower or faster movement through an area. One reason Hurricane Mitch was so deadly was that it slowed down and lingered over a small portion of Central America for nearly a week, dropping huge amounts of rain (Figure 5.2.2).

As good as the models are, scientists are unable to forecast the path of a hurricane more than 72 hours in advance. Within 24 hours of a hurricane's reaching land, forecasts of its path and where it might make landfall improve considerably, but the margin of error still is about 100 miles (160 kilometers). When one considers the difficulty and expense of evacuating a heavily settled coastal area, this degree of error is far from ideal. Within 6 hours of landfall, the immediate path of a hurricane is much more predictable, but by then the time for taking precautions is limited.

Atmospheric scientists and geographers believe that we have entered a more active hurricane period. Since 1995, a change in the multiyear cycle of sea surface temperatures in the Atlantic Ocean may be contributing to more intense tropical depressions. With settlement and population growth intensifying in coastal zones of North America and the Caribbean, major storms are likely to do more damage. Apart from deaths resulting from Hurricane Mitch, fatalities from hurricanes over the last 30 years have declined because of better forecasting of hurricane movement, resulting in initiation of evacuation procedures at least 24 hours ahead of a storm's landing. But forecasting cannot reduce the damage to crops, forests, homes, or infrastructure. An unfortunately timed storm can destroy a banana harvest or shut down a resort for a season or more. The sheer force of a storm can radically change the landscape as well, turning a palm-lined sandy beach into a barren rocky shore covered with debris.

Source: Adapted from Edward Aguado and James E. Burt, 2004, *Understanding Weather and Climate*, 3rd ed., Upper Saddle River, NJ: Prentice Hall, pp. 367–74.

Figure 5.2.2 Hurricane Mitch Satellite imagery shows the immense scale of one of the most destructive hurricanes to hit the region. Mitch achieved wind speeds of over 200 mph and unleashed torrential rains. The storm and subsequent floods killed more than 10,000 people, left 2.5 million people homeless, and caused over $5 billion in damage. This image was taken on October 26, 1998, three days before Mitch made landfall. *(Hal Pierce/Laboratory of Atmospheres/NASA/ Goddard Space Flight Center)*

Hurricanes Each year several **hurricanes** form, pounding the Caribbean as well as Central and North America with heavy rains and fierce winds. In July through October, westward-moving low-pressure disturbances form off the coast of West Africa and pick up moisture and speed as they move across the Atlantic. These tropical disturbances are usually no more than 100 miles (160 kilometers) across, but to achieve hurricane status they must reach velocities of more than 75 miles (121 kilometers) per hour. Hurricanes may take several paths through the region, but they typically enter through the Lesser Antilles. They then arc north or northwest and collide with the Greater Antilles, Central America, Mexico, or southern North America before moving to the northeast and dissipating in

Figure 5.8 Grenada After Hurricane Ivan
A resident of St. Georges in Grenada sits by the remains of his home after Hurricane Ivan destroyed it in 2004. One of the most destructive tropical storms to hit the Antilles in years, most of Grenada's infrastructure was damaged by the storm. *(Lucas Oleniuk/Zuma)*

the Atlantic Ocean. The hurricane zone lies just north of the equator on both the Pacific and Atlantic sides of the Americas. Typically a half-dozen to a dozen hurricanes form each season and move through the region; most of them never hit land or, if they do, inflict little damage (see "Geographic Tools: Tracking Hurricanes").

Most long-time residents of the Caribbean have felt the full force of at least one major tropical storm in their lifetimes. The destruction caused by these storms is not just from the high winds, but also from the heavy downpours that can cause severe flooding and deadly coastal tidal surges (Figure 5.8). In 1998, the torrential rains of Hurricane Mitch, the most deadly hurricane in a century, resulted in the deaths of at least 10,000 people in Honduras, Nicaragua, and El Salvador. Mud slides and flooding ravaged structures and roads, leaving an estimated 2.5 million people homeless. In 2004 the strongest storm to hit the Antilles in a decade was Hurricane Ivan. The island nation of Grenada took a direct hit and 85 percent of the island's infrastructure was destroyed. Fewer than half of the country's hotels were open by January 2005 and exports of nutmeg—the island's principle cash crop—were down 90 percent. Barbados, Tobago, the Netherlands Antilles, Cuba, and Jamaica also suffered damage from Ivan. In total, about 70 lives and some $2 billion in insurance losses were claimed.

Forests, Savannas, and Mangroves Tropical forests, once common throughout the region, are today found almost exclusively in the rimland, the Guianas, and Belize, and even these areas are increasingly under pressure. As for the rest of the Caribbean, small pockets of forest remain broken by fields and savannas. On the island of Puerto Rico, for example, the small El Yunque rain forest offers a glimpse of the Caribbean's forested past. Protected by the Spanish in the nineteenth century, today El Yunque is the smallest national forest in the U.S. system, but it is also one of the most biologically diverse (Figure 5.9).

The palm savannas, found mostly in the tropical savanna (Aw) zones, are important biomes in the Caribbean. The palm savannas of Hispaniola and Cuba are quite fertile and easily adapted to agriculture. In central Cuba, the natural grasslands studded with palms have the best soils in the entire country and are planted mostly in sugarcane and citrus. The savanna soils of southern Guyana, however, are acidic with little agricultural potential.

For centuries the coastal mangrove swamps of the region were largely left undisturbed. These wet environments are poorly suited to human settlement, although they are vital nurseries for young crustaceans and fish. Found throughout the Caribbean, especially on the calmer leeward shores, the tangled stands of woody mangrove are now often cleared to create open beaches. Removal of mangroves, besides eliminating vital marine habitat, exposes coasts to increased erosion. In addition, mangroves decline when disturbances such as increased soil erosion and runoff from areas upstream increase the silt load in the water. Moreover, mangrove swamps and other coastal wetlands are increasingly vulnerable to the long-term effects of global warming, especially rises in sea level.

Figure 5.9 El Yunque National Forest
A tourist takes in the scene of a waterfall surrounded by lush forest in El Yunque National Forest. An important tourist destination for Puerto Rico, this small national forest is a protected remnant of the extensive tropical forests that once existed across the Antilles. *(Robert Fried/Stock Boston)*

POPULATION AND SETTLEMENT: Densely Settled Islands and Rimland Frontiers

In the Caribbean, the population density is generally quite high and, as in neighboring Latin America, increasingly urban. Eighty-six percent of the region's population is concentrated on the four islands of the Greater Antilles (Figure 5.10). Add to this Trinidad and Tobago's 1.4 million and Guyana's 800,000, and most of the population of the Caribbean is accounted for by six countries and one U.S. territory (Puerto Rico). Of these, Puerto Rico has the greatest population density, with 1,157 people per square mile (445 people per square kilometer), whereas Jamaica has 634 people per square mile (244 people per square kilometer) and Hispaniola averages more than 625 people per square mile (240 people per square kilometer) (Table 5.1).

In absolute numbers, few people inhabit the Lesser Antilles; nevertheless, some of these microstates are densely settled. The small island of Barbados is an extreme example. With only 166 square miles (430 square kilometers) of territory, it has 1,682 people per square mile (647 people per square kilometer). Bermuda, which is one-third of the size of the District of Columbia, has over 3,000 people per square mile (1,200 people per square kilometer). Population densities on St. Vincent, Martinique, and Grenada, while not as high, are still more than 700 people per square mile (270 people per square kilometer). If one considers the scarcity of arable land on some of these islands, it is clear that access to land is a basic resource problem for many inhabitants of the Caribbean. The growth in the region's population coupled with its scarcity of land has forced many people into the cities or abroad. It also has forced many Caribbean states to be net importers of food.

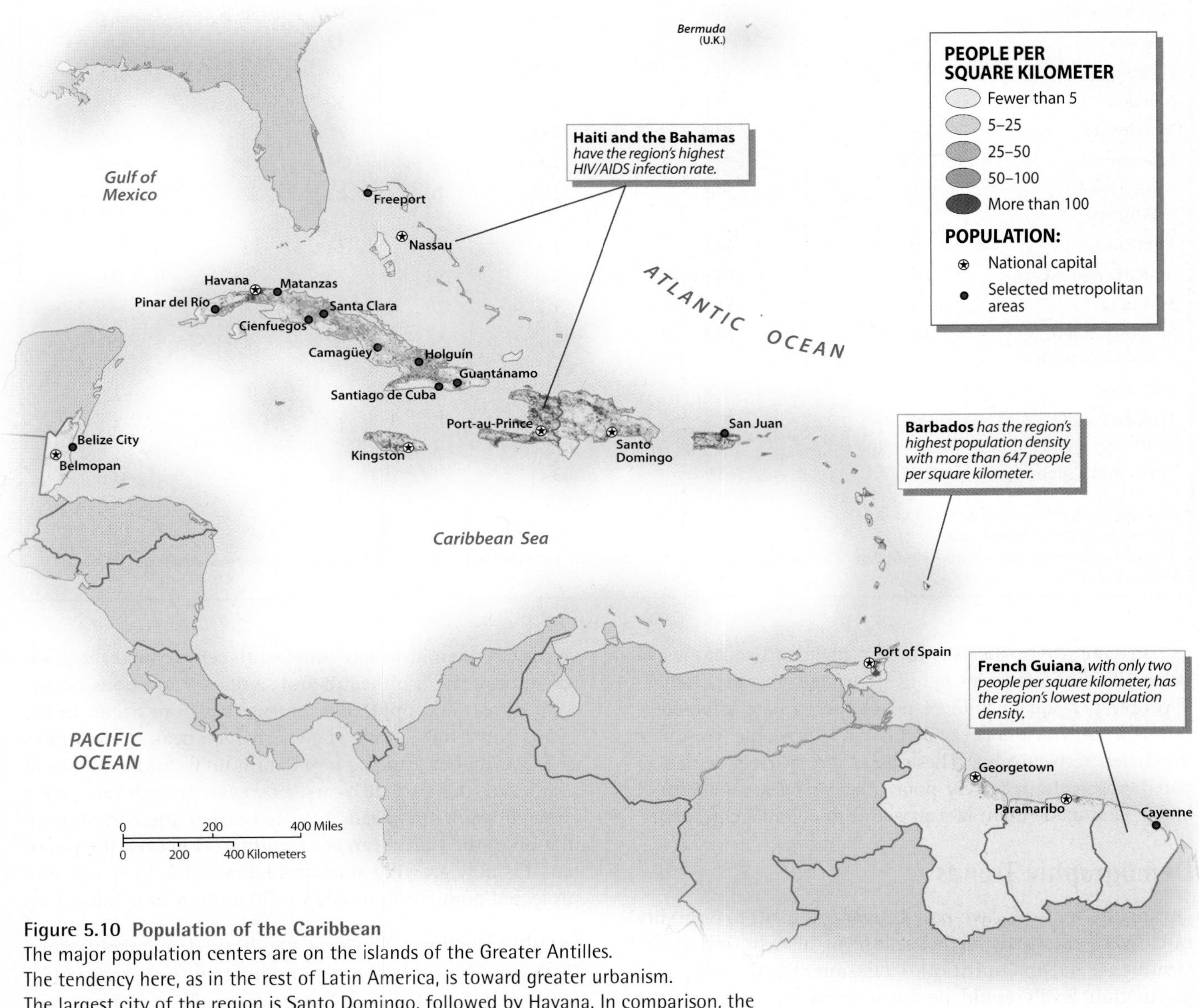

Figure 5.10 Population of the Caribbean
The major population centers are on the islands of the Greater Antilles. The tendency here, as in the rest of Latin America, is toward greater urbanism. The largest city of the region is Santo Domingo, followed by Havana. In comparison, the rimland states are very lightly settled.

TABLE 5.1 Population Indicators

Country	Population (Millions, 2007)	Population Density (per Square Kilometer)	Total Fertility Rate	Percent Urban	Percent <15[a]	Percent >65[b]	Net Migration (per 1,000, 2000–05)
Anguilla	0.01	134	1.7		22	7	2.4
Antigua and Barbuda	0.1	195	2.3	39	28	7	0.0
Bahamas	0.3	24	1.9	90	28	6	1.3
Barbados	0.3	647	1.9	53	22	12	
Belize	0.3	14	3.3	50	41	4	−0.8
Bermuda	0.07	1,241	1.9		18	13	2.3
Cayman	0.05	177	1.9		20	9	16.5
Cuba	11.2	101	1.5	76	19	11	
Dominica	0.1	93	3.0	73	29	10	−9.4
Dominican Republic	9.4	192	2.9	65	33	6	−3.3
French Guiana	0.2	2	4.0	76	35	4	4.6
Grenada	0.1	288	2.1	31	32	5	
Guadeloupe	0.5	274	2.2	100	26	8	−1.6
Guyana	0.8	4	2.7	28	32	5	
Haiti	9.0	323	4.0	36	39	4	−2.6
Jamaica	2.7	244	2.1	49	30	8	−7.6
Martinique	0.4	368	2.0	98	22	12	
Montserrat	0.01	93	1.8		24	11	14.2
Netherlands Antilles	0.2	248	1.8	70	23	10	0.0
Puerto Rico	3.9	445	1.7	94	21	13	
St. Kitts and Nevis	0.05	180	2.3	32	28	8	0.0
St. Lucia	0.2	315	1.7	28	28	7	
St. Vincent and the Grenadines	0.1	286	2.0	45	29	7	−8.5
Suriname	0.5	3	2.5	74	30	6	−7.2
Trinidad and Tobago	1.4	270	1.6	12	25	6	−3.1
Turks and Caicos	0.02	51	3.0		31	4	43.8

[a]*Percentage of population younger than 15.*
[b]*Percentage of population older than 65.*
Sources: Population Reference Bureau, World Population Data Sheet, 2007; *net migration rate data from* UN International Migration, 2006; *Bermuda data from* CIA World Factbook, 2007.

In contrast to the islands, the mainland territories of Belize and the Guianas are lightly populated; Guyana averages 9 people per square mile (3 people per square kilometer), Suriname only 8, and Belize 36 (2.7 and 14 people per square kilometer, respectively). These areas are sparsely settled in part because the relatively poor quality and accessibility of arable land made them less attractive to colonial enterprises.

Demographic Trends

During the years of slave-based sugar production, mortality rates were extremely high because of disease, inhumane treatment, and malnutrition. Consequently, the only way population levels could be maintained was through the continual importation of African slaves. With the end of slavery in the mid-to-late nineteenth century and the gradual improvement of health and sanitary conditions on the islands, natural population increase began to occur. In the 1950s and 1960s, many states achieved peak growth rates of 3.0 or higher, causing population totals and densities to soar. Over the past 20 years, however, growth rates have come down or stabilized. As noted earlier, the current population of the Caribbean is 41 million. However, the population is now growing at an annual rate of 1.2 percent, and projected population in 2025 is 50 million (see Table 5.1).

Fertility Decline The most significant demographic trend in the Caribbean is the decline in fertility (Figure 5.11). Cuba and Trinidad and Tobago have the region's lowest total fertility rates (TFRs). In socialist Cuba, the education

Figure 5.11 Smaller Caribbean Families
A family in Bermuda enjoys a day at the beach. The average Caribbean family is far smaller now than 30 years ago. Higher levels of education, improved availability of contraception, an increase in urban living, and a larger percentage of women in the labor force contribute to slower population growth rates in many countries. *(Tony Arruza/Tony Arruza Photography)*

of women combined with the availability of birth control and abortion has resulted in 1.5 children for the average woman (compared to 2.0 in the United States). In capitalist Trinidad and Tobago, similar results have been achieved (a TFR of 1.6). Population pressures in Trinidad and Tobago also have been eased by a steady out-migration of islanders overseas, especially to the United Kingdom and North America. The decision to emigrate and a preference for smaller families have contributed to slower growth rates throughout the region. Even states with relatively high TFRs, such as Haiti, have seen a decline in family size. Haiti's TFR fell from 6.0 in 1980 to 4.0 in 2007.

The Rise of HIV/AIDS The rate of HIV/AIDS infection in the Caribbean is more than three times that in North America, making the disease an important regional issue. Although nowhere near the infection rates in Sub-Saharan Africa (see Chapter 6), 1.6 percent of the Caribbean population between the ages of 15 and 49 has HIV/AIDS. The highest rates are in some of the most populated islands. In Haiti, one of the earliest locations where AIDS was detected, more than 3.8 percent of the population between the ages of 15 and 49 is infected with the virus. The infection rate in Guyana has climbed to 2.4 percent, and Trinidad and Tobago have a 2.6 percent infection rate as of 2006. AIDS is already the largest single cause of death among young men in the English-speaking Caribbean.

Many Caribbean countries have taken steps to educate their populations about the spread of HIV/AIDS in an effort to lower infection rates. The Bahamas, for example, had an estimated HIV/AIDS infection rate of 4 percent among the 15- to 49-year-old population in 2001. Officials there have developed educational campaigns and introduced drug treatments to help prevent mother-to-child transmission; efforts have paid off, and the infection rate fell to 3.3 percent in 2006. Cuba, which witnessed a surge in both tourism and prostitution that typically contributes the spread of HIV/AIDS, claims to have an infection rate of only 0.1 percent among its 15- to 49-year-old population. Education programs and an effective screening and reporting system have apparently kept the infection rate down in Cuba.

Emigration Driven by the region's limited economic opportunities, a pattern of emigration to other Caribbean islands, North America, and Europe began in the 1950s. For more than 50 years, a Caribbean **diaspora**—the economic flight of Caribbean peoples across the globe—has defined existence and identity for much of the region (Figure 5.12). Barbadians generally choose England, most settling in the London suburb of Brixton with other Caribbean immigrants. In contrast, one out of every three Surinamese has moved to the Netherlands, with most residing in Amsterdam. As for Puerto Ricans, only slightly more live on the island than reside on the U.S. mainland. In the 1980s, roughly 10 percent of Jamaica's population legally immigrated to North America (some 200,000 to the United States and 35,000 to Canada). Cubans have made the city of Miami their destination of choice since the 1960s. Today they make up the majority of that city's population.

Intraregional movements also are important. Perhaps one-fifth of all Haitians do not live in their country of birth, with their most common destination being the neighboring Dominican Republic, followed by the United States, Canada, and French Guiana. Dominicans are also on the move; the vast majority come to the United States, settling in New York City, where they are the single largest immigrant group. Others, however, simply cross the Mona Passage and settle in Puerto Rico. As a region, the Caribbean has one of the highest annual rates of net migration in the world at −2.8 per thousand. That means for every 1,000 people in the region, 2.8 annually leave. Individual countries have much higher rates, such as Guyana at −10.7 per 1,000 or Dominica at −9.4. Of course, some lightly populated but prosperous territories such as the Cayman Islands and the Turks and Caicos have high positive annual rates of net migration (16.5 and 43.8 per 1,000, respectively). The economic implications of this labor-related migration are significant and will be discussed later.

Most migrants, with the exception of the Cubans, are part of a **circular migration** flow. In this type of migration, a man or woman typically leaves children behind with

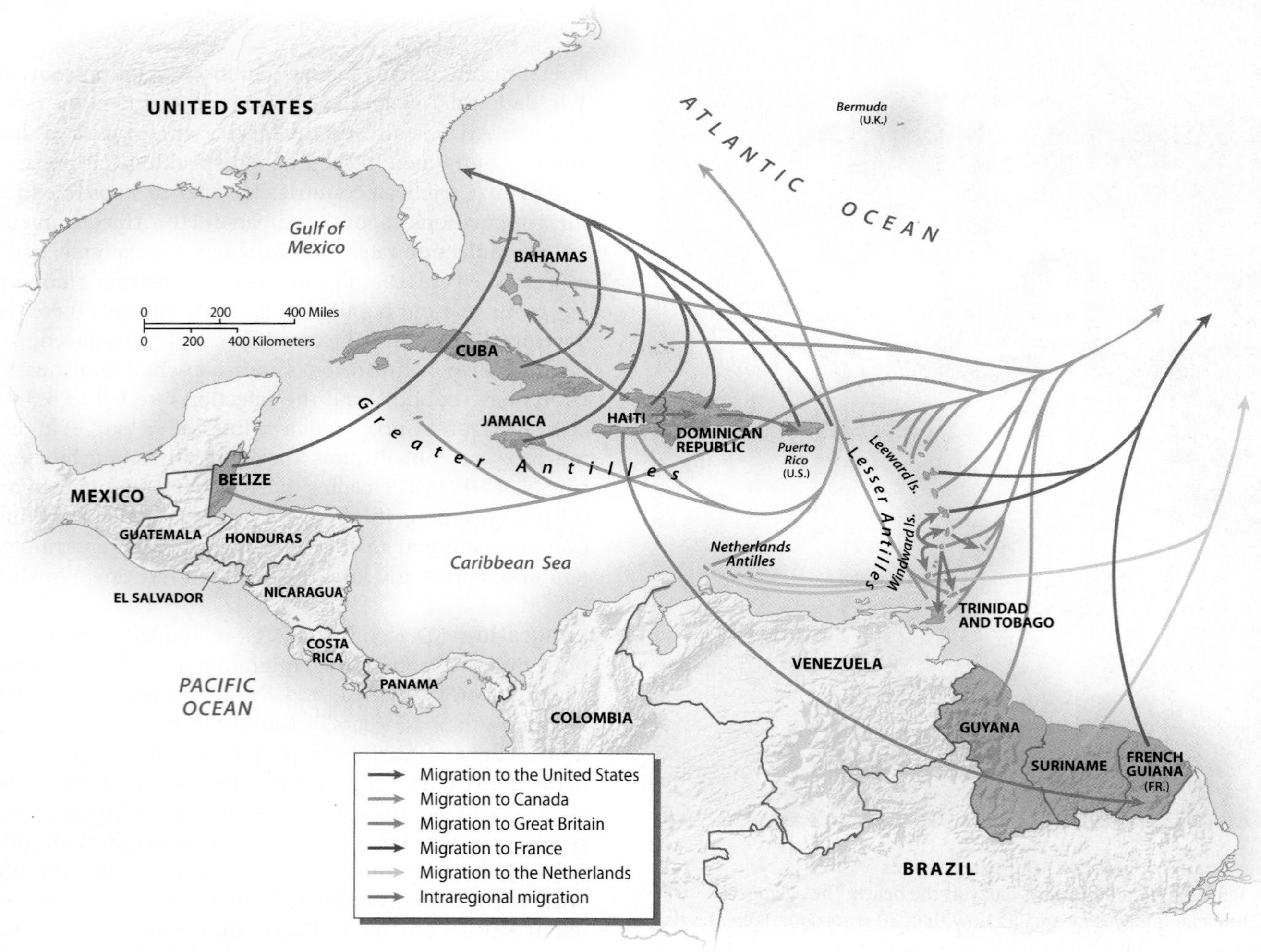

Figure 5.12 Caribbean Diaspora
Emigration has long been a way of life for Caribbean peoples. With relatively high education levels but limited professional opportunities, migrants from the region head to North America, Great Britain, France, and the Netherlands. Intraregional migrations between Haiti and the Dominican Republic or the Dominican Republic and Puerto Rico also occur. *(Data from Levin, 1987,* Caribbean Exodus, *Westport, CT: Praeger)*

relatives in order to work hard, save money, and return home. Other times a **chain migration** begins, in which one family member at a time is brought over to the new country. In some cases, large numbers of residents from a Caribbean town or district send migrants to a particular locality in North America. Thus, chain migration can account for the formation of immigrant enclaves. Caribbean immigrants have increasingly practiced **transnational migration**—the straddling of livelihoods and households between two countries. Dominicans are probably the most transnational of all the Caribbean groups. They regularly move back and forth between two islands, Hispaniola and Manhattan. Dominican President Leonel Fernandez (1996–2000 and 2004–08) grew up in New York City, still holds a green card, and has said he intends to return when his term is finished.

The Rural–Urban Continuum

Initially, plantation agriculture and subsistence farming shaped Caribbean settlement patterns. Low-lying arable lands were dedicated to export agriculture and controlled by wealthy colonial landowners. Only small amounts of land were set aside for subsistence production. Over time, villages of freed or runaway slaves were established, especially in remote areas of the interior. But the vast majority of people continued to live on estates as owners, managers, or slaves. Cities were formed to serve the administrative and social needs of the colonizers, but most were small, containing a small fraction of a colony's population. The colonists who linked the Caribbean to the world economy saw no need to develop major urban centers.

Even today the structure of Caribbean communities reflects the plantation legacy. Many of the region's subsistence farmers are ancestors of former slaves who continue to work their small plots and seek seasonal wage labor on estates. The social and economic patterns generated by slavery still mark the landscape. Rural communities tend to be loosely organized; labor is transient; and small farms are scattered wherever pockets of land are available. Traditionally, men tended to leave home for seasonal labor, so that matriarchal family structures and female-headed households are common.

Caribbean Cities Since the 1960s, the mechanization of agriculture, offshore industrialization, and population growth have caused a surge in rural-to-urban migration. Cities have grown accordingly, and today 64 percent of the region is classified as urban. Of the large countries, Cuba is the most urban (76 percent) and Haiti the least (36 percent). Caribbean cities are not large by world standards, as only four have 1 million or more residents: Santo Domingo (2.7 million), Havana (2.2 million), Port-au-Prince (1.5 million), and San Juan (1.0 million). All but Port-au-Prince were laid out by the Spanish.

Like their counterparts in Latin America, the Spanish Caribbean cities were laid out on a grid with a central plaza. Vulnerable to raids by rival European powers and pirates, these cities were usually walled and extensively fortified. The oldest continually occupied European city in the Americas is Santo Domingo in the Dominican Republic, settled in 1496, and today the largest city in the region. Havana emerged as the most important colonial city in the region, serving as a port for all incoming and outgoing Spanish galleons. Strategically situated on Cuba's north coast at a narrow opening to a natural deep-water harbor, Havana became an essential city for the Spanish empire. Consequently, Havana possesses a handsome collection of colonial architecture, especially from the eighteenth and nineteenth centuries. Only recently did Santo Domingo edge out Havana as the Caribbean's largest city (see "Cityscapes: In the Streets of Havana").

Other colonial powers left their mark on the region's cities. For example, Paramaribo, the capital of Suriname, has been described as a tropical, tulipless extension of Holland. In the British colonies, a preference for wooden whitewashed cottages with shutters is evident. Yet the British and French colonial cities tended to be unplanned afterthoughts; these port cities were built to serve the rural estates rather than serve the needs of all residents. Most of them have grown dramatically over the last 40 years. No longer small ports for agricultural exports, increasingly these cities are oriented to welcoming cruise ships and sun-seeking tourists.

Caribbean cities and towns do have their charms and reflect a variety of cultural influences. Most people still get around by foot, bicycle, or public transportation; neighborhoods are filled with small shops and services that are within easy walking distance. Streets are narrow, and the pace of life is markedly slower than in North America and Europe. Even when space is tight in town, most settlements are close to the sea and its cooling breezes. An afternoon or evening stroll along the waterfront is a common activity. Flowering shrubs and swaying palms add to the tropical ambiance.

Housing Throughout the region, houses are often simple structures (made of wood, brick, or stucco), sometimes raised off the ground a few feet to avoid flooding, and painted in soft pastels (Figure 5.13). The sudden surge in urbanization in the Caribbean is best explained by an erosion of rural jobs, rather than a rise in urban opportunities. Thousands poured into the cities as economic migrants, erecting shantytowns and filling the ranks of the informal sector. Squatter settlements in Port-au-Prince and Santo Domingo are especially bad, with residents living in wretched housing without the benefit of sewers and running water. Electricity is usually pirated from nearby power lines.

The one place that dramatically breaks from this pattern is Cuba. Forged in a socialist mode, Cubans are housed in uniform government-built apartment blocks like those seen throughout Russia and eastern Europe. These unimaginative complexes contain hundreds of identical one- and two-bedroom apartments where basic modern amenities (plumbing, electricity, and sewage) are provided. Compared with other large cities of the developing world, the lack of squatter settlements makes Havana unusual.

Figure 5.13 Belize City Cottage
Residents of Belize City build their wooden cottages on stilts as protection against flooding. Shuttered cottages are typical throughout the British Caribbean. *(Rob Crandall/www.robcrandall.com)*

CULTURAL COHERENCE AND DIVERSITY: A Neo-Africa in the Americas

Linguistic, religious, and ethnic differences abound in the Caribbean. A score of former European colonies, millions of descendents of ethnically distinct Africans and indentured workers from India, and isolated Amerindian communities on the mainland challenge any notion of cultural coherence.

Within such diversity, common historical and social processes provide the glue for the region. European colonies, with their plantation-based economies, reproduced similar social structures throughout the region. The imprint of more than 7 million African slaves, creating a neo-Africa in the Americas, is the focus of continuing scholarly research to show the important linkages between the Caribbean and the wider Atlantic world. Last, in a process called **creolization**, African and European cultures were blended in the Caribbean. Through this mixing, European languages were transformed into vibrant local dialects, and at times entirely new languages were created (Papiamento

CITYSCAPES In the Streets of Havana

Nearly a century before the Jamestown and Plymouth colonies in the United States, Havana was settled in 1519. Called the "Pearl of the Caribbean," the city's deep and protected harbor, mild climate, and ideal location for catching trade winds back to Europe made it a setting of economic and political importance for the Spanish Empire. In fact, Spain held on to this island colony for nearly 400 years, until losing it during the Spanish-American War in 1898.

Havana is special for its architectural splendor and regional significance as the Caribbean's historic center of trade (Figure 5.3.1). Old Havana was actually a walled city, defendable from intruders, especially pirates or European rivals. Unlike the typical provincial cities found in the Caribbean and Latin America, Havana attracted travelers from around the world. Sailors, merchants, slaves, soldiers, and government officials circulated through the city's port.

By the 1950s, Havana was the Antillean metropolis, bringing in more tourists and investment than any other city in the region. The famous promenade, the Malecón, emerged as the city's social center. The beach resorts, casinos, and several hundred brothels cemented Havana's reputation as "Sin City" in

Figure 5.3.1 Old Havana
A Cuban family strolls down a narrow cobblestoned street near the cathedral in Old Havana, Cuba. The best examples of eighteenth- and nineteenth-century colonial architecture in the Caribbean are found in Havana. *(Rob Crandall/www.robcrandall.com)*

and French Creole). This melding also produced the rich and diverse musical traditions now heard throughout the world: reggae, salsa, merengue, *rara*, and calypso.

The Cultural Imprint of Colonialism

The formation of European colonies in the Caribbean destroyed indigenous societies and imposed completely different social systems and cultures. Plantation-based agriculture dependent on forced and indentured labor was the clearest expression of how the colonists ordered space and society. Unlike colonies that relied on indigenous labor, Caribbean plantations depended on labor, both forced and indentured, from Africa and Asia.

The arrival of Columbus in 1492 triggered a devastating chain of events that depopulated the region within 50 years. A combination of Spanish brutality, enslavement, warfare, and disease reduced the densely settled islands, which supported up to 3 million Caribs and Arawaks, into an uninhabited territory ready for the colonizer's hand. The demographic collapse of Amerindian populations occurred throughout the Americas (see Chapter 4), but the death rates were highest in the Caribbean. Only fragments of Amerindian communities survive, mostly on the rimland.

By the mid-sixteenth century, as rival European states vied for Caribbean territory, the lands they fought for were virtually uninhabited. In many ways, this simplified their task, as they did not have to acknowledge indigenous land claims or work amid Amerindian societies. Instead, the Caribbean territories were reorganized to serve a plantation-based production system. The critical missing element was labor. Once slave labor from Africa, and later indentured labor from Asia, was secured, the small Caribbean colonies became surprisingly profitable. Much of Caribbean culture and society today can be traced to the same processes that created plantation America.

Plantation America The term **plantation America** was coined by anthropologist Charles Wagley to designate a cultural region that extends from midway up the coast of Brazil through the Guianas and the Caribbean into the southeastern United States. Ruled by a European elite dependent on an African labor force, this society was primarily coastal and produced agricultural exports. Other characteristics

the mid-twentieth century. It seemed destined to be a nexus in the Las Vegas and Miami triangle of gambling and tourism. Yet the Cuban Revolution of 1959 derailed multi-million-dollar tourism investments, led to an exodus of tens of thousands of Cubans, and shut down tourism and gambling. Consequently, the great city was oddly frozen in time as Castro's revolution focused more on rural areas than Old Havana.

Because of fears that this Caribbean treasure would be lost, Old Havana was declared a UN World Heritage Site in 1982, and slowly funds began to trickle into the city to restore its architectural treasures. At that time, the buildings of the old city had been converted mostly to residences, and were in serious need of renovation. The turbulent rebirth of Havana as a major tourist destination began in the 1990s as Cuba coped with the demise of its economic benefactor, the former Soviet Union. Desperate for hard currency, the country reluctantly turned to tourism to earn foreign exchange. Today nearly 2 million visitors a year come to the island, most spending some time in Havana.

The streets of Havana buzz with small private enterprises, including flea markets, street vendors, and privately owned family restaurants, all of which are licensed and sanctioned by the government. Whereas having U.S. dollars was once a punishable offense, today Cubans freely carry and exchange them. Foreign investment has poured into the city, with more than $60 million spent in 2000 on historic preservation in Havana. Resources, especially fuel, still are scarce for most Cubans. In an effort to avert a transportation crisis, more than 2 million bicycles were imported from China in the 1990s. The city of Havana, with 2 million residents, boasts more than 1 million bicycles (Figure 5.3.2). The city's infrastructure has improved, but new tensions have emerged. New tourist buildings are shoddily built and typically out-of-scale with historic ones. More important are the social changes that new people and investments have brought. While Cubans embrace their new economic opportunities, they have growing concerns about rising social inequities that their revolution had tried to eliminate.

Figure 5.3.2 Bicycle Transport in Havana
A Cuban couple shows off their new Chinese bicycle on Havana's seaside promenade, the Malecón. When subsidized oil from the former Soviet Union stopped flowing into Cuba in the early 1990s, bicycles became an essential component of the city's transportation system. *(Rob Crandall/www.robcrandall.com)*

Source: Adapted from Joseph L. Scarpaci, Roberto Segre, and Mario Coyula, 2002, *Havana: Two Faces of the Antillean Metropolis*, Chapel Hill: University of North Carolina Press.

included a reliance on **monocrop production** (a single commodity, such as sugar) under a plantation system that concentrated land in the hands of elite families. Such a system created rigid class lines, as well as the formation of a multiracial society in which people with lighter skin were privileged. The term *plantation America* is not meant to describe a race-based division of the Americas, but rather a production system that relied upon export commodities, coerced labor, and limited access to land (Figure 5.14).

Asian Immigration Before detailing the pervasive influence of Africans on the Caribbean, the lesser-known Asian presence deserves discussion. By the mid-nineteenth century, most colonial governments in the Caribbean had begun to free their slaves. Fearful of labor shortages, they sought **indentured labor** (workers contracted to labor on estates for a set period of time, often several years) from South and Southeast Asia.

The legacy of these indentured arrangements is clearest in Suriname, Guyana, and Trinidad and Tobago. In Suriname, a former Dutch colony, more than one-third of the population is of South Asian descent, and 16 percent are Javanese (from Indonesia). Guyana and Trinidad were British colonies, and most of their contract labor came from India. Today half of Guyana's population and 40 percent of Trinidad and Tobago's claim South Asian ancestry. Hindu temples are found in the cities and villages, and many families speak Hindi in the home (Figure 5.15). Such racial diversity has led to tension. During Guyana's general election in 2001, tensions between Afro- and Indo-Guyanese were heightened, and international election observers were called in to monitor the vote. In the end, an Indo-Guyanese candidate was peacefully elected. Most of the former English colonies have Chinese populations of not more than 2 percent. Once these East Asian immigrants fulfilled their agricultural contracts, they often became merchants and small-business owners, positions they still hold in Caribbean society.

Creating a Neo-Africa

African slaves were first introduced to the Americas in the sixteenth century, partly in response to the demographic collapse of the Amerindians. The flow of slaves continued

Figure 5.14 Sugar Plantation A historical illustration (1823) of slaves harvesting sugarcane on a plantation in Antigua. Sugar production was profitable but arduous work. Several million Africans were enslaved and forcibly relocated into the region. *(Michael Holford/Michael Holford Photographs)*

into the nineteenth century. This forced migration of mostly West Africans to the Americas was only part of a much more complex African diaspora—the forced removal of Africans from their native area. The slave trade also crossed the Sahara to include North Africa and linked East Africa with a slave trade in the Middle East (see Chapter 6). The best documented slave route is the transatlantic one; at least 10 million Africans landed in the Americas, and it is estimated that another 2 million died en route. More than half of these slaves were sent to the Caribbean (Figure 5.16).

This influx of slaves, combined with the extermination of nearly all the native inhabitants, recast the Caribbean as the area with the greatest concentration of African transfers in the Americas. The African source areas extended from Senegal to Angola, and slave purchasers intentionally mixed tribal groups in order to dilute ethnic identities. Consequently, intact transfer of African religions and languages into the Caribbean did not occur; instead languages, customs, and beliefs were blended.

Figure 5.15 South Asian Indians in Trinidad Trinidadians of Indian ancestry celebrate a Hindu festival by sending burning boats down the Marianne River in Trinidad. About 40 percent of Trinidad and Tobago's population is ethnically India and adhere to the Hindu faith. *(Steve Raymer/Corbis)*

Maroon Societies Communities of runaway slaves—termed **maroons**—offer the most compelling examples of African cultural diffusion across the Atlantic. Hidden settlements of escaped slaves existed wherever slavery was practiced. Called *maroons* in English, *palenques* in Spanish, and *quilombos* in Portuguese, many of these settlements were short-lived, but others have endured and allowed for the survival of African traditions, especially farming practices, house designs, community organization, and language.

The maroons of Suriname and French Guiana still manifest clear links to West Africa. They form the largest maroon population in the Western Hemisphere. Whereas other maroon societies gradually were assimilated into local populations, to this day the maroons of Suriname maintain a distinct identity. These runaways fled the Dutch coastal plantations in the seventeenth and eighteenth centuries, forming riverine settlements amid the interior rain forest. Six distinct maroon tribes formed, ranging in size from a few hundred to 20,000. Clear manifestations of West African cultural traditions persist, including religious practices, crafts, patterns of social organization, agricultural systems, and even dress (Figure 5.17). Living relatively undisturbed for 200 years, these rainforest inhabitants fashioned a rich ritual life for themselves involving oracles, spirit possession, and witch doctors. Yet pressures to modernize and extract resources have placed the maroons in direct conflict with the state and private businesses. The maroons in Suriname have been directly affected by the construction of dams, gold

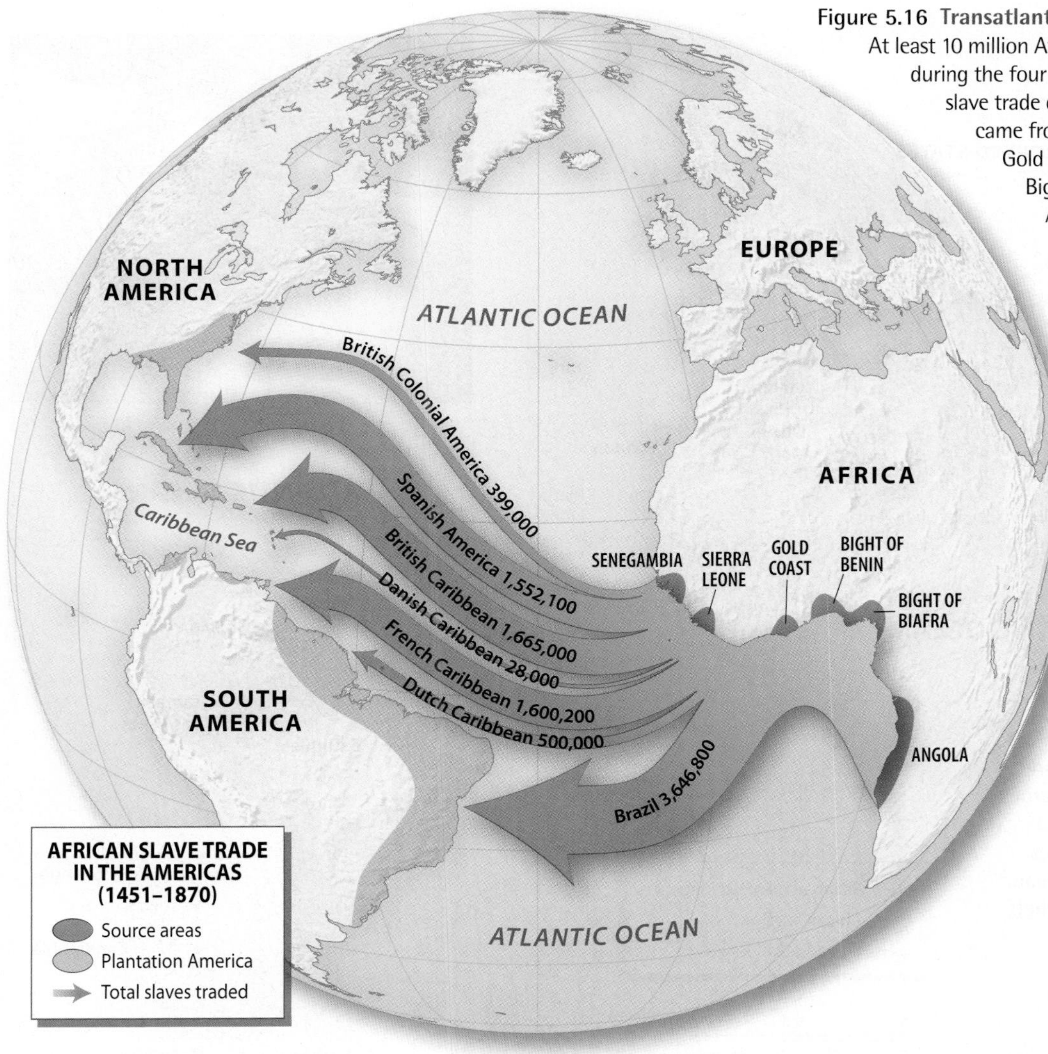

Figure 5.16 Transatlantic Slave Trade
At least 10 million Africans landed in the Americas during the four centuries in which the Atlantic slave trade operated. Most of the slaves came from West Africa, especially the Gold Coast (now Ghana) and the Bight of Biafra (now Nigeria). Angola, in southern Africa, was also an important source area. *(Data based on Curtin, 1969,* The Atlantic Slave Trade, A Census, *Madison: University of Wisconsin Press, p. 268)*

mining operations, and logging concessions. From 1986 until 1992 there was a civil war between the maroons and the Creole-run military in which hundreds of maroons were killed and villages destroyed. Although peace was brokered in 1992, the maroons continue to fight for legal recognition of ancestral claims to land and resources.

African Religions Linked to maroon societies, but more widely diffused, is the transfer of African religious and magical systems to the Caribbean. These patterns are another reflection of neo-Africa in the Americas and are most closely associated with northeastern Brazil and the Caribbean. In Chapter 4 we discussed how millions of Brazilians practice the African-based religions of Umbanda, Macuba, and Candomblé, along with Catholicism. Likewise, Afro-religious traditions in the Caribbean have evolved into unique forms that have clear ties to West Africa. The most widely practiced are Voodoo (also Vodoun) in Haiti, Santería in Cuba, and Obeah in Jamaica. These religions have their own priesthood and unique patterns of worship. Their impact is considerable; the father and son dictators of Haiti, the Duvaliers, were

Figure 5.17 Maroon Village in Suriname
A maroon woman carries a pail of water to her home in the village of Stonuku, Suriname. Maroon communities existed throughout the Caribbean as slaves ran away from plantations and formed villages in remote localities. The maroon communities in Suriname still retain many African traditions. *(Robert Caputo/Aurora/Getty Images)*

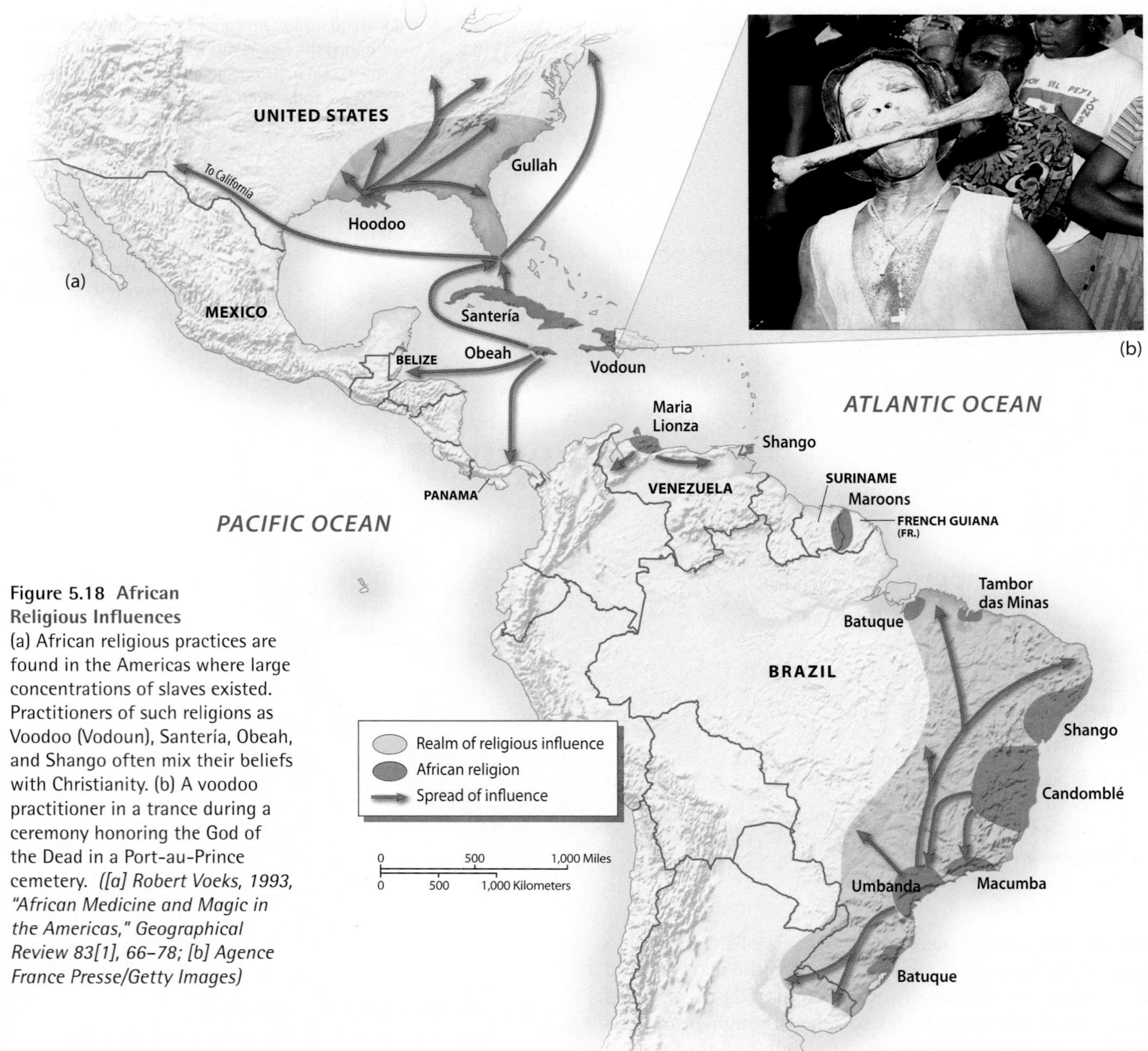

Figure 5.18 African Religious Influences
(a) African religious practices are found in the Americas where large concentrations of slaves existed. Practitioners of such religions as Voodoo (Vodoun), Santería, Obeah, and Shango often mix their beliefs with Christianity. (b) A voodoo practitioner in a trance during a ceremony honoring the God of the Dead in a Port-au-Prince cemetery. *([a] Robert Voeks, 1993, "African Medicine and Magic in the Americas," Geographical Review 83[1], 66–78; [b] Agence France Presse/Getty Images)*

known to hire Voodoo priests to scare off government opposition during their rule from 1957 to 1986. Moreover, as Figure 5.18 shows, many of these religions have diffused from their areas of origin. Santería is practiced in Florida and New York by some Cuban immigrants. Likewise, belief in Obeah diffused when Jamaicans migrated to Panama and Los Angeles.

Creolization and Caribbean Identity

Creolization refers to the blending of African, European, and even some Amerindian cultural elements into the unique sociocultural systems found in the Caribbean. Early anthropologists tended to shun the Caribbean as "cultureless," since an indigenous society was replaced with fragments of European and African traditions. Yet the Creole identities that have formed over time are complex; they illustrate the cultural and national identities of the region. Today Caribbean writers (V. S. Naipaul, Derek Walcott, and Jamaica Kinkaid), musicians (Bob Marley, Celia Cruz, and Juan Luís Guerra), and artists (Trinidadian costume designer Peter Minshall) are internationally regarded. Collectively, these artists are representative of their individual islands and of Caribbean culture as a whole.

The story of the Garifuna people illustrates creolization at work. Settled along the Caribbean rimland from the southern coast of Belize to the northern coast of Honduras, the Garifuna (formerly called the *Black Carib*) are descendants of African slaves who speak an Amerindian language.

Figure 5.19 Garifuna Girl
A Garifuna girl sits on fishing nets in Batalla Village on the north coast of Honduras. The Garifuna are mostly of African ancestry, but they speak an Amerindian language. *(Rob Crandall/www.robcrandall.com)*

Unions between Africans and Carib Indians on the island of St. Vincent produced an ethnic group that was predominantly African but spoke an Indian language. In the late eighteenth century, Britain forcibly resettled some 5,000 Garifuna from St. Vincent to the Bay Islands in the Gulf of Honduras. Over time, the Garifuna settled along the Caribbean coast of Central America, living in isolated fishing communities from Honduras to Belize (Figure 5.19). In addition to maintaining an Indian language, the Garifuna are the only group in Central America who regularly eat bitter manioc—a root crop common in lowland tropical South America. It is assumed that they acquired their taste for manioc from their exposure to Carib culture. The Afro-Indian blend that the Garifuna manifest is unique, but the process of creolization is recognizable throughout the Caribbean, especially in language and music.

Language The dominant languages in the region are European: Spanish (24 million speakers), French (9 million), English (6 million), and about half a million Dutch (Figure 5.20). Yet these figures tell only part of the story.

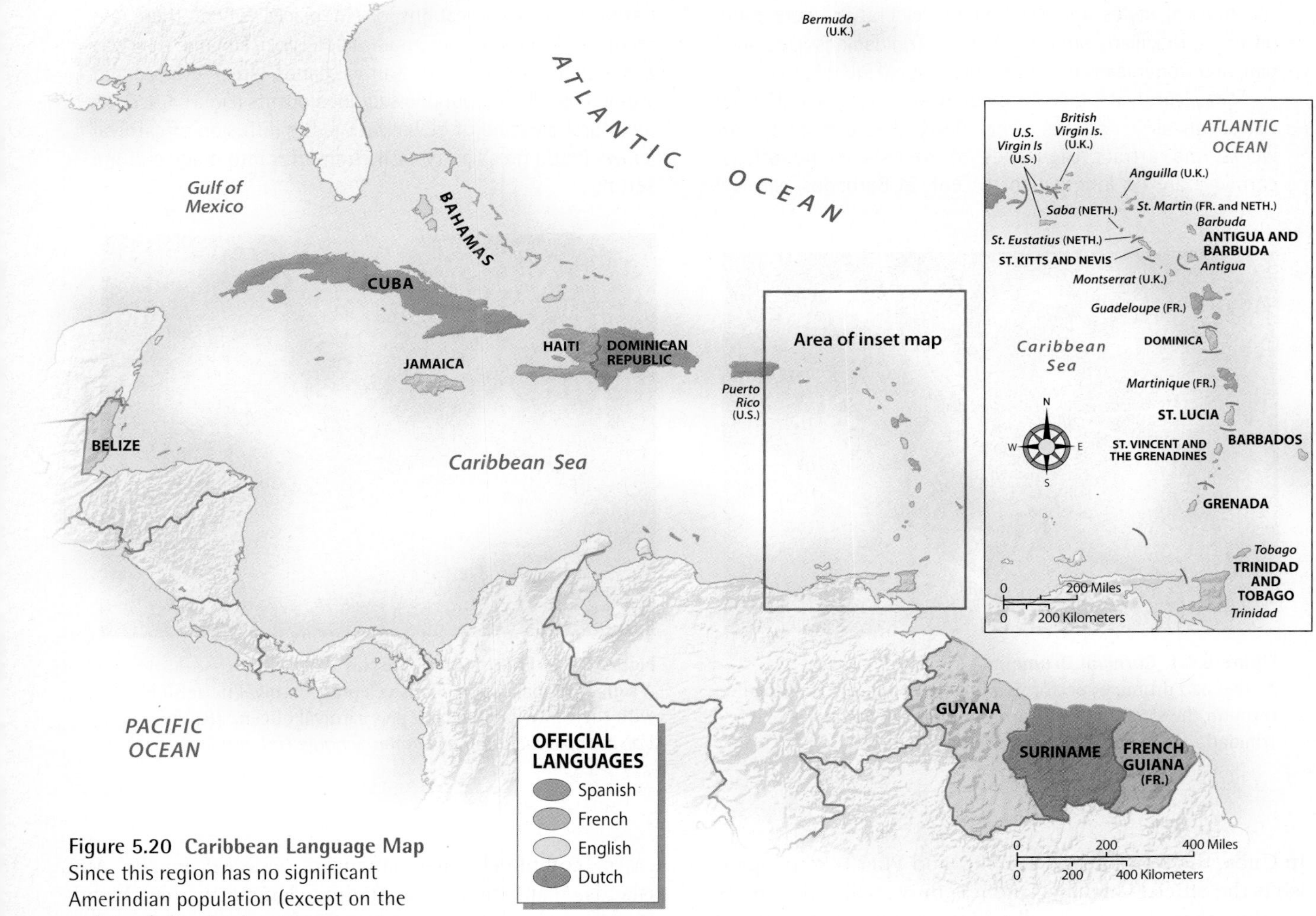

Figure 5.20 Caribbean Language Map
Since this region has no significant Amerindian population (except on the mainland), the dominant languages are European: Spanish (24 million), French (8 million), English (6 million), and Dutch (0.5 million). However, many of these languages have been creolized, making it difficult for outsiders to understand them.

GLOBAL TO LOCAL Caribbean Carnival Around the World

The origins of modern carnival are in Christian Lenten beliefs and African musical traditions. The term is Italian in origin, a reference to giving up eating meat (*carnivale* means to put meat away) in observance of Lent (the 40-day period prior to Easter). In the context of the Caribbean and Brazil, former slaves imbued carnival with special meaning because it became their opportunity to break the monotony of their daily lives as well as provide a chance to mock the white elite while appearing to observe Christian traditions. In the Americas, carnival was always associated with African rhythms, dancing, and costuming. Afro-Trinidadians, for example, might dress as European royalty and mawkishly put on white face while dancing to rhythmic music. In the nineteenth century, local governments made several attempts to ban carnival, but the music and the celebration were hard to resist; over time carnival celebrations became part of the national identity for Brazil and Trinidad, and to a lesser extent for the Dominican Republic and Cuba. In each of these cases, the musical styles and method of celebration were quite different. Brazilians preferred Samba, Trinidadians liked steel pan, and Dominicans moved to merengue (Figure 5.4.1).

Today, carnival is celebrated on nearly every island of the Caribbean as a national street party that can go on for weeks and attract thousands of tourists. Many official carnivals are no longer tied to Lent. In Barbados, carnival runs from late May to early August; in the Cayman Islands, it takes place in May; and in Cuba, it is in July. Stripped of its religious significance, carnival is now a celebration of life in the Caribbean in which people express themselves through music, costumes, dancing, beauty contests, and parades. Over time, the celebrations have become more and more alike, with popular musical styles and elaborate but revealing costumes.

Carnival celebrations have followed the Caribbean diaspora to new settings in North America and Europe. More than a dozen U.S. cities such as Atlanta, Baltimore, and Boston have carnivals that last a day or two in the summer. One of the biggest carnivals in North America is in Toronto, where a large Caribbean immigrant population maintains the tradition every July. In London, Birmingham, and Leicester, Caribbean carnivals are celebrated annually. The same is true for the Netherlands, where a large number of Surinamese have introduced carnival to Amsterdam and Rotterdam. Carnival has even caught on in places where there are relatively few Caribbean migrants. Helsinki, Finland, now has a summer carnival with samba bands, steel drums, and women parading around in sequined outfits (Figure 5.4.2). As a cultural measure of globalization, the diffusion of carnival shows that a great party easily translates into many cultural settings.

Figure 5.4.1 Carnival Drummer
A steel pan drummer performs while his drum cart is pushed through the streets during carnival in Port of Spain, Trinidad. *(Rob Crandall/www.robcrandall.com)*

Figure 5.4.2 Finnish Samba Band
A sign of globalization, Finns celebrate carnival in Helsinki with their own samba band and carnival queen. *(Paivi Vaisanen/Association of Samba Schools Finland)*

In Cuba, the Dominican Republic, and Puerto Rico, Spanish is the official language, and it is universally spoken. As for the other countries, colloquial variants of the official language exist, especially in spoken form, that can be difficult for a nonnative speaker to understand. In some cases, completely new languages emerge; in the ABC islands, Papiamento (a trading language that blends Dutch, Spanish, Portuguese, English, and African languages) is the *lingua franca*, with usage of Dutch declining. In Suriname the vernacular language is Sranan Tongo (an amalgam of Dutch and English with many African words). Similarly, French Creole or *patois* in Haiti has constitutional status as a distinct language. In practice, French is used in higher education, government, and the courts, but patois (with clear African influences) is the language of the street, the home, and oral tradition. Most Haitians speak patois, but only the formally educated know French.

With independence in the 1960s, Creole languages became politically and culturally charged with national meaning. During the colonial years, Creole was negatively viewed as a corruption of standard European forms. Those who spoke Creole were considered uneducated or backward. As linguists began to study these languages, they found that while the vocabulary came from Europe, the syntax or semantic structure had other origins, notably from the African language families. Although most formal education is taught using standard language forms, the richness of vernacular expression and its ability to instill a sense of identity are appreciated. Locals rely on their ability to switch from standard to vernacular forms of speech. Thus, a Jamaican can converse with a tourist in standard English and then switch to a Creole variant when a friend walks by, effectively excluding the outsider from the conversation. While this ability to switch is evident in many cultures, it is widely used in the Caribbean.

Music The rhythmic beats of the Caribbean might be the region's best-known product. This small area is the hearth of reggae, calypso, merengue, rumba, zouk, and scores of other musical forms. The roots of modern Caribbean music reflect a combination of African rhythms with European forms of melody and verse. These diverse influences coupled with a long period of relative isolation sparked distinct local sounds. As circulation among Caribbean inhabitants increased, especially during the twentieth century, musical traditions were grafted onto each other, but characteristic sounds remained (see "Global to Local: Caribbean Carnival Around the World").

The famed steel pan drums of Trinidad were created from oil drums discarded from a U.S. military base there in the 1940s. The bottoms of the cans are pounded with a sledge hammer to create a concave surface that produces different tones. During carnival, racks of steel pans are pushed through the streets by dancers while the panmen sound off. So skilled are these musicians that they even perform classical music, and government agencies encourage troubled teens to learn steel pan.

The eclectic sound and the ingenious rhythms make Caribbean music very popular. It is much more than good dancing music; the music is closely tied to Afro-Caribbean religions and is a popular form of political protest. In Haiti, rara music mixes percussion instruments, saxophones, and bamboo trumpets, while weaving in funk and reggae basslines (Figure 5.21). The songs always are performed in French Creole and typically celebrate Haiti's African ancestry and the use of Voodoo. The lyrics address difficult issues, such as political oppression or poverty. Consequently, rara groups and other musicians have been banned from performing and even forced into exile—most notably, folk singer Manno Charlemange, who later returned to Haiti and was elected mayor of Port-au-Prince in the 1990s.

The music of the late Bob Marley also takes a political stand. Jamaican-born Marley sang of his life in the Kingston ghetto of Trenchtown. He was a devout Rastafarian who believed that Jah was the living force, that New World Africans should look to Africa for a prince to emerge (determined to be Haile Selassie of Ethiopia), and that *ganja* (marijuana) should be consumed regularly. It was Marley's political voice as peacemaker, however, that touched so many lives. His first hit, "Simmer Down," was written to quell street violence that had erupted in Kingstown in 1964. Other songs, such as "Stand-up" and "No Woman No Cry," had a message of social unity and freedom from oppression that resonated in the 1970s. Commercial success never dulled Marley's political edge. He was wildly popular in Africa, and before his death in 1981, one of his last concerts was in Zimbabwe to mark its independence.

Figure 5.21 Haiti's Rara Music
Performed in procession, rara music is sung in patois. It is considered the music of the poor and used to express risky social commentary. This rara band performs at a folk festival in Washington, DC. *(Rob Crandall/www.robcrandall.com)*

GEOPOLITICAL FRAMEWORK: Colonialism, Neocolonialism, and Independence

Caribbean colonial history is a patchwork of rival powers dueling over profitable tropical territories. By the seventeenth century, the Caribbean had become an important

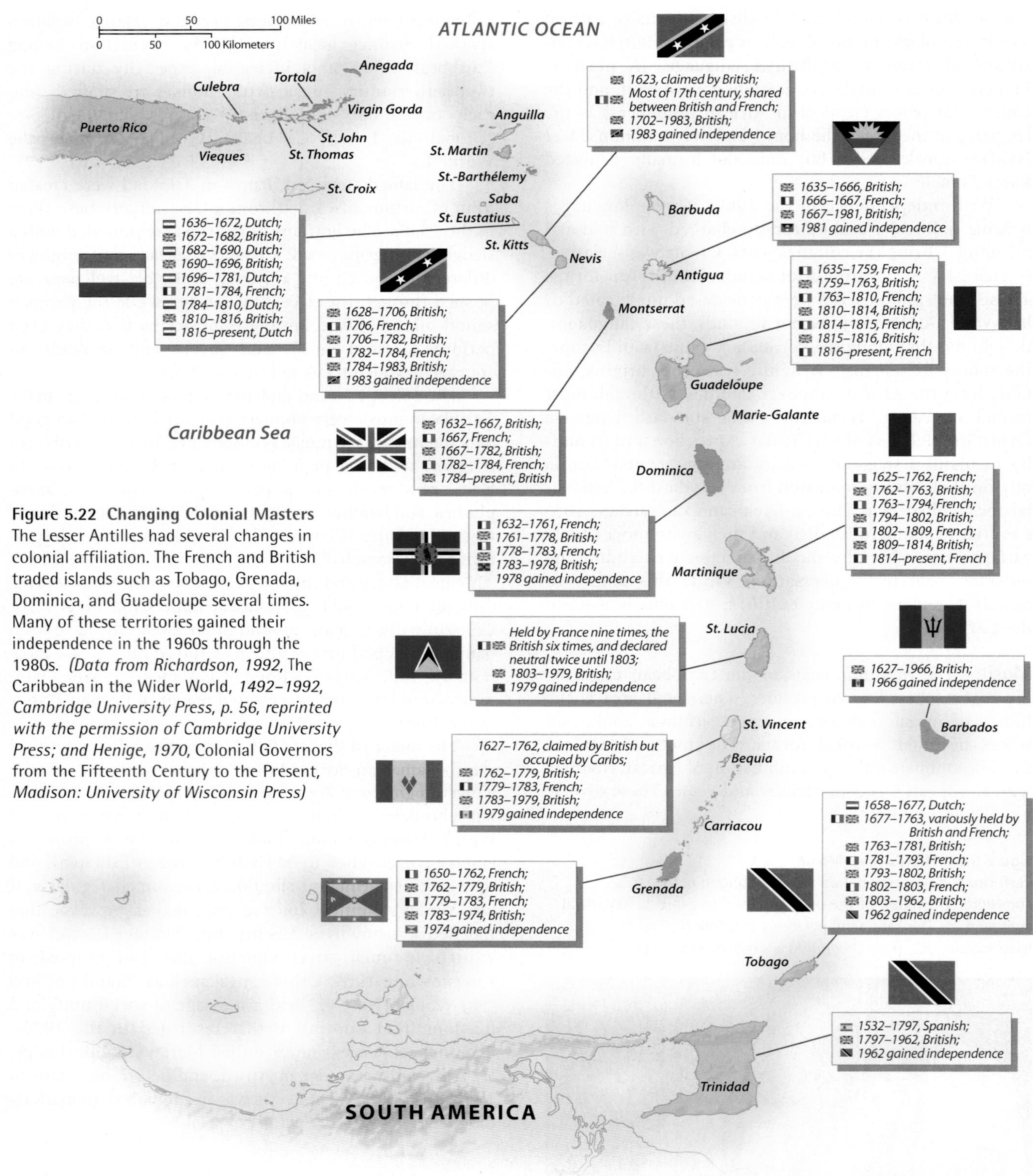

Figure 5.22 Changing Colonial Masters The Lesser Antilles had several changes in colonial affiliation. The French and British traded islands such as Tobago, Grenada, Dominica, and Guadeloupe several times. Many of these territories gained their independence in the 1960s through the 1980s. *(Data from Richardson, 1992,* The Caribbean in the Wider World, *1492–1992, Cambridge University Press, p. 56, reprinted with the permission of Cambridge University Press; and Henige, 1970,* Colonial Governors from the Fifteenth Century to the Present, *Madison: University of Wisconsin Press)*

proving ground for European colonial ambitions. Spain's grip on the region was tentative, and rivals felt confident that they could win territory by gradually moving from the eastern edge of the sea to the west. Many territories, especially islands in the Lesser Antilles, changed hands several times (Figure 5.22). In a few instances, contested colonial holdings have produced contemporary border disputes. Only recently did the Guatemalan government give up its claim to Belize, arguing that the British ignored Spanish claims to the area and illegally acquired it. Also, there are several long-standing border disputes among the Guianas. Yet there are indications of new-found regional cooperation, especially with the expansion of the Caribbean Community membership beyond the English-speaking countries to include Haiti and Suriname.

In the seventeenth and eighteenth centuries, Europeans viewed the Caribbean as a strategic and profitable region in which to produce sugar, rum, and spices. Geopolitically, rival European powers also felt that their presence in the Caribbean limited Spanish authority there. Yet Europe's geopolitical dominance in the Caribbean began to wane by the mid-nineteenth century just as the U.S. presence increased. Inspired by the **Monroe Doctrine**, which claimed that the United States would not tolerate European military involvement in the Western Hemisphere, the U.S. government made it clear that it considered the Caribbean to be within its sphere of influence. Even though several English, Dutch, and French colonies persisted after this date, the United States indirectly (and sometimes directly) asserted its control over the region, ushering in a period of **neocolonialism**.

Life in the "American Backyard"

For much of the twentieth century the United States maintained a proprietary attitude toward the Caribbean, referring to it as the "American backyard." Initial U.S. foreign policy objectives were to free it from European tyranny and foster democratic governance. Yet time and again, American political and economic ambitions betrayed those goals. President Theodore Roosevelt made his priorities clear with imperialistic policies that emphasized the construction of the Panama Canal and the maintenance of open sea-lanes. The United States later offered benign-sounding development packages such as the Good Neighbor Policy (1930s), the Alliance for Progress (1960s), and the Caribbean Basin Initiative (1980s). The Caribbean view of such initiatives has been guarded at best. Rather than feeling liberated, many residents believe that one kind of political dependence was being traded for another: colonialism for neocolonialism.

In the early 1900s, the role of the United States in the Caribbean was overtly military and political. The Spanish- American War (1898) secured Cuba's freedom from Spain and also resulted in Spain's ceding the Philippines, Puerto Rico, and Guam to the United States; the latter two still are U.S. territories. The U.S. government also purchased the Danish Virgin Islands in 1917, renaming them the U.S. Virgin Islands and developing the harbor of St. Thomas. French, English, and Dutch colonies were tolerated as long as these allies recognized the supremacy of the United States in the region. Avowedly against colonialism, the United States had become much like an imperial force.

One of the requirements of an empire is the ability to impose one's will, by force if necessary. Marines were landed and U.S.-backed governments were installed throughout the Caribbean basin. These were not short-term engagements; U.S. troops occupied the Dominican Republic from 1916 to 1924, Haiti from 1913 to 1934, and Cuba from 1906 to 1909 and 1917 to 1922 (see Figure 5.23). Even today several important military bases are in the region, including Guantánamo in eastern Cuba. There is greater reluctance to commit troops in the area now, but as recently as 1994 and 2004 U.S. troops were sent to Haiti to suppress political violence and prevent a mass exodus of Florida-bound refugees.

Many critics of U.S. policy in the Caribbean complain that business interests overshadow democratic principles in the determination of that policy. U.S. banana companies settled the coastal plain of the Caribbean rimland and operated as if they were independent states. Sugar and rum manufacturers from the United States bought the best lands in Cuba, Haiti, and Puerto Rico. Meanwhile, truly democratic institutions remained weak, and social development saw little improvement. True, exports increased, railroads were built, and port facilities improved; but levels of income, education, and health remained abysmally low throughout the first half of the twentieth century.

The Commonwealth of Puerto Rico Puerto Rico is both within the Caribbean and apart from it because of its status as a commonwealth of the United States. Throughout the twentieth century, various Puerto Rican independence movements sought to uncouple the island from the United States. Even today, residents of the island are divided about their island's political future. At the same time, Puerto Rico depends on U.S. investment and welfare programs; U.S. food stamps are a major source of income for many Puerto Rican families. Commonwealth status also means that Puerto Ricans can freely move between the island and the U.S. mainland, a right they actively assert. In other ways, Puerto Ricans symbolically manifest their independence; for example, they support their own "national" sports teams and send a Miss Puerto Rico to international beauty pageants. The dispute over Vieques Island also has showcased a current of Puerto Rican independence. The small island off Puerto Rico's east coast has been used by the U.S. Navy for live bombing exercises. In 2000, Puerto Rico elected its first female governor, Sila Calderon, on a platform calling for the end of Navy-led live-fire exercises on Vieques. After years of protest that included Puerto Rican activists' placing themselves in the line of fire, President George W. Bush agreed to find a new space for naval exercises in 2003.

Puerto Rico led the Caribbean in the transition from an agrarian economy to an industrial one beginning in the 1950s, and became the model for the rest of the region. Puerto Rican President Muñoz Marín championed an industrialization program called "Operation Bootstrap." Through tax incentives and cheap labor, hundreds of U.S. textile and apparel firms relocated to Puerto Rico. Over the next two decades, 140,000 industrial jobs were added, resulting in a marked increase in per capita gross national product (GNP). In the 1970s, when Puerto Rico faced stiff competition from Asian apparel manufacturers, the government encouraged petrochemical and pharmaceutical

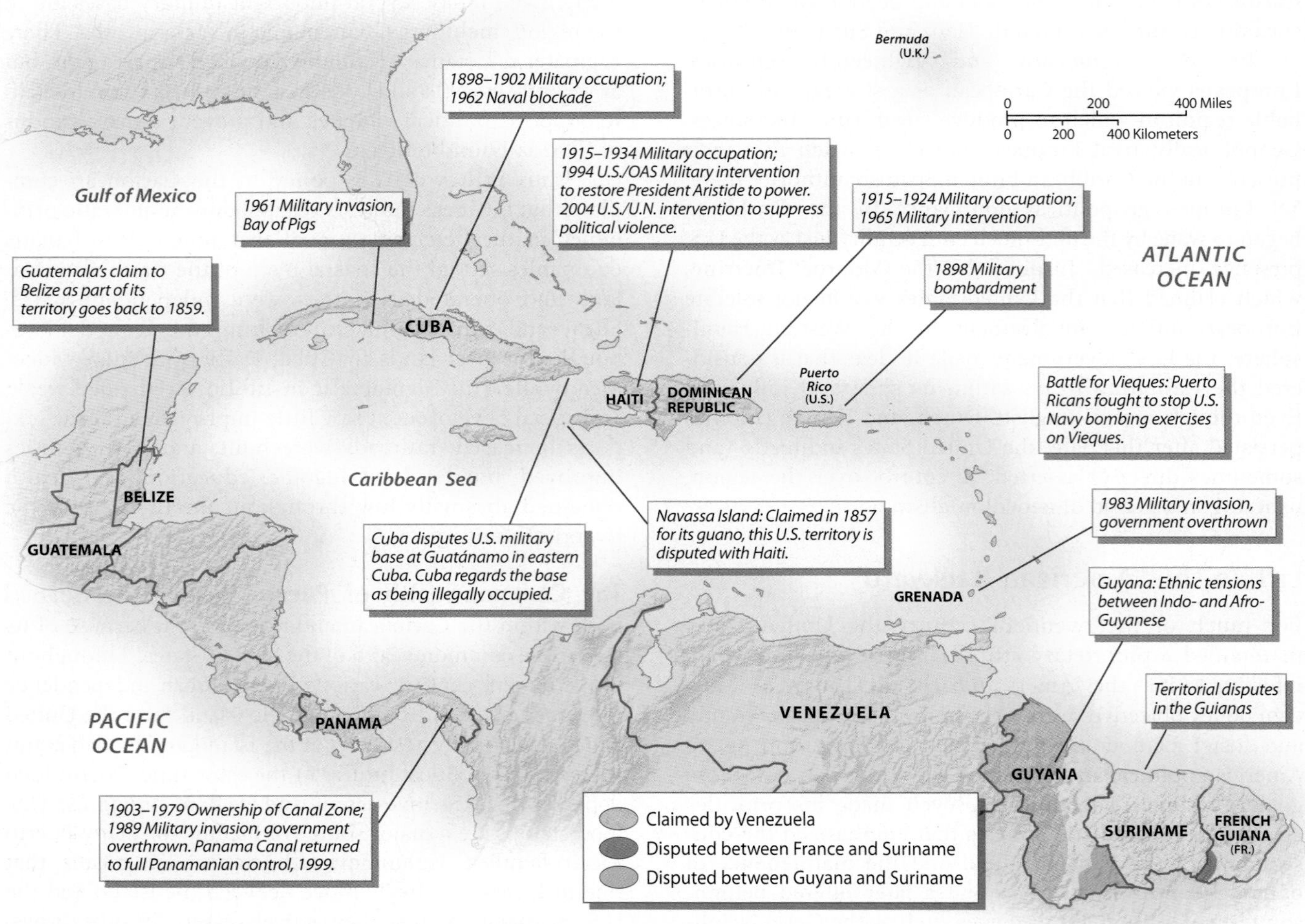

Figure 5.23 Caribbean Geopolitics: U.S. Military Involvement and Regional Disputes
The Caribbean was labeled the geopolitical backyard of the United States, and U.S. military occupation was a common occurrence in the first half of the twentieth century. Border and ethnic conflicts also exist, most notably in the Guianas. *(Data from Tenenbaum, ed., 1996,* 1996 Encyclopedia of Latin American History and Culture, *vol. 5, p. 296, with permission of Charles Scribner's Sons; and Allcock, 1992,* Border and Territorial Disputes, *3rd ed., Harlow, Essex, UK: Longman Group)*

plants to relocate to the island (Figure 5.24). Puerto Rico is the most industrialized locality in the region, with a purchasing power parity (PPP) of $19,300 per capita. Yet it still shows many signs of underdevelopment, including rampant out-migration, low rates of educational attainment, and widespread poverty and crime.

Cuba and Regional Politics The most profound challenge to U.S. authority in the region came from Cuba and its superpower ally, the former Soviet Union. In the 1950s, a revolutionary effort led by Fidel Castro began in Cuba against the pro-American Batista government. Cuba's economic productivity had soared but its people were still poor, uneducated, and increasingly angry. The contrast between the lives of average cane workers and the foreign elite was stark and inescapable. Castro tapped a deep vein of Cuban resentment against six decades of American neocolonialism. After Castro's government nationalized American industries and took ownership of all foreign-owned properties, the United States responded by refusing to buy Cuban sugar and ultimately ending diplomatic relations with the state. Various U.S. trade embargoes against Cuba have existed for more than four decades. What sealed Cuba's fate as a geopolitical enemy was its establishment of diplomatic relations with the USSR in 1960 during the height of the Cold War. With the Soviet Union financially and militarily backing Castro, a direct U.S. invasion of Cuba was too risky. The fall of 1962 produced one of the most dangerous episodes of the Cold War when Soviet missiles were discovered on Cuban soil. Ultimately, the Soviet Union removed its weapons; in return, the United States promised not to invade Cuba.

Castro wanted to extend his socialist vision abroad, supporting at different times revolutionaries in Colombia,

Venezuela, Guyana, Suriname, Bolivia, and across the Atlantic in Angola (Figure 5.25). Other Caribbean states watched the Cuba socialist experiment with interest, especially its remarkable strides in literacy and public health. By the mid-1970s, Cuba had the lowest infant mortality rate in all of the Caribbean and Latin America. Yet, Cuban-style socialism did not readily transfer to other countries, in part, because the United States was determined to prevent it, and Soviet support of Cuba proved too expensive. Even with the end of the Cold War, when Cuba lost its financial support from the Soviet Union, it managed to reinvent itself by growing its tourism sector and courting foreign investment, especially from Spain. More recently, Castro and Venezuelan President Hugo Chavez have become close political allies, signing an important trade agreement in 2004 to exchange medical resources (from Cuba) and petroleum (from Venezuela). All the while, the United States maintains its tough trade sanctions against Cuba and forbids U.S. tourists from visiting the island.

A new political era for Cuba appears imminent. In 2006 Fidel Castro turned 80 years old and his health began to fail. After suffering grave complications after intestinal surgery, Fidel asked his brother, Raul Castro, to assume the duties of President. Raul is widely thought to be willing to

Figure 5.25 Fidel Castro
A world political figure for nearly five decades, Fidel Castro came to power in 1959. Since then he has promoted his vision of socialism within Cuba and beyond. His autocratic rule, Cold War alliance with the Soviet Union, and defiant attitude toward the United States has long challenged U.S. authority in the region. In 2006, Castro turned 80 years old and his health began to fail. Raul Castro, his brother, temporarily assumed the duties of president. As of 2008, the political leadership of Cuba is poised for a major transition. *(Rob Crandal/www.robcrandall.com)*

Figure 5.24 Puerto Rican Pharmaceutical Plant
Women at work in a pharmaceutical plant in Puerto Rico. Since the 1980s US pharmaceutical companies have located plants in Puerto Rico due to lower costs and tax advantages offered by the US government. *(David Frazier/The Image Works)*

encourage more private enterprise in Cuba but as of 2008 no major economic or political changes have occurred.

Independence and Integration

Given the repressive colonial history of the Caribbean, it is no wonder that the struggle for political independence began more than 200 years ago. Haiti was the second colony in the Americas to gain independence, in 1804; the United States was the first, in 1776. However, the achievement of political independence by many states in the region has not guaranteed economic independence. Many Caribbean states struggle to meet the basic needs of their people. Surprisingly, today some Caribbean territories maintain their colonial status as an economic asset. For example, the French territories of Martinique, Guadeloupe, and French Guiana are overseas departments of France; residents have full French citizenship and social welfare benefits.

Perhaps the hardest task facing the Caribbean is promoting regional integration. Scattered islands, a divided rimland, different languages, and limited economic resources inhibit the formation of a meaningful regional trade block. Yet over the years CARICOM (the Caribbean Community and Common Market) has steadily expanded beyond its network of former English colonies to include Suriname and Haiti, thus becoming more representative of the region as a whole.

Independence Movements Haiti's revolutionary war began in 1791 and ended in 1804. Spanish, French, and British forces were involved, as well as factions within Haiti that had formed along racial lines. In the end, a full-blown race war ensued. The island's population was halved through casualties and emigration; ultimately, the former slaves became the rulers. Independence, however, did not allow this crown of the French Caribbean to prosper. Plantation America watched in horror as Haitian slaves used guerrilla tactics to gain their freedom. Fearing that other colonies might follow Haiti's lead, plantation owners in other countries were on guard for the slightest hint of revolt. For its part, Haiti did not become a leader in liberation. Mired in economic and political problems, it was shunned by the European powers and never embraced by the states of the Spanish mainland when they became independent in the 1820s.

Several revolutionary pulses followed in the nineteenth century. In the Greater Antilles, the Dominican Republic finally gained independence in 1844 after wresting control of the territory from Spain and Haiti. Cuba and Puerto Rico were freed from Spanish colonialism in 1898, but their independence was compromised by greater U.S. involvement. The British colonies also faced revolts, especially in the 1930s, yet it was not until the 1960s that independent states emerged from the English Caribbean. First the larger colonies of Jamaica, Trinidad and Tobago, Guyana, and Barbados gained their independence. Other British colonies followed throughout the 1970s and early 1980s: the Bahamas in 1973, Grenada in 1974, Dominica in 1978, St. Vincent and the Grenadines in 1979, St. Lucia in 1979, Antigua and Barbuda in 1981, Belize in 1981, and St. Kitts and Nevis in 1983. Suriname, the only Dutch colony on the rimland, became an autonomous territory in 1954 but remained part of the Kingdom of the Netherlands until 1975, when it declared itself an independent republic.

Present-Day Colonies Britain still maintains several crown colonies in the region: the Cayman Islands, the Turks and Caicos, Anguilla, Montserrat, and Bermuda. The combined population of these islands is about 120,000 people, yet their standard of living is high, due in part to their specialization in the recently developed industry of offshore financial services (see "Geography in the Making: Bermuda's Economic Rise and Political Uncertainty"). French Guiana, Martinique, and Guadeloupe are each departments of France and thus, technically speaking, not colonies. Together they total 1 million people. The Dutch islands in the Caribbean are considered autonomous countries that are part of the Kingdom of the Netherlands. Curaçao, Bonaire, St. Martin, Saba, and St. Eustatius make up the federation of the Netherlands Antilles. Aruba left the federation in 1986 and governs without its influence. Together, the population of the Dutch islands is a quarter million people.

Regional Integration Experimentation with regional trade associations as a means to improve the economic competitiveness of the Caribbean began in the 1960s. The goal of regional cooperation was to improve employment rates, increase intraregional trade, and ultimately reduce external dependence. The countries of the English Caribbean took the lead in this development strategy. In 1963, Guyana proposed an economic integration plan with Barbados and Antigua. In 1972, the integration process intensified with the formation of the **Caribbean Community and Common Market (CARICOM)**, referred to earlier. Representing the former English colonies, CARICOM proposed an ambitious regional industrialization plan and the creation of the Caribbean Development Bank to assist the poorer states. CARICOM also oversees the University of the West Indies, with campuses in Trinidad, Jamaica, and Barbados. As important as this trade group is as an institutional symbol of collective identity, it has produced limited improvements in intraregional trade.

Today there are 15 full-member states in CARICOM—all of the English Caribbean, French-speaking Haiti, and Dutch-speaking Suriname. Other dependencies, such as Anguilla, Bermuda, Turks and Caicos, Cayman Islands, and the British Virgin Islands, are associate members. The Netherlands Antilles and the Dominican Republic have observer status but have proposed switching to associate or full status. In 2007 the U.S. Virgin Islands' government announced that it would seek ties with CARICOM. As CARICOM's membership grows, so does its services. In 2005 CARICOM passports began to be issued to facilitate travel between member nations. And, during the 2007

GEOGRAPHY IN THE MAKING

Bermuda's Economic Rise and Political Uncertainty

Bermuda's splendid isolation, its status as an overseas territory of the United Kingdom, and its talent for attracting financial services and rich homeowners has brought great wealth to this small archipelago 600 miles (960 kilometers) due east of South Carolina. To call Bermuda part of the Caribbean is a stretch, due to its remote location in the North Atlantic. Yet in 2003 it became an associate member of CARICOM, underscoring its solidarity with the Caribbean Community. With a population of nearly 60,000, it has much in common with the Cayman Islands and the Turks and Caicos.

By some measures, Bermudans boast the highest per capita income in the world at $70,000 (PPP). The island, first settled in 1609 by shipwrecked English colonists on their way to Virginia, has been a major upscale tourist destination for Americans since the 1950s. With its manicured golf courses, cricket fields, coral sands, and the occasional man in Bermuda shorts, the island epitomizes a Caribbean ideal. Its economic boom was created, however, by overseas financial services: first offshore banking and more recently the insurance industry. Bermuda is now a global leader in the reinsurance business. Reinsurance underwrites part of the risk of other insurance companies. After the terrorist attacks in 2001 and Hurricane Katrina in 2005, Bermuda received an even larger portion of the reinsurance business. Consequently, in the capital city of Hamilton many highly skilled professionals are employed to administer this dominant financial service (Figure 5.5.1).

Yet amid such prosperity, new rumblings for independence have stirred. In 1995, a referendum for independence from the United Kingdom was soundly defeated. In 2006, Bermuda's newly elected premier, Dr. Ewart Brown, once again raised the issue. Continued support for independence is explained, in part, by the island's racial divide. White Bermudans (one-third of the population) generally oppose independence; many, though not all, black Bermudans support the idea. As Bermuda's economy transitions further into financial services and away from tourism, one of its fundamental assets is the political stability that is tied to its status as an overseas colony.

Figure 5.5.1 Hamilton, Bermuda
Front Street in Hamilton, Bermuda is a reflection of the territory's ties to the United Kingdom and its prosperity. Tourism and financial services explain Bermuda's wealth. *(Redux Pictures)*

The irony for Bermuda is that the features that make it a candidate for sovereignty—prosperity, high education levels, and stability—are also easily undermined by introducing any political change. Moreover, in the highly competitive world of offshore financial services, there are always competitors waiting in the wings.

Cricket World Cup a CARICOM Visa was issued to facilitate travel between the nine host countries for this global sporting event.

One scholar of the region argues that a limiting factor to this regional integration is a "small-islandist ideology." Islanders tend to keep their backs to the sea, oblivious to the needs of neighbors. At times, such isolationism results in suspicion, distrust, and even hostility toward nearby states. But in spite of the desire to remain inward-looking, economic necessity dictates engagement with partners both inside and outside the region. And so this peculiar status of isolated proximity unfolds in the Caribbean, expressing itself in diverse global linkages and wildly uneven levels of social and economic development.

ECONOMIC AND SOCIAL DEVELOPMENT: From Cane Fields to Cruise Ships

Collectively, the population of the Caribbean, albeit poor by U.S. standards, is economically better off than most of Sub-Saharan Africa, South Asia, and China. Most Caribbean countries fall in the lower-middle-income range set by the

World Bank. With the exception of Haiti where the majority of the population is living on less than $2 per day, extreme poverty is relatively rare. Despite periods of economic stagnation in the Caribbean, social gains in education, health, and life expectancy are significant (Table 5.2). Historically, the Caribbean's links to the world economy were tropical agricultural exports, yet several specialized industries, such as tourism, offshore financial services, and assembly plants, have challenged the dominance of agriculture. These industries grew because of the region's proximity to North America and Europe, the availability of cheap labor, and the implementation of policies that created a nearly tax-free environment for foreign-owned companies. Unfortunately, growth in these sectors does not employ all the region's workers, so the lure of jobs in North America and Europe is still strong.

From Fields to Factories and Resorts

Agriculture used to dominate the economic life of the Caribbean. Decades of turbulent commodity prices and decline in preferential trade agreements with former colonial masters have produced more hardship than prosperity. Ecologically, the soils are overworked, and no frontier areas remain in which to expand production, save for areas of the rimland. Moreover, agricultural prices have not kept

TABLE 5.2 Development Indicators

Country	GNI[a] per Capita, PPP[b] (2005)	GDP[c] Average Annual Percent Growth (2000–05)	Life Expectancy (2007)	Percent Population Living on Less than $2 a Day	Under Age 5 Mortality Rate 1990	Under Age 5 Mortality Rate 2005	Gender Equity[d]
Anguilla	8,800		77				
Antigua and Barbuda	10,900		72				
Bahamas	21,600		71				
Barbados	18,400		76				
Belize	6,740		70				
Bermuda	70,000		78				
Cayman	43,800		80				
Cuba	4,100	3.4	77		13	7	110
Dominica	5,560		74				
Dominican Republic	7,150	2.8	72	11	65	31	111
French Guiana			75				
Grenada	7,260		65				
Guadeloupe			79				
Guyana	4,230		65				
Haiti	1,840	−0.5	58	78	150	120	
Jamaica	4,110	1.8	72	13	20	20	104
Martinique			79				
Montserrat	3,400		79				
Netherlands Antilles	16,000		76				
Puerto Rico	19,300		78				
St. Kitts and Nevis	12,500		70				
St. Lucia	5,980		74				
St. Vincent and the Grenadines	6,460		71				
Suriname	7,100		69				
Trinidad and Tobago	13,170	8.3	69	39	33	19	104
Turks and Caicos	11,500		75				

[a]*Gross national income.*
[b]*Purchasing power parity.*
[c]*Gross domestic product.*
[d]*Ratio of female-to-male enrollments in primary and secondary school percentage. Numbers below 100 have more males in primary/secondary school; numbers above 100 have more females in primary/secondary schools.*

Sources: World Bank, World Development Indicators, 2007; *life expectancy 2007 and percentage of population living on $2 a day data from Population Reference Bureau,* World Data Sheet, 2007; *Bermuda data from* CIA World Factbook, 2007.

pace with rising production costs, so wages and profits remain low. With the exception of a few mineral-rich territories, such as Trinidad, Guyana, Suriname, and Jamaica, most countries have systematically tried to diversify their economies, relying less on their soils and more on manufacturing and services.

Comparing export figures over time demonstrates the shift away from monocrop dependence. In 1955, Haiti earned more than 70 percent of its foreign exchange through the export of coffee; by 1990, coffee accounted for only 11 percent of its export earnings. Similarly, in 1955 the Dominican Republic earned close to 60 percent of its foreign exchange through sugar, but 35 years later sugar earned less than 20 percent of the country's foreign exchange, and pig iron exports nearly equaled that of sugar.

Sugar and Coffee The economic history of the Caribbean cannot be separated from the production of sugarcane. Even relatively small territories such as Antigua and Barbados yielded fabulous profits because there was no limit to the demand for sugar in the eighteenth century. Once considered a luxury crop, it became a popular necessity for European and North American laborers by the 1750s. It sweetened tea and coffee and made jams a popular spread for stale bread. In short, it made the meager and bland diets of ordinary people tolerable, and it also boosted caloric intake. Distilled into rum, sugar produced a popular intoxicant. Though it is hard to imagine today, consumption of a pint of rum a day was not uncommon in the 1800s.

Sugarcane is still grown throughout the region for domestic consumption and export. Its economic importance has declined, however, mostly because of increased competition from sugar beets grown in the midlatitudes. The Caribbean and Brazil are the world's major sugar exporters. Until 1990, Cuba alone accounted for more than 60 percent of the value of world sugar exports. The country earned 80 percent of its foreign exchange through sugar production. Cuba's dominance in sugar exports had more to do with its subsidized and guaranteed markets in eastern Europe and the Soviet Union than with exceptional productivity. Since 1990, the value of the Cuban sugar harvest has declined by two-thirds.

Coffee is planted in the mountains of the Greater Antilles. Haiti has been the most dependent on coffee, relying on peasant sharecroppers to tend the plants and harvest the beans. For other countries, coffee is a valued specialty commodity. Beans harvested in the Blue Mountains of Jamaica, for example, fetch two to three times the going price for similar highland coffee grown in Colombia. Puerto Rico and Cuba also are trying to develop a niche in the gourmet coffee market. An important production distinction with coffee, in contrast to sugar, is that it is mostly grown on small farms and sold to buyers or delivered to cooperatives. Typically, farmers plant other crops between the coffee bushes so that they can meet their subsistence needs as well as produce a cash crop. Even with this self-provisioning system, peasants often seasonally abandon their farms . work elsewhere as laborers. The instability of coffee prices, which were especially low in 2000 and 2001, makes the economics of growing coffee hard to predict.

The Banana Wars The major banana exporters are in Latin America, not the Caribbean. In fact, the success of banana plantations is mixed in this region, as banana plants are especially vulnerable to hurricanes. Still, several small states in the Lesser Antilles, most notably Dominica, St. Vincent, and St. Lucia, have become dependent on bananas, gaining as much as 60 percent of their export earnings from the yellow fruit. Bananas have not made people rich, yet their production for export has fostered greater economic and social development. In the eastern Caribbean, where most bananas are grown on small farms of five acres, the landowners are the laborers and thus earn two to four times more than banana plantation workers in Ecuador and Central America. Moreover, for these small states, banana exports are their link with the global economy. Pressures on the European Union to drop the preferential treatment given to banana growers from the former colonies has undermined the economic viability of this crop in the Caribbean (Figure 5.26). Where once a market and a minimum price were guaranteed, under the new global order neither is certain.

The case of the eastern Caribbean underscores what happens to the losers in globalization. In 1996, the United States, Ecuador (the world's leading banana exporter), Mexico, Guatemala, and Honduras took the European Union's (EU) restrictive banana trade agreement to the World Trade Organization court. The agreement was denounced as unfair, and the EU was told to eliminate it by 1998. To make matters worse, consumer tastes had changed, giving preference to a uniformly large and unblemished yellow banana typical of the Latin American plantations but not always grown in places like St. Lucia. The growers of the eastern Caribbean, to survive in this newly competitive environment, will have to produce a more standardized fruit and increase their yield per acre. While island governments are trying to aid in this transition, local farmers have experimented with new crops, such as okra, tomatoes, avocados, and marijuana. A few well-tended marijuana plants will earn 30 times more per pound than bananas. Just what will happen to family-run banana farms is hard to tell, but many Caribbean growers fear that the days of the banana economy are numbered. The combination of increased global competition and changing consumer tastes has forced many farmers and rural laborers to find employment elsewhere.

Assembly-Plant Industrialization One regional strategy to deal with the unemployed was to invite foreign investors to set up assembly plants and thus create jobs. This strategy first succeeded in Puerto Rico in the 1950s and was copied throughout the region. In Puerto Rico's "Operation Bootstrap," island leaders encouraged U.S. investment by offering cheap labor, local tax breaks, and, most important, federal tax exemptions (something only Puerto Rico can do because of its special status as a commonwealth of the United States). Initially the program was so successful that by 1970 nearly 40 percent of the island's gross domestic product (GDP) came from manufacturing. Today, 22 percent of the Puerto Rican labor force is employed in industry, and this sector accounts for nearly half of the island's GDP. Yet competition from other states with even lower wages and the U.S. Congress's decision in 1996 to phase out many of the tax exemptions may threaten Puerto Rico's ability to maintain its specialized industrial base.

Through the creation of **free trade zones (FTZs)**, duty free and tax-exempt industrial parks for foreign corporations, the Caribbean is an increasingly attractive location to assemble goods for North American consumers. The Dominican Republic developed its FTZs by taking advantage of tax incentives and guaranteed access to the U.S. market first offered through the Caribbean Basin Initiative. The number of operational FTZs in the Dominican Republic in 2001 was 16, including an FTZ on the island's north shore, near the Haitian border, optimistically named "Hong Kong of the Caribbean." Plans exist to double the number of FTZs. Firms from the United States and Canada are the most frequent investors in these zones, followed by Dominican, South Korean, and Taiwanese firms. Traditional manufacturing on the island was tied to sugar refining, whereas up to 60 percent of production in the FTZs is in garments and textiles (Figure 5.27). Manufacturing is the largest economic sector in the Dominican Republic, accounting for nearly one-fifth of the country's GDP.

In the Dominican Republic as well as Jamaica and Haiti, the growth in manufacturing depended on national and international development policies that supported export-led growth through direct foreign investment. Although new jobs are being created and national economies are diversifying in the process, critics believe that foreign investors gain more than the host countries. Since most goods are assembled from imported materials, there is little integration with national suppliers. Often higher than local averages, wages are still miserable compared with those in the developed world—sometimes just \$2 to \$3 a day. Moreover, as other developing countries compete with the Caribbean for the establishment of FTZs, this strategy may become less beneficial over time.

Offshore Banking The rise of offshore banking in the Caribbean is most closely associated with the Bahamas, which began this industry back in the 1920s. **Offshore banking** centers appeal to foreign banks and corporations by offering specialized services that are confidential and tax-exempt. Localities that provide offshore banking make money through registration fees, not taxes. The Bahamas were so successful in developing this sector that by 1976 the country was the 3rd largest banking center in the world. Its dominance began to decline because of competitors from the Caribbean, Hong Kong, and Singapore and because there was no longer a tax advantage in booking large international loans offshore. Concerns about corruption and laundering of drug money also hurt the islands' financial status in the 1980s, and major reforms were introduced to remove illegal funds. By 1998, the Bahamas' ranking among global financial centers had dropped to 15th. Still, offshore banking remains an important part of the Bahamian economy. In the 1990s, the Cayman Islands emerged as the region's leader in financial services. With a population of 40,000, this crown colony of Britain has some 70,000 registered companies and one of the highest per capita PPP of the region at \$43,800. In 2007, it was estimated that the Caymans were the 5th largest banking center in the world after New York, London, Hong Kong, and Tokyo.

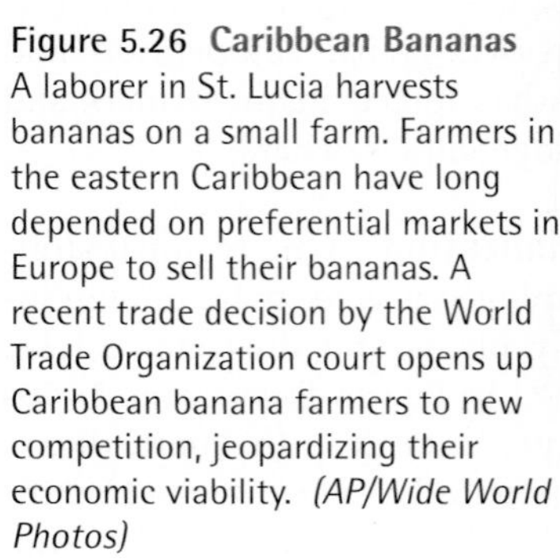

Figure 5.26 Caribbean Bananas A laborer in St. Lucia harvests bananas on a small farm. Farmers in the eastern Caribbean have long depended on preferential markets in Europe to sell their bananas. A recent trade decision by the World Trade Organization court opens up Caribbean banana farmers to new competition, jeopardizing their economic viability. *(AP/Wide World Photos)*

Figure 5.27 Free Trade Zones in the Dominican Republic Another sign of globalization is the proliferation of duty-free and tax-exempt industrial parks in the Caribbean. Currently 16 FTZs are operational, with foreign investors from the United States, Canada, South Korea, and Taiwan. *(Modified from Warf, 1995, "Information Services in the Dominican Republic,"* Yearbook, Conference of Latin Americanist Geographers *21, 15)*

Each of the offshore banking centers in the Caribbean (Figure 5.28) tries to develop special financial niches to attract clients, such as banking, credit card accounts, functional operations, insurance, or trusts. The Caribbean is an attractive locality for such services: it is close to the United States, which is home to many of the registered firms; it offers a variety of locations to satisfy client demand; and it is easily accessible by telecommunications. On the supply side, the resource-poor islands of the region see financial services as a way to attract hard currency. Envious of the economic success of the Bahamas, Bermuda, and the Cayman Islands, countries such as Antigua, Aruba, Barbados, and Belize hope that greater prosperity will come from establishing close ties to international finance.

The growth of offshore financial centers is indicative of the forces of globalization. Multinational corporations are accustomed to moving physical and financial assets across borders. Yet taxation is based on national boundaries. The rise of offshore financial centers is a time-tested vehicle for large corporations to minimize their taxes. As money moves with greater speed and ease around the planet, so too have offshore financial centers blossomed, especially in the Caribbean, Southeast Asia, the Pacific, and Europe (Liechtenstein and Jersey).

Ironically, banking does not employ large numbers of people and can quickly unravel with the slightest sign of political instability. Offshore banking also attracts money tied to the drug trade. A major transshipment area for cocaine bound for North America and Europe, the Caribbean is also a major money laundering center. This brings such problems to the region as increases in drug consumption, corruption of local officials, and even drug-related murders. International efforts in the 1990s sought to curb money laundering by threatening the suspension of privacy provisions whenever criminal activities are suspected. After the terrorist attacks on the United States in September 2001, interest in promoting know-your-customer laws as a means of tracking down the offshore assets of terrorist organizations was renewed. The upshot has been a more stringent regulatory environment that is making offshore banking less attractive to newcomers. Grenada recently announced that it is no longer a site for offshore banking. The major banking centers, such as the Cayman Islands and the Bahamas, are likely to endure while smaller operations such as those on St. Kitts or Barbados will have a more difficult time surviving.

Online gambling is the newest industry for the microstates of the Caribbean. Antigua and St. Kitts began legal online gambling services in 1999. Other states soon followed, and as of 2003 Dominica, Grenada, Belize, and the Cayman Islands had gambling domain sites. Even though online gambling began only in 1996, it expanded rapidly and, like offshore banking, took root in the Caribbean. Restrictions imposed upon overseas Internet gambling sites by the United States were deemed illegal by the World Trade Organization (WTO) in 2007. The tiny nation of Antigua is currently seeking compensation from the United States for millions of dollars in lost revenue due to the illegal restrictions placed on Antigua's online business. While the WTO sided with Antigua, the future of Internet gambling remains uncertain.

Tourism Environmental, locational, and economic factors converge to strengthen tourism in this part of the world. The earliest visitors to this tropical sea admired its clear and sparkling turquoise waters. By the nineteenth century, elite North Americans were fleeing winter to enjoy the restorative warmth of the Caribbean during its dry season. By the twentieth century, tourism was well established with both destination resorts and cruise lines. By the 1950s, the leader in tourism was Cuba, and the Bahamas was a distant second. Castro's rise to power, however, eliminated this sector of the island's economy for nearly three decades and opened the door for other islands to step in.

Five islands hosted two-thirds of the 18 million international tourists who came to the Caribbean in 2004: Puerto Rico, the Dominican Republic, Cuba, the Bahamas, and Jamaica (Figure 5.29). Puerto Rico saw its tourist sector begin to grow with commonwealth status in 1952. San Juan is now the largest home port for cruise lines and the second largest cruise-ship port in the world in terms of total visitors. The cruise business, in combination with stayover visitors, accounted for more than 3.5 million arrivals in 2004. The Bahamas attributes most of its economic development and high per capita income to tourism. With more than 1.5 million stayover visitors in 2004 and 1.7 million cruise-ship passengers, the Bahamas is another major hub for tourism in the region. Some 30 percent of the Bahamian population is employed in tourism, and tourism represents nearly half the country's GDP (Figure 5.30). Because the Bahamas is closer to the United States than most Caribbean destinations, the majority of its international clientele is American (about 80 percent), followed by European and Canadian.

The Dominican Republic also sees more than 3.5 million visitors annually, many of them Dominican nationals who live overseas. Since 1980, tourist receipts have increased 20-fold, making tourism the leading foreign-exchange earner at more than \$3.0 billion. The Dominican Republic earned more tourism dollars in 2004 than any other Caribbean country, although Puerto Rico

Figure 5.28 Caribbean Offshore Banking Centers
Another development strategy used by the smaller islands is to offer specialized banking services to international corporations. These services are both confidential and tax exempt. The Bahamas, Cayman Islands, and Panama dominate offshore banking in the region. *(From Roberts, 1995, "Small Place, Big Money: The Cayman Islands and the International Financial System,"* Economic Geography *71[3], 237–56)*

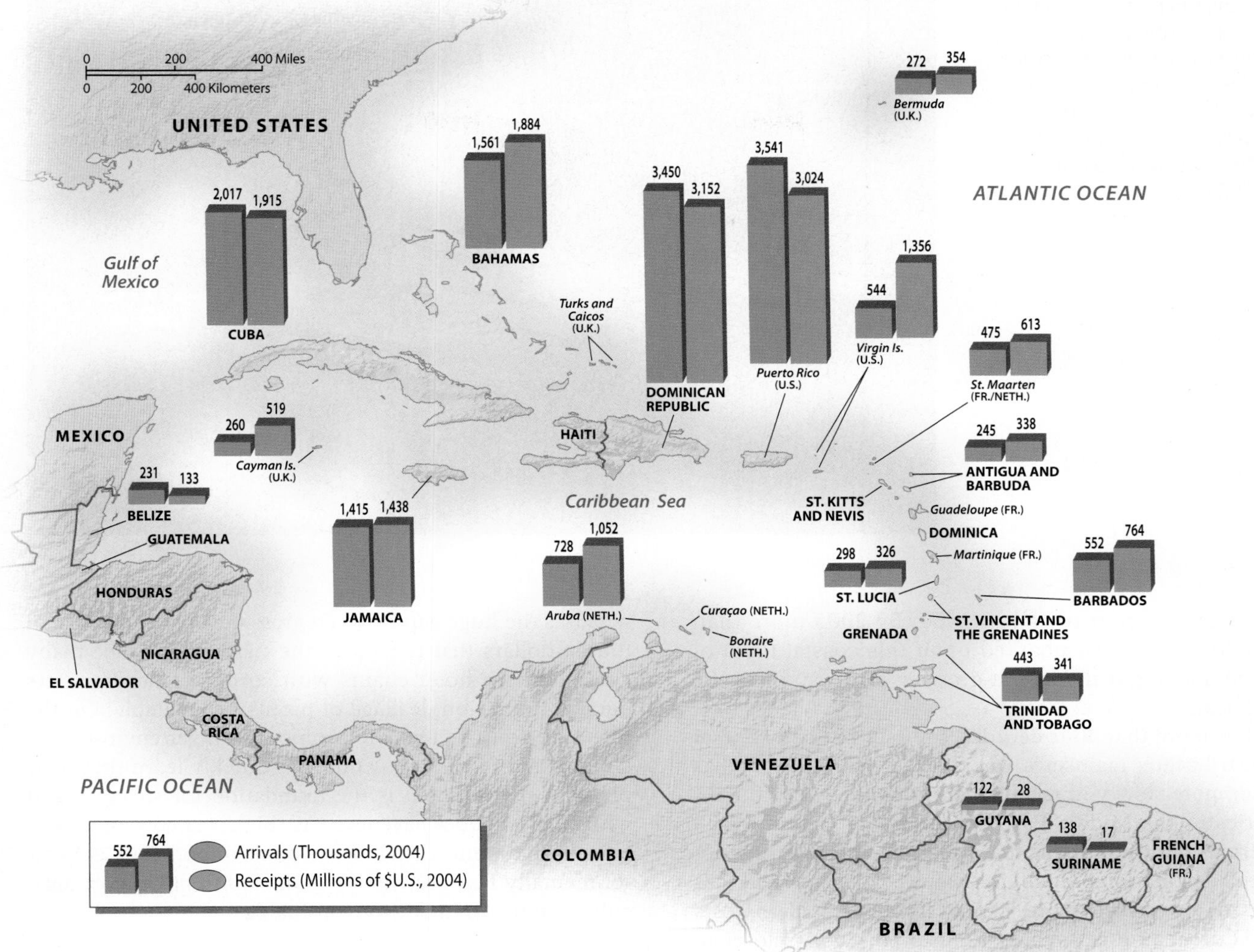

Figure 5.29 Global Linkages: International Tourism in the Caribbean
Tourism directly links the Caribbean to the global economy. Each year 15 million tourists come to the islands, mostly from North America, Latin America, and Europe. The most popular destinations are the Dominican Republic, Puerto Rico, Cuba, the Bahamas, and Jamaica. *(From* Tourism Market Trends, *2005 edition)*

was a close second. Jamaica has become similarly dependent on tourism for hard currency, earning more than $1.5 billion in tourist receipts in 2005. The vast majority of tourists to Jamaica are from the United States and the United Kingdom. While drawing 1.4 million stayovers in 2004, Jamaica was a port of call for another half-million cruise-ship passengers (Figure 5.31).

After years of neglect, tourism is being revived in Cuba in an attempt to earn badly needed hard currency. Tourism represented less than 1 percent of the national economy in the early 1980s. By 2004, more than 2 million tourists (mostly Canadians and Europeans) yielded gross receipts approaching $2 billion. Conspicuous in their absence are travelers from the United States, forbidden to travel to Cuba because of the U.S.-imposed sanctions. With U.S. investors out of the picture, Spanish and other European investors are busy building up Cuba's tourist capacity, anticipating that someday the U.S. ban will be lifted.

As important as tourism is for the larger islands, it is often the principal source of income for smaller ones. The Virgin Islands, Barbados, Turks and Caicos, and, recently, Belize all greatly depend on international tourists. To show how quickly this sector can grow, consider this example: Belize began promoting tourism in the early 1980s, when just 30,000 arrivals came each year. Close to North America and English-speaking, it specialized in a more rustic ecotourism that showcased its interior tropical forests and coastal barrier reef. In the mid-1990s, the number of land-based tourists peaked at 300,000, and tourism was credited with employing one-fifth of the workforce. By 2000 the arrival figure had dropped by two-thirds. Yet today Belize City is reeling from an influx of day visitors from cruise ships. Belize

Figure 5.30 Caribbean Cruise Ships Cruise ships in port at Nassau, the Bahamas. Tourism is vital to many Caribbean states, but most cruise ships are owned by companies outside the region and offer relatively little employment for Caribbean workers. *(Galen Rowell/Corbis)*

City became a port of call in 2000. By 2005 more than 800,000 visitors disembarked upon this coastal town of 60,000, making it the fastest-growing tourist port in the Caribbean.

For more than four decades, tourism has been one of the Caribbean's mainstays. Yet this regional industry has grown more slowly in recent years compared with the industries in the Middle East, southern Europe, and even Central America. It seems that Americans are favoring domestic destinations, such as Hawaii, Florida, and Las Vegas, or are going to more "exotic" localities, such as Costa Rica. European tourists also seem to be staying closer to home or venturing to new locations, such as Dubai on the Persian Gulf or Goa in India. A destination's reputation can suddenly deteriorate as a result of social or natural forces. In the 1980s, a wave of crimes against tourists in Jamaica caused a drop in visitors. It took a locally owned hotel chain to market all-inclusive trips on private beaches to restore the island's image as a safe place to visit. Hurricane Ivan in 2004 virtually shut down Grenada to visitors for an entire season, but islanders managed to rebuild most of their infrastructure by the following year.

Tourism-led growth has detractors for other reasons. It is subject to the overall health of the world economy and current political affairs. Thus, if North America experiences a recession or international tourism declines because of heightened fears of terrorism, the flow of tourist dollars to the Caribbean dries up. Where tourism is on the rise, local resentment may build as residents confront the disparity between their own lives and those of the tourists. There is also a serious problem of **capital leakage**, which is the huge gap between gross receipts and the total tourist dollars that remain in the Caribbean. Since many guests stay in hotel chains with corporate headquarters outside the region, leakage of profits is inevitable. On the plus side, tourism tends to promote stronger environmental laws and regulation. Countries quickly learn that their physical environment is the foundation for success. And while tourism does have its costs (higher energy and water consumption and demand for more imports), it is environmentally less destructive than traditional export agriculture and currently more profitable.

Social Development

Although the record for economic growth in the region is inconsistent, measures of social development are stronger, in part due to Cuba's accomplishments in health care and education. With the exception of Haiti's abysmally low life expectancy of 58 years, most Caribbean peoples have an average life expectancy of more than 70 years (Table 5.2). Literacy levels are high, and in most countries there is near parity in terms of school enrollment by gender. Indeed, higher levels of educational attainment and outmigration have contributed to a marked decline in natural increase rates over the last 30 years, so that the average Caribbean woman has two or three children. Despite real social gains, many inhabitants are chronically underemployed, poorly housed, and dependent on foreign remittances. For rich and poor alike, the temptation to leave the region in search of better opportunities is always present.

Status of Women The matriarchal basis of Caribbean households is a distinguishing characteristic of the region. The rural custom of men leaving home for seasonal employment tends to nurture strong and self-sufficient female networks. Women typically run the local street markets. With men absent for long periods of time, women tend to make household and community decisions. While giving women local power, this position does not always confer status. In rural areas, female status is often undermined by the relative exclusion of women from the cash economy; men earn wages while women provide subsistence.

As Caribbean society urbanizes, more women are being employed in assembly plants (the garment industry, in particular, prefers to hire women), in data-entry firms, and in tourism. With new employment opportunities, female labor-force participation has surged; in countries such as the Bahamas, Barbados, Jamaica, and Puerto Rico more than 40 percent of the workforce is female. Increasingly, women are the principal earners of cash, which challenges established gender roles. Also, there are signs of greater political involvement by women. In March 2006 Portia Simpson-Miller was elected prime minister of Jamaica, making her the first female head of government for that country. Educated in the United States, Simpson-Miller is the third woman to lead a country in the English Caribbean. Both Dominica and Guyana have had women prime ministers.

Education Many Caribbean states have excelled in educating their citizens. Literacy is the norm, and the expectation is for most people to receive at least a high school degree. In many respects, Cuba's educational accomplishments are the most impressive, given the size of the country and its high illiteracy rates in the 1960s. Today nearly all adults are literate. Hispaniola is the obvious contrast to Cuba's success. While the Dominican Republic has made strides in improving adult literacy (87 percent of adults are literate), half of all Haitian adults are illiterate. Political stability and economic growth have helped the Dominican Republic better its social conditions over the past decade. In fact, many Haitians have fled to the Dominican Republic because conditions, although far from ideal, are much better than in their homeland.

Education is expensive for these nations, but it is considered essential for development. Ironically, many states express frustration about training professionals for the benefit of developed countries in a phenomenon called **brain drain**. In the early 1980s, the prime minister of Jamaica complained that 60 percent of his country's college-educated workers left for the United States, Canada, and Britain, representing a subsidy to these economies far greater than the foreign aid Jamaica received from them. A 2007 World Bank study of skilled migrants revealed that 40 percent of Caribbean immigrants living abroad were college educated. No other region in the world has this many educated people leaving. To be fair, some of these migrants moved when they were young and received their education in North America and Europe. Still, other immigrants, especially health professionals, received their educations in the Caribbean and were recruited abroad because of higher wages and better opportunities. Brain drain occurs throughout the developing world, especially between former colonies and the mother countries. Given the small population of many Caribbean territories, each professional person lost to emigration can negatively impact local health care, education, and enterprise. While the outflow of professionals continues to be high, many countries report on return migration of Caribbean peoples from North America and Europe and the potential of returnees to contribute to the social and economic development of the region.

Labor-Related Migration The combination of inequitable economies and limited resources force many residents to migrate. Besides influencing regional

Figure 5.31 Resort Life
Swimmers enjoy the turquoise waters of Jamaica's famous Montego Bay on the island's northern coast. A major destination for sun-seeking North Americans and Europeans, Jamaica faces stiff competition from its other Caribbean neighbors to maintain tourist revenues. *(Richard Cummins/Corbis/Bettmann)*

economies, this strategy impacts community and household structures as well. Historically, labor circulated within the Caribbean in pursuit of the sugar harvest—Haitians went to the Dominican Republic; residents of the Lesser Antilles journeyed to Trinidad. During the construction of the Panama Canal in the early 1900s, thousands of Jamaicans and Barbadians journeyed to the Canal Zone, and many remained.

After World War II, better transportation and political developments in the Caribbean produced a surge of migrants to North America. This trend began with Puerto Ricans going to New York in the early 1950s and intensified in the 1960s with the arrival of nearly half a million Cubans. Since then, large numbers of Dominicans, Haitians, Jamaicans, Trinidadians, and Guyanese also have migrated to North America, typically settling in Miami, New York, Los Angeles, and Toronto.

Crucial in this exchange of labor from south to north is the flow of cash remittances (monies sent back home). Immigrants are expected to send something back, especially when immediate family members are left behind. Collectively, remittances add up; it is estimated that $3 billion was sent to the Dominican Republic by immigrants in 2006, making remittance income the country's second leading industry. Governments and individuals alike depend on these transnational family networks. Families carefully select the member most likely to succeed abroad in the hope that money will flow back and a base for future immigrants will be established. A Caribbean nation in which no one left would soon face a crisis. Economic

PEOPLE ON THE MOVE Migrants and Remittances

For many Caribbean residents, remittances are an essential component for everyday survival. Sent by immigrants working abroad to family members and friends, remittances tend to be directed to lower-income groups and thus are vital in reducing poverty and improving social equity in the region. Dependence upon remittances encourages labor migration, both within the region and beyond. Plus, new technologies and increased competition have reduced the costs of money transfers. What use to be an informal flow of cash is now formalized through the use of banks, ATM cards, and money transfer services such as Western Union. Thus, a Jamaican in New York can deposit money in a local bank and have a family member withdraw it from an ATM machine in a rural convenience store in Jamaica.

The Inter-American Development Bank estimated that $62.3 billion in remittances reached Latin American and the Caribbean in 2006—one-third of the monies going to Mexico alone. Over $7 billion of these remittances went to eight Caribbean states: Dominican Republic, Haiti, Jamaica, Guyana, Barbados, Suriname, Belize, and Trinidad and Tobago. When considering the small size of these states, remittance dollars are incredibly important (Table 5.6.1). In Haiti, the poorest nation in the region, remittances grew from $800 million to $1.65 billion from 2001 to 2006. Moreover, remittances equaled one-third of the country's GDP in 2006. Jamaica also saw its remittances nearly double in the same time frame, reaching $1.8 billion in 2006 and accounting for 18 percent of its GDP. Jamaica also has the distinction of receiving the most remittances on a per capita basis for all the Americas—$623 per year. The next highest country is Barbados, with a per capita figure of $418. Not surprisingly, all of these large remittance recipients have negative rates of net migration (again, see Table 5.1), which means that more people leave the country than arrive each year. The decision to emigrate and remit money is usually a family matter. Yet, as data from the Caribbean reveal, it has become a standard practice for tens of thousands of households in the region, as well as a clear expression of the globalization of labor flows.

TABLE 5.6.1 **Remittances and Key Economic Indicators in the Caribbean, 2006**

Country (Ordered by Percent of GDP)	Remittances as Percent of GDP	Remittances per Capita $	Annual Transfer of Remittances ($ millions)
Guyana	37	360	270
Haiti	35	116	1,650
Jamaica	18	623	1,770
Dominican Republic	13	271	2,900
Barbados	4	418	300
Suriname	4	122	102
Belize	4	149	93
Trinidad and Tabago	1	71	110
			7,195 Total

Source: Adapted from Making the Most of Family Remittances, Second Report of the Inter-American Dialogue Task Force on Remittances, *May 2007.*

opportunities are limited, and the populations have expanded far faster than the economies. Many of the people who leave would be jobless at home, or would force someone else to be jobless (see "People on the Move: Migrants and Remittances").

A common strategy for many immigrants is to work abroad and retire in the Caribbean with a pension and savings that ensure a comfortable old age. Returnees often bring cash, new skills, and increased expectations. There is anecdotal evidence that this community of migrants can introduce positive economic and political changes in particular localities. Still, it is unlikely that migrants and their remittances can challenge the larger forces that make migration attractive in the first place. Their impact is experienced primarily at the household level, and is too fragmented to represent a national development force.

Summary

- The Caribbean is more integrated into the global economy than most areas in the developing world, even though it remains on the economic periphery. Although small in area, the region offers some of the clearest examples of the long-term effects of globalization from plantation agriculture to offshore banking.
- This tropical region has been exploited to produce export commodities such as sugar and bananas. The region's warm waters and mild climate attract millions of tourists. Yet serious problems with deforestation, soil erosion, and water contamination have degraded urban and rural environments. Global climate change poses a serious threat to the region.
- Population growth in the Caribbean has slowed over the past two decades. The average woman now has two or three children. Many people leave the region for employment opportunities in North America and Europe. While life expectancy is quite high, health officials are alarmed by the increases in HIV infection rates.
- The Caribbean was forged through European colonialism and the labor of millions of Africans. The blending of African and European elements, referred to as creolization, has resulted in many unique cultural expressions in music, language, and religion.
- Today the region contains 20 independent countries and several dependent territories. With the end of the Cold War, many microstates in the region fear that their lack of strategic significance may result in neglect by the United States and Europe, which may limit their ability to participate in world trade.
- In terms of economic development, employment in assembly plants, tourism, and offshore banking has steadily replaced agricultural jobs in the Caribbean. The region's positive strides in social development, namely in education, health, and the status of women, distinguish it from other developing areas.

Key Terms

brain drain *(page 217)*
capital leakage *(page 216)*
Caribbean Community and Common Market (CARICOM) *(page 208)*
chain migration *(page 194)*
circular migration *(page 193)*
creolization *(page 195)*
diaspora *(page 193)*
free trade zone (FTZ) *(page 212)*
Greater Antilles *(page 185)*
hurricanes *(page 189)*
indentured labor *(page 197)*
isolated proximity *(page 182)*
Lesser Antilles *(page 186)*
maroons *(page 198)*
monocrop production *(page 197)*
Monroe Doctrine *(page 205)*
neocolonialism *(page 205)*
offshore banking *(page 212)*
plantation America *(page 196)*
rimland *(page 183)*
transnational migration *(page 194)*

Questions for Review

1. What characteristics define the Caribbean as a world region?
2. What environmental, economic, and locational factors contributed to the growth of tourism in the Caribbean?
3. What is the typical path of a hurricane in the Caribbean, how are they tracked, and when are these meteorological disturbances likely to occur?
4. How will global climate change impact the Caribbean over the next century?
5. What factors contributed to the deforestation of the Caribbean islands? Why are the rimland forests, in comparison, still intact?
6. What happened to the Amerindian population in the Caribbean?
7. What cultural traditions transferred from Africa to the Caribbean via the slave trade?
8. What is the history of carnival in the Caribbean? How is it important today? Why are carnival traditions spreading around the world?
9. Which territories in the region are still colonies? Why have they not sought independence?
10. What are the banana wars, and how do they illustrate the tensions brought on by globalization?

Thinking Geographically

1. After looking at the map of offshore banking in the Caribbean, what conclusions can you draw about where this occurs?
2. Contrast the historical African diaspora to the contemporary Caribbean one. What patterns are formed by these two distinct population movements? What social and economic forces are behind them? Are they comparable?
3. What are the advantages and disadvantages of tourism as a development strategy?
4. What advantages might Caribbean free trade zones retain over their competitors in Southeast and East Asia? What disadvantages might they face?
5. Why have U.S. actions in the region been considered neocolonial? Are there other regions in the world where the United States exerts neocolonial tendencies?
6. Compare and contrast the industrial development of Cuba with Puerto Rico. How were these two islands integrated into the world economy during the twentieth century?
7. Why does chain migration occur between the Caribbean and North America or Europe? How does a migrant benefit from being part of a chain migration? How does chain migration affect sending communities?
8. In the twenty-first century, will agricultural exports continue to be important for Caribbean economies?
9. How might the September 2001 terrorist attacks on the United States impact the regional economy of the Caribbean?
10. Are remittances a sign of the Caribbean's isolation or integration with the global economy? How might remittances be used to foster development?

Regional Novels and Films

Novels

Reinaldo Arenas, *Farewell to the Sea: A Novel of Cuba* (1994, Penguin)

Aimé Césaire, *Notebook on a Return to the Native Land* (2001, Wesleyan University Press)

Patrick Chamoiseau, *Texaco* (1988, Vintage International)

Edwidge Danticat, ed., *The Butterfly's Way: Voices from the Haitian Dyaspora in the United States* (2001, Soho)

Zee Edgell, *Beka Lamb* (1982, Heinemann)

Cristina Garcia, *Dreaming in Cuba* (1992, Random House)

Jamaica Kincaid, *A Small Place* (1988, Penguin)

William Luis, ed., *Dance Between Two Cultures: Latino Caribbean Literature Written in the United States* (1997, Vanderbilt)

V. S. Naipaul, *The Middle Passage* (1963, Macmillan)

Derek Walcott, *Omeros* (1990, Farrar, Straus & Giroux)

Films

Bitter Cane (1986, Haiti)

Buena Vista Social Club (1999, Cuba)

Circle of Voodoo (2002, Haiti)

The Emperor's Birthday: The Rastafarians Celebrate (1997, Jamaica)

The Harder They Come (1972, Jamaica)

I Am Cuba (1995, Russia and U.S.)

Inside Castro's Cuba (1995, Cuba)

The King Does Not Lie: The Initiation of a Shango Priest (1993, Trinidad)

No Seed (2002, Antigua and Barbuda)

Sugar Cane Alley (1983, Martinique)

Bibliography

Duval, David Timothy, ed. 2004. *Tourism in the Caribbean: Trends, Development, Prospects*. London: Routledge.

Grossman, Lawrence. 1998. *The Political Ecology of Bananas: Contract Farming, Peasants, and Agrarian Change in the Eastern Caribbean*. Chapel Hill: University of North Carolina.

Hillstrom, Kevin, and Hillstrom, Laurie Collier. 2004. *Latin America and the Caribbean: A Continental Overview of Environmental Issues*. Santa Barbara, CA: ABC Clio.

Klak, Thomas, ed. 1998. *Globalization and Neoliberalism: The Caribbean Context*. Lanham, MD: Rowman and Littlefield.

Mintz, Sidney. 1985. *Sweetness and Power: The Place of Sugar in Modern History*. New York: Viking Penguin.

Richardson, Bonham C. 1992. *The Caribbean in the Wider World, 1492–1992*. Cambridge, UK: Cambridge University Press.

Roberts, Susan. 1995. "Small Place, Big Money: The Cayman Islands and the International Financial System." *Economic Geography* 71(3), 237–56.

Scarpaci, Joseph, Segre, Roberto, and Coyula, Mario. 2002. *Havana, Two Faces of the Antillean Metropolis*. Chapel Hill: University of North Carolina Press.

Voeks, Robert. 1993. "African Medicine and Magic in the Americas." *Geographical Review* 83(1), 66–78.

Wilson, Mark. 2003. "Chips, Bits and the Law: An Economic Geography of Internet Gambling." *Environment and Planning* 35(7), 1245–60.

Additional bibliographic resources are at the *Diversity Amid Globalization* website: http://wwwprenhall.com/rowntree/.

6

Sub-Saharan Africa

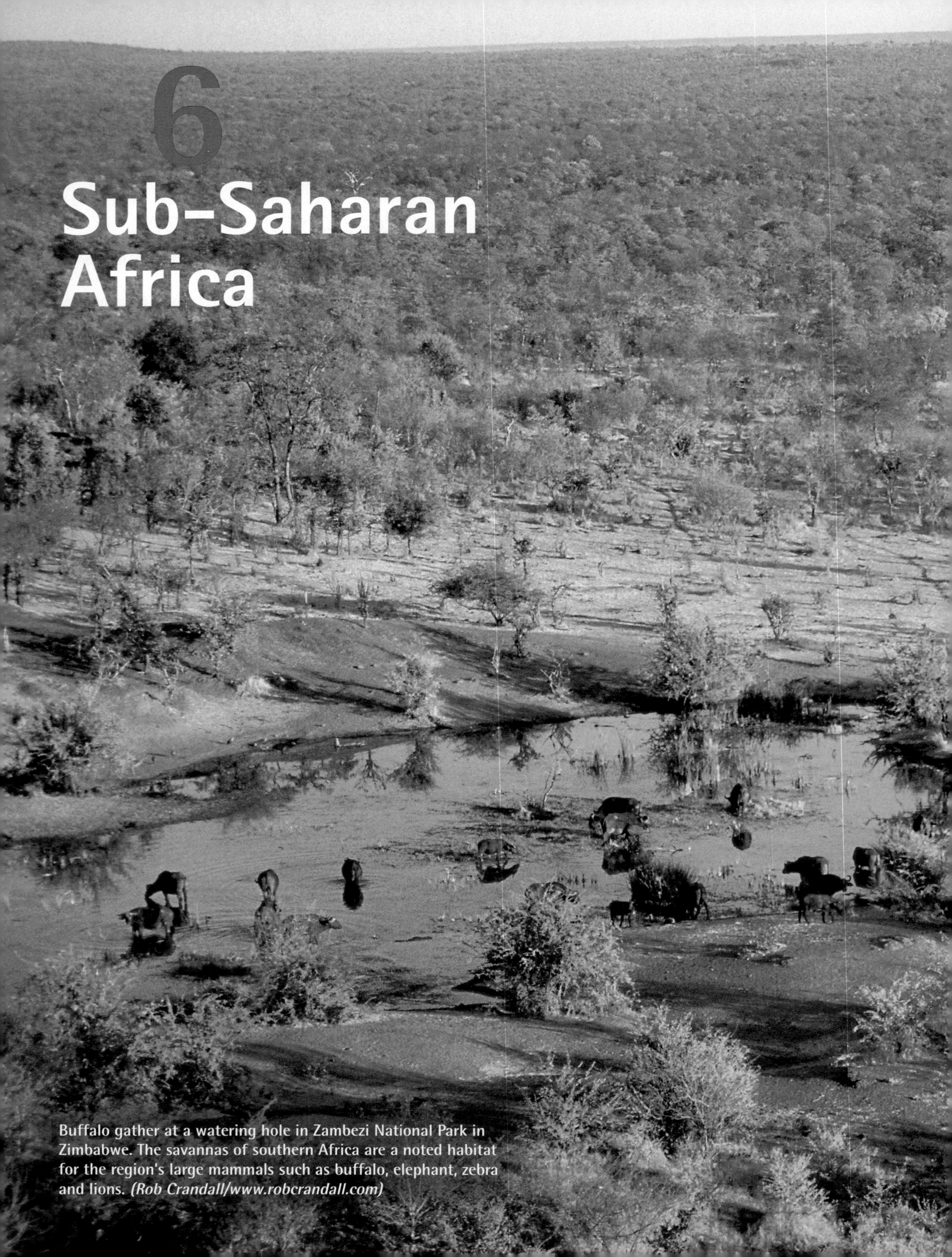

Buffalo gather at a watering hole in Zambezi National Park in Zimbabwe. The savannas of southern Africa are a noted habitat for the region's large mammals such as buffalo, elephant, zebra and lions. *(Rob Crandall/www.robcrandall.com)*

ENVIRONMENTAL GEOGRAPHY

Wood is a main source of energy for this region. Kenyan professor Wangari Maathai won the Noble Peace Prize for her Greenbelt Movement, which led to the planting of millions of trees by rural women.

POPULATION AND SETTLEMENT

As a region, Sub-Saharan Africa has the highest infection rates of HIV/AIDS. As a result of this disease, many countries have seen life expectancy rates plummet into the 40s. Yet the population continues to grow due to high fertility rates.

CULTURAL COHERENCE AND DIVERSITY

This is a region with large and growing numbers of Muslims, Christians, and animists. With a few exceptions, religious diversity and tolerance has been a distinctive feature of the region.

GEOPOLITICAL FRAMEWORK

Most countries gained their independence in the 1960s. Since then, many ethnic conflicts have resulted as governments struggle for a sense of national unity within the boundaries drawn by European colonialists.

ECONOMIC AND SOCIAL DEVELOPMENT

In terms of global trade, Sub-Saharan Africa's connection with the world is limited. Yet as global demand for natural resources grows, investors from Asia, Europe, and North America are hoping to extract more of the region's metals and fossil fuels.

Compared with Latin America and the Caribbean, Africa south of the Sahara is poorer and more rural, and its population is very young. Nearly 800 million people reside in this region, which includes 48 states and 1 territory. Demographically, this is the world's fastest-growing region (2.5 percent rate of natural increase); in most countries, nearly half the population (43 percent) is younger than 15 years old. Income levels are low and over half of the region's population lives on less than $1 per day. Life expectancy is only 48 years. It is the only major world region that has grown poorer over the past 30 years. For many people, this part of the world has become synonymous with poverty, disease, violence, and refugees. Overlooked in the all-too-frequent negative headlines are initiatives by private, local, and state groups to improve the region's quality of life. Many states have reduced infant mortality, expanded basic education, and increased food production in the past two decades despite the profound economic and political crises in this region.

Sub-Saharan Africa—that portion of the African continent lying south of the Sahara Desert—is a commonly accepted world region (Figure 6.1). It is historically referred to as "Black Africa" because of the skin color of its indigenous inhabitants, but such a race-based division falls apart under closer inspection. The coherence of this region has less to do with skin color than with similar livelihood systems and a shared colonial experience. No common religion, language, philosophy, or political system ever united the area. Instead, diffuse cultural bonds developed from a variety of lifestyles and idea systems that evolved here. The impact of outsiders also helped determine the region's identity. Slave traders from Europe, North Africa, and Southwest Asia treated Africans as chattel; up until the mid-1800s millions of Africans were taken from the region and sold into slavery. In the late 1800s, the entire African continent was divided by European colonial powers, imposing political boundaries that remain to this day. In the postcolonial period, which began in the 1960s, Sub-Saharan African countries faced many of the same economic and political challenges (see "Setting the Boundaries").

The region is culturally complex; 20 or more languages may be spoken in one large state. Consequently, most Africans understand and speak several languages. Ethnic identities do not conform to the political divisions of Africa, sometimes resulting in bloody ethnic warfare, as witnessed in Rwanda in the mid-1990s and in Sudan today. Nevertheless, throughout the region peaceful coexistence between distinct ethnic groups is the norm. Although the cultural significance of European colonizers cannot be ignored, especially the adoption of European languages, religions, educational systems, and political ideas, the daily rhythms of African life are far removed from the industrial world. Most Africans still engage in subsistence agriculture. Women in particular are charged with tending crops and procuring household necessities (Figure 6.2). Male roles revolve around tending livestock and the public life of the village and market. As is happening elsewhere in the developing world, cities are growing rapidly, although just one-third of the region's population is classified as urban.

The influence of African peoples outside the region is profound, the legacy of a slave trade that resulted in the transfer of African peoples, religious systems, and musical traditions throughout the Western Hemisphere. Even today, African-based religious systems are widely practiced in the Caribbean and Latin America, especially in Brazil. American jazz, Brazilian samba, and Cuban rumba would not exist but for the influence of Africans. The popularity of "world music" has created a large international audience for performers such as South Africa's Ladysmith Black Mambazo. Through music and the arts, Sub-Saharan Africa has exerted a significant influence on global culture.

Setting the Boundaries

Sub-Saharan Africa includes 43 mainland states plus the island nations of Madagascar, Cape Verde, São Tomé and Principe, Seychelles, Mauritius, and the French territory of Reunion. When setting this particular regional boundary, the major question one faces is how to treat North Africa.

Many scholars argue for Africa's continental integrity as a culture region. The Sahara has never formed a complete barrier between the Mediterranean north and the remainder of the African land mass. Moreover, the Nile River forms a corridor several thousand miles long of continuous settlement linking North Africa directly to the center of the continent. There is no obvious place to divide the watershed between northern and Sub-Saharan Africa. Political units, such as the African Union, are modern examples of the continent's indivisibility.

The lack of a clear divide across Africa does not nullify its division into two world regions. North Africa is generally considered more closely linked, both culturally and physically, to Southwest Asia. Arabic is the dominant language and Islam the dominant religion of North Africa. Consequently, North Africans feel more closely connected to the Arab hearth in Southwest Asia than to the Sub-Saharan world.

The decision to view Sub-Saharan Africa as a world region still presents the problem of where to divide it from Africa north of the Sahara. We draw the line along political boundaries so that the Mediterranean states of North Africa are discussed with Southwest Asia. Sudan is discussed in both Chapters 6 and 7 because it shares characteristics common to both regions. Sudan is Africa's largest state in terms of area; it is one-fourth the size of the United States. In the more populous and powerful north, Muslim leaders have forged an Islamic state that is culturally and politically oriented toward North Africa and Southwest Asia. Southern Sudan, however, has more in common with the animist and Christian groups of the Sub-Saharan region.

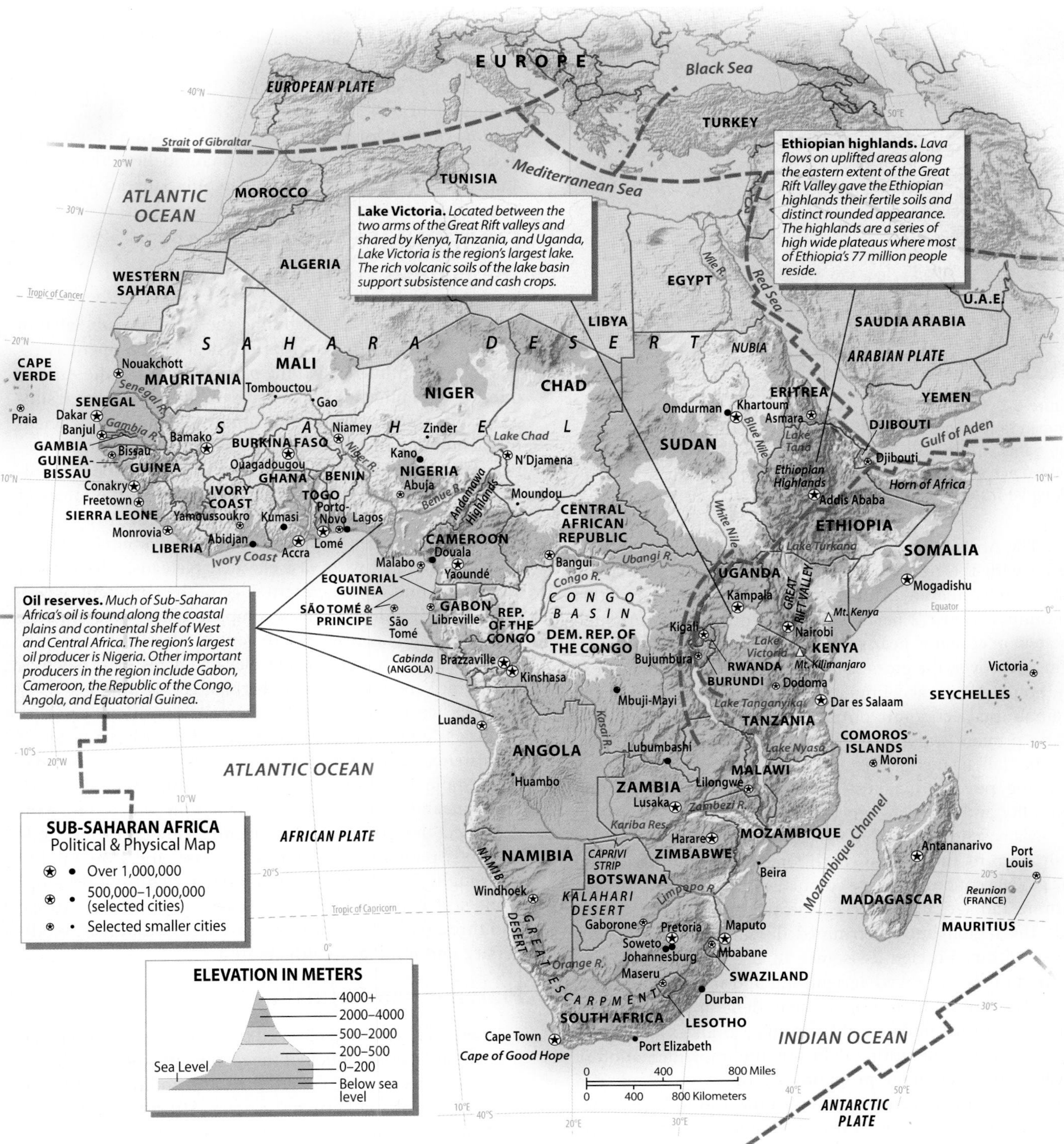

Figure 6.1 Sub-Saharan Africa
Africa south of the Sahara includes 48 states and 1 territory. This vast region of rain forest, tropical savanna, and desert is home to nearly 800 million people. Much of the region consists of broad plateaus ranging from 1,600 to 6,500 feet (500 to 2,000 meters) in elevation. Although the population is growing rapidly, the overall population density of Sub-Saharan Africa is low. Considered one of the least developed regions of the world, it remains an area rich in natural resources.

Figure 6.2 Women Farmers in Senegal
Women farmers grow manioc in Tattaguine, Senegal. Throughout Sub-Saharan Africa, subsistence agriculture is still widely practiced and women are often responsible for tending these crops. *(Dung Vo Trung/Sygma/Corbis)*

The African economy, however, is marginal when compared with the rest of the world. According to the World Bank, Sub-Saharan Africa's economic output in 2004 amounted to just 2 percent of global output, even though the region contains nearly 12 percent of the world's population. Moreover, the gross national product of just one country, South Africa, accounts for more than one-third of the region's total economic output. Even the region's share of world primary exports (such as copper, cocoa, oil, and coffee) has fallen since 1980. Of greatest concern is the region's grinding poverty manifested by malnutrition, high child mortality, and short life expectancies.

Many scholars feel that Sub-Saharan Africa has benefited little from its integration (both forced and voluntary) into the global economy. Slavery, colonialism, and export-oriented mining and agriculture served the needs of consumers outside the region but undermined domestic production for those within. Foreign assistance in the postindependence years initially improved agricultural and industrial output, but also led to mounting foreign debt and corruption, which over time undercut the region's economic gains. One consequence of mounting debt was the implementation of **structural adjustment programs** by the World Bank and the International Monetary Fund in the 1980s. These controversial economic measures are designed to reduce government spending and encourage private sector initiatives. Yet they typically trigger drastic cutbacks in government-supported services and food subsidies, which disproportionately affect the poor. The idea of debt forgiveness for Africa's poorest states is steadily being applied as a strategy to reduce human suffering in the region.

The past few years have witnessed a surge in philanthropic outreach to the region. Rock star Bono of U2 and the Bill and Melinda Gates Foundation have led the One Campaign, directing millions of dollars to support health care, disease prevention, education, and poverty reduction in Sub-Saharan Africa. Private capital investment in Sub-Saharan Africa lags far behind other developing regions but foreign investment is on the rise, especially in oil-rich states such as Angola, Chad, and Equatorial Guinea. Through a combination of internal reforms, better governance, foreign assistance, and foreign investment in infrastructure and technology, many believe that social and economic gains are possible in Sub-Saharan Africa.

ENVIRONMENTAL GEOGRAPHY: The Plateau Continent

The largest land mass straddling the equator, Sub-Saharan Africa is vast in scale and its physical environment is remarkably beautiful. Called the *plateau continent*, the African interior is dominated by extensive uplifted areas that resulted from the breakup of **Gondwanaland**, an ancient megacontinent that included Africa, South America, Antarctica, Australia, Madagascar, and Saudi Arabia. Some 250 million years ago, it began to split apart through the forces of continental drift. As this process unfolded, the African land mass experienced a series of continental uplifts that left much of the area with vast elevated plateaus. The highest areas are found on the eastern edge of the continent, where the Great Rift Valley forms a complex upland area of lakes, volcanoes, and deep valleys. In contrast, lowlands prevail in West Africa, although smaller elevated areas exist, such as the Guinea and the Adamawa highlands (Figure 6.1).

The landscape of Sub-Saharan Africa offers a palette of intense colors: deep red soils studded with plantings of subsistence crops; the blue tropical sky; golden savannas that ripple with the movement of animal herds; dark rivers meandering through towering rain forests; and sun-drenched deserts. Amid this beauty, however, one finds relatively poor soils, widespread disease, and vulnerability to drought. Large areas for potential agricultural development still exist, however, especially in southern Africa; and throughout the continent vast water resources, biodiversity, and mineral wealth abound.

Plateaus and Basins

A series of plateaus and elevated basins dominate the African interior and explain much of the region's uniqueness. Generally, elevations increase toward the south and east of the continent. Most of southern and eastern Africa lies well above 2,000 feet (600 meters), and sizable areas sit above 5,000 feet (1,500 meters). These areas are typically referred to as High Africa; Low Africa includes West Africa and much of Central Africa. The higher plateaus of Kenya, Zimbabwe, and Angola are noted for their cooler climates and relatively abundant moisture. Steep escarpments form where plateaus abruptly end, as illustrated by the majestic Victoria Falls on the Zambezi River (Figure 6.3). Much of southern Africa is rimmed by a landform called the **Great Escarpment**, which begins in southwestern Angola and ends in northeastern South Africa, creating a natural barrier to coastal settlement. South Africa's Drakensberg Mountains (with elevations reaching 10,000

Figure 6.3 Victoria Falls
The Zambezi River descends over Victoria Falls. A fault zone in the African plateau explains the existence of a 360-foot (110-meter) drop. The Zambezi has never been important for navigation, but it is a vital supply of hydroelectricity for Zimbabwe, Zambia, and Mozambique. *(Rob Crandall/www.robcrandall.com)*

feet, or 3,100 meters) rise up from the Great Escarpment. Because of this landform, coastal plains tend to be narrow, with few natural harbors, and river navigation is impeded by a series of falls. Fairly wide coastal lowlands are found in parts of western Africa, Mozambique, and Somalia, but even these lack extensive areas of fertile alluvial soil (see Figure 6.1).

Though Sub-Saharan Africa is an elevated land mass, it has few significant mountain ranges. The one extensive area of mountainous topography is in Ethiopia, which lies in the northern portion of the Rift Valley zone. Yet even there the dominant features are high plateaus intercut with deep valleys rather than actual mountain ranges. Receiving heavy rains in the wet season, the Ethiopian Plateau forms the headwaters of several important rivers, most notably the Blue Nile, which joins the White Nile at Khartoum, Sudan.

A discontinuous series of volcanic mountains, some of them quite tall, are associated with the southern half of the Rift Valley. Kilimanjaro at 19,000 feet (5,900 meters) is the continent's tallest mountain, and nearby Mount Kenya (17,000 feet, or 5,200 meters) is the second tallest. The Rift Valley itself reveals the slow but inexorable progress of geological forces (Figure 6.4). Eastern Africa is slowly being torn away from the rest of the continent, and within some tens of millions of years it will form a separate land mass. Such motion has already produced a great gash across the uplands of eastern Africa, much of which is occupied by elongated and extremely deep lakes (most notably Nyasa, Malawi, and Tanganyika). In central eastern Africa this rift zone splits into two separate valleys, each of which is flanked by volcanic uplands. Between the eastern and western rifts lies a bowl-shaped depression, the center of which is filled by Lake Victoria—Africa's largest body of water. Not surprisingly, some of the densest areas of settlement are found amid the fertile and well-watered soils that border the Rift Valley.

Watersheds Africa south of the Sahara conspicuously lacks the broad, alluvial lowlands that influence patterns of settlement throughout other regions. The four major river systems are the Congo, Nile, Niger, and Zambezi. Smaller rivers, such as the Orange in South Africa; the Senegal, which divides Mauritania and Senegal; and the Limpopo in Mozambique, are locally important but drain much smaller areas. Ironically, most people think of Africa south of the Sahara as suffering from water scarcity and tend to discount the size and importance of the watersheds, or catchment areas, that these river systems drain.

The Congo River (or Zaire) is the largest watershed in terms of drainage and flow volume in the region. It is second only to South America's Amazon River in annual volume of flow. The Congo flows across a relatively flat basin that lies more than 1,000 feet (300 meters) above sea level, meandering through Africa's largest tropical forest, the Ituri (Figure 6.5). Entry from the Atlantic into the Congo Basin is prevented by a series of rapids and falls, making the Congo River only partially navigable. Despite these limitations, the Congo River has been the major corridor for travel within the Republic of the Congo and the Democratic Republic of the Congo (formerly Zaire); the capitals of both countries, Brazzaville and Kinshasa, rest on opposite sides of the river.

The Nile River, the world's longest, is discussed in Chapter 7 as the lifeblood of Egypt and Sudan. Yet this river originates in the highlands of the Rift Valley zone, and is an important link between North and Sub-Saharan Africa. The Nile begins in the lakes of the rift zone (Victoria and Edward) before descending into a vast wetland in southern

Figure 6.4 The Rift Valley
An aerial view of a portion of the Rift Valley in Kenya. This is one of the continent's most dramatic landforms, extending nearly 5,000 miles (8,000 kilometers) from Lake Nyasa to the Red Sea. The Rift Valley zone is a series of faults that have created volcanoes, escarpments, elongated lakes, and valleys. *(Altitude/Y. Arthus-B./Peter Arnold, Inc.)*

Figure 6.5 Congo River Fishermen Congo River fishermen in a canoe ply the river as a source of fish. Behind them is their village and the immense Ituri rain forest. *(Robert Caputo/Aurora & Quanta Productions, Inc.)*

Sudan known as the Sudd. Agricultural development projects in the 1970s greatly increased the agricultural potential of the Sudd, especially its peanut crop. Unfortunately, the past three decades of civil war in Sudan ravaged this area, turning farmers and herders into refugees and undermining the productive capacity of this important ecosystem.

Like the Nile, the Niger River is the critical source of water for two otherwise arid countries: Mali and Niger. Originating in the humid Guinea highlands, the Niger flows first to the northeast and then spreads out to form a huge inland delta in Mali before making a great bend southward at the margins of the Sahara near Gao. On the banks of the Niger River are the capitals of Mali (Bamako) and Niger (Niamey), as well as the historic city of Tombouctou (Timbuktu). After flowing through the desert, the Niger River returns to the humid lowlands of Nigeria where the Kainji Reservoir temporarily blocks its flow to produce electricity for Africa's most populous state.

The considerably smaller Zambezi River originates in Angola and flows east, spilling over an escarpment at Victoria Falls, and finally reaching Mozambique and the Indian Ocean. More than other rivers in the region, the Zambezi is a major supplier of commercial energy. Sub-Saharan Africa's two largest hydroelectric installations, the Kariba on the border of Zambia and Zimbabwe and the Cabora Bassa in Mozambique, are on this river. Flooding in the lower Zambezi and Limpopo rivers devastated Mozambique in 2000, afflicting nearly 2 million people with failed crops, property damage, and disease. It seems that the worst flooding in 50 years was aggravated when countries upstream released additional waters from their locks to keep dams from bursting.

Soils With a few major exceptions, Sub-Saharan Africa's soils are relatively infertile and thus cannot easily support the intensive agriculture needed to feed large populations. Generally speaking, fertile soils are young soils, those deposited in recent geological time by rivers, volcanoes, glaciers, or windstorms. In older soils—especially those located in moist tropical environments—natural processes tend to wash out most plant nutrients over time. Over most of Sub-Saharan Africa, the agents of soil renewal have largely been absent; the region has few alluvial lowlands where rivers periodically deposit fertile silt, and it did not experience significant glaciation in the last ice age, as did North America.

Portions of Sub-Saharan Africa are, however, noted for their natural soil fertility, and not surprisingly these areas support denser settlement. Some of the most fertile soils are in the Rift Valley, enhanced by the volcanic activity associated with the area. The population densities of rural Rwanda and Burundi, for example, are partially explained by the highly productive volcanic soils. The same can be said for highland Ethiopia, which supports the region's second largest population, one of more than 70 million people. The Lake Victoria lowlands and central highlands of Kenya also are noted for their sizable populations and productive agricultural bases.

In the drier grasslands and semidesert areas one finds a soil type called *alfisols*. High in aluminum and iron, these red soils have greater fertility than comparable soils found in wetter zones. This helps explain the tendency of farmers to plant in drier areas, such as the Sahel, even though they risk exposure to drought. With irrigation, many agronomists suggest that the southern African countries of Zambia and Zimbabwe could greatly increase commercial grain production on these soils.

Climate and Vegetation

Sub-Saharan Africa lies in the tropical latitudes. Beginning just north of the Tropic of Cancer, crossing the equator, and extending past the Tropic of Capricorn in the south, it is the largest tropical land mass on the planet. Only the far south

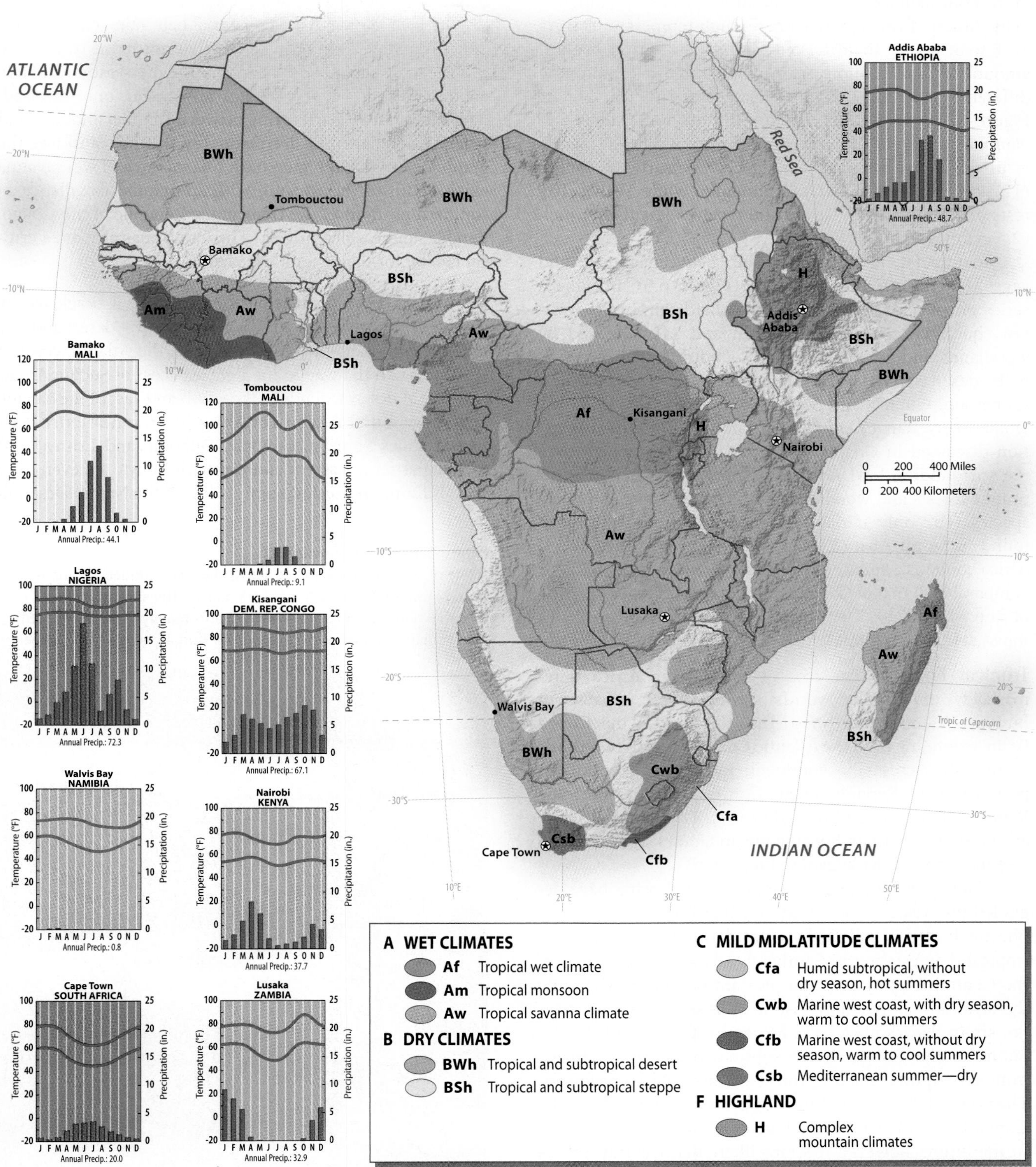

Figure 6.6 Climate Map of Sub-Saharan Africa

Much of the region lies within the tropical humid and tropical dry climatic zones; thus, the seasonal temperature changes are not great. Precipitation, however, varies significantly. Compare the distinct rainy seasons in Lusaka and Lagos: Lagos is wettest in June and Lusaka receives most of its rain in January. Although there are important tropical forests in West and Central Africa, much of the territory is tropical savanna. *(Temperature and precipitation data from E. A. Pearce and C. G. Smith, 1984,* The World Weather Guide, *London: Hutchinson)*

of the continent extends into the subtropical and temperate belts. Much of the region averages high temperatures from 70°F to 80°F (22° to 28°C) year-round. The seasonality and amount of rainfall, more than temperature, determine the different vegetation belts that characterize the region. As Figure 6.6 shows, Addis Ababa, Ethiopia, and Walvis Bay, Namibia, have similar average temperatures, but the former is in the moist highlands and receives nearly 50 inches (127 centimeters) of rainfall annually, while Walvis Bay rests on the Namibian Desert and receives less than 1 inch (2.5 centimeters).

To understand the relationship between climate and vegetation in Sub-Saharan Africa (Figure 6.6), imagine a series of concentric vegetation belts that begin in the western equatorial zone as forest (Am and Af), followed by woodlands and grasslands (Aw), semidesert (BSh), and finally desert (BWh). The montane zones of East Africa, Cameroon, Guinea, and South Africa (especially along the Drakensberg Range) exhibit the forces of altitudinal zonation discussed in Chapter 4. Capturing more rainfall than the surrounding lowlands, these mountains often support unique forests and woodlands with large numbers of native species. The only midlatitude climates of the region are found in South Africa. The southwestern corner of South Africa contains a small zone of Mediterranean climate that is comparable to that of California and noted for its production of fine wine (Csb). The eastern coast of South Africa has a moist subtropical climate, not unlike that of Florida (Cfa).

Tropical Forests The core of Sub-Saharan Africa is remarkably moist. The world's second-largest expanse of humid equatorial rain forest, the Ituri, lies in the Congo Basin, extending from the Atlantic Coast of Gabon two-thirds of the way across the continent, including the northern portions of the Republic of the Congo and the Democratic Republic of the Congo (formerly Zaire). The conditions here are constantly warm to hot, and precipitation falls year-round (see the graph for Kisangani in Figure 6.6).

Commercial logging and agricultural clearing have degraded the western and southern fringes of this vast forest, but much of it is still intact. Considering the high rates of tropical deforestation in Southeast Asia and Latin America, the Central African case is a pleasant exception. Certainly the area's low population base of subsistence farmers does not strain the resource base. Moreover, Gabon and the Republic of the Congo derive substantial foreign exchange from oil exports, making forest extraction less necessary. The political chaos engulfing the Democratic Republic of the Congo has made large-scale logging a difficult proposition. In the future, however, it seems likely that Central Africa's rain forests could suffer the same kind of degradation experienced in other equatorial areas. As deforestation proceeds elsewhere in the world, the trees of equatorial Africa become increasingly valuable, hence, more vulnerable.

Savannas Wrapped around the Central African rainforest belt in a great arc lie Africa's vast tropical wet and dry savannas. Savannas are dominated by a mixture of trees and tall grasses in the wetter zones immediately adjacent to the forest belt and shorter grasses with fewer trees in the drier zones (Figure 6.7). North of the equatorial belt, rain generally falls only from May to October. The farther north one travels, the less the total rainfall and the longer the dry season. Climatic conditions south of the equator are similar, only reversed, with the wet season occurring between October and May and precipitation generally decreasing toward the south (see the graph for Lusaka in Figure 6.6). A larger area of wet savanna exists south of the equator, with substantial woodlands in southern portions of the Democratic Republic of the Congo, Zambia, and eastern Angola. These savannas also are a critical habitat for the region's large fauna.

Deserts The vast extent of tropical Africa is bracketed by several deserts. The Sahara, the world's largest desert and one of its driest, spans the land mass from the Atlantic coast of Mauritania all the way to the Red Sea coast of Sudan. A narrow belt of desert extends to the south and east of the Sahara, wrapping around the **Horn of Africa** (the northeastern corner that includes Somalia, Ethiopia, Djibouti, and Eritrea) and pushing as far as eastern and northern Kenya. An even drier zone is found in southwestern Africa. In the Namib Desert of

Figure 6.7 African Savannas
A line of wildebeests marches across the tree-studded savannas found in Masai Mara National Park, Kenya. The savannas, an essential habitat for Africa's wildlife, are steadily being converted into agricultural and pastoral lands for human needs. *(Nik Wheeler)*

Figure 6.8 Kalahari Desert
A San man teaches his four-year-old son to hunt with a bow and arrow. The Kalahari, though not a true desert, does not get enough rain to support agriculture. Yet this ecosystem is home to abundant wildlife and hunter-gatherers, such as the San (formerly known as the Bushmen). *(Louis Gubb/The Image Works)*

coastal Namibia, rainfall is a rare event, although temperatures are usually mild (see the graph for Walvis Bay in Figure 6.6). Inland from the Namib lies the Kalahari Desert. Most of the Kalahari is not dry enough to be classified as a true desert, because it receives slightly more than 10 inches (25 centimeters) of rain a year (Figure 6.8). Its rainy season, however, is brief. Most of the precipitation is immediately absorbed by the underlying sands. Surface water is thus scarce, giving the Kalahari a desertlike aspect for most of the year.

Africa's Environmental Issues

The prevailing perception of Africa south of the Sahara is one of environmental scarcity and degradation, no doubt fostered by televised images of drought-ravaged regions and starving children. Single explanations such as rapid population growth or colonial exploitation cannot fully capture the complexity of Africa's environmental issues or the ingenious ways that people have adapted to living in marginal ecosystems. Because much of Sub-Saharan Africa's population is rural and poor, earning its livelihood directly from the land, sudden environmental changes are keenly felt and can cause mass migrations, famine, and even death. As Figure 6.9 illustrates, **desertification**, the expansion of desertlike conditions as a result of human-induced degradation, and deforestation are commonplace. Sub-Saharan African also is vulnerable to drought, most notably in the Horn of Africa, parts of southern Africa, and the Sahel. Many scientists fear that droughts will come more often and be prolonged under global climate change scenarios. At the same time, wildlife tourism is an increasingly important source of foreign exchange for many African states. Throughout the region, national parks have been created in an effort to strike a balance between humans' and animals' competing demands for land.

The Sahel and Desertification In the 1970s, the **Sahel** became an emblem for the dangers of unchecked population growth and human-induced environmental degradation when a relatively wet period came to an abrupt end and 6 years of drought (1968–74) ravaged the land. The Sahel is a zone of ecological transition between the Sahara to the north and wetter savannas and forest in the south. Life depends on a delicate balance of limited rain, drought-resistant plants, and a pattern of animal **transhumance**, the movement of animals between wet-season and dry-season pasture. Parts of the Sahel were important areas of settlement long before European colonization. What appears to be desert wasteland in April or May is transformed into a lush garden of millet, sorghum, and peanuts after the drenching rains of June. Relatively free of the tropical diseases found in the wetter zones to the south, Sahelian soils also are quite fertile, which helps explain why people continue to live there despite the unreliable rainfall patterns (Figure 6.10). During the drought of the early 1970s, rivers in the area diminished and desertlike conditions began to move south. Tens of millions of people inhabited this area, and farmers and **pastoralists** whose livelihoods had come to depend on the more abundant precipitation of the relatively wet period were temporarily forced out.

The main culprits in Sahelian desertification are the expansion of agriculture and overgrazing, leading to the loss of natural vegetation and declines in soil fertility. Through most of the Sahel, French colonial authorities forced villagers to grow peanuts as an export crop, a policy continued by the newly independent states of the region. However, peanuts tend to deplete several key soil nutrients, which means that peanut farms are often abandoned after a few years as cultivators move on to fresh sites. Since peanuts grow underground, moreover, the soil must be overturned at harvest time. This typically occurs at the onset of the dry season, when dry winds from the Sahara can carry away the fine dirt of the newly harvested field. With the loss of topsoil, the ground no longer absorbs the precipitation that does fall—often in drenching torrents—during the brief rainy season. The end result of this grim process is the spread of desertlike conditions regardless of actual changes in precipitation.

Overgrazing also has been implicated in Sahelian desertification, as is true in other regions of the world. Livestock is a traditional product of the region, and animal production dramatically expanded after World War II as the population swelled and the need for export earnings grew. In many areas, natural pasture was greatly reduced by intensified grazing pressure, leading to increased wind

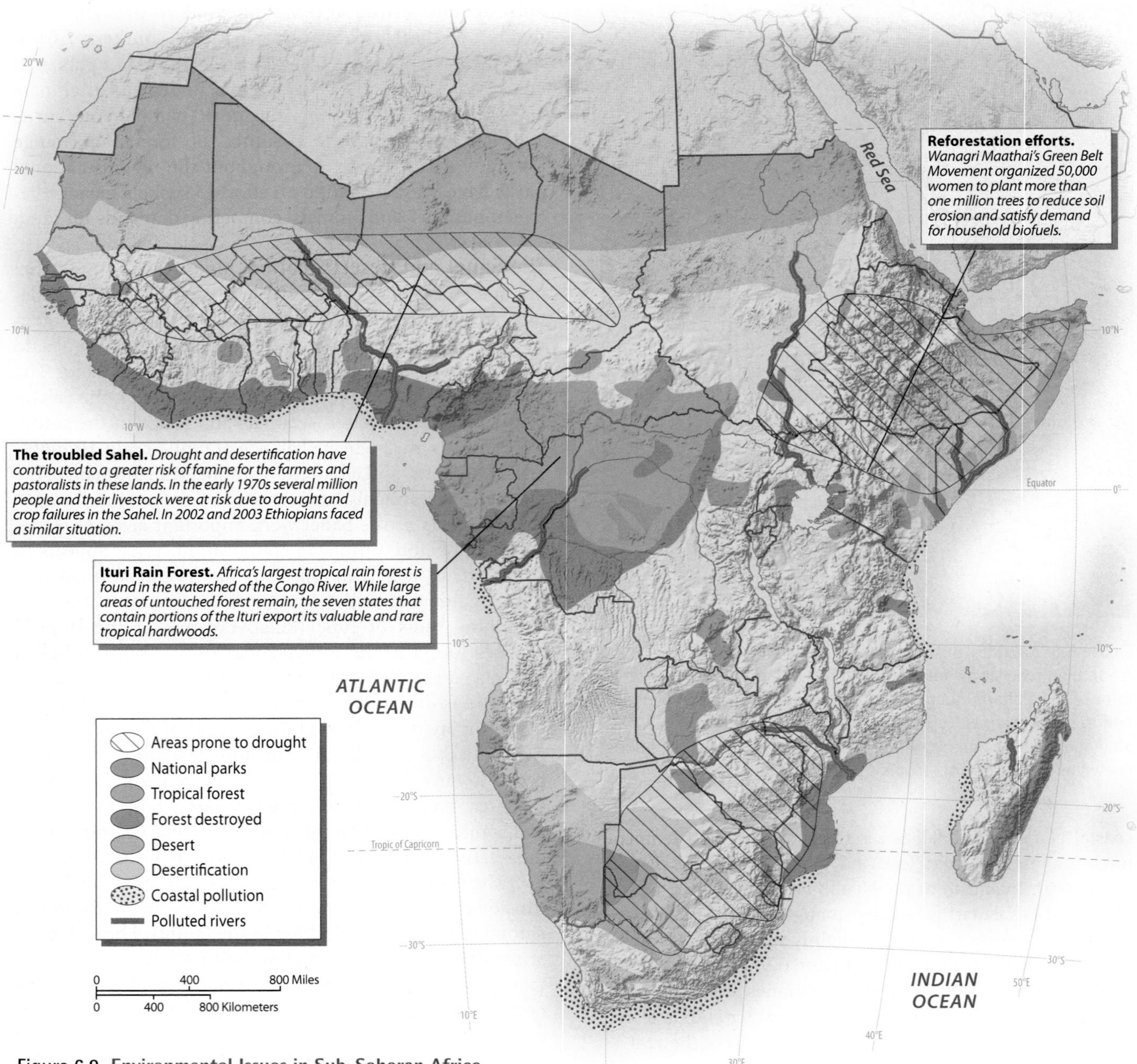

Figure 6.9 Environmental Issues in Sub-Saharan Africa
Given the immense size of Sub-Saharan Africa, it is difficult to generalize about environmental problems. Dependence on trees for fuel places strains on forests and wooded savannas throughout the region. In semiarid regions, such as the Sahel and Horn of Africa, population pressures, climate change, and land-use practices seem to have exacerbated desertification. Yet Sub-Saharan Africa also supports the most impressive array of wildlife, especially large mammals, on Earth. *(Adapted from* DK World Atlas, *1997, London: DK Publishing, p. 75)*

erosion. National and international developmental agencies, hoping to increase production further, began to dig deep wells in areas that previously had been unused by herders through most of the year. The new supplies of water, in turn, allowed year-round grazing in places that, over time, could not withstand it. Large barren circles around each new well began to appear even on satellite images.

Pessimists view the process of Sahelian degradation as virtually irreversible and argue that the region can no longer support the millions of people who presently reside there; a few foresee a future of mass starvation exacerbated by global climate change. Optimists contend that the land can quickly recover if appropriate practices are followed, pointing out that when the rains returned in the late 1970s, so too did many farmers and herders (see "Geographic Tools: Monitoring Land Cover and Conservation Changes in Niger"). Debates also surround the issue of climatic fluctuation itself and whether the future for the Sahel

Figure 6.10 Sahel in Bloom After the drenching rains of summer, the Sahel can produce both subsistence and export crops. In southern Chad, beans are grown for local consumption. *(Frank Kroenke/Das Fotoarchiv/Peter Arnold, Inc.)*

will be wetter or drier as a result of global climate change. Regardless of the causes, many of the areas experiencing desertification are the same ones most vulnerable to drought and famine. These conditions may become more common with global climate change.

Deforestation Although Sub-Saharan Africa still contains extensive forests in some areas, much of the region has relatively little tree cover. Unlike tropical America, forest clearance in the wet and dry savanna is of greater local concern than the limited commercial logging of the rain forest. North of the equator, for example, only a few wooded areas remain in a landscape dominated by grasslands, savanna (grassland with scattered trees and shrubs), and cropland. Highland Ethiopia once was covered with lush forests, but these have long since been reduced to a few remnant patches. Loss of woody vegetation has resulted in extensive hardship, especially for women and children who must spend many hours a day scrounging for wood (Figure 6.11). Deforestation, especially in the woodlands of the savannas, aggravates problems of increased runoff, soil erosion, and shortages of **biofuels** (wood and charcoal used for household energy needs, especially cooking).

In the southern tropical savanna, where human population has historically been lighter, extensive tracts of dry woodland remain intact. Zambia and sections of Mozambique, Angola, and Tanzania still are extensively wooded, although the actual tree coverage is not dense. The trees of the dry forest have little commercial value compared to some of the species found in the rain forest, but they provide vital subsistence resources for local people as well as wildlife habitat. Even in these more abundant areas, biofuel scarcity is common around the larger towns and villages. In some countries, village women have organized into community-based nongovernmental organizations (NGOs) to plant trees and create green belts to meet future fuel needs. One of the most successful efforts is in Kenya under the leadership of Wangari Maathai. Maathai's Green Belt Movement has more than 50,000 members, mostly women, organized into 2,000 local community groups. Since the movement's inception in 1977, these local groups have successfully planted millions of trees. In those areas, village women now spend less time collecting fuel, and local environments have improved. Maathai's success has drawn interest from other African countries,

Figure 6.11 Wood Fuel in Ethiopia A women walks to Addis Ababa with a large bundle of firewood to sell in the city market. Collecting firewood is the work of women, both for household subsistence needs and for sale in local markets. *(Louis Gubb/Corbis/Bettmann)*

GEOGRAPHIC TOOLS

Monitoring Land Cover and Conservation Changes in Niger

Much to everyone's surprise, Niger is becoming greener. Through satellite imagery and field inventories at the village level, local agronomists have documented that tree-covered areas have increased by 7 million acres (Figure 6.2.1). These basic geographic tools were used to reveal unexpected improvements in tree cover. It appears that better local conservation practices and improved rainfall have led to a dramatic increase in vegetation over the past 30 years. More interesting still, increases in tree cover have occurred in some of the most densely populated areas. Given that the Sahel is an area noted for drought, famine, and desertification, documenting land cover change and explaining how such changes happen offers valuable lessons for other drought-prone areas.

The greening of Niger was not the result of a large-scale government effort or a foreign assistance program. The impetus for change came from farmers themselves who realized after the drought of 1984 that maintaining tree cover was important for the long-term sustainability of farming practices. Farmers began actively protecting saplings instead of clearing them from their fields, including the nitrogen-fixing *goa* tree which had disappeared from many villages. During the rainy season, the goa tree loses its leaves, so it does not compete with crops for water or sun. The leaves themselves are added fertilizer to the soil. In general, trees and shrubs reduce wind erosion, and their presence also reduces runoff and erosion when the rains do come. Sahelian farmers also use branches, pods, and leaves from trees for fuel and animal fodder. Any excess can be sold at market.

Given the utility of native trees, why were farmers removing them? Much vegetation disappeared in the 1970s when the region suffered prolonged drought. Until the 1990s, all trees were considered property of the state of Niger, thus giving farmers little incentive to protect them. Since then, the government has recognized the value of allowing individuals to own trees. Not only can farmers sell branches, pods, fruit, and bark, they also conserve them to ensure their agricultural livelihood.

For forest regeneration to work it requires the effort of the whole community. If farmers don't take action themselves and the community does not support them, forest regeneration will not be successful. Consequently, some villages in Niger are much greener than others. In the village of Moussa Bara, where regeneration has been a success, not one child died of malnutrition in the famine of 2005, largely because farmers were able to earn income from selling firewood when their crops failed.

The Sahel is a poor region that has always been vulnerable to drought. Yet it is also an area of fertile soils and few tropical diseases. It is uncertain how global climate changes will impact this region; some years may be wetter than average but there is the possibility of more prolonged droughts. For now, at least, Niger offers a provocative example of how small and relatively simple conservation practices can have a positive impact.

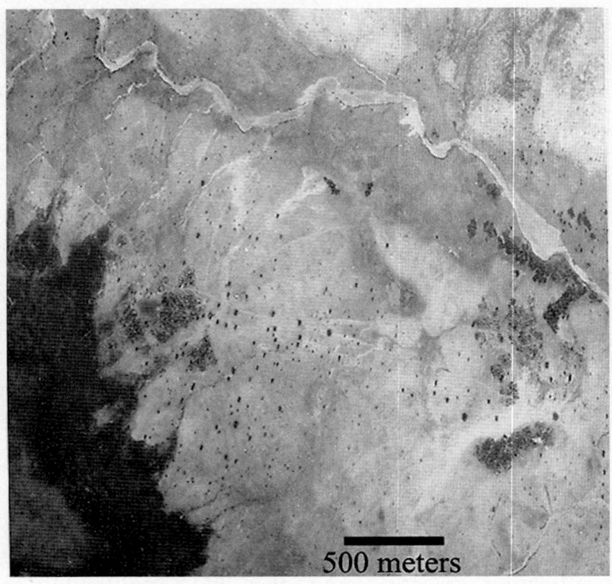

(a)

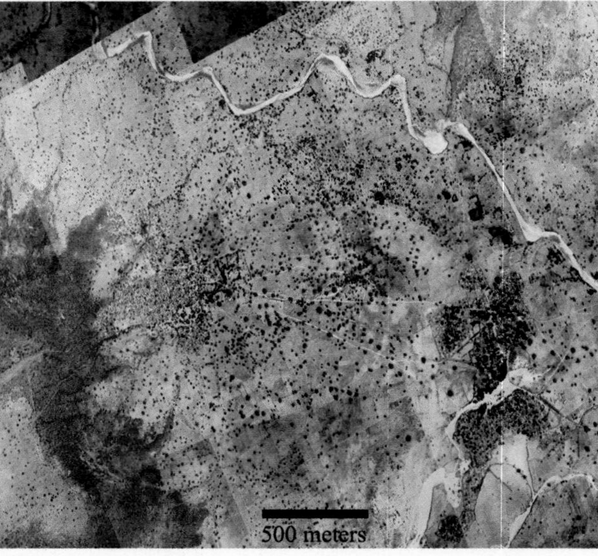

(b)

Figure 6.2.1 More Trees in Southern Niger
These two satellite images demonstrate an increase in tree cover in Niger. In 1975 (a), there were relatively few trees in this part of the Sahel. By 2003 (b), tree cover had inrcreased substantially.
(USGS Center for Earth Resources Observation and Science)

Source: Adapted from Lydia Polgreen, "In Niger, Trees and Crops Turn Back the Desert," *New York Times*, February 11, 2007.

Figure 6.12 Deforestation in Madagascar
Forest clearing for agriculture and fuel has denuded much of the countryside in Madagascar. Serious problems with soil erosion exist throughout the country, limiting agricultural productivity. During the rainy season, the discharge of topsoil into the Indian Ocean is visible from space. *(Bios/M. Gunther/Peter Arnold, Inc.)*

spurring a Pan-African Green Belt Movement largely organized through nongovernmental organizations interested in biofuel generation, protection of the environment, and the empowerment of women. In 2004 Professor Maathai was awarded a Nobel Peace Prize for her contribution to sustainable development, democracy, and peace.

Destruction of rain forest for logging is most evident in the southern fringes of Central Africa's Ituri forest (see Figure 6.9). Given the vastness of this forest and the relatively small number of people living there, however, it is less threatened than other forest areas. Two smaller rain forests, one along the Atlantic coast from Sierra Leone to western Ghana and the other along the eastern coast of the island of Madagascar, have nearly disappeared. These rain forests have been severely degraded by commercial logging and agricultural clearance. At present rates, several countries in West Africa will be completely deforested early in this century. Madagascar's eastern rain forests, as well as its western dry forests, have suffered serious degradation in the past three decades (Figure 6.12). Deforestation in Madagascar is especially worrisome because the island forms a unique environment with a large number of native species. Many species of fauna (including birds, reptiles, and lemurs) now are endangered, and several face imminent extinction.

Wildlife Conservation Sub-Saharan Africa is justly famous for its wildlife. In no other region of the world can one find such abundance and diversity, especially of large mammals. The survival of wildlife here reflects, to some extent, the historically low human population density and the fact that sleeping sickness (transferred by the tsetse fly) and other diseases have kept people and their livestock out of many areas. In addition, many African peoples have developed various means of successfully coexisting with wildlife.

But as is true elsewhere in the world, wildlife is quickly declining in much of Sub-Saharan Africa. West Africa now has few areas of prime habitat, and what remains is rapidly shrinking. The most noted wildlife reserves are in East Africa; in Kenya and Tanzania these reserves are major tourist attractions and are thus economically important. Even there, however, population pressure, political instability, and poverty make the maintenance of large wildlife reserves difficult. Poaching is a major problem, particularly for rhinoceroses and elephants; the price of a single horn or tusk in distant markets represents several years' wages for most Africans. Ivory is avidly sought in East Asia, especially in Japan. In China, powdered rhino horn is used as a traditional medicine, whereas in Yemen rhino horn is prized for dagger handles. During the 1980s, the region's elephant population fell by more than half, to 600,000, in part due to the ivory trade. Rhino populations are far smaller, and the black rhino is an endangered species.

The most secure wildlife reserves now are located in southern Africa (Figure 6.13). Zoologists and park guides have employed handheld Global Positioning Systems, called *CyberTrackers*, to record wildlife observations and map their findings. In Botswana, even the San bushmen have been given these devices to record the animal movements that only these expert hunters usually observe. In fact, elephant populations are considered to be too large in countries such as Botswana and South Africa. Some wildlife experts here contend that herds should be culled to prevent overgrazing and that the ivory should be legally sold in the international market in order to generate revenue for further conservation. Only such a market-oriented approach, they argue, will give African countries a long-term incentive to preserve habitat. Many environmentalists, not surprisingly, disagree strongly.

Figure 6.13 South African Wildlife
An African elephant traversing the savannas of Kruger National Park in South Africa, one of the region's oldest wildlife parks. South Africa, Botswana, Namibia, and Zimbabwe have the largest elephant populations. Other countries have seen a troubling decline in elephant numbers. *(Rob Crandall/www.robcrandall.com)*

In 1989, a worldwide ban on the legal ivory trade was imposed as part of the Convention on International Trade in Endangered Species (CITES). While several African states, such as Kenya, lobbied hard for the ban, others, such as Zimbabwe, Namibia, and Botswana, complained that their herds were growing and the sale of ivory had helped pay for conservation efforts. Conservationists feared that lifting the ban would bring on a new wave of poaching and illegal trade. In the late 1990s, the ban was lifted so that some southern African states could sell down their substantial inventories of ivory and continue with limited sales. The long-term repercussions for elephant survival are not yet known, although political leaders from Kenya and India have voiced strong opposition to all ivory sales. The ivory controversy shows how global markets and international policies impact elephant survival in Sub-Saharan Africa. At the same time, differences within the region make it difficult to come up with a consistent conservation strategy.

Climate Change and Vulnerability in Sub-Saharan Africa

Global climate change poses extreme risks for this region due to its widespread poverty, recurrent droughts, and overdependence on rain-fed agriculture. Sub-Saharan Africa is the lowest emitter of greenhouse gases in the world, but it is likely to experience greater human vulnerability to global warming because of the region's limited resources to both respond and adapt to environmental change. The areas most vulnerable are arid and semiarid regions such as the Sahel and Horn of Africa, some grassland areas, and the coastal lowlands of West Africa and Angola.

Climate change models suggest that parts of highland East Africa and equatorial Central Africa may receive more rainfall. Thus some lands that are currently marginal for farming might be made more productive. These effects are likely to be negated by the decline in agricultural productivity in the Sahelian belt as well as the grasslands of southern Africa, especially in Zambia and Zimbabwe. Drier grassland areas could deplete wildlife populations, which are a major source behind the growing tourist economy. As in Latin America, higher temperatures in the tropics may also result in the expansion of vector-borne diseases such as malaria and dengue fever into the highlands where they were once relatively rare. Given the relatively high elevations in the region, the negative consequences of sea level rising would be mostly felt on the West African coast (Senegal, Gambia, Sierra Leone, Nigeria, Cameroon, and Gabon).

Even without the threat of climate change, famine stalks many areas of Africa. During 2002 and 2003, the rains did not fall in Ethiopia, and agronomists estimated that 15 million people would require food aid or face death by starvation. In this recent battle with famine, early warnings were taken seriously and international food aid was aggressively sought and distributed. This was not the case in the early 1980s, when nearly 1 million Ethiopians perished, many of them needlessly because aid was not delivered to certain areas of the country due to ethnic fighting. In 2005 crops in Niger failed due to drought and an estimated 2.5 million people were vulnerable to famine but relief efforts resulted in relatively low mortality rates. Areas of southern Africa experienced drought in 2003, making 14 million people in six states vulnerable to food shortages. In this part of Africa, with the highest HIV/AIDS infection rate in the world, food shortages are aggravated by a large percentage of adults too weak to grow food because of infection. Many people, while not technically dying from famine, are succumbing to AIDS prematurely because they are weakened by hunger (Figure 6.14).

The World Food Program estimates that millions of Africans struggle against starvation each year. Famine occurs when there is a surge in hunger-related mortality. Most global warming scenarios show that prolonged drought will become more common in parts of the region

Figure 6.14 Food for Drought Victims
Food aid was delivered to drought victims in central Niger in 2005. With adequate drought warning and international support, famine-related deaths can be avoided in the region. *(Issouf Sanogo/AFP/Getty Images)*

over time. As critical resources such as water and food become scarce, resource-driven conflicts and human displacement become more likely. Some scholars argue that the ethnic conflict in Sudan's Darfur region is driven, in part, by resource competition between pastoralists and agriculturalists in an ecosystem that is experiencing water scarcity.

POPULATION AND SETTLEMENT: Young and Restless

Sub-Saharan Africa's population is growing quickly. Although the global population is projected to increase by nearly 40 percent by 2050, the projected population growth for Africa south of the Sahara is 128 percent over the same time period. The population is very young, with 43 percent younger than age 15, compared to just 17 percent for more developed countries (Table 6.1). Only 3 percent of the region's population is over 65 years of age, whereas in Europe 16 percent is over 65. Families tend to be large, with a woman having an average of five or six children. Yet child and maternal mortality rates are high, reflecting disturbingly low access to basic health services. The most troubling indicator for the region is its stubbornly low life expectancy, which dropped from 52 years in 1995 to 49 years in 2007, in part due to the AIDS epidemic. Life expectancy in other developing nations is much better; India's is 63 years and China's 72 years. The growth of cities also is a major trend. In 1980 it was estimated that 23 percent of the population lived in cities; now the figure is 34 percent.

Behind these demographic facts lie complex differences in settlement patterns, livelihoods, belief systems, and access to health care. Although the region is experiencing rapid population growth, Sub-Saharan Africa is not densely populated. The entire region holds some 788 million persons—roughly half the population that is crowded into the much smaller land area of South Asia. In fact, the overall population density of the region (83 people per square mile, or 32 people per square kilometer) is similar to that of the United States at 80 people per square mile (31 people per square kilometer). Just six states account for half of the region's population: Nigeria, Ethiopia, the Democratic Republic of Congo, South Africa, Sudan, and Tanzania (see Table 6.1). True, there are states with very high population densities: Rwanda has 344 per square kilometer and the island nation of Mauritius has a population density of 618. Yet there are many sparsely inhabited states. Chad has just 20 people per square mile (8 people per square kilometer) and Namibia has only 6 people per square mile (3 people per square kilometer). Such states do not consider population growth a serious issue. Many of the governments of the more densely settled territories, however, began to seriously promote family planning policies in the 1980s.

Crude population density is an imperfect indicator of whether a country is overpopulated. Geographers are often more interested in **physiological density**, the number of people per unit of arable land. The physiological density in Chad, where only 3 percent of the land is arable, is much higher than its crude population density. Perhaps a more telling indicator of population pressure and potential food shortages is **agricultural density**, the number of farmers per unit of arable land. Because the majority of people in Sub-Saharan Africa earn their livings from agriculture, agricultural density indicates the number of people who directly depend on each arable square mile. The agricultural density of many Sub-Saharan countries is 10 times greater than their crude population density.

Population Trends and Demographic Debates

The growth in particular areas of Sub-Saharan Africa with high agricultural densities and rates of natural increase has combined with reversals in some economic and social indicators to make demographers concerned about the region's overall well-being. The Sahel, for example, is not crowded by European or Asian standards, but it may already contain too many people for the land to support, given the unpredictability of its limited rainfall.

This assertion remains controversial, however—as does the entire field of African demography. Some believe that the region could support many more people than it presently does; these people even argue that Sub-Saharan Africa as a whole is still underpopulated and therefore will benefit from continued high population growth. Others argue that the region is a demographic time bomb and that unless fertility is quickly reduced, Sub-Saharan Africa will face massive famines in the near future. The majority of African states officially support lowering rates of natural increase and are slowly promoting modern contraception practices both to reduce family size and to protect people from sexually transmitted diseases. One thing is certain: the region will continue to grow in population.

Family Size A preference for large families is the basis for the region's demographic growth. In the 1960s, many areas in the developing world had comparable total fertility rates (TFR) of 5.0 or higher, but by the mid-1990s only people in Sub-Saharan Africa continued to have such large families. In Sub-Saharan Africa, a combination of cultural practices, rural lifestyles, and economic realities encourage large families (Figure 6.15).

Throughout the region large families are considered prestigious and guarantee a family's lineage and status. Even now, most women marry young, typically when they are teenagers, which increases their opportunity to have children. Demographers often point to the limited formal education available to women as another factor contributing to high fertility. Ethnic rivalries also may encourage certain groups to have more children. In ethnically divided states, such as Kenya and Nigeria, resources often are divided according to ethnic affiliation. An ethnic group's decision to reduce its numbers could, over time, weaken its political influence and thus be seen as unsound. Also, stubbornly high child mortality rates may encourage

TABLE 6.1 Population Indicators

Country	Population (Millions, 2007)	Population Density (per Square Kilometer)	Total Fertility Rate	Percent Urban	Percent <15[a]	Percent >65[b]	Net Migration (per 1,000, 2000–05)
Angola	16.3	13	6.8	40	47	2	1.9
Benin	9.0	80	5.7	39	44	3	2.5
Botswana	1.8	3	3.1	54	38	3	−0.7
Burkina Faso	14.8	54	6.2	16	47	3	1.6
Burundi	8.5	306	6.8	10	45	3	5.5
Cameroon	18.1	38	4.9	53	42	3	0.2
Cape Verde	0.5	122	3.5	56	38	6	−2.1
Central African Republic	4.3	7	5.0	38	43	4	−2.3
Chad	10.8	8	6.5	21	46	3	6.0
Comoros	0.7	318	4.9	37	42	3	−2.7
Congo	3.8	11	5.3	60	42	3	−0.8
Dem. Rep. of Congo	62.6	27	6.7	32	47	3	−1.2
Djibouti	0.8	36	4.2	82	39	3	−2.6
Equatorial Guinea	0.5	18	5.6	39	42	4	0.0
Eritrea	4.9	41	5.3	19	43	2	14.1
Ethiopia	77.1	70	5.4	16	43	3	−0.4
Gabon	1.3	5	3.4	84	26	5	−2.3
Gambia	1.5	134	5.1	50	42	3	4.4
Ghana	23.0	96	4.4	44	40	4	0.1
Guinea	10.1	41	5.7	30	46	3	−6.7
Guinea-Bissau	1.7	47	7.1	30	48	3	0.2
Ivory Coast	20.2	63	5.0	47	41	3	−4.3
Kenya	36.9	64	4.9	19	42	2	−1.3
Lesotho	1.8	59	3.5	13	39	5	−4.0
Liberia	3.8	34	6.8	58	47	2	−15.4
Madagascar	18.3	31	5.2	26	45	3	0.0
Malawi	13.1	110	6.3	17	46	3	−0.3
Mali	12.3	10	6.6	31	48	4	−2.1
Mauritania	3.1	3	4.8	40	40	4	2.1
Mauritius	1.3	618	1.7	42	24	7	0.0
Mozambique	20.4	25	5.4	35	43	3	−0.2
Namibia	2.1	3	3.6	33	41	3	−0.6
Niger	14.2	11	7.1	17	48	3	−0.2
Nigeria	144.4	156	5.9	44	45	3	−0.3
Reunion	0.8	319	2.4	89	27	7	1.3
Rwanda	9.3	355	6.1	17	46	3	1.1
São Tomé and Principe	0.2	161	4.1	58	42	4	−2.7
Senegal	12.4	63	5.3	41	44	4	−1.8
Seychelles	0.1	190	2.1	53	25	8	0.0
Sierra Leone	5.3	74	6.1	36	42	4	17.5
Somalia	9.1	14	6.8	34	45	3	4.5
South Africa	47.9	39	2.7	53	32	4	0.2
Sudan	38.6	15	4.5	41	41	4	−3.0
Swaziland	1.1	65	3.6	23	41	4	−1.2
Tanzania ·	38.7	41	5.4	23	44	3	−1.7
Togo	6.6	116	5.1	40	43	3	−0.1
Uganda	28.5	118	6.7	12	50	3	−0.1
Zambia	11.5	15	5.5	35	46	2	−1.2
Zimbabwe	13.3	34	3.8	36	41	3	−0.8

[a]*Percentage of population younger than 15.*
[b]*Percentage of population older than 65.*
Sources: Population Reference Bureau, World Population Data Sheet, *2007; net migration rate data from* UN International Migration, *2006.*

Figure 6.15 Large Families
This family in Senegal includes five children. Large families are still common in Sub-Saharan Africa. The average total fertility rate for the region is 5.6 children per woman. *(Nomi Baumgartl/ Bilderberg/Peter Arnold, Inc.)*

women to have more children. In some of the poorest and war-torn states, one in five children die before their fifth birthday. Religious affiliation has little bearing on the region's fertility rates; Muslim, Christian, and animist communities all have similarly high birthrates.

The everyday realities of rural life make large families an asset. Children are an important source of labor; from tending crops and livestock to gathering fuelwood, they add more to the household economy than they take. Also, for the poorest places in the developing world, such as Sub-Saharan Africa, children are seen as social security. When the parent's health falters, they expect their grown children to care for them.

Government policies toward family size have shifted dramatically in the past three decades. During the 1970s, population growth was not perceived as a problem by many African governments; in fact, many equated limiting population size with a neocolonial attempt to slow regional development. By the 1980s, a shift in national policies occurred. For the first time, government officials argued that smaller families and slower population growth were needed for social and economic development. Following the United Nations International Conference on Population and Development in Cairo, Egypt, in 1994, organizers announced the following ambitious goals: to bring the natural increase rate down to 2.0 by the year 2020 and to increase the rate of contraceptive use to 40 percent. They may reach their goal: as of 2007, the regional rate of natural increase was down to 2.5 percent, and nearly 22 percent of married women used some method of contraception. Migration patterns within the region and out of it are also influencing growth rates. Africans are mobile people and movement across borders within the region has long been a part of livelihood systems and survival strategies during times of conflict. As Table 6.1 shows, some African states such as Burundi, Chad, Eritrea, and Sierra Leone have very high rates of net migration either due to receiving refugees or former refugees returning home. Similarly, the states with some of the highest negative net migration rates are places experiencing conflict such as Liberia or the Ivory Coast. In general, many of the international flows in Africa are from one poor country to another and most are not recorded. Sub-Saharan Africans are also leaving the region in small but substantial numbers.

Other factors are converging to slow the growth rate. As African states slowly become more urban, there is a corresponding decline in family size—a pattern seen throughout the world. Tragically, declines in natural increase also are occurring as a result of AIDS.

The Impact of AIDS on Africa Now in its third decade, HIV/AIDS may become one of the deadliest epidemics in modern human history. As of 2006, two-thirds of the HIV/AIDS cases in the world were found in Sub-Saharan Africa, and it was estimated that in 2006 alone, there were 2 million AIDS-related deaths in the region. The virus is thought to have originated in the forests of the Congo, possibly crossing over from chimpanzees to humans sometime in the 1950s. Yet it was not until the 1980s that the impact of the disease was widely felt. Women bear a disproportionate burden of the HIV/AIDS epidemic, as they are more likely than men to be infected with HIV and they are usually the caregivers for those who are infected. In Sub-Saharan Africa, as in much of the developing world, the disease is transmitted by unprotected heterosexual encounters. Until the late 1990s, many African governments were unwilling to acknowledge publicly the severity of the situation or to discuss frankly the measures necessary for prevention. The consequences of this inaction have been deadly.

Southern Africa is ground zero for the AIDS epidemic that is ravaging the region (Figure 6.16). In South Africa, the most populous state in southern Africa, 5.5 million people (nearly one in five people aged 15–49) are infected with HIV/AIDS. The rate of infection in neighboring Botswana is 24 percent (down from a staggering 36 percent in 2001) for the same age group. Although the rate of infection is highest in southern Africa, East Africa and West Africa are seeing increases. In Nigeria, it is estimated that 4 percent of the 15–49 age group has HIV or AIDS. For Kenya the figure is 6 percent. By comparison, only 0.6 percent of the same age group in North America is infected. Infection rates are so high that demographers anticipate a slower population growth rate for the entire region. Rates of natural increase have plummeted in southern Africa; Botswana now has a negative rate of increase at −0.1, which means that it is losing population.

Sadly, the social and economic implications of this epidemic are profound. Life expectancy rates have tumbled in the last decade, in a few places dropping to below 40 years. AIDS typically hits the portion of the population that is most active economically. Time lost to care for sick family members and the outlay of workers' compensation benefits has reduced economic productivity and overwhelmed public services in hard-hit areas. The disease makes no class distinctions, so that countries are losing peasant farmers and educated professionals (doctors, engineers, and teachers). Infection rates among newborns are high but several countries have been successful at using antiviral medicines to

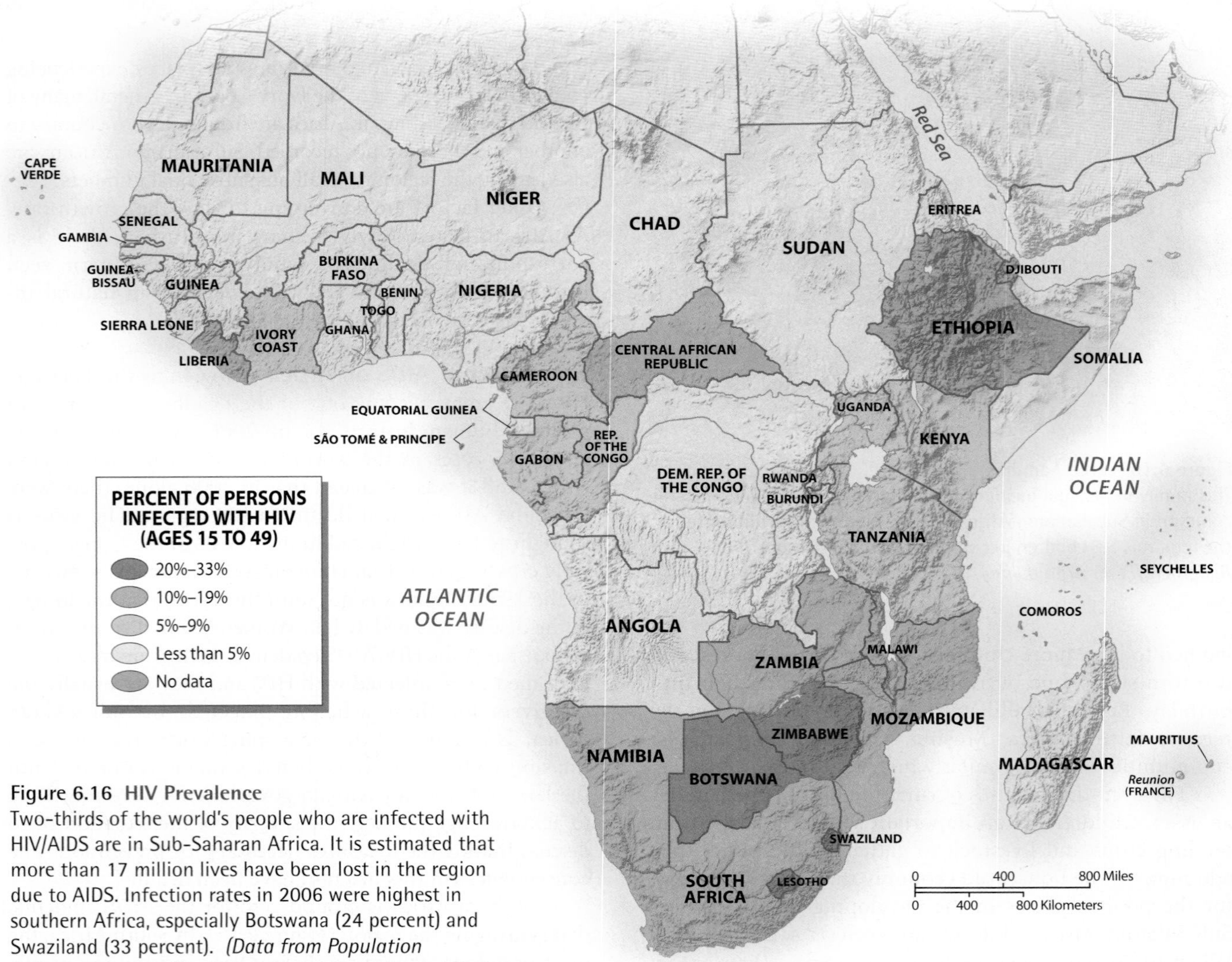

PERCENT OF PERSONS INFECTED WITH HIV (AGES 15 TO 49)

- 20%–33%
- 10%–19%
- 5%–9%
- Less than 5%
- No data

Figure 6.16 HIV Prevalence
Two-thirds of the world's people who are infected with HIV/AIDS are in Sub-Saharan Africa. It is estimated that more than 17 million lives have been lost in the region due to AIDS. Infection rates in 2006 were highest in southern Africa, especially Botswana (24 percent) and Swaziland (33 percent). *(Data from Population Reference Bureau,* World Population Data Sheet, *2006)*

block the transmission of HIV from mothers to children at birth. Many countries are struggling to care for millions of children orphaned by AIDS.

Antiretroviral drug therapies have prolonged the lives of people with HIV/AIDS for nearly two decades in the developed world, yet governments in Sub-Saharan Africa are just beginning to disseminate these treatments. The government of South Africa insisted on its right to use cheap generic drugs to prolong the lives of those with HIV/AIDS, even though this violated patent laws. The world's major drug companies promptly sued South Africa but backed down in 2001 in the face of growing international pressure (and the realization that pharmaceutical companies in developing countries such as India would happily provide generic alternatives). Through a UN-led agreement, generic drugs are now available, and the cost of a year's supply of a three-drug cocktail has dropped to just $300 per person annually, although even this is still beyond the financial reach of most Africans. In 2003, the South African government announced that it would supply antiretroviral drugs to every South African who needs them; their goal was to be treating 3 million people by 2005 (Figure 6.17).

Figure 6.17 HIV/AIDS Clinic in South Africa
HIV-positive patients wait to receive doses of an antiretroviral medication at a rural clinic in the Lusikisiki District, Eastern Cape. Such medications made available by the South African government can prolong the lives of HIV/AIDS patients. Programs such as these offer some hope from the ravages of this disease. *(Gideon Mendel/Corbis)*

For now, the surest way to stem the epidemic is through prevention, mostly through educating people about how the virus is spread and persuading them to change their sexual behavior. In Uganda, state agencies, along with NGOs, began a national no-nonsense campaign for AIDS awareness in the 1980s that focused on the schools. President Museveni of Uganda pioneered the catchy phrase "Abstain-Be faithful-use a Condom" (ABC). As a result of explicit materials, role-playing games, and frank discussion, the prevalence of HIV among women in prenatal clinics had declined by the late 1990s. In Tanzanian villages along Lake Victoria, local governments have tried to curb risky behavior associated with the spread of HIV/AIDS by fining villagers. Dancing after dark, for example, may result in the fine of one chicken. Although there is much more awareness about the disease, and some countries have seen infection rates decline, the battle against AIDS is only beginning.

Patterns of Settlement and Land Use

Because of the dominance of rural settlements in Sub-Saharan Africa, people are widely scattered throughout the region (Figure 6.18). Population concentrations are the highest in West Africa, highland East Africa, and the eastern half of South Africa. The first two areas have some of the region's best soils, and indigenous systems of permanent agriculture developed there. In South Africa, the more densely settled east results from an urbanized economy based on mining, as well as the forced concentration of black South Africans into eastern homelands.

Figure 6.18 Population Distribution
The majority of people in Sub-Saharan Africa live in rural areas. Some of these rural zones, however, are densely settled, such as West Africa and the East African highlands. Major urban centers, especially in South Africa and Nigeria, support millions. There is only one megacity in the region with more than 10 million (Lagos, Nigeria), but over two dozen cities have more than 1 million residents.

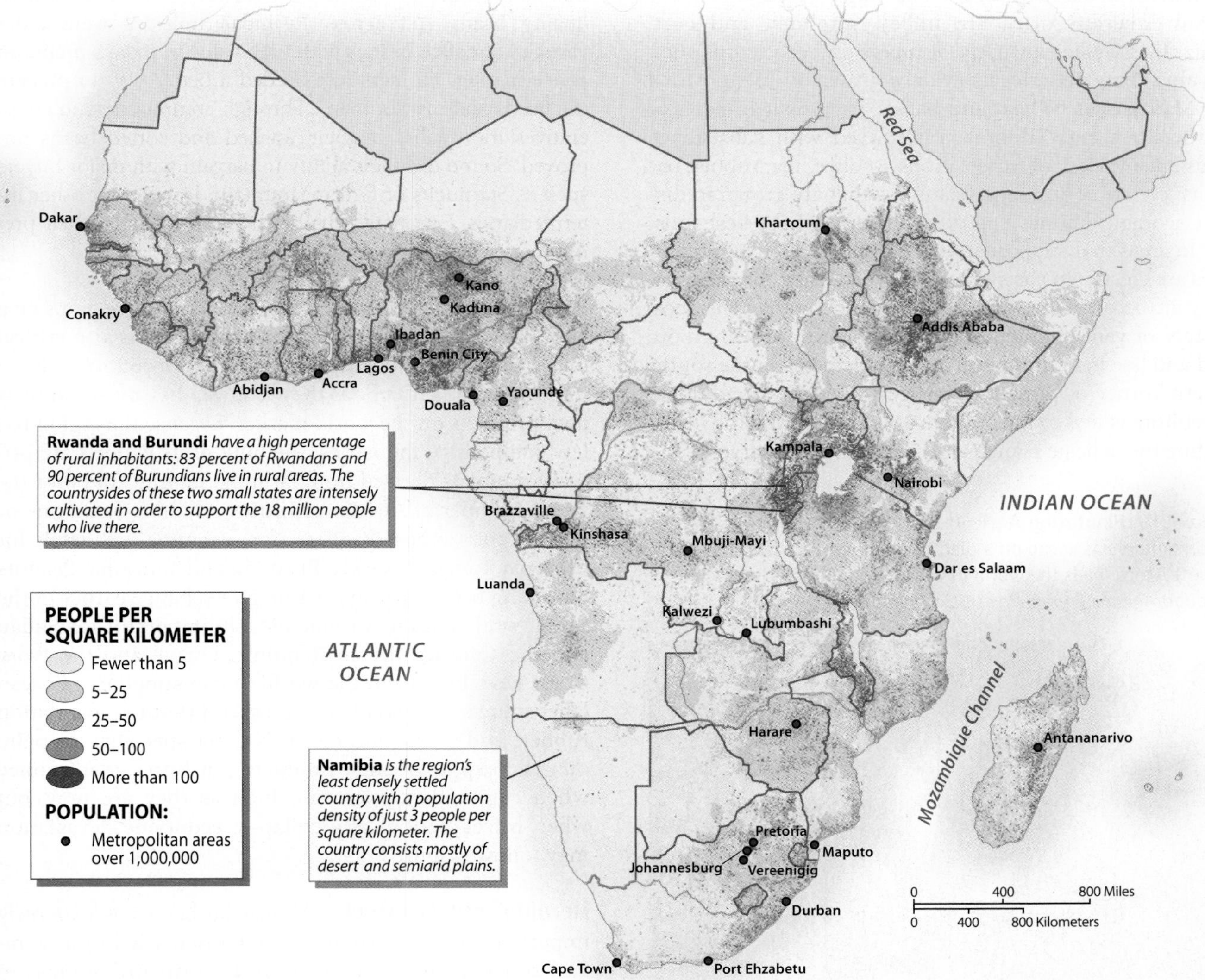

West Africa is more heavily populated than most of Sub-Saharan Africa, although the actual distribution pattern is patchy. Density in the far west, from Senegal to Liberia, is moderate; this area is characterized by broad lowlands with decent soils, and in many areas the cultivation of wet rice has enhanced agricultural productivity. Greater concentrations of people are found along the Gulf of Guinea, from southern Ghana through southern Nigeria, and again in northern Nigeria along the southern fringe of the Sahel. Nigeria is moderately to densely settled through most of its extensive territory; with 144 million inhabitants, it stands as the demographic core of Sub-Saharan Africa. The next largest country, Ethiopia, has a population of 77 million.

As more Africans move to cities, patterns of settlement are evolving into clusters of higher concentration. Localities that were once small administrative centers for colonial elites have mushroomed into major cities. The region even has its own megacity; greater Lagos has some 14 million inhabitants. Throughout the continent, African cities are growing faster than rural areas. But before examining the Sub-Saharan urban scene, a more detailed discussion of rural subsistence is needed.

Agricultural Subsistence The staple crops over most of Sub-Saharan Africa are millet, sorghum, and corn (maize), as well as a variety of tubers and root crops such as yams. Irrigated rice is widely grown in West Africa and Madagascar. Wheat and barley are grown in parts of South Africa and Ethiopia. Intermixed with subsistence foods are a variety of export crops—coffee, tea, rubber, bananas, cocoa, cotton, and peanuts—that are grown in distinct ecological zones and often in some of the best soils.

In areas that support annual cropping, population densities are greater. In parts of humid West Africa, for example, the yam became the king of subsistence crops. The Ibos' mastery of yam production allowed them to procure more food and live in denser permanent settlements in the southeastern corner of present-day Nigeria. Much of traditional Ibo culture is tied to the arduous tasks of clearing the fields, tending the delicate plants, and celebrating the harvest.

Figure 6.19 Plantation Agriculture
A worker harvests cacao on a plantation in Ivory Coast. This area of West Africa leads the world in cacao production. *(Issouf Sanaogo/Agence France Presse/Getty Images)*

Over much of the continent, African agriculture remains relatively unproductive, and rural population densities tend to be low. Amid the poorer tropical soils, cropping usually entails shifting cultivation, or **swidden**. This process involves burning the natural vegetation to release fertilizing ash and planting crops such as maize, beans, sweet potatoes, banana, papaya, manioc, yams, melon, and squash. Each plot is temporarily abandoned once its source of nutrients has been exhausted. Swidden cultivation often is a very finely tuned adaptation to local environmental conditions, but it is unable to support high population densities.

Women are the subsistence farmers of the region, producing for their household needs as well as for local and foreign markets. In the densely settled country of Rwanda, most farms are small but the mountainous volcanic soil is ideal for growing coffee. In an effort to rebuild the country after the ethnic genocide in the 1990s, the government targeted improving the quality of the country's coffee through the formation of cooperatives that would bring ethnic groups together and improve farmers' incomes. The cooperatives have taken off and farmers, many of them war widows, have seen their incomes improve. For decades the country's coffee production languished, and farmers earned very little for their low-quality beans. Yet across Rwanda's hillsides were older varieties of coffee bushes with high value in today's premium coffee market. Farmers just needed a better way to prepare the beans and market them. Through community-run cooperatives the quality of their washed and sorted beans improved. So too did their ability to bargain with major buyers such as Starbucks or Green Mountain, leaving out other intermediaries. For many small farmers in Rwanda, their premium coffee is now a source of pride.

Plantation Agriculture Designed to produce crops on a large scale for export, plantation agriculture is also critical to the economies of many states. If African countries are to import the modern goods they require, they must sell their own products on the world market. Because the region has few competitive industries, the bulk of its exports are primary products derived from farming, mining, and forestry.

A number of African countries rely heavily on one or two export crops. Coffee, for example, is vital for Ethiopia, Kenya, Rwanda, Burundi, and Tanzania. Peanuts have long been the primary foreign exchange earner in the Sahel, while cotton is tremendously important for Sudan and the Central African Republic. Ghana and the Ivory Coast have long been the world's main suppliers of cacao (the source of chocolate); Liberia produces plantation rubber; and many farmers in Nigeria specialize in palm oil. The export of such products can bring good money when commodity prices are high as they are now, but when prices periodically collapse, economic devastation may follow (Figure 6.19).

Herding and Livestock Animal husbandry is extremely important in Sub-Saharan Africa, particularly in the semiarid zones. Camels and goats are the principal animals in

Figure 6.20 Masai Pastoralists
A Masai man holds a cow steady while a woman collects blood being drained from the animal's neck. Pastoral groups such as the Masai live in the drier areas of Sub-Saharan Africa. The Masai live in Kenya and Tanzania. Other pastoral groups are found in the Sahel and the Horn of Africa. *(Kennan Ward/Corbis)*

the Sahel and the Horn of Africa, but farther south cattle are primary. Many African peoples have traditionally specialized in cattle raising and are often tied into mutually beneficial relationships with neighboring farmers. Such pastoralists typically graze their stock on the stubble of harvested fields during the dry season and then move them to drier uncultivated areas during the wet season when the pastures turn green. Farmers thus have their fields fertilized by the manure of the pastoralists' stock, while the pastoralists find good dry-season grazing. At the same time, the nomads can trade their animal products for grain and other goods of the sedentary world. Several pastoral peoples of East Africa, however, are noted for their extreme reliance on cattle and general (but never complete) independence from agriculture. The Masai of the Tanzanian–Kenyan borderlands traditionally derive a large percentage of their nutrition from drinking a mixture of milk and blood (Figure 6.20). The blood is obtained by periodically tapping the animal's jugular veins, a procedure that evidently causes little harm.

Large expanses of Sub-Saharan Africa have been off-limits to cattle because of infestations of **tsetse flies**, which spread sleeping sickness to cattle, humans, and some wildlife. Where wild animals, which harbor the disease but are immune to it, were present in large numbers, especially in environments containing brush or woodland (which are necessary for tsetse fly survival), cattle simply could not be raised. Some evidence suggests that tsetse fly infestations dramatically increased in the late 1800s, greatly harming African societies dependent on livestock but benefiting wildlife populations. At present, tsetse fly eradication programs are reducing the threat, and cattle raising is spreading into areas that were previously forbidden. When people and their stock move into new areas in sizable numbers, wildlife almost inevitably declines.

Sub-Saharan Africa thus presents a difficult environment for raising livestock because of the virulence of its animal diseases. In the tropical rain forests of Central Africa, cattle have never survived well and the only domestic animal that thrives is the goat. Raising horses, moreover, has historically been feasible only in the Sahel and in South Africa. As we shall see later, the disease environment of tropical Africa has presented a variety of problems for humans as well.

Figure 6.21 Urban Slums of Luanda, Angola
The Roche Santiero market on the outskirts of Luanda serves the city's growing poor population. Like many Sub-Saharan African cities, Luanda is growing faster than the surrounding rural areas. *(Rob Crandall/www.robcrandall.com)*

Urban Life

Although Sub-Saharan Africa is considered the least urbanized region in the developing world, most Sub-Saharan cities are growing at twice the national growth rates. If present trends continue, half of the region's population may well be living in cities by 2025. One of the consequences of this surge in city living is urban sprawl. Rural-to-urban migration, industrialization, and refugee flows are forcing the cities of the region to absorb more people and use more resources. As in Latin America, the tendency is toward urban primacy, the condition in which one major city is dominant and at least three times larger than the next largest city. Luanda, the capital of Angola, was built for half a million people but now has at least 4 million people (many of them displaced during the country's long years of war). Most of the city's residents live in the growing slums that surround the colonial core (Figure 6.21).

European colonialism greatly influenced urban form and development in the region. Africans, however, had an urban tradition prior to the colonial era, although a very small percentage of the population lived in cities. Ancient cities, such as Axum in Ethiopia, thrived 2,000 years ago. Similarly, in the Sahel, prominent trans-Saharan trade centers, such as Timbuktu (Tombouctou) and Gao, have existed for more than a millennium. In East Africa, an urban mercantile culture emerged that was rooted in Islam and the Swahili language. The prominent cities of Zanzibar, Tanzania, and Mombasa, Kenya, flourished by supporting a trade network that linked the East African highlands with the Persian Gulf. The stone ruins of Great Zimbabwe in southern Africa are testimony to the achievements of stone working, metallurgy, and religion achieved by Bantu groups in the fourteenth century. West Africa, however, had the most developed precolonial urban network, reflecting both indigenous and Islamic traditions. Today it supports some of the region's largest cities.

West African Urban Traditions The West African coastline is dotted with cities, from Dakar, Senegal, in the far north to Lagos, Nigeria, in the east. Nearly half of Nigerians live in cities, and half a dozen of Nigeria's metropolitan areas have populations of more than 1 million. Historically, the Yoruba cities in southwestern Nigeria are the best documented. Developed in the twelfth century, cities such as Ibadan were walled and gated, with a palace encircled by large rectangular courtyards at the city center. An important center of trade for an extensive hinterland, Ibadan also was a religious and political center. Lagos was another Yoruba settlement. Founded on a coastal island on the Bight of Benin, most of the modern city has spread onto the nearby mainland. Its coastal setting and natural harbor made this relatively small, indigenous city attractive to colonial powers. When the British took control in the mid-nineteenth century, the city's size and importance grew.

Today, Lagos has some 14 million inhabitants (Figure 6.22). In 1960, it was a city of only 1 million. Unable to keep up with the surge of rural migrants, Lagos's streets are clogged with traffic; for those living on the city's periphery, three- and four-hour commutes (one way) are common. City officials struggle to build enough roads and provide electricity, water, sewage service, and employment for all of these people. For many, the informal sector (unregulated services and trade) provides urban employment and services. Crime is another major problem for Lagos. The chances of being attacked on the streets or robbed in one's home are quite high, even though most windows are barred and houses are fenced. To deal with this problem, Lagos has the highest density of police in Nigeria, but the

Figure 6.22 Downtown Lagos, Nigeria
Lagos is home to 14 million people, making it the region's largest city. A classic primate city, its streets teem with buses, there are modern highrises, thousands of pedestrians, and street vendors. *(AP/Wide World Photos)*

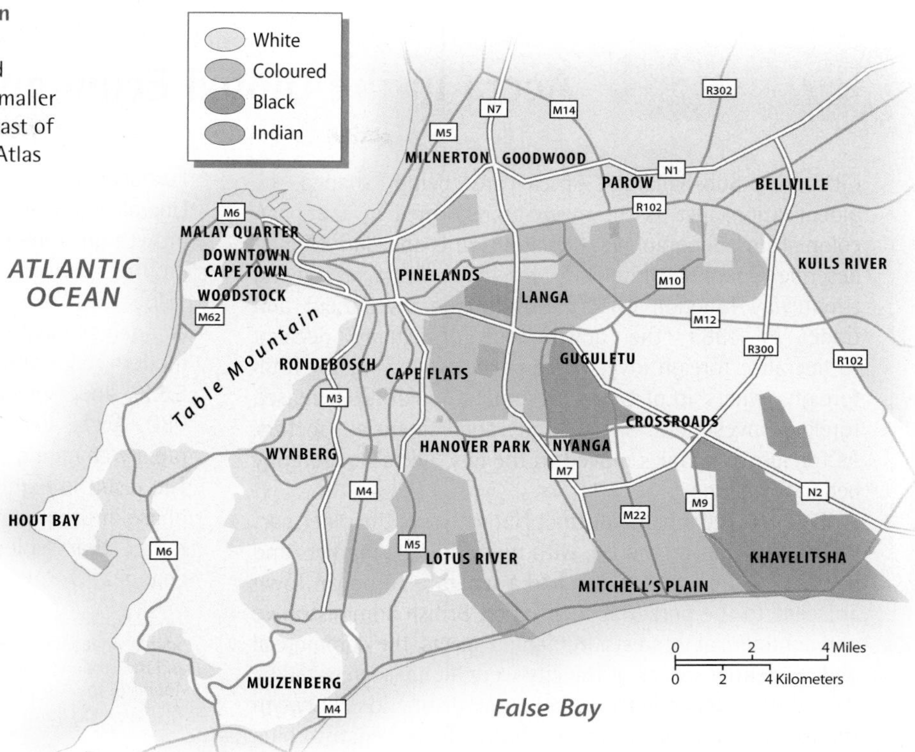

Figure 6.23 Racial Segregation in Cape Town Under apartheid, most land in Cape Town was designated for white use. Coloureds, blacks, and Indians (South Asians) were crammed into far smaller areas on the less desirable flat and sandy soils east of Table Mountain. *(From Christopher, 1994,* The Atlas of Apartheid, *London: Routledge, p. 156)*

weak social bonds between urban migrants, widespread poverty, and the gap between rich and poor seem to encourage lawlessness.

Most West African cities are hybrids, combining Islamic, European, and national elements such as mosques, Victorian architecture, and streets named after independence leaders. Accra is the capital city of Ghana and home to nearly 2 million people. Originally settled by the Ga people in the sixteenth century, it became a British colonial administrative center by the late 1800s. Here again, the modern city is being transformed through neoliberal policies geared toward attracting international corporations (see "Cityscapes: Accra in the Global Economy").

Urban Industrial South Africa The major cities of southern Africa, unlike those of West Africa, are colonial in origin. Cities such as Lusaka, Zambia, and Harare, Zimbabwe, grew as administrative or mining centers. South Africa is one of the most urbanized states in the entire region, and it is certainly the most industrialized. The foundations of South Africa's urban economy rest largely on its incredibly rich mineral resources (diamonds, gold, chromium, platinum, tin, uranium, coal, iron ore, and manganese). Eight of its metropolitan areas have more than 1 million people; the largest of them are Johannesburg, Durban, and Cape Town.

The form of South African cities continues to be imprinted by the legacy of **apartheid**, an official policy of racial segregation that shaped social relations in South Africa for nearly 50 years. Even though apartheid was abolished in 1994, it is still evident in the landscape. Under apartheid rules, all cities were divided into residential areas according to racial categories: white, **coloured** (a South African term describing people of mixed African and European ancestry), Indian (South Asian), and African (black). In Cape Town, whites occupied the largest and most desirable portions of the city, especially the scenic areas below Table Mountain (Figure 6.23). Blacks were crowded into the least desired areas, forming squatter settlements called *townships* in places such as Gugulethu. Today blacks, coloureds, and Indians are legally allowed to live anywhere they want. Yet the economic differences between racial groups, as well as deep-rooted animosity, hinder residential integration. Postapartheid settlements, such as Cape Town's Delpht South, promote residential integration of black and coloured households, something that was forbidden during the apartheid years (Figure 6.24). South African cities, such as Cape Town and Johannesburg, are also home to thousands of immigrants from other African states who have settled there in the last decade in search of economic opportunities not available in their countries of origin.

Figure 6.24 Elite Cape Town Landscape Europeans settled in this prize farmland to the east of Table Mountain outside of Cape Town. Today, South Africa's top vineyards are found here. *(Rob Crandall/www.robcrandall.com)*

CITYSCAPES

Accra in the Global Economy

Cities in Sub-Saharan Africa are being shaped by globalization, just as they were transformed by European colonialism. Geographers Richard Grant and Jan Nijman describe Ghana's capital city and port, Accra, as having a gateway function in terms of foreign investment and trade. In 1983, the government of Ghana decided to liberalize foreign investment policies, allowing for total foreign ownership of companies and stimulating increased foreign investment in mining and financial sectors. As foreign companies moved to the city, Accra's geography began to change.

Colonial Accra had a distinct Native Town, the older part of the city built by the Ga, with its traditional markets and bazaars (Figure 6.3.1). It also had a separate European Town adjacent to the port that focused on British administrative and commercial interests. In many respects, the commercial and residential spaces of the city were highly segregated by race. With independence in 1957, the distinctions between the former European Town and Native Town began to blur, and, increasingly, the buildings and structures of the European Town took on the symbolism of the new nation. Liberalization of investment policies in the 1980s led to a growth in foreign investment in Ghana's financial and producer services. This growth stimulated the creation of a "Global CBD" on the east side of the city, where foreign companies clustered in areas with secure land title, new modern roads and parking, and access to the airport (Figure 6.3.2). Upper-income housing also is found near the Global CBD. Accra, like many cities in the developing world, is rapidly changing as a result of an influx of foreign capital. The result is highly segregated spaces that reflect a global phase in urban development in which world market forces, rather than colonial or national ones, are driving the changes.

Source: Adapted from Richard Grant and Jan Nijman, 2002, "Globalization and the Corporate Geography of Cities in the Less-Developed World," *Annals of the Association of American Geographers 92*(2), 320–40.

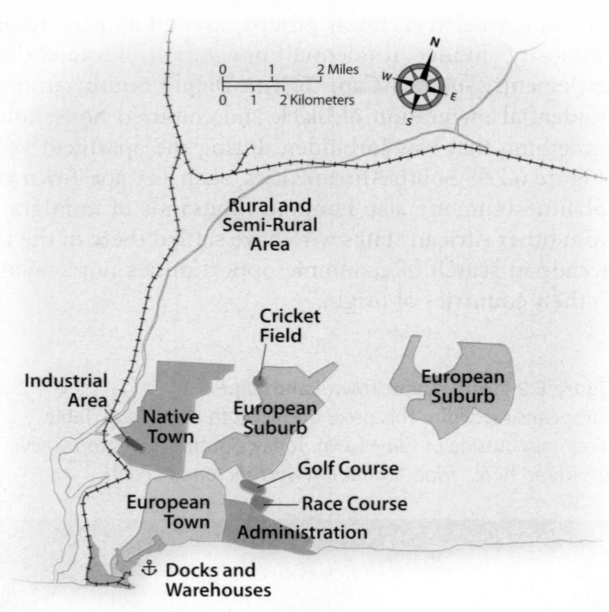

Figure 6.3.1 Colonial Accra
In this model the colonial city is centered on a port with warehouse and docks. Nearby, the European Town is well developed and separate from the commercial activities of the Native Town. *(From Grant and Nijman, 2002, "Globalization and the Corporate Geography of Cities in the Less-Developed World,"* Annals of the Association of American Geographers *92[2], 320–40)*

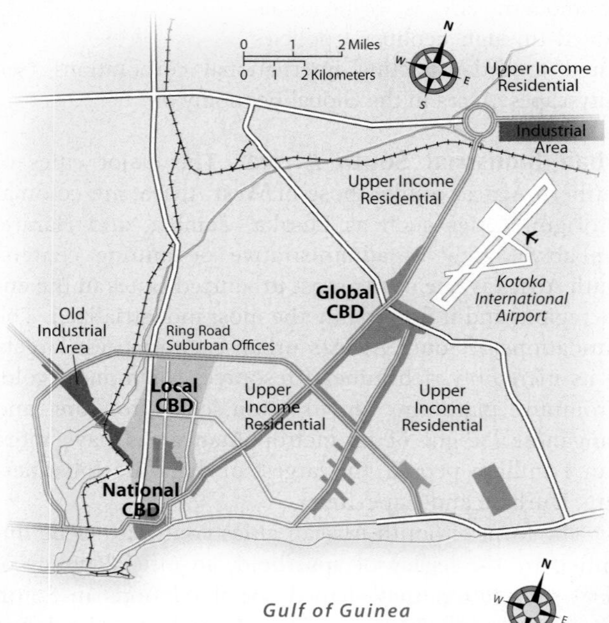

Figure 6.3.2 Contemporary Accra
This model represents the global phase of Accra's development with three distinct CBDs. The national CBD was developed from the former European Town. The local CBD is nearby and formed from the Native Town. The global CBD, fostered by international investment, is removed from the other two and close to the international airport. *(From Grant and Nijman, 2002, "Globalization and the Corporate Geography of Cities in the Less-Developed World,"* Annals of the Association of American Geographers *92[2], 320–40)*

CULTURAL COHERENCE AND DIVERSITY: Unity Through Adversity

No world region is culturally homogeneous, but most have been partially unified in the past by widespread systems of belief and communication. Traditional African religions, however, were largely limited to local areas, and the religions that did become widespread, Islam and Christianity, are primarily associated with other world regions. A handful of African trade languages have long been understood over vast territories (Swahili in East Africa, Mandingo and Hausa in West Africa), but none spans the entire Sub-Saharan region. Sub-Saharan Africa also lacks a history of widespread political union or even of an indigenous system of political relations. The powerful African kingdoms and empires of past centuries all were limited to distinct subregions of the land mass.

The lack of traditional cultural and political coherence across Sub-Saharan Africa is not surprising if one considers the region's vast scale. Sub-Saharan Africa is more than four times larger than Europe or South Asia. Had foreign imperialism not impinged on the region, it is quite possible that West Africa and southern Africa would have developed into their own distinct world regions.

An African identity south of the Sahara was forged through a common history of slavery and colonialism as well as struggles for independence and development. More telling, the people of the region often define themselves as African, especially to the outside world. That Sub-Saharan Africa is poor by most economic measures, no one will argue. And yet the cultural expressions of its people—its music, dance, and art—are joyous. Africans share an extraordinary resilience and optimism that visitors to the region often comment on. The cultural diversity of the region is obvious, yet there seems to be a unity drawn from surviving adversity.

Language Patterns

In most Sub-Saharan countries, as in other former colonies, the persistence of multiple languages reflects the layers of ethnic, colonial, and national identity. Indigenous languages, many from the Bantu subfamily, often are localized to relatively small rural areas. More widely spoken African trade languages, such as Swahili or Hausa, serve as a lingua franca over broader areas. Overlaying native languages are Indo-European and Afro-Asiatic ones (French and English; Arabic and Somali). Figure 6.25 illustrates the complex pattern of language families and major languages found in Africa today. Contrast the larger map with the inset that shows current "official" languages. A comparison of the two shows that most African countries are multilingual, which can be a source of tension within states. In Nigeria, for example, the official language is English, yet there are millions of Hausa, Yoruba, Igbo (or Ibo), Ful (or Fulani), and Efik speakers, as well as speakers of dozens of other languages.

African Language Groups Three of the six language groups mapped in Figure 6.25 are unique to the region (Niger-Congo, Nilo-Saharan, and Khoisan), while the other three (Afro-Asiatic, Austronesian, and Indo-European) are more closely associated with other parts of the world. Afro-Asiatic languages, especially Arabic, dominate North Africa and are understood in Islamic areas of Sub-Saharan Africa as well. Amharic in Ethiopia and Somali of Somalia are also Afro-Asiatic languages. The Malayo-Polynesian language family is limited to the island of Madagascar, which many believe was first settled by seafarers from Indonesia some 1,500 years ago. Indo-European languages, especially French, English, Portuguese, and Afrikaans, are a legacy of colonialism and widely used today.

Of the three language groups found exclusively in the region, the Niger-Congo language group is by far the most influential. This linguistic group originated in West Africa and includes Mandingo, Yoruba, Ful(ani), and Igbo, among others. Around 3,000 years ago a people of the Niger-Congo stock began to expand out of western Africa into the equatorial zone. This group, called the Bantu, commenced one of the most far-ranging migrations in human history, which introduced agriculture into large areas of central and southern Africa. One Bantu group migrated east across the fringes of the rain forest to settle in the Lake Victoria basin in East Africa, where they formed an eastern Bantu core that later pushed south all the way to South Africa (Figure 6.26). Another group moved south, into the rain forest proper. The equatorial rainforest belt immediately adjacent to the original Bantu homeland had been very sparsely settled by the ancestors of the modern pygmies (a distinct people noted for their short stature and hunting skills). Pygmy groups, having entered into close trading relations with the Bantu newcomers, eventually came to speak Bantu languages as well. Although several pygmy populations have persisted to the present, their original languages disappeared long ago.

Once the Bantu migrants had advanced beyond the rain forest into the savannas and woodlands, their agricultural techniques proved highly successful and their influence expanded (see Figure 6.26). Around 650 CE, Bantu-speaking peoples reached South Africa. Over the centuries, the various languages and dialects of the many Bantu-speaking groups, which often were separated from each other by considerable distances, gradually diverged from each other. Today there are several hundred distinct languages in the Bantu subfamily of the great Niger-Congo group. All Bantu languages, however, remain closely related to each other, and a speaker of one can generally learn any other without undue difficulty.

Most individual Sub-Saharan languages are limited to relatively small areas and are significant only at the local scale. One language in the Bantu subfamily, Swahili, eventually became the most widely spoken Sub-Saharan language. Swahili originated as a trade language on the East African coast, where a number of merchant colonies from Arabia were established around 1100 CE. A hybrid society grew up

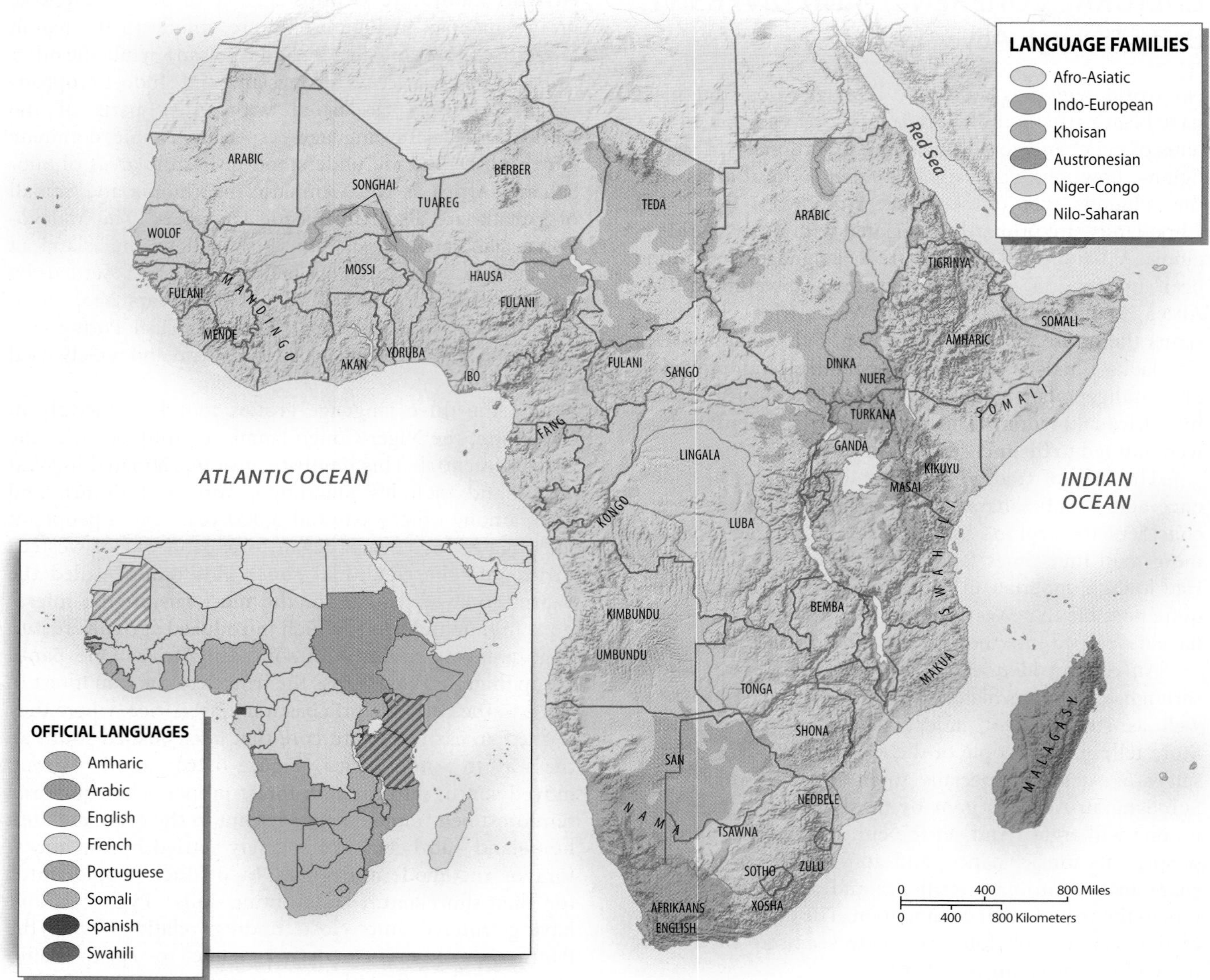

Figure 6.25 African Language Groups and Official Languages
Mapping language is a complex task for Sub-Saharan Africa. There are languages with millions of speakers, such as Swahili, and there are languages spoken by a few hundred people living in isolated areas. Six language families are represented in the region. Among these families are scores of individual languages (see the labels on the map). Because most modern states have many indigenous languages, the colonial language often became the "official" language because it was less controversial than picking from among several indigenous languages. English and French are the most common official languages in the region (see inset).

in a narrow coastal band of modern Kenya and Tanzania, one speaking a language of Bantu structure enriched with many Arabic words. While Swahili became the primary language only in the narrow coastal belt, it spread far into the interior as the language of trade. After independence, both Kenya and Tanzania adopted Swahili as an official language. Swahili, with some 60 million speakers, is the lingua franca of East Africa. It has generated a fairly extensive literature and is often studied in other regions of the world.

Language and Identity Ethnic identity as well as linguistic affiliation have historically been highly unstable over much of the region. The tendency was for new groups to form when people threatened by war fled to less-settled areas, where they often mixed with refugees from other places. In such circumstances, new languages arise quickly and divisions between linguistically distinct groups are blurred. Nevertheless, distinct **tribes** formed that consisted of a group of families or clans with a common kinship, language, and definable territory. The impetus to formalize tribal boundaries came from European colonial administrators who were eager to establish a fixed indigenous social order to better control native peoples. In this process, a cultural map of Sub-Saharan Africa evolved, albeit flawed. Some tribes were artificially

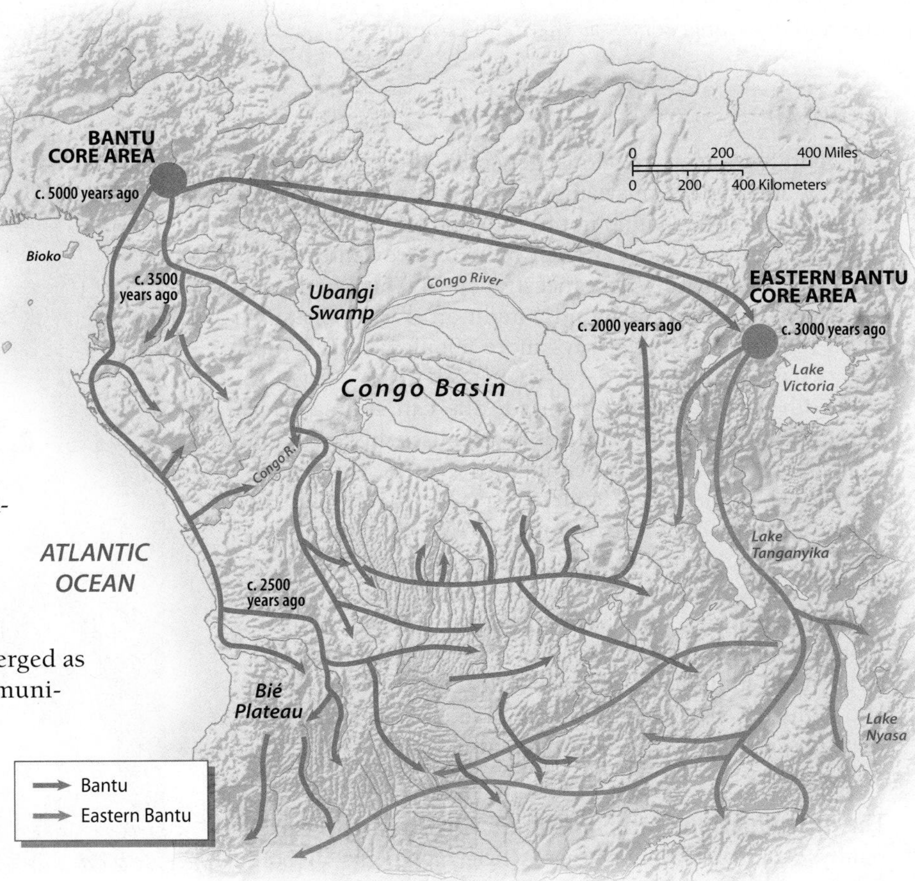

Figure 6.26 Bantu Migrations
Bantu languages, a subfamily of the Niger-Congo language family, are widely spoken throughout Sub-Saharan Africa. The out-migration of Bantu tribes from an original core in West Africa and a secondary core in East Africa helps explain the diffusion of Bantu languages, which include Zulu, Swahili, Bemba, Shona, Lingala, and Kikuyu, among others. *(From Newman, 1995,* The Peopling of Africa, *New Haven, CT: Yale University Press, p. 141)*

divided; meaningless names were applied; and territorial boundaries were often misinterpreted.

Social boundaries between different ethnic and linguistic groups have become more stable in recent years, and a number of individual languages have emerged as particularly important vehicles of communication at the national scale. Wolof in Senegal; Mandingo and other closely related Mande languages in Mali; Mossi in Burkina Faso; Yoruba, Hausa, and Igbo in Nigeria; Kikuyu in central Kenya; and Zulu, Xhosha, and Sotho in South Africa all are nationally significant languages spoken by millions of people (Figure 6.27). None, however, has the status of being the official language of any country. Indeed, a single language has a clear majority status in only a handful of Sub-Saharan countries, which is partially explained by the arbitrary territorial boundaries superimposed by Europeans. The more linguistically homogeneous countries of the region include Somalia (where virtually everyone speaks Somali) and the very small states of Rwanda, Burundi, Swaziland, and Lesotho.

European Languages In the colonial period, European countries used their own languages for administrative purposes in their African empires. Education in the colonial period also stressed literacy in the language of the imperial power. In the postindependence period, most Sub-Saharan African countries have continued to use the languages of their former colonizers for government and higher education. Few of these new states had a clear majority language that they could employ, and picking any minority tongue would have aroused the opposition of other peoples. The one exception is Ethiopia, which maintained its independence during the colonial era. The official language is Amharic, although other indigenous languages are used, especially in the southwestern corner of the country.

Two vast blocks of European languages exist in Africa today: Francophone Africa, encompassing the former colonies of France and Belgium, where French serves as the main language of administration; and Anglophone Africa, where the use of English prevails (see inset, Figure 6.25). Early Dutch settlement in South Africa resulted in the use of Afrikaans (a Dutch-based language) by several million South Africans. Portuguese is spoken in Angola and

Figure 6.27 Multilingual South Africa
A South African sign warns pedestrians not to cross a highway in three languages: English, Afrikaans, and Zulu. Throughout Sub-Saharan Africa, many people are multilingual because of the diversity of languages that exist in each state. *(Rob Crandall/ www.robcrandall.com)*

Mozambique, former colonies of Portugal. In Mauritania and Sudan, Arabic serves as the main language.

Religion

Indigenous African religions generally are classified as *animist*, a somewhat misleading catchall term used to classify all local faiths that do not fit into one of the handful of "world religions." Most animist religions are centered on the worship of nature and ancestral spirits, but the internal diversity within the animist tradition is vast. Classifying a religion as animist says more about what it is not than what it actually is.

With such warnings in mind, one can still say that much of Sub-Saharan Africa has an animist religious tradition. If one goes back far enough, the entire region was animist—but that is, of course, true for the rest of the world as well. Both Christianity and Islam actually entered the region early in their histories, but they advanced slowly for many centuries. Since the beginning of the twentieth century, both religions have spread rapidly—more rapidly, in fact, than in any other part of the world. But tens of millions of Africans still hold animist beliefs, and many others combine animist practices and ideas with their observances of Christianity and Islam.

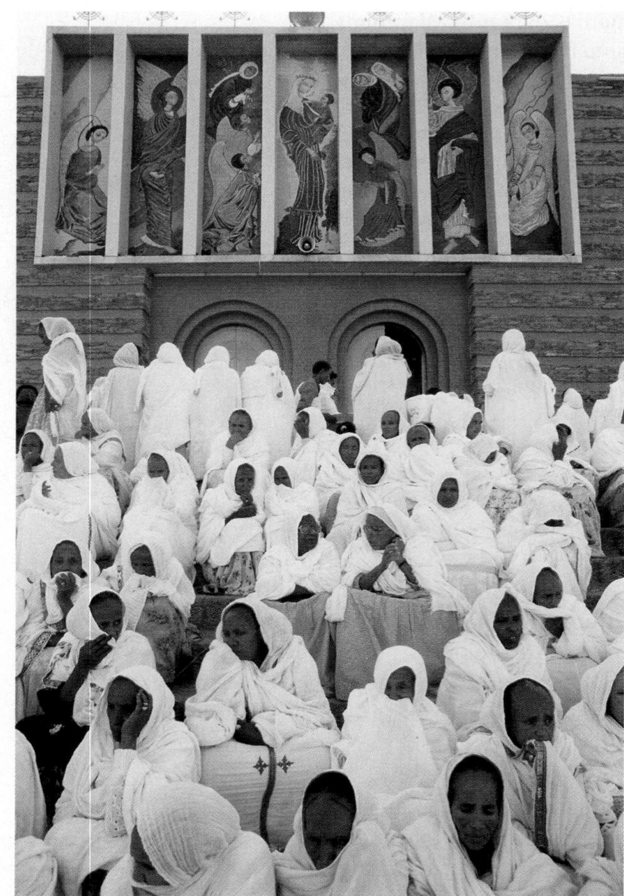

Figure 6.28 Eritrean Christians at Prayer
Coptic Christians gather in front of St. Mary's Church in Asmara, Eritrea, on Good Friday, a holy day for Christians. Half of the populations in Eritrea and Ethiopia belong to the Coptic Church, which has ties to the Christian minority in Egypt. *(Ed Kashi/Corbis/Bettmann)*

The Introduction and Spread of Christianity Christianity came first to northeast Africa. Kingdoms in both Ethiopia and central Sudan were converted by 300 CE—the earliest conversions outside the Roman Empire. The peoples of Ethiopia adopted the Coptic form of Christianity and have thus historically looked to Egypt's Christian minority for their religious leadership (Figure 6.28). Today, roughly half of the population of both Ethiopia and Eritrea follow Coptic Christianity; most of the rest are Muslim, but there are still some animist communities, especially in Ethiopia's western lowlands.

European settlers and missionaries introduced Christianity to other parts of Sub-Saharan Africa beginning in the 1600s. The Dutch, who began to colonize South Africa at this time, brought their Calvinist Protestant faith. Today most of their descendants, the Afrikaners as well as many coloureds, still practice this rather puritanical faith (other coloureds are Muslims). Later European immigrants to South Africa brought Anglicanism and other Protestant creeds, as well as Catholicism. A substantial Jewish community also emerged, concentrated in the Johannesburg area. Most black South Africans eventually converted to one or another form of Christianity as well. In fact, churches in South Africa were instrumental in the long fight against white racial supremacy. Religious leaders such as Bishop Desmond Tutu were outspoken critics of the injustices of apartheid and worked to bring down the system.

Elsewhere in Africa, Christianity came with European missionaries, most of whom arrived after the mid-1800s. As was true in the rest of the world, missionaries had little success where Islam had preceded them, but they eventually made numerous conversions in animist areas. As a general rule, Protestant Christianity prevails in areas of former British colonization, whereas Catholicism is important where France, Belgium, and Portugal had staked their empires. In the postcolonial era, African Christianity has diversified, at times taking on a life of its own independent from foreign missionary efforts. Increasingly active in the region are various Pentecostal, Evangelical, and Mormon missionary groups, mostly from the United States. Yet many areas also have seen the emergence of syncretic faiths, in which Christianity is complexly intertwined with traditional belief systems. It is difficult to map the distribution of Christianity in Africa, however, because it has spread irregularly across the entire non-Islamic portion of the region.

The Introduction and Spread of Islam Islam began to advance into Sub-Saharan Africa 1,000 years ago (Figure 6.29). Berber traders from North Africa and the Sahara introduced the religion to the Sahel, and by 1050

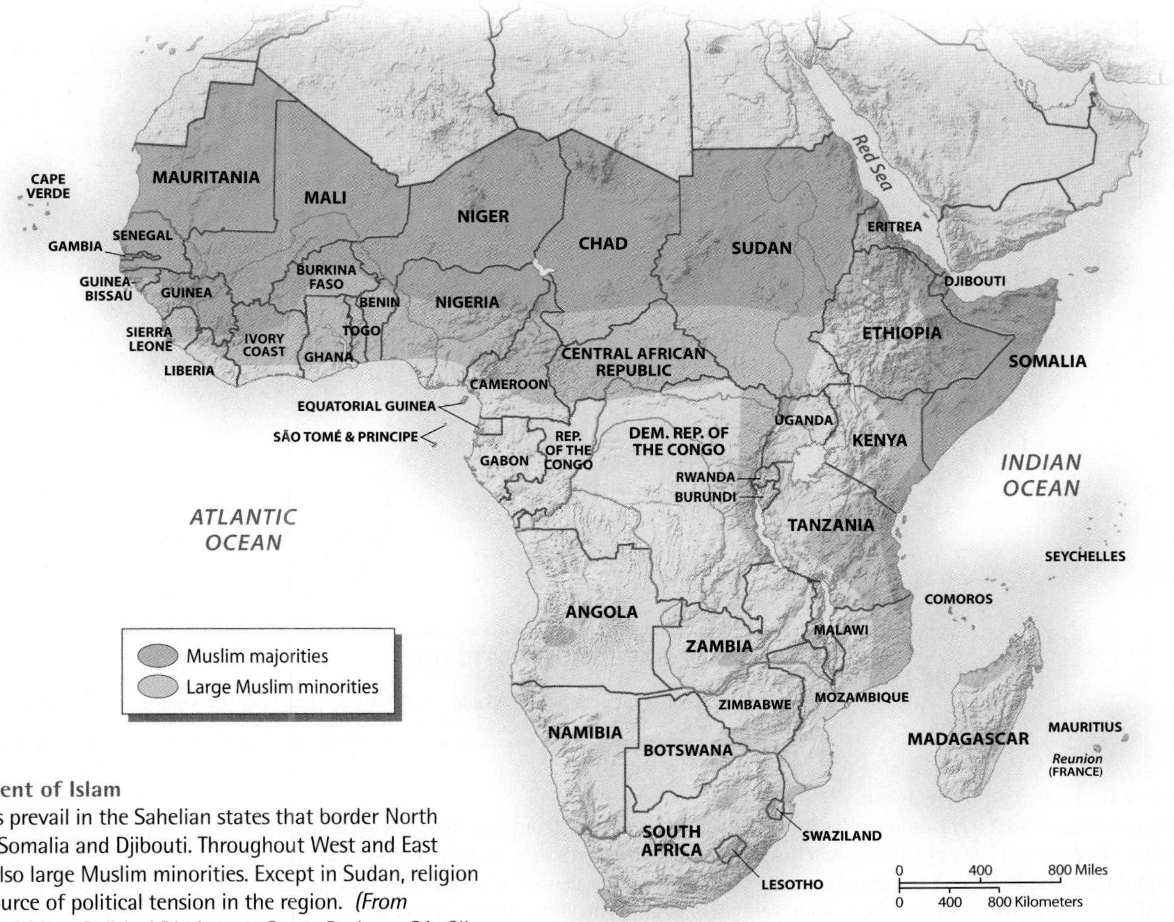

Figure 6.29 Extent of Islam
Muslim majorities prevail in the Sahelian states that border North Africa, as well as Somalia and Djibouti. Throughout West and East Africa there are also large Muslim minorities. Except in Sudan, religion has not been a source of political tension in the region. *(From Phillips, 1984,* The African Political Dictionary, *Santa Barbara, CA: Clio Press Ltd., p. 196)*

the Kingdom of Tokolor in modern Senegal emerged as the first Sub-Saharan Muslim state. Somewhat later, the ruling class of the powerful Mande-speaking mercantile empires of Ghana and Mali converted as well. In the fourteenth century, the emperor of Mali astounded the Muslim world when he and his huge entourage made the pilgrimage to Mecca, bringing with them so much gold that they set off an inflationary spiral throughout Southwest Asia.

Mande-speaking traders, whose networks spanned the area from the Sahel to the Gulf of Guinea, gradually introduced the religion to other areas of West Africa. There, however, many peoples remained committed to animism, and Islam made slow and fitful progress. Even in the Sahel, syncretic forms of Islam prevailed through the 1700s. In the early 1800s, however, the pastoral Fulani people launched a series of successful holy wars designed to shear away animist practices and to establish pure Islam. Today orthodox Islam prevails through most of the Sahel. Farther south Muslims are mixed with Christians and animists, but their numbers continue to grow and their practices tend to be orthodox as well (Figure 6.30).

Figure 6.30 West African Mosque
A mosque rises above the houses in the town of Mankono, Ivory Coast. Islam entered this part of West Africa more than 600 years ago, but conversions were limited, with many Ivorians retaining their animist religious practices. Today Christian, Muslim, and animist faiths are practiced in the Ivory Coast, which reflects a pattern of religious tolerance seen throughout much of Sub-Saharan Africa. *(Victor Englebert/Englebert Photography, Inc.)*

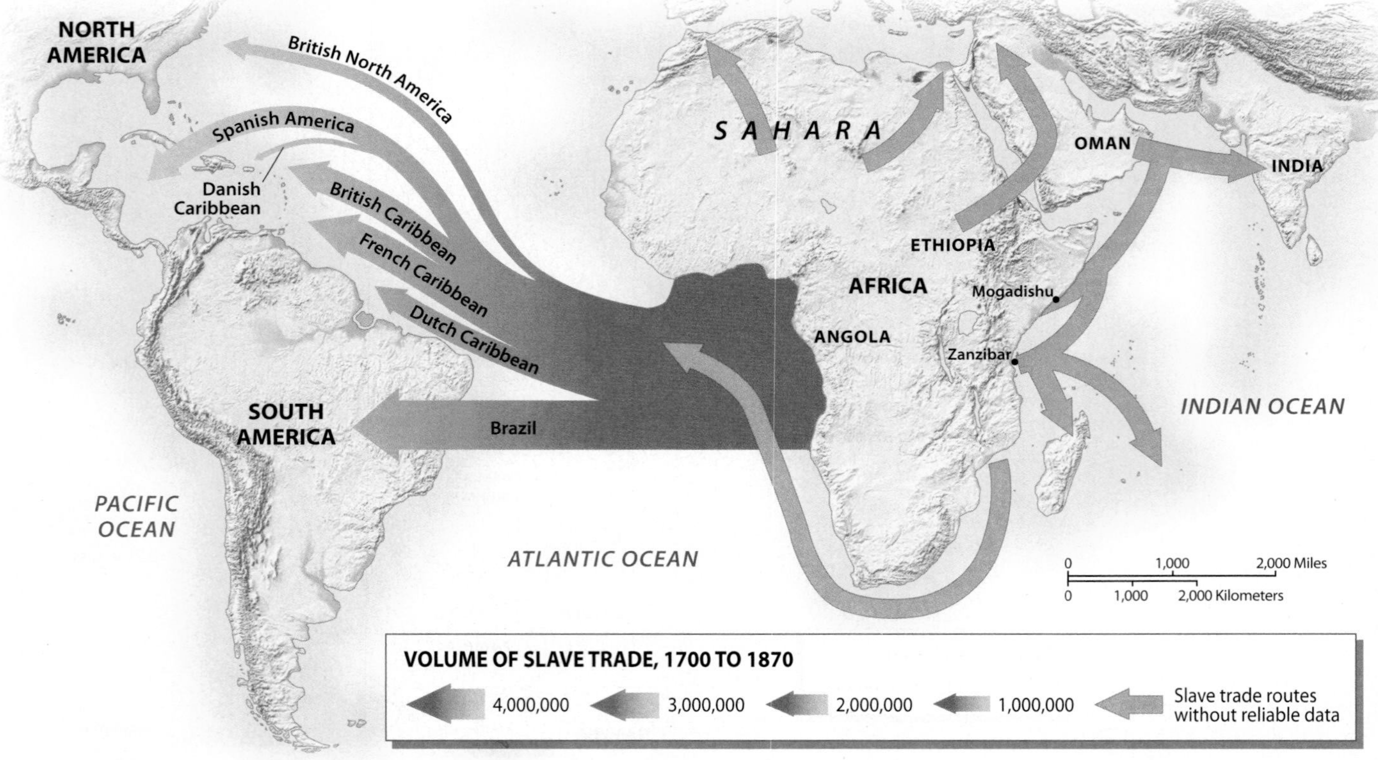

Figure 6.31 African Slave Trade
The slave trade had a devastating impact on Sub-Saharan societies. From ship logs, it is estimated that 12 million Africans were shipped to the Americas to work as slaves on sugar, cotton, and rice plantations; the majority went to Brazil and the Caribbean. Yet other slave routes existed, although the data are less reliable. Africans from south of the Sahara were used as slaves in North Africa. Others were traded across the Indian Ocean into Southwest Asia and South Asia.

Interaction Between Religious Traditions The southward spread of Islam from the Sahel, coupled with the northward dissemination of Christianity from the port cities, has generated a complex religious frontier across much of West Africa. In Nigeria, the Hausa are firmly Muslim, while the southeastern Igbo are largely Christians. The Yoruba of the southwest are divided between Christians and Muslims. In the more remote parts of Nigeria, moreover, animist traditions remain vital. But despite this religious diversity, there has been little overt religious animosity in Nigeria—or elsewhere in West Africa, for that matter. This may change; Human Rights Watch estimates that more than 11,000 Nigerians have died in several hundred separate incidents of violence across religious and ethnic lines since 1999. In 2000, seven Nigerian states imposed Muslim sharia laws, which triggered much violence, especially in the northern city of Kaduna. Still, most of West Africa's regional conflicts continue to be framed more along ethnic terms, rather than religious ones.

Religious conflict historically has been far more acute in northeastern Africa, where Muslims and Christians have struggled against each other for centuries. Islam came early to the coastal areas of the Horn, and soon it had virtually isolated the Ethiopian highlands from the rest of the Christian world. In recent years, religious conflict in the Horn has remained muted and has been largely replaced by ethnic and ideological struggles. The new country of Eritrea, which is roughly half Christian and half Muslim, has in the past few years emerged as something of a model of peaceful coexistence between members of different faiths. Sudan, on the other hand, is currently the scene of an intense conflict that is both religious and ethnic in origin. In the 1970s, the Arabic-speaking Muslims of north and central Sudan emerged as the country's dominant group and began to build an Islamic state. Experiencing both religious discrimination and economic exploitation, the Christian and animist peoples of the south subsequently launched a massive rebellion. Fighting since the 1980s has been intense, with the government generally controlling the main towns and roads and the rebels maintaining power in the countryside. Southern Sudan has excellent soils and oil reserves, which have been only partially exploited because of the years of conflict. In 2003 and 2004, just as a peace agreement was reached in the south between the Sudan People's Liberation Army (SPLA) and the government of Khartoum, a brutal ethnic conflict broke out in the western province of Darfur. This conflict, which will be discussed in more detail later, is between Muslims from different ethnic groups. Since the Darfur conflict began, an estimated 300,000 Sudanese have died and 2 million have been displaced. Overall, the years of religious and ethnic conflict have extracted a heavy toll on the Sudanese; in 2005, more than 6 million internally displaced people were living in Sudan, more than in any other country.

Sub-Saharan Africa is a land of profound religious vitality. Both Christianity and Islam are spreading rapidly, but animism continues to hold widespread appeal. Many new and syncretic forms of religious expression also are emerging. With such a diversity of faiths, it is fortunate that religion has seldom been the cause of overt conflict.

Globalization and African Culture

The slave triangle that linked Africa to the Americas and Europe set in process patterns of cultural diffusion that transferred African peoples and practices across the Atlantic. Tragically, slavery undermined the demographic and political strength of African societies, especially in West Africa, from where the most slaves were taken. An estimated 12 million Africans were shipped to the Americas as slaves from the 1500s until 1870. As Figure 6.31 shows, slavery impacted the entire region, sending Africans not just to the Americas, but to Europe, North Africa, and Southwest Asia. The vast majority, however, worked on plantations across the Americas.

Out of this tragic diaspora came a melding of African cultures with Amerindian and European ones. Rumba, jazz, bossa nova, the blues, and rock and roll have African rhythms at their core. Cultural exchanges are never one-way. Foreign language, religion, and dress were absorbed by Africans who remained in the region. With independence in the 1960s and 1970s, several states sought to rediscover their ancestral roots by openly rejecting European cultural elements. Ironically, the search for traditional religions found African scholars traveling to the Caribbean and Brazil to consult with Afro-religious practitioners about ceremonial elements lost to West Africa.

Popular culture in Africa, like everywhere else in the world, is a dynamic mixture of global and local influences. Kwaito, the latest musical craze in South Africa, sounds a lot like rap from the United States. A closer listen, however, reveals an incorporation of local rhythms, lyrics in Zulu and Xhosa, and themes about life in the postapartheid townships. Lagos, Nigeria has emerged as Africa's film capital. Referred to as Nollywood because over 500 films a year are made there, most are shot in a few days and with budgets of $10,000. These pulp movies are often criticized for being exploitive imitations of Hollywood B-movies. Yet the African movies touch on themes that resonate with Africans such as religion, ethnicity, corruption, witchcraft, and ritual killings. They can be seen all over English-speaking Africa. Recently, television viewers in Addis Ababa were able to phone in their votes for "Ethiopian Idols," inspired by similar shows in the United Kingdom and United States.

Music in West Africa Nigeria is the musical center of West Africa, with a well-developed and cosmopolitan recording industry. Modern Nigerian styles such as juju, highlife, and Afro-beat are influenced by jazz, rock, reggae, and gospel, but they are driven by an easily recognizable African sound. Further up the Niger River, lies the country of Mali. Bamako, the capital, is also a music center that has produced scores of recording artists. Many Malian musicians descend from a traditional caste of musical storytellers performing on either the traditional kora (a cross between a harp and a lute) or the guitar. The musical style is strikingly similar to that of blues from the Mississippi Delta, so much so that Ali Farka Touré, from northern Mali, was referred to as the Bluesman of Africa because of his distinctive, yet familiar, guitar work. Each January, not far from Timbuktu, music fans from West Africa and Europe gather for the Festival in the Desert. In this remote Saharan locale, a celebration of Malian music and Touareg nomadic culture draws together western tourists, African musicians, and nomads (Figure 6.32).

Figure 6.32 Festival in the Desert
Every January thousands of Touareg nomads, Malian musicians, and western tourists gather in the oasis town of Essankane, Mali to attend the Festival in the Desert. In this image, a western tourist sits among the Touareg at the festival. *(Alamy Images)*

Contemporary African music can be both commercially and politically important. Nigerian singer Fela Kuti became a voice of political conscience for Nigerians struggling for true democracy. From an elite family and educated in England, Kuti borrowed from jazz, traditional, and popular music to produce the Afro-beat sound in the 1970s. Yet it was his searing and angry lyrics that attracted the most attention. Acutely critical of the military government, he sang of police harassment, the inequities of the international economic order, and even Lagos's infamous traffic. Singing in English and Yoruba, his message was transmitted to a larger audience, and he became a target of state harassment. At times self-exiled in Ghana, in the 1980s Kuti was jailed briefly by the Nigerian government, but widespread protests eventually led to his early release. Though Fela Kuti's protest music is unpopular with the state, it has been copied by other groups, making music an important form of political expression in Nigeria as well as in other Sub-Saharan states. Sadly, Kuti died in Lagos in 1997 from complications of AIDS; he was 58 years old. His son, however, is now an important recording artist in Nigeria.

For many African emigrants, music can also be deeply nostalgic. No singer expresses this longing for home better than Cesária Evora from Cape Verde. A Grammy-winning singer, Evora is known for a style of singing called the *mornas*, ballads that express sadness and yearning. Her need to sing is rooted in the emotion of *sodade*—a longing to find a better life elsewhere combined with the hope of returning to live among one's people. This need to leave one's home in order to support one's family is an experience that has shaped Cape Verdeans since the late 1800s when mass emigration began (see "People on the Move: Cape Verde's Culture of Migration"). As more Africans emigrate from their countries of origin in response to dire economic or political conditions, their music is reflective of this experience.

Figure 6.33 Ethiopian Distance Runners
Haile Gebrselassie (left) races with two other Ethiopian runners, Kenenisa Bekele (right) and Sileshi Sihine, in a 10,000-meter race in France. The sport of distance running is passionately followed in Ethiopia, and champions such as Gebrselassie are national heroes. *(Francois Xavier Marit/Agence France Presse/Getty Images)*

The Pride of Runners Ethiopia and Kenya have produced many of the world's greatest distance runners. Abebe Dikila won Ethiopia's and Africa's first Olympic gold medal running barefoot at the Rome games in 1960. Since then nearly every Olympic games has yielded medals for these two states. Running is a national pastime in countries where elevation—Addis Ababa sits at 7,300 feet (2,200 meters) and Nairobi at 5,300 feet (1,600 meters)—increases oxygen-carrying capacity. Ethiopia is one of the world's poorest countries, yet its runners won eight medals at the Sydney Olympics in 2000 and seven medals at the 2004 Games in Athens. Past medalists Haile Gebrselassie and Derartu Tulu are national celebrities in Ethiopia who are idealized by the country's youth. Tulu, the first black African woman to win a gold medal in distance running, is a forceful voice for women's rights in a country where women are discouraged from putting on running shorts. There is talk of a political career for Gebrselassie after he hangs up his running shoes (Figure 6.33).

East Africa's running talent has not gone unnoticed. Qatar, in the Persian Gulf, has enticed several world-class runners from East Africa to change citizenship by guaranteeing income and other bonuses. A Kenyan steeplechase runner, Stephen Cherono, now has a Qatar passport and a new name, Saif Saeed Shaheen. His native Kenyan town also has a new running track.

Sub-Saharan Africa has shared many of its cultural practices with the world, and the creative and athletic vitality of its people continues to be recognized. In contrast to the fairly peaceful coexistence of diverse religious traditions in Sub-Saharan Africa, ethnic tensions have often boiled over into violent disputes. To understand the roots of ethnic conflict in the region, it is necessary first to examine the region's political history, paying particular attention to the legacy of European colonization.

GEOPOLITICAL FRAMEWORK: Legacies of Colonialism and Conflict

The duration of human settlement in Sub-Saharan Africa is unmatched by any other region. After all, humankind originated there, evidently evolving from a rather apelike

PEOPLE ON THE MOVE Cape Verde's Culture of Migration

A string of islands some 400 miles off the coast of Senegal, Cape Verde had its origins as a colonial outpost for the Portuguese. This rocky, arid, and chronically poor country in the North Atlantic has less than half a million people (Figure 6.4.1). By necessity, the residents of Cape Verde have been outward looking. Cape Verdeans are a blend of Portuguese and West Africans, linked through four centuries of colonialism and the slave trade. The islands themselves were valued as provisioning sites by the Portuguese, connecting their homeland with African and American colonial territories. It was not until 1975 that Cape Verde gained its independence from Portugal. Today, the Cape Verdean diaspora is estimated to be larger than the country's population.

Migration as a livelihood strategy has been reinforced time and again when droughts drove people off the islands to opportunities elsewhere, be it Providence, São Paulo, Lisbon, or Rotterdam. It is not unusual for an elderly mother to have all her grown children living abroad while raising grandchildren or even great-grandchildren in Cape Verde. Remittances account for 12 percent of the country's gross domestic product. These same remittances have pushed per capita income figures to relatively high levels for West Africa. Consequently, the country has embraced its emigrants, allowing them to vote in national elections and elect a representative to the National Assembly.

Figure 6.4.1 Map of Cape Verde

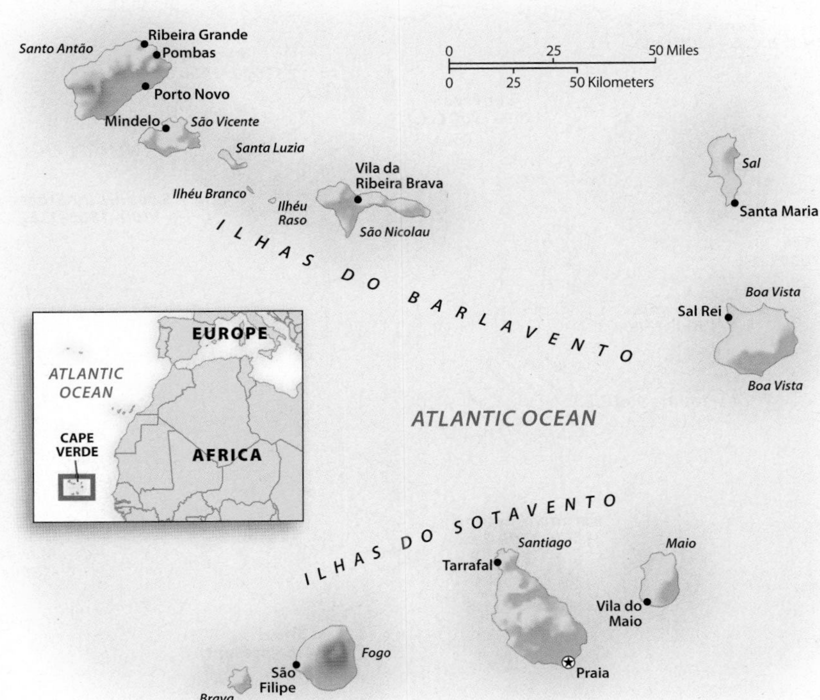

Figure 6.4.2 Mindelo, Capital City of Cape Verde *(Bruno Barbier/Robert Harding World Imagery/Corbis)*

Yet just when emigration seems indispensable to Cape Verdeans, the doors to potential host countries are closing. Holland has long been a popular destination for Cape Verdeans, but the Dutch have recently made it more difficult to enter because applicants must pass new language tests and pay higher visa fees. A popular Cape Verdean song warns that "Holland belongs to the Dutch. . . . Watch out. . . . Because they can make you go back swimming . . . and you'll get home with seaweed in your teeth." Similarly, opportunities to settle in the United States, Portugal, and Brazil are waning. Despite the rising barriers, many Cape Verdeans remain confident they will leave.

At the same time, Cape Verde has become a country of destination for even poorer immigrants from West Africa. For the same reasons that the islands were a transit point in the past, immigrants arrive here with the hope of eventually moving on to the Canary Islands, which are part of Spain, in hopes of gaining a foothold in Europe. While immigrants from the West African mainland risk their lives on rickety boats, there are also Chinese and South Asian immigrants arriving with the hopes of using Cape Verde as a springboard to Europe (Figure 6.4.2). This too is characteristic of an age of migration; today most sending countries are also receiving immigrants.

Source: Adapted from Jason Deparle, "Border Crossings; In a World on the Move, a Tiny Land Strains to Cope," *New York Times*, June 24, 2007.

Australopithicus all the way to modern *Homo sapiens*. Over the millennia, many diverse ethnic groups formed that defy simple classification. Conflicts among these groups existed, with certain groups (the Bantu, for example) overwhelming others (the Khoisan). But different peoples also cooperated and coexisted.

With the arrival of Europeans, patterns of human relatedness and ethnic relations were changed forever. As Europeans rushed to carve up the continent to serve their imperial ambitions, they instituted various policies that heightened ethnic tensions and promoted hostility. Many of the region's modern conflicts can trace their roots back to the colonial era, especially the arbitrary drawing of political boundaries. Others are attributed to struggles over national identity and political control among different ethnic groups.

Indigenous Kingdoms and European Encounters

The first significant state to emerge in Sub-Saharan Africa was Nubia, which controlled a large territory in central and northern Sudan some 3,000 years ago (Figure 6.34); 1,000 years later the Kingdom of Axum arose in northern Ethiopia and Eritrea. Both of these states were strongly influenced by political models derived from Egypt and Arabia. The first wholly indigenous African states were founded in the Sahel around 700 CE. Kingdoms such as

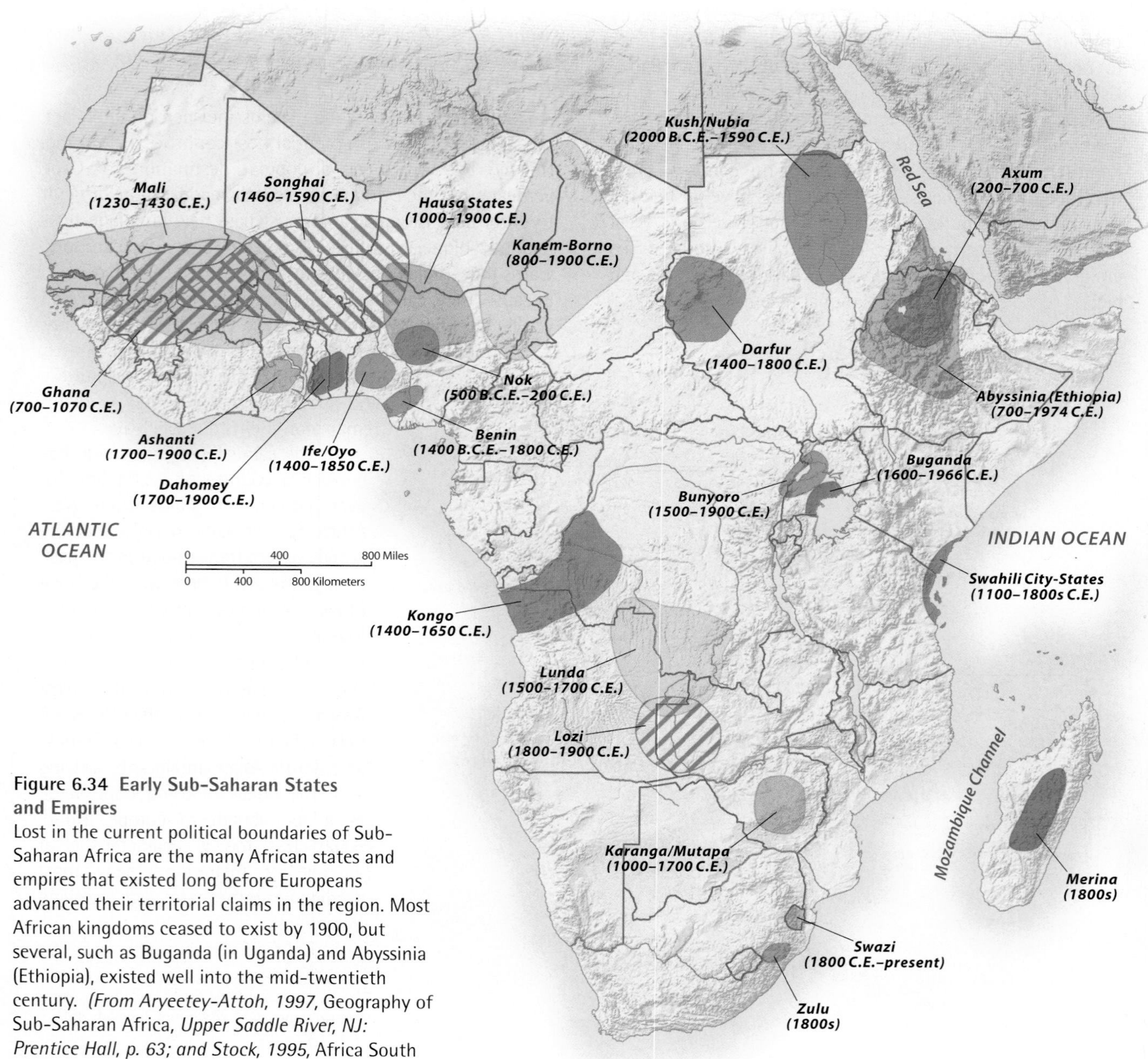

Figure 6.34 Early Sub-Saharan States and Empires
Lost in the current political boundaries of Sub-Saharan Africa are the many African states and empires that existed long before Europeans advanced their territorial claims in the region. Most African kingdoms ceased to exist by 1900, but several, such as Buganda (in Uganda) and Abyssinia (Ethiopia), existed well into the mid-twentieth century. *(From Aryeetey-Attoh, 1997,* Geography of Sub-Saharan Africa, *Upper Saddle River, NJ: Prentice Hall, p. 63; and Stock, 1995,* Africa South of the Sahara, *New York: Guilford Press, p. 62)*

Ghana, Mali, Songhai, and Kanem-Bornu grew rich by exporting gold to the Mediterranean and importing salt from the Sahara, and they maintained power over lands to the south by monopolizing horse breeding and mastering cavalry warfare.

Over the next several centuries, a variety of other states emerged in West Africa. Some were large but diffuse empires organized through elaborate hierarchies of local kings and chiefs; others were centralized states focused on small centers of power. The Yoruba of southwestern Nigeria, for example, developed a city-state form of government, and their homeland is still one of the most urbanized and densely populated parts of Africa. The most powerful Sub-Saharan states continued to be located in the Sahel until the 1600s, when European coastal trade undercut the lucrative trans-Saharan networks. Subsequently, the focus of power moved to the Gulf of Guinea, where well-organized African states (such as Dahomey and Ashanti) as well as Europeans took advantage of the lucrative opportunities presented by the slave trade and in the process increased their military and economic power.

Early European Encounters Unlike the relatively rapid colonization of the Americas, Europeans needed centuries to gain effective control of Sub-Saharan Africa. Portuguese traders arrived along the coast of West Africa in the 1400s, and by the 1500s they were well established in East Africa as well. Initially, the Portuguese made large profits, converted a few local rulers to Christianity, established several fortified trading posts, and acquired dominion over the Swahili trading cities of the east. They stretched themselves too thin, however, and had little ultimate success in any of their endeavors. Only where a sizable population of mixed African and Portuguese descent emerged, as along the coasts of modern Angola and Mozambique as well as the islands of Cape Verde, could Portugal maintain power. Along the Swahili, or eastern, coast they were eventually expelled by Arabs from Oman, who subsequently established their own mercantile empire in the area.

The Disease Factor One of the main reasons for the Portuguese failure was the disease environment of Africa. With no resistance to malaria and other tropical diseases, roughly half of all Europeans who remained on the African mainland died within a year. Protected both by their formidable armies and by the diseases of their native lands, African states were able to maintain an upper hand over European traders and adventurers well into the 1800s. Unlike the Americas, where European conquest was facilitated by the introduction of Old World diseases that devastated native populations (see Chapters 4 and 5), Sub-Saharan Africa's native diseases limited European settlement until the mid-nineteenth century.

The hazards of malaria and other tropical diseases such as sleeping sickness were compensated by the lure of profit, and soon other European traders followed the Portuguese. By the 1600s, Dutch, British, and French firms dominated the lucrative export of slaves, gold, and ivory from the Gulf of Guinea. The Dutch also established a settler colony in South Africa, safely outside the tropical disease zone, to supply their ships bound for Indonesia. For the next 200 years, European traders came and went, occasionally building fortified coastal posts, but they almost never ventured inland and seldom had any real influence on African rulers. By exporting millions of slaves, however, they had a profoundly negative impact on African society.

In the 1850s, European doctors discovered that a daily dose of quinine would offer protection against malaria, radically changing the balance of power in Africa. Explorers immediately began to penetrate the interior of the continent, while merchants and expeditionary forces began to move inland from the coast. The first imperial claims soon followed. The French quickly grabbed power along the easily navigated Senegal River, whereas the British established **protectorates** over the indigenous states of the Gold Coast (modern-day Ghana), in which the British guaranteed protection for these territories in exchange for various trade preferences.

European Colonization

In the 1880s, European colonization of the region quickly accelerated, leading to the so-called scramble for Africa. By this time, after the invention of the machine gun, no African state could long resist European force. The exact reasons for the abrupt division of Africa among the colonial powers remain controversial, but several developments seem to have been crucial. One was the British seizure in 1882 of Egypt, a territory that the French had long coveted. In return, the infuriated French began to seize additional lands in West Africa, equatorial Africa, and Madagascar.

Another contributing factor was the desire of several new European countries to join the game of empire-building. Because Asia was either occupied by established European powers or controlled by still-formidable indigenous empires, Africa emerged as the main arena of rivalry and expansion. Even though Belgium had been a country only since 1830, its king quickly began to carve out a personal empire along the Congo River, using particularly brutal techniques. The German government—which itself dated back only to 1871—began to claim territories wherever German missionaries were active, and it soon staked out the colonies of Togo, Cameroon, Namibia, and Tanganyika (modern Tanzania minus the island of Zanzibar). The Italians eyed the Horn of Africa, while Spain acquired a small coastal foothold in equatorial West Africa. Alarmed by such activity, the Portuguese began to push inland from their coastal possessions in Angola and Mozambique.

Also in the early 1800s, two small territories were established in West Africa so that freed and runaway slaves would have a place to return to in Africa. The territory that was to become Liberia was set up by the American Colonization Society in 1822 to settle former African American slaves. By 1847, it was the independent and free state of Liberia. Sierra Leone served a similar function for ex-slaves from the British Caribbean, but it remained a protectorate of Britain until the 1960s. Despite the good intentions behind the creation of these territories, they too were colonies. Liberia, in particular, was imposed on existing indigenous groups who viewed their new "African" leaders with contempt.

The Berlin Conference As the scramble intensified, tensions among the participating countries mounted. Rather than risk war, 13 countries convened in Berlin at the invitation of the German chancellor Bismarck in 1884 in a gathering known as the **Berlin Conference**. During the conference, which no African leaders attended, rules were established as to what constituted "effective control" of a territory, and Sub-Saharan Africa was carved up and traded like properties in a game of Monopoly (Figure 6.35). Exact boundaries in the interior, which was still poorly known, were not determined, and a decade of "orderly" competition remained as imperial armies marched inland. While European arms were by the 1880s far superior to anything found in Africa, several indigenous states did mount effective resistance campaigns. In central Sudan, an Islamic-inspired force held out against the British until 1900, and as late as 1914, the Darfur region of western Sudan maintained tenuous independence.

Eventually European forces prevailed everywhere, with one major exception: Ethiopia. The Italians had conquered

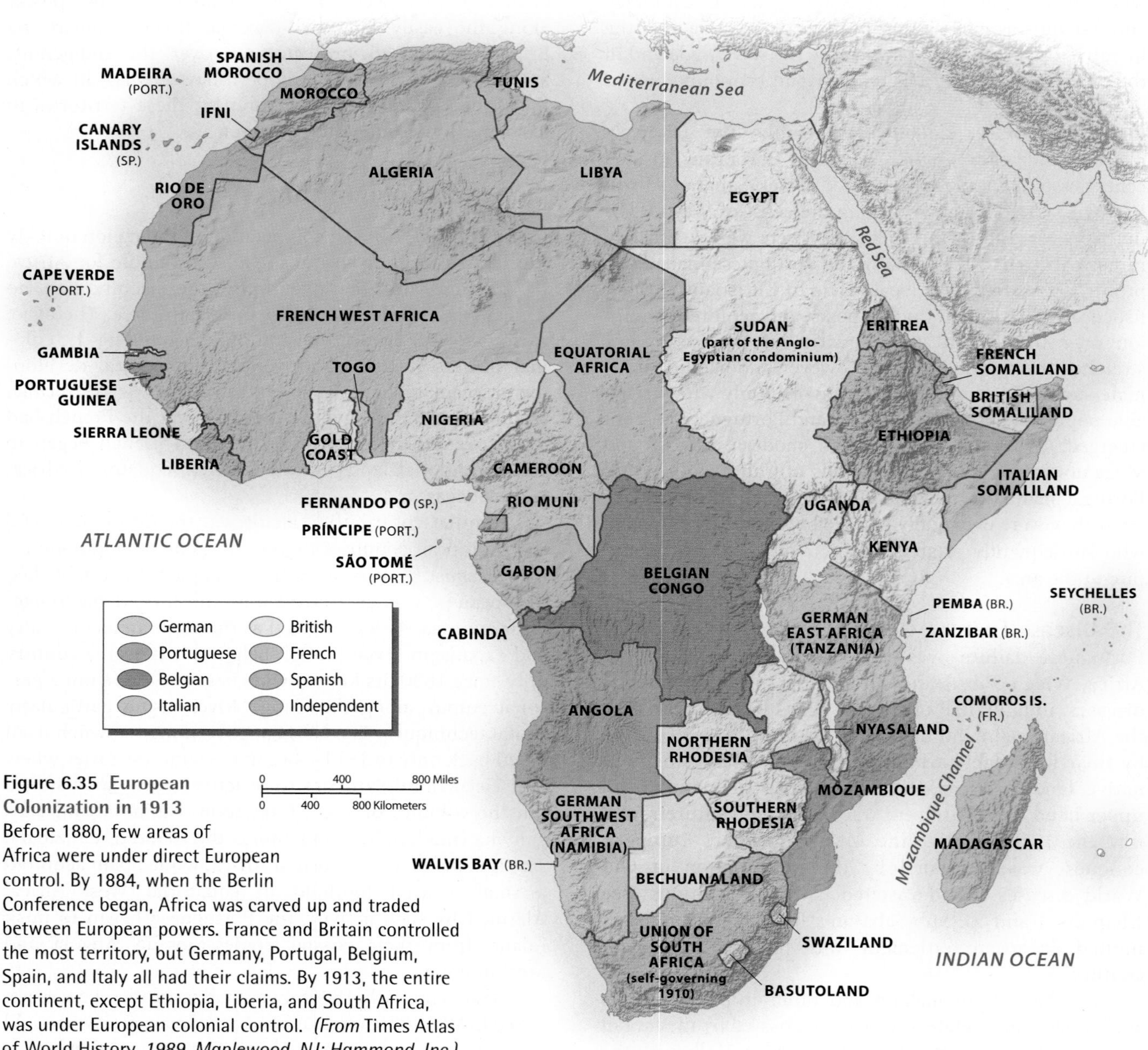

Figure 6.35 European Colonization in 1913 Before 1880, few areas of Africa were under direct European control. By 1884, when the Berlin Conference began, Africa was carved up and traded between European powers. France and Britain controlled the most territory, but Germany, Portugal, Belgium, Spain, and Italy all had their claims. By 1913, the entire continent, except Ethiopia, Liberia, and South Africa, was under European colonial control. *(From* Times Atlas of World History, *1989, Maplewood, NJ: Hammond, Inc.)*

the Red Sea coast and the far northern highlands (modern Eritrea) by 1890, and they quickly set their sights on the large Ethiopian kingdom, called *Abyssinia*, which had itself been vigorously expanding for several decades. In 1896, however, Abyssinia defeated the invading Italian army, earning the respect of and recognition by the European powers. In the 1930s, fascist Italy launched a major invasion of the country, now renamed Ethiopia, to redeem its earlier defeat, and with the help of poison gas and aerial bombardment, it quickly prevailed. By 1942, Ethiopia had regained its freedom.

Although Germany was a principal instigator of the scramble for Africa, it lost its own colonies after suffering defeat in World War I. Britain and France then partitioned most of Germany's African empire between them. Figure 6.35 shows the colonial status of the region in 1913, prior to Germany's territorial loss. The French held most of West Africa, but the British controlled populous Nigeria and several other coastal territories. The French also colonized Gabon in western equatorial Africa, Madagascar, and the small but strategic enclave of Djibouti at the southern end of the Red Sea. The British holdings were larger still, covering a continuous swath of territory in the east from Sudan to South Africa. Belgium and Portugal were the other main colonial powers. The government of Belgium had taken direct control over the personal domain of King Leopold II and had extended it to the south, eventually reaching the mineral-rich copper belt. Portugal, the weakest European power, controlled huge territories in southwestern and southeastern Africa (Angola and Mozambique). Britain's drive to the north from South Africa, organized by imperial dreamer and diamond magnate Cecil Rhodes, thwarted the Portuguese effort to bridge the continent.

Establishment of South Africa While the Europeans were cementing their rule over Africa after World War I, South Africa was inching toward political freedom, at least for its white population. South Africa was one of the oldest colonies in Sub-Saharan Africa, and it became the first to obtain its political independence from Europe in 1910. Its economy is the most productive and influential of the region, yet the legacy of apartheid made it an international outcast, especially within Africa.

The original body of Dutch settlers in South Africa grew slowly, expanding to the north and east of its original nucleus around Cape Town. As a farming and pastoral people largely isolated from the European world, these Afrikaners, or Boers, developed an extremely conservative cultural outlook marked by an intensifying belief in their racial superiority over the native population. In 1806 the British, then the world's unchallenged maritime power, seized the Cape district from the Dutch. Relations between the British rulers and their new Afrikaner subjects quickly soured, in part because the British attempted to restrict the enslavement of Africans as part of a larger abolitionist movement. In the 1830s, the bulk of the Afrikaner community opted to leave British territory and strike out for new lands. After being rebuffed by the powerful Zulu state in the Natal area, they settled on the northeastern plateau, known as the *high veldt*. By the 1850s, they had established two "republics" in the area: the South African Republic (commonly called the *Transvaal*) and the Orange Free State (Figure 6.36).

Once relocated on the high veldt, the Afrikaners found themselves threatened by Zulu armies. The Zulus had been a relatively minor group until a Zulu king introduced centralized rule and new military techniques in the 1820s. From then on Zulu armies were virtually invincible. After a series of inconclusive wars, the British intervened in 1878 on behalf of the Dutch settlers. The first British army sent into Zululand was annihilated; the second, equipped with machine guns, prevailed. By 1900, the British had incorporated the Zulu into South Africa's Natal province.

The British had more difficulty subduing the Afrikaners. By the late 1800s, it had become clear that the two Boer republics were sitting above one of the world's greatest troves of mineral wealth. The British were increasingly drawn to the area, and they grew resentful of Afrikaner power. The result was the Boer War, which turned into a protracted and brutal guerrilla struggle between the British Army and mobile Afrikaner bands. By 1905, the Boers relented, after which the British joined the two former republics to Cape Province and Natal to form the Union of South Africa. Five years later South Africa was given its independence from Britain. Britons and other Europeans continued to settle there, but the Afrikaners remained the majority white group. Black and coloured South Africans, however, greatly outnumbered whites.

It wasn't until 1948, when the Afrikaner's National Party gained control of the government, that they introduced the policy of "separateness" known as *apartheid*. British South Africans had enacted a series of laws that were prejudicial to nonwhite groups, but it was under Afrikaner leadership that racial separation become more formalized and systematic. Operating on three scales—petite, meso-, and grand—apartheid managed social interaction by controlling space. Petite apartheid, like Jim Crow laws in the United States, created separate service entrances for government buildings, bus stops, and restrooms based on skin color. Mesoapartheid divided the city into residential sectors by race, giving whites the best and largest urban zones. Finally, grand apartheid was the construction of black **homelands** by ethnic group. Technically, blacks were to become citizens of the nominally independent homelands (see the 1970 map in Figure 6.36). Likened to the reservations created for Native Americans in the United States, homelands were rural, overcrowded, and on marginal land. Moreover, to ensure the notion that every black had a homeland, some 3 million blacks were forcibly relocated into homelands during apartheid, and residence outside the homelands was strictly regulated.

Granted its independence in 1910, South Africa was the first state in the region freed from colonial rule. Yet because of its formalized system of discrimination and racism, it was hardly a symbol of liberty. Ironically, at the

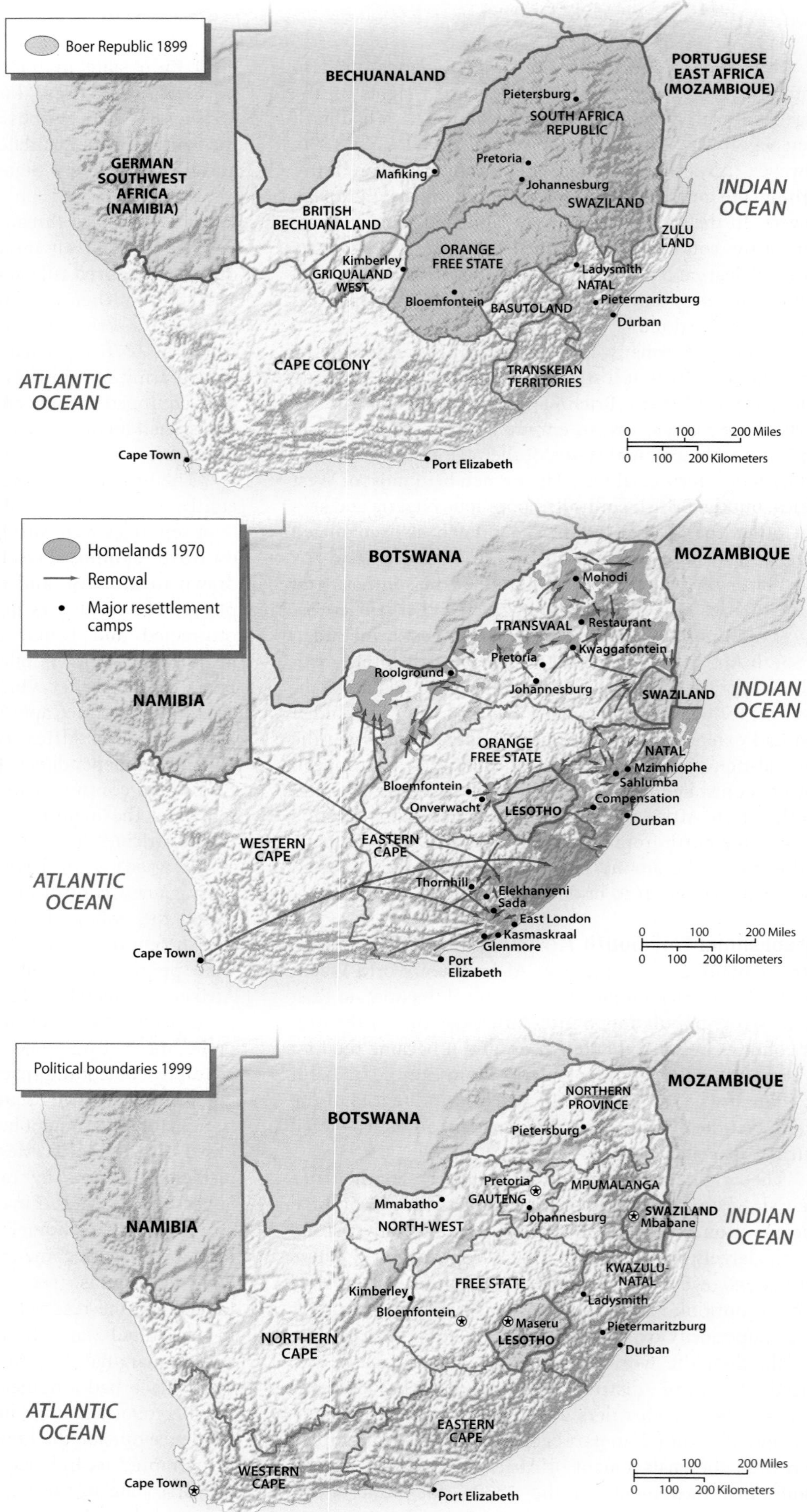

Figure 6.36 The Evolution of South African Political Boundaries
South Africa's internal political boundaries have been redrawn several times in the past century. To appease the Afrikaans (Dutch) settlers, a Boer republic was established in the late nineteenth century with British approval. It was later reabsorbed into the Union of South Africa when gold and diamonds were discovered. During the apartheid years, homelands were created with the intent of having all black South Africans live in these ethnic-based rural settings. In order to accomplish this, 3 million black South Africans were forcibly relocated. Lastly, in the new South Africa, homelands were eliminated and new province boundaries were drawn. *(From Christopher, 1994,* The Atlas of Apartheid, *London: Routledge, pp. 16, 83)*

same time that the Afrikaners tightened their political and social control over the nonwhite population, the rest of the continent was preparing for political independence from Europe.

Decolonization and Independence

Decolonization of the region happened rather quickly and peacefully beginning in 1957. Independence movements, however, had sprung up throughout the continent, some dating back to the early 1900s. Workers' unions and independent newspapers became voices for African discontent and the hope for freedom. Black intellectuals, who had typically studied abroad, were influenced by the ideas of the **Pan-African Movement** led by W. E. B. Du Bois and Marcus Garvey in the United States. Founded in 1900, the movement's slogan of "Africa for Africans" encouraged a trans-Atlantic liberation effort. Nevertheless, Europe's hold on Africa remained secure through the 1940s and early 1950s, even as other colonies in South and Southeast Asia gained their independence.

By the late 1950s, Britain, France, and Belgium decided that they could no longer maintain their African empires and began to withdraw. (Italy had already lost its colonies during World War II, and Britain gained Somalia and Eritrea.) Once started, the decolonization process moved rapidly. By the mid-1960s, virtually the entire region had achieved independence. In most cases, the transition was relatively peaceful and smooth, with the exception of southern Africa.

Portugal refused to relinquish its colonies in the 1960s. As a result, the people of Angola and Mozambique turned to armed resistance. The most powerful rebel movements adopted a socialist orientation and received support from the Soviet Union and Cuba. A new Portuguese government came to power in 1974 and withdrew abruptly from its African colonies. At this point, Marxist regimes quickly came to power in both Angola and Mozambique. The United States, and especially South Africa, responded to this perceived threat by supplying arms to rebel groups that opposed the new governments. As was common in much of Sub-Saharan Africa, the resulting struggles turned out to be more firmly grounded in ethnic loyalty than in Cold War ideology. Fighting dragged on for nearly three decades in Angola and Mozambique. The countryside in both states was heavily riddled with land mines. Efforts to clear the mines and make the land usable are ongoing but sporadic. With the end of the Cold War, however, outsiders lost their interest in perpetuating these conflicts. Mozambique has been at peace since the mid-1990s. After several failed attempts at peace in Angola, the Angolan army signed a peace treaty with Unita rebels in 2002 that ended a 27-year conflict in which more than 300,000 people died and more than 3 million Angolans were displaced.

African Unity to the African Union With the exception of southern Africa, the 1960s were a time of hope for Sub-Saharan Africa, and dynamic African leaders put their mark on the region during the early independence period. Men such as Kenya's Jomo Kenyatta, Ivory Coast's Felix Houphuët-Boigney, Julius Nyerere of Tanzania, and Ghana's Kwame Nkrumah became powerful father figures who molded their new nations (Figure 6.37). President Nkrumah's vision for Africa was the most expansive. After helping secure independence for Ghana in 1957, his ultimate aspiration was the political unity of Africa. While his dream was never realized, it set the stage for the founding of the Organization of African Unity (OAU) in 1963, which was renamed the **African Union (AU)** in 2002.

The AU is a continent-wide organization headquartered in Addis Ababa, Ethiopia. Originally its purpose was to mediate disputes between neighbors, although its record is mixed. Certainly in the 1970s and 1980s it was a constant voice of opposition to South Africa's minority rule, and it intervened in some of the more violent independence movements in southern Africa. With the renaming of the organization, an ambitious continental development plan called the New Partnership for Africa's Development (Nepad) was initiated. The partnership stresses economic development, regional security, governmental reform, and

Figure 6.37 A Monument to Kwame Nkrumah
Charismatic independence leader Kwame Nkrumah is remembered with this monument in Accra, Ghana. Nkrumah led Ghana to an early independence in 1957; he was also a founder of the Organization of African Unity (OAU), now called the African Union. *(Victor Englebert/Englebert Photography, Inc.)*

accountability. The first chair of the AU, South African president Thabo Mbeki, called for new initiatives to end the "marginalization of Africa" and for the continent to take its "rightful place" in global affairs.

Given the scale of the African continent, it is not surprising that groups of states formed regional organizations to facilitate intraregional exchange and development. The two most active regional organizations are the Southern African Development Community (SADC) and the Economic Community of West African States (ECOWAS). Both were founded in the 1970s but became more prominent in the 1990s (Figure 6.38). SADC and ECOWAS are anchored by the region's two largest economies: South Africa and Nigeria. The Economic Community of Central African States (ECCAS) was founded in the mid-1980s and is headquartered in Libreville, Gabon. Its effectiveness has been hampered by the political instability of the area. Several ECCAS members also are members of SADC.

Apartheid's Demise in South Africa While fighting continued in the former Portuguese zone, South Africa underwent a remarkable transformation. Through the 1980s, its government had remained firmly committed to white supremacy. Under apartheid only whites enjoyed real political freedom, whereas blacks were denied even citizenship in their own country—technically, they were citizens of homelands. Yet because labor was needed in South African cities and mines, nonwhites were allowed to

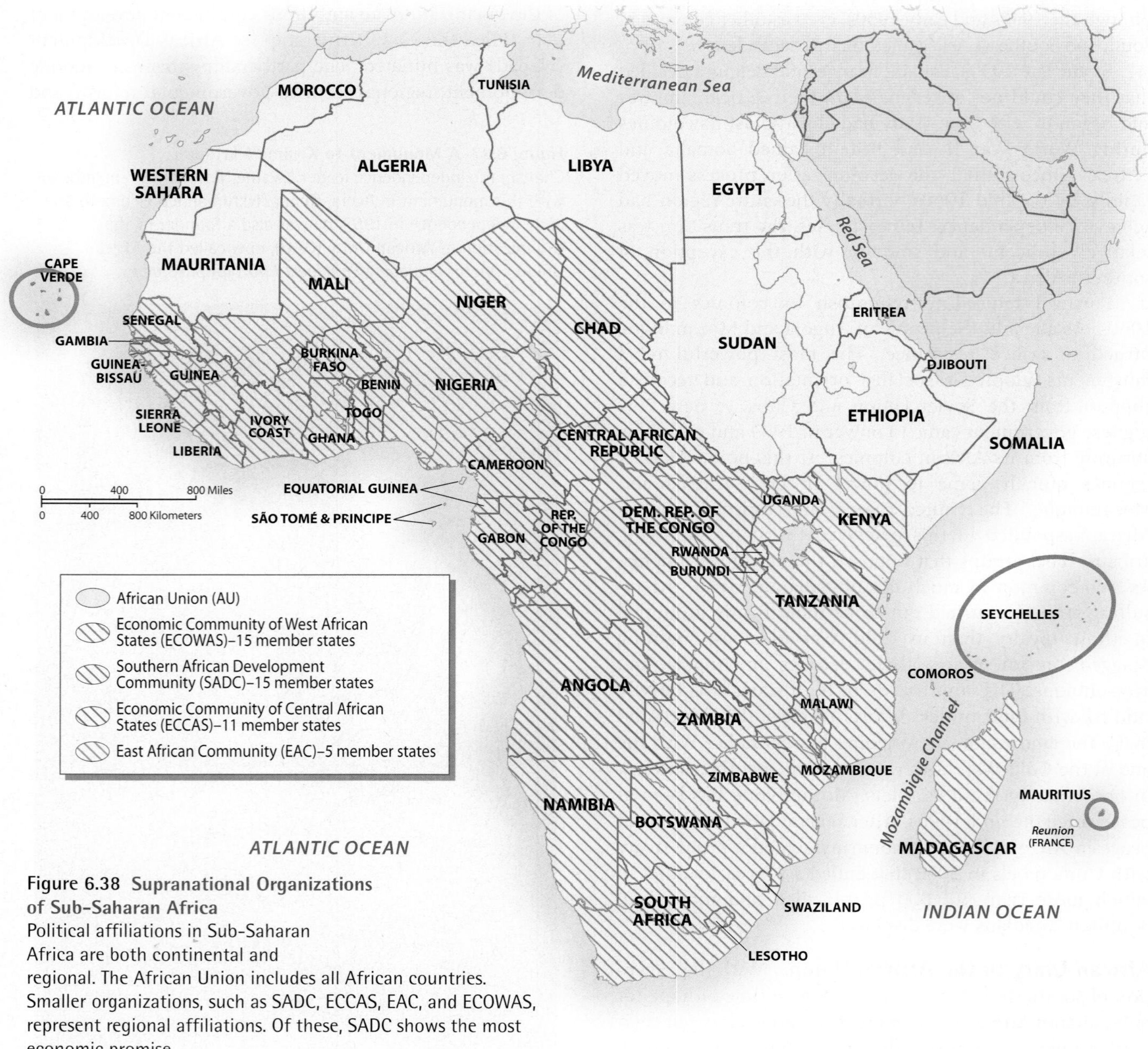

Figure 6.38 Supranational Organizations of Sub-Saharan Africa
Political affiliations in Sub-Saharan Africa are both continental and regional. The African Union includes all African countries. Smaller organizations, such as SADC, ECCAS, EAC, and ECOWAS, represent regional affiliations. Of these, SADC shows the most economic promise.

reside in segregated neighborhoods on the outskirts of cities, called **townships**. Some 20 miles southwest of Johannesburg, for example, is the famous black township of Soweto. Home to Nelson Mandela and Desmond Tutu, it was the principal home for many blacks working in Johannesburg and a center for political resistance to apartheid. In the postapartheid era, Soweto is a mixture of lean-tos and middle-class housing that still attracts mostly black migrants from former homelands and neighboring countries. Planners estimate that nearly 2 million people reside in Soweto (Figure 6.39).

Opposition to apartheid began in the 1960s, intensifying and becoming more violent by the 1980s. Blacks led the opposition, but coloureds and Asians (who suffered severe, but less extreme, discrimination) also opposed the Afrikaner government. As international pressure mounted, white South Africans found themselves ostracized. Many corporations refused to do business there, and South African athletes (regardless of color) were banned from most international competitions, such as the Olympics and World Cup Soccer. Increasing numbers of whites also opposed the apartheid system, and many businesspeople began to believe that apartheid threatened to undermine their economic endeavors.

The first major change came in 1990, when South Africa withdrew from Namibia, which it had controlled as a protectorate since the end of World War I. South Africa now stood alone as the single white-dominated state in Africa. A few years later, the leaders of the Afrikaner-dominated political party decided they could no longer resist the pressure for change. In 1994, free elections were held in which Nelson Mandela, a black leader who had been imprisoned for 27 years by the old regime, emerged as the new president. Black and white leaders pledged to put the past behind them and work together to build a new, multiracial South Africa. The homelands themselves were the first to be eliminated from the political map of the new South Africa (see the 1999 map in Figure 6.36). In 1999, peaceful elections were held, and Thabo Mbeki replaced Nelson Mandela as the country's leader. Mbeki was reelected to a second term in 2004. Unfortunately, the legacy of apartheid is not so easily erased. Residential segregation is officially illegal, but neighborhoods are still sharply divided along race lines. Under the new system a black middle class is rapidly emerging, but most blacks remain extremely poor (and most whites remain quite prosperous). Violent crime has increased in many parts of the country, and rural migrants have poured into South African cities in search of employment. Because the political change was not matched by a significant economic transformation, the hopes of many people are frustrated. The AIDS crisis has placed additional strains on the government's ability to foster development.

Enduring Political Conflict

Although most Sub-Saharan countries made a relatively peaceful transition to independence, virtually all of them immediately faced a difficult set of institutional and political problems. In several cases, the old authorities had done virtually nothing to prepare their colonies for independence. Lacking an institutional framework for independent government, countries such as the Democratic Republic of the Congo confronted a chaotic situation from the beginning. Only a handful of Congolese had received higher education, let alone been trained for administrative posts. The indigenous African political framework had been essentially destroyed by colonization, and in most cases very little had been built in its place. Even more problematic was the fact that the European colonial powers had essentially ignored indigenous cultural and political patterns, both in dividing Africa among themselves and in creating administrative subdivisions within their own imperial territories.

The Tyranny of the Map All over Africa, different ethnic groups found themselves forced into the same state with peoples of different linguistic and religious backgrounds, many of whom had recently been their enemies. At the same time, a number of the larger ethnic groups of the region found their territories split between two or more countries. The Hausa people of West Africa, for example, were divided between Niger (formerly French) and Nigeria (formerly British), each of which they had to share with several former enemy groups.

Given the imposed political boundaries, many African countries struggled to generate a common sense of national identity and establish stable political institutions. **Tribalism**, or loyalty to the ethnic group rather than to the state, has emerged as the bane of African political life. Especially in rural areas, tribal identities usually supersede national ones. Because virtually all of Africa's countries inherited an inappropriate set of colonial borders, one

Figure 6.39 Nelson Mandela's Home
In the township of Soweto, outside of Johannesburg, is the former home of President Nelson Mandela. A typical "matchbox" house built for black laborers, the building is now a museum that commemorates Mandela's struggle against apartheid.
(Rob Crandall/www.robcrandall.com)

might assume that they would have been better off scrapping the system altogether and drawing a new political map based on indigenous identities. Such a strategy was impossible, as all the leaders of the newly independent states realized. Any new territorial divisions would have created winners and losers, and thus would have proved tremendously contentious. Moreover, because ethnicity in Sub-Saharan Africa was traditionally fluid, and because many groups were territorially interspersed among their neighbors, it would have been difficult to generate a clear-cut system of division. Finally, most African ethnic groups were considered too small to form viable countries. With such complications in mind, the new African leaders in the 1960s agreed that colonial boundaries should remain. The violation of this principle, they argued, would lead to pointless wars between states and interminable civil struggles within them.

Despite the determination of Africa's leaders to build their new nations within existing boundaries, challenges to the states began soon after independence. Figure 6.40 maps the ethnic and political conflicts that have disabled parts of Africa since 1995. The human cost of this turmoil is several million refugees and internally displaced persons. **Refugees** are people who flee their state because of a well-founded fear of persecution based on race, ethnicity, religion, or political orientation. Nearly 3 million Africans were considered refugees in 2007. Added to this figure are another 13 million internally displaced persons. **Internally displaced persons** have fled from conflict but still reside in their country of origin. At the end of 2005,

Figure 6.40 Geopolitical Issues—Conflict and Refugees
Many Sub-Saharan countries have experienced wars or serious insurrections since 1995. These same states are also likely to produce refugees (red circles) and internally displaced persons (yellow circles). As of 2007, 3 million Africans were refugees and 11 million were internally displaced. *(Data from U.S. Committee for Refugees, 2007,* World Refugee Survey*)*

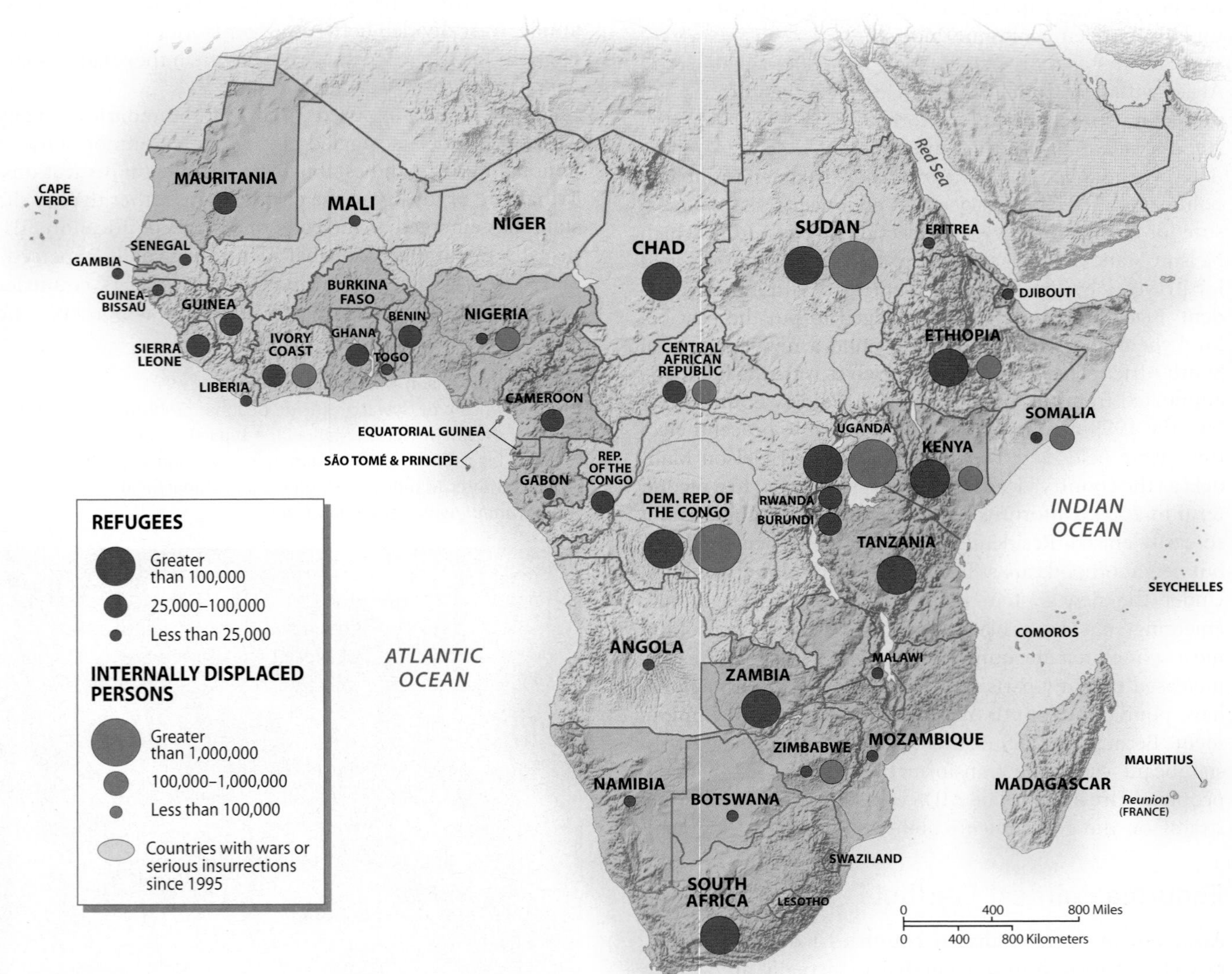

Sudan had the largest number of internally displaced people (5 million), followed by the Democratic Republic of the Congo (1.7 million) and Uganda (1.7 million). These populations are not technically considered refugees, making it difficult for humanitarian nongovernmental organizations and the United Nations to assist them.

Over the past decade, the number of internally displaced people has risen, while the total number of refugees has declined. This decline is due, in part, to the reluctance of neighboring states to take on the burden of hosting refugees. Tanzania had more than half a million refugees (mostly from Burundi) and Chad had 200,000 (mostly from Sudan) in 2004. With international aid on the decline, it is understandable that poor African states that struggle to serve their own people are disinclined to respond to the needs of refugees.

Ethnic Conflicts As Figure 6.40 suggests, more than half of the states in the region have experienced wars or serious insurrections since 1995. Fortunately, in the past few years peace has returned to Sierra Leone, Liberia, and Angola, states that produced large number of refugees in the 1990s. In the Ivory Coast, where conflict began as violence spilled over from Liberia in 2002, a fragile peace exists as of 2007 between the New Forces rebel group in the north and the government-controlled south. And, after a decade of conflict and nearly 4 million deaths, the Democratic Republic of the Congo held peaceful elections in 2006. Yet as peace was advancing in parts of West and Central Africa, violent ethnic conflict erupted in Darfur.

Darfur, Sudan, is a Sahelian province the size of France. In 2003, an ethnic conflict between Arab nomads and herders and non-Arab "black" farmers ignited. Since the fighting began, at least 300,000 people have died and more than 2 million have been displaced. The conflict is ethnically driven and, in 2004, the U.S. government accused the Sudanese government of **genocide**—the deliberate and systematic killing of a racial, political, or cultural group. The African Union has sent some troops and the UN has been trying to force the Sudanese to accept a large peacekeeping force. But in the name of nonintervention, China has supported Sudan's effort to block the deployment of UN troops to Darfur by vetoing or abstaining from Security Council resolutions. China has invested heavily in Sudan's oil fields and, as of 2007, continues to support the Sudanese government in Khartoum, despite considerable diplomatic pressure to use its influence to force the Sudanese government to stop the killing.

The conflict began when a non-Arab rebel group (later called the Sudan Liberation Movement) attacked a Sudanese military base. The SLM complained of economic marginalization and wanted a power-sharing arrangement with the Sudanese government. The Sudanese government responded to the rebels by encouraging Arab militiamen (called *janjaweed*—an Arabic term meaning horse and gun) to attack non-Arab civilians in Darfur. The janjaweed storm villages on horseback and camelback, destroying grain supplies and livestock, killing villagers, and setting buildings on fire. Human rights groups are especially alarmed by the systematic use of rape by the janjaweed as a weapon of ethnic cleansing and terror. The government of Khartoum claims it has been unable to stop the janjaweed atrocities, yet international observers continue to produce evidence that the government is supporting the Arab militiamen through aerial bombardment and attacks on civilians. Figure 6.41 shows the more than 500 villages that have been

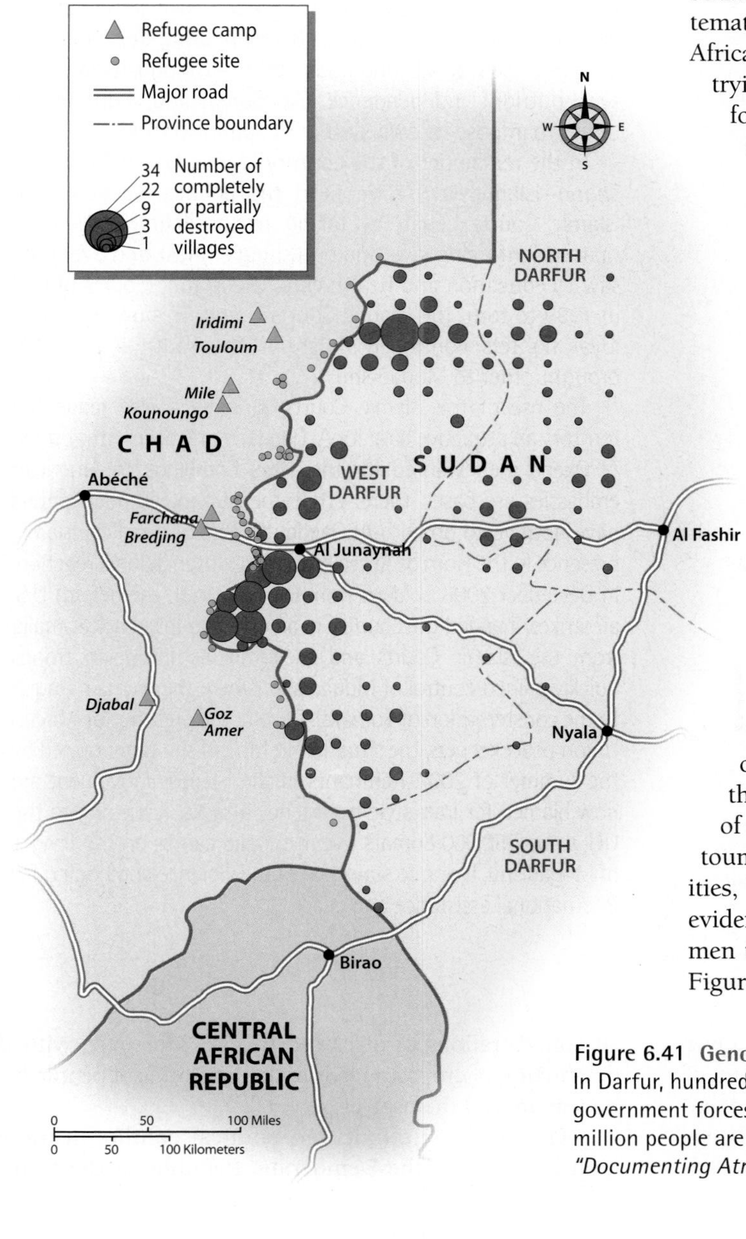

Figure 6.41 Genocide in Darfur
In Darfur, hundreds of villages have been totally or partially destroyed by government forces and militia fighters. More than 2 million of Darfur's 6 million people are either internally displaced or refugees in Chad. *(From "Documenting Atrocities in Darfur," Department of State Publication 11182)*

GEOGRAPHY IN THE MAKING The Battle Over Somalia

Somalia's political collapse was somewhat surprising; most residents are ethnic Somalis, and the country is one of the few ethnically homogeneous states of Sub-Saharan Africa. Somalis, however, have long been divided into fiercely independent clans. As clan warfare intensified in the 1980s, food could no longer be easily produced or distributed, and a massive famine followed. In 1992, UN forces, led by the United States, arrived to impose order and distribute food. They were successful at the latter task, but they were utterly ineffective at the former, and they soon were forced to withdraw in 1993 after 18 U.S. troops were killed.

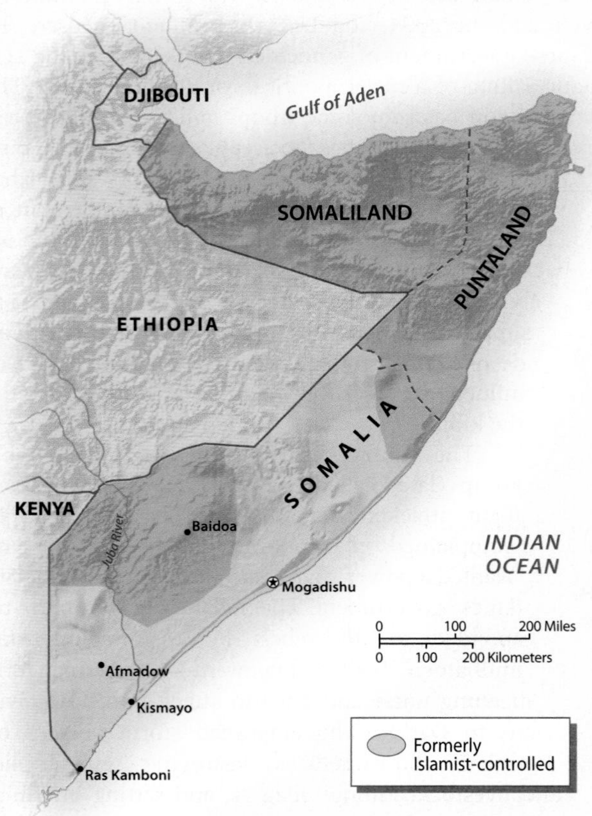

Figure 6.5.1 Somalia Divided
A state in name only, Somalia has had no central govenment since 1991. Two areas in the north seek political autonomy or independence. Meanwhile, Islamic courts have controlled areas in the south.

Although no functioning central government has existed since 1991, Somalia has been informally divided into several clan territories controlled by major warlords and their militias. Several alternative political territories have formed in response to the need for order (Figure 6.5.1).

In the northern half of the country, the territory of Somaliland declared its independence early in 1991. Once a British protectorate, Somaliland has a constitution, a functioning parliament, government ministries, a police force, a judiciary, and a president. The territory produces its own currency and passports. Although it has few natural resources, Somaliland has survived through a structure of power sharing among the clans and the support of the Somalian diaspora. Yet no country has recognized this territory, in part because no government exists in Somalia to negotiate the secession. In 1998, neighboring Puntland also declared itself an autonomous state. Although it does not seek outright independence like Somaliland, Puntland is trying to impose its own system of administration.

In the remainder of the country Islamic Courts (based on *sharia*—Islamic law) have filled the political vacuum. The Islamic Courts began by taking responsibility for judicial matters (enforcing laws and punishing crimes), and over time saw to education and health care. The Islamic Courts united in 1999 to form the Islamic Courts Union. This union and its loyal fighters controlled much of the south and briefly brought order to Mogadishu.

The rise of the Islamic Courts in Somalia also made the territory an attractive base for Al Qaeda; many believe that three Al Qaeda men wanted for the 1998 bombings of American embassies are based there. While the bulk of Islamic fighters have nothing to do with Al Qaeda, the rise of a radical Islamic presence in the Horn of Africa brought an international reaction. In December 2006, soldiers from Ethiopia, with the help of U.S. air strikes, invaded the country in an effort to take back Somalia from the Islamic Courts and their militias. Ethiopian troops quickly gained control of Mogadishu, where they remain (much to the consternation of Somalis). Despite the presence of African Union peacekeepers, the situation in Mogadishu deteriorated by the summer of 2007. Remnants of the Islamist movement are now blamed for Iraq-style insurgency attacks. According to the UN, some 250,000 Somalis live in squalid camps on the fringes of Mogadishu, trying to stay clear of the violence and hoping for international assistance and peace.

completely or substantially destroyed in Darfur. Over 250,000 refugees are living in Chad. The government of Sudan has resisted UN intervention but has allowed some African Union soldiers to assist the internally displaced people. However, the food program has not been able to reach all concentrations of displaced people. Moreover, with the destruction of crops and livestock, thousands of people have no way to feed themselves.

The deadliest ethnic and political conflict in the region has been in the Democratic Republic of the Congo.

Between 1998 and 2004, an estimated 3.8 million people died, although many of the deaths were from war-induced starvation and diseases rather than bullets or machetes. In 1996 and 1997, a loose alliance of armed groups from Rwanda (led by Tutsis) and Uganda joined forces with other groups in the Congo and marched their way across the country, installing Laurent Kabila as president. Under Kabila's rocky and ruthless leadership, which ended in assassination in 2001, rebel groups again invaded from Uganda and Rwanda and soon controlled the northern and eastern portion of the country while the Kinshasa-based government loosely controlled the western and southern portions.

With Kabila's death in 2001, his son Joseph took power and signed a peace accord with the rebels in 2002. In 2003 rebel leaders were part of a transitional government and an unsteady peace was in place with help from the UN, the African Union, and support by Western donors. Remarkably, elections were held in the summer of 2006 and Joseph Kabila was elected president. The country has no real experience with democracy, a civil service that barely functions, and virtually no roads or working infrastructure for the estimated 63 million people who live there. In 2005, foreign aid accounted for about one-quarter of the country's gross national income. For now, continuing peace and the ability of the Congo's people to improvise and cope under such adversity may gradually bring about social and economic improvements.

Secessionist Movements Problematic African political boundaries have occasionally led to attempts by territories to secede and form new states. The Shaba (or Katanga) province in what was then the state of Zaire (now the Democratic Republic of the Congo) tried to leave its national territory soon after independence. The rebellion was crushed a couple of years after it started with the help of France and Belgium. Similarly, the Igbo in the oil-rich Niger Delta of Nigeria proclaimed an independent state of Biafra in 1967. After a short but brutal war during which Biafra was essentially starved into submission, Nigeria was reunited.

In 1991, the government of Somalia disintegrated. The territory has since been ruled by clan-based warlords and their militias, who have informally divided the country amongst themselves. **Clans** are social units that are branches of a tribe or ethnic group larger than a family. While clans are able to maintain order in limited areas of the country, the lack of overall governance and associated violence has produced many refugees. Currently some 400,000 Somalis are recognized refugees, mostly living in Kenya and Ethiopia. Another 400,000 of the country's 9 million people are internally displaced persons. To complicate matters, the northern portion of Somalia has declared its independence and has sought international recognition as a sovereign state. Meanwhile Islamic courts with their well-armed militias control the south, where Al Qaeda operatives are believed to reside (see "Geography in the Making: The Battle Over Somalia").

Only one territory in the region has successfully seceded. In 1993, Eritrea gained independence from Ethiopia after two decades of civil conflict. This territorial secession is striking because Ethiopia forfeited its access to the Red Sea, making it landlocked. Yet the creation of Eritrea still did not bring about peace. After years of fighting, the transition to Eritrean independence began remarkably well. Unfortunately, border disputes between the two countries erupted in 1998, resulting in some 100,000 troop deaths. In 2000, a peace accord was reached and the fighting stopped.

ECONOMIC AND SOCIAL DEVELOPMENT: The Struggle to Rebuild

By almost any measure, Sub-Saharan Africa is the poorest and least-developed world region. Whereas other regions have experienced significant improvements in life expectancy and per capita income since 1990, Sub-Saharan Africa's life expectancy has declined and its income levels are either the same or lower. The average GNI per capita figure using purchasing power parity is about $1,900. Some states such as Botswana, Equatorial Guinea, Mauritius, Namibia, Seychelles, and South Africa have much higher per capita GNI-PPP (see Table 6.2). Purchasing power parity is a relative income figure; when absolute income is measured many states have average per capita incomes of less than $1 a day. Nearly all the states in the region are ranked at the bottom of most lists comparing per capita GNI or purchasing power parity. Using other development indicators such as life expectancy, percentage of population living on less than $2 a day, and child mortality, the region also fares poorly.

In the 1990s, most countries in Sub-Saharan Africa saw their economic productivity decline, with low or negative growth rates from 1990 to 1999. The economic and debt crisis of the 1980s and 1990s prompted the introduction of structural adjustment programs. Promoted by the International Monetary Fund (IMF) and the World Bank, structural adjustment programs typically reduce government spending, cut food subsidies, and encourage private-sector initiatives. Yet these same policies cause immediate hardships for the poor, especially women and children, and lead to social protest, most notably in the cities. The idea of debt forgiveness for Africa's poorest states, argued for by the unlikely duo of economist Jeffrey Sachs and rock star Bono from U2, has gradually gained acceptance as a strategy to reduce human suffering by redirecting monies that would have gone to debt repayment to the provision of services and the building of infrastructure. Sachs argues that in order for the region to get out of the poverty trap it will need substantial sums of new foreign aid and investment.

The implementation of the **Millennium Development Goals** is one example of programs aimed at fostering development in the region. The Millennium Development Goals

TABLE 6.2 Development Indicators

Country	GNI[a] per Capita, PPP[b] (2005)	GDP[c] Average Annual % Growth (2000–05)	Life Expectancy (2007)	Percent of Population Living on Less Than $2 a Day (2006)	Under Age 5 Mortality Rate		Gender Equity[d]
					1990	2005	
Angola	2,210	9.9	41		260	260	260
Benin	1,110	4.0	56	74	185	150	73
Botswana	10,250	5.9	34	50	58	120	102
Burkina Faso	1,220	5.1	51	72	210	191	77
Burundi	640	2.2	49	88	190	190	83
Cameroon	2,150	3.7	50	51	139	149	83
Cape Verde	6,000		71				
Central African Republic	1,140	−1.4	43	84	168	193	65
Chad	1,470	14.5	51		201	208	60
Comoros	2,000		64				
Congo	810	3.9	52		110	108	89
Dem. Rep. of Congo	720	4.4	45		205	205	73
Djibouti	2,240		54				
Equatorial Guinea	7,580		49				
Eritrea	1,010	3.5	57		147	78	70
Ethiopia	1,000	4.2	49	78	204	127	76
Gabon	5,890	1.7	57		92	91	
Gambia	1,920	3.7	58	83	151	137	97
Ghana	2,370	5.1	59	79	122	112	91
Guinea	2,240	2.9	54		234	160	74
Guinea-Bissau	700	−0.5	46		253	200	
Ivory Coast	1,490	−0.1	51	49	157	195	67
Kenya	1,170	3.4	53	58	97	120	94
Lesotho	3,410	2.9	36	56	101	132	103
Liberia		−6.8	45		235	235	
Madagascar	880	2.0	57	85	168	119	96
Malawi	650	3.4	40	76	221	125	98
Mali	1,000	5.9	53	91	250	218	75
Mauritania	2,150	4.0	62	63	133	125	96
Mauritius	12,450	4.0	72		23	15	98
Mozambique	1,270	8.6	43	78	235	145	82
Namibia	7,910	4.6	52	56	86	62	101
Niger	800	3.7	56	86	320	256	72
Nigeria	1,040	5.9	47	92	230	194	82
Reunion			76				
Rwanda	1,320	5.1	47	84	173	203	99
São Tomé and Principe	1,200	4.4	64				
Senegal	1,770	4.7	62	63	149	119	90
Seychelles	15,940		72				
Sierra Leone	780	13.7	48	75	302	282	71
Somalia	600	2.6	48		225	225	
South Africa	12,120	3.7	51	34	60	68	101
Sudan	2,000	6.1	58		120	90	89
Swaziland	5,190	2.3	33		110	160	94
Tanzania	730	6.9	50	90	161	122	95
Togo	1,550	2.7	58		152	139	72
Uganda	1,500	5.6	47		160	136	96
Zambia	950	4.7	38	94	180	182	92
Zimbabwe	1,940	−5.9	37	83	80	132	95

[a] *Gross national income.*
[b] *Purchasing power parity.*
[c] *Gross domestic product.*
[d] *Ratio of female-to-male enrollments in primary and secondary school percentage. Numbers below 100 have more males in primary/secondary school; numbers above 100 have more females in primary/secondary schools.*

Sources: World Bank, World Development Indicators, 2007; *life expectancy 2007 and percentage of population living on $2 a day data from* Population Reference Bureau, World Data Sheet, 2007; *gender equity data from* Millennium Development Goals.

(MDG) are part of a global United Nations effort to reduce extreme poverty by 2015. The goals are to:

- Halve the proportion of people living on incomes of less than $1 a day between 1990 and 2015;
- Ensure universal primary schooling and promote gender equity in school enrollment by 2015;
- Reduce child mortality by two-thirds from 1990 to 2015;
- Reduce by three-quarters the maternal mortality rate between 1990 and 2015;
- Halt the growing prevalence of HIV/AIDS and malaria, and begin to reduce their spread by 2015;
- Halve the proportion of the population without sustainable access to safe drinking water and sanitation by 2015.

As of 2007, several regions of the world are on track to meet many of these goals, although it appears that Sub-Saharan Africa will not meet any of the targets by 2015. That said, some individual countries have made impressive gains toward achieving some of these goals.

International interest in Africa's struggles seems to be growing; between 2002 and 2005 foreign aid to the region doubled from $11.5 billion to $23 billion. Much of this aid is in the form of fighting disease, improving health care, and providing basic education. There are still pressing needs for infrastructural development. Concern also exists that foreign aid, which was lavished on the region in the past, often fails to foster real development. Now, as in the past, many donated funds end up in the hands of a few corrupt officials.

On the positive side, there are signs of economic growth since 2000. For most states, the average annual growth rates from 2000 to 2005 are positive. In Angola, Botswana, Chad, Mali, Mozambique, Nigeria, Sierra Leone, Rwanda, Sudan, and Tanzania, average annual growth rates exceeded 5 percent (see Table 6.2). Such dazzling figures, while good news overall, can be deceptive. Some are due to soaring oil prices (notably so for Angola, Chad, Nigeria, and Sudan), whereas others are from countries that began from a very low base after years of conflict (Mozambique, Sierra Leone, and Rwanda). Still, for the first time in many years, Sub-Saharan African economies are growing at a faster rate than their populations.

Roots of African Poverty

In the past, outside observers often attributed Africa's poverty to its environment. Favored explanations included the infertility of its soils, the erratic patterns of its rainfall, the lack of navigable rivers, and the virulence of its tropical diseases. Most contemporary scholars, however, argue that such handicaps are not prevalent throughout the region and that even where they do exist they can be—and have often been—overcome by human labor and ingenuity. The favored explanations for African poverty today look much more to historical and institutional factors than to environmental circumstances.

Numerous scholars have singled out the slave trade for its debilitating effect on Sub-Saharan African economic life. Large areas of the region were effectively depopulated, and many people were forced to flee into poor, inaccessible refuges. Colonization was another blow to Africa's economy. European powers invested little in infrastructure, education, and public health and instead were interested mainly in extracting mineral and agricultural resources for their own benefit. Several plantation and mining zones did achieve a degree of prosperity under colonial regimes, but dynamic national economies failed to develop. In almost all cases, the rudimentary transport and communications systems were designed to link administration centers and zones of extraction directly to the colonial powers, rather than to their own hinterlands or neighboring areas. On achieving independence, Sub-Saharan African countries thus faced economic and infrastructural challenges that were as daunting as their political problems (Figure 6.42). Today, the average amount of paved road per person in countries such as Ghana, Ethiopia, Kenya, Tanzania, and Uganda is just .01 kilometers per person. The average for the non-African developing world is 2.8 miles (4.5 kilometers) per person.

Failed Development Policies The first decade or so of independence was a time of relative prosperity and optimism for many African countries. Most of them relied heavily on the export of mineral and agricultural products, and through the 1970s commodity prices generally remained high. Some foreign capital was attracted to the region, and in many cases the European economic presence actually increased after decolonization. (This is one reason critics refer to the period as one of "neocolonialism.")

The relatively buoyant economies of the 1960s and early 1970s disappeared in the 1980s as most commodity

Figure 6.42 Lack of Infrastructure
Young boys work to free an overloaded truck stuck in a muddy West African road. Throughout the region, a critical lack of infrastructure (especially good roads) impedes trade and communication. *(James Strachan/Robert Harding World Imagery)*

prices began to decline. Sizable foreign debt began to weigh down many Sub-Saharan countries. By the end of the 1980s, most of the region entered a virtual economic tailspin. By the early 1990s, the region's foreign debt was around $200 billion. Although low compared to that of other developing regions (such as Latin America), Sub-Saharan Africa's debt was the highest in the world as a percentage of its economic output.

Many economists contend that Sub-Saharan African governments enacted counterproductive economic policies and thus brought some of their misery on themselves. Keen to build their own economies and reduce their dependency on the former colonial powers, most African countries followed a course of economic nationalism. Thus, they set about building steel mills and other forms of heavy industry in which they were simply not competitive (Figure 6.43). In many cases, inefficient and often corrupt government ministries took over large segments of the economy. Local currencies also were often maintained at artificially elevated levels, which benefited the elite who consume imported products but undercut exports. Some former French colonies kept their currencies pegged to the franc, which made successful export strategies almost impossible.

Food Policies The largest blunders made by Sub-Saharan leaders were in agricultural and food policies. The main objective was to retain a cheap supply of staple foods in the urban areas. A modern industrial economy could emerge, or so it was argued, only if manufacturing wages remained low, which in turn was feasible only if food could remain inexpensive. Another reason for keeping food inexpensive was that it would help maintain political stability in urban areas. The main problem with this policy was that the majority of Africans were farmers, who could not make money from their crops because prices were kept artificially low. Thus they opted to grow food mainly for subsistence, rather than to sell—often at a loss—to national marketing boards. At the same time, they were encouraged to shift into export crops, such as coffee, cocoa, peanuts, and cotton. The end result was the failure to meet staple food needs at a time when the population was growing explosively. In the 1980s, famine occurred in 22 African states, partly as a result of failed agricultural policies combined with drought.

Although Sub-Saharan Africa is poorer than South Asia in terms of per capita GNI, its people are generally better nourished and often better housed. Starvation has followed war and drought in Africa, but day-to-day food intake is adequate. Such "nutritional prosperity" can be explained in part by the region's low population density and by the fact that many people still rely largely on subsistence production, which does not figure into official economic statistics.

Many African countries could vastly increase their production of staple foods, according to prevailing economic wisdom, if cultivators could be offered adequate incentives and have access to both supplies and markets. Most countries in the region began to reform their agricultural policies in the 1980s, but reform has been difficult because, as explained previously, the urban poor have grown dependent on low prices of staple goods.

Corruption Although prevalent through most of the world, corruption also seems to have been particularly rampant in several African countries. Civil servants typically are not paid a living wage and thus are virtually forced to solicit bribes. Teachers solicit informal fees from students in order to get paid. According to a recent poll of international businesspeople, Nigeria ranks as the world's most corrupt country. (Skeptical observers, however, point out that several Asian nations with highly successful economies, such as China, also are noted for high levels of corruption, so that corruption alone may not be an explanation for Africa's economic problems.)

With millions of dollars in loans and aid pouring into the region, officials at various levels were tempted to skim from the top. Some African states, such as the Democratic Republic of the Congo (DRC), were dubbed kleptocracies. A **kleptocracy** is a state in which corruption is so institutionalized that politicians and government bureaucrats siphon off a huge percentage of the country's

Figure 6.43 Industrialization Heavy industry, such as this chemical plant in Kafue, Zambia, has failed to deliver Sub-Saharan Africa from poverty. In the worst cases, these industrial enterprises were unable to produce competitive products for world and domestic markets. *(Marc & Evelyne Bernheim/Woodfin Camp & Associates)*

wealth. President Mobutu of the DRC was a legendary kleptocrat. While his country was saddled with an enormous foreign debt, he reportedly skimmed several billion dollars and deposited them in Belgian banks.

Links to the World Economy

Sub-Saharan Africa's trade connection with the world is limited, accounting for about 2 percent of global trade. The level of overall trade is low both within the region and outside it. Traditionally, most exports went to the European Union, especially the former colonial powers. The United States was the second most common destination. This trade pattern is changing, especially as trade with China and India grows. As of 2005, exports to the EU were only slightly larger than those to Asia and the United States, and Asia's share of Africa's commodity exports (oil, gas, wood, and food) is growing the fastest. The patterns of imports are also changing. The majority of African countries still turn to Europe for imports but Asia (particularly China) now accounts for one-third of all imports. Imports from the United States are only about 10 percent of the region's total. And, the movement of goods between countries in the region is steadily growing.

By most measures of connectivity, Sub-Saharan Africa lags behind other developing regions. Telephone lines are scarce; Nigeria has only 8 lines per 1,000 persons. The regional average is slightly better, at 16 lines per 1,000, but still far behind North Africa and Southwest Asia. About 15 percent of region's households have televisions, whereas nearly all households have televisions in high-income countries. One hopeful change is the expansion of mobile telephones in Africa. No longer reliant on expensive fixed telephone lines for communication, multinational providers now are competing for mobile-phone customers. There are 78 mobile subscribers per 1,000 people in the region. Subscriber growth in the region was 40 percent in 2006 and mobile operators are among the biggest companies and largest taxpayers. Plus, it is a market that is still young. The growth of mobile phones is a powerful example of the desire for Africans to acquire modern telecommunication systems long denied them due to limited land lines (Figure 6.44).

Figure 6.44 Mobile Phones for Africa
A shopkeeper has steady customers for mobile phones in Abidjan, Ivory Coast. Mobile telephones are quickly becoming the preferred instrument of communication in Sub-Saharan Africa. They do not require expensive fixed lines, and multinational firms are competing for the region's business. *(Andre Ramasore/Galbe.com)*

Africa lacks the infrastructure and goods to facilitate more trade. Only southern Africa has a telecommunications and road network of any note. Of the few roads that exist in Central and West Africa, the majority are not paved. This, too, is a legacy of colonialism, and one that keeps Africans from nurturing economic linkages with each other and with the outside world. Trade groups, such as the Southern African Development Community, have the potential to build up regional economies and infrastructure but have yet to achieve the influence of similar groups such as Mercosur in South America.

Aid Versus Investment As a poor region, Sub-Saharan Africa is linked to the global economy more through the flow of financial aid and loans than through the flow of goods. As Figure 6.45 reveals, for several states aid accounts for more than 20 percent of the GNI. In the extreme case of Liberia, economic assistance in 2005 equaled 54 percent of the GNI and in Burundi it was 47 percent. Most of this aid comes from a handful of developed countries (see the inset in Figure 6.45). The United States, France, and the United Kingdom provided the most aid in 2005, followed by Germany, the Netherlands, and Japan. Foreign direct investment in Sub-Saharan Africa increased 10-fold from 1990 to 2004. Optimists interpret this as a sign of Africa's economic renewal, but even this increase represents less than 2 percent of global foreign investment. The largest recipients of foreign investment have been the region's major oil producers: Nigeria, Angola, Sudan, Chad, and Equatorial Guinea. The poorer states—without minerals or oil to export—were largely bypassed by foreign investors. The relative lack of capital underscored the fact that all of the foreign direct investment in Sub-Saharan Africa in 2004 was just two-thirds of the foreign investment in Mexico alone.

The reasons foreign investors shun Africa are fairly obvious. The region is generally perceived to be too poor and unstable to merit much attention, and most foreign investors eager to take advantage of low wages put their money in Asia or Latin America. Sub-Saharan Africa, some economists contend, is thus starved for capital. Other scholars, however, see foreign investment as more of a trap—and one that the region would be wise to avoid if it seeks broad-based and lasting economic gains. The countries of the region tried to build more autonomous and less capitalistic economic systems in the first decades of independence. Today, most states are opting for greater integration into the global system. Many African countries have begun to privatize sectors of their economies, a process that often leads to faster economic growth but also tends to heighten income disparities and negative

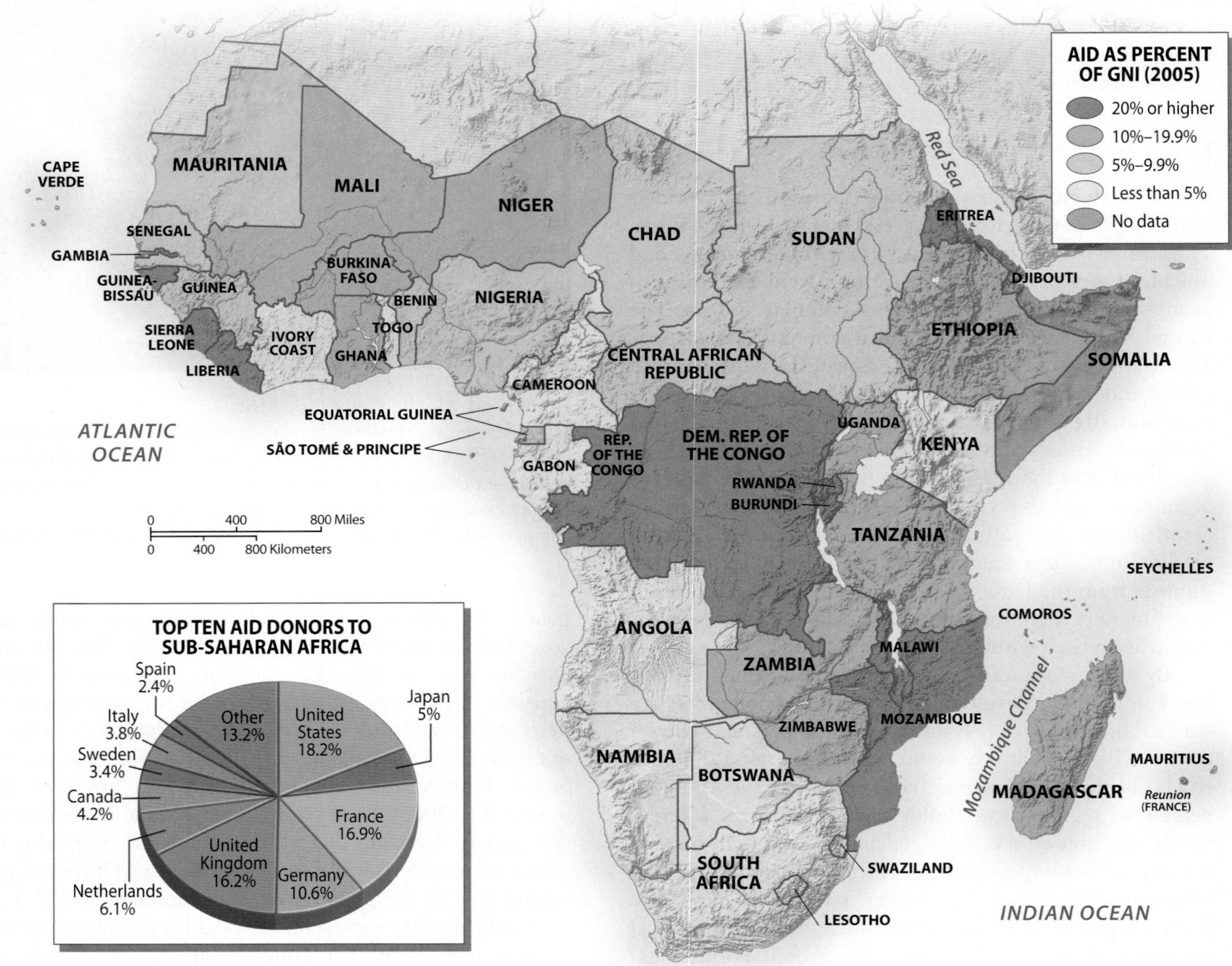

Figure 6.45 Global Linkages: Aid Dependency
Many states in Sub-Saharan Africa are dependent on foreign aid as their primary link to the global economy. This figure maps aid as a percentage of GNI, which ranges from less than 1 percent to 54 percent for Liberia and 47 percent for Burundi. *(Data from the World Bank,* World Development Indicators, *2007)*

environmental impacts. The negotiation of debt relief has also freed up capital for social investment.

Debt Relief The World Bank and the IMF proposed in 1996 to reduce debt levels for heavily indebted poor countries, many of which are in Sub-Saharan Africa. Most Sub-Saharan states are indebted to official creditors such as the World Bank, not to commercial banks (as is the case in Latin America and Southeast Asia). Under this new World Bank/IMF program, substantial debt reduction will be given to Sub-Saharan countries that are determined to have "unsustainable" debt burdens. Mauritania, for example, spends six times more money on repaying its debts than it does on health care. States qualify for different levels of debt relief provided they present a poverty reduction strategy. Uganda was the first state to qualify for the program, using the money it saved on debt repayment to expand primary schooling. Ghana qualified for debt relief in 2004 and received a $3.5 billion relief package. Other countries that have benefited from debt reduction are Tanzania, Mozambique, Ethiopia, Mauritania, Mali, Niger, Nigeria, Senegal, Burkina Faso, and Benin. With luck, more African countries will be liberated from burdensome foreign debt payments and redirect those same resources toward improving services and infrastructure.

China's Growing Role in the Region Beijing hosted nearly all the African heads of state at the China-African Forum in November 2006. The event signaled China's growing economic trade and investment in the region at a time when U.S. and European interests focus on proffering aid or fighting terrorism. China wants to secure the oil and ore it needs for its massive industrial economy. In exchange, it offers African nations money for roads, railways and schools with relatively few strings attached. China is interested in investing in Africa to improve its export commodity production; it has no interest in interfering in African countries' domestic affairs. Some African leaders see China as a new kind of global partner, one that wants straight commercial relations without an ideological or political agenda. China's trade with Africa is growing faster than with any other region except Southwest Asia, increasing ten-fold in the past decade. Angola, which China invested heavily in, is now China's largest foreign supplier of oil—beating out Saudi Arabia. Sudan, with its largely untapped oil reserves, has also received substantial Chinese investment despite international pressure not to negotiate with the Sudanese government due to the genocide in Darfur.

China's engagement with Africa is not new. In the 1960s and 1970s it supported the socialist governments of postcolonial Africa with development assistance, especially Tanzania. In the 1970s Beijing funded and built the Tan-Zam Railroad that gave landlocked Zambia an outlet to the sea via Tanzania, bypassing the need for Zambia to use apartheid South Africa's ports. China's investment today is strictly commercial and substantially larger. In addition to buying up oil, China purchases copper from Zambia and wood from the Democratic Republic of the Congo. It also is selling cheap manufactured goods and clothing in record quantities.

The biggest complaint about Chinese investment in Africa comes from the West. There is frustration that China readily invests in states such as Sudan and Zimbabwe with wretched human rights records. China mostly ignores the lending standards that have been employed in the region to curb corruption. Some Africans complain China is only interested in exporting raw materials and importing manufactured goods back to Africa. The garment industry in Lesotho, for example, is contracting due to the increased competition from cheaper Chinese imports. In turn China has gained diplomatic friends and new sources for raw materials from Sub-Saharan Africa. Interestingly, Chinese officials see themselves as the developing world's biggest beneficiary of globalization, and they are partnering with the region most ignored by globalization.

Economic Differentiation Within Africa

As in most regions, considerable differences in levels of economic and social development persist. In many respects, the small island nations of Mauritius and the Seychelles have little in common with the mainland. With a high per capita GNI, life expectancies averaging in the low 70s, and economies built on tourism, they could more easily fit into the Caribbean were it not for their Indian Ocean location. In contrast, Malawi has a per capita GNI of $650 and 76 percent of the population lives on less than $2 per day. In the developed world, where a latte at Starbucks costs $3, it is hard to imagine existence on less than $1 or $2 a day, but such is the case for nearly half of all Africans living south of the Sahara.

Life for the Poorest Rose Shanzi, a mother of five, lives in the town of Maramba, Zambia, not far from the city of Livingstone, Zimbabwe, site of Victoria Falls. Rose sells tomatoes in the local market, and if she sells enough, she and her children eat that day. Her daily goal is to earn 75 cents, which will enable her to purchase the vegetables and ground corn (called *mealie meal*) needed to make the evening supper. Yet there are not many buyers and competition is stiff because few people have money.

Livingstone and Maramba used to have three dozen clothing manufacturers that employed hundreds of people, like Rose's husband. The deluge of used clothing from the West killed this industry during the 1990s and plunged families like Rose's into stomach-tightening poverty (see "Global to Local: The Used Clothing Trade"). Then Rose's husband died. With no other job prospects, she joined other vendors in the market to earn enough to feed herself and her family. Rose has a high school education and was a housewife until her husband lost his job. Now she sits in her vegetable stand among the neatly piled mounds of tomatoes for 12 hours a day. Rose earns between $12 and $18 a month from her stand. Half of her earnings go to food. She pays $2 a month for school fees, $2 a month for water, $1.50 for property tax, and 50 cents for government health insurance. Water, education, and health care were free before structural adjustment policies forced payment for these basic services. Rose used to be able to buy meat occasionally with her earnings, but no longer can because of her new expenses. There are luckless days when she returns home with no food. "You don't know what suffering is until you have watched your babies go hungry," says Rose. "I have suffered many times." There are millions of Africans living in extreme poverty in every state in the region. In rural areas many of the very poor still practice subsistence farming and are barely a part of the formal economy. In the rapidly growing urban areas, massive slums are homes to millions with inadequate shelter and no regular access to clean water. Most make their living through the informal economy. These are the people that the Millennium Development Goals hope to reach, although securing the resources and finding the right programs to ease this poverty is both a difficult and long-term proposition.

Oil and Mineral Producers Another group of relatively prosperous Sub-Saharan countries benefits from substantial oil and mineral reserves and small populations (Figure 6.46). The prime example is Gabon, a country of noted oil wealth that is inhabited by 1.4 million people. Its neighbors, the Republic of the Congo and Cameroon, also benefit from oil, but both have experienced economic declines in recent years, with that of Cameroon being particularly dramatic. Equatorial Guinea, a former Spanish colony, began producing significant amounts of oil in 1998. American investment is behind much of the offshore drilling, most notably by Exxon Mobil. As this small country of just half a million people begins to earn substantial oil revenues, observers hope that the lives of average citizens will

Figure 6.46 Chad–Cameroon Pipeline
Workers unload pipe destined for the 600-mile (960 kilometers) long pipeline that pumps oil from Chad to Cameroon's Atlantic port of Kribi. The pipeline was finished in 2003. *(Hans-Jurgen Burkard/Bilderberg/Peter Arnold, Inc.)*

improve; yet, after several record years of production the oil revenues have not been invested in the country or its citizens. Farther south, Namibia and Botswana also have the advantage of small populations and abundant mineral resources, especially diamonds. Over the past few years, both countries have enjoyed sound government and experienced significant economic growth.

South Africa is the region's unchallenged economic powerhouse and one of the world's mining superpowers. The country leads the world in gold production, and is a major producer of other strategic metals and gems as well. Only South Africa has a well-developed and well-balanced industrial economy. It also boasts a healthy agricultural sector. But while South Africa is undeniably a wealthy country by African standards, the country is beset by severe and widespread poverty. Despite the end of apartheid, the country maintains one of the worst distributions of income in the world.

The Leaders of ECOWAS Nigeria has the largest oil reserves in the region, but its huge population has kept its per capita GNI-PPP at a relatively low $1,040. Oil money also has helped make Nigeria notoriously inefficient. A small minority of its population has grown fantastically wealthy, more by manipulating the system than by engaging in productive activities. Most Nigerians, however, remain trapped in poverty. Oil money also led to the explosive growth of the former capital of Lagos, which is Africa's only megacity at 14 million and still growing. As a result, the Nigerian government opted to build a new capital city in Abuja, located near the country's center, a move that has proved tremendously expensive. The southwestern corner of the country, in the Niger delta where most of the oil is produced, has seen little of the benefits of oil production but bears the burden of the environmental costs.

Ivory Coast and Senegal, formerly the core territories of the French Sub-Saharan empire, still function as commercial centers, but they also suffered economic downturns. In the mid-1990s, the Ivorian economy again began to grow. Boosters within the country called it an emerging "African elephant" (comparing it to the successful "economic tigers" of eastern Asia). Yet a destructive civil war began in 2002 that still divides the country and undermines the economy. Ghana, a former British colony and ECOWAS member, also began an economic recovery in the 1990s. In 2001, it negotiated with the IMF and World Bank for debt relief and in 2004 it qualified for $3.5 billion in relief. The country's average annual growth from 2000 to 2005 was an impressive 5.1 percent.

East Africa Long the commercial and communications center of East Africa, Kenya experienced economic decline and political tension throughout the 1990s. In recent years the economy has grown and per capita GNI-PPP is at $1,170. Kenya boasts good infrastructure by African standards, and about 1 million foreign tourists come each year to marvel at its wildlife. Agricultural exports of coffee, tea, and *khat* (a leaf that is chewed as a mild stimulant throughout the Horn of Africa) dominate the economy. As for social indicators, Kenyan women are having fewer babies now than in the late 1970s (down from an average of 8 to 5 children per woman), and those children are better educated. If Kenya can avoid political implosion, it could lead East Africa into better economic integration with the southern and western parts of the continent. The political and economic indicators for Kenya's neighbor, Uganda, also are looking good with a per capita GNI-PPP of $1,500 and an average annual growth rate of over 5 percent (see Table 6.2).

Measuring Social Development

By world standards, measures of social development in Africa are extremely low. Yet there are some positive trends, especially with regard to child survival, education, and gender equity. Also, many governments in the region have reached out to the modern African diaspora (economic migrants who have left the region for employment in Europe and North America). As in other parts of the

GLOBAL TO LOCAL The Used Clothing Trade

What begins as charity in the West—giving away old clothing—blossoms into thousands of small businesses in the villages and cities across Africa. In Uganda it's called *mivumba* and in Zambia it's *salaula*; throughout Africa one can go to any market and find the vendors of used clothing. In a region where many people live on less than $2 a day, most people rely on secondhand clothing imported from overseas. This is one of the reasons why children in the Sahel or the Congo wear t-shirts emblazoned with Western movies, pop stars, and sports teams or why a farmer in Uganda wears a wool sports coat.

Much of the used clothing comes from the United States. Charitable organizations in the West that receive clothing donations sell them to exporters who sort and press the clothing into bales and ship them to Sub-Saharan Africa. Once in cities like Lusaka or Kampala, the bales of dresses, jackets, pants, and t-shirts become part of the informal economy, purchased by the pound and sold by vendors in local markets and city shops and by traders on bicycles (Figure 6.6.1). The used clothing section of most markets is much larger than the food section. In the morning, bales are opened publicly and carefully examined for their contents. The quality of the clothing within often determines whether or not a seller will make his or her money back. There is much recycling of fabric: sweaters are unraveled and the yarn used to knit baby blankets; curtains can be turned into dresses. Some reports estimate that one-third of Sub-Saharan Africans are wearing cast-off clothing from Europe and North America.

Yet several African textile manufacturers are crying foul, arguing that something must be done to stop the flow of used clothing. They believe that as long as used clothes are in the markets, no textile industry can survive in the region. In some countries, most notably South Africa, the direct import of used clothing has been banned since 1999. Nigeria, Ethiopia, and Eritrea have imposed their own prohibitions. In Uganda, the import duty on mivumba was raised in 2003 to appease local textile manufacturers, and the importation of used undergarments was banned. In the 1970s, Uganda had a vibrant textile industry based on the cotton plantations set up by the British. Economic and political turmoil in the 1980s destroyed it. By the time the mills were ready to operate in the late 1980s, secondhand clothes had arrived, undermining local demand. Yet, the used clothing business employs many people and supplies consumers with a less expensive product. The reason this market exists is that many people live on too little money. In today's globalized economy, a shirt sewn in a Honduras sweat shop and worn by a New Jersey teenager gets repackaged and sold to a Ugandan farmer.

Figure 6.6.1 Used Clothing Vendor
A vendor carries a 100-pound (45 kilo) bale of used clothing through the streets of Kampala, Uganda. The clothing is sold at local markets and is a vital part of the informal economy. *(Marco Longari/Galbe.com)*

Source: Adapted from Carter Dougherty, "Trade Theory vs. Used Clothes in Africa," *New York Times*, June 3, 2004.

world, African immigrant organizations have worked to improve schools and health care, and immigrants themselves have returned to invest in businesses and real estate. The economic impact of remittances on this region is relatively small when compared to Latin America or the Caribbean, but it is growing.

Understandably, reducing child mortality is one of the Millennium Development Goals. In a country where the child mortality rate is above 200, it means that one-out-of-five children die before their fifth birthday. As Table 6.2 shows, about half of the states in the region saw modest improvements in child survival between 1990 and 2005. Ethiopia and Malawi actually experienced dramatic gains in child survival rates. Yet in countries undergoing continued conflict (Ivory Coast) or with high HIV/AIDS infection rates (Botswana), child mortality increased. Hence the region is a long way from cutting child mortality by two-thirds, as stated in the MDG goals.

Meeting Educational Needs Basic education is another obstacle for the poorest people in the region who struggle with the small fees and lost labor associated with sending children to school. Still, there has been a tremendous push by states and donors to educate more African children. The goal of universal access to primary education is a daunting one for a region in which 44 percent of the population is less than 15 years old. The UN reports

Figure 6.47 Ugandan School Children
Cheering at a school event, these children in rural Bumwalukani, Uganda are receiving a good education as a result of efforts by a US-based NGO founded by a Ugandan emigrant from this village. Educating the region's children is one of the region's most pressing challenges. *(Washington Post Writers Group)*

that as of 2005, some 70 percent of African children were enrolled in primary school (up from just 54 percent in 1990). The region is home to one-sixth of the world's children under 15 but half of the world's uneducated children. Girls are still less likely than boys to attend school. In West African countries such as Chad, Niger, or Ivory Coast, girls are decidedly underrepresented—in Chad, for every 100 boys there are 60 girls in school (see Table 6.2). Yet many states in southern Africa, such as South Africa, Botswana, and Namibia, have achieved gender equity among students.

UN data show that of all the world regions, Sub-Saharan Africa has the highest average student-to-teacher ratio and the lowest percentage of trained teachers. In fact, a single teacher may be in charge of 100 students, and may lack basic teaching resources such as books, chalk, or paper. The renewed focus on education since 2000 has resulted in many more resources being directed to schools, from the state and private donors and organizations. One example of the transformative influence of better resourced schools comes from the village of Bumwalukani, Uganda. This remote village in eastern Uganda has government schools that accommodate hundreds of children but classes are crowded, teachers are poorly trained and paid, and educational materials are scarce. Malaria and hunger keep overall attendance low. A new school in Bumawalukani, the Arlington Academy of Hope, was built in 2003 and funded by a U.S. NGO started by a Ugandan emigrant who grew up in this rural mountain village. Students pay a small fee to attend the school and are given hot lunches, uniforms, and incomparable educational resources compared to the government schools (Figure 6.47). Teachers are recruited from the best teaching schools in Kampala and are well paid by Ugandan standards. All classes are in English. In Uganda, students who excel receive scholarships to attend residential high schools in the capital. Many of the students from Arlington Academy are able to continue their education because of the foundation received in their model village school.

Life Expectancy and Health Issues Sub-Saharan Africa's figures on life expectancy are, overall, the world's lowest. Only the poorest Asian countries, such as Afghanistan, Nepal, and Laos, compare with the average African level. Despite these dismal figures, some progress has been made in enhancing life expectancy in Sub-Saharan Africa. While the region's average life expectancy of 49 years may seem incredibly short, it must be remembered that infant and childhood mortality figures depress these numbers; average life expectancy for adults is substantially higher. States such as Kenya, Botswana, and Zimbabwe established relatively high life expectancy figures in the 1980s, but the growing AIDS epidemic eroded their accomplishments.

The causes of short life expectancy generally are related to extreme poverty, environmental hazards (such as drought), and various environmental and infectious diseases (cholera, measles, malaria, schistosomiasis, and AIDS). Often these factors work in combination. The effects of disease are exacerbated by poverty, with undernourished children being the most vulnerable to the effects of high fevers. For example, cholera outbreaks occur in crowded slums and villages where food or water is contaminated by the feces of infected persons and basic infrastructure is lacking. Tragically, diseases that are preventable, such as measles, occur when people have no access to or cannot afford vaccines. Funding from the Gates Foundation in particular has targeted fighting tropical diseases as a critical first step in overall development, and some diseases have seen dramatic declines in incidence due to the availability of relatively simple treatments.

The scarcity of doctors and health facilities, especially in rural areas, also helps explain Sub-Saharan Africa's short life expectancies and high child mortality levels. The more successful countries in the region have discovered that inexpensive rural clinics dispensing basic treatments, such as oral rehydration therapy for infants with severe diarrhea, can substantially improve survival rates. Another problem is the severity of the African disease environment itself. Malaria, which kills a half-million African children each year, is reappearing in many areas as the disease-causing organisms develop resistance to common drugs and the mosquito carriers develop resistance to insecticides. Although most Africans have partial immunity to malaria, it remains a major killer, and many who survive infections remain debilitated. Schistosomiasis, carried by freshwater snails, causes chronic diarrhea and cramping, and affects 200 million Africans and can lead to dehydration and susceptibility to other diseases. Collectively, national and international health agencies, along with local NGOs, have improved the availability of basic health care in Sub-Saharan Africa, but much work remains to be done. Moreover, the HIV/AIDS epidemic is overwhelming the region's health-care capabilities.

Women and Development

Development gains cannot be made in Africa unless the economic contributions of African women are recognized. Officially, women are the invisible contributors to local and national economies. In agriculture, women account for 75 percent of the labor that produces more than half the food consumed in the region. Tending subsistence plots, taking in extra laundry, and selling surplus produce in local markets all contribute to household income. Yet because many of these activities are considered informal economic activities, they are not counted. For many of Africa's poorest people, however, the informal sector is the economy. Some research suggests that it can account for 30 to 50 percent of the gross domestic product in certain states. Within this sector, women dominate.

Status of Women The social position of women is difficult to gauge for Sub-Saharan Africa. Women traders in West Africa, for example, have considerable political and economic power. By such measures as female labor force participation, many Sub-Saharan African countries show relative gender equality. And women in most Sub-Saharan societies do not suffer the kinds of traditional social liabilities encountered in much of South Asia, Southwest Asia, and North Africa. In 2006, the first African woman president, Mrs. Johnson-Sirleaf, took power in Liberia. A former World Bank economist and grandmother, she beat out a Liberian football star in a runoff election.

By other measures, however, such as the prevalence of polygamy, the practice of "bride-price," and the denial of property inheritance, African women do suffer discrimination. Perhaps the most controversial issue regarding women's status is the practice of female circumcision, or genital mutilation. In Sudan, Ethiopia, Somalia, and Eritrea, as well as parts of West Africa, almost 80 percent of girls are subjected to this practice, which is extremely painful and can have serious health consequences. Yet because the practice is considered traditional, most African states are unwilling to ban it.

Regardless of their social position, most African women still live in remote villages where educational and wage-earning opportunities remain limited and bearing numerous children remains a major economic contribution to the family. As educational levels increase and urban society expands—and as reduced infant mortality provides greater security—one can expect fertility in the region to gradually decrease. Governments can greatly quicken the process by providing birth control information and cheap contraceptives—and by investing more money in health and educational efforts aimed at women. As the economic importance of women receives greater attention from national and international organizations, more programs are being directed exclusively toward them.

Building From Within Major shifts in the way development agencies view women and women view themselves have the potential to transform the region. All across the continent, support groups and networks have formed, raising women's consciousness, offering women microcredit loans for small businesses, and harnessing their economic power. From farm-labor groups to women's market associations, investment in the organization of women has paid off. In Kenya, for example, hundreds of women's groups organize tree plantings to prevent soil erosion and ensure future fuel supplies.

Whether inspired by feminism, African socialism, or the free market, community organizations have made a difference in meeting basic needs in sustainable ways. No doubt the majority of the groups fall short of all of their objectives. Yet for many people, especially women, the message of creating local networks to solve community problems is an empowering one.

Summary

- The largest land mass straddling the equator, Africa is called the plateau continent because it is dominated by extensive uplifted plains. Key environmental issues facing this tropical region are desertification, deforestation, and drought. Global climate change is likely to make these problems worse. At the same time, the region supports a tremendous diversity of wildlife, especially large mammals.
- With three-quarters of a billion people, Sub-Saharan Africa is the fastest-growing region in terms of population. Yet it is also the poorest region, with two-thirds of the population living on less than $2 a day and the lowest average life expectancy of 48 years. Disease, especially the scourge of HIV/AIDS, has driven down life expectancy in the region.

- Culturally, this is an extremely diverse region, where multiethnic and multireligious societies are the norm. With a few exceptions, religious diversity and tolerance has been a distinctive feature of the region. In the nearly 50 years since independence, most states have forged pluralistic but distinct national identities. Many African cultural expressions such as music, dance, religion, and film have had an influence well beyond the region.
- Since 1995 there have been numerous bloody ethnic and political conflicts in the region. Fortunately, peace now exists in many conflict-ridden states such as Angola, Liberia, Sierra Leone, and the Democratic Republic of the Congo. Still, on-going ethnic and territorial disputes in Darfur, Sudan; northern Uganda; and Somalia have produced millions of internally displaced persons and refugees.
- In terms of contemporary economic globalization, Sub-Saharan Africa's connections to the global economy are weak. With 12 percent of the world's population, the region accounts for only about 2 percent of the world's economic activity. Most of the region's economic ties are through international aid and loans rather than trade. Foreign direct investment, especially from China, is beginning to increase in the region.
- Poverty is the region's most pressing issue. Since 2000, economic growth has occurred, led in part by debt forgiveness policies, the end of some long-running political conflicts, and an increase in commodity prices (mainly oil and metals). Foreign aid to the region doubled between 2002 and 2005, in part due to the targets set by the Millennium Development Goals. It remains to be seen if the international community is truly committed to helping millions of Africans improve their lives.

Key Terms

African Union (AU) *(page 261)*
agricultural density *(page 237)*
apartheid *(page 245)*
Berlin Conference *(page 258)*
biofuels *(page 233)*
clan *(page 267)*
coloured *(page 245)*
desertification *(page 231)*
genocide (page 265)
Gondwanaland *(page 226)*
Great Escarpment *(page 226)*
homelands *(page 259)*
Horn of Africa *(page 230)*
internally displaced persons *(page 264)*
kleptocracy *(page 270)*
Millennium Development Goals *(page 267)*
Pan-African Movement *(page 261)*
pastoralists *(page 231)*
physiological density *(page 237)*
protectorates *(page 257)*
refugee *(page 264)*
Sahel *(page 231)*
structural adjustment programs (page 226)
swidden *(page 242)*
township *(page 263)*
transhumance *(page 231)*
tribalism *(page 263)*
tribe *(page 248)*
tsetse fly *(page 243)*

Questions for Review

1. Describe the relationship between soil types and patterns of settlement in Sub-Saharan Africa.
2. Is desertification a natural or a human-induced process? Where does it occur? How might global climate change impact desertification trends?
3. What are the demographic, social, and economic consequences of AIDS in Sub-Saharan Africa?
4. Explain the factors that contribute to high population growth rates in the region.
5. What was the political significance of the 1884 Berlin Conference for Africa?
6. Why have there been relatively few boundary changes since African nations won their independence?
7. How are private and grassroots organizations shaping African development?
8. Explain the history and policies that make South Africa distinct from other states in the region.
9. How might the food policies of African governments have undermined the region's ability to feed itself?
10. What are the region's key economic resources that may facilitate its integration into the global economy?

Thinking Geographically

1. What factors might explain why European conquest and settlement occurred much earlier in tropical America than in tropical Africa?
2. What are some of the cultural and political ties that unite Africa with the Americas? With Southwest Asia? With South Asia?
3. Discuss how increasing urbanization in the twenty-first century might affect the overall structure of the population of Sub-Saharan Africa.
4. More than any other region, Sub-Saharan Africa is noted for its wildlife, especially large mammals. What environmental and historical processes explain the existence of so much fauna? Why are there relatively fewer large mammals in other world regions?
5. Compare and contrast the role of tribalism in Sub-Saharan Africa with that of nationalism in Europe.
6. Historically, how was Sub-Saharan Africa integrated into the global economy? Was its role similar to or different from that of other developing regions?
7. Sub-Saharan Africa is a particularly problematic world region because of its relationship to North Africa. Should the

Sahara be considered a cultural divide between these two regions? Why or why not?

8. Compare the contrasting development model put forward by the United States and Europe with that of China. Will Chinese influence in the region alter the course of development for the region?
9. Is desertification a natural or a human-induced process? Where does it occur? How might global climate change impact desertification trends?
10. How useful are the Millennium Development Goals in measuring development? What are the problems in tracking Sub-Saharan Africa's relative progress given the lack of infrastructure and data?

Regional Novels and Films

Novels

Chinua Achebe, *Things Fall Apart* (1959, Fawcett Crest)
Ishmael Beah, *A Long Way Gone* (2007, Sarah Crichton Books)
Nuruddin Farah, *Knots* (2007, Riverhead Books)
Alexandra Fuller, *Don't Let's Go to the Dogs Tonight: An African Childhood* (2001, Random House)
Nadine Gordimer, *Burger's Daughter* (1979, Viking Press)
Barbara Kingsolver, *The Poisonwood Bible* (1998, Harper)
Doris Lessing, *African Laughter: Four Visits to Zimbabwe* (1992, Harper)
Alan Paton, *Cry, The Beloved Country* (1948, Scribner's)
Ngugi Wa Thiong'o, *A Grain of Wheat* (1967, Heinemann)
Wole Soyinka, *You Must Set Forth at Dawn* (2006, Random House)
Ngugi Wa Thiong'o, *Petals of Blood* (1977, Heinemann)

Films

Daresalam (2000, Chad)
The Gods Must Be Crazy (1981, Botswana/South Africa)
Hotel Rwanda (2004, U.S.)
The Last King of Scotland (2006, U.S./UK)
Live and Become (2005, France/Israel)
Lumumba (2000, France)
Masai: The Rain Warriors (2005, Kenya/France)
Thunderbolt (2000, Nigeria)
Tsotsi (2005, South Africa)
Youssou N'Dour: Return to Gorée (2006, Senegal/Switzerland)

Bibliography

Aryeetey-Attoh, Samuel, ed. 1997. *Geography of Sub-Saharan Africa*. Upper Saddle River, NJ: Prentice Hall.

Christopher, A. J. 1984. *Colonial Africa: A Historical Geography*. Totowa, NJ: Barnes & Noble.

Curtin, Philip D. 1969. *The Atlantic Slave Trade*. Madison: University of Wisconsin Press.

Grove, Alfred T. 1994. *The Changing Geography of Africa*. Oxford: Oxford University Press.

Manning, Patrick. 1990. *Slavery and African Life: Occidental, Oriental, and African Slave Trades*. Cambridge, UK: Cambridge University Press.

Ndegwa, Stephen N. 1996. *The Two Faces of Civil Society: NGOs and Politics in Africa*. West Hartford, CT: Kumarian Press.

Newman, James L. 1995. *The Peopling of Africa: A Geographical Interpretation*. New Haven, CT: Yale University Press.

Rain, David. 1999. *Eaters of the Dry Season: Circular Labor Migration in the West African Sahel*. Boulder, CO: Westview Press.

Reader, John. 1997. *Africa: A Biography of the Continent*. London: Penguin Books.

Stock, Robert. 2004. *Africa South of the Sahara: A Geographical Interpretation*. 2nd ed. New York: Guilford Press.

Additional bibliographic resources are at the *Diversity Amid Globalization* Website: http//www.prenhall.com/rowntree/.

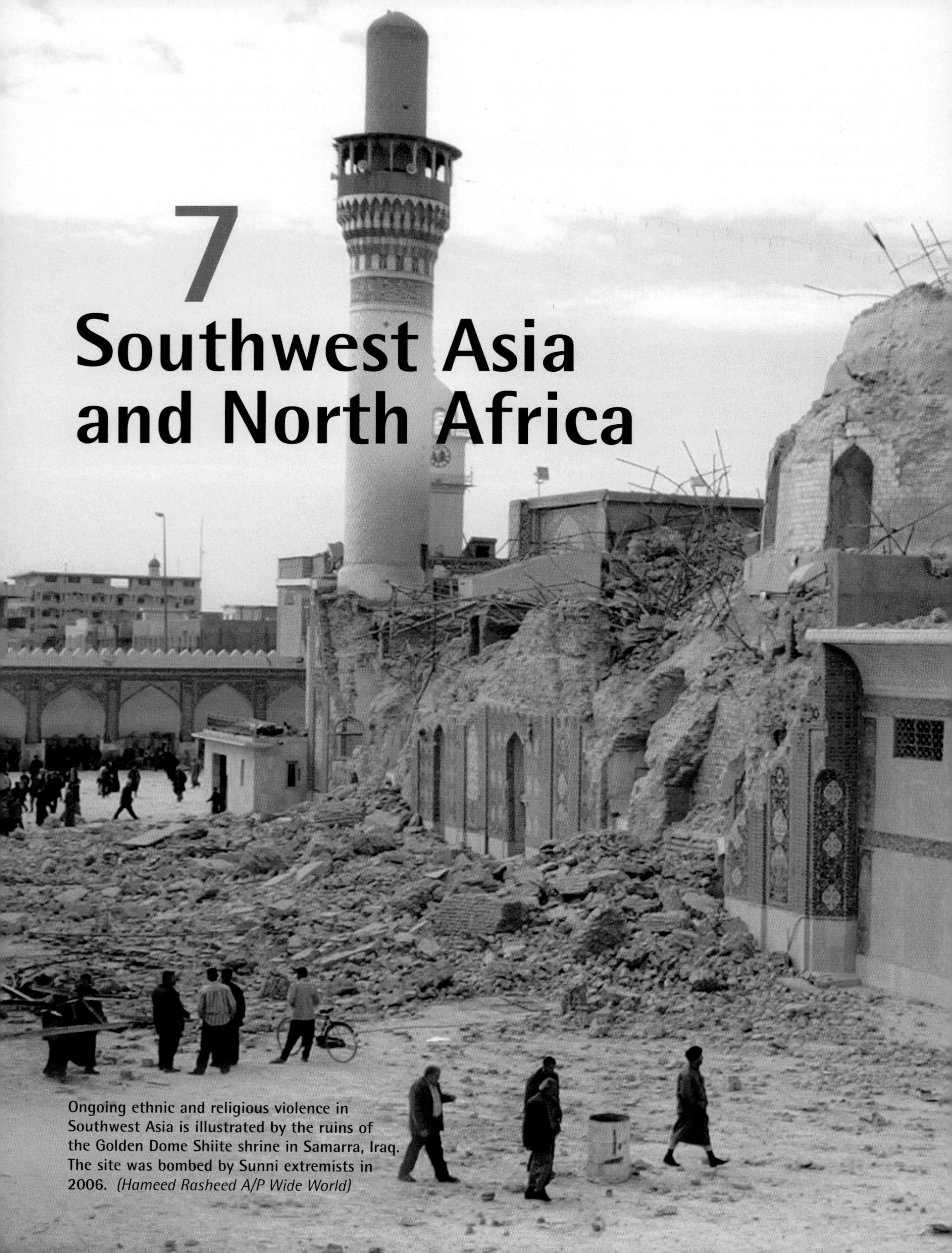

7

Southwest Asia and North Africa

Ongoing ethnic and religious violence in Southwest Asia is illustrated by the ruins of the Golden Dome Shiite shrine in Samarra, Iraq. The site was bombed by Sunni extremists in 2006. *(Hameed Rasheed A/P Wide World)*

ENVIRONMENTAL GEOGRAPHY

Water shortages are likely to increase across this arid region in the early twenty-first century as growing populations, rapid urbanization, and increasing demands for agricultural land strain already limited water supplies.

POPULATION AND SETTLEMENT

Rapid population growth in North African cities such as Algiers and Cairo is far outpacing the ability of these urban places to supply adequate housing and services.

CULTURAL COHERENCE AND DIVERSITY

The heart of the Islamic world, this region finds itself at the center of the global rise of Islamist movements that often come into conflict with Western values and traditions.

GEOPOLITICAL FRAMEWORK

Ongoing political instability is a fact of life across much of the region. Religious and ethnic differences, shifting political allegiances, and persisting economic problems all contribute to the region's geopolitical challenges.

ECONOMIC AND SOCIAL DEVELOPMENT

Changing world oil prices have a tremendous economic impact on this region, which holds some two-thirds of all petroleum reserves on the planet.

Climate, culture, and oil all help define the complex Southwest Asia and North Africa world region (see "Setting the Boundaries"). Straddling the historic meeting ground between Europe, Asia, and Africa, the region sprawls across thousands of miles of parched deserts, rugged plateaus, and oasis-like river valleys. It extends 4,000 miles (6,400 kilometers) between Morocco's Atlantic coastline and Iran's eastern boundary with Pakistan. More than two dozen nations are included within its borders, with the largest populations found in Egypt, Turkey, and Iran (Figure 7.1). Generally, its climates are arid, although the region's diverse physical geography causes precipitation to vary considerably.

Culturally, diverse languages, religions, and ethnic identities have molded land and life within the region for centuries. One traditional zone of conflict is the Middle East, where Jewish, Christian, and Islamic peoples have yet to resolve long-standing cultural tensions and political differences, particularly as they relate to the state of Israel and various Palestinian groups within the region. Iraq also has been a setting for violence as an American-led invasion toppled the regime of Saddam Hussein and sharpened religious and ethnic differences dividing the country. Nearby Iran and its development of a nuclear capability have also increased tensions. In addition, Southwest Asia and North Africa's extraordinary petroleum resources place the area in the global economic spotlight. The strategic value of oil has combined with ongoing ethnic and religious conflicts to produce one of the world's least-stable political settings, one prone to geopolitical conflict both within and between the countries of the region.

No world region has better exemplified the theme of globalization throughout history than Southwest Asia and North Africa. A key global **culture hearth**, the region witnessed many cultural innovations that subsequently diffused widely to other portions of the world. As an early center for agriculture, several great civilizations, and three major world religions, the region has been a key human crossroads for thousands of years. Important trade routes have connected North Africa with the Mediterranean and Sub-Saharan Africa. Southwest Asia also has had historical ties to Europe, the Indian subcontinent, and Central Asia. As a result, innovations within the region have often spread far beyond its bounds. For example, the domestication of wheat and cattle in Southwest Asia had far-reaching global impacts. In addition, religions born within the region (Judaism, Christianity, and Islam) have shaped many other parts of the world.

Particularly within the past century, globalization also has operated in the opposite direction: the region's strategic importance has made it increasingly vulnerable to

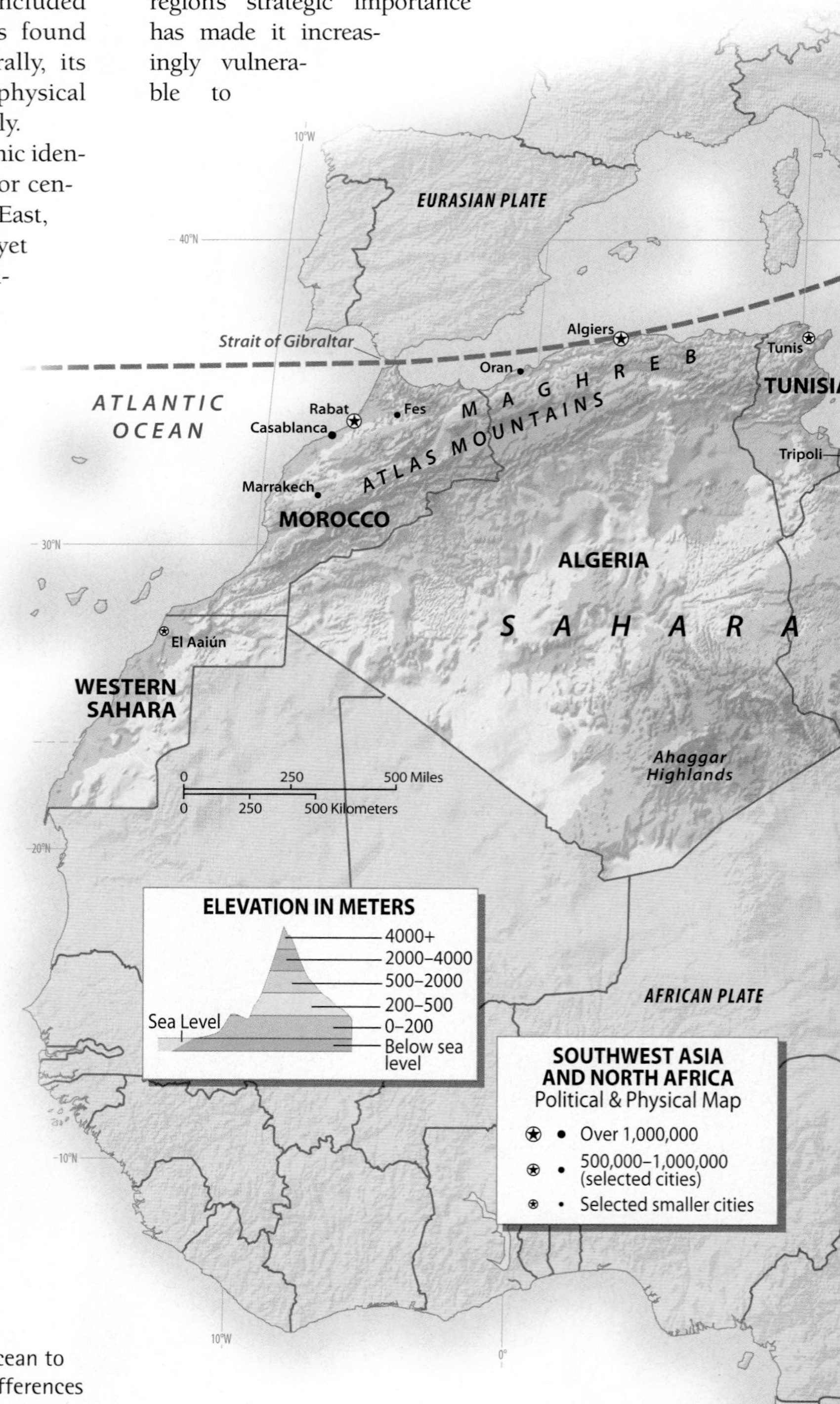

Figure 7.1 Southwest Asia and North Africa
This vast region extends from the shores of the Atlantic Ocean to the Caspian Sea. Within its boundaries, striking cultural differences and globally important petroleum reserves have contributed to recent political tensions. The region's economic geography reveals great differences in wealth, as it includes some of the world's richest and poorest nations. In addition, its fragile arid environments and growing populations pose twenty-first-century challenges.

outside influences. Traditional lifeways have been transformed (Figure 7.2). The twentieth-century development of the petroleum industry, largely initiated by U.S. and European investment, has had enormous, but selective, consequences for economic development. Global demand for oil and natural gas has powered rapid industrial change within the region, defining its pivotal role in world trade (Figure 7.3). Many key members of the **Organization of Petroleum Exporting Countries (OPEC)** are found within the region, and these countries strongly influence global prices and production levels for petroleum. Still, the regional patterns of petroleum resources are complex. Oil-rich nations such as Saudi Arabia, Kuwait, and Iran have been fundamentally transformed, while petroleum-poor neighbors such as Jordan, Lebanon, and Tunisia have seen less dramatic impacts. Politically, petroleum resources also have elevated the region's strategic importance, adding to many traditional cultural tensions within the realm.

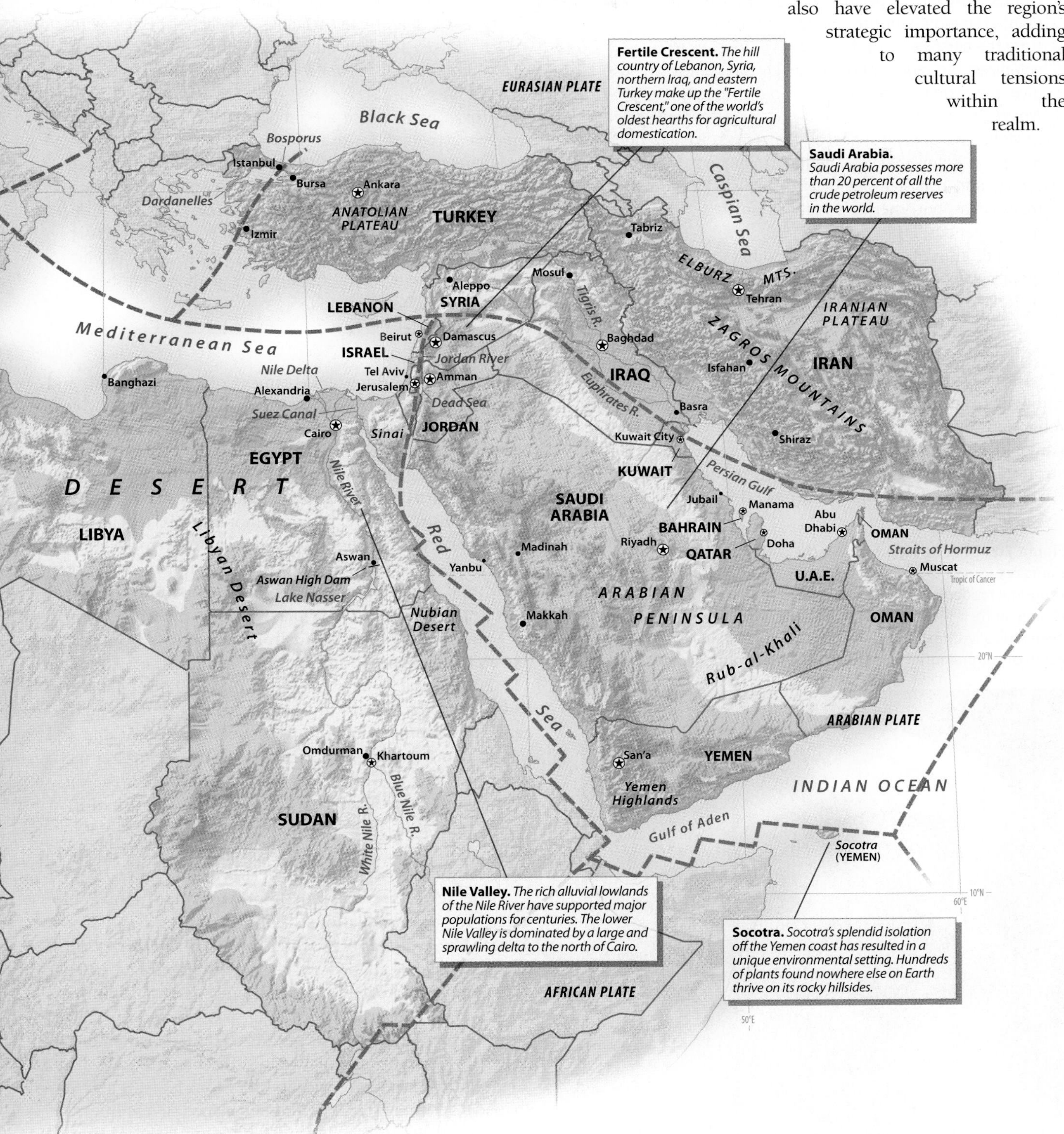

Setting the Boundaries

Southwest Asia and North Africa is both a cumbersome term and a complex region. Often the same area is simply called the *Middle East*, but some experts would exclude the western parts of North Africa as well as Turkey and Iran from such a region. In addition, the *Middle East* carries with it a peculiarly European vantage point—Lebanon is in the "middle of the east" only from the perspective of the western Europeans who colonized the region and still shape the names we give the world today. Instead, *Southwest Asia and North Africa* offers a useful and straightforward way to describe the general limits of the region. Largely arid climates, an Islamic religious heritage, abundant fossil fuel reserves, and persistent political instability are dominant in many areas, but internal variations on these themes make generalizations difficult.

There also are problems with simply defining the geographical limits of the region. To the northwest, a small piece of Turkey actually sits west of the Bosporus Strait, generally considered to be the dividing line between Europe and Asia (see Figure 7.1). The addition of Cyprus (off the coast of Turkey) to the European Union in 2004 effectively removed it from the region, and that country is now treated in the chapter on Europe (see Chapter 8). To the northeast, the largely Islamic peoples of Central Asia share many religious and linguistic ties with Turkey and Iran, but we have chosen to treat these groups in a separate chapter on Central Asia (see Chapter 10).

African borders also are problematic. The conventional regional division of "North Africa" from "Sub-Saharan Africa" often cuts directly through the middle of modern Mauritania, Mali, Niger, Chad, and Sudan. We place all of these transitional countries in Chapter 6, Sub-Saharan Africa, but also include some discussion of Sudan within our Southwest Asia and North Africa material because of its status as an "Islamic republic," a political and cultural designation that ties it strongly to the Muslim world.

More recently, **Islamic fundamentalism** has advocated a return to more traditional practices within the Muslim religion. Fundamentalists in any religion advocate a conservative loyalty to enduring beliefs within their faith and they strongly resist change. A related political movement within Islam known as **Islamism** challenges the encroachment of global popular culture and blames colonial, imperial, and Western elements for many of the region's political, economic, and social problems. Islamists resent the role they claim the West has played in creating poverty in their world, and many Islamists advocate merging civil and religious authority and rejecting modern, Western-style consumer culture. While often demonized in the Western press, particularly when associated with terrorist acts of violence, Islamic fundamentalism and Islamism can be seen as cultural and political reactions to the disruptive role that forces of globalization have played across the region.

Figure 7.2 Man at Livestock Market, Al Ayn, United Arab Emirates
Longtime residents of Southwest Asia have witnessed incredible change in their lifetimes. In many localities, traditional ways have been radically transformed by urbanization, foreign investment, and external cultural influences. *(Rob Crandall/www.robcrandall.com)*

The globally contentious role played by Al Qaeda, an extremist Islamist group, particularly since the September 11, 2001, attacks against the United States, has been linked to the region in multiple ways. Many Al Qaeda recruits have come from countries such as Saudi Arabia and Egypt, and money used to finance global terrorist attacks often has been traced back to the region. Indeed, the early twenty-first century will no doubt see the continuation of these cultural and political tensions as traditional, sometimes extremist, values are juxtaposed with the ever-increasing influence of the modern world (Figure 7.3).

The region's environment provides additional challenges for the populations of Southwest Asia and North Africa. The availability of water in this largely dry portion of the world has shaped the region's physical and human geographies. Biologically, the region's plants and animals must adapt to the aridity of long dry seasons and short, often unpredictable rainy periods. Human settlement is linked to water in similar ways. Whether it comes from precipitation, underground aquifers, or rivers, water has shaped settlement patterns and has placed severe limits on agricultural development across huge portions of the region. In the future, the region's growing population of

Figure 7.3 Saudi Arabian Oil Refinery
Eastern Saudi Arabia's Ras Tanuna Oil Refinery links the oil-rich country to the world beyond. Huge foreign and domestic investments since 1960 have dramatically transformed many other settings in the region. *(Minosa/Scorpio/Corbis/Sygma)*

more than 400 million people will stress these available resources even further and water issues will no doubt increase economic and political instability.

ENVIRONMENTAL GEOGRAPHY: Life in a Fragile World

In the popular imagination, much of Southwest Asia and North Africa is a land of shifting sand dunes, searing heat, and scattered oases. Although examples of those stereotypes certainly can be found across the region, the actual physical setting, in both landforms and climate, is considerably more complex (Figure 7.1). In reality, the regional terrain varies greatly, with rocky plateaus and mountain ranges more common than sandy deserts. Even the climate, although dominated by aridity, varies remarkably from the dry heart of North Africa's Sahara Desert to the well-watered highlands of northern Morocco, coastal Turkey, and western Iran. One theme is pervasive, however: a lengthy legacy of human settlement has left its mark on a fragile environment, and the entire region will be faced with increasingly daunting ecological problems in the decades ahead.

Legacies of a Vulnerable Landscape

The island of Socotra illustrates the region's fragile and vulnerable environment and suggests how processes of globalization may threaten the area's long-term ecological health. Socotra's stony slopes rise out of the shimmering waters of the Indian Ocean about 230 miles (368 kilometers) southeast of Yemen. The island's unique natural and cultural history has recently caught the world's attention. Separated for millions of years from the mainland of the Arabian Peninsula, Socotra's environment evolved in isolation. More than 30 percent of the island's 850 plants are found nowhere else on Earth. Exotic dragon's blood trees dot many of the island's dry and rocky hillsides, and dozens of other species have only recently been catalogued by botanists (Figure 7.4). Offshore, equally rare coral reefs and unusual fish populations have evolved as a part of the island's unique environmental setting in the wet–dry tropics of the northwestern Indian Ocean.

Socotra's unique environmental heritage now hangs in the balance. Global participants in the 1992 environmental summit at Rio de Janeiro established a $5 million international biodiversity project on Socotra, and Edinburgh's Royal Botanic Garden has conducted extensive studies on the island. Some of the island's unique plants have already been commercially developed as cosmetics and herbal remedies, and in 1998 a secretive French expedition, financed by a large pharmaceutical company, gathered hundreds of rare plants for laboratory testing. At the same time, the Yemeni government has invited international petroleum companies to explore the island's offshore oil and gas potential, and it has considered a grand scheme for developing a series of luxury tourist hotels that would forever change the island's character. In 2003, a local group of environmentally conscious residents helped establish the Socotra Eco-tourism Society, dedicated to developing sustainable tourism on the island. Only time will tell how Socotra, which is often cited as the Galápagos of the Indian Ocean, will fare in the twenty-first century. A recently developed mixed-use zoning plan and a newly paved airstrip suggest development is inevitable.

The larger environmental history of Southwest Asia and North Africa contains many examples of the clever but short-sighted legacies of its human occupants. Lengthy human settlement in a marginal land has resulted in deforestation, soil salinization and erosion, and depleted water resources (Figure 7.5). Indeed, the pace of population growth and technological change during the past century suggests that the region's vulnerable environment is destined to face even greater challenges in the twenty-first century.

Figure 7.4 Socotra's Dragon's Blood Tree
Unique to Socotra, the rare dragon's blood tree reflects the island's environmental isolation. Evolving as a part of the island's ecosystem, the tree survives in the region's dry tropical climate. *(Chris Hellier/Corbis/Bettmann)*

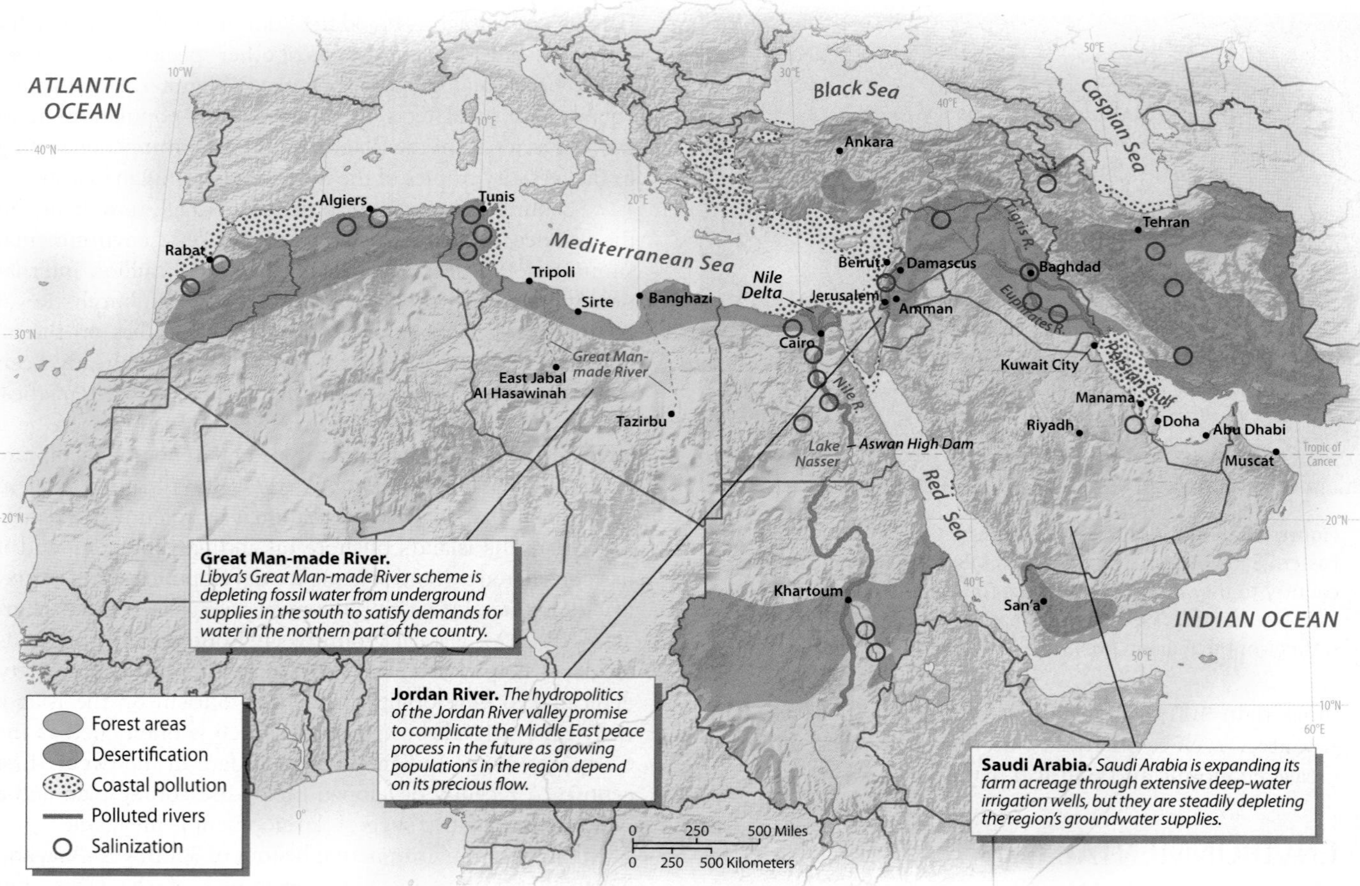

Figure 7.5 Environmental Issues in Southwest Asia and North Africa
Growing populations, pressures for economic development, and pervasive aridity combine to create environmental hazards across the region. Long human occupance has contributed to deforestation, irrigation-induced salinization, and expanding desertification. Saudi Arabia's deep-water wells, Egypt's Aswan High Dam, and Libya's Great Man-made River are all recent technological attempts to expand settlement, but they may carry a high long-term environmental price tag.

Deforestation and Overgrazing Deforestation is an ancient problem in Southwest Asia and North Africa. Although much of the region is too dry for trees, the more humid and elevated lands that ring the Mediterranean once supported heavy forests. Included in these woodlands are the cedars of Lebanon, cut during ancient times and now reduced to a few scattered groves in a largely denuded landscape. In many settings, growing demands for agricultural land caused upland forests to be removed and replaced with grain fields, orchards, and pastures.

Human activities have conspired with natural conditions to reduce most of the region's forests to grass and scrub. Mediterranean forests grow slowly, are highly vulnerable to fire, and usually fare poorly if subjected to heavy grazing. Browsing by goats in particular often has been blamed for much of the region's forest loss, but other livestock have also impacted the vegetative cover, especially in steeply sloping semiarid settings that are slow to recover from grazing. Deforestation also has resulted in a millennia-long deterioration of the region's water supplies and in accelerated soil erosion.

Some forests survive in the mountains of northern and southern Turkey, and northern Iran also retains considerable tree cover. Scattered forests can be found in western Iran, the Levant, and the Atlas Mountains. Moreover, several governments have launched reforestation drives and forest preservation efforts in recent years. For example, both Israel and Syria have expanded their coverage of wooded lands since the 1980s, and more than 5 percent of Lebanon's total area is now part of the Chouf Cedar Reserve formed in 1996 to protect old-growth cedars.

Salinization Salinization, or the buildup of toxic salts in the soil, is another ancient environmental issue in a region where irrigation has been practiced for centuries (Figure 7.5). The accumulation of salt in the topsoil is a common problem wherever desert lands are subjected to extensive irrigation. All freshwater contains a small amount of dissolved salt, and when water is diverted from streams into fields, salt remains in the soil after the moisture is absorbed by the plants and evaporated by the sun. In humid climates, accumulated salts are washed away by saturating rains, but in arid climates this rarely occurs. Where irrigation is practiced, salt concentrations build up over time, leading to lower crop yields and eventually to land abandonment.

Hundreds of thousands of acres of once-fertile farmland within the region have been destroyed or degraded by salinization. The problem has been particularly acute in Iraq, where centuries of canal irrigation along the Tigris and Euphrates rivers has seriously degraded land quality. Similar conditions plague central Iran, Egypt, and irrigated portions of the Maghreb.

Managing Water Residents of the region are continually challenged by many other problems related to managing water in one of the driest portions of Earth. As technological change has accelerated, the scale and impact of water management schemes have had a growing effect on the region's environment. Sometimes the costs are justified if large new areas are brought into productive and sustainable agricultural use. In other cases, however, the long-term environmental price will far outweigh any immediate and perhaps short-lived gains in agricultural production.

Regional populations have been modifying drainage systems and water flows for thousands of years. The Iranian **qanat system** of tapping into groundwater through a series of gently sloping tunnels was widely replicated on the Arabian Peninsula and in North Africa. With simple technology, farmers directed the underground flow to fields and villages where it could be efficiently utilized.

In the past half century, however, the scope of environmental change has been greatly magnified. Water management schemes have become huge engineering projects, major budget expenditures, and even political issues, both within and between countries of the region. One remarkable example is Egypt's Aswan High Dam, completed in 1970 on the Nile River south of Cairo (Figure 7.5). Many benefits came with the dam's completion. Most important, greatly increased storage capacity in the upstream reservoir promoted more year-round cropping and an expansion of cultivated lands along the Nile, critical changes in a country that continues to experience rapid population growth. The dam also generates large amounts of clean electricity for the region. But the environmental costs have been high. The dam has changed methods of irrigation along the Nile, greatly increasing salinization. While fresh sediments and soil nutrients once annually washed across the valley floor with high river flows, more controlled irrigation has meant greatly increased inputs of costly fertilizers. Sediments that used to move downstream now accumulate behind the dam, infilling Lake Nasser. Other problems with the dam include an increased incidence of schistosomiasis (a debilitating parasitic disease spread by waterborne snails in irrigation canals) and the collapse of the Mediterranean fishing industry near the Nile Delta, an area previously nourished by the river's silt.

Israel's "Peace Corridor" project is another massive engineering venture within the region. Designed to bring Red Sea water north into the Dead Sea, the effort received increased government backing in 2007. The ambitious plan calls for a 120-mile (200-kilometer) conduit to be built from near Aqaba to the southern end of the Dead Sea (Figure 7.6). The water could be desalinated, generate hydroelectricity, and help save the Dead Sea and its major tourist resorts from literally drying up.

Elsewhere, **fossil water**, or water supplies stored underground during earlier and wetter climatic periods, has also been put to use by modern technology. Libya's "Great Man-made River" scheme taps underground water in the southern part of the country, transports it 600 miles (965 kilometers) to

Figure 7.6 Water Issues in the Middle East
Many water-related issues complicate the geopolitical setting in the Middle East. The Jordan River system has been a particular focus of conflict. *(Modified from Soffer, 1999,* Rivers of Fire: The Conflict over Water in the Middle East, *p. 180. Reprinted by permission of Rowman and Littlefield Publishers, Inc.)*

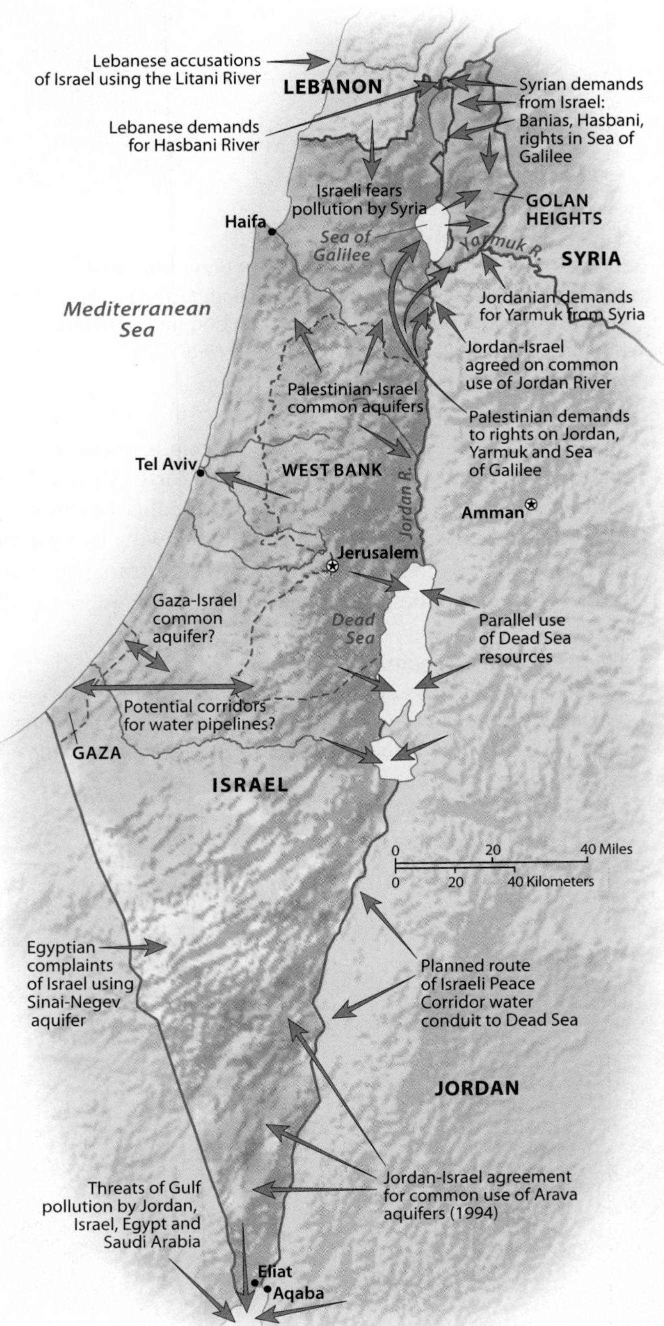

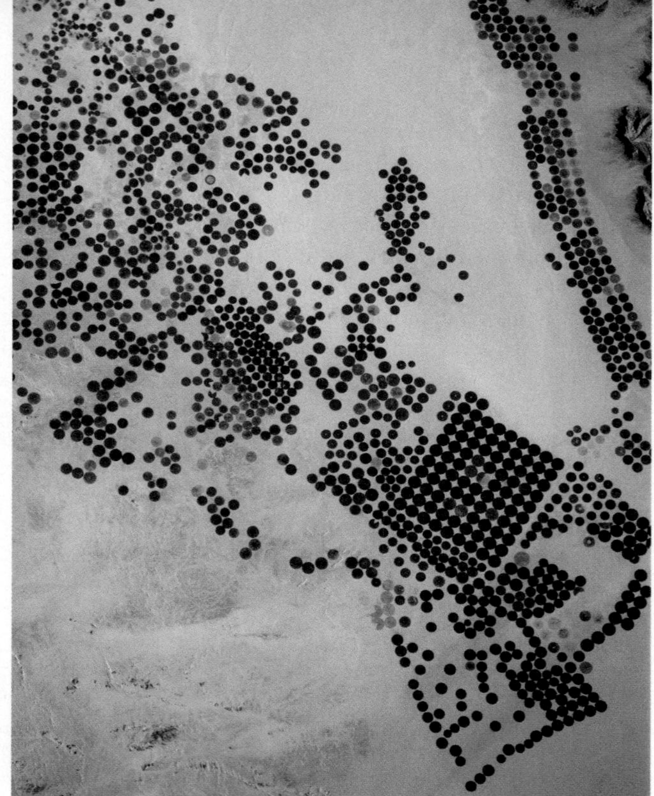

Figure 7.7 **Saudi Arabian Irrigation**
These irrigated fields in the Saudi Desert draw from wells over 4,000 feet deep. While significantly expanding the country's food production, such efforts are rapidly depleting underground supplies of fossil water. *(Digital image © 1996 Corbis; original image courtesy of NASA)*

northern coastal zones, and uses the precious resource to expand agricultural production in one of North Africa's driest countries (Figure 7.5). Similarly, Saudi Arabia has invested huge sums to develop deep-water wells, allowing it to expand its food output greatly (Figure 7.7). Unfortunately, these underground supplies are being depleted much more rapidly than they are being recharged, thus limiting the long-term sustainability of such ventures.

Most dramatically, **hydropolitics**, or the interplay of water resource issues and politics, has raised tensions between countries that share drainage basins. For example, both Sudan's plans to expand its irrigation networks along the upper Nile and Ethiopia's Blue Nile Dam project are causes of concern in Egypt. To the north, Turkey's growing development of the upper Tigris and Euphrates rivers (the Southeast Anatolian Project) has raised issues with Iraq and Syria, who argue that capturing "their" water might be considered a provocative political act. Hydropolitics also has played into negotiations between Israel, the Palestinians, and other neighboring states, particularly in the valuable Jordan River drainage, which runs through the center of the area's most hotly disputed lands (Figure 7.6). Israelis fear Palestinian and Syrian pollution; nearby Jordanians argue for more water from Syria; and all regional residents must deal with the uncomfortable reality that, regardless of their political differences, they must drink from the same limited supplies of freshwater.

Regional Landforms

A quick tour of the region's physical setting reveals a surprising diversity of environments and landforms. In North Africa, the **Maghreb** region (meaning "western island") includes the nations of Morocco, Algeria, and Tunisia and is dominated near the Mediterranean coastline by the Atlas Mountains. The rugged flanks of the Atlas rise like a series of islands above the narrow coastal plains to the north and the vast stretches of the lower Saharan deserts to the south (Figure 7.8). They reach heights of more than 13,000 feet (3,965 meters) in central Morocco and sweep eastward to dominate the physical settings of northern Algeria and Tunisia. South and east of the Atlas Mountains, interior North Africa varies between rocky plateaus and extensive

Figure 7.8 **Atlas Mountains**
Residents of Morocco's Atlas Mountains adapt to the region's steep slopes, and their Mediterranean-style agriculture takes advantage of higher rates of annual precipitation than are found in the arid deserts below. *(Abdelijalil Bounhar/AP Wide World)*

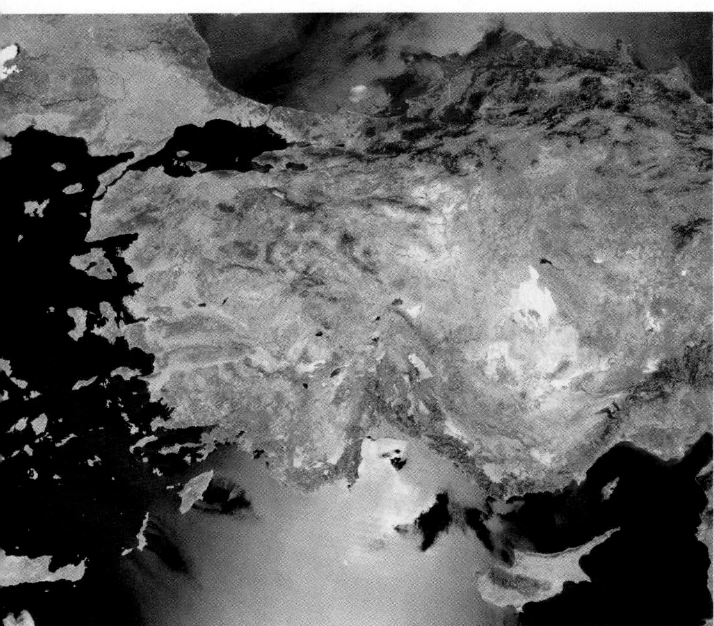

Figure 7.9 **Satellite View of Turkey**
This satellite image of Turkey suggests the varied, quake-prone terrain encountered across the Anatolian Plateau. The Black Sea coastline is visible near the top of the image and the island-studded Aegean Sea borders Turkey on the west. *(NASA/MODIS)*

lowlands. Even here, however, the desert topography is diverse: the bare and steeply sloping mountains of southern Algeria contrast markedly with the sandier reaches of the Libyan Desert and the river-carved lowlands of the Nile as it flows northward through Sudan and Egypt.

Southwest Asia is generally more mountainous than North Africa. In the **Levant**, or eastern Mediterranean region, mountains rise within 20 miles (32 kilometers) of the sea, and the highlands of Lebanon reach heights of more than 10,000 feet (3,048 meters). Farther south, the Arabian Peninsula forms a massive tilted plateau, with western highlands higher than 5,000 feet (1,524 meters) gradually sloping eastward to extensive lowlands in the Persian Gulf area. The Persian Gulf separates the Arabian Peninsula from nearby Iran, reaching its narrowest point at the strategically located Straits of Hormuz, through which pass vast amounts of the region's petroleum exports.

North and east of the Arabian Peninsula lie the two great upland areas of Southwest Asia: the Iranian and Anatolian plateaus (*Anatolia* refers to the large peninsula of Turkey, sometimes called *Asia Minor*) (Figure 7.9). Both of these plateaus average between 3,000 and 5,000 feet (915 to 1,524 meters) in elevation and are associated with tectonically active mountain uplifts and fault zones that are prone to earthquakes. One dramatic quake in western Turkey in August 1999 measured 7.4 on the Richter scale, killed more than 17,000 people, and left several hundred thousand residents homeless (see "Geographic Tools: Assessing Earthquake Hazards in Turkey"). Another quake near the Iranian city of Bam in late 2003 measured 6.8 on the Richter scale and claimed more than 40,000 lives.

Smaller lowlands characterize other portions of Southwest Asia. Narrow coastal strips are common in the Levant, along both the southern (Mediterranean) and northern (Black Sea) Turkish coastlines, and north of Iran's Elburz Mountains near the Caspian Sea. Iraq contains the most extensive alluvial lowlands in Southwest Asia, dominated by the Tigris and Euphrates rivers, which flow southeast to empty into the Persian Gulf. Although much smaller, the distinctive Jordan River Valley also is a notable lowland region that straddles the strategic borderlands of Israel, Jordan, and Syria and drains southward to the Dead Sea (Figure 7.10).

Patterns of Climate

Although often termed the *dry world*, a closer look at Southwest Asia and North Africa reveals a more complex climatic pattern (Figure 7.11). Both latitude and altitude come into play. Aridity dominates large portions of the region. A nearly continuous belt of desert lands stretches eastward from the Atlantic coast of southern Morocco across the continent of Africa, through the Arabian Peninsula, and into central and eastern Iran. Throughout this vast dry zone, plant and animal life have adapted to extreme conditions. Deep or extensive root systems and rapid life cycles allow desert plants to benefit from the limited moisture they receive. Similarly, animals adjust by efficiently storing water, hunting nocturnally, or migrating seasonally to avoid the worst of the dry cycle.

Some of the driest conditions are found across North Africa. Away from the Atlas Mountains, the Sahara Desert dominates much of the region. Much of the central Sahara receives less than 1 inch (2.5 centimeters) of rain a year and can thus support only the most meager forms of vegetation. Saharan summers feature extremely hot days and warm evenings (the world's record high, 136°F [58°C] was recorded in Libya), but winters are generally pleasant, even cool at night. Only in the tropical latitudes of southern Sudan do precipitation levels rise significantly. Here,

Figure 7.10 **Jordan Valley**
This view of the Jordan Valley shows a fertile mix of irrigated vineyards and date palm plantations. *(Corbis)*

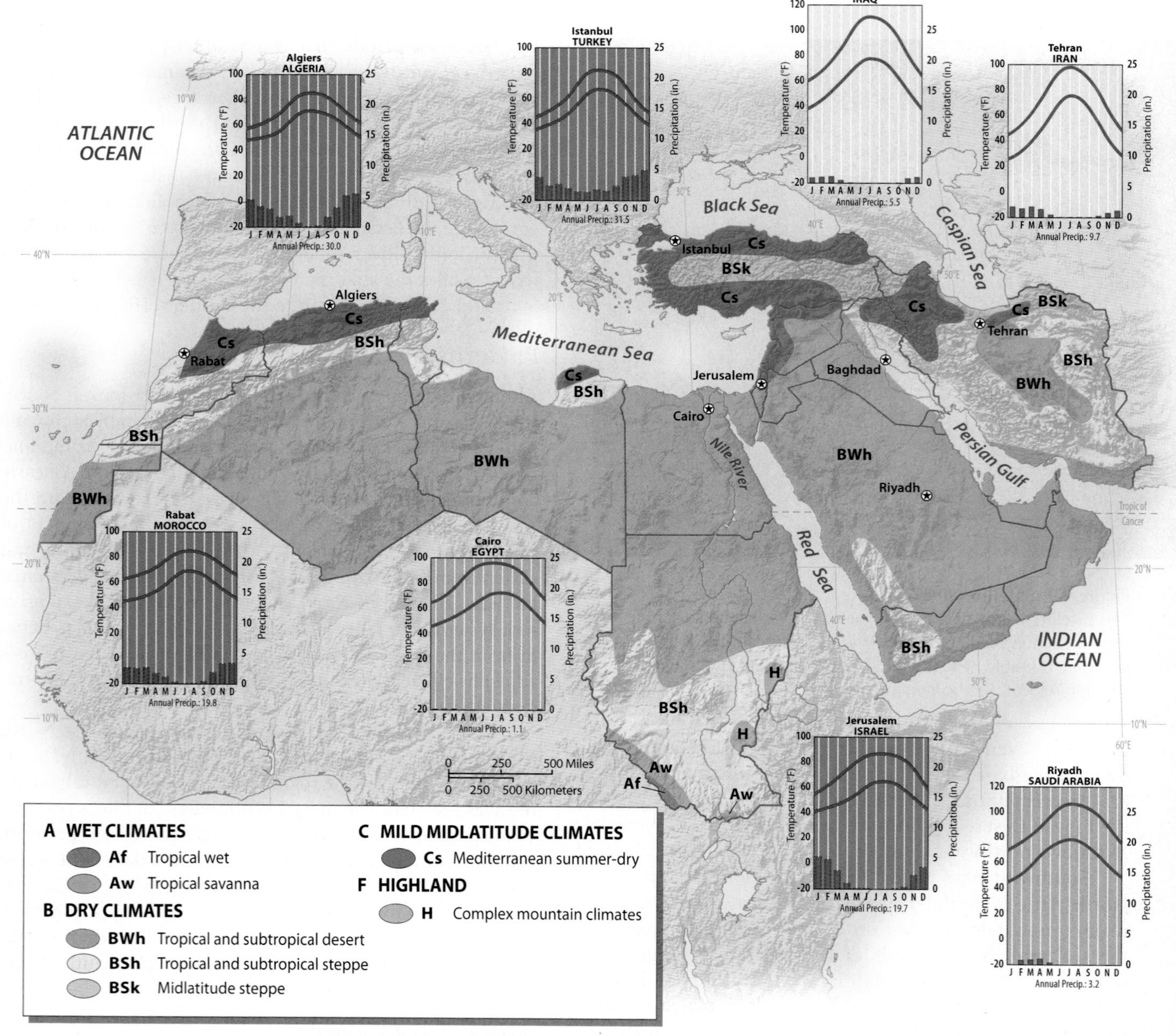

Figure 7.11 Climate Map of Southwest Asia and North Africa
Dry climates dominate from western Morocco to eastern Iran. Within these zones, persistent subtropical high-pressure systems offer only limited opportunities for precipitation. Elsewhere, mild midlatitude climates with wet winters are found near the Mediterranean Basin and Black Sea. To the south, tropical savanna climates provide summer moisture to southern Sudan.

summer rains produce between 20 and 50 inches (51 to 127 centimeters) of precipitation, and tropical savannas and woodlands replace the desert vegetation to the north.

Across the Red Sea, deserts also dominate Southwest Asia. Most of the Arabian Desert is not quite as dry as the Sahara proper, although the Rub-al-Khali along Saudi Arabia's southern border is one of the world's most desolate areas. On the fringe of the summer-monsoon belt, the Yemen Highlands along the southwestern edge of the Arabian Peninsula receive much more rain than the rest of the region and thus form a more favorable site for human habitation. To the northwest, precipitation also increases slightly in the central Tigris and Euphrates river valleys, with Baghdad, Iraq, averaging just over 5 inches (13 centimeters) of rain annually. Another major arid zone lies across Iran, a country divided into a series of minor mountain ranges and desert basins (Figure 7.12).

Elsewhere, altitude and latitude dramatically alter the desert environment and produce a surprising amount of climatic variety. For example, the Atlas Mountains and the nearby lowlands of northern Morocco, Algeria, and Tunisia experience a distinctly Mediterranean climate in which hot, dry summers alternate with cooler, relatively wet winters. In these areas, the landscape resembles that found in nearby southern Spain or Italy (Figure 7.13). A second zone of Mediterranean climate extends along the Levant coastline into the nearby mountains and northward across large portions of northern Syria, Turkey, and northwestern Iran.

GEOGRAPHIC TOOLS Assessing Earthquake Hazards in Turkey

Western Turkey was rocked by a devastating earthquake (magnitude 7.4) on August 17, 1999. Striking at 3 A.M. local time, the quake killed more than 17,000 people, injured 40,000, and collapsed more than 20,000 reinforced concrete buildings (Figure 7.2.1). While it was the worst quake to hit Turkey since 1939, the temblor offered both Turkish and American scientists an invaluable opportunity to gain knowledge and share information on both the behavior of earthquake faults and how human structures survive or fail during major quakes. Within hours of the initial quake, a team of scientists from the United States Geological Survey (USGS) were on the ground assessing damage and studying aftershocks. The USGS venture was part of a 30-year history of cooperation between the two countries. Both nations face significant earthquake hazards, and they have pooled their scientific information in order to understand the similar geology and consequences of large earthquake events. In particular, California scientists were interested in the striking geological parallels between the North Anatolian fault system in Turkey and the San Andreas fault that runs through the western portion of the Golden State.

Figure 7.2.1 Earthquake Damage in Turkey
Rescuers work to uncover victims in Golcuk just after the devastating August 1999 earthquake in western Turkey. The city, located about 60 miles (100 kilometers) east of Istanbul, experienced hundreds of collapsed buildings and thousands of casualties. *(AP/Wide World Photos)*

Once on the scene, the USGS scientists gathered information from digital seismographs stationed on site in Turkey. They also studied aerial photographs of quake-affected areas and conducted extensive field work in the worst-hit regions. They quickly assembled their own highly sensitive data-gathering equipment in order to monitor the impact of the dozens of major aftershocks that hit the area. They used soil and tectonic maps to understand how varying ground motion and acceleration seemed to hit some areas much worse than others. Their findings will help Turkey's government plan for future quakes, a certainty in this hazard-prone part of the world, as well as offer critical information for North American scientists interested in understanding earthquakes in their own backyard.

In particular, the USGS initiatives in Turkey provided key data on how major-magnitude earthquakes impact large urban areas. In the United States, San Diego, Los Angeles, San Francisco, Seattle, and Salt Lake City are seen as especially vulnerable to similar large quakes. In addition, the loss of more than 20,000 reinforced concrete buildings in Turkey has a chilling parallel in the United States. Many buildings in the United States are built in a similar fashion, particularly older structures. Even newer buildings that are up to code may not be able to withstand the strong ground movements witnessed in the Turkish quake. Finally, the Turkish example (in which repeated quakes occur along a single fault system) may help scientists in the United States predict future movements along similar systems, such as the San Andreas. In Turkey, seismic data suggest that one great quake sets the stage for the next quake, and such forecasting models are increasingly being applied to the similar California setting.

The Uncertainties of Regional Climate Change

Projected changes in global climate will tend to aggravate already-existing environmental issues within North Africa and Southwest Asia. Temperature changes are predicted to have a greater impact on the region than changes in precipitation. The already arid and semiarid region will likely remain relatively dry, but warmer average temperatures are likely to have several major consequences. First, higher overall evaporation rates and lower overall soil moisture across the region will more likely stress crops, grasslands, and other

Figure 7.12 Arid Iran
Large portions of Iran are arid. Settlements are strongly oriented around zones of higher moisture or opportunities for irrigated agriculture. This view of a remote Iranian desert resembles the dry, basin-and-range landscape of the arid American West. *(Cory Langley)*

Figure 7.13 Algerian Orange Harvest
The Mediterranean moisture in northern Algeria produces an agricultural landscape similar to that of southern Spain or Italy. Winter rains create a scene that contrasts sharply with deserts found elsewhere in the region. *(Dott. Giorgio Gualco/Bruce Coleman Inc.)*

vegetation. Semiarid lands are particularly vulnerable, especially dryland cropping systems that cannot depend on irrigation. Even in irrigated zones, higher temperatures are likely to reduce yields for crops such as wheat and maize. Second, warmer temperatures will likely reduce net runoff into the region's already stressed streams and rivers, potentially reducing hydroelectric potential and water that is available for the region's increasingly urban population. Third, there will be a higher likelihood of extreme weather events, such as record-setting summertime temperatures. These extreme events will undoubtedly lead to more heat-related deaths as residents struggle to adapt, particularly in urban settings where many people cannot afford air conditioning.

Sea level changes will pose special threats to the Nile Delta (Figure 7.14). This portion of northern Egypt is a vast, low-lying landscape of settlements, farms, and marshland. Studies that simulate rising sea levels reveal that much of the region will be lost, either to inundation, erosion, or salinization. Farmland losses of more than 250,000 acres (100,000 hectares) are quite possible with even modest sea level changes. In the Egyptian city of Alexandria, $30 billion in losses have been projected because sea level changes will devastate the city's huge resort industry as well as nearby residential and commercial areas. Overall, experts suggest that a rise of one meter (3.3 feet) in sea level could displace 8 million Egyptians in coastal and delta settings.

Some experts have also attempted to estimate the broader political and economic costs associated with potential climate changes within the region. For example, given the political instability of the Middle East, even relatively small changes in water supplies, particularly where they might involve several nations, could significantly add to the potential for conflict within the region. Projected economic implications of climate change may depend on the relative importance of agriculture within different nations. In addition, wealthier nations such as Israel and

Figure 7.14 Nile Valley
This satellite image of the Nile Valley dramatically reveals the impact of water on the North African desert. Cairo lies at the southern end of the delta, where it begins to widen toward the Mediterranean Sea. The coastal city of Alexandria sits on the northwest edge of the low-lying Nile Delta. *(NASA/MODIS)*

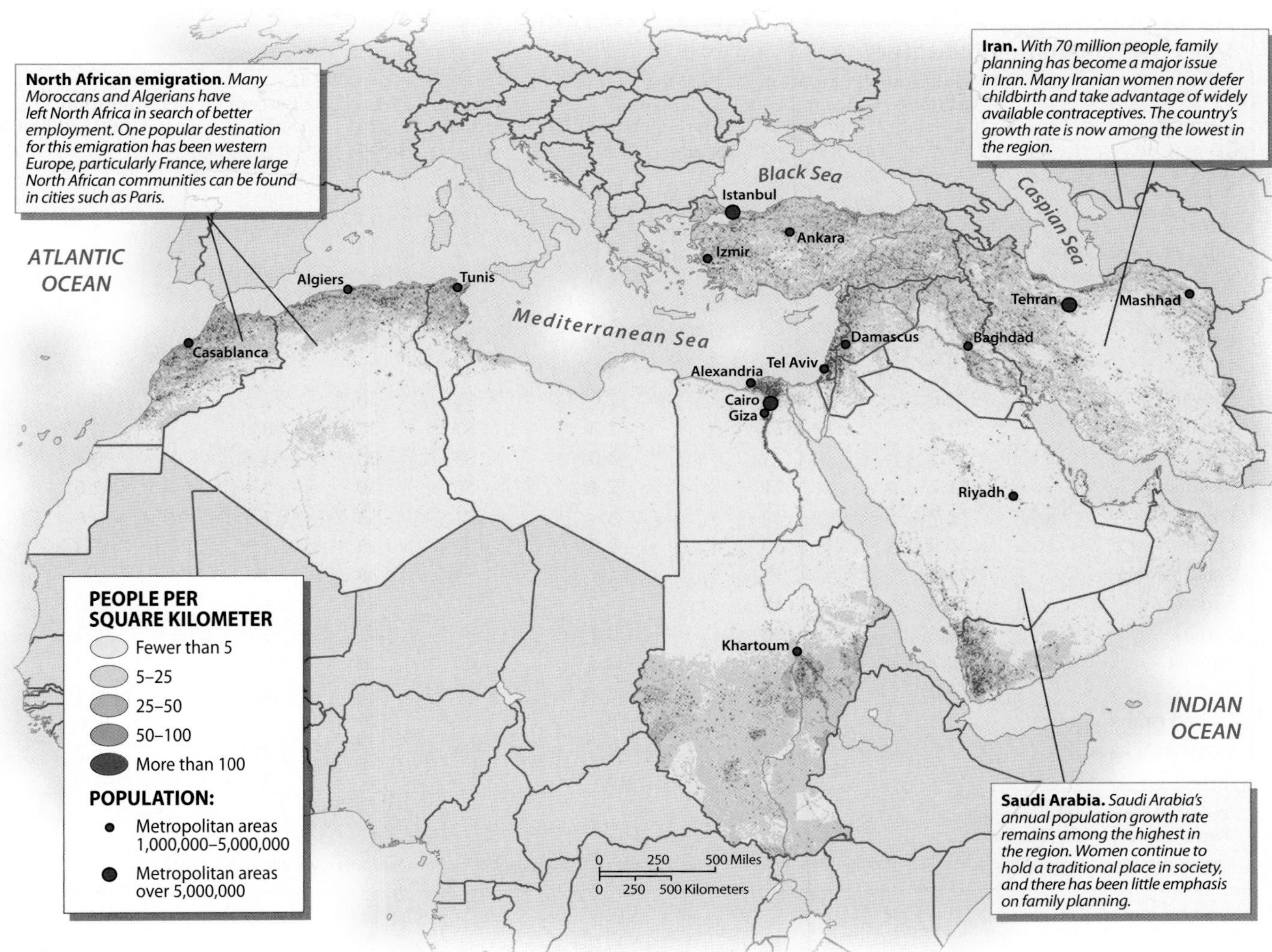

Figure 7.15 Population Map of Southwest Asia and North Africa
The striking contrasts between large, sparsely occupied desert zones and much more densely settled regions where water is available are clearly evident. The Nile Valley and the Maghreb region contain most of North Africa's people, while Southwest Asian populations cluster in the highlands and along the better-watered shores of the Mediterranean.

Saudi Arabia may have more available resources to plan, adjust, and adapt to climate shifts and extreme events versus poorer, less-developed countries such as Yemen, Syria, or Sudan.

POPULATION AND SETTLEMENT: Changing Rural and Urban Worlds

The geography of human population across Southwest Asia and North Africa demonstrates the intimate tie between water and life in this part of the world. The pattern is complex: large areas of the population map remain almost devoid of permanent settlement, while more moisture-favored lands suffer increasingly from problems of crowding and overpopulation (Figure 7.15). Almost everywhere across the region, humans have uniquely adapted themselves to living within arid or semiarid settings.

The Geography of Population

Today, more than 400 million people live in Southwest Asia and North Africa (Table 7.1). The distribution of that population is strikingly varied. In countries such as Egypt, large zones of almost empty desert land stand in sharp contrast to crowded, well-watered locations, such as those along the Nile River. While the overall population density in such countries appears modest, the **physiological density**, which is the number of people per unit of arable land, is among the highest on Earth. Patterns of urban geography also are highly uneven: although less than two-thirds of the overall population is urban, many nations are overwhelmingly dominated by huge and sprawling cities that produce the same problems of urban crowding found elsewhere in the developing world. Rates of recent urban growth have been phenomenal: Cairo, a modest-sized city of 3.5 million people in the 1960s, has quadrupled in population in the past 40 years.

TABLE 7.1 Population Indicators

Country	Population (Millions, 2007)	Population Density (per Square Kilometer)	Total Fertility Rate	Percent Urban	Percent <15[a]	Percent >65[b]	Net Migration (per 1,000, 2000–05)
Algeria	34.1	14	2.4	58	30	5	–0.6
Bahrain	0.8	1,098	2.6	100	27	3	0.0
Egypt	73.4	73	3.1	43	33	5	–1.3
Gaza and West Bank	4.0	666	4.6	72	44	3	–2.3
Iran	71.2	43	2.0	67	29	5	–4.1
Iraq	29.0	66	4.9	67	42	3	1.8
Israel	7.3	332	2.8	92	28	10	4.9
Jordan	5.7	64	3.5	82	37	3	3.7
Kuwait	2.8	156	2.6	98	26	2	19.5
Lebanon	3.9	377	2.3	87	27	8	–2.0
Libya	6.2	4	3.0	85	30	4	0.4
Morocco	31.7	71	2.4	55	30	5	–2.6
Oman	2.7	9	3.4	71	33	3	–12.8
Qatar	0.9	80	2.8	100	23	1	42.3
Saudi Arabia	27.6	13	4.1	81	38	2	2.2
Sudan	38.6	15	4.5	41	41	4	–3.0
Syria	19.9	108	3.5	50	37	3	–0.3
Tunisia	10.2	62	2.0	65	26	6	–0.4
Turkey	74.0	94	2.2	66	28	6	–0.7
United Arab Emirates	4.4	53	2.7	74	20	1	49.6
Western Sahara	0.5	2	2.9	92	31	2	6.2
Yemen	22.4	42	6.2	26	47	2	–1.0

[a]*Percentage of population younger than 15.*
[b]*Percentage of population older than 65.*
Sources: Population Reference Bureau, World Population Data Sheet, 2007; *net migration rate data from* UN International Migration, 2006.

Across North Africa, two dominant clusters of settlement, both shaped by the availability of water, account for most of the region's population (see Figure 7.15). In the Maghreb, the moister slopes of the Atlas Mountains and nearby better-watered coastal districts have accommodated denser populations for centuries. Today, concentrations of both rural and urban settlement extend from south of Casablanca in Morocco to Algiers and Tunis on the shores of the southern Mediterranean. Indeed, most of the populations of Morocco, Algeria, and Tunisia crowd into this crescent, a stark contrast to the almost empty lands south and east of the Atlas Mountains. Casablanca and Algiers are the largest cities in the Maghreb. They have rapidly growing metropolitan populations of 3 to 4 million residents each. Farther east, much of Libya and western Egypt is very thinly settled. Egypt's Nile Valley, however, is home to the other great North African population cluster. The vast majority of Egypt's 75 million people live within 10 miles of the river. Optimistic politicians and planners are creating a second corridor of denser settlement in Egypt's "New Valley" west of the Nile by diverting water from Lake Nasser into the Western Desert.

Most Southwest Asian residents are clustered in favored coastal zones, moister highland settings, and desert localities where water is available from nearby rivers or subsurface aquifers. High population densities are found in better-watered portions of the eastern Mediterranean (Israel, Lebanon, and Syria) and Turkey. Nearby Iran is home to 70 million residents, but population densities vary considerably from thinly occupied deserts in the east to more concentrated settlements near the Caspian Sea and across the more humid highlands of the northwest. Turkey's Istanbul (formerly Constantinople) and Iran's Tehran are Southwest Asia's largest urban areas, and both have grown in recent years as rural populations gravitate toward the economic opportunities of these cities. Elsewhere, sizable populations are scattered through the Tigris and Euphrates valley, the Yemen Highlands, and near oases where groundwater can be tapped to support agricultural or industrial activities.

Water and Life: Rural Settlement Patterns

Water and life are closely linked across the rural settlement landscapes of Southwest Asia and North Africa. Indeed, the diverse environments of Southwest Asia, in particular, are home to one of the world's earliest hearths of **domestication**, where plants and animals were purposefully selected and bred for their desirable characteristics. Beginning around 10,000 years ago, increased experimentation with wild varieties of wheat and barley led to agricultural settlements that later included domesticated animals, such as cattle, sheep, and goats. Much of the early agricultural activity fo-

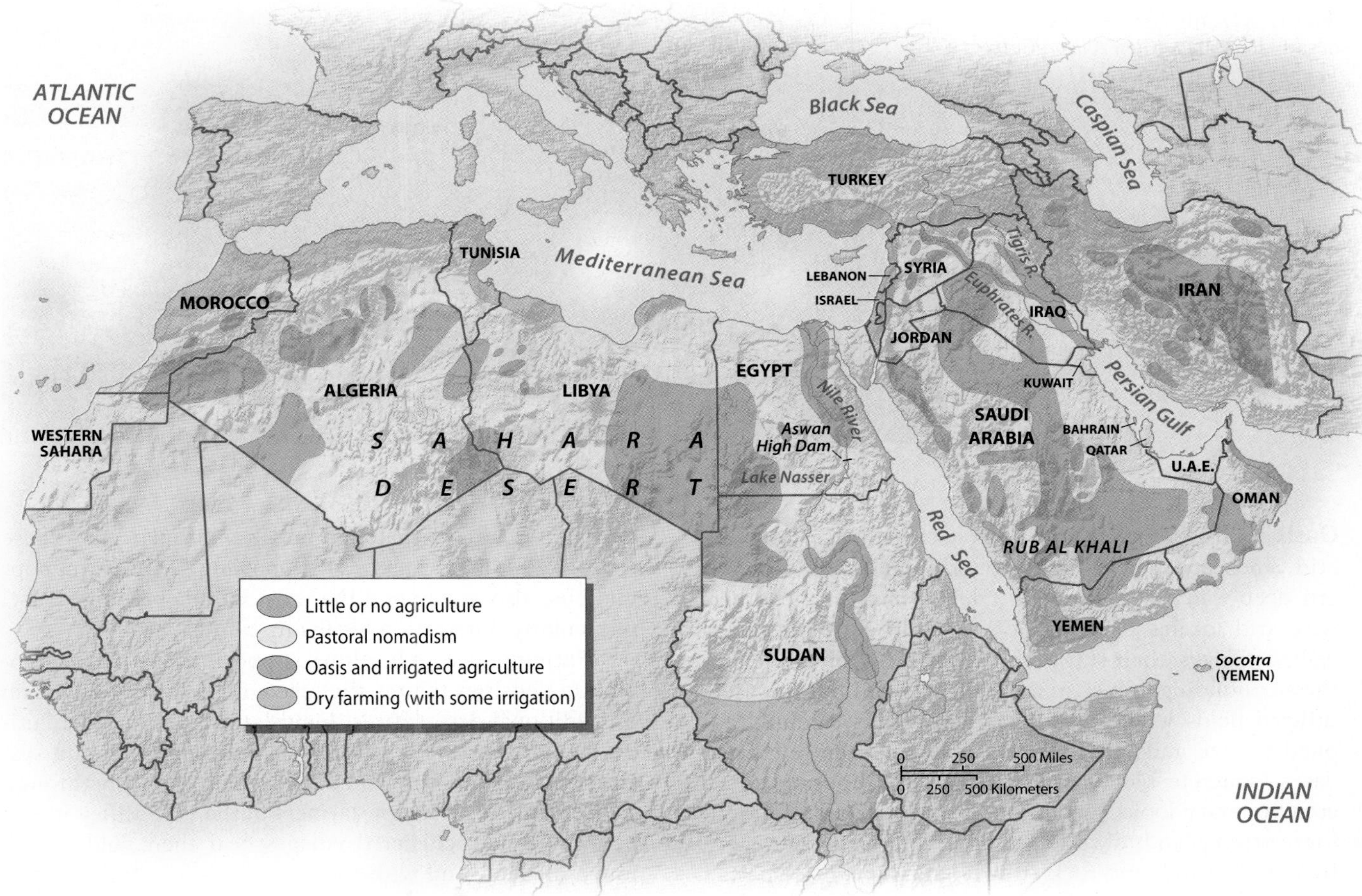

Figure 7.16 Agricultural Regions of Southwest Asia and North Africa
Important agricultural zones include oases and irrigated farming where water is available. Elsewhere, dry farming supplemented with irrigation is practiced in midlatitude settings. *(Modified from Clawson and Fisher, 2004,* World Regional Geography, *8th ed., Upper Saddle River, NJ: Prentice Hall, and Bergman and Renwick, 1999,* Introduction to Geography, *Upper Saddle River, NJ: Prentice Hall)*

cused on the **Fertile Crescent**, an ecologically diverse zone that stretches from the Levant inland through the fertile hill country of northern Syria into Iraq. Between 5,000 and 6,000 years ago, better knowledge of irrigation techniques and increasingly centralized political states promoted the spread of agriculture into nearby valleys such as the Tigris and Euphrates (Mesopotamia) and North Africa's Nile Valley. Since then, different peoples of the region have adapted to its environmental diversity and limitations in distinctive ways. In the process, they have practiced forms of agriculture appropriate to their settings and have left their own unique imprints upon the landscape (Figure 7.16).

Pastoral Nomadism Most common in the drier portions of the region, **pastoral nomadism** is a traditional form of subsistence agriculture in which practitioners depend on the seasonal movement of livestock for a large part of their livelihood. Arabian Bedouins, North African Berbers, and Iranian Bakhtiaris provide surviving examples of nomadism within the realm. Today, however, with fewer than 10 million nomads remaining, the lifestyle is in decline, the victim of constricting political borders, a reduced demand for traditional beasts of burden such as camels, competing land uses, and selective overgrazing. In addition, government resettlement programs in Saudi Arabia, Syria, Egypt, and elsewhere are actively promoting a more settled lifestyle for many nomadic groups.

The settlement landscape of pastoral nomads reflects their need for mobility and flexibility as they seasonally move camels, sheep, and goats from place to place. Near highland zones such as the Atlas Mountains or the Anatolian Plateau, nomads practice **transhumance** (the seasonal movement of animals between wet-season and dry-season pastures) by herding their livestock to cooler, greener high country pastures in the summer and then returning them to valley and lowland settings for fall and winter grazing. Elsewhere, seasonal movements often involve huge territories of desert to support small groups of a few dozen families. In addition, nomads trade with sedentary agricultural populations, a mutually beneficial relationship in which they exchange meat, milk, hides, and wool for cereal and orchard crops available at desert oases.

Figure 7.17 Oasis Agriculture The green fields and trees of Morocco's Tinghir oasis contrast dramatically with the surrounding desert landscape. *(Glen Allison/Getty Images Inc.–Stone Allstock)*

Oasis Life Permanent oasis settlements dot the arid landscape where high groundwater levels or modern deep-water wells provide reliable moisture in otherwise arid locales (Figure 7.17). Tightly clustered, often walled villages, their sun-baked mud houses blending into the surrounding scene, sit adjacent to small but intensely utilized fields where underground water is carefully applied to tree and cereal crops. In more recently created oases, concrete blocks and prefabricated housing add a contemporary look. Surrounded by large zones of desert, these green islands of rural activity stand out in sharp contrast to the sand- and rock-strewn landscape.

Traditional oasis settlements are composed of close-knit families who work their own irrigated plots or, more commonly, work for absentee landowners. Although oases are usually small, eastern Saudi Arabia's Hofuf oasis covers more than 30,000 acres (12,150 hectares). While some crops are raised for local consumption, commercial trade always has played a role in such settings. In the past century, the expanding world demand for products such as figs and dates has included even these remote oases in the global economy, and many products end up on the tables of hungry Europeans or North Americans. New drilling and pumping technologies, particularly in Saudi Arabia, have added to the size and number of oasis settlements. But oasis life across the region faces major challenges. Population growth, groundwater depletion, and the pressures of global cultural change threaten the economic and social integrity of these settlements.

Settlement Along Exotic Rivers For centuries, the region's densest rural settlement has been tied to its great river valleys and their seasonal floods of water and enriching nutrients. In such settings, **exotic rivers** transport precious water and nutrients from distant, more humid lands into drier regions, where the resources are utilized for irrigated farming. The Nile and the Tigris and Euphrates rivers are the largest regional examples of such activity, and both have large, densely settled deltas. Other linear irrigated settlements can be found near the Jordan River in Israel and Jordan, along short streams originating in North Africa's Atlas Mountains, and on the more arid peripheries of the Anatolian and Iranian plateaus. These settings, although capable of supporting sizable rural populations, also are among the most vulnerable to overuse, particularly if irrigation results in salinization.

Farming in such localities supports much higher population densities than is the case with pastoral nomadism or traditional desert oases. Fields are small, intensely utilized, and connected with closely managed irrigation systems designed to store and move water efficiently through the settlement. In Egypt, farmers of the Nile Valley live in densely settled, clustered villages near their fields; work much of their land with the same tools and technologies their ancestors used; and grow a mix of cotton, rice, wheat, and forage crops. The scale and persistence of their agricultural imprint is dramatically visible from space (Figure 7.14). But rural life is also changing in such settings. New dam- and canal-building schemes in Egypt, Israel, Syria, Turkey, and elsewhere are increasing the storage capacity of river systems, which allows for more year-round agricultural activity. Higher-yielding rice and wheat varieties and more mechanized agricultural methods also have raised food output, particularly in places such as Egypt and Israel. Some of the most efficient farms in the region are associated with Israeli **kibbutzes**, collectively worked settlements that produce grain, vegetable, and orchard crops irrigated by waters from the Jordan River and from the country's elaborate feeder canals. For example, Israel's National Water Carrier system takes water from the Sea of Galilee and moves it south and west, where intensively worked agricultural operations produce key food crops for nearby Tel Aviv and Jerusalem.

The Challenge of Dryland Agriculture Mediterranean climates in portions of the region permit varied forms of dryland agriculture that depend largely on seasonal moisture to support farming. These zones include the better-watered valleys and coastal lowlands of the northern Maghreb, lands along the shore of the eastern Mediterranean, and favored uplands across the Anatolian and Iranian plateaus. A regional variant of dry farming is practiced across Yemen's terraced highlands in the moister corners of the southern Arabian Peninsula, and in a smaller zone of Oman just east of the Persian Gulf.

Often, a variety of crops and livestock surrounds the Mediterranean villages of the region. Drought-resistant tree crops are common, with olive groves, almond trees, and citrus orchards producing output for both local consumption and commercial sale. Elsewhere, favored locations support grape vineyards, while more marginal settings are used to grow wheat and barley or to raise forage crops to feed cattle, sheep, and goats. Vulnerability to drought and the availability of more sophisticated water management strategies are leading some Mediterranean farmers to utilize more irrigated cropping, thus improving production of cotton, wheat, citrus fruits, and tobacco across portions of the region. More mechanization, crop specialization, and fertilizer use are also transforming such agricultural settings, following a pattern set earlier in nearby areas of southern Europe. One commercial adaptation of growing regional and global importance is Morocco's flourishing hashish crop. More than 200,000 acres (80,000 hectares) of cannabis are cultivated in the hill country near Ketama in northern Morocco, generating more than $2 billion annually in illegal exports (mostly to Europe).

Figure 7.18 The Islamic Landscape
Iranian mullahs discuss the religious questions of the day beneath the minarets of the Moussavi Mosque in Qom. Islam has left a widespread mark on the region's cultural landscapes. *(S. Franklin/Corbis/Sygma)*

Figure 7.19 The Old City of Fes, Morocco
These narrow streets in the old city of Fes are lined with shops and commercial stalls. The winding labyrinth of alleyways and cul-de-sacs forms an almost impenetrable maze to visitors, and it exemplifies the complex urban landscapes of the traditional Islamic city. *(Dave Bartruff/Corbis/Bettmann)*

Many-Layered Landscapes: The Urban Imprint

Cities have played a pivotal role in the region's human geography. Indeed, some of the world's oldest urban places are located in the region. Today, enduring political, religious, and economic ties link the city and countryside.

A Long Urban Legacy Cities in the region traditionally have played important functional roles as centers of political and religious authority, as well as key focal points of local and long-distance trade. Urbanization in Mesopotamia (modern Iraq) began by 3500 BCE, and cities such as Eridu and Ur reached populations of 25,000 to 35,000 residents. Similar centers appeared in Egypt by 3000 BCE, with Memphis and Thebes assuming major importance amid the dense populations of the middle Nile Valley. These ancient cities were key centers of political and religious control. Temples, palaces, tombs, and public buildings dominated the urban landscapes of such settlements, and surrounding walls (particularly in Mesopotamia) offered protection from outside invasion. By 2000 BCE, however, a different kind of city was emerging, particularly along the shores of the eastern Mediterranean and at the junction points of important caravan routes. Centers such as Beirut, Tyre, and Sidon, all in modern Lebanon, as well as Damascus in nearby Syria, exemplified the growing role of trade in creating urban landscapes. Expanding port facilities, warehouse districts, and commercial thoroughfares suggested how trade and commerce shaped these urban settlements, and many of these early Middle Eastern trading towns have survived to the present.

Islam also left a lasting mark because cities traditionally served as centers of Islamic religious power and education. By the eighth century, Baghdad had emerged as a religious center, followed soon thereafter by the appearance of Cairo as a seat of religious authority and expansion. Urban settlements from North Africa to Turkey felt the influences of Islam. Indeed, the Islamic Moors carried its characteristic Muslim signature to Spain, where it shaped urban centers such as Córdoba and Málaga. Islam's impact upon the settlement landscape merged with older urban traditions across the region and established a characteristic Islamic cityscape that exists to this day. Its traditional attributes include a walled urban core, or **medina**, dominated by the central mosque and its associated religious, educational, and administrative functions (Figure 7.18). A nearby bazaar, or *suq*, functions as a marketplace where products from city and countryside are traded (Figure 7.19). Housing districts feature an intricate maze of narrow, twisting streets that maximize shade and accentuate the privacy of residents, particularly women. Houses have small windows, frequently are situated on dead-end streets, and typically open inward to private courtyards often shared by extended families with similar ethnic or occupational backgrounds.

More recently, European colonialism added another layer of urban landscape features in selected cities. Particularly in North Africa, coastal and administrative centers during the late nineteenth century added dozens of architectural features from Britain and France. Victorian building blocks, French mansard roofs, suburban housing districts, and wide European-style commercial boulevards complicated the settlement landscapes of dozens of cities, both old and new. Centers such as Algiers (French), Fes (French) and Cairo (British) vividly displayed the effects of colonial control, and many of these urban signatures remain today.

Signatures of Globalization

Since 1950, dramatic new forces have transformed the urban landscape. Cities have become key gateways to the global economy. As the region has been opened to new investment, industrialization, and tourism, the urban landscape reflects the fundamental changes taking place. Expanded airports, commercial and financial districts, industrial parks, and luxury tourist facilities all mark the impress of the global economy.

Further, as urban centers become focal points of economic growth, surrounding rural populations are drawn to the new employment opportunities, thus fueling rapid population increases. The results are both impressive and problematic. Many traditional urban centers, such as Algiers and Istanbul, have more than doubled in size in recent years. Booming demand for homes has produced ugly, cramped high-rise apartment houses in some government-planned neighborhoods, while elsewhere sprawling squatter settlements provide little in the way of quality housing or municipal services.

Crowded Cairo, now with 15 million people, exemplifies the pattern of urban expansion. The city is now much more densely settled, as well as larger in size (Figure 7.20). Its legendary "City of the Dead" neighborhood, home to almost 1 million residents, is an urban cemetery intermingled with homes and apartment houses, a vivid acknowledgment of the premium put on living space. Many central-city neighborhoods have become so crowded and congested that wealthier urbanites have left, moving to outlying suburbs with large villas and lower population densities. Other suburbs are sites for major industrial expansion, often financed through foreign investment. In some cases, entirely

Figure 7.20 The Growth of Cairo
Today a city of more than 15 million people, Cairo has grown immensely in area since its Islamic origins more than 12 centuries ago. Just since 1930, the city has sprawled in all directions from its roots along the Nile, and now the metropolitan area includes a growing number of industrial and residential suburbs. *(Modified from Max Rodenbeck, 1999,* Cairo: The City Victorious, *p. x. Reprinted by permission of Alfred A. Knopf, Inc.)*

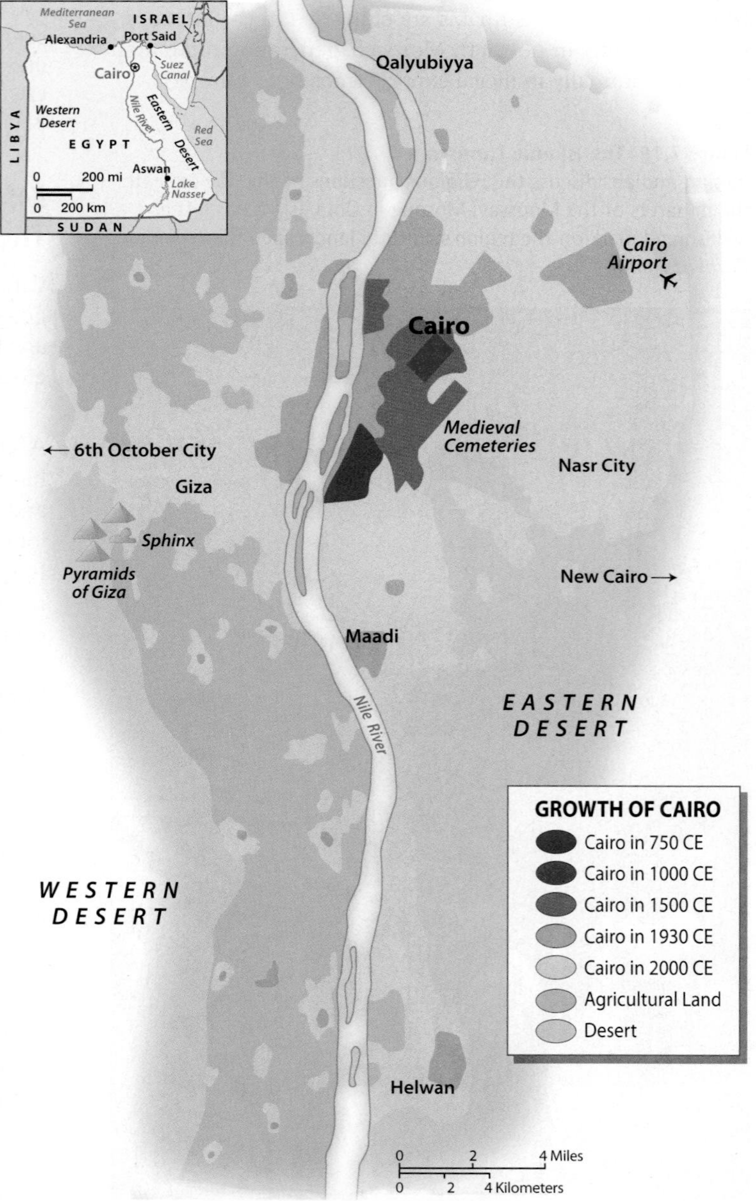

new industrial centers are being established, part of Egypt's ambitious New-Town Program.

Certainly, the oil-rich states of the Persian Gulf display the most extraordinary changes in the urban landscape (Figure 7.21). Before the twentieth century, urban traditions were relatively weak in the area, and even as late as 1950 only 18 percent of Saudi Arabia's population lived in cities. All that changed, however, as the global economy's demand for petroleum mushroomed. Today, the Saudi Arabian population is more urban than those of many industrialized nations, including the United States, and the capital city of Riyadh has grown to more than 3 million people. Particularly after 1970, other cities, such as Abu Dhabi (United Arab Emirates), Doha (Qatar), and Kuwait City (Kuwait), grew in size and took on modern Western characteristics including futuristic architecture and new transportation infrastructure. In addition, investments in petrochemical industries have fueled the creation of new urban centers, such as Jubail along Saudi Arabia's Persian Gulf coastline. The result is an urban settlement landscape in which traditional and global influences intermingle, producing cityscapes where domed mosques, resort complexes, mirrored bank buildings, and oil refineries coexist beneath the dusty skies and desert sun.

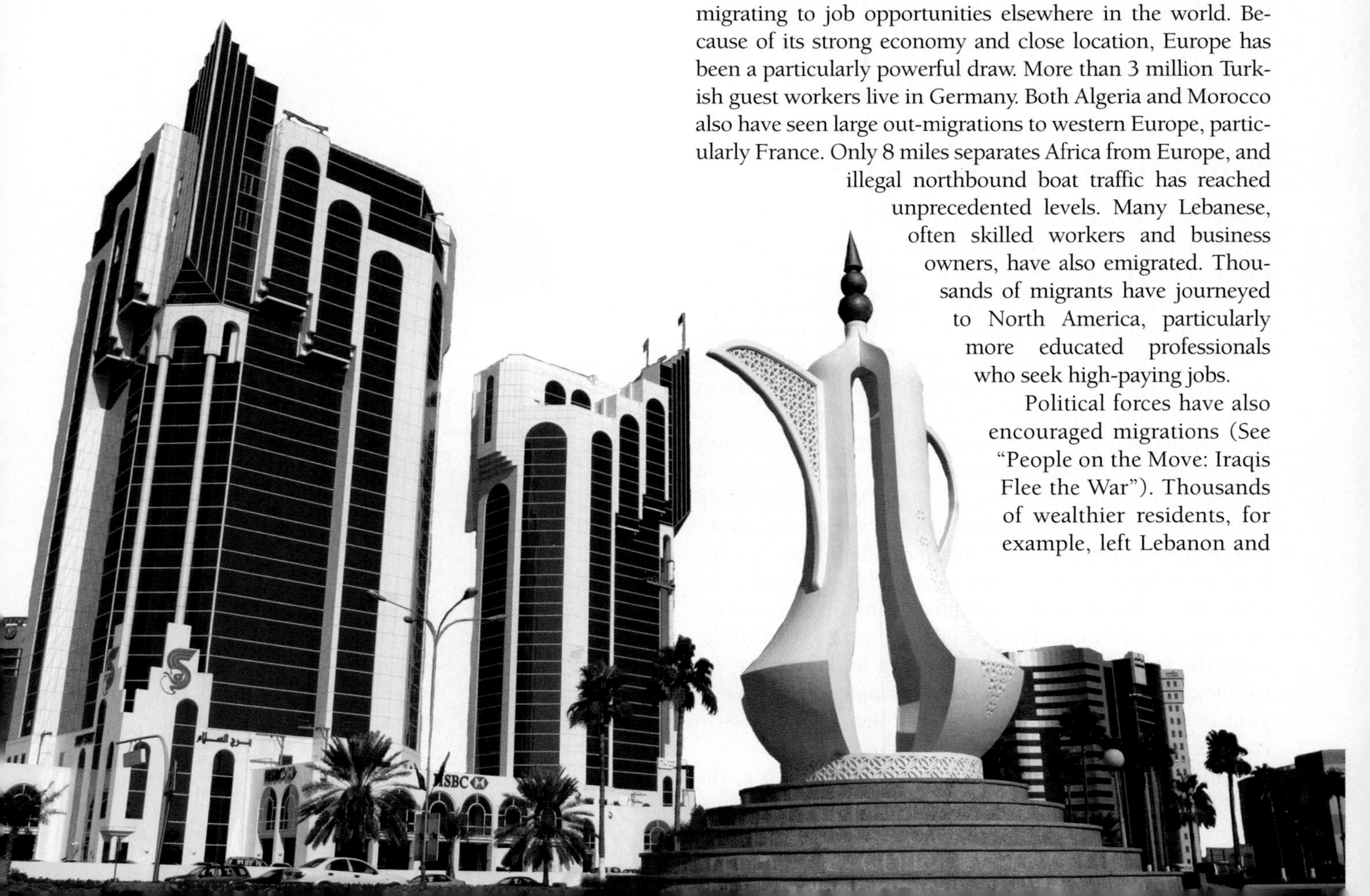

Figure 7.21 Modern Doha, Qatar
The rapidly changing urban landscapes of the oil-rich Persian Gulf region are illustrated in the modern architecture seen in Doha's West Bay district. *(Michel Setboun/Corbis)*

Recent Migration Patterns

While pastoral nomads have crisscrossed the region for ages, entirely new patterns of migration have been sparked by the dynamic global economy and by recent political events. Several major migration streams have profoundly reshaped the human geography of the region. First, the rural-to-urban shift seen so widely in many parts of the less-developed world is also reworking population patterns across Southwest Asia and North Africa. The Saudi Arabian example is echoed in many other countries within the region. Cities from Casablanca to Tehran are experiencing phenomenal growth rates, spurred by immigration.

Second, large numbers of workers are migrating within the region to areas with growing job opportunities. Labor-poor countries such as Saudi Arabia, Kuwait, and the United Arab Emirates (UAE), for example, have attracted thousands of Jordanians, Palestinians, Yemenis, and Iranians to construction and service jobs within their borders. Indeed, the demand for workers has been so strong that large numbers of South Asians, Southeast Asians, and East Africans, particularly from Muslim nations, have flooded into the region. Today, for instance, more than 85 percent of Saudi Arabia's labor force is foreign and they annually export more than $20 billion in wages. In Dubai (UAE), Pakistani cab drivers, Filipino nannies, and Indian shop clerks typify a foreign workforce that currently makes up 80 percent of the city's 1 million inhabitants (Figure 7.22).

A third major pattern of movement involves residents migrating to job opportunities elsewhere in the world. Because of its strong economy and close location, Europe has been a particularly powerful draw. More than 3 million Turkish guest workers live in Germany. Both Algeria and Morocco also have seen large out-migrations to western Europe, particularly France. Only 8 miles separates Africa from Europe, and illegal northbound boat traffic has reached unprecedented levels. Many Lebanese, often skilled workers and business owners, have also emigrated. Thousands of migrants have journeyed to North America, particularly more educated professionals who seek high-paying jobs.

Political forces have also encouraged migrations (See "People on the Move: Iraqis Flee the War"). Thousands of wealthier residents, for example, left Lebanon and

Figure 7.22 Labor Camp in Shariah City, United Arab Emirates These South Asian workers live a few miles outside wealthy Dubai and enjoy a card game during their time off from work. Over 80 percent of the country's population is foreign-born and many immigrants work in the fast-growing construction and service sectors of the economy. *(Rob Crandall/www.robcrandall.com)*

Iran during the political turmoil of the 1980s and are today living in cities such as Toronto, Los Angeles, and Paris. Elsewhere, more recent political instability has provoked other refugee movements. Thousands of Afghans remain in Iran. Huge numbers of refugees have been forced from their land and settlements in Darfur in western Sudan. The movement of more than 500,000 Russian Jewish immigrants to Israel after the 1991 breakup of the Soviet Union also reshaped the cultural and political geography of that country.

Shifting Demographic Patterns

While high population growth remains a critical issue throughout the region, the demographic picture is shifting. Uniformly high rates of population growth in the 1950s and 1960s have been replaced by more varied regional patterns, and many nations are seeing birthrates fall fairly rapidly. For example, women in Tunisia, Iran, and Turkey are now averaging fewer than three births, representing a large decline in the total fertility rate (see Table 7.1). Indeed, some experts now predict that fertility rates on the European and North African sides of the Mediterranean will almost converge (at just over two births per woman) by 2030. Many factors help explain the changes. More urban, consumer-oriented populations have opted for fewer children. Many Arab women now are delaying marriage into the middle 20s and even early 30s. Even in more traditional Bahrain, the average marriage age of women has risen from 15 to 23 years of age since the early 1970s. Family planning initiatives are expanding in many countries. For example, programs in Tunisia, Egypt, and Iran have greatly increased access to contraceptive pills, IUDs, and condoms since 1980.

Still, demographic challenges loom. Areas such as the West Bank, Gaza, and Yemen experience some of the highest rates of natural increase on the planet. In some localities, persisting patterns of poverty and traditional ways of rural life contribute to the large rates of population increase, and even in more urban and industrialized Saudi Arabia, annual growth rates remain very high. Such growth rates result from the combination of high birthrates and low death rates.

Elsewhere, even as new births are declining, already existing populations pose problems. For most countries in the region, more than 35 percent of the present population is younger than 15 years old, promising another large generation of people in need of food, jobs, and housing. In Egypt, for example, even though birthrates are likely to decline, the labor market will need to absorb more than 500,000 new workers annually over the next 10 to 15 years just to keep up with the country's large, youthful population. As that population ages in the mid-twenty-first century, it will continue to demand jobs, housing, and social services (Figure 7.23).

Figure 7.23 Egypt's Changing Population While Egypt's annual population growth rates are slowing, its 2000 population pyramid reveals a large number of young people already in the population. Future (2050) implications are clear: Egypt's population will grow larger and older in the years to come. *(U.S. Census Bureau, International Data Base)*

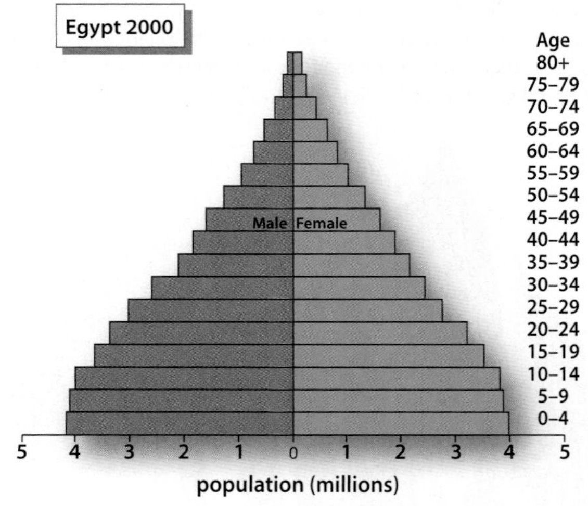

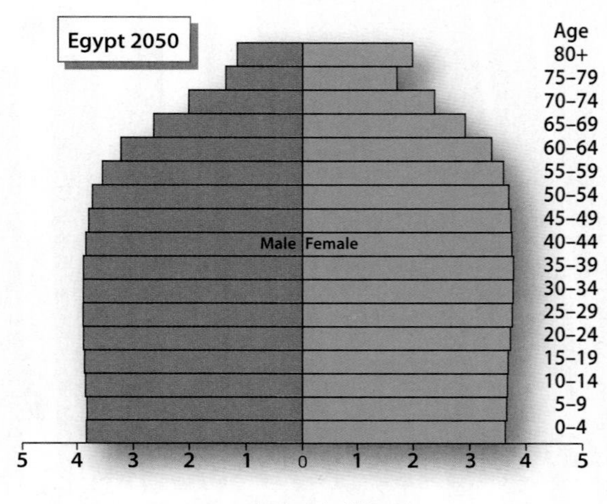

PEOPLE ON THE MOVE Iraqis Flee the War

Since 2005, huge numbers of Iraqis have left their homes, fleeing neighborhoods and communities that are no longer safe. Some move to other portions of the war-torn nation, while others have simply left the country (Figure 7.3.1). By any measure, it is the largest refugee crisis in the Middle East since 1948. By 2007, about 8 percent of the nation's population had left the country. Approximately 1.2 million Iraqis fled to Syria, many of them ending up in the vicinity of Damascus. Another 750,000 are in Jordan, 100,000 in Egypt, 50,000 in Iran, and 40,000 in Lebanon. On the road into Syria, UN aid workers recently witnessed more than 2,000 people per day flooding into the country. Iraqis are now also the largest national group claiming political asylum in Europe.

The political instability that followed the American invasion of Iraq in 2003 is the driving force behind these massive movements of refugees. Further, sectarian violence between Sunni and Shiite Muslims expanded greatly after the February 2006 attack on the Golden Mosque, a Shiite Muslim holy shrine in the Iraqi city of Samarra. That attack sparked an escalation in the violence, forcing a growing number of people from their homes. Urban violence in Baghdad has also reshaped neighborhoods as Sunnis and Shiites have quarreled over which zones of the city they control. The result is that many residents must leave their homes and look for nearby housing in what they consider to be "safer" areas. Many people take the more dramatic step and head for the border. In large portions of western Iraq, almost the entire Shiite population has left, while many Sunnis in southern Iraq or in Shiite-dominated eastern Baghdad have similarly moved on. More generally, economic misery is encouraging emigration. While some Iraqis have benefited economically since the overthrow of Saddam Hussein, many others have lost their jobs as local economies are devastated by violence. Indeed, many of the migrants are well-educated professionals who are taking their skills and expertise with them as they leave their country.

Figure 7.3.1 Iraqis in Exile
These Iraqis, recently displaced by violence in their home country, presently live in Damascus, Syria. *(Bassem Tellawi/AP Wide World)*

Meanwhile, host nations such as Syria and Jordan have done little to make life easier for these migrants. The Syrians, although sympathetic to the refugees, have forbidden them to buy property and have limited their access to health care. Many refugee neighborhoods consist of rough concrete- and cinder-block buildings, hastily assembled on the outskirts of Damascus. While there are strict Syrian limits put on employment opportunities, thousands of young Iraqi women have informally become "casino girls," prostitutes who serve wealthy Arab clients. In nearby Jordan, there is even less support for the refugees. Almost 200,000 refugees are Iraqi children of school age, with only about 14,000 attending classes. Jordanian officials have compared the "problem" to illegal Mexican immigration into the United States. As patience for Iraqi refugees ebbs, the long-term outlook for this floating Middle Eastern population remains clouded. Will they ultimately return to a more peaceful Iraq? Will they migrate to Europe and beyond? Will they become simply another refugee population in the region, a nation in exile without a place to call home?

Given recent trends in job growth and migration, the region's large cities probably will bear the brunt of future population increases. Growing populations will also impose new demands on the region's already limited water resources.

CULTURAL COHERENCE AND DIVERSITY: Signatures of Complexity

While Southwest Asia and North Africa clearly define the heart of the Islamic and Arab worlds, a surprising degree of cultural diversity characterizes the region. Muslims practice their religion in varied ways, often disagreeing profoundly on basic religious views as well as on how much of the modern world and its mass consumer culture should be incorporated into their daily lives. In addition, diverse religious minorities complicate the region's contemporary cultural geography. Linguistically, Arabic languages form an important cultural core historically centered on the region. Still, many non-Arab peoples, including Persians, Kurds, and Turks, also populate important homelands. Understanding these varied patterns of cultural geography is essential to comprehending many of the region's political tensions, as well as appreciating why many of its residents resist processes of globalization and celebrate the lasting cultural identity of their home neighborhoods and communities.

Patterns of Religion

Religion permeates the lives of most people within the region. Its centrality stands in sharp contrast to largely secular cultures in many other parts of the world. Whether it is the quiet ritual of morning prayers or profound discussions about contemporary political and social issues, religion is part of the daily routine of most regional residents from Casablanca to Tehran. The geographies of religion—their points of origin, paths of diffusion, and patterns of modern regional distribution—are essential elements in understanding cultural and political conflicts within the region.

Hearth of the Judeo-Christian Tradition Both Jews and Christians trace their religious roots to the eastern Mediterranean, and while neither group is numerically dominant across the area, each plays a key cultural role. The roots of Judaism lie deep in the past: Abraham, an early patriarch in the Jewish tradition, lived some 4,000 years ago and led his people from Mesopotamia to Canaan (modern-day Israel), near the shores of the Mediterranean. From Jewish history, recounted in the Old Testament of the Holy Bible, springs a rich religious heritage focused on a belief in one God (or **monotheism**), a strong code of ethical conduct, and a powerful ethnic identity that continues to the present. Around 70 CE, during the time of the Roman Empire, most Jews were forced to leave the eastern Mediterranean after they challenged Roman authority. The resulting forced migration, or diaspora, of the Jews took them to the far corners of Europe and North Africa. Only in the past century have many of the world's far-flung Jewish populations returned to the religion's hearth area, a process that accelerated greatly with the formation of the Jewish state of Israel in 1948.

Christianity also emerged in the vicinity of modern-day Israel and has left a lasting legacy across the region. An outgrowth of Judaism, Christianity was based on the teachings of Jesus and his disciples, who lived and traveled in the eastern Mediterranean about 2,000 years ago. While many Christian traditions became associated with European history, some forms of early Christianity remained strong near the religion's original hearth. To the south, one stream of Christian influences linked with the Coptic Church diffused into northern Africa, shaping the culture of places such as Egypt and Ethiopia. In the Levant, another group of early Christians, known as the Maronites, retained a separate cultural identity that sur-

Figure 7.24 Diffusion of Islam
The rapid expansion of Islam that followed its birth is shown here. From Spain to Southeast Asia, Islam's legacy remains strongest nearest its Southwest Asian hearth. In some settings, its influence has ebbed or has come into conflict with other religions, such as Christianity, Judaism, and Hinduism. *(Modified from Rubenstein, 2005,* An Introduction to Human Geography, *8th ed., Upper Saddle River, NJ: Prentice Hall)*

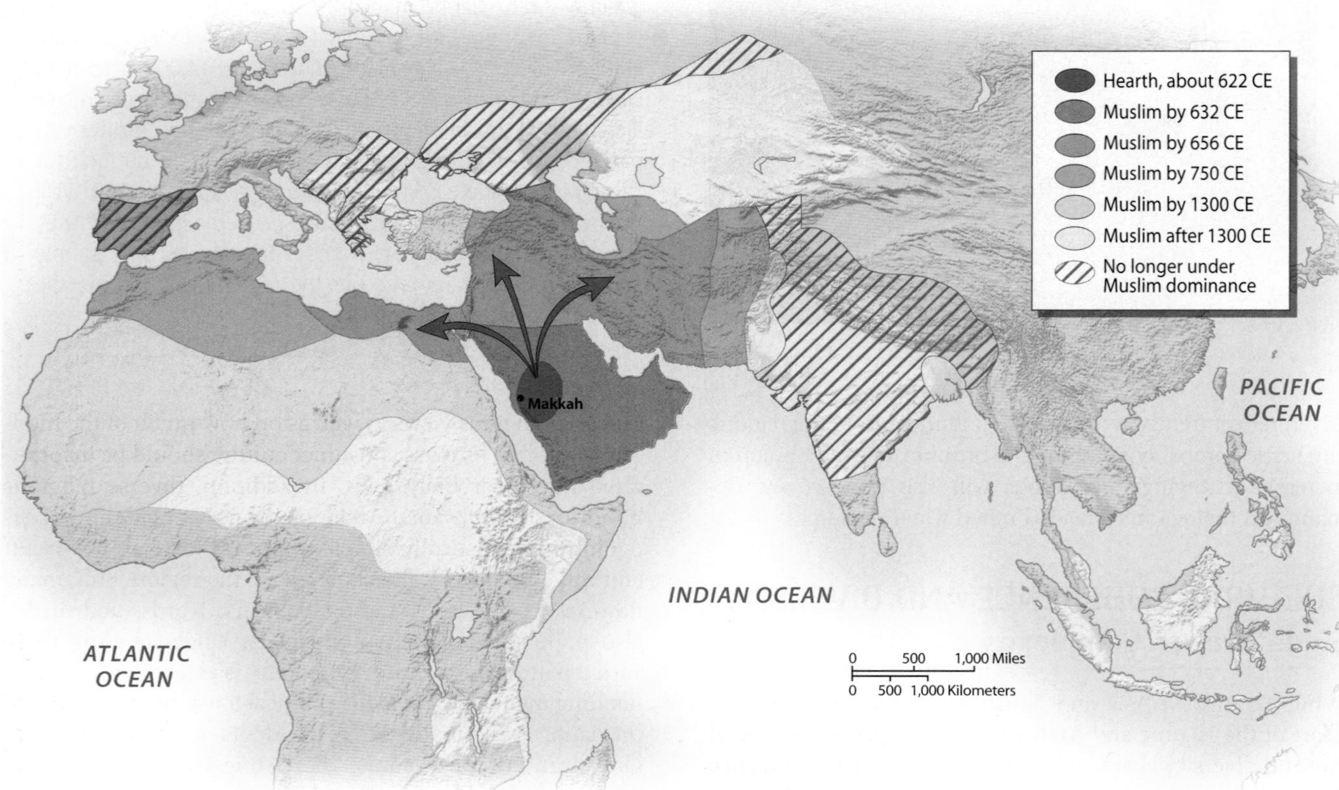

vives today. The region remains the "Holy Land" in the Christian tradition and attracts millions of the faithful annually as they visit the most sacred sites in the Christian world.

The Emergence of Islam Islam originated in Southwest Asia in 622 CE, forming yet another cultural hearth of global significance. Muslims can be found today from North America to the southern Philippines; however, the Islamic world is still centered on its Southwest Asian origins. Most Southwest Asian and North African peoples still follow its religious teachings and moral doctrines. Muhammad, the founder of Islam, was born in Makkah (Mecca) in 570 CE and taught in nearby Medinah (Medina) (Figure 7.24). In many respects, his creed parallels the Judeo-Christian tradition. Muslims believe that both Moses and Jesus were true prophets and that both the Hebrew Bible (or Old Testament) and the Christian New Testament, while incomplete, are basically accurate. Ultimately, however, Muslims hold that the **Quran** (or Koran), a book of revelations received by Muhammad from Allah (God), represents God's highest religious and moral revelations to humankind.

The basic teachings of Islam offer an elaborate blueprint for leading an ethical and religious life. Islam literally means "submission to the will of God," and the creed rests on five essential pillars: (1) repeating the basic creed ("There is no God but God, and Muhammad is his prophet"); (2) praying facing Makkah five times daily; (3) giving charitable contributions; (4) fasting between sunup and sundown during the month of Ramadan; and (5) making at least one religious pilgrimage, or **Hajj**, to Muhammad's birthplace of Makkah (Figure 7.25). Islam is a more austere religion than most forms of Christianity, and its modes of worship and forms of organization are generally less ornate. Muslims avoid the use of religious images, and the strictest interpretations even forbid the depiction of the human form. Followers of Islam are prohibited from drinking alcohol and are instructed to lead moderate lives, avoiding excess. Many Islamic fundamentalists still argue for a **theocratic state**, such as modern-day Iran, in which religious leaders (ayatollahs) guide policy.

A major religious schism divided Islam early on, and the differences endure to the present. The breakup occurred almost immediately after the death of Muhammad in 632 CE. Key questions surrounded the succession of religious power. One group, now called the **Shiites**, favored passing on power within Muhammad's own family, specifically to Ali, his son-in-law. Most Muslims, later known as **Sunnis**, advocated passing power down through the established clergy. This group emerged victorious. Ali was killed, and his Shiite supporters went underground. Ever since, Sunni Islam has formed the mainstream branch of the religion, to which Shiite Islam has presented a recurring, and sometimes powerful, challenge. The Shiites argue that a successor to Ali will someday return to reestablish the pure, original form of Islam. The Shiites also are more hierarchically organized than the Sunnis. Iran's ayatollahs, for example, have overriding religious and political power.

Figure 7.25 Makkah
Thousands of faithful Muslims gather at the Grand Mosque in central Makkah, part of the pilgrimage to this sacred place that draws several million visitors annually. Portions of the city's commercial district are in the distance. *(Awad Awad/AFP/Getty Images)*

In a very short period of time, Islam diffused widely from its Arabian hearth, often following camel caravan routes and Arab military campaigns as it expanded its geographical range and converted thousands to its beliefs. By the time of Muhammad's death in 632 CE, the peoples of the Arabian Peninsula were united under its banner. Shortly thereafter, the Persian Empire fell to Muslim forces and the Eastern Roman (or Byzantine) Empire lost most of its territory to Islamic influences. By 750 CE, Arab armies had swept across North Africa, conquered most of Spain and Portugal, and established footholds in Central and South Asia. At first, only the Arab conquerors followed Islam, but the diverse inhabitants of Southwest Asia and North Africa gradually were absorbed into the religion, although with many distinct local variants. By the thirteenth century, most people in the region were Muslims, while older religions such as Christianity and Judaism became minority faiths or disappeared altogether.

Between 1200 and 1500, Islamic influences expanded in some areas and contracted in others. The Iberian Peninsula (Spain and Portugal) returned to Christianity by 1492, although many Moorish (Islamic) cultural and architectural features remained behind and still shape the

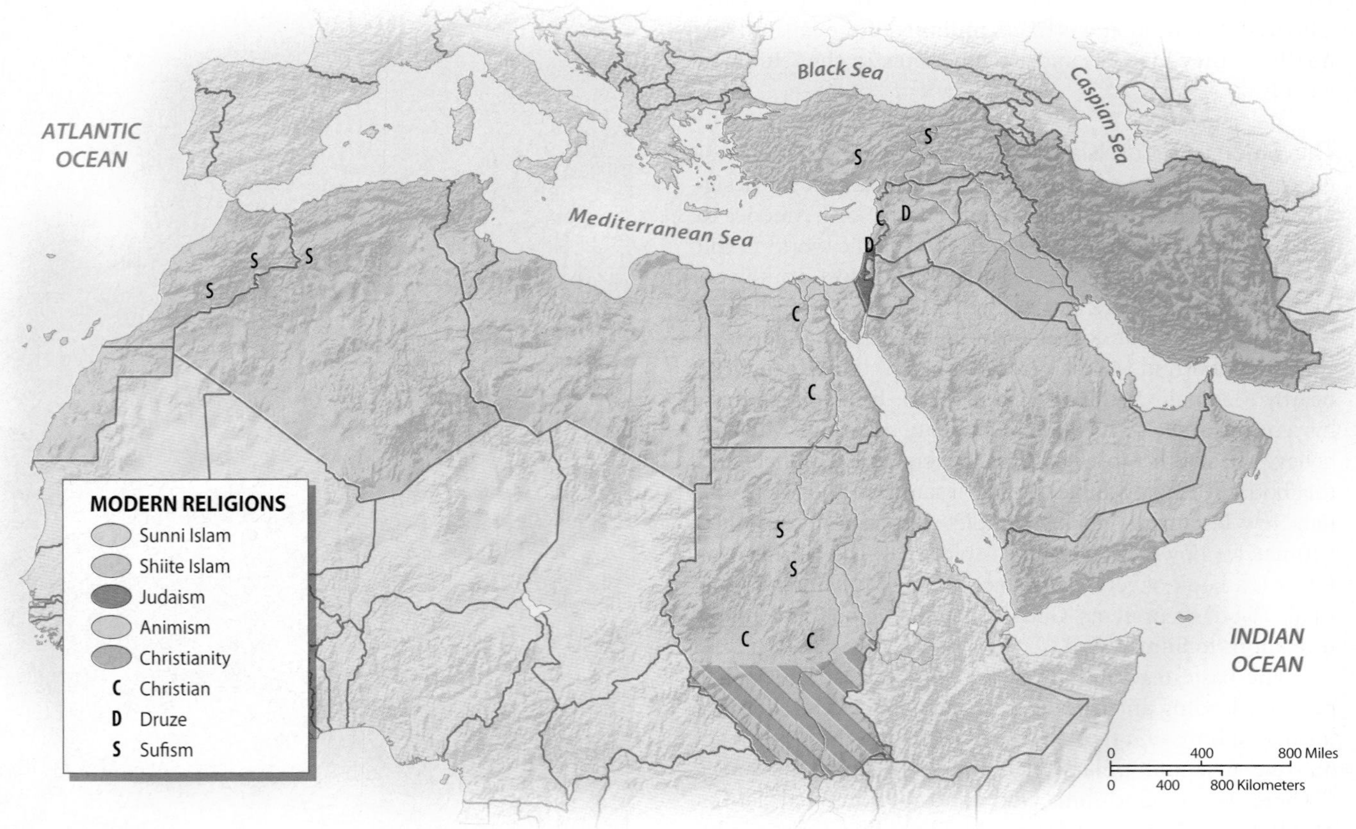

Figure 7.26 **Modern Religions**
Islam persists as the dominant religion across the region. Most Muslims are tied to the Sunni branch, while Shiites are found in places such as Iran and southern Iraq. In some locales, however, Christianity and Judaism remain important. African animism is found in southern portions of Sudan. *(Modified from Rubenstein, 2008,* An Introduction to Human Geography, *9th ed., Upper Saddle River, NJ: Prentice Hall)*

region today. At the same time, Muslims expanded their influence southward and eastward into Africa. In addition, Muslim Turks largely replaced Christian Greek influences in Southwest Asia after 1100. One group of Turks moved into the Anatolian Plateau and finally conquered the last vestiges of the Byzantine Empire in 1453. These Turks soon created the vast **Ottoman Empire** (named after one of its leaders, Osman), which included southeastern Europe (including modern-day Albania, Bosnia, and Serbia) and most of Southwest Asia and North Africa. The legacy of the Ottoman Empire was considerable. It offered a new, distinctly Turkish interpretation of Islam, and it provided a focus of Muslim political power within the region until the empire's disintegration in the late nineteenth and early twentieth centuries.

Modern Religious Diversity Today, Muslims form the majority population in all of the countries of Southwest Asia and North Africa except Israel, where Judaism is the dominant religion (Figure 7.26). Still, divisions within Islam have created key cultural differences within the region. While most (73 percent) of the region is dominated by Sunni Muslims, the Shiites (23 percent) remain an important element in the contemporary cultural mix. In Iraq, for example, the majority Shiite population in the southern portion of the country (around Najaf, Karbala, and Basra) set its own cultural and political course following the fall of Saddam Hussein. Many Shiites saw the departure of Hussein (who gathered most of his support from Sunnis in the central and western parts of Iraq) as an opportunity for increasing their cultural and political influence within the country. The result has been the bitter sectarian violence evident since 2006. Shiites claim a majority in nearby Iran, as well, where their religious fervor has fundamentally shaped government policies since the late 1970s. In addition, they form a major religious minority in Lebanon, Bahrain, Algeria, Egypt, and Yemen. Strongly associated with the recent flowering of Islamist movements, the Shiites also have benefited from rapid growth rates because their brand of Islam is particularly appealing to the poorer, powerless, and more rural populations of the region. While some Sunnis have been attracted to Islamist tendencies as well, many reject its more radical cultural and political teachings and argue for a more modern Islam that accommodates some Western values and traditions.

nities are found as far to the east as Egypt, but Morocco is the center of this language group, where it plays an important role in shaping that nation's cultural identity.

Persians and Kurds Although Arabic spread readily through portions of Southwest Asia, much of the Iranian Plateau and nearby mountains are dominated by older Indo-European languages. Here the principal tongue remains Persian, although, since the tenth century, the language has been enriched with Arabic words and written in the Arabic script. Persian, like other languages, developed distinct local dialects. Today Iran's official language is called *Farsi*, which denotes the form of Persian spoken in Fars, the area around the city of Shiraz. Thus, although both Iran and neighboring Iraq are Islamic nations, their ethnic identities spring from quite different linguistic and cultural traditions.

The Kurdish speakers of northern Iraq, northwest Iran, and eastern Turkey add further complexity to the regional pattern of languages. Kurdish, also an Indo-European language, is spoken by 10 to 15 million people in the region. Kurdish has not historically been a written language, but the Kurds do have a strong sense of shared cultural identity. Indeed, "Kurdistan" has sometimes been called the world's largest nation without its own political state because the group remains a minority in several countries of the region. In the 2003 war in Iraq, the Kurds emerged as a cohesive group in the northern part of the country that was opposed to Saddam Hussein. Iraqi Kurds have gained more political autonomy in a post-Saddam Iraq state and have escaped much of the sectarian violence that has destabilized the remainder of the country. Their leading city of Kirkuk has been called the "Kurdish Jerusalem" because its history and settlement so richly capture the group's cultural identity (Figure 7.29). Nearby Kurds in eastern Turkey, however, still complain that their ethnic identity is frequently challenged by the majority Turks, leaving some to wonder if they will attempt to join forces with their Iraqi neighbors.

The Turkish Imprint Turkish languages provide more variety across much of modern Turkey and in portions of far northern Iran. The Turkish languages are a part of the larger Altaic language family that originated in Central Asia. Turkey remains the largest nation in Southwest Asia dominated by that family. Tens of millions of people in other countries of Southwest and Central Asia speak related Altaic languages, such as Azeri, Uzbek, and Uighur. During the era of the Ottoman Empire, Turkish speakers ruled much of Southwest Asia and North Africa, but Iran is the only other large country in the region today where the language persists.

Figure 7.29 Kirkuk
The northern Iraqi city of Kirkuk remains a focus of Kurdish cultural identity. Here, Iraquis shop for fruits and vegetables in a commercial district of the city. *(Brennan Linsley/AP Wide World)*

Regional Cultures in Global Context

Many cultural connections tie the region with the world beyond (see "Global to Local: The Egypt of Naguib Mahfouz"). Islam links the region with a Muslim population that now lives in many different settings around the world. In addition, European and American cultural influences have multiplied greatly across the region since the mid-nineteenth century. Colonialism, the boom in petroleum investment, and the growing presence of Western-style popular culture have had enduring impacts from the Atlas Mountains to the Indian Ocean.

Islamic Internationalism While Islam is geographically and theologically divided, all Muslims recognize the fundamental unity of their religion. This religious unity extends far beyond Southwest Asia and North Africa. Islamic communities are well established in such distant places as central China, European Russia, central Africa, and the southern Philippines. Today, Muslim congregations also are expanding rapidly in the major urban areas of western Europe and North America, largely through migration but also through local conversions. Islam is thus emerging as a truly global religion. Even with its global reach, however, Islam remains centered on Southwest Asia and North Africa, the site of its origins and its most holy places. As Islam expands in number of followers and geographical scope, the religion's tradition of pilgrimage ensures that Makkah will become a city of increasing global significance in the twenty-first century. The global growth of Islamist fundamentalism and Islamism also focuses attention on the region, where these contemporary movements recently burst upon the scene. In addition, the oil wealth accumulated by many Islamic nations will be used to sustain and promote the religion. Countries such as Saudi Arabia and Libya invest in Islamic banks and economic ventures and make donations to Islamic cultural causes, colleges, and hospitals worldwide.

Globalization and Cultural Change The region also is struggling with how its growing role in the global economy is changing traditional cultural values. European colonialism left its own cultural legacy not only in the architectural landscapes still found in the old colonial centers, but also in the widespread use of English and French among the Western-educated elite across the region. In oil-rich countries, huge capital investments have had important cultural implications as the number of foreign workers has grown and as more affluent young people have embraced elements of Western-style music, literature, and clothing. The expansion of Islamic fundamentalism and Islamism are in many ways

GLOBAL TO LOCAL The Egypt of Naguib Mahfouz

When Egyptian novelist Naguib Mahfouz died in 2006, Egypt's religious affairs minister declared that "he transformed Arabic literature from the local to the universal and he enriched it with his ideas." Indeed, part of the special genius of Mahfouz, who lived to the age of 94, was to capture the intimate, local flavor of Egyptian geography and then translate that to a larger global audience. He accomplished this with an effective use of characters and larger universal themes that resonated with readers from around the world. Mahfouz eagerly studied Western literary practices, blending them with the Arab and Islamic traditions he knew so well. Many experts compared his work with that of Balzac, Dickens, and Kafka because Mahfouz was so adept at addressing human themes such as suffering, the alienation of the individual, and the clash between tradition and modernity. More than any other Arab writer of the twentieth century, Mahfouz managed to transcend his cultural world by telling its story simply and vividly to a global audience (Figure 7.4.1).

That audience grew over the years. He was the first and only Arab to win the Nobel Prize for Literature in 1988. His famed Cairo Trilogy of novels (*Palace Walk, Palace of Desire*, and *Sugar Street)*, set on the streets of that famed Egyptian city, were originally published in Arabic in the 1950s, but were in later decades translated into many other languages, including English.

Mahfouz's long familiarity with Cairo's many streets and neighborhoods enables him to take the reader into the city with an intimate look at the interplay between people and place. In *Palace of Desire*, young Yasin returns to his boyhood neighborhood:

> When his feet brought him to al-Gamaliya Street, he was so choked up he felt he would die. He had not been there for eleven years.... yet it remained exactly the way it had been when he was growing up. Nothing had changed. The street was still so narrow a handcart would almost block it when passing by. The protruding balconies of the houses almost touched each other overhead. The small shops resembled the cells of a beehive, they were so close together and crowded with patrons, so noisy and humming. The street was unpaved, with gaping holes full of mud. The boys who swarmed along the sides of the street made footprints in the dirt with their bare feet. There was the same never-ending stream of pedestrian traffic. Uncle Hasan's snack shop and Uncle Sulayman's restaurant too remained just as he had known them.... His heart pounded so strongly it almost deafened his ears.

Even in translation, Mahfouz's sense of place captures something local and timeless, and his genius for writing made that vision accessible far beyond the bounds of the Arab World.

Sources: Edward W. Said, "The Cruelty of Memory," *The New York Review of Books*, November 30, 2000; "Naguib Mahfouz, Obituary," *The Economist*, September 2, 2006; and Naguib Mahfouz, 1991, *Palace of Desire*, New York: Doubleday.

Figure 7.4.1 Naguib Mahfouz
Egypt's great novelist captures a unique sense of place and time in the stories he tells. Mahfouz's Cairo Trilogy has been widely translated into English and other languages. *(Thomas Hartwell/TimePix)*

reactions to the threat posed by external cultural influences, particularly the supposed evils of European colonialism, American and Israeli power, and local governments and cultural institutions that are seen as selling out to the West.

Technology has also contributed to cultural change. Educated and more affluent residents in the region have been embracing the Internet and its power to access a global wealth of information and entertainment. Cell phone use also has increased rapidly. In Morocco, for example, while less than 6 percent of the country's population is served by fixed phone lines, more than 85 percent is within the reach of cellular phones. Widespread television-viewing has transformed the region, with viewers offered everything from Islamist religious programming to American-style reality TV and talent shows. Even in conservative Iran, where satellite dishes are officially banned, millions of people have access to them, beaming in multicultural programming from around the globe. Some fundamentalists in Islamic republics, such as Iran and Sudan, would like to impose a complete technological and cultural wall between themselves and the West; such actions remain problematic in today's world. Indeed, many middle-class and well-to-do Iranians privately enjoy their wealth, including the pleasures of Western-style food, drink, fashion, and popular culture.

In contrast, conservative cultural influences are gaining strength and visibility in many Islamic settings. For example, in Tunisia, long known for its more Western-leaning, modernized brand of Islam, the growing use of headscarves and veils among conservative women has caused a stir. A ban on such traditional practices was enacted in 1981 by a Tunisian government that described it as a "sectarian form of dress." More recently, such clothing has been seen by some as promoting the political agenda of extremist Islamist elements within the country. Interestingly, European human rights advocates have come to the aid of the women as they are increasingly harassed by police to remove the scarves and sign pledges that they will not go back to wearing them. Greater public displays of religion are also changing Turkey in subtle ways. A growing traditional Islamist presence in the Turkish government and in the civil service bureaucracy has allowed for more religious study and prayers in the public school system. Science textbooks are also being changed, deemphasizing Western interpretations in favor of more traditional religiously based explanations.

GEOPOLITICAL FRAMEWORK: Never-Ending Tensions

Geopolitical tensions remain high in Southwest Asia and North Africa (Figure 7.30). Some of the tensions surround the struggle of different ethnic, religious, and

Figure 7.30 Geopolitical Issues in Southwest Asia and North Africa
Political tensions continue across much of the region, including those surrounding events in Iraq. Although regional conflict remains oriented around Israel, its neighboring states, and the rights of resident Palestinians, other regional trouble spots periodically erupt into violence. Islamist political movements have challenged political stability in settings from Algeria to Saudi Arabia. Elsewhere, ongoing ethnic conflicts shape daily life in western Sudan's Darfur region. Several large American military bases are found in the region.

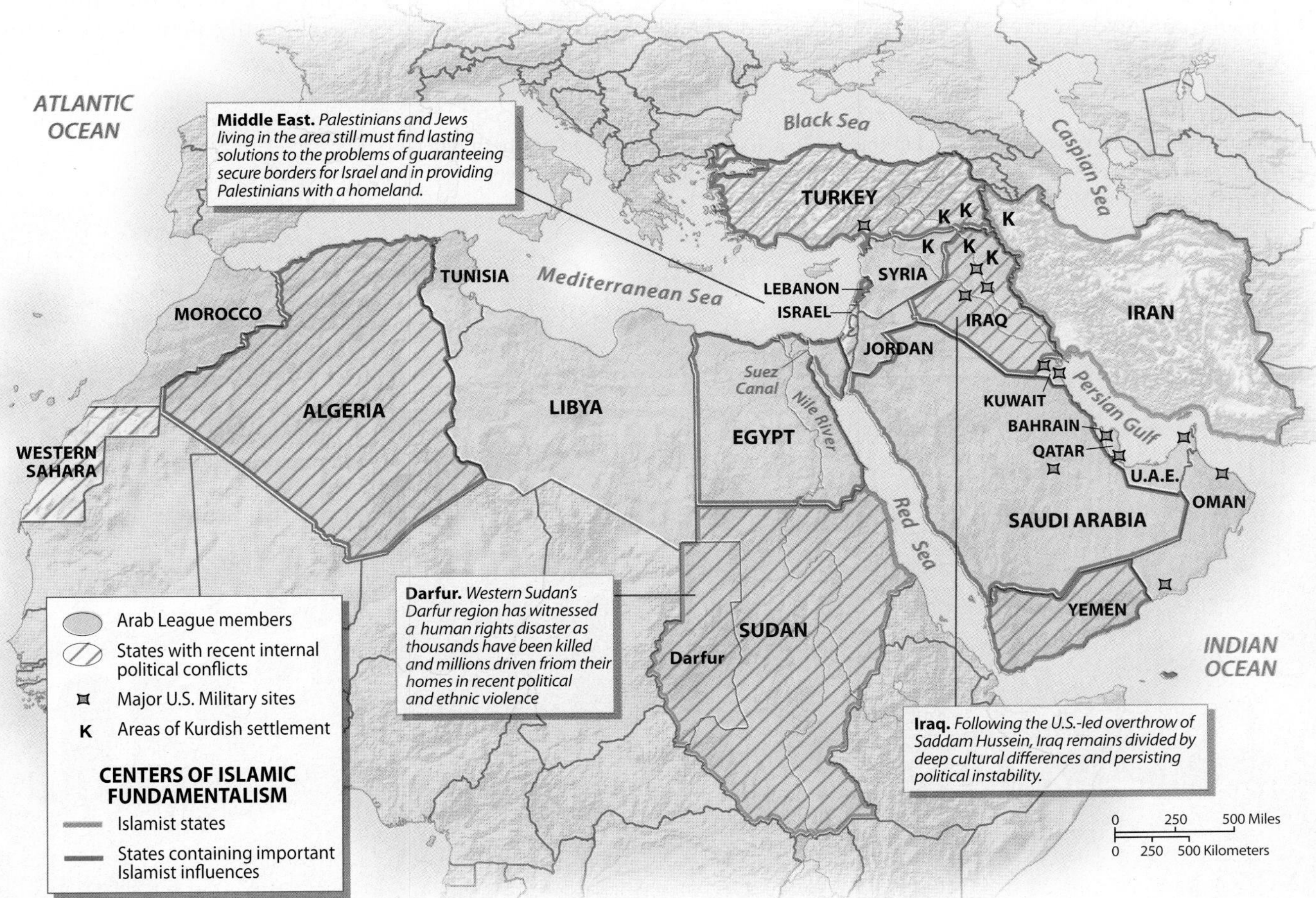

linguistic groups to live with one another in a rapidly changing world of political relationships. The region's brief but complex ties to the era of European colonialism also contribute to present difficulties, because the boundaries of many countries were imposed by colonial powers. American political power has been repeatedly exercised throughout the region, evident in its multiple military bases and in its recent reentry to Iraq. Geographies of wealth and poverty also enter the geopolitical mix: some residents profit from petroleum resources and industrial expansion, while others struggle just to feed their families. Islamist elements in many countries such as Iran, Egypt, Algeria, Turkey, and Saudi Arabia have changed the political atmosphere dramatically in the past 20 years. The result is a political climate charged with tension, a region in which the sounds of bomb blasts and gunfire have become a common characteristic of everyday life.

The Colonial Legacy

European colonialism arrived relatively late in Southwest Asia and North Africa, but the era left an important impact upon the modern political landscape. The Turks were one reason for Europe's late participation in imposing colonial rule. Between 1550 and 1850, much of the region was dominated by the Turkish Ottoman Empire, which expanded from its Anatolian hearth to engulf much of North Africa as well as nearby areas of the Levant, the western Arabian Peninsula, and modern-day Iraq. Ottoman rule imposed an outside political order and economic framework that had lasting consequences. The tide began to turn, however, in the early nineteenth century as the European presence increased and Ottoman power ebbed. Still, it took a century for Ottoman influences to be replaced with largely European colonial dominance after World War I (1918). While much of that direct European control ended by the 1950s, old colonial ties persist in a variety of economic, political, and cultural contexts. Indeed, it is still common to encounter British English on the streets of Cairo, whereas French still can be heard in Algiers and Beirut.

Imposing European Power French colonial ties have long been a part of the region's history. Beginning around 1800, France became committed to a colonial presence in North Africa which included an expedition to Egypt by Napoleon. By the 1830s, Algeria moved into the French sphere of influence. In the next 120 years, several million French, Italian, and other European immigrants poured into the country, taking the best lands from the Algerian people. The French government expected this territory to become an integral part of France, dominated by the growing French-speaking immigrant population (Figure 7.31). Later, France established its influence in Tunisia (1881) and Morocco (1912), ensuring a lasting French political and cultural presence in the Maghreb. These **protectorates** in Tunisia and Morocco retained some political autonomy, but remained under a broader sphere of French influence and protection from other competing colonial powers. Finally, France's victory over

Figure 7.31 **Algiers**
European colonial influences still abound in the old French capital of Algiers in northern Algeria. The city's modern streets have witnessed periodic political and religious tensions since 1992. *(Francoise Perri/Woodfin Camp & Associates)*

the German–Ottoman Turk alliance in World War I produced additional territorial gains in the Levant, as France gained control of the northern zone encompassing the modern nations of Syria and Lebanon.

Great Britain's colonial fortunes also grew within the region before 1900. To control its sea-lanes to India, Britain established a series of nineteenth-century protectorates over the small coastal states of the southern Arabian Peninsula and the Persian Gulf. In this manner, such places as Kuwait, Bahrain, Qatar, the United Arab Emirates, and Aden (in southern Yemen) were loosely incorporated into the British Empire. Nearby Egypt also caught Britain's attention. Once the British-engineered **Suez Canal** linked the Mediterranean and Red seas in 1869, European banks and trading companies gained more influence over the Egyptian economy. The British took more direct control in 1883. In the process, Britain also inherited a direct stake in Sudan, since Egyptian soldiers and traders had been pushing south along the Nile for decades.

Another series of British colonial gains within the region came at the close of World War I. In Southwest Asia, British and Arab forces had joined to expel the Turks during the war. To obtain Arab trust, Britain promised that an independent Arab state would be created in the former Ottoman territories. At roughly the same time, however, Britain and France signed a secret agreement to partition the area. When the war ended, Britain opted to slight its Arab allies and honor its treaty with France, with one exception: the Saud family convinced the British that a smaller country (Saudi Arabia) should be established, focused on the desert wastes of the Arabian Peninsula. Saudi Arabia became fully independent in 1932. Elsewhere, however, Britain carried out its plan to partition lands with France. Britain divided its new territories into three entities: Palestine (now Israel and Gaza) along the Mediterranean coast; Transjordan to the east of the Jordan River (now Jordan); and a third zone that later became Iraq. Iraq, in particular, was a territory contrived by the British that combined three dissimilar former Ottoman provinces. It included the centers of Basra in the south (an Arabic-speaking Shiite area), Baghdad in the center (an Arabic-speaking Sunni area), and Mosul in the north (a Kurd-dominated zone).

Other settings within the region felt more marginal colonial impacts. Libya, for example, was long regarded by Europeans as a desert wasteland. Italy, never a dominant colonial power, expelled Turkish forces from the coastal districts by 1911 but did not subdue the Saharan oases until the 1930s. Spain also carved out a territorial stake within the region, gaining control over southern Morocco (now Western Sahara) in 1912. To the east, Persia and Turkey never were directly occupied by European powers. In Persia, the British and Russians agreed to establish mutual spheres of economic influence (the British in the south, the Russians in the north), while respecting Persian independence. In 1935, Persia's modernizing ruler, Reza Shah, changed the country's name to Iran.

In nearby Turkey, the old core of the Ottoman Empire was almost partitioned by European powers following World War I. After several years of fighting, however, the Turks expelled the French from southern Turkey and the Greeks from western Turkey. The key to the successful Turkish resistance was the spread of a new modern, nationalist ideology under the leadership of Kemal Ataturk. Ataturk decided to emulate the European countries and establish a culturally unified and resolutely secular state. He was successful, and Turkey was quickly able to stand up to European power.

Decolonization and Independence European colonial powers began their withdrawal from several Southwest Asian and North African colonies before World War II. By the 1950s, most of the countries in the region were independent, although many maintained political and economic ties with their former colonial rulers. In North Africa, Britain finally withdrew its troops from Sudan and Egypt in 1956. Libya (1951), Tunisia (1956), and Morocco (1956) achieved independence peacefully during the same era, but the French colony of Algeria became a major problem. Since several million French citizens resided there, France had no intention of simply withdrawing. A bloody war for independence began in 1954, and France agreed to an independent Algeria in 1962, but the two nations continued to share a close—if not always harmonious—relationship thereafter.

Southwest Asia also lost its colonial status between 1930 and 1960, although many of the imposed colonial-era boundaries continue to shape the regional geopolitical setting. While Iraq became independent from Britain in 1932, its later instability in part resulted from its artificial borders, which never recognized much of its cultural diversity. Similarly, the French division of its Levant territories into the two independent states of Syria and Lebanon (1946) greatly angered local Arab populations and set the stage for future political instability in the region. As a favor to its small Maronite Christian majority, France carved out a separate Lebanese state from largely Arab Syria, even guaranteeing the Maronites constitutional control of the national government. The action created a culturally divided Lebanon as well as a Syrian state that repeatedly has asserted its influence over its Lebanese neighbors.

Modern Geopolitical Issues

The geopolitical instability in Southwest Asia and North Africa will continue in the twenty-first century. It remains difficult to predict political boundaries that seem certain to change as a result of negotiated settlements or political conflict. A quick regional transect from the shores of the Atlantic to the borders of Central Asia suggests how these forces are playing out in different settings early in the new century.

Across North Africa Varied North African settings threaten the region's political stability. Libya's leader, Colonel Muammar al-Qaddafi, has intermittently provoked regional tensions since he took power in 1969. Libya has financed violent political movements directed against Israel, western Europe, and the United States. Its relationships with nearby Egypt and Chad also have been strained, although Qaddafi has recently campaigned for greater African unity. Since 2004, he has vowed to disarm and work to open his country to more Western investment, winning praise and growing diplomatic recognition from both the European Union and the United States.

Elsewhere in North Africa, Islamist political movements have reshaped the political landscape in several states. Most notably, Algeria was plunged into an escalating cycle of Islamist-led violence and protests for much of the 1990s, and more than 150,000 Algerians and foreign visitors were killed in the process. Since 1999, an amnesty has encouraged some rebel groups to lay down their arms, but the government still faces active opposition from Islamist extremists who have officially linked themselves with the terrorist group, Al Qaeda. Nearby Egypt has found itself ensnared in Islamist-initiated instability, and groups such as the Muslim Brotherhood have pushed for more radical political change. Egyptian president Hosni Mubarak, who maintains close ties to the United States, has successfully used the "war on terror" to crack down on many of his domestic political enemies. Still, several dissident Islamist groups remain in Egypt, and Egyptians have played an active role in the larger Al Qaeda movement worldwide.

Sudan faces some of the most daunting political issues in North Africa. Some cut across religious divides while others seem to focus more on ethnic and racial differences within the country. A Sunni Islamist state since a military coup in 1989, Sudan imposed Islamic law across the country and in the process antagonized both moderate Sunni Muslims as well as the nation's large non-Muslim (mostly Christian and animist) population in the south. A long civil war between the Muslim north and the Christian and animist south also proved disastrous, producing more than 2 million casualties (mostly in the south) in the past 20 years. Although the carnage in the south has lessened since 2001 (a tentative peace agreement was signed in 2004), a newer conflict erupted in the Darfur region in the western portion of the country (Figure 7.30). Ethnicity, race, and control of territory seem to be at the center of the struggle in the largely Muslim region as a well-armed Arab-led militia group (with many ties to the central government in Khartoum) has attacked hundreds of black-populated villages, killing over 200,000 people and driving 2.5 million more from their homes.

The Arab–Israeli Conflict The 1948 creation of the Jewish state of Israel produced another enduring zone of cultural and political tensions within the eastern Mediterranean. Jewish migration to the area accelerated after the British took Palestine from the defeated Ottoman Empire after World War I. In 1917, Britain issued the Balfour Declaration, considered by some to be a pledge to encourage the "establishment of Palestine as a home for the Jewish people." After World War II, the British agreed to withdraw from the area, and the United Nations divided the region into two states, one to be predominantly Jewish, the other primarily Muslim (Figure 7.32). Indigenous Arab Palestinians rejected the deal because they saw it as an unfair partitioning of land in favor of a relatively small number of Jewish settlers. War erupted as the British departed. Jewish forces proved victorious, and by 1949 Israel had gained a larger share of land than it had originally been allotted. The remainder of Palestine, including the West Bank and the Gaza Strip, passed to Jordan and Egypt, respectively. Hundreds of thousands of Palestinian refugees fled from Israel to nearby countries, such as Egypt, Jordan, and Lebanon, where many of them remained in makeshift camps. Under these difficult conditions, the Palestinians nurtured the idea of creating their own state in the land that had become Israel.

Israel's relations with neighboring countries were bitter from the beginning. The supporters of Arab unity and Muslim solidarity sympathized with the Palestinians, while their antipathy toward Israel grew. Israel and its Arabic-speaking neighbors fought major wars in 1956, 1967, and 1973. In territorial terms, the Six-Day War of 1967 was the most important conflict (Figure 7.32). In this struggle against Egypt, Syria, and Jordan, Israel occupied substantial new territories in the Sinai Peninsula, the Gaza Strip, the West Bank, and the Golan Heights. Israel annexed the eastern part of the formerly divided city of Jerusalem, arousing particular bitterness among the Palestinians, since Jerusalem is a sacred city in the Muslim tradition (it contains the Dome of the Rock and the holy Al-Aqsa Mosque) (Figure 7.27). Jerusalem is sacred in Judaism as well (it contains the Temple Mount and the holy Western Wall of the ancient Jewish temple), and Israel remains adamant in its claims to the entire city. A peace treaty with Egypt later resulted in the return of the Sinai Peninsula in 1982, but tensions then focused on the other occupied territories that remained under Israeli control. To strengthen its geopolitical claims, Israel also built additional Jewish settlements in the West Bank and in the Golan Heights, further angering Palestinian residents and nearby Syrians.

Palestinians and Israelis began to negotiate a settlement in 1993. Preliminary agreements called for the construction of a quasi-independent Palestinian state in the Gaza Strip and across much of the West Bank. Radical factions on both sides denounced the settlement as a sellout, and political polarization among both the Palestinians and the Israelis heightened tensions across the region. A tentative agreement late in 1998 strengthened the potential control of the ruling **Palestinian Authority (PA)** in the Gaza Strip and portions of the West Bank (Figure 7.33).

A new cycle of heightened violence erupted late in 2000, however, as Palestinian attacks against Jews increased and as the Israelis continued with the construction of new settlements in occupied lands (especially in the West Bank), resisting PA moves toward more autonomy (Figure 7.34). Palestinian leaders harshly criticized continuing construction of these new Jewish settlements, particularly in the vicinity of Jerusalem. Even more controversial has been the continuing construction of the Israeli security barrier, a partially completed 26-foot (8-meter) high wall destined to separate the Israelis from Palestinians across

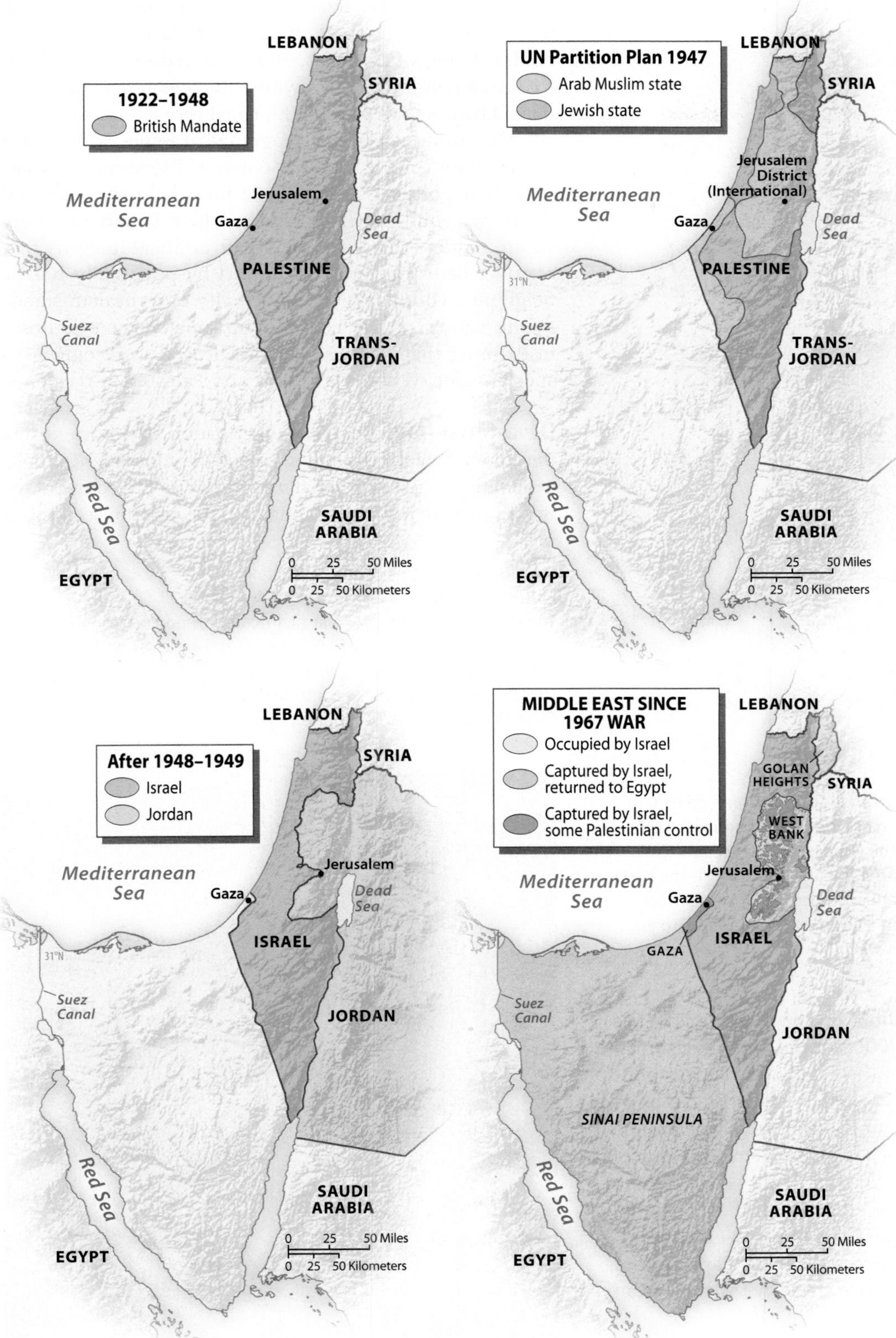

Figure 7.32 Evolution of Israel
Modern Israel's complex evolution began with an earlier British colonial presence and a UN partition plan in the late 1940s. Thereafter, multiple wars with nearby Arab states produced Israeli territorial victories in settings such as Gaza, the West Bank, and the Golan Heights. Each of these regions continues to figure importantly in the country's recent relations with nearby states and with resident Palestinian populations. *(Modified from Rubenstein, 2008,* An Introduction to Human Geography, *9th ed., Upper Saddle River, NJ: Prentice Hall)*

much of the West Bank region (see "Geography in the Making: Walling in Bethlehem and the Israeli Security Barrier"). Israeli supporters of the barrier (to be more than 400 miles [640 kilometers] long when completed) see it as the only way they can protect their citizens from suicide bombings and more terrorist attacks. Palestinians see it as a land grab, an "apartheid wall" designed to socially and economically isolate many of their settlements along the Israeli border. To the west, Israel withdrew troops and settlers from the Gaza Strip in 2005, but political instability thereafter has resulted in repeated Israeli incursions and reoccupations of the small, war-torn region.

Figure 7.33 West Bank
Portions of the West Bank were returned to Palestinian control in the 1990s, but Israel has partially reasserted its authority in some of these areas since 2000, citing the increased violence in the region. New Israeli settlements also are scattered through the West Bank in areas still under their nominal control. *(Reprinted from Rubenstein, 2005,* An Introduction to Human Geography, *8th ed., Upper Saddle River, NJ: Prentice Hall)*

The political fragmentation of the Palestinians has added further uncertainty. In 2006, control of the Palestinian government was split between the Fatah and Hamas political parties. Hamas has long been seen by many Israelis as an extremist and violent Palestinian political party, whereas the Fatah party has shown more of a willingness to work peacefully with the state of Israel. In 2007, the split between these rival Palestinian factions became violent with Hamas gaining effective control of the PA within Gaza and with Fatah maintaining its greatest influence across the West Bank region. The split makes an eventual peace treaty with Israel seem even more difficult to achieve.

Instability also continues along Israel's northern border with Lebanon. In 2006, a Shiite militia group known as Hezbollah increased its rocket attacks into northern Israel, prompting an armed response by the Israelis, including selective bombing of suspected Hezbollah strongholds in Lebanon as well as limited ground movements into portions of southern Lebanon. Well-funded by anti-Israeli Islamist elements in both Syria and Iran, Hezbollah supports expanded Palestinian territorial rights and threatens Israel's political stability, particularly along its vulnerable northern border. Along with Hamas, Hezbollah represents a radicalized Arab political element within the region that shows little inclination to negotiate with Israel (or vice versa).

One thing is certain: geographical issues will remain at the center of the regional conflict. Palestinians, whatever their political stripes, hope for a land they can call their own, and Israelis continue their search for more secure borders that guarantee their political integrity in a region where they are surrounded by potentially hostile neighbors. Ultimately, the sacred city of Jerusalem stands at the center of the conflict. Imaginative compromises in defining that political space will need to recognize its special value to both its Jewish and Palestinian residents.

Devastated Iraq Iraq is another nation state born during the colonial era that has yet to escape the consequences of its geopolitical origins. When the country was carved out of the British Empire in 1932, it contained the cultural seeds of its

Figure 7.34 Jewish Settlement, West Bank
The new houses and well-planned neighborhoods of the West Bank Jewish settlement of Eli (foreground) contrast with the older Palestinian settlement in the distance. Many Palestinians resent the construction of these controversial Israeli settlements. *(Nir Elias/Reuters/Corbis/Bettmann)*

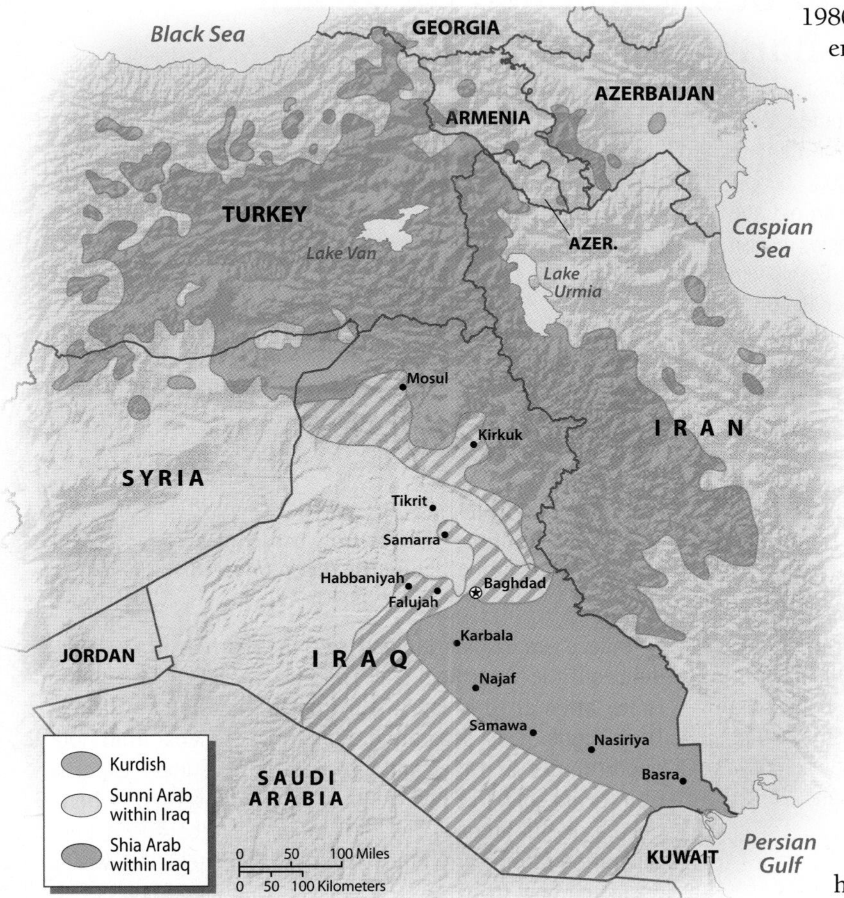

Figure 7.35 Multicultural Iraq
Iraq's complex colonial origins produced a state with varying ethnic characteristics. Shiites dominate south of Baghdad, Sunnis hold sway in the western triangle zone, and Kurds are most numerous in the north near oil-rich Kirkuk and Mosul.

later troubles decades before Saddam Hussein came to power. Iraq remains culturally complex today (Figure 7.35). Most of the country's Shiite population lives in the lower Tigris and Euphrates valleys near and south of Baghdad. Indeed, the region focused around the city of Basra contains some of the holiest Shiite shrines in the world. In northern Iraq, the culturally distinctive Kurds have their own ethnic identity and political aspirations. Many Kurds want complete independence from Baghdad and they have managed to establish a federal region that already enjoys a great deal of autonomy from the central Iraqi government. A third major subregion is dominated by the Sunnis and includes part of the Baghdad area as well as a triangle of territory to the north and west that includes Sunni strongholds such as Falujah and Tikrit. After enjoying a great deal of power under Saddam Hussein, many Sunnis feel they may get the short end of any future power-sharing agreement within the country.

Led by Saddam Hussein, Iraq created a major source of instability in the region before he was removed in 2003. In 1980, Hussein invaded oil-rich but politically weakened Iran to gain a better foothold in the Persian Gulf. Eight years of bloody fighting resulted in a stalemate, and the conflict left Iraq's finances in disarray. Iraq then invaded and overran Kuwait in 1990, claiming it as an Iraqi province. A U.S.-led UN coalition, receiving substantial support from Saudi Arabia, expelled Iraq from Kuwait in early 1991. Twelve years later, the 2003 American-led invasion of the country replaced one set of uncertainties with another.

When Iraqi leaders assumed control of their new state on June 28, 2004, more than 135,000 American troops remained in the country. Thereafter, growing religious and ethnic violence between different Iraqi factions has thrown portions of the nation into civil war. Rival Sunni and Shiite groups have forced many Iraqis from their communities. Entire neighborhoods and villages have witnessed dramatic ethnic shifts as residents react to changing local conditions. Often, local religious factions as well as elements of Al Qaeda have also directed violence at remaining American troops within the country. These attacks have led to ongoing political debates concerning the American role in the region and the correct path toward solving the crisis within Iraq. Some have suggested partitioning the country into three parts (Shiite, Kurd, and Sunni), but this poses its own difficulties, both with complex distributions of people and resources within Iraq as well as with neighboring states that might see such divisions as further destabilizing the region. A final geopolitical solution has yet to take shape, but it will need to recognize the cultural differences that divide the country, as well as allow for a revival of the country's crippled oil industry.

Instability in Saudi Arabia The region's greatest oil power also finds itself in the throes of geopolitical changes that are complex and difficult to predict. A conservative monarchy (the Saud family) controls the country and has been unwilling to promote much in the way of democratic reforms. On the surface, it has supported U.S. efforts in the region to provide stable flows of petroleum, but beneath the surface certain elements of the regime may have financed radically anti-American groups such as Al Qaeda. The Saudi people themselves, largely Sunni Arabs, are torn between an allegiance to their royal family (and the economic stability it brings), the lure of a more democratic, open Saudi society, and an enduring distrust of foreigners, particularly Westerners. Furthermore, the Sunni majority includes Wahhabi sect members, whose radical Islamist philosophy has fostered anti-American sentiment and encouraged Al Qaeda (15 of the 19 hijackers in the September 2001 terrorist attacks were Saudis). Add to this a large number of foreign laborers and a persisting American military and economic

Walling in Bethlehem and the Israeli Security Barrier

Located southwest of Jerusalem in the much-disputed West Bank region of the Middle East, Bethlehem is a modern Palestinian and Christian community of 60,000 people (Figure 7.5.1). It is home to a Muslim majority and an old and important Christian minority. Nearby, several new Jewish settlements add further ethnic complexity. Today, Bethlehem finds itself encircled by the new Israeli security barrier. Begun in 2002, this barrier now snakes its way across much of the northwest border zone between Israel and West Bank Palestinian settlements. A shorter segment also stretches south of Jerusalem to the area around Bethlehem (Figure 7.5.2). The planned 440-mile (700-kilometer) wall is Israel's response to repeated suicide bombings in the country since 2000. Barrier advocates suggest that walling out the Palestinian threat is the only way to guarantee Israeli security. They have watched with grim satisfaction as the reinforced concrete wall (26 feet, or 8 meters, high) has gone up, complete with razor wire, ditches, electronic sensors, and manned watchtowers. Large portions of the barrier were completed by 2007, with future additions planned across southern portions of the West Bank as well as in the Jordan Valley farther east (see Figure 7.5.1). Ironically, cash-strapped Palestinian laborers often are used to build the very wall critics claim is destined to doom their cause and take their land.

Figure 7.5.1 Route of the Israeli Security Barrier
The map shows the completed and planned portions of the Israeli security barrier as well as many of the Israeli settlements located in the West Bank region.

Figure 7.5.2 Israeli Security Barrier
This segment of the Israeli security barrier passes near Bethlehem, south of Jerusalem. *(Kevin Unger–KPA/Zuma)*

Indeed, most Palestinians see the barrier (they have dubbed it the "Apartheid Wall") as a blatant land grab and an Israeli attempt to isolate them from the rest of the world. They argue it is separating many Palestinians from job opportunities in Israel and preventing Palestinian exports from being sold. Even worse, they cite many settings where the barrier is literally dividing village neighborhoods and separating farmers from their fields. In Bethlehem, the wall is encircling the town on several sides, harming the tourist industry and vitality of the local economy. A growing series of Israeli-monitored checkpoints in the area and throughout the West Bank is making Palestinian movement more difficult outside their own communities. Palestinians claim that the location of the barrier, often running well inside the West Bank territory rather than along the internationally recognized border with Israel, is nothing less than a grab for more real estate. Israelis counter that a decline in the frequency of suicide bombings demonstrates that the barrier is working. In particular, they cite Bethlehem as a traditional source area of recent terrorist attacks within Israel.

Whether it is a line of safety and security in the sand or a cruel, inhumane barricade reminiscent of the Cold War, the barrier is not likely to disappear anytime soon. While Israeli courts have ordered portions of the barrier realigned, the Israeli government seems firmly committed to its extension. A United Nations–supported ruling by the International Court of Justice in 2004 proclaimed the barrier illegal and ordered it removed. The court's ruling is not enforceable and therefore construction is likely to continue, although portions of the barrier may be moved to accommodate local Palestinian needs. As Palestinians painfully adjust to their predicament, Israelis hope the barrier will pressure moderates to bargain more aggressively for a solution to the longer-term crisis. Meanwhile, Israel's enhanced security comes at a high price and the residents of Bethlehem and elsewhere in the West Bank appear likely to continue paying for it until the larger conflict is resolved.

presence within the country (one of the chief complaints of Al Qaeda leader Osama bin Laden) and you have a setting ripe for political instability. High world oil prices have recently helped keep the level of protests muted, however, as a healthy Saudi economy has made for fatter wallets and a more satisfied, complacent population.

The Non-Arab Fringe: Turkey and Iran Turkey and Iran, two large non-Arab states, define the northern rim of Southwest Asia. While these two countries have not seen the degree of political instability evident in neighboring Iraq, both nations find themselves strategically positioned between diverse, often contradictory geopolitical forces. Many pro-Westerners within Turkey, for example, are committed to joining the European Union. To do so, the country has embarked on an active agenda of reforms designed to demonstrate its commitment to democracy. Press freedoms and multiparty elections have been permitted, the role of the military downplayed, and minority groups (particularly the Kurds) have gained greater recognition from the central government. At the same time, anti-Western Islamist political elements have been on the rise, linked to a growing number of terrorist bombings in the country since 2003. Moderate Turks are hoping that its road to the European Union will not be interrupted with the violence, but Islamists and ultra nationalists are suspicious of the EU pledge to move the country in a radically different direction. Turkey also has continuing tensions with nearby Greece and Cyprus, two countries that are now both in the EU. Final approval of Turkey's admission still remains clouded by all of these concerns.

Nearby Iran increasingly garners international attention. Islamic fundamentalism dramatically appeared on the political scene in 1978 and 1979 as Shiite Muslim clerics overthrew the shah, an authoritarian, pro-Western ruler friendly to U.S. political and economic interests. The Ayatollah Khomeni, a once-exiled religious leader, took power in 1979 and proclaimed an Islamic republic in which religious officials ruled both clerical and political affairs. Building on considerable domestic distrust of both the shah and the United States, Khomeni led a revolution that continues to shape Iranian politics today.

Iran now finds itself at the center of a growing regional power struggle. Iran supports Shiite Islamist elements throughout the region (such as Hezbollah). The United States, Russia, and China all have a keen eye on its rich global reserves of oil (10 percent) and natural gas (15 percent). Even more problematic is Iran's ongoing nuclear development program, an initiative its government claims is solely related to the peaceful construction of power plants. Many in the West, however, remain unconvinced of the government's motives and demand that its program be stopped or opened to international inspections before allowing for greater international cooperation with the Iranian regime. Meanwhile, an educated Iranian middle class that never fully embraced the fundamentalist cause has quietly advocated more ties with the West. Recently, however, those calls for moderation have been drowned out by an Islamist Iranian government that continues to play toward more radical, often anti-Western, anti-Israeli elements in Iranian society.

An Uncertain Political Future

Few areas of the world pose more geopolitical questions than Southwest Asia and North Africa. Twenty years from now, the region's political map could look quite different from the one today. The region's strategic global importance increased greatly after World War II, propelled into the international spotlight by the creation of Israel, the tremendous growth in the world's petroleum economy, Cold War tensions between the United States and the Soviet Union, and, more recently, the rise of Islamic fundamentalism and Islamist political movements. Now that the Cold War between global superpowers has ended, the region is experiencing a geopolitical reorientation.

One key question that remains is what geopolitical role should the United States play in Southwest Asia? That simple question continues to confound policymakers in the United States as well as residents of that ever-troubled part of the world. Since the United States embarked on its war on terror, its geopolitical presence in the region has inevitably grown. The 2003 Iraqi invasion put large numbers of American troops in the region. But many questions remain. How does the United States identify and deal with hostile elements? Should the United States go it alone in these efforts ("unilateralism"), or depend on sharing the responsibilities with other countries or international organizations ("multilateralism")? Regional resistance movements directed against the United States have grown more powerful in many countries within the region, drawing on growing popular suspicion of American motives and on the increasingly heated religious rhetoric of Islamist elements.

Four interrelated settings appear destined to hold continuing U.S. attention within the region. First, the U.S. involvement in Iraq remains in flux, with religious and ethnic tensions within the country further complicating a continuing U.S. presence in the country. A second question mark is oil-rich Saudi Arabia. Terrorist links to Saudi Arabia have renewed friction between the two countries. Domestic challenges to the all-powerful al-Saud royal family also appear to be on the rise. U.S. observers fret that a sudden change in Saudi Arabian politics would have global ramifications, particularly if oil exports are disrupted. Third, the growing geopolitical presence of Iran challenges American, Israeli and many moderate Arab (particularly Saudi Arabian) interests within the region. Iran's regional aspirations are thus yet another key question mark for this portion of the world. Finally, recent events in Israel, Gaza, and the West Bank continue to make global headlines. Palestinian infighting, continuing attacks on Israel, and Israeli reprisals (including the construction of its security barrier) have further enflamed the region. What is an appropriate role for the United States in tactically improving short-term relationships between Israel and its Palestinian populations and in crafting a longer-term strategic policy aimed at achieving a permanent peace in the region? All of these issues will command a great deal of attention within the United States. How the United States responds to Southwest Asia's shifting geopolitics seems destined to shape the region for decades to come.

ECONOMIC AND SOCIAL DEVELOPMENT: Lands of Wealth and Poverty

Southwest Asia and North Africa is a region of both incredible wealth and disheartening poverty. While a few of its countries enjoy great prosperity, due mainly to rich reserves of petroleum and natural gas, other nations within the region are among the least developed in the world (Table 7.2). Overall, recent economic growth rates have lagged behind those of the more-developed world. Persisting political instability has contributed greatly to the regional economic malaise. These economic stumbling blocks have had profound social consequences: investments in education, health care, and new employment opportunities have slowed considerably in many countries from the heady gains of the late 1970s and 1980s. Petroleum will no doubt figure prominently into the region's future economic relationships with the rest of the world, but many countries in the area also have focused on increasing agricultural output, investing in new industries, and promoting tourism as important ways to broaden the regional economic base.

The Geography of Fossil Fuels

The striking global geographies of oil and natural gas reveal the region's persisting importance in the world oil economy, as well as the extremely uneven distribution of these resources within the region (Figure 7.36). Higher oil prices early in the new century have once again highlighted the economic clout of this region of the world. Saudi Arabia remains one of the major producers of petroleum in the world, and Iran, the United Arab Emirates, Libya, and Algeria also contribute significantly. The region plays an important though less dominant role in natural gas production. The distribution of fossil fuel reserves suggests that regional supplies will not be exhausted anytime soon. Overall, with only 7 percent of the world's

TABLE 7.2 Development Indicators

Country	GNI[a] per Capita, PPP[b] (2005)	GDP[c] Average Annual % Growth (2000–05)	Life Expectancy (2007)	% Population Living on Less Than $2 a Day	Under Age 5 Mortality Rate 1990	Under Age 5 Mortality Rate 2005	Gender Equity[d]
Algeria	6,770	5.2	72	29	69	39	102
Bahrain	21,290		74				
Egypt	4,440	3.7	71	15	104	33	
Gaza and West Bank		−0.9	72		40	23	104
Iran	8,050	5.8	70	7	72	36	99
Iraq		−11.4	57		50		76
Israel	25,280	1.9	80		12	6	105
Jordan	5,280	6.1	72	7	40	26	102
Kuwait	24,010	7.3	78		16	11	110
Lebanon	5,740	4.0	71		37	30	104
Libya		5.3	73		41	19	106
Morocco	4,360	4.3	70	14	89	40	88
Oman	14,680	3.0	74		32	12	99
Qatar			73				
Saudi Arabia	14,740	4.2	75		44	26	101
Sudan	2,000	6.1	58		120	90	89
Syria	3,740	3.7	73		39	15	94
Tunisia	7,900	4.5	74	7	52	24	105
Turkey	8,420	5.2	72	19	82	29	84
United Arab Emirates	24,090	8.2	79		15	9	126
Western Sahara			65				
Yemen	920	3.3	60	45	139	102	61

[a]*Gross national income.*

[b]*Purchasing power parity.*

[c]*Gross domestic product.*

[d]*Ratio of female-to-male enrollments in primary and secondary school percentage. Numbers below 100 have more males in primary/secondary school; numbers above 100 have more females in primary/secondary schools.*

Sources: World Bank, World Development Indicators, *2007; life expectancy 2007 and percentage of population living on less than $2 a day data from* Population Reference Bureau, World Data Sheet, *2007; gender equity data from* Millennium Development Goals.

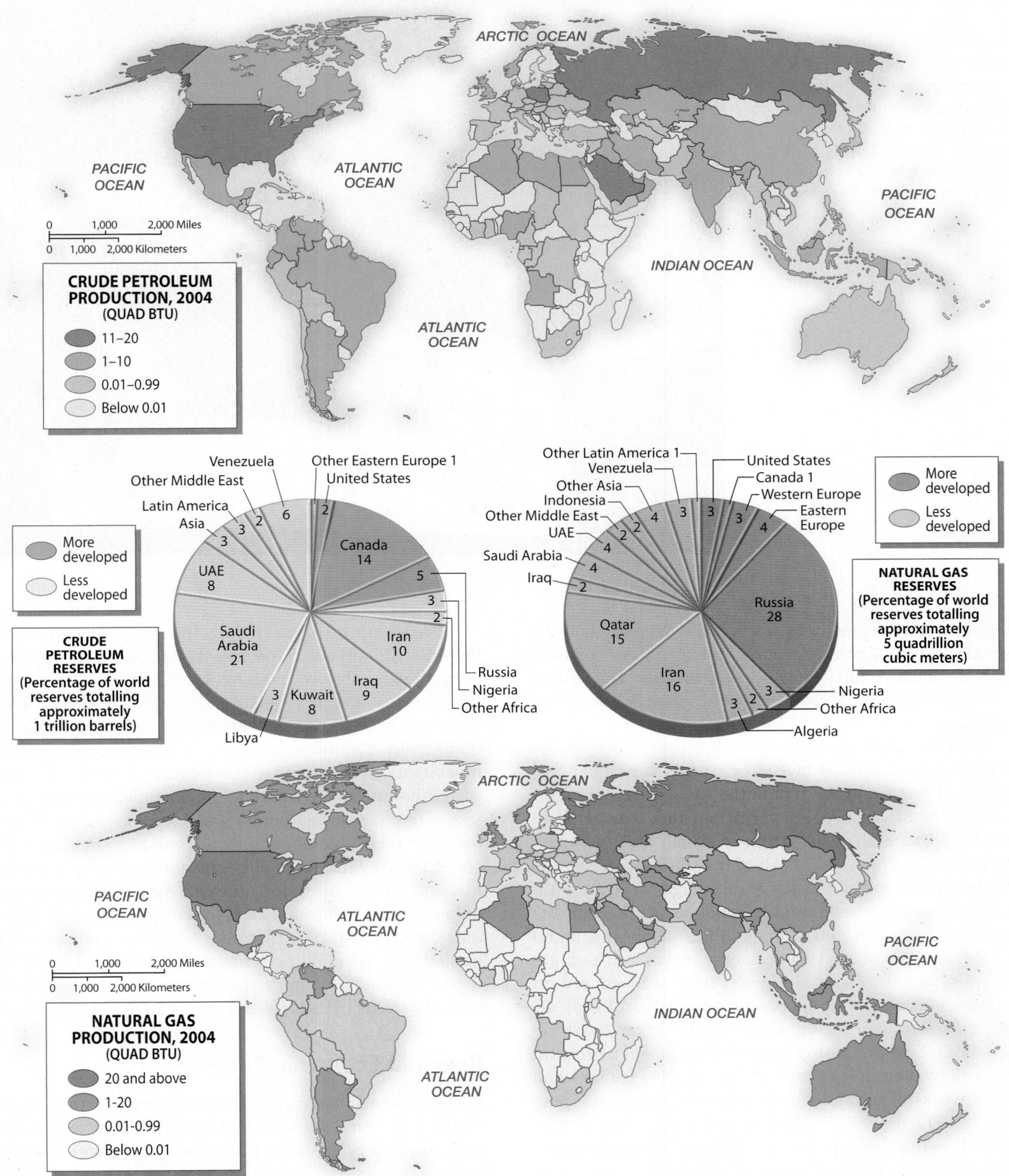

Figure 7.36 Crude Petroleum and Natural Gas Production and Reserves
The region plays a central role in the global geography of fossil fuels. Abundant regional reserves suggest that the pattern will continue. *(Modified from Rubenstein, 2008,* An Introduction to Human Geography, *9th ed., Upper Saddle River, NJ: Prentice Hall)*

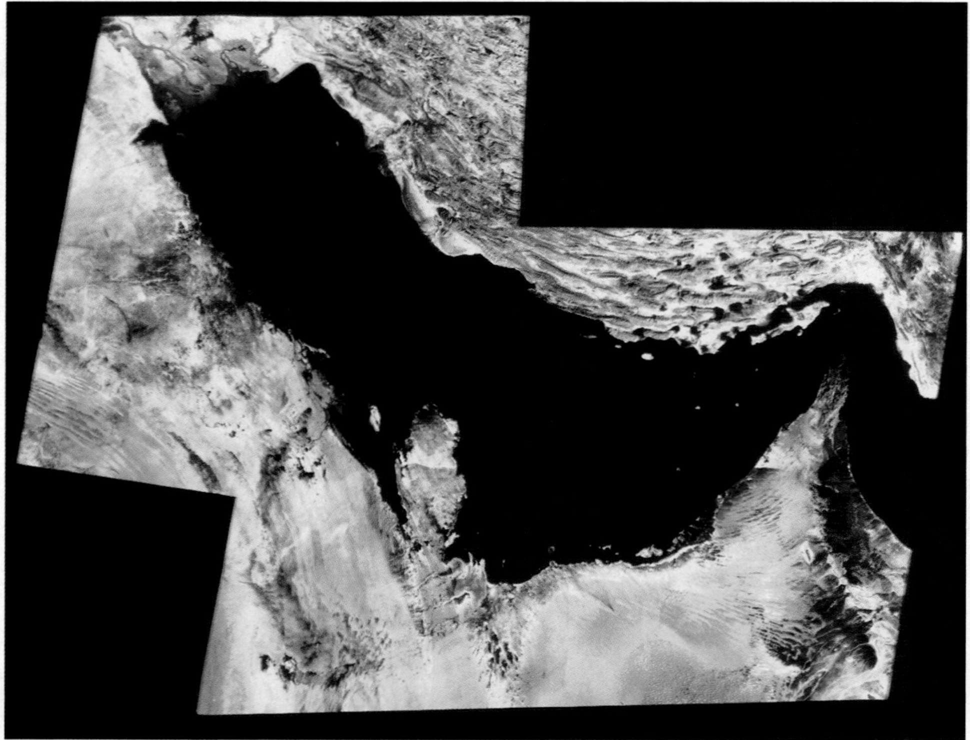

Figure 7.37 Persian Gulf This satellite view of the Persian Gulf reveals one of the world's richest sources of petroleum. Sedimentary rocks, both on land and offshore, contain additional reserves that can sustain production for decades. *(Earth Satellite Corporation/Science Photo Library/Photo Researchers, Inc.)*

population, the region holds a staggering 69 percent of the world's proven oil reserves. Saudi Arabia's pivotal position, both regionally and globally, is clear: its 28 million residents live atop 21 percent of the planet's known oil supplies.

Several major geological zones supply much of the region's output of fossil fuels. The world's largest concentration of petroleum lies within the Arabian–Iranian sedimentary basin, a geological formation that extends from northern Iraq and western Iran to Oman and the lower Persian Gulf (Figure 7.37). All of the states bordering the Persian Gulf reap the benefits of oil and gas deposits within this geological basin, and it is not surprising that the world's densest concentration of OPEC members is found in the area. A second important zone of oil and gas deposits includes eastern Algeria, northern and central Libya, and scattered developments in northern Egypt. As in the Persian Gulf region, these North African fields are tied to regional processing points and to global petroleum markets by a complex series of oil and gas pipelines and by networks of technologically sophisticated oil shipping facilities (Figure 7.38). Third, a zone of growing importance is being developed in the African nation of Sudan, including areas near and south of Khartoum. Near Khartoum, the commercial center of Alsunut has become Africa's largest commercial construction site. Home to regional oil giants Petrodar and the Greater Nile Petroleum Operating Company, Alsunut is being built on Sudan's petroleum riches and is financed by a global set of investors from the Gulf region, Malaysia, China, Pakistan, and elsewhere.

Even with all these riches, the geography of fossil fuels is strikingly uneven across the region. Some states—even those with tiny populations (Bahrain, Qatar, and Kuwait, for example)—contain incredible fossil fuel reserves, especially when considered on a per capita basis. Many other countries, however, and millions of regional inhabitants, reap relatively few benefits from the oil and gas economy. In North Africa, for example, Morocco possesses few developed petroleum reserves, and even the fruits of Sudan's blossoming wealth will likely remain very concentrated within a small segment of a large, poor population. Even in oil-rich Southwest Asia, the distribution of fossil fuels is uneven: Israel, Jordan, and Lebanon all lie outside favored geological zones for either petroleum or natural gas. While Turkey has some developed fields in the far southeast, it must import substantial supplies to meet the needs of its large and industrializing population.

Regional Economic Patterns

Remarkable economic differences characterize the region (Table 7.2). Some oil-rich countries have prospered greatly since the early 1970s, but in many cases fluctuating oil prices, political disruptions, and rapidly growing populations have reduced prospects for economic growth. Other nations, although poor in oil and gas reserves, have seen brighter prospects through moves toward greater economic diversification. Finally, some countries in the region are subject to persisting poverty, where rapid population growth and the basic challenges of economic development combine with political instability to produce very low standards of living.

Higher-Income Oil Exporters The most prosperous countries of Southwest Asia and North Africa owe their wealth to massive oil reserves. States such as Saudi Arabia, Kuwait (after recovering from the Gulf War), Qatar, Bahrain, and the United Arab Emirates benefit from fossil fuel production, as well as from their relatively small populations. Since the oil-price hikes of the 1970s, these countries have reaped billions in revenues that have had a fundamental impact on their economies. For example, rising oil prices in the past four years have nearly doubled Saudi Arabia's economic output (Table 7.2). By 2007, the Saudi government had an accumulated surplus of more than $250 billion, thanks to foreign oil sales.

Nations such as Saudi Arabia, Kuwait, Qatar, and the United Arab Emirates, have made huge investments in transportation infrastructure, urban commercial and financial centers, and other petroleum-related industries. The Saudi petroleum-processing and shipping centers of Jubail (on the Persian Gulf) and Yanbu (on the Red Sea) exemplify this commitment to expand their economic base beyond the simple extraction of crude oil. With these industrial hubs, the Saudis participate more broadly in the benefits of the oil economy by refining and manufacturing petroleum-related products for export. Other oil-rich cities have cultivated an even more cosmopolitan image, with globalized Dubai in the United Arab Emirates appealing to both international tourists and business interests (see: "Cityscapes: The Instant Global City of Dubai").

While much of the oil wealth in these nations remains concentrated among the ruling elite, petrodollars also have provided real improvements for the larger population. Billions of dollars have poured into new schools, medical facilities, low-cost housing, and modernized agriculture, fundamentally raising the standard of living in the past 40 years. In addition, the Saudis have opened their economy to more foreign investment both within and beyond the petroleum sector.

Still, problems remain, even in these centers of relative wealth. Poor people do reside in Saudi Arabia and elsewhere in the oil-rich region. For example, the Shiite minority in eastern Saudi Arabia has a standard of living far below that of urban populations elsewhere in the country. In addition, large foreign workforces are common across the oil-rich zone, and they typically are paid far less than domestic laborers. Foreign men usually work in construction and the oil industry, whereas foreign women mostly work as domestic servants. Foreign laborers typically live and work under tough conditions, but they receive wages that are far higher than they could earn at home.

Another challenge relates to the shifting fortunes of the oil economy itself. When prices rise, as they did in the 1970s, early 1980s, and early this decade, these oil-rich nations gain an economic windfall. Recently, GNI per capita figures in major producing states such as Kuwait and the United Arab Emirates rival those of many European nations (see Table 7.2). But dependence on oil and gas revenues clearly has a darker side: falling prices, such as those seen in the mid-1980s or in the late 1990s, created immediate economic pain for the major Middle East producers. Such fluctuations in world oil markets will inevitably continue in the future. While OPEC continues to be a significant global economic force, many non-OPEC producers also compete in the world marketplace (including Europe, Mexico, and Russia), thus diminishing OPEC's power to set prices. As a result, future economic growth in these wealthier oil-dependent states will likely be moderate, at best. Although Saudi Arabia, blessed with abundant reserves of easily accessible oil, can survive with high or low prices in the future, countries such as Bahrain and Oman are faced with the additional problem of rapidly depleting their reserves over the next 20 to 30 years.

Lower-Income Oil Exporters Other areas of the region are important secondary players in the oil trade, but different political and economic variables often have hampered sustained economic growth. In North Africa,

Figure 7.38 Bejaia, Algeria
The northern Algeria port of Bejaia sits at the terminus of a major oil pipeline in this portion of North Africa. *(Sebastian Cailleux/Corbis)*

CITYSCAPES

The Instant Global City of Dubai

Studded with eye-popping skyscrapers and famous for its fabricated islands in the shape of palms, Dubai got its start with oil-dollars. This bustling city on the Persian Gulf has since diversified into a major financial and commercial hub of the Middle East. Its leaders, the ruling al-Maktoum family, aspire to make it a global city that competes with London, Tokyo, New York, and Hong Kong. In less than a lifetime, Dubai has grown from a small fishing village of 40,000 in 1960, to a glittering metropolis of 1.3 million residents. Clustered developments such as Internet City, Media City, Healthcare City (partnered with Harvard University), and the Dubai International Financial Center are poised to serve twenty-first-century-needs of the Middle East. Foreigners can purchase property in Dubai; thus real estate development and speculation drive much of the city's economy. The city's showpiece project, the Burj Dubai, will be the world's tallest building (Figure 7.6.1).

Tourists from the Middle East, Russia, and Europe pour into this city, some 5 million a year. The Burj al Arab, a seven-star hotel in the shape of a sail, serves a jet-set elite from Moscow to Riyadh (Figure 7.6.2). International tourists, glitzy sporting events, and major corporation headquarters (most recently, Haliburton) find this Arab boomtown irresistible. In 2005, Dubai added indoor skiing to the Mall of the Emirates (the largest mall in the Middle East, naturally). City billboards proclaim, in both English and Arabic, "If you can dream it, it can happen."

Figure 7.6.1 Burj Dubai, Dubai, United Arab Emirates South Asian contract laborers from Bangladesh, India, and Pakistan work at the construction site of Burj Dubai, or "Dubai Tower" in Dubai. The 2,600-foot (800-meter)-high structure will vie to be the world's tallest building when completed. *(Rob Crandall/www.robcrandall.com)*

Algeria, Libya, and portions of Sudan illustrate this scenario. Algerian oil and natural gas overwhelmingly dominate its exports, but the past 15 years have also brought political instability and increasing shortages of consumer goods. While the country contains some excellent agricultural lands in the north, the overall amount of arable land has increased little over the past 25 years, even as the country's population has grown by more than 50 percent. The pressure on the rural economy has accelerated movement to increasingly crowded cities, and many of the nation's upwardly mobile workers have left the country for employment opportunities in western Europe. Nearby Libya also remains a major exporter of oil and natural gas, but foreign investment beyond the petroleum industry has remained limited. Still, Qaddafi has undertaken ambitious efforts to expand the agricultural economy and improve water availability in his desert land, and he recently has pushed for expanding the country's tourism industry in the hope of attracting more outsiders. In Sudan, recent political strife in the south as well as ongoing upheaval in the west (Darfur) has slowed that country's move toward developing its considerable petroleum resources. At the very least, western European and North American participation in these ventures has been severely limited due to the hostile political environment. Much of the nation remains mired in poverty. The country's larger infrastructure has seen little new investment; settlement remains overwhelmingly rural; and secondary school enrollments stand at less than 25 percent of the school-age population

In Southwest Asia, huge economic and political challenges face Iraq. The American invasion and subsequent civil war crippled much of Iraq's already deteriorated infrastructure, and ongoing political instability has made the task of rebuilding its economy even more difficult. Health care in the country has deteriorated greatly since 1990, and food shortages are much more common in many areas than they were 20 years ago. Iraq certainly has the potential to be a prosperous country. Its population density is moderate and its petroleum resource base is

Figure 7.6.2 Landscapes of Luxury, Dubai, United Arab Emirates
The canals of the elite Madinat Jumeirah Arabian Resort are visible in the foreground. In the background, the telltale sail of the luxurious Burj Al Arab Hotel is one of the most dramatic visual symbols of affluence in the region. The hotel bills itself as the world's first seven-star hotel. *(Rob Crandall/www.robcrandall.com)*

Dubai's rise is emblematic of the integrating yet uneven forces of globalization. Oil and ambition created the city; yet it also benefited indirectly from post-September 11 geopolitics. Arab investors flush with cash began to invest more in the Middle East—Dubai was ready and waiting. Yet, the fact that four-out-of-five residents in Dubai are foreign-born, mostly from South Asia, usually gets overlooked in the promotional hype about this Gulf city. There is no other major city in the world that is so dependent upon foreign labor. Almost all are guest workers with no access to citizenship. Due to their sheer numbers, immigrants are Dubai. South Asians call it the "best run Indian city." Yet the foreign-born are socially and spatially isolated from the Emiratis—a separation that is intensified by housing policies that benefit only Emiratis. Similarly, most companies are obligated to house foreign workers, but these accommodations are hostels on the urban periphery, far from where most Emirati live.

Visitors are repeatedly told that Dubai is a more open and tolerant Arab city, which is true. Openness, however, has its limits. Although Emiratis enthusiastically embrace the trappings of Western capitalism and consumerism, they are increasingly concerned about the negative cultural influences of foreigners on their identities and values. The minority status of Emiratis in their own country does not go unnoticed. As Professor Abdallah of the UAE University observed, "usually minorities assimilate into the majority but we don't want to assimilate into the majority. We want to preserve the localness, the Emiratiness of this city."

substantial. Yet as long as political instability plagues the country, sustained economic expansion is unlikely.

The situation in Iran also is challenging. The country is large and populous and has a relatively diverse economy. Iran's oil reserves are huge and have seen active commercial development since 1912. The country also has a sizable industrial base, much of it built in the 20 years prior to the fundamentalist revolution of 1979. In addition, considerable capital flowed into modernizing agriculture during the same period, bringing the benefits of increased mechanization and new irrigation projects. Still, most observers agree that today Iran is relatively poor, burdened with a stagnating if not declining standard of living. Since 1980, the country's fundamentalist leaders have downplayed the role of international trade in consumer goods and services, fearing they would import unwanted cultural influences from abroad. Making matters worse were the twin challenges of a costly and bloody war with Iraq in the 1980s, followed by the struggling oil economy of the 1990s. Recently, however, the country's economic prospects have brightened somewhat with new economic links to Central Asia and with a recovery in oil prices. In addition, Iran's literacy rate (particularly for women) has risen, reflecting a new emphasis on rural education in the country. Still, the political differences with potential Western trading partners are vast, and Iran's ongoing nuclear development program has further isolated the country from many nations, including the United States.

Prospering Without Oil Some countries, while lacking petroleum resources, have nevertheless found paths to increasing economic prosperity. Israel, for example, supports one of the highest standards of living in the region, even with its tremendous political challenges (see Table 7.2). The nation's sparse natural resource base would hardly seem conducive to economic development. But the Israelis and many foreigners have invested large amounts of capital to create a highly productive agricultural and industrial base. The country also is emerging as a global center for high-tech computer and telecommunications products.

Figure 7.39 Israeli High-Tech Industry
These Israeli workers are employees of the Telrad factory, a high-tech division of the Koor conglomerate. *(Ricki Rosen/Corbis SABA)*

Many U.S. and European companies maintain development and production centers there, and the country is increasingly becoming known for its fast-paced and highly entrepreneurial business culture, which resembles California's Silicon Valley (Figure 7.39). Tourism is another important sector of the economy, as the country attracts global visitors interested in the region's rich cultural heritage. Even so, Israel has daunting economic problems. Its persisting struggles with the Palestinians and with neighboring states have sapped much of its potential vitality. Defense spending absorbs a large share of total gross national income, necessitating high tax rates. Poverty among the Palestinians is widespread, and the gap between rich and poor within the country has widened considerably.

Turkey also has a diversified economy. Whereas its per capita income is modest even by regional standards, it has shown far greater economic dynamism in recent years than have many of the major oil producers in the region. Lacking petroleum, Turkey produces varied agricultural and industrial goods for export. Almost half of the population remains employed in agriculture, and the country's principal commercial products include cotton, tobacco, wheat, and fruit. The industrial economy has grown since 1980, including exports of textiles, food, and chemicals. Turkey remains the most important tourist destination in the region as well, attracting more than 6 million visitors annually in recent years. In addition, Istanbul enjoys increasing prominence as a regional finance and investment center. Many Turkish leaders hope to strengthen the economy by eventually joining the European Union. Turkey already has closer links to the West than any other country of the region except Israel (it is, for example, a member of NATO), and its trade potential with Europe is great. Still, economic integration with the West will not be easy. Some European leaders are wary of extending their economic union to what they consider a politically unstable non-European state in which anti-Western Islamist forces seem to be gaining political power.

Other oil-poor countries in the region aspire to economic prosperity. In North Africa, Tunisia has fostered relative economic stability without the benefit of large oil reserves. Although the country possesses a few important oil fields south of Tunis, most of its workforce is in agriculture, services (particularly tourism), and manufacturing. Recent government policies have favored economic reforms, increased private investment, and encouraged participation in global markets, such as those that bring the region's citrus fruits to grocery stores in the United States.

Regional Patterns of Poverty The poorer countries of the region share the problems of much of the less-developed world. In North Africa, large areas of Morocco and Egypt face distinctive economic challenges. Morocco remains poorer than either Algeria or Tunisia (Table 7.2). Poverty is especially widespread in the Atlas Mountains, where many Berber communities have little access to modern services or infrastructure. Elsewhere in the country, economic reforms have spurred privatization, encouraged investment by foreign banks, and slowed inflation. Still, illiteracy is widespread, and the country suffers from the **brain drain** phenomenon as some of its brightest young people leave for better jobs in western Europe. Rural population pressures and a stagnating agricultural economy also have spurred internal movements from the countryside to major cities, including Casablanca and Rabat, but migrants often are frustrated by ongoing problems of underemployment even in these rapidly expanding metropolitan areas.

Egypt's economic prospects also are unclear. On the one hand, the country has experienced real economic growth in the last decade as President Hosni Mubarak pushed for smaller government deficits and a multibillion-dollar privatization program to put government-controlled assets under more efficient management. Egypt also actively courts foreign investment in its expanding industrial sector, and the streets of Cairo are increasingly sprinkled with the flash of Rolls-Royces, Porsches, and Lamborghinis. Even so, many Egyptians still live in poverty, and the gap between rich and poor continues to widen. Millions of farmers till tiny plots of land along the Nile, while others scramble for living wages in low-paying service and manufacturing jobs in Cairo and Alexandria. Furthermore, the demographic clock is ticking: Egypt's 75 million people already make it the region's most populous state, and recent efforts to expand the nation's farmland have met with numerous environmental, economic, and political problems. Future prosperity hinges on the country's ability to expand its economy faster than its population.

In Southwest Asia, Yemen remains the poorest country on the Arabian Peninsula. Positioned far from most of the region's principal oil fields, Yemen's low per capita GNI puts it on par with many nations in impoverished Sub-Saharan Africa or South Asia. The largely rural country relies mostly on marginally productive subsistence agriculture, and much of its mountain and desert interior lacks effective links to the outside world (Figure 7.40). The present state emerged in 1990 with the political union of North and South Yemen. Coffee, cotton, and fruits are commercial agricultural products and modest oil exports bring in needed foreign currency. Overall, however, high unemployment and marginal subsistence farming remain widespread across the country.

Unique problems afflict the Palestinian populations of Gaza and the West Bank. Continued declines in employment and income have devastated the economy as political disruptions discourage investment and conflicts among the Palestinians and with the Israelis destroy infrastructure (see Table 7.2). Recent violence among the Palestinians in Gaza further decimated the economy of that war-torn region. In the nearby West Bank, when Israeli forces suspect political dissidents and terrorist elements in a given Palestinian town or urban neighborhood, their response is to simply destroy the settlement, leveling houses and shops in the process. Poverty now grips two-thirds of the Palestinian population; unemployment hovers above 40 percent; and the ongoing construction of the Israeli security barrier promises to further disrupt the Palestinian economy in the years to come.

A Woman's Changing World

The role of women in the largely Islamic region also remains a major social issue. Female labor participation rates in the workforce are among the lowest in the world, and large gaps typically exist between male and female literacy (see Table 7.2). In the most conservative parts of the region, few women are allowed to work outside the home. Even in parts of Turkey, where Western influences are widespread, it is rare to see rural women selling goods in the marketplace or driving cars in the street. More orthodox Islamic states impose legal restrictions on the activities of women. In Saudi Arabia, for example, women are not allowed to drive. In Iran, full veiling remains mandatory in more conservative parts of the country. Generally, Islamic women lead more private lives than men: much of their domestic space is shielded from the world by walls and shuttered windows, and their public appearances are filtered through the use of the face veil or chador (full-body veil).

Yet in some places women's roles are changing, even within the norms of more conservative, Islamist societies. Many young Algerian women demonstrate the pattern (Figure 7.41). Most studies suggest the younger generation are more religious than their parents and are more likely to cover their heads and drape their bodies with traditional religious clothing. At the same time, they are more likely to be educated and employed than ever before. Today, 70 percent of Algeria's lawyers and 60 percent of its judges are women. A majority of university students are women and women dominate the health-care field. These new social and economic roles also help explain why the birthrates in such settings are declining and are likely to continue doing so. The Algerian case also demonstrates how complex social processes actually unfold, rarely following simple models for how "traditional" societies might evolve.

Women's roles are shifting elsewhere in the region. In Sudan and Saudi Arabia, a growing number of women pursue high-level careers. Education may be segregated, but it is available. Libya's Qaddafi has singled out the modernization of women as a high priority, and today more women than men graduate from the country's university system. In the Western Sahara, Saharawi women play a leading role in the country's political fight for independence from Morocco and their broader educational backgrounds and social freedoms (including the right to divorce their husbands) further separate them from Moroccan women. Women also have a more visible

Figure 7.40 Rural Yemen
This isolated hilltop village in the rugged mountains of northern Yemen is surrounded by small, terraced fields and livestock pastures. *(Monique Jacot/All Rights Reserved/Woodfin Camp & Associates)*

Figure 7.41 Algerian Women Even within the norms of a more conservative Islamist society, Algerian women are increasingly visible and highly productive contributors to that nation's work force. *(Francois Perry/Woodfin Camp)*

social position in Israel, except in fundamentalist Jewish communities where conservative social customs require more traditional domestic roles.

Global Economic Relationships

Southwest Asia and North Africa share close economic ties with the world. While oil and gas remain critical commodities that cement these international linkages, the growth of manufacturing and tourism exemplify the region's changing role in the world. New intraregional and interregional ties promise to further rework the ways in which this dynamic part of the world functions in the twenty-first century's global economy. Consumers within the region are also driving new demand for European, Asian, and North American goods and services (Figure 7.42).

OPEC's Changing Fortunes OPEC has not gone away, and the region's enduring role in oil and gas production ensures that fossil fuels will continue to be a major international export for many nations in Southwest Asia and North Africa (see Figure 7.36). While OPEC no longer can control oil and gas prices globally, it still influences the cost and availability of these critical products. Western Europe, the United States, Japan, and many less industrialized countries depend on the region's fossil fuels. In the case of Saudi Arabia, for example, petroleum and its related products make up more than 90 percent of its exports. One recent trend evident in many major oil-producing countries (such as Saudi Arabia) is their increasing willingness to form partnerships with foreign corporations, a pattern that will accelerate the economic integration of the region with the rest of the world.

Figure 7.42 Iranian Consumers Three Iranian teenagers ponder the latest computers at the Paytakht mall, a popular shopping spot in northern Tehran. *(Kaveh Kazemi/Corbis)*

Another energy-related aspect of globalization relates to the flow of investment capital out of the region, a process long dominated by petrodollars. As oil prices increased 10-fold between 1970 and 1980, many of those investments ended up in foreign bank accounts, stock markets, and real estate. Although the region's role in the global oil economy has declined some since 1980, recent estimates still value Gulf Arab foreign investments at more than $800 billion. These investments influence everything from the cost of office buildings in downtown Singapore to the yields on Brazilian bonds.

Beyond the key OPEC producers, other countries within the region are less dependent on oil-related exports and economic diversification has generally contributed to their greater integration in the global economy. Turkey, for example, ships textiles, food products, and manufactured goods to its principal trading partners, Germany, the United States, Italy, France, and Russia. Tunisia sends more than 60 percent of its exports (mostly clothing, food products, and petroleum) to nearby France and Italy. Israeli exports emphasize the country's highly skilled workforce: products such as cut diamonds, electronics, and machinery parts are exported to the United States, western Europe, and Japan.

Regional and International Linkages Future interconnections with the global economy may depend increasingly on cooperative economic initiatives far beyond OPEC. Relations with the European Union are critical. Since 1996, Turkey has enjoyed closer economic ties with the EU, but recent attempts at full membership in the organization have failed. Other so-called Euro-Med agreements also have been signed between the EU and countries across North Africa and Southwest Asia that border the Mediterranean Sea. Supporters argue that these agreements will bring more export-oriented industries to the region and encourage free trade between member states.

Many Arab countries, however, are wary of too much European dominance. They formed a regional political organization known as the Arab League in 1945. Seventeen league members established the Greater Arab Free-Trade Area (GAFTA) in 2005, designed to eliminate all intraregional trade barriers and spur economic cooperation within the region. Smaller organizations tie together other regional constituencies. For example, the six-member Gulf Cooperation Council provides major oil producers in the area with a forum in which to discuss shared security and trade-related issues, and the five-member Union of the Arab Maghreb is dedicated to promoting economic integration and a free-trade zone across North Africa. In addition, Saudi Arabia has played a particularly key role in regional economic development through organizations such as the Islamic Development Bank and the Arab Fund for Economic and Social Development.

The Geography of Tourism Tourists are another link to the global economy. Traditional magnets such as ancient historical sites and globally significant religious areas (for a multitude of faiths) draw millions of visitors annually. As the developed world becomes wealthier, there is also a growing global demand for recreational spots that can offer beaches, sunshine, and novel entertainment. Indeed, many miles of the Mediterranean, Black, and Red Sea coastlines are now lined with the upscale, but often ticky-tacky, landscapes of resort hotels and condominiums dedicated to serving travelers' needs. More adventurous travelers seek ecotourist activities such as snorkeling in Naama Bay on Egypt's Sinai Coast or four-wheeling among the Berbers in the Moroccan backcountry. Endangered wildlife also beckons photographers and poachers hoping to catch a glimpse of a South Arabian grey wolf, Nubian ibex, or a darting Persian squirrel.

All of this activity means big business to many countries in the region, and the economic impacts of tourism seem likely to grow during the twenty-first century. Tourism already is a huge part of the regional economy in settings such as Turkey, Israel, and Egypt. Elsewhere, smaller inflows of tourist dollars still have an extraordinary impact on the economies of, for example, northern Morocco and Tunisia, Lebanon (near Beirut), and the holy Saudi Arabian cities of Makkah and Medinah. In some cases, tourism remains much less developed, but the small existing inflows in such places as Libya and Jordan may be the beginning of a much larger economic trend in future years, depending on the region's political stability.

The growth of tourism, while further connecting the region to the global economy, has come at a considerable price. In addition to the visual blight of high-rise hotels, real environmental damage is increasingly the product of the growing human presence in many fragile regional settings. Various archeological and sacred sites have been negatively impacted across the region. More broadly, the economic realities of the tourist economy have produced a local underclass of poorly paid service workers, a growing sense of social distance between the haves and have-nots in tourist settings, and a consumption-oriented lifestyle that contrasts sharply with many traditional ways of life. Still, these changes appear inevitable for a region that, even in uncertain political times, seems to irresistibly attract tourists from every corner of the globe.

Summary

- Positioned at the meeting ground of Earth's largest land masses, the Southwest Asia and North Africa region historically has played a central role in world history and in processes of globalization. In ancient times, the region's inhabitants were early contributors to the agricultural revolution, a long process of plant and animal domestication destined to reshape world cultures. The region also served as a birthplace for some of the world's earliest urban civilizations, offering in the process a fundamentally new way of human settlement that continues to reshape the distribution of global populations today. Three of the world's great religions—Judaism, Christianity, and Islam—emerged beneath its desert skies.
- Many nations within the region suffer from significant environmental challenges. Twentieth-century population

growth across the region was dramatic. At the same time, it has been difficult and costly to expand the region's limited supplies of agricultural land and water resources. The results, apparent from the eroded soils of the Atlas Mountains to overworked garden plots along the Nile, are a classic illustration of the environmental price paid when population growth outstrips the ability of the land to support it.

- Culturally, the region remains the hearth of Christianity, the spatial and spiritual core of Islam, and the political and territorial salvation of modern Judaism. Muslims worldwide are influenced by its cultural and political evolution, and the Arab–Israeli conflict will continue to be a critical element in global geopolitics.
- Political conflicts have disrupted economic development across the region. Civil wars, conflicts between states, and regional tensions have worked against initiatives for greater cooperation and trade. Perhaps most important, the region must deal with the conflict between modernity and more fundamentalist interpretations of Islam. One thing is certain: future cultural change will be guided by a complex response to Western influences.
- Abundant reserves of oil and natural gas, coupled with the global economy's reliance on fossil fuels, dictate that the region will remain prominent in world petroleum markets. Also likely are moves toward economic diversification and integration, which will gradually draw the region closer to Europe and other participants in the global economy.

Key Terms

brain drain *(page 324)*
culture hearth *(page 282)*
domestication *(page 294)*
exotic rivers *(page 296)*
Fertile Crescent *(page 295)*
fossil water *(page 287)*
Hajj *(page 303)*
hydropolitics *(page 288)*
Islamic fundamentalism *(page 284)*
Islamism *(page 284)*
kibbutzes *(page 296)*
Levant *(page 289)*
Maghreb *(page 288)*
medina *(page 298)*
monotheism *(page 302)*
Organization of Petroleum Exporting Countries (OPEC) *(page 283)*
Ottoman Empire *(page 304)*
Palestinian Authority (PA) *(page 312)*
pastoral nomadism *(page 295)*
physiological density *(page 293)*
protectorates *(page 310)*
qanat system *(page 287)*
Quran *(page 303)*
Shiites *(page 303)*
Suez Canal *(page 311)*
Sunnis *(page 303)*
theocratic state *(page 303)*
transhumance *(page 295)*

Questions for Review

1. Why do Southwest Asia and North Africa form a useful world region? What are some of the problems associated with defining the region?
2. Describe the climatic changes you might experience as you traveled from the eastern Mediterranean coast to the highlands of Yemen. What are some of the key climatic variables that explain these variations?
3. Discuss five important human modifications of the Southwest Asian and North African environment, and assess whether these changes have benefited the region.
4. Discuss how pastoral nomadism, oasis agriculture, and dryland wheat farming represent distinctive adaptations to the regional environments of Southwest Asia and North Africa. How do these rural lifeways create distinctive patterns of settlement?
5. Compare the modern maps of religion and language for the region, and identify three major examples where Islam dominates non-Arabic-speaking areas. Explain why that is the case.
6. Describe the role played by the French and British in shaping the modern political map of Southwest Asia and North Africa. Provide specific examples of their lasting legacy.
7. Define Islamic fundamentalism and Islamism. Outline three regional examples where Islamic fundamentalism or Islamism redefined the domestic geopolitical setting since the late 1970s.
8. Explain how ethnic differences have shaped Iraq's political conflicts in the past 50 years.
9. Describe the basic geography of oil reserves across the region, and compare the pattern with the geography of natural gas reserves.
10. What strategies for economic development have recently been employed by nations such as Turkey, Israel, Egypt, and Morocco? How successful have they been, and how do they relate to the theme of globalization?

Thinking Geographically

1. How might a major project for transferring water from Turkey to the Arabian Peninsula affect the development of Saudi Arabia? What would be some of the potential political and ecological ramifications of such a project?
2. Why are birthrates declining in selected countries within the region? Despite the cultural differences with North America, what common processes seem to be at work in both regions that have contributed to this demographic transition?
3. What economic changes could occur if Israel and the Palestinians were to reach a lasting peace? What kinds of general connections might be found between political conflict and economic conditions throughout the region?

4. Are relations between the region's Muslim societies and the United States likely to improve or worsen? What factors would you cite to defend your answer?
5. What might be some of the reasons for Southwest Asia and North Africa's general failure to match the rates of economic and industrial growth found in North America?
6. Why has the idea of Arab nationalism failed to achieve any lasting geopolitical changes in the region?
7. Imagine you are a ruler of a conservative, Islamist Arab state. What might be the advantages and disadvantages of opening up your country to the Internet?

Regional Novels and Films

Novels

Shmuel Agnon, *Shira* (1989, Schocken)

Hanan Al-Shaykh, *Beirut Blues: A Novel* (1995, Doubleday)

Murid Barghuthi, *I Saw Ramallah* (2003, Doubleday)

Simin Daneshvar, *Savushun: A Novel About Modern Iran* (1990, Mage)

Kahlil Gibran, *Broken Wings: A Novel* (1957, Citadel)

Barbara Hodgson, *The Tattooed Map* (1995, Chronicle Books)

Yasar Kemal, *Salman the Solitary* (1998, Harvill)

Naguib Mahfouz, *Palace of Desire* (1991, Doubleday)

Naguib Mahfouz, *Palace Walk* (1991, Doubleday)

Abdelrahman Munif, *Cities of Salt: A Novel* (1987, Random House)

Films

Casablanca (1942, U.S.)

The Children of Heaven (1999, Iran)

The English Patient (1996, U.S.)

The House on Chelouche Street (1973, Israel)

Kadosh (1999, Israel)

Lawrence of Arabia (1962, U.S.)

Lion of the Desert (1981, U.K.)

The Lizard (Marmoulak) (2004, Iran)

The Ten Commandments (1956, U.S.)

Time of Favor (2000, Israel)

Bibliography

Cramer, Richard Ben. 2004. *How Israel Lost: The Four Questions*. New York: Simon and Schuster.

Esposito, John L., ed. 1999. *The Oxford History of Islam*. Oxford, UK: Oxford University Press.

Ibrahim, Fouad, and Ibrahim, Barbara. 2003. *Egypt: An Economic Geography*. New York: Macmillan.

Keyder, Caglar, ed. 1999. *Istanbul: Between the Global and the Local*. Lanham, MD: Rowman and Littlefield.

Lewis, Bernard. 2004. *The Crisis of Islam: Holy War and Unholy Terror*. New York: Random House.

Mostyn, Trevor, ed. 1988. *The Cambridge Encyclopedia of the Middle East and North Africa*. New York: Cambridge University Press.

Rodenbeck, Max. 1999. *Cairo: The City Victorious*. New York: Knopf.

Soffer, Arnon. 1999. *Rivers of Fire: The Conflict Over Water in the Middle East*. Lanham, MD: Rowman and Littlefield.

Swearingen, Will D., and Bencherifa, Abdellatif, eds. 1996. *The North African Environment at Risk*. Boulder, CO: Westview Press.

United States Geological Survey. 2000. "Implications for Earthquake Risk Reduction in the United States from the Kocaeli, Turkey, Earthquake of August 17, 1999." USGS Circular 1193. Washington, DC: Government Printing Office.

Additional bibliographic resources are at the *Diversity Amid Globalization* Website: http://www.prenhall.com/rowntree/.

8 Europe

Under the midnight sun of arctic Norway, the many troublesome issues facing Europe seem far away, at least for the moment. *(Les Rowntree)*

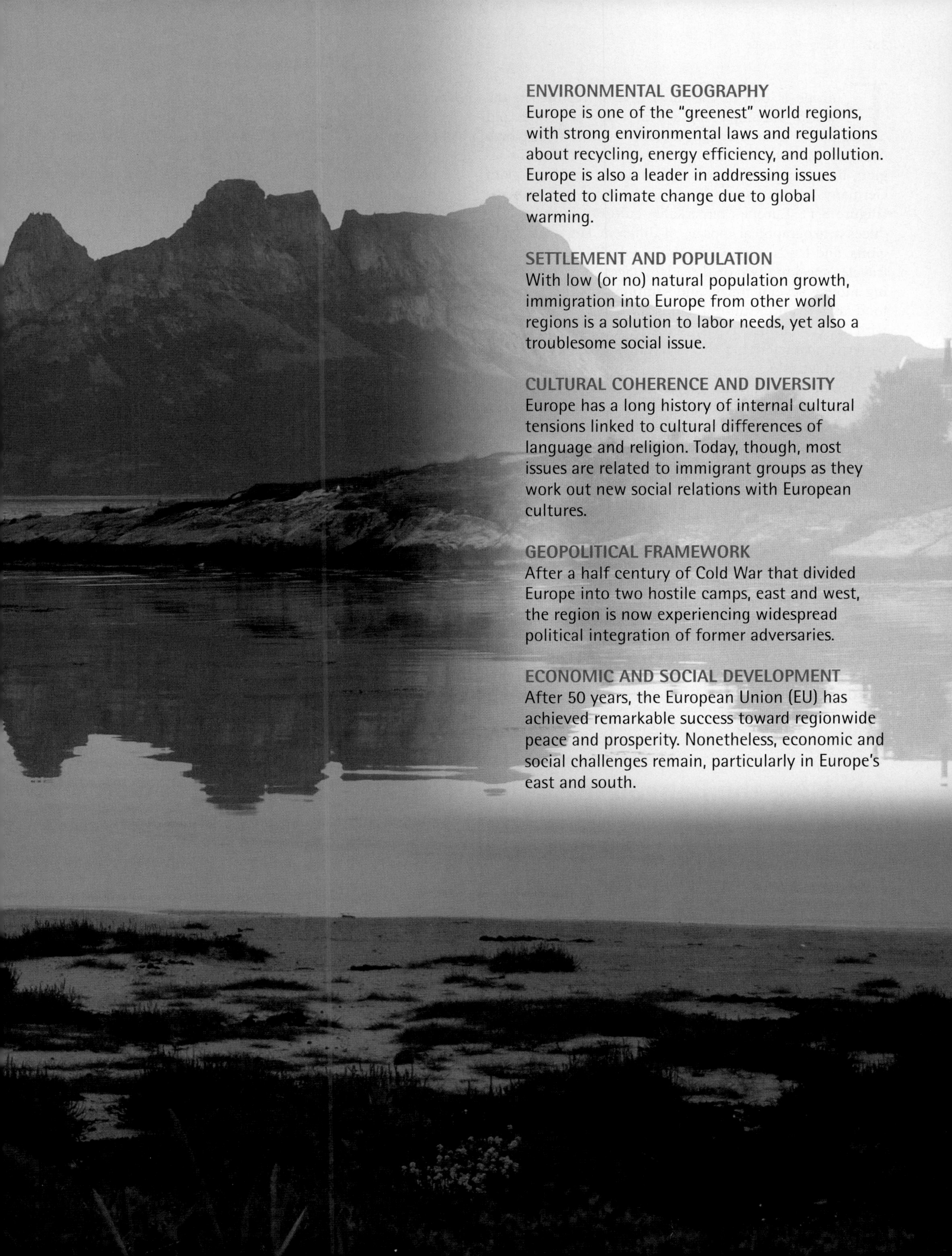

ENVIRONMENTAL GEOGRAPHY
Europe is one of the "greenest" world regions, with strong environmental laws and regulations about recycling, energy efficiency, and pollution. Europe is also a leader in addressing issues related to climate change due to global warming.

SETTLEMENT AND POPULATION
With low (or no) natural population growth, immigration into Europe from other world regions is a solution to labor needs, yet also a troublesome social issue.

CULTURAL COHERENCE AND DIVERSITY
Europe has a long history of internal cultural tensions linked to cultural differences of language and religion. Today, though, most issues are related to immigrant groups as they work out new social relations with European cultures.

GEOPOLITICAL FRAMEWORK
After a half century of Cold War that divided Europe into two hostile camps, east and west, the region is now experiencing widespread political integration of former adversaries.

ECONOMIC AND SOCIAL DEVELOPMENT
After 50 years, the European Union (EU) has achieved remarkable success toward regionwide peace and prosperity. Nonetheless, economic and social challenges remain, particularly in Europe's east and south.

Europe is one of the most diverse regions in the world, encompassing a wide array of people and places in an area considerably smaller than North America. More than half a billion people reside in this region, living in 40 countries that range in size from giant Germany to microstates such as Andorra and Monaco (Figure 8.1). Europe's remarkable cultural diversity produces a geographical mosaic of different languages, religions, and landscapes. Commonly, a day's journey finds a traveler speaking two or three languages, possibly changing money several times, and sampling distinct regional food and drink. Cultural landscapes also vary widely, with diverse house types, settlement forms, and field patterns characterizing the everyday scene (see "Setting the Boundaries").

Though the traveler may revel in Europe's cultural and environmental diversity, these regional differences also have entangled Europe in a troubled past, with neighbors warring with each other. In the twentieth century alone, Europe was the principal battleground of two world wars, followed by a 45-year **Cold War** that divided the continent into two hostile and highly armed camps—Europe (along with the United States) against the former Soviet Union.

Today, however, a spirit of cooperation prevails as Europe sets aside nationalistic agendas and works toward regional economic, political, and cultural integration through the **European Union (EU)**. This supranational organization is made up of 27 countries, anchored by the western European states of the United Kingdom, France, and Germany, and now extending to eastern Europe and the Baltic. Undoubtedly, the social, environmental, and economic policies of the EU will continue to transform all of Europe during the next decades. But not all Europeans are convinced that the EU is the answer to the region's problems. In Norway, for example, the government has voted to join the EU on two occasions only to have its population twice reject the idea in national referenda. Opponents of the EU argue that individual countries lose control over important sovereign matters; for some people—and a few states—EU membership carries more costs than benefits (Figure 8.2). Thus, a major theme in Europe today is the tension between large-scale regional integration and smaller-scale impulses toward independence and autonomy; between a new and shared sense of "Europeanness" and traditional national and regional identities. Additionally, there is a new layer of tension as millions of new immigrants work out their own relationships with Europe.

ENVIRONMENTAL GEOGRAPHY: Human Transformation of a Diverse Landscape

Despite its small size, Europe's environmental diversity is extraordinary. Within its borders are found a startling array of landscapes, from the Arctic tundra of northern Scandinavia to the barren hillsides of the Mediterranean islands, and from the explosive volcanoes of southern Italy to the glaciated seacoasts of western Norway and Iceland (Figure 8.4).

Setting the Boundaries

The European region is small compared to the United States. In fact, Europe from Iceland to the Black Sea would fit easily into the eastern two-thirds of North America. A more apt comparison would be to Canada, because Europe, too, is a northern region; more than half of Europe lies north of the 49th parallel, which in North America is the line of latitude forming much of the border between the United States and Canada (see Figure 8.3).

Europe currently contains 40 countries that range in size from superstates such as France and Germany, to microstates like Liechtenstein, Andorra, Monaco, and San Marino. Currently, the population for these combined states totals about 531 million people.

The notion that Europe is a continent with clearly defined boundaries is a misconception rooted in history. The Greeks and Romans divided their worlds into the three continents of Europe, Asia, and Africa separated by the Mediterranean Sea, the Red Sea, and the Bosporus Straits. A northward extension of the Black Sea was thought to separate Europe from Asia, and only in the sixteenth century was this proven false. Instead, explorers and cartographers discovered that the "continent" of Europe was firmly attached to the western portion of Asia.

Since that time, geographers have not agreed on the eastern boundary of Europe. During the existence of the Soviet Union, some took the continent's border all the way east to the Ural Mountains, while others drew the line at the western boundary of the Soviet Union. However, with the disintegration of the Soviet Union in 1990, the eastern boundary of Europe became even more problematic. Now some geography textbooks extend Europe to the border with Russia, which places the three countries of Moldova, Ukraine, and Belarus, former Soviet republics, in eastern Europe. We reject this definition, however, since these countries are still aligned culturally and economically with Russia and, instead, we base our definition on the expanded membership of the European Union into eastern Europe.

Figure 8.1 Europe
Stretching from Iceland in the Atlantic to the Black Sea, Europe includes 40 countries, ranging in size from large states, such as France and Germany, to the microstates of Liechtenstein, Andorra, San Marino, and Monaco. Currently, the population of the region is about 531 million. Europe is highly urbanized and, for the most part, affluent, particularly the western portion. However, economic and social disparities between eastern and western Europe remain a problem.

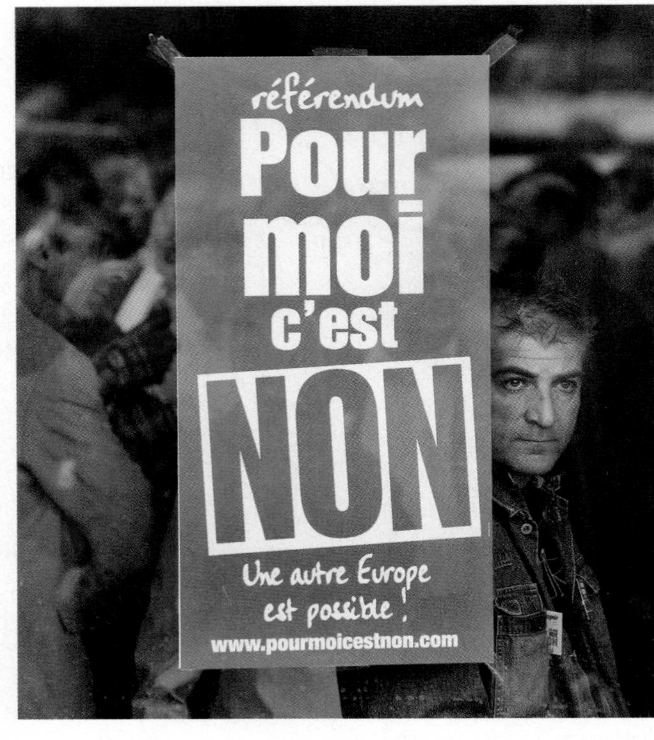

Figure 8.2 Anti-EU Feeling
Not everyone is happy with the European Union, for many groups feel its regulations trample on the uniqueness of individual nations. Here, a Frechman lobbies for a no vote on a recent EU issue. *(AP/Wide World Photos)*

Four factors explain this environmental diversity:

- The complex geology of this western extension of the Eurasian land mass has produced some of the newest, as well as the oldest, landscapes in the world.
- Europe's latitudinal extent from the Arctic to the Mediterranean subtropics affects climate, vegetation, and many human activities (Figure 8.3).
- The interaction of land and sea (the Atlantic, Black, and Mediterranean) shape regional climates and bioregions.
- Last, the long history of human settlement has transformed and modified Europe's natural landscapes in fundamental ways over thousands of years (see "Geographic Tools: Landscape Restoration in England").

Landform and Landscape Regions

European landscapes can be organized into four general topographic regions: the European Lowland, forming an arc from southern France to the northeast plains of Poland, but also including southeastern England; the alpine mountain system, extending from the Pyrenees in the west to the Balkan Mountains of southeastern Europe; the Central Uplands, positioned between the Alps and the European Lowland; and the Western Highlands, which include mountains in Spain, portions of the British Isles, and the highlands of Scandinavia.

The European Lowland This lowland, also known as the North European Plain, is the unquestionable economic focus of western Europe because of its high population density, intensive agriculture, large cities, and major industrial regions. Though not completely flat by any means, most of this lowland lies below 500 feet (150 meters) in elevation. In places, however, it is broken by rolling hills, plateaus, and uplands (such as in Brittany,

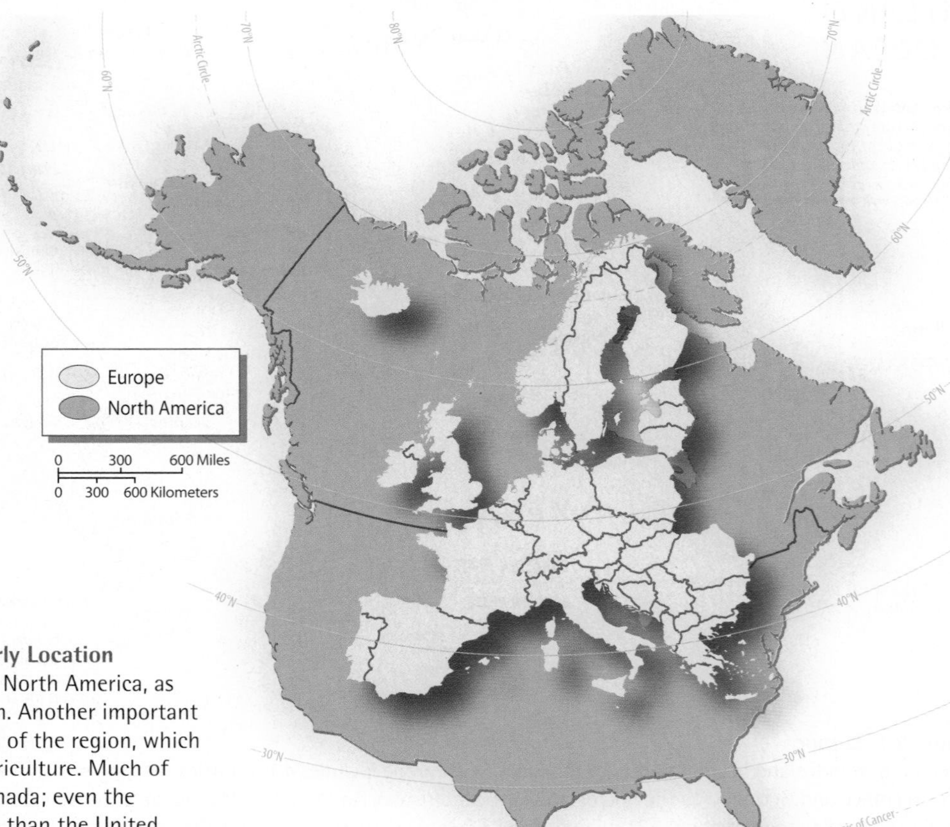

Figure 8.3 Europe: Size and Northerly Location
Europe is about two-thirds the size of North America, as shown in this cartographic comparison. Another important characteristic is the northerly location of the region, which affects its climate, vegetation, and agriculture. Much of Europe lies at the same latitude as Canada; even the Mediterranean lands are farther north than the United States–Mexico border.

France), where elevations exceed 1,000 feet (300 meters). Many of Europe's major rivers (the Rhine, the Loire, the Thames, and the Elbe) meander across this lowland and form broad estuaries before emptying into the Atlantic. Several of Europe's great ports inhabit these strategic lowland settings, namely London, Le Havre, Rotterdam, and Hamburg.

The Rhine River delta conveniently divides the unglaciated lowland in the south from the glaciated plains to the north, areas that were covered by a Pleistocene ice sheet until about 15,000 years ago. Because of these continental glaciers, the North European Lowland, including the Netherlands, Germany, Denmark, and Poland, is far less fertile for agriculture than the unglaciated portion in Belgium and France (Figure 8.5). Rocky clay materials in Scandinavia were eroded and transported south by glaciers. As the glaciers later retreated with a warming climate, piles of glacial debris known as **moraines** were left on the plains of Germany and Poland. Elsewhere in the north, glacial meltwater created infertile outwash plains that have limited agricultural potential.

Figure 8.5 The European Lowland
Also known as the North European Plain, this large lowland extends from southwestern France to the plains of northern Germany and into Poland. Although this landform region has some rolling hills, most of it is less than 500 feet (150 meters) in elevation. *(P. Vauthey/Corbis/Sygma Photo News)*

The Alpine Mountain System

The alpine mountain system forms the topographic spine of Europe. It consists of a series of mountains running east to west from the Atlantic to the Black Sea and the southeastern Mediterranean. Though these mountain ranges carry distinct regional names, such as the Pyrenees, Alps, Appenines, Carpathians, Dinaric Alps, and Balkan Ranges, they have similar geologic traits. All were created more recently (about 20 million years ago) than other upland areas of Europe, and all are constructed from a complex arrangement of rock types.

The *Pyrenees* form the political border between Spain and France (including the microstate of Andorra). This rugged range extends almost 300 miles (480 kilometers), stretching from the Atlantic to the Mediterranean. Within the mountain range, glaciated peaks reaching to 11,000 feet (3,350 meters) alternate with broad glacier-carved valleys. The Pyrenees are home to the Basque people in its western reaches, as well as to the distinctive Catalan-speaking minorities in the east. Both groups have strong separatist traditions.

Figure 8.4 Iceland Landscape
Sitting astride the Mid-Atlantic ridge where two tectonic plates move apart, Iceland is a fascinating mixture of volcanoes, glaciers, and spectacular waterfalls. This is a farmstead in the northeast coast. *(Les Rowntree)*

The centerpiece of Europe's geologic system is the prototypical mountain range the Alps, reaching more than 500 miles (800 kilometers) from France to eastern Austria. These impressive mountains are highest in the west, ascending more than 15,000 feet (4,575 meters) in Mt. Blanc on the French–Italian border; in Austria, to the east, few peaks exceed 10,000 feet (3,050 meters). Though easily crossed today by car or train through a system of long tunnels and valley-spanning bridges, these mountains have historically formed an important cultural divide between the Mediterranean lands to the south and central and western Europe in the north.

The Appenine Mountains are located south of the Alps; however, the two ranges are physically connected by the hilly coastline of the French and Italian Riviera. Forming the mountainous spine of Italy, the Appenines are generally lower and lack the spectacular glaciated peaks and valleys of the true Alps. Farther to the south, the Appenines take on their own distinct character with the impressive and explosive volcanoes of Mt. Vesuvius (just over 4,000 feet, or 1,200 meters) outside Naples and the much higher (almost 11,000 feet, or 3,350 meters) Mt. Etna off Italy's toe on the island of Sicily. These two active volcanoes result from their location on the meeting ground of the African and Eurasian tectonic plates.

To the east, the Carpathian Mountains define the limits of the alpine system in eastern Europe. They are a plow-shaped upland area that extends from eastern Austria to the Iron Gate gorge, a narrow passage for river traffic where the borders of Romania and Serbia intersect. While about the same length as the main alpine chain, the Carpathians are not nearly as high. The highest summits in Slovakia and southern Poland are less than 9,000 feet (2,780 meters).

Central Uplands In western Europe, a much older highland region occupies an arc between the Alps and the European Lowland in France and Germany. These mountains are much lower in elevation than the alpine system, with their highest peaks at 6,000 feet (1,830 meters). Dated to about 100 million years ago, much of this upland region is characterized by rolling landscapes of about 3,000 feet (less than 1,000 meters).

Their importance to western Europe is great because they contain the raw materials for Europe's industrial areas. In both Germany and France, for example, these uplands have provided the iron and coal necessary for each country's steel industry; in the eastern reaches, mineral resources from the Bohemian highlands have also fueled major industrial areas in Germany, Poland, and the Czech Republic.

Western Highlands The Western Highlands define the western edge of the European subcontinent, extending from Portugal in the south, through the portions of the British Isles in the northwest, to the highland backbone of Norway, Sweden, and Finland in the far north. These are Europe's oldest mountains, formed about 300 million years ago.

As with other upland areas that traverse many separate countries, specific place-names for these mountains differ from country to country. A portion of the Western Highlands form the highland spine of England, Wales, and Scotland, where picturesque glaciated landscapes are found at modest elevations of 4,000 feet (1,220 meters) or less. These U-shaped glaciated valleys also appear in Norway's uplands, where they produce a spectacular coastline of **fjords**, or flooded valley inlets similar to the coastlines of Alaska and New Zealand.

Though less elevated, the Fenno-Scandian Shield of Sweden and northern Finland is noteworthy because it is made up of some of the oldest rock formations in the world, dated conservatively at 600 million years. This **shield landscape** was eroded to bedrock by Pleistocene glaciers and, because of the cold climate and sparse vegetation, still has extremely thin soils that severely limit agricultural activity (Figure 8.6). As with other heavily glaciated areas like the Canadian Shield of North America, numerous small lakes dot the countryside, giving clues to the impressive erosional power of ice sheets.

Europe's Climates

Three principal climates characterize Europe (Figure 8.7). Along the Atlantic coast, a moderate and moist maritime climate dominates, modified by oceanic influences. Farther inland, continental climates prevail, with hotter summers and colder winters. Finally, dry-summer Mediterranean climates are found in southern Europe, from Spain to Greece.

One of the most important climate controls is that of the Atlantic Ocean. Though most of Europe is at a relatively high latitude (London, England, for example, is slightly farther north than Vancouver, British Columbia), the mild North Atlantic current, which is a continuation of the warm Atlantic Gulf Stream, moderates coastal temperatures from Iceland and Norway to Portugal and even inland to the western reaches of Germany. As a result, this maritime influence gives Europe a climate 5 to 10°F (2.8 to 5.7°C) warmer than comparable latitudes without this oceanic effect. In the **marine west-coast climate** region, no winter months average below freezing, though cold rain, sleet, and an occasional blizzard are common winter visitors. Summers are often cloudy and overcast with frequent drizzle and rain as moisture flows in from the ocean. Ireland, the Emerald Isle, offers an appropriate snapshot of this maritime climate.

With increasing distance from the ocean (or where a mountain chain limits the maritime influence, as in Scandinavia), land-mass heating and cooling becomes a strong climatic control, producing hotter summers and colder winters. Indeed, all **continental climates** average at least one month below freezing during the winter.

Figure 8.6 Northern Landscapes Northern Europe is a harsh land characterized by landscapes with little soil, glaciated rock expanses, sparse vegetation, and thousands of lakes. Pleistocene glaciers sculpted this region 20,000 years ago, and its high-latitude location maintains an Ice Age character that includes a very short growing season. *(Macduff Everton/Corbis/Bettmann)*

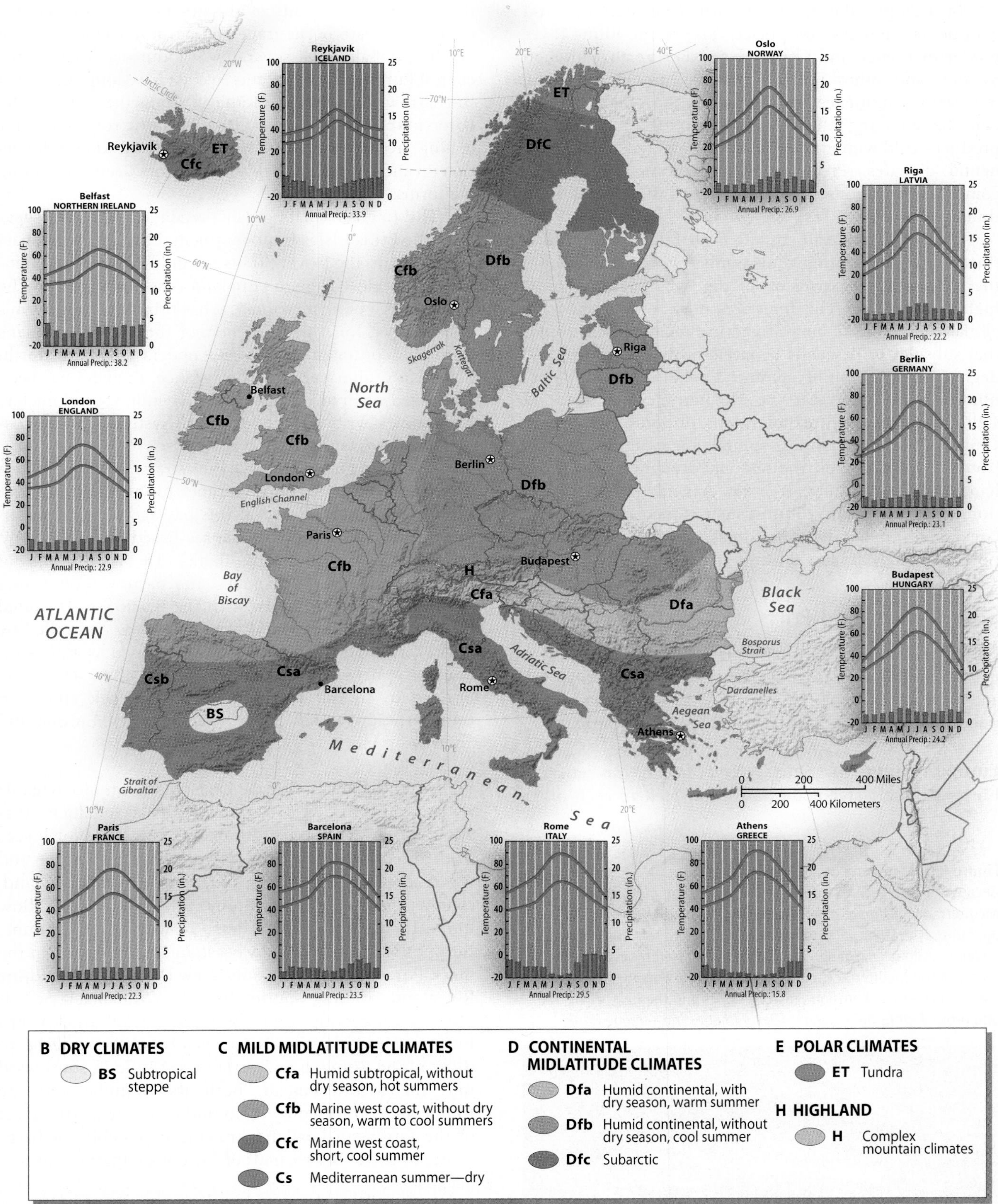

Figure 8.7 Climate Map of Europe
Three major climate zones dominate Europe. Close to the Atlantic Ocean the marine west-coast climate has cool seasons and steady rainfall throughout the year. Farther inland, continental climates have at least one month averaging below freezing, as well as hot summers, with a precipitation maximum falling during the warm season. The dry-summer Mediterranean climate is found in southern Europe.

In Europe, the transition between maritime and continental climates takes place close to the Rhine River border of France and Germany. Farther north, although Sweden and other nearby countries are close to the moderating influence of the Baltic Sea, high latitude and the blocking effect of the Norwegian mountains produces cold winter temperatures characteristic of continental climates. Precipitation in continental climates comes as rain from summer frontal systems and local thundershowers.

The **Mediterranean climate** is characterized by a distinct dry season during the summer, which results from the warm-season expansion of the Atlantic (or Azores) high-pressure area. This high pressure is produced by the global circulation of air warmed in the equatorial tropics that subsides or descends between latitudes of 30 and 40 degrees, thus inhibiting summer rainfall. This same phenomenon also produces the Mediterranean climates of California, western Australia, parts of South Africa, and Chile. While these rainless summers may attract tourists from northern Europe, the seasonal drought can be problematic for agriculture. It is no coincidence that traditional Mediterranean cultures, such as the Arab, Moorish, Greek, and Roman, have been major innovators of irrigation technology (Figure 8.8).

Seas, Rivers, and Ports

Europe remains a maritime region with strong ties to its surrounding seas. Even landlocked countries like Austria and the Czech Republic have access to the ocean through an interconnected network of navigable rivers and canals.

Europe's Ring of Seas Five major seas encircle Europe; these water bodies are connected to each other through narrow straits with strategic importance for controlling waterborne trade and naval movement. In the north, the Baltic Sea separates Scandinavia from north-central Europe. Denmark and Sweden have long controlled the narrow Skagerrak and Kattegat straits that connect the Baltic to the North Sea. Besides its historic role as a major fishing ground, the North Sea is now well known for its rich oil and natural gas fields mined from deep-sea drilling platforms.

The English Channel (in French, *La Manche*) separates the British Isles from continental Europe. At its narrowest point, the Dover Straits are only 20 miles (32 kilometers) wide. Though England has thought of the Channel as a protective moat, it has primarily been a symbolic barrier, for it deterred neither the French Normans from the continent nor Viking raiders from the north. Since 1993, after decades of resistance, England has been connected to France through the 31-mile (50-kilometer) Eurotunnel, with its high-speed rail system carrying passengers, autos, and freight.

Gibraltar guards the narrow straits between Africa and Europe at the western entrance to the Mediterranean Sea. Britain's stewardship of this passage remains an enduring symbol of its once great sea-based empire. Finally, on Europe's southeastern flanks are the Straits of Bosporus and the Dardanelles, the narrows connecting the eastern Mediterranean with the Black Sea. Disputed for centuries, these pivotal waters are now controlled by Turkey. Though these straits are often used as a physical boundary between Europe and Asia, they are easily bridged in several places to facilitate truck and train transportation within Turkey and between Europe and Southwest Asia.

Rivers and Ports Europe is also a region of navigable rivers connected by a system of canals and locks that allow inland barge travel from the Baltic and North seas to the Mediterranean, and between western Europe and the Black Sea. Many rivers on the European Lowland, such as the Loire, Seine, Rhine, Elbe, and Vistula, flow into Atlantic or Baltic waters. However, the Danube, Europe's longest river, flows east and south, rising in the Black Forest of Germany only a few miles from the Rhine River and running southeastward to the Black Sea. It offers a connecting artery between central and eastern Europe. Similarly, the Rhône headwaters rise close to those of the Rhine in Switzerland, yet it flows southward into the Mediterranean. Both the Danube and the Rhône are connected by locks and canals with the rivers of the European Lowland, making it possible for barge traffic to travel between all of Europe's fringing seas and oceans.

Major ports are found at the mouths of most western European rivers, serving as transshipment points for inland waterways as well as focal points for rail and truck networks. From south to north, these ports include Bordeaux at the mouth of the Garonne, Le Havre on the Seine, London on the Thames, Rotterdam (the world's largest port

Figure 8.8 Mediterranean Agriculture
Because of water scarcity during the hot, dry summers in the Mediterranean climate region, local farmers have evolved agricultural strategies for making the best use of soil and water resources. Global warming is projected to make water even more scarce in the Mediterranean. In this photo from Portugal, the terraces used to prevent soil erosion on steep slopes and the mixture of tree and ground crops are evident. *(Getty Images Inc.—Image Bank)*

in terms of tonnage) at the mouth of the Rhine, Hamburg on the Elbe River, and, to the east in Poland, Szczezin on the Oder and Gdansk on the Vistula.

Europe's Environmental Issues, East and West

Because of its long history of agriculture, resource-extraction, industrial manufacturing, and urbanization, Europe has its share of serious environmental problems. Compounding the situation is the fact that pollution rarely respects political boundaries. Air pollution from England, for example, creates serious acid rain problems in Sweden, and water pollution by factories in Switzerland creates major problems for the Netherlands where Rhine River water is used for municipal drinking supplies. Clearly, when environmental problems cross national boundaries, solutions must come from intergovernmental cooperation (Figure 8.9).

However, since the 1970s, when the European Union (EU) added environmental issues to its economic and political agenda, western Europe has been increasingly effective in addressing its varied environmental problems with regional solutions. In addition to the more obvious environmental problems of air and water pollution, the EU also has taken the lead on matters of recycling, waste

GEOGRAPHIC TOOLS Landscape Restoration in England

Because thousands of years of European settlement have altered the landscape, there is great interest today in restoring certain landscapes to their earlier state. In England, landscape restoration efforts are focused on two environments, ancestral woodlands and the heath lands. To accomplish this, a number of geographic tools are used in these projects.

A first step in landscape restoration is to recreate the ecology of the lost environment. This is often done by using a combination of methods such as palynology (pollen analysis), sediment dating, and archival sources with historical descriptions to reconstruct the earlier landscape.

Once a baseline is established and restoration efforts begin, other tools are used to monitor progress. Botanical and wildlife surveys are done regularly to tabulate plant and animal species. Often these data are located and mapped with a Global Positioning Systems (GPS) and then fed into geographic information system (GIS) files. Using these computerized maps, geographers and ecologists can document how landscape changes over time. In addition, both land and aerial photography is used to track environmental change in the restored landscapes.

One of the restoration efforts in England focuses on the original or ancestral woodlands of England, which were deciduous forests of oak, beech, elm, and similar broadleaf trees. Drawing upon historical land-use maps and local tax records, a restoration group called the Woodland Trust calculated that these forests once covered more than a million acres (335,000 hectares). Today, however, the broadleaf woodland is about half that amount, primarily because it has been replaced by planted nonnative conifer trees. Another restoration effort in England focuses on the treasured—yet disappearing—lowland heath land, a tundralike environment composed primarily of grasses, heather, and gorse. Unlike England's ancestral forest that is primarily natural, heath lands were created over thousands of years by human activity. More specifically, they resulted from a long history of grazing sheep and cattle on these environments. Because these animals tend to eat young saplings, tree growth was inhibited and the heath lands expanded as treeless meadows.

However, in the last 200 years these heath lands have disappeared as livestock grazing has become more specialized and localized on more productive lands. As a result, trees have taken over many former heath lands (Figure 8.2.1). Using historical maps from 1800 to recreate England's landscape of that period, restoration groups calculate that fully 82 percent of the heath lands have been lost to forests. As a result, restoration efforts focus on removing trees to allow native heath to regrow.

Figure 8.2.1 Oxshott Heath in England
This heath landscape west of London embodies both the beauty and the problems of heath restoration. Note the willow trees (to the right) encroaching on the heath. In the background are ancestral woodlands fringing the heath. *(Allison Green)*

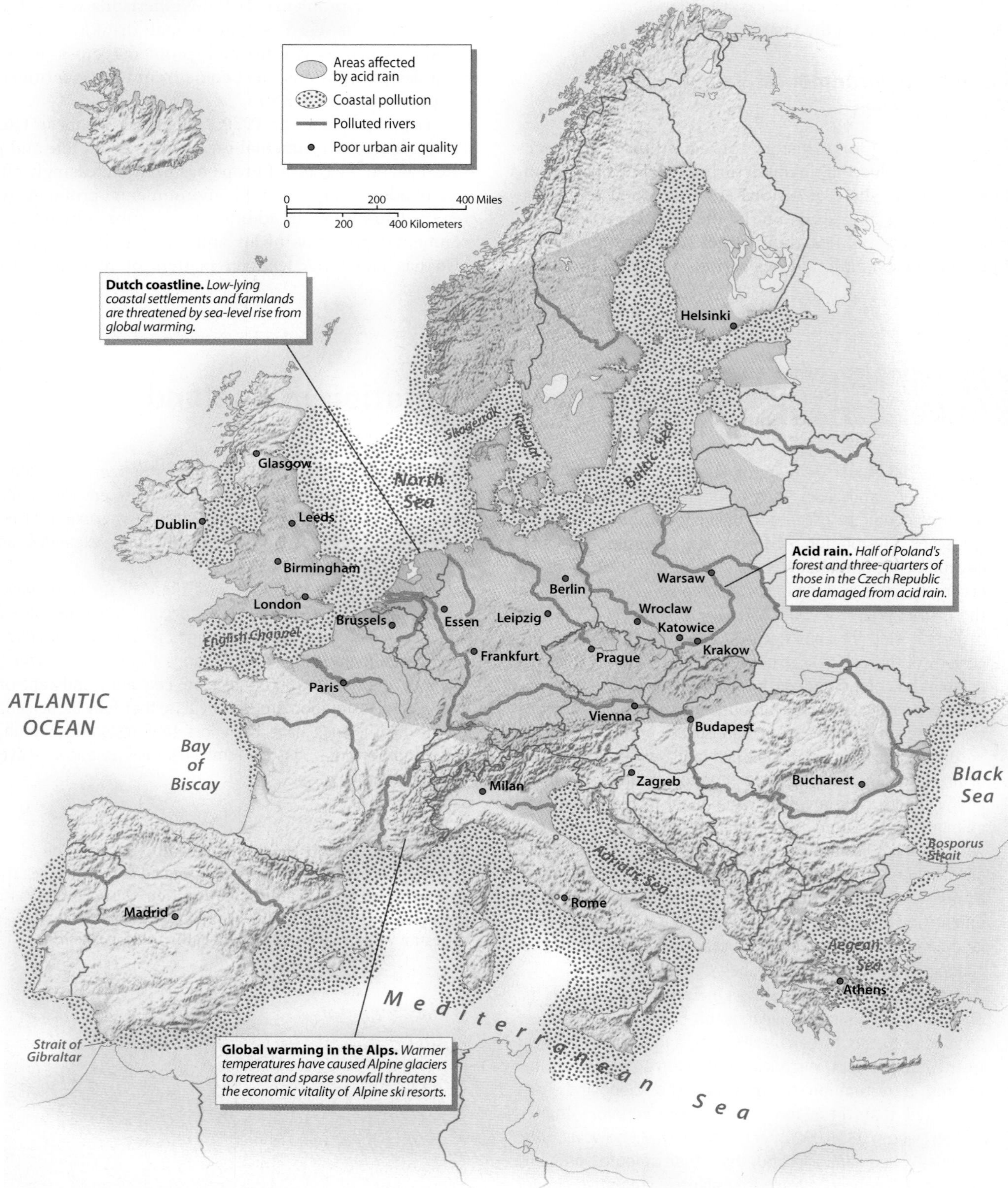

Figure 8.9 Environmental Issues in Europe
While western Europe has worked energetically over the last 50 years to solve environmental problems such as air and water pollution, eastern Europe lags behind somewhat because environmental protection was not a high priority during the postwar communist period. Current efforts, though, show great promise.

Figure 8.10 Acid Rain and Forest Death
Acid precipitation has taken a devastating toll on eastern European forests, such as those shown here in Bohemia, Czech Republic. In this country, three-quarters of the forests are damaged from acid precipitation, which is caused by industrial and auto emissions both in and outside the country. *(Karol Kallay/Bilderberg Archiv der Fotografen)*

management, reduced energy usage, and sustainable resource use. As a result, western Europe is probably the greenest of the major world regions.

The situation in eastern Europe, though, has been decidedly grimmer because of its communist past. During the period of Soviet economic planning (1945–90), little attention was paid to environmental issues because of the communist government's emphasis on short-term industrial output. As a result, the environmental costs of that historical period have been high. Until recently, 90 percent of Poland's rivers had no aquatic or plant life, and more than 50 percent of the country's forest trees showed signs of damage from air pollution (Figure 8.10). Human illness is also high. According to recent reports, one-third of Poland's population is expected to suffer from an environmentally induced disease such as cancer or respiratory illness. The future, however, appears more positive with the expansion of EU environmental policies and cleanup funds for new member states in eastern Europe (Figure 8.11). Even before EU membership was granted, in fact, countries such as Poland, Romania, and Slovakia had to enact environmental legislation comparable to that of western European states.

Global Warming in Europe: Problems and Prospects

The fingerprints of global warming are everywhere in Europe, from dwindling sea ice and sparse snow cover in arctic Scandinavia to more frequent droughts in the water-starved Mediterranean (Figure 8.12). Furthermore, the projections for future climate change are ominous—world-class ski resorts in the Alps will wither away with snowless winters while summertime heat waves could devastate Europe's cities. Also, higher seas will threaten the Low Countries where much settlement lies below sea level.

Because of these possibilities Europe has become a world leader in addressing global warming with numerous policies and programs to reduce greenhouse gas (GHG) emissions and adapt to changing environmental conditions. While most attention is understandably given to those negative effects that threaten human lives and activities, there are also several projected changes from climate change that might be considered positive, even beneficial. Examples are a possibly longer agricultural growing season in northern Europe and year-round ship travel in the ice-free arctic seas. However, these changes are not thought to compensate for the problems caused by environmental changes in other parts of Europe.

Europe and the Kyoto Protocol The EU entered the 1997 Kyoto negotiations with an innovative scheme underscoring its philosophy that regional action to solving environmental problems was superior to that taken by individual countries, at least in those parts of the world

Figure 8.11 Toxic Landscape in Romania
A troublesome legacy of Soviet communism in eastern Europe is the numerous toxic dump sites and polluted landscapes found in the region, such as this site in Romania. Money from the EU is helping to clean up these sites now that Romania is a member. *(Filip Horvat/Corbis/SABA Press Photos, Inc.)*

Figure 8.12 Retreating Glacier in Iceland
Glaciers throughout Europe are melting as a result of global warming. In Iceland, glacial retreat is a bit more complicated since sub-surface volcanic activity warms the bottom layers of ice sheets. *(Les Rowntree)*

where there was a mosaic of small states. More specifically, the EU proposed setting a collective goal of an 8 percent reduction below 1990 GHG emission levels as an umbrella for its individual member states.

Under this umbrella, or bubble as it's also called, some European states would be asked to make larger reductions while others were actually allowed an increase in GHG emissions. The point of this was to allow growth and industrial development in poorer countries like Spain, Greece, and Portugal, all of which were allowed GHG increases while demanding greater reductions from the more developed states such as Germany, Denmark, England, and France. The result of this differential umbrella approach was that regionally, for Europe as a whole, there would be an 8 percent reduction in GHG emissions.

This umbrella approach has remained a working assumption of the EU as it grew from 15 members at Kyoto to its current 27 states. Relevant to note is that many of the eastern Europe EU members are well below 1990 levels of GHG emissions because of changes in their economy, coupled with EU emission standards far more rigorous than those during the Soviet period.

The EU's Emission Trading Scheme In 2005, the EU inaugurated the world's largest GHG emission trading scheme, a plan intended to

Figure 8.13 Wind Power in Northern Europe
Not only is Europe trying to reduce its CO_2 emissions, but it is also a world leader in generating renewable energy from wind, sun, and biofuels. This large wind farm is in Denmark. *(National Geographic Society)*

become a main pillar of the EU global warming policy. Under this scheme, each participating country specifies emission caps or limits for individual sources of GHG emissions such as large industrial or power facilities. Each plant gets a maximum amount of emissions allowed for a set period and to comply with this target, the facility can either reduce its emissions or, instead, purchase credits from those plants whose emissions are lower than its GHG allowance. This, then, sets up a trading market where GHG emission allowances (in tons of carbon) are bought and sold.

But not all has gone well with this trading scheme. In April 2006, the price was about $40 a ton, then dropped dramatically to $14 a ton in May 2006 when it became clear to traders that individual countries had been too generous with their allowances and there was no need for plants to reduce their emissions. At this writing (2007) a complete lack of confidence characterizes the market with the price of a ton of carbon hovering around $1. However, major revisions on emission allocations are planned for the second trading period beginning in January 2008.

Results to Date Unfortunately, despite its good intentions, current data show the EU is not on target to meet its 2012 goal of an 8 percent reduction as required by the Kyoto Protocol. Instead, so far the EU bubble has reduced GHG emissions by 0.6 percent below its 1990 baseline, which is only a tenth of the way to the Kyoto target. The main reason for this failure, EU experts say, is the unanticipated growth in truck transport over the last decade. Additionally, Spain and other less-developed countries that were allowed GHG emission increases overshot their targets, largely because of greater than anticipated industrial development. This minuscule overall EU reduction results despite dramatic reductions of around 12 percent from the industrial powers of England and Germany.

To compensate for this slow start, as well as to reaffirm Europe's commitment as a world leader in reducing GHG emissions, in early 2007 the EU agreed to set a new target of a 20 percent reduction over 1990 to be achieved by 2020. Whether this is realistic and achievable or not, only time will tell (Figure 8.13).

POPULATION AND SETTLEMENT: Slow Growth and Rapid Migration

The map of Europe's population distribution (Figure 8.14) shows that population densities are higher in the historical industrial core areas of western Europe (England, the Netherlands, northern France, northern Italy, and western Germany) than in the periphery to the east and north. While this generalization overlooks important urban clusters in Mediterranean Europe, it does convey an accurate sense of a densely settled European core set apart from a more rural, agricultural periphery.

While this population pattern is closely linked to historic areas of industrialization, there are many contemporary consequences of this core-periphery distribution. For example, the EU's development policies have for several decades directed economic subsidies from the affluent, highly urbanized core to the less affluent, agricultural periphery. Further, while western Europe's urban-industrial core is characterized by extremely low natural growth rates, it is the target area for migrants—both legal and illegal—from Europe's peripheral countries, as well as from outside Europe (see "People on the Move: Smuggling Illegal Immigrants into Europe").

Natural Growth: Beyond the Demographic Transition

Probably the most striking characteristic of Europe's population is its slow or even negative natural growth (Table 8.1). More to the point, in many European countries the death rate exceeds the birthrate, meaning that there is simply no natural growth at all. Were it not for in-migration from other countries and other world regions, these no-growth countries would record a decline in population over the next few decades.

There seem to be several causes for zero population growth in western Europe. First of all, recall from Chapter 1 that the concept of the demographic transition was formulated from the historical change in European growth rates as the population moved from rural settings to more urban and industrial locations. What we see today is an extension of that model, namely the continued expression of the low fertility–low mortality of the fourth stage of the demographic transformation. Some demographers suggest adding a fifth stage to the model—a "postindustrial" phase in which population falls below replacement levels. Evidence for this comes from the highly urbanized and industrialized populations of Germany, France, and England, all of which are below zero population growth when immigration is excluded.

These low-birthrate countries are all attempting to increase their rate of natural growth. In Germany, for example, couples are given an outright monetary gift if they produce a child. In Austria, the government gives a modest cash award to couples when they marry, and still more follows if they give birth. France provides monthly payments to families with children that last until the children reach school age. Additionally, most EU countries have liberal maternity benefits to guarantee both parents time off from their jobs without penalty. Yet despite these incentives, most European countries still have natural growth rates below replacement levels and several, most notably in the former communist countries of eastern Europe, actually show negative growth. In these countries the political and economic turmoil associated with the transition from communism to capitalism may help explain very low birthrates.

PEOPLE PER SQUARE KILOMETER
Fewer than 5
5–25
25–50
50–100
More than 100

POPULATION:
Metropolitan areas 1,000,000–5,000,000
Metropolitan areas over 5,000,000

0 150 300 Miles
0 150 300 Kilometers

Negative growth and migration. *Negative natural growth in Germany is offset by relatively high rates of in-migration from eastern and southern Europe, former Soviet Union lands, and even Asia.*

High densities. *Dense concentrations of people in both cities and rural areas produce the highest densities in Europe—1,266 people per square mile in the Netherlands and 889 per square mile in Belgium.*

No growth. *Many eastern European countries have negative natural growth. This plight is worsened by out-migration to more affluent countries of western Europe.*

Declining population. *Because of low natural growth rates, population projections for Italy show a decline of 7.7 million people or about 13 percent by the year 2050.*

ATLANTIC OCEAN
North Sea
Skagerrak
Kattegat
Baltic Sea
English Channel
Hamburg
Berlin
Warsaw
London
Prague
Paris
Munich
Vienna
Budapest
Milan
Turin
Bucharest
Belgrade
Black Sea
Sofia
Bosporus Strait
Adriatic Sea
Madrid
Barcelona
Rome
Naples
Dardanelles
Aegean Sea
Mediterranean Sea
Strait of Gibraltar

Figure 8.14 Population Map of Europe
The European region includes more than 531 million people, many of them clustered in large cities in both western and eastern Europe. As can be seen on this map, the most densely populated areas are in England, the Netherlands, Belgium, western Germany, northern France, and south across the Alps to northern Italy. Because most European countries have very little, if any, natural growth, in-migration is a major issue.

TABLE 8.1 Population Indicators

Country	Population (Millions, 2007)	Population Density (per Square Kilometer)	Total Fertility Rate	Percent Urban	Percent <15[a]	Percent >65[b]	Net Migration (per 1,000, 2000–05)
Western Europe							
Austria	8.3	99	1.4	67	16	17	2.5
Belgium	10.6	348	1.7	97	17	17	1.3
France	61.7	112	2.0	77	18	16	1.0
Germany	82.3	230	1.3	75	14	19	2.7
Liechtenstein	0.04	221	1.4	15	17	12	5.9
Luxembourg	0.5	180	1.7	83	19	14	8.7
Monaco	0.1	33,104		100	13	22	10.5
Netherlands	16.4	394	1.7	65	18	14	1.9
Switzerland	7.5	182	1.4	68	16	16	1.1
United Kingdom	61.0	251	1.8	90	18	16	2.3
Eastern Europe							
Bulgaria	7.7	69	1.4	71	13	17	−0.2
Czech Republic	10.3	131	1.3	74	15	14	1.0
Hungary	10.1	108	1.3	65	15	16	1.0
Poland	38.1	122	1.3	62	16	13	−0.4
Romania	21.6	90	1.3	55	16	15	−1.4
Slovakia	5.4	110	1.3	56	16	12	0.2
Southern Europe							
Albania	3.2	110	1.8	45	27	8	−6.5
Bosnia & Herzegovina	3.8	75	1.2	46	18	13	2.1
Croatia	4.4	79	1.4	56	16	17	4.4
Cyprus	1.0	111	1.5	62	19	11	
Greece	11.2	85	1.3	59	14	19	3.2
Italy	59.3	197	1.4	68	14	20	2.1
Macedonia	2.0	80	1.4	59	20	11	−1.0
Malta	0.4	1,288	1.4	95	17	13	2.8
Montenegro	0.6	45	1.6	64	20	13	−1.9
Portugal	10.7	116	1.4	55	16	17	4.8
San Marino	0.03	508	1.2	84	15	16	7.3
Serbia	9.3	108	1.8	52	20	15	−1.9
Slovenia	2.0	99	1.3	49	14	16	1.0
Spain	45.3	90	1.4	77	14	17	9.7
Northern Europe							
Denmark	5.5	127	1.9	72	19	15	2.3
Estonia	1.3	30	1.6	69	15	17	−1.5
Finland	5.3	16	1.8	62	17	16	1.6
Iceland	0.3	3	2.1	93	21	12	1.2
Ireland	4.4	62	1.9	60	20	11	9.8
Latvia	2.3	35	1.4	68	14	17	−1.0
Lithuania	3.4	52	1.3	67	16	16	−1.2
Norway	4.7	12	1.9	78	19	15	2.6
Sweden	9.1	20	1.9	84	17	17	3.5

Sources: Population Reference Bureau, World Population Data Sheet, 2007; *net migration rate data from* UN International Migration, *2006.*

Figure 8.15 Migration into Europe
Historically, Europe opened its doors to migrants to help solve post–World War II labor shortages, but now immigration is a contentious and controversial issue as culturally homogeneous countries like France, England, and Germany confront tensions resulting from large numbers of immigrants, legal and illegal.

Migration to and Within Europe

Migration is one of the most challenging population issues facing Europe today because the region is caught in a web of conflicting policies and values (Figure 8.15). Currently, there is widespread resistance to unlimited migration into Europe, ostensibly because of high unemployment in the European industrial countries and a concern that scarce jobs should go first to European citizens, not to immigrants, but also because of an array of social and political concerns, ranging from foreign terrorism to dilution of a country's dominant culture.

Historically, the region has witnessed a massive influx of migrants from former European colonies in Asia, Africa, and the Caribbean and, as a result, the former colonial powers of England, France, and the Netherlands have been the major recipients of this immigration. England, for example, has inherited large numbers of former colonials from India, Pakistan, Jamaica, and Hong Kong. Indonesians (from the former Dutch East Indies) are common in the Netherlands, while emigrants in France are often from former colonies in both northern and Sub-Saharan Africa (Figure 8.16).

More recently, political and economic changes in eastern Europe and the former Soviet Union have generated a new wave of migrants to Europe. Within Germany, for example, thousands of former East Germans have moved to the more prosperous and dynamic western parts of unified Germany. Although the German government has attempted to redress the economic differences within the country, recent data suggest that the unemployment rate is still more than twice as high in former East Germany than in the west.

As a result of these different migration streams, coupled with the country's postwar reliance on foreign workers from Turkey and the Balkans, Germany has become a reluctant land of immigration that receives 400,000 newcomers each year. Currently, about 7.5 million foreigners live in Germany, making up about 9 percent of the population. France has a similar percentage of foreign-born, which is about the same percentage as the foreign-born share of the U.S. population. However, while the United States celebrates its immigration heritage, Germany, France, and England struggle with this new cultural diversity.

The Geography of "Fortress Europe"

An important and perhaps even aggravating ingredient in European concerns about foreign migration is the agreement between the heartland countries of Europe to facilitate free movement for its citizens across borders without passport inspections and checks. This agreement has reshaped the political geography of Europe by creating divisions between insiders and outsiders. To those on the inside, it creates a long-dreamed-of "Europe without borders," while to those on the outside it presents a "Fortress Europe," a defensive perimeter perceived as hostile to migrants from Asia, Africa, and the Russian domain. At the "hard" borders on EU's perimeter, particularly in Spain, Italy, Slovakia, and Poland, foreigners and foreign goods are subject to lengthy passport checks, visa requirements and searches, whereas crossing a "soft" border within the European heartland usually involves little more than a visual check and a wave from the border police, if there is even a border station.

Taking its name from the city of Schengen, Luxembourg, where the original declaration of intent was signed in 1985, the EU's **Schengen Agreement** has as one of its goals the gradual reduction of border formalities for travelers moving between EU countries (Figure 8.17). Though reduced border formalities within western Europe seemed like reasonable and desirable goals in 1985, the situation is much more complicated today because of the large number of illegal migrants from the former Soviet lands, Africa, and south Asia. This concern about illegal immigrants was elevated by the September 2001 terrorist attacks in New York City and more recent incidents within Europe itself.

The Landscapes of Urban Europe

One of the major characteristics of Europe's population is its high level of urbanization. All but several Balkan countries have more than half their population in cities,

Figure 8.16 Immigrant Riots in France
With the influx of immigrants from former colonial lands, social tensions have increased considerably in France, England, and Germany. Expressions of these tensions are the riots and unrest that periodically break out in these countries. Commonly, immigrants are ghettoized into large public housing projects. *(Getty)*

Figure 8.17 Czech Republic Border Station
This border station between Germany and the Czech Republic was once a highly fortified Iron Curtain checkpoint. But today, with the Czech Republic's membership in the EU, most people pass through with a cursory wave after showing their passport. *(Les Rowntree)*

and several countries, namely the United Kingdom and Belgium, are more than 90 percent urbanized. Despite the different levels of urbanization, few question that Europe's cities today—as they have been throughout history—are central to its politics, culture, and economy.

The Past in the Present Three historical eras dominate most European city landscapes. The medieval (roughly 900–1500 CE), Renaissance–Baroque (1500–1800 CE), and industrial (1800–present) periods each left characteristic marks on the European urban scene. Learning to recognize these stages of historical growth provides visitors to Europe's cities with fascinating insights into both past and present landscapes (Figure 8.18).

The **medieval landscape** is one of narrow, winding streets, crowded with three- or four-story masonry buildings with little setback from the street. This is a dense landscape with few open spaces, except around churches or public buildings. Here and there, public squares or parks are clues about earlier open-air marketplaces where medieval commerce was transacted.

As picturesque as we find medieval-era districts today, they nevertheless present challenges to modernization because of their narrow, congested streets and antiquated housing. Often such housing lacks modern plumbing and heating and has small, cramped rooms and hallways in contrast with modern housing. These medieval sectors are often inhabited by low- and fixed-income residents. As a result, in many areas these quarters are populated by elderly people, university students, and ethnic migrants.

In contrast to the cramped and dense medieval landscape, those areas of cities built during the **Renaissance–Baroque** period (1500–1800 CE) have a landscape that is much more open and spacious, with expansive ceremonial buildings and squares, monuments, ornamental gardens, and wide boulevards lined with elaborate residences. During this period, a new aesthetic and sense of urban planning arose in Europe that resulted in the restructuring of many European cities, particularly the large capitals. These changes were primarily for the benefit of the new urban elite: the royalty, aristocrats, and successful merchants. City dwellers of lesser means remained in the medieval quarters, which became increasingly crowded and cramped as more of the city was devoted to the space-extensive pleasures of the ruling classes.

Aggravating crowding during this time was the fact that defensive structures, such as city walls and other fortifications, constrained the outward spread of growing cities. With the advent of high-powered assault artillery, European cities were forced to build an extensive system of defensive walls. Once girdled by these walls, the cities could not expand laterally; instead, densities had to increase within their confines. A common solution was to add several new stories to the medieval houses.

Figure 8.18 Urban Landscapes
This aerial view of central Paris shows how the contemporary landscape expresses different historical periods. The tightly clustered houses and streets are of the medieval period, while the broad boulevards and large public buildings with gardens are examples of later Renaissance-Baroque city planning. *(Corbis/Bettmann)*

PEOPLE ON THE MOVE Smuggling Illegal Immigrants into Europe

Europe is the destination of choice for many thousands of immigrants each year, both legal and illegal. As is the case in so many other parts of the world, the plight of illegal immigrants can be a tragic one. Often these people are taken advantage of by smugglers who promise much and deliver little, charging exorbitant rates for uncertain outcomes.

Worldwide, people smuggling is now big business and, as a result, organized crime has become increasingly involved. European immigration officials estimate that fully 90 percent of those attempting to enter EU countries were helped by international crime networks.

Commonly immigrants pay around $15,000 to be smuggled into Europe, life savings for most, and organized crime appears to be more interested in accumulating riches than the immigrant's success. Many immigrants end up broke, sick, and hopelessly stranded in a jail or detention center in Africa rather than Europe. Or worse, they drown or die in refugee camps.

A common smuggling route takes immigrants from all parts of the world to the West African coast where they are loaded on decrepit boats hoping to make landfall on the Canary Islands, an autonomous community of Spain. If the boats make it that far, immigrants are usually told to swim ashore and ask for political asylum. If successful, the immigrants may be taken to a refugee center in mainland Spain where, after a period of questioning, they could be released and ordered out of the country. Instead, though, most immigrants would begin an underground and illegal life in Europe, moving from country to country, across the soft borders of the EU.

In 2006 the Canary Islands were overwhelmed by almost 7,000 illegal immigrants who made it to shore. Uncounted others were not so lucky, either ending up back in Africa or, worse, dead from drowning or exposure in ill-equipped boats. Those reaching the Canary Islands are not just from African countries but from countries such as Sri Lanka, Afghanistan, Myanmar, even India and Pakistan.

The fact that many of these immigrants come from great distances and from a mix of countries is taken as another indication that organized crime systems now control the people-smuggling business in Europe. Further evidence comes from the fact that the ships smuggling these immigrants fly a variety of foreign flags and have international crews. North Korean ships with Georgian or Russian crews seem particularly popular with the smugglers since North Korea has not signed international treaties on human trafficking.

Figure 8.3.1 Illegal Migration to Europe
This boatload of would-be immigrants to Europe was intercepted by Spanish authorities as it approached the Canary Islands. *(AP/Wide World Photos)*

These Renaissance–Baroque landscapes are still very much a part of the contemporary city and are, in fact, often the most prominent landmarks sought out by tourists. For example, after its devastating fire in 1666, London was completely rebuilt in the Baroque style with showpiece parks, boulevards, and monumental buildings such as St. Paul's Cathedral. In Paris, although major boulevard-building took place in the mid-nineteenth century, it was still very much an expression of the Baroque mentality. In Vienna, the grand boulevard of the Ringstrasse, or ring-street, that encircles the inner city was built in the late nineteenth century when city fortifications were declared obsolete and torn down.

The Industrial Period Industrialization dramatically altered the landscape of European cities. Historically, factories clustered together in cities beginning in the early nineteenth century, drawn by their large markets and labor force and supplied by raw materials shipped by barge and railroad. Full-blown industrial districts of factories and worker tenements grew up around these transportation lines.

In continental Europe, where many cities retained their defensive walls until the late nineteenth century, the new industrial districts were often located outside the fortifications, removed somewhat from the historical central city. In Paris, for example, when the railroad was constructed in the 1850s, it was not allowed to penetrate the city walls. As a result, terminals and train stations for the network of tracks were located beyond the original fortifications. Although the walls of Paris are long gone, this pattern of outlying train stations persists today, delimiting the historical placement of the inner fortifications. The historical industrial areas that

CITYSCAPES Berlin Reinvents Itself

Berlin is in many ways a microcosm of Europe, showing considerable scar tissue from the wars and political divisions of the twentieth century, yet now healing by reinventing itself for the twenty-first century with its promise of political and economic integration. As a result, the city's landscape is a fascinating mosaic of the past and present, along with hopes for the future.

Berlin has been the historical capital of the north German region since the fifteenth century. In 1871, with the formation of the German state, it became the capital of Germany until the end of World War II in 1945. At that time, Berlin, along with all of Germany, was occupied by the four Allied powers—the United States, France, Britain, and the Soviet Union. Berlin was in the Soviet zone, and, as Cold War tensions increased, the city was increasingly isolated from western Germany.

Today, however, Berlin is a growing city of about 3.5 million. Of that population, about 2.4 million live in what was West Berlin during the Cold War, with 1.1 million in the former communist east zone. Dividing the two areas was the infamous Berlin Wall, built in 1961 to prevent the East Berlin population from fleeing East Germany, and torn

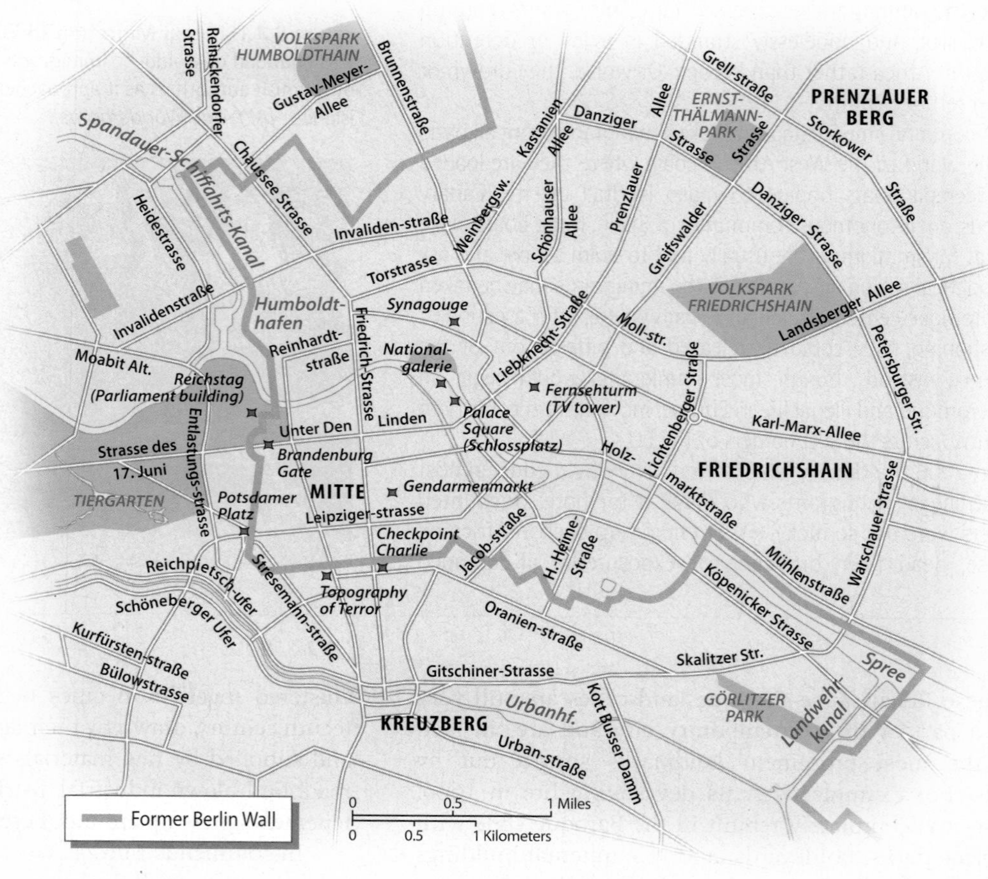

Figure 8.4.1 Map of Berlin
This map shows the former extent of the infamous Berlin Wall that divided the city into two distinct political units until 1989. Note that the traditional city center, Mitte, was in former East Berlin.

were outside the walls in what were essentially suburban locations also were along these tracks. This arrangement contrasts with the geography of industrial areas in North American cities, where factories and working-class housing often occupied central-city locations since there were no urban fortifications or walls to restrict the railroads.

In London, factories and workers located in the eastern part of the city, close to the busy docks of the Thames River. Early city planning (some would call it "social engineering") reinforced the distinction between the middle- and upper-class West End and London's impoverished blue-collar, industrial east side. This historical

down in 1989 when the East Berliners revolted against their government.

Today, the remnants of the Wall are major attractions that remind inhabitants and tourists alike of the city's divided past (Figure 8.3.1). The entire course of the Wall is marked in the streets by a continuous trace of inlaid bricks. In some places, actual remnants of the Wall remain, such as along Muehlenstrasse, where the concrete slabs have been painted by artists from throughout the world and form the extensive open-air "East Side Gallery." In other areas, vacant lots and boarded-up buildings provide mute testimony to the Wall's ominous presence (Figure 8.3.2).

The communist past also is found in the huge (1,207 feet, or 368 meters) TV tower on Alexanderplatz that dominates the city. It was built in the 1970s to block out Western television programs and broadcast state propaganda. The large square itself is ringed by huge, unattractive buildings that were formerly state department stores, mixed in with the gigantic multistory apartment houses that were the communist solution to postwar housing problems.

The historical city center (Mitte) of Berlin was in the former east zone and languished under the communists; now, with the Wall's demise and the city unified, this area is once more the vibrant heart of Berlin. Today, restored historic landmarks include the parliament, the Brandenburg Gate, the Berlin cathedral, and several opera houses, along with numerous museums. These buildings now form an attractive backdrop to the dynamic street life of this lively area. Rather surprisingly, within two years after communism fell, world-class department stores and hotels returned to the Mitte, proving wrong the many skeptics who thought the former East Berlin would never recover from its communist past.

Berlin's future is seen downtown around Potsdamer Square, where skyscrapers designed by world-famous architects blossom. These skyscrapers house corporate headquarters for companies that predict Berlin will become the economic capital of the newly expanded EU.

Figure 8.4.2 The Wall Zone Still today, the former Wall zone is marked by fences, vacant lots, weeds, and boarded-up buildings. *(Les Rowntree)*

west–east difference has lessened in postwar decades with massive urban renewal projects of office buildings, shopping malls, and middle-class housing. Much of the impetus for these recent projects came from the need to rebuild London's east side, which was heavily damaged by bombing in World War II. Postwar rebuilding has been a mixed blessing for some other European cities that suffered heavy war damage. In many cities, such as Berlin, Rotterdam, and Warsaw, the historic landscape was largely destroyed. In its place today is a landscape of modern office buildings, apartment houses, downtown highways, and shopping malls (see "Cityscapes: Berlin Reinvents Itself").

Figure 8.19 Protecting the Paris Skyline
The Parisians so value their traditional skyline and its landmarks (such as the Eiffel Tower, shown here) that they have passed regulations banning high-rises in the central city. Instead, skyscrapers are clustered just outside the historic city center. *(Les Rowntree)*

Protecting the Sense of Place Europeans take a strong sense of cultural identity from their cities and, as a result, have legislated a number of innovative programs designed to protect their urban landscapes. The unique skylines of many cities, which have become unique signatures and symbols, are now protected by law. The best example can be seen in contrasting the skylines of Paris and London. In Paris, strong building limitations protect the inner-city skyline so that the historic bell towers of Notre Dame and the more recent Eiffel Tower remain visually prominent. As a result, the traditional skyline remains largely intact (Figure 8.19). New high-rise buildings are clustered in outlying nodal points, as in the high-density office and residential neighborhood of La Défense just to the west of the inner city. London, in contrast, has done little to protect its skyline, and now modern high-rise office buildings obscure the landmark cathedral of St. Paul's and the spires of Westminster.

At the street level in European cities, it is common to find that historic preservation measures have protected and renovated medieval neighborhoods while at the same time maintaining the traditional urban fabric and sense of place. These are often expensive projects, because restoring old buildings often costs more than demolition and replacement with modern structures. Still, most European cultures have decided that the benefits outweigh the costs. Once again, Paris was a leader in this historic preservation movement; in the 1960s, it allocated public funds to protect its medieval district. By the late 1970s, similar historic preservation measures were found throughout Europe; examples include Salzburg, Austria, and Regensburg, Germany (Figure 8.20).

Figure 8.20 Historic Preservation of Cities
Most European countries have laws and regulations designed to preserve and protect their cultural heritage as embodied in historical cities. One of the earliest historic preservation projects in Germany was the inner city of Regensburg, on the Danube River. *(Les Rowntree)*

CULTURAL COHERENCE AND DIVERSITY: A Mosaic of Differences

The rich cultural geography of Europe demands our attention for several reasons. First, the highly varied and fascinating mosaic of languages, customs, religions, ways of life, and landscapes that characterize Europe has also shaped strong local and regional identities that have often stoked the fires of separatism and nationalism. Examples are found today in Northern Ireland and the Balkans.

Second, European cultures have played leading roles in processes of globalization. Through European colonialism and imperialism, languages, religion, economies, and values have transformed every corner of the globe. If you question this, consider cricket games in Pakistan, high tea in India, Dutch architecture in South Africa, and the millions of French-speaking inhabitants of equatorial Africa. Even without such modern technologies as satellite TV, the Internet, and Hollywood films and video, European culture spread across the world, changing the speech, religion, belief systems, dress, and habits of millions of people on every continent.

Today, though, many European countries resist global culture (Figure 8.21). France, for example, struggles against both U.S.-dominated popular culture as well as the multicultural influences of its large migrant population. The cultural geography of contemporary Europe is also complicated by the fact that while Europe fends off global culture with one hand, with the other it creates its own unique supranational culture through the EU integration. At times it seems as though the supranational economic and political integration of the EU is offset by the tensions of local and regional cultures for autonomy and independence.

Figure 8.21 Global Culture in Europe
U.S. popular culture is received with great ambivalence in Europe. While many people embrace everything from fast food to Hollywood movies, at the same time they protest the loss of Europe's traditional regional cultures. This photo was taken in Prague, Czech Republic. *(Les Rowntree)*

Geographies of Language

Language has always been an important component of nationalism and group identity in Europe (Figure 8.22). While some small ethnic groups like the Irish or the Bretons work hard to preserve their local language in order to reinforce their own cultural identity, millions of Europeans are busy learning multiple languages so they can communicate across cultural and national boundaries. The EU itself, while primarily an integrating force in Europe, honors this linguistic mosaic by having 19 official languages.

As their first language, 90 percent of Europe's population speaks a Germanic, Romance, or Slavic language, all of which belong to Western linguistic groups of the Indo-European family. Germanic and Romance speakers each number almost 200 million in the European region, while there are about 80 million Slavic speakers in Europe west of the former Soviet Union.

Germanic Languages Germanic languages dominate Europe north of the Alps. This linguistic region includes Germany, Scandinavia, two Baltic republics (Latvia and Lithuania), and the English-speaking areas of the United Kingdom. Approximately 5,000 years ago, the earliest Germanic tribes were centered along the Baltic coast of what is today Germany and Denmark. From this prehistoric cultural hearth, Germanic peoples spread south, east, and west over the course of thousands of years, and with this diffusion, regional dialects evolved into distinct languages, such as German, English, and Dutch.

Today about 90 million people speak German as their first language. This is the dominant language of Germany, Austria, Liechtenstein, Luxembourg, eastern Switzerland, and several small areas in Alpine Italy. As is true of most languages, there are very strong regional dialects in German that set apart German-speaking Swiss, for example, from the Letzeburgish German spoken in Luxembourg. All dialects, though, are mutually intelligible.

English is the second largest Germanic language, with about 60 million native speakers. Additionally, uncounted millions learn English as a second language, particularly in the Netherlands and Scandinavia, where many are as fluent as native speakers. Linguistically, English is closest to the Low German spoken along the coastline of the North Sea, which reinforces the notion that an early form of English evolved in the British Isles through contact with the coastal peoples of northern Europe. However, one of the distinctive traits of English that sets it apart from German is that almost a third of the English vocabulary is composed of Romance words brought to England during the Norman French conquest of the eleventh century.

Elsewhere in this region, Dutch (Netherlands) and Flemish (northern Belgium) account for another 20 million

Figure 8.22 Language Map of Europe
Ninety percent of Europeans speak an Indo-European language. These languages can be grouped into the major categories of Germanic, Romance, and Slavic languages. Ninety million Europeans speak German as a first language, which places it ahead of the 60 million who list English as their native language. However, given the large number of Europeans who speak fluent English as a second language, one could make the case that English is the dominant language of modern Europe.

people, and roughly the same number of Scandinavians speak the closely related languages of Danish, Norwegian, and Swedish. Icelandic, though, is a distinct language because of its long separation from its Scandinavian roots.

Romance Languages Romance languages, namely French, Spanish, and Italian, evolved as regional dialects from their common ancestry in the vulgar (or everyday) Latin used within the Roman Empire. Today about

Figure 8.23 The Alphabet of Ethnic Tension
With the departure of many of Kosovo's Serbs during recent ethnic unrest, signs in the Cyrillic alphabet were taken off many stores, shops, and restaurants as Kosovars of Albanian ethnicity restated their claims to the region. Here the Albanian owner of a restaurant in the capital city of Pristina scrapes off Cyrillic letters. *(David Brauchli/AP/Wide World Photos)*

60 million Europeans speak Italian as their first language. Italian is also an official language of Switzerland and is spoken on the French island of Corsica.

French is spoken in France, western Switzerland, and southern Belgium, where it is known as Walloon. Today there are about 55 million native French speakers in Europe. As with other languages, French also has very strong regional dialects, so strong that linguists differentiate between two forms of French in France itself, that spoken in the north (the official form because of the dominance of Paris) and the language of the south, or *langue d'oc*. This linguistic divide expresses long-standing tensions between Paris and southern France. In the last decade, the strong regional consciousness of the southwest (centered on Toulouse and the Pyrenees) has led to a resurrection of its own distinct language, Occitanian.

Spanish also has very strong regional variations. About 25 million people speak Castillian Spanish, the country's official language, which dominates the interior and northern areas of that large country. However, the Catalan form, which some argue is a completely separate language, is found along the eastern coastal fringe, centered on Barcelona, Spain's major city in terms of population and the economy. This distinct language reinforces such a strong sense of cultural separateness that the state of Catalonia has been given autonomous status within Spain.

Portuguese is spoken by 12 million speakers in Portugal and in the northwestern corner of Spain. Portugal's colonial legacy is imprinted by the fact that far more people speak this language in its former colony of Brazil than in Europe. Finally, Romanian represents the most eastern extent of the Romance language family where it is spoken by 24 million people. Though unquestionably a Romance language, Romanian also contains many Slavic words.

The Slavic Language Family Slavic is the largest Western subfamily of the Indo-European language family if one includes European Russia and its neighboring states. Traditionally, Slavic speakers are separated into northern and southern groups, divided by the non-Slavic speakers of Hungary and Romania.

To the north, Polish has 35 million speakers, and Czech and Slovakian about 14 million each. These numbers pale in comparison, however, with the number of northern Slav speakers in nearby Ukraine, Belarus, and Russia, where one can easily count more than 150 million. Southern Slav languages include three groups: 14 million Serbo-Croatian speakers (now considered separate languages because of the political troubles between Serbs and Croats), 11 million Bulgarian-Macedonian, and 2 million Slovenian.

The use of two alphabets further complicates the geography of Slavic languages (Figure 8.23). In countries with a strong Roman Catholic heritage, such as Poland and the Czech Republic, the Latin alphabet is used in writing. In contrast, countries with close ties to the Orthodox church use the Greek-derived **Cyrillic alphabet**, as is the case in Bulgaria, Macedonia, Croatia, and Serbia.

Cultural Diversity: Minor Indo-European Languages

One of the most interesting minor language families is the Celtic group, which includes languages native to Ireland, Scotland, and the Brittany region of France. Though Celtic languages were probably dominant in much of Europe several thousand years ago, today Celtic speakers number just over 1 million. The prehistoric cultural hearth for Celts, dating from 1500 BCE, is identified from archeological materials stretching along the northern foothills of the Alps, including portions of Austria, Switzerland, and France.

One theory suggests that Celts fled from this hearth as they were caught between the northward movement of Romans and the southward advance of Germanic peoples in 200 or 300 CE. As a result, Celts moved to the western fringes of Europe, reestablishing their culture on the Brittany peninsula as well as in the highlands of what are now Wales, Scotland, and Ireland. These fringe areas then served as the new hearth area for Celtic cultures that flourished for more than 1,000 years.

Modern Celtic languages, however, are struggling for survival. In western France, the Celtic Breton language counts fewer than half a million speakers, about one-third of its numbers in 1900. A major decline took place after World War II when the Catholic clergy in Celtic Brittany switched to French. In Wales, about the same number of people speak Welsh, though several groups are trying to preserve the language and culture. In the highlands and islands of Scotland, the situation is even more desperate. Here the number of Gaelic speakers is thought to be less than 100,000. In Ireland, the government has tried various strategies for promoting the use of Irish, but to little avail. Until 1980, the teaching of Irish was mandatory in public schools, but that was dropped because it was thought to be highly impractical. Land grants and other forms of settlement subsidies were offered to Irish speakers, but this did not stem the increasing use of English. Today only about 60,000 people use Gaelic in everyday conversation.

In southeastern Europe, modern residents trace their linguistic heritage to the Hellenic language of ancient Greece. About 10 million people in Greece and Cyprus speak this tongue. Much of the vitality of Hellenic languages comes from its use in the Greek Orthodox Church. It is said that modern Greek has the same relationship to classical Greek as contemporary Italian does to Latin. Finally, in the northeastern corner of Europe, the Baltic languages include Lettish (spoken in Latvia) and Lithuanian. Together they have about 4 million speakers.

Non-Indo-European Languages Only 5 percent of Europe's population speak non-Indo-European languages, and, with the exception of the Hungarians, they are found primarily on the outer reaches of the region. Magyar counts some 13 million, mainly in Hungary, but this linguistic region also spills over into nearby Romania, Slovakia, and northern Serbia.

Magyar tradition maintains that their ancestors were a herding people who entered Europe from the Asian steppes around 900 CE. Once settled in the central Danube basin, they created an important culture that, despite its linguistic differences, has played a major role in Europe's affairs. The Austro-Hungarian Empire, for example, dominated eastern Europe until World War I.

Far to the north, the Uralic linguistic family includes Estonian speakers, an even smaller number of Sami people (often called Lapps), and a larger number of Finns. In total, about 5 million people speak these related languages, which possibly originated in the forest and tundra lands of northern Eurasia.

On the southeastern periphery of Europe, the Altaic language family includes Turkish. Because of the long rule of the Ottoman Turks in southeastern Europe, significant Turkish minorities remain in countries such as Bulgaria, where, despite discrimination and attempts to expel them, they constitute almost 10 percent of the nation's population. Also important are the more than 3.5 million Turks currently in Germany, **guest workers**, or immigrant workers, invited to that country to solve its postwar labor problems, but now the focus of considerable discrimination and even blatant racism.

Last, the Basque people on the Atlantic border of France and Spain at the western end of the Pyrenees mountain chain speak Euskara, a language unlike any other in Europe. Some argue that the Basques are ancestors of the original prehistoric Europeans. Perhaps more important than this uncertain past is the violent terrorism that today pits many Spanish Basque people against the Madrid government as they seek political autonomy for their unique culture. Even though a large number of Basques have emigrated to North and South America, there are still more than 600,000 speakers in this part of Europe.

Geographies of Religion, Past and Present

Religion is absolutely inseparable from the geography of cultural coherence and diversity in Europe because so many of today's cultural tensions are embedded in historical events. To illustrate, strong cultural borders in the Balkans and eastern Europe are drawn based upon the eleventh-century split into Christianity's Eastern and Western churches, while in Northern Ireland blood is still shed over the tensions between the seventeenth-century division into Catholicism and Protestantism. Additionally, there is considerable tension today regarding the large Muslim migrant populations in England, France, and Germany. Understanding these important contemporary issues thus involves a brief look back at the historical geography of Europe's religious complexity (Figure 8.24).

The spread of early Christianity into Roman Europe was slow until the Edict of Tolerance by the emperor Constantine in 313 CE. After that, the new religion appears to have spread more rapidly both within and outside the Empire. The Celtic peoples of Britain adopted Christianity by 600 CE, and their missionaries played a central role in spreading this religion to the Germanic tribes of central Europe by 800, and then even farther east and north by the tenth century. This spread appears to have taken place by **hierarchical diffusion**, with local peoples converting because of the adoption of Christianity by a tribal leader or ruler. Centuries later, this same sort of hierarchical diffusion was important during the conflict between Catholics and Protestants.

The Schism Between Western and Eastern Christianity

In southeastern Europe, early Greek missionaries spread Christianity through the Balkans and into the lower reaches of the Danube. Progress was slower than in western Europe, perhaps because of continued invasions by peoples from the Asian steppes. There were other problems as well, primarily the refusal of these Greek missionaries to accept the growing church hierarchy and control of Roman bishops.

This tension with Western Christianity led in 1054 to an official split of the Eastern church from Rome. This Eastern church subsequently splintered into sects closely linked to specific nations and states. Today, for example, we find Greek Orthodox, Bulgarian Orthodox, and Russian Orthodox churches, all of which have different rites and rituals, yet share cultural affinity. The current political ties between Russians and Serbians, for example, grow from their shared language and religion.

Another factor that distinguished Eastern Christianity from Western was the Orthodox use of the Cyrillic alphabet instead of the Latin. Because Greek missionaries were primarily responsible for the spread of early Christianity in southeastern Europe, it is not surprising that they used an alphabet based on Greek characters. More precisely, this alphabet is attributed to the missionary work of

Figure 8.24 Religions of Europe
This map shows the divide in western Europe between the Protestant north and the Roman Catholic south. Historically, this distinction was much more important than it is today. Note the location of the former Jewish Pale, which was devastated by the Nazis during World War II. Today ethnic tensions with religious overtones are found primarily in the Balkans, where adherents of Roman Catholicism, Eastern Orthodoxy, and Islam are found in close proximity to one another.

St. Cyril in the ninth century. As a result, the division between Western and Eastern churches, and between the two alphabets, remains one of the most problematic cultural boundaries in Europe.

The Protestant Revolt Besides the division between Western and Eastern churches, the other great split within Christianity occurred between Catholics and Protestants. This division arose in Europe during the sixteenth century and has divided the region ever since. With the exception of the troubles in Northern Ireland, however, tensions today between these two major groups are far less damaging than in the past when the European landscape was scarred by religious battles.

Perhaps the low point came during the Thirty Years' War (1618–48), when central and western Europe were laid waste by war, famine, and the plague. Archival evidence suggests that more than 50 percent of the affected population was either killed or died from starvation or sickness. Fields lay fallow, forests expanded, and villages were deserted as this religious struggle devastated Europe from Bohemia (now the Czech Republic) to Belgium on the Atlantic shore. Today, this contested landscape still delineates the north–south boundary between Protestant and Catholic Europe.

Conflicts with Islam Both Eastern and Western Christian churches also struggled with incursions from Islamic empires to Europe's south and east. Even though historical Islam was reasonably tolerant of Christianity in its conquered lands, Christian Europe did not reciprocate by accepting Muslim imperialism. The first crusade to reclaim Jerusalem from the Turks took place in 1095 CE. After the Ottoman Turks took Constantinople in 1453 and gained control over the Straits of Bosporus and the Black Sea, they moved rapidly to spread their Muslim empire throughout the Balkans and arrived at the gates of Vienna in the middle of the sixteenth century. There, Christian Europe made its stand and turned back the Turks.

However, Ottoman control of southeastern Europe lasted until the empire's demise in the early twentieth century. This historical presence of Islam explains the current troublesome mosaic of religions in the Balkans, with intermixed areas of Muslims, Orthodox, and Roman Catholics. Commonly, ethnic boundaries in that part of Europe are drawn along religious lines.

Muslim incursions into Spain from North Africa left an enduring Moorish imprint throughout the Iberian Peninsula in terms of architecture and technology. Irrigation and water-control techniques and strategies brought by the Muslims, for example, benefited not just the Spanish, but also those parts of the world colonized by Spain and Portugal.

A Geography of Judaism Europe became a problematic homeland for Jews after their forced dispersal from Palestine during the Roman Empire. At that time, small Jewish settlements were found in cities throughout the Mediterranean. Later, by 900 CE, about 20 percent of the Jewish population was clustered in the Muslim lands of the Iberian Peninsula where Islam showed greater tolerance than Christianity had for Judaism. After the Christian reconquest of Iberia, however, Jews once more faced severe persecution and fled from Spain to more tolerant countries in western and central Europe.

One focus for migration was the area in eastern Europe that became known as the Jewish Pale. In the late Middle Ages, at the invitation of the Kingdom of Poland, which offered an edict of tolerance, Jews settled in cities and small villages in what is now eastern Poland, Belarus, western Ukraine, and northern Romania (see Figure 8.25). Jews collected in this region for several centuries in the hope of establishing a true European homeland, despite the poor natural resources of this marshy, marginal agricultural landscape.

Until emigration to North America began in the 1890s, 90 percent of the world's Jewish population lived in Europe and most were clustered in the Pale. Even though many emigrants to the United States and Canada came from this area, the Pale remained the largest aggregation of Jews in Europe until World War II. Tragically, Nazi Germany was able to use this ethnic clustering to its advantage by focusing extermination activities on this area.

In 1939, on the eve of World War II, there were approximately 9.5 million Jews in Europe, or about 60 percent of the world's Jewish population. During the war, German Nazis murdered some 6 million Jews during the horror of the Holocaust. Today fewer than 2 million Jews live in Europe. Since 1990 and the lifting of constraints on Jewish emigration from Russia, Belarus, and Ukraine, more than 100,000 Jews have emigrated to Germany, giving it the fastest-growing Jewish population outside Israel (Figure 8.25).

The Patterns of Contemporary Religion In Europe today, there are about 250 million Roman Catholics and fewer than 100 million Protestants. Generally, Catholics are found in the southern half of the region, except for significant numbers in Ireland and Poland, while Protestants dominate in the north. Additionally, since World War II interest in organized religion has declined in western

Figure 8.25 Jewish Synagogue in Berlin
Before World War II, Berlin had a large and thriving Jewish population, as attested to by this synagogue built in 1866. However, during the Nazi period, the Jewish population was forcibly removed and largely exterminated. This synagogue was then used by the Nazis to store military clothing until it was heavily damaged by Allied bombing. It was restored and opened again in 1995 as both a synagogue and museum. *(Les Rowntree)*

Europe, resulting in plummeting church attendance. This trend is so marked that the term **secularization** is used to refer to the widespread movement away from the historically prominent organized religions of Europe.

Not to be overlooked are the 13 million Muslims in Europe. Most are migrants and their families from Africa and southwestern Europe, although some are recent converts to Islam. More than half of the Muslim population (7.5 million) is found in France, with the second largest number (around 4 million) in Germany. While most Muslims worship in mosques converted from existing buildings, several new—and rather grandiose—mosques have been built lately in France, Germany, and Britain.

Catholicism dominates the religious geography and cultural landscapes of Italy, Spain, France, Austria, Ireland, and southern Germany. In these areas, large cathedrals, monasteries, iconic monuments to Christian saints, and religious place-names draw heavily on pre-Reformation Christian culture. Because visible beauty is part of the Catholic tradition, elaborate religious structures and monuments offer a more ornate landscape than in Protestant lands.

Protestantism is most widespread in northern Germany, the Scandinavian countries, and England and is intermixed with Catholicism in the Netherlands, Belgium, and Switzerland. Because of its reaction against the perceived aesthetic excesses of the Catholic Church, the landscape of Protestantism is much more sedate and subdued. Large cathedrals and religious monuments in Protestant countries are associated primarily with the Church of England, which has strong historical ties to Catholicism; St. Paul's Cathedral and Westminster Abbey in London are examples.

Tragically, another sort of religious landscape has emerged in Northern Ireland, where barbed-wire fences and concrete barriers separate Protestant and Catholic neighborhoods in an attempt to reduce violence between warring populations (Figure 8.26). Religious affiliation is a major social force that influences where people live, work, attend school, shop, and so on. Of the 1.6 million inhabitants of Northern Ireland, 54 percent are Protestant and 42 percent Roman Catholic. The Catholics feel they have been discriminated against by the Protestant majority. As a result, they want a closer relationship with the Republic of Ireland across the border to the south. The Protestant reaction is to forge even stronger ties with the United Kingdom. Unfortunately, these differences have been expressed in prolonged violence that has led to about 4,000 deaths since the 1960s, resulting largely from paramilitary groups on both sides who promote their political and social agendas through terrorist activities.

The emerging geography of religion in eastern Europe remains highly differentiated and complex. In Poland, where the Catholic Church was strong and remained an active opponent of communism during the Cold War years, religion seems to be playing a prominent role in contemporary life. More specifically, three-quarters of the Polish people consider themselves practicing Catholics. This reemergence of religious commitment far exceeds church attendance in some of the historically Catholic countries of western Europe. Historically, Catholicism also dominated in Hungary, and recent census figures suggest that more than two-thirds of the population characterizes itself as Catholic in today's postcommunism environment. This is also the case in nearby Slovakia. In contrast, religious adherence in the Czech Republic appears to be evenly divided between Catholics and those who consider themselves atheist.

Figure 8.26 Religious Tensions in Northern Ireland
Despite considerable and continual efforts to forge peace between Protestants and Catholics in Northern Ireland, outbreaks of violence are still common. Here, in Belfast, sectarian fighting broke out only hours after groups from both religions met to discuss ways to end the fighting. *(Cathal McNaughton/Getty Images, Inc.—Liaison)*

European Culture in a Global Context

Europe, like all world regions, is currently caught up in a period of profound cultural change; in fact, many would argue that the pace of cultural change in Europe is accelerated because of the complicated interactions between globalization and Europe's internal agenda of political and economic integration. While newspaper pundits celebrate the "New Europe" of integration and unification, other critics refer to a more tension-filled Europe of foreign migrants and guest workers troubled by ethnic discrimination and racism (Figure 8.27).

Globalization and Cultural Nationalism Since World War II, Europe has been defensive about, and even hostile to, cultural contamination from North America. Some countries are more reactionary than others. While a few large countries, such as England and Italy, seem to accept and even embrace the onrush of American popular culture, other countries have expressed outright indignation over the corrupting impact of U.S. popular culture on speech, music, food, and fashion. France, for example, is often thought of as the poster child for the antiglobalization struggle as it fights to preserve local foods and culture against the homogenizing onslaught of fast-food outlets and genetically engineered crops.

France has taken cultural nationalism to new levels with protective legislation against foreign influences. Radio stations, for example, must devote at least 40 percent of their air play to French songs and musicians, and French filmmakers are subsidized and supported by government

Figure 8.27 Turkish Store in Germany
This Turkish store in the Kreuzberg district of Berlin, sometimes referred to as "Little Istanbul," serves not only the large Turkish population, but also German shoppers who have found this store an appealing aspect of the country's broadening ethnicity. Unfortunately, to many, Europe's migrant cultures are more problematic. *(Embassy of the Federal Republic of Germany)*

subventions in the hope of staving off a complete takeover by Hollywood. The official body of the French Academy, whose job it is to legislate proper usage of the French language, has a long list of English and American words and phrases that are banned from use in official publications and from highly visible advertisements or billboards. Though the French commonly use "le weekend" or "le software" in their everyday speech, these words will never be found in official governmental speeches or publications.

While France sets the model for cultural nationalism, these ideas seem to be spreading to other European countries. Germany, for example, recently started public debate over banning certain English phrases and words in commercial advertisements and "proper" speech. There is a clear irony here, given that Europe's languages have always borrowed heavily from each other. Examples include the large number of French and German cognates in the English language, the similarities between Romance languages, and the shared vocabulary of Scandinavian tongues.

Migrants and Culture Migration patterns also are influencing the cultural mosaic of Europe. Historically, Europe spread its cultures worldwide through aggressive colonialization and imperialism. Today, however, the region is experiencing a reverse flow as millions of migrants move into Europe, bringing their own distinct cultures from the far-flung countries of Africa, Asia, and Latin America. Unfortunately, in some areas of Europe, as mentioned earlier, the products of this cultural exchange are highly problematic.

Ethnic clustering, even ghettoization, is now common in the cities and towns of western Europe. The high-density apartment buildings of suburban Paris, for example, are home to large numbers of French-speaking Africans and Arab Muslims caught in the cultural crossfire of high unemployment, poverty, and racial discrimination. Recent riots in these areas underscore the alienation and frustration felt by the immigrant population, while traditional French culture struggles with how to accept this new dimension of globalization. The political landscape of many European countries now contains far-right, nationalistic parties with an exclusionary and discriminatory political agenda that is distinctly anti-immigrant. More to the point, these political parties celebrate traditional customs, symbols, and icons associated with an idealized, nationalistic past, one that clearly does not include migrant foreigners. In Germany, for example, conservative politicians have ignited a firestorm by promoting the notion of a German *leitkultur*, or "guiding culture," that immigrants should subscribe to before being granted visas or citizenship. Although the full content of this guiding culture remains elusive and coded, it most certainly includes the German language, Christianity, and adherence to traditional forms of social behavior. Most noxious, perhaps, is the reaction of young German skinheads who purport to defend with terrorism and violence "true and pure" Aryan values historically associated with Nazi Germany (Figure 8.28).

GEOPOLITICAL FRAMEWORK: A Dynamic Map

One of Europe's unique characteristics is its dense fabric of 40 independent states within a relatively small area. No other world region demonstrates the same kind of geopolitical fragmentation.

But Europe's varied geopolitical landscape has provided as much problem as promise. Twice in the last century Europe shed blood to redraw its political borders. Within the last two decades alone, eight new states appeared in eastern Europe, more than half through war. The map of geopolitical troubles suggests that still more political fragmentation may take place in the near future, perhaps even violently (Figure 8.29) (see "Geography in the Making: Terrorism in Europe").

Figure 8.28 Neo-Nazis in Germany
Purporting to embrace "pure and true" Aryan values, these neo-Nazi Germans provoke police at a rally against foreigners. A rise in extreme forms of nationalism has become increasingly problematic in Germany, with anger often directed against Turks and other immigrants. *(Joanna B. Pinneo/Aurora & Quanta Productions)*

Figure 8.29 Geopolitical Issues in Europe
While the major geopolitical issue of the early twenty-first century remains the integration of eastern and western Europe into the European Union, numerous issues of micro- and ethnic nationalism also engender geopolitical fragmentation. In other parts of Europe, such as Spain, France, and Great Britain, questions of local ethnic autonomy within the nation-state structure challenge central governments.

Most of Europe, however, sees a brighter geopolitical future. For many, this optimism is based on a widespread spirit of cooperation through unification rather than continued fragmentation. Now empowered by the European Union's recent successes at political and economic integration, many Europeans take solace from the sudden and unanticipated end of the Cold War in 1989. Many argue that the disasters of the twentieth century were of Europe's own making. If true, this region seems determined to avoid those mistakes in the twenty-first century by giving the world a new geopolitical model for peace and prosperity.

Redrawing the Map of Europe Through War

Two world wars redrew the geopolitical maps of twentieth-century Europe (Figure 8.30). Because of these conflicts, empires, nation-states, and even remnant feudal territories have appeared and disappeared within the last 100 years. By the early twentieth century, Europe was divided into two opposing and highly armed camps with exaggerated national egos that tested each other for a decade before the tragic outbreak of World War I in 1914. France, Britain, and Russia were allied against the new nation-states of Italy and Germany, along with the ineffectual Austro-Hungarian or Hapsburg Empire, which controlled a complex mosaic of ethnic groups in central Europe and the Balkans.

Though at the time World War I was referred to as the "war to end all wars," it fell far short of solving Europe's geopolitical problems. Additionally, American entry into the war in 1917 ensured that country's continued involvement in European geopolitical affairs for the rest of the century.

When Germany and Austria-Hungary surrendered in 1918, the Treaty of Versailles redrew the map of Europe with two goals in mind: to punish the losers through loss of territory and severe fiscal reparations, and to recognize the nationalistic aspirations of certain peoples by creating new nation-states. As a result, the new states of Czechoslovakia and Yugoslavia were born. Additionally, Poland was reestablished, as were the Baltic states of Finland, Estonia, Latvia, and Lithuania.

Though the goals were lofty, the redrawn map created a fragmented Europe in which few states were satisfied. New states were generally resentful when kinfolk were left outside the new borders and became minorities in other new states. This created an epidemic of interwar **irredentism**, or state policies for reclaiming lost territory (real or imagined) beyond borders inhabited by people of the same ethnicity. Examples include the large German population in the western portion of the new state of Czechoslovakia and the Hungarians in western Romania.

This imperfect geopolitical solution was aggravated greatly by the global economic depression of the 1930s that brought high unemployment, food shortages, and political unrest to Europe. Three competing ideologies promoted their own solutions to Europe's pressing problems: Western democracy and capitalism; communism from the Soviet revolution to the east; and a fascist totalitarianism promoted by Mussolini in Italy and Hitler in Germany. With industrial unemployment commonly approaching 25 percent in western Europe, public opinion fluctuated wildly between extremist solutions. In 1936, Italy and Germany once again joined forces through the Rome–Berlin "axis" agreement, and this alignment was countered with mutual protection treaties between France, Britain, and the Soviet Union. When an imperialist Japan signed a pact with Germany, the scene was set for a second global war.

Nazi Germany tested Western resolve in 1938 by first annexing Austria, the country of Hitler's birth, and then Czechoslovakia under the guise of providing protection for Germans located there. After signing a nonaggression pact with the Soviet Union, Hitler invaded Poland on September 1, 1939. Two days later, France and Britain declared war on Germany. Within a month, the Soviet Union moved into eastern Poland, the Baltic states, and Finland to reclaim territories lost through the peace treaties of World War I. Nazi Germany then moved westward and occupied Denmark, the Netherlands, Belgium, and France and began preparations to invade England. In 1941, the war took several startling new turns. In June, Hitler broke the nonaggression pact with the Soviet Union and, catching the Red Army by surprise, took the Baltic states and drove deep into Soviet territory. When Japan attacked Pearl Harbor, Hawaii, in December, the United States entered the war in both the Pacific and Europe.

By early 1944, the Soviet army had recovered most of its territorial losses and moved against the Germans in eastern Europe, beginning a long communist hegemony in that region. By agreement with the Western powers, the Red Army stopped when it reached Berlin in April 1945. At that time, Allied forces crossed the Rhine River and began their occupation of Germany. With Hitler's suicide, Germany signed an unconditional surrender on May 8, 1945, ending the war in Europe. But with Soviet forces firmly entrenched in the Baltics, Poland, Czechoslovakia, Bulgaria, Romania, Hungary, Austria, and eastern Germany, the military battles of World War II were quickly replaced by the geopolitical reality of an ideological Cold War that lasted 45 years, until 1990.

A Divided Europe, East and West

From 1945 until 1990, Europe was divided into two geopolitical and economic blocs, east and west, separated by the infamous **Iron Curtain** that descended shortly after the armistice of World War II. In 1961, this division took concrete expression when the Soviets built the Berlin Wall to stop Germans from fleeing to the west. East of this border, the Soviet Union imposed the heavy imprint of communism on all activities—political, economic, military, and cultural. To the west, as Europe rebuilt from the ravages of the war, new alliances and institutions were created to counter the Soviet presence in Europe.

Cold War Geography The seeds of the Cold War are commonly thought to have been planted at the Yalta Conference

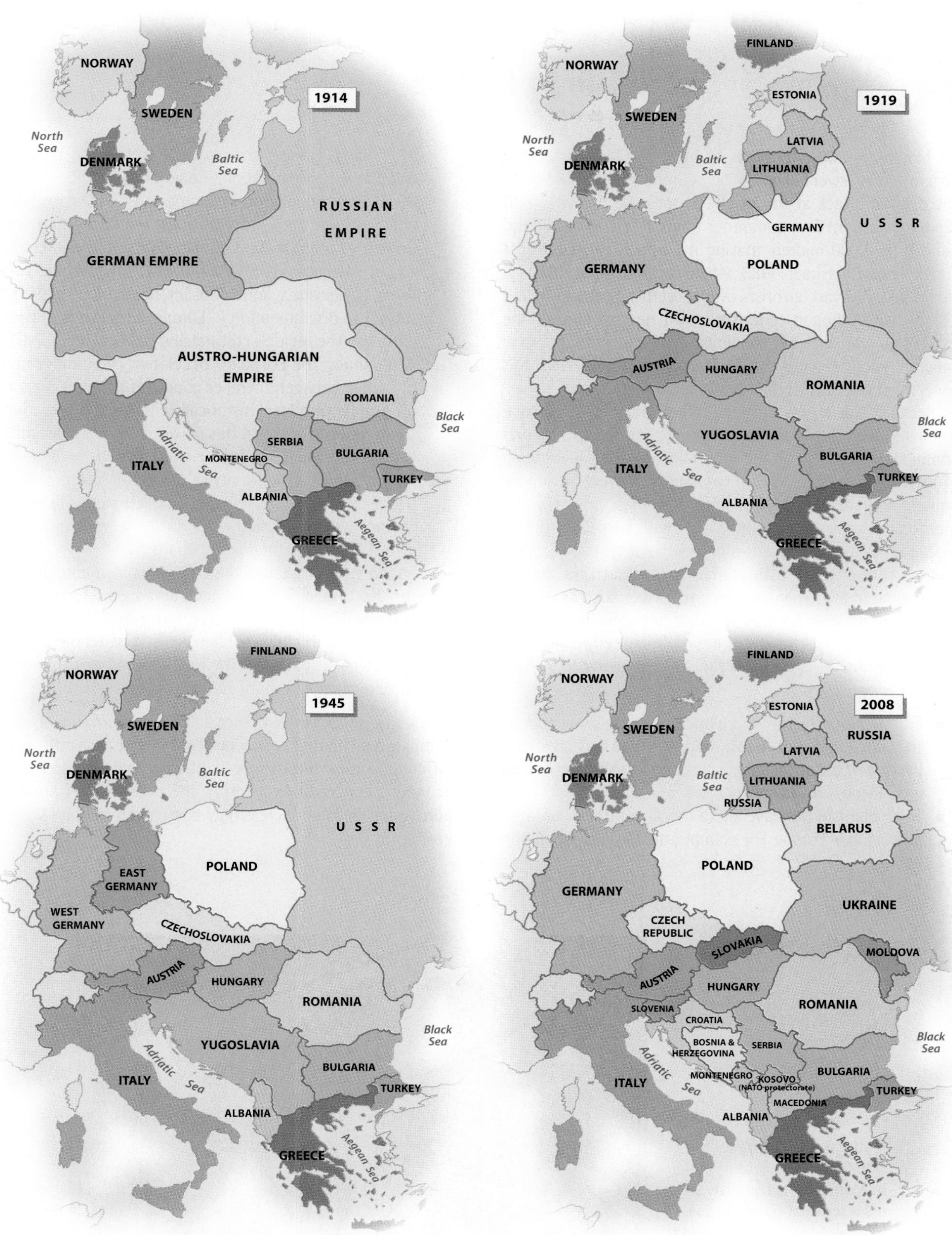

Figure 8.30 A Century of Geopolitical Change
At the outset of the twentieth century, central Europe was dominated by the German, Austro-Hungarian (or Hapsburg), and Russian empires. Following World War I, these empires were largely replaced by a mosaic of nation-states. More border changes followed World War II, largely as a result of the Soviet Union's turning that area into a buffer zone between itself and western Europe. With the demise of Soviet hegemony in 1989, further political change took place.

GEOGRAPHY IN THE MAKING

Terrorism in Europe

On March 11, 2004, terrorists bombed the Madrid subway system. That event—known in Europe as 3/11—was a coordinated attack at the height of the rush hour, with 10 explosions aboard four commuter trains. It left 191 dead and more than 1,800 injured, making it Europe's worst terrorist attack since December 1988, when Pan Am flight 103 was blown up by Libyan terrorists over Lockerbie, Scotland, killing 270 people. Although terrorism is not new to Europe, the region's growing web of geopolitical involvements aggravates concern about attacks on civilian populations, and this concern escalated with the Madrid subway bombing.

Because Spain has suffered many attacks by Basque terrorists over the years, immediate blame for the Madrid bombings was placed on the notorious ETA, the violent wing of Basque separatism. Within two weeks, however, evidence pointed toward extremist Islamist terrorists, primarily the Moroccan Islamic Combatant Group, with suspected ties to Al Qaeda. As a result, Spain—along with the rest of Europe—had to confront the reality that terrorist attacks are no longer limited to homegrown internal issues such as Basque separatism or Irish-English religious tension. Instead, the region has become a target for international terrorism.

The reasons for this new reality are complex and include the following:

- Europe's long history of colonialism with Islamic countries such as Saudi Arabia, Iraq, and Iran has created a backlog of historical animosities, as well as a complicated web of modern-day allegiances.
- These colonial ties have led to the large number of Muslims living in Europe. For example, the Muslim population in England, France, and Spain is largely the result of colonial ties. Colonialism, however, does not explain the Muslim population in Germany. There the explanation lies with the history of guest workers, along with Germany's amnesty and human rights policies. Regardless of background, however, the sensitivities and concerns of Europe's large Muslim population, a people subject to discrimination in Europe, adds layers of complexity to the region's cultural and political life.
- Additionally, the EU policy of relatively free movement of people between member countries makes it possible for groups who were historically clustered in one country to move to another. This makes it easier for terrorists to travel between, say, France and England.
- Heightening the sensitivity of Europe's Muslim population is the festering religious and ethnic strife in the Balkans where Muslims still suffer a wide range of atrocities from Orthodox Christians.
- Support for or against the U.S.-led war in Iraq, or even for the United States in general, appears to influence vulnerability to international terrorism. Many experts explain the Madrid bombings as a reaction against Spain's participation in the Iraq War and an attempt to force a change in that country's policies.

Clearly, terrorism, whether local or international, is a complicated matter involving the intersection of many different, often intertwined cultural, political, and even economic concerns. As a result, the threat of terrorism in Europe has become increasingly complex in the last several years.

Figure 8.5.1 Terrorist Bombing in Madrid In March 2004, terrorists set off a series of bombs in the Madrid, Spain, subway during rush hour, killing 191 people and injuring more than 1800. *(Kai Pfaffenbech/ Corbis)*

of February 1945 when Britain, the Soviet Union, and the United States met to plan the shape of postwar Europe. Because the Red Army was already in eastern Europe and moving quickly on Berlin, Britain and the United States agreed that the Soviet Union would occupy eastern Europe and the Western allies would occupy parts of Germany.

The larger geopolitical issue, though, was the Soviet desire for a **buffer zone** between its own territory and western Europe. This buffer zone consisted of an extensive bloc of satellite countries that would cushion the Soviet heartland against possible attack from western Europe. Some scholars argue that this concern goes back to the Russian civil war (1918–21), when western states attempted to intervene in the hope of influencing the outcome of the communist revolution. Regardless of its historical roots, the result was that former military allies ceased to be friends in 1945. Instead, Europe fell into two hostile blocs serving the interests of the two new global superpowers, the United States and the Soviet Union.

In the east the Soviet Union took control of the Baltic states, Poland, Czechoslovakia, Hungary, Bulgaria, Romania, Albania, and Yugoslavia. The last two states later distanced themselves from the Soviet Union by taking different approaches to communism. Austria and Germany were divided into occupied sectors by the four (former) allied powers. In both cases, the Soviet Union dominated the eastern portion of each country, areas that contained the capital cities of Berlin and Vienna. Both capital cities, in turn, were divided into French, British, U.S., and Soviet sectors. In 1955, with the creation of an independent and neutral Austria, the Soviets withdrew from their sector, effectively moving the Iron Curtain eastward to the Hungary– Austria border. This was not the case with Germany, however, which quickly evolved into two separate states (West Germany and East Germany) that remained apart until 1990.

Along the border between east and west, two hostile military forces faced each other for almost half a century. Both sides prepared for and expected an invasion by the other across the barbed wire of a divided Europe (Figure 8.31). In the west, NATO (the North Atlantic Treaty Organization) forces, including the United States, were encamped from West Germany south to Turkey. To the east, the Warsaw Pact forces were anchored by the Soviets but also included token armies from most satellite countries. Both NATO and the Warsaw Pact countries were armed with nuclear weapons, making Europe a tinderbox for what could be a third world war.

Berlin was the flashpoint that brought these forces close to a fighting war on two occasions. In winter 1948, the Soviets imposed a blockade on the city by denying Western powers access to Berlin across its East German military sector. This attempt to starve the city into submission was thwarted by a nonstop airlift of food and coal by NATO. In August 1961, the Soviets built the Berlin Wall to curb the flow of East Germans seeking political asylum in the west. The Wall became the concrete-and-mortar symbol of a firmly divided postwar Europe. For several days while the Wall was being built and the West agonized over destroying it, NATO and Warsaw Pact tanks and soldiers faced each other with loaded weapons at point-blank range. Though war was averted, the Wall stood for 28 years, until November 1989.

The Cold War Thaw The symbolic end of the Cold War in Europe came on November 9, 1989, when East and West Berliners joined forces to rip apart the Wall with jackhammers and hand tools (Figure 8.32). Within a year, East and West Germany were officially reunified into a single nation-state, and during the following decade all other Soviet satellite states from the Baltic to the Black Sea also underwent major geopolitical changes.

In August 1989, Poland elected the first noncommunist government to lead an eastern European state since World War II. Following this, with just one exception, peaceful revolutions with free elections spread throughout eastern Europe as communist governments renamed themselves and broke with doctrines of the past. In Romania, though, street fighting between citizens and military resulted in the violent overthrow and execution of the communist dictator, Nicolae Ceausescu.

Figure 8.31 The Iron Curtain
During the Cold War when Germany was divided into east and west, villages were oftern split in half by the Iron Curtain border with communist East Germany. Here, the village is truncated by the communist "kill zone" in the foreground where those who attempted to flee East Germany were shot on sight. *(James Blair/NGS Image Collection)*

(a)

(b)

Figure 8.32 The Berlin Wall
In August 1961, the East German and Soviet armies built a concrete and barbed-wire structure, known simply as "the Wall," to stem the flow of East Germans leaving the Soviet zone. It was the most visible symbol of the Cold War until November 1989, when Berliners physically dismantled it after the Soviet Union renounced its control over eastern Europe. *([a] Corbis/Bettmann; [b] Anthony Suau/Getty Images, Inc.–Liaison)*

In late 1989, as all Europeans nervously awaited a Soviet response to developments in eastern Europe, President Mikhail Gorbachev said that the Soviet Union had no moral or political right to intervene in the domestic affairs of eastern Europe. Following this statement, Hungary and Czechoslovakia opened their borders to the west, and eastern Europeans freely visited and migrated to western Europe for the first time in 50 years. Estimates are that more than 100,000 East Germans fled their country during this time.

As a result of the Cold War thaw, the map of Europe began changing once again. Germany reunified in 1990. Elsewhere, separatism and ethnic nationalism, long suppressed by the Soviets, were unleashed in southeastern Europe. Yugoslavia fragmented into the independent states of Slovenia, Croatia, and Bosnia. More recently, Macedonia and Montenegro have become independent states. In central Europe, Czechoslovakia was replaced by two separate states, the Czech Republic and Slovakia in 1993.

A New Geopolitical Stability? While the final shape of post–Cold War Europe is still far from clear, there is little question that geopolitical tensions have been reduced between former adversaries as major steps have been taken to foster stability in the historically treacherous region of eastern Europe. In early 1999, three former Warsaw Pact enemies—Poland, the Czech Republic, and Hungary—became members of NATO. Russia resisted this encroachment into its former sphere of influence and buffer zone, so much so that NATO offered Moscow a special relationship of its own within the alliance. In a historic document, NATO affirmed that it no longer viewed Russia as an enemy, had no intention of deploying nuclear weapons on the territory of its new members, and, further, would not station permanent combat forces in those countries. Despite Russia's continued unease, NATO added seven more former communist countries in May 2004: Bulgaria, Romania, Slovenia, Slovakia, Estonia, Latvia, and Lithuania.

The Balkans: Europe's Geopolitical Nightmare

The former Yugoslavia tests the mettle of the new, unified Europe. More specifically, both NATO and the EU are haunted by their inability to solve the complexities of ethnic conflict, micronationalism, and state disintegration. In 1918, the world responded to Balkan factionalism by creating the new state of Yugoslavia. However, by the last decade of the twentieth century, it was painfully apparent that this was not a solution at all. Again and again, ethnic fault lines in the Balkans have threatened global stability (Figure 8.33).

That the problems in the Balkans are global—not merely local or regional—needs to be emphasized. Indeed, a complicating geopolitical dimension of the Balkan issue is that these struggles and tensions play out on four different and complicated scales. These are:

- First, there are local and regional tensions within former Yugoslavia (now called Serbia) itself, which consists of two provinces, Serbia and Kosovo.
- As well there are the tensions between the neighboring states of Macedonia, Albania, Bosnia, Croatia, and Slovenia.
- Third, there are relations with the rest of Europe, with NATO, and with the EU, that range from military and peacekeeping actions to EU offers of asylum for refugees from the war-torn Balkans.
- Finally, and perhaps most vexing, there are global implications of the crisis. Elements beyond Europe

include the United Nations, which has committed peacekeeping monitors and observers; Russia, with its strong pro-Serb bias; the Islamic countries (primarily Iraq, Iran, and Saudi Arabia), which support Muslim populations in the Balkans; and the United States, which as a NATO and UN member has committed military forces to enforce Balkan peace agreements.

Pundits commonly take refuge in Balkan history to explain the current troubles, pointing out centuries of regional rivalries, clan feuds, religious intolerance, and ethnic cleansing. Implied in this perspective is the belief that the Balkans is a region unlike any other part of the world, thus we should accept as inevitable the violence and terrorism of ethnic separatism. Indeed, the term **balkanization** is used to describe the fragmented geopolitical processes involved with small-scale independence movements and the phenomenon of mini-nationalism as it develops along ethnic fault lines.

Historical Background As World War I ended in 1918, so did the Austro-Hungarian and Ottoman Turk empires' control of the Balkans. The independent kingdom of Serbia became the magnet for the creation of the new state of Yugoslavia in 1929. Attached to Serbia were a collection of other Hapsburg lands, including Slovenia, Croatia, Bosnia and Herzegovina, Montenegro, Macedonia, and Kosovo. The result was a new state of three major religions (Islam, Catholicism, and Eastern Orthodoxy), a handful of languages, at least eight different ethnic groups, and two distinct alphabets. From the outset, a Serbian king ruled over this new country and Serbs dominated political and economic activity. Belgrade and the Serbian core amassed wealth at the expense of the ethnic periphery. As a result, Yugoslavia was marked by internal hostility for two decades until it was invaded by Hitler in 1941.

German troops were welcomed as liberators in Croatia, a region that was particularly restless under Serbian rule. Major resistance to Nazi occupation, however, came from two fronts: Serbian royalists who supported the king and opposed Hitler, and partisans who were mainly pan-Yugoslav communists. The latter were led by Marshall Tito, a Croat, and supported by the Allies in his battle against Nazi Germany.

Following World War II, Britain, the United States, and

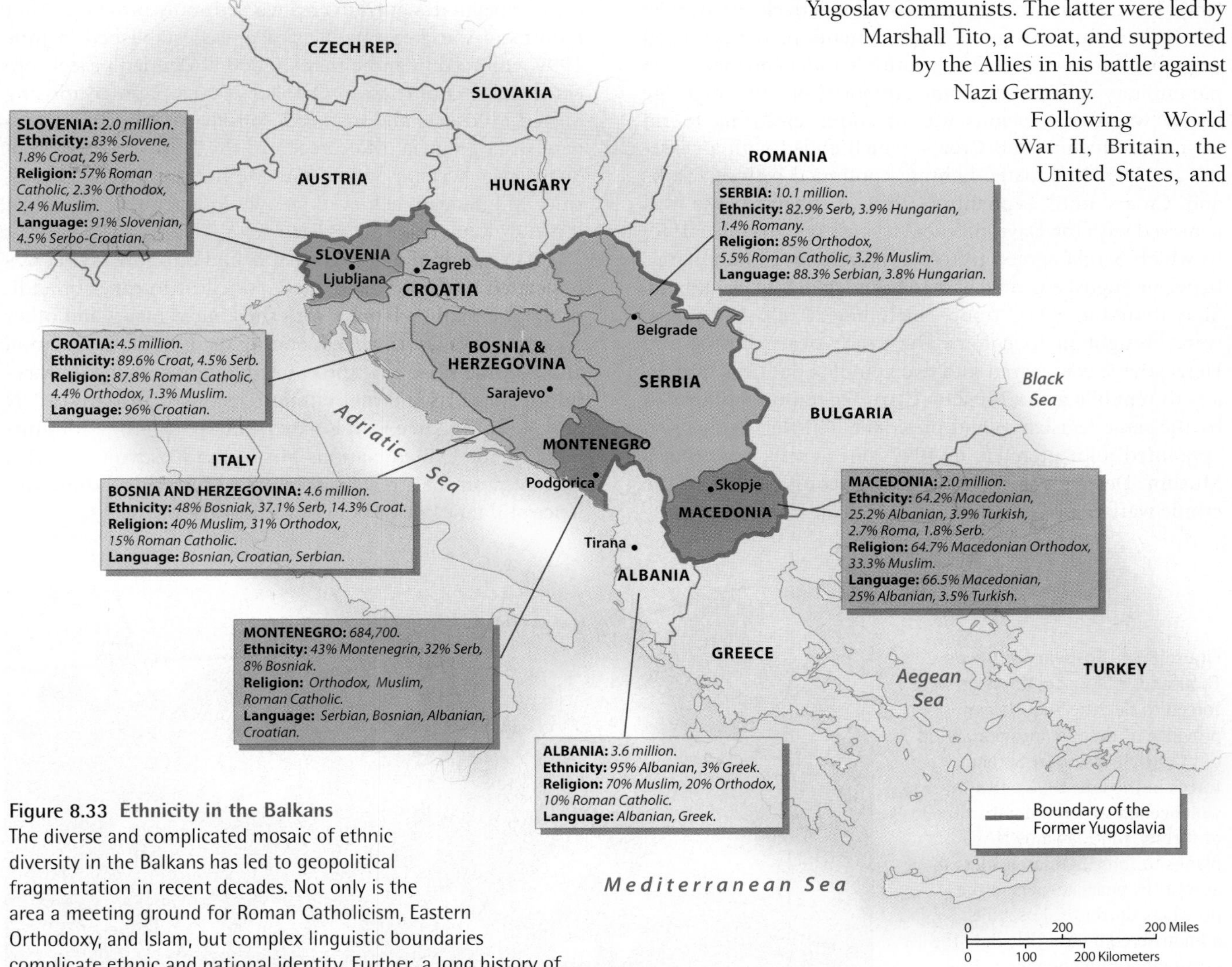

Figure 8.33 **Ethnicity in the Balkans**
The diverse and complicated mosaic of ethnic diversity in the Balkans has led to geopolitical fragmentation in recent decades. Not only is the area a meeting ground for Roman Catholicism, Eastern Orthodoxy, and Islam, but complex linguistic boundaries complicate ethnic and national identity. Further, a long history of discrimination and retaliation between ethnic groups is embedded in ethnic consciousness. *(Data from CIA World Factbook, 2007)*

the Soviet Union rewarded Tito for his efforts by backing him as the leader of a new socialist Yugoslavia. Conventional wisdom posits that the three major ethnic groups—the Serbs, Croats, and Muslims—coexisted under Tito's strong leadership by subordinating their separatist agendas to the larger goal of a communist state independent of the Soviet Union. This relatively peaceful coexistence continued for almost a decade after Tito's death in 1980. During that period, leaders of the various Yugoslavian states took turns serving as the country's president.

More Recent Events The region's geopolitical stability began to unravel in 1990 in concert with the lessening of Soviet control in eastern Europe. In that year, elections held in the different republics brought the issue of secession from Yugoslavia to a head. While secessionist parties gained control in Slovenia and Croatia, Serbian voters (in what observers consider to be a government-controlled election) opted for continued unification.

Slovenia and Croatia declared independence in 1991, Macedonia in January 1992, Bosnia and Herzegovina in April 1992. When the Yugoslavian army attacked Slovenia, the Balkan situation got Europe's attention. A negotiated settlement resulted in that country's independence. Serb paramilitary units (with some support from the Yugoslav army) waged a ferocious war of ethnic cleansing to rid Bosnia of Muslims and Croats, which ended with a cease-fire in 1995. In Croatia, fighting continued between Serbs and Croats until September 1995. A fragile peace was achieved with the Dayton Peace Accords of November 1995 in which Serbia agreed to respect a new set of boundaries between Yugoslavia and its newly independent neighbors. Sixty thousand NATO troops, including 20,000 Americans, were brought in to enforce the peace treaty. Bosnia and Herzegovina was carved into two complex political entities, a Serb republic and a Muslim–Croat federation, both ruled by the same legislature and president. This legislature also appointed joint premiers in 1997, one a Serb, the other a Muslim. Despite the Dayton Peace Accords, however, the ethnic warfare plaguing the region was far from over.

Kosovo, to the south of Serbia, is another trouble spot (Figure 8.34). This predominantly Muslim Albanian-populated province of 2 million lost its autonomous status shortly after the 1990 elections, and Kosovar rebels proclaimed an independent republic in 1997. With the effective collapse of the Albanian government in 1997, looting of military bases in that country provided the Kosovo insurgents with a new and enlarged supply of arms, which resulted in escalating attacks on the Serb-dominated Yugoslav army and police. Serb authorities responded by expelling Albanian Kosovars. In an attempt to end the conflict, the UN sent 2,000 unarmed monitors into Kosovo in November 1998; however, this UN effort did not work, nor did a diplomatic solution that would have guaranteed autonomy for Kosovo within Yugoslavia because of rejection by then President Milosevic.

As a result of this diplomatic failure and increased violence by Serbs against the Kosovars, NATO began an intensive air war against Yugoslavia in March 1999. In retaliation, the Serbian army and police killed or exiled hundreds of thousands of ethnic Kosovars in a violent ethnic cleansing campaign designed to create a Serb-only province. After months of warfare, a fragile peace was established in June 1999, enforced by more than 50,000 NATO-led peacekeepers from 30 countries. In October 2000 a popular uprising ousted Milosevic and installed Vojislav Kostunica as president of Yugoslavia. More recently, the Federal Republic of Yugoslavia became Serbia and Montenegro in 2003 and, since Montenegro achieved its own independence in 2006, it is now known simply as Serbia.

As a result of a new governmental outlook, Serbia was reinstated in the UN and the Council of Europe. In addition, diplomatic relations with the United States and other countries were normalized, ending its diplomatic isolation and offering hope that the country might somehow peacefully resolve its internal ethnic tensions. Currently, a UN peacekeeping force is in Kosovo and the region is administered by the United Nations Mission in Kosovo (UNMIK). However, on the streets and in the countryside, ethnic violence still takes place between Kosovars and Serbs.

Figure 8.34 Kosovar Refugees Thousands of Muslim Kosovars were forced to flee the Yugoslavian province of Kosovo during a period of ethnic cleansing by Serbian forces in the late 1990s. This violence led to aerial bombardment of former Yugoslavia by NATO planes to force Serbian leaders to accept UN peacekeepers in the province. Currently, Kosovo is administered by the UN while the possibility of independence from Serbia is discussed *(Buu/Getty Images, Inc.–Liaison)*

ECONOMIC AND SOCIAL DEVELOPMENT: Integration and Transition

As the acknowledged birthplace of the Industrial Revolution, Europe in many ways invented the modern economic system. Though Europe was the world's industrial leader in the early twentieth century, it was eclipsed by Japan and the United States as Europe struggled to cope with the legacy of two world wars, a decade of global depression, and the Cold War.

In the last 50 years, however, the European Union has guided an increasingly successful economic integration. In fact, western Europe's success at blending national economies has given the world a new model for regional cooperation, an approach that could well be imitated in Latin America and Asia in the twenty-first century. Eastern Europe, however, has not fared so well. The results of four decades of Soviet economic planning were, at best, mixed. The total collapse of that system in 1990 cast eastern Europe into a period of chaotic economic, political, and social transition from which the region is only now emerging.

Accompanying western Europe's economic boom has been an unprecedented level of social development as measured by worker benefits, health services, education, literacy, and gender equality. Though the improved social services set a laudable standard for the world, they are currently under attack by cost-cutting business people and politicians who argue they add so much to the cost of business that European goods cannot compete in the global marketplace. As a result, today the European economy is at a crossroads with several different yet undefined pathways possible.

Europe's Industrial Revolution

Europe is the cradle of modern industrialism. Two fundamental changes were associated with this transformation. First, machines replaced human labor in many manufacturing processes, and, second, inanimate energy sources, such as water, steam, electricity, and petroleum, powered the new machines. Though we commonly apply the term *Industrial Revolution* to this transformation, implying rapid change, it took more than a century for the interdependent pieces of the industrial puzzle to come together. This happened first in England between 1730 and 1850. Once it occurred, however, rapid change ensued as industrialization spread to other parts of Europe and the world.

Centers of Change England's textile industry, located on the flanks of the Pennine Mountains, was the center of early industrial innovation. More specifically, innovation and change took place in the small towns and villages of Yorkshire and Lancashire, away from the rigid control of the urban guilds. Yorkshire, on the eastern side of the Pennines, had been a center of woolen textile making since medieval times, drawing raw materials from the extensive sheep herds of that region and using the clean mountain waters to wash the wool before it was spun in rural cottages. On the western flank of the mountains, Lancashire workers used cotton from England's overseas colonies (including the United States) as the basis for their newer textile industry. By the 1730s, water wheels drove mechanized looms at the rapids and waterfalls of the Pennine streams (Figure 8.35).

Figure 8.35 Water Power and Textiles
Europe's industrial revolution began on the flanks of England's Pennine Mountains, where swift-running streams and rivers were used to power large cotton and wool looms. Later many of these textile plants switched to coal power. *(James Marshall/Corbis/Bettmann)*

Locational Factors of Early Industrial Areas By the 1790s, the steam engine had become the preferred source of energy to drive the new looms because waterpower was hindered by low stream flow during dry weather. However, the steam engine required vast amounts of wood fuel, and England's forests had been depleted since the 1600s because of shipbuilding. While Great Britain possessed large amounts of coal, it was expensive to transport long distances by horse-drawn cart. Only with the advent of the railroad after 1820 could coal be moved long distances at reasonable cost.

Another fundamental innovation was the change from charcoal to coke in the manufacture of iron and steel. Until the early nineteenth century, iron foundries throughout Europe used wood charcoal in the smelting process, which created a pattern of small iron mills dispersed throughout rural forest areas, close to sources of charcoal. In the 1790s, coal-based coke was first substituted for charcoal in the Midlands iron and steel industry. Not only was coke cheaper than charcoal in timber-poor England, it also permitted the use of lower-grade iron ore. Following this change, iron and steel foundries relocated from rural forest areas to urban clusters adjacent to coalfields. By the 1820s, the world's first industrial landscapes had appeared in the English Midlands and Scottish Lowlands. Europe's industrial era had begun.

London was an important exception to the rule that the British Industrial Revolution developed near the

Figure 8.36 Industrial Regions of Europe
From England, the Industrial Revolution spread to continental Europe, starting with the Sambre-Meuse region on the French–Belgian border, then diffusing to the Ruhr area in Germany. Readily accessible surface coal deposits powered these new industrial areas. Early on, iron ore for steel manufacture came from local deposits, but later it was imported from Sweden and other areas in the shield country of Scandinavia. Most of the newer industrial areas are closely linked to urban areas.

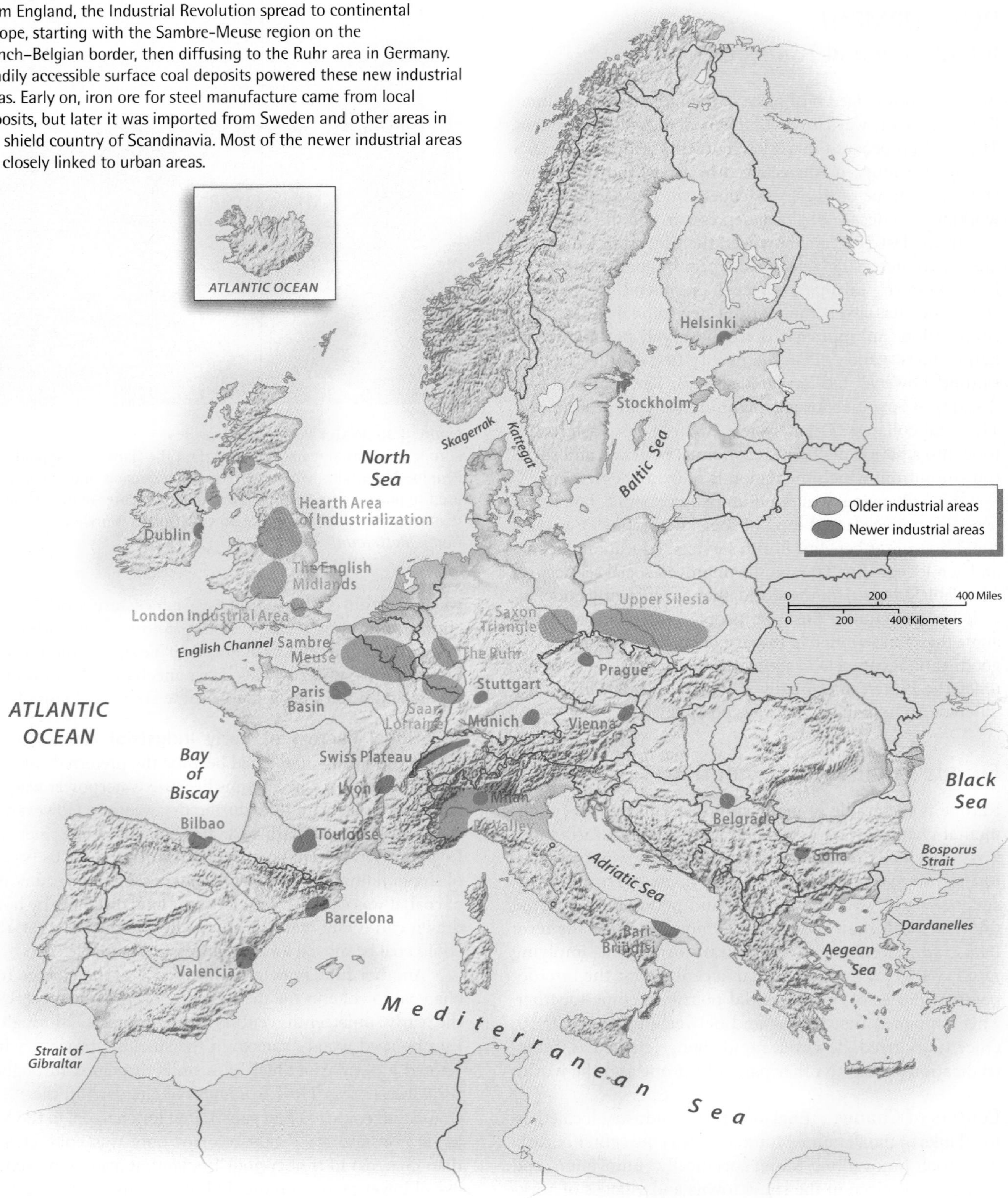

coalfields. Despite being far removed from natural resources, this world city developed into the largest industrial center in the United Kingdom because it had access to a large market, a large supply of skilled labor, a port for overseas trade, and a merchant class with access to capital and proximity to banks, insurance firms, investors, and other institutional support systems needed by industry.

Development of Industrial Regions in Continental Europe By the 1820s, the first industrial districts had begun appearing in continental Europe. These hearth areas were, and still are, near coalfields (Figure 8.36). The first area outside Britain was the Sambre-Meuse region, named for the two river valleys straddling the French–Belgian border. Like the English Midlands, it had a long history of cottage-based wool textile manufacturing that quickly converted to the new technology of steam-powered mechanized looms. Additionally, a metalworking tradition drew on charcoal-based iron forges in the nearby forests of the Ardenne Mountains. Coal was also found in these mountains; in 1823 the first coke blast furnace outside Britain started operation in Liege, Belgium.

By the second half of the nineteenth century, the dominant industrial area in all Europe (including England) was the Ruhr district in northwestern Germany, near the Rhine River. Rich coal deposits close to the surface fueled the Ruhr's transformation from a small textile region to one oriented around heavy industry, particularly iron and steel manufacturing. By the early 1900s, the Ruhr had exhausted its modest iron ore deposits and was importing ore from Sweden, Spain, and France. Several decades later the Ruhr industrial region became synonymous with the industrial strength behind Nazi Germany's war machine and was bombed heavily in World War II (Figure 8.37).

To the southwest of the Ruhr, the industrial region of Saar-Lorraine straddles the border between Germany and France. Though this region contains rich deposits of both coal and iron ore, a much-contested political boundary separated these resources until after World War II. Most of the coal lies in the German Saarland, while iron ore is found primarily in the French Lorraine district. As the industrial strength of this region grew in the late nineteenth century, so did political strife over its control. Twice Germany annexed the iron deposits in France, and twice France attempted unsuccessfully to occupy the coal areas of the Saarland. Only since 1953 have these resources been integrated through trade agreements.

Historical boundary shifts and political tensions between states also plagued the Upper Silesia industrial district, which now lies in southern Poland west of Krakow. Historically a center of charcoal-iron smelting, this region was originally part of Prussia before German unification in the 1870s. German and Prussian financial interests developed the region in the late nineteenth century, combining the rich coalfields with imported coke and coal so that by 1900 Upper Silesia provided almost 10 percent of Germany's steel. After World War I, however, Germany lost half the region as the Polish border was adjusted westward. The remaining German half was awarded to Poland in 1945 with yet another boundary shift following World War II.

While most early industrial development took place on or near coalfields, a noteworthy exception is the Po Plain in northern Italy, where hydroelectric power generated in the nearby Alps provided energy for industrialization. Because Italy lacked extensive coal deposits, the government subsidized the generation and transmission of electrical energy. Indeed, the northern Italian city of Milan became the first European city to have electric lights in 1883.

Rebuilding Postwar Europe: Economic Integration in the West

Because of its historical momentum, Europe was unquestionably the leader of the industrial world in the early twentieth century. More specifically, before World War I, European industry was estimated to produce 90 percent of the world's manufactured output. However, four decades of political and economic chaos, punctuated by two devastating world wars, left Europe divided and in shambles. By 1945, its cities were in ruins; industrial areas were destroyed; vast populations were dispirited, hungry, and homeless; and millions of refugees moved about Europe looking for solace and stability.

In 1947, the European Recovery Program (ERP) was established to rebuild the region. Through the Marshall Plan, the United States was a major contributor to ERP. In eastern Europe, however, then controlled by the Soviet Union, this aid was rejected because of U.S. involvement. In its place the Soviets created in 1949 the **Council for Mutual Economic Assistance (CMEA**, or, alternatively, Comecon), which linked eastern European aid and recovery to the centralized, command economies of communism.

ECSC and EEC In 1950, western Europe began discussing a new form of economic integration that would avoid both the historical pattern of nationalistic independence through tariff protection as well as the economic inefficiencies resulting from the duplication of industrial effort. Instead, Robert Schuman, France's foreign minister, proposed that German and French coal and steel production be coordinated by a new supranational authority. In May 1952, France, Germany, Italy, the Netherlands,

Figure 8.37 The Ruhr Industrial Landscape
Long the dominant industrial region in Europe, the Ruhr region was bombed heavily during World War II because of its central role in providing heavy arms to the German military. It has been rebuilt, has modernized, and now is competitive with newer industrial regions. *(Andrej Reiser/Bilderberg Archiv der Fotografen)*

Belgium, and Luxembourg ratified a treaty that joined them in the European Coal and Steel Community (ECSC). Because of the immediate success of the ECSC, these six states soon agreed to work toward further integration by creating a larger European common market that would foster the free movement of goods, labor, and capital. In March 1957, the Treaty of Rome was signed, establishing the European Economic Community (EEC), popularly called the *Common Market*, and the forerunner of today's European Union.

Britain remained outside the new EEC for several reasons, primarily because of its economic obligations to former colonies (such as Canada) through Commonwealth agreements and an ambivalence about losing its historic independence from continental Europe, but also because of French objection to English participation. Throughout the 1960s, EEC membership for the United Kingdom was resisted by France for fear the English economy was simply a Trojan horse for American interests. Shut out of the EEC by France, Britain instead joined with Switzerland, Austria, Portugal, Denmark, Sweden, and Norway in 1958 to form the short-lived European Free Trade Association (EFTA).

European Community and Union The EEC reinvented itself in 1965 with the Brussels Treaty that laid the groundwork for adding a political union to the successful economic community. In this "second Treaty of Rome," aspirations for more than economic integration were clearly stated with the creation of an EEC council, court, parliament, and commission. The EEC also changed its name to the European Community (EC). To be fully successful, though, the EC had to expand its membership, and this occurred eight years later when the French dropped their opposition to the admission of the United Kingdom. In 1973, the EC added not just the United Kingdom, but also Denmark, Ireland, and Norway, though voters in that Scandinavian country subsequently rejected admission in a referendum. The EFTA, as a result, disintegrated. During the 1980s, Greece, Portugal, and Spain were added to the EC, with Austria, Finland, and Sweden joining in 1995.

In 1991, the EC again changed its name, this time to the European Union (EU), and expanded its goals once again with the Treaty of Maastricht (named after the town in the Netherlands in which delegates met). While economic integration remains an underlying theme in the new constitution, the EU has also moved into intergovernmental and supranational affairs with an agenda for common foreign policies and mutual security agreements (see "Global to Local: Bananas, Beef, and U.S.–EU Trade Wars").

More recently, in May 2004 the EU added 10 new states including a core group of former Soviet-controlled communist countries from eastern Europe. These new members—Latvia, Estonia, Lithuania, Poland, Slovakia, the Czech Republic, Hungary, Slovenia, Malta, and Cyprus—brought the total to 25; Bulgaria and Romania were then admitted in January 2007 resulting in the current total of 27 EU countries (Figure 8.38). Several Balkan countries have applied for EU membership, as has Turkey. No date is set for their admission.

Euroland: The European Monetary Union States usually have their own monetary systems since coining money is a fundamental component of political sovereignty. As a result, in many parts of the world crossing a political border usually means changing money and becoming familiar with a new system of bills and coins. Until recently this was also the case in Europe, which meant buying pounds sterling in England, marks in Germany, francs in France, lira in Italy, and so on. However, this mosaic of coinage systems has become an artifact of the past as Europe moves from individual state monetary systems to a common currency.

On January 1, 1999, in what many saw as a historical advance for the notion of a united Europe, 12 of the then 15 EU member states created the European Monetary Union (EMU). As of that date, cross-border business and trade transactions began taking place in a new monetary unit, the *euro*. While the euro remained a fiscal abstraction for three years, on January 1, 2002, euro coins and bills became available for everyday use and by July 2002, the new monetary system had replaced the old national currencies in everyday life.

Although the United Kingdom and several other EU member countries are still uncertain about joining EMU, the euro is generally judged to be highly successful. In fact, given the strength of the euro against other global currencies, the euro is starting to replace the dollar as the international currency of preference.

Economic Integration, Disintegration, and Transition in Eastern Europe

Historically, eastern Europe has been less developed economically than its western counterpart (Table 8.2). Partial explanations come from the fact that eastern Europe is not particularly rich in natural resources, for it has only modest amounts of coal, less iron ore, and little oil and natural gas. Furthermore, the few resources found in the region have been historically exploited by outside interests, including the Ottomans, Hapsburgs, Germans, and, more recently, the Soviet Union.

While the Soviet-dominated economic planning of the postwar period attempted to develop eastern Europe through a coordinated regional economy that also served Soviet goals, the collapse of that centralized system in 1991 cast many eastern European countries into deep economic, political, and social chaos from which they are only now emerging. As a result, one cannot understand the present-day problems of eastern Europe without background on the earlier period of Soviet-bloc economic planning.

The Soviet Plan In 1949, the Soviet Union took control of eastern Europe's economies. Its goals were complete

Figure 8.38 The European Union
The driving force behind Europe's economic and political integration has been the European Union (EU), which was formed in the 1950s as an organization focused solely on rebuilding the region's coal and steel industries. After initial success, the Common Market (as it was then called) expanded into a wider range of economic activities. Today, the EU has 27 members.

economic, political, and social integration through a **command economy**, one that was centrally planned and controlled. The methods were straightforward: in most countries the state owned all means of production so the economy could be managed for the benefit of the people as a whole, not for private interests. Put differently, this meant that most economic activity was nationalized.

Further, supply and demand were coordinated at a regional level so that states would not compete with each other. Natural resources, such as coal, iron, natural gas,

TABLE 8.2 Development Indicator

Country	GNI[a] per Capita, PPP[b] (2005)	GDP[c] Average Annual % Growth (2000–05)	Life Expectancy (2007)	Population Living on Less Than $2 a Day (2006)	Under Age 5 Mortality Rate		Gender Equity[d]
					1990	2005	
Western Europe							
Austria	33,140	1.5	80		10	5	102
Belgium	32,640	1.5	79		10	5	103
France	30,540	1.5	81		9	5	105
Germany	29,210	0.7	79		9	5	
Liechtenstein			80				
Luxembourg	65,340		78				
Monaco							
Netherlands	32,480	0.7	80	0.5	9	5	99
Switzerland	37,080	0.9	81	0.6	9	5	94
United Kingdom	32,690	2.4	79	2.3	10	6	107
Eastern Europe							
Bulgaria	8,630	5.0	73	<2	19	15	100
Czech Republic	20,140	3.5	76	<2	13	4	101
Hungary	16,940	4.1	73	<2	17	8	107
Poland	13,490	3.2	75	<2	18	7	109
Romania	8,940	5.8	71	13	31	19	105
Slovakia	15,760	4.9	74	3	14	8	104
Southern Europe							
Albania	5,420	5.3	75	12	45	18	99
Bosnia & Herzegovina	7,790	5.0	74	<2	22	15	
Croatia	12,750	4.7	75		12	7	104
Cyprus	22,230		78				
Greece	23,620	4.4	79		11	5	105
Italy	28,840	0.6	81		9	4	106
Macedonia	7,080	1.7	74	<2	38	17	103
Malta	18,960		80				
Montenegro		5.1	73		15	11	107
Portugal	19,730	0.5	78	<2	14	5	108
San Marino			81				
Serbia		5.1	72		15	11	107
Slovenia	22,160	3.4	78	<2	10	4	109
Spain	25,820	3.1	80		9	5	107
Northern Europe							
Denmark	33,570	1.2	78		9	5	109
Estonia	15,420	7.5	73	8	16	7	114
Finland	31,170	2.4	79		7	4	107
Iceland	34,760		81				
Ireland	34,720	5.2	78		9	6	103
Latvia	13,480	7.9	72	5	18	11	115
Lithuania	14,220	7.8	71	8	13	9	110
Norway	40,420	2.0	80		9	4	109
Sweden	31,420	2.3	81	2	7	4	112

[a] *Gross national income.*
[b] *Purchasing power parity.*
[c] *Gross domestic product.*
[d] *Ratio of female-to-male enrollments in primary and secondary school percentage. Numbers below 100 have more males in primary/secondary school; numbers above 100 have more females in primary/secondary schools.*

Sources: World Bank, World Development Indicators, *2007; percentage of population living on $2 a day data from Population Reference Bureau,* World Data Sheet, *2007; gender equity data from* Millennium Development Goals.

GLOBAL TO LOCAL Bananas, Beef, and U.S.–EU Trade Wars

A premise of economic globalization—for better or worse—is the notion of free trade, that all countries should have equal access to all others as potential trading partners, unhindered by preferential agreements or protectionist quotas, tariffs, or taxes. While both the United States and the European Union pay lip service to free trade, a good deal of hypocrisy lies beneath this agreement.

The reality of U.S.-EU trade is that it is characterized by protection of special interests, cutthroat trade wars, and retaliatory tariffs. But these measures are more than abstract legal issues argued in the world's courtrooms, for each decision has local implications that can spell boom or bust for local farmers and consumers. Two examples make the case.

- **The U.S.–EU Banana War.** Several EU countries gave preferential trade agreements to former colonies or overseas territories that produced bananas. France, for example, favored bananas from its Caribbean territory of Guadeloupe by setting a quota on "dollar" banana imports from U.S. companies. Chiquita Banana (the company, not the banana) took offense at this barrier to European markets and persuaded the U.S. government to argue its case against the EU.

  The "banana war" erupted in 1999 when the United States (prodded by Chiquita) slapped a 100 percent tax on a range of European goods from Scottish cashmere sweaters to French designer handbags and Roquefort cheese. As a result, the Scottish woolen industry (from sheep farmers to factory workers) was on the verge of collapse. Likewise, thousands of France's farmers and cheesemakers faced financial ruin.

  The EU counterattacked by asking the World Trade Organization (WTO) to convene a panel to review the whole gamut of U.S. trade policy. This action brought the United States to the negotiating table and, as a result, in 2001 a truce was called to the banana wars when the United States rescinded its tariffs and the EU dismantled its policy of preferential treatment for Caribbean and African bananas. Ironically, Chiquita Banana was still a loser as it was brought to the brink of bankruptcy by a consumer boycott, resulting in a 50 percent loss of market share in Europe.

Figure 8.6.1 Banana War Casualty
A Caribbean banana farmer says goodbye to his lucrative crop of "green gold" that previously enjoyed guaranteed markets in France. *(Alamy)*

- **The U.S.–EU Beef War.** Before the banana war was settled, the United States and the EU were at it again, this time over the issue of beef imports. Since 1988, the EU had banned U.S. beef that contained growth hormones, but in 1999 the EU upped the ante by banning all beef because of health concerns about other growth additives commonly used in the U.S. cattle industry. Many analysts linked this ban to the then-unsettled banana war.

In response to the EU ban, the United States once again put a 100 percent tax on a wide range of European imports, including truffles, pork, French mustard, cheese, onions, and chocolate. These retaliatory tariffs were sanctioned by the WTO because it ruled the EU ban on beef had cost U.S. ranchers about $7 million. U.S. officials had sought penalties worth $900 million, which was their estimate of the damage to ranchers.

Currently, the "beef war" goes on. Although the WTO essentially ruled against the EU ban, Europe refuses to lift it and is presenting the world with more scientific evidence that growth hormones are harmful.

and petroleum, were moved across national boundaries to serve the greater good. The Soviet Union also exported to eastern Europe those resources in short supply and, in return, guaranteed markets for the region's products. Additionally, each individual state subsidized its industry to facilitate overall economic development and provide manufacturing jobs for a reduced agricultural labor force. Finally, to increase food supplies, the agricultural sector was reshaped through **collectivization** (state ownership), mechanization, and centralized planning with the goal of increasing food production.

The Results After 40 years of communist economic planning, the results were mixed within eastern Europe. In Poland and Yugoslavia, many farmers strongly resisted the nationalization of agriculture, thus, most arable and productive land remained in private hands. However, in Romania, Bulgaria, Hungary, and Czechoslovakia, 80 to

90 percent of the agricultural sector was converted to state and collective farms. But despite these radical changes, food production did not increase dramatically, and food shortages, particularly for urban dwellers, were common across eastern Europe (Figure 8.39).

Perhaps most notable during the Soviet period were the dramatic changes to the industrial landscape with many new factories built in both rural and urban areas and fueled by cheap energy and raw materials imported from the Soviet Union. In defining the pathway to economic development in eastern Europe, planners chose heavy industry, such as steel plants and truck manufacturing, over consumer goods. And as in the Soviet Union, the bare shelves of retail outlets in eastern Europe became both a sign of communist shortcomings and, perhaps more important, a source of considerable public tension. As western Europeans enjoyed an increasingly high standard of living with an abundance of consumer goods, eastern Europeans struggled to make ends meet as the utopian vision promised by Soviet communists became increasingly elusive.

Transition and Changes Since 1991 As Soviet political hegemony over eastern Europe disintegrated in 1990, so did the integrated economies of communist planning. In place of the Soviet domination came a painful period of economic transition that was close to outright chaos in some eastern European countries.

The causes for this economic pain were complex. As the Soviet Union moved to address its own economic and political turmoil, it stopped exporting subsidized natural gas and petroleum to eastern Europe and instead sold it on the open global market to gain hard currency. Without cheap energy, many eastern European industries were unable to operate and shut down, laying off thousands of workers. In the first two years of the transition (1990–92), industrial production fell 35 percent in Poland and 45 percent in Bulgaria (Figure 8.40). In addition, the markets guaranteed under a command economy, many of them in the Soviet Union, dissolved overnight, causing many factories and services to close.

Given this chaos, in the 1990s eastern European countries began redirecting their economies away from Russia and toward western Europe with the goal of joining the EU. This change, though, meant moving quickly from a socialist-based economy of state ownership and control to a capitalist economy predicated on private ownership and free markets. To achieve this, states such as the Czech Republic, Hungary, and Poland had to go through a period of **privatization**, which is the transfer to private ownership of those firms and industries previously owned and run by state governments.

A good deal of hardship came with this transition. Price supports, tariff protection, and subsidies were removed from consumer goods as countries moved to a free market. For the first time, goods from western Europe and other parts of the world were now plentiful and common in eastern European retail stores. The irony, though, is that this prolonged period of economic transition has taken its toll on many eastern European consumers for few could afford these long-dreamed-of products. As well, unemployment rose to more than 15 percent, and underemployment was common.

Still today financial security is elusive for many in eastern Europe because of irregular paychecks for those

Figure 8.39 Polish Agriculture
One of the most problematic issues facing eastern European countries is the modernization of the agricultural sector. This is particularly pressing now that eastern European countries have gained membership in the EU and their farmers must now compete with western European agriculture. *(Hans Madej/Bilderberg Archiv der Fotografen)*

Figure 8.40 Post-1989 Hardship
With the fall of communism in eastern Europe after 1989, economic subsidies and support from the Soviet Union ended. Accompanying this transition has been a high unemployment rate resulting from the closure of many industries, such as this plant in Bulgaria. *(Rob Crandall/www.robcrandall.com)*

with jobs and uncertain welfare benefits for those without. Furthermore, for many the basic costs of food, rent, and utilities are higher today under a free-market system than under the subsidized economies of communism. Nonetheless, despite these problems, it is quite clear that the preferred economic and political trajectory of eastern Europe is toward western Europe and full integration with the EU.

Regional Disparities Within Eastern Europe

To many observers, a troublesome pattern of dual economies is emerging in eastern Europe, with some countries speeding ahead to reap the rewards of successful change and integration with the EU while others appear to be falling ever farther behind their neighbors (Figure 8.41).

By some accounts, successful economic transitions have taken place in the Czech Republic, Slovenia, Hungary, and Poland. Though still far behind affluent western European states, these four countries have all forged strong trade ties with western Europe, are receiving investment from the West, and are well along the path to privatization. As a result, a commercial boom has recently added a globalized dimension to landscapes in these countries. In Prague, for example, a handful of Western-style mega malls now ring the outskirts of that city, while billboards advertising Western goods and services blight the Czech landscape in both city and countryside. In Warsaw, new high-rise office buildings have transformed the stodgy communist-era central city; outside Wroclaw, like the area outside Prague, new mega malls are found everywhere. However, not everyone is happy with this blossoming globalization. For example, many Czechs feel as though they have lost their beloved capital city of Prague to international tourism and Western commercialism,

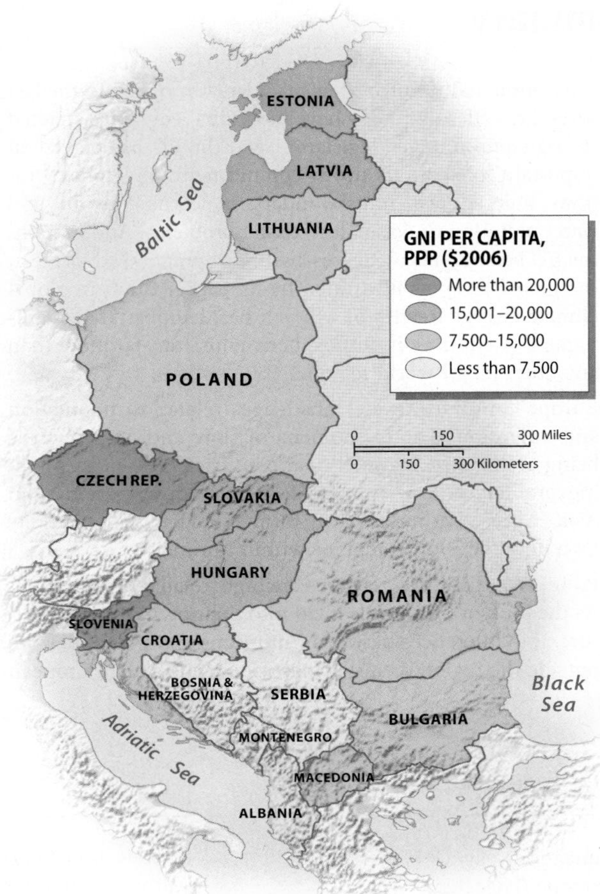

Figure 8.41 Economic Disparities in Eastern Europe
While some countries in eastern Europe are doing fairly well economically, others are lagging far behind. At the head of the pack are Slovenia and the Czech Republic, followed by Hungary and Slovakia. See Table 8.2 for more complete data.

with prices that are far beyond the reach of the Czech middle class.

At the other end of the spectrum, several Balkan countries such as Macedonia and Montenegro rank along with Albania as lowest in economic and social development measures. Common to these countries is widespread political and ethnic strife that has inhibited development. Eastern European countries that make up the middle ground of economic and social change include the three Baltic countries—Latvia, Estonia, and Lithuania—all formerly part of the Soviet Union and now members of the EU, along with Slovakia, separated from the Czech Republic in the "Velvet Divorce" of 1993. Struggling somewhat are Bosnia and Herzegovina, Croatia, and Serbia, weakened and distracted by mutual animosities and internal strife. The two Black Sea countries of Bulgaria and Romania, both highly dependent on the Soviet Union for economic stability during the postwar period, are the newest members of the EU and, while the fact of their membership indicates they have already made much progress, analysts are uncertain about their future prospects.

Summary

- Environmentally, Europe has made great progress in the last several decades. Not only have individual countries enacted strong environmental legislation, but the EU has played an important role with its strong commitment to regional solutions that address transboundary concerns like air and water pollution or ocean and coastal problems. Additionally, the EU has become an aggressive environmental advocate at the global level, as illustrated by its participation in global climate change treaties in which it has demonstrated a willingness to cut back its atmospheric emissions far more than any other major world region.
- Europe continues to face challenges related to population and migration. The phenomena of slow and no growth is being confronted by individual countries. The most pressing problem is how Europe deals with immigration from Asia, Africa, Latin America, former Soviet lands, and its own underdeveloped regions within Europe.
- Political and cultural tensions are high related to migration. As the bulk of Europeans meld into their own common culture, the differences between Europeans and migrants from outside the region seems to increase. While some European countries, namely Germany, France, and England, appear committed to becoming multicultural societies, the pathway to this goal is still unclear despite explicit legal actions against ethnic and racial discrimination.
- With the end of the Cold War, Europe's geopolitical issues are continually being redefined. The role and future direction of NATO is one of the important topics facing Europe, for it has brought together historic enemies and softened long-standing political animosities. Many European states, however, would like to replace NATO with a new military alliance, a European security force without a strong and dominating U.S. presence.
- The EU celebrated its 50th birthday in March 2007 in a self-congratulatory mood. Despite continuing criticisms over minor bureaucratic matters, there is little question about its important role in rebuilding Europe from the devastation of World War II; uniting former enemies with a shared economic, political, and cultural agenda, and delineating a peaceful and prosperous future. Challenges remain, without a doubt, particularly in ensuring that the new member countries in eastern Europe and the Balkans become full partners in the EU's success.

Key Terms

balkanization *(page 367)*
buffer zone *(page 365)*
Cold War *(page 332)*
collectivization *(page 375)*
command economy *(page 373)*
continental climate *(page 336)*
Council for Mutual Economic Assistance (CMEA) *(page 371)*
Cyrillic alphabet *(page 355)*
European Union (EU) *(page 332)*
fjord *(page 336)*
guest worker *(page 356)*
hierarchical diffusion *(page 356)*
Iron Curtain *(page 362)*
irredentism *(page 362)*
marine west-coast climate *(page 336)*
medieval landscape *(page 348)*
Mediterranean climate *(page 338)*
moraines *(page 335)*
privatization *(page 376)*
Renaissance–Baroque landscape *(page 348)*
Schengen Agreement *(page 347)*
secularization *(page 359)*
shield landscape *(page 336)*

Questions for Review

1. How does the European Lowland differ north and south of the Rhine River?
2. Describe the different upland and mountain regions of Europe.
3. What are the major factors influencing Europe's weather and climate?
4. List those European countries that are below replacement in natural population increase and those that will grow in the next 20 years. Explain the factors behind these differences.
5. Explain what is meant by "Schengenland" and "Fortress Europe." What are the geographic advantages and disadvantages of the new arrangement behind these two terms?
6. What are the major tongues of the Germanic language family? Which have the most speakers?
7. Map the interface boundaries, both historical and modern, between Christianity and Islam in different parts of Europe. Where is this interface still a problem? Why?
8. List the new nation-states that appeared on Europe's map in 1919, 1945, and 1992.
9. Trace the evolution of the EU since 1952, noting how its goals and membership have changed. Be prepared to make a map of its membership at different points in time.

Thinking Geographically

1. To compare the scale of Europe to North America, draw a circle 500 miles (800 kilometers) in diameter, which is a day's journey, around Frankfurt, Germany. Then do the same for Chicago. What does this exercise tell you about the differences in scale between the two regions?
2. Investigate the implications of sea level rise from global warming for different parts of Europe, including the Dutch coastline. How might this influence the EU's policy on control of atmospheric emissions?
3. How and why has western Europe become more energy efficient than North America?
4. Map the different rates of natural increase in Europe, noting which countries will grow and which will decline in the next 20 years. Then, link this map to a discussion of migration in Europe. Given these two factors, how might the population map change in 20 years?
5. Find good maps of several mid-sized European cities, and deduce the historical development of the cities based on their street patterns, arrangement of open spaces, boulevards, parks, and so on. Draw on the differences between the medieval and Renaissance–Baroque periods for your interpretation.
6. Look at the ads in a French or German news magazine and list the "globalized" English words used. Discuss your findings in terms of what sectors of society seem to use these foreign terms.
7. Critically examine the geographic aspects of the Dayton Peace Accords, and explain whether you think the boundaries were drawn wisely.
8. Research the fiscal, political, and social changes that had to take place in Hungary and Poland before they were admitted to the EU.

Regional Novels and Films

Novels

John Berger, *Once in Europa* (1983, Random House)

Emilie Carles, *A Life of Her Own: A Countrywoman in Twentieth-Century France* (1991, Rutgers University Press)

James Joyce, *The Dubliners* (1967, Viking Press)

Milan Kundera, *Ignorance* (2002, Harper & Row)

V. S. Naipal, *The Enigma of Arrival* (1987, Knopf)

Charles Power, *In the Memory of the Forest* (1997, Scribner's)

Peter Schneider, *The Wall Jumper: A Berlin Story* (1998, University of Chicago Press)

Dimitur Talev, *The Iron Candlestick* (1964, Foreign Language Press)

Films

Au Revoir, Les Enfants (1987, France)

Before the Rain (1995, Macedonia)

Bend It Like Beckham (2002, U.S., UK)

Goodbye, Lenin (2003, Germany)

L'Auberge Espanole (2003, France)

The Lives of Others (Germany, 2006)

The Pianist (2002, U.S., Poland)

The Postman (1994, Italy)

The Remains of the Day (1994, UK)

Underground (1995, Yugoslavia [Bosnia])

Bibliography

Allcock, John. 2000. *Explaining Yugoslavia.* New York: Columbia University Press.

Barnes, Ian, and Hudson, Robert. 1998. *The Historical Atlas of Europe: From Tribal Societies to a New European Unity.* New York: Macmillan.

Denitch, Bogdan. 1996. *Ethnic Nationalism: The Tragic Death of Yugoslavia.* Minneapolis: University of Minnesota Press.

Doherty, Paul, and Poole, Michael. 1997. "Ethnic Residential Segregation in Belfast, Northern Ireland, 1971–1991." *Geographical Review* 87(4), 520–36.

Hamilton, Kimberly. 1994. *Migration and the New Europe.* Boulder, CO: Westview Press.

Jordan, Terry, and Jordan, Bella Bychova. 2001. *The European Culture Area.* 3rd ed. Lanham, MD: Rowman and Littlefield.

Martin, Philip L. 1998. *Germany: Reluctant Land of Immigration.* Washington, DC: American Insitute for Contemporary German Studies.

Pells, Richard. 1997. *Not Like Us: How Europeans Have Loved, Hated, and Transformed American Culture Since World War II.* New York: Basic Books.

"Poverty in Eastern Europe: The Land That Time Forgot." 2000. *The Economist,* September 23, pp. 27–30.

Additional bibliographic references are at the *Diversity Amid Globalization* Website: http://www.prenhall.com/rowntree/.

9
The Russian Domain

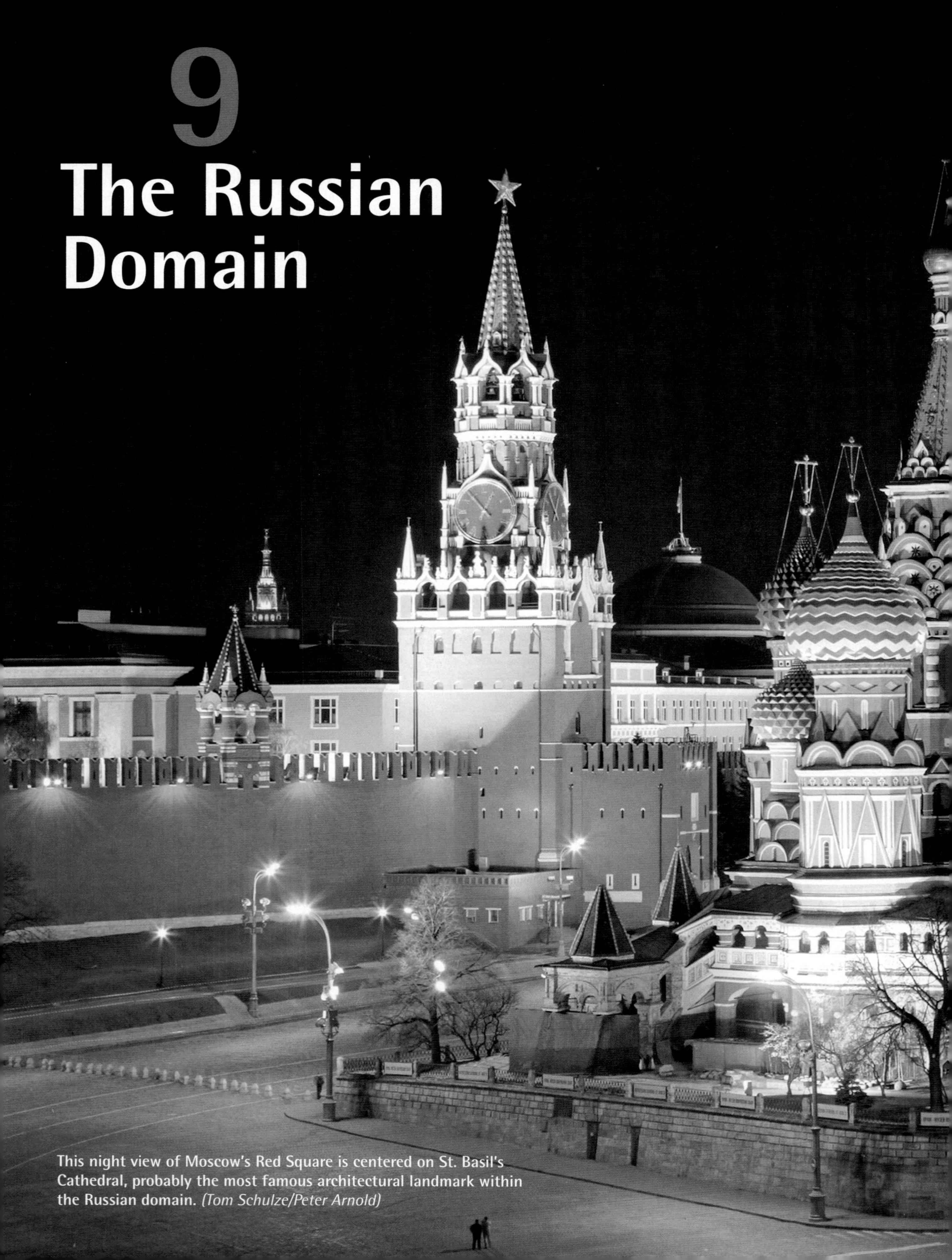

This night view of Moscow's Red Square is centered on St. Basil's Cathedral, probably the most famous architectural landmark within the Russian domain. *(Tom Schulze/Peter Arnold)*

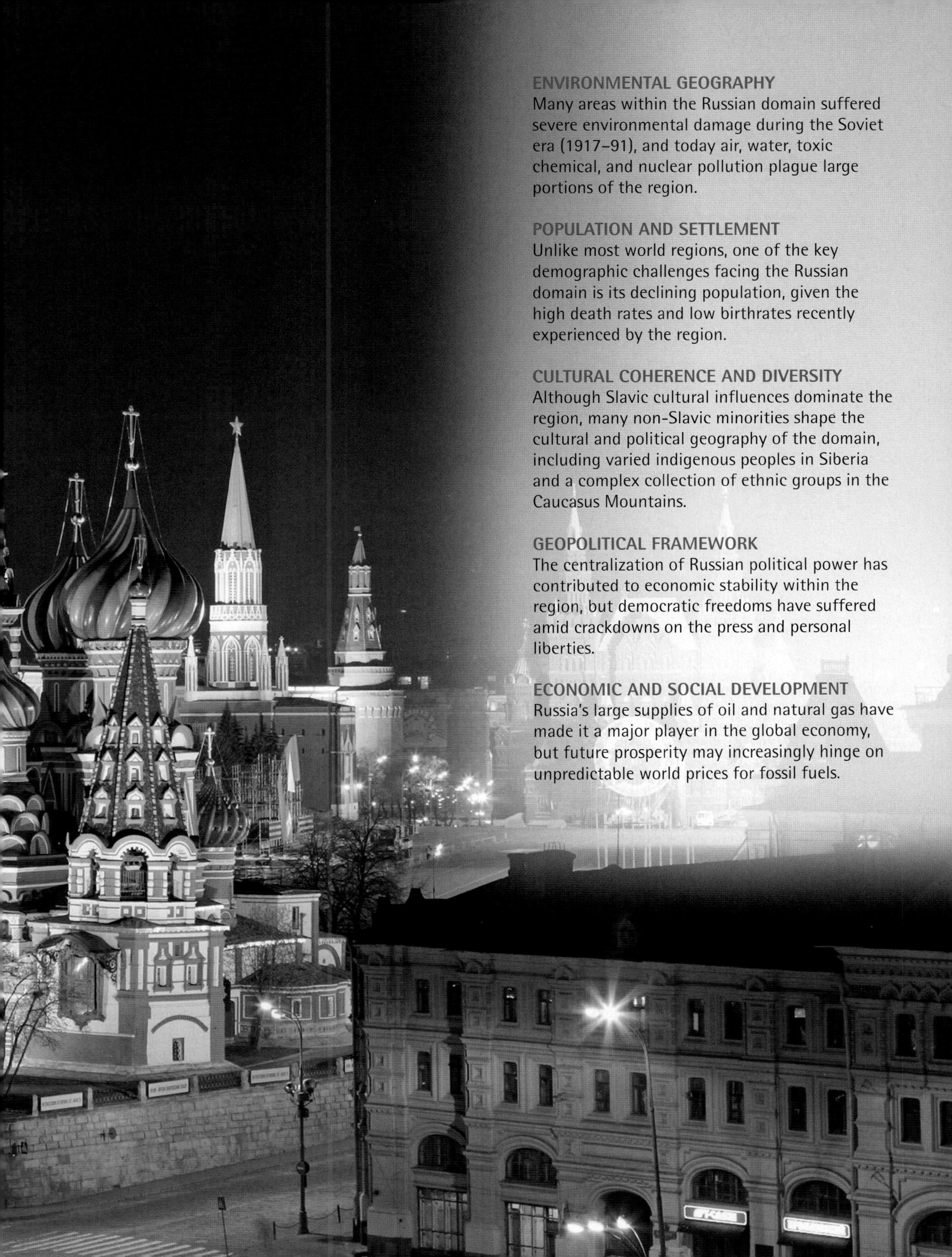

ENVIRONMENTAL GEOGRAPHY

Many areas within the Russian domain suffered severe environmental damage during the Soviet era (1917–91), and today air, water, toxic chemical, and nuclear pollution plague large portions of the region.

POPULATION AND SETTLEMENT

Unlike most world regions, one of the key demographic challenges facing the Russian domain is its declining population, given the high death rates and low birthrates recently experienced by the region.

CULTURAL COHERENCE AND DIVERSITY

Although Slavic cultural influences dominate the region, many non-Slavic minorities shape the cultural and political geography of the domain, including varied indigenous peoples in Siberia and a complex collection of ethnic groups in the Caucasus Mountains.

GEOPOLITICAL FRAMEWORK

The centralization of Russian political power has contributed to economic stability within the region, but democratic freedoms have suffered amid crackdowns on the press and personal liberties.

ECONOMIC AND SOCIAL DEVELOPMENT

Russia's large supplies of oil and natural gas have made it a major player in the global economy, but future prosperity may increasingly hinge on unpredictable world prices for fossil fuels.

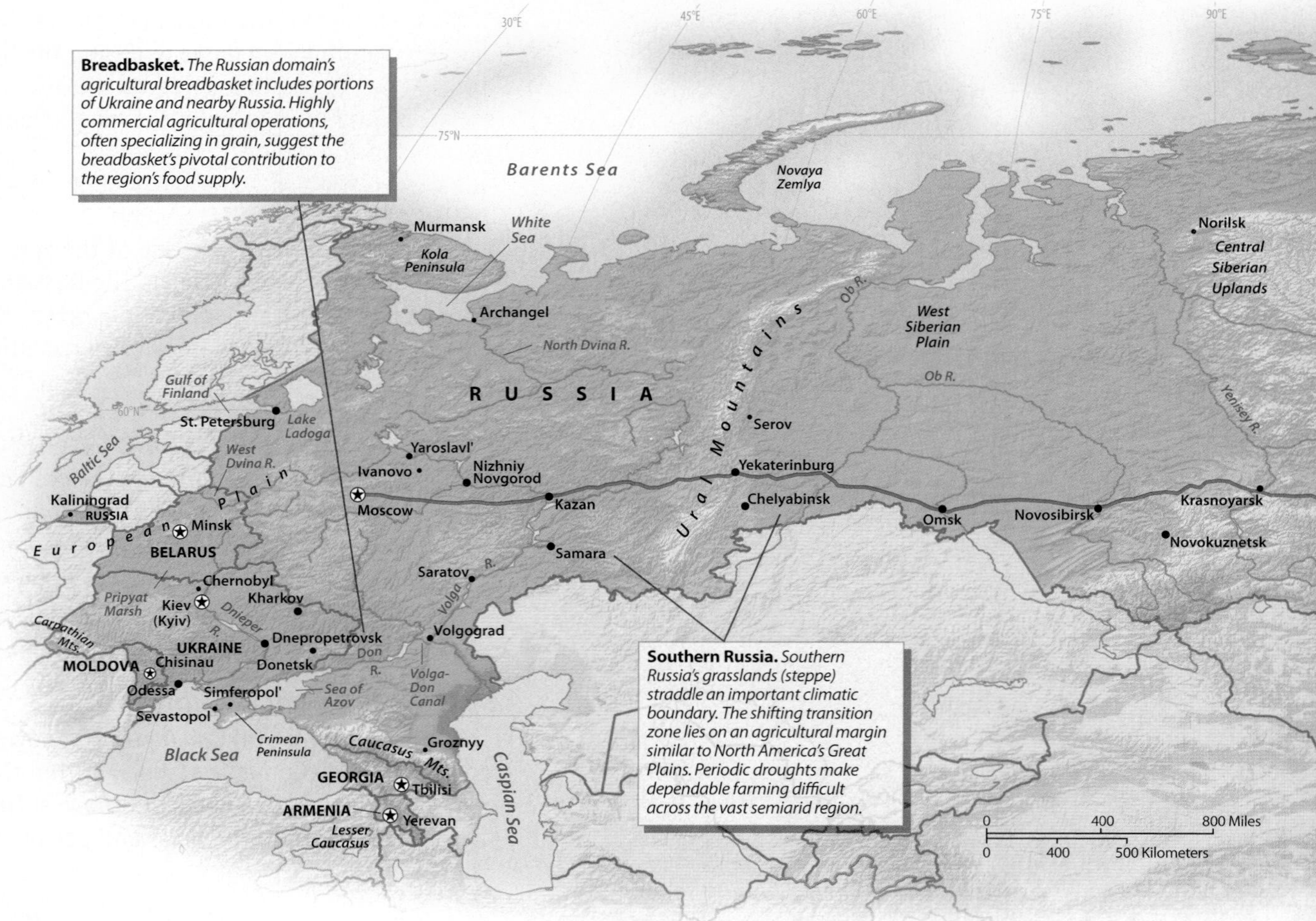

Figure 9.1 The Russian Domain
Russia and its neighboring states of Belarus, Ukraine, Moldova, Georgia, and Armenia make up a dynamic and unpredictable world region that has experienced tremendous change in the past decade. Sprawling from the Baltic Sea to the Pacific, the region includes huge industrial centers, vast farmlands, and almost-empty stretches of tundra.

The Russian domain sprawls across the vast northern half of Eurasia and includes not only Russia itself, but also the nations of Ukraine, Belarus, Moldova, Georgia, and Armenia (see Figure 9.1; see also "Setting the Boundaries"). The land is rich with superlatives: endless Siberian spaces, unlimited natural resources, legends of ruthless Cossack warriors, and tales of epic wars and revolutions are all part of the region's geographical and historical mythology (Figure 9.2). Indeed, the rise of Russian civilization remarkably parallels the story of the United States. Both cultures grew from small beginnings to become imperial powers that benefited from the fur trade, gold rushes, and transcontinental railroads during the nineteenth century. In addition, both countries experienced dramatic change resulting from industrialization in the twentieth century.

Recently, however, the Russian domain has witnessed breathtaking change. With the fall of the Soviet Union in 1991, new political and economic institutions dramatically reshaped everyday life. Russia, in particular, opened its society to democratic reforms. But economic collapse in the late 1990s produced steep declines in living standards throughout the region. Political instability also grew between neighboring states as well as within countries.

Since 2000, strong and increasingly centralized leadership within Russia has set the region on a different course. With the help of rising energy prices (Russia is a major exporter of oil and natural gas), real economic improvements have benefited the region, helping stabilize its economy and its larger political role in the world. At the same time, many of the democratic reforms welcomed with the fall of communism in 1991 are now being eroded away by a powerful Russian government committed to more direct state control. This new assertiveness on the part of the Russian government is increasing tensions within that country as well as between Russia and its neighboring states.

Globalization is shaping the Russian domain in complex ways. The region's relationship with the rest of the

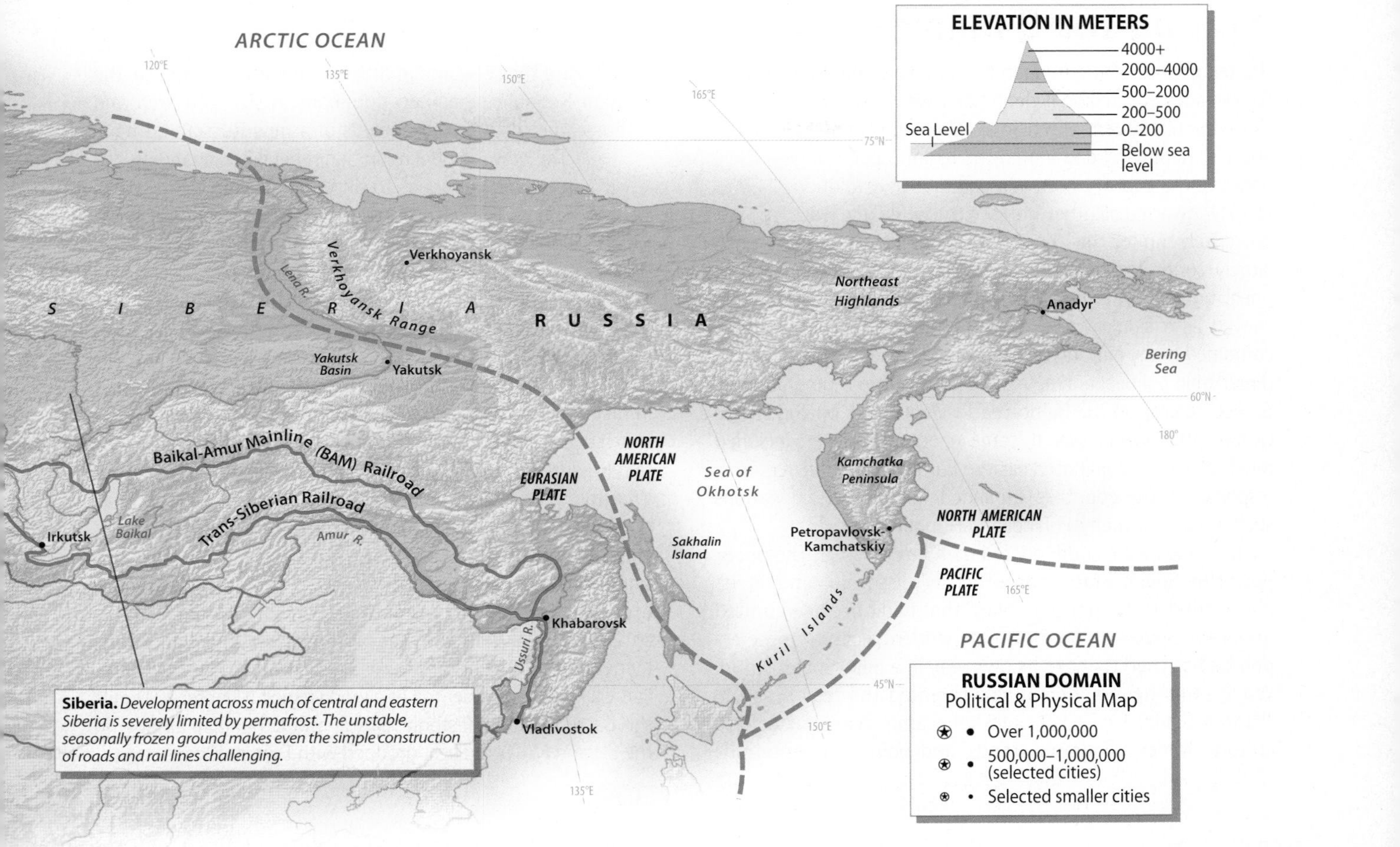

ELEVATION IN METERS
4000+
2000–4000
500–2000
200–500
0–200
Below sea level
Sea Level

RUSSIAN DOMAIN
Political & Physical Map
Over 1,000,000
500,000–1,000,000 (selected cities)
Selected smaller cities

Siberia. *Development across much of central and eastern Siberia is severely limited by permafrost. The unstable, seasonally frozen ground makes even the simple construction of roads and rail lines challenging.*

world shifted dramatically during the last 10 years of the twentieth century. Until the end of 1991, all six countries belonged to the Soviet Union, the world's most powerful communist state. Under Soviet control, the region's twentieth-century economy saw large increases in industrial output that made the nation a major global producer of steel, weaponry, and petroleum products. Its communist system offered a powerful set of ideas that promised economic prosperity and hope to residents within the region and around the world. Indeed, the political and military reach of the Soviet Union spanned the globe, making it a superpower on par with the United States. The Soviet presence dominated many eastern European countries, and nations from Cuba to Vietnam enjoyed close strategic ties with the country.

Suddenly, as the old communist order evaporated early in the 1990s, the now-independent republics of Russia, Ukraine, Belarus, Moldova, Georgia, and Armenia had to carve out new regional and global relationships. With the breakdown of Soviet control, the region also felt the growing presence of western European and American influences. Westernized popular culture combined with radical economic changes to transform the region. These new global relationships have not been easy. Social tensions have risen within the region as people grapple with fundamental economic changes. Political relationships with neighboring regions in Europe and Asia remain uncertain. The Russian domain has also been more fully exposed to both the opportunities and the competitive pressures of the global economy. The result is a world region that has

Figure 9.2 Cossack
Natives of the Ukrainian and Russian steppe, the highly mobile Cossacks played a pivotal role in aiding Russian expansion into Siberia during the sixteenth century. Many modern descendants retain their skills of horsemanship and are proud of their distinctive ethnic heritage. *(Julien Chatelin/ Getty Images, Inc.—Liaison)*

Setting the Boundaries

The boundaries of the Russian domain have shifted over time. For decades the regional definitions were relatively easy because the highly centralized Soviet Union (or Union of Soviet Socialist Republics) dominated the region's political geography. The country, born in a communist revolution in 1917, dwarfed all other states in the world and was powerfully united by the Soviet government and largely controlled by ethnic Russians. Most geographers agreed that the Soviet Union could thus be viewed as a single world region. Although the nation contained many cultural minorities, the Soviet Union wielded remarkable political and economic power from Leningrad (now St. Petersburg) on the Baltic Sea to Vladivostok on the Pacific Ocean. After World War II, some geographers even included much of Soviet-dominated eastern Europe within the region in response to the country's expanded military role in nations such as East Germany, Poland, and Hungary.

The maps were suddenly redrawn late in 1991. The once-powerful Soviet state was officially dissolved, and in its place stood 15 former "republics" that had once been united under the Soviet Union. Now independent, each of these republics has tried to make its own way in a post-Soviet world. While some geographers initially treated the region as the "Former Soviet Union," it quickly became clear that diverse cultural forces, economic trends, and political orientations were taking the republics in different directions. Even so, the Russian Republic remained dominant in size and area and thus came to form the nucleus of a new Russian domain.

The new regional definition reflects the changing political and cultural map since the breakup of the Soviet Union. The term *domain* suggests persisting Russian influence within the five other nations included in the region. Russia, Ukraine, and Belarus make up the core of the region. Enduring cultural and economic ties closely connect these three countries. In addition, Moldova, Georgia, and Armenia, while more culturally distinctive, are still best classified within a zone of Russian influence. In fact, Armenia and Russia recently signed a treaty of "friendship, cooperation, and mutual understanding" that will maintain Russian military bases in the country for another 25 years. Whereas Georgia and Moldova have had more stressful relations with their giant neighbor, Russian influence also shapes almost every aspect of life in these former Soviet republics. Two significant areas that were once a part of the Soviet Union have been eliminated from the domain. The mostly Muslim republics of Central Asia and the Caucasus (Kazakhstan, Uzbekistan, Kyrgyzstan, Turkmenistan, Tajikistan, and Azerbaijan) have become aligned with a Central Asia world region (Chapter 10), while the Baltic republics (Estonia, Latvia, and Lithuania), all now in the EU and NATO, are best grouped with Europe (Chapter 8).

seen its global linkages redefined in the recent past. Today, fluctuating world oil markets, shifting patterns of foreign investment, new patterns of migration, and the shadowy flows of illegal drugs and Russian mafia money all demonstrate the unpredictable nature of the Russian domain's global connections.

Slavic Russia (population 140 million) dominates the region. Although only about three-quarters the size of the former Soviet Union, Russia's dimensions still make it the largest state on Earth. West of Moscow, the country's European front borders Finland and Poland, while far to the east Mongolia and China share a sparsely populated boundary with sprawling Russian Siberia. Its area of 6.6 million square miles (17 million square kilometers) dwarfs even Canada, and its 11 time zones are a reminder that dawn in Vladivostok on the Pacific Ocean is still only dinnertime in Moscow.

With the demise of the Soviet Union, the Russians ended almost 75 years of Marxist rule. After a decade of political and economic instability (1991–2000), Russia has made impressive progress in the new century. Russian president Vladimir Putin built a reputation for strong leadership (2000–08). Much of the economic growth has come to Russian cities, where expanding middle and professional classes are enjoying better living standards. Many rural areas, however, remain deeply mired in poverty. Another concern is that Putin's desire for wealth and power has been matched by his need for more centralized political control. Many Russians do not want to lose hard-won democratic freedoms gained in the 1990s, while others suggest that the price of economic progress may be a more authoritarian state. What comes next is difficult to predict, and the country currently stands at an economic and political crossroads between the old ways and the new. A 2003 agreement to form a "Common Economic Space" between Russia, Ukraine, Belarus, and Kazakhstan suggests that the Russian desire for more centralized authority within the region did not end with the demise of the Soviet Union. Such an organization may also act as a counterbalance to a rapidly expanding European Union (EU) to the west.

The bordering states of Ukraine, Belarus, Moldova, Georgia, and Armenia will inevitably be linked to the evolution of their giant neighbor, even as they attempt to make their own way as independent nations. Emerging from the shadows of Soviet dominance has been difficult. Ukraine, in particular, has the size, population, and resource base to become a major European nation, but it has struggled to create real political and economic change since independence. With 45 million people and a rich storehouse of resources, Ukraine's size of 233,000 square miles (604,000 square kilometers) is similar to that of France. Nearby Belarus is smaller (80,000 square miles or 208,000 square kilometers), and its population of 10 million is likely to remain more closely tied economically and politically to Russia. Presently, its strikingly authoritarian and antiforeign leadership reflects many aspects of the old Soviet empire.

Moldova, with 4 million people, shares many cultural links with Romania, but its economic and political connections have kept it more closely tied to the Russian domain than to nearby portions of central Europe. South of Russia and beyond the bordering Caucasus Mountains, the Transcaucasian countries of Armenia and Georgia are similar in size to Moldova. Their populations differ culturally from their Slavic neighbor to the north. In addition, these two nations face significant political challenges: Armenia shares a hostile border with Azerbaijan (see Chapter 10), and Georgia's ethnic diversity and contentious relations with neighboring Russia threaten its political stability.

ENVIRONMENTAL GEOGRAPHY: A Vast and Challenging Land

The Kamchatka Peninsula dangles dramatically into the waters of the North Pacific Ocean, far from the hectic streets of St. Petersburg or Moscow (Figure 9.1). Unlike much of the Russian domain, which has been poisoned by almost a century of reckless development and environmental exploitation, the Kamchatka region remains relatively untouched by the outside world (Figure 9.3). But that may be changing soon and Russia's ability to preserve the region will say a great deal about that country's future commitment to preserving and restoring its environmental health.

Six native species of Pacific salmon thrive in the region, spawning in the millions in the area's free-flowing rivers. A move to preserve the salmon habitat is growing among both Russian and global environmental organizations. The salmon are at the center of a complex environmental web: they provide food for the brown bears, seals, and Stellar's sea eagles that are abundant here; they also remain part of the subsistence diet of the Koryak and other native peoples of these coastal zones; and they offer ecotourism and sports fishing opportunities found nowhere else in the world. But outside threats are rapidly appearing. Canadian and South Korean oil companies have acquired drilling rights in the region and the huge Russian oil consortium, Rosneft, has similar leases in the offshore waters. Illegal poaching of salmon and salmon eggs is also on the rise. A concerted global effort is now underway to prevent development in seven sensitive spawning areas of Kamachatka wilderness. One Russian-supported plan calls for protecting an area nearly triple the size of Yellowstone National Park. If such initiatives on the Kamchatka Peninsula are successful, it will mark a new era of environmental consciousness in a part of the world that has produced some of the most toxic settings on the planet. It will also mark an important marriage between a fragile local setting and a larger collection of national and international environmental organizations dedicated to seeing the wild salmon continue to run.

A Devastated Environment

The Russian domain has no shortage of environmental disasters. The breakup of the Soviet Union and subsequent opening of the region to international public scrutiny

Figure 9.3 Kamchatka Peninsula
The Kamchatka Peninsula offers many spectacular natural settings as well as a habitat for a variety of Pacific salmon. *(Wolfgang Kaehler/Corbis; inset, Dmitry Beliakov/The New York Times/Redux Pictures)*

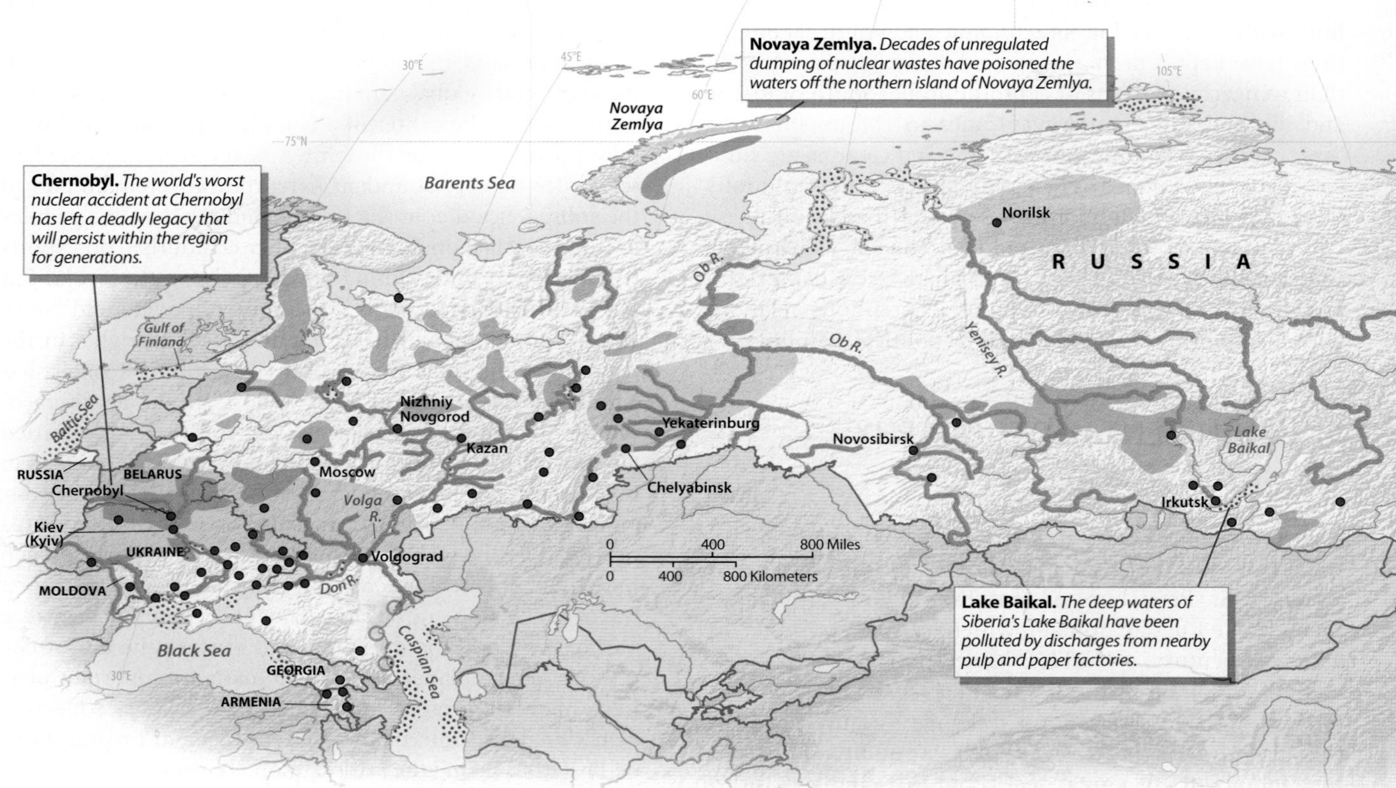

Figure 9.4 Environmental Issues in the Russian Domain
Varied environmental hazards have left a devastating legacy across the region. The landscape has been littered with nuclear waste, heavy metals, and air pollution. Fouled lakes and rivers pose additional problems in many localities. Present economic difficulties and political uncertainties only add to the costly challenge of improving the region's environmental quality in the twenty-first century.

revealed some of the world's most severe environmental degradation (Figure 9.4). Even official studies commissioned by the Russian government estimate that almost two-thirds of the Russian population live in an environment harmful to their health and that the country's environmental problems continue to worsen. The studies also suggest that nearly 65 million Russians live in areas of chronically poor air quality and that the drinking water is unsafe in half of the country.

The frenetic pace of seven decades of Soviet industrialization took its toll across the region. Even in some of the most remote reaches of Russia, careless mining and oil drilling, the spread of nuclear contamination, and rampant forest cutting have resulted in frightening environmental damage. New Russian environmental and antinuclear movements have protested these ecological disasters, but to date these movements remain a minor political voice in a region dominated by the desire for economic growth. Indeed, the magnitude of many of these environmental challenges is so great that they have global implications and may affect world climate patterns, water quality, and nuclear safety. For example, since the 1980s, the global environmental costs of Siberian forests lost to lumbering and pollution may have exceeded the more widely publicized destruction of the Brazilian rain forest.

Air and Water Pollution Poor air quality plagues hundreds of cities and industrial complexes throughout the region (see Figure 9.4). The traditional Soviet practice of building large clusters of industrial processing and manufacturing plants in concentrated areas, often with minimal environmental controls, has produced an ongoing legacy of fouled air that stretches from Belarus to Russian Siberia. A traditional reliance on abundant but low-quality coal also contributes to pollution problems. The air quality in dozens of cities within the region typically fails to meet health standards, particularly in the winter when cold-air inversions trap the polluted atmosphere for days on end. Large numbers of urban residents across the region suffer from chronic respiratory problems. Siberia's northern mining and smelting city of Norilsk is one of Russia's most polluted urban areas. In addition, a large swath of larch-dominated forest has died in a huge zone of contamination that stretches more than 75 miles (120 kilometers) southeast of the city. Elsewhere, growing rates of private car ownership have greatly increased automobile-related pollution, especially because many vehicles lack catalytic converters and still use leaded gasoline. Today, 90 percent of Moscow's air pollution has been linked to the city's growing automobile traffic.

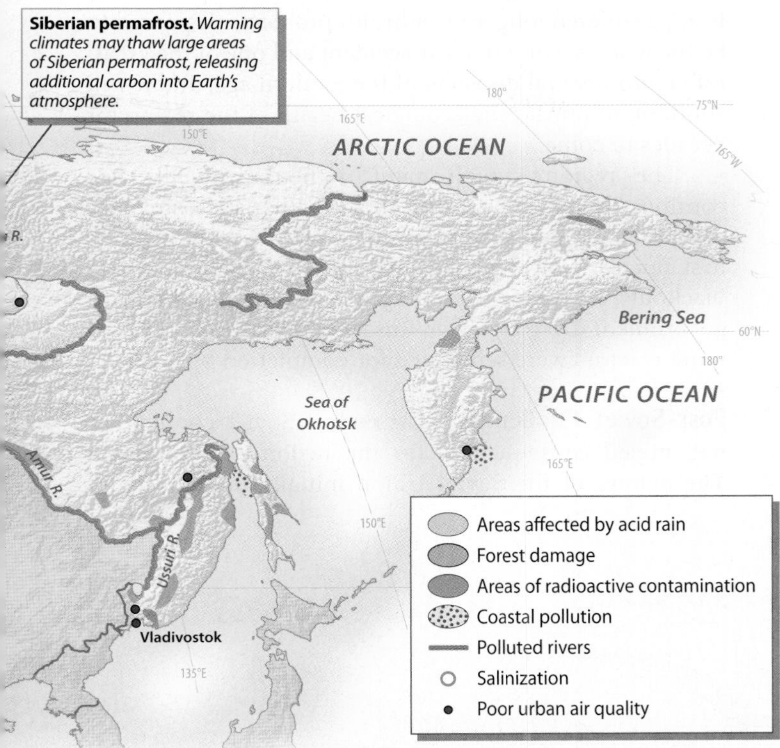

Degraded water is another hazard that residents of the region must cope with daily (see Figure 9.4). Municipal water supplies are constantly vulnerable to industrial pollution, flows of raw sewage, and demands that increasingly exceed capacity. For example, the Baltic Sea near the city of St. Petersburg has reached a critical level of pollution that has killed fish and threatens to permanently damage the region's ecosystem. The biggest problem is that 30 percent of all the residential and industrial waste that enters the sea via the nearby Neva River is unfiltered raw sewage, a toxic mix of heavy metals and human waste that is rapidly killing the Baltic. Elsewhere, the extensive industrialization and dam-building along Russia's Volga Valley has produced a corridor of degraded water that stretches for hundreds of miles. Water pollution has also affected much of the northern Black Sea, large portions of the Caspian Sea shoreline, and even Arctic waters off Russia's northern coast.

The pollution of Siberia's Lake Baikal attracted international attention in the 1970s and 1980s. Lake Baikal, the world's largest reserve of freshwater, is a remarkable natural feature. Not only is the lake almost 400 miles (644 kilometers) long, but it is also 5,300 feet (1,615 meters) deep, occupying a structural rift in the continental crust (Figure 9.5). Lake Baikal is home to a large array of unique species, including the world's only freshwater seal. Until the mid-twentieth century, Baikal's water was exceptionally pure. In the 1950s and 1960s, however, the Soviet government built pulp and paper factories along its shores, attracted by both the lake's clean water (useful for producing high-quality wood fibers) and the abundant forests in the surrounding uplands. With factory discharges, the lake's purity rapidly declined. By the 1970s, Russian ecologists warned that Baikal's entire ecosystem was threatened. Owing in part to the resulting international pressure, pollution from the paper mills was reduced.

Recently, Lake Baikal also became the center of attention as Russia planned to expand a major Siberian oil pipeline linking Russian resources to East Asian markets. High oil prices have encouraged the Russians to make large new investments in their petroleum industry, but many environmentalists feared that these growing global demands for oil might have destructive local consequences for Lake Baikal. In 2006, major protests and petition-signing drives opposed the planned pipeline's close proximity to the north shore of the lake. The initiative caught the attention of Russian president Putin who dramatically ordered that the pipeline be directed further away from the lake's fragile ecosystem.

The Nuclear Threat The nuclear era brought its own particularly deadly dangers to the region. The Soviet Union's aggressive nuclear weapons and nuclear energy programs expanded greatly after 1950, and issues of environmental safety were often

Figure 9.5 Lake Baikal
Southern Siberia's Lake Baikal is one of the world's largest deep-water lakes. Industrialization devastated water quality after 1950 as pulp and paper factories poured wastes into the lake. Recent cleanup efforts have helped, but many environmental threats remain. *(Digital Image © 1996 Corbis)*

ignored. Northeast Siberia's Sakha region, for example, suffered regular nuclear fallout in the era of above-ground nuclear testing. In other areas, nuclear explosions were widely utilized for seismic experiments, oil exploration, and dam-building projects. The once-pristine Russian Arctic has also been poisoned. During the Soviet era, the area around the northern island of Novaya Zemlya served as a huge and unregulated dumping ground for nuclear wastes. Nearby, dozens of atomic submarines have been abandoned to rust away among the fjords of the Kola Peninsula. Aging nuclear reactors also dot the region's landscape, often contaminating nearby rivers with plutonium leaks.

Nuclear pollution is particularly pronounced in northern Ukraine, where the Chernobyl nuclear power plant suffered a catastrophic meltdown in 1986. Large areas of nearby Belarus were also devastated in the Chernobyl disaster. Fallout contaminated fertile agricultural lands across much of the southern part of the country and rendered about 20 percent of the nation unhealthy to even live in. The reactor burned for 16 days, pouring smoke 2 miles into the sky and spreading nuclear contaminants from southern Russia to northern Norway. Thousands died directly from the accident, and millions of other residents across the region and in nearby affected portions of Europe suffered longer-term health problems. It proved to be the world's worst nuclear accident and one of the greatest environmental disasters of the modern age, and it will continue to impact the ecological health of the region for decades to come.

The region's involvement with the nuclear age continues to take new forms. In 2001 Russia's Rostov Nuclear Energy Station went on-line, making it the region's first nuclear power plant since the Soviet era. Growing blackouts and energy shortages have produced a new government drive to revive Russia's nuclear industry, and more reactors were scheduled for completion by 2010.

Post-Soviet Challenges The end of Soviet control has had mixed consequences for the region's environment. The demise of the Soviet Union initially brought about

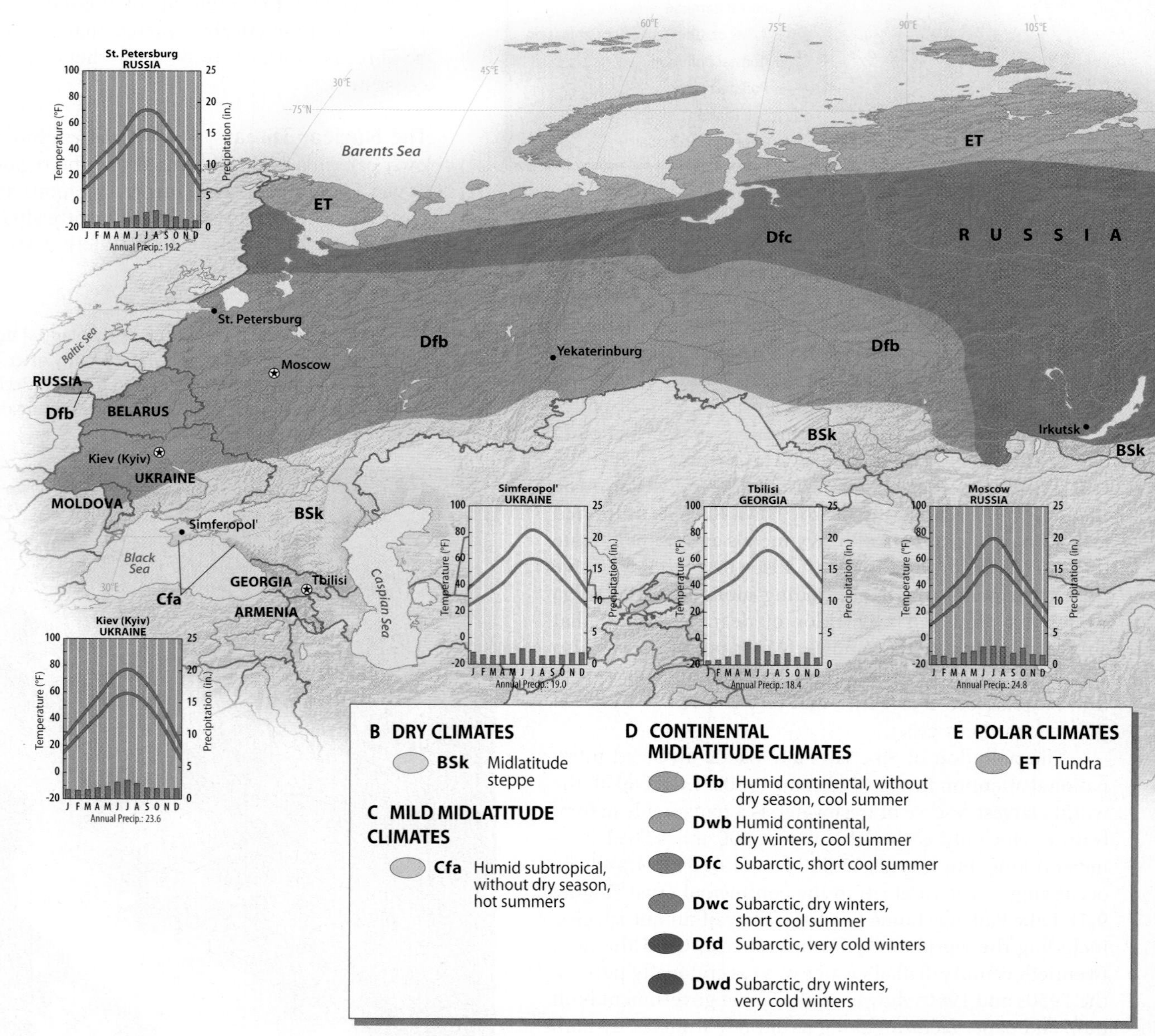

Figure 9.6 **Siberian Lumber** A growing supply of Siberian lumber is flowing toward markets in both Japan and China, threatening the environmental quality and sustainability of Russia's forest resources. *(Bodryashkin Valery/ Itar-Tass/Landov)*

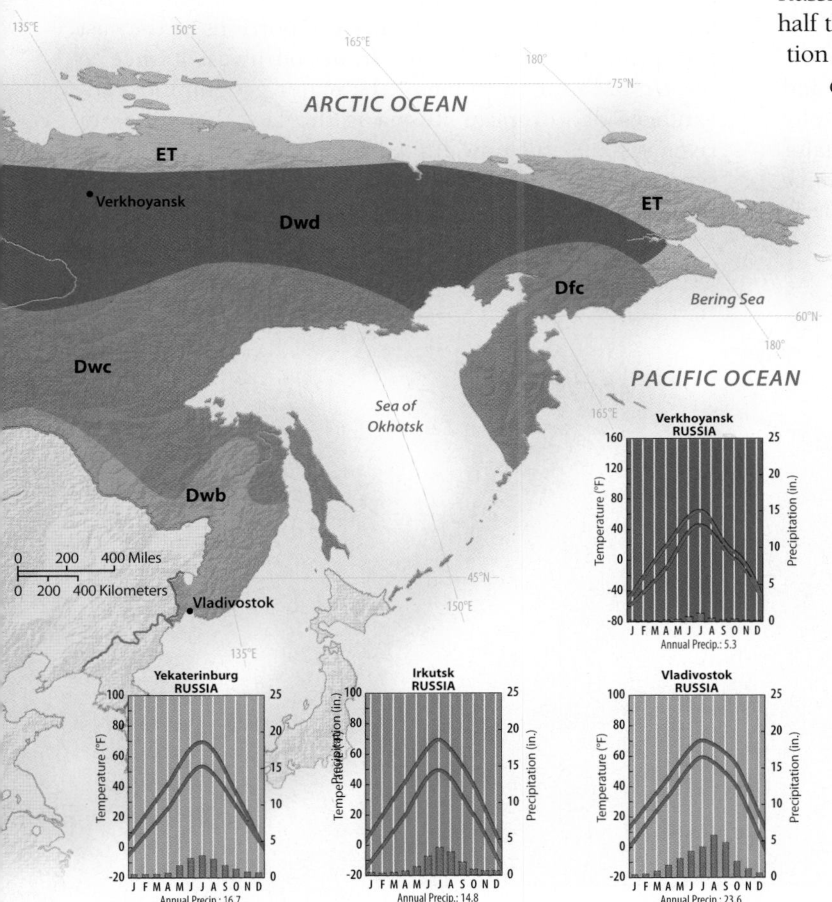

Figure 9.7 **Climate Map of the Russian Domain** The region's northern latitude and large landmass suggest that continental climates dominate. Indeed, farming is greatly limited by short growing seasons across much of the region. Aridity imposes limits elsewhere. Only a few small zones of mild midlatitude climates are found on the warming shores of the Black Sea in the far southwestern corner of the region, producing subtropical conditions in western Georgia.

environmental improvement in some areas. Many factories shut down because they were no longer economically viable, which itself reduced pollution. The huge steel mills in the Russian city of Magnitogorsk, for example, produce less than half the raw steel they did a decade ago, and global competition threatens to reduce demand further. Ironically, this decreased production has resulted in cleaner air. Although costly, advanced pollution control equipment is also beginning to be imported from western Europe. Elsewhere in Russia, nuclear warhead storage facilities have been consolidated, and government authorities are responsible for maintaining control over the nation's nuclear weapons. An environmental consciousness also is growing among young, educated Russians and Ukrainians.

On the other hand, official spending (in Russia) to protect the environment is today only a tiny percentage of levels seen in the final years of the Soviet Union. Most cities in the region have been left to finance their own cleanup projects. For example, industrial Volgograd (in the southern Volga Valley) needs low-cost antipollution technology that could save hundreds of lives annually, but the city cannot afford the $100,000 price tag. An especially vexing problem is the disposal of nuclear byproducts, now sometimes smuggled out of the country and possibly used in illegal weapons manufacturing.

Russia has also depended heavily on its natural resources to finance its return to economic health, but this response to growing global demand has produced new environmental problems for the region. State-run petroleum and natural-gas companies, for example, have tremendous power and freedom to produce profits as quickly as possible, regardless of the consequences. A similar mentality shapes the timber industry. With huge demand for lumber from nearby Japan and China, there has been widespread timber poaching reported in eastern Siberia (Figure 9.6). At present rates of cutting, much of this natural resource will vanish in the next 10 to 20 years. Thus, growing East Asian appetites for new homes and furniture are directly contributing to the destruction of Siberia's forests. The forests of northwestern Russia also continue to disappear at an alarming rate, responding to similar market-driven pressures from Europe (see "Geographic Tools: Mapping Russian Forests").

Russia also faces an impending crisis of wildlife extinction. The old Soviet regime had some success in protecting both endangered species and sizable areas of natural habitat noted for their biological diversity. Unregulated hunting and trapping now are on the rise, however, and in some areas are virtually uncontrolled. It is questionable whether such animals as the Siberian tiger (only 330–370 still survive in the wild) and the Amur leopard (25–40 wild animals remain), both inhabiting the forests of Russia's Far East, will survive the next few decades.

A Diverse Physical Setting

The Russian domain sprawls from the Baltic Sea in the west to the Amur River in the Far East. Size, latitude, and topography create a distinctive physical geography (Figure 9.1). The fact that the region occupies a major portion of the world's largest land mass means that huge distances separate people and resources. Particularly within Russia, thousands of miles and many days of road or rail travel add immeasurable human and economic costs to the movement of people, goods, and services within the country.

The region's northern latitudinal position also is critical to understanding its climate, vegetation, and agriculture. Indeed, the Russian domain provides the world's largest example of a high-latitude continental climate where seasonal temperature extremes and short growing seasons profoundly limit opportunities for human settlement (Figure 9.7). In terms of latitude, Moscow is positioned as far north as Ketchikan, Alaska, and even the Ukrainian capital of Kiev (Kyiv) would sit north of the Great Lakes in Canada. Thus, apart from a subtropical zone near the Black Sea, much of the region experiences a classic continental climate with hard, cold winters and marginal agricultural potential.

The European West An airplane flight over the western portions of the Russian domain would reveal a vast, barely changing landscape below. European Russia, Belarus, and Ukraine cover the eastern portions of the vast European Plain, which runs from southwest France to the Ural Mountains. One of the major geographical advantages of European Russia is the fact that different river systems, all now linked by canals, flow into four separate drainages. The Dnieper and Don rivers flow into the Black Sea; the West and North Dvina rivers

drain into the Baltic and White seas, respectively; and the Volga River runs to the Caspian Sea (Figures 9.1 and 9.8).

Even though European Russia has a milder climate than Siberia, most of it experiences cold winters and cool summers by North American standards. Moscow (see Figure 9.7), for example, is about as cold as Minneapolis in January, yet not nearly as warm in July. In Ukraine, Kiev is milder, however, and Simferopol', near the Black Sea, offers wintertime temperatures that average more than 20°F (11°C) higher than those of Moscow.

Three distinctive environments shape agricultural potential in the European West (Figure 9.9). North of Moscow and St. Petersburg, poor soils and cold temperatures severely limit farming, and much of the land remains in coniferous forest. Belarus and central portions of European Russia possess longer growing seasons, but acidic **podzol soils**, typical of northern forest environments, limit agricultural output and the ability of the region to support a highly productive farm economy. Still, diversified agriculture includes grain (rye, oats, and wheat) and potato cultivation, swine and meat production, and dairying. South of 50 degrees latitude, forests gradually give way to steppe environments dominated by grasslands and by fertile "black earth" **chernozem soils**, which have proven valuable for commercial wheat, corn, and sugar beet cultivation and for commercial meat production (Figure 9.10).

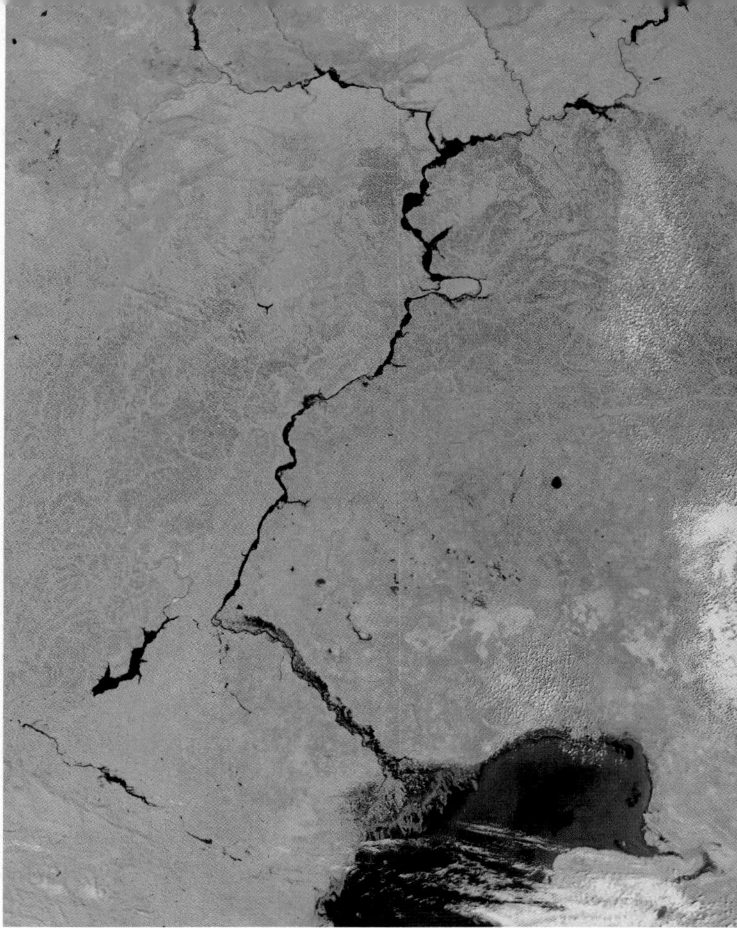

Figure 9.8 Volga Valley
This satellite image shows the lower Volga River flowing through southwestern Russia. In the lower portion of the image, the city of Volgograd sits along the western edge of the river where it bends sharply to the southeast on its journey to the Caspian Sea. *(NASA/MODIS)*

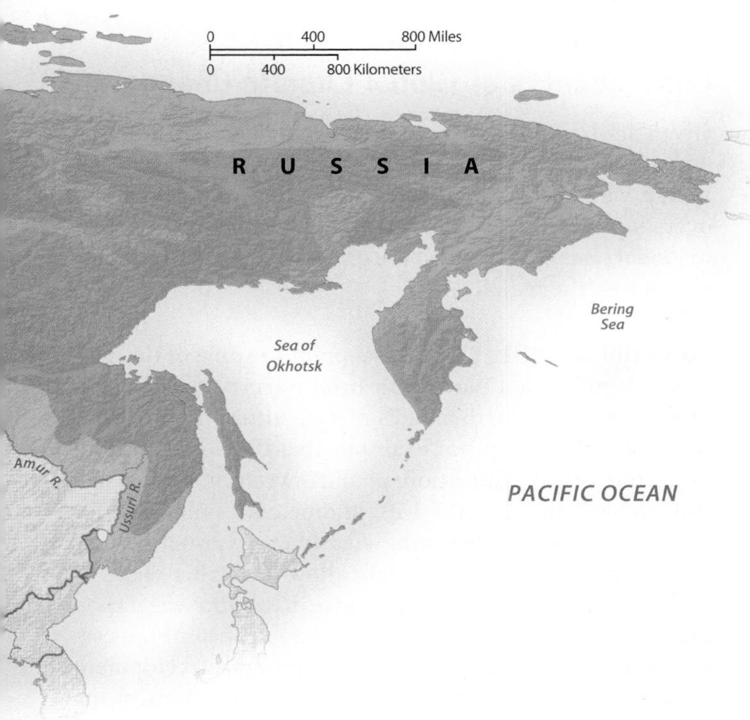

The Ural Mountains and Siberia The Ural Mountains (see Figure 9.1) mark European Russia's eastern edge, separating it from Siberia, or Asian Russia. Despite their geographical significance as the traditional division between continents, the Urals are not a particularly impressive range; several of their southern passes are less than 1,000 feet (305 meters) high. Although not agriculturally productive, the range is still significant for two reasons: its ancient rocks are heavily mineralized, and it once marked Russia's eastern cultural boundary.

East of the Urals, the vastness of Russian Siberia unfolds across the landscape for thousands of miles.

Figure 9.9 Agricultural Regions
Harsh climate and poor soils combine to limit agriculture across much of the Russian domain. Better farmlands are found in Ukraine and in European Russia south of Moscow. Portions of southern Siberia support wheat production but yield marginal results. In the Russian Far East, warmer climates and better soils translate into higher agricultural productivity. *(Modified from Clawson and Fisher, 2004,* World Regional Geography, *8th ed., Upper Saddle River, NJ: Prentice Hall)*

The great Arctic-bound Ob, Yenisey, and Lena rivers (see Figure 9.1) drain millions of square miles of northern country that includes the level and poorly drained West Siberian Plain, the hills and plateaus of the Central Siberian Uplands, and the rugged and isolated Northeast Highlands. Wintertime climatic conditions are severe across the region.

Siberian vegetation and agriculture reflect the climatic setting. The northern portion of the region is too cold for tree growth and instead supports tundra vegetation, which is characterized by mosses, lichens, and a few ground-hugging flowering plants. South of the tundra, the Siberian **taiga**, or coniferous forest zone, dominates a large portion of the Russian interior. Only localized agriculture (marginal wheat farming and potato cultivation) is possible even within favored portions of the zone. In the east, much of the tundra and taiga regions also are associated with **permafrost**, a cold-climate condition of unstable, seasonally frozen ground that limits the growth of vegetation and makes problematic the construction of even simple railroad tracks. Conditions moderate across southwestern Siberia: longer growing seasons and better soils offer more agricultural opportunities, although precipitation decreases and becomes less dependable along the border with Kazakhstan.

The Russian Far East Proximity to the Pacific Ocean, a more southerly latitude, and a pair of fertile river valleys create a distinctive subregion within the Russian Far East. About the same latitude as North America's New England, the region features longer growing seasons and milder climates than those found to the west or north (see Figure 9.9). Here, the continental climates of the Siberian interior meet the seasonal monsoon rains of East Asia. It is also a fascinating zone of ecological mixing: conifers of the taiga mingle with Asian hardwoods, and reindeer, Siberian tigers, and leopards find common ground. Natural hazards also haunt the region as part of the tectonically active Pacific Rim of Fire. The Kuril Islands and the Kamchatka Peninsula both experience earthquakes and volcanic eruptions, but low population densities limit the human impact of this tectonic activity.

Figure 9.10 Commercial Wheat Production
This wheat field in southern Ukraine typifies some of the region's most productive farmland. Longer growing seasons and good soils are important assets in this portion of the Russian domain. *(Art Directors & TRIP Photo Library)*

The Caucasus and Transcaucasia In European Russia's extreme south, flat terrain gives way first to hills and then to the Caucasus Mountains, a large earthquake-prone range stretching between the Black and Caspian seas (see Figure 9.1). The Caucasus mark Russia's southern boundary. Farther south lies Transcaucasia and the distinctive natural settings of Georgia and Armenia. Extensive lowlands and low plateaus are found here, and a second, less-formidable mountain range, aptly named the Lesser Caucasus, runs through the southern part of Georgia and along the boundary between Armenia and Azerbaijan.

Patterns of both climate and terrain in the Caucasus and Transcaucasia are tremendously complex. Rainfall is generally high in the western zone, with some slopes supporting dense forests. The area's eastern valleys, on the other hand, are semiarid or arid. In areas of adequate rainfall or where irrigation is possible, agriculture can be quite productive. Georgia in particular has long been a noted producer of fruits, vegetables, flowers, and wines (Figure 9.11). The Georgian capital of Tbilisi enjoys a rare subtropical climate within the region that averages 30°F (17°C) higher in the winter than does bone-chilling Moscow.

Consequences of Global Climate Change

Given its latitude and continental climates, the Russian domain is often cited as a world region that would benefit from a warmer global climate. But such an interpretation oversimplifies the complex natural and human responses to global climate change, some of which are already occurring across the region.

Potential Benefits Optimists point to some of the economic benefits that may result from warmer Eurasian climates. Consider the dynamics of agriculture (Figure 9.9). Some models predict, for example, that the northern limit of spring cereal cultivation in northwestern Russia will shift 60–90 miles (100–150 kilometers) poleward for every 1°C (1.8°F) of warming. Elsewhere, warmer springs in settings such as Belarus might allow for higher yields of maize and sunflowers. Areas once in tundra vegetation may be more suitable for boreal forest expansion. Less severe winters may make energy and mineral development in subarctic settings less costly. In nearby portions of the Arctic Ocean and Barents Sea, less sea ice will translate to easier navigation, more high-latitude commerce, and more

Figure 9.11 Subtropical Georgia
The modifying influences of the Black Sea and a more southern latitude produce a small zone of humid subtropical agriculture in Georgia. The lush landscapes of these tea plantations offer a sharp contrast to the colder country found north of the Caucasus. *(Sovfoto/Eastfoto)*

ice-free days in northern Russian ports. Less ice on northward-flowing Siberian rivers such as the Ob and Yenisey may also make these waterways more valuable as corridors of commerce.

Potential Hazards Even with such rosy scenarios within the Russian domain, might the long-term regional and global costs outweigh the benefits? Climate experts point to three areas of particular concern. First, anticipated rises in global sea levels will hit low-lying areas of the Black and Baltic seas particularly hard. Officials in St. Petersburg, Russia's second largest city, are already contemplating significant costs associated with the challenges of controlling the Baltic's rising waters. The city's natural setting is essentially in a low-lying coastal estuary along the Baltic. Similar areas along the Black Sea coast are likely to threaten residents in southern Ukraine.

Second, changes in ecologically sensitive arctic and subarctic ecosystems are already leading to major disruptions in wildlife and indigenous human populations in those settings of northern Russia. Take the example of the polar bear (Figure 9.12). The shrinking volume of arctic sea ice habitat for the bears has meant that they are forced to widen their search for food. This has brought them into closer contact with arctic villages and also disrupted traditional hunting practices. Poachers have also profited, increasing their illegal harvests. Paradoxically, Russian officials and a number of wildlife management experts have argued that increasing legal and controlled hunting (versus illegal poaching) of the bears may be the best way to more effectively maintain their populations in this period of environmental flux.

Finally, the largest potential change within the region, with truly global implications, relates to the thawing of the Siberian permafrost. Substantial areas of northern Russia are covered with permafrost that is already close to thawing. Thus, even minor increases in temperature could have large and irreversible consequences for the region. Major changes in topography (mud flows, slumping, and erosion), drainage (lake coverage, rivers), and vegetation will be the result. Existing fish and wildlife populations will need to adjust in order to survive. Human infrastructure such as buildings, roads, and pipelines will also require substantial modification. But the greatest potential global impact may come with the huge release of carbon that is currently stored in existing permafrost environments. The soils frozen in permafrost contain large amounts of organic material that decomposes quickly when thawed. Most of the planet's permafrost could release its carbon reservoir within the next century, the equivalent of 80 years of burning fossil fuels. Such a contribution to the world's carbon budget, which would likely further warm Earth, is only beginning to be incorporated into models of global climate change. Thus, the survival of the Siberian permafrost may hold one of the keys to slowing or quickening further global warming.

Figure 9.12 Russian Polar Bears
Reduced ice flows in the arctic are disrupting the habitat of Russia's polar bears, forcing them to range closer to many human settlements. *(Galen Rowell/Corbis)*

GEOGRAPHIC TOOLS Mapping Russian Forests

Greenpeace Russia and the World Resources Institute recently sponsored one of the most ambitious satellite mapping efforts ever. Their goal was to accurately locate and assess the extent of remaining intact boreal forest lands in northwestern Russia (Figure 9.2.1). It was the largest project of its kind ever in Russia and it demonstrated how satellite imagery could be efficiently used to assess forest health and disturbance in high-latitude environments where existing maps are sketchy and where fieldwork is expensive and time-consuming.

Figure 9.2.1 Boreal Forest in Northwestern Russia
The once endless boreal forests of northern Russia are quickly being harvested for commercial lumber markets in nearby western Europe. Only a small percentage of roadless forest remains today. *(Konrad Wothe/Minden Pictures)*

The results of the study were sobering. Only the most remote portions (14 percent of the total) of the vast 563-million-acre (228-million-hectare) region, stretching from Finland to the Urals, retained their status as "intact forest landscapes." Such intact areas, critical for forest health and wildlife preservation, were defined as roadless natural areas of at least 123,000 acres (50,000 hectares) that were basically undisturbed by development. Most of the region, however, has felt the impact of a commercial timber industry that is busily supplying the never-ending demands for lumber in nearby western Europe.

Measuring and mapping the remaining intact forests involved five different sources. Geographers on the research team accumulated general topographic maps of the region to get basic information on human infrastructure and they consulted Soviet-era thematic maps of general regional forest cover. Next, they collected middle resolution (with a moderate amount of detail) satellite imagery from a MSU-SK scanner on an orbiting Russian satellite. These images were used to identify larger clear-cuts, agricultural fields, and areas of obvious second-growth forest. Then, researchers used more costly, but higher-resolution (more detailed) Landsat images to make final decisions on the extent of smaller-scale human disturbances. Finally, several dozen research teams were sent to make field observations across the zone and verify the accuracy of the satellite data. Final maps of the intact and disturbed forests were then completed using an ArcView GIS software package that matched the images with base maps and other layers showing drainage and human activity.

The study demonstrated that popular images of unending Russian forests are more myth than reality. Only a fragment of the great boreal forests of the Russian north remain. Authors of the study hope that their findings will prompt Russian officials to preserve much of the surviving intact forest, particularly because many of these tracts contain only marginally valuable timber.

Source: Alexey Yaroshenko, Peter Potapov, and Svetlana Turubanova, 2001, *The Last Intact Forest Landscapes of Northern European Russia*, Moscow: Greenpeace Russia.

POPULATION AND SETTLEMENT: An Urban Domain

The six states of the Russian domain are home to about 200 million residents (Table 9.1). While they are widely dispersed across a vast Eurasian land mass, most live in cities. The region's distinctive distributions of natural resources, and changing migration patterns continue to shape its population geography. Although government policies have encouraged migration into the eastern portions of the domain, the population remains strongly concentrated in the traditional centers of the European West. Low birthrates and higher death rates also remain a critical concern as these trends threaten to shrink future regional populations.

Population Distribution

Striking differences in population densities exist between European and Asian portions of the Russian domain. The more favorable agricultural setting of the European West historically encouraged higher densities of population

TABLE 9.1 Population Indicators

Country	Population (Millions, 2007)	Population Density (per Square Kilometer)	Total Fertility Rate	Percent Urban	Percent <15[a]	Percent >65[b]	Net Migration (per 1,000), 2000–05
Armenia	3.0	101	1.7	64	22	11	–6.6
Belarus	9.7	47	1.2	73	16	14	–0.2
Georgia	4.5	65	1.3	52	18	13	–10.8
Moldova	4.0	118	1.3	45	20	10	–1.9
Russia	141.7	8	1.3	73	15	14	0.6
Ukraine	46.5	77	1.3	68	14	16	–2.9

[a]*Percentage of population younger than 15.*
[b]*Percentage of population older than 65.*
Source: Population Reference Bureau, World Population Data Sheet, 2007; *net migration rate data from* UN International Migration, 2006.

than did the more inhospitable conditions found across central and northern Siberia. Although Russian efforts over the past century have encouraged a wider dispersal of the population, it remains heavily concentrated in the west (Figure 9.13). European Russia is home to more than 100 million people, while Siberia, although far larger, holds only some 35 million. When one adds the 60 million inhabitants of Belarus, Moldova, and Ukraine, the imbalance between east and west becomes even more striking.

The European Core The region's largest cities, biggest industrial complexes, and most productive farms are located in the European Core, a subregion that includes Belarus, much of Ukraine, and Russia west of the Urals. The Core supports population densities that greatly exceed those to the east, although thesy are modest by northwest European or East Asian standards. The sprawling city of Moscow and its nearby satellite centers clearly dominate the settlement landscape with a metropolitan area of more than 10 million people (Figure 9.14). Within 250 miles (400 kilometers), a series of other major urban centers are closely linked to Moscow. Largest of these is the industrial city of Nizhniy Novgorod (formerly named Gor'kiy, in the Soviet period) (1.3 million), traditionally oriented around automobile and heavy equipment manufacturing.

Beyond Moscow's immediate orbit, three other areas of concentrated urban settlement dominate European Russia. On the shores of the Baltic Sea, St. Petersburg (4.7 million) offers a major trading window to the West. Between 1712 and 1917 it served as the capital of the Russian Empire, and it acquired a rich skyline of baroque architecture and beautiful churches. Many have compared the city's urban landscape to the great cities of western Europe (see "Cityscapes: St. Petersburg").

Far southeast of Moscow, a second urban focus is oriented along the lower and middle stretches of the Volga River. Industrialization along this part of the Volga accelerated greatly during World War II, as the region lay somewhat removed from German advances in the west. Today, the highly commercialized river corridor, also blessed with nearby petroleum reserves, supports a diverse industrial base strategically located to serve the large populations of the European Core. From north to south, the four Volga Valley cities of Kazan (1.1 million), Samara (1.2 million), Saratov (900,000), and Volgograd (1 million; Stalingrad in the Soviet period) are the largest settlements, and each possesses a sizable industrial infrastructure.

A third constellation of key population centers on the eastern edge of the Core is anchored along the resource-rich alignment of the Ural Mountains. From the gritty industrial landscapes of Serov (1.0 million) in the north to Chelyabinsk (1.1 million) and Magnitogorsk (425,000) in the south, the Urals region specializes in iron and steel manufacturing, metals smelting and refining, and heavy machinery construction. Outside of these industrial centers, however, the rural population density of the Urals is lower than in the better agricultural lands found farther west within the Core.

Beyond Russia, major population clusters within the European Core also are found in Belarus and Ukraine. The Belorussian capital of Minsk (1.7 million) is the dominant urban center in that country, and its landscape recalls the drab Soviet-style architecture of an earlier era (Figure 9.15). In nearby Ukraine, the capital of Kiev (Kyiv, 2.7 million) straddles the Dnieper River, and the city's rich architectural heritage is a reminder of its historic role in the political and economic geography of the European interior. Other large industrial centers, such as Kharkov (1.5 million), Dnepropetrovsk (1.1 million), and Donetsk (1.1 million), are located in resource-rich eastern Ukraine and benefit from their proximity to deposits of coal and iron ore. Southwest of Ukraine, Moldova's largest city is the capital of Chisinau (Kishinev, 750,000). Smaller population clusters also are found in Georgia's fertile Black Sea lowlands and around the capital of Tbilisi (1.3 million). To the southeast, the Armenian capital of Yerevan (1.2 million) contains almost one-third of that nation's population.

Siberian Hinterlands Leaving the southern Urals city of Yekaterinburg on a Siberia-bound train, one

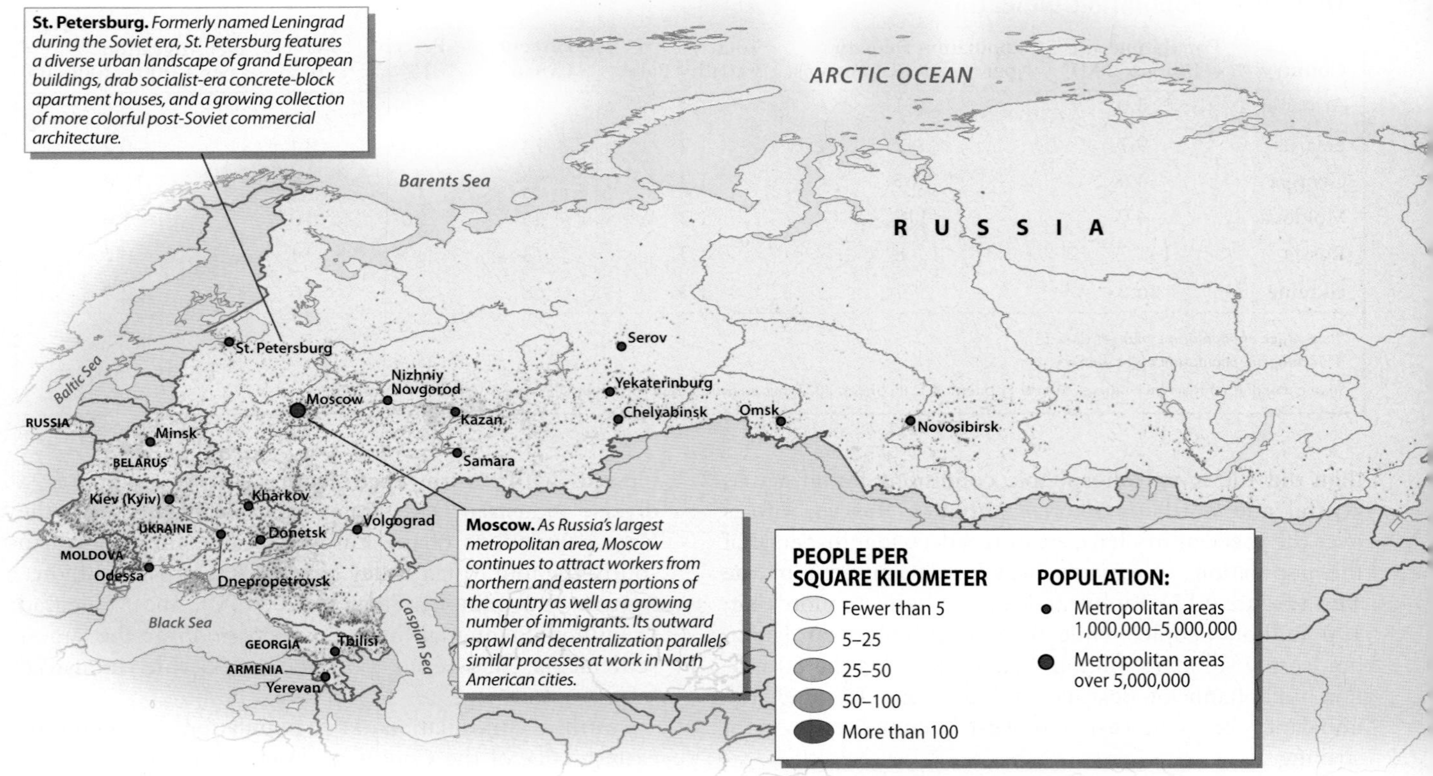

St. Petersburg. *Formerly named Leningrad during the Soviet era, St. Petersburg features a diverse urban landscape of grand European buildings, drab socialist-era concrete-block apartment houses, and a growing collection of more colorful post-Soviet commercial architecture.*

ARCTIC OCEAN
Barents Sea
RUSSIA
Baltic Sea
St. Petersburg
Nizhniy Novgorod
Moscow
Kazan
Samara
Serov
Yekaterinburg
Chelyabinsk
Omsk
Novosibirsk
Minsk
BELARUS
Kiev (Kyiv)
Kharkov
UKRAINE
Donetsk
Volgograd
MOLDOVA
Odessa
Dnepropetrovsk
Black Sea
Caspian Sea
GEORGIA
Tbilisi
ARMENIA
Yerevan

Moscow. *As Russia's largest metropolitan area, Moscow continues to attract workers from northern and eastern portions of the country as well as a growing number of immigrants. Its outward sprawl and decentralization parallels similar processes at work in North American cities.*

PEOPLE PER SQUARE KILOMETER
Fewer than 5
5–25
25–50
50–100
More than 100

POPULATION:
Metropolitan areas 1,000,000–5,000,000
Metropolitan areas over 5,000,000

Figure 9.13 Population Map of the Russian Domain
Population within the region is strongly clustered west of the Ural Mountains. Dense agricultural settlements, extensive industrialization, and large urban centers are found in Ukraine, much of Belarus, and across western Russia south of St. Petersburg and Moscow. A narrower chain of settlements follows the better lands and transportation corridors of southern Siberia, but most of Russia east of the Urals remains sparsely settled.

is acutely aware that the land ahead is ever more sparsely settled (see Figure 9.13). The distance between cities grows, and the intervening countryside reveals a landscape shifting gradually from farms to forest. The Siberian hinterland is divided into two characteristic zones of settlement. To the south, an alignment of isolated but sizable urban centers follows the **Trans-Siberian Railroad**, a key railroad corridor to the Pacific completed in 1904 (see Figure 9.1). The industrial settlements along the route benefit from good east–west connections as well as from proximity to natural resources such as oil, natural gas, coal, and iron ore. The eastbound traveler encounters Omsk (1.1 million) as the rail line crosses the Irtysh River, Novosibirsk (1.4 million) at its junction with the Ob River, and Irkutsk (600,000) near the southwest corner of Lake Baikal. More than 1,500 miles (2,415 kilometers) beyond, the better agricultural lands and industrial opportunities of the Russian Far East contribute to higher population densities. Khabarovsk (620,000) is the leading settlement of the Amur Valley, while the port city of Vladivostok (640,000) provides the principal

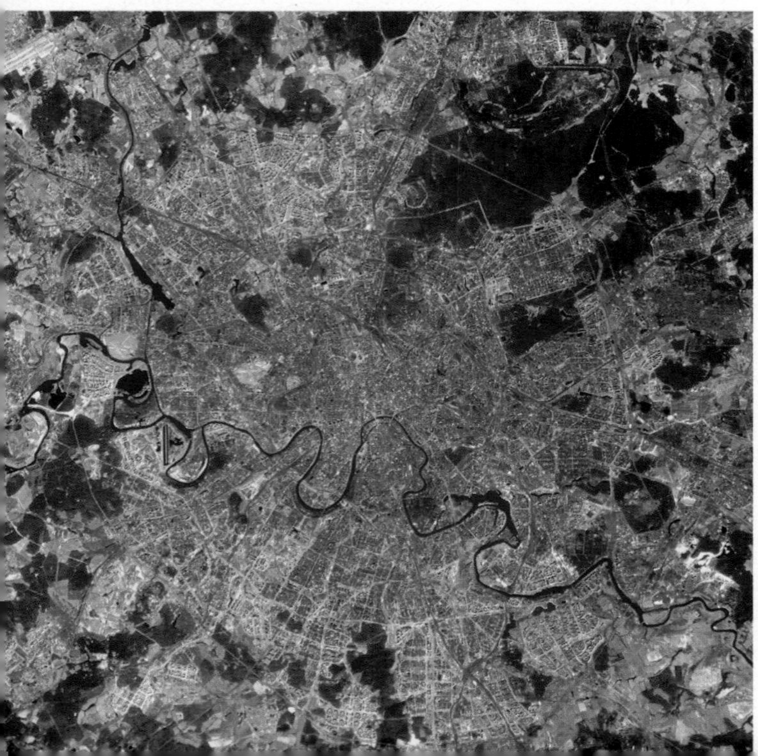

Figure 9.14 Metropolitan Moscow
Sprawling Moscow extends more than 50 miles (80 kilometers) beyond the city center. The city is home to more than 10 million people, and the relative strength of its urban economy continues to attract migrants from elsewhere in the country, thus putting more pressure on its infrastructure. *(CNES/Spot Image/Photo Researchers, Inc.)*

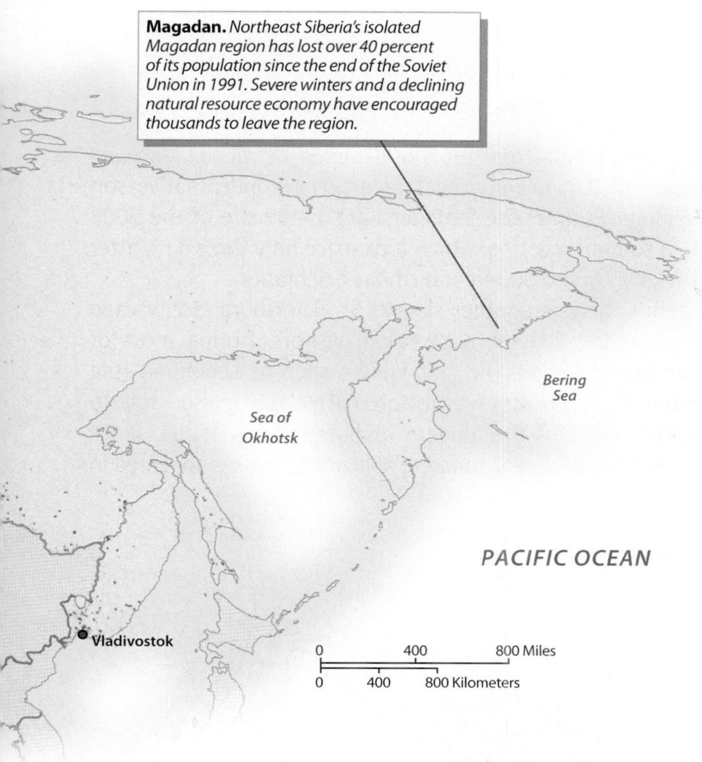

Russian access to the Pacific. A thinner sprinkling of settlement appears along the more recently completed (1984) **Baikal-Amur Mainline (BAM) Railroad**, which parallels the older line, but runs north of Lake Baikal to the Amur River. North of the BAM line, however, the seemingly empty vastness of the Siberian hinterland dominates the scene. Settlements are few and far between, and larger urban areas are usually either regional administrative capitals, such as Yakutsk (230,000), or resource extraction and processing centers, such as Norilsk (240,000).

Regional Migration Patterns

Over the past 150 years, millions of people within the Russian domain have been on the move. These major migrations, both forced and voluntary, reveal sweeping examples of human mobility that rival the great movements from Europe and Africa or the transcontinental spread of settlement across North America.

Eastward Movement Just as settlers of European descent moved west across North America, exploiting natural resources and displacing native peoples, European Russians moved east across the vast Siberian frontier. Although the deeper historical roots of the movement extend back several centuries, the pace and volume of the eastward drift accelerated in the late nineteenth century once the Trans-Siberian Railroad was completed. Peasants were attracted to the region by its agricultural opportunities (in the south) and by greater political freedoms than they traditionally enjoyed under the **tsars** (or czars; Russian for "Caesar"), the authoritarian leaders who dominated politics during the pre-1917 Russian Empire. Almost 1 million Russian settlers moved into the Siberian hinterland between 1860 and 1914.

The eastward migration continued during the Soviet period, once communist leaders consolidated power during the late 1920s and saw the economic advantages of developing the region's rich resource base. The German invasion of European Russia during World War II demonstrated that there were also strategic reasons for settling the eastern frontier, and this propelled further migrations during and following the war. Indeed, by the end of the communist era, 95 percent of Siberia's population was classified as Russian (including Ukrainians and other western immigrants). With the completion of the BAM Railroad in the 1980s, yet another corridor of settlement opened in Siberia, prompting new migrations into a region once remote from the outside world.

Figure 9.15 Minsk
This row of mass-produced apartment houses is found along the banks of the Svisloch River in Minsk, Belarus. *(Nik Wheeler/ Corbis/Bettmann)*

CITYSCAPES St. Petersburg

It was an improbable place for one of Europe's great cities. Literally rising from swamplands in 1703, St. Petersburg grew from the creative imagination of its founder, Tsar Peter the Great. Seven feet tall, Peter towered above the swamps he envisioned to become Russia's great window to the West. While it took time, St. Petersburg gradually assumed that role in later years. It became the westward-looking tsarist capital, anchored on a protected arm of the Baltic Sea, until the Soviet takeover in 1917 (when the capital returned to Moscow). Peter's imagination helped create a Renaissance-style city built around grand avenues, palaces, and a network of canals that still draws comparisons with Venice and Amsterdam (Figure 9.3.1). Water remains a central element in the cityscape today. Miles of canals, crossed by a latticework of scenic bridges, connect the Neva, Fontanka, and Moyka rivers (Figure 9.3.2).

Near the city center, elaborate baroque and neoclassical mansions (many now converted to other uses) and commercial buildings recall earlier days when Russians such as Tchaikovsky and Dostoyevsky walked the streets. Cutting through the heart of the city and bordered by the bright lights of shops, cafes, and theaters, Nevsky Prospekt ends near the Neva River, just two blocks from the world-famed Hermitage Museum, home to 2.8 million pieces of art. No wonder native son Vladimir Putin chose St. Petersburg as the site of the 2003 G-8 summit meeting, which took place only three days after the city's grand 300-year birthday celebration.

But there is another side to St. Petersburg, a cityscape beyond the ken of most casual visitors. During most of the Soviet period, the city, known then as Leningrad (the name St. Petersburg was restored after 1991), also came to reflect some of the modern drabness of Soviet life. Block after block of dreary concrete apartment houses appeared in

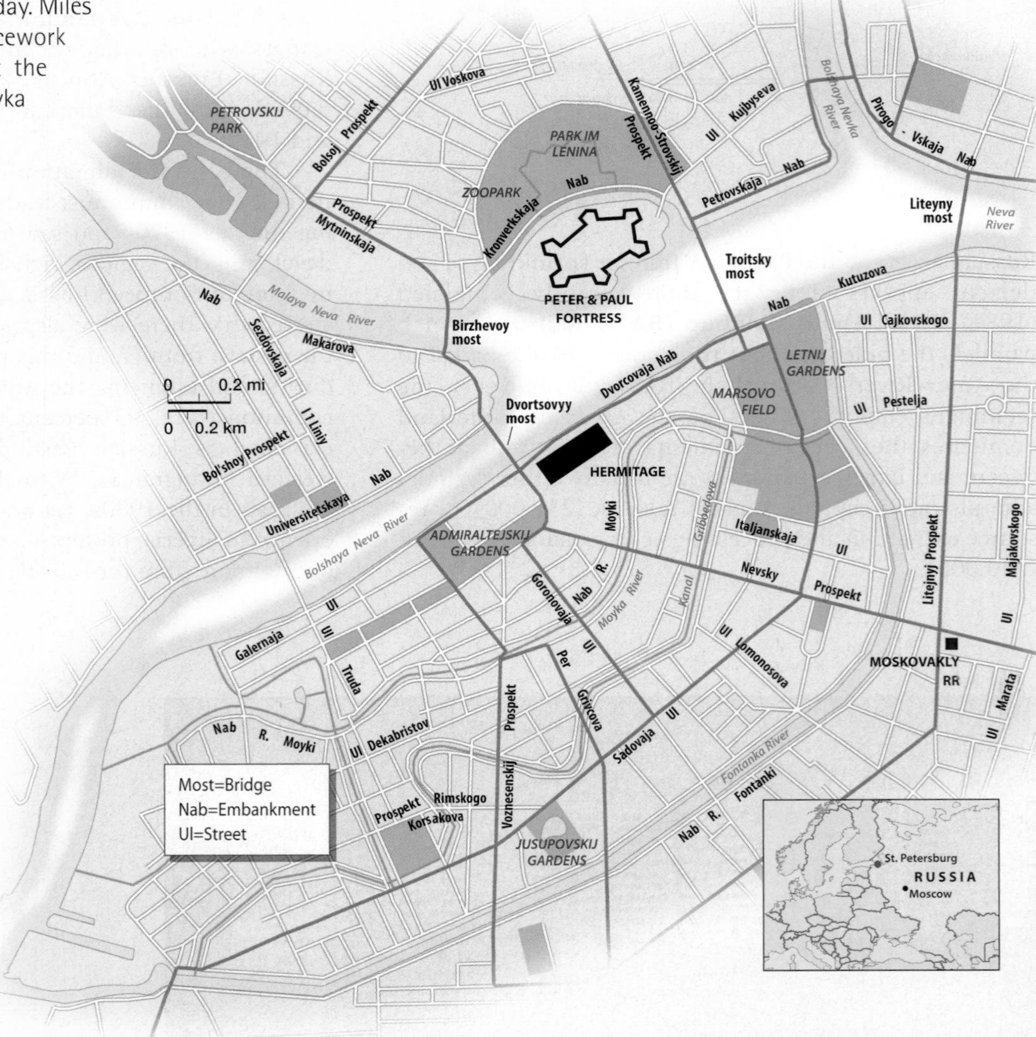

Figure 9.3.1 Map of St. Petersburg
The grand avenues and commercial thoroughfares of downtown St. Petersburg are laid out amid nearby rivers and canals, creating an urban framework that many have compared to Venice and Amsterdam.

many of the city's suburban districts. Many of the city's graceful old buildings fell into disrepair, victims of a Soviet mindset that valued modern progress over tsarist architecture. Worsening economic times in the 1980s also did little for the city's appearance as its 4.7 million residents struggled with the chores of daily life. Recently, however, investments in urban preservation and an inflow of foreign investment have perked up the city's economy and freshened its urban landscape. Downtown cafes are bustling, theaters are getting a facelift, and it is rumored that some central government ministries may relocate to the city from Moscow. Gradually, a new day seems to be dawning on an urban landscape that still manages to capture the full breadth and imagination of a people who were willing to build their fortunes in a swamp.

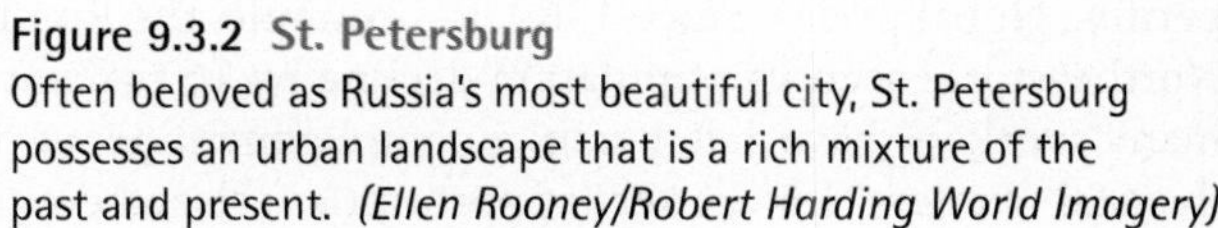

Figure 9.3.2 St. Petersburg
Often beloved as Russia's most beautiful city, St. Petersburg possesses an urban landscape that is a rich mixture of the past and present. *(Ellen Rooney/Robert Harding World Imagery)*

Political Motives Political motives also have shaped migration patterns. Particularly in the case of Russia, leaders from both the imperial and Soviet eras saw advantages in moving selective populations to new locations. The infilling of the southern Siberian hinterland had a political as well as an economic rationale. Both the tsars and the Soviet leaders saw their political power grow as Russians moved into the resource-rich Eurasian interior. For some, however, the move to Siberia was not voluntary as the region became a repository for political dissidents and troublemakers. Especially in the Soviet period, uncounted millions were forcibly relocated to the region's infamous **Gulag Archipelago**, a vast collection of political prisons in which inmates often disappeared or spent years far removed from their families and home communities. The communist regime of Joseph Stalin (1928–53) was particularly noted for its forced migrations, including the removal of thousands of Jews to the Russian Far East after 1928. More recently, political instability within Russia's internal republic of Chechnya produced large-scale refugee migrations into nearby regions (Figure 9.16).

Russification, the Soviet policy of resettling Russians into non-Russian portions of the Soviet Union, also changed the region's human geography. Millions of Russians were given economic and political incentives to move elsewhere in the Soviet Union in order to increase Russian dominance in many of the outlying portions of the country. The migrations were geographically selective in that most of the Russians moved either to administrative centers where they took government positions or to industrial complexes focused on natural resource extraction. As a result, by the end of the Soviet period, Russians made up significant minorities within former Soviet republics (now independent nations) such as Kazakhstan (30 percent Russian), Latvia (30 percent), and Estonia (26 percent). Among its Slavic neighbors, Belarus remains 11 percent Russian and Ukraine more than 17 percent Russian, with concentrations particularly high in the eastern portions of the countries.

New International Movements In the post-Soviet era, Russification has often been reversed (see Figure 9.16). Several of the newly independent non-Russian countries imposed rigid language and citizenship requirements, which encouraged many Russian residents to leave. In other settings, ethnic Russians simply experienced varied forms of social and economic discrimination. By 2007, more than 6 million Russians had left former Soviet republics and returned to their homeland. As a result, the Central Asia and Baltic regions, once a part of the Soviet Union, have seen their Russian populations decline significantly since 1991, often by 20 to 35 percent. Russia's economic growth has also produced an increasing number of illegal immigrants who have simply come to the country looking for employment and a better life for their families. Central Asia and China have been important source areas for this flow of illegal migrants (see "People on the Move: Russia's Illegal Immigrants").

The domain's more open borders also have made it easier for other residents to leave the region (see Figure 9.16). Poor economic conditions and the region's unpredictable politics have encouraged many to emigrate.

The "brain drain" of young, well-educated, upwardly mobile Russians has been considerable. Sometimes, ethnic links play a part in migration patterns. For example, many Russian-born ethnic Finns have moved to nearby Finland, much to the consternation of the Finnish government. Russia's Jewish population also continues to fall, a pattern begun late in the Soviet period. These emigrants have flocked mostly to Israel or the United States, where they locate in familiar Jewish neighborhoods such as Miami, Los Angeles, and New York City's Brighton Beach district.

The Urban Attraction Regional residents also have been bound for the cities. The Marxist philosophy embraced by Soviet planners encouraged urbanization. In 1917, the Russian Empire was still overwhelmingly rural and agrarian; 50 years later, the Soviet Union was primarily urban. Planners saw great economic and political advantages in efficiently clustering the population, and Soviet policies dedicated to large-scale industrialization obviously favored an urban orientation. Today, Russian, Ukrainian, and Belorussian rates of urbanization are comparable to those of the industrialized capitalist countries (Table 9.1).

Soviet cities grew according to strict governmental plans. Planners selected different cities for different purposes. Some were designed for specific industries, while others had primarily administrative roles. A system of internal passports prohibited people from moving freely from city to city. Instead, people generally went where the government assigned them jobs. Moscow, the country's leading administrative city, thrived under the Soviet regime. It formed the undisputed core of Soviet bureaucratic power as well as the center of education, research, and the media. Specialized industrial cities also grew at a rapid pace. In the mining and metallurgical zone of the southern Urals, centers such as Yekaterinburg and Chelyabinsk mushroomed into major urban centers. Another cluster of specialized industrial cities, including Kharkov and Donetsk, emerged in the coal districts of eastern Ukraine.

With the end of the Soviet Union, however, people gained basic freedoms of mobility. In addition, with a more open economy, especially in Russia, shifting urban employment opportunities increasingly reflect how effectively local economies could compete in the global market. This new economic reality, for example, has led to the depopulating of many older industrial areas, as they simply cannot produce raw materials or finished industrial goods at competitive global prices. Since 1989, for example, the Russian Northeast has seen its population decline by 16 percent as many workers have left harsh, unemployment-prone industrial centers for better opportunities elsewhere. In many cases, people are freely gravitating toward growth opportunities in locations of new foreign investment, principally in larger urban areas in western and southern Russia. Some also are receiving relocation assistance from a Russian government that admits many of these outlying settlements simply cannot be justified in a more market-oriented economy.

Inside the Russian City

Large Russian cities possess a core area, or center, that features superior transportation connections; the best-stocked, upscale department stores and shops; the most desirable housing; and the most important offices (both governmental and private). At the top of the urban hierarchy, large cities such as Moscow and St. Petersburg also feature extensive public spaces and examples of monumental architecture at the city center. Inner-city decay, so characteristic of the United States, is not a feature of the Russian central city.

Within the city, there is usually a distinctive pattern of concentric land-use zones, each of which was built at a later date as one moves outward from the center. Such a ringlike urban morphology is not a unique phenomenon, but it is probably more highly developed here than in most parts of the world, owing to the extensive power of government planners during the Soviet period.

At their very center, the cores of many older cities predate the Soviet Union. Pre-1900 stone buildings often dominate city centers. Some of these are former private mansions that were turned into government offices or subdivided into apartments during the communist period but are now being sold to private investors and residents. But pressures for inner-city growth are immense. Many of these older buildings are being leveled in rapidly growing urban settings such as downtown Moscow. Construction cranes are a common site as central city space is highly prized by commercial developers (Figure 9.17). Since 1992, for example, urban preservation experts estimate that more than 2,000 historic buildings have been destroyed in that city, including dozens of structures that had supposedly received official protection. Sometimes the preservation laws are simply ignored, while in other cases convenient fires or building condemnations eliminate barriers to new commercial construction. One large new luxury retailing mall just off Red Square in downtown Moscow generates millions in revenues as cash-carrying residents scoop up Armani Casa furniture, Brioni suits, and Italian jewelry. Nightclubs and bars are filled with pleasure seekers as the city's growing professional elite mingles with foreign visitors and tourists. Meanwhile, the nearby "Moscow City" business zone is expanding into older industrial areas, offering new office space for Russian and international firms.

Beyond the dynamic urban core, one often can find a ring of public housing projects and fully planned *sotzgorods*, or socialist neighborhoods. These projects became most common in the heavy industrial cities that mushroomed during the early decades of the Soviet Union. Another common urban zone, removed some distance from city cores, is the *chermoyuski*. Chermoyuski are large, uniform apartment blocks built during the 1950s and 1960s. At that time a more prosperous Soviet Union could give each family its own small, private flat.

Farther out from the city centers are the ***mikrorayons***, the much-larger housing projects of the 1970s and 1980s (Figure 9.18). Mikrorayons are typically composed of massed blocks of standardized apartment buildings, ranging from 9 to 24 stories in height. Each mikrorayon was to form a self-contained community, with grocery stores and other basic services located within walking distance of each apartment building. The largest of these supercomplexes contain up to 100,000 residents. While planners hoped that mikrorayons would foster a sense of community, most now serve largely as suburban bedroom communities for larger metropolitan areas. Although public transportation is well developed in most cities of the region, commuting time between city centers and these outer apartment complexes is usually 45 minutes or more.

Some of Russia's most rapid urban growth in recent years has occurred on the metropolitan periphery, paralleling the North American experience. Moscow, for example, has seen its urban reach expand far beyond the city center. The surrounding administrative district (the Moscow Oblast) contains more than

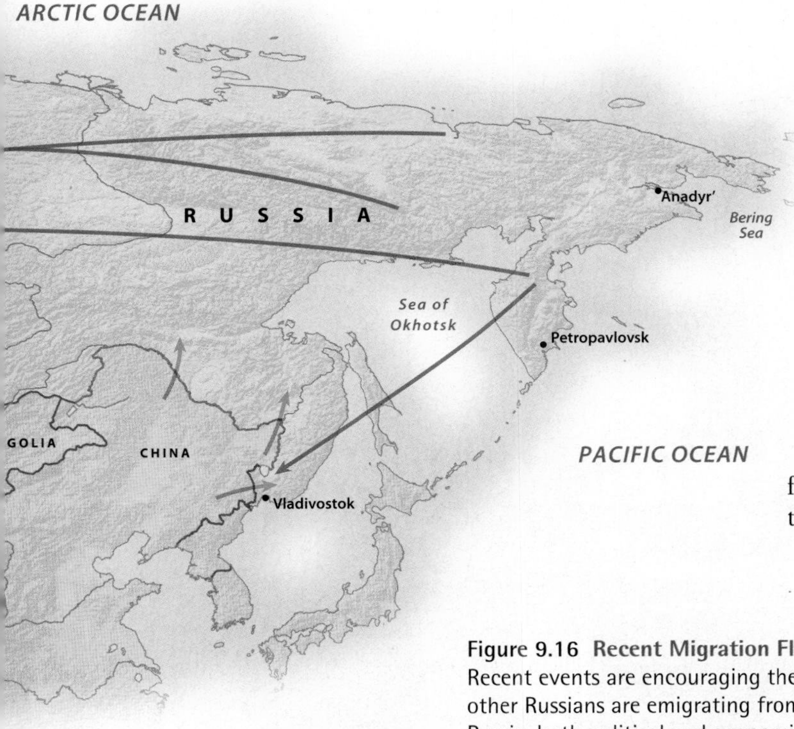

Figure 9.16 Recent Migration Flows in the Russian Domain
Recent events are encouraging the return of ethnic Russians from former Soviet republics, while other Russians are emigrating from the domain for economic, cultural, and political reasons. Within Russia, both political and economic forces also are at work encouraging people to be on the move.

7 million residents. Already home to hundreds of international companies, the outer city has been attracting many global corporate offices, including those of PepsiCo, Procter and Gamble, and Bristol-Myers. Land prices and tax rates are lower than in the central city, and the transportation and telecommunications infrastructure is relatively new. New suburban shopping malls, paralleling the North American model, also are popping up on the urban fringe, allowing residents to shop and be entertained without having to visit the city center.

Elsewhere on Moscow's urban fringe, elite ***dacha***, or cottage communities, appeal to many more well-to-do residents, particularly during the summer months. The tradition of rural retreats, dating back to the Russian Empire, also thrived during the Soviet era as Communist Party officials sought an escape from the dreary bureaucratic chores of the city. Today, there are about 300 cottage settlements on the Moscow periphery, including many new elite privatized developments northwest and southeast of the city that cater to the country's business class. The older dacha belt to the west of the city along the Moskva River also offers summer homes, spa retreats, and holiday hotels in a zone dubbed the "Sub-Moscow Switzerland." The dachas illustrate how contemporary settlement patterns blend landscape elements inherited from the Soviet period with those reflecting new forces at work across the Russian domain.

Figure 9.18 Moscow Housing
For many residents in larger Russian cities, home is a high-rise apartment house. Many of these satellite centers were built in the Soviet era. Poor construction and a lack of landscaping often yield a bleak suburban scene, but nearby stores, entertainment, and public transportation offer important amenities. *(Itar-Tass/Sovfoto/Eastfoto)*

The Demographic Crisis

Recently, Russia's Ministry of Labor and Social Development acknowledged that the country faced a crisis of persistently declining populations. Government officials

Figure 9.17 Downtown Moscow
The city's downtown landscape has been rapidly transformed in the recent economic boom. It remains a complex and fascinating mix of imperial, Soviet, and post-Soviet influences. *(Grigory Dukor/Reuters/Corbis)*

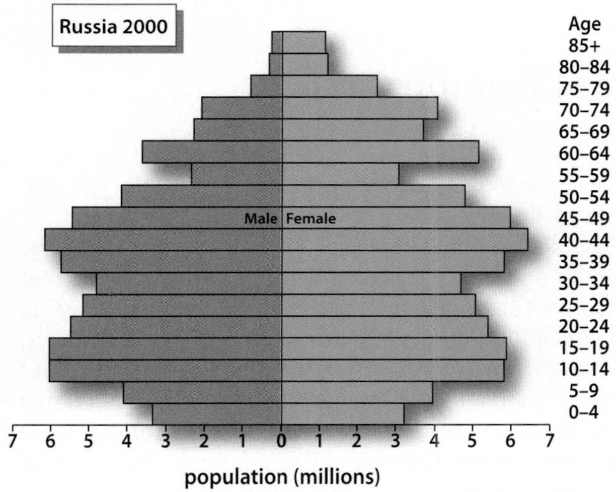

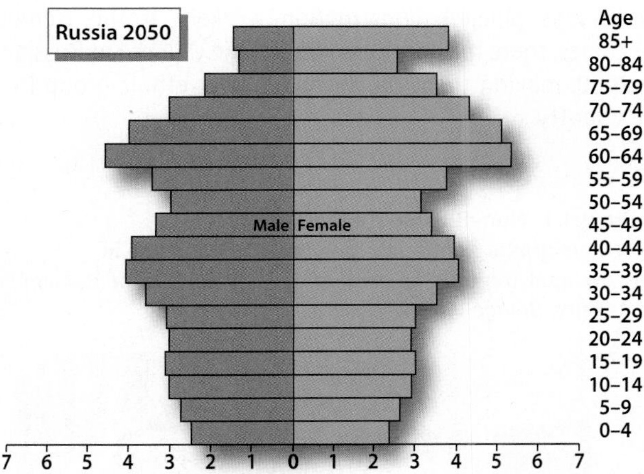

Figure 9.19 Russia's Changing Population
These two population pyramids provide a recent glimpse (2000) as well as predicted patterns (2050) of Russia's population structure. Present trends suggest that Russia's population will continue to age, with relatively fewer young people supporting a relatively large elderly population. Also note the impact of earlier wars and higher death rates on older adult Russian males. *(U.S. Census Bureau, International Data Base)*

pledged to improve the nation's health-care system, provide more incentives for increasing the birthrate, and foster immigration. The success of those initiatives remains to be seen, however, and government studies in 2000 predicted that Russia's population could fall by a startling 25 million by 2030. Similar conditions are affecting the other countries within the region (see Table 9.1). Declining populations, low birthrates, and rising mortality, particularly among middle-aged males, all are symptoms of the grim demographic scenario that has engulfed the region. During World War II, large numbers of deaths combined with low birthrates to produce sizable population losses. While population increases accelerated in the 1950s, growth slowed by 1970, and death rates began exceeding birthrates in the early 1990s. Two population pyramids tell the troubling tale (Figure 9.19). The first shows that, whereas adult-age populations are prominently represented (with the exception of older males and the birth crash of World War II), few children are being added. Looking ahead to 2050, today's shrinking numbers take an even bigger toll as the pattern of small families is likely to continue.

In 2006, President Putin recognized that demographic decline was Russia's "most acute problem." He has pushed for a long-term program aimed at raising birthrates. Under the plan, mothers of multiple children receive cash payments, extended maternity leave, and extensive day-care subsidies. Recently, there have been increases in Russian birthrates, perhaps a function of changing government policies as well as a better economy and more optimism about the future. Russian birthrates are now higher than those of Germany and Japan and there are long waiting lists for urban kindergarten spots in Moscow (Figure 9.20). Employers are also offering more benefits to both mothers and fathers, making parenthood more appealing for some couples.

Even with rising birthrates, the other challenge throughout the region will be to stabilize death rates, particularly for older males. Especially among Russian men, low life expectancy appears related to many stress-related conditions, such as alcoholism and heart disease. Murder and suicide rates have climbed rapidly in the past 20 years. In addition, perhaps 20 to 30 percent of the increase in death rates may be attributed to the region's increasingly toxic environment. Whatever the complex causes, the average lifespan for Russians has fallen appreciably since the late 1980s, and for men the drop has been truly unprecedented in the developed world. Potentially, improved economic performance might lead toward better health care and to less stress-related disease. Only time will tell. Meanwhile the region's population totals will likely continue their slow decline.

Figure 9.20 Young Russian Parents
A slight rise in Russian birthrates may help slow that country's population decline. Still, high death rates are a major issue, particularly for males throughout the region, and they are the product of many environmental and lifestyle variables. *(Yuri Belinsky/Itar-Tass/Landov Photo)*

PEOPLE ON THE MOVE Russia's Illegal Immigrants

The story has a familiar ring to it. More than 10 million illegal immigrants are suspected in the country. Most are young, upwardly mobile people, predominantly male, who come for better-paying jobs. Recently, the government has implemented tighter controls to restrict border crossings and enacted tougher penalties against businesses that hire illegal immigrants. There is a growing national debate concerning how many legal foreign workers should be allowed in the country and whether or not illegal immigrants should be granted amnesty. Indeed, Russia's story is similar to that of the United States. In fact, as a percentage of total population, there are more illegal immigrants in Russia than in the United States.

Even with the parallels, there are also real differences. Russia faces major declines in its native-born population, due to high death rates and low birthrates. In that sense, immigrants can be rapidly absorbed into a growing Russian economy and many migration experts within the country suggest Russia should further expand its immigrant population. In fact, Russian leaders have openly called for ethnic Russians living in foreign countries to return to their native cultural homeland.

But the actual flows of illegal immigrants are more complex. While 80 percent of Russia's immigrants—legal and illegal—come from portions of the former Soviet Union, a growing number of workers are from ethnically non-Slavic regions such as Central Asia (Figure 9.16). Street cleaners from Kyrgyzstan and shopkeepers from Uzbekistan are increasingly common sights in Moscow, where almost one-third of the country's illegal immigrants may live (Figure 9.4.1). One recent estimate suggests that about 20 percent of Tajikistan's economy is now based on funds sent to the country from émigrés living and working in Russia. But there is growing Russian resistance to these non-Slavic arrivals. For example, a new law in Moscow limits the number of non-Russians who can operate market stalls on the streets of the city. Growing violence against these non-Russian immigrants has also erupted in Moscow and elsewhere.

Another stream of illegal migrants is reshaping the human and cultural geography of the Russian Far East (Figure 9.16). Large numbers of ethnic Chinese are crossing the Amur River border into Russia every day to trade, work, and live. Many Chinese cross the border as "tourists" but quickly melt into the countryside. Others are migrating into eastern Siberian cities, setting up shop as small business owners. Recent proposals for completing an oil pipeline between Russia and China have also included plans to use tens of thousands of additional Chinese laborers as pipeline construction workers. If the trend continues, there may be 10 million ethnic Chinese in Russia by 2050, making them the second largest ethnic group in the country.

Figure 9.4.1 Non-Russian Immigrants
These immigrants from Tajikistan are standing in line in Moscow, applying for employment visas to work in the Russian capital city. *(Image Works)*

CULTURAL COHERENCE AND DIVERSITY: The Legacy of Slavic Dominance

For hundreds of years, Slavic peoples speaking the Russian language expanded their influence from an early homeland in central European Russia. Eventually, this Slavic cultural imprint spread north to the Arctic Sea, south to the Black Sea and Caucasus, west to the shores of the Baltic, and east to the Pacific Ocean. In this process of diffusion, Russian cultural patterns and social institutions spread widely, influencing many non-Russian ethnic groups that continued to live under the rule of the Russian Empire. The legacy of that Slavic expansion continues today. It offers Russians a rich historical identity and sense of nationhood. It also provides a meaningful context in which to understand the way present-day Russians are dealing with forces of globalization and how non-Russian cultures have evolved within the region.

The Heritage of the Russian Empire

The expansion of the Russian Empire paralleled similar events in western Europe. As Spain, Portugal, France, and Britain carved out empires in the Americas, Africa, and

Asia, the Russians expanded eastward and southward across Eurasia. Unlike other European empires, however, the Russians formed one single territory, uninterrupted by oceans or seas. Only with the fall of the Soviet Union in 1991 did this transformed empire finally begin to dissolve.

Origins of the Russian State The origin of the Russian state lies in the early history of the **Slavic peoples**, defined linguistically as a distinctive northern branch of the Indo-European language family. The Slavs originated in or near the Pripyat marshes of modern Belarus. Some 2,000 years ago they began to migrate to the east, reaching as far as modern Moscow by 200 CE. Slavic political power grew by 900 CE as Slavs intermarried with southward-moving warriors from Sweden known as *Varangians*, or *Rus*. Within a century, the state of Rus extended from Kiev (the capital) in modern Ukraine, to Lake Ladoga near St. Petersburg. The new Kiev-Rus state interacted with the rich and powerful Byzantine Empire of the Greeks, and this influence brought Christianity to the Russian realm by 1000 CE. Along with the new religion came many other aspects of Greek culture, including the Cyrillic alphabet. Even as groups such as the Russians and Serbs converted to **Eastern Orthodox Christianity**, a form of Christianity historically linked to eastern Europe and church leaders in Constantinople (modern Istanbul), their Slavic neighbors to the west (the Poles, Czechs, Slovaks, Slovenians, and Croatians) accepted Catholicism. The resulting religious division split the Slavic-speaking world into two groups, one oriented to the west, the other to the east and south. This early Russian state soon faltered and split into several principalities that were then ruled by invading Mongols and Tatars (a group of Turkish-speaking peoples).

Growth of the Russian Empire By the fourteenth century, however, northern Slavic peoples overthrew Tatar rule and established a new and expanding Slavic state (Figure 9.21). The core of the new Russian Empire lay near the eastern fringe of the old state of Rus. The former center around Kiev was now a war-torn borderland (or "Ukraine" in Russian) contested by the Orthodox Russians, the Catholic Poles, and the Muslim Turks. Gradually this area's language diverged from that spoken in the new Russian core, and *Ukrainians* and Russians developed into two separate peoples. A similar development took place among the northwestern Russians, who experienced several centuries of Polish rule and over time were transformed into a distinctive group known as the *Belorussians*.

The Russian Empire expanded remarkably in the sixteenth and seventeenth centuries (see Figure 9.21). Former Tatar territories in the Volga Valley (near Kazan) were incorporated into the Russian state in the mid-1500s. The Russians also allied with the seminomadic **Cossacks**, Slavic-speaking Christians who had earlier migrated to the region to seek freedom in the ungoverned steppes (Figure 9.2). The Russian Empire granted them considerable privileges in exchange for their military service, an alliance that facilitated Russian expansion into Siberia during the seventeenth century. Premium furs were the chief lure of this immense northern territory. By the 1630s, Russian power was entrenched in central Siberia, and, by the end of the century, it had reached the Pacific Ocean. Chinese resistance, however, delayed Russian occupation of the Far East region until 1858, and the imperial designs of the Japanese halted further expansion to the southeast when the Russians lost the Russo-Japanese War in 1905.

While the Russian Empire expanded to the east with great rapidity, its westward expansion was slow and halting. In the 1600s, Russia still faced formidable enemies in Sweden, Poland, and the Ottoman (Turkish) Empire. By the 1700s, however, all three of these states had weakened, allowing the Russian Empire to gain substantial territories. After defeating Sweden in the early 1700s, Tsar Peter the Great (1682–1725) obtained a foothold on the Baltic, where he built the new capital city of St. Petersburg. Later in the eighteenth century, Russia defeated both the Poles and the Turks and gained all of modern-day Belarus and Ukraine. Tsarina Catherine the Great (1762–96) was particularly pivotal in colonizing Ukraine and bringing the Russian Empire to the warm-water shores of the Black Sea (Figure 9.22).

The nineteenth century witnessed the Russian Empire's final expansion. Large gains were made in Central Asia, where a group of once-powerful Muslim states was no longer able to resist the Russian army. The mountainous Caucasus region proved a greater challenge, as the peoples of this area had the advantage of rugged terrain in defending their lands. South of the Caucasus, however, the Christian Armenians and Georgians accepted Russian power with little struggle, because they found it preferable to rule by the Persian or Ottoman empires.

The Legacy of Empire The expansion of the Russian people was one of the greatest human movements Earth has ever witnessed. By 1900, a traveler going from St. Petersburg on the Baltic to Vladivostok on the Sea of Japan would everywhere encounter Russian peoples speaking the same language, following the same religion, and living under the rule of the same government. Nowhere else in the world did such a tightly integrated cultural region cover such a vast space.

The history of the Russian Empire also reveals points of ongoing tension with the world beyond. One of these tensions centers on Russia's ambivalent relationship with western Europe. Russia shares with the West the historical legacy of Greek culture and Christianity. Since the time of Peter the Great, Russia has undergone several waves of intentional Westernization. At the same time, however, Russia has long been suspicious of—even hostile to—European culture and social institutions. Although elements of this debate were transformed during the Soviet period, the central tension remains, and it influences Russia to this day. As Europe further unifies through the

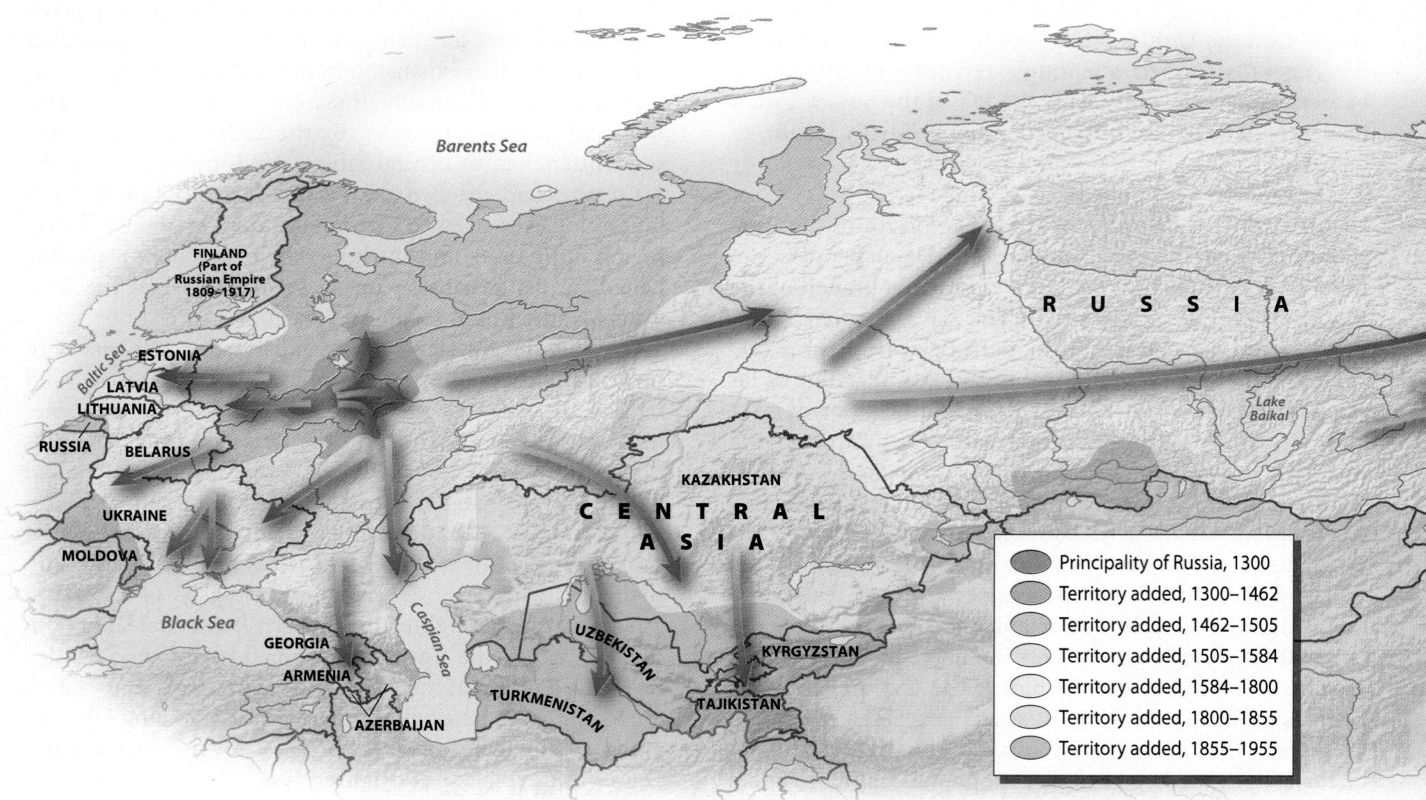

Figure 9.21 Growth of the Russian Empire
Beginning as a small principality in the vicinity of modern Moscow, the Russian Empire took shape between the fourteenth and sixteenth centuries. After 1600, Russian influence stretched from eastern Europe to the Pacific Ocean. Later, portions of the empire were added in the Far East, Central Asia, and near the Baltic and Black seas. *(Modified from Bergman and Renwick, 1999,* Introduction to Geography, *Upper Saddle River, NJ: Prentice Hall)*

expansion of the European Union (EU), this gap between east and west may take new forms, but it is unlikely to disappear anytime soon.

Geographies of Language

Slavic languages dominate the region (Figure 9.23). The distribution of Russian-speaking populations is complicated. Russian, Belorussian, and Ukrainian are closely related languages. Some linguists argue that they ought to be considered separate dialects of a single Russian language, as they are all mutually intelligible. Most Ukrainians, however, insist that Ukrainian is a distinct language in its own right, and there is a well-developed sense of national distinction between Russians and Ukrainians. Belorussians, on the other hand, are more inclined to stress their close kinship with the Russians.

Patterns in Belarus, Ukraine, and Moldova The geographic pattern of the Belorussian people is relatively simple. The vast majority of Belorussians reside in Belarus, and most people in Belarus are Belorussians (see Figure 9.23). The country does, however, contain scattered Polish and Russian minorities. For the Russians, this presents few problems, since Russians and Belorussians can relatively easily assume each other's ethnic identity.

The situation in Ukraine, however, is more complex. Because of different historical patterns of territorial conquest, Russian speakers dominate large parts of eastern Ukraine, while they make up a much smaller portion of western Ukraine's population (Figure 9.24). Similarly, the Crimean Peninsula, now a part of Ukraine, has long ethnic and political connections to Russia. As a result, there are many Ukrainian citizens in the eastern and southern parts of the country who rarely speak Ukrainian. Conversely, many Ukrainians born in the west never learn Russian. Kiev (Kyiv), the national capital, has been described as bilingual, "a Russian-speaking city whose people know how to speak Ukrainian." Since 2004, Ukrainian speakers have increased their efforts to enforce Ukrainian cultural and linguistic traditions, a campaign that has provoked popular and political resistance in the eastern portion of the country.

In nearby Moldova, Romanian (a Romance language) speakers are dominant, although ethnic Russians and Ukrainians each make up about 13 percent of the country's population. Use of the official Romanian language has been criticized by many Slavic speakers, particularly in the region east of the Dniester River (Transdniester) that borders

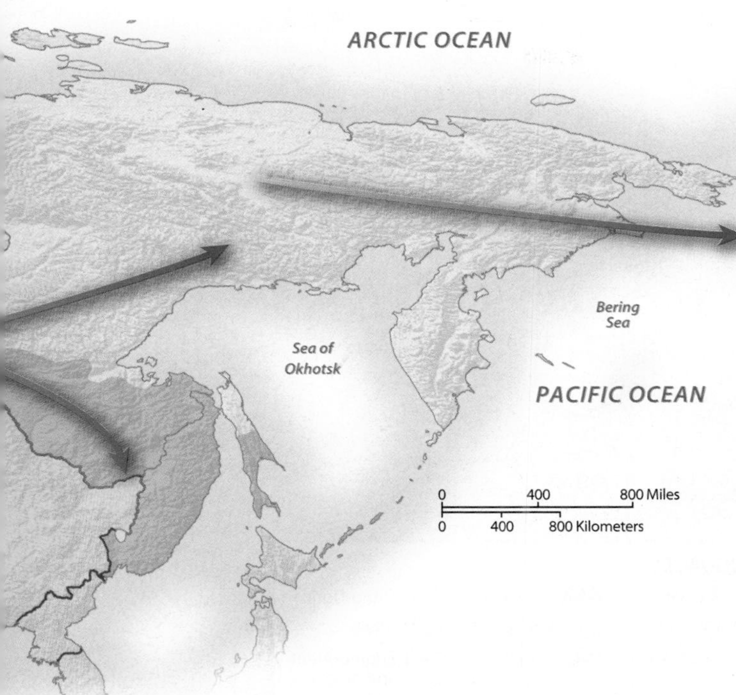

Ukraine. During the Soviet period, such geographical complexities had little consequence, because the distinction between Russians, Ukrainians, and Moldavians was not viewed as important in official circles. Now that Russia, Ukraine, and Moldova are separate countries with a heightened sense of national distinction, this issue has emerged as a significant source of tension across the region.

Patterns Within Russia Approximately 80 percent of Russia's population claims a Russian linguistic identity. Russian speakers inhabit most of European Russia, but there are large enclaves of other peoples. The Russian linguistic zone extends across southern Siberia to the Sea of Japan. In sparsely settled lands of central and northern Siberia, Russians are numerically dominant in many areas, but they share territory with varied indigenous peoples.

Finno-Ugric (Finnish-speaking peoples), though small in number, dominate sizable portions of the non-Russian north. The Finno-Ugric (or Uralic) speakers make up an entirely different language family from the Indo-European Russians. While many have been culturally Russified, distinct Finnish-speaking peoples such as the Karelians, Komi, and Mordvinians remain a part of Russia's modern cultural geography. Altaic speakers also complicate the country's linguistic geography. This language family includes the Volga Tatars, whose territory is centered on the city of Kazan in the middle Volga Valley. While retaining their ethnic identity, the Turkish-speaking Tatars have extensively intermarried with and borrowed from their Russian neighbors. Bilingual education is now common in the schools, with ethnic Tatars free to use their traditional language. Yakut peoples of northeast Siberia also represent Turkish speakers within the Altaic family. Other Altaic speakers in Russia belong to the Mongol group. In the west, examples include the Kalmyk-speaking peoples of the lower Volga Valley who migrated to the region in the 1600s. Far to the east, the Buryats live in the vicinity of Lake Baikal and represent an indigenous Siberian group closely tied to the cultures and history of Central Asia.

The plight of many native peoples in central and northern Siberia parallels the situation in the United States, Canada, and Australia. Rural indigenous peoples in each of these settings remain distinct from dominant European cultures. Such groups also are internally diverse and often are divided into a number of unrelated linguistic clusters. One entire linguistic grouping of Eskimo-Aleut speakers is limited to approximately 25,000 people, who are widely dispersed through northeast Siberia. Another indigenous Siberian group is the Altaic-speaking Evenki, whose traditional territory covers a large portion of central and eastern Siberia (Figure 9.25). Many of these Siberian peoples have seen their traditional ways challenged by the pressures of Russification, just as indigenous peoples elsewhere in the world have been subjected to similar pressures of cultural and political assimilation. Unfortunately, other common traits seen within such settings are low levels of education, high rates of alcoholism, and widespread poverty.

Figure 9.22 Catherine the Great
Tsarina Catherine the Great ruled the Russian Empire between 1762 and 1796 and expanded imperial influence over portions of southern Russia, Ukraine, and the warm-water coastline of the Black Sea. *(Getty)*

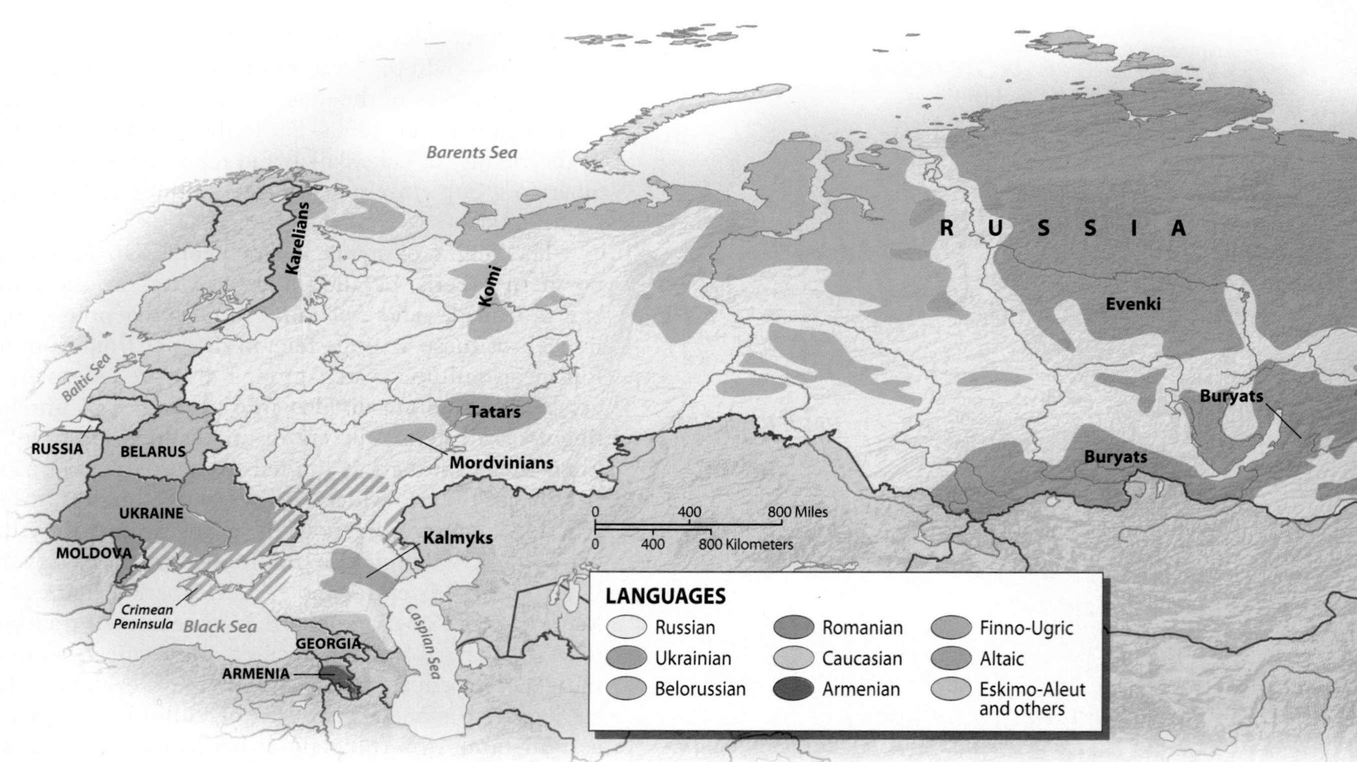

Figure 9.23 Languages of the Russian Domain
Slavic Russians dominate the realm, although many linguistic minorities are present. Siberia's diverse native peoples add cultural variety in that area. To the southwest, the Caucasus Mountains and the lands beyond contain the region's most complex linguistic geography. Ukrainians and Belorussians, while sharing a Slavic heritage with their Russian neighbors, add further variety in the west.

Transcaucasian Languages Although small in size, Transcaucasia offers a bewildering variety of languages (Figure 9.26). From Russia, along the north slopes of the Caucasus, and to Georgia and Armenia east of the Black Sea, a complex history and a fractured physical setting have combined to produce some of the most complicated language patterns in the world. No fewer than three language families

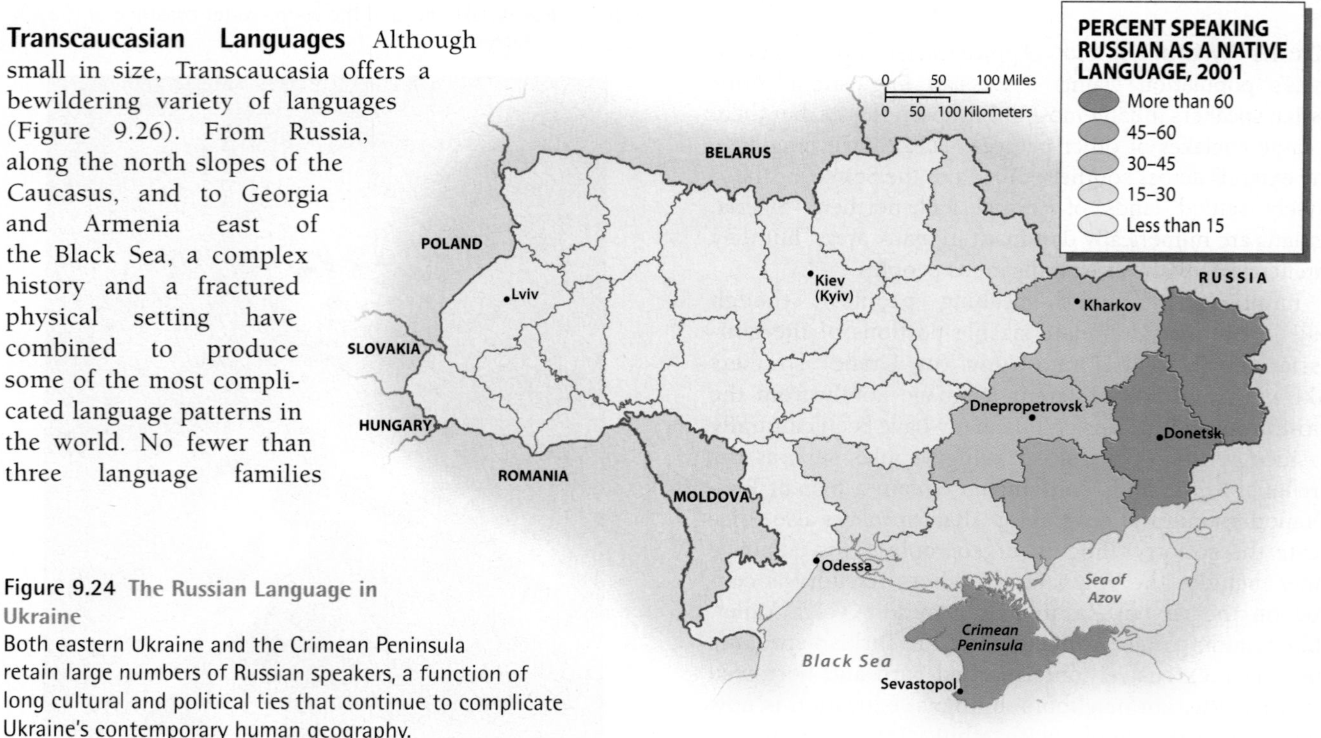

Figure 9.24 The Russian Language in Ukraine
Both eastern Ukraine and the Crimean Peninsula retain large numbers of Russian speakers, a function of long cultural and political ties that continue to complicate Ukraine's contemporary human geography.

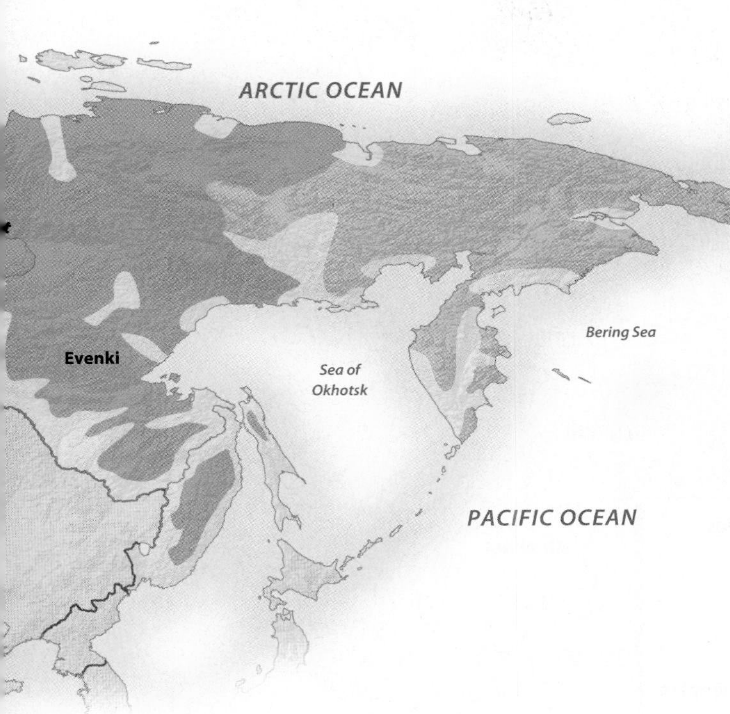

(Caucasian, Altaic, and Indo-European) are spoken within a region smaller than Ohio, and many individual languages are represented by small, isolated cultural groups. Not surprisingly, language remains a pivotal cultural and political issue within the fragmented Transcaucasian subregion.

Geographies of Religion

Most Russians, Belorussians, and Ukrainians share a religious heritage of Eastern Orthodox Christianity. For hundreds of years, Eastern Orthodoxy served as a central cultural presence within the Russian Empire (Figure 9.27). Indeed, church and state were tightly fused, until the demise of the empire in 1917. Under the Soviet Union, however, religion in all forms was severely discouraged and actively persecuted. Most monasteries and many churches were converted into museums or other kinds of public buildings, and schools disseminated the doctrine of atheism.

Contemporary Christianity With the downfall of the Soviet Union, however, a religious revival has swept much of the Russian domain. In the past 15 years, more than 12,000 Orthodox churches have been returned to religious uses. Now, an estimated 75 million Russians are members of the Orthodox Church, including almost 500 monastic orders dispersed across the country. Within Russia, the Orthodox Church appears headed toward a return to its former role as state church. The government increasingly is using church officials to sanction various state activities, which is particularly ironic because many of Russia's current leaders played roles in the earlier Soviet period when religious observances were banned. For example, when he became president in 2000, former Soviet-era KGB agent Vladimir Putin professed his Orthodox beliefs and claimed that his mother baptized him during the height of Soviet repression. In turn, the Church has supported recent Russian political campaigns, both domestically and internationally.

Other forms of Western Christianity also are present in the region. For example, the people of western Ukraine, who experienced several hundred years of Polish rule, eventually joined the Catholic Church. Eastern Ukraine, on the other hand, remained fully within the Orthodox framework. This religious split reinforces the cultural differences between eastern and western Ukrainians. Western Ukrainians have generally been far more nationalistic, hence, more firmly opposed to Russian influence than eastern Ukrainians.

Elsewhere, Christianity came early to the Caucasus, but modern Armenian forms—their roots dating to the fourth century CE—differ slightly from both Eastern Orthodox and Catholic traditions. Georgian Christianity, however, is more closely tied to the Orthodox faith. Evangelical Protestantism has also been on the rise since the demise of the Soviet Union.

Figure 9.25 Minority Evenki
Russia's indigenous Evenki population speaks an Altaic language and they share many of the economic and social problems of native peoples in North America and Australia. *(Dean Conger/Corbis)*

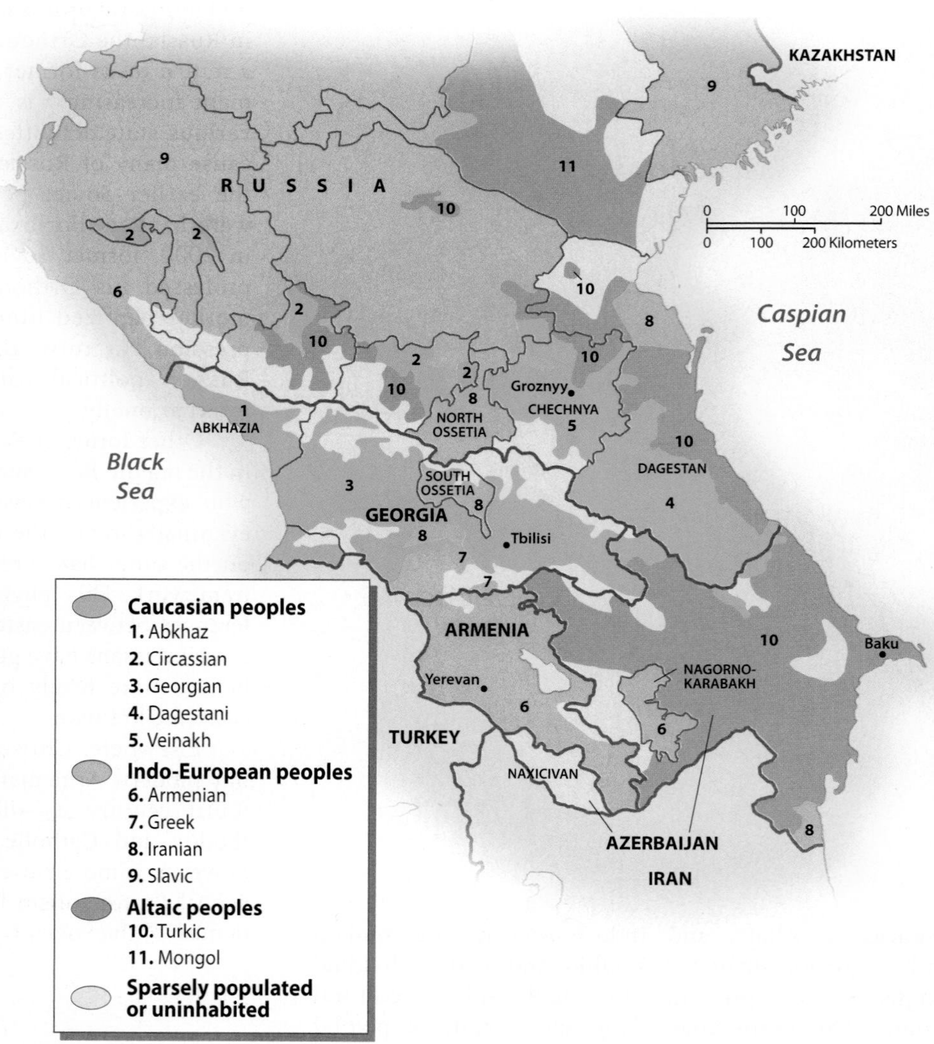

Figure 9.26 Languages of the Caucasus Region
A complicated mosaic of Caucasian, Indo-European, and Altaic languages characterizes the Caucasus region of southern Russia and nearby Georgia and Armenia. Persisting political problems have erupted in the region as local populations struggle for more autonomy. Recent examples include independence movements in Chechnya and in nearby Dagestan. *(Modified from Martin Glassner,* Political Geography, *2nd ed., 1996, p. 599. Reprinted by permission of John Wiley & Sons, Inc.)*

Non-Christian Traditions Non-Christian religions also appear and, along with language, shape ethnic identities and tensions within the region. Islam is the largest non-Christian religion. In Russia, there are some 7,000 mosques and approximately 20 million adherents (Figure 9.28). Most are Sunni Muslims, and they include peoples in the North Caucasus, the Volga Tatars, and Central Asian peoples near the Kazakhstan border. Growth rates among Russia's Muslim population are three times that of the non-Muslim population. In addition, Moscow's economic boom has attracted many Muslim immigrants. Recent estimates, including illegal immigrants, suggest that 20 percent of Moscow's population is now Islamic. To date, Islamic fundamentalism has not become a major cultural or political issue in the region, although Russian military involvement in Muslim-dominated regions such as Chechnya has at times had cultural and religious overtones. As in many other parts of the world, a growing Islamic political consciousness is present, particularly in Russia.

Russia, Belarus, and Ukraine are also home to more than 1 million Jews, who are especially numerous in the larger cities of the European West. Jews suffered severe persecution under both the tsars and the communists. Recent out-migrations, prompted by new political freedoms, have further reduced their numbers in Russia, Belarus, and Ukraine. Buddhists also are represented in the region, associated with the Kalmyk and Buryat peoples of the Russian interior. Indeed, Buddhism has witnessed a recent renaissance and now claims approximately 1 million practitioners, mostly in Asiatic Russia.

Russian Culture in Global Context

Russian culture has interacted in varied ways with the world beyond. Russian cultural norms have for centuries embodied both an inward orientation toward traditional forms of expression and an outward orientation directed primarily to western Europe. By the nineteenth century, even as Russian peasants interacted rarely with the outside world, Russian high culture had become thoroughly Westernized, and Russian composers, novelists, and playwrights gained considerable fame in Europe and the United States. By the turn of the twentieth century, Russian avant-garde artists were among the international leaders in devising the experimental styles of modernism.

Soviet Days During the Soviet period, a new mixture of cultural relationships unfolded within the socialist state. Initially, European-style modern art flourished in the Soviet Union, encouraged by the radical ideas of the new rulers. By the late 1920s, however, Soviet leaders turned against modernism, which they viewed as the decadent expression of a declining capitalist world. Many Soviet artists fled to the West, and others were exiled to Siberian labor camps. Increasingly, state-sponsored Soviet artistic productions centered on **socialist realism**, a style devoted to the realistic depiction of workers harnessing the forces of nature or struggling against capitalism. Still, traditional high arts, such as classical music and ballet, continued to receive lavish state subsidies, and to this day Russian artists regularly achieve worldwide fame.

Turn to the West By the 1980s, it was clear that the attempt to fashion a new Soviet culture based on socialist realism and working-class solidarity had failed. The younger generation instead adopted a rebellious stance, turning for inspiration to fashion and rock music from the West. The mass-consumer culture of the United States proved immensely popular, symbolized above all by brand-name chewing gum, jeans, and cigarettes. The Soviet government attempted to ward off this perceived cultural onslaught, but with little success. Soviet officials could more easily censor books and other forms of written expression, but even here they were increasingly frustrated. By the end of the Soviet period, the persistent secrecy that separated the Soviet Union from the West had broken down, thus enabling more people and influence to flow into the region.

After the fall of the Soviet Union in 1991, basic freedoms brought an inrush of global cultural influences, particularly to the region's larger urban areas such as Moscow. Shops were quickly flooded with Western books and magazines; people pondered the financial mysteries of home mortgages and condominium purchases; and they reveled in the new-found pleasures of fake Chanel handbags and McDonald's hamburgers. English-language classes became even more popular in cities such as Moscow, where Russians hurried to embrace the world their former leaders had warned them about for generations. Cultural influences streaming into the country were not all Western in inspiration. Films from Hong Kong and Mumbai (Bombay), as well as the televised romance novels *(telenovelas)* of Latin America, for example, proved far

Figure 9.27 Russian Orthodox Church, Siberia
A revival of interest in the Russian Orthodox Church followed the collapse of the Soviet Union in the early 1990s. The newly built Znamensky Cathedral in Kemerovo displays many of the faith's characteristic cultural landscape signatures. *(Novosti/Sovfoto/Eastfoto)*

Figure 9.28 Moscow Mosque
Moscow's growing immigrant population has included many Muslims from portions of the former Soviet Union, particularly from Central Asia. Today, their presence in the capital is an increasingly visible element in the city's cultural landscape. *(AP/Wide World Photo)*

more popular in the Russian domain than in the United States. This onrush of global cultural influences, however, has not equally spread throughout the region. While urban residents have access and money to explore such options, rural life across the Russian domain remains far more wedded to traditional cultural institutions and values.

The Music Scene Younger residents of the region also have embraced the world of popular music, and their enthusiasm for American and European performers, as well as their support of a budding home-grown music industry, symbolizes the changing values of an increasingly post-Soviet generation. MTV Russia went on the air at midnight on September 26, 1998. A year later, the network sponsored a huge open-air concert on Red Square. Headlined by the Red Hot Chili Peppers, the event was a far cry from Soviet-era tank parades in front of Communist Party officials. By 2005, Russian MTV reached more than 60 million viewers, particularly in larger urban areas. Sony Music Entertainment also established its Russian operations in December 1999. It has been followed by several of the world's other major recording companies. Sony, BMG, and other labels have signed multiple Russian acts for domestic markets, helping boost a native pop-music culture. Universal also has opened the way for Russian performers to go global. For example, it sponsored the English-language debut of Russian pop singer Alsou in 2001. Three years later, Ukrainian singer Ruslana won the coveted Eurovision 2004 award for her song "Wild Dance" (Figure 9.29). In a superb example of cultural globalization, the high-tech performance of the song at the awards show in Istanbul, Turkey, was seen by more than 100 million television viewers worldwide, and it featured a mix of Ukrainian folk music traditions, leather-clad female dancers, and on-stage acrobatics.

A Revival of Russian Nationalism Even as the Russian domain finds itself increasingly swept up in external cultural influences, countertrends that emphasize Russian nationalism persist. Extremely conservative nationalists resist foreign influences such as Western-style music, and they often have been supported by more traditional elements of the Russian Orthodox Church. Even President Putin and the Russian government have carefully resurrected numerous symbols from the Russian past in order to cultivate a renewed sense of Russian identity within the region. Recently, Russian leaders proposed a return to the tsarist-era coat of arms (the double-headed eagle) to symbolize the ongoing connections between Russian society today and the earlier culture of Dostoyevsky and Pushkin. Even the glory of Soviet communist days (1917–91) has been revived, with a strong Russian tone. The Soviet-era national anthem was reintroduced (with new lyrics) in 2001, and a number of Soviet holidays, celebrations, and national awards have been reinstated.

Figure 9.29 Ruslana
For her song "Wild Dance," Ukrainian singer Ruslana won top honors at the 2004 Eurovision Awards in Turkey. *(Ethan Miller/Reuters/Corbis/Bettmann)*

GEOPOLITICAL FRAMEWORK: Transformation of a Global Superpower

The geopolitical legacy of the former Soviet Union still weighs profoundly upon the Russian domain. After all, the bold lettering of the "Union of Soviet Socialist Republics" dominated the Eurasian map for much of the twentieth century, and the country's global political reach left no corner of the world untouched. Former Soviet republics still struggle to define new geopolitical identities for themselves. Neighboring states continue to view the region with nervousness left over from the days of Soviet political and military power. Present demands for more local and regional political control within countries such as Russia, Ukraine, and Georgia can still be understood in the context of the Soviet era, when a highly centralized government gave little voice to regional dissent. At the same time, Russia's new global visibility and its desire to recentralize authority in Moscow have raised concerns elsewhere within the domain, in Europe, and in the United States.

Geopolitical Structure of the Former Soviet Union

The Soviet Union rose from the ashes of the Russian Empire, which collapsed abruptly in 1917. The ultraconservative policies of the Russian tsar generated opposition among businesspeople and workers, while the peasants, who formed the majority of the population, had always resisted the powerful land-owning aristocracy. After the fall of the tsar and the aristocracy, a broad-based coalition government assumed authority. Several months later, however, the **Bolsheviks**, a faction of Russian communists representing the interests of the industrial workers, seized power within the country. The leader of these Russian communists was Vladimir Ilyich Ulyanov, usually known by his self-selected name, Lenin. Lenin was a key architect of the Soviet Union, which fully emerged between 1917 and 1922. It became the Russian Empire's successor state.

The new socialist state reconfigured Eurasian political geography. Although it resembled the territorial contours of the Russian Empire and centralized authority continued to be concentrated in European Russia, the political and economic structure of the country was radically transformed. Lenin and the other communist leaders were aware that they faced a major challenge in organizing the new state.

The Soviet Republics and Autonomous Areas Soviet leaders designed a geopolitical solution that maintained their country's territorial boundaries and acknowledged, at least theoretically, the rights of its non-Russian citizens. Each major nationality was to receive its own "union republic," provided it was situated on one of the nation's external borders (Figure 9.30). Eventually 15 such republics were established, thus creating the Soviet Union. A number of these republics were quite small, while the massive Russian Republic sprawled over roughly three-quarters of the Soviet terrain. Each republic was to have considerable political autonomy, vested with the right to withdraw from the union if it so desired. In practice, however, the Soviet Union remained a centralized state, with important decisions made in the capital of Moscow.

The Soviets came up with different geopolitical solutions to acknowledge smaller ethnic groups and nationalities that were not situated on the country's external borders. Indeed, dozens of significant minority groups pressed for recognition. One solution to this problem was the creation of **autonomous areas** of varying sizes that recognized special ethnic homelands but did so within the structure of existing republics. Thus, within the Russian Republic, the larger nationalities, such as the Yakut and the Volga Tatars, were granted their own "autonomous republics" (not to be confused with the 15 higher-level union republics, such as Russia or Ukraine). The system's main deficiency was that the autonomy it was designed to provide proved to be more of a charade than a reality.

Centralization and Expansion of the Soviet State In the early Soviet era, it appeared that the framework of separate republics and autonomous areas might allow non-Russian peoples to protect their own cultures and establish their own social and economic policies (provided, of course, that such policies embodied Marxist principles). From the beginning, however, it was clear that such self-determination would be a temporary measure. According to the official ideology of Marxist beliefs, the gradual development of a communist society would see the withering away of all significant ethnic differences and the disappearance of religion. In the future, a new classless Soviet society was supposed to emerge.

By the 1930s, it was clear that national autonomy would not have any real significance within an increasingly centralized Soviet state. The chief architect of this political consolidation was Soviet leader Joseph Stalin, who did everything he could to centralize power in Moscow and to assert Russian authority. Stalin launched a ruthless plan of state-controlled agricultural production and industrialization. Although many people initially resisted his policies, Stalin never hesitated to use force to bring about his vision of a purer socialist revolution within the Soviet Union. Stalin also utilized more subtle methods of control in his quest to produce a new communist society. The Soviet landscape itself came to reflect communist ideas. Soviet-style architecture, paintings, cinema, and many other visual elements in the everyday world came to symbolize the new world view.

The Stalin period also saw the geographical enlargement of the Soviet Union. As a victorious power in World War II, the country acquired southern Sakhalin and the Kuril Islands from Japan. It regained the Baltic republics (Lithuania, Latvia, and Estonia—independent between 1917 and 1940), as well as substantial territories formerly belonging to Poland, Romania, and Czechoslovakia.

After World War II, the Soviet Union also gained significant authority, although not actual sovereignty, over a broad swath of eastern Europe. As they pushed the German army west toward the end of the war, Soviet troops advanced across much of the region, actively working to establish communist regimes thereafter. In the words of British leader Winston Churchill, the Soviets extended an **"Iron Curtain"** between their eastern European allies and the more democratic nations of western Europe by restricting the free flow of people and information.

As eastern Europe disappeared behind the Iron Curtain, the Soviet Union and the United States, while

Figure 9.30 Soviet Geopolitical System
During the Soviet period, the boundaries of the country's 15 internal republics often reflected major ethnic divisions. Ultimately, however, many of the ethnically non-Russian republics pressured the Soviet government for more political power. As the Soviet Empire disintegrated, the former republics became politically independent states and now form an uneasy ring of satellite nations around Russia. *(Modified from Rubenstein, 1999,* Introduction to Human Geography, *Upper Saddle River, NJ: Prentice Hall)*

allies during World War II, became antagonists in a global **Cold War** of escalating military competition that lasted from 1948 to 1991. In its post–World War II heyday, the Soviet Union became a global superpower, one of only two countries equipped with enough nuclear weapons to ensure global destruction. At the height of its power in the late 1970s, it enjoyed close military and economic alliances not only with eastern Europe, but also with certain countries in Asia (Mongolia, North Korea, Vietnam, Laos, and Cambodia), the Caribbean region (Cuba and Nicaragua), and Africa (Angola, Somalia, and several others).

End of the Soviet System Ironically, Lenin's system of culturally defined republics helped sow the seeds of the Soviet Union's downfall. Even though the nationally based republics and autonomous areas of the Soviet Union were never allowed real freedom, they did provide a lasting political framework for the maintenance of distinct cultural identities. Indeed, contrary to the expectations of Soviet leaders, ethnic nationalism intensified in the post–World War II era as the Soviet system grew less repressive. When Soviet president Mikhail Gorbachev initiated his policy of **glasnost**, or greater openness, during the 1980s, several republics—most notably the Baltic states of Lithuania, Latvia, and Estonia—demanded outright independence.

Other forces also worked toward the political end of the Soviet regime. A failed war in Afghanistan in the early 1980s frustrated both the Soviet leaders and population. In eastern Europe, multiple protests over Soviet dominance emerged after World War II from Hungary, Czechoslovakia, and Poland. Worsening domestic economic conditions, increasing food shortages, and a declining quality of life led to fundamental questions concerning the value of centralized planning within the country. In response, President Gorbachev introduced **perestroika**, or planned economic restructuring, aimed at making production more efficient and more responsive to the needs of Soviet citizens. In 1991, however, Gorbachev saw his authority slip away amid rising pressures for political decentralization and more dramatic economic reforms. During the summer, Gorbachev's regime was further imperiled by the popular election of reform-minded Boris Yeltsin as the head of the Russian Republic and by a failed military coup by commu-

nist hard-liners. By late December, all of the country's 15 constituent republics had become independent states, and the Soviet Union ceased to exist.

Current Geopolitical Setting

Post-Soviet Russia and the nearby independent republics have radically rearranged their political relationships since the collapse of the Soviet Union in 1991. All of the former republics still struggle to establish stable political relations with their neighbors, and tensions within Russia continue as recent leaders have pushed for more centralized control and more limits on media independence and personal liberties (Figure 9.31). For a time it seemed that a looser political union of most of the former republics, called the **Commonwealth of Independent States (CIS)**, would emerge from the ruins of the Soviet Union. All the former republics, with the exception of the three Baltic states, joined the CIS soon after the dismemberment of the old union (see Figure 9.31). By the early twenty-first century, however, the CIS had developed into little more than a forum for discussion, without real economic or political power. That situation may change in the future, however, as a reinvigorated Russian state reasserts its broader regional influence.

Russia continues to maintain a military presence in many of the former Soviet republics. **Denuclearization**, the return of nuclear weapons from outlying republics to Russian control and their partial dismantling, was completed during the 1990s. The Soviet-era nuclear arsenals of Kazakhstan, Ukraine, and Belarus were removed in the process. Elsewhere, Tajikistan invited the Russian army in during the early 1990s to quell its own internal ethnic struggles, and Armenia has maintained close military ties with Moscow. Russia also maintains naval bases in Ukraine's Crimea region (Figure 9.32). The Soviet-era Black Sea fleet was traditionally based in the Crimean port of Sevastopol, now in Ukraine. After lengthy negotiations, Russia and Ukraine agreed in 1997 to share the naval base. Russia retains some 80 vessels and 15,000 service members in Ukraine. But the longer-term political sovereignty of the entire Crimean Peninsula remains in doubt. The peninsula has been contested real estate within the Russian domain for centuries. Some Ukrainians would like nothing better than to see the Russians leave. Many ethnic Russians living in the area, however, see the continuing Russian military presence as protecting their long-term interests.

Geopolitics in the South and West While Russia shares many cultural ties with Belarus and Ukraine, growing political tensions have threatened recent relationships between these two countries. Leaders in Belarus have been extremely slow to open up their economy to outside investment. They have also resisted too much control from nearby Russia. The two countries quarreled recently over the price of natural gas that Russia has sold to Belarus. The Russians have also accused their western neighbor of stealing oil from pipelines that cross Belarus on their way to western Europe. While the two nations have pledged a political union sometime in the future, it appears unlikely that such a harmonious marriage will occur anytime soon.

Russia's rocky relations with Ukraine have also transformed the geopolitical landscape in the western portion of the domain. Russia and Ukraine signed an agreement in 1998 to develop closer trade relations and foster economic linkages. Indeed, Russian investments in Ukraine have grown sharply since 2000. Still, tensions between the two countries have increased. Ukraine remains highly dependent upon Russian energy supplies. In addition, major Russian pipelines to central and western Europe must pass through Ukraine. The result is that the two nations have often wrangled over energy prices and the availability of oil and natural gas supplies.

Ukraine's internal politics have also riled Russia. In 2004, Viktor Yushchenko, a reformist leader and Ukrainian nationalist, came to power after a controversial election that even included an attempt to poison the winning candidate. Yushchenko often criticized Russian leaders and excessive Russian interference with Ukraine's affairs. On the other hand, Viktor Yanukovich, the opposition candidate, received the political blessings of Russian president Putin. Not surprisingly, Yanukovich also received wide support from the ethnic Russian areas in the southern and eastern portions of Ukraine. Since 2004, several

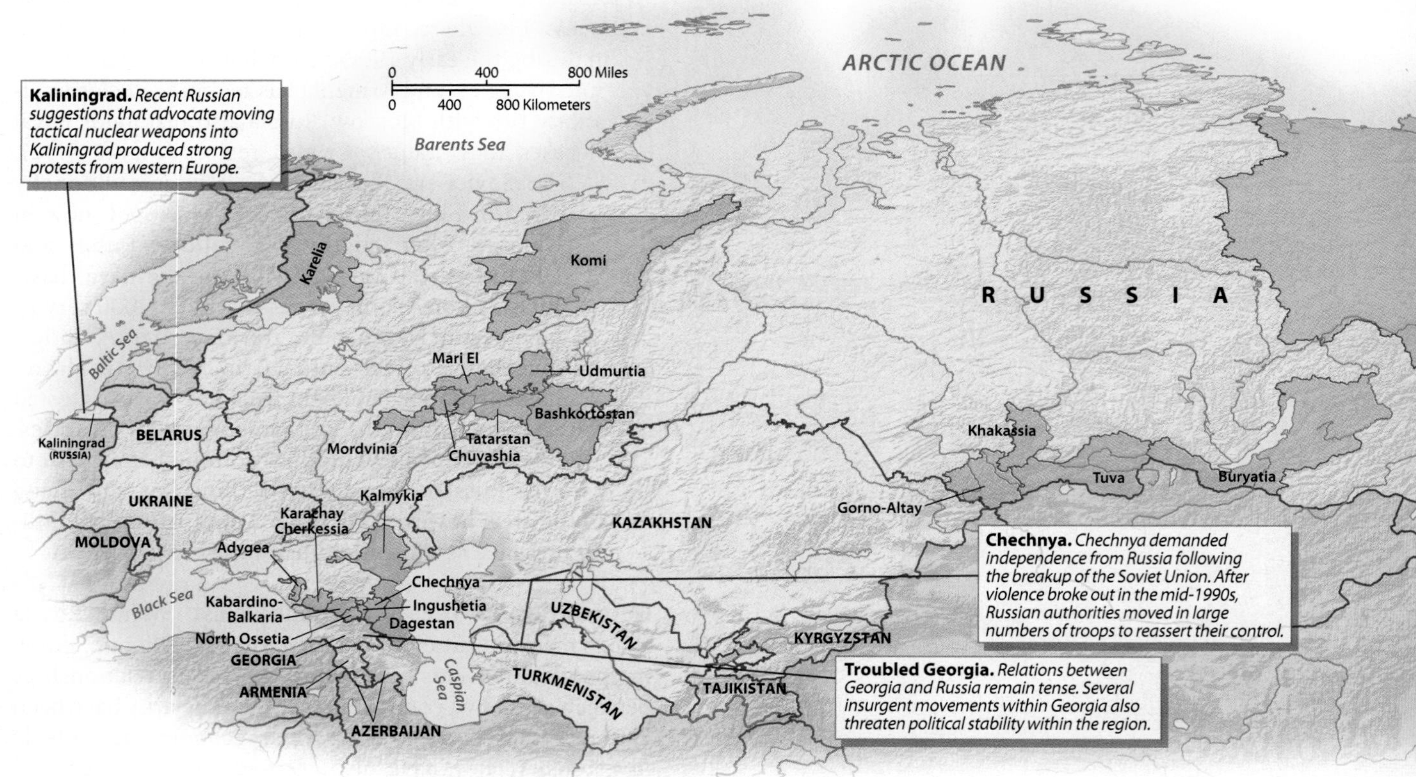

Figure 9.31 Geopolitical Issues in the Russian Domain
The Russian Federation Treaty of 1992 created a new internal political framework that acknowledged many of the country's ethnic minorities. Recently, however, Russian authorities have moved to centralize power and limit regional dissent. Russia's relations with several nearby states remain strained. *(Modified from Bergman and Renwick, 1999,* Introduction to Geography, *Upper Saddle River, NJ: Prentice Hall)*

major political upheavals have gripped the country, including one period where Yushchenko served as Ukraine's president and Yanukovich served as its prime minister. It seems likely that neighboring Russia will continue to watch its neighbor closely and attempt to influence Ukrainian politics in any way it can to serve its own political and economic interests.

Tiny Moldova also has witnessed political tensions in the post-Soviet era. Conflict has repeatedly flared in the Transdniester region in the eastern part of the country where Slavic separatists have pushed for independence from a central government dominated by Romanian-speaking Moldavians. Russia has encouraged the case for "independence," seeing it as a way to increase their influence in the region. On the other hand, a growing number of Romanians have suggested that Moldavians join with Romania and dump troublesome Transdniester in the process. Meanwhile, Russian troops remain in the country.

Transcaucasia also remains unstable. Georgia experienced a bloodless coup in 2003. Since then, the Georgian government has moved toward closer ties with the United States, even suggesting that it might join NATO. This has greatly annoyed the Russians. In addition, Georgia's arrest of several Russian officers suspected of being spies contributed to escalating tensions between the two nations. Recently, Russia severed all trade and transportation links between the two countries. North Georgia remains particularly unstable. The Abkhazia and South Ossetia regions that border Russia have threatened to secede and have had considerable covert support from their Russian neighbors. Strategic oil and gas pipelines also are being completed through the region (from Caspian Sea fields), and this will surely increase Georgia's geopolitical importance in the coming decades.

In nearby Armenia, the territories of the Christian Armenians and the Muslim Azeris interpenetrate one other in a complex fashion. The far southwestern portion of Azerbaijan (Naxicivan) is actually separated from the rest of the country by Armenia, while the important Armenian-speaking district of Nagorno-Karabakh is officially an autonomous portion of Azerbaijan. Almost immediately after independence in 1991, Armenia and Azerbaijan went to war over these territories. After Armenia successfully occupied much of Nagorno-Karabakh in 1994, the fighting between the countries diminished. No final peace treaty has been signed, however, and Azerbaijan demands the return of

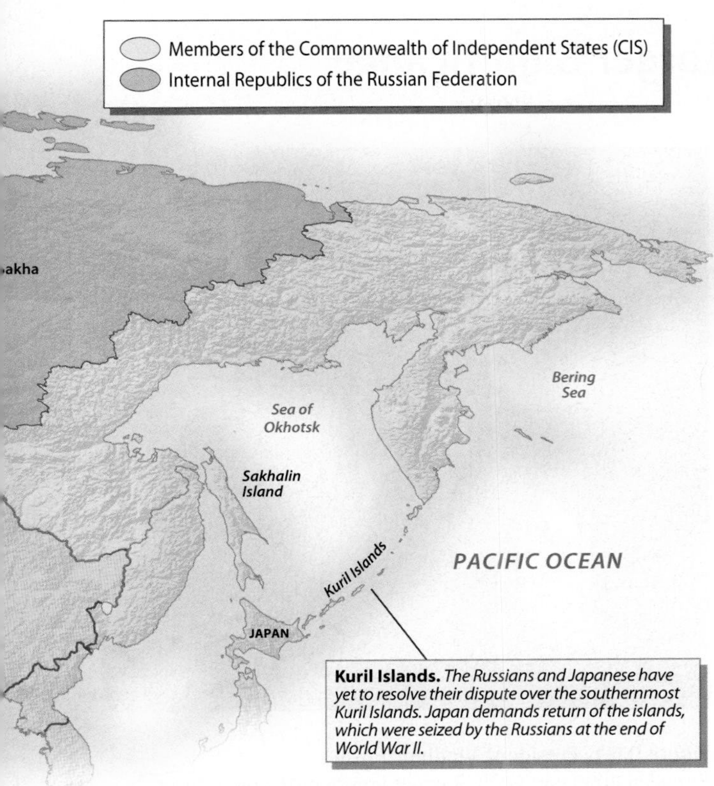

Kuril Islands. *The Russians and Japanese have yet to resolve their dispute over the southernmost Kuril Islands. Japan demands return of the islands, which were seized by the Russians at the end of World War II.*

the territory. Meanwhile, Russian political and economic power in Armenia is increasing and the tiny Transcaucasian country seems likely to remain within the influence of its giant neighbor.

Geopolitics Within Russia Within Russia, further pressures for devolution, or more localized political control, produced the March 1992 signing of the Russian Federation Treaty. The treaty granted Russia's internal autonomous republics and its lesser administrative units greater political, economic, and cultural freedoms, including more control of their natural resources and foreign trade. Conversely, it weakened Moscow's centralized authority to collect taxes and to shape policies within its varied hinterlands. Defined essentially along ethnic lines, 21 regions possess status as republics within the federation and now have constitutions that often run counter to national mandates. Scattered from central Siberia to the north-facing slopes of the Caucasus, these republics reflect much of the linguistic and religious diversity of the nation (see Figure 9.31).

But Russian president Putin and others in Moscow argue that devolution comes with a price. They fear that Russia could fragment, resulting in further political instability. Indeed, many of the country's local and regional governments have been plagued by incompetence and widespread corruption. Political fragmentation makes it more difficult to introduce national reforms, to make large-scale investments in infrastructure, and to engage in comprehensive long-term planning. Leaders in Moscow certainly used their more centralized power to quell the recent instability in the Russian republic of Chechnya (See "Geography in the Making: Chechnya's Larger Significance"). In addition, advocates argue that a more centralized state can deal more effectively with global

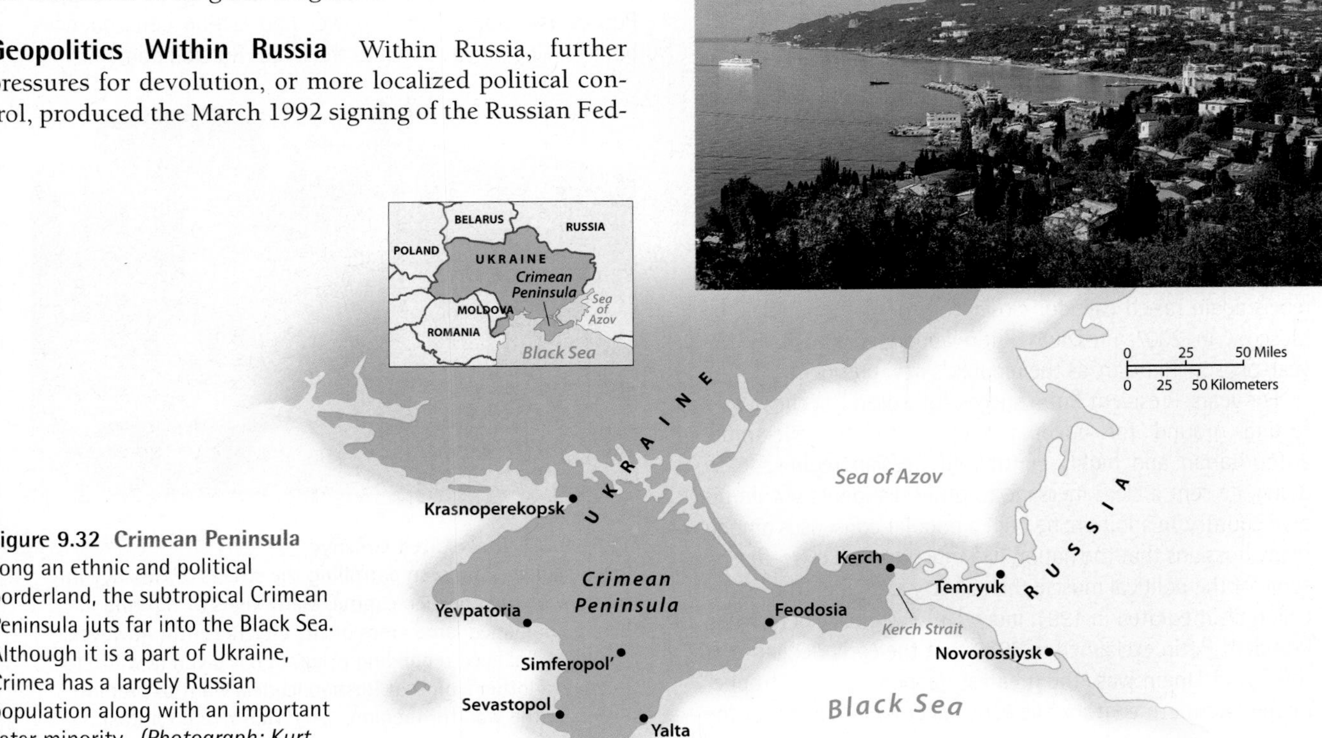

Figure 9.32 Crimean Peninsula Long an ethnic and political borderland, the subtropical Crimean Peninsula juts far into the Black Sea. Although it is a part of Ukraine, Crimea has a largely Russian population along with an important Tatar minority. *(Photograph: Kurt Scholz/SuperStock, Inc.)*

GEOGRAPHY IN THE MAKING

Chechnya's Larger Significance

Since the disintegration of the Soviet Union, Russia's Caucasus Mountains region has been a particularly explosive setting for internal political tensions and rapid change (see Figure 9.26). The area has also been an important geopolitical laboratory in which Russian leaders have learned how to exercise greater military and political authority.

In 1994, leaders of the internal Chechnyan Republic vowed to establish a genuinely independent state. At the time, they challenged Russian president Yeltsin who was struggling with many competing forces within the fledgling Russian republic. Russian troops were sent in to quiet the rebellion, but many insurgents simply retreated to the rugged mountains in the southern portion of the republic. The rebels increased their armed attacks in the late 1990s.

As Russian president Vladimir Putin came to power (acting prime minister in 1999; president between 2000 and 2008), he launched a massive military invasion of Chechnya (Figure 9.5.1). The Russian army took firm control of Chechnya's capital of Groznyy, located in the plains just to the north of the Caucasus (Figure 9.5.2). But the Russian military force of 100,000 soldiers did more than quiet the rebels. It also helped Putin secure domestic support for his policies from the Russian people. After the terrorist attacks upon the United States in 2001, Putin once again played the Chechen card, citing their large, insurgent Muslim populations as potentially dangerous "terrorists" within his own country. Indeed, some Chechen extremists participated in violence against Russians, making a crackdown politically popular, not only in Russia but also in the eyes of Western political leaders.

Since their return, the Russians have been accused of looting Chechnyan natural resources (especially metals and oil) and trying to grab a share of the lucrative global drug trade that flows through the remote mountainous region. Meanwhile, about half the republic's population fled the fighting, causing massive refugee movements and dislocations in neighboring areas, particularly to nearby Ingushetia. The 2004 assassination of Chechen president Kadyrov by Chechen rebels again raised calls for closer control of the republic by Moscow. In 2007, President Putin appointed Kadyrov's 30-year-old son, Ramzan, as the republic's new president.

For years, President Putin successfully used Chechnya as a testing ground for building support for his ever-more authoritarian and highly centralized Russian regime. In so doing, he sent a clear message to other dissidents within his own country. In addition, he built a popular consensus among many Russians that the nation had much to gain by re-flexing some of the political muscle the country lost when the Soviet Union disintegrated in 1991. Indeed, it was no accident when President Putin exclaimed in 2005 that the earlier collapse of the Soviet Union was "the greatest geopolitical catastrophe" of the twentieth century. His early successes in crushing the Chechen rebels proved pivotal in reversing that collapse. Putin's assertion of state power had lasting consequences, both for Chechnya as well as the larger Russian domain.

Figure 9.5.1 President Vladimir Putin
Between 2000 and 2008, Russian president Vladimir Putin worked to centralize political power within Russia and rebuild Russia's political and economic presence around the world. His authoritarian approach was exemplified by his dealings with the insurgent Russian republic of Chechnya. *(Sergei Chirikov/epa/Corbis)*

Figure 9.5.2 Devastated Groznyy
Russian soldiers are seen patrolling the streets of Groznyy in this view of the Chechen capital. Many years of fighting in the area damaged large areas of the city. Recently, however, a Russian-financed rebuilding program has given new life to Groznyy, another sign that Russian leaders want to promote success in the war-torn region. *(AP/Wide World Photos)*

political and economic affairs. Since 2002, these forces of centralization have gathered strength, particularly under President Putin's watchful eye.

Russian Challenge to Civil Liberties Citizens of the Russian Republic enjoyed a genuine flowering of democratic freedoms during the initial post-Soviet years. A multiparty political system, independent media, and a growing array of locally and regionally elected political officials all signaled real change from the authoritarian legacy of the Soviet period.

Since 2002, however, many of these hard-won civil liberties have slipped away, victims of President Putin's campaign to consolidate political power, increase the authority of the central government, and silence critics who disagreed with his policies. Many of these losses came amidst President Putin's own declared "war on terror" within Russia and the increasingly antagonistic rhetoric he has leveled against the international community, including the United States. While some Russians vehemently protested these losses of political liberties as well as human rights abuses, many Russians argue that such adjustments are a tolerable price to be paid for a much improved economy, a better standard of living, and the country's newly elevated prestige on the global geopolitical stage.

Several events illustrate the changes afoot. First, a bill signed by President Putin in 2004 eliminated the popular election of dozens of Russian governorships and mayoral positions. Now, the Russian president nominates these officials, almost always receiving approval from weakened regional legislatures and municipal governments. Second, many of the country's media outlets—both television and print—have returned to state control or are now controlled by state-run companies such as Gazprom. This has silenced many critics of the government. Third, some of the country's most outspoken critics have simply been taken prisoner or mysteriously murdered. In 2003, the head of one of Russia's largest oil companies, Mikhail Khodorkovsky, was imprisoned for alleged tax violations. The assets of his company, Yukos Oil, were later auctioned off with most of the energy reserves ending up in the hands of state-controlled corporations. Three years later, one of Russia's leading journalists, Anna Politkovskaya, was murdered outside her apartment. Politkovskaya was a highly visible critic of President Putin, the central government, and Russia's harsh treatment of Chechnya. She is 1 of 13 journalists who were murdered or died under suspicious circumstances between 1999 and 2007. More print and television journalists have quit their jobs or been fired from key media positions. Recently, Russian officials have also increased their surveillance and regulation of the Internet, the latest chapter in a move to silence opposition. One study of global media freedoms in 2007 ranked Russia alongside regimes such as Burma and North Korea when it came to its increasingly repressive policies.

The Shifting Global Setting

Historically, much of Russia's larger geopolitical history has been focused on its creation of a land-based empire across much of Eurasia. As British geographer Halford Mackinder argued in 1904, Russia's control of the heartland, the huge interior land mass of Eurasia, gave it a global advantage over maritime powers such as Great Britain. The collapse of the Soviet Union, however, resulted in a simultaneous crash of Russian global power. Russia, with only a little more than half of the population of the former Soviet Union and with a teetering economy, could no longer afford to support and subsidize allies such as Cuba. In retrospect, Mackinder's thesis on the dominance of the heartland no longer seems supportable. The Russian example suggests that power in the modern world flows much more from economic productivity and technological prowess than from the mere control of continental territory.

Regional Tensions Since the fall of the Soviet Union, regional political tensions continue to challenge the Russians, in both the east and west. In East Asia, the boundary between Russia and China was imposed by the Russian Empire in 1858 and has never been fully accepted by Beijing. The two countries battled over the area in 1969, but since then relations have improved. The leaders of both Russia and China now stress the need for cooperation, but the potential for renewed conflict remains. The Chinese want Russian oil, gas, and lumber and Chinese workers are increasingly slipping across the border to work in the Russian Far East.

Territorial disagreements also complicate Russia's relationship with Japan. Japan has long demanded the return of the four southernmost islands of the Kuril Archipelago seized by the Soviet Union in 1945. Although Russia could gain major financial benefits in any agreement, it has refused to return any territory toJapan. Growing economic ties between the two countries, however, are contributing to better political relationships.

To the west, Russia worries about the expansion of the North Atlantic Treaty Organization (NATO). Although most Russian leaders accepted the inevitable inclusion of Poland, Hungary, and the Czech Republic into NATO, they strongly opposed the recent addition of the Baltic republics (Estonia, Latvia, and Lithuania) to the increasingly powerful organization. In fact, the move provoked some Russian military leaders to recommend new troop and weapon deployments on the country's western borders. The plight of Russian Kaliningrad is a related sore point (Figure 9.33). After World War II, the former Soviet Union added a small but strategic territory on the Baltic Sea. It was the northern portion of East Prussia (now the port of Kaliningrad), previously part of Germany. It still forms a small but highly strategic Russian **exclave**, which is defined as a portion of a country's territory that lies outside its contiguous land area. Kaliningrad (population 950,000) remains a part of the Russian

Figure 9.33 Kaliningrad Kaliningrad, a Russian exclave on the Baltic Sea, is now surrounded by Poland and Lithuania, both EU- and NATO-member states. The multistory Hotel Kaliningrad (right) borders the Lenin Prospekt in this downtown view of the Russian exclave. *(Tom Schulze/Sovfoto/Eastfoto)*

Federation, but since 2004, neighboring Lithuania and Poland have both joined NATO and the European Union (EU). Russian officials fear that being in the midst of NATO and EU nations may encourage some residents of Kaliningrad to think about independence. In response, the Russians have beefed up their military presence in the exclave, making it clear that they have no intention of giving up one of their valuable port cities on the Baltic Sea.

A Growing Global Presence Today, Russian leaders appear determined to reassert the nation's global political status. Russian officials have increasingly expressed their displeasure at the notion of seeing the United States as the world's sole superpower. They have also bristled at NATO's expansion into what they consider their own "post-Soviet sphere of influence," even accusing the alliance of sparking a new arms race in Europe. The country's nuclear arsenal, while reduced in size, remains a powerful counterpoint to American, European, and Chinese interests. Russia also retains a permanent seat on the United Nations Security Council, arguably the world's most important geopolitical body. Its inclusion in the G-8 economic meetings signifies its enduring international clout. High energy prices have recently increased Russia's economic importance on the world stage.

And despite their nominal independence from direct Russian control, the other nations within the Russian domain clearly feel the presence of their mammoth neighbor. Indeed, the long geopolitical history of the domain suggests that Russia's recent reemergence as a more powerful, centralized state upon the global stage is a sign of things to come.

ECONOMIC AND SOCIAL DEVELOPMENT: An Era of Ongoing Adjustment

The economic future of the Russian domain remains difficult to predict. Economic declines devastated the entire region for much of the 1990s following the collapse of the Soviet Union. It marked the most abrupt economic contraction within the industrialized world since the Great Depression of the 1930s. Since 2002, however, greater foreign investment and better economic conditions have brought improvement to much of the region (Table 9.2). Russia in particular, has benefited tremendously from its natural gas and oil exports, particularly in an era of higher world energy prices and strengthening global demand.

The true economic potential of the Russian domain always has been difficult to gauge. Optimists point to the vast size, abundant natural resources, and well-educated, urbanized populations of the region as significant assets. Indeed, in its heyday, the Soviet Union rose to become one of the great industrial powers in the world, and it did so in a remarkably short period of time.

Skeptics note that size also brings its disadvantages, particularly by raising transportation costs within the region's economy. They also point out that Russia's northern

TABLE 9.2 Development Indicators

Country	GNI[a] per Capita, PPP[b]	GDP[c] Average Annual % Growth (2000–05)	Life Expectancy	Percent of Population Living on Less Than $2 a Day	Under Age 5 Mortality Rate 1990	Under Age 5 Mortality Rate 2005	Gender Equity[d]
Armenia	5,060	12.4	71	31	54	29	108
Belarus	7,890	7.5	70	2	19	12	105
Georgia	3,270	7.4	73	25	47	45	103
Moldova	2,150	7.1	69	64	35	16	109
Russia	10,640	6.2	65	12	27	18	110
Ukraine	6,720	8.0	68	5	26	17	102

[a]*Gross national income.*
[b]*Purchasing power parity.*
[c]*Gross domestic product.*
[d]*Ratio of female-to-male enrollments in primary and secondary school percentage. Numbers below 100 have more males in primary/secondary school; numbers above 100 have more females in primary/secondary schools.*

Source: World Bank, World Development Indicators, 2007; *percentage of population living on $2 a day data from* Population Reference Bureau, World Data Sheet, 2007; *gender equity data from* Millennium Development Goals.

location always has made food production problematic. Most notably, since the breakup of the Soviet Union, most economies within the region have struggled to evolve in a stable and predictable fashion toward greater productivity and output. Thus, the region's future path to prosperity remains clouded as it spends the early years of the twenty-first century reconfiguring an economy that was created largely during the Soviet period.

The Legacy of the Soviet Economy

The birth of the Soviet Union in 1917 initiated a radical change within the region's economy. Under the Russian Empire, most people were peasants, farming the land much as they had done for centuries. Following the revolution, however, the Soviet Union quickly emerged to rival, and even surpass, many of the most powerful economies on Earth. During that era of unparalleled growth, much of the region's present economic infrastructure was established, including new urban centers and industrial developments, as well as a modern network of transportation and communication linkages. Thus, even though the Soviet Union has been dissolved, much of its economic legacy remains to shape the progress and the challenges of the region's contemporary economy.

As communist leaders such as Stalin consolidated power in the 1920s and 1930s, they nationalized Russian industries, creating a system of **centralized economic planning** in which the state controlled production targets and industrial output. The Soviets stressed heavy, basic industries (steel, machinery, chemicals, and electricity generation), deferring demand for consumer goods to the future. Huge increases in production were realized, but at a significant human cost.

The Soviets also nationalized agriculture. By the late 1920s, communist leader Stalin was shifting agricultural land into large-scale collectives and state farms that organized production around state-mandated production goals. Marxist ideology reconfigured the countryside. As collectivization spread, the agricultural landscape was reorganized. Field boundaries and cropping complexes reflected state-established production goals; rapid mechanization replaced human and animal labor with tractors; and entirely new areas (such as the Virgin and Idle Lands Project east of the Volga River) were put into agricultural production to boost total food output. Still, the agricultural sector struggled. Most peasants initially resisted collectivization, and many slaughtered their animals rather than turn them over to state ownership.

Soviet-era industrial expansion was far more successful than any gains seen in agriculture. The nation's development of heavy industries during the 1930s proved vitally important in World War II. Although the German army advanced to within miles of Moscow, new centers of heavy industry in the resource-rich Urals and in Siberia's Kuznetsk Basin in the vicinity of Novosibirsk remained untouched, supplying the Russian army with the materials necessary to conduct the war. The industrial economy boomed in the 1950s and 1960s, and Soviet leaders confidently predicted that their country would be the world's most highly developed nation by 1980. It was a pivotal era in establishing the industrial complexes that remain dominant today (Figure 9.34).

Much of the Russian domain's basic infrastructure—its roads, rail lines, canals, dams, and communications networks—also originated during the Soviet period and reflect the planning necessities of that era. Indeed, massive projects to improve Soviet infrastructure were pursued with patriotic fervor. While some proved to be a waste of resources, many have been excellent long-term investments that form the backbone of the post-Soviet Union's economic geography. Dam and canal construction,

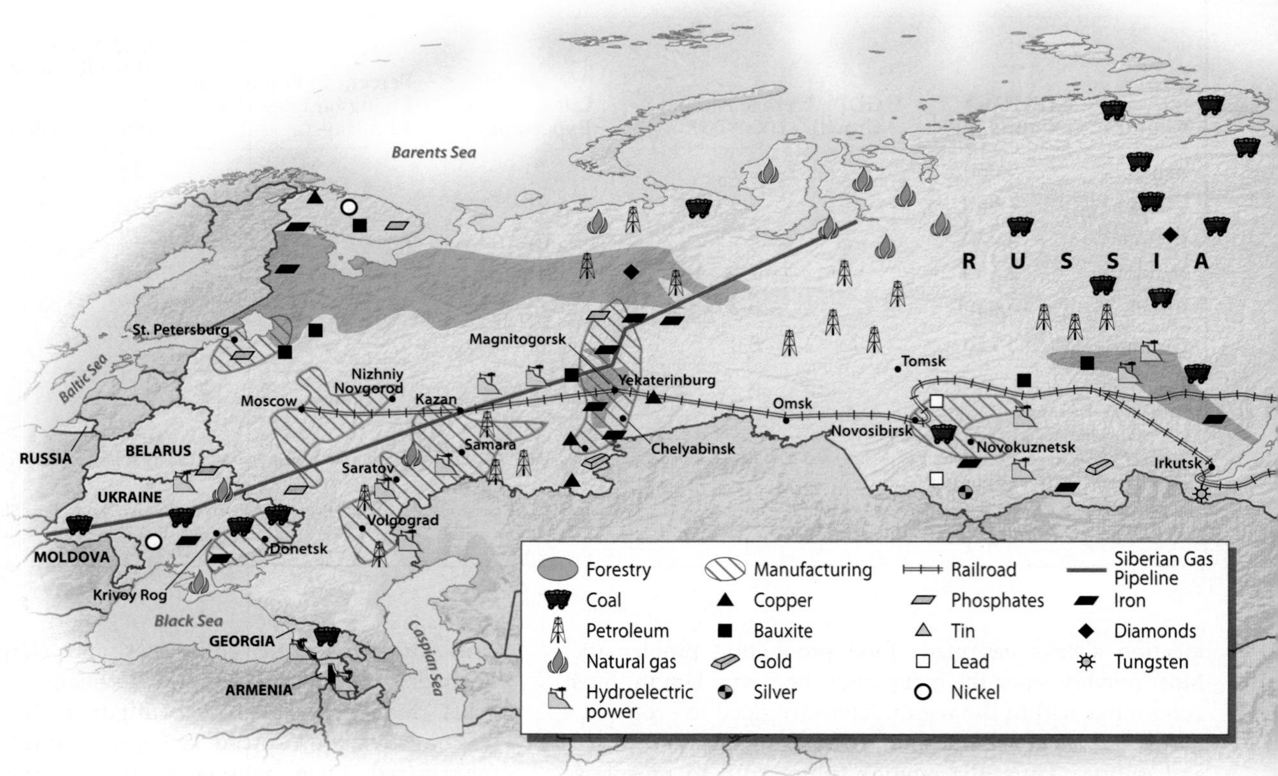

for example, turned the main rivers of European Russia into a virtual network of interconnected reservoirs. Invaluable links such as the Volga-Don Canal (completed in 1952), which connected those two key river systems, have greatly facilitated the movement of industrial raw materials and manufactured goods within the country (Figure 9.35). The Soviets also improved the country's railroad network. Thousands of miles of new track were added in the European west, and the Trans-Siberian Line was modernized and then complemented by the addition of the BAM link across central Siberia. Farther north, the Siberian Gas Pipeline was built to link the energy-rich fields of the Soviet arctic with growing demand in Europe.

Overall, the postwar period saw real economic improvements for the Soviet people. Russia and the other Soviet republics were increasingly industrialized and largely urbanized. A massive housing campaign in the 1950s and 1960s ensured most families a private flat. Consumer goods, such as televisions, telephones, and even private automobiles, began to spread more widely through society. The Soviet social system also achieved some notable successes. Literacy was virtually universal; quality health care was readily available; and every capable person was guaranteed employment.

Despite these successes, however, economic and social problems increased during the 1970s and 1980s. Soviet agriculture was still inefficient, and the country increasingly relied on grain imports. Manufacturing efficiency and quality failed to match the standards of the West, particularly in regard to consumer goods. Soviet citizens were burdened by recurrent shortages and usually had to spend many hours each week waiting in line to purchase basic necessities. Rigid centralized planning simply did not have the flexibility necessary to provide adequate goods and services. Equally troubling was the fact that the Soviet Union was failing to participate fully in the technological revolutions that were transforming the United States, Europe, and Japan, a failure that had serious military consequences. Disparities also visibly grew between the Soviet elite and an everyday citizenry that still enjoyed few personal freedoms. Political leaders vacationed in luxurious dachas, shopped in private stores, and avoided the long lines that most people faced to buy shoddier goods in the state-managed retail outlets. Such disparities of wealth further undermined faith in the socialist system. By the late 1980s, the Soviet Union had reached both an economic and a political impasse.

The Post-Soviet Economy

Fundamental economic changes have transformed the Russian domain since the demise of the Soviet Union. Particularly within Russia itself, much of the highly centralized state-controlled economy has been modified, replaced by a mixed economy of state-run operations and private enterprise. The changeover has been very difficult.

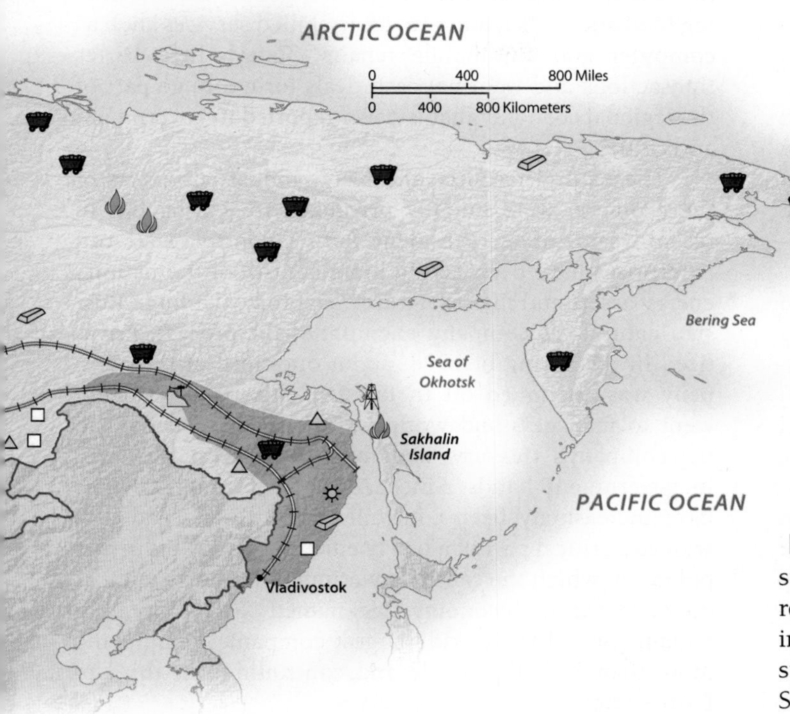

Figure 9.34 **Major Natural Resources and Industrial Zones**
The region's varied natural resources and chief industrial zones are widely distributed. Fossil fuels are in abundance, although their distance from markets often imposes special costs. In southern Siberia, rail corridors offer access to many mineral resources. In the mineral-rich Urals and eastern Ukraine, proximity to natural resources sparked industrial expansion, while Moscow's industrial power is related to its proximity to markets and capital. *(Modified from Bergman and Renwick, 1999,* Introduction to Geography, *Upper Saddle River, NJ: Prentice Hall; and Rubenstein, 2008,* An Introduction to Human Geography, *9th ed., Upper Saddle River, NJ: Prentice Hall)*

Fundamental problems of unstable currencies, corruption, and changing government policies plagued the system for much of the 1990s. In Russia, for example, steel output declined from almost 70 million tons in 1992 to less than 50 million tons at the end of the decade. Higher oil and natural gas prices have contributed to an impressive recovery in the Russian economy in the last few years, and selective economic growth has also improved conditions elsewhere in the region (see Table 9.2).

Figure 9.35 **Volga-Don Canal**
This view near Volgograd suggests the enduring economic importance of the Volga-Don Canal. Built during the Soviet era, the canal remains a key commercial link that facilitates the economic integration of southern Russia. *(Carl Wolinsky/Stock Boston)*

Redefining Regional Economic Ties Economic ties still link the six nations of the Russian domain, but the region has seen many of those relationships redefined in the post-Soviet era. The collapse of the communist state transformed planned internal exchanges between Soviet republics into less-predictable flows of foreign trade. Under the old regime, for example, government officials made certain that Ukraine received adequate supplies of oil from elsewhere in the country, just as Russia obtained cotton for its textile industry from Uzbekistan. Under the new economy, Ukraine must buy oil, or barter other products for it, from Russia, just as Russia suddenly must purchase more cotton from abroad.

Even with these disruptions, Russia retains an overwhelming economic presence within the region. Russian dominance is likely to continue, as suggested by that country's leading role in the creation of a "Common Economic Space" with neighboring Ukraine, Belarus, and Kazakhstan. It still is unclear how effective this Russian-led response to the growth of the EU will be in coming years. Meanwhile, however, regional economic ties remain strong. Belarus, for example, is still dependent upon Russian economic assistance, particularly in the form of cheap energy exports. Similarly, Ukraine imports a great deal of its raw materials (particularly oil) from Russia, and its exports (principally metals, food, and machinery) are led by return flows to the north. Recently, cash-starved Ukrainian companies also have looked to Russia for help, and the result has been a Russian-led buying binge of cheap Ukrainian assets. Ukrainian aluminum smelters, television stations, and oil refineries now are controlled by large Russian corporations, certainly a new post-Soviet twist on a long-established Russian presence within the country.

Privatization and State Control Even as hopes initially ran high for a smooth restructuring of the post-Soviet economy, the early years were difficult. A dramatic move came in October 1993, when the government initiated a massive program to privatize the Russian economy. Millions of Russians were given the option to buy into newly privatized agricultural lands and industrial companies. These initiatives greatly opened the economy to more private initiative and investment. Unfortunately, few legal and financial safeguards invited many abuses and often resulted in mismanagement and corruption within the new system. Elsewhere in the region, privatization proceeded more slowly. Much of the Belorussian economy, for example, remains mired in inflexible, corrupt state-controlled companies.

The agricultural sector continues to struggle. No matter its economic system, much of the region always will be challenged by short growing seasons, poor soils, and moisture deficiencies. Russia's best farmlands remain limited to a small slice of southern territory wedged between the Black and Caspian seas, leaving most of the rest of the country on the agricultural margins. In addition, the recent drastic economic changes have not been easy for the nation's farmers. Almost 90 percent of the country's farmland was privatized by 2003, with many farmers forming voluntary cooperatives or joint-stock associations to work the same acreage they did under the Soviet system. While crop prices have risen, costs have gone up even faster, and many farmers are not skilled in dealing with the uncertainties of a market-driven agricultural economy. Elsewhere, the government has held some agricultural land in the form of state farms, employing workers less willing to take risks in the fickle farm economy. Overall, agriculture employs about 11 percent of Russia's workforce, and the basic distribution of crops remains little changed from the Soviet era.

Russia has encouraged the rapid privatization of the service sector. Thousands of privatized retailing establishments have appeared, and they now dominate that portion of the economy. In addition, the long-established "informal economy" continues to flourish. Even during the Soviet era, millions of citizens earned extra money by informally selling Western consumer goods, manufacturing food and vodka, and providing skilled services such as computer and automobile repairs. Today, these barter transactions and informal cash deals form a huge part of the regional economy that is never reported to government authorities.

The natural resource and heavy industrial sectors of the economy were initially privatized in Russia, but in recent years, under President Putin's control, state-run enterprises have been taking more control of the nation's energy assets and infrastructure. Gazprom, the huge Russian natural gas company, exemplifies the process. Privatized in the spring of 1994, about one-third of the company was auctioned off to Russian citizens; 15 percent went to managers and workers; 10 percent remained in the company's treasury; and most of the rest was kept in government hands. Since 2005, Gazprom's activities have increasingly been controlled by the state and it is seen as a critical part of a newly emerging "state industrial policy" in which the central government is playing a more direct role in the economy. Nicknamed "Russia, Inc.," it remains one of the world's largest companies, employing more than 300,000 people and controlling one-third of Earth's known natural gas reserves.

Particularly in Russia, the successes of the new economy are increasingly visible on the landscape, especially in the country's major metropolitan areas. Luxury malls, office buildings, and more fashionable housing subdivisions are now part of the urban scene as the middle class grows in settings such as Moscow. Stroll the area near that city's Red Square today and you will encounter Planet Sushi; the Moscow Bentley, Ferrari, and Maserati dealership; as well as a growing number of trendy night clubs (Figure 9.36). Much of the wealth is highly concentrated. A 2007 survey by *Forbes Magazine* suggests that Moscow

Figure 9.36 Moscow Night Club Affluent young people in Moscow party late into the night as they enjoy the material fruits of selective economic growth that has arrived in post-Soviet Russia. *(Hans-Jurgen Burkard/Bilderberg/ Peter Arnold, Inc.)*

can now claim more billionaires than any other city in the world. A cruise through the elite Rublyevka housing district reveals luxury homes on par with anything in Beverly Hills, complete with in-home theaters, swimming pools, and ice skating rinks. On the other hand, the gap between increasing urban affluence and grinding rural poverty is growing dramatically. For example, in many villages across southern Siberia, there are no telephones, jobs, or money. Vodka is easier to find than running water. Closed shops and a continued lack of services continue to be a part of everyday life in such rural settings, which find themselves outside the bright glow and better economic opportunities of the cities.

The Challenge of Corruption Throughout the Russian domain, corruption also remains widespread. Doing business often means lining the pockets of government officials, company insiders, or trade union representatives. Much of the country's wealth has been funneled into private Swiss bank accounts. Cumbersome national and regional tax policies also prevent efficient revenue collections from most private companies and individuals involved in the informal economy, and thus the government sees little benefit from such operations.

Organized crime remains pervasive in Russia and controls many aspects of the economy. The government's own interior ministry estimates that the Russian mafia may control about 40 percent of the private economy and 60 percent of state-run enterprises. While organized crime certainly existed in the Soviet era, the more liberal economic and political environment that followed has allowed it to flourish even more widely and openly. The mafia provided critical capital and jobs to many unemployed young men in the unstable months immediately following the collapse of communism. Many close ties also remain between organized crime and Russian intelligence-collecting agencies. Various local and regional crime organizations have divided up much of the economy. One group might control the construction business in a Moscow suburb, whereas another syndicate oversees drug dealing and prostitution, and still another helps funnel illegal DVDs to eager consumers. The Russian mafia has also gone global. It has been implicated in huge money-laundering schemes that involve Russian, British, and U.S. banks as well as the flow of International Monetary Fund investments into the region. Banks in Switzerland, Liechtenstein, and Cyprus have been involved in similar schemes.

Ongoing Social Problems Even with better economic times, many residents of the Russian domain still struggle to make ends meet. Street crime has escalated with growing urban poverty and insecurity. Unemployment, rising housing costs, and declining social welfare expenditures have hit many families hard. Often, both husband and wife work multiple low-paying jobs with few benefits and long hours.

In many settings, women pay an especially heavy price. Increasing domestic violence has been widely reported in the post-Soviet era. Beatings and rapes are common. A survey in Moscow suggested that one-third of divorced women had experienced domestic violence, while a women's rights group in Ukraine reported that rape was an all-too-common crime in many villages. International organizations have sharply criticized government authorities for doing little to change the situation. Many women are compelled to sell homemade products or services on city streets for extremely low wages (Figure 9.37). Politically, women in Russia, Belarus, and Ukraine serve in fewer governmental positions than their counterparts in western Europe or the United States, suggesting that women are not able to play a high-profile role in shaping major social and economic policies within the region. Still, reforms undertaken during the communist era produced high female literacy rates and educational achievements (Table 9.2) that have persisted in all of the countries within the Russian domain.

Figure 9.37 Life on the Street
This young street vendor at the Moscow Food Market hopes to sell her pickled vegetables to passing consumers. Many Russians have not fully participated in the country's economic expansion since 2002. *(John Egan/The Hutchison Library)*

The region also suffers from many significant healthcare issues. Conditions in Russia illustrate the challenges. A shortage of vaccines and medical services has contributed to the return of such diseases as cholera, typhus, and the bubonic plague. Chronic illnesses, often related to lifestyle, produce the highest mortality rates in the developed world for cardiovascular disease, alcoholism, and smoking, particularly for Russian men. One recent study suggests that half of all Russian men and one-third of Russian women are plagued by long-term drinking problems. Many Russian men also smoke, creating a growth industry for multinational tobacco companies eager to increase sales. AIDS is rapidly on the rise, spreading especially among the region's young drug-using population. Russia and Ukraine have some of the fastest growth rates for HIV infection in the world, particularly among people between 20 and 30 years of age. Sixty percent of Russian children also are unhealthy, and infant mortality rates in rural parts of the country are higher than in many less-developed regions of the world. In addition, toxic environmental conditions have extracted a huge price from the region, although precise estimates of their impacts are difficult to make.

Growing Economic Globalization

The relationship between the Russian domain and the world beyond has shifted greatly since the end of communism. During much of the Soviet era, the region was relatively isolated from the world economic system. By the 1970s, however, the Soviet Union had begun to export large quantities of fossil fuels to the West while importing more food products. Connections with the global economy quickly intensified with the downfall of the Soviet Union. As Russia and its neighbors selectively embraced globalization, the region has intersected with the world economy in a variety of ways.

A New Day for the Consumer Most visibly, a barrage of new consumer imports now reaches many residents of the Russian domain, particularly those living in its larger cities. All of the symbols of global capitalism are visible in the heart of Moscow and, increasingly, in many other settings throughout the region. Luxury goods from the West also have found a small but enthusiastic market among the Russian elite, a group noted for its devotion to BMW automobiles, Rolex watches, and other status emblems. Most Russians find such luxuries beyond their limited budgets, but they are interested in purchasing basic foods, cheap technology, and popular clothing and media from their Western neighbors or from eastern and southern Asia.

Attracting Foreign Investment Despite political and economic uncertainties, most countries of the Russian domain are successfully attracting a growing inflow of foreign investment. For example, Russia brought in more than $40 billion in the first half of 2007, easily eclipsing the pace of inflows from earlier years. As measured by total foreign investment, the strongest global ties by far have been with the United States, Japan and western Europe, particularly Germany and Great Britain. The success of Russian equity markets and the relative stability of its financial sector have encouraged some. Still, many potential investors are put off by continuing uncertainties with Russian legal frameworks, lingering problems with slow bureaucracies and red tape, and growing concerns about the ultimate political aims of Russian leaders. Large outside investments have been made in Russia's oil and gas economy, but recently some of those opportunities have cooled amid the country's desire to limit foreign ownership of its energy resources and infrastructure.

Elsewhere in the region, connections with the global economy vary. In Belarus, a lack of economic reforms continues to slow new investment. About 80 percent of the nation's industrial sector is controlled by the state and the political climate seems opposed to foreign investment. Neighboring Ukraine, however, has succeeded in attracting more investment since 2004, but it is still early in the process of fully opening its economy to outside invest-

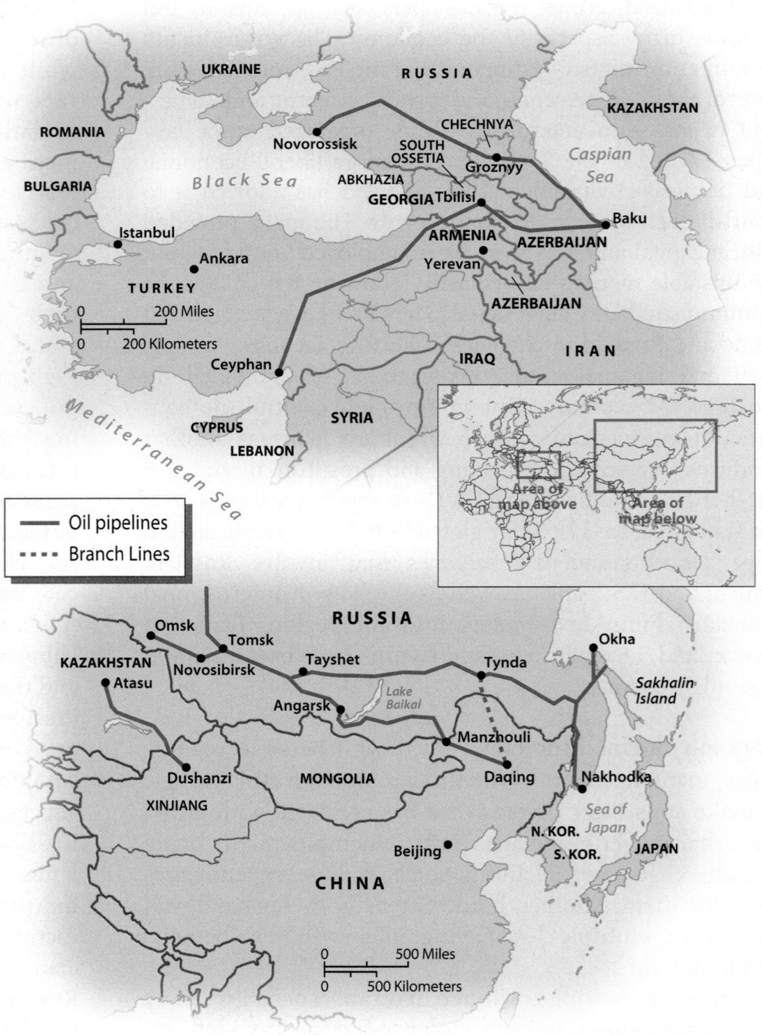

Figure 9.38 Russia's Expanding Pipelines These two maps show new and planned oil pipelines that are designed to expand Russia's presence in the global petroleum economy. Different pipeline projects in the Far East would benefit nearby China or Japan while projects near the Caspian Sea take pipelines through politically unstable portions of the region.

ment. Similarly, Moldova, Georgia, and Armenia have all made some progress in economic reforms that help reduce barriers to foreign investment, but the ultimate economic success of these nations remains closely linked to policies and conditions that unfold within Russia.

Globalization and Russia's Petroleum Economy

Russia's oil and gas industry remains one of the strongest economic links between the region and the global economy, and the diverse international connections it has forged suggest the increasing importance of the sector to the region's future. The statistics are impressive: Russia's energy production now makes up more than 25 percent of its entire economic output. Russia has 35 percent of the world's natural gas reserves (mostly in Siberia) and is the world's largest gas exporter. As for oil, Russia is by far the world's largest non-OPEC producer, and it is the second largest oil exporter in the world (behind Saudi Arabia). It far outpaces the United States in annual output (major oilfields are in Siberia, the Volga Valley, the Far East, and the Caspian Sea region), and it possesses more than 60 billion barrels of proven reserves.

The dynamic nature of the Russian oil and gas business exemplifies the forces of globalization and the region's changing economy. Prior to the breakup of the Soviet Union, about half of Russia's oil and gas exports went to other Soviet republics, such as Ukraine and Belarus. While these two nations still depend on Russian supplies, the primary destination for Russian petroleum products has overwhelmingly shifted to western Europe. Russia now supplies that region with more than 25 percent of its natural gas and 16 percent of its crude oil, and those linkages are likely to grow even stronger. An agreement between Russia and the EU in 2000 aimed at the rapid expansion of these East–West linkages. The Siberian Gas Pipeline already weds distant Asian fields with western Europe via Ukraine, and those connections are being supplemented by new lines through Belarus (the Yamal-Europe Pipeline) and Turkey (the Blue Stream Pipeline). Similar expansions are dramatically refashioning the geography of oil exports. The huge Druzhba Pipeline as well as the expanding oil port facility at Primorsk (near St. Petersburg) serves western Europe and other global markets.

Pipeline politics continues to play a growing role in the Russian domain (Figure 9.38). Many new and planned oil and gas pipelines pass through strategic, sometimes politically unstable regions, and these no doubt will be in the global spotlight in the years to come as new pipeline investments are made. For example, a large export terminal opened at Novorossisk (on the Black Sea) in 2001, delivering Caspian Sea supplies to the world market via a pipeline passing through troubled Chechnya. To the south, another new oil pipeline between Baku (on the Caspian Sea) and Ceyphan (on the Turkish Mediterranean) crosses both Azerbaijan and Georgia. In the Russian Far East, both China and Japan are lobbying hard for other pipeline projects. The Russians are building a large new Siberian Pacific Pipeline to link its Siberian fields to Asian markets. The Chinese want Russian oil to flow to Daqing, where it could be refined for national and regional markets. Japan prefers a large new facility at the Pacific port of Nakhodka, well-positioned to supply Japan and offering Russia easy access to global markets via the Pacific Ocean. Other links connect the system with developments on Sakhalin Island, where several major energy projects are currently being completed.

Figure 9.39 Sakhalin Island
Large-scale investments by both foreign and Russian interests have concentrated on energy-rich Sakhalin Island. The area promises to be a producer of both oil and natural gas in the years to come. *(AP/Wide World Photos)*

While many global companies have participated in the development of Russia's energy infrastructure, recent events suggest that state-controlled Russian companies will play a larger role in the future. Some private companies that were based in Russia, but that also utilized foreign capital, have simply been nationalized. Other North American, European, and Japanese energy companies have seen their role in Russian projects limited or eliminated. On Sakhalin Island, for example, Gazprom has been allowed to take control of several ventures originally dominated by companies such as Royal Dutch Shell (Figure 9.39). Ironically, Russian authorities have stripped control of these projects from foreign companies by suggesting they failed to follow environmental regulations. Once in Russian hands, environmental concerns have faded away.

Local Impacts of Globalization As the geography of the Russian petroleum industry suggests, the local impacts of globalization are highly selective. Obviously, portions of the region that are close to oil and gas wells, pipeline and refinery infrastructure, and key petroleum shipping points are greatly affected (both economically and environmentally) by the changing global oil economy. The same is true more broadly: globalization has affected different locations within the Russian domain in very distinctive ways. Within Russia, capitalism has brought its most dramatic, though selective, benefits to Moscow. Luxury goods, fancy cars, and upscale restaurants cater to the city's newly rich. Indeed, much of the foreign investment in the country has gone to the Moscow area.

Beyond Moscow, St. Petersburg and the Siberian city of Omsk have seen growing global investment, while new oil and gas prospects in Siberia, the Far East (Sakhalin Island), and near the Caspian Sea have attracted other investments. Cities such as Nizhniy Novgorod and Samara also have succeeded in attracting global capital

GLOBAL TO LOCAL The Tale of Two Siberias

As the global economy has evolved over the past 50 years, many of Russia's mining and manufacturing towns have had either to change or die. Today, in the post-Soviet era, they are especially exposed to both the challenges and the opportunities of the global economy. The result is a dramatically varied geography of impacts between the winners and the losers. In some parts of Siberia, areas oriented around noncompetitive extractive activities and outdated heavy industries are seeing a mass exodus of jobs and population. In fact, Russians living in several northern industrial settlements recently qualified for a World Bank program designed to assist governments with "urban liquidation." The gold-mining center of Magadan, for example, will receive liquidation assistance from both the World Bank and the Russian government. Eleven settlements and 6,500 people will be affected, leaving only four towns occupied within the region. Many residents need little incentive to leave. Indeed, since 1991, more than 40 percent of Magadan's population emigrated from the region. The landscape is bleak, winters are long and brutal, and the program will offer a modest beginning to those lucky enough to receive assistance (Figure 9.6.1). Overall, more than 500,000 people may be relocated from such outdated settlements.

On the other hand, Siberian cities that have proven their competitive edge in the global economy have shown new life. Novosibirsk, a southern Siberian city of 1.4 million people, has witnessed real economic growth and sharply increased international trade since 1999 (Figure 9.6.2). The city serves important regional political functions and is home to the Siberian Branch of the Russian Academy of Sciences. In addition, it has attracted a lengthening list of foreign corporate offices, including representatives of PepsiCo, Pfizer, Honeywell, Volvo, and Hoffman-La Roche. The results are impressive: foreign exports (chemicals, machinery, processed food) more than doubled in the past five years, and imports (machinery and chemicals) have risen by more than 70 percent. Ukraine and Kazakhstan remain the city's largest trading partners, but Germany, the United States, and even China are busily establishing new links with the city. Such localities, in Siberia and elsewhere, will reap persisting rewards as they find their profitable niche on the global stage, while other places lose people and close up shop, victims of a world economy that has passed them by.

Figure 9.6.1 Northeastern Siberia
Bleak settlements in northeast Siberia have often lost population since the 1990s as migrants seek out better job opportunities elsewhere. *(Heidi Bradner/The Image Works)*

Figure 9.6.2 Novosibirsk
The Russian city of Novosibirsk has become a major focus of investment and urban growth within Siberia since the disintegration of the Soviet Union. *(Vladimir Zinin/Sovfoto/Eastfoto)*

Figure 9.40 Vladivostok
The busy harbor of Vladivostok remains Russia's leading trade center in the Far East. With easy access to markets in Japan, China, and even the United States, this Pacific port is poised to grow as Russia's economy recovers. *(Itar-Tass/Sovfoto/Eastfoto)*

by building a probusiness economic environment. In addition, port cities such as Vladivostok are well positioned to take advantage of their accessibility to nearby markets, even though political red tape and corruption have hampered much growth in Pacific Basin trade (Figure 9.40).

Elsewhere, globalization clearly has imposed penalties (see "Global to Local: The Tale of Two Siberias"). Older, less globally competitive localities have been hit hard. Aging steel plants, for example, no longer have guaranteed markets for their high-cost, low-quality products as they did in the days of the planned Soviet economy. Instead, they must compete on the global market, a market that is increasingly prone to lower prices and weakening demand for many of the industrial goods the region produces. Many of the region's extractive centers are similarly vulnerable in a global economy of rapidly changing commodity prices.

Summary

- The Russian domain's environmental setting is rich in energy, metals, and timber resources, but its northern latitude limits the region's agricultural potential. Future climate change will introduce new environmental uncertainties, particularly in the arctic and subarctic portions of the region. Regional populations must also deal with the Soviet-era legacy of a toxic industrial and nuclear environment across many portions of the domain.
- Declining and aging populations are part of the sobering reality for much of the region. While some localities see modest population growth related to in-migration (mostly toward expanding urban areas), many rural areas and less competitive industrial zones are likely to see continued outflows of people and very low birthrates.
- The region's traditional cultural geography developed centuries ago, the complex product of Slavic languages, Orthodox Christianity, and numerous ethnic minorities that continue to complicate the scene today. Russia remains the most culturally diverse nation within the region, although Moldova, Georgia, and Ukraine each contain important cultural minorities. Further changing the region are new global influences, a set of products, technologies, and attitudes that often clash with traditional cultural values.
- Much of the region's political legacy is rooted in the Russian Empire, a land-based system of colonial expansion that greatly enlarged Russian influence after 1600 and was then reasserted in somewhat different form during the Soviet period. Inside Russia, increasing pressures for more centralized political power are likely, given forces set in motion by President Putin. Within the region, Russia's growing geopolitical influence is being met by resistance in localities such as Ukraine and Georgia. Beyond the region, Russia's growing visibility on the international stage may signal its reemergence as a truly global political power.
- The region's economic geography remains defined by its abundant resources and varied natural setting, as well as by the Marxist-guided industrialization and urbanization of the twentieth century. In the region's post-Soviet economy, Russia's oil and gas industry is its strongest link to the global economy, while other countries struggle to attain economic stability.

Key Terms

autonomous areas *(page 413)*
Baikal-Amur Mainline (BAM) Railroad *(page 397)*
Bolsheviks *(page 413)*
centralized economic planning *(page 421)*
chernozem soils *(page 391)*
Cold War *(page 414)*
Commonwealth of Independent States (CIS) *(page 415)*
Cossacks *(page 405)*
dacha *(page 401)*
denuclearization *(page 415)*
Eastern Orthodox Christianity *(page 405)*
exclave *(page 419)*
glasnost *(page 414)*
Gulag Archipelago *(page 399)*
Iron Curtain *(page 413)*
mikrorayons *(page 401)*
perestroika *(page 414)*
permafrost *(page 392)*
podzol soils *(page 391)*
Russification *(page 399)*
Slavic peoples *(page 405)*
socialist realism *(page 411)*
taiga *(page 392)*
Trans-Siberian Railroad *(page 396)*
tsars *(page 397)*

Questions for Review

1. Compare the climate, vegetation, and agricultural conditions of Russia's European west with those of Siberia and the Russian Far East.
2. Describe some of the high environmental costs of industrialization within the Russian domain.
3. Discuss how major river and rail corridors have shaped the geography of population and economic development in the region. Provide specific examples.
4. Contrast Soviet and post-Soviet migration patterns within the Russian domain, and discuss the changing forces at work.
5. Describe some of the major land-use zones in the modern Russian city, and suggest why it is important to understand the impact of Soviet-era planning within such settings.
6. What were the key phases of colonial expansion during the rise of the Russian Empire, and how did each enlarge the reach of the Russian state?
7. What are some of the key ethnic minority groups within Russia and the neighboring states, and how have they been recognized in the region's geopolitical structure?
8. Describe how centralized planning created a new economic geography across the former Soviet Union. What is its lasting impact?
9. Briefly summarize the key strengths and weaknesses of the post-Soviet Russian economy and suggest how globalization has shaped its evolution.

Thinking Geographically

1. How might it be argued that Russia's natural environment is one of its greatest assets as well as one of its greatest liabilities?
2. In the future, how might the forces of capitalism and free markets reshape the landscapes and land uses of cities within the Russian domain?
3. What options do peoples such as the Volga Tatars or the Siberian Buryats have for preserving their cultural autonomy? What might be some of the advantages and disadvantages of such peoples' pressing for greater political independence?
4. On a base map of the Russian domain, suggest possible political boundaries 20 years from now. What forces will work for a larger Russian state? A smaller Russian state?
5. What were some of the greatest strengths and weaknesses of centralized Soviet-style planning between 1917 and 1991? How were Ukraine and Belarus impacted? Why did the system ultimately fail?
6. Why is organized crime such a critical problem in this part of the world?
7. How have the growing forces of globalization affected Russian culture?
8. From the perspective of a 22-year-old Russian college student and resident of Moscow, write a short essay that suggests how the economic and political changes over the past 10 to 15 years have changed your life.

Regional Novels and Films

Novels

Fyodor Dostoyevsky, *The Brothers Karamazov* (orig. pub. 1879; 1995, Bantam Books)

Fyodor Dostoyevsky, *Crime and Punishment* (orig. pub. 1866; 1984, Bantam Classics)

Eugenia Ginzburg, *Journey into the Whirlwind* (orig. pub. 1967; 1975, Harcourt Brace)

Nikolai Gogol, *Dead Souls* (orig. pub. 1842; 1961, Viking Press)

Mikhail Lermantov, *A Hero of Our Time* (orig. pub. 1840; 1966, Penguin)

Andrei Makine, *Dreams of My Russian Summers* (1998, Scribner's)

Andrei Makine, *Once Upon the River Love* (1999, Penguin)

Aleksandr Solzhenitsyn, *One Day in the Life of Ivan Denisovich* (orig. pub. 1962; 1998, Signet Classics)

Leo Tolstoy, *War and Peace* (orig. pub. 1869; 1982, Viking Press)

Irene Zabytko, *The Sky Unwashed: A Novel* (2000, Algonquin Books)

Films

Brother (1997, Russia)

Burnt by the Sun (1994, France/Russia)

Dr. Zhivago (1965, U.S.)

Gorky Park (1983, U.S.)

Leo Tolstoy's Anna Karenina (1997, U.S.)

Prisoner of the Mountains (1996, Russia)

Reds (1981, U.S.)

Russia House (1990, U.S.)

Russian Ark (2002, Russia)

The Thief (1997, Russia)

Bibliography

Bater, James H. 1996. *Russia and the Post-Soviet Scene: A Geographical Perspective*. London: Arnold.

Billington, James. 2004. *Russia in Search of Itself*. Baltimore: Johns Hopkins University Press.

Crews, Robert D. 2006. *For Prophet and Tsar: Islam and Empire in Russia and Central Asia*. Cambridge, MA: Harvard University Press.

Dobrenko, Evgeny, and Naiman, Eric, eds. 2003. *The Landscape of Stalinism: The Art and Ideology of Soviet Space*. Seattle: University of Washington Press.

Griffin, Nicholas. 2001. *Caucasus: A Journey to the Land Between Christianity and Islam*. Chicago: University of Chicago Press.

Hosking, Geoffrey. 2001. *Russia and the Russians*. Cambridge, MA: Harvard University Press.

Karny, Yo'av. 2000. *Highlanders: A Journey to the Caucasus in Quest of Memory*. New York: Farrar, Straus & Giroux.

Shaw, Denis, 1999. *Russia in the Modern World: A New Geography*. Oxford, UK: Blackwell.

Thornton, Judith, and Ziegler, Charles, eds. 2002. *Russia's Far East: A Region at Risk*. Seattle: University of Washington Press.

Wilson, Andrew. 2000. *The Ukrainians: Unexpected Nation*. New Haven, CT: Yale University Press.

Additional bibliographic resources are at the *Diversity Amid Globalization* Website: http://www. prenhall.com/rowntree.

10
Central Asia

A buried gas pipeline in Tukmenistan has been exposed by shifting desert sands. *(Reza/Getty Images)*

ENVIRONMENTAL GEOGRAPHY
Intensive agriculture along the rivers that flow into the deserts of Central Asia has resulted in water shortages, leading to the desiccation of many of the region's lakes and wetlands.

POPULATION AND SETTLEMENT
Pastoral nomadism, the traditional way of life across much of Central Asia, is gradually disappearing as people settle in towns and cities.

CULTURAL COHERENCE AND DIVERSITY
In much of eastern Central Asia, the growing Han Chinese population threatens the indigenous culture of the Tibetan and Uyghur people; in the west, the role of Islam in social and political life remains a major issue.

GEOPOLITICAL FRAMEWORK
Due to its rich resource base and strategic position, Central Asia has become a zone of geopolitical rivalry in which the United States, Russia, China, India, Pakistan, and Iran vie for influence.

ECONOMIC AND SOCIAL DEVELOPMENT
Despite its abundant resources, Central Asia remains a relatively poor region, although much of it does enjoy relatively high levels of social development.

Central Asia does not appear in most books on world regional geography (Figure 10.1). Although it covers a larger expanse than the United States, it is a remote and lightly populated area dominated by high mountains, barren deserts, and semiarid steppes (grasslands). Geopolitically, Central Asia has been long dominated by external forces. Until 1991 it contained only two independent countries, Mongolia and Afghanistan. The rest of the region was at that time divided between the Soviet Union and China. Central Asia began to reappear in discussions of global geography following the breakup of the Soviet Union. Suddenly a number of new countries appeared on the international scene, prompting scholars to reexamine the position of Central Asia in human affairs (see "Setting the Boundaries").

Central Asia was more firmly established on the map of the world after September 11, 2001, when it became evident that the attack on the World Trade Center and the Pentagon were planned and organized by Osama bin Laden and his Al Qaeda organization operating out of Afghanistan and receiving support from that country's extreme Islamic government known as the **Taliban**. Although the subsequent U.S.-led war against the Taliban quickly proved successful, Afghanistan has remained mired in fighting and political chaos. Tension has intensified in many other parts of Central Asia, moreover, as repressive governments seek to maintain power and forestall radical Islamic fundamentalism, and as competition intensifies over the region's abundant natural resources.

One reason for Central Asia's former lack of prominence is that it was poorly integrated into international trade networks. This situation began to change in the 1990s as large oil and gas reserves were found, especially in Kazakhstan, Turkmenistan, and Azerbaijan. Moreover, a number of external countries are seeking to exert geopolitical influence over Central Asia, including Iran, Pakistan, India, the United States, and Russia. China's strict control and periodic repression of its Central Asian lands also highlight the significance of the region.

Central Asia forms a large, compact region in the center of the Eurasian landmass. Alone among the world regions, it lacks ocean access. Owing to its continental position in the

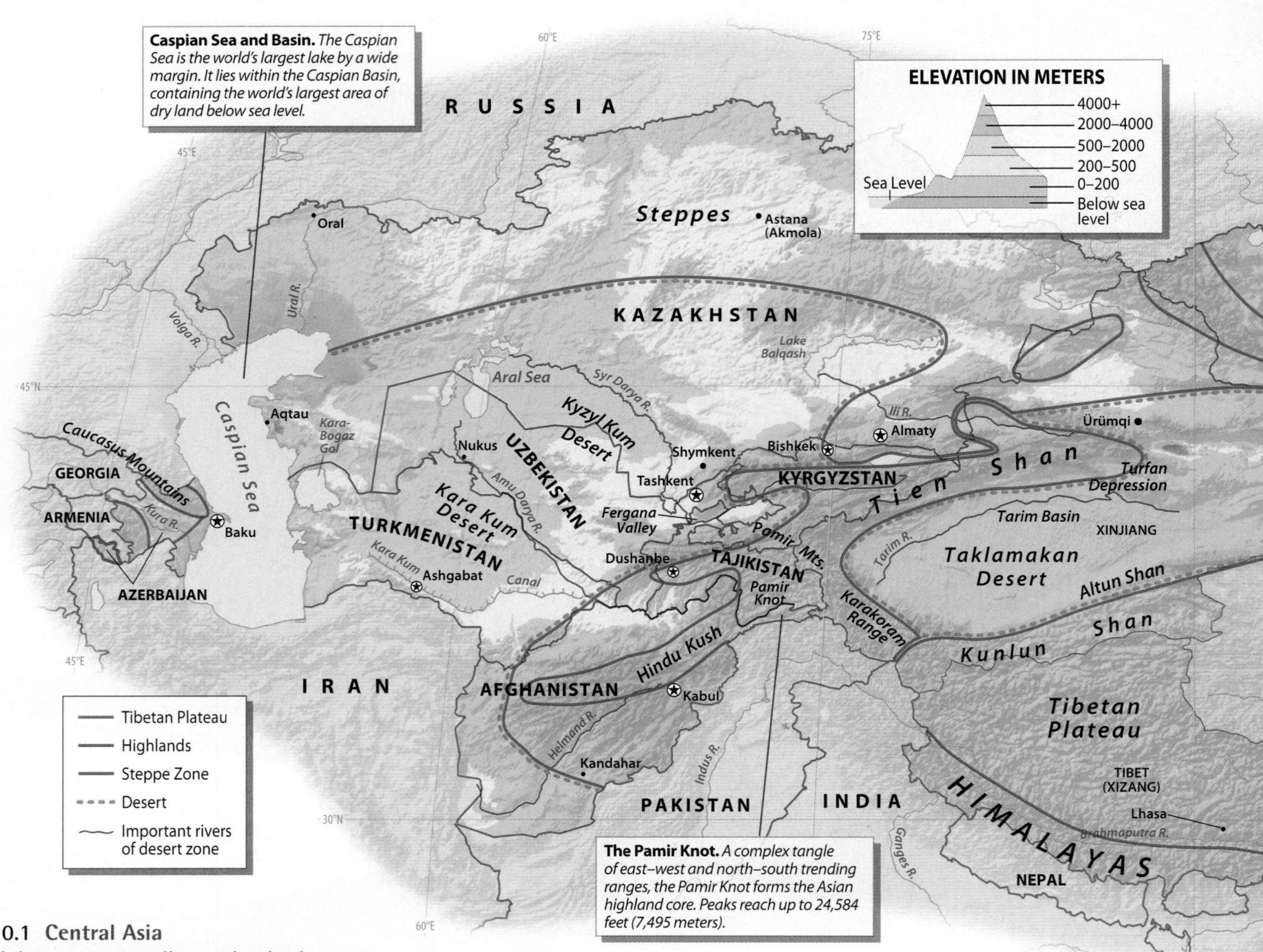

Figure 10.1 Central Asia
Central Asia, a vast, sprawling region in the center of the Eurasian continent, is dominated by arid plains and basins and lofty mountain ranges and plateaus. Eight independent countries—Kazakhstan, Turkmenistan, Uzbekistan, Kyrgyzstan, Tajikistan, Azerbaijan, Afghanistan, and Mongolia—form Central Asia's core. China's lightly populated far west and north are often placed within Central Asia as well, due to patterns of cultural and physical geography.

Setting the Boundaries

The term *Central Asia* is defined differently by different writers. Most authorities agree that it includes five newly independent former-Soviet republics: Kazakhstan, Kyrgyzstan, Uzbekistan, Tajikistan, and Turkmenistan. This chapter also includes other territories—Mongolia; Afghanistan; another post-Soviet state, Azerbaijan; and the autonomous regions of western China, Tibet, Xinjiang, and Nei Mongol (Inner Mongolia). The large Chinese province of Qinghai is also occasionally discussed, as it lies on the Tibetan Plateau and most of its indigenous peoples are either Tibetans or Mongolians.

The inclusion of these additional territories within Central Asia is controversial. Azerbaijan is often classified with its neighbors in the Caucasus region (Georgia and Armenia); western China is obviously part of East Asia by political criteria; and Mongolia is also often placed within East Asia because of both its location and its historical connections with China. Afghanistan can alternatively be located within either South Asia or Southwest Asia. Indeed, some writers would include the entire former Soviet zone within Southwest Asia, although others link it instead with a Russia-centered post-Soviet world region.

But considering Central Asia's historical unity, its common environmental circumstances, and its reentry onto the stage of global geopolitics, we think that it deserves consideration in its own right. It also makes sense to define its limits rather broadly. For example, Azerbaijan is linked both culturally (language and religion) and economically (oil) more to Central Asia than to Armenia and Georgia.

At the same time, however, any unity that Central Asia possesses is far from stable. Continuing Chinese political control over southeastern Central Asia and Han Chinese (meaning people of Chinese cultural heritage) migration into that region threatens any claims for regional coherence. Moreover, Central Asia itself remains deeply divided along cultural lines. Most of the region is Muslim in religious orientation and Turkic in language, but both the northeastern and southeastern sections (Mongolia and Tibet) are firmly Buddhist. Only time will tell whether Central Asia will continue to merit recognition as a distinct world region in its own right.

CENTRAL ASIA
Political & Physical Map
Over 1,000,000
500,000–1,000,000 (selected cities)
Selected smaller cities

The gorge country of eastern Tibet. *Several extremely steep canyons alternate with lofty ridges, making eastern Tibet one of the most topographically forbidding places in the world.*

center of the world's largest landmass, Central Asia is noted for its rigorous climate. High mountains, deep basins, and extensive plateaus magnify its climatic extremes. The aridity of the region, as we shall see, has also contributed to some of the most severe environmental problems in the world.

ENVIRONMENTAL GEOGRAPHY: Steppes, Deserts, and Threatened Lakes of the Eurasian Heartland

Much of Central Asia has a relatively clean environment, owing largely to its generally low population density. Industrial pollution is a serious problem only in the larger cities, such as Uzbekistan's Tashkent and Azerbaijan's Baku. Some parts of Central Asia, such as northwestern Tibet, remain practically pristine, with little human impact of any kind. Elsewhere, however, the typical environmental dilemmas of arid environments plague the region: desertification (the spread of deserts), salinization (the accumulation of salt in the soil), and desiccation (the drying up of lakes and wetlands). The destruction of the Aral Sea has been particularly tragic.

The Shrinking Aral Sea

One of the great environmental disasters of the twentieth century was the virtual destruction of the Aral Sea, a vast lake located on the boundary of Kazakhstan and Uzbekistan in western Central Asia. The Aral's only sources of water are the Amu Darya and Syr Darya rivers, which flow out of the Pamir Mountains, some 600 miles (960 kilometers) to the southeast. For thousands of years, water diversions for irrigation have lowered flows in both of these

Figure 10.2 The Northern Aral Sea Due to water diversions from the Amu Darya and Syr Darya Rivers, the Aral Sea has lost most of its volume and now forms two separate lakes. The government of Kazakhstan has recently built dikes to protect the water of the smaller Northern Aral, which is fed by the flow of the Syr Darya River. *(Ogoniok Lovchine/ Corbis & Sygma)*

rivers, but after 1950 the scale of diversion vastly expanded. The valleys of the two rivers formed the southernmost farming districts of the Soviet Union and thus the rivers became vital suppliers of water for warm-season crops, especially cotton. The largest of these projects was the Kara Kum Canal, which carries water from the Amu Darya across the deserts of southern Turkmenistan.

Unfortunately, the more river water was diverted for crop production, the less freshwater was available for the Aral Sea. The Aral proved to be particularly sensitive because it is relatively shallow. By the 1970s the shoreline was retreating at an unprecedented rate; eventually a number of "seaside" villages found themselves stranded more than 40 miles (64 kilometers) inland. As salinity levels increased, most fish species disappeared. New islands began to emerge, and by the 1990s the Aral Sea consisted of two separate lakes (see "Geographical Tools: Remote Sensing, GIS, and the Tragedy of the Aral Sea").

The destruction of the Aral Sea resulted in economic and cultural damage, as well as ecological devastation. Fisheries were wiped out and even local agriculture suffered. The retreating lake left large salt flats on its exposed beds; windstorms pick up the salt, along with the agricultural chemicals that have accumulated in the lake's shallows, and deposit it in nearby fields. Crop yields have thus declined; desertification has accelerated; and public health has been undermined. Particularly hard hit by this destruction are the Karakalpak people, members of a relatively powerless ethnic group who inhabit the formerly rich delta of the Amu Darya River. Recent reports indicate that the Karakalpak have the highest levels of infant and maternal death in the former Soviet Union.

In 2001, the oil-rich government of Kazakhstan decided that it had to save what was left of its portion of the Aral Sea. By reconstructing canals, sluices, and other waterworks along the Syr Darya River, it doubled the flow of water reaching the northern Aral, which correspondingly began to rise (Figure 10.2). A large dam built across the lakebed prevented the extra water from flowing south into the much larger but more saline southern Aral Sea. As a result, the salt level of the northern lake dropped significantly by 2007, allowing the return of fish and other forms of wildlife. The southern Aral Sea, meanwhile, continues to deteriorate. Thus far, the oil-poor, farming-dependent government of Uzbekistan has done little to reverse the course of its destruction.

Other Environmental Issues

Although the desiccation of the Aral Sea represents Central Asia's worst environmental crisis, other lakes in the region are also endangered. In addition, large areas of Central Asia suffer from extensive desertification and deforestation.

Shrinking and Expanding Lakes Western Central Asia supports large lakes because it forms a low-lying basin, without drainage to the ocean, that is virtually surrounded by mountains and other more humid areas. The world's largest lake, by a huge margin, is the Caspian Sea, located along the region's western boundary; the 4th largest was until recently the Aral Sea, and the 15th largest is Lake Balkhash in eastern Kazakhstan. Several other large lakes, such as Kyrgyzstan's spectacular Ysyk-Kol, are also located in the region.

Neither the Aral nor the Caspian are true seas because they are not connected with the ocean, but they are extremely large and salty. The Caspian is less salty than the ocean (particularly in the north), while until the 1970s the Aral was only slightly brackish (salty). Lake Balqash is almost fresh in the west but is quite salty in its long eastern extension. Because none of these lakes are drained by rivers, all naturally fluctuate in size, depending on how much precipitation falls in their drainage basin in a given year. Like the Aral Sea, a number of Central Asia's lakes have suffered from reduced water flow and hence increasing salinity; Balqash, for example, has recently shrunk by some 770 square miles (2,000 square kilometers).

The story of the Caspian is more complicated. The Caspian Sea receives most of its water from the large rivers of the north, the Ural and the Volga, which drain much of

GEOGRAPHIC TOOLS

Remote Sensing, GIS, and the Tragedy of the Aral Sea

Remote sensing technology is directly relevant to the study of the Aral Sea; satellite images clearly and immediately reveal the extensive and rapid reduction of the lake (Figure 10.2.1). But remote sensing and other technologically sophisticated geographical tools, such as geographic information system (or GIS, which allows spatially organized databases to generate numerous maps and map overlays), have also allowed geographers and other environmental scientists to more fully understand both the processes that have caused the tragedy and the broader social and environmental effects. Such tools are also invaluable for designing plans to reduce the damage.

Geographers Elena Lioubimtseva (Grand Valley State University) and Grigoriy Kapustin (Geography Institute, Russian Academy of Sciences) have recently used several sequences of medium-resolution satellite images to determine the spatial patterns and the underlying processes behind desertification in the Amu Darya Delta along the southern edge of the lake—one of the most devastated environments of the world. Their work shows the significance of overgrazing, wind erosion, and the salinization of soils, revealing as well local changes in microclimate caused by the shrinkage of the lake.

An even more comprehensive project, "GIS of the Aral Sea," is led by Rainer Ressl (German Aerospace Center), Andrei Ptichnikov (Institute of Geography, Russian Academy of Sciences), and Philip Micklin (USAID advisor on water management policy for the Uzbek government). As its Website (http://igras.geonet.ru/igras_e/project/aral/index.htm) explains, this project, under development since 1993, uses "dozens of georeferenced cartographic layers and a huge data base. It covers topics of physical geography and topography, social, demographic, health, water management and irrigation, ecological problems, desertification and biodiversity of the region around the Aral Sea, and the Sea itself. Satellite imagery of different resolution ... is widely used to update the map layers. Models of optimal water and land use of irrigated lands, of ecosystem transformation and desertification are under construction."

If the problems of the Aral Sea are to be genuinely addressed at any level, the concerted efforts of a number of both national and international agencies will be required. Despite the overwhelming severity of those problems, it is heartening to learn that geographers have responded by devising such internationally focused projects, using the best and most up-to-date technologies at their disposal.

Figure 10.2.1 The Shrinking of the Aral Sea
These three satellite images—the first taken in May 1973, the second in August 1987, and the third in July 2000—show the dramatic shrinkage of the Aral Sea. A 60 percent reduction of the lake's size has occurred since the 1960s, resulting in severe economic damage and environmental degradation. *(EROS Data Center, U.S. Geological Survey)*

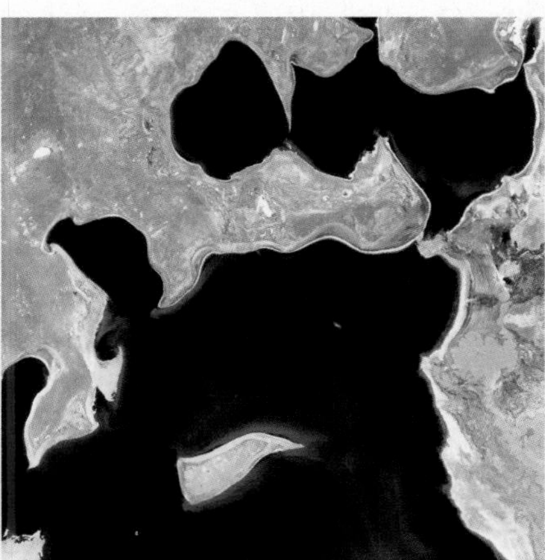

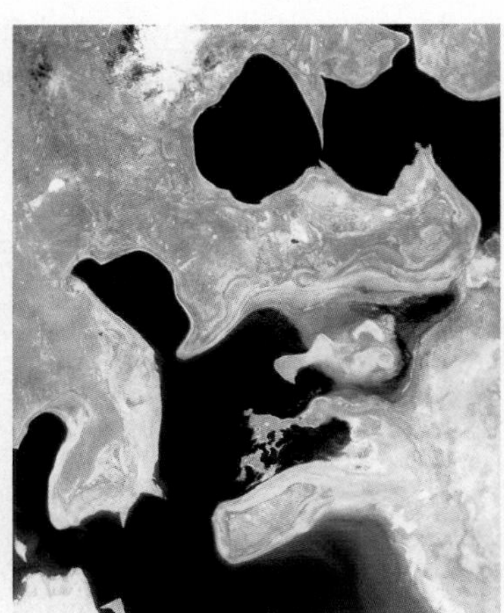

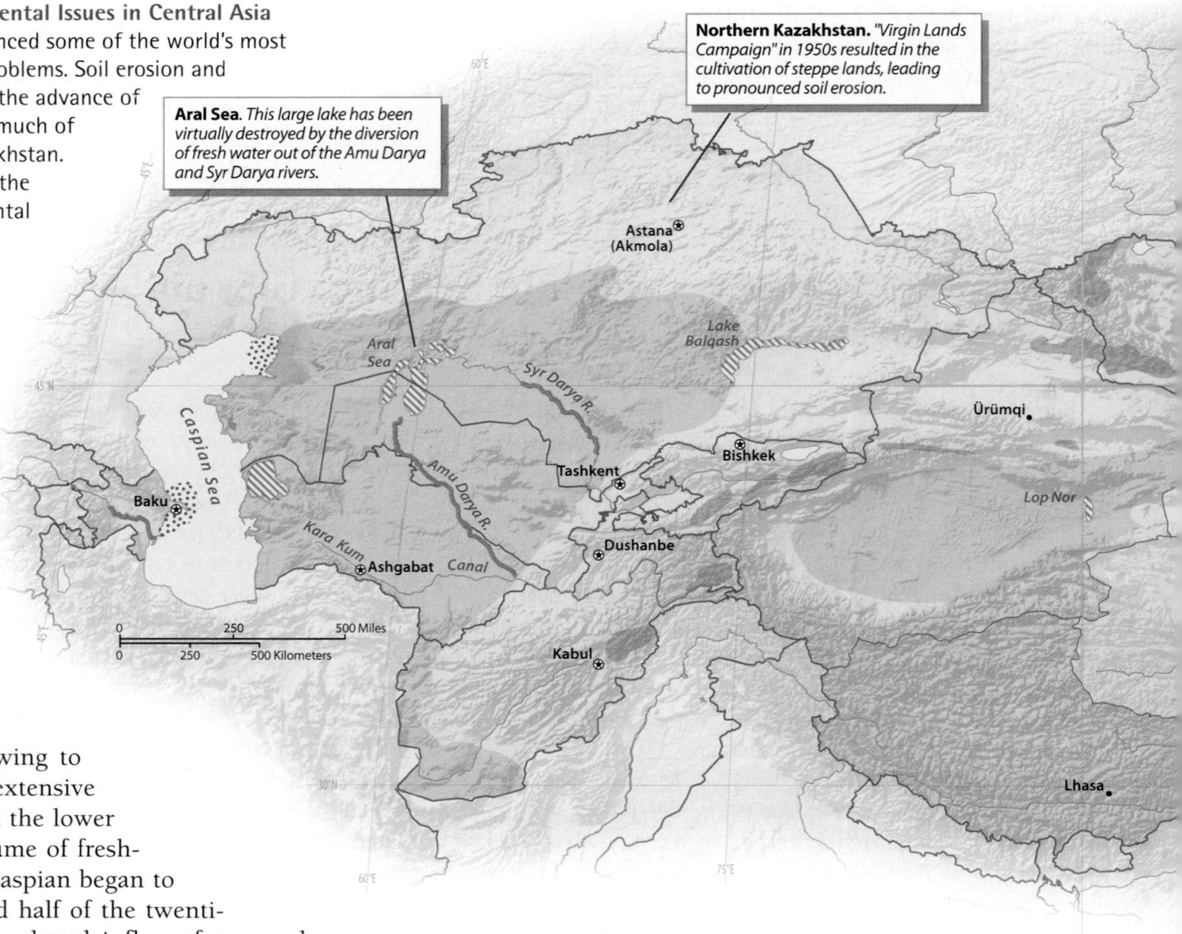

Figure 10.3 Environmental Issues in Central Asia Central Asia has experienced some of the world's most severe desertification problems. Soil erosion and overgrazing have led to the advance of desertlike conditions in much of western China and Kazakhstan. In western Central Asia, the most serious environmental problems are associated with the diversion of rivers for irrigation and the corresponding desiccation of lakes. Oil pollution is a particularly serious issue in the Caspian Sea area.

European Russia. Owing to the development of extensive irrigation facilities in the lower Volga basin, the volume of freshwater reaching the Caspian began to decline in the second half of the twentieth century. With a reduced influx of water, the level of the great lake dropped, exposing as much as 15,000 square miles (39,000 square kilometers) of former lake bed. A reduced volume of water resulted in increased salinity levels, undermining fisheries. The Russian caviar industry, centered in the northern Caspian, suffered extensive damage.

After reaching a low point in the late 1970s, the Caspian began to rise, presumably because of higher than normal precipitation in its drainage basin. By the late 1990s it had risen some 8.2 feet (2.5 meters). This enlargement, too, has caused problems, inundating, for example, some of the newly reclaimed farmlands in the Volga Delta. The most serious current environmental threat to the Caspian, however, is pollution from the oil industry rather than fluctuation in size. In early 2007, the sudden, unexplained death of hundreds of Caspian seals raised major concerns about the health of this vast lake.

Water development projects have also begun to impact the Tibetan Plateau. Concerned about water shortages in the North China Plain, China is now diverting part of the flow of the headwaters of the Yangtze River into headwaters of the Huang He (or Yellow River). This and other transfer schemes now threaten several of the Tibetan Plateau's large wetlands, havens for many rare species of migrating waterfowl.

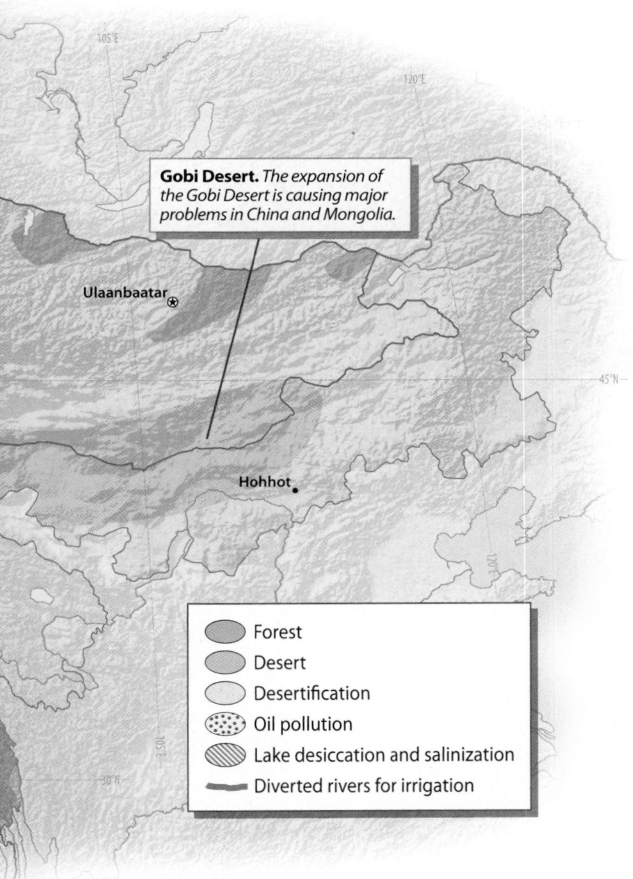

Desertification Desertification is another major concern in Central Asia (Figure 10.3). In the eastern part of the region, the Gobi and Taklamakan deserts have gradually spread southward, encroaching on settled lands in northeastern China. Farmers sometimes have to relocate their houses three or four times over the course of their lives in order to avoid the sands. The Chinese government has tried to stabilize dune fields with massive tree- and grass-planting campaigns, but such efforts have been only partially successful.

Northern Kazakhstan has also seen extensive desertification. This area was one of the main sites of the ambitious Soviet "Virgin Lands Campaign" of the 1950s, in which semiarid grasslands were plowed and planted with wheat. Many of these lands have since returned to native grasses, but not before erosion stripped away much of their productivity.

Deforestation has also harmed the region, resulting in wood shortages and reduced dry-season water supplies. Although most of Central Asia is too dry to support forests, many of its mountains were once well wooded. Today extensive forests can be found only in the wild gorge country of the eastern Tibetan Plateau; in some of the more remote west- and north-facing slopes of the Tien Shan, Altai, and Pamir mountains; and in the mountains of northern Mongolia.

Central Asia's Physical Regions

To understand why Central Asia suffers from such environmental problems as lake desiccation, an examination of the region's physical geography is necessary. In general, Central Asia is dominated by grassland plains (or steppes) in the north, desert basins in the southwestern and central areas, and high plateaus and mountains in the south-center and southeast (Figure 10.4). Several mountain ranges also extend into the heart of the region, dividing the desert zone into a series of separate basins and giving rise to the rivers that flow into the deserts and hence into the imperiled lakes.

The Central Asian Highlands The highlands of Central Asia originated in one of the great tectonic events of Earth's history: the collision of the Indian subcontinent into the Asian mainland. This ongoing impact has created the highest mountains in the world, the Himalayas, located along the boundary of South Asia and Central Asia.

Yet the Himalayas are merely one portion of a much larger network of high mountains and plateaus. To the northwest they merge with the Karakoram Range and then the Pamir Mountains. From the so-called Pamir Knot, a complex tangle of mountains situated where Pakistan, Afghanistan, China, and Tajikistan converge, other towering ranges radiate outward in several directions. The Hindu Kush sweeps to the southwest through central Afghanistan; the Kunlun Shan extends to the east (along the northern border of the Tibetan Plateau); and the Tien Shan swings out to the northeast into China's Xinjiang province. All of these

Figure 10.4 Tibetan Plateau
The Tibetan Plateau is dominated by alpine grasslands and tundra interspersed with rugged mountains and saline lakes. In summer the sparse vegetation offers forage for the herds of nomadic Tibetan pastoralists. Much of the northern part of Tibet, however, is too high to sustain even low-intensity land use. *(Michel Peissel/ SIPA Press)*

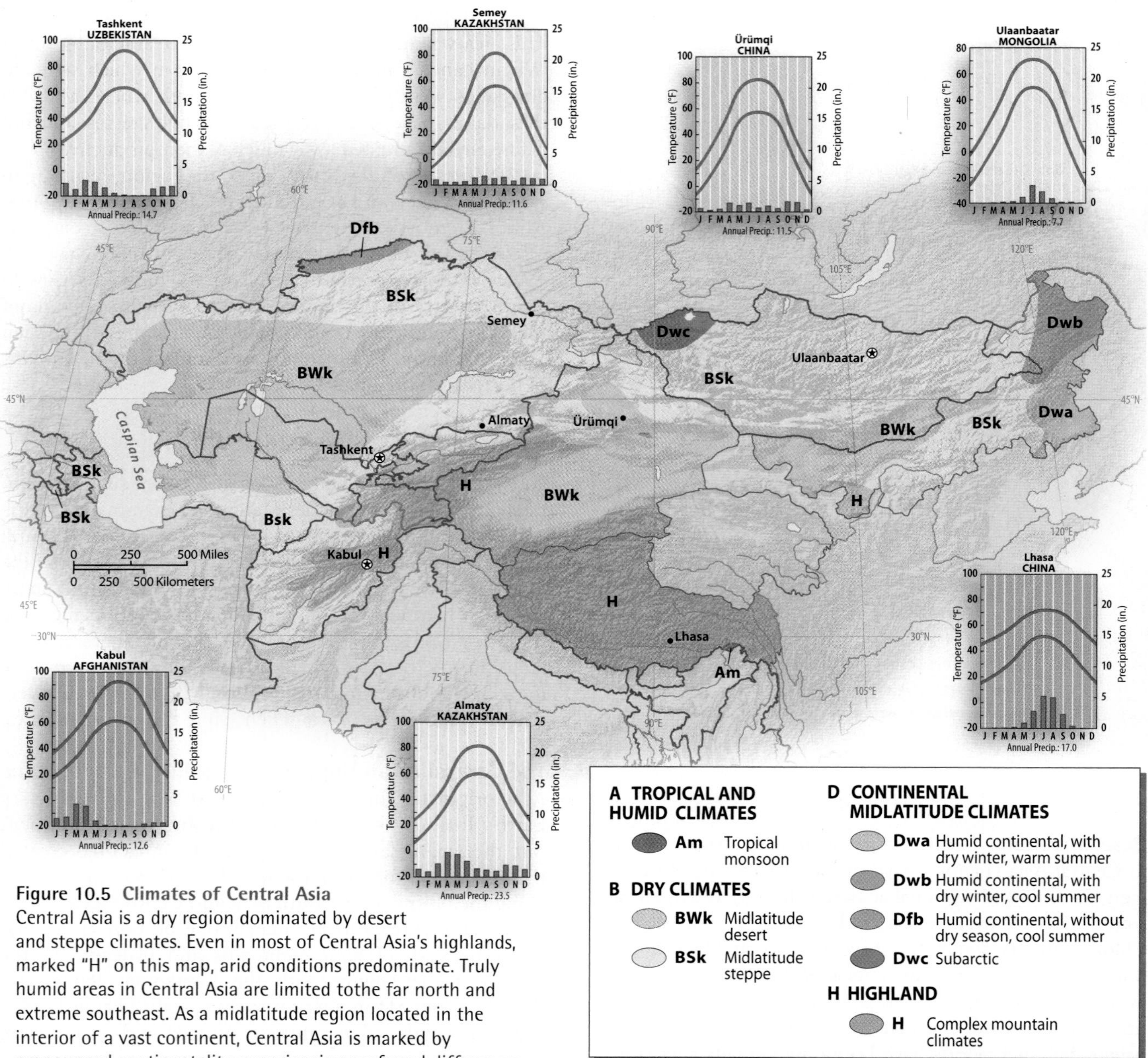

Figure 10.5 Climates of Central Asia
Central Asia is a dry region dominated by desert and steppe climates. Even in most of Central Asia's highlands, marked "H" on this map, arid conditions predominate. Truly humid areas in Central Asia are limited tothe far north and extreme southeast. As a midlatitude region located in the interior of a vast continent, Central Asia is marked by pronounced continentality, experiencing profound differences between winter and summer temperatures.

ranges have peaks higher than 20,000 feet (6,000 meters) in elevation. Much lower but still significant ranges are found along Turkmenistan's boundary with Iran and Azerbaijan's boundaries with Russia, Armenia, and Iran.

More extensive than these mountain ranges, however, is the Tibetan Plateau (Figure 10.5). This massive upland extends some 1,250 miles (2,000 kilometers) from east to west and 750 miles (1,200 kilometers) from north to south. More remarkable than its size is its elevation; virtually the entire area is higher than 12,000 feet (3,700 meters) above sea level, and its average height is about 15,000 feet (4,600 meters).

Most of the large rivers of South, Southeast, and East Asia originate on the Tibetan Plateau and adjoining mountains, including the Indus, Ganges, Brahmaputra, Mekong, Yangtze, and Huang He. These rivers pass either through the canyon-lands of eastern Tibet or through gaps in the Himalayas to the south. Most of the Tibetan Plateau, however, drains internally. Runoff from mountain snowfields trickles down to evaporate from innumerable lakes and marshes at the lowest plateau elevations.

The greater part of the Tibetan Plateau, at about 15,000 feet (4,600 meters) of elevation, lies near the maximum height at which human life can exist. Rather than forming a flat, tablelike surface, the plateau is punctuated with east-west-running ranges alternating with undrained basins. Although the southeastern sections of the plateau receive ample precipitation, most of Tibet is arid. Cut off by high ranges from any source of moisture, large areas of the plateau receive only a few inches of rain a year (Figure 10.5). Winters on the Tibetan Plateau are cold, and while summer afternoons can be warm, summer nights remain chilly.

The Plains and Basins Although the mountains of Central Asia are higher and more extensive than those found anywhere else in the world, most of the region is characterized by plains and basins of low and intermediate elevation. This lower-lying zone can be divided into two main areas: a central belt of deserts punctuated by lush river valleys and a northern swath of semiarid steppe.

Central Asia's desert belt is itself divided into two segments by the Tien Shan and Pamir mountains. To the west lie the arid plains of the Caspian and Aral sea basins, located primarily in Turkmenistan, Uzbekistan, and southern Kazakhstan. Most of this area is relatively flat and very low, with the surface of the Caspian Sea lying 92 feet (28 meters) below sea level. The climate of this region is strikingly continental; summers are dry and hot, whereas winter temperatures average well below freezing. Central Asia's eastern desert belt extends for almost 2,000 miles (3,200 kilometers) from the extreme west of China to the southeastern edge of Inner Mongolia. It is conventionally divided into two deserts: the Taklamakan, found in the Tarim Basin of Xinjiang, and the Gobi, which runs along the border between Mongolia proper and the Chinese region of Inner Mongolia.

The environment of western Central Asia is distinguished from that of eastern Central Asia in part by its larger rivers. More snow falls on the western than the eastern slopes of the Pamir Mountains, giving rise to more abundant runoff. The largest of these rivers, as we have seen, flow into the Aral Sea. Others, such as the Helmand of Afghanistan, terminate in shallow lakes or extensive marshes and salt flats. In Xinjiang, one substantial river, the Tarim, does flow out of the highlands onto the basin floor, where it frequently shifts course across the sandy lowlands. Before the completion of irrigation projects in the 1960s, it terminated in a salty lake called Lop Nor. Subsequently, the newly dried-out Lop Nor salt flat was used periodically by China for testing nuclear weapons.

North of the desert zone, rainfall gradually increases and desert eventually gives way to the great grasslands, or steppe, of northern Central Asia. Near the region's northern boundary, trees begin to appear in favored locales, outliers of the great Siberian taiga (coniferous forest) of the north. A nearly continuous swath of grasslands extends some 4,000 miles (6,400 kilometers) east to west across the entire region, only partially broken by Mongolia's Altai Mountains. Summers on the northern steppe are usually pleasant, but winters can be brutally cold.

Global Warming and Central Asia

Most climate change experts expect Central Asia to be hard hit by global warming. The Tibetan Plateau has already seen marked increases in temperature, resulting in the reduction of permafrost and the rapid drawback of mountain glaciers. Eighty percent of Tibet's glaciers are presently retreating, some at rates of up to 7 percent a year, leading to predictions that the region will lose up to half of its ice fields within the next 50 years. Similar levels of glacial retreat are occurring throughout the Central Asia highlands, including the Pamir and Tien Shan mountains. Tree-ring data from Mongolia indicate that the last century has been the warmest in more than a thousand years. Overall, Central Asia has experienced a 3.6°F (2°C) temperate increase since the early 1900s.

The retreat of Central Asia's glaciers is especially worrisome because of their role in providing dependable flows of water for the region's rivers. In some areas, the melting of ice has resulted in the temporary flooding of lowland basins, but the long-term result will be to further reduce freshwater resources in a region already hampered by aridity. Climate change from global warming is also likely to reduce precipitation in the arid lowlands of western Central Asia, compounding the problem. Prolonged and devastating droughts have recently struck Afghanistan and several of its neighbors, indicating a possible shift to a drier climate. As a result, the UN Intergovernmental Panel on Climate Change (IPCC) has predicted a 30 percent crop decline for Central Asia as a whole by mid-century. Efforts to save the region's endangered lakes, moreover, will likely be undermined by the increased aridity.

But as is true elsewhere in the world, global warming will not affect all parts of Central Asia in a uniform manner. Some areas, including the Gobi Desert and the Tibetan Plateau, could possibly see a long-term increase in precipitation. Such forecasts are scant consolation, however, in a region that is now experiencing rampant desertification. Some global climate models also indicate increased precipitation in the boreal forest zone of Siberia that extends into the mountains of northern Mongolia. But as the rivers coming out of these mountains generally flow north, few benefits would be reaped by the rest of Central Asia.

Overall, Central Asia is not a large producer of greenhouse gasses, although the oil-rich economies of Kazakhstan and Azerbaijan are noted for their inefficient energy use. The traditional forms of livelihood found across Central Asia, based on nomadic pastoralism and peasant agriculture, generate little excess carbon dioxide. Greenhouse gas emissions from the region are rapidly increasing, however, due mainly to the expansion of the fossil fuel industry and the growth of cities and modern transportation systems, as we shall see in the following section.

POPULATION AND SETTLEMENT: Densely Settled Oases amid Vacant Lands

Most of Central Asia is sparsely populated (Figure 10.6). Large areas are essentially uninhabited, either too arid or too high to support human life, while vast expanses are populated only by widely scattered groups of nomadic **pastoralists** (people who raise livestock for subsistence

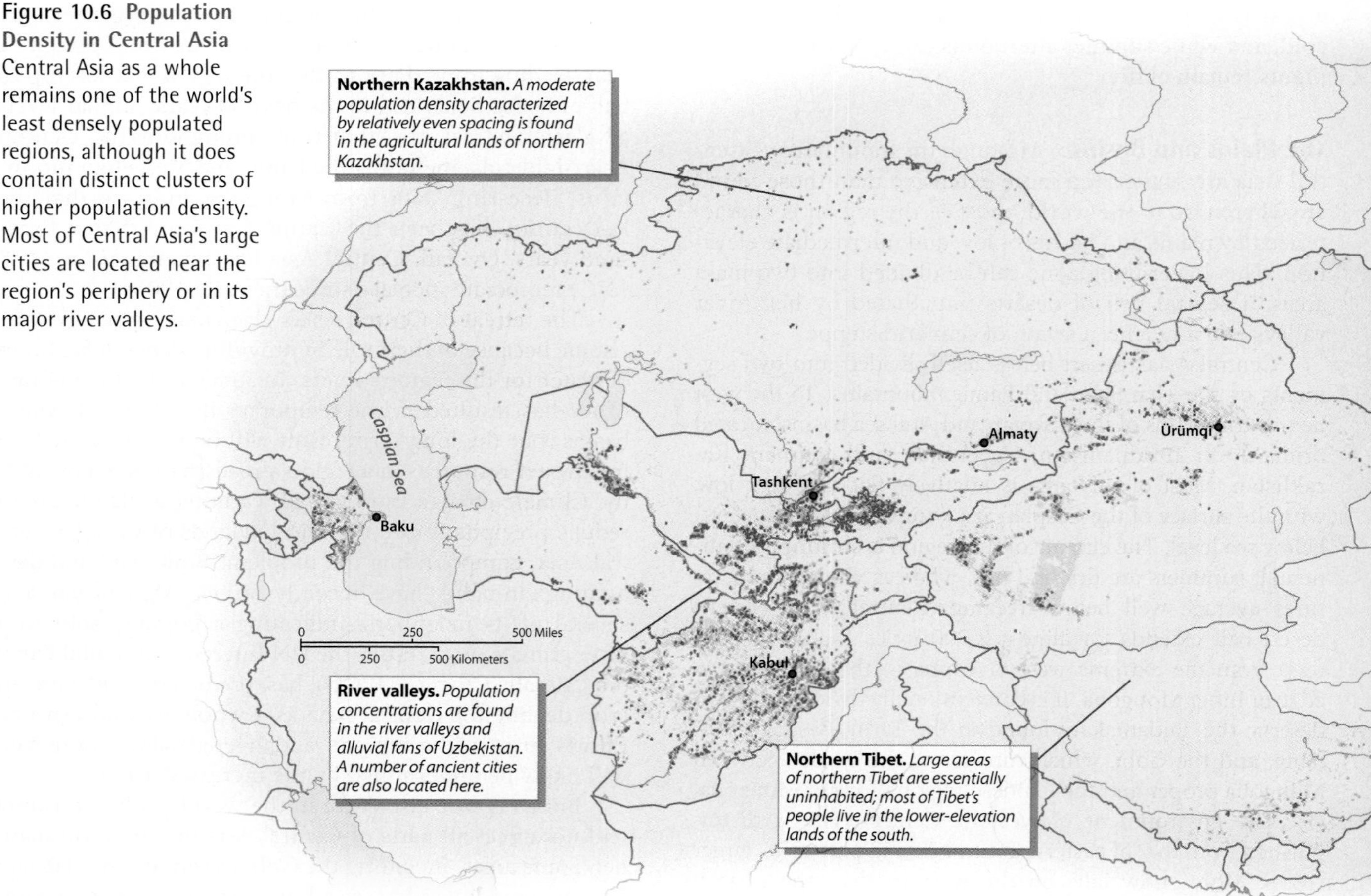

Figure 10.6 Population Density in Central Asia Central Asia as a whole remains one of the world's least densely populated regions, although it does contain distinct clusters of higher population density. Most of Central Asia's large cities are located near the region's periphery or in its major river valleys.

purposes). Mongolia, which is more than twice the size of Texas, has only 2.6 million inhabitants. But as is common in arid environments, those few lowland areas with good soil and dependable water supplies can be thickly settled. Despite its overall aridity, Central Asia is well endowed with perennial rivers and fertile oases. While the nomadic pastoralists of the steppe and desert zones have dominated the history of Central Asia, the sedentary peoples of the river valleys have always been more numerous. Even the forbidding Tibetan Plateau has been demographically dominated by settled peasant farmers for thousands of years.

Highland Population and Subsistence Patterns

The environment of the Tibetan Plateau is particularly harsh. Not only is the climate cold and water often scarce or brackish, but ultraviolet radiation in this high-elevation area is always pronounced. Only sparse grasses and herbaceous plants—so-called mountain tundra—can survive, and human subsistence is obviously difficult under such conditions. The only feasible way of life over much of the Tibetan Plateau is nomadic pastoralism based on the yak, an altitude-adapted relative of the cow. Several hundred thousand people manage to make a living in such a manner, roaming with their herds over vast distances.

Sedentary farming in Tibet is possible only in a few favorable locations, generally those that are *relatively* low in elevation and that have good soils and either adequate rain or a dependable irrigation system. The main zone of agricultural settlement lies in the far south, where protected valleys offer favorable conditions. The population of Tibet proper (the Chinese autonomous region of Xizang) is only 2.8 million, a small number indeed considering the vast size of the area.

Population densities are also low in the other highland areas of Central Asia, although settled agricultural communities can be found in protected valleys. The complex topography of the Pamir Range in particular offers a large array of small and nearly isolated valleys that are suitable for agriculture and intensive human settlement. Not surprisingly, this area is marked by great cultural and linguistic diversity. Many villages here are noted for their agricultural terraces and well-tended orchards.

Central Asia's mountains are vitally important for people living in the adjacent lowlands, whether they are settled farmers or migratory pastoralists. Many herders use the highlands for summer pasture; when the lowlands are parched, the high meadows provide rich grazing. The Kyrgyz (of Kyrgyzstan) are noted for their traditional economy based on **transhumance**, moving their flocks from lowland pastures in the winter to highland meadows in the summer. The farmers of Central Asia rely on the highlands for their wood supplies and, more important, for their water. Settled agricultural life in most of Central Asia is possible only because of the rivers and streams flowing out of the region's mountains.

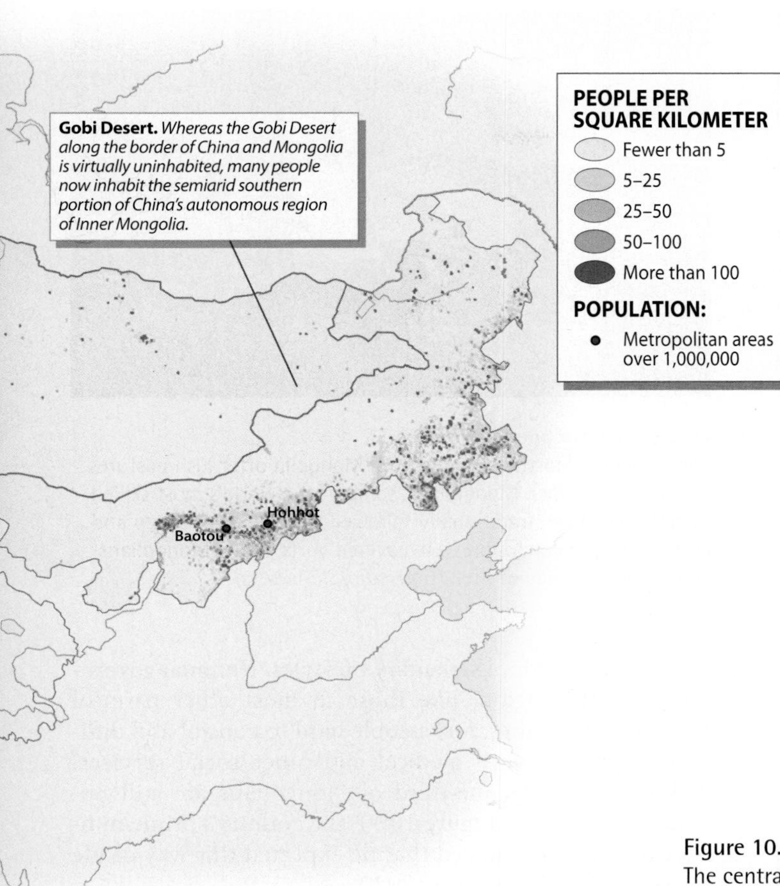

Lowland Population and Subsistence Patterns

Most of the inhabitants of Central Asia's deserts live in the narrow belt where the mountains meet the basins and plains. Here water supplies are adequate and soils are not contaminated with salt or alkali, as is often the case in the basin interiors. The population distribution pattern of China's Tarim Basin forms an almost perfect ringlike structure (Figure 10.7). Streams flowing out of the mountains are diverted to irrigate fields and orchards in the narrow fertile band situated between the steep slopes of the mountains and the arid wasteland of the central basin.

The population west of the Pamir Range, in former Soviet Central Asia, is also concentrated in the transitional zone nestled between the highlands and the plains. A series of **alluvial fans**, fan-shaped deposits of sediments dropped by streams flowing out of the mountains, have long been devoted to intensive cultivation. Fertile **loess**, a silty soil deposited by the wind, is widespread, and in a few favored areas winter precipitation is high enough to allow rain-fed agriculture. Several large valleys in this area also offer fertile and easily irrigated farmland

Figure 10.7 Population Patterns in Xinjiang's Tarim Basin
The central portion of the Tarim Basin is a virtually uninhabited expanse of sand dunes and salt flats. Along the edge of the basin, however, dense agricultural and urban settlements are located where streams running out of the surrounding mountains allow for intensive irrigation. The largest of these oasis communities are found along the southwestern fringe of the basin.

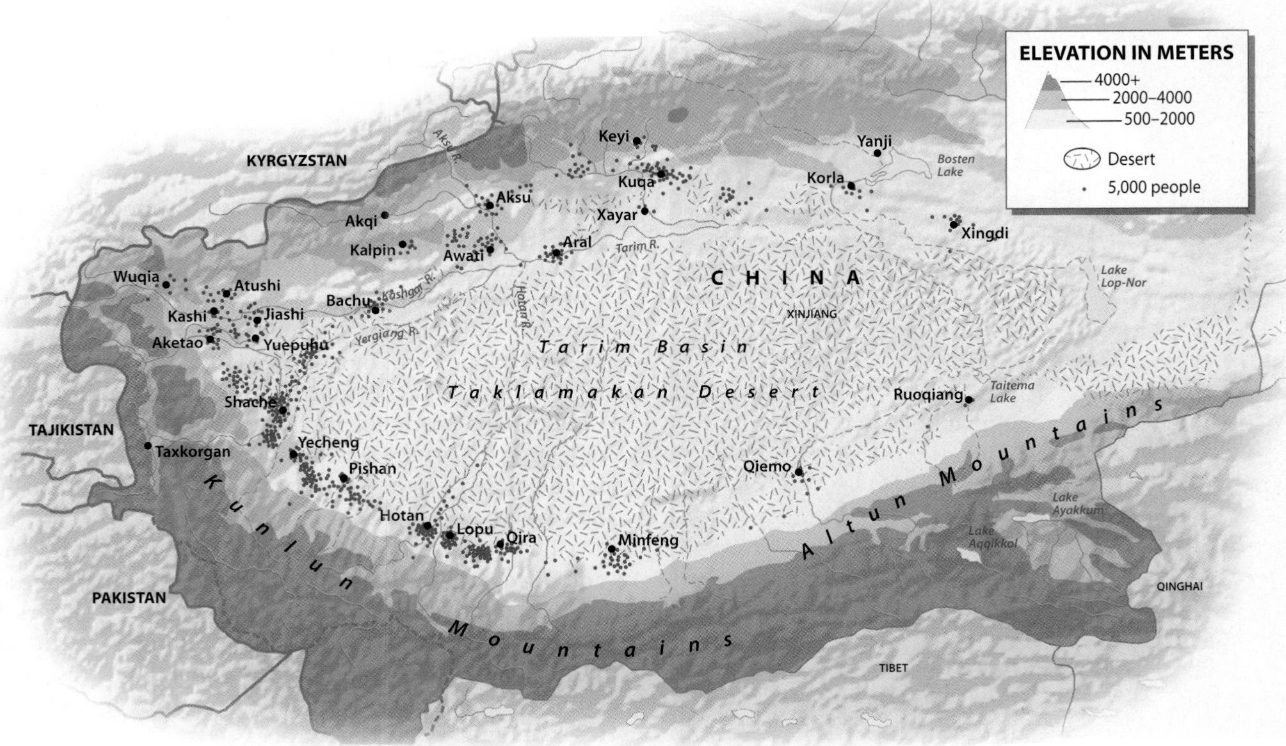

(Figure 10.8). The densely populated Fergana Valley of the upper Syr Darya River is shared by three countries: Uzbekistan, Kyrgyzstan, and Tajikistan. In the far west of the region, Azerbaijan's Kura River Basin is another area of intensive agriculture and concentrated settlement.

Unlike the other deserts of the region, the Gobi has few sources of permanent water. Rivers draining Mongolia's highlands flow to the north or terminate in interior basins, while only a few of the larger streams from the Tibetan Plateau reach the Gobi proper. (The Huang He, however, does swing north to reach the desert edge before turning south to flow through the Loess Plateau.) Owing to this scarcity of **exotic rivers** (those originating in more humid areas) and to its own aridity, the Gobi remains one of Asia's least-populated areas.

The steppes of northern Central Asia are the classical land of nomadic pastoralism. Until the 1900s, virtually none of this area had ever been plowed and farmed. To this day, pastoralism remains a common way of life across the grasslands. In northwestern China and the former Soviet republics, however, many pastoral peoples have been forced to adopt sedentary lifestyles. National governments in the region, like those in most other parts of the world, find migratory people hard to control and difficult to provide with medical and other social services. Although roughly one-third of Mongolians are still nomadic pastoralists (Figure 10.9), the country's prime minister recently announced that he expected this way of life to essentially disappear by 2020.

Figure 10.8 **Farmland in Uzbekistan**
The fertile river valleys of Uzbekistan have been intensely cultivated for many centuries, producing large harvests of fruits and vegetables in addition to cotton and grain. *(David Turnely/ Corbis/Bettmann)*

Figure 10.9 **Steppe Pastoralism**
The steppes of northern and central Mongolia offer lush pastures during the summer. Mongolians, some of the world's most skilled horse-riders, have traditionally followed their herds of sheep and cattle, living in collapsible, felt-covered yurts. Many Mongolians still follow this way of life. *(Goussard/SIPA Press)*

In northern Kazakhstan, the Soviet regime converted the most productive pastures into farmland in the mid-1900s in order to increase the country's supply of grain. Some of these lands have since reverted to steppe, but large areas remain under the plow, and Kazakhstan is a major producer of spring wheat. Consequently, northern Kazakhstan has the highest population density of the steppe belt.

Population Issues

Although Central Asia remains a low-density environment, some portions are growing at a moderately rapid pace. The demographic data on Central Asia show overall fertility rates near the middle of those for the developing world as a whole, while exhibiting pronounced variation between different areas (Table 10.1). Afghanistan, the least-developed and most male-dominated country of the region, has the highest birthrate by a substantial margin. Kazakhstan's birthrate, on the other hand, is relatively low, reflecting in part the small families of the Russian speakers in the north of the country. Mongolia and Azerbaijan, however, have even lower birthrates.

Low fertility, however, can be counteracted by immigration, as is the case in much of western China. Here substantial population growth over the past 30 years has stemmed from the migration of Han Chinese into the area. While many of the indigenous inhabitants resent this influx, the Chinese government claims that it is necessary for economic development.

TABLE 10.1 Population Indicators

Country	Population (Millions, 2007)	Population Density (per Square Kilometer)	Total Fertility Rate	Percent Urban	Percent <15[a]	Percent >65[b]	Net Migration (per 1,000, 2000–05)
Afghanistan	31.9	49	6.8	20	45	2	16.0
Azerbaijan	8.6	99	2.1	52	24	7	−2.4
Kazakhstan	15.5	6	2.5	57	27	8	−8.0
Kyrgyzstan	5.2	26	2.8	35	32	6	−2.9
Mongolia	2.6	2	2.0	59	29	4	−3.9
Tajikistan	7.1	50	3.4	26	32	3	−10.9
Turkmenistan	5.4	11	2.9	47	34	5	−0.4
Uzbekistan	26.5	59	2.7	36	35	5	−2.3

[a]*Percentage of population younger than 15.*
[b]*Percentage of population older than 65.*
Source: Population Reference Bureau, World Population Data Sheet, *2007; net migration rate data from* UN International Migration, *2006.*

Population growth in most parts of the former Soviet zone has stemmed largely from moderately high levels of fertility. During the final years of the Soviet Union, birthrates among Central Asian Muslims were substantially higher than those elsewhere in the country, perhaps contributing to the breakup of the Soviet Union.

Migration has also complicated demographic patterns in the former Soviet zone. After the breakup of the Soviet Union, millions of Russians and Ukrainians left Central Asia to return to their homelands. Kazakhstan, which in the 1950s had more Russians than Kazakhs, witnessed a particularly sharp population decline in the early 1990s (Figure 10.10). Kazakhstan's ethnic German population,

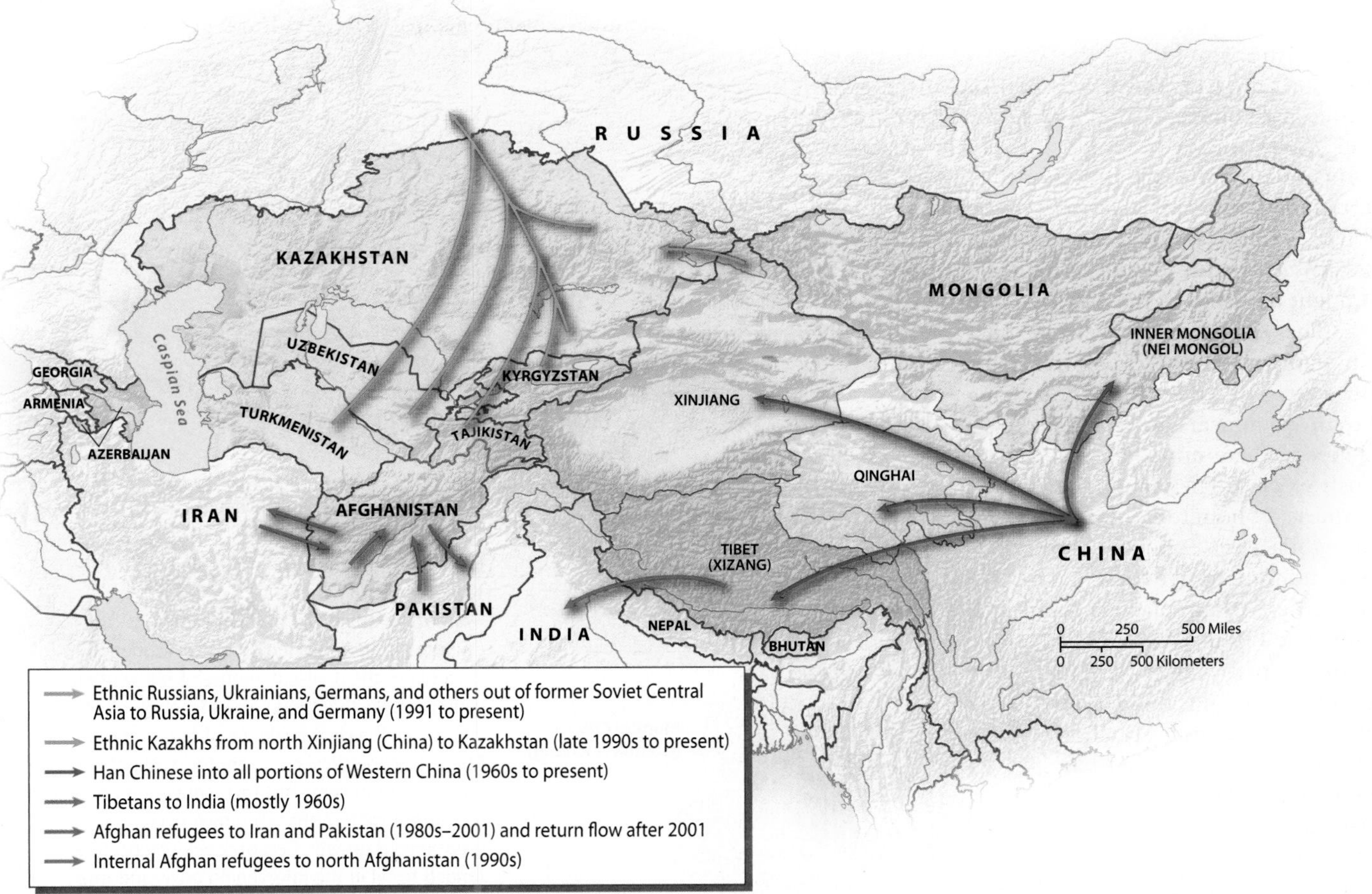

Figure 10.10 Recent Migration and Refugee Flows
Since the early 1990s, large numbers of people have been moving out of the former Soviet Central Asia and into the Chinese portions of Central Asia. Afghanistan has one of the largest refugee problems in the world.

once almost a million strong, virtually disappeared. This migration stream particularly affected the region's urban areas, where many of the Europeans were concentrated. More recently, Russia's economic boom has led many Azerbaijanis, Uzbeks, and other Central Asians to seek employment in Moscow and St. Petersburg, even though they face substantial discrimination and even violence. In 2006 alone, racially motivated attacks in Russia, directed mostly against Central Asians and people from the Caucasus, resulted in 54 deaths and 520 serious injuries.

In the oil-rich countries of Central Asia, the current energy boom has begun to reverse the flow of migrants. Although people of European background continue to leave Kazakhstan, economic migrants from elsewhere in Central Asia—and from China—are taking their place. Over 400,000 illegal immigrants were thought to live in Kazakhstan as of early 2007. The Kazak government has also encouraged the immigration of ethnic Kazakhs from northwestern China, and several hundred thousand have responded.

Urbanization in Central Asia

Although the steppes of northern Central Asia had no real cities before the modern age, the river valleys and oases have been partially urbanized for millennia. Cities such as Samarkand and Bukhara in Uzbekistan were famous even in medieval Europe for their riches and their lavish architecture (see "Cityscapes: Life on the Streets of Samarkand"). This early urban development was built upon the region's economic and political position. The Amu Darya and Syr Darya valleys lay near the midpoint of the trans-Eurasian silk route, and they formed the core of a number of empires based on the cavalry forces of the steppe. The modern era of steamships and oceanic trade, however, brought hardship to the cities of Central Asia.

The conquest of Central Asia by the Russian and Chinese empires ushered in a new wave of urban formation. Cities slowly began to appear on the Kazakh steppes, where none had previously existed. The Manchu and Chinese conquerors of Inner Mongolia and Xinjiang built new administrative and garrison cities, often placing them only a few miles from indigenous urban sites (Figure 10.11). The old indigenous cities of the region are characterized by complex and almost mazelike networks of streets and alleyways, while the neighboring Chinese cities were constructed according to a strict geometrical order. The recent growth of urban populations in the area has obscured this old dualism, but several new "twin cities" have emerged in Xinjiang, where Han Chinese immigrants have built settlements next to indigenous cities.

Although it is possible to distinguish Russian/Soviet cities from indigenous cities in the former Soviet zone, this dichotomy is not clear-cut. In Uzbekistan, for example, Tashkent is largely a Soviet creation, while many parts of Bukhara and Samarkand still reflect the older urban patterns. Several major cities, such as Kazakhstan's former capital of Almaty, did not exist before Russian colonization. In Azerbaijan, Baku emerged as a major city in the early twentieth century as the first Caspian oil fields began to be intensively exploited. Moreover, almost everywhere one can see the effects of centralized Soviet urban planning and design.

Urbanization is gradually but unevenly spreading across Central Asia. North-central Kazakhstan recently witnessed the rise of a major city, Astana, the country's new, centrally located capital. Even Mongolia, long a land virtually without permanent settlements, now has more people living in cities than in the countryside. In some parts of the region, however, cities remain relatively few and far between. Only 26 percent of the people of Tajikistan, for example, are urban residents. Tibet similarly remains a predominantly rural society, but the influx of Han Chinese into the region is creating a larger urban system. To understand this movement and its broader ramifications, one must examine Central Asia's patterns of cultural geography.

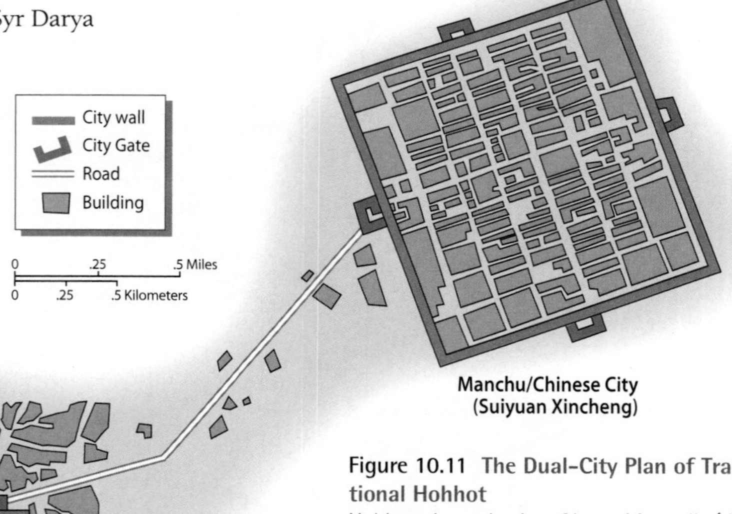

Figure 10.11 The Dual-City Plan of Traditional Hohhot
Hohhot, the main city of Inner Mongolia (the Chinese autonomous region of Nei Monggol), exemplifies the dual-city nature of many urban centers of western China. The Manchu/Chinese city shows the clear evidence of traditional Chinese geometrical city planning, while the older Mongol city, at some distance, shows a more haphazard pattern that stemmed from spontaneous growth. Recent developments have tended to fill in the intervening areas and thus obscure these traditional distinctions. *(From Piper R. Gaubatz, 1996, Beyond the Great Wall: Urban Form and Transformation on the Chinese Frontiers, Stanford: Stanford University Press, p. 67)*

CITYSCAPES Life on the Streets of Samarkand

Like all cities of the former Soviet Union, Uzbekistan's Samarkand has its share of drab concrete buildings designed for public housing and government offices. And like all of the older cities of Central Asia, it has its share of crowded traditional neighborhoods with narrow streets (Figure 10.3.1) and houses facing inward toward private courtyards. But in other ways, Samarkand is unique; it contains more sites of historical and religious significance than any other city in the region. Life in the streets of central Samarkand, a city that poets once dubbed the "Rome of the East," forces one to confront and appreciate the past with an unusual intensity. In recognition of its significance, Samarkand was declared a United Nations Educational, Scientific and Cultural Organization (UNESCO) world heritage site in 2001.

Figure 10.3.1 The Narrow Streets of Samarkand
The older districts of Samarkand are noted for their tightly packed buildings and narrow streets and alleys. Many of the old city's roads are too narrow for wheeled vehicles. *(Zafer Kizikaya/Coral Planet)*

Samarkand's historical riches derive in part from its physical geography. Located in a fertile valley at roughly the midway point of the Silk Road that once connected China to the Mediterranean, it was an important city as far back as the time of Alexander the Great. Many centuries later, it became the capital city of the famous and feared Turkic-Mongol conqueror, Tamerlane (Timur-i-leng; 1336–1405). Tamerlane and his descendents endeavored to beautify the city, building mosques, tombs, madrassahs (Islamic colleges), and a host of other public buildings. Many of these architectural marvels remain. Muslims from throughout Central Asia and beyond often make pilgrimages to visit the sacred sites of the city, some viewing the trip as second in importance only to the hajj to Mecca.

At the core of historical Samarkand sits Registan Square (Figure 10.3.2), described in the nineteenth century by Lord George Curzon (Britain's Viceroy of India) as simply "the best central city square in the world." Majestic buildings, including the famous Ulugbek Madrassah, line three sides of the square. Nearby, sitting in the midst of a crowded and colorful market, is the Bibi-Khanum Mosque, which was the largest in the world when built in the fifteenth century. But it is Tamerlane's final resting place, the Gur-Emir Mausoleum, that is considered by many as the most spectacular example of traditional Central Asian architecture.

Figure 10.3.2 Traditional Architecture in Samarkand
Samarkand is famous for its lavish Islamic architecture, some of it dating to the 1400s. This photograph shows some of the buildings arrayed around Registan Square, the city's traditional core. *(Haley/SIPA Press)*

CULTURAL COHERENCE AND DIVERSITY: A Meeting Ground of Different Traditions

Although Central Asia has a certain environmental unity, its cultural coherence is more questionable. The western half of the region is largely Muslim and is often classified as part of Southwest Asia. Northeastern and southeastern Central Asia—Mongolia and Tibet—are characterized by a distinctive form of Tibetan Buddhism. Tibet is culturally linked to both South and East Asia, and Mongolia is intimately associated with China, but neither fits easily within any world region. Such complexity can best be understood by examining the region's historical geography.

Historical Overview: Steppe Nomads and Silk Road Traders

The river valleys and oases of Central Asia were early sites of sedentary, agricultural communities. Archaeologists have discovered abundant evidence of farming villages dating back to the Neolithic period (beginning circa 8000 BCE) in the Amu Darya and Syr Darya valleys and along the rim of the Tarim Basin. After the domestication of the horse around 4000 BCE, nomadic pastoralism emerged in the steppe belt as a new human adaptation. Eventually pastoral peoples gained power over the entire region, transforming not only the history of Central Asia but also that of virtually all of Eurasia. In the premodern period, pastoral nomads enjoyed major military advantages over sedentary societies. They had access to large numbers of horses at a time when cavalry almost always held the edge over infantry. Not until the age of gunpowder were the benefits of pastoralism offset by the demographic and economic advantages held by the more populous agriculturally based states.

The earliest recorded languages of Central Asia, spoken both in the oasis communities and among some of the pastoralists, were members of the Indo-European linguistic family. Through the first millennium CE, the inhabitants of the Amu Darya and Syr Darya valleys spoke languages closely related to Persian. Many aspects of Persian heritage are still found in southwestern Central Asia.

Indo-European languages began to be replaced on the steppe roughly 2,000 years ago by languages in the Altaic family, which include Mongolian and Turkic. By the second century BCE, a powerful nomadic empire of Turkic-speaking peoples arose in what today is Mongolia, forcing the Chinese to begin building the Great Wall as a defensive measure. As Turkic power spread through most of Central Asia, Turkic languages gradually began to replace Indo-European tongues in the oasis communities. This process was never completed, however, and southwestern Central Asia remains a meeting ground of the Persian and Turkic languages. Uzbek, for example, may be classified within the Turkic language family, but a great deal of its basic vocabulary is of Persian origin.

The Turks were eventually replaced on the eastern steppes by another group of Altaic speakers, the Mongols. In the late 1100s the Mongols, under the leadership of Genghis Khan, united the pastoral peoples of Central Asia and used the resulting force to conquer nearby sedentary societies. By the late 1200s the Mongol Empire had grown into the largest contiguous empire the world had ever seen, stretching from Korea and southern China in the east to the Carpathian Mountains and the Euphrates River in the west (Figure 10.12). In today's Mongolia, Genghis Khan has reemerged as the country's national hero. China is also trying to claim the same legacy, having recently built a Genghis Khan theme park in Inner Mongolia.

Protected by mountain barriers and by the rigorous conditions of the plateau, Tibet has taken a different course from the rest of Central Asia. Tibet emerged as a strong, unified kingdom around 700 CE. Tibetan unity and power did not persist, however, and the region reverted to its former state of semi-isolation. Tibet was incorporated for a short period in the 1200s into the Mongol Empire, and in later centuries other Mongol states occasionally enjoyed limited powers over the Tibetans. These interactions resulted in the establishment of Mongolian communities in the northeastern portion of the plateau and in the eventual conversion of the Mongolian people to Tibetan Buddhism.

Contemporary Linguistic and Ethnic Geography

Today most of Central Asia is inhabited by peoples speaking Altaic languages, but patterns of linguistic geography remain complex (Figure 10.13). A few indigenous Indo-European languages are confined to the southwest, while Tibetan remains the main language of the plateau. Russian is also widely spoken in the west and in most former Soviet countries retains an official status, whereas Mandarin Chinese is increasingly important in the east. In both Tibet and Xinjiang, Mandarin Chinese is now the basic language of higher education. In Tibet, most urban merchants are now Chinese-speaking immigrants, causing much concern among the indigenous people.

Tibetan Tibetan is divided into a number of distinct dialects that are spoken over almost the entire inhabited portion of the Tibetan Plateau. Approximately 6 million people speak Tibetan; of these, roughly 2.5 million live in Tibet proper, while most of the rest reside in China's provinces of Qinghai and Sichuan. Tibetan is usually placed in the Sino-Tibetan family, implying a shared linguistic ancestry between the Chinese and the Tibetan peoples, but many scholars argue that no definite relationship between the two has been established. Tibetan has an extensive literature written in its own script, most of which is devoted to religious topics.

Mongolian Mongolian forms a cluster of closely related dialects spoken by approximately 5 million people. The standard Mongolian of both the independent country of Mongolia and China's Inner Mongolia is called Khalkha; other Mongolian dialects include Buryat (found in southern Siberia) and Kalmyk (found to the northwest of the Caspian Sea). Mongolian has its own distinctive script, which dates back some 800 years, but Mongolia itself adopted the Cyrillic alphabet of Russia in 1941. Efforts are now being made to revive the old script.

Although Mongolian speakers form about 90 percent of the population of Mongolia, in China's Inner Mongolian Autonomous Province they have been almost submerged by a wave of Han Chinese migrants over the past 50 years. Today only about 2 million out of the 25 million residents of Inner Mongolia speak Mongolian.

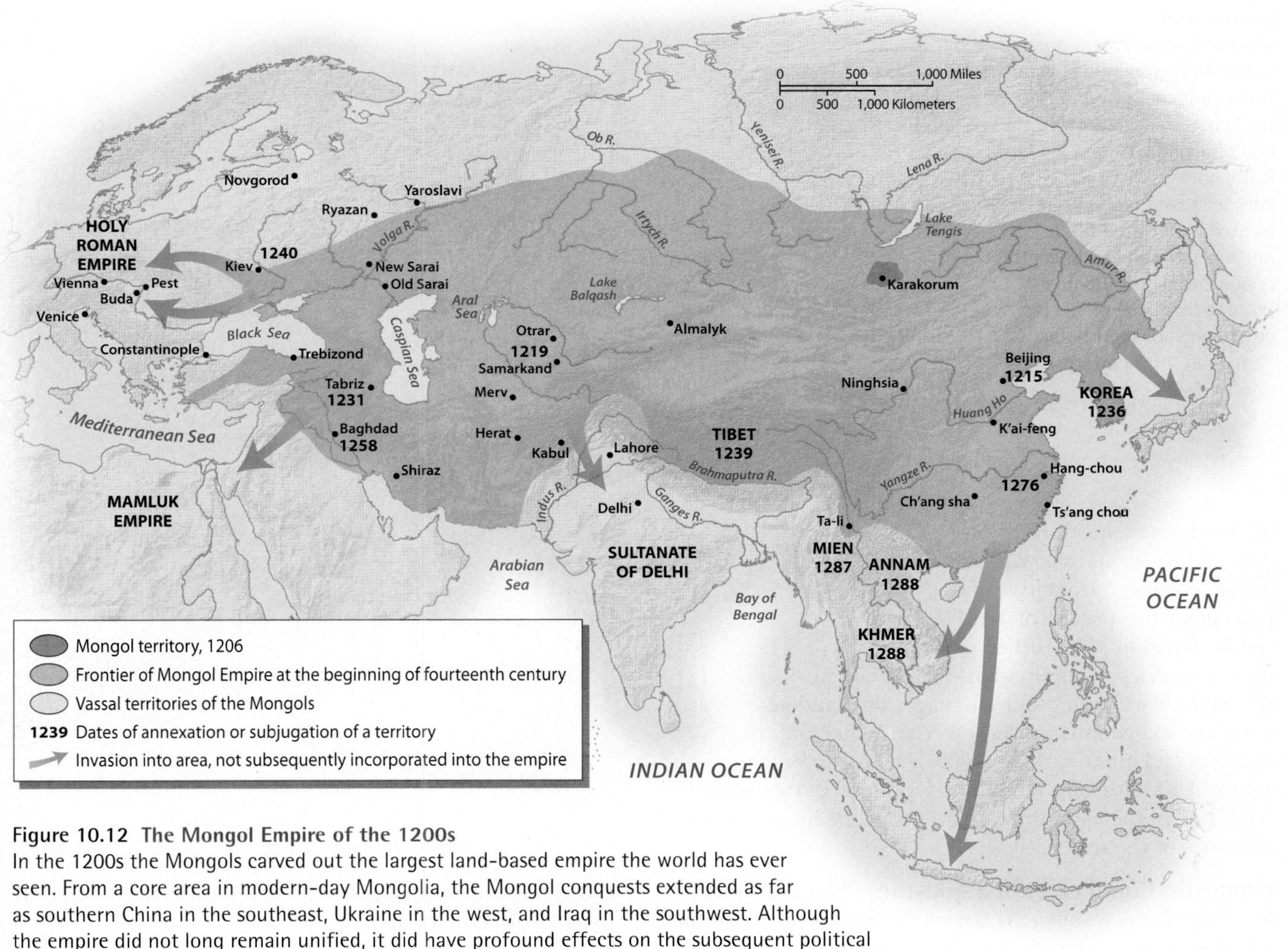

Figure 10.12 The Mongol Empire of the 1200s
In the 1200s the Mongols carved out the largest land-based empire the world has ever seen. From a core area in modern-day Mongolia, the Mongol conquests extended as far as southern China in the southeast, Ukraine in the west, and Iraq in the southwest. Although the empire did not long remain unified, it did have profound effects on the subsequent political and economic history of Eurasia.

Turkic Languages Far more Central Asians speak Turkic languages than Mongolian and Tibetan combined. Central Asia's Turkic linguistic sphere extends from Azerbaijan in the west through China's Xinjiang province in the east. The various Turkic languages are not as closely related to each other as are the dialects of Mongolian, but they are still obviously kindred tongues. Six main Turkic languages are found in Central Asia; five are associated with newly independent (former Soviet) republics of the west, while the sixth, Uyghur, is the main indigenous language of northwestern China.

Uyghur is an old and important language, dating back almost 2,000 years. The Uyghur number about 8 million, almost all of whom live in Xinjiang. As recently as 1949, the Uyghur formed about 90 percent of the population of Xinjiang; now, because of Han Chinese immigration, that figure has dropped to 45 percent. Eastern and northern Xinjiang are now largely Chinese-speaking. There are also about 1 million Kazakh speakers in far northern Xinjiang.

Of the six countries of the former Soviet Central Asia, five—Kazakhstan, Uzbekistan, Turkmenistan, Kyrgyzstan, and Azerbaijan—are named after the Turkic languages of their dominant populations. In three of these countries the indigenous people form a substantial majority. Some 82 percent of the people of Azerbaijan speak Azeri (there are more Azeris in northern Iran, however, than there are in Azerbaijan); some 80 percent of the people of Uzbekistan speak Uzbek as their native tongue; and some 77 percent of the people in Turkmenistan speak Turkmen. With more than 23 million speakers, Uzbek is the most widely spoken Central Asian language. In the Amu Darya Delta in the far north of Uzbekistan, however, most people speak a different Turkic language called Karakalpak, while the country's older cities, such as Samarkand, are still partly Tajik- (or Persian-) speaking. Kazakh speakers, moreover, are found in the sparsely populated Uzbek deserts. In Azerbaijan, non-Azeris, including such peoples as the Talysh who speak a language related to Persian, have been under pressure to conform to the linguistic and cultural norms of the national majority.

In the two other Turkic republics, the nationality for which the country is named forms less than two-thirds of the total population. Some 65 percent of the inhabitants

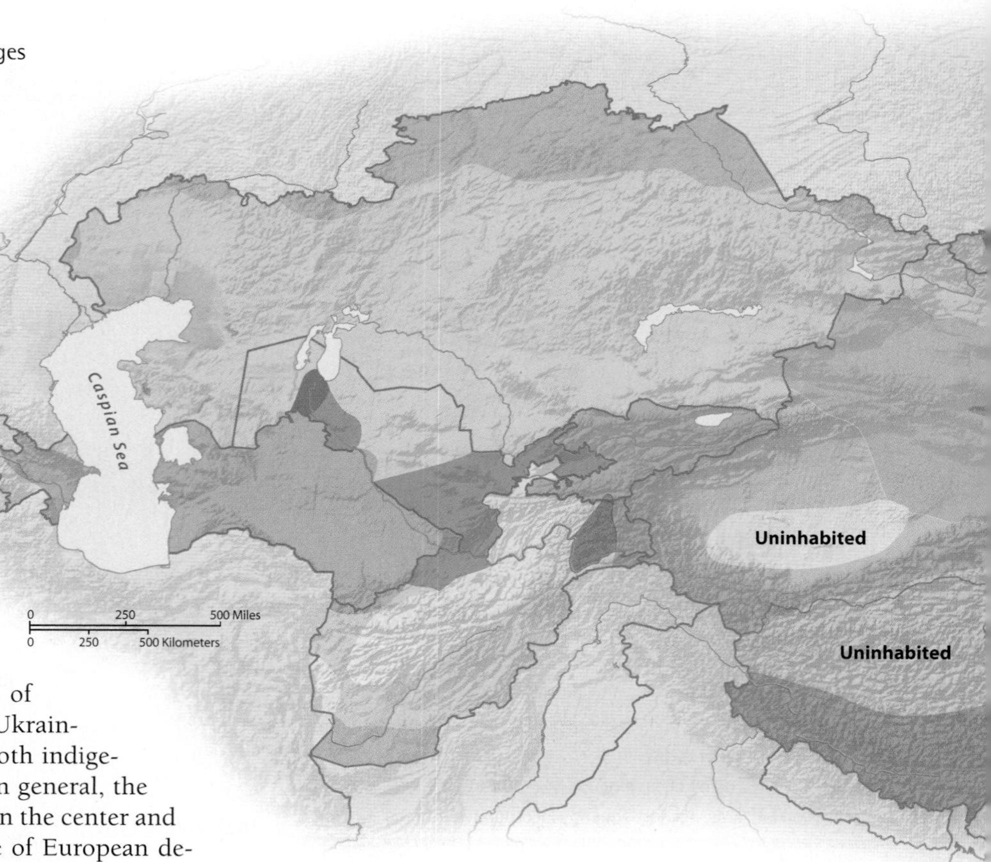

Figure 10.13 Linguistic Geography of Central Asia
Most of Central Asia is dominated by languages in the Altaic family, which includes both the Turkic languages (found through most of the center and the west of the region) and Mongolian (found in Central Asia's northeast). Several Indo-European languages, however, are located in both the far northwestern and southcentral regions, while the Tibeto-Burman language of Tibetan covers most of the Tibetan Plateau in southeastern Central Asia.

of Kyrgyzstan speak Kyrgyz as their native language, while roughly 60 percent of the people of Kazakhstan speak Kazakh. The other residents of these countries speak Russian, Uzbek, Ukrainian, and a variety of other languages both indigenous and exogenous to Central Asia. In general, the Kazakhs and other Turkic peoples live in the center and south of Kazakhstan, while the people of European descent live in the agricultural districts of the north and in the cities of the southeast.

Linguistic Complexity in Tajikistan The sixth republic of the former Soviet Central Asia, Tajikistan, is dominated by people who speak an Indo-European rather than a Turkic language. Tajik is so closely related to Persian that it is considered to be a Persian (or Farsi) dialect. Roughly 5.8 million people in Tajikistan, about 80 percent of the total population, speak Tajik as their main language. The remote mountains of eastern Tajikistan are populated by peoples speaking a variety of distinctive Indo-European languages, sometimes collectively referred to as "Mountain Tajik."

Tajikistan, like much of the rest of former Soviet portion of Central Asia, is noted for its complex mixture of languages. About a 15 percent of its people, for example, are Uzbeks, resulting in a certain degree of ethnic animosity. Such linguistic mixing was even more pronounced in earlier centuries, as Soviet policy encouraged gradual ethnic homogenization. The Soviet authorities also

Figure 10.14 Uyghur Mosque
A small mosque, illustrating traditional Uyghur architecture, survives amid blocks of modern apartments in Urumchi, Xinjiang. Traditional forms of housing and urban design can still be found in Uyghur communities in northwestern China, but they are gradually disappearing. *(Chris Stowers/Panos Pictures)*

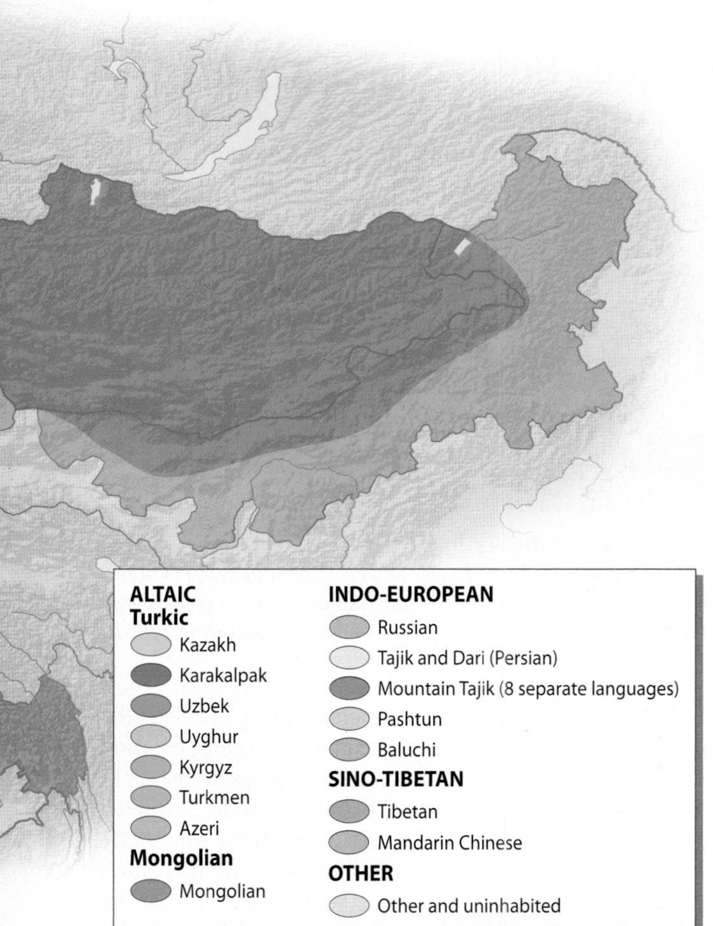

devised complex political boundaries among the different peoples of Central Asia, partly in order to play one group against another and thus bolster their own authority.

Language and Ethnicity in Afghanistan The linguistic geography of Afghanistan is even more complex than that of Tajikistan (Figure 10.15). Afghanistan was never colonized by outside powers, and it is one of the few countries of the world to have inherited the boundaries of a premodern, indigenous kingdom. This kingdom emerged in the 1700s on traditional dynastic lines that did not reflect ethnic or linguistic divisions ("dynastic" linkages are those based on the family of the monarch). The modern nation-state ideal—that each country should be identified with a particular national group—never had any currency in Afghanistan. (See "People on the Move: Afghan Refugees Return Home.")

The eighteenth-century creators of Afghanistan were mostly members of the Pashtun ethnic group, but they did not attempt to build a nation-state around Pashtun identity. Indeed, more Pashtuns live in Pakistan than in Afghanistan. In Afghanistan, estimates of the proportion of the populace speaking Pashtun vary from 40 to 60 percent. Most of them live to the south of the Hindu Kush.

Approximately half of the people of Afghanistan speak Dari, Afghanistan's variant of Persian, as their first language. Dari speakers are concentrated in the north. Three separate ethnicities are ascribed to the Dari-speaking people: the settled farmers and townspeople in the west and north are Tajik; the traditionally seminomadic people of the west-center are Aimaks; and the East Asian–appearing villagers of the central mountains are Hazaras, reputed to be descendants of medieval Mongol conquerors. Although

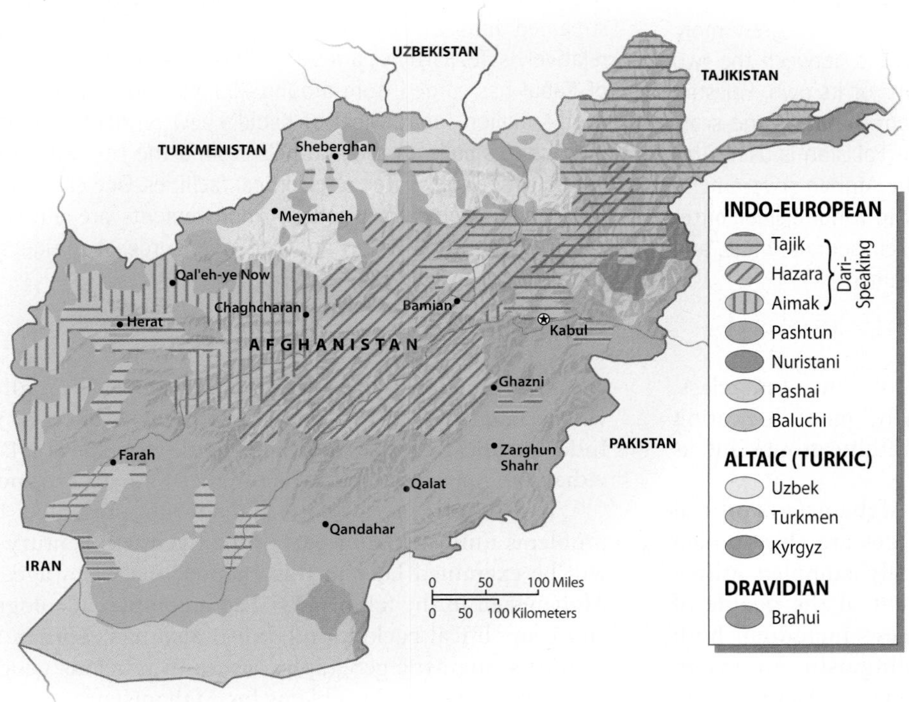

Figure 10.15 Afghanistan's Ethnolinguistic Patchwork
Afghanistan is one of the world's more ethnically complex countries. Its largest ethnic group is that of the Pashtuns, a people who inhabit most of the southern portion of the country as well as the adjoining borderlands of Pakistan. Northern Afghanistan is inhabited mostly by Uzbeks, Tajiks, and Turkmens—whose main population centers are located in Uzbekistan, Tajikistan, and Turkmenistan, respectively. The Hazaras of Afghanistan's central mountains, like the Tajiks, speak a form of Persian but are considered to be a separate ethnic group in part because they, unlike other Afghans, follow Shiite rather than Sunni Islam. Smaller groups are found elsewhere in the country.

PEOPLE ON THE MOVE

Afghan Refugees Return Home

On April 15, 2007, over 9,000 Afghan refugees returned to their home country from refugee camps in Pakistan. Before leaving, most received $100 from United Nations High Commissioner for Refugees (UNHCR) "encashment centers" where sophisticated procedures, including iris scanning of individuals above age five, were instituted to prevent fraud (Figure 10.4.1). Many refugees returned with more than cash. Some had dismantled their houses and were literally porting their beams back to Afghanistan.

This one-day rush into Afghanistan was not entirely unprecedented. In the four preceding months, more than 200,000 Afghans had returned home, bringing the total number of repatriations since 2002 to well over 3 million. But April 15 was significant as it marked the end of a "grace period"; after that date, Pakistan would begin prosecuting undocumented Afghans residing in the country. The over 2 million *registered* Afghans in Pakistan had little to fear, but only for the time being. By December 2009, Pakistan's government insists, they too must return to their country of origin.

The story of the Afghan refugees dates back to the late 1970s when the Soviet Union invaded Afghanistan and the country entered a period of prolonged bloodshed. At first, most Afghan refugees streaming into Pakistan were relatively well treated, as large amounts of money poured into the refugee camps that became centers of anti-Soviet military resistance. By the mid-1990s, however, funds for camp schools and clinics had largely vanished. The fall of Afghanistan's Taliban government in late 2001 was followed by the return of many refugees, but as the security situation in Afghanistan again deteriorated, many Afghans remaining in Pakistan decided to stay put. Their situation grew more vulnerable, however, as tensions mounted between the two countries. With deep political problems of its own, Pakistan decided to address the refugee problem through large-scale expulsions. Some observers think that Pakistan is using this threat as a tool for bargaining with the Afghan government. The two governments remain locked in a number of bitter disputes, including one focused on their boundary line, and attempts at negotiation have been strained.

Figure 10.4.1 Refugees
In March 2007, Afghan refugees in a camp in Peshawar, Pakistan were frantically loading their belongings on trucks in preparation for returning home to Afghanistan. *(Arshad Arbab/epa/Corbis)*

The Afghan refugees are themselves split over the planned repatriation. A sizable segment of the population was born or grew up in Pakistan and thus has no real links to Afghanistan and no desire to return. Some refugees, however, long for home and lost loved ones; others are eager to seek economic opportunities in Afghanistan's cities. One prospective returnee told reporters that he could make $2.80 (U.S.) a day in the Afghan city of Jalalabad, as compared to his $1.65 daily wage in the refugee camp.

Because of the dismal security situation in southern Afghanistan, many if not most returning refugees have streamed into the capital city of Kabul, which remains relatively safe. Partly as a result of this flow, the population of Kabul has surged from around 700,000 in the 1970s to over 4 million today. Most of Kabul's new residents live in mud houses built on the arid hills around the city, with no access to sewage, water, or electrical facilities. One can only hope that some infrastructural improvements are put in place by late 2009, when another huge surge of refugees could enter the city.

traditionally forming something of an underclass, Afghanistan's Hazaras are now showing more dedication to education than the country's other ethnic groups and as a result are beginning to prosper.

Although Pashtun and Dari are Afghanistan's official languages, a variety of other languages are also spoken in the country, creating an especially complex ethnic patchwork. Approximately 11 percent of the people of Afghanistan speak Turkic languages, including both Uzbek and Turkmen. Other ethnolinguistic groups in Afghanistan include the Baluchis, the Nuristanis (who speak five separate languages), and the Pashai, all of whose languages fall in the Indo-Iranian branch of the Indo-European family, and the Brahui, who speak a Dravidian language related to the tongues of southern India.

The ethnic mixture in Afghanistan presented few problems until the latter part of the twentieth century. As will be examined later in this chapter, the war waged in Afghanistan in the fall of 2001 had a complex ideological and geopolitical background, but it also had roots in the country's linguistic geography. Issues of religious geography have also caused problems for Afghanistan.

Geography of Religion

An assortment of religions characterized ancient and medieval Central Asia. The major overland trading routes of premodern Eurasia crossed the region, giving easy access to both merchants and missionaries. Several varieties of Buddhism, Islam, Christianity, and Judaism, as well as several minor religions, have all thrived at various times and places within the region. Eventually, however, religious lines hardened, and Central Asia was divided into two opposed spiritual camps: Islam triumphed in the west and center, while Tibetan Buddhism prevailed in Tibet and Mongolia.

Islam in Central Asia As is true elsewhere in the Muslim world, different Central Asian peoples are known for their different interpretations of Islamic orthodoxy. The Pashtuns of Afghanistan are noted for their strict Islamic ideals—although critics contend that Pashtun religious practices, such as forbidding women's faces to be seen in public, are based more on their own customs (Figure 10.16). By contrast, the traditionally nomadic groups of the northern steppes, such as the Kazakhs, are often considered lax in their religious observances. Although most of the region's Muslims are Sunnis, Shiism is dominant among the Hazaras of central Afghanistan, the Azeris of Azerbaijan, and the mountain dwellers of eastern Tajikistan.

Figure 10.16 Afghan Women in Public
Especially in the Pashtun areas of Afghanistan, women have traditionally been forced to cover their entire bodies when in public areas. In areas that were controlled by the Taliban, such dress codes were strictly enforced. Extremely modest dress still prevails in Afghanistan. *(Lemoyne/Getty Images, Inc.–Liaison)*

Under the communist rule of China, the Soviet Union, and Mongolia, all forms of religion were discouraged. Chinese authorities and student radicals suppressed Islam in Xinjiang during the Cultural Revolution of the late 1960s and early 1970s. Mosques were destroyed or converted to museums and religious schools were closed. Periodic persecution of Islam also occurred in Soviet Central Asia, and until the 1970s many observers thought that that religion was slowly disappearing from the region.

Religious expression was not so easily discouraged, however. Interest in Islam began to grow in former Soviet Central Asia in the 1980s. In the post-Soviet period, Islam continues to revive as people reassert their indigenous heritage and identity. Thus far, however, secularism remains widespread and signs of mass Islamic political fundamentalism are few. Still, most Central Asian leaders remain wary of any signal that Islamic fundamentalism is on the rise. In 2005, for example, Tajikistan banned the wearing of Islamic head-scarves in public schools. In response to widespread criticism, the country refined its dress code in 2007 to also ban revealing Western styles—as well as graduation parties and the use of mobile phones and private cars at the country's secondary schools.

In China, Muslims now enjoy basic freedom of worship, but the state still closely monitors religious expression out of fear that it is entangled with political separatism and terrorism. Although some evidence suggests that Islam has served as a focal point of an anti-Chinese movement among the Uyghur people, most Uyghur leaders insist that their program is not radically fundamentalist.

In Afghanistan, however, Islamism (or politically radical Islamic fundamentalism) did emerge as a powerful political movement. From the mid-1990s until 2001, most of the country was controlled by the Taliban, an extremist organization that insisted that all aspects of society conform to its own harsh version of Islamic orthodoxy. This commitment was demonstrated in early 2001 when Afghanistan's religious authorities oversaw the destruction of Buddhist statues in the country—including some of the world's largest and most magnificent works of art—as symbols of a non-Islamic past. Although the Taliban regime fell in late 2001, Islamism remains a powerful force in Afghanistan.

Islam is not the only religion represented in western Central Asia. Many Russians belong to the Russian Orthodox Church, and Uzbekistan has a small Jewish population. Kazakhstan also counts some 100,000 Catholics, mostly people of Polish or German descent.

Tibetan Buddhism Mongolia and Tibet stand apart from the rest of Central Asia in the adherence of their people to Tibetan Buddhism. Buddhism entered Tibet from India many centuries ago, where it merged with the indigenous religion of the area, called Bon. The resulting faith is more oriented toward mysticism than are other forms of Buddhism, and it is more hierarchically organized. Standing at the apex of Lamaist society is the Dalai Lama, considered to be a reincarnation of the **Bodhisattva** of Compassion (a Bodhisattva is a spiritual being who helps others attain enlightenment). Ranking below him is the Panchen Lama, followed by other religious officials. Until the Chinese conquest, Tibet was essentially a **theocracy**, or religious state, with the Dalai Lama enjoying political as well as religious authority. A substantial proportion of Tibet's male population has traditionally become monks, and monasteries once wielded both economic and political power (Figure 10.17).

Tibetan Buddhism suffered severe persecution in the 1960s. The Dalai Lama fled to India with many of his followers after China invaded Tibet in 1959, and has since been a powerful advocate for the Tibetan cause in international circles. During the 1960s and 1970s, an estimated 6,000 Tibetan Buddhist monasteries were destroyed and thousands of monks were killed. The number of active monks today is only about 5 percent of what it was before the Chinese occupation. Many monasteries have been allowed to reopen, although their activities are severely limited. Still, the Buddhist faith continues to form the basis of Tibetan identity, and in so doing helps keep alive the dream of independence or at least of real autonomy within China.

In Mongolia the downfall of communism has allowed the Tibetan Buddhist faith to experience a renaissance. Several monasteries have been refurbished, and many people are returning to their national religion. The intensity of Buddhist belief, however, is not as strong in Mongolia as it is in Tibet.

Central Asian Culture in International Context

During the Soviet period, the Russian language spread widely through western Central Asia. Russian served both as a common language and as a means of instruction in higher education. One had to be fluent in Russian in order to reach any position of responsibility. The Cyrillic (or Russian) script, moreover, replaced the modified Arabic script that was previously used for the indigenous languages. Russian speakers settled in all of the major cities, and many became influential. After the fall of the Soviet Union, however, most Russians migrated back to Russia, especially from the region's poorer countries. As a result, the use of Russian in education, government, business, and the media has declined in favor of local languages.

But despite its recent decline, Russian remains the common tongue of the former Soviet zone of Central Asia. Even many of the leaders of these countries speak their own languages imperfectly, and Russian is the only language that can be effectively used for communication across Central Asia's national boundaries. The continuing prominence of Russia, however, irritates many people. Kazakh nationalists, for example, are upset that Russian remains the country's "official language of communication between different ethnic groups," and they recently lobbied successfully to switch from the Cyrillic to the Roman alphabet for writing Kazakh. In 2007, Tajikistan went so far as to order its citizens to change their names if they sounded too Russian; President Emomali Rakhmonov thus adopted the most Tajik-sounding surname of "Rakhmon."

Although Central Asia is remote and poorly integrated into global cultural circuits, it is hardly immune to the forces of globalization. Even the tensions existing throughout the region between religious and secular orientations, and between ethnic nationalism and multiethnic inclusion, are aspects of the current global condition. So, too, the increased usage of English throughout Central Asia shows that this part of the world is not cut off from global culture. Such influences are especially marked in the oil cities of the Caspian Basin, such as Azerbaijan's Baku.

In late 2006, an internationally focused cultural uproar hit Kazakhstan in regard to Sacha Baron Cohen's hit film *Borat*. Hard-line Kazakh nationalists, furious at the film's misleading portrayal of their country, were able to limit local distribution and shut down a Kazakhstan-based *Borat* Website. Others in Kazakhstan argued that such actions would only discredit their government and encourage Cohen to continue mocking their country. They also noted that a large increase in inquiries from abroad about tourism in Kazakhstan followed the release of the film, which in turn encouraged Kazakhstan's Ministry of Tourism and Sports in its efforts to build major new ski resorts in the mountains along the border with Kyrgyzstan.

Figure 10.17 Lamaist Buddhist Monastery
Tibet is well known for its large Buddhist monasteries, buildings that in earlier years served as seats of political as well as religious authority. The Potala Palace in Lhasa, traditional seat of the Dalai Lama, is the largest, most important, and most famous of such monastic establishments. *(Alain le Garsmeur/Panos Pictures)*

GEOPOLITICAL FRAMEWORK: Political Reawakening in a Power Void

Central Asia has played a minor role in global political affairs for the past several hundred years. Before 1991 the entire region, except Mongolia and Afghanistan, lay under direct Soviet or Chinese control. Mongolia, moreover, was a Soviet satellite, and even Afghanistan came under Soviet domination in the late 1970s. Although the breakup of the Soviet Union saw the emergence of six new Central Asian countries, all of them remain geopolitically insecure (Figure 10.18).

Partitioning of the Steppes

Although Central Asia through most of the twentieth century was largely controlled by outside powers, this was not always the case. Before 1500, the region was a power center whose mobile armies threatened the far more populous, sedentary states of the Eurasian rim. Military developments, however, changed the balance of power, allowing the wealthier agricultural states to defeat the nomadic pastoralists and take over their lands by the 1700s. The winners in this struggle were the two largest states bordering the steppes: Russia and China.

The Manchu conquest of China in 1644 undercut the autonomy of the peoples of the eastern steppe. The Manchus came from Manchuria, the eastern borderlands of Central Asia, and were themselves skilled in the arts of cavalry warfare. By the mid-1700s the Chinese empire, now ruled by the Manchus, stood at its greatest territorial extent. Within its grasp lay not only Mongolia and Xinjiang but also Tibet and a slice of modern Kazakhstan.

Although Manchu-ruled China declined rapidly after 1800, it was still able to retain most of its Central Asian territories through the 1800s. But by the early 1900s, Chinese authority had begun to diminish in Central Asia as well. When the Manchu (also called Ch'ing or Qing) dynasty fell in 1912, Mongolia became independent, although China did manage to keep the extensive borderlands of Inner Mongolia (Nei Mongol). Tibet had earlier gained de

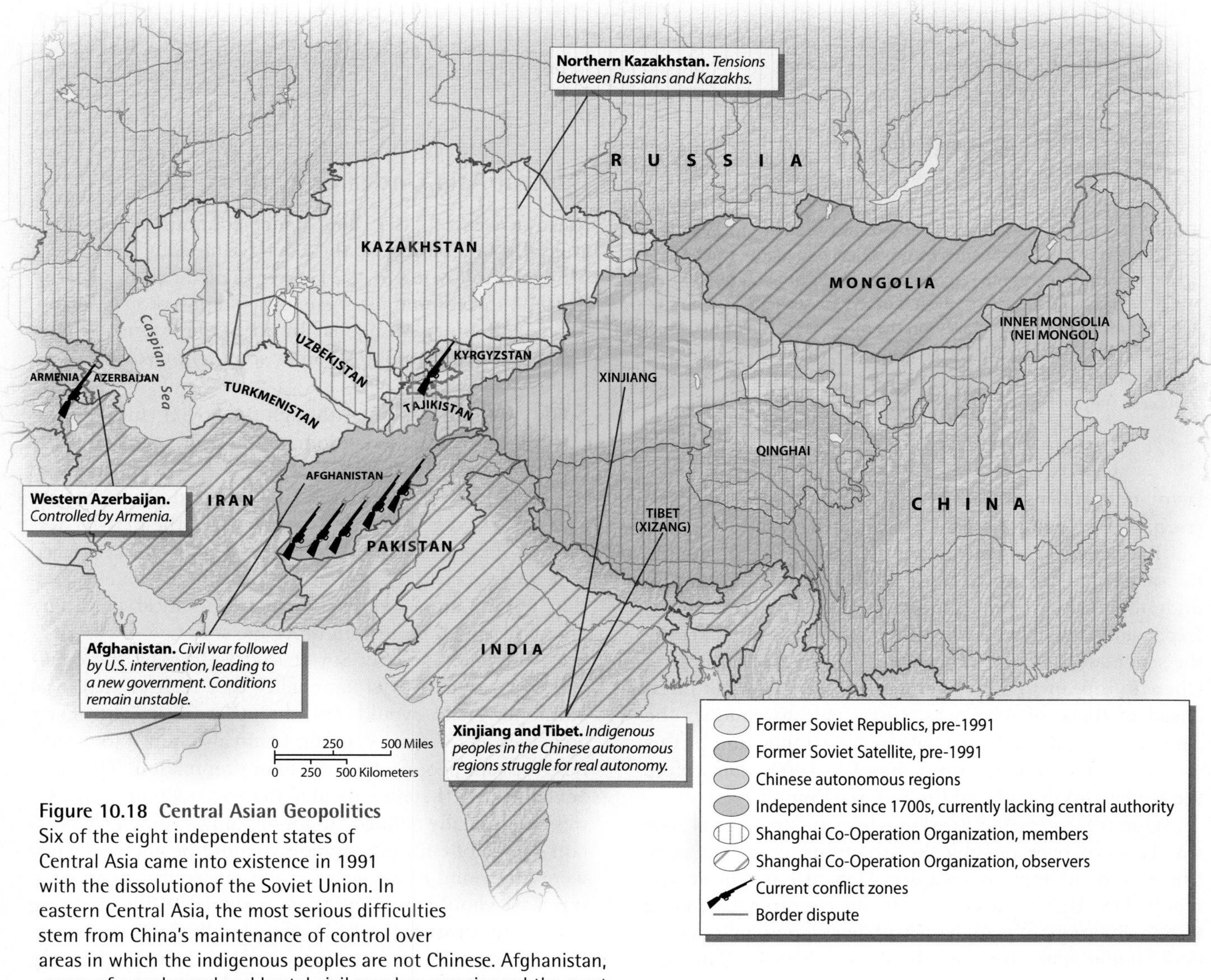

Figure 10.18 Central Asian Geopolitics Six of the eight independent states of Central Asia came into existence in 1991 with the dissolutionof the Soviet Union. In eastern Central Asia, the most serious difficulties stem from China's maintenance of control over areas in which the indigenous peoples are not Chinese. Afghanistan, scene of a prolonged and brutal civil war, has experienced the most extreme forms of geopolitical tension in the region.

Figure 10.19 **Political Boundaries Around the Fergana Valley** Some of the world's most convoluted political boundaries can be found in the vicinity of the Fergana Valley. The central portion of the valley belongs to Uzbekistan, which is otherwise separated from it by high mountains. The lower valley, on the other hand, is part of Tajikistan, the core area of which is likewise separated from the valley by highlands. The Fergana's upper periphery belongs to Kyrgyzstan. Note also the small exclaves of Uzbekistan within Kyrgyzstan.

facto independence, and even Xinjiang lay beyond the reach of effective Chinese authority during the 1920s.

Russia began to advance into Central Asia at roughly the same time as China. In the 1700s the Russian Empire undertook the systematic conquest of the Kazakh steppes. Expansion farther to the south, however, was blocked by the sedentary states of Uzbekistan. Only in the late 1800s, when European military techniques and materials raced ahead of those of Asia, was Russia able to conquer the Amu Darya and Syr Darya valleys. Its subjugation of this area was not completed until the early 1900s, just before the Soviet Union replaced the Russian Empire.

One reason for the Russian advance into Central Asia was concern over possible British influence in the area. Britain did attempt to conquer Afghanistan, but it was rebuffed by Afghan forces—and by the country's forbidding terrain. Subsequently, Afghanistan's position as an independent "buffer state" between the Russian Empire (later the Soviet Union) and the British Empire in South Asia remained secure. The British also sent a major military expedition to Tibet in the early twentieth century and came close to creating an autonomous Tibetan state under British "protection." This move heightened China's determination to regain control over Tibet.

Central Asia Under Communist Rule

Western Central Asia came under communist rule soon after the foundation of the Soviet Union in 1917, with Mongolia following in 1924. After the Chinese revolution of 1949, the communist system was imposed on Xinjiang, Tibet, and Inner Mongolia. In all of these areas, major changes in the geopolitical order soon followed.

Soviet Central Asia Although the Soviet Union inherited the Russian imperial domain in Central Asia virtually intact, it enacted new policies. The Soviet regime sought to create a socialist economy and to build a new society that would eventually knit together all of the massive territories of the country. Central Asia's leaders were replaced by Communist Party officials loyal to the new state; Russian immigration was encouraged; and local languages could no longer be written in Arabic script.

Although the early Soviet leaders foresaw the emergence of a single Soviet nationality, they realized that local ethnic diversity would not disappear overnight. Early Soviet leaders such as Vladimir Lenin also hoped to protect non-Russian peoples from Russian domination. The early Soviet leaders therefore divided the Soviet Union into a series of nationally defined "union republics" in which a certain degree of cultural autonomy would be allowed. They were uncertain, however, what the relevant units in Central Asia should be. Was there a single Turkic-speaking nationality, or were the Turkic peoples divided into a number of separate nationalities, with the Tajiks forming yet another? For several years boundaries shifted as new "republics" appeared on the map. Finally, in the 1920s the modern republics of Kazakhstan, Kyrgyzstan, Tajikistan, Uzbekistan, Turkmenistan, and Azerbaijan assumed their present configurations. In certain areas, such as the fertile Fergana Valley, the political boundaries so drawn remained extremely complex and contested (Figure 10.19).

Some scholars argue that the Soviet policy backfired severely. Rather than forming a transitional step on the way to a Soviet identity, the constituent republics of the Soviet Union instead nurtured local nationalisms that ultimately undermined the Soviet system. Identities such as Turkmen, Uzbek, and Tajik that had been somewhat vague in the pre-Soviet period were now given real political significance. Another problem undercutting Soviet unity was the fact that the cultural and economic gaps separating Central Asians from Russians did not diminish as much as planned. Islam remained entrenched in some areas and began to revive elsewhere in the 1980s.

The Chinese Geopolitical Order After China reemerged as a united country in 1949, it too was able to reclaim most of its old Central Asia territories. China's communist leaders promised the non-Chinese peoples a significant degree of political self-determination as well as cultural autonomy, and thus they found much local support in Xinjiang. Tibet, isolated behind its mountain walls and virtually independent for many years, presented a greater obstacle. China occupied Tibet in 1950, but the Tibetans launched a rebellion in 1959. When the rebellion was crushed, the Dalai Lama and some 100,000 followers found refuge in India (Figure 10.20).

Loosely following the Soviet nationalities model, China established autonomous regions in areas occupied primarily by non-Han Chinese peoples, including Xinjiang, Tibet proper (called Xizang in Chinese), and Inner Mongolia. Such autonomy, however, often turned out to be more theoretical than real, and it did not prevent the massive immigration of Han Chinese into these areas. Nor were all parts of Chinese Central Asia granted autonomous status. The large and historically Tibetan and Mongolian province of Qinghai, for example, remained an ordinary Chinese province.

Current Geopolitical Tension

Although the former Soviet portion of Central Asia weathered the post-1991 transition to independence relatively smoothly, the region still suffers from a number of actual and potential ethnic conflicts as well as from the struggle between radical Islam and secular government. Much of China's Central Asian territory is troubled, but China retains a firm grip. Afghanistan experienced a particularly brutal civil war that was temporarily brought to an end by U.S. occupation following September 11, 2001. By 2003, however, Afghanistan's seemingly interminable war had again been reignited.

Independence in Former Soviet Lands The breakup of the Soviet Union in 1991 generally proceeded peacefully in Central Asia. The six newly independent countries had been dependent on the Soviet system, however, and it was no simple matter for them to chart their own courses. They still had to cooperate with Russia on security issues, and all initially opted to remain part of the Commonwealth of Independent States, the rather hollow successor of the Soviet Union. In most cases authoritarian rulers, rooted in the old order, retained power and sought to undermine opposition groups. All told, democracy made less progress in Central Asia than in other parts of the former Soviet Union. Kyrgyzstan seemed to form an exception to this rule, especially in 2005 when citizens protesting corruption and authoritarianism forced President Askar Akayev out of office. By 2007, however, the new Kyrgyzstan government appeared to be following in the footsteps of the old, resulting in new rounds of ineffective protest and widespread political unrest.

Kazakhstan, Tajikistan, and Azerbaijan faced more difficult post-Soviet transitions. Many Kazakhs wanted to create a national state centered on Kazakh identity, and they resented the presence and power of Russians, Ukrainians, Germans, and others of European background. These Europeans feared what they considered alien Central Asian cultural standards, and for a time there was some concern that the Russians of northern Kazakhstan, the country's breadbasket, might attempt to join their land with Russia. In the late 1990s, Kazakhstan moved its capital from Almaty in the south to Astana in the Russian-dominated north, partly to forestall any plans for secession. The continued out-migration of Russians and other Europeans from Kazakhstan, however, has gradually made this issue irrelevant.

In Tajikistan war broke out almost immediately after independence in 1991. Many members of the smaller ethnic groups living in the mountainous east resented the authority of the lowland Tajiks and thus rebelled. They were joined by Islamist groups seeking to overthrow the secular state. The civil war was officially ended in 1997 when the Islamists agreed to work through normal political channels. However, political tensions remain pronounced,

Figure 10.20 Chinese Invasion of Tibet
In 1959 China launched a massive invasion of Tibet. Pursued by the Chinese army, the 23-year-old Dalai Lama (second from the lead) is shown here escaping over the Zsagola Pass, ultimately to find refuge in India. *(Hg/AP/Wide World Photos)*

and much of the mountainous area is still beyond the effective control of the central government. By 2007, continuing violence led Tajikistan's government to increase its pressure on the Islamic Renaissance Party, the only legal Islamist party in former Soviet Central Asia.

Azerbaijan also experienced strife following the breakup of the Soviet Union. Armenia invaded and occupied a portion of far-western Azerbaijan, allowing the Armenian-speaking highlands in the western portion of the country to form the "breakaway republic" of Nagorno-Karabakh. Hostile relations with Armenia make it difficult for Azerbaijan to control its **exclave** of Naxcivan, a piece of Azerbaijani territory separated from the rest of the country by Armenia and Iran. Although the situation remains tense, Armenia and Azerbaijan announced in 2007 that they were willing to consider peace proposals put forth by international mediators.

Although Uzbekistan and Turkmenistan had relatively easy transitions out of the Soviet Union, both countries are ruled by repressive governments, allowing little room for personal freedom or opposition movements. In 2007, Uzbekistan increased its censorship of the news, blocked many Internet sites, and cracked down on human rights. Until his death in late 2006, Turkmenistan's president, Saparmurat Niyazov—who called himself simply "Turkmenbashi" ("father of the Turkmens")—maintained a lavish personality cult, forcing all students in his country to read and honor his lengthy book on Turkmen history, philosophy, and mythology. Upholding strict Turkmen nationalism, Niyazov kept his country out of most regional associations while forcing non-Turkmen workers out of public employment. Turkmenistan continues to be run by people loyal to the former dictator, and thus seems likely to continue following this highly nationalistic path. Indeed, the man named acting president after Niyazov's death is widely rumored to be his illegitimate son.

Strife in Western China Local opposition to Chinese rule in Central Asia increased during the 1990s, although without success. Any form of protest in Tibet has been severely repressed, but the Tibetans have brought the attention of the world to their struggle. China maintains several hundred thousand troops in a region that has only 2.8 million civilian inhabitants. Such an overwhelming military presence is considered necessary because of both Tibetan resistance and the strategic importance of the region. In early 2007, the Dalai Lama stated in several interviews that he had come to accept Chinese sovereignty over Tibet, leading many to hope for reduced tensions, but his announcement several weeks later that Tibet is "ruled by terror" shows that the conflict is far from being settled.

China's control of Xinjiang is more secure than its hold on Tibet. Although fragmented by deserts and mountains, Xinjiang's terrain is less forbidding than that of Tibet and it is now the home of millions of Han Chinese immigrants. Xinjiang is also more vital to China than is Tibet. It contains a variety of mineral deposits (including oil) essential for Chinese industry, and it has been the site of nuclear weapons tests. Many Uyghurs, not surprisingly, oppose such uses of their homeland, and they resent the periodic suppression of their religion by the communist regime (Figure 10.21).

China's position is that all of its Central Asian lands are integral portions of its national territory. Those who advocate independence are viewed as traitors and are sometimes considered to be in league with Western political forces that have sought for the past 200 years to keep China weak and divided. Chinese officials have also linked separatist elements in Xinjiang to a global movement of radical Islamic fundamentalists. While most Uyghur nationalist leaders claim that their movement is not founded on religion, the involvement of a number of

Figure 10.21 Ethnic Tension in Xinjiang
Relations between Han Chinese officials and Muslim Central Asians have grown increasingly tense in recent years. Here a Chinese official levies a trading tax on merchants in Kashgar's Sunday market. *(Chris Stowers/Panos Pictures)*

Uyghur separatists in Afghanistan's Taliban movement and in Al Qaeda has not helped their cause.

War in Afghanistan None of the conflicts in western China or in the former Soviet republics compares in intensity to the situation in Afghanistan. Afghanistan's troubles began in 1978 when a Soviet-supported military "revolutionary council" seized power. The new Marxist-oriented government began to suppress religion, which led almost immediately to widespread rebellion. When the government was about to collapse, the Soviet Union responded with a massive invasion. Despite its power, the Soviet military was never able to gain control of the more rugged parts of the country. Moreover, Pakistan, Saudi Arabia, and the United States ensured that the anti-Soviet forces remained well armed.

The exhausted Soviets finally withdrew their troops in 1989. The puppet government that they installed remained to face the insurgents alone. It managed to hold the country's core around the city of Kabul for a few years, largely because the opposition forces were themselves divided. Local warlords grabbed power in the countryside, destroying any semblance of central authority and committing a number of atrocities, including the shelling of civilian areas, the abduction of noncombatants, and the mass raping of women from enemy communities.

In 1995–96, the Taliban arrived on the Afghan scene. The Taliban was founded by young Muslim religious students disgusted with the anarchy that was consuming their country. They were convinced that only the firm imposition of Islamic law could end corruption, quell the disputes among the country's different ethnic groups, and ultimately bring peace to Afghanistan. Large numbers of soldiers flocked to the Taliban standard, and, with major military support from Pakistan, they won control of most of Afghanistan by the late 1990s. The Taliban also offered refuge in Afghanistan to the radical Islamist organization Al Qaeda, led by Osama bin Laden.

The Taliban acquired power not only from its religious nature, but also from the ethnic divisions of Afghanistan. It gained most of its strength from the Pashtuns of the southern half of the country. The Pashtuns have a reputation for militarism, as well as good connections for military supplies among their relatives across the Pakistan border. The main opposition to the Taliban came from the country's other ethnic groups, especially the Uzbeks and Tajiks of the north and the Shiite Hazaras of the central mountains.

By the end of the century, most of the people of Afghanistan lost faith in the Taliban, largely because of the severe restrictions that it imposed on daily life. Such constraints were most pronounced for women, but even men were compelled, by both whippings and imprisonment, to obey the Taliban's numerous decrees. Most forms of recreation were outlawed, including television, films, music, and even kite-flying. Little escaped the watchful eyes of the Taliban's religious police.

The aftermath of September 11, 2001, completely changed the balance of power in the region. The United States and Britain, working closely with anti-Taliban groups concentrated in northern Afghanistan, launched a major war against Al Qaeda and the Taliban government. Within a few months, formal Taliban power collapsed. By early 2002, the emphasis in Afghanistan shifted to rebuilding the country's institutions and establishing an effective central government. An interim administration emerged under Hamid Karzai, while a traditional "grand assembly" agreed upon a new constitution in 2004. Successful general elections allowed Karzai to keep his office, offering hope to the beleaguered country.

Unfortunately, the coming of democracy to Afghanistan did not solve the country's numerous political problems. By 2003, the Taliban had successfully regrouped in new bases in northeastern Pakistan, from which they reestablished partial control over large swaths of the Pashtun-speaking parts of Afghanistan. Small but bloody battles continue to break out between the Taliban movement and the Afghan government and the U.S. and NATO military forces that support it (Figure 10.22). Although the non-Pashtun northern and central portions of Afghanistan remain far more stable, local warlords still command sizable forces while the authority of the central government remains minimal.

Global Dimensions of Central Asian Geopolitics

As the previous discussion indicates, Central Asia has recently emerged as a key arena of geopolitical tension. Vying for power and influence in the region are a number of important countries, including China, Russia, Pakistan, India, Iran, and the United States. In addition, the revival of Islam has generated international geopolitical repercussions. Relations among the various countries of Central Asia are generally good, although Kyrgyzstan still has several minor border conflicts with its neighbors, particularly in regard to Uzbekistan's **exclaves** within its territory (Figure 10.20).

A Continuing U.S. Role After the downfall of the Taliban regime in 2001, the U.S. military retained a significant position in Afghanistan. As the fighting intensified after 2003, additional troops were eventually necessary; in early 2006 a NATO International Security Assistance Force was authorized by the UN Security Council to attempt to restore order in the southern half of the country. As of early 2007, roughly 27,000 U.S. troops were stationed in Afghanistan, along with an additional 24,000 troops from other NATO countries. While most of these forces stay in secure military bases, smaller units of Special Forces deploy in remote areas where the Taliban remains active. Achieving peace has not been easy, however, and in late 2006 the UN Security Council warned that Afghanistan remained at high risk of becoming a failed state.

Afghanistan is not the only Central Asian country to have experienced a substantial U.S. armed presence. After 9/11, the United States established military operations in Uzbekistan, Tajikistan, and Kyrgyzstan. These countries were eager to host the U.S. military because they feared that Islamic fundamentalist movements could overthrow their governments. Already by the late 1990s, the Islamic Movement in Uzbekistan (IMU), a group dedicated initially to the creation of an Islamic state in the Fergana Valley, was engaging the armies of Uzbekistan, Tajikistan, and Kyrgyzstan. As a result, military budgets increased throughout the region, straining governmental finances but also allowing Central Asian governments to get an upper hand on the IMU. The IMU suffered a major blow in early 2007 when a contingent of its fighters operating out of northern Pakistan had a falling-out with local Pashtun tribes, resulting in battles that left over 300 Uzbek militants dead.

The United States' military presence in Central Asia has generated a number of controversies. In 2005, Uzbekistan ordered the United States to abandon its local military base after the U.S. government harshly criticized the massacre of several hundred nonviolently protesting citizens by the Uzbekistan government. Evidence suggests that the influence of the United States is currently declining across much of Central Asia as Russia and China court the region's leaders. Mongolia, however, has sought closer ties with the United States (as well as with Japan), due largely to its concerns about falling under the influence of China or Russia. The United States also continues to maintain a military air base in Kyrgyzstan.

Relations with China and Russia After the breakup of the Soviet Union, relations between the newly independent countries of Central Asia and China remained tense for a number of years. China objected to the boundary lines separating it from Tajikistan and Kyrgyzstan and asked for adjustments. These boundary disputes were peacefully settled in the early years of the new millennium, with Kyrgyzstan ceding 222,400 acres (90,000 hectares) of disputed land to China. China subsequently initiated a series of diplomatic maneuvers designed to quell tensions, fight Islamic separatism, and gain greater access to the natural resources of Central Asia.

After the fall of the Soviet Union, Russia continued to regard all of the former Soviet territories of Central Asia as lying within its sphere of geopolitical influence, and as such has resented U.S. military initiatives. Russia has, moreover, retained or built military bases of its own in both Tajikistan and Kyrgyzstan. Economics and infrastructure also tie Russia to the former Soviet Central Asian sphere. The main transportation line linking European Russia to central Siberia, for example, cuts across northern Kazakhstan, while former Soviet Central Asia's rail and pipeline links to the outside world are still largely oriented toward Russia. Kazakhstan also contains the huge Baikonur Cosmodrome, Russia's main site for rocket launches.

By the early years of the new millennium, most Central Asian leaders concluded that the potential economic and political advantages to be gained by cooperation with Russia and China outweighed the disadvantages. The major consequence of this attitude was the formation of the **Shanghai Cooperation Organization** (or SCO, commonly referred to as the "Shanghai Six"), composed of China, Russia, Kazakhstan, Kyrgyzstan, Tajikistan, and Uzbekistan (Figure 10.23). The SCO seeks cooperation on such security issues as terrorism and separatism, aims to enhance trade, and serves as a counterbalance against the United States. Although not a member, Mongolia does have "observer" status in the SCO

Figure 10.22 War in Afghanistan Warfare continues in Afghanistan as troops from the United States and other countries attempt to root out remnants of the Taliban and maintain order. Tense encounters with local residents often result when foreign soldiers search for weapons and insurgent fighters. *(AP/Wide World Photos)*

Figure 10.23 **Meeting of the Shanghai Cooperation Organization** The leaders of the six countries that make up the Shanghai Cooperation Organization have gathered together to discuss their common concerns. Pictured here are the leaders (or former leaders) of Russia, China, Kazakhstan, Kyrgyzstan, Tajikistan, and Uzbekistan. *(ITAR-Tass/Konstantin Zavrashin, Presidential Press Services/AP Wide World Photos)*

(as do Iran, Pakistan, and India); Turkmenistan, on the other hand, has rejected membership not only in the SCO but in most other regional associations as well.

The Roles of Iran, Pakistan, India, and Turkey

Iran is also interested in the new republics of Central Asia. It is a major trading partner, and it offers a good potential route to the ocean. Since the completion of a rail link between Iran and Turkmenistan in the late 1990s, some of Central Asia's global trade has been reoriented toward Iran's ports. Iran's cultural links with the region are old and deep, particularly in Tajikistan and northern Afghanistan, where linguistic ties are strong. Iran remained a major enemy of Afghanistan's Taliban regime and offered military help to the Northern Alliance. It thus gained a good measure of influence in Afghanistan, especially in the western region of Herat and among the Shiite Hazaras of the central mountains. Iran's relations with Azerbaijan, on the other hand, remain tense, in part because many Azerbaijanis refer to northwestern Iran—an Azeri-speaking region—as "Southern Azerbaijan." Some analysts believe that Iran, which runs a large spy network

GEOGRAPHY IN THE MAKING Afghanistan's Never-Ending War

On November 10, 2007, six U.S. troops were killed in an ambush in eastern Afghanistan, bringing the total U.S. death count for the year to 108, and making 2007 the deadliest year for American forces since the beginning of the war in 2001. 2007 also saw increased fighting in areas of northern and western Afghanistan, such as the province of Herat, that had previously been relatively peaceful. Suicide bombings increased as well; one blast in November 2007 killed six members of Afghanistan's parliament. At the same time, the production of opium and heroin continued to surge, accounting by 2007 for over half of Afghanistan's total economic output, according to the United Nations Office on Drugs and Crime. Such figures and events seemed to support the U.N. Security Council's 2006 warning that Afghanistan was at risk of becoming a failed state due to the Taliban resurgence, drug production and trafficking, and weak institutions.

Not all the news coming out of Afghanistan in 2007, however, indicated failure on the part of NATO forces and the Afghan government. In the same year, several key Taliban leaders were killed or captured, including Mullah Dadullah, who had been in charge of military operations in southern Afghanistan. In October 2007, Canadian troops surrounded and defeated a group of approximately 300 insurgents, thus stopping a potential Taliban offensive on the important city of Qandahar. The U.S. military has responded to the worsening situation in Afghanistan by enhancing the role of the Marine Corps, which some observers think will improve security. Overall, Taliban casualties continue to be far higher than those of the United States and its allies.

Overall, however, the situation in Afghanistan can only be described as grim. Young Afghans continue to sign up with the Taliban, a movement that enjoys substantial support in the Pashtun-speaking parts of the country. High levels of unemployment, pervasive crime and insecurity, and civilian casualties at the hands of U.S. and NATO forces—as well as infighting among Pashtun tribal groups—continue to send men into the ranks for the Taliban. The Taliban also enjoys high levels of support, as well as vital sanctuaries, across the border in Pakistan's Federally Administered Tribal Districts. Finally, the Taliban has increasingly recruited foreign fighters, counting among its ranks soldiers from Uzbekistan, Chechnya, several Arabic-speaking counties, Turkey, and China. It can field as many as 10,000 insurgents at a given time, making it a potent military force.

Considering the combination of Taliban insurgency, global networks of militant Islam, rugged topography, and massive opium and heroin production, Afghanistan is likely to remain a war-wracked, impoverished, and insecure country for many years to come. As a result, it seems unlikely that U.S. military forces will be pulled out of the country any time soon.

in Azerbaijan, might bomb the oil pipeline that runs from Baku to the Mediterranean Sea in southern Turkey if Azerbaijan were to cooperate with the United States in any kind of attack on Iran's nuclear facilities.

Pakistan has also strived to gain influence in Central Asia, hoping that pipelines will eventually carry Central Asian oil and natural gas to its new deep-water port at Gwadar. Any such pipelines, however, would have to pass through the rugged topography and perilous political environment of Afghanistan. During the Taliban period, Pakistan enjoyed very close relations with Afghanistan's government. The aftermath of 9/11 thus put Pakistan in a serious bind. In response to both U.S pressure and promises, Pakistan joined the anti-Taliban coalition and supplied the U.S. military with valuable intelligence. The subsequent retreat of the Taliban into northwestern Pakistan, however, has resulted in extremely tense relations between the two countries, partially blocking off Pakistan's access into the rest of Central Asia.

Pakistan's interest in Central Asia ensures that India remains concerned about the region as well. India enjoys good relations with the government of Afghanistan (to which it has promised some $750 million in aid), and, like Pakistan, hopes to tap into the natural gas reserves of Central Asia. In 2006, India opened a military base in Tajikistan—its first in a foreign country—where it now maintains two squadrons of fighter aircraft.

Turkey's connections with Central Asia are potentially close. Most Central Asians speak Turkic languages, while Turkey also offers itself as the model—contrary to those of Iran and Pakistan—of the modern, secular state of Muslim heritage. It is unclear, however, whether the Turkish system can be exported easily, even to other Turkic-speaking states. But if Central Asia were to experience substantial Western-oriented economic growth, it seems likely that its relations with Turkey would strengthen.

ECONOMIC AND SOCIAL DEVELOPMENT: Abundant Resources, Devastated Economies

By most measures, Central Asia is one of the poorer regions of the world. Afghanistan in particular stands near the bottom of almost every list of economic and social indicators. Other Central Asian countries, however, enjoy relatively high levels of health and education, a legacy of the social programs enacted by their former communist regimes. These same governments, however, built inefficient economic systems, and after the fall of the Soviet Union, western Central Asia experienced a spectacular economic decline. Although much of the region is now experiencing a major economic expansion, growth has been largely based on the extraction of natural resources, particularly oil and natural gas. If fossil fuel prices decline significantly in the near future, Central Asia will be hard hit.

Economic Development in Central Asia

Soviet economic planners sought to spread the benefits of economic development widely across their country. This required building large factories even in remote areas such as Tajikistan regardless of the costs involved. Such Central Asian industries relied heavily on subsidies from the Soviet central government. When those subsidies ended, the industrial base of the region began to collapse, leading to plummeting living standards that have only recently begun to recover. But as is true elsewhere in the former Soviet Union, certain well-connected individuals have grown very wealthy since the fall of communism. The growing gap between the rich and the poor remains a major concern across most of Central Asia.

TABLE 10.2 Development Indicators

Country	GNI[a] per Capita, PPP[b] (2005)	GDP[c] Average Annual % Growth (2000–05)	Life Expectancy	Percent of Population Living on Less Than $2 a Day (2006)	Under Age 5 Mortality Rate		Gender Equity[d]
					1990	2005	
Afghanistan		12.0	42				55
Azerbaijan	4,890	12.7	72	1.5	105	89	98
Kazakhstan	7,730	10.1	66	16	63	73	106
Kyrgyzstan	1,870	4.0	66	21	80	67	105
Mongolia	2,190	5.8	66	75	108	49	116
Tajikistan	1,260	9.6	64	43	115	71	84
Turkmenistan			62		97	104	
Uzbekistan	2,020	5.3	67		79	68	96

[a]*Gross national income.*
[b]*Purchasing power parity.*
[c]*Gross domestic product.*
[d]*Ratio of female-to-male enrollments in primary and secondary school percentage. Numbers below 100 have more males in primary/secondary school; numbers above 100 have more females in primary/secondary schools.*

Source: World Bank, World Development Indicators, 2007; *percentage of population living on $2 a day data Population Reference Bureau,* World Data Sheet, 2007; *gender equity data from* Millennium Development Goals.

The Postcommunist Economies As Table 10.2 shows, no Central Asian country or region could be considered prosperous by global standards. Kazakhstan generally stands as the most developed, and it probably has the best prospects. Since Kazakhstan's agricultural base is potentially productive and its population density is low, it could emerge as a major food exporter. More important, it has two of the world's largest underutilized deposits of oil and natural gas, the vast Tengiz and Kashagan fields in the northeastern Caspian Basin (Figure 10.24), as well as sizable deposits of other minerals. Kazakhstan has signed agreements with Western oil companies to exploit its oil reserves. Since 2002, Kazakhstan's oil and gas wealth has led to a burgeoning economy.

Owing to its sizable population, Uzbekistan has the second largest economy in the region. Uzbekistan did not decline as sharply as its neighbors after the fall of the Soviet Union, largely because it retained many aspects of the old command economy (that is, an economy run by governmental planners rather than by private firms responding to the market). Uzbekistan remains the world's second largest exporter of cotton, and it has significant gold and natural gas deposits. However, environmental degradation threatens cotton production, and both inefficient state control and widespread corruption continue to hamper production in many economic sectors.

Kyrgyzstan, in contrast, moved aggressively after the fall of the Soviet Union to privatize former state-run industries. Along with Mongolia, it is Central Asia's only member of the World Trade Organization (WTO). But Kyrgyzstan's economy is heavily agricultural, few of its industries are competitive, and political strife has discouraged foreign investment. Kyrgyzstan does, however, enjoy the largest supply of freshwater in Central Asia, a resource that will likely become increasingly valuable in years to come, and its mineral reserves are substantial. Gold exports in particular are so vital to the Kyrgyz economy that a decline in production from a single mine in 2002 resulted in a 0.5 percent drop in the country's total economic output.

Figure 10.24 Oil and Gas Pipelines
Central Asia has some of the world's largest oil and gas deposits and recently has emerged as a major center for drilling and exploration. Because of its landlocked location, Central Asia cannot easily export its petroleum products. Pipelines have been built to solve this problem, and a number of others are currently being planned. Pipeline construction is a contentious issue, however, as several of the potential pathways lie across Iran, a country that remains under U.S. sanctions.

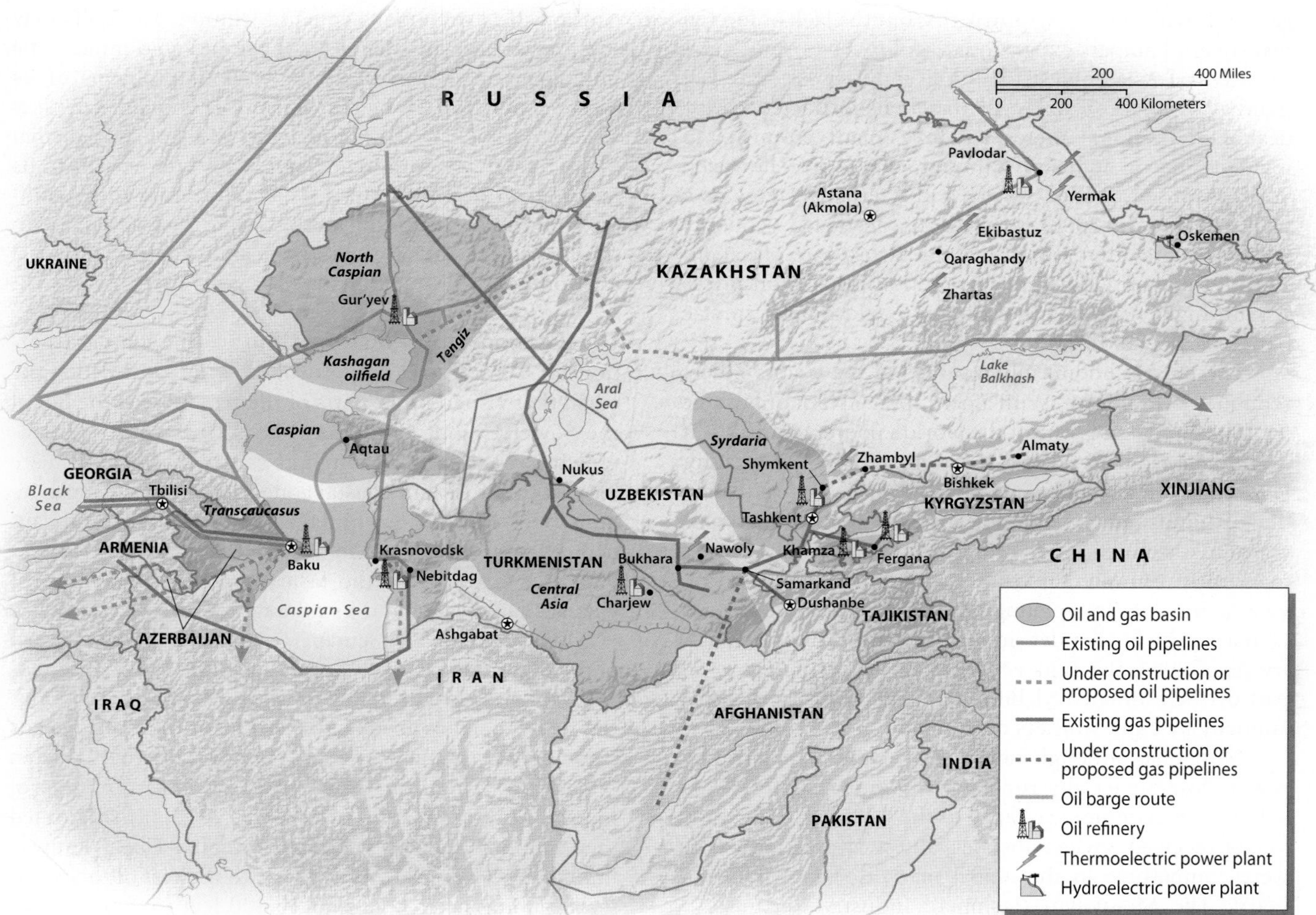

Figure 10.25 Oil Development in Azerbaijan
Although oil has brought a certain amount of wealth to Azerbaijan, it has also resulted in extensive pollution and visual blight. Because most of the petroleum is located either near or under the Caspian Sea, this sea—actually the world's largest lake—is now ecologically endangered. *(John Spaull/Panos Pictures)*

Turkmenistan also has a substantial agricultural base, due mainly to Soviet irrigation projects, and it remains a major cotton exporter. It retains a state-run economy and has resisted pressure for economic liberalization. Government planners hope that the development of new oil and gas fields will soon bring prosperity, but the lack of foreign investment hampers expansion.

Central Asia's oldest fossil fuel industry is located in Azerbaijan (Figure 10.25). Azerbaijan has attracted a great deal of international interest and investment, promising to revitalize its oil industry. Through the 1990s, however, its economy responded slowly, and Azerbaijan remains a poor country. In 2006, however, its oil-fueled economy reportedly grew by 34 percent, the fastest rate of expansion in the world. Some observers think that if oil prices remain high, Azerbaijan's economy could double in size between 2006 and 2010.

The most economically troubled of the former Soviet republics is Tajikistan. With a per capita GNI of only some $1,260, Tajikistan is one of the world's poorer countries. It is burdened by its remote location, rugged topography, lack of natural resources, and political strife. Unfortunately, one of Tajikistan's most important enterprises seems to be the transshipment of heroin from Afghanistan toward markets in Europe. Tajikistan's government hopes to revive its fortunes by opening new transportation routes to China and by further developing its hydroelectrical industry. As of 2007, the Tajik economy was supported in good part by the estimated $1 billion in remittances sent home annually by Tajik workers residing in Russia.

Mongolia, although never part of the Soviet Union, was a close Soviet ally run by a communist party. It too suffered an economic collapse in the 1990s. Mongolia no longer received Soviet subsidies, and few of its industries were competitive in the global market. (See "Global to Local: The Mongolian Cashmere Industry Readjusts.") Mongolia thus emerged in the postcommunist period as a poor country, and its economy continued to decline into the late 1990s. By the early years of the new century, however, major investments by both Chinese and Western firms in the mining sector resulted in rapid economic expansion. If current plans come to fruition, Mongolia may soon have the world's largest copper and gold mines. The mining boom, however, is not without problems of its own. As much of Mongolia's current economic growth is linked to the Chinese market, many Mongolians fear that their country is becoming economically subservient to its much larger neighbor. Small-scale, unlicensed "ninja miners," moreover, have suffered severely from mercury poisoning while causing a significant amount of local environmental degradation.

The Economy of Western China The Chinese portions of Central Asia did not suffered the same economic crash that visited other parts of the region with the fall of the Soviet Union. China has one of the world's fastest-growing economies, although its centers of dynamism are located in the distant coastal zone. Still, even the more remote parts of the country have experienced a significant degree of development in recent years. By 2007 Inner Mongolia had one of the fastest-growing economies of China, thanks largely to its mineral resources.

Tibet in particular remains burdened by poverty. Most of the plateau is relatively cut off from the Chinese economy and is even more isolated from the global economy. But many Tibetans are able to provide for their basic needs through subsistence activities. Hoping to reduce Tibetan separatism, China has been investing large amounts of money in infrastructural projects. In 2006, it inaugurated the monumentally expensive Qinghai-Tibet Railway linking Tibet's capital city of Lhasa to the rest of the country. Reaching heights of over 16,400 feet (5,000 meters),

GLOBAL TO LOCAL

The Mongolian Cashmere Industry Readjusts

For centuries, Mongolia has produced high-quality cashmere textiles derived from the soft undercoat of a particular variety of goat. Areas with particularly cold winters—such as Mongolia—produce especially soft and luxurious fibers. Herders in southern Mongolia have relied for hundreds of years on cashmere for their cash needs, raising small herds of goats in a sustainable manner and laboriously combing out the underfur of their animals each spring (Figure 10.6.1).

From the 1920s until 1990, most of Mongolia's goat herds were collectively owned, the cashmere fibers being transformed into textiles in state-owned factories. Everything changed, however, when communism collapsed. As unemployment skyrocketed, tens of thousands of urban Mongolians returned to the countryside to herd goats, figuring that they could always earn cash by selling cashmere. Lacking experience, however, many of these new herders stocked too many goats, leading to serious overgrazing hence the degradation of pasturelands. When a series of dry summers followed by brutally cold winters hit in 2000 and 2001, millions of goats died, forcing many of the new cashmere producers back into the urban economy.

Economic as well as environmental disorder has plagued Mongolia's cashmere industry after the fall of communism. In the early 1990s, Mongolia's government sought to earn more foreign exchange by developing a larger textile industry, and was distressed that many herders were sending their cashmere combings to Chinese mills for processing. In 1994, Mongolia therefore banned the export of raw cashmere, resulting in the rapid opening of 50 new Mongolian textile factories. In the following year, however, the Asian Development Bank informed Mongolia that essential new loans would be held back until the ban—regarded by the bank as anticompetitive—was dropped. Mongolia soon complied, putting its newly revived cashmere textile industry in deep trouble.

To revive its cashmere trade, Mongolia needed a new strategy. One problem was that its own textile styles were out-of-date, tending toward neon colors and stodgy designs from the 1950s. In response, a number of local firms joined together under the banner of the Mongolian Fibermark Society and sought assistance from international development organizations. When the United States Agency for International Development (USAID) began an economic reform initiative in Mongolia in 2003, the cashmere producers eagerly sought its support. By 2005, USAID was helping connect Mongolian producers with buyers in wealthy countries and providing information about the kinds of cashmere textiles that would satisfy demanding consumers in North America, Europe, and East Asia.

Figure 10.6.1
Mongolian girl holding a young goat in southern Mongolia. *(Gavriel Vecan/OnAsia.com)*

Within a year, these efforts appeared to be bearing fruit. In early 2006, the Mongolian Fibermark Society's exhibition at a Las Vegas textile trade fair generated considerable interest, as did an exhibit in New York in early 2007. It is still difficult, however, for Mongolia's cashmere textile manufacturers to compete with those of China, just as it is still hard for Mongolia's goat herders to earn a consistent income in an area noted for its periodic droughts and harsh winters. But then again, it is precisely the severe climate of Mongolia that produce such high-quality cashmere in the first place.

Figure 10.26 Chinese Business in Lhasa, Tibet
An elderly Tibetan woman peers into a hair salon in Lhasa, Tibet, in 2006. As Han Chinese immigrants stream into Tibet's cities, rapid cultural change and economic modernization leave many Tibetans behind. *(Paula Bronstein/Getty Images)*

the trains on this railway have to be equipped with supplemental oxygen. As a result of this and other transportation projects, Tibet's tourism economy is booming. Many Tibetans, however, argue that most of the benefits are flowing to Han Chinese immigrants rather than to the indigenous peoples of the region (Figure 10.26).

Most observers agree that Xinjiang has tremendous economic potential. It boasts significant mineral wealth, including a substantial portion of China's oil and natural gas reserves, and it has benefited from major infrastructural development projects over the past several decades. But as is the case in Tibet, many of the indigenous peoples of Xinjiang believe that the wealth of their region is being monopolized by the Chinese state and the Han Chinese immigrants. Some point to the Xinjiang Production and Construction Corps, a quasi-military firm that runs hundreds of industrial enterprises, giant farms, schools, and hospitals. This huge company was initially formed by the army that rejoined Xinjiang to China in 1949, and it is still run almost entirely by Han Chinese.

Economic Misery in Afghanistan Afghanistan, with one of the weakest economies in the world, is Central Asia's poorest country. Since the late 1970s, Afghanistan has suffered nearly continuous war, undermining virtually all economic endeavors. Even before the war it was an impoverished country with little industrial or commercial development. Its only significant legitimate exports are animal products, handwoven carpets, and a few fruits, nuts, and semiprecious gemstones.

Through all of these years of chaos, Afghanistan remained economically competitive in one area: the production of illicit drugs for the global market. According to the CIA, it had emerged by 1999 as the world's largest producer of opium. Although opium production declined in the final years of Taliban rule, it again surged after the fall of the regime in 2001. In late 2006, the United Nations drug agency announced that the country's opium harvest had increased 49 percent over that of the previous year. The country's drug economy has also grown more sophisticated, as Afghan opium is increasingly processed into heroin within the country.

Illegal drug production does little to help Afghanistan's government, whose financial situation remains dire. Many Afghan provinces are controlled by local warlords, who often feud over the opium trade, and thus contribute little if any taxes to the central government. As a result, Afghanistan must rely substantially on foreign assistance.

Central Asian Economies in Global Context

Afghanistan— despite its poverty and relative isolation— is thoroughly embedded in the global economy, albeit through illicit products. The overland flow of Afghan opium and heroin to Europe implicates several other Central Asian countries in the international drug economy. Countries such as Uzbekistan and Turkmenistan also have more regular connections with the global economy, but overall the extent of globalization in Central Asia remains low (Figure 10.27).

In the former Soviet area, the most important international connections remain with Russia, although economic ties with China are rapidly expanding. In the early post-Soviet period, both Russia and the newly independent Central Asian countries were ambivalent about their continuing economic relationships, while border conflicts hampered ties with China. Disentangling Central Asian economies from that of Russia, however, proved difficult,

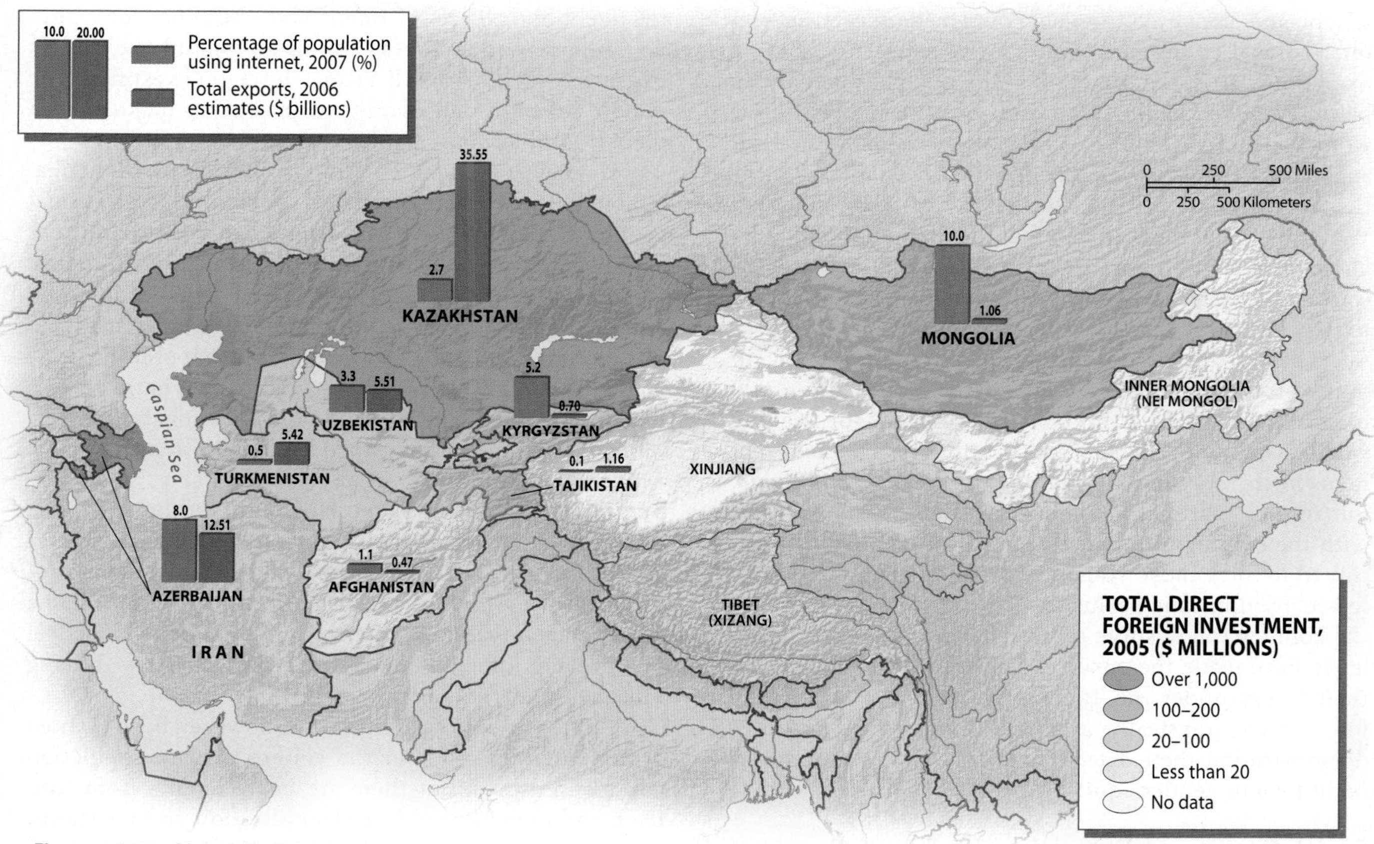

Figure 10.27 Global Linkages
Although Azerbaijan and Kazakhstan are relatively well connected to the rest of the world through their exports of oil and natural gas, the other countries of Central Asia remain somewhat isolated from global economy.

largely because of infrastructural linkages inherited from the Soviet period. As of 2007, two-thirds of Turkmenistan's natural gas exports flowed through Russia's state-owned Gazprom pipelines. Since the creation of the Shanghai Cooperation Organization, moreover, Uzbekistan, Kazakhstan, Kyrgyzstan, and Tajikistan have eagerly sought closer linkages with the booming Russian and Chinese economies. Connections to China were enhanced in 2006 with the opening of the Kazakhstan–China oil pipeline, which carries petroleum from Kazakhstan to Xinjiang.

The United States and other Western countries, as well as India, are also drawn to the area by its oil and natural gas deposits. Most large oil companies have established operations in Azerbaijan and Kazakhstan. The delivery of Central Asian oil to the west was enhanced in 2006 with the opening of the extraordinarily expensive Baku-Tbilsi-Ceyhan Pipeline—the second longest in the world—which runs from the Caspian Sea through Georgia, hence across much of Turkey, before reaching the Mediterranean at the port of Ceyhan near Turkey's border with Syria. Western goods, including luxury items such as sport utility vehicles, are now readily available to those profiting from oil wealth in cities such as Kazakhstan's Almaty and Azerbaijan's Baku. Although Turkmenistan also has significant fossil fuel deposits—including the fifth largest reserves of natural gas in the world—foreign investment has been slowed by the actions of its authoritarian and erratic government.

Social Development in Central Asia

Social conditions in Central Asia vary more than economic conditions. In the former Soviet territories, levels of health and education remain fairly high but are generally declining. Not surprisingly, Afghanistan stands at the bottom of the scale by every measure.

Social Conditions and the Status of Women in Afghanistan Social conditions in Afghanistan are no better than economic circumstances, although reliable information is difficult to obtain. As can be seen in Table 10.2, the average life expectancy in the country is a mere 42 years, one of the lowest figures in the world. Infant and childhood mortality levels remain extremely high. Not only does Afghanistan suffer from constant warfare, but its rugged topography and ongoing warfare hinder the provision of basic social and medical services. Illiteracy is commonplace and is notably gender-biased. Afghanistan's 21 percent adult female literacy figure is one of the lowest in the world.

Women in most parts of Afghanistan—and especially in Pashtun areas—lead highly constrained lives. In many areas they must completely conceal their bodies, including their faces, when venturing into public. Such restrictions intensified in the 1990s. Dress controls were strictly enforced where the Taliban gained authority. In most areas, Taliban forces prevented women from working, attending school, and often even from obtaining medical care. Several international aid agencies were forced to scale back their operations in Afghanistan after their female employees began to suffer from constant harassment. The assault on women by the Taliban did not go uncontested in Afghanistan. Many women risked their lives to document the horrific abuses that Afghan females were suffering. While Taliban leaders argued that they were upholding Islamic orthodoxy, their opponents contended that they were actually enforcing an extreme version of Pashtun customary law.

The fall of the Taliban brought temporary joy to most of the women of Afghanistan. Many promptly uncovered their faces and began working—or seeking work. The new Afghan constitution proclaimed that henceforth men and women would enjoy equal rights. But in most parts of the country, little actually changed with the coming of the new regime. Social customs continued to force most women into arranged marriages followed by domestic seclusion. Indeed, in some areas of the country the upsurge in both crime and political violence have made the position of women even less secure than it was under the Taliban. In recent years, Taliban forces have physically attacked several girls' schools, demonstrating their determination to thwart any movement toward gender equity.

Social Conditions in the Former Soviet Republics and Mongolia Compared to women in Afghanistan, those in the former Soviet portion of Central Asia enjoy a high social position. Traditionally, women had much more autonomy among the northern pastoral peoples (especially the Kazakh) than among the Uzbek and Tajik oasis dwellers of the south, but under Soviet rule the position of women everywhere improved. In the former Soviet republics today, adult women's educational rates are comparable to those of men, and women are well represented in the workplace. Some evidence suggests, however, that the position of women has been declining in many areas since the fall of the Soviet Union. In Kazakhstan several leaders have blamed feminism for the country's low birthrate, and in much of the region newly rich men are increasingly practicing polygamy partly in order to have large numbers of children. As education becomes more expensive throughout the area, moreover, girls are increasingly staying at home.

The figures in Table 10.2 reveal generally favorable levels of social welfare overall for the former Soviet zone. This is especially notable when one considers the dismal state of the region's economy—or contrasts conditions with those of neighboring Afghanistan. Tajikistan, with a per capita GNI of only $1,260, has a female life expectancy of 68 years as well as almost universal adult literacy. The social successes of Tajikistan and its northern and western neighbors reflect investments made during the Soviet period. It is unclear, however, whether health and educational facilities can be maintained in the face of economic and political turmoil. Certainly the region's relatively high levels of infant and childhood mortality do not bode well. Whereas Tajikistan might hope that its educated workforce will attract foreign investment, its remote location and continued instability make such a scenario unlikely.

Somewhat similar conditions prevail in Mongolia, although the preponderance of women in professional positions is striking. This is a legacy of both the former socialist system and of traditional Mongol culture, in which women have long had a good deal of domestic authority. In 2000, one survey found that women made up 60 percent of Mongolia's lawyers, 77 percent of its doctors, and 84 percent of its university graduates (Figure 10.28). But despite such educational and professional advances, Mongolia's women have gained little political power.

Social Conditions in Western China Although China as a whole has made significant progress in health and education, many reports suggest that the indigenous peoples of Tibet and Xinjiang have been left behind. Some sources claim that up to 60 percent of the non-Han people of Xinjiang are illiterate, and that illiteracy among Tibetans stands at approximately 55 percent.

China's minority peoples have, however, been granted certain exemptions from the country's strict population control measures. But there are many reports from Xinjiang and elsewhere of forced sterilization and abortions, adding to the local political tensions.

Figure 10.28 Mongolian Women
A female technician examines slides under a microscope at the Agricultural Institute in Ulan Bator, Mongolia. *(Dean Conger/Corbis)*

Summary

- Central Asia, long hidden by Russian and Chinese domination, has recently reappeared on the map of the world. Environmental problems, among others, have brought the region to global attention. The destruction of the Aral Sea is one of the worst environmental disasters the world has experienced, and many other lakes in the region have experienced similar problems. Desertification has devastated many areas, and the booming oil and gas industries have created their own environmental disasters.
- Large movements of people in Central Asia have also attracted global attention. The biggest issue at present is the migration of Han Chinese into Tibet and Xinjiang, which is steadily turning the local peoples of these areas into minority groups. The migration of Russian speakers out of the former Soviet areas of Central Asia is another important demographic issue. In Afghanistan, war and continuing chaos have generated large refugee populations.
- Religious tension has recently emerged as a major cultural issue through much of western Central Asia. Radical Islamic fundamentalism remains a potent force in southern and western Afghanistan, and to a somewhat lesser extent in the Fergana Valley of Uzbekistan, Tajikistan, and Kyrgyzstan. Although nonradical forms of Islam are prevalent through most of the region, Central Asian leaders have used the threat of religiously inspired violence to maintain repressive and antidemocratic policies.
- While China maintains a firm grip on Tibet and Xinjiang, the rest of Central Asia has emerged as a key area of geopolitical competition. Russia, the United States, China, Iran, Pakistan, and India all contend for influence. Central Asian countries have attempted to play these powers against each other in order to maintain or increase their world status. Overall, political structures throughout the region remain largely authoritarian, and in recent years a number of Central Asian countries have moved closer to China and Russia, in part because of the development of the Shanghai Cooperation Organization.
- The economies of Central Asia are gradually opening up to global connections, largely because of their substantial fossil fuel reserves. However, Central Asia is likely to face serious economic difficulties for some time, especially since the region is not a significant participant in global trade and has attracted little foreign investment outside the fossil-fuel sector. The growing of opium and the manufacture of heroin remain a serious problem in Afghanistan. Drug trafficking, moreover, has weakened legitimate economic activities in other countries of the region.

Key Terms

alluvial fan (*page 443*)
Bodhisattva (*page 453*)
exclave (*page 458*)
exotic river (*page 444*)
loess (*page 443*)
pastoralist (*page 441*)
Shanghai Cooperation Organization (SCO); "Shanghai Six" (*page 460*)
Taliban (*page 434*)
theocracy (*page 453*)
transhumance (*page 442*)

Questions for Review

1. Describe the three major regions of Central Asia as defined by physical geography.
2. Why is lake level a particularly serious concern in Central Asia?
3. Why is the distribution of population in Central Asia so uneven? Why are certain areas virtually uninhabited?
4. Describe the different agricultural patterns that are found on (a) the plains of northern Kazakhstan, (b) the major river valleys of Uzbekistan, and (c) the Tibetan Plateau.
5. Why is much of Central Asia often called "Turkestan"? Why does this label not apply to the entire region?
6. How does religion act as a political force in Central Asia? How does this situation vary in different parts of the region?
7. Why is the United States concerned about the fossil fuel resources of Central Asia? How do transportation routes play into this concern?
8. What role does Russia currently play in Central Asian geopolitics? How has Russian power historically influenced the region?
9. How do social conditions (health, longevity, literacy, etc.) vary across the border between Afghanistan and Uzbekistan? Why are the disparities so pronounced?
10. If western Central Asia is so rich in natural resources, why did the region experience such pronounced economic decline in the 1990s?

Thinking Geographically

1. To what extent will Central Asia's continental location be a disadvantage in years to come? Is coastal access truly significant in determining a country's competitive position in the global economy? Explain.
2. Will the countries of former Soviet Central Asia be able to maintain their high levels of education now that they are independent? If so, will they be able to use education to their own economic advantage? Explain.
3. Is Russia's influence in Central Asia bound to decline now that Russia has no direct political authority in the area? Explain.
4. Which external connection will prove most important for Central Asia in years to come—those based on religion, language, economic ties, or geopolitical connections?
5. Is China's political control over eastern Central Asia justified? Should Han Chinese migration to the area be a concern to the United States? Explain.
6. Is Chinese control over Tibet a legitimate concern of U.S. foreign policy? Should the United States use trade sanctions to attempt to influence Chinese policy in the area? Explain.
7. How might the desiccation of Central Asia's great lakes best be addressed?

Regional Novels and Films

Novels

Rinjing Dorje, *Tales of Uncle Tompa, The Legendary Rascal of Tibet* (1997, Barrytown Ltd.)

Mark Frutkin, *Invading Tibet* (1993, Soho Press)

James Hilton, *Lost Horizon* (1996, Morrow)

Khaled Hosseini, *The Kite Runner* (2003, Riverhead Books)

Amin Maalouf, *Samarkand: A Novel* (1998, Interlink)

Hilary Roe Metternich, ed., *Mongolian Folktales* (1996, Avery Press)

Films

Kandahar (2001, Iran)

Kundun (1997, U.S.)

Osama (2003, Afghanistan)

Pure Coolness (2007, Kyrgyzstan)

The Road under the Heavens (2006, Uzbekistan)

Seven Years in Tibet (1997, U.S.)

The Story of the Weeping Camel (2003, Mongolia)

Urga: Close to Eden (1991, Mongolia)

Bibliography

Adshead, S. A. M. 1993. *Central Asia in World History*. New York: St. Martin's Press.

Bissell, Tom. 2003. *Chasing the Sea: Lost Among the Ghosts of Empire in Central Asia*. New York: Pantheon.

Drompp, Michael. 1989. "Centrifugal Forces in the Inner Asian 'Heartland': History *Versus* Geography." *Journal of Asian History* 23, 135–55.

Ewans, Martin. 2002. *Afghanistan: A Short History of Its People and Politics*. New York: Perennial.

Kobori, Iwao, and Glantz, Michael. 1998. *Central Eurasian Water Crisis: Caspian, Aral, and Dead Sea*. Tokyo: United Nations University Press.

Mahnovski, Sergei. 2007. *Economic Dimension of Security in Central Asia*. Santa Monica, CA: RAND Corporation.

Rashid, Ahmed. 2001. *Taliban: Militant Islam, Oil, and Fundamentalism in Central Asia*. New Haven, CT: Yale University Press.

Rumer, Boris, ed. 2000. *Central Asia and the New Global Economy*. New York: M. E. Sharpe.

Sabloff, Paula. 2002. *Modern Mongolia: Reclaiming Genghis Khan*. Philadelphia: University Museum Publications.

Sinor, Denis, ed. 1990. *The Cambridge History of Early Inner Asia*. Cambridge, UK: Cambridge University Press.

Additional bibliographic resources are at the *Diversity Amid Globalization* Website: http//www.prenhall.com/rowntree/.

11
East Asia

A Chinese worker examines progress on the massive Three Gorges Dam built across the Yangtze River. This single hydroelectric facility provides some roughly three percent of China's total electricity demand. (*Tischler Fotografen/Peter Arnold*)

ENVIRONMENTAL GEOGRAPHY

China has long experienced severe deforestation and soil erosion, and its current economic boom is generating the worst pollution problems in the world.

POPULATION AND SETTLEMENT

China is currently undergoing a major transformation as tens of millions of peasants move from poor villages in the interior to booming cities of the coastal region.

CULTURAL COHERENCE AND DIVERSITY

Despite the presence of several unifying cultural features, East Asia in general and China in particular are divided along several striking cultural lines.

GEOPOLITICAL FRAMEWORK

Both Korea and China remain bitterly divided nations, with Taiwan pitted against the People's Republic of China and South Korea set in opposition against North Korea.

ECONOMIC AND SOCIAL DEVELOPMENT

Over the past several decades, East Asia has emerged as a core area of the world economy, with China experiencing one of the most rapid economic expansions the world has ever seen.

East Asia, composed of China, Japan, South Korea, North Korea, and Taiwan (Figure 11.1), is the most populous region of the world. China alone is inhabited by more than 1.3 billion people, more than live in any other region of the world except South Asia. Although East Asia is historically unified by cultural features, in the second half of the twentieth century it was divided ideologically and politically, with the capitalist economies of Japan, South Korea, Taiwan, and Hong Kong separated from the communist bloc of China and North Korea. Differences in levels of economic development also remained pronounced. As Japan reached the pinnacle of the global economy, much of China remained mired in extreme poverty.

Since the 1990s, however, divisions with East Asia have been reduced. While China is still governed by the Communist Party, it has embarked on a path of capitalist development. Business ties between its booming coastal zone and Japan, South Korea, and Taiwan have quickly strengthened. Animosity still exists between North Korea and South Korea and between China and Taiwan, but East Asia as a whole has witnessed a gradual reduction in political tensions (see "Setting the Boundaries").

A thousand years ago, East Asia was not only the most populous but also the most economically advanced part of the world. In the 1800s, however, its fortunes declined abruptly. China suffered repeated famines and revolutions and was exploited by European imperial powers. Japan managed to retain full independence and to build a powerful military, but it remained a poor country until the second half of the twentieth century. In 1945, at the end of World War II, Japan was utterly defeated, while China remained gripped in a brutal civil war between nationalist and communist forces. The future of the region looked bleak.

Today, East Asia must be recognized as one of the core areas of the world economy, and it is emerging as a center of political power as well. Japan, South Korea, and Taiwan are among the world's key trading states. Tokyo, Japan's largest city, stands alongside New York and London as one of the financial centers of the globe, and is noted as both a consumer and producer of global culture. China has more recently emerged as a global trading power. Its coastal zone, particularly the former British colony of Hong Kong, is now tightly integrated within global networks of information, commerce, and entertainment. China is also becoming a major regional military power. This development has caused much concern among its neighbors—and among distant countries, such as the United States.

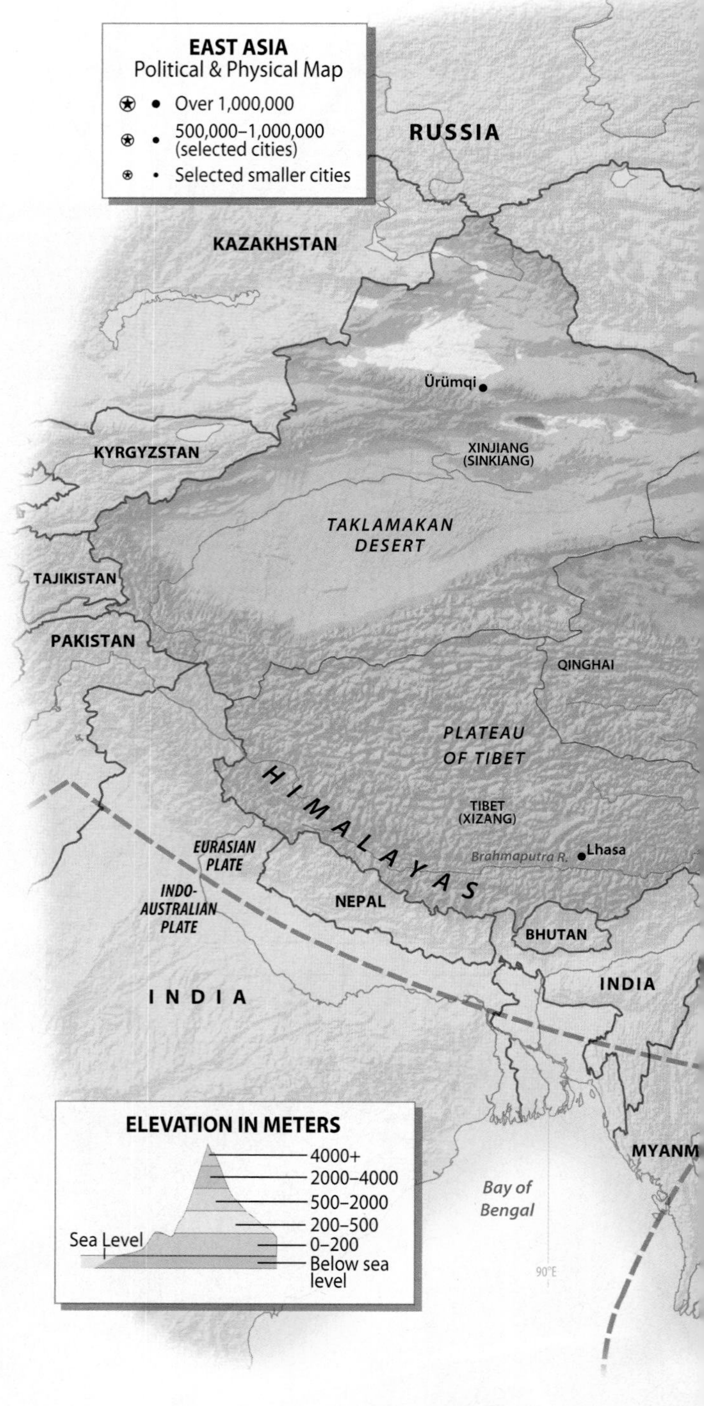

ENVIRONMENTAL GEOGRAPHY: Resource Pressures in a Crowded Land

Environmental problems in East Asia are particularly severe owing to a combination of the region's large population, its rapid industrial development, and its physical geography (Figure 11.2). The wealthier countries of the region, particularly Japan, have been able to invest heavily in environmental protection. China, in contrast, not only has less money available for conservation, but also has large-scale environmental problems that began centuries ago. Our discussion begins with one of China's most controversial environmental issues, the building of the Three Gorges Dam on the Yangtze River.

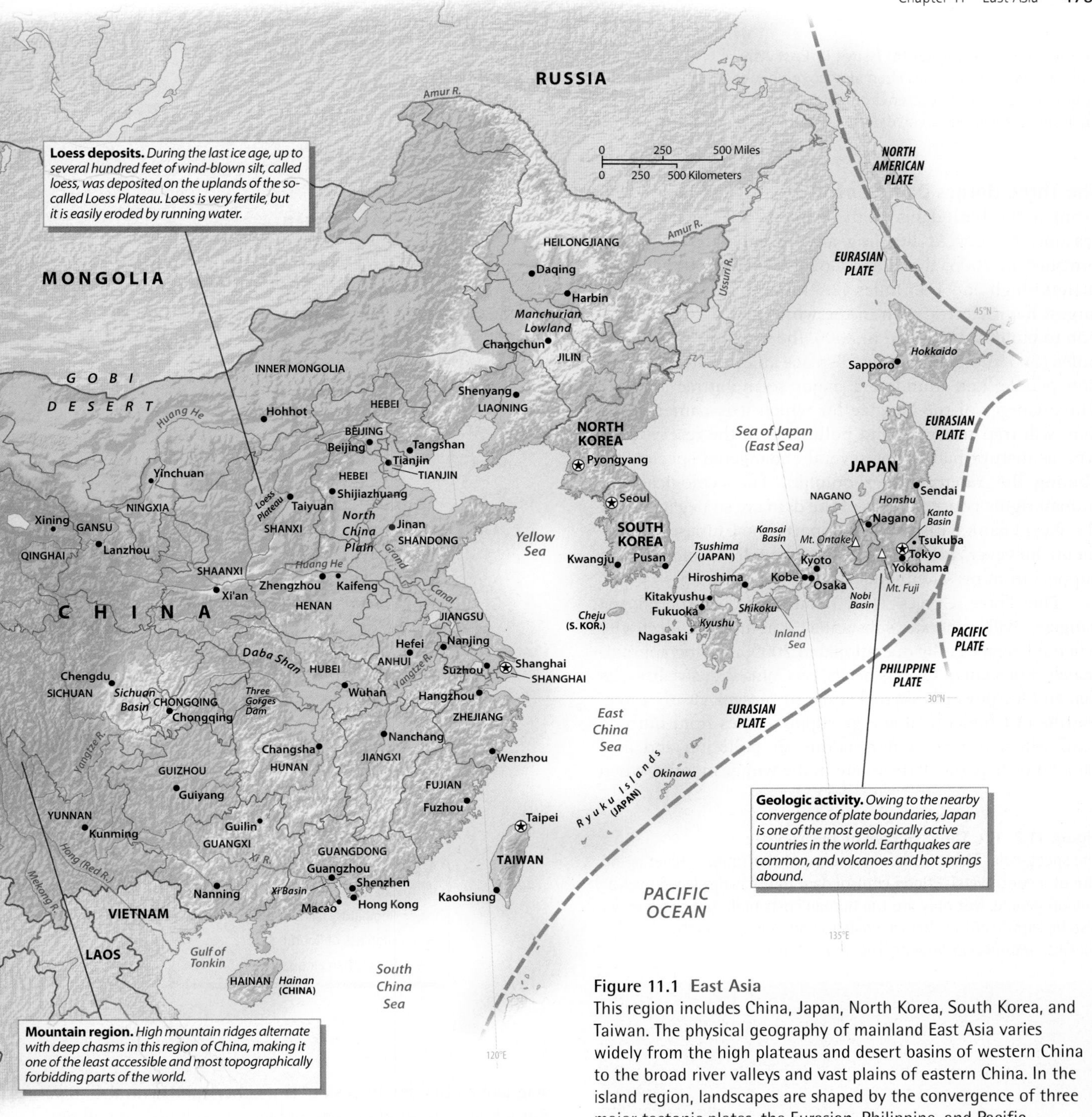

Figure 11.1 East Asia
This region includes China, Japan, North Korea, South Korea, and Taiwan. The physical geography of mainland East Asia varies widely from the high plateaus and desert basins of western China to the broad river valleys and vast plains of eastern China. In the island region, landscapes are shaped by the convergence of three major tectonic plates, the Eurasian, Philippine, and Pacific.

Dams, Flooding, and Soil Erosion in China

The Yangtze River (also called the Chang Jiang) is one of the most important physical features of East Asia. This river, the third largest (by volume) in the world, emerges from the Tibetan highlands onto the rolling lands of the Sichuan Basin, passes through a magnificent canyon in the Three Gorges area (Figure 11.3), and then meanders across the lowlands of central China before entering into the sea near the city of Shanghai. The Yangtze has historically been the main avenue of entry into the interior of China, and is celebrated in Chinese literature for its beauty and power. Since the 1990s, however, it has become the focal point of an environmental controversy of global proportions.

Figure 11.2 **Environmental Issues in East Asia**
This vast world region has been almost completely transformed from its natural state and continues to have serious environmental problems. In China, some of the more pressing environmental issues involve deforestation, flooding, water control, and soil erosion.

The Three Gorges Dam Controversy The Chinese government has built dams on the Yangtze River for two main reasons: to prevent flooding and to generate electricity. Finished in 2006, the Three Gorges Dam—600 feet (180 meters) high and 1.45 miles (2.3 kilometers) long—is the largest hydroelectric dam in the world and cost $22.5 billion to build. When filled in 2009, the reservoir will be 350 miles (563 kilometers) long, displace an estimated 1.2 million people, and inundate a major scenic attraction (the Three Gorges of the Yangtze for which it is named). The dam will trap sediment and pollutants in the reservoir, as well as disrupt habitat for several endangered species, including the Yangtze River dolphin. The ecological and human rights consequences of the dam are so negative that the World Bank withdrew its support. The Chinese government, however, has continued to provide financing with support from private financial firms.

The Three Gorges Dam, along with others on the Yangtze, will generate substantial amounts of electricity when it becomes fully operational in 2009, meeting approximately 3 percent of China's needs. As China industrializes, its demand for power is skyrocketing. At present, nearly three-fourths of China's total energy supply comes from burning coal, which results in horrendous air pollution and acid rain. Smog is particularly severe in the winter months, when stagnant air masses often sit over northern and central China. But while dam-building may reduce air pollution, it will not be nearly adequate to supply China's vast energy needs. Most environmentalists therefore argue that the damage caused by the dams is greater than the benefits they confer.

Figure 11.3 **The Three Gorges of the Yangzte**
The spectacular Three Gorges landscape of the Yangzte River is the site of a controversial flood control dam that has displaced over a million people. Not only are the human costs high, but there may also be significant ecological costs to endangered aquatic species. *(Panarama Images/The Image Works)*

Chinese planners also praise the water-control benefits provided by dams on the Yangtze. The lower and middle stretches of the Yangtze periodically suffer from devastating flooding, and planners hope that the Three Gorges Dam will store enough water to help protect 15 million people and over 3 million acres (1.2 million hectares) of farmland on the lower Yangtze floodplains. However, flooding has been exacerbated over the past several decades by the declining area of lakes on the Yangtze floodplain, which function as overflow areas during

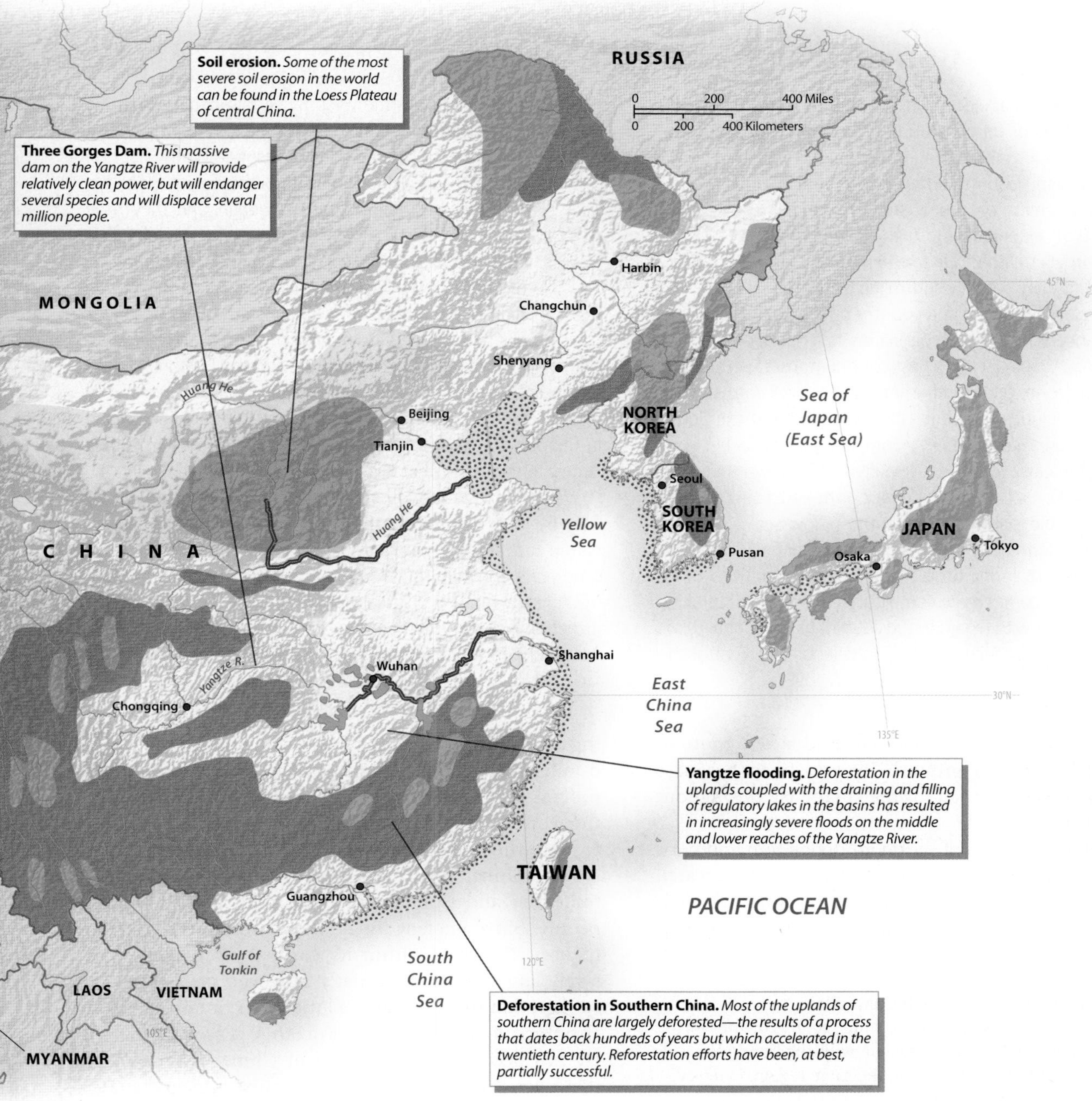

floods. East of Sichuan, the Yangtze passes through several basins (in Hunan, Hubei, and Jiangxi provinces), each containing a group of lakes. During periods of high water, river flows are diverted into these **regulatory lakes**, thus reducing the flow downstream. After the flood season is over, water drains from the lakes and therefore helps maintain the water level in the river downstream. In the 1950s and 1960s, however, the Chinese government artificially drained many of these lakes in order to convert the lake beds to farmland. As a result, flooding downstream has intensified.

Flooding in Northern China The North China Plain, which has been deforested for thousands of years, is plagued by both drought and flood. This area is dry most of the year, yet often experiences heavy downpours in summer. Since ancient times, large-scale hydraulic engineering projects have both controlled floods and allowed irrigation. But no matter how much effort has been put into water control, disastrous flooding has never been completely prevented (Figure 11.4).

The worst floods in northern China are caused by the Huang He, or Yellow River, which cuts across the North China Plain. Owing to upstream erosion, the Huang He carries a huge **sediment load**, or suspended clay, silt, and sand, making it the world's muddiest major river. When the river enters the low-lying plain, its velocity slows and its sediments begin to settle and accumulate in the riverbed. As a result, the level of the riverbed gradually rises above that of the surrounding lands. Eventually the

Setting the Boundaries

East Asia is easily marked off on the map of the world by the territorial extent of its constituent countries: China, Taiwan, North Korea, South Korea, and Japan (see Figure 11.1). Japan is an island country composed of four main islands and a chain of much smaller islands (the Ryukyus) that extends almost to Taiwan. To the north, Japan claims the four southernmost Kuril Islands, but these are actually controlled by Russia. Taiwan is basically a single-island country, but its political status is uncertain since China claims it as part of its own territory. China itself is a vast continental state—the third most extensive in the world—that reaches well into Central Asia. It also includes the large island of Hainan in the South China Sea. Korea constitutes a compact peninsula between northern China and Japan, but it is divided into two sovereign states: North Korea and South Korea. South Korea also encompasses the sizable island of Cheju in the Yellow Sea.

In political terms, such a straightforward definition of East Asia is appropriate. If one turns to cultural considerations, however, the issue becomes more complicated. The main cultural inconsistency arises with regard to the western half of China. This is a huge but lightly populated space; only about 5 percent of the residents of China live in the west. By certain criteria, only the populous eastern half of China (often called **"China proper"**) really fits into the East Asian world region (see Figure 11.12 on p. 484). The indigenous inhabitants of western China are not Chinese by culture and language, and they have never accepted the religious and philosophical beliefs that give historical unity to East Asian civilization. In the northwestern quarter of China, called *Xinjiang* in Chinese, most indigenous inhabitants speak Turkic languages and are Muslim in religion. Tibet, in southwestern China, has a highly distinctive culture, and the Tibetans in general resent Chinese authority. Xinjiang and Tibet are thus appropriately classified within both East Asia and Central Asia (covered in Chapter 10). In this chapter we will examine western China largely to the extent that it is politically part of China.

Vietnam presents another anomaly for the definition of East Asia. By most cultural and historical criteria, Vietnam belongs within East Asia. Vietnam was part of China for almost a thousand years, and many of its cultural patterns are similar to those of China. By modern political measures, however—as well as by its actual location—Vietnam fits much better within Southeast Asia. Vietnam will therefore be covered in Chapter 13.

river must break free of its course to find a new route to the sea over lower-lying ground. Twenty-six such course changes have been recorded for the Huang He throughout Chinese history.

Through the process of sediment deposition and periodic course changes, the Huang He has actually created the vast North China Plain. In prehistoric times, the Yellow Sea extended far inland. Even today the sea is retreating as the Huang He's delta expands; one study revealed a 6-mile advance of the land in a three-year period. Such a process occurs in other alluvial plains, but nowhere else in the world is it so pronounced.

Figure 11.4 Flooding on the North China Plain
Major flooding, sometimes inundating large sections of the North China Plain, has been a historical problem with the Huang He River. Severe droughts can also plague the same region. Extensive dikes have been built along much of the river to protect the countryside from flooding, as seen in this photo taken near the historical city of Kaifeng. *(Yang Xiuyun/ChinaStock Photo Library)*

Nowhere else, moreover, is it so destructive. The North China Plain has been densely populated for millennia and is now home to more than 300 million people. Since ancient times, the Chinese have attempted to keep the river within its banks by building progressively larger dikes. Eventually, however, the riverbed rises so high that the flow can no longer be contained and catastrophic flooding results. For this reason, the Huang He has been called "the river of China's sorrow." While the river has not changed its course since the 1930s, most geographers agree that another course correction is inevitable.

Erosion on the Loess Plateau The Huang He's sediment burden is derived from the eroding soils of the Loess Plateau, located to the west of the North China Plain. **Loess** consists of fine, windblown sediment that was deposited on this upland area during the last Ice Age, accumulating in some places to depths of several hundred feet.

Loess forms fertile soil, but it washes away easily when exposed to running water. At the dawn of Chinese civilization, the semiarid Loess Plateau was covered with tough grasses and scrubby forests that retained the soil. Chinese farmers, however, began to clear the land for the abundant crops that it yields when rainfall is adequate. Cultivation required plowing, which, by exposing the soil to water and wind, exacerbated erosion. As the population of the region gradually increased, the remaining areas of woodland diminished, leading to faster rates of soil loss. As the erosion process continued, great gullies cut across the plateau, steadily reducing the extent of arable land.

Today, the Loess Plateau is one of the poorest parts of China. Population is only moderately dense by Chinese standards, but good farmland is limited and drought is common. The Chinese government encourages the construction of terraces to conserve the soil, but such efforts have not been effective everywhere. Campaigns to plant woody vegetation on the most severely eroded lands have achieved only limited success. Seedlings require careful attention if they are to survive the harsh climate, and local farmers do not always have adequate knowledge or incentives to care for them.

Other East Asian Environmental Problems

Although the problems associated with the Yangtze and Huang He rivers are among the most spectacular environmental issues in East Asia, they are hardly the only ones. Deforestation, urban pollution, and the loss of wildlife are increasing problems in much of the region.

Forests and Deforestation Most uplands of China and South Korea support only grass, meager scrub, and stunted trees (Figure 11.5). China lacks the historical tradition of forest conservation that characterizes Japan. During earlier periods of Chinese history, hillsides were often cleared for fuelwood, and in some instances entire forests were burned for ash that could be used as fertilizer. In much of southern China, sweet potatoes, maize, and other crops have been grown on steep and easily eroded hillsides for hundreds of years. After centuries of exploitation, many uplands are now so degraded that they cannot easily regenerate forests.

Over the past 50 years, China has initiated a number of reforestation efforts. One of the most ambitious projects involved planting more than 300 million trees along a 3,000-mile (4,800-kilometer)-long swath of northern China south of the Gobi Desert. This "great green wall" was designed to stop the southward expansion of the desert and reduce the dust storms that often hit northern China. Like other Chinese reforestation projects, however, it has only been partially successful. At present, substantial forests are found only in China's far northeast, where forests are now being protected from logging in the province of Heilongjiang, and along the eastern slopes of the Tibetan Plateau, where rugged terrain restricts commercial forestry. As a result, China suffers a severe shortage of forest resources.

Mounting Pollution As China's industrial base expands, other environmental problems, such as water pollution and toxic-waste dumping, are growing more acute, particularly in the booming coastal areas. The burning of high-sulfur coal has resulted in particularly severe air pollution, a problem aggravated by the increasing number of automobiles driven by the Chinese. China now has the most serious air pollution problem in the world. According to one World Health Organization (WHO) report, air pollution kills some 656,000 Chinese citizens each year, while another 95,000 are killed by polluted drinking water. Although Chinese citizens have organized thousands of protests against pollution and other forms of environmental destruction, the government insists that economic growth remain the country's main priority.

Considering its large population and intensive industrialization, Japan's environment is relatively clean. The very density of its population gives certain environmental advantages, allowing, for example, a highly efficient public transportation system. In the 1950s and 1960s, Japan did suffer from some of the world's worst pollution. Soon afterward the Japanese government passed stringent environmental laws.

Japan's cleanup was aided by its insular location, because winds usually carry smog-forming chemicals out to sea. Equally important has been the phenomenon of **pollution exporting**. Because of Japan's high cost of production and its strict environmental laws, many Japanese companies have moved their dirtier factories overseas. In effect, Japan's pollution has been partially displaced to poorer countries. Of course, the same argument can also be made about the United States and western Europe, but to a somewhat lesser extent.

Serious air and water pollution problems also emerged early in other East Asian countries. But Taiwan and South Korea, which have large chemical, steel, and other heavy industries, responded by imposing more stringent environmental controls as they grew wealthier. These two countries have also followed Japan's footsteps by setting up new factories in poorer countries—including China—that have less strict environmental standards.

Figure 11.5 Denuded Hillslopes in China
Because of the need to clear forests for wood products and agricultural lands, China's mountain slopes have long been deforested. Without forest cover, soil erosion is a serious issue. *(Bob Sacha/Corbis)*

Figure 11.6 Panda Habitat
A large panda is shown here in its native habitat in Wolong Valley, China. China has made great efforts to protect panda habitat, which is limited to high-altitude forests and bamboo thickets in western Sichuan and neighboring provinces. *(Fritz Polking/The Image Works)*

Endangered Species The growing number of endangered species in East Asia has been linked to the region's economic rise. Many forms of traditional Chinese medicine are based on products derived from rare and exotic animals. Deer antlers, bear gallbladders, snake blood, tiger penises, and rhinoceros horns are believed to have medical effectiveness, and certain individuals will pay fantastic sums of money to purchase them. As wealth has accumulated, trade in such substances has expanded. China itself has relatively little remaining wildlife, but other areas of the world help supply its demand for wildlife products.

China has, however, made reasonable efforts to protect some of its remaining areas of habitat. Among the most important of these are the high-altitude forests and bamboo thickets of western Sichuan province, home of the panda bear (Figure 11.6). Efforts are also being made to preserve wildlands in northern Manchuria, where evidence of a surviving population of Siberian tigers was discovered in 1997, and in the canyonlands of northwestern Yunnan, which were declared off-limits to loggers in 1999.

Wildlife is also scarce in Korea. Ironically, however, the so-called demilitarized zone that separates North from South Korea functions as an unintentional wildlife sanctuary, supporting populations of several endangered species, including the Asiatic black bear.

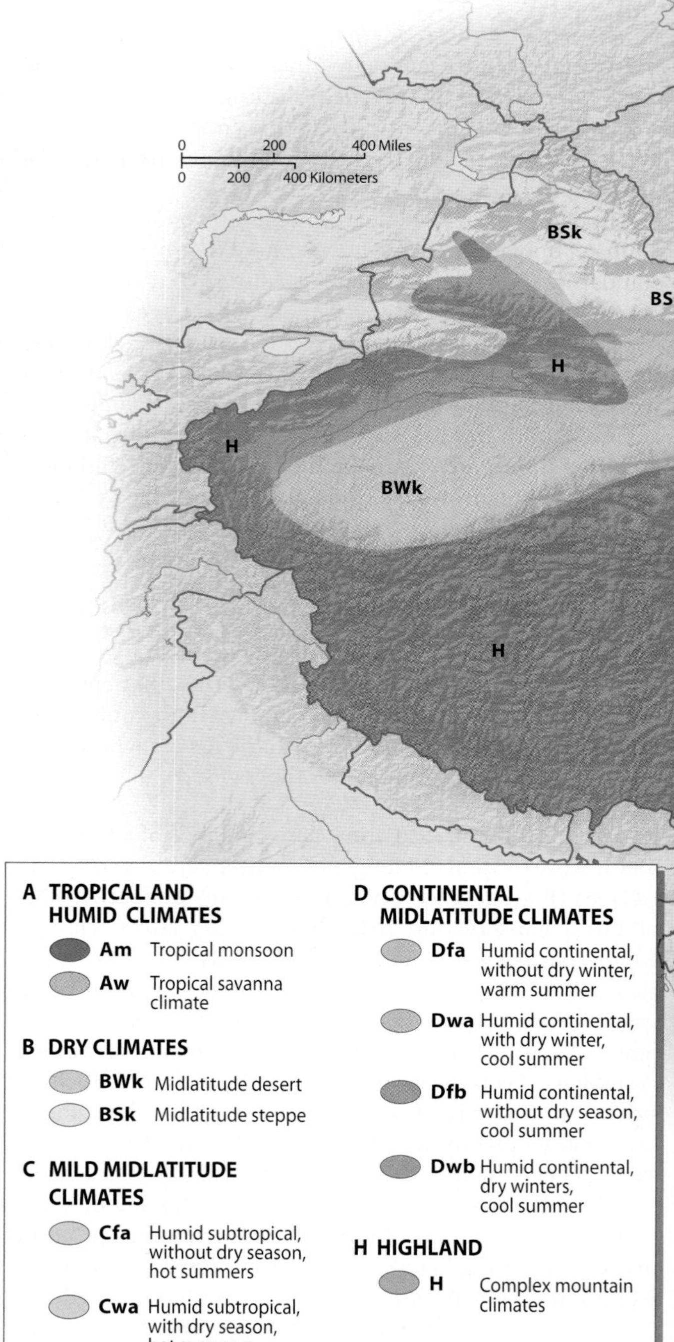

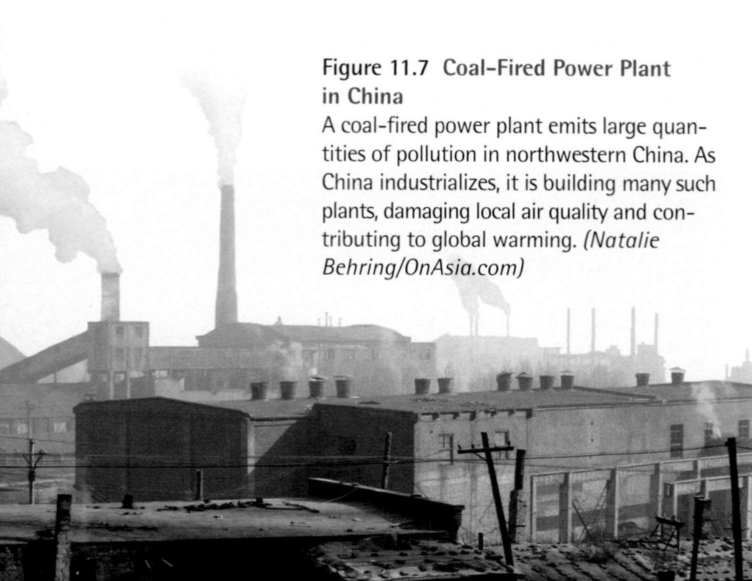

Figure 11.7 Coal-Fired Power Plant in China
A coal-fired power plant emits large quantities of pollution in northwestern China. As China industrializes, it is building many such plants, damaging local air quality and contributing to global warming. *(Natalie Behring/OnAsia.com)*

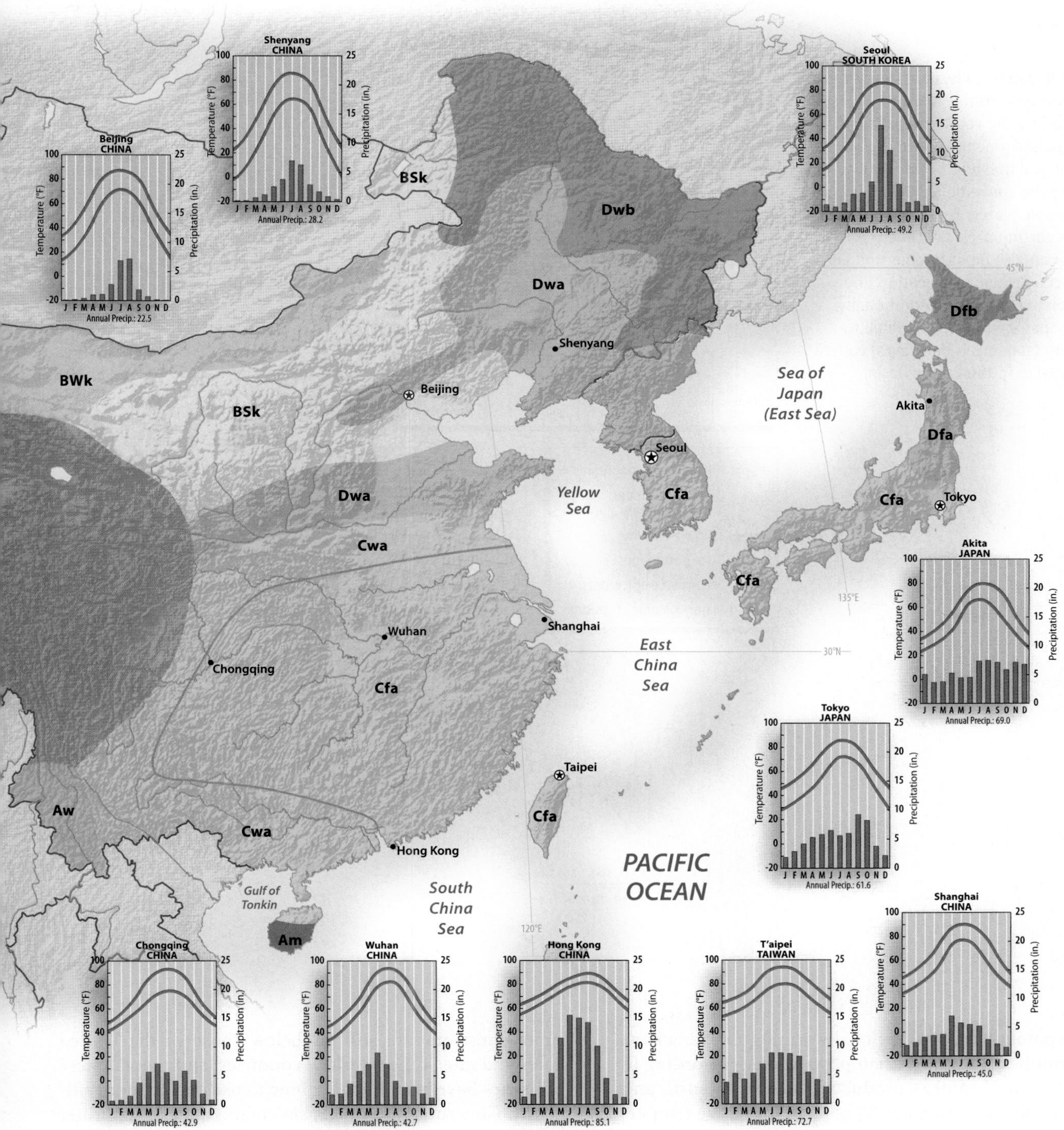

Figure 11.8 Climate Map of East Asia
East Asia is located in roughly the same latitudinal zone as North America, and climatic parallels exist between the two world regions. The northernmost tip of China lies at about the same latitude as Quebec and shares a similar climate, whereas southern China approximates the climate of Florida. In Japan, maritime influences produce a milder climate.

Global Warming and East Asia

East Asia has come to occupy a central position in global warming debates, largely because of China's extremely rapid increases in carbon emissions. From levels only about half those of the United States in 2000, China's total production of greenhouse gases probably surpassed those of the United States in 2007. This staggering rise has been caused both by China's explosive economic growth and by the fact that it relies on burning coal to meet most of its energy needs. China is currently building one coal-fired power plant every week on average, and it plans to continue doing so for the foreseeable future (Figure 11.7).

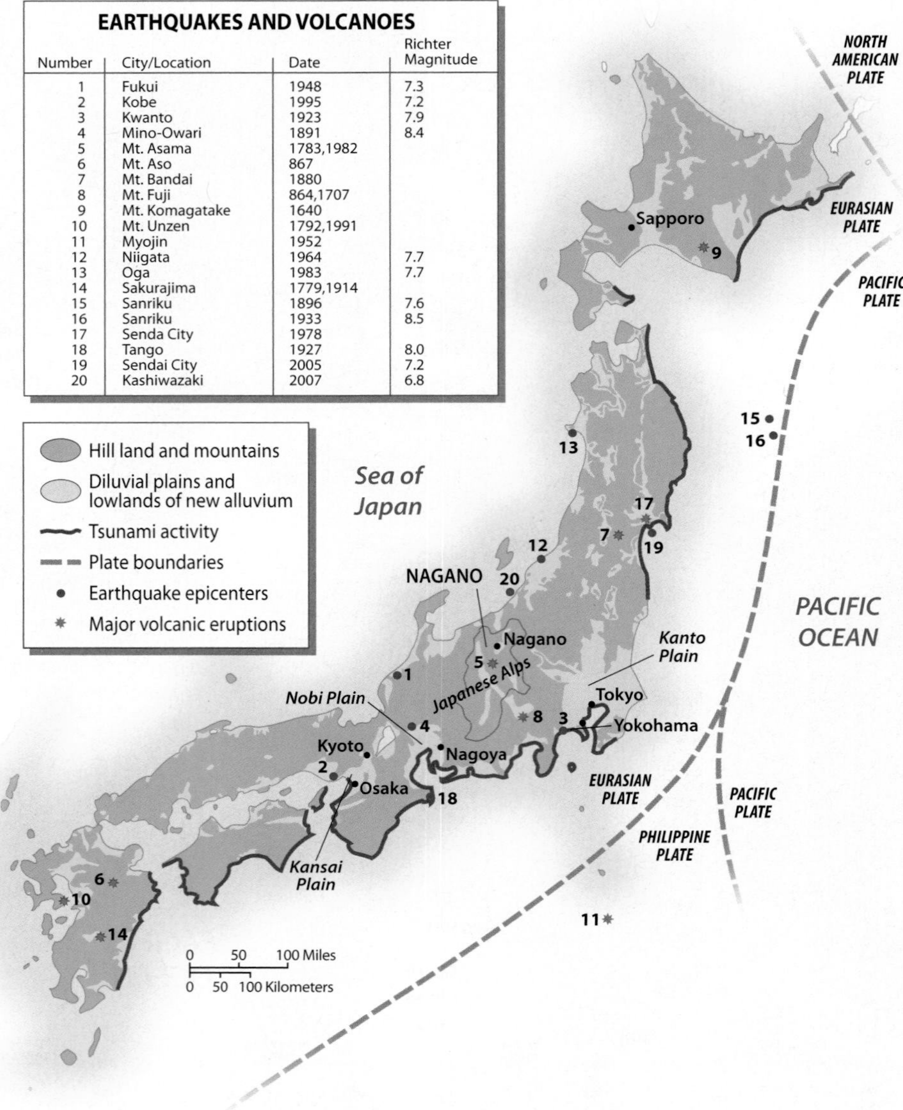

EARTHQUAKES AND VOLCANOES

Number	City/Location	Date	Richter Magnitude
1	Fukui	1948	7.3
2	Kobe	1995	7.2
3	Kwanto	1923	7.9
4	Mino-Owari	1891	8.4
5	Mt. Asama	1783,1982	
6	Mt. Aso	867	
7	Mt. Bandai	1880	
8	Mt. Fuji	864,1707	
9	Mt. Komagatake	1640	
10	Mt. Unzen	1792,1991	
11	Myojin	1952	
12	Niigata	1964	7.7
13	Oga	1983	7.7
14	Sakurajima	1779,1914	
15	Sanriku	1896	7.6
16	Sanriku	1933	8.5
17	Senda City	1978	
18	Tango	1927	8.0
19	Sendai City	2005	7.2
20	Kashiwazaki	2007	6.8

Figure 11.9 Japan's Physical Geography Japan has several sizable lowland plains, primarily along the coastline, but they are interspersed among rugged mountains and uplands. Because of its location at the convergence of three major tectonic plates, Japan experiences numerous earthquakes. Volcanic eruptions can also be hazardous and are linked directly to Japan's location on tectonic plate boundaries. Additionally, much of Japan's coast is vulnerable to devastating tsunamis (tidal waves) caused by earthquakes in the Pacific Basin.

The potential effects of global warming in China have serious implications for human populations. According to one report, the country's production of wheat, corn, and rice could fall by as much as 37 percent if average temperatures increase by 3.5 to 5.5°F (2 or 3°C) over the next 50 to 80 years. Increased evaporation rates, coupled with the melting of glaciers on the Tibetan Plateau, could greatly intensify the water shortages that already plague much of northern China. Officials believe that by the middle of the twenty-first century, even Shanghai, located in the humid Yangtze Delta, will either have to use desalinized water or build an expensive network of canals, dams, and pumps to import water from south-central China for municipal use. In the wet zones of southern China, global warming concerns center on the probability of more intense storms in this already flood-prone region.

Despite these concerns, the Chinese government has insisted that economic growth take priority over reducing greenhouse gas emissions. Officials argue that China's per capita emissions are still far below those of the United States and most other wealthy countries, and that Chinese industries have been learning how to use energy more efficiently. Government officials also maintain that wealthier countries should take the lead on reducing emissions and figuring out other ways to combat global warming.

In June 2007, China released its first national plan on climate change, which calls for a 20 percent gain in energy efficiency by 2010. China's new strategy also includes a major expansion of nuclear power, increased use of renewable energy sources, and the continuation of ambitious reforestation efforts. While environmentalists were pleased to see China addressing these issues, most feel that the country has not gone nearly far enough to curb its extraordinarily fast rise in greenhouse gas emissions.

China's contribution to the causes of global warming has tended to overshadow that of other East Asian countries. Japan, South Korea, and Taiwan are all major emitters of greenhouse gases, but all also have energy-efficient economies, releasing far less carbon than the United States on a per capita basis. In fact, in 2003 the United States produced roughly six times the amount of greenhouse gases as Japan. Overall, Japan has been a strong proponent of international treaties, such as the Kyoto Protocol, designed to force reductions in greenhouse gas emissions. But while its total emissions have been relatively flat in recent years, Japan has not achieved the reductions that it promised in Kyoto. Increasingly, however, Japanese high-tech companies, like those of South Korea and Taiwan, realize that it could be very good business to develop new energy technologies to reduce the threat of climate change.

East Asia's Physical Geography

East Asia is situated in the same general latitudinal range as the United States, although it extends considerably farther north and south. The northernmost tip of China lies as far north as central Quebec, while China's southernmost point is at the same latitude as Mexico City. The climate of southern China is thus roughly comparable to southern Florida and the Caribbean, whereas that of northern China is more similar to south-central Canada (Figure 11.8, see page 481).

The island belt of East Asia, extending from northern Japan through Taiwan, is situated at the intersection of three tectonic plates (the basic building blocks of Earth's crust): the Eurasian, the Pacific, and the Philippine. This area, particularly Japan, is therefore geologically active, experiencing numerous earthquakes and dotted with volcanoes (Figure 11.9).

Japan's Physical Environment Although slightly smaller than California, Japan extends farther north and south. As a result, Japan's extreme south, in southern Kyushu and the Ryukyu Archipelago, is subtropical, while northern Hokkaido is almost subarctic. Most of the country, however, is temperate. The climate of Tokyo is not unlike that of Washington, DC—although Tokyo is distinctly rainier.

Japan's climate varies not only from north to south, but also from southeast to northwest across the main axis of the archipelago. In the winter, the area facing the Sea of Japan receives much more snow than the Pacific Ocean coastline. During this time of the year, cold winds from the Asian mainland blow across the relatively warm waters of the Sea of Japan. The air picks up moisture over the sea and deposits it, usually as snow, when it hits land (Figure 11.10). The Pacific coast of Japan, on the other hand, is far more vulnerable to typhoons (hurricanes), which frequently strike the country.

The Pacific coast of Japan is separated from the Sea of Japan coast by a series of mountain ranges. Japan is one of the world's most rugged countries, with mountainous terrain covering some 85 percent of its territory. Most of these uplands are thickly wooded, making Japan also one of the world's most heavily forested countries (Figure 11.11). Japan owes its lush forests both to its mild, rainy climate and to its long history of forest conservation. For hundreds of years, both the Japanese state and its village communities have enforced strict conservation rules, ensuring that timber and firewood extraction would be balanced by tree growth.

Figure 11.10 Heavy Snow in Japan's Mountains
Cold, moist air moving off the Sea of Japan produces heavy snows in northwestern Japan and along the country's mountainous spine. Numerous major ski areas dot the Japanese Alps, several of which have hosted world-class sports competitions. *(George Mobley/NGS Image Collection)*

Along Japan's coastline and interspersed among its mountains are limited areas of alluvial plains (see Figure 11.9). Although these lowlands were once covered by forests and wetlands, they have long since been cleared and drained for intensive agriculture. The largest Japanese lowland is the Kanto Plain to the north of Tokyo, but even it is only some 80 miles wide and 100 miles long

Figure 11.11 Forested Landscapes of Japan
Although much of Japan is heavily forested and supports a viable wood products industry, the yield is not large enough to satisfy demand. As a result, Japan imports timber extensively from North America, Southeast Asia, and Latin America. *(Robert Holmes/Corbis/Bettmann)*

(130 by 160 kilometers). The country's other main lowland basins are the Kansai, located around Osaka, and the Nobi, centered on Nagoya. In the mountainous province of Nagano, smaller basins are sandwiched between the imposing peaks of the Japanese Alps (see Figure 11.9).

Taiwan's Environment Taiwan, an island about the size of Maryland, sits at the edge of the continental land mass. To the west, the Taiwan Strait is only about 200 feet (60 meters) deep; to the east, ocean depths of many thousands of feet are found 10 to 20 miles (16 to 32 kilometers) offshore.

Taiwan itself forms a large tilted block. Its central and eastern regions are rugged and mountainous, while the west is dominated by an alluvial plain. Bisected by the tropic of Cancer, Taiwan has a mild winter climate, but it is sometimes battered by typhoons in the early autumn. Unlike nearby areas of China proper, Taiwan still has extensive forests, concentrated in its eastern upland areas.

Chinese Environments Even if one excludes its Central Asian provinces of Tibet and Xinjiang, China is a vast country with diverse environmental regions. For the sake of convenience, China proper can be divided into two main areas, one lying to the north of the Yangtze River valley, the other including the Yangtze and all areas to the south. As Figure 11.12 shows, each of these divisions can be subdivided into a number of distinctive regions.

Southern China is a land of rugged mountains and hills interspersed with lowland basins. The lowlands of southern China are far larger than those of Japan. One of the most distinctive is the former lake bed of central Sichuan (in the central Upper Yangtze region of Figure 11.12). Protected by imposing mountains, Sichuan is noted for its mild winter climate. Eastward from Sichuan, the Yangtze River passes through several other broad basins (in the Middle Yangtze region), partially separated from each other by hills and low mountains, before flowing into a large delta near Shanghai.

South of the Yangtze Valley, the mountains are higher (up to 7,000 feet, or 2,150 meters, in elevation), but they are still interspersed with alluvial lowlands. Sizable valleys are found in the far south, the Xi Basin in Guangdong province being the largest. Here the climate is truly tropical, free of frost. West of Guangdong lie the moderate-elevation plateaus of Yunnan and Guizhou, the former noted for its perennial springlike weather. Finally, to the northeast of Guangdong lies the rugged coastal province of Fujian. With its narrow coastal plain and deeply indented

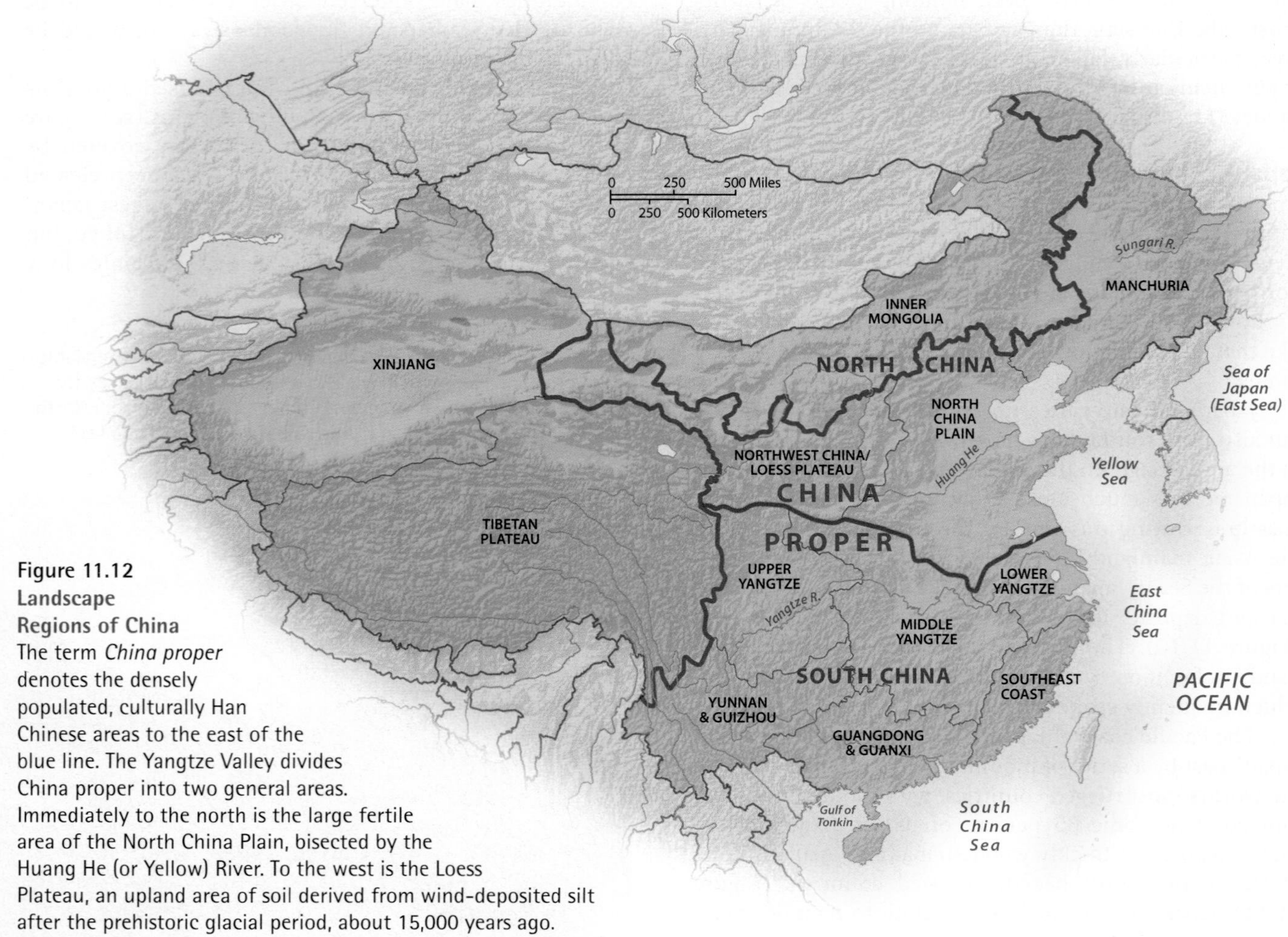

Figure 11.12 Landscape Regions of China The term *China proper* denotes the densely populated, culturally Han Chinese areas to the east of the blue line. The Yangtze Valley divides China proper into two general areas. Immediately to the north is the large fertile area of the North China Plain, bisected by the Huang He (or Yellow) River. To the west is the Loess Plateau, an upland area of soil derived from wind-deposited silt after the prehistoric glacial period, about 15,000 years ago.

Figure 11.13 The Fujian Coast of China
In southeastern China lies the rugged coastal province of Fujian. Here the coastal plain is narrow and the shoreline deeply indented, producing a picturesque landscape. Because of limited agricultural opportunities along this rugged coastline, many Fujianese people work in maritime activities. *(Dean Conger/Corbis/Bettmann)*

coastline, Fujian's landscape does not support a productive agricultural economy; as a result, the Fujianese people have often sought a maritime way of life, many of them working in the fishing and shipping industries (Figure 11.13).

North of the Yangtze Valley, the climate is both colder and drier than it is to the south. Summer rainfall is generally abundant except along the edge of the Gobi Desert, but the other seasons are dry. Desertification is a major threat in parts of the North China Plain, which have experienced prolonged droughts in recent years. With the exception of a few low mountains in Shandong province, the entire area is a virtually flat plain. Seasonal water shortages in this area are growing increasingly severe as withdrawals for irrigation and industry increase, leading to concerns about an impending water crisis.

West of the North China Plain sits the Loess Plateau. This is a fairly rough upland of moderate elevation and uncertain precipitation. It does, however, have fertile soil—as well as huge coal deposits. Farther west one finds the semiarid plains and uplands of Gansu province, situated at the foot of the great Tibetan Plateau.

China's far northeastern region is called *Dongbei* in Chinese and *Manchuria* in English. Manchuria is dominated by a broad, fertile lowland basin sandwiched between mountains stretching along China's borders with North Korea, Russia, and Mongolia. Although winters here can be brutal, summers are usually warm and moist. Manchuria's peripheral uplands have some of China's best-preserved forests and wildlife refuges.

Korean Landscapes Korea forms a well-demarcated peninsula, partially cut off from Manchuria by rugged mountains and sizable rivers (Figure 11.14). Like Japan, its latitudinal range is pronounced. The far north, which just touches Russia's Far East, has a climate not unlike that of Maine, whereas the southern tip is more reminiscent of the Carolinas. Korea is a mountainous country with scattered

Figure 11.14 The Mountains and Lowlands of Korea
Like Japan, Korea has extensive uplands interspersed with lowland plains. The country's highest mountains are in the north, whereas its most extensive alluvial plains are in the south. South Korea's provinces, shown here, are culturally as well as physically distinctive.

alluvial basins, a landscape that has, as in many other parts of East Asia, deeply influenced its demographic and agricultural development. The lowlands of the southern portion of the peninsula are more extensive than those of the north, giving South Korea a distinct agricultural advantage over North Korea. However, North Korea has more abundant mineral deposits, forests, and hydroelectric resources.

POPULATION AND SETTLEMENT: A Realm of Crowded Lowland Basins

East Asia, along with South Asia, is the most densely populated region of the world (Figure 11.15). The lowlands of Japan, Korea, and China are among the most intensely

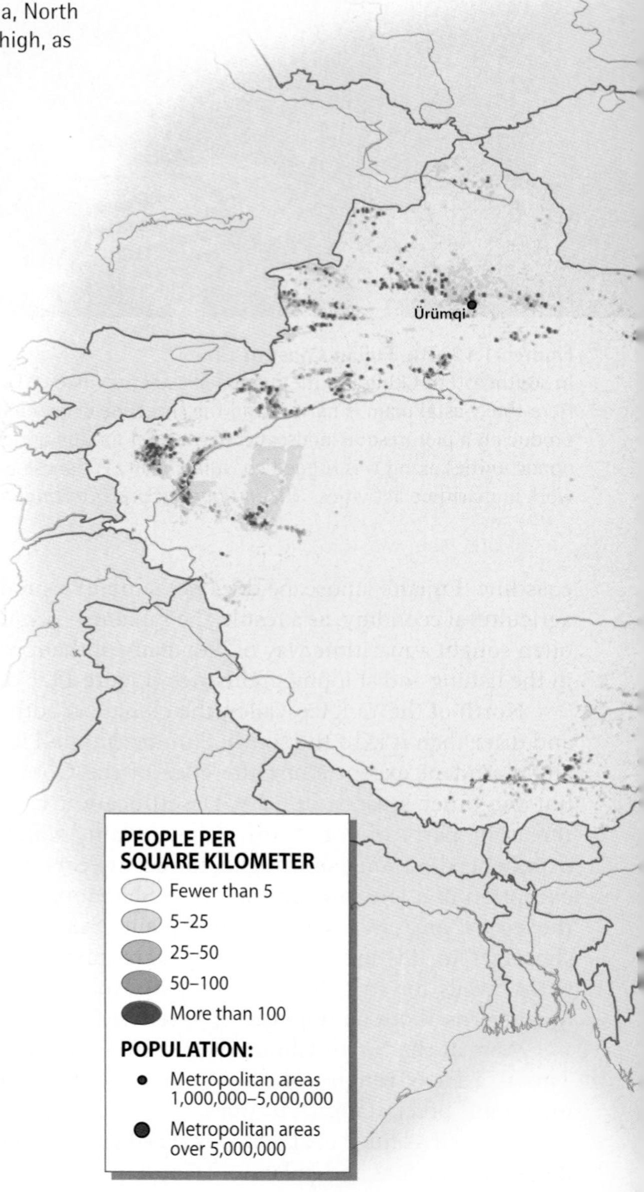

Figure 11.15 Population Map of East Asia
Parts of East Asia are extraordinarily densely settled, particularly in the coastal lowlands of China and Japan. This contrasts with the sparsely settled lands of western China, North Korea, and northern Japan. Although the total population of this world region is high, as is the overall density, the rate of natural population increase has slowed rather dramatically in the last several decades.

used portions of Earth, containing East Asia's major cities and most of its agricultural lands.

Although the population density of East Asia is extremely high, the region's demographic growth rate has declined dramatically since the 1970s. While China's population is still expanding, its rate of growth is now relatively low (Table 11.1). In Japan especially, the current concern is impending population loss, hence an aging population that will need to be supported by a shrinking number of younger workers.

Agriculture and Settlement in Japan

Japan is a highly urbanized country, supporting two of the largest urban agglomerations in the world. Yet it is also one of the world's most mountainous countries, with lightly inhabited uplands. Agriculture must therefore share the limited lowlands with cities and suburbs, resulting in extremely intensive farming practices.

Japan's Agriculture Lands Japanese agriculture is largely limited to the country's coastal plains and interior basins. Japanese rice farming has long been one of the most productive forms of agriculture in the world, helping support a population of 127 million people on a relatively small and rugged land. Although rice is grown in almost all Japanese lowlands, the country's premier rice-growing districts lie along the Sea of Japan coast in Honshu. Vegetables are also grown intensively in all of Japan's lowland basins, even on tiny patches within urban neighborhoods (Figure 11.16). The valleys of central and northern

TABLE 11.1 Population Indicators

Country	Population (Millions, 2007)	Population Density (per Square Kilometer)	Total Fertility Rate	Percent Urban	Percent <15[a]	Percent >65[b]	Net Migration (per 1000, 2000–05)
China	1,318.0	138	1.6	44	20	8	−0.3
Hong Kong	6.9	6,308	1.0	100	14	12	8.8
Japan	127.7	338	1.3	79	14	21	0.4
North Korea	23.3	193	2.0	60	27	8	0.0
South Korea	48.5	487	1.1	82	18	10	−0.3
Taiwan	22.9	636	1.1	78	18	10	

[a]*Percentage of population younger than 15.*
[b]*Percentage of population older than 65.*

Sources: Population Reference Bureau, World Population Data Sheet, *2007; net migration rate data from* UN International Migration, *2006.*

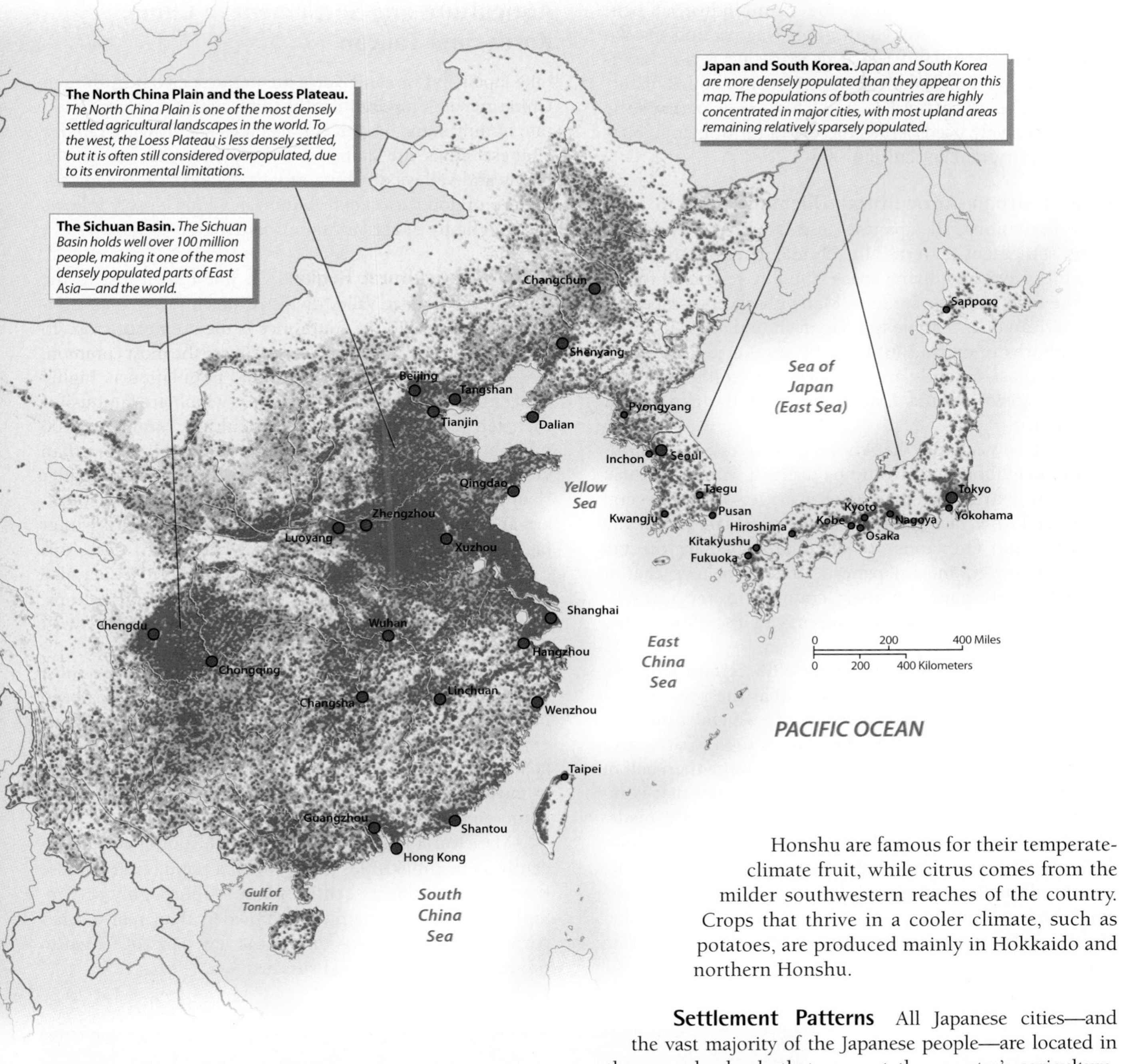

Honshu are famous for their temperate-climate fruit, while citrus comes from the milder southwestern reaches of the country. Crops that thrive in a cooler climate, such as potatoes, are produced mainly in Hokkaido and northern Honshu.

Settlement Patterns All Japanese cities—and the vast majority of the Japanese people—are located in the same lowlands that support the country's agriculture. Not surprisingly, the three largest metropolitan areas—Tokyo, Osaka, and Nagoya—sit near the centers of the three largest plains.

The fact that Japan's settlements are largely restricted to roughly 15 percent of its land area means that the country's effective population density—the actual crowding that it experiences—is one of the highest in the world.

Figure 11.16 Japanese Urban Farm
Japanese landscapes often combine dense urban settlement with small patches of intensively farmed agriculture. Here a cabbage farm coexists with an urban neighborhood. *(Kyodo News International, Inc.)*

This is especially notable in the main industrial belt, which extends from Tokyo through Nagoya and Osaka, hence along the Inland Sea (the maritime region sandwiched between Shikoku, western Honshu, and Kyushu) to the northern coast of Kyushu. This area is perhaps the most intensively used part of the world for human habitation, industry, and agriculture.

Japan's Urban–Agricultural Dilemma Due to such space limitations, all Japanese cities are characterized by dense settlement patterns. In the major urban areas, the amount of available living space is highly restricted for all but the most affluent families. Many observers—especially American ones—argue that Japan should allow its cities and suburbs to expand into nearby rural areas. However, because most uplands are too steep for residential use, such expansion would have to come at the expense of farmland.

As it now stands, the wholesale conversion of farmland to neighborhoods would be difficult. Although most farms are extremely small (the average is only several acres) and only marginally viable, farmers are politically powerful and croplands are often protected by the tax code. Moreover, most Japanese citizens believe that it is vitally important for their country to remain self-sufficient at least in rice. Rice imports are therefore highly restricted.

U.S. economic interests, along with certain Japanese consumer advocates, stress the fact that the present system forces Japanese consumers to pay several times the world market price for rice, their staple food. Critics also point out that Japan relies on imports for more than half of its overall food needs as well as for the energy that is required to grow rice. But whatever arguments are made, change will only come gradually, if at all. For the moment, Japan must live with the tensions resulting from the fact that its affluent population is forced to live in tight proximity and to pay high prices for basic staples. Somewhat similar patterns, moreover, are found in other East Asian countries.

Agriculture and Settlement in China, Korea, and Taiwan

Like Japan, Taiwan and Korea are urban societies. Although China remains predominantly rural, with an urbanization rate of only some 44 percent, its cities are growing rapidly. Chinese cities are rather evenly distributed across the plains and valleys of China proper. As a result, the overall pattern of population distribution in China closely follows the geography of agricultural productivity.

China's Agricultural Regions A line drawn just to the north of the Yangtze Valley divides China into two main agricultural regions. To the south, rice is the dominant crop. To the north, wheat, millet, and sorghum are the most common.

In southern and central China, population is highly concentrated in the broad lowlands, which are famous for their fertile soil and intensive agriculture. More than 100 million people live in the Sichuan Basin, while more than 70 million reside in deltaic Jiangsu, a province smaller than Ohio. Cropping occurs year-round in most of southern and central China; summer rice alternates with winter barley or vegetables in the north, whereas two rice crops can be harvested in the far south. Southern China also produces a wide variety of tropical and subtropical crops, and moderate slopes throughout the area supply sweet potatoes, corn, and other upland crops.

The North China Plain is one of the world's most thoroughly **anthropogenic landscapes**; that is, it has been heavily transformed by human activities. Virtually its entire extent is either cultivated or occupied by houses, factories, and other structures of human society. Too dry in most areas for rice, the North China Plain supports the main crops of wheat, millet, and sorghum.

Manchuria was a lightly populated frontier zone as recently as the mid-1800s. Today, with a population of more than 100 million, its central plain is thoroughly settled. Still, Manchuria remains less crowded than many other parts of China, and it is one of the few parts of China to produce a consistent food surplus.

Figure 11.17 **Loess Settlement** A typical subterranean dwelling carved out the soft loess sediment in central China. Approximately 70 million people live similarly in the Loess Plateau region. Unfortunately, this region is prone to major earthquakes that take a high toll on the local population because of dwelling collapse. *(Christopher Liu/ChinaStock Photo Library)*

The Loess Plateau is more thinly settled yet, supporting only some 70 million inhabitants. But considering the area's aridity and widespread soil erosion, this is a high figure. Like other portions of northern China, the Loess Plateau produces wheat and millet. One of its unique settlement features is subterranean housing. Although loess erodes quickly under the impact of running water, it holds together well in other circumstances. For millennia, villagers have excavated pits on the surface of the plateau, from which they have tunneled into the earth to form underground houses (Figure 11.17). These subterranean dwellings are cool in the summer and warm in the winter. Unfortunately, they tend to collapse during earthquakes. One quake in 1920 killed an estimated 100,000 persons; another in 1932 killed some 70,000.

Patterns in Korea and Taiwan Like China and Japan, Korea is a densely populated country. It contains some 71 million persons (23 million in the north and 48 million in the south) in an area smaller than Minnesota. South Korea's population density is approximately 1,200 per square mile (463 per square kilometer), a higher density than Japan's. Most of South Korea's people are crowded into the alluvial plains and basins of the west and south. The highland spine, extending from the far north to northeastern South Korea, remains relatively sparsely settled. South Korean agriculture is dominated by rice. North Korea, in contrast, relies heavily on corn and other upland crops that do not require irrigation.

Taiwan is the most densely populated state in East Asia. Roughly the size of the Netherlands, it contains more than 22 million inhabitants. Its overall population density is more than 1,500 people per square mile (627 per square kilometer), one of the highest figures in the world. Because mountains cover most of central and eastern Taiwan, virtually the entire population is concentrated in the narrow lowland belt in the north and west. Large cities and numerous factories are scattered here amid lush farmlands. Despite its highly productive agriculture, Taiwan, like other East Asian countries, is forced to obtain much of its food from abroad.

Agriculture and Resources in Global Context

Japan and South Korea, like Taiwan, are major food importers, and China has recently moved in the same direction. Other resources also are being drawn in from all quarters of the world by the powerful economies of East Asia.

Global Dimensions of Japanese Agriculture and Forestry Japan may be mostly self-sufficient in rice, but it is still one of the world's largest food importers. As the Japanese have grown more prosperous over the past 60 years, their diet has grown more diverse. That diversity is made possible largely by importation of food.

Japan imports food from a wide array of other countries. It procures both meat and the feed used in its domestic livestock industry from the United States, Canada, and Australia. These same countries supply wheat needed to produce bread and noodles. Even soybeans, long a staple of the Japanese diet, must be purchased from Brazil and the United States. Japan has one of the highest rates of fish consumption in the world, and the Japanese fishing fleet scours the world's oceans to supply the demand (Figure 11.18). Japan also purchases prawns and other seafood, much of it farm-raised in former mangrove swamps, from Southeast Asia and Latin America.

Figure 11.18 Japanese Fishing Boat
A Japanese whaling ship leaves the port of Shimonoseki in southwestern Japan. While Japanese whaling is particularly controversial, the larger Japanese fishing fleet covers much of the world's ocean surface in search of high-quality seafood. *(AP Photo/Itsuo Inouye)*

Japan also depends on imports to supply its demand for forest resources. While its own forests produce high-quality cedar and cypress logs, it obtains most of its construction lumber and pulp (for papermaking) from western North America and Southeast Asia. As the rain forests of Malaysia, Indonesia, and the Philippines diminish, Japanese interests are beginning to turn to Latin America and Africa as sources of tropical hardwoods. Japanese and South Korean firms also are looking to eastern Russia, a nearby and previously little-exploited forest zone.

Japan is able to support its large and prosperous population on such a restricted land base because it can purchase resources from abroad. Certainly all countries engage in trade, but Japan's basic resource dependence is pronounced. Almost all of the oil, coal, and other minerals that it consumes are imported. If forced to rely on its own resources, Japan would find itself in an economic and ecological bind.

Although geographers once looked at resource endowments as a main support of each country's economy, such a view is no longer supportable in the increasingly interconnected world. As long as a country can export items of value, it can obtain whatever imports it requires. This also means, however, that the environmental degradation generated by a successful economy, such as that of Japan, becomes globalized as well.

Global Dimensions of Chinese Agriculture Until the late 1990s, China was essentially self-sufficient in food, despite its huge population and crowded lands. But rapid ecoomic growth has resulted in an increased consumption of meat, which requires large amounts of feed grain. Growth also has brought about the loss of agricultural lands to residential and industrial development. Between 2002 and 2004, China is estimated to have lost more than 2 percent of its farmland. As a result, in 2003 the country had to import 40 million tons of grain.

Whether China will ever again be self-sufficient in food is an open question, but it is clear that the country's current demands for agricultural products—as well as for a wide array of other resources—are reordering patterns of global trade. Many Latin American, African, and other Asian countries now rely heavily on the Chinese market. Mineral producers in particular have made large profits as Chinese demand has generated soaring prices for copper, nickel, and other basic mineral commodities

Although China remains a net food importer, it also exports a number of specialty crops and processed foods. Increasingly, however, consumers in foreign markets are concerned about the safety of Chinese exports. In 2007, contaminated pet-food ingredients from China resulted in many animal deaths in the United States, and in the same year the United States banned the import of a number of Chinese seafood products until they could be proved safe. China responded by banning chicken and pork products from Tyson Foods, a U.S. company. At the same time, however, the Chinese government shut down more than 100 food-processing plants, and went so far as to execute the former head of its own food and drug agency after he was convicted of corruption.

Korean Agriculture in a Global Context South Korea has also made the transition, like Japan and Taiwan, to a global food and resource procurement pattern. It is now one of the world's main importers of wheat, corn, and soybeans. North Korea, on the other hand, has pursued a goal of strict self-sufficiency. While relatively successful for a number of years, in the mid-1990s this policy resulted in widespread famine after a series of floods, followed by drought, destroyed most of the country's rice and corn crops. Since then, North Korea has relied heavily on international aid—much of it coming from South Korea—to feed its people, but malnutrition and even starvation remain widespread.

Urbanization in East Asia

China has one of the world's oldest urban foundations, dating back more than 3,500 years. In medieval and early modern times, East Asia as a whole possessed a well-developed system of cities. In the early 1700s, Tokyo, then called Edo, probably overshadowed all other cities, with a population of more than 1 million.

But despite this early start, East Asia was overwhelmingly rural at the end of World War II. Some 90 percent of China's people then lived in the countryside, and even Japan was only about 50 percent urbanized. But as the region's economy began to grow after the war, so did its cities. Japan, Taiwan, and South Korea are now between 70 and 85 percent urban, which is typical for advanced industrial countries. Some 56 percent of China's people still live in rural areas, but this figure is decreasing steadily.

Chinese Cities Traditional Chinese cities were clearly separated from the countryside by defensive walls. Most were planned in accordance with strict geometrical principles that were thought to reflect the cosmic order. The old-style Chinese city was dominated by low buildings and characterized by straight streets. Houses were typically built around courtyards, and narrow alleyways served both commercial and residential functions.

China's urban fabric began to change during the colonial period. A group of port cities was taken over by European interests, which proceeded to build Western-style buildings and modern business districts. By far the

Figure 11.19 Contemporary Shanghai
This vibrant city of approximately 20 million embodies the new China with its massive high-rise apartments, industrial developments, and office towers. *(Hiroshi Harada/DUNQ/Photo Researchers, Inc.)*

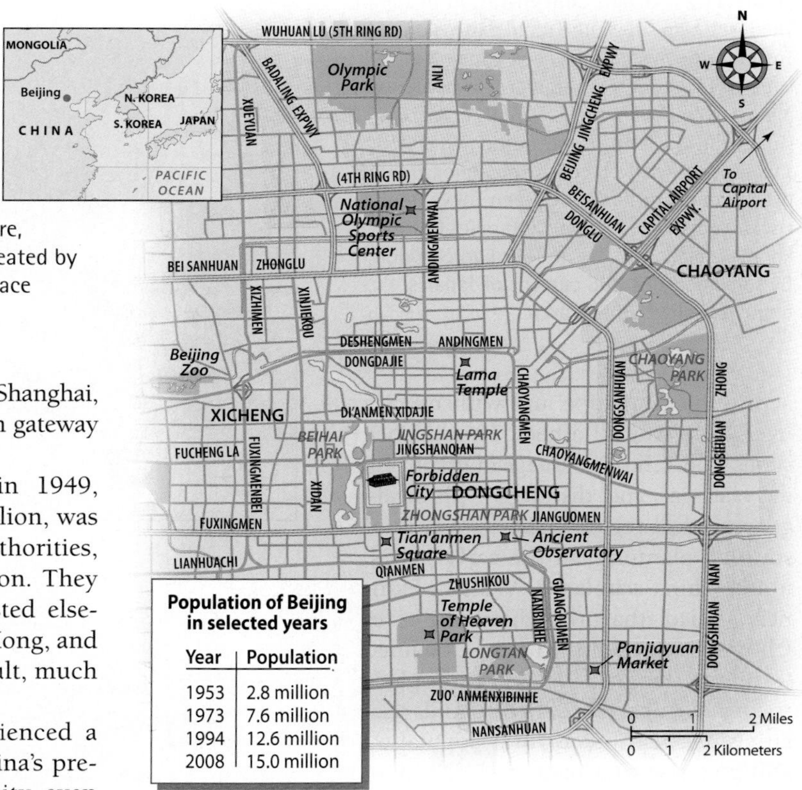

Figure 11.20 Beijing, China
The historic capital during the Manchu period (1644–1912), Beijing regained its status as capital city in 1949. Under communist rule, much of Beijing's historical landscape was razed and replaced with large blocks of government offices and massive apartment buildings. Only the historically significant Forbidden City, home to the Manchu rulers, was saved from this transformation of the urban landscape. Tienanmen Square, reputed to be the largest open square in any city, was created by clearing buildings away from the area in front of the Palace Museum.

Population of Beijing in selected years

Year	Population
1953	2.8 million
1973	7.6 million
1994	12.6 million
2008	15.0 million

most important of these semicolonial cities was Shanghai, built near the mouth of the Yangtze River, the main gateway to interior China.

When the communists came to power in 1949, Shanghai, with a population of more than 10 million, was the second-largest city in the world. The new authorities, however, viewed it as a decadent, foreign creation. They therefore milked it for taxes, which they invested elsewhere. Most of the wealthy people fled to Hong Kong, and relatively few migrants were allowed in. As a result, much of the city began to decay.

Since the late 1980s, Shanghai has experienced a major revival and is again in many respects China's premier city. Migrants are now pouring into the city, even though the state still tries to restrict the flow, and building cranes crowd the skyline. Official statistics now put the population of the metropolitan area at roughly 18 million, but the actual number, including temporary migrants, is estimated to be more than 20 million—including a 3-million-strong "floating" population of undocumented migrants. The new Shanghai is a city of massive high-rise apartments and concentrated industrial developments (Figure 11.19). Shanghai has the third busiest port in the world (after Singapore and Hong Kong) and boasts the world's fastest train. But despite Shanghai's revived economic fortunes, the city remains politically secondary to Beijing, China's capital.

Beijing was China's capital during the Manchu period (1644–1912), a status it regained in 1949. Under communist rule, Beijing was radically transformed; old buildings were razed and broad avenues were plowed through neighborhoods (Figure 11.20). Crowded residential districts gave way to large blocks of apartment buildings and massive government offices. Some historically significant structures were saved; the buildings of the Forbidden City, for example, where the Manchu rulers once lived, survived as a complex of museums. The area immediately in front of the Palace Museum, however, was cleared. The resulting plaza, Tienanmen Square, is reputed to be the largest open square in any city of the world. In 2007 and 2008, the Chinese government invested large amounts of money into Beijing's urban infrastructure in preparation for the city's 2008 Summer Olympics.

The Chinese Urban System China's urban system as a whole is fairly well balanced, with sizable cities relatively evenly spaced

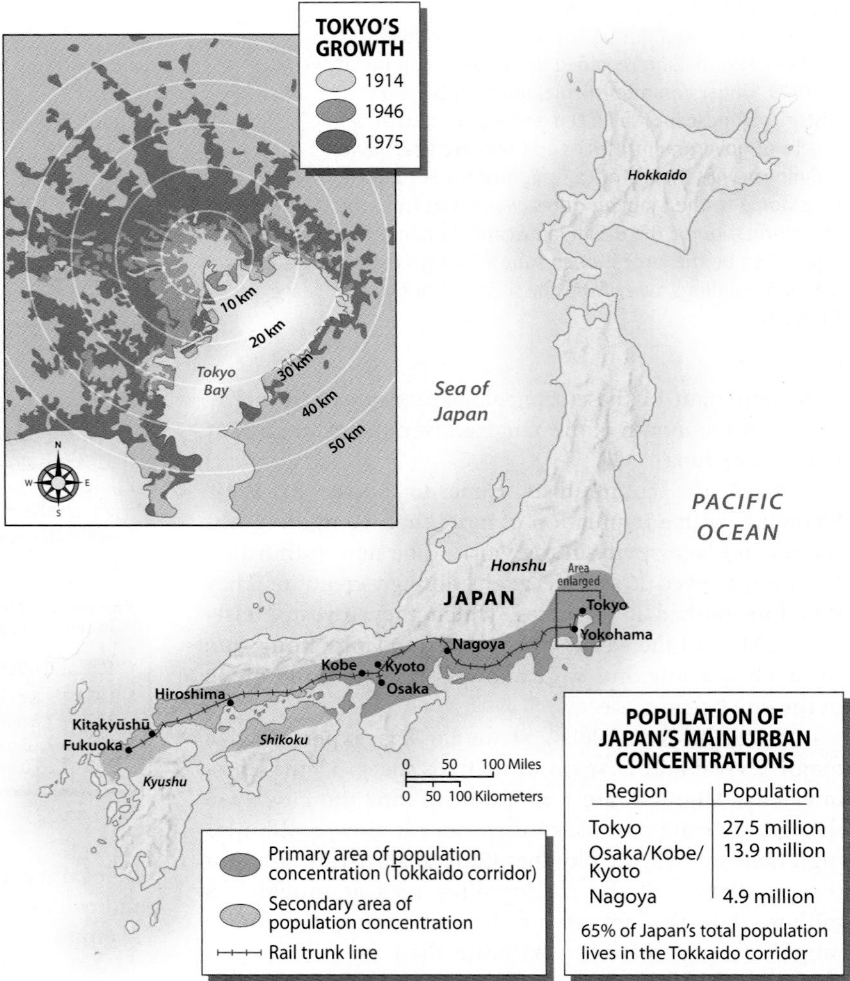

Figure 11.21 Urban Concentration in Japan The inset map shows the rapid expansion of Tokyo in the postwar decades. Today the Greater Tokyo metropolitan area is home to almost 30 million people. The larger map shows the cluster of urban settlements along Japan's southeastern coast. The major area of urban concentration is between Tokyo and Osaka, a distance of some 300 miles (482 kilometers), known as the Tokkaido corridor. By some accounts, 65 percent of Japan's population lives in this area.

across the landscape and with no city overshadowing all others. This balance stems from China's heritage of urbanism, its vast size and distinctive physical–geographical regions, and its legacy of socialist planning. In the early 1960s, noted China scholar G. William Skinner argued that the distribution of Chinese cities is best explained through the use of **central place theory**. Central place theory holds that an evenly distributed rural population will give rise to a regular hierarchy of urban places, with uniformly spaced larger cities surrounded by constellations of smaller cities, each of which, in turn, will be surrounded by smaller towns.

In the 1990s, Beijing and Shanghai vied for the first position among Chinese cities, with Tianjin, serving as Beijing's port, coming in a close third. All three of these cities, along with Chongqing in the Sichuan Basin, have been removed from the regular provincial structure of the country and granted their own metropolitan governments. In 1997, another major city, Hong Kong, passed from British to Chinese control. Rather than becoming an ordinary part of China, Hong Kong was granted a unique status as an autonomous "special administrative region," and allowed to largely manage its own affairs. While not as populous as Beijing or Shanghai, Hong Kong is far wealthier. The greater metropolitan area of the Xi Delta, composed of Hong Kong, Shenzhen, and Guangzhou (called *Canton* in the West), is now in some respects China's premier urban area.

In addition to these four primary Chinese cities are several dozen large and growing urban centers. More than 30 of them have more than 1 million inhabitants, and in the near future many more will pass that milestone. Yet for all of this, China remains a predominantly rural society. But industrial development is increasingly moving into the countryside, as factories sprout in a seemingly haphazard fashion amid rice fields and vegetable plots.

Other Regional Urban Patterns The urban structures of East Asia's maritime fringe are different from those of China. Both South Korea and Taiwan are noted for their pronounced **urban primacy**, a condition that occurs when the urban population is concentrated in a single city. Japan supports a new urban phenomenon called a **superconurbation**, or megalopolis, which is a huge zone of coalesced metropolitan areas (Figure 11.21).

Seoul, the capital of South Korea, overwhelms all other cities in the country. Seoul itself is home to more than 10 million people, and its greater metropolitan area contains some 40 percent of South Korea's total population. All of South Korea's major governmental, economic, and cultural institutions are concentrated there. Seoul's explosive and generally unplanned growth has resulted in serious congestion. The South Korean government is promoting industrial growth in other cities, but none of its actions has yet challenged the primacy of the capital.

Taiwan is similarly characterized by a high degree of urban primacy. The capital city of Taipei, located in the far north, mushroomed from some 300,000 people during the Japanese colonial period (from 1895 to 1945) to more than 6 million in the metropolitan area today.

CITYSCAPES Life on the Streets of Tokyo

Although Tokyo is a huge city located at the center of one of the world's largest metropolitan areas, it can feel like a small, intimate place. This is partly because it is divided into 23 districts (or *ku*, in Japanese), each of which has its own distinctive character and specializes in specific goods and services (Figure 11.2.1). Fortunately, Tokyo has a superb system of public transportation, making it relatively easy to get from one neighborhood to another (Figure 11.2.2).

Several Tokyo neighborhoods are noted as entertainment districts. For people looking for exciting nightlife and a cosmopolitan atmosphere, Roppongi is the place to go. Emerging around a now-defunct U.S. military base, Roppongi is the most Westernized part of Japan, and thus appeals to many tourists and expatriates. Younger Japanese looking for fun and fashion more often head for Harajuku, famous for its boutiques and sidewalk cafes. Well-heeled business executives, on the other hand, tend to favor the Akasaka nightlife district, celebrated for its expensive hotels and "hostess bars."

Other Tokyo neighborhoods are better known for their shopping potential. Some are highly specialized. Akihabara, for example, has more than 600 stores focused on electronics. Tsukiji, on the other hand, boasts the largest fish market in Japan and perhaps the world. The most expensive shops—and some of the world's most expensive real estate—are found in the Ginza, a swanky neighborhood also noted for its art galleries and exclusive clubs. Tokyo's department store center is found in Shibuya, which also has an active and rather hip nightlife scene. One also can shop, and do much more, in Shinjuku, Tokyo's main business district. Reminiscent of midtown Manhattan, Shinjuku boasts a collection of impressive skyscrapers.

Other Tokyo neighborhoods are better known for their historical significance. Although Hibiya is an important financial district, it is more famous as the site of the Imperial Palace and the center of the old city. Ueno, also part of the old downtown, supports the National Museum, which has the world's largest collection of Japanese art.

Tourists less often venture into Tokyo's other districts, but many are worth the effort. If one wants to see the rougher side of the city—which, by U.S. standards, is not very rough at all—head for Ikebukuro, which has aptly been described as "working man's Tokyo." For the geographically inclined, one can do no better than simply walk the narrow back streets and encounter the unexpected. Turn one corner, and you might find an entire street of small-scale machine shops, each of which is relatively open to visual inspection; turn another, and you might come across a large assemblage of tiny printing establishments. For a pedestrian's encounter of diversity amid globalization, few places are more interesting and accessible than Tokyo.

Figure 11.2.2 Tokyo Train Station
Tokyo boasts an extremely efficient public transportation system, based largely on subways and commuter rail lines. Trains and train stations, however, can be very crowded during rush hours. *(Hiroshi Harada/Dumo/Photo Researchers, Inc.)*

Figure 11.2.1 Neighborhoods and Districts of Tokyo
Tokyo is divided into numerous administrative districts, each of which has an individual personality. Each district, in turn, contains a number of distinctive neighborhoods.

Japan has traditionally been characterized by urban "bipolarity" rather than urban primacy. Until the 1960s, Tokyo, the capital and main business and educational center, together with the neighboring port of Yokohama, was balanced by the mercantile center of Osaka and its port of Kobe. Kyoto, the former imperial capital and the traditional center of elite culture, is also situated in the Osaka region. A host of secondary and tertiary cities balance Japan's urban structure. Nagoya, with a metropolitan area of some 5 million persons, remains the center of the automobile industry and is one of the few large Japanese cities in which it is more efficient to travel by car than by public transportation.

As Japan's economy boomed in the 1960s, 1970s, and 1980s, so did Tokyo (see "Cityscapes: Life on the Streets of Tokyo"). The capital city then outpaced all other urban areas in almost every urban function. The Greater Tokyo metropolitan area now contains 25 to 28 million people, depending on how it is defined. The Osaka–Kobe metropolitan area stands at a distant second, with 10 to 14 million inhabitants. Concerned about the increasing primacy of Tokyo, the Japanese government has been steering new developments to other parts of the country. Such efforts, however, have been only moderately successful.

Most of Japan's other main cities have seen modest population gains over the past several decades. Urban growth, in general, has been supported by rural depopulation. Metropolitan expansion has been particularly pronounced in the cities linking Tokyo to Osaka, an area known as the Tokkaido corridor. Transportation connections are superb along this route, and proximity to Tokyo, Osaka, and Nagoya encourages development. The result has been the creation of a superconurbation (Figure 11.21). One can travel from Tokyo to Osaka on the main rail line, a distance of almost 300 miles (480 kilometers), and never leave the urbanized area. By some accounts, 65 percent of Japan's people are crowded into this narrow supercity.

Japanese cities sometimes strike foreign visitors as rather gray and monotonous places, lacking historical interest. Little of the country's premodern architecture remains intact. Traditional Japanese buildings were made of wood, which survives earthquakes much better than stone or brick. Fires have therefore been a long-standing hazard, and in World War II the U.S. Air Force fire-bombed most Japanese cities, virtually obliterating them. In addition, Hiroshima and Nagasaki were completely destroyed by atomic bombs. The one exception was Kyoto, the old imperial capital, which was spared devastation. As a result, Kyoto is famous for its beautiful (wooden) Buddhist monasteries and Shinto temples, which ring the basin in which central Kyoto lies.

Other Japanese cities were largely reconstructed in the late 1940s and 1950s, a period when Japan was still relatively poor and could afford only inexpensive concrete buildings. In the boom years of the 1980s, however, modern skyscrapers rose in many of the larger cities and postmodernist architecture began to lend variety, especially to the wealthier neighborhoods. Critics sometimes contend that Japanese cities still have relatively few features that show a distinctive cultural flavor. Others disagree, noting that if one looks closely, one may see aspects of Japanese and, more generally, East Asian culture everywhere.

CULTURAL COHERENCE AND DIVERSITY: A Confucian Realm?

East Asia is in some respects one of the world's most unified cultural regions. Although different parts of East Asia have their own unique cultural features, the entire region shares certain historically rooted ways of life and systems of ideas.

Most of these East Asian commonalties can be traced back to ancient Chinese civilization. Chinese civilization emerged roughly 4,000 years ago, largely in isolation from the Eastern Hemisphere's other early centers of civilization in the valleys of the Indus, Tigris-Euphrates, and Nile rivers. As a result, East Asian civilization developed along several unique lines.

Unifying Cultural Characteristics

The most important unifying cultural characteristics of East Asia are related to religious and philosophical beliefs. Throughout the region, Buddhism and Confucianism have shaped not only individual beliefs but also social and political structures. Although the role of traditional belief systems has been seriously challenged over the past 50 years, especially in China, historically grounded cultural patterns have not disappeared.

The Chinese Writing System The clearest distinction between East Asia and the world's other cultural regions is found in written language. Existing writing systems elsewhere in the world are based on the alphabetic principle, in which each symbol represents a distinct sound. East Asia, in contrast, evolved an entirely different system, **ideographic writing**. In ideographic writing, each symbol (commonly called a *character*) primarily represents an idea rather than a sound (although the symbols can denote sounds in certain circumstances). As a result, ideographic writing requires the use of a large number of distinct symbols.

The East Asian writing system can be traced to the dawn of Chinese civilization, some 3,200 years ago. As the Chinese Empire expanded and the prestige of Chinese civilization carried its culture to other lands, the Chinese writing system spread. Japan, Korea, and Vietnam all came to use the same system, although in Japan it was substantially modified, while in Korea and Vietnam it was much later largely replaced by alphabetic systems.

The Chinese ideographic writing system has one major disadvantage and one major advantage when compared with alphabetic systems, both of which stem from the fact that it is largely divorced from spoken language. The disadvantage is that it is difficult to learn; to be literate, a person must memorize thousands of characters. The advantage is that two literate people do not have to speak the same language to be able to communicate, because the written symbols that they use to express their ideas are the same.

This advantage was tremendously important for the creation of a unified Chinese culture. When the Chinese Empire expanded south of the Yangtze River beginning in about 200 BCE, ethnic groups speaking a variety of languages were suddenly brought into the same political and cultural system. Because they could adopt the Chinese writing system, the peoples of southern China were able to integrate fully into the country without necessarily adopting its spoken language.

Korean Modifications In Korea, Chinese characters were adopted at an early date and were used exclusively for hundreds of years. In the 1400s, however, Korean officials decided that Korea needed its own alphabet. They wanted to allow more widespread literacy, hoping also to more clearly differentiate Korean culture from that of China. The use of the new script spread quickly through the country. Korean scholars and officials, however, continued to use Chinese characters, regarding their own script as suitable only for popular writings. Today the Korean script is used for almost all purposes, but scholarly works still contain scattered Chinese characters.

Japanese Modifications The writing system of Japan is even more complex (Figure 11.22). Initially the Japanese simply borrowed Chinese characters, referred to in Japanese as *kanji*. Owing to the grammatical differences between Japanese and Chinese, the exclusive use of kanji resulted in awkward sentence construction. The Japanese solved this quandary by developing a quasi-alphabet known as *hiragana* that allowed the expression of words and parts of speech not easily represented by Chinese characters. In hiragana, each symbol represents a distinct syllable, or combination of a consonant and a vowel sound. (A different but essentially parallel system, called *katakana*, is used in Japan for spelling words of foreign origin.) Eventually the two styles of writing merged, and written Japanese came to employ a complex mixture of symbols. Japanese kanji today differ slightly from Chinese characters. Even so, it is still relatively easy for anyone literate in Japanese to learn how to read Chinese—but not to speak it!

Unlike Chinese, Japanese is easily written in the Roman alphabet. The resulting *romanji* style of writing is especially important in advertisements and for computer use. Computer keyboards using roman letters are not only much less cumbersome than Japanese keyboards, but they are also a component of modern global culture. Computer technology, however, allows romanji entries to be easily converted into the traditional Japanese mixture of kanji, hiragana, and katakana symbols.

The Confucian Legacy Just as the use of a common writing system helped forge cultural linkages throughout East Asia, so too **Confucianism** (the philosophy developed by Confucius) came to occupy a significant position in all of the societies of the region. So strong is the heritage of Confucius that some writers refer to East Asia as the "Confucian world." In Japan, however, Confucianism never had the influence that it did in China and Korea.

The premier philosopher of Chinese history, Confucius (or *Kongfuzi*, in Mandarin Chinese) lived during the sixth century BCE, a period of marked political instability. Confucius's goal was to create a philosophy that could generate stability. Although Confucianism is often considered to be a religion, Confucius himself was mostly interested in the "here and now," focusing his attention on how to lead a correct life and organize a proper society. Confucian thought does not deny the existence of a deity or an afterlife, but neither does it give them much consideration.

Confucius stressed deference to the proper authority figures, but he also thought that authority has a responsibility to act in a benevolent manner. The most basic level of the traditional Confucian moral order is the family unit, considered the bedrock of society. The ideal family structure

Figure 11.22 Japanese Writing
The writing system of Japan was originally based on Chinese characters, known in Japan as *kanji*. Because of grammatical differences, however, the Japanese developed two unique "alphabets" of syllables, known as *katakana* and *hiragana*. Here kanji and katakana symbols are visible. *(Hiroshi Harada/DUNQ/Photo Researchers, Inc.)*

is patriarchal, and children are told to obey and respect their parents—especially their fathers—as well as their elder brothers.

Confucian philosophy also stresses the need for a well-rounded and broadly humanistic education. To a certain extent, Confucianism advocates a kind of meritocracy, holding that an individual should be judged on the basis of behavior and education, rather than on family background. The high officials of Imperial China (pre-1912)—the powerful **Mandarins**—were thus selected by competitive examinations. Only wealthy families, however, could afford to give their sons the education needed for success on those grueling tests.

Confucianism in Japan In Japan, Confucianism was never as important as it was on the mainland. Japanese officials were actually able to exclude certain Confucian beliefs that they considered dangerous. The most important of these was the revocable "mandate of heaven." According to this notion, the emperor of China derived his authority from the principle of cosmic harmony, but such a mandate could be withdrawn if he failed to fulfill his duties. This idea was used both to explain and to legitimize the rebellions that occasionally resulted in a change of China's ruling dynasty. In Japan, on the other hand, a single imperial dynasty has persisted throughout the entire period of written history. Although the emperor of Japan has had little real power for more than a thousand years, the sanctity of his family lineage continues to form a basic principle of Japanese society.

The Modern Role of Confucian Ideology The significance of Confucianism in East Asian development has been hotly debated for the past hundred years. In the early 1900s, many observers believed that the conservatism of the philosophy, derived from its respect for tradition and authority, was responsible for the economically backward position of China and Korea. But because East Asia has enjoyed the world's fastest rates of economic growth over the past several decades, such a position is no longer supportable. Some scholars now argue that Confucianism's respect for education and the social stability that it generates give East Asia an advantage in international competition.

Over the past 100 years, Confucianism has lost much of the hold that it once had on public morality throughout East Asia, and especially in China. Under the communist system, China's rulers sought for decades to discourage if not eliminate Confucian thought. Currently, however, some Chinese officials are pushing for a revival of the philosophy, hoping that it will lead to enhanced social stability.

Religious Unity and Diversity

Certain religious beliefs have worked alongside Confucianism to cement together cultures of the East Asian region. The most important culturally unifying beliefs are associated with Mahayana Buddhism. Other religious practices, however, have had a more divisive role.

Mahayana Buddhism Originating in India in the sixth century BCE, Buddhism stresses escape from an endless cycle of rebirths to reach union with the divine cosmic principle (or nirvana). By the second century CE, Buddhism had reached China, and within a few hundred years it had spread throughout East Asia. Today Buddhism remains widespread everywhere in the region, although it is far more significant in mainland Southeast Asia and Sri Lanka.

The variety of Buddhism practiced in East Asia—Mahayana, or Greater Vehicle—is distinct from the Theravada Buddhism of Sri Lanka and Southeast Asia (Figure 11.23). Most important, Mahayana Buddhism simplifies the quest for nirvana, in part by suggesting that entities *(boddhisatvas)* exist who refuse divine union for themselves in order to help others spiritually. Mahayana Buddhism is also nonexclusive; in other words, one may follow it while simultaneously practicing other faiths. Thus, many Chinese consider themselves to be both Buddhists and Taoists (as well as Confucianists), whereas most Japanese are at some level both Buddhists and followers of Shinto.

As Mahayana Buddhism spread through East Asia in the medieval period, many different sects emerged. Probably the best known is Japanese Zen, which demands that its followers engage in the rigorous practice of "mind emptying." At one time, Buddhist monasteries associated with Zen and other sects were rich and powerful. In all East Asian countries, however, periodic reactions against Buddhism resulted in the persecution of monks and the suppression of monasteries. One reason for this opposition was the fact that government officials often viewed Buddhism as a foreign religion that placed India—rather than China—at the center of the world.

Figure 11.23 The Buddhist Landscape
Mahayana Buddhism has been traditionally practiced throughout East Asia. This Golden Buddha statue is located in Baomo Park in Chi Lei Village, Guangdong Province, China. *(Rob Crandall/www.robcrandall. com)*

Despite such hardships, East Asian Buddhism was never extinguished. But it also never became the focal point of society that it did in mainland Southeast Asia—or as Islam did in Southwest Asia and Christianity in Europe. In Japan, that position was partially captured by a different religion, Shinto.

Shinto The religious practice of Shinto is so closely bound to the idea of Japanese nationality that it is questionable whether a non-Japanese person can follow it. Shinto began as the animistic worship of nature spirits, but it was gradually refined into a subtle set of beliefs about the harmony of nature and its connections with human existence. Until the late 1800s, Buddhism and Shinto were complexly intertwined. Subsequently, the Japanese government began to disentangle the two faiths while elevating Shinto into a nationalistic cult focused on the divinity of the Japanese imperial family. After World War II, the more excessive aspects of nationalism were removed from the religion.

Shinto is still a place- and nature-centered religion. Certain mountains, particularly the volcanic Mount Fuji, are considered sacred and are thus climbed by large numbers of people (Figure 11.24). Major Shinto shrines, often located in scenic places, attract numerous pilgrims; most notable is the Ise Shrine south of Nagoya, site of the cult of the emperor. Local Shinto shrines, as well as Buddhist temples, offer leafy oases in otherwise largely treeless Japanese urban neighborhoods.

Taoism and Other Chinese Belief Systems The Chinese religion of Taoism (or Daoism) is similarly rooted in nature worship. Like Shinto, it stresses the acquisition of spiritual harmony and the pursuit of a balanced life. Taoism is indirectly associated with *feng shui*, also called **geomancy**, the Chinese practice of designing buildings in accordance with the spiritual powers that supposedly course through the local topography. Even in hypermodern Hong Kong, skyscrapers worth millions of dollars have occasionally gone unoccupied because their construction failed to accord with geomantic principles.

Despite the fact that both Taoism and Buddhism were historically followed throughout the country, traditional religious practice in China has always embraced local particularism; in other words, it has focused on the unique attributes of particular places. Many minor gods were traditionally associated with single cities or other specific areas. In modern-day rural China, village gods are often still honored.

Minority Religions Small numbers of followers of virtually all world religions can be found in the increasingly cosmopolitan cities of East Asia. Millions of Chinese and Japanese belong to Christian churches, although Christians are still only a small percent of the population of either country. South Korea, on the other hand, is now roughly 30 percent Christian (mostly Protestant), and reportedly sends more missionaries abroad than any country except the United States. Some reports indicate that Christianity is growing rapidly in China—despite persecution—but reliable information is scarce.

Larger than China's Christian population is its Muslim community. Several tens of millions of Chinese-speaking Muslims, called *Hui*, are concentrated in Gansu and Ningxia in the northwest and in Yunnan province in the south. Smaller clusters of Hui, often segregated in their own villages, live in almost every province of China. The only Muslim congregations in Japan and South Korea, on the other hand, are associated with recent and probably temporary immigrants from South, Southeast, and Southwest Asia.

Secularism in East Asia For all of these varied forms of religious expression, East Asia is still one of the most secular regions of the world. In Japan, most people occasionally observe Shinto or Buddhist rituals and maintain a small shrine for their ancestors, although only a small segment of the population is deeply religious. Japan also has a number of "new religions," sometimes called cults,

Figure 11.24 Mt. Fuji, Japan This picturesque volcanic mountain, sacred to Japan's Shinto religion, is climbed by large numbers of religious pilgrims each year. In the foreground are tea fields. *(Pacific Stock)*

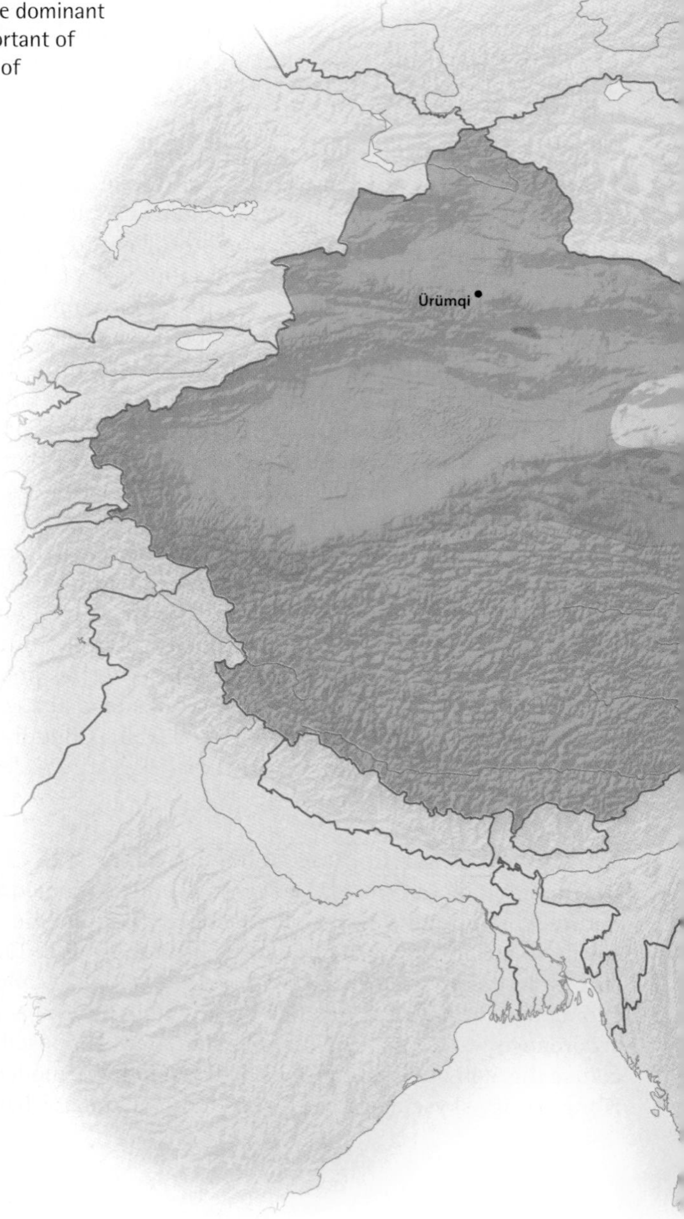

Figure 11.25 The Language Geography of East Asia
The linguistic geography of Korea and Japan is very straightforward, as the vast majority of people in those countries speak Korean and Japanese, respectively. In China, the dominant Han Chinese speak a variety of closely related Sinitic languages, the most important of which is Mandarin Chinese. In the peripheral regions of China, a large number of languages—belonging to several different linguistic families—can be found.

a few of which are noted for their fanaticism. But for Japanese society as a whole, religion is not particularly important.

After the communist regime took power over China in 1949, all forms of religion and traditional philosophy were discouraged and sometimes severely repressed. Under the new regime, atheistic **Marxist** philosophy—the communistic belief system developed by Karl Marx—became the official ideology. In the 1960s, many observers thought that traditional forms of Chinese religion would survive only in overseas Chinese communities. With the easing of Marxist orthodoxy during the 1980s and 1990s, however, many forms of spiritual expression began to return, especially in China's more prosperous coastal areas. In North Korea, on the other hand, Marxist orthodoxy is rigidly enforced. North Korea is also noted for its official ideology of *juche*, or "self-reliance." Ironically, juche demands absolute loyalty to North Korea's political leaders.

Figure 11.26 Ainu Man
The indigenous Ainu people of northern Japan are much reduced in population, but they still maintain a number of their cultural traditions. In this photograph, Ainu men participate in the Marimo Festival on the northern Japanese island of Hokkaido. *(Masa Uemura/Alamy Images)*

Linguistic and Ethnic Diversity

Japanese and Mandarin Chinese may partially share a system of writing, but the two languages bear no direct relationship (Figure 11.25). In their grammatical structures, Chinese and Japanese are more different from each other than are Chinese and English. Japanese, however, has adopted many words of Chinese origin.

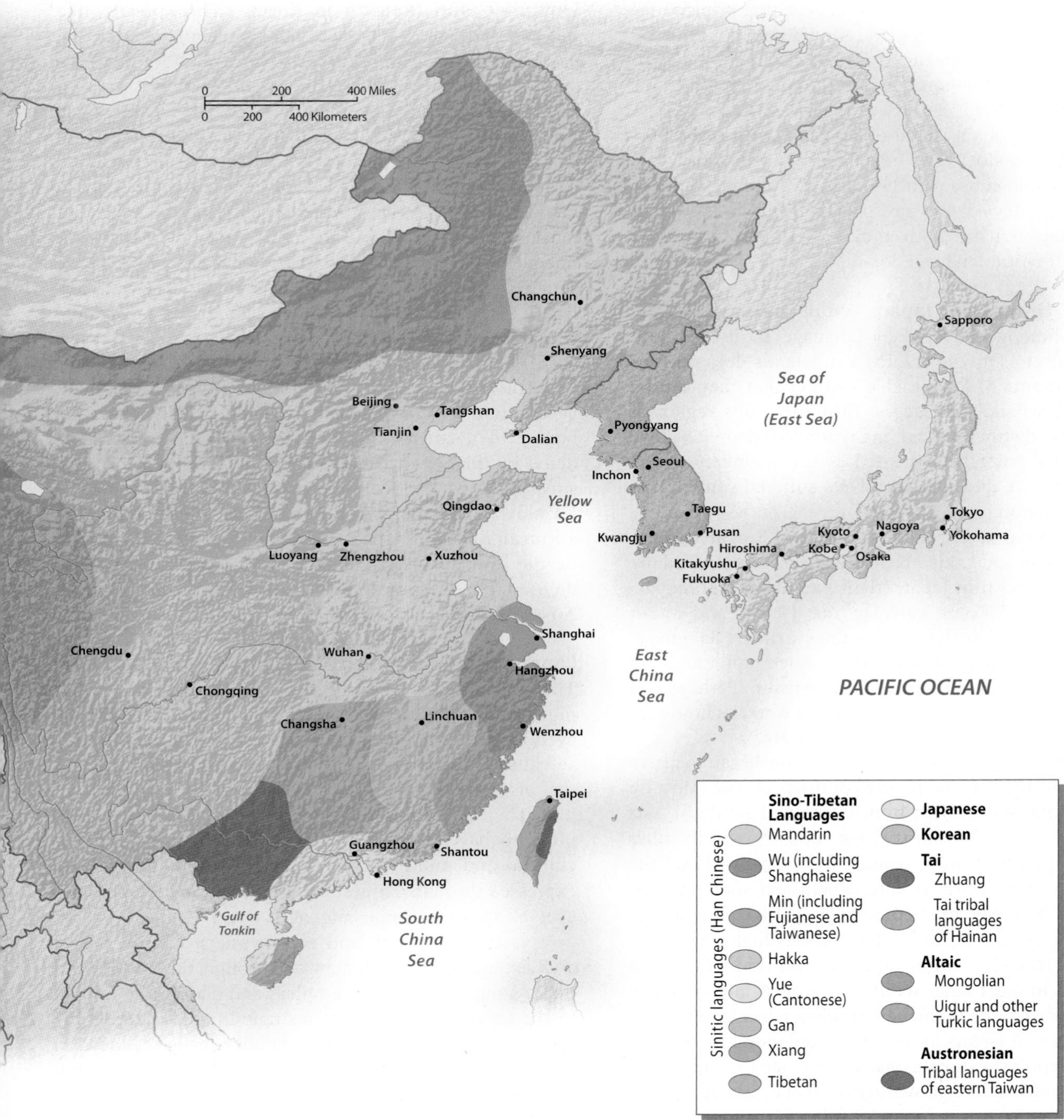

Language and National Identity in Japan According to most scholars, Japanese is not related to any other language. Korean is also usually classified as the only member of its language family. Many linguists, however, think that Japanese and Korean should be grouped together because they share many grammatical features. Only the mutual distrust between the Japanese and the Koreans, some suggest, has prevented this linguistic relationship from being acknowledged.

From several perspectives, the Japanese form one of the world's most homogeneous peoples, and they tend to regard themselves in such a manner. To be sure, minor cultural and linguistic distinctions are noted between the people of western Japan (centered on Osaka) and eastern Japan (centered on Tokyo), and many individual regions have distinctive customs and lifeways. Overall, however, such differences are of little significance.

In earlier centuries, the Japanese archipelago was divided between two very different peoples: the Japanese living to the south, and the Ainu inhabiting the north. The Ainu are completely distinct from the Japanese. They possess their own language and have a different physical appearance (Figure 11.26). Owing to their facial features, the Ainu were once categorized as members of the "Caucasian race." However, few scholars now believe that humankind is divided into separate races, and Ainu do not appear to be closely related to Europeans by genetic criteria.

For centuries the Japanese and the Ainu competed for land, and by the tenth century CE the Ainu had been largely driven off the main island of Honshu. Until the 1800s, however, Hokkaido largely remained Ainu territory. The Japanese people subsequently began to colonize Hokkaido, putting renewed pressure on the Ainu. Today, about 15,000 Ainu remain, with only 15 people supposedly speaking the Ainu language.

Minority Groups in Japan While the Japanese are relatively homogeneous, their language is divided into several dialects. Only in the Ryukyu Islands does one encounter variants of Japanese so distinct that they can be considered separate languages. Ethnically, this region of Japan is also distinct. In fact, many Ryukyu people believe that they have not been treated as full members of the Japanese nation, and that they have suffered some forms of discrimination.

The approximately 700,000 people of Korean descent living in Japan today also have felt discrimination. Many of them were born in Japan (their parents and grandparents having left Korea early in the 1900s) and speak Japanese rather than Korean as their primary language. But despite their deep bonds to Japan, such individuals are not easily able to obtain Japanese citizenship. Perhaps as a result of such treatment, many Japanese Koreans hold radical political views and support North Korea.

Starting in the 1980s, other immigrants began to arrive in Japan, mostly from the poorer countries of Asia. Most do not have legal status. Men from China and southern Asia typically work in the construction industry and in other dirty and dangerous jobs; women from Thailand and the Philippines often work as entertainers and sometimes as prostitutes. Roughly 200,000 Brazilians of Japanese ancestry have returned to Japan for the relatively high wages they can earn. However, immigration is less pronounced in Japan than in most other wealthy countries, and relatively few migrants acquire permanent residency, let alone citizenship. Because of Japan's impending population loss, however, some scholars estimate that it may need to import as many as 5 million foreign workers over the next decade.

The most victimized people in Japan may well be the **Burakumin**, or *Eta*, an outcast group of Japanese whose ancestors worked in "polluting" industries such as leathercraft. While discrimination is now illegal, the Burakumin are still among the poorest and least-educated people in Japan. Private detective agencies do a brisk business checking prospective marriage partners and employees for possible Burakumin ancestries. The Burakumin, however, have banded together to demand their rights; the Buraku Liberation League is politically powerful and is reputed to have close connections with the Yakuza (the Japanese mafia). The Burakumin usually live in separate neighborhoods and are concentrated in the Osaka region of western Japan.

Language and Identity in Korea The Koreans, like the Japanese, are a relatively homogeneous people. The vast majority of people in both North and South Korea speak Korean and unquestioningly consider themselves to be members of the Korean nation.

South Korea does, however, have a strong sense of regional identity, which can be traced back to the medieval period when the peninsula was divided into three separate kingdoms. The people of southwestern South Korea, centered on the city of Kwangju, tend to be viewed as distinctive, and many southwesterners believe that they have suffered periodic discrimination. In contract, the Kyongsang region of southeastern South Korea has supplied most of the country's political leaders and has received more than its share of development funds.

Korean identity also has an international component. Several million Korean-speakers reside directly across the border in northern China. Because of deportations ordered by the Soviet Union in the mid-twentieth century, substantial Korean communities also can be found in Kazakhstan in Central Asia. Over the past several decades, a Korean **diaspora**—the scattering of a particular group of people over a vast geographical area—has brought hundreds of thousands of Koreans to the United States, Canada, Australia, and New Zealand.

Language and Ethnicity Among the Han Chinese The geography of language and ethnicity in China is far more complex than that of Korea or Japan. This is true even if one considers only the eastern half of the country, so-called China proper. The most important distinction is that of separating the Han Chinese from the non-Han peoples. The Han, who form the vast majority, are those people who have long been incorporated into the Chinese cultural and political systems and whose languages are expressed through Chinese writing. The Han do not, however, all speak the same language.

Northern, central, and southwestern China—a vast area extending from Manchuria through the middle and upper Yangtze Valley to the valleys and plateaus of Yunnan in the far south—constitute a single linguistic zone. The spoken language here is generally called *Mandarin Chinese*. Mandarin is divided into a number of dialects, but the standard dialect of the Beijing area is gradually spreading. Standard Mandarin—called *Pŭtōnghuà* locally—is China's official national language.

In southeastern China, from the Yangtze Delta to China's border with Vietnam, a number of separate languages are spoken. Peoples speaking these languages are Han Chinese, but they are not native Mandarin speakers. Traveling from south to north, one encounters Cantonese (or Yue) spoken in Guangdong, Fujianese (alternatively Hokkienese, or Min, locally) spoken in Fujian, and Shanghaiese (or Wu), spoken in and around the city of Shanghai and in Zhejiang province. These are true languages, not dialects, since they are not mutually intelligible. They are usually called dialects, however, because they have no distinctive written form.

The Hakka are one group of people speaking a southern Chinese language that are occasionally not considered to be Han Chinese. Evidently, their ancestors fled northern China roughly a thousand years ago to settle in the rough upland area where Guangdong, Fujian, and Jiangxi provinces meet. Later migrations took them throughout southern China, where they typically settled in hilly wastelands (Figure 11.27). The Hakka traditionally made their living by growing upland crops such as sweet potatoes and by working as loggers, stonecutters, and metalworkers. Today they form one of the poorest communities of southern China.

Despite their many differences, all of the languages of the Han Chinese (including Hakka) are closely related to each other and belong to the same Sinitic language subfamily. Since their basic grammars and sound systems are similar, a person speaking one of these languages has little difficulty learning another. It is usually difficult, however, for speakers of European languages—or of Japanese or Korean—to gain fluency in these tongues. All Sinitic languages are **tonal** and monosyllabic; their words are all composed of a single syllable (although compound words can be formed from several syllables), and the meaning of each basic syllable changes completely according to the pitch in which it is uttered.

The Non-Han Peoples

Many of the more remote upland districts of China proper are inhabited by groups of non-Han peoples speaking non-Sinitic languages. Such peoples are usually classified as tribal, implying that they have a traditional social order based on self-governing village communities. Such a view is not entirely accurate, however, because some of these groups once had their own kingdoms. What they do have in common is a heritage of cultural and sometimes political tension with the Han Chinese (Figure 11.28).

Over the course of many centuries, the territory occupied by these non-Han communities has steadily declined in size, caused by the continued expansion of the Han as well as by non-Han emigration. Acculturation into Chinese society and intermarriage with the Han also have reduced many non-Han groups. Their main concentrations today are in the rougher lands of the far north and the far south.

As many as 11 million Manchus live in the more remote portions of Manchuria. The Manchu language is related to those of the tribal peoples of central and southeastern Siberia. Only about 60 Manchus, however, still speak their own language, the rest having abandoned it for Mandarin Chinese. This is an ironic situation, because the Manchus ruled the Chinese Empire from 1644

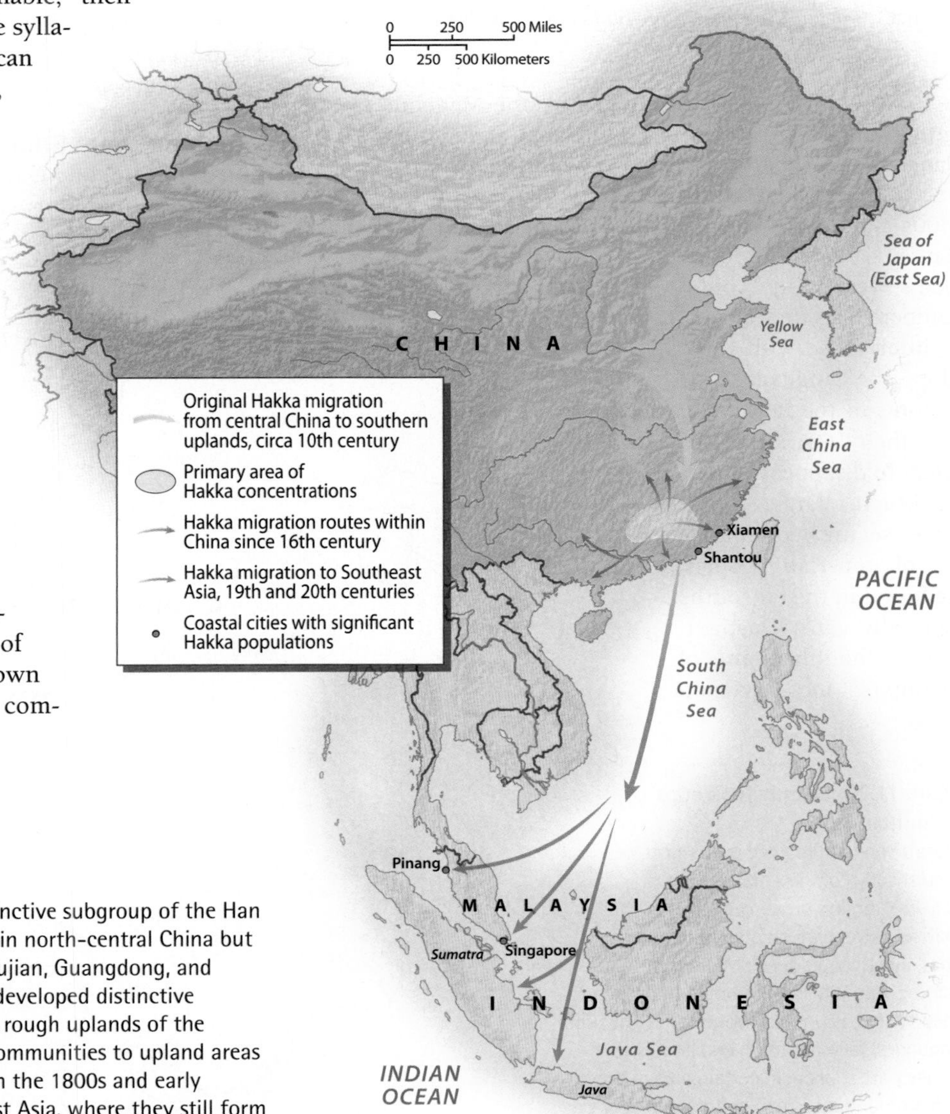

Figure 11.27 The Hakka Diaspora
The Hakka form an important and distinctive subgroup of the Han Chinese. They seem to have originated in north-central China but long ago migrated to the area where Fujian, Guangdong, and Jiangxi provinces converge. Here they developed distinctive agricultural patterns well suited to the rough uplands of the region. Later movements took Hakka communities to upland areas throughout much of southern China. In the 1800s and early 1900s, many Hakka moved to Southeast Asia, where they still form important communities.

Figure 11.28 Tribal Villages in South China
Non-Han people are usually classified as "tribal" in China, which assumes they have a traditional social order based upon autonomous village communities. Shown are Yi people at an open-air market in the village of Xhanghe in Yunnan province. *(Michael S. Yamashita/NGS Image Collection)*

to 1912. Until the end of this period, the Manchus prevented the Han from settling in central and northern Manchuria, which they hoped to preserve as their homeland. Once Chinese were allowed to settle in Manchuria in the 1800s—in part to prevent Russian expansion—the Manchus soon found themselves vastly outnumbered. As they began to intermarry with and adopt the language and customs of the newcomers, their own culture began to disappear.

Much larger and more secure communities of non-Han peoples are found in the far south, especially in Guangxi. Most of the inhabitants of Guangxi's more remote areas speak languages of the Tai family, closely related to those of Thailand. Because there are as many as 18 million non-Han people in Guangxi, it has been designated an **autonomous region**. Such autonomy was designed to allow non-Han peoples to experience "socialist modernization" at a different pace from that expected of the rest of the country. Critics contend that very little real autonomy has ever existed. (In addition to Guangxi, there are four other autonomous regions in China. Three of these, Xizang [Tibet], Nei Monggol [Inner Mongolia], and Xinjiang, are located in Central Asia and are discussed in Chapter 10. The final autonomous region, Ningxia, located in northwestern China, is distinguished by its large concentration of Hui [Mandarin-speaking Muslims].)

Other areas with sizable numbers of non-Han peoples are Yunnan and Guizhou, in southwestern China, and western Sichuan. Most tribal peoples here practice swidden agriculture (also called "slash and burn"; see Chapters 6 and 13) on rough slopes; flatter lands are generally occupied by rice-growing Han Chinese. A wide variety of separate languages, falling into several linguistic families, are found among the ethnic groups living in the uplands. Figure 11.29 shows that in Yunnan, the resulting ethnic mosaic is staggeringly complex.

Language and Ethnicity in Taiwan Taiwan is also noted for its linguistic and ethnic complexity. In the island's mountainous eastern region, a few small groups

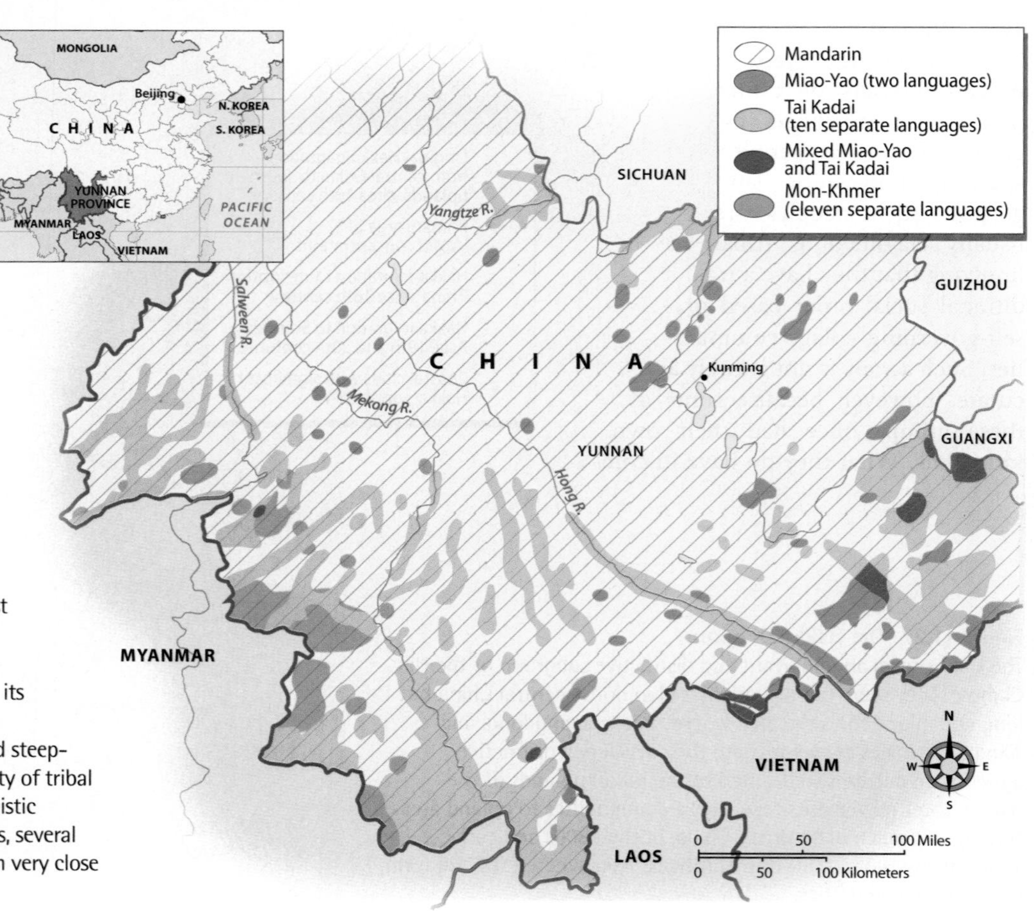

Figure 11.29 Language Groups in Yunnan
China's Yunnan Province is the most linguistically complex area in East Asia. In Yunnan's broad valleys and relatively level plateau areas and in its cities, most people speak Mandarin Chinese. In the hills, mountains, and steep-sided valleys, however, a wide variety of tribal languages, falling into several linguistic families, are spoken. In certain areas, several different languages can be found in very close proximity.

of "tribal" peoples speak languages related to those of Indonesia (belonging to the Austronesian language family). These peoples resided throughout Taiwan before the sixteenth century. At that time, however, Han migrants began to arrive in large numbers. Most of the newcomers spoke the Fujianese dialect, which evolved into the distinctive language of Taiwanese.

Taiwan was transformed almost overnight in 1949, when China's nationalist forces, defeated by the communists, sought refuge on the island. Most of the nationalist leaders spoke Mandarin, which they made the official language. Taiwan's new leadership discouraged Taiwanese, viewing it as a mere local dialect. As a result, tension developed between the Taiwanese and the Mandarin communities. Only in the 1990s did Taiwanese speakers begin to reassert their language rights, a movement that accelerated after 2000. Interestingly, as Taiwan has become more fully global, its own cultural framework has grown more locally specific.

GLOBAL TO LOCAL The Korean Wave

A century ago, Korea was commonly referred to as a "hermit kingdom," in reference to its relative isolation from global cultural and political currents. Today, North Korea has taken the ideal of isolation several steps further, largely cutting itself off from the rest of the world. South Korea has taken the opposite approach, and is today one of Asia's most globalized countries. Cultural products as well as technological devices from all over the world are readily available in the streets of Seoul and other large South Korean cities.

In recent years, however, South Korea's most notable contribution to cultural globalization has been as an exporter rather than importer. In the 1990s, South Korean music, movies, television shows and even cuisine began to follow the country's electronic exports into markets throughout East and Southeast Asia and beyond. The Chinese media was so taken by the phenomenon that they dubbed it the *Hallyu*, or "Korean Wave." By 2004, some two-thirds of all television viewers in Japan were tuning into the popular Korean television drama, *Winter Sonata*. By 2006, Chinese television stations devoted more time to South Korean shows than to those of all other foreign countries put together. The Hallyu movement has in turn prompted massive tourism into South Korea, with the number of tourist arrivals going from 2.8 million in 2003 to 3.7 million in 2004.

Not surprisingly, the Korean Wave has generated resentment in other Asian countries. Both Vietnam and Taiwan have recently threatened to ban or at least limit Korean broadcasting, while in Japan a comic book called "Halting the Korean Wave" has been a minor hit. Still, considering the animosity that so often marks relations between Japan and Korea, it is remarkable that Korean popular culture is so well-liked and has generated so little overt opposition in Japan.

Scholars are not sure how to explain the Hallyu phenomenon. Some argue that Korean films and TV shows are popular because they tend to deal with themes of family life and obligations during a period of technological change, incorporating Confucian values that resonate strongly in other East Asian countries. Others look more to issues of business and showmanship. In the 1900s, Korean shows tended to be much cheaper than those produced in North America, Europe, or Japan, yet were almost as sophisticated in terms of production techniques. Since then, Korean cultural products have grown more technologically advanced, enhancing their appeal.

But whatever the explanation, it is undeniable that South Korean popular culture is big business. South Korean stars are now among the highest paid entertainers in the world, with at least 10 actors earning more than $10 million a year (Figure 11.3.1).

Figure 11.3.1 South Korean Film Star
South Korea's noted film star Yi-Hyun So attends a press conference in regard to her new film "Jung" at the Pusan International Movie Festival. *(Photo by Yuan Hongyan/ ChinaFotoPress/Zuma Press. © Copyright 2006 by ChinaFotoPress)*

East Asian Cultures in Global Context

Not just Taiwan but all of East Asia has long exhibited tensions between an internal orientation and tendencies toward cosmopolitanism. This dichotomy has both a cultural and an economic dimension. Until the mid-1800s, East Asian countries attempted to insulate themselves from Western cultural influences. Japan subsequently opened its doors but remained ambivalent about foreign ideas. Only after its defeat in 1945 did Japan really opt for a globalist orientation. It was followed in this regard by South Korea, Taiwan, and Hong Kong (then a British colony). The Chinese and North Korean governments, to the contrary, decided during the early **Cold War** decades of the 1950s and 1960s to isolate themselves as much as possible from Western and global culture.

The Cosmopolitan Fringe The capitalist countries of East Asia are characterized by a vibrant internationalism, especially in the large cities, which coexists with strong national and local cultural identity. Virtually all Japanese, for example, study English for 6 to 10 years, and, although relatively few learn to speak it fluently, most can read and understand a good deal. Business meetings among Japanese, Chinese, and Korean firms are sometimes conducted in English. Relatively large numbers of advanced students, especially from Taiwan, study in the United States and other English-speaking countries, and thus bring home a kind of cultural bilingualism. Internet usage, with its wide global connections, is widespread in East Asia's cosmopolitan fringe.

The current cultural flow is not merely from a globalist West to a previously isolated East Asia. Instead, the exchange is growing more reciprocal. Hong Kong's action films are popular throughout most of the world and have influenced filmmaking techniques in Hollywood. Over the past few years, South Korean popular culture has spread throughout East Asia and in much of the rest of the world (see "Global to Local: The Korean Wave"). Japan almost dominates the world market in video games, and its ubiquitous comic-book culture and animation techniques are now following its karaoke bars in their overseas march.

Cultural globalization is, of course, as controversial in Japan as it is elsewhere. Japanese ultranationalists are few but vocal, calling their fellow citizens to resist the decadence of the West and to return to the military traditions of the **samurai**, the warrior class of premodern Japan. Many other Japanese people, however, contend that their country is too insular, worrying that they do not possess the English-language and global cultural skills necessary to operate effectively in the world economy.

Figure 11.30 Chinese Theme Park
As China's economy grows, its people are spending increasing amounts of money on entertainment. Theme parks, which now number over 2,000, are particularly popular. Happy Valley Theme Park in Beijing, shown here, was China's largest when opened in 2006. *(China Photos/Getty Images)*

The Chinese Heartland In one sense, Japan is more culturally predisposed to cosmopolitanism than is China. The Japanese have always borrowed heavily from other cultures (particularly from China itself), whereas the Chinese have historically been more self-sufficient. However, the southern coastal Chinese have long tended to be more international in their orientation, with close linkages to the Chinese diaspora communities of Southeast Asia and ultimately to maritime trading circuits extending over much of the globe.

In most periods of Chinese history, the interior orientation of the center prevailed over the external orientation of the southern coast. After the communist victory of 1949, only the small British enclave of Hong Kong was able to pursue international cultural connections. In the rest of the country, a dour and puritanical cultural order was rigidly enforced. While this culture was largely founded on the norms of Chinese peasant society, it was also influenced by the communist-oriented cultural system that had emerged in the Soviet Union.

After China began to liberalize its economy and open its doors to foreign influences in the late 1970s and early 1980s, the southern coastal region suddenly assumed a new prominence. Through this gateway, global cultural patterns began to penetrate the rest of the country. The result has been the emergence of a vibrant but somewhat showy urban popular culture throughout China that is replete with such global features as nightclubs, karaoke bars, fast-food franchises, and theme parks (Figure 11.30).

The liberalization of Chinese culture has allowed the reemergence of regional identities. From the late 1960s to the late 1970s, all forms of localism were rigorously suppressed as China's leaders sought to build a nationally uniform, working-class culture. Today, markers of local ethnicity are on the rise, especially in southeastern China, where most people do not speak Mandarin and have remained culturally distinctive. Southern Chinese culture, particularly Cantonese culture, is now widely considered to be attractive precisely because it is identified as cosmopolitan. The Guangdong region, including Hong Kong, has been the main gateway for foreign culture to enter China, and is thus viewed as being at the forefront of globalization. An interesting if unanswerable question is whether China's coastal resurgence will have any impact on its geopolitical orientation.

GEOPOLITICAL FRAMEWORK: Enduring Cold War Tensions

Much of the political history of East Asia revolves around the centrality of China and the ability of Japan to remain outside China's grasp. The traditional Chinese conception of geopolitics was based on the idea of a universal empire: all territories were either supposed to be a part of the Chinese Empire, pay tribute to it and acknowledge its supremacy, or stand outside the system altogether. Until the 1800s, the Chinese government would not recognize any other as its diplomatic equal. When China could no longer maintain its power in the face of European aggression, the East Asian political system fell into disarray. As European power declined in the 1900s, China and Japan contended for regional leadership. After World War II, East Asia was split by larger Cold War rivalries (Figure 11.31).

Figure 11.31 Geopolitical Issues in East Asia
East Asia remains one of the world's geopolitical hot spots. Tensions are particularly severe between capitalist, democratic South Korea and the isolated communist regime of North Korea, and between China and Taiwan. China has had several border disputes, whereas Japan and Russia have not been able to resolve their quarrel over the southern Kuril Islands.

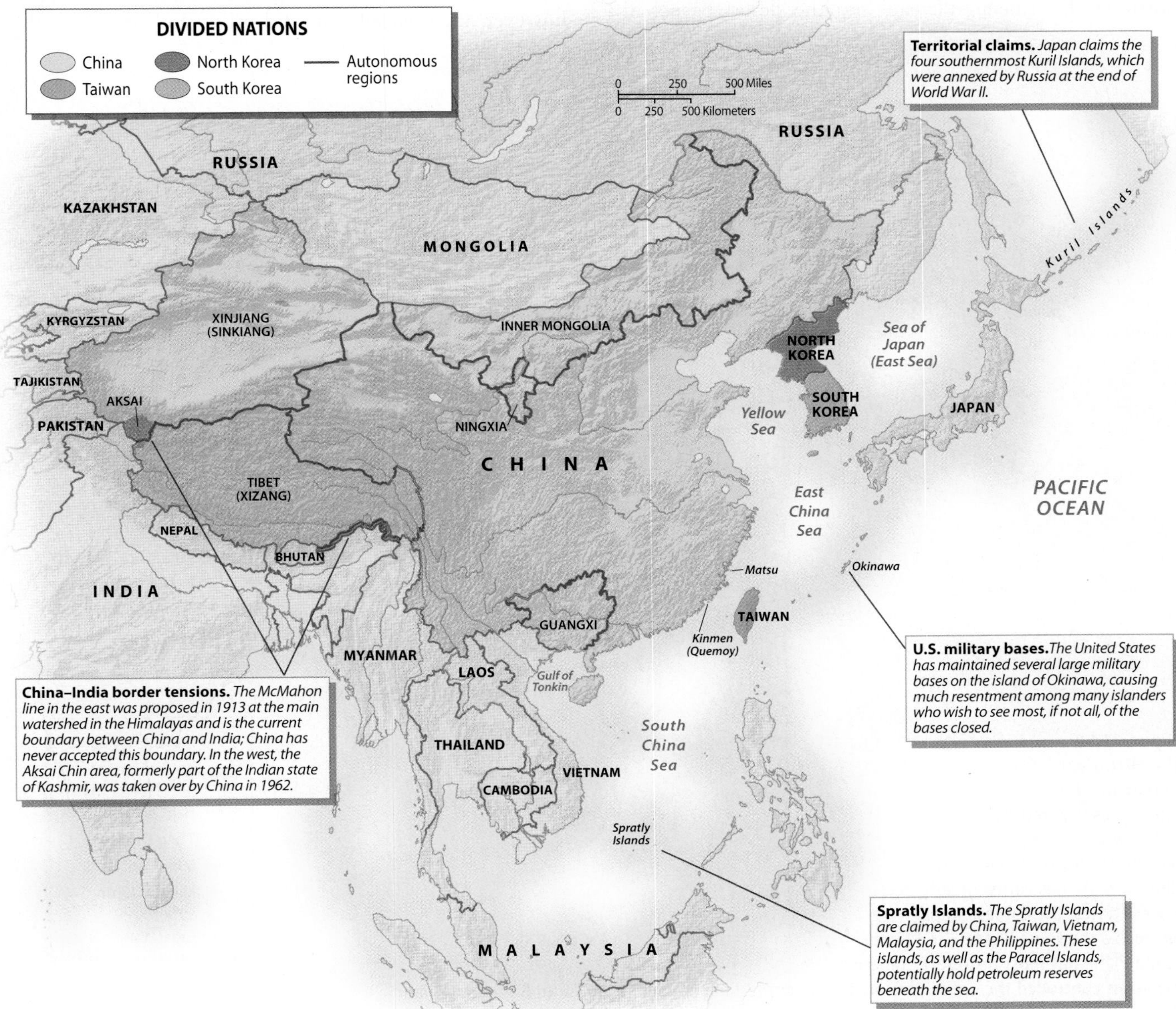

The Evolution of China

The original core of Chinese civilization was the North China Plain and the Loess Plateau. For many centuries, periods of unification alternated with times of division into competing states. The most important episode of unification occurred in the third century BCE. Once political unity was achieved, the Chinese Empire began to expand vigorously to the south of the Yangtze Valley. Subsequently, the ideal of the imperial unity of China triumphed, with periods of division seen as indicating disorder. This ideology helped cement the Han Chinese into a single people.

Several Chinese dynasties rose and fell between 219 BCE and 1912, most of them controlling roughly the same territory (Figure 11.32). The core of the Chinese Empire remained China proper, excluding Manchuria. Other lands, however, were sometimes ruled as well. The most important of these was a western projection extending north of the Tibetan Plateau into the desert basins of Central Asia (modern Xinjiang). China valued this area primarily because the vital trading route to western Eurasia (the "silk road") passed through it.

Various Chinese dynasties attempted to conquer Korea, but the Koreans resisted. Eventually China and Korea worked out an arrangement whereby Korea paid token tribute and acknowledged the supremacy of the Chinese Empire, and in return Korea received trading privileges and retained independence. When foreign armies invaded Korea—as did those of Japan in the late 1500s—China sent troops to support its "vassal kingdom."

For most of the past 2,000 years, the Chinese Empire was Earth's wealthiest and most powerful state. Its only real threat came from the pastoral peoples of Mongolia and Manchuria. Although vastly outnumbered by China, these societies were organized on a highly effective military basis. Usually the Chinese and the Mongols enjoyed a mutually beneficial trading relationship. Periodically, however, they waged war, and on several occasions the

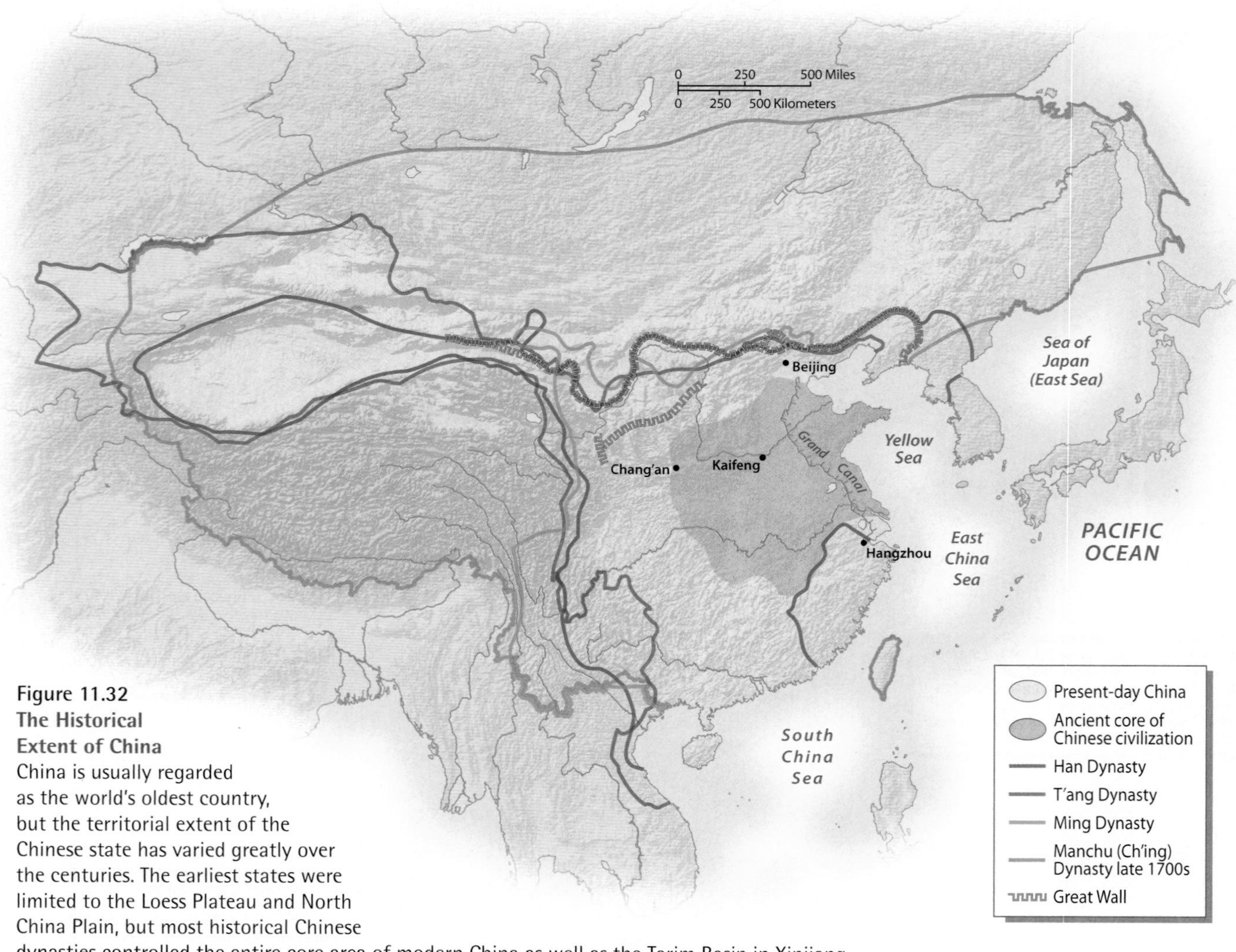

Figure 11.32
The Historical Extent of China
China is usually regarded as the world's oldest country, but the territorial extent of the Chinese state has varied greatly over the centuries. The earliest states were limited to the Loess Plateau and North China Plain, but most historical Chinese dynasties controlled the entire core area of modern China as well as the Tarim Basin in Xinjiang. Before the 1600s, however, China seldom held control of Tibet, Inner Mongolia, or central and northern Manchuria.

Figure 11.33 The Great Wall of China
The Great Wall of China runs 1,500 miles (2,400 kilometers) east to west from the Yellow Sea to deep within Central Asia. The first parts of the wall were built in the fourth century BCE; however, most of the wall was either rebuilt or finished in later times, mainly in the fifteenth and sixteenth centuries. *(Michael Howell/Pacific Stock)*

northern nomads conquered China (see Chapter 10). (The Great Wall along China's border did not, in other words, provide adequate defense; see Figure 11.33.) In time, the conquering armies adopted Chinese customs in order to govern the far more numerous Han people.

The Manchu Qing Dynasty The final and most significant conquest of China occurred in 1644, when the Manchus toppled the Ming Dynasty and replaced it with the Qing (also spelled Ch'ing) Dynasty. Like earlier conquerors, the Manchus retained the Chinese bureaucracy and made few institutional changes. Their strategy was to adapt themselves to Chinese culture, yet at the same time to preserve their own identity as an elite military group. Their system functioned well until the mid-nineteenth century, when the empire began to crumble before the onslaught of European and, later, Japanese power.

China's most significant legacy from the Manchu Qing Dynasty was the extension of its territory to include much of Central Asia. The Manchus subdued the Mongols and eventually established control over eastern Central Asia, including Tibet. Even the states of mainland Southeast Asia sent tribute and acknowledged Chinese supremacy. Never before had the Chinese Empire been so extensive or so powerful.

The Modern Era From its height of power and extent in the 1700s, the Chinese Empire descended rapidly in the 1800s as it failed to keep pace with the technological progress of Europe. Threats to the empire had always come from the north, and Chinese and Manchu officials saw little peril from European merchants operating along their coastline. The Europeans were distressed by the amount of silver needed to obtain Chinese silk, tea, and other products, and by the fact that the Chinese disdained their manufactured goods. In response, the British began to sell opium, which Chinese authorities viewed as a threat. When the imperial government tried to suppress the opium trade in the 1840s, the British attacked and quickly prevailed.

This first "opium war" ushered in a century of political and economic chaos in China. The British demanded free trade in selected Chinese ports, and in the process overturned the traditional policy of managed trade based on the acknowledgment of Chinese supremacy. As European enterprises penetrated China and undermined local economic interests, anti-Manchu rebellions began to break out. At first, all such uprisings were crushed, but not before causing tremendous destruction. Meanwhile, European power continued to advance. In 1858, Russia annexed the northernmost reaches of Manchuria, and by 1900 China had been divided—as shown in Figure 11.34—into separate "**spheres of influence**" where the interests of different European countries prevailed. (In a sphere of influence, the colonial power had no formal political authority but did enjoy informal influence and tremendous economic clout.)

A successful rebellion launched in 1911 finally toppled the Manchus and destroyed the empire, but subsequent efforts to establish a unified Chinese Republic were not successful. In many parts of the country, local military leaders, or "warlords," grabbed power for themselves. By the 1920s, it appeared that China might be completely dismembered. The Tibetans had gained autonomy; Xinjiang was under Russian influence; and in China proper Europeans and local warlords vied with the weak Chinese Republic for power. In addition, Japan was increasing its demands and seeking to expand its territory.

The Rise of Japan

Japan did not emerge as a unified state until the seventh century. From its earliest days, Japan looked to China (and, at first, to Korea as well) for intellectual and political models. Its offshore location insulated Japan from the threat of rule by the Chinese Empire. At the same time, the Japanese conceptualized their islands as a separate empire, equivalent in certain respects to China. Between 1000 and 1580, however, Japan was divided into a number of mutually hostile feudal realms and had no real unity.

The Closing and Opening of Japan Around 1600, Japan was reunited by the armies of the Tokugawa **Shogunate** (a shogun was a supreme military leader who theoretically remained under the emperor). At this time,

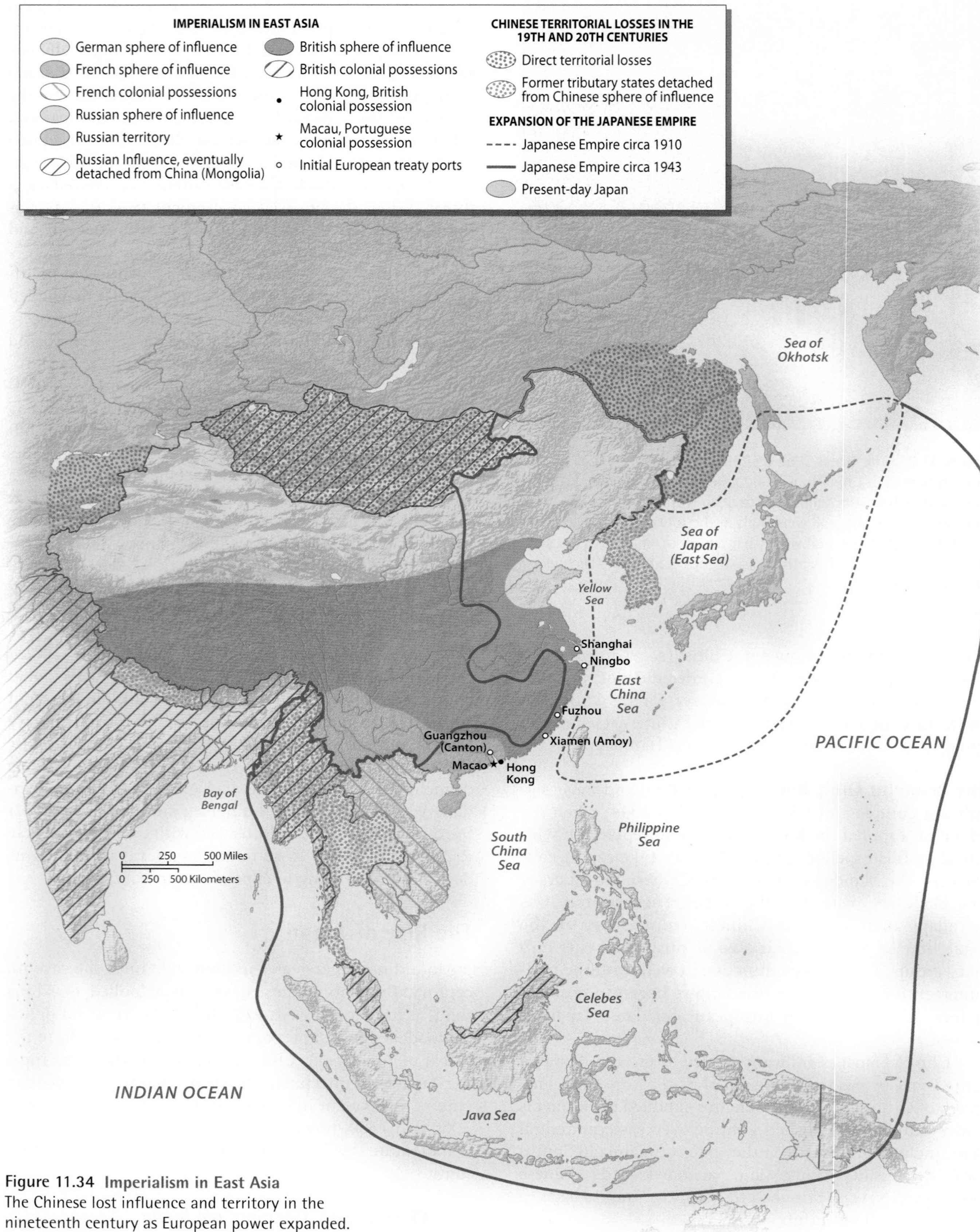

Figure 11.34 Imperialism in East Asia
The Chinese lost influence and territory in the nineteenth century as European power expanded. Although China regained its autonomy and most of its territory in the 1900s, Russia retained large areas that were formerly under Chinese control. The first half of the twentieth century saw the rapid expansion of the Japanese Empire, which ended with the defeat of Japan in World War II.

Japan attempted to isolate itself from the rest of the world. Until the 1850s, Japan traded with China mostly through the Ryukyu islanders and with Russia through Ainu intermediaries. The only Westerners allowed to trade in Japan were the Dutch, and their activities were strictly limited.

Japan remained largely closed to foreign commerce and influence until U.S. gunboats sailed into Tokyo Bay to demand access in 1853. Aware that they could no longer keep the Westerners out, Japanese leaders set about modernizing their economic, administrative, and military systems. This effort accelerated when the Tokugawa Shogunate was toppled in 1868 by the Meiji Restoration. (It is called a *restoration* because it was carried out in the emperor's name, but it did not give the emperor any real power.) Unlike China, Japan successfully accomplished most of its reform efforts.

The Japanese Empire Japan's new rulers realized that their country remained threatened by European imperial power. They therefore nurtured the development of a silk export industry, which gave them the funds needed to buy modern equipment (see "Geographic Tools: The East Asian Gazetteer and the Reconstruction of Historical Geography"). They also decided that the only way to meet the European challenge was to become expansionistic themselves. Japan therefore took control over Hokkaido and began to move farther north into the Kuril Islands and Sakhalin.

In 1895, the Japanese government tested its newly modernized army against China, winning a quick and profitable victory that gave it authority over Taiwan. Tensions then mounted with Russia as the two countries vied for power in Manchuria and Korea. The Japanese defeated the Russians in 1905, giving them considerable influence in northern China. With no strong rival in the area, Japan annexed Korea in 1910. Alliance with Britain, France, and the United States during World War I brought further gains, as Japan was awarded Germany's island colonies in Micronesia.

The 1930s brought a global depression, putting a resource-dependent Japan in a difficult situation. The country's leaders sought a military solution, and in 1931 Japan conquered Manchuria. In 1937, Japanese armies moved south, occupying the North China Plain and the coastal cities of southern China. The Chinese government withdrew to the relatively inaccessible Sichuan Basin to continue the struggle. During this period, Japan's relations with the United States steadily deteriorated. When the United States cut off the export of scrap iron, Japan began to experience a resource crunch.

In 1941, Japan's leaders decided to destroy the American Pacific fleet in order to clear the way for the conquest of resource-rich Southeast Asia. Their grand strategy was to unite East and Southeast Asia into a "Greater East Asia Co-Prosperity Sphere." This "sphere" was to be ruled by Japan, however, and was designed to keep the Americans and Europeans out. Japanese forces elsewhere in East Asia sometimes engaged in brutal acts. In Korea, the colonial government evidently planned to extinguish the Korean language in favor of Japanese. As a result, serious tensions emerged between the Japanese and the other East Asians—tensions that persist to this day.

Postwar Geopolitics

With the defeat of Japan at the end of World War II, East Asia became an arena of rivalry between the United States and the Soviet Union. Initially, American interests prevailed in the maritime fringe, while Soviet interests advanced on the mainland. Soon, however, East Asia began to experience its own revival.

Japan's Revival Japan lost its colonial empire when it lost World War II. Its territory was reduced to the four main islands plus the Ryukyu Archipelago and a few minor outliers. In general, the Japanese government accepted this loss of land. The only remaining territorial conflict concerns the four southernmost islands of the Kuril chain, which were taken by the Soviet Union in 1945. Although Japan still claims these islands, Russia refuses even to discuss relinquishing control, causing strains in Russo-Japanese relations.

After losing its overseas possessions, Japan was forced to rely on trade to obtain the resources needed for its economy. Japan's military power was strictly limited by the constitution imposed on it by the United States, forcing Japan to rely on the U.S. military for much of its defense needs. Even today, the U.S. Navy patrols many of Japan's vital sea-lanes, and U.S. armed forces maintain several bases within the country. This U.S. military presence, however, has become controversial. Particularly contentious are the U.S. bases on Okinawa, which take up much of the island's territory.

Slowly but steadily, however, Japan's own military has emerged as a strong regional force despite the constitutional limits imposed on it. In recent years, fears about North Korea's nuclear program and China's military growth have led the Japanese government to reconsider its constitutionally imposed limitations on military spending. In 2007, Japan announced that it would make international peacekeeping a military priority, strengthen its missile defenses, and enhance its coordination with U.S. troops. In the same year, Japan spent $1.28 billion on ballistic missile defense.

Other East Asian countries are concerned about the potential threat posed by a remilitarized Japan. Such a perception was strengthened in the early years of the new millennium when Japan's former prime minister visited the Yasukuni Shrine, which contains a military cemetery in which several war criminals from World War II are buried. Anti-Japanese sentiments in China—which have occasionally boiled over into huge public protests—have also been encouraged by the publication of Japanese textbooks that minimize the country's atrocities during the war.

GEOGRAPHIC TOOLS

The East Asian Gazetteer and the Reconstruction of Historical Geography

East Asia is a particularly good place in which to conduct research in historical geography because of its gazetteer tradition. In common usage, a gazetteer is simply a kind of index: a list of place-names and their spatial coordinates found at the back of most comprehensive atlases. But a gazetteer can also be a geographical encyclopedia, providing detailed information about a large array of particular places. Gazetteers of this sort have historically been used most extensively in East Asia, as they were believed to be essential tools for government. Using gazetteer data, geographers and historians literally can map out landscape transformations in East Asia over periods of hundreds of years.

One geographer who has used gazetteer data extensively is Kären Wigen, especially in her award-winning book *The Making of a Japanese Periphery, 1750–1920* (University of California Press, 1995). Here she examined how the region around the city of Ina, located in a sizable valley in the heart of the Japanese Alps, has repeatedly been transformed. In the 1700s, Ina was a minor feudal domain that formed a key node in the packhorse trains that carried goods across the mountainous central region of Japan. It also supported its own major handicraft industry focused on high-quality paper goods. As Wigen discovered by mapping out gazetteer data (see Figure 11.4.1), the paper industry entailed a complex spatial organization; although the entire Ina region supplied the paper-mulberry bark from which paper was made, different villages specialized in the manufacture of different paper-craft items.

As Wigen discovered by examining gazetteers from later decades, the economic geography of the Ina region was completely transformed during the mid- and late 1800s. In part because of changing fashions, decorative paper goods no longer commanded high prices, and as a result the industry underwent a rapid decline. Simultaneously, the regional economy was subordinated to that of the rapidly centralizing Japanese state. To obtain foreign exchange, the state began to export silk thread. As a result, Ina's paper mulberries were ripped out, to be replaced by a different species of mulberry used to feed the voracious silkworms.

The prosperity silk brought to Ina did not prove sustainable. The depression of the 1930s, followed by World War II, combined to destroy the industry. In the postwar period, mulberries were again uprooted, this time to be replaced by apple and other fruit trees. By the turn of the century, however, fruit growing was no longer very profitable, and the farmers of Ina were again looking for alternative crops.

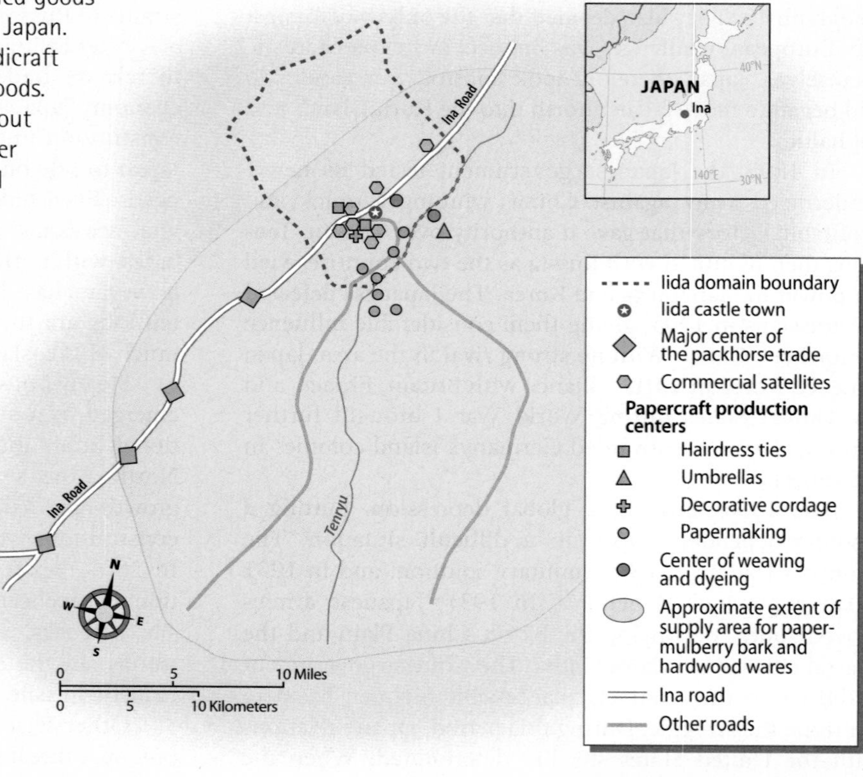

Figure 11.4.1 Ina's Paper Industry
In the early 1800s, before Japan's modernization, the area around Ina in central Japan supported a sizable and highly specialized papercraft industry. Different villages focused on the production of different paper objects.

The Division of Korea The aftermath of World War II brought much greater changes to Korea than to Japan. As the end of the war approached, the Soviet Union and the United States agreed to divide Korea; Soviet forces were to occupy the area north of the 38th parallel, and U.S. troops would occupy the south. Soon two separate regimes were established. In 1950, North Korea invaded South Korea, seeking to reunify the country. The United States, with support from the United Nations, supported the south, while China aided the north. The war ended in a stalemate, and Korea has remained a divided country, its two governments locked in a continuing Cold War (Figure 11.35).

Large numbers of U.S. troops remained in the south after the war. South Korea in the 1960s was a poor, agrarian country that could not defend itself. Over the past 30 years, however, the south has emerged as a wealthy trading nation while the fortunes of the north have plummeted. The southerners' fear of the north gradually lessened, especially among the members of the younger generation. Many South Korean students have resented the presence of U.S. forces, seeking instead to build friendly relations with North Korea. Their periodic and sometimes violent demonstrations have been a significant force in South Korean politics. Changing military priorities, meanwhile, led the United States to reduce its presence in Korea, dropping its troop strength in 2004 from roughly 37,000 to some 25,000 soldiers.

In the late 1990s, the South Korean government began to pursue better relations with North Korea, giving it large amounts of food aid in exchange for reduced hostility. It is not clear, however, whether such policies have been at all successful. Seoul's new approach toward the north, called the "sunshine policy," also resulted in a slight cooling of relations between South Korea and the United States.

The Division of China World War II brought tremendous destruction and loss of life to China. Even before the war began, China had already been engaged in a civil conflict between nationalists (who favored an authoritarian capitalist economy) and communists. The communists had originally been based in the middle Yangtze region, but in 1934 nationalist pressure forced them out. Under the leadership of Mao Zedong, they retreated in the "Long March," which took them to the Loess Plateau, an area close to both the traditional power center of northern China and the industrialized zones of Manchuria. After the Japanese invaded China proper in 1937, the two camps cooperated, but as soon as Japan was defeated, China again found itself embroiled in civil war. In 1949, the communists proved victorious, forcing the nationalists to retreat to Taiwan.

A latent state of war has persisted ever since between China and Taiwan. Although no battles have been fought, gunfire has periodically been exchanged over Kinmen and Matsu, two small Taiwanese islands just off the mainland. The Beijing government still claims Taiwan as an integral part of China and vows eventually to reclaim it. The nationalists who gained power in Taiwan long maintained that they represented the true government of China. It was actually made a crime in Taiwan to advocate Taiwanese independence, because the fiction had to be maintained that Taiwan was merely one province of a temporarily divided China.

The idea of the intrinsic unity of China continues to be influential. In the 1950s and 1960s, the United States recognized Taiwan as the only legitimate government of China, but its policy changed in the 1970s after U.S. leaders decided that it would be more useful—and more realistic—to recognize mainland China. Soon China entered the United Nations, and Taiwan found itself diplomatically isolated, virtually without international recognition. (As of 2007, only 24 countries, mostly small African, Pacific, and Caribbean states, continued to recognize Taiwan and, in return, to receive Taiwanese aid.) In reality, however, Taiwan is a fully separate country, and many of its citizens would like to proclaim it an independent republic. In 2000, Taiwan elected a former advocate of Taiwanese independence. China, however, immediately threatened to invade if Taiwan were to declare itself a separate country. In 2001, both China and Taiwan entered the World Trade Organization (WTO), but Taiwan was forced to join under the awkward name of "Chinese Taipei," in order to avoid the suggestion that it is a sovereign country.

In 2007, a public opinion poll showed that only 8.2 percent of the citizens of Taiwan favored an immediate proclamation of independence. But while the Taiwanese people and government are content to maintain the island's uncertain political situation, the ideal of eventual reunification is losing ground. 2007 also saw the systematic purging of words and phrases suggesting unity with China in Taiwanese textbooks. Thus the common term "cross-strait ties" was replaced with "China-Taiwan ties." Another 2007 public opinion poll found that a full 74 percent of the island's people agree that in actuality Taiwan is a sovereign, independent state.

Some observers think that China might eventually try to reclaim Taiwan, pointing to its overwhelming military

Figure 11.35 The Demilitarized Zone in Korea
North and South Korea were divided along the 38th parallel after World War II. Today, even after the conflict of the early 1950s, the demilitarized zone, or DMZ (which runs near the parallel), separates these two states. U.S. armed forces are active in patrolling the DMZ. *(Yonhap/AP/Wide World Photos)*

advantages. However, many others view such a scenario as highly unlikely, pointing instead to the more than $100 billion annual trade flow between the island and the mainland.

Chinese Territorial Issues Despite the fact that it has been unable to regain Taiwan, China has been successful in retaining the Manchu territorial legacy. In the case of Tibet, this has required considerable force; resistance by the Tibetans compelled China to launch a full-scale invasion in 1959. The Tibetans, however, have continued to struggle for real autonomy if not actual independence, as they fear that the Han Chinese now moving to Tibet will eventually outnumber them and undermine their culture (Figure 11.36; also see Chapter 10).

The postwar Chinese government also retained control over Xinjiang in the northwest, as well as Inner Mongolia (or Nei Monggol), a vast territory stretching along the Mongolian border. Like Tibet, Nei Monggol and Xinjiang are classified as autonomous regions. The peoples of Xinjiang are asserting their religious and ethnic identities, and separatist sentiments are widespread. However, most Han Chinese regard Nei Monggol and Xinjiang as integral parts of their country, and they regard any talk of succession as treasonous.

China claims several other areas that it does not control. It contends, for example, that former Chinese territories in the Himalayas were illegally annexed by Britain when it controlled South Asia, resulting in several unresolved border disputes with India. The two countries went to war in 1962 when China occupied an uninhabited highland district in northeastern Kashmir. Tensions between India and China have eased over the past decade, but their territorial disputes remain unresolved. China also asserts its rights over a number of tiny islands—most of them submerged at high tide—in the South China Sea. The Paracel Islands, however, are also claimed by Taiwan and Vietnam, whereas the Spratly Islands are claimed by Taiwan, Vietnam, Malaysia, and the Philippines (see Chapter 13).

One territorial issue was finally resolved in 1997 when China reclaimed Hong Kong. In the isolationist 1950s, 1960s, and 1970s, Hong Kong acted as China's window on the outside world, and it grew prosperous as a capitalist enclave. As Chinese relations with the outer world opened in the 1980s, Britain decided to honor its treaty provisions and return Hong Kong to China. China in turn promised that Hong Kong would retain its fully capitalist economic system for at least 50 years. Civil liberties not enjoyed in China itself were also to remain protected in Hong Kong (Figure 11.37). Wealthy citizens

Figure 11.36 Chinese Soldiers in Tibet
Following a full-scale invasion of Tibet in 1959, China continues to increase its presence through its military forces, the relocation of migrants into the area from other parts of China, and rebuilding programs that mask the traditional Tibetan landscape. Here Chinese soldiers observe a praying Tibetan while eating ice cream bars. *(Galen Rowell/Corbis/Bettmann)*

Figure 11.37 Distinctive Hong Kong
Although part of China since 1997, Hong Kong remains both politically and culturally distinctive from the rest of China. In this Hong Kong scene, a bride is posing for photographs in front of a sexually suggestive advertisement. *(Rob Crandall/www.robcrandall.com)*

of Hong Kong, however, grew nervous about the transfer of sovereignty, and many acquired residency rights in Canada or Australia as a fallback.

It is unclear how much autonomy will actually be allowed for Hong Kong. Over the past several years, China has acted firmly against efforts by its people to install a more democratic government. As a result of such incidents, hundreds of thousands of people have taken to the streets of Hong Kong in protest.

In 1999 Macao, the last colonial territory in East Asia, was returned to China (Figure 11.38). This small Portuguese enclave, located across the estuary from Hong Kong, has functioned largely as a gambling refuge. Macao represents a remnant of an earlier episode of political globalization in East Asia; contemporary aspects of globalization are generating different geographical patterns and problems.

Global Dimensions of East Asian Geopolitics

In the early 1950s, East Asia was divided into two hostile Cold War camps: China and North Korea were allied with the Soviet Union, while Japan, Taiwan, and South Korea were linked to the United States. The Chinese–Soviet alliance soon deteriorated into mutual hostility, and in the 1970s China and the United States found that they could accommodate each other, sharing an enemy in the Soviet Union. Subsequently, growing economic ties between the United States and China resulted in a further easing of hostility. In contrast, North Korea's relations with the United States and many other countries have only grown more heated during the same period.

Figure 11.38 Macao Casino
Macao, reclaimed by China from Portugal in 1999, retains a unique mixture of Chinese and Portuguese cultural influences. Its economic mainstay remains gambling, which is prohibited in the rest of China. *(Rob Crandall/www.robcrandall. com)*

The North Korean Crisis In 1994, North Korea refused to allow international inspections of its nuclear power facilities—which many countries believed were being used for weapons production—provoking an international crisis. It eventually relented, however, in exchange for energy assistance from the United States. But by 2002 the agreement had fallen apart, as evidence mounted that North Korea was continuing to pursue nuclear weapons, focusing now on enriched uranium rather than plutonium. North Korea also continued to develop missiles capable of carrying nuclear warheads, although its 2005 test of an advanced model that could potentially reach the western United States was not successful. Other complaints by the international community have focused on North Korea's trafficking in illegal drugs and the fact that it reportedly holds more than 200,000 of its own citizens in brutal labor-camp prisons (see "Geography in the Making: North Korea in Crisis?").

In response to North Korea's nuclear ambitions, the United States has advocated multilateral negotiations, involving South Korea, China, Japan, and Russia. North Korea, however, insisted for years that it would negotiate only with the United States. Owing in part to pressure from China, its main trading partner, North Korea began to relent in 2004, although progress in the ensuing "six nation talks" remained maddeningly slow. The process appeared to have failed completely in October 2006, when North Korea detonated a small nuclear device. Negotiations continued, however, and in March 2007 North Korea announced that it had shut down its main nuclear facility, a claim verified by international inspectors several months later. In return, North Korea is expected to receive 1 million tons of fuel oil in addition to other concessions. Even with its reactors shut down, however, North Korea is still estimated to have enough enriched uranium to build six more nuclear bombs.

China on the Global Stage The end of the Cold War, coupled with the rapid economic growth of China, has reconfigured the balance of power in East Asia. The United States no longer needs China to offset the Soviet Union, and the U.S. military has become increasingly worried about the growing power of the Chinese armed forces. China's neighbors also have become more concerned. China now has the largest army in the world, as well as nuclear capability, and sophisticated missile technology. China is also rapidly building up its navy, buying advanced ships from Russia and beginning to construct its own. Like Japan, China is concerned about the vulnerability generated by its reliance on foreign oil, and as a result some of its planners have advocated developing a naval force in the Indian Ocean.

GEOGRAPHY IN THE MAKING

North Korea in Crisis?

For many years, North Korea has been engaged in a complex struggle with the United States and other members of the international community in regard to its nuclear weapons program. Despite several rounds of negotiations, little if any progress was made until 2007. In 1994, North Korea signed on to the "Agreed Framework" that called on it to freeze its plutonium production in exchange for economic aid and help in building two modern nuclear power plants. International nuclear inspectors, however, grew frustrated with North Korea's refusal to offer full cooperation, and in 1998 both the United States and Japan were angered by Pyongyang's testing of advanced missiles.

In 2002, the Agreed Framework collapsed completely when the United States discovered evidence that North Korea was carrying out a secret uranium-enriching program, also designed to produce nuclear weapons. When the U.S. halted oil shipments in retaliation, North Korea repaired and reopened its Yongbyon nuclear facility, raising fears of renewed plutonium production. North Korea continued to test advanced missiles, and in October 2006 it detonated a small nuclear device. The United Nations Security Council responded to this nuclear blast by imposing heightened sanctions; special efforts were made to keep luxury goods, including French wines and jet skis, out of the hands of North Korea's top leaders.

Negotiations continued as North Korea began to feel the pinch of international sanctions. In early 2007, North Korea agreed to shut down the Yongbyon reactors in exchange for 50,000 metric tons of fuel aid. The United States further agreed to seek more normal relations with North Korea and to remove the country from its terrorism blacklists. In July 2007, international inspectors confirmed that the reactors had indeed been shuttered, and the promised aid began to flow into the country. In September, a U.S. delegation announced that it was satisfied with the situation at Yongbyon.

Despite North Korea's efforts at dismantling its nuclear weapons programs, many observers remain suspicious about its ultimate goals. Many experts believe that North Korea has stockpiled enough plutonium to produce six to eight nuclear weapons. The deal signed in early 2007, however, requires the Pyongyang government to account for all of its fissile materials and to provide the U.S. with a list of all of its nuclear stockpiles.

On December 19, 2007, the U.S. State Department's top Korea expert traveled to North Korea to review the country's disarmament efforts. On the same day, China's top nuclear envoy left North Korea after visiting its main nuclear facilities. A week earlier, a U.S.-led team began moving fuel rods from the Yongbyon reactor to water pools, a key step in the disabling process.

Relations between North and South Korea also showed signs of improvement in late 2007. Leaders of the two countries signed an eight-point agreement on issues ranging from discussing the eventual establishment of permanent peace to economic cooperation to the expansion of transportation links. On a lighter note, the two leaders also talked about the creation of a joint Olympic cheering squad.

China is thus coming of age as a major force in global politics. Whether it is a force to be feared by other countries is a matter of considerable debate. Chinese leaders insist that they have no expansionistic designs and no intention of interfering in the internal affairs of other countries. They regard concerns expressed by the United States and other countries about their human rights record, as well as their activities in Tibet, as undue meddling in their own internal affairs.

Current opinions on China in the United States vary tremendously. Many U.S. leaders, particularly those in the business community, contend that the two countries should ignore their political differences and develop closer economic and cultural ties. Some East Asian experts similarly argue that the United States must respect China's sovereignty more carefully—or risk a future conflict in the region. Critics, on the other hand, think that China's trade practices are unfair, its labor, human-rights, and environmental records appalling, and its actions in Tibet unsupportable. China's growing military strength is another cause for concern, leading the United States to seek stronger military ties with Japan, Australia, and India.

Regardless of what one thinks about East Asia's international relations, its economic ascent is undeniable. Yet East Asia does have a number of serious economic problems.

ECONOMIC AND SOCIAL DEVELOPMENT:
An Emerging Core of the Global Economy

East Asia exhibits extreme differences in economic and social development. Japan's urban belt contains one of the world's greatest concentrations of wealth, whereas many interior districts of China remain extremely poor. Overall, however, East Asia has experienced rapid economic growth since the 1970s, with two of its economies, those of Taiwan and South Korea, jumping from the ranks of underdeveloped to developed (Table 11.2). Over the past two decades, China has experienced extraordinarily rapid

TABLE 11.2 Development Indicators

Country	GNI[a] per Capita, PPP[b] (2005)	GDP[c] Average Annual % Growth (2000–05)	Life Expectancy (2007)	Percent of Population Living on Less than $2 a Day (2006)	Under Age 5 Mortality Rate 1990	Under Age 5 Mortality Rate 2005	Gender Equity[d]
China	6,600	9.6	72	47	49	27	98
Hong Kong	34,670	4.3	82				93
Japan	31,410	1.4	82		6	4	98
North Korea			71		55	55	
South Korea	21,850	4.6	79		9	5	87
Taiwan			77				

[a]*Gross national income.*
[b]*Purchasing power parity.*
[c]*Gross domestic product.*
[d]*Ratio of female-to-male enrollments in primary and secondary school percentage. Numbers below 100 have more males in primary/secondary school; numbers above 100 have more females in primary/secondary schools.*

Sources: World Bank, World Development Indicators, *2007; life expectancy 2007 and percentage of population living on $2 a day data from* Population Reference Bureau, World Data Sheet, *2007; gender equity data from* Millennium Development Goals.

economic growth, and now has the world's second largest economy when calculated on the basis of purchasing power parity. Increasingly, East Asia functions as a global economic core (Figure 11.39). But again, growth has not been experienced everywhere. North Korea, for example, has experienced a desperate decline over the past several decades.

In terms of social development, the picture is brighter. Even in the poorer parts of China, most people have

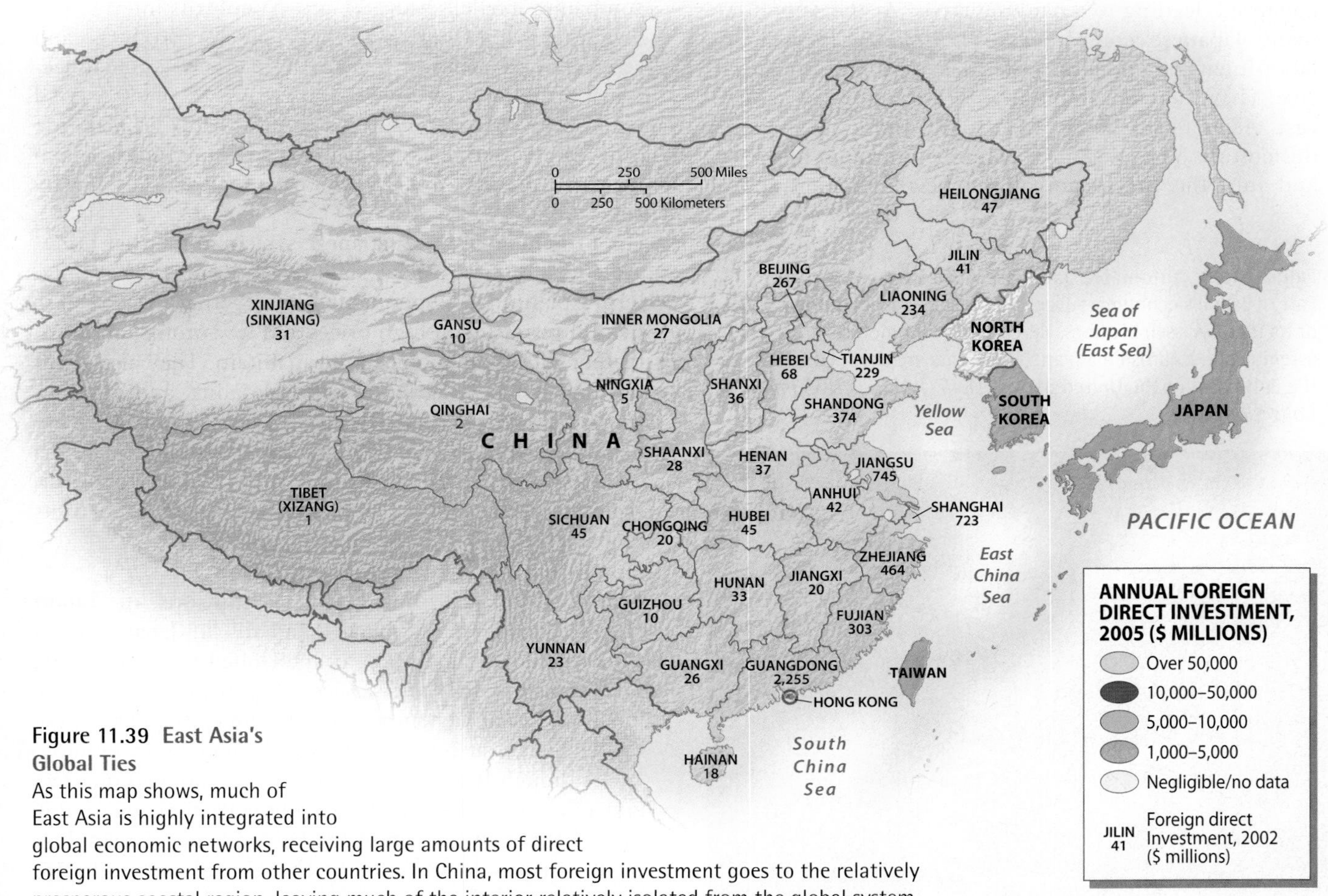

Figure 11.39 East Asia's Global Ties
As this map shows, much of East Asia is highly integrated into global economic networks, receiving large amounts of direct foreign investment from other countries. In China, most foreign investment goes to the relatively prosperous coastal region, leaving much of the interior relatively isolated from the global system.

enjoyed relatively good access to education and health care. As China moves to a market economy, however, such "modern" troubles as unemployment and homelessness have begun to appear. In general, East Asia has experienced fewer such problems than most other parts of the world.

Japan's Economy and Society

Japan was the pacesetter of the world economy in the 1960s, 1970s, and 1980s. In the early 1990s, however, the Japanese economy experienced a major setback, and growth has been slow ever since. But despite its recent problems, Japan is still by some measurements the world's second largest economic power.

Japan's Boom and Bust Although Japan's heavy industrialization began in the late 1800s, most of its people remained poor. The 1950s, however, saw the beginnings of the Japanese "economic miracle." Shorn of its empire, Japan was forced to export manufactured products. Beginning with inexpensive consumer goods, Japanese industry moved to more sophisticated materials, including automobiles, cameras, electronics, machine tools, and computer equipment (Figure 11.40). By the 1980s, it was the leader in many segments of the global high-tech economy.

In the early 1990s, Japan's inflated real estate market collapsed, leading to a banking crisis. At the same time, many Japanese companies discovered that producing labor-intensive goods at home had become too expensive. They therefore began to relocate factories to Southeast Asia and China. Because of these and related difficulties, Japan's economy slumped through the 1990s and into the first years of the new millennium. The Japanese government made several attempts to revitalize the economy through massive state spending, resulting in a huge government debt, totaling 176 percent of the country's gross domestic product. By 2007, however, the Japanese economy apparently had turned a corner, exhibiting a modest but sustainable growth rate of about 2.6 percent. One of the main reasons for this turnaround was the rapid expansion of Japanese industrial exports to China.

Figure 11.40 Automated Japanese Auto Factory
Part of Japan's economic success has resulted from automation of its factory assembly lines. Here Mazda automobiles are assembled in a Hiroshima plant. These cars are destined for the east coast of the United States. *(Jodi Cobbings/NGS Image Collection)*

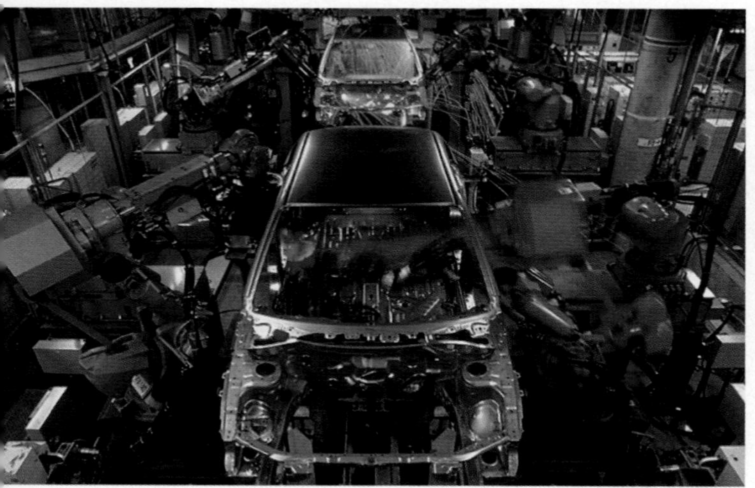

In every sense, Japan remains a core country of the global economic system. Its economic system increasingly spans the globe, as Japanese multinational firms invest heavily in production facilities in North America and Europe, as well as in poorer countries. Japan is a world leader in a large array of high-tech fields, including robotics, optics, and machine tools for the semiconductor industry. It is also one of the world's largest creditor nations, owning a large percentage of U.S. government bonds.

The Japanese Economic System The business environments in Japan and the United States are quite different, reflecting distinct versions of the capitalist economic system. In Japan, the bureaucracy maintains far greater control over the economy than it does in the United States. Japanese corporations are also structured differently from those of the United States. Large groups of companies (called *keiretsu*) are complexly intertwined, owning each other's stock and buying products and services from each other. Because of these interconnections, Japanese firms are less influenced by investors and stockbrokers than are those of the United States. Another difference is that Japanese workers seldom switch companies, and the core workforce of each corporation is rarely subjected to layoffs.

Proponents of the Japanese system argue that it creates business and social stability and encourages long-term planning. Opponents argue that it reduces flexibility, results in high prices and low profits, and ultimately will prove too costly to maintain. They also point out that Japanese agriculture, wholesaling, and distribution remain rather inefficient. Japan has only recently begun to see the rise of large-scale discount stores, which offer lower prices but also threaten the viability of small-scale merchants as well as downtown shopping districts in small cities.

Living Standards and Social Conditions in Japan Despite its affluence, living standards in Japan remain somewhat lower than those of the United States. Housing, food, transportation, and services are particularly expensive in Japan. Certain amenities that are almost standard in the United States, such as central heating, remain uncommon.

Although the Japanese may live in cramped quarters and pay high prices for basic products, they also enjoy many benefits unknown in the United States. Unemployment remains lower than in the United States; health care is provided by the government; and crime rates are extremely low. By such social measures as literacy, infant mortality, and average longevity, Japan surpasses the United States by a comfortable margin. Japan also lacks the extreme poverty found in certain pockets of American society.

Japan, of course, has its share of social problems. Koreans and alien residents from other Asian countries suffer discrimination, as do members of the indigenous Japanese underclass, the Burakumin. Japan's more remote rural areas have few jobs, and many have seen population decreases. In many small villages, most of the remaining people are elderly. Farming is an increasingly marginal occupation, and many farm families survive only because one family member works in a factory or office. Professional and managerial occupations in Japan's cities are noted for their long hours and high levels of stress.

Women in Japanese Society Critics often contend that Japanese women have not shared the benefits of their country's success. Advanced career opportunities remain limited, especially for those who marry and have children. Mothers are still expected to devote themselves to their families and to their children's education. (The Japanese educational system is hierarchically organized, and poor performance in secondary school usually precludes career success.) Japanese business executives often work, or socialize in bars with their coworkers, late every evening, and thus contribute little to child care. Japan's recent economic problems seem to have further reduced career opportunities for women.

One response to the conditions faced by Japanese women has been a drop in the marriage rate. Many Japanese women are delaying marriage, and a sizable number may be abandoning it altogether. Japan has seen an even more dramatic decline in its fertility rate; beginning in 2005, Japan has seen more deaths than births, and its population is now declining. Whether this is due to the domestic difficulties of Japanese women or merely the result of the pressures of a postindustrial society is an open question. Fertility rates have, after all, dropped even lower in many parts of Europe.

Regardless of the cause, Japanese economic planners are concerned about the increasing dependency burden associated with a shrinking population. As Japan's population ages, increasing numbers of retirees will have to be supported by declining numbers of workers. Some argue that only immigration can solve this problem, but most Japanese remain wary of allowing in large numbers of foreigners.

The Newly Industrialized Countries

In the 1960s, 1970s, and 1980s, the Japanese path to development was successfully followed by its former colonies, South Korea and Taiwan. Hong Kong also emerged as a newly industrialized economy in this period, although its economic and political systems remained distinctive.

The Rise of South Korea The postwar rise of South Korea was even more remarkable than that of Japan. During the period of Japanese occupation, Korean industrial development was concentrated in the north, which is rich in natural resources. The south, in contrast, remained a densely populated, poor, agrarian region. South Korea emerged from the bloody Korean War as one of the world's least-developed countries.

In the 1960s, the South Korean government initiated a program of export-led economic growth. It guided the economy with a heavy hand and denied basic political freedom to the Korean people. By the 1970s, such policies had proved highly successful in the economic realm. Huge Korean industrial conglomerates, known as *chaebol*, moved from manufacturing inexpensive consumer goods to heavy industrial products and then to high-tech equipment.

At first South Korean firms remained dependent on the United States and Japan for basic technology. By the 1990s, however, South Korea emerged as one of the world's main producers of semiconductors. South Korean wages also rose at a rapid clip. The country has invested heavily in education (by some measures it has the world's most intensive educational system), which has served it well in the global high-tech economy. Large South Korean companies are themselves now strongly multinational, building new factories in the low-wage countries of Southeast Asia and Latin America, as well as in the United States and Europe.

Contemporary South Korea The political and social development of South Korea has not been nearly as smooth as its economic progress. Throughout the 1960s and 1970s, student-led protests against the dictatorial government were brutally repressed. As the South Korean middle class expanded and prospered, pressure for democratization grew, and by the late 1980s it could no longer be denied. But even though democratization has been successful, political tension has not disappeared (Figure 11.41).

South Korea's periodic political crises have been accompanied by economic difficulties. In 1997, for example, with the country's banking system in chaos, its economy entered a deep recession. Although the South Korean economy soon recovered, it remained unstable for several years, with periods of fast growth alternating with slowdowns. Critics contend that the South Korean economy needs substantial reforms, in particular the breaking up of the large conglomerates (chaebol). By 2007, however, the South Korean economy was expanding at a solid 4 percent a year, and was characterized by moderate inflation, low unemployment, and large export surpluses.

South Korea has attempted to use its own economic success to help North Korea, encouraging Korean companies to invest in the North. Such ventures, however, have not proved very successful. North Korean industry, suffering from parts shortages and decaying infrastructure, continues to decline. Political complications arising from

Figure 11.41 Protests in South Korea
South Korean students have long been noted for their militant protests against their government, international organizations such as the IMF, and, on occasion, the United States. This rally was in the capital city, Seoul, in 1997, to protest government financial and trade policies. *(Yun Jai-hyoung/AP/Wide World Photos)*

North Korea's nuclear program, moreover, have stalled economic cooperation between the two Koreas. In 2007, however, South Korea began delivering 400,000 tons of rice as promised to the North, hoping to prevent another round of famine in the impoverished country.

Taiwan and Hong Kong Like South Korea, Taiwan and Hong Kong also have experienced rapid economic growth since the 1960s. The Taiwanese government, like that of South Korea and Japan, has guided the economic development of the country. Taiwan's economy, however, is organized not around large conglomerates and linked business firms, but rather around small to midsized family firms. This characteristically Chinese form of business organization is sometimes said to give Taiwan greater economic flexibility than its northern neighbors, but it has prevented it from entering certain industries that require huge concentrations of capital. Unlike the other capitalist economies of the region, however, Taiwan suffered little from the "Asian crisis" of the late 1990s.

Hong Kong, unlike its neighbors, has been characterized by one of the most **laissez-faire** economic systems in the world, one with a great amount of market freedom with little governmental control (*laissez-faire* is a French term meaning "let it be"). State involvement has been minimal, which is one reason the city's business elite was nervous about the transition to Chinese rule. Hong Kong traditionally functioned as a trading center, but in the 1960s and 1970s it emerged as a major producer of textiles, toys, and other consumer goods. By the 1980s, however, such cheap products could no longer be made in such an expensive city. Hong Kong's industrialists subsequently began to move their plants to nearby areas in southern China, while Hong Kong itself increasingly specialized in business services, banking, telecommunications, and entertainment. Fears that its economy would falter after the Chinese takeover in 1997 were not realized. Since 2005, Hong Kong's economy has been further strengthened by a number of Chinese companies listing their initial public offerings on the booming Hong Kong stock exchange.

Both Taiwan and Hong Kong have close overseas economic connections. Linkages are particularly tight with Chinese-owned firms located in Southeast Asia. Taiwan's high-technology businesses are also intertwined with those of the United States; there is a constant back-and-forth flow of talent, technology, and money between Taipei and Silicon Valley. Hong Kong's economy also is closely bound with that of the United States (as well as those of Canada and Britain), but its closest connections, not surprisingly, are with the rest of China. The same is also true in regard to Taiwan. China is now Taiwan's largest export market and, since 2006, its second-largest source of imports after Japan.

Chinese Development

China dwarfs all of the rest of East Asia in both physical size and population. Its economic takeoff is thus reconfiguring the economy of the entire region and, to some extent, that of the world as a whole. But despite its recent growth, China's economy has a number of weaknesses. Parts of its vast interior remain trapped in poverty, and some of its largest industries are not competitive. The future of the Chinese economy is thus one of the biggest uncertainties facing both the East Asian and the global economies.

China Under Communism More than a century of war, invasion, and near-chaos in China ended in 1949 when the communist forces led by Mao Zedong seized power. The new government, inheriting a weak economy, set about nationalizing private firms and building heavy industries. Certain successes were realized, especially in Manchuria, where a large amount of heavy industrial equipment was inherited from the Japanese colonial regime.

In the late 1950s, however, China experienced an economic disaster ironically called the "Great Leap Forward." One of the main ideas behind this scheme was that small-scale village workshops could produce the large quantities of iron needed for sustained industrial growth. Communist Party officials demanded that these inefficient workshops meet unreasonably high production quotas. In some cases the only way they could do so was to melt peasants' agricultural tools. Peasants also were forced to contribute such a large percentage of their crops to the state that many went hungry. The result was a horrific famine that may have killed 20 million people.

The early 1960s saw a return to more pragmatic policies, but toward the end of the decade a new wave of radicalism swept through China. This "Cultural Revolution" was aimed at mobilizing young people to stamp out the remaining vestiges of capitalism. Thousands of experienced industrial managers and college professors were expelled from their positions. Many were sent to villages to be "reeducated" through hard physical labor; others were simply killed. The economic consequences of such policies were devastating.

Toward a Postcommunist Economy When Mao Zedong, revered as an almost superhuman being, died in 1976, China faced a crucial turning point. Its economy was nearly stagnant and its people were desperately poor. A political struggle ensued between pragmatists hoping for change and dedicated communists. The pragmatists emerged victorious, and by the late 1970s it was clear that China would embark on a different economic path. The new China would seek closer connections with the world economy and take a modified capitalist road to development (Figure 11.42).

China did not, however, transform itself into a fully capitalist country. The state continued to run most heavy industries, and the Communist Party retained a monopoly on political power. Instead of suddenly abandoning the communist model, as the former Soviet Union later did, China allowed cracks to appear in which capitalist ventures could take root and thrive. By 2005, 70 percent of the Chinese economy was controlled by the private sector.

One of China's first capitalist openings, in the late 1970s, was in agriculture, which had previously been dominated by large-scale communal farms. Individuals were suddenly allowed to act as agricultural entrepreneurs, selling produce in the open market. Owing to this change, the income of many farmers rose dramatically. By the late 1980s, however, the focus of growth had shifted to the urban-industrial sector. As the government became concerned about inflation, it placed price caps on agricultural products and increased taxes on farmers. By the early years of the new millennium, many rural areas in China's interior provinces were experiencing economic distress, prompting many people to relocate to China's coastal cities or to foreign countries (see "People on the Move: The Chinese Global Diaspora").

Industrial Reform One of China's most important early industrial reforms involved opening **Special Economic Zones (SEZs)** in which foreign investment was welcome and state interference minimal. The Shenzhen SEZ, adjacent to Hong Kong, proved particularly successful after Hong Kong manufacturers found it a convenient source of cheap land and labor. Additional SEZs soon were opened, mostly in the coastal region. The basic strategy was to attract foreign investment that could generate exports, the income from which would supply China with the capital it needed to build its infrastructure and thus achieve conditions for sustained economic growth.

Other market-oriented reforms followed. Former agricultural cooperatives were allowed to transform themselves into capitalist entities. Many of these "township and village enterprises" proved highly successful. By the early 1990s, the Chinese economy was growing at nearly 10 percent a year, perhaps the fastest rate of expansion the world has ever seen. China emerged as a major trading nation, and by the mid-1990s it had amassed large quantities of foreign reserves; as of 2006, its total trading surplus was estimated at a whopping $180 billion. Between 2001 and 2006, China was estimated to have accounted for roughly one third of the world's total economic growth.

China's economy has grown so quickly in recent years that many economists have feared that it would overheat, leading to unsustainably high rates of inflation. Thus far, however, inflation has remained moderate. Other critics contend that China must fully abandon centralized planning if its expansion is to be sustainable over the long run. China's leadership, however, has made it clear that reform will be a gradual process, maintaining

Figure 11.42 Capitalism in China
Although China is officially a communist country, its economy is actually highly capitalistic. As China develops, banking has become an increasingly important segment of its economy. *(Rob Crandall/ www.robcrandall.com)*

PEOPLE ON THE MOVE The Chinese Global Diaspora

One of the largest waves of migration the world has ever seen is currently underway in East Asia. Since the coming of economic reform in the late 1970s, some 200 million Chinese have uprooted themselves to seek greater opportunities elsewhere. Most of the migrants are rural Chinese from the interior of the country moving to the booming cities of the coastal zone. However, an estimated 18 million people have left China altogether.

Emigration from China is nothing new. In the 1800s, millions of Chinese moved to Southeast Asia, North America, and South America. During that period, almost all Chinese emigrants came from the southern coastal zone. Today the southern coastal province of Guangdong gains more migrants than it loses, whereas the less economically successful areas of interior and northern China now export a disproportionate number of people. The current wave also differs from that of the 1800s in that it sends migrants to all parts of the world. Chinese immigrants are now streaming out to Europe, Central Asia, Russia, Africa, and elsewhere, ending up in some 150 countries. In another departure from the past, the Chinese government now often encourages and even assists emigrants, in part because it values the $20 billion in remittances returned to the country annually.

In different parts of the world, Chinese immigrants enter different economic niches. In oil-rich African and Central Asian countries, many work on construction brigades run by Chinese companies building basic infrastructure. In Europe, migrants from China often work in factories and restaurants, often under poor conditions. In Romania, for example, thousands of Chinese women, earning on average $260 a month, have taken textile-factory jobs that were abandoned by Romanians moving to western and central Europe in search of better pay. In countries rich and poor, Chinese migrants often run shops and other small businesses.

The influx of Chinese workers sometimes generates economic and cultural tensions. In 2006, citizens of Tonga in the Pacific, angry that local Chinese businesses were recruiting workers from China rather than from the local population, looted several dozen Chinese-run shops. In a 2007 incident that received global attention and infuriated the Chinese government, militant female religious students in Pakistan seized nine Chinese women from a local massage parlor, accusing them of prostitution. In Russia, extreme nationalists worry that the 100,000 Chinese who have recently moved to the Russian Far East threaten the territorial integrity of their country. Moscow, which now has six Chinese language newspapers, has seen the recent rise of pronounced anti-Chinese sentiments.

Figure 11.6.1 China's Contemporary Global Diaspora Officials in Half Moon Bay, California, seize a ship containing roughly 100 Chinese nationals being smuggled into the United States. Despite China's booming economy, hundreds of thousands of Chinese citizens migrate overseas every year. *(Lov Dematteis/The Image Works)*

In Europe and the United States, increasingly strict restrictions on immigration have resulted in a profitable Chinese "people smuggling" business. By 2007, it typically cost $30,000 to illegally transport a Chinese worker to Britain and about $70,000 to get one into the United States—figures approximately double what they had been in the mid-1990s (Figure 11.6.1). As a result of such exorbitant prices, illegal Chinese immigrants usually take many years to pay off those who smuggled them across the border, often suffering severe exploitation in the meantime.

that it has proved quite successful at managing its economic expansion.

China's economic growth has resulted in increased tensions with the United States. China exports far more to the United States than it imports, leading some U.S. politicians to demand that China allow its currency to rise against the dollar. In 2005, China did let its currency appreciate, but only by a modest 2.1 percent. Other U.S. critics accuse China of unfairly keeping the price of its labor low in order to enhance exports. China's large and growing holdings of U.S. treasury bonds, however, make it tricky for the United States to exert pressure.

Social and Regional Differentiation The Chinese economic surge unleashed by the reforms of the late 1970s and 1980s resulted in growing **social and regional differentiation**. In other words, certain groups of people—and certain portions of the country—prospered far more than others. Despite its official socialism, the Chinese state encouraged the formation of an economic elite, having

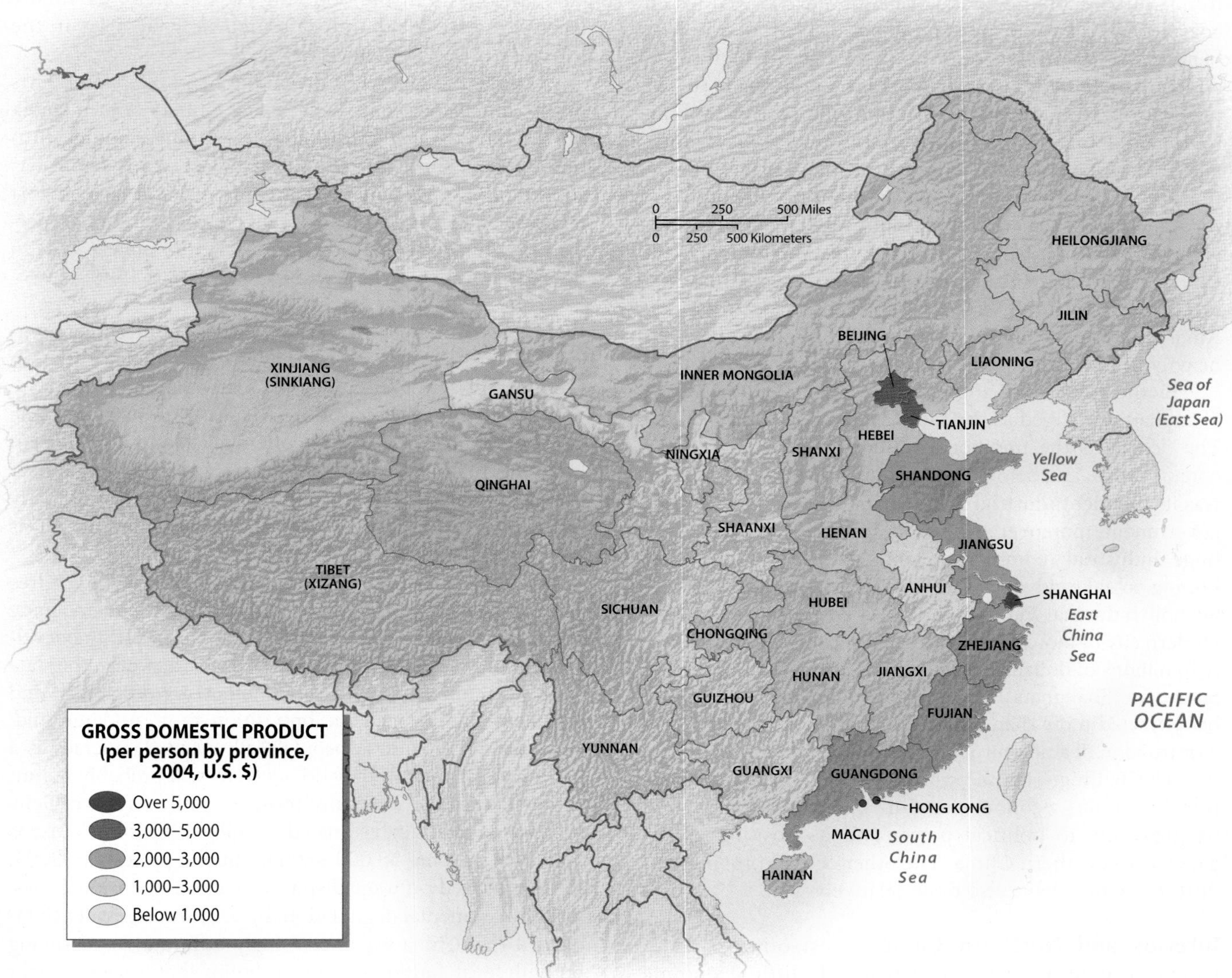

Figure 11.43 Economic Differentiation in China
Although China has seen rapid economic expansion since the late 1970s, the benefits of growth have not been evenly distributed throughout the country. Economic prosperity and social development are concentrated on and near the coast. Most of the interior remains mired in poverty. The poorest part of China is the upland region of Guizhou in the south-central part of the country.

concluded that wealthy individuals are necessary to transform the economy. The least-fortunate Chinese citizens were sometimes left without work, and many millions have migrated to the booming coastal cities. The government has attempted to control the transfer of population, but with only partial success. Shantytowns, as well as homeless populations, began to emerge around China's cities. Many migrant children, owing to their lack of official residence, have trouble enrolling in school, undermining their future economic opportunities.

As China's economic boom accelerated, so did economic differences. The rapid growth of the elite population made China the world's fastest-growing market for luxury automobiles and even for golf-course developments. At the same time, vulnerable state-owned enterprises were increasingly abandoning their provision of housing, medical care, and other social services, leaving many people destitute—particularly the elderly.

Economic disparities in China, as in other countries, are geographically structured. Before the reform period, the communist government attempted to equalize the fortunes of the different regions, giving special privileges to individuals from poor places. Such efforts were not wholly successful, and some provinces continued to be deprived. Since the coming of market reforms, moreover, the process of regional economic differentiation has accelerated (Figure 11.43).

The Booming Coastal Region Most of the benefits from China's economic transformation have flowed to the coastal region and to the capital city of Beijing. The first beneficiaries were the southern provinces of Guangdong and Fujian. This region was perhaps predisposed to the new economy, because the southern Chinese have long been noted for engaging in overseas trade. Guangdong and Fujian have also benefited from their close connections with the overseas Chinese communities of Southeast Asia and North America. Proximity to Taiwan and especially Hong Kong also proved helpful. Vast amounts of capital have flowed to the south coastal region from foreign (and Hong Kong-based) Chinese business networks. American, Japanese, and European firms also have invested heavily in the region.

By the 1990s, the Yangtze Delta, centered on the city of Shanghai, reemerged as the most dynamic region of China. The delta was the traditional economic (and intellectual) core of China, and before the communist takeover Shanghai was its premier industrial and financial center. The Chinese government, moreover, has encouraged the development of huge industrial, commercial, and residential complexes, hoping to take advantage of the region's dynamism. The Suzhou Industrial Park in Jiangsu is emerging as a hypermodern city of more than a half-million people, thanks largely to billions of dollars of foreign investment, most of it from Singapore. Shanghai's Pudong industrial development zone has attracted more than $10 billion, much of it going to the construction of a new airport and subway system.

The Beijing–Tianjin region also has played a major role in China's economic boom. Its main advantage is its proximity to political power and its position as the gateway to northern China. The other coastal provinces of northern China have also done relatively well.

Interior and Northern China Most other parts of China, in contrast, have seen relatively little economic expansion. Central and northern Manchuria were formerly quite prosperous, owing to fertile soils and early industrialization, but have not participated much in the boom. Many of the state-owned heavy industries of the Manchurian "**rust belt**," or zone of decaying factories, are not efficient. Manchuria's once-productive oil wells, moreover, are largely exhausted. China and Russia have, however, agreed to begin building an oil pipeline from Siberia to Daqing in Manchuria in 2008, hoping to revive the economy of this decaying oil town.

Most of the interior provinces of China likewise largely missed the wave of growth in the 1980s. In many areas, rural populations continue to grow while the natural environment deteriorates. One consequence is high levels of underemployment and out-migration. By most measures, poverty increases with distance from the coast. As a result of such discrepancies, China is encouraging development in the west (including Tibet and Xinjiang as well as the western provinces of China proper), but thus far its efforts have been largely limited to transportation improvement and natural resource extraction. Evidence suggests, however, that mounting labor shortages in the booming coastal areas are finally leading industrialists to locate their new factories in the interior portions of the country.

Scholars debate the true conditions in the poorer parts of interior China. Some believe that China's official statistics are too positive, hiding significant hunger and destitution. Others think that the country's economic boom has substantially raised living standards even in the poorest districts. According to official statistics, fewer than 10 percent of China's people now fall below the poverty line, a figure that was over 50 percent as recently as 1980.

Rising Tensions China's explosive but uneven economic growth has generated a number of problems. Inflation has made planning difficult and created hardship for those on fixed incomes. Corruption by state officials is by some accounts rampant; success often seems to depend on knowing the right people and having the right connections. China's crime rate, moreover, which was extremely low, is rising rapidly. Organized crime in particular is emerging as a major problem in many areas.

A more significant issue has been the struggle for free expression and political openness. The desire for democratic reform has been enhanced by rising incomes and the development of a sizable middle class. In 1989, however, the state crushed a movement for government accountability and democracy, and forced the opposition to go underground. Whether most Chinese people really want democracy is a controversial issue. If they do, tensions will probably mount as long as China's economy prospers while its rulers deny basic freedoms. An estimated 3 million Chinese citizens took part in some 58,000 separate public protests in 2004, mostly focused on wage disputes, evictions, pension issues, and environmental degradation. In 2007, much evidence indicated that China was increasingly infiltrating and spying on both local and foreign organizations that it feared might disrupt the 2008 Beijing Olympics.

China's political and human-rights policies have complicated its international relations. Sources of tension with the United States and other wealthy countries are also economic in nature. Not only China's large and growing trade surplus, but also its reluctance to enforce copyright and patent law irritate many of its trading partners. Several U.S. firms have accused Chinese businesses of pirating music, software, and brand names. China has made efforts to stop such activities, but critics contend that its actions have not been adequate. As the global market grows and as popular culture becomes globalized, copyright and trademark infringement are becoming increasingly profitable and hence increasingly difficult to control.

Social Conditions in China

Despite its pockets of persistent poverty, China has achieved significant progress in social development. Since coming to power in 1949, the communist government has made large investments in medical care and education, and today China boasts fairly impressive health and longevity figures.

Human well-being in China is also geographically structured. The literacy rate, for example, remains relatively low in many of the poorer parts of China, including the uplands of Yunnan and Guizhou and the interior portions of the North China Plain. Literacy is much more widespread in Manchuria, the Yangtze Delta, and most major urban areas. Such regional disparities may increase as the gap between the wealthy and the poor grows larger.

China's Population Quandary Population policy also remains an unsettling issue for China. With 1.3 billion people highly concentrated in less than half of its territory, China has one of the world's highest effective population densities. By the 1980s, its government had become so concerned that it instituted the famous "one-child policy." Under this plan, couples in normal circumstances are expected to have only a single offspring and can suffer financial and other penalties if they do not comply (Figure 11.44). Although the one-child policy was never fully implemented in many rural areas, China's demographic strategy has been generally successful; the average fertility level is now only 1.6, and the population is growing at the relatively slow rate of 0.6 percent a year. China will likely reach roughly 1.5 billion persons before stabilization occurs.

Figure 11.44 China's Population Policies
One aspect of China's population policy is the expansion of child-care facilities so that mothers can be near their children while at work. This enables women to resume participating in the workforce soon after giving birth. This photo shows a typical day-care center attached to an industrial plant in Guangdon province in coastal China. *(Xinhua/Getty Images, Inc.—Liaison)*

The decline in Chinese fertility, however, has brought about problems of its own. China is now worried about impending labor shortages as well as its aging population. Thus far, China has not provided adequate facilities for caring for its elderly population. As a result, the government is increasingly allowing couples to have more than one child, hoping that the additional children will care for their parents as they grow old.

China's population policy has also has generated social tensions and human-rights abuses. Particularly troubling is the growing gender imbalance in the Chinese population. Baby boys in the country now far outnumber baby girls. This asymmetry reflects the practice of honoring one's ancestors; because family lines are traced through male offspring, one must produce a male heir to maintain one's lineage. Many couples are therefore desperate to produce a son. Some opt to bear more than one child regardless of penalties. Another option is gender-selective abortion; if ultrasound reveals a female fetus, the pregnancy is sometimes terminated. Baby girls also are commonly abandoned, and well-substantiated rumors of female infanticide circulate. International women's organizations, as well as antiabortion groups, are concerned about these effects of China's population policy. Most environmentalists, however, applaud China for lowering its birthrate.

The Position of Women Women have historically had a relatively low position in Chinese society, as is true in most other civilizations. One particularly blatant traditional expression of this was the practice of foot binding: the feet of elite girls were usually deformed by breaking and binding them in order to produce a dainty appearance. This crippling and painful practice was eliminated only in the twentieth century. In certain areas of southern China it was also common in earlier times for girls to be married, hence to leave their own families, when they were mere toddlers (such marriages, of course, would not be consummated for many years).

Not all women suffered such disabilities in premodern China. Some individuals achieved fame and fortune—a few even through military service. Among the Hakka, women enjoyed a relatively high social position and were seldom subjected to foot binding. Both the nationalist and communist governments have, moreover, sought to begin equalizing the relations between the sexes. Many of their measures have been successful, and women now have a relatively high level of participation in the Chinese workforce. But it is still true that throughout East Asia—in Japan no less than in China—few women have achieved positions of power in either business or government.

Summary

- The economic success of East Asia has been accompanied by severe environmental degradation. Japan, South Korea, and Taiwan have responded by enacting strict environmental laws and by moving many of their most polluting industries overseas. The major environmental issue in the region today concerns the rapid growth of the Chinese economy. Pollution in Chinese cities is so serious that it has major negative effects on human health, while much of the Chinese countryside suffers from such problems as soil erosion and desertification.
- Although East Asia is a densely populated region, its birthrates have plummeted in recent decades. Japan is now experiencing population decline, which will put severe pressure on the Japanese economy. In China, the biggest demographic challenge results from the massive movement of people from the interior to the coast and from rural villages to the rapidly expanding cities. China has thus been trying to redirect development toward the interior, but thus far has had little success.
- East Asia is united by deep cultural and historical bonds. China has the largest influence because, at one time or another, it ruled most of the region. Although Japan has never been under Chinese rule, it still has profound historical connections to Chinese civilization. The more prosperous parts of East Asia have welcomed cultural globalization over the several past decades.
- Geopolitically, East Asia remains characterized by strife. China and Korea are still suspicious of Japan, and they worry that it might rebuild a strong military force. Japan is concerned about the growing military power of China and especially about the nuclear arms and missiles of North Korea. While relations between North and South Korea have improved in recent years, they remain tense—as do those between Taiwan and China. North Korean weapons development, moreover, is a global concern.
- With the notable exception of North Korea, all East Asian countries have experienced major economic growth since the late 1900s. The most important story is the rapid rise of China. China's economic expansion has reduced poverty nationwide, but has also generated serious tensions between the wealthier, more globally oriented coastal regions and the less prosperous interior provinces. The rise of China also has global implications, as many countries in all regions of the world have profited by exporting the raw materials needed by China's booming industries.

Key Terms

anthropogenic landscape *(page 488)*
autonomous region *(page 502)*
Burakumin *(page 500)*
central place theory *(page 492)*
China proper *(page 478)*
Cold War *(page 504)*
Confucianism *(page 495)*
diaspora *(page 500)*
geomancy *(page 497)*
ideographic writing *(page 494)*
laissez-faire *(page 518)*
loess *(page 478)*
Mandarin *(page 496)*
Marxism *(page 498)*
pollution exporting *(page 479)*
regulatory lakes *(page 477)*
rust belt *(page 522)*
samurai *(page 504)*
sediment load *(page 477)*
Shogunate *(page 507)*
social and regional differentiation *(page 520)*
Special Economic Zones (SEZs) *(page 519)*
spheres of influence *(page 507)*
superconurbation *(page 492)*
tonal language *(page 501)*
urban primacy *(page 492)*

Questions for Review

1. How does the physiography of the Huang He River valley differ from that of the valley of the Yangtze?
2. Why is Japan so much more heavily forested than China?
3. Why has Shanghai emerged as the most populous urban area in China?
4. What are the major ways in which Japanese cities differ from those of the United States?
5. Where are the non-Han peoples of China concentrated? Why are they concentrated in these areas?
6. What have been the main consequences of the geographical division of Korea into two states?
7. Historically speaking, how did China and Japan act differently as imperial powers?
8. How have the different countries of East Asia followed different paths to economic development?
9. Where in China would one find the most rapid economic development, and why would one find it there?
10. How does the position of women in Japan compare to the position of women in other wealthy, industrialized countries?

Thinking Geographically

1. Discuss the advantages and disadvantages of China's building of major dams.
2. Discuss the ramifications, both positive and negative, of Japan's allowing the importation of rice and opening its agricultural lands to urban development.
3. What might be the consequences of China's granting true autonomy to the Tibetans and other non-Han peoples? Independence?
4. What are the potential implications of Taiwan's declaring itself an independent country?
5. Discuss the potential ramifications of the United States' restricting the importation of Chinese goods in order to put pressure on the Chinese government for human-rights reforms.
6. Discuss the advantages and disadvantages of China's current population policies.
7. Do you think that East Asia will emerge as the center of the world economy in the next century?

Regional Novels and Films

Novels

Sawako Ariyoshi, *The River Ki* (1982, Kodansha)

Xingjian Gao, *Soul Mountain* (2000, Flamingo)

Kazuo Ishigura, *An Artist of the Floating World* (1989, Vintage Books)

Peter H. Lee, *Flowers of Fire* (1986, University of Hawaii Press)

Wang Shuo, *Playing for Thrills* (1998, Penguin)

Hsueh-Chin Tsao, *The Dream of the Red Chamber* (1958, Doubleday)

Films

Farewell, My Concubine (1993, China)

The Gate of Heavenly Peace (1995, China)

Little Red Flowers (2006, China)

Nomugi Pass (1979, Japan)

The Old Garden (2006, South Korea)

Princess Mononoke (1997, Japan)

Seven Samurai (1954, Japan)

A Taxing Woman (1987, Japan)

301, 302 (1995, South Korea)

Twilight Samurai (2002, Japan)

Why Has Bodi-Darma Left for the East? (1989, Korea)

Yellow Earth (1984, China)

Bibliography

Brown, Melissa. 2004. *Is Taiwan Chinese?: The Impact of Culture, Power, and Migration on Changing Identities*. Berkeley: University of California Press.

The Contemporary Atlas of China. 1988. Boston: Houghton Mifflin.

Cybriwsky, Roman A. 1998. *Tokyo: The Shogun's City and the Twenty-First Century*. New York: Wiley.

Elvin, Mark. 2004. *The Retreat of the Elephant: An Environmental History of China*. New Haven, CT: Yale University Press.

Harris, Mark Edward. 2007. *Inside North Korea*. San Francisco: Chronicle Books.

Knapp, Ronald G., ed. 1992. *Chinese Landscapes: The Village as Place*. Honolulu: University of Hawaii Press.

Kynge, James. 2006. *China Shakes the World: A Titan's Rise and Troubled Future—and the Challenge for America*. Boston: Houghton Mifflin.

Ryu, Je-Hun. 2004. *Reading the Korean Cultural Landscape*. Elizabeth, NJ: Hollym International.

Smil, Vaclav. 2004. *China's Past, China's Future: Energy, Food, Environment*. London: Curzon Press.

Totman, Conrad. 1989. *The Green Archipelago: Forestry in Preindustrial Japan*. Berkeley: University of California Press.

Additional bibliographic resources are at the *Diversity Amid Globalization* Website: http//www.prenhall.com/rowntree/.

12
South Asia

Slum settlements sprout in front of new high-rise buildings in Bangalore, India. Rapid economic growth in the poor countries of South Asia generates striking disparities as well as serious social and political tensions. *(Dinodia)*

ENVIRONMENTAL GEOGRAPHY

While the arid parts of South Asia suffer from water shortages and the salinization of the soil, the humid areas often experience devastating floods.

POPULATION AND SETTLEMENT

South Asia will soon become the most populous region in the world; although birthrates have come down substantially in India and Bangladesh, they remain high in Pakistan.

CULTURAL COHERENCE AND DIVERSITY

South Asia is one of the most culturally diverse regions of the world, with India alone having more than a dozen official languages as well as numerous followers of most major religions.

GEOPOLITICAL FRAMEWORK

South Asia is burdened not only by a large number of violent secession movements, but also by the struggle between the nuclear-armed countries of India and Pakistan, which threatens both regional and global stability.

ECONOMIC AND SOCIAL DEVELOPMENT

Although South Asia as a whole is one of the poorest regions of the world, parts of India are experiencing rapid economic development based on the high-tech skills of the educated segment of its population.

South Asia, a land of deep historical and cultural interconnections, has recently experienced intense political conflict. Since independence from Britain and the political division of the region in 1947, India and Pakistan, South Asia's two largest countries, have fought several wars and remain locked in a bitter conflict concerning possession of the disputed territory of Kashmir in the northern reaches of the region. Religious divisions are linked to this geopolitical turmoil, for India is primarily a Hindu country (with a large Muslim minority), while neighboring Pakistan and Bangladesh are both predominantly Muslim (see Figure 12.1 and "Setting the Boundaries").

Parallel to these geopolitical tensions are demographic concerns. Given its current rate of growth, South Asia soon will surpass East Asia as the world's most populous region. The underlying issue, however, is not simply the gigantic population of this region, but whether it can support these people given its economy and resource base. Although agricultural production has increased slightly faster than population over the past three decades, some experts think that these improvements are approaching their limit. Compounding this serious situation is the widespread poverty of South Asia; it is, along with Sub-Saharan Africa, the poorest part of the world. Roughly one-third of India's people subsist on less than $1 a day.

South Asia remains less connected to the contemporary globalized world than East or Southeast Asia. Until recently, South Asia was noted for its slow rate of economic growth and inward orientation. India, however, has begun to experience rapid economic growth, and South Asia as a whole is now gaining an important global role, based in part on the impressive levels of scientific and technical skills found within the labor force. Parts of India, for example, have emerged as major players in information technology (IT), and are now tightly linked to California's Silicon Valley and other centers of the global IT industry.

ENVIRONMENTAL GEOGRAPHY: Diverse Landscapes, from Tropical Islands to Mountain Rim

South Asia's environmental geography covers a wide spectrum that ranges from the highest mountains in the world to densely populated delta islands barely above sea level; from one of the wettest places on Earth to dry, scorching deserts; from tropical rain forests to degraded scrublands to coral reefs (Figure 12.1). All of these ecological zones have their own distinct and complex environmental problems. To illustrate the complexity of South Asian

Setting the Boundaries

Formidable mountains separate this region from the rest of Asia, although one of the main ranges, the Hindu Kush, lies to the northwest of South Asia in central Afghanistan. Sometimes called the Indian subcontinent, mainland South Asia includes India, Pakistan, and Bangladesh as well as the smaller countries of Nepal and Bhutan. Also included in the region are a number of islands in the Indian Ocean, including the countries of Sri Lanka and the Maldives, as well as the Indian territories of the Lakshadweep, and the Andaman and Nicobar islands.

The unity of this region comes more from a shared history than from any contemporary political or cultural coherence. In the past, the region could be characterized as a meeting ground between the major religions of Hinduism, Islam, and, to a lesser degree, Buddhism. Additionally, South Asia shares a colonial history dominated by British administration. Today, however, the region is torn by ethnic, political, and religious tensions that have surfaced more intensely since independence in 1947.

India is by far the largest South Asian country, both in size and in population. Covering more than 1 million square miles (2,590,000 square kilometers), India is the world's seventh largest country in area and, with more than 1 billion inhabitants, second only to China in population. Although it has a Hindu majority, India contains a tremendous amount of religious, ethnic, linguistic, and political diversity.

Pakistan and Bangladesh are next largest in size and population. Until independence in 1947, Pakistan was part of a larger, undivided British colonial realm simply called *India*. Because of its strong ties to Islam, however, Pakistan is now sometimes considered more closely connected to its Muslim neighbors in Southwest Asia than it is to the rest of South Asia. Bangladesh, on India's eastern shoulder, also is a predominantly Muslim country. Although a relatively small country in area, Bangladesh is one of the most densely populated in the world—and also one of the poorest—with a population of 149 million in an area about the size of Wisconsin. Bangladesh has a short border with Burma (Myanmar), but it is otherwise virtually engulfed by India, which wraps around the country to the north and the northeast.

Nepal and Bhutan are both located in the Himalayan Mountains, sandwiched between India and the Tibetan Plateau of China. Nepal is much the larger of the two in both area and population, and is far more open to and engaged with the contemporary world. Bhutan, on the other hand, has purposely remained a relatively isolated Buddhist kingdom, although it is beginning to transition to democratic rule.

The two island countries of Sri Lanka (formerly Ceylon) and the Maldives round out South Asia. Sri Lanka has since 1983 been mired in a civil war that has taken a high toll on its potentially strong economy. The Maldives, in contrast, has a vibrant tourist economy, but it is deeply threatened by the prospect of global warming and the associated rise in sea level.

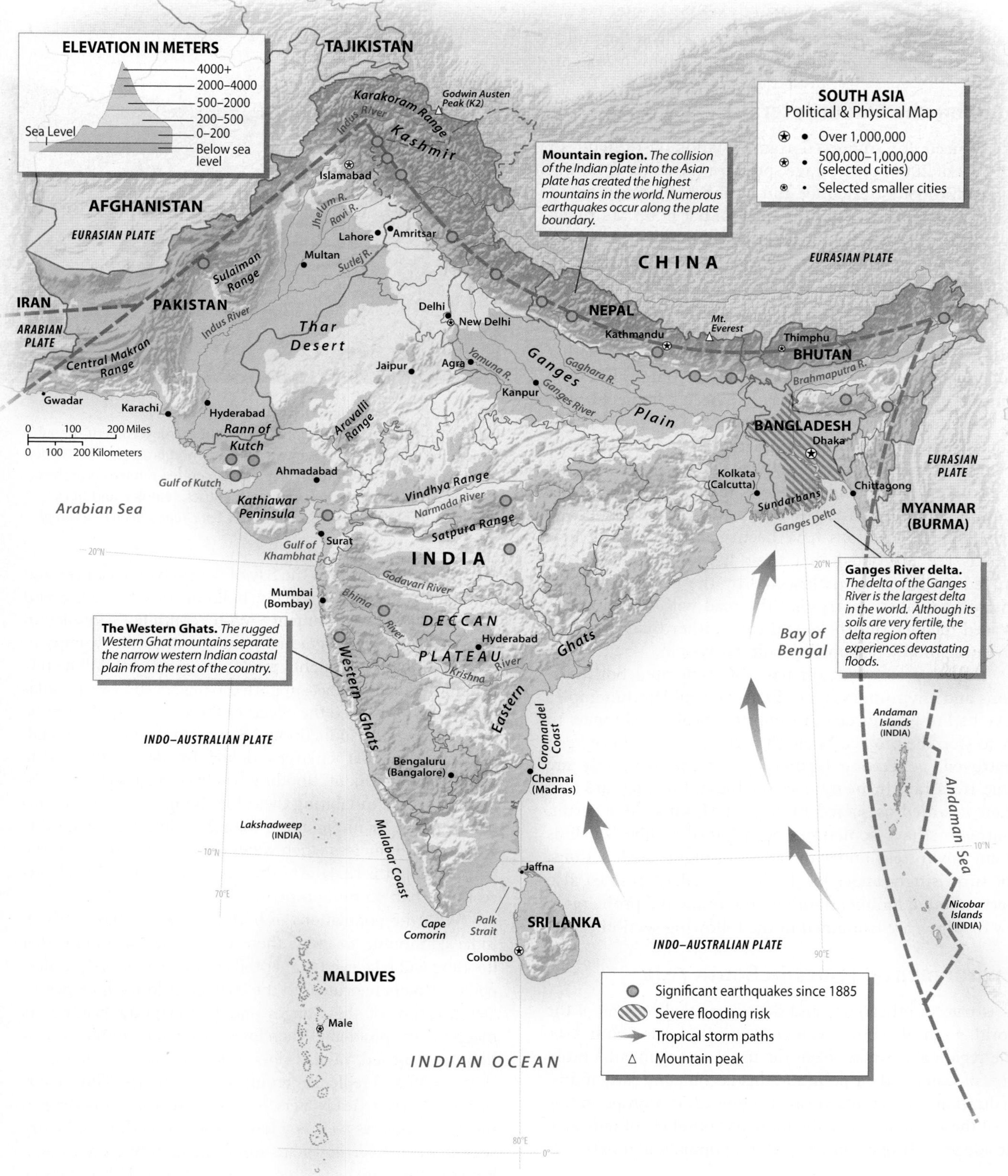

Figure 12.1 South Asia
The second most populous region in the world, South Asia is dominated by India, Pakistan, and Bangladesh. The two Himalayan countries of Nepal and Bhutan, along with the island nations of Sri Lanka and the Maldives, round out the region. Four extensive physical subregions cover most of South Asia: the high Himalayan mountains in the north; the expansive Indus–Ganges lowland that reaches from Pakistan to the delta lands of Bangladesh; peninsular India, dominated by the Deccan Plateau; and the island realm that includes Sri Lanka and the Maldives. Many of this seismically active region's landscapes are products of the slow northward movement of the Indo-Australian tectonic plate against the Eurasian plate. Typhoons from the Bay of Bengal pose a significant risk to parts of eastern South Asia.

environmental issues, let us begin by looking at the building of India's new "Golden Quadrilateral" highway system.

Building the Quadrilateral Highway

India's need for improved transportation is difficult to deny. As of 2006, the average speed for a trucker traveling between Kolkata (Calcutta) and Mumbai (Bombay) could be as low as 6.8 miles (11 kilometers) per hour. Multilane highways are rare, while checkpoints and tollbooths, typically demanding long waits, abound. To address the transport needs generated by its booming economy, India has undertaken a massive $12.3 billion road project designed to connect its four largest cities, New Delhi, Kolkata (Calcutta), Chennai (Madras), and Mumbai (Bombay), with a modern highway (Figure 12.2). As of March 1, 2007, some 95 percent of the highway had been completed.

Building the new highway system has not proved easy. Truckers have protested the higher taxes and tolls that are needed to finance it, while citizen groups have stopped construction on several occasions with large protests demanding more underpasses, overpasses, and cattle crossings. Some concessions have been made, but many rural Indians are infuriated that their homes have been destroyed and their farms bisected by the massive project. Environmentalists worry that the road will lead to an increased reliance on trucks and cars, rather than the more environmentally responsible railroad system.

Highway construction has also generated both religious and environmental conflicts. Several Hindu temples have had to be relocated, prompting local worshippers to try to stop the project. Many thousands of trees have been destroyed, generating further opposition. Not only are large trees rare in many parts of India, but they are also often regarded as sacred by devout Hindus. As a result, contractors have sometimes been forced to hire Muslims to cut down trees under the cover of night. Such actions, not surprisingly, anger local villagers. Villagers must also worry about a number of other environmental problems, a few of which are examined in the following section.

Figure 12.2 The Golden Quadrilateral Highway
India's infrastructure is notoriously poor, but the government is now responding with a massive highway construction program. The Golden Quadrilateral Highway, shown here, has generated numerous protests, as villagers object to the destruction of houses, temples, and trees that lie in its path. *(Tyler Hicks/The New York Times/Redux Pictures)*

Environmental Issues in South Asia

As is true in other poor and densely settled regions of the world, a raft of serious ecological issues plague South Asia. The region also suffers from the usual environmental problems of water and air pollution that accompany early industrialization and manufacturing (Figure 12.3). Compounding all of these problems are the immense numbers of new people added each year through natural population growth.

Natural Hazards in Bangladesh The link between population pressure and environmental problems is nowhere clearer than in the delta area of Bangladesh, where the search for fertile land has driven people into hazardous areas, putting millions at risk from seasonal flooding as well as from the powerful **cyclones** that form over the Bay of Bengal. For millennia, drenching monsoon rains have eroded and transported huge quantities of sediment from the Himalayan slopes to the Bay of Bengal by the Ganges and Brahmaputra rivers, gradually building this low-lying and fertile deltaic environment. With continual population growth, people have gradually moved into the swamps to transform them into highly productive rice fields. While this agricultural activity has supported Bangladesh's vast population, it has intensified the effects of the area's natural hazards.

Although periodic floods are a natural, even beneficial, phenomenon that enlarges deltas by depositing fertile river-borne sediment, flooding has become a serious problem for people inhabiting these low-lying areas. In September 1998, for example, more than 22 million Bangladeshis were made homeless when water covered two-thirds of the country (Figure 12.4). Another major flood in 2007 affected more than 35 million people.

With the populations of both Bangladesh and northern India continuing to grow, there is a strong possibility that flooding will take even higher tolls in the next decade as desperate farmers relocate into the hazardous lower floodplains. Deforestation of the Ganges and Brahmaputra headwaters magnifies the problem. When forest cover and ground vegetation are removed, rainfall does not soak into the ground to slow runoff and replenish groundwater supplies. Thus deforestation in river headwaters results in increased flooding in the wet season, as well as a lower water level during the dry season when river flows are supplemented by groundwater. Both problems are increasing in parts of Bangladesh and India. Bangladesh's water problems are further magnified by the fact that many of its aquifers are contaminated with arsenic, threatening the health of as many as 80 million people.

Forests and Deforestation Forests and woodlands once covered most of South Asia, except for the desert areas in the northeast, but in most places tree cover has vanished as a result of human activities. The Ganges Valley

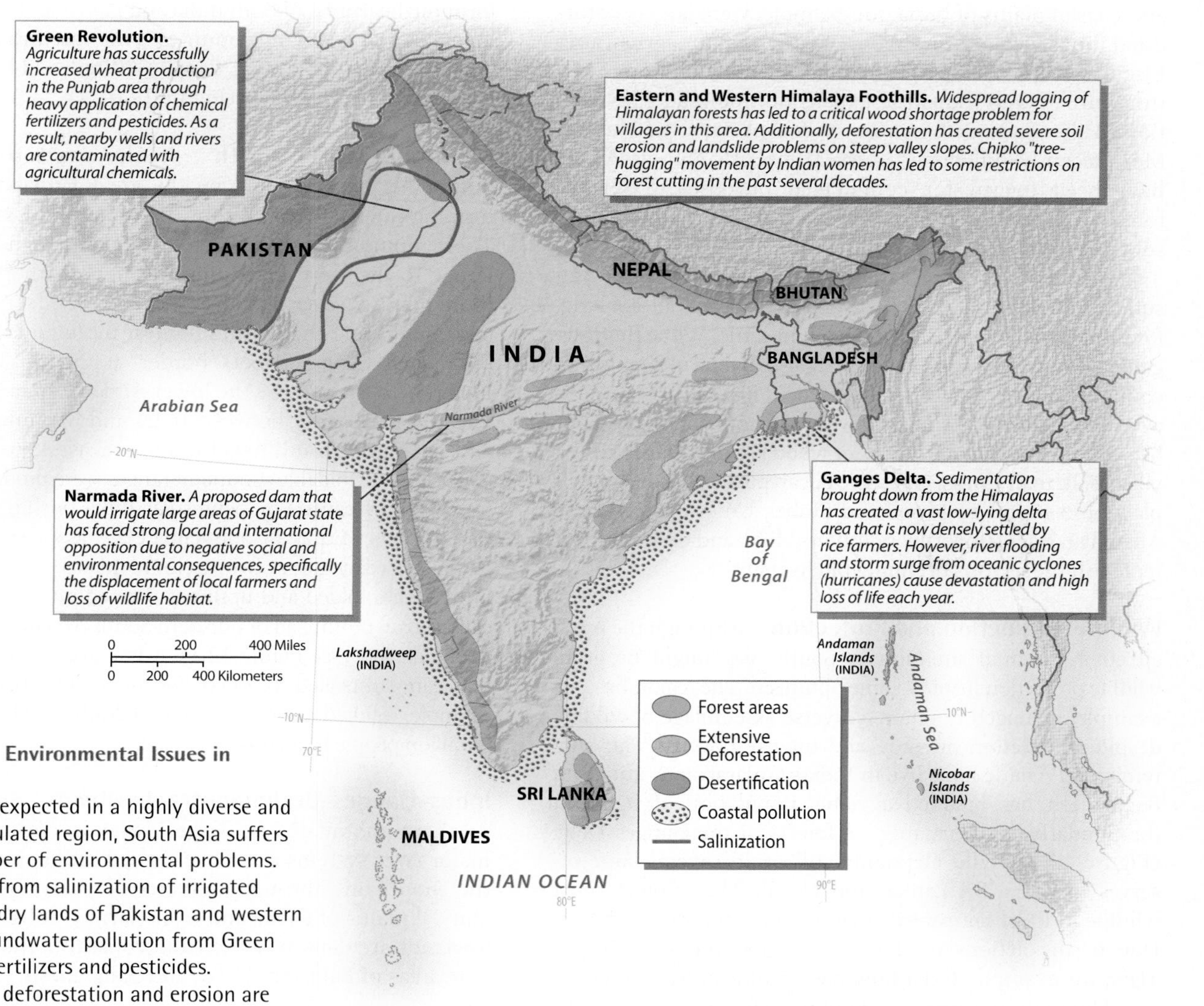

Green Revolution. *Agriculture has successfully increased wheat production in the Punjab area through heavy application of chemical fertilizers and pesticides. As a result, nearby wells and rivers are contaminated with agricultural chemicals.*

Eastern and Western Himalaya Foothills. *Widespread logging of Himalayan forests has led to a critical wood shortage problem for villagers in this area. Additionally, deforestation has created severe soil erosion and landslide problems on steep valley slopes. Chipko "tree-hugging" movement by Indian women has led to some restrictions on forest cutting in the past several decades.*

PAKISTAN
NEPAL
BHUTAN
INDIA
BANGLADESH
Arabian Sea
Narmada River
20°N

Narmada River. *A proposed dam that would irrigate large areas of Gujarat state has faced strong local and international opposition due to negative social and environmental consequences, specifically the displacement of local farmers and loss of wildlife habitat.*

Bay of Bengal

Ganges Delta. *Sedimentation brought down from the Himalayas has created a vast low-lying delta area that is now densely settled by rice farmers. However, river flooding and storm surge from oceanic cyclones (hurricanes) cause devastation and high loss of life each year.*

0 200 400 Miles
0 200 400 Kilometers
Lakshadweep (INDIA)
Andaman Islands (INDIA)
Andaman Sea
10°N
70°E
SRI LANKA
MALDIVES
INDIAN OCEAN
80°E
90°E
Nicobar Islands (INDIA)

Forest areas
Extensive Deforestation
Desertification
Coastal pollution
Salinization

Figure 12.3 Environmental Issues in South Asia
As might be expected in a highly diverse and densely populated region, South Asia suffers from a number of environmental problems. These range from salinization of irrigated lands in the dry lands of Pakistan and western India to groundwater pollution from Green Revolution fertilizers and pesticides. Additionally, deforestation and erosion are widespread in upland areas.

Figure 12.4 Flooding in Bangladesh
Devastating floods are common in the low-lying delta lands of Bangladesh. Heavy rains come with the southwest monsoon, especially to the Himalayas, and powerful cyclones often develop over the Bay of Bengal. *(Baldev/Corbis/Sygma)*

and coastal plains of India, for example, were largely deforested thousands of years ago to make room for agriculture. Elsewhere, forests were cleared more gradually for agricultural, urban, and industrial expansion. Railroad construction in the nineteenth century was especially destructive. More recently, hill slopes in the Himalayas and elsewhere have been logged for commercial purposes. Extensive forests still can be found, however, in the far northern, southwestern, and east-central areas of South Asia.

As a result of deforestation, many villages of South Asia suffer from a shortage of fuelwood for household cooking, forcing people to burn dung cakes from cattle. While this low-grade fuel provides adequate heat, it also diverts nutrients that could be used as fertilizers to household fires. Where wood is available, collecting it may involve many hours of female labor because the remaining sources of wood are often far from the villages. In many areas, extensive eucalyptus stands have been planted to supply fuelwood and timber. Unfortunately, these Australian trees support very little wildlife, and do nothing to replenish the region's declining biodiversity.

Wildlife: Extinction and Protection Although the overall environmental situation in South Asia might be grim, wildlife protection inspires some optimism. The region has, for example, managed to retain a diverse assemblage of wildlife despite population pressure and intense poverty. The only remaining Asiatic lions live in India's Gujarat state, and even Bangladesh has retained a viable population of tigers in the Sundarbans, the mangrove forests of the Ganges Delta (Figure 12.5). Wild elephants still roam several large reserves in India, Sri Lanka, and Nepal. This protection of wildlife in India far exceeds that in most other parts of Asia. Due to the actions of the conservation program Project Tiger, for example, India now has 28 official tiger reserves.

Although the current state of wildlife protection may be positive, the future is uncertain as pressure mounts to convert wildlands to farmlands. The best remaining zone of extensive wildlife habitat is in India's far northeast, an area subjected to rapid immigration and political unrest. Moreover, wild animals, particularly tigers and elephants, threaten crops, livestock, and even people living near the reserves. When a rogue elephant herd ruins a crop or a tiger kills livestock, government agents are usually forced to destroy the animal. As a result, several key species now seem to be in decline, although conservation groups struggle valiantly on their behalf.

The Four Subregions of South Asia

South Asia forms a distinct landmass separated from the rest of the Eurasian continent by a series of sweeping mountain ranges, including the Himalayas—the highest in the world. For this reason it is often called the Indian **subcontinent**, in reference to its largest country. To better understand environmental conditions in this vast region, the Indian subcontinent can be broken down into four physical subregions, starting with the high mountain ranges of its northern fringe and extending to the tropical islands of the far south. Lying south of the mountains are the extensive river lowlands that form the heartland of both India and Pakistan. Between river lowlands and the island countries is the vast area of peninsular India, extending more than 1,000 miles (1,600 kilometers) from north to south (Figure 12.6).

Mountains of the North South Asia's northern rim of mountains is dominated by the great Himalayan Range, forming the northern borders of India, Nepal, and Bhutan. These mountains are linked to the equally high Karakoram Range to the west, extending through northern Pakistan. More than two dozen peaks exceed 25,000 feet (7,620 meters), including the world's highest mountain, Everest, on the Nepal–China (Tibet) border at 29,028 feet (8,848 meters). To the east are the lower Arakan Yoma Mountains, forming the border between India and Myanmar (Burma) and separating South Asia from Southeast Asia.

These formidable mountain ranges were produced by tectonic activity caused by peninsular India pushing northward into the larger Eurasia Continental Plate; as a result of the collision between these two tectonic plates, great mountain ranges were folded and upthrust. The entire region is seismically active, putting all of northern South Asia in serious earthquake danger (see Figure 12.1). A massive earthquake in the Pakistani-controlled section of Kashmir on October 8, 2005, for example, resulted in roughly 100,000 deaths and left more than 3 million people homeless.

Indus-Ganges-Brahmaputra Lowlands South of the northern mountains lie large lowlands created by three major river systems that have carried sediments eroded off the mountains through millions of years, building vast alluvial plains of fertile and easily farmed soils. These river lowlands are densely settled and constitute the population core areas of Pakistan, India, and Bangladesh.

Figure 12.5 Indian Wildlife
The only remaining Asiatic lions live in Gir National Park in the Indian state of Gujarat. Despite its large population and extreme poverty, India has managed to protect a significant amount of natural habitat. *(Philippe Caron/Sygma/Corbis)*

Of these three rivers, the Indus is the longest, covering more than 1,800 miles (2,880 kilometers) as it flows southward from the Himalayas through Pakistan to the Arabian Sea, providing much-needed irrigation waters to the desert areas in the southern portions of that country. The broad band of cultivated land in the desert zone of central and southern Pakistan that is made possible by the Indus is clearly visible on satellite images of the area (Figure 12.6).

Even more densely settled is the vast lowland of the Ganges, which, after flowing out of the Himalayas, travels southeasterly some 1,500 miles (2,400 kilometers) to empty into the Bay of Bengal. The Ganges not only provides the fertile alluvial soil that has made northern India one of the world's most densely settled areas but has also long served as a transportation corridor. Given the central role of this important river in South Asia's past and present, it is understandable that Hindus consider the Ganges sacred.

Although this large South Asian lowland often is referred to as the Indus–Ganges Plain, this term neglects the Brahmaputra River. This river rises on the Tibetan Plateau and flows easterly, then southward and westerly over 1,700 miles (2,720 kilometers), joining the Ganges in central Bangladesh and spreading out over the vast delta, the largest in the world. Unlike the sparsely populated Indus Delta in Pakistan, the Ganges–Brahmaputra Delta is very densely settled, containing more than 3,000 people per square mile (1,200 per square kilometer).

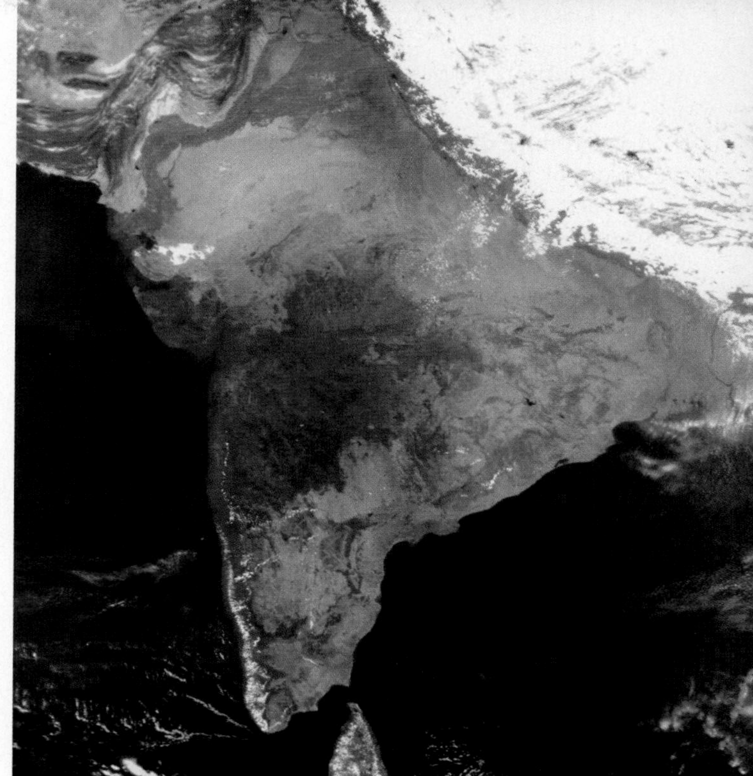

Figure 12.6 South Asia from Space
The four physical subregions of South Asia are clearly seen in this satellite photograph, from the snow-clad Himalayas in the north to the islands of the south. The Deccan Plateau is dark, fringed by white clouds as moist air is lifted over the uplands of the Western Ghats. *(Earth Satellite Corporation/Science Photo Library/Photo Researchers, Inc.)*

Peninsular India Jutting southward is the familiar shape of peninsular India, made up primarily of the Deccan Plateau, which is bordered on each coast by narrow coastal plains backed by elongated north–south mountain ranges. On the west are the higher Western Ghats, which are generally about 5,000 feet (1,524 meters) but reach higher than 8,000 feet (2,438 meters) near the peninsula's southern tip; to the east, the Eastern Ghats are lower and discontinuous, thus forming less of a transportation barrier to the broader eastern coastal plain of peninsular India. On both coastal plains, fertile soils and an adequate water supply support population densities comparable to the Ganges lowland to the north.

Soils are poor or average over much of the Deccan, but in Maharashtra state basaltic lava flows have produced fertile black soils. A reliable water supply for agriculture is a major problem in most areas. Much of the western plateau lies in the rain shadow of the Western Ghats, giving it a somewhat dry climate. Small reservoirs or tanks have been the traditional method for collecting monsoon rainfall for use during the dry season. More recently, deep wells and powerful pumps have mined groundwater to support irrigated crops and village water needs.

Partly because of the overuse of these aquifers, the Indian government is building a series of large dams to allow more extensive irrigation. These projects, however, are highly controversial, largely because the reservoirs created by the dams will result in the displacement of hundreds of thousands of rural residents. A case in point is the Narmada Dam project, which involves the construction of 30 major dams along the Narmada River. One of these dams, Sardar Sarovar, could displace more than 100,000 people. Local residents and activists throughout India have joined forces in opposition, but farmers in neighboring Gujarat, the main beneficiaries, strongly support the project. In 2006, official permission was given to raise the height of this massive dam from 361 feet (110 meters) to 400 feet (122 meters).

The Southern Islands At the southern tip of peninsular India lies the island country of Sri Lanka, which is almost linked to India by a series of small islands called Adam's Bridge. Sri Lanka is ringed by extensive coastal plains and low hills, but mountains reaching more than 8,000 feet (2,438 meters) occupy the southern interior, providing a cool, moist climate. Because the main winds arrive from the southwest, that portion of the island is much wetter than the rain-shadow area of the north and east.

Forming a separate country are the Maldives, a chain of more than 1,200 islands stretching south to the equator some 400 miles (640 kilometers) off the southwestern tip of India. The combined land area of these islands is only about 116 square miles (290 square kilometers), and only a quarter of the islands are actually inhabited. The islands of the Maldives are flat, low coral atolls, with a maximum elevation of just over 6 feet (2 meters) above sea level.

South Asia's Monsoon Climates

The dominant climatic factor for most of South Asia is the **monsoon**, the distinct seasonal change of wind direction, which corresponds to wet and dry periods. Most of South

Asia has three distinct seasons. First is the warm and rainy season of the southwest monsoon from June through October. This is followed by a relatively cool and dry season, extending from November until February, when the dominant winds are from the northeast. Only a few areas in far northwestern and southeastern South Asia get substantial rainfall during this otherwise dry time. The third season is the hot period from March to late May, which builds up with great heat and humidity until the monsoon's much anticipated and rather sudden "burst" in early June.

This monsoon pattern is caused by large-scale meteorological processes that affect much of Asia (Figure 12.7). During the Northern Hemisphere's winter, a large high-pressure system forms over the cold Asian landmass. Cold, dry winds flow outward from the interior of this high-pressure cell, over the Himalayas and down across South Asia. As winter turns to spring, these winds diminish, resulting in the hot, dry season of March through May. Eventually this buildup of heat over South and Southwest Asia produces a large thermal low-pressure cell. By early June this low-pressure cell is strong enough to draw in warm, moist air from the Indian Ocean.

Orographic rainfall results from the uplifting and cooling of moist monsoon winds over the Western Ghats. As a result, some stations receive more than 200 inches (508 centimeters) of rain during the four-month wet season. On the climate map (Figure 12.8), these are the areas of Am, or tropical monsoon, climate. Inland, however, a strong rain-shadow effect dramatically reduces rainfall on the Deccan Plateau. Farther north, as the monsoon winds are forced up and over the mountains, copious amounts of rainfall are characteristic. Cherrapunji, India, at 4,000 feet (1,220 meters), is a strong contender for the title of world's wettest place, with an average rainfall of 451 inches (1,128 centimeters).

Not all of South Asia, however, receives high rainfall totals from the southwest monsoon. In Pakistan and the Indian state of Rajasthan, precipitation is low and variable enough to result in steppe and desert climates. In Karachi, the annual total is less than 10 inches (25 centimeters).

Regardless of whether rainfall is heavy or light, the monsoon rhythm affects all of South Asia in many different ways, from the delivery of much-needed water for crops and villages, to the mood of millions of people as they eagerly await relief from oppressive heat (Figure 12.9). Some years the monsoon delivers its promise with abundant moisture; in other years, though, it brings scant rainfall, resulting in crop failure and hardship. In the near future, the monsoonal patterns may change due to global warming.

Global Warming and South Asia

Many areas of South Asia are particularly vulnerable to the effects of global warming. Even a minor rise in sea level will inundate large areas of the Ganges–Brahmaputra Delta in Bangladesh. Already, over 18,500 acres (7,500 hectares) of swampland in the Sunderbans have been submerged. Several small islands in this area have recently

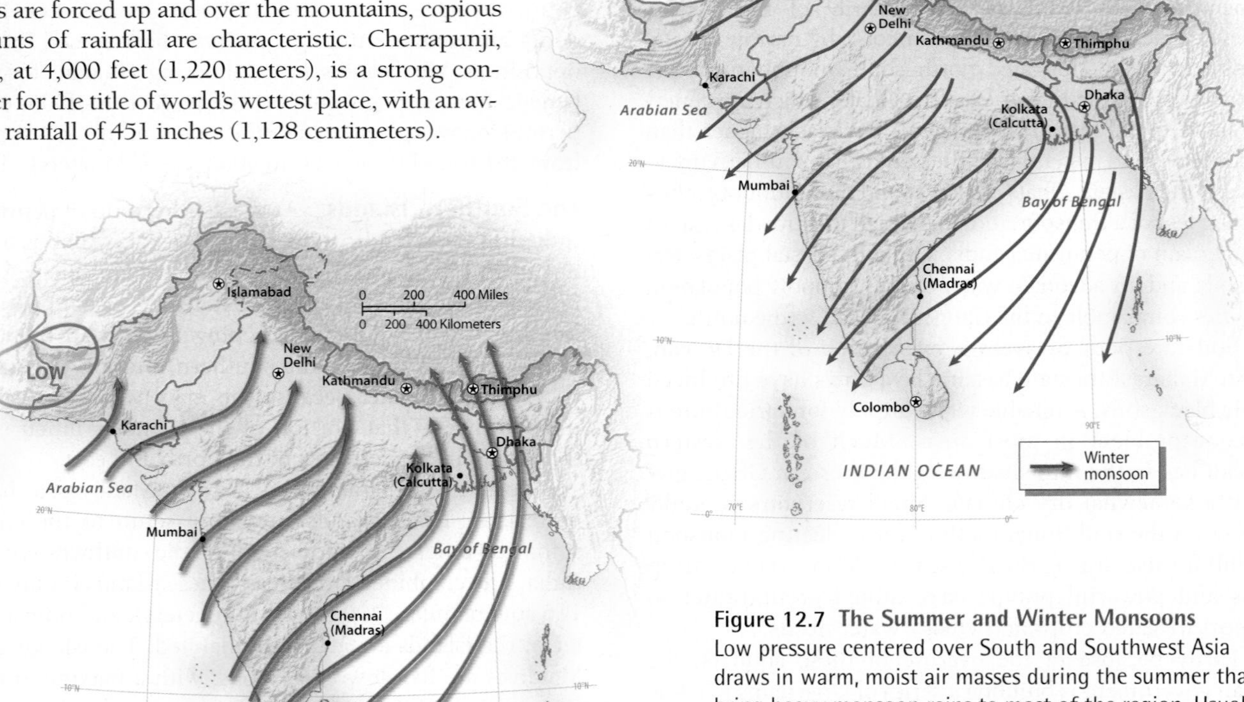

Figure 12.7 **The Summer and Winter Monsoons** Low pressure centered over South and Southwest Asia draws in warm, moist air masses during the summer that bring heavy monsoon rains to most of the region. Usually these rains begin in June and last for several months. During the winter, high pressure forms over northern Asia. As a result, winds are reversed from those of the summer. During this season, only a few coastal locations along India's east coast and in eastern Sri Lanka receive substantial rain.

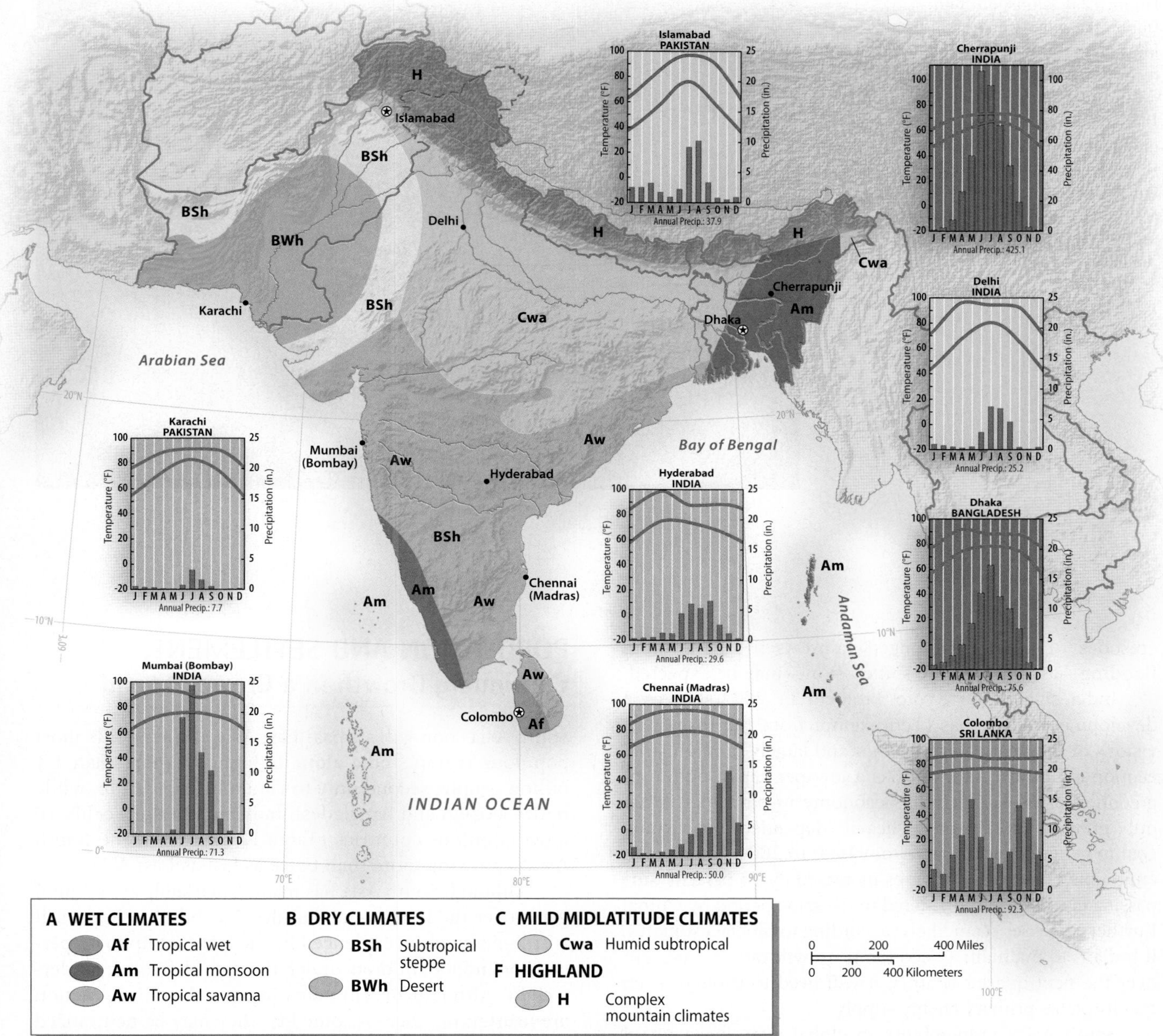

Figure 12.8 Climates of South Asia
Except for the extensive Himalayas, South Asia is dominated by tropical and subtropical climates. Many of these climates show a distinct summer rainfall season that is associated with the southwest monsoon. The climographs for Mumbai (Bombay) and Delhi are excellent illustrations. However, the climographs for east-coast locations such as Madras, India, and Colombo, Sri Lanka, show how some locations also receive rains from the northeast monsoon of the winter.

disappeared, due to a combination of sea level rise and subsidence. A 2007 report from the India government suggests that up to 7 million people could be displaced from coastal areas by the end of the century due to a predicted 3.3-foot (1-meter) rise in sea level. If the most severe sea level forecasts come to pass, the atoll nation of the Maldives would simply vanish beneath the waves.

South Asian agriculture is likely to suffer from a number of problems linked to global climate change. Most Himalayan glaciers are rapidly retreating, threatening the dry-season water supplies of the Indus–Ganges Plain, an area that already suffers from overuse of groundwater resources. Increased winter temperatures of up to 6.4°F (3°C) could destroy the vital wheat crop of Pakistan and northwestern India, undermining the food security of both countries. In parts of South Asia, however, global warming could result in increased rainfall due to an intensification of the summer monsoon. Unfortunately, such a new precipitation regime will likely be characterized by more intense cloudbursts coupled with fewer

Figure 12.9 Monsoon Rain During the summer monsoon, some Indian cities such as Mumbai (Bombay) receive more than 70 inches (178 centimeters) of rain in just three months. These daily torrents cause floods, power outages, and daily inconvenience. However, the monsoon rains are crucial to India's agriculture. If the rains are late or abnormally weak, crop failure often results. *(Sharad J. Devare/Dinodia Picture Agency)*

episodes of gentle, prolonged rain. As a result, more flooding—as well as more soil erosion—may be expected.

India signed the Kyoto Protocol in 2002, but as a developing country has been exempt from the main provisions of the treaty. With its poor and largely nonindustrial economies, South Asia still has a low per capita output of greenhouse gases. But India's economy in particular is not only growing rapidly, but is heavily dependent on burning coal to generate electricity. From 1990 to 2004, India's total emissions of greenhouse gases increased by 55 percent, surpassing even the gain registered in the same period by China. Further increases seem likely; according to official estimates, if India is to maintain an economic growth rate of 8 percent over the next quarter century, it will need to triple or even quadruple its primary energy supply.

South Asia's vulnerability to global warming coupled with its rapidly growing emissions of greenhouse gases have prompted numerous calls to action. Experts were heartened in early 2007, when India's government signed an agreement with Japan aimed at encouraging Japanese companies to invest in India's energy sector and to transfer energy-saving technology. Many South Asian officials, however, argue that the responsibility for addressing global climate change lies with the already industrialized countries, and that local attempts to reduce emissions could undermine economic growth. Several Indian scientists have gone further, arguing that if and when global warming affects South Asia, its consequences could be as much positive as negative. Overall, however, South Asia's scientific and public policy communities are very concerned about the effects of global climate change, especially since rises in sea level and changes in precipitation regimes will have such potentially catastrophic impacts on populations in the region.

POPULATION AND SETTLEMENT: Continuing Growth in a Crowded Land

South Asia soon will surpass East Asia as the world's most populous region. India alone is home to more than 1.1 billion people, second only to China in population, while both Pakistan and Bangladesh rank among the world's 10 most populous countries (Table 12.1). At present, South Asia is growing more than twice as fast as East Asia.

Although South Asia has made remarkable agricultural gains over the last several decades, its ability to feed itself is still marginal, as evidenced by the fact that up to 47 percent of India's children under the age of five are underweight. Although recent gains in agricultural production are heartening, there is some fear they may be neutralized by continuing population growth.

The Geography of Family Planning

While all South Asian countries have family planning programs, the commitment to these policies—along with the results—varies widely from place to place. A brief overview of population policies in the three largest South Asian countries provides important background for further discussion.

India Widespread concern over India's population growth began in the 1960s. To some extent, the measures taken over the last 40 years have been successful; the total fertility rate (TFR) dropped from 6 in the 1950s to the current rate of 2.9. Fertility rates vary widely within India, from below 2.0 in the states of Goa, Kerala, and Tamil Nadu to a problematic high of 4.4 in Uttar Pradesh, which, with more than 166 million people, would be the sixth largest country in the world were

TABLE 12.1 Population Indicators

Country	Population (Millions, 2007)	Population Density (per Square Kilometer)	Total Fertility Rate	Percent Urban	Percent <15[a]	Percent >65[b]	Net Migration (per 1,000, 2000–05)
Bangladesh	149.0	1,035	3.0	23	33	4	−0.5
Bhutan	0.9	19	2.9	31	33	5	0.0
India	1,131.9	344	2.9	28	33	5	−0.3
Maldives	0.3	1,020	2.8	27	34	4	0.0
Nepal	27.8	189	3.1	14	41	4	−0.8
Pakistan	169.3	213	4.1	34	40	4	−2.4
Sri Lanka	20.1	306	2.0	15	27	6	−1.6

[a]*Percentage of population younger than 15.*
[b]*Percentage of population older than 65.*
Sources: Population Reference Bureau, World Population Data Sheet, 2007; *net migration rate data from* UN International Migration, 2006.

it independent. Because of the high fertility of Uttar Pradesh and its neighboring states in northern India, much of the progress made in other areas of the country has been effectively canceled. A strong relationship is evident between women's education and family planning; where women's literacy has increased most dramatically, fertility levels have rapidly declined (see "Geographic Tools: Survey Work on Contraceptive Use in West Bengal").

As is the case in China, a distinct cultural preference for male children is found in most of South Asia, a tradition that further complicates family planning. Where allowed, sex determination clinics provide couples with information about the sex of the fetus, resulting in a higher rate of abortion for female fetuses. Consequently, especially in northern India, a lower TFR is accompanied by a higher ratio of male-to-female infants. In southern South Asia (particularly Sri Lanka and the Indian state of Kerala), where women have a much higher social position, sex ratios are balanced and birthrates are much lower. Southern India, at least, is now undergoing demographic stabilization.

Pakistan and Bangladesh Pakistan, with a population of more than 169 million, appears to have an ambivalent attitude toward family planning. Although the government's official position is that the birthrate is excessive, the country still lacks an effective, coordinated family planning program. As a result, the TFR remains very high at 4.1. In 2005, Pakistan's government began urging Muslim scholars to promote birth control, but this program has met much resistance, especially in the more conservative, rural portions of the country.

Densely settled Bangladesh, with a population about half that of the United States packed into an area smaller than the state of Wisconsin, has made significant strides in family planning. As recently as 1975, the TFR was 6.3, but it had dropped to 3.0 by 2006. The success of family planning can be attributed to strong support from the Bangladesh government for advertising through radio and billboards. Also important are the more than 35,000 women fieldworkers who take information about family planning into every village in the country (Figure 12.10). Village outreach is vitally important because Bangladesh—like the rest of South Asia—remains a predominantly rural society.

Migration and the Settlement Landscape

The most densely settled areas of South Asia still coincide with areas of fertile soils and dependable water supplies (Figure 12.11). The largest rural populations are found in the core area of the Ganges and Indus river valleys and on the coastal plains of India. Settlement is less dense on the Deccan Plateau and is relatively sparse in the arid lands of the northwest. While most of South Asia's northern mountains are too rugged and high to support dense human settlement, there are major population clusters in the Katmandu Valley of Nepal, situated at 4,400 feet (1,341 meters), and the Valley, or Vale, of Kashmir in northern India, at 5,200 feet (1,585 meters).

As is true in most other parts of the world, South Asians have migrated for hundreds of years from poor and

Figure 12.10 Family Planning in Bangladesh
Bangladesh has been one of the most successful nations in South Asia in reducing its fertility rate through family planning. Many women in Bangladesh use oral contraceptives. This photo shows a woman receiving a contraceptive implant. *(Peter Barker/Panos Pictures)*

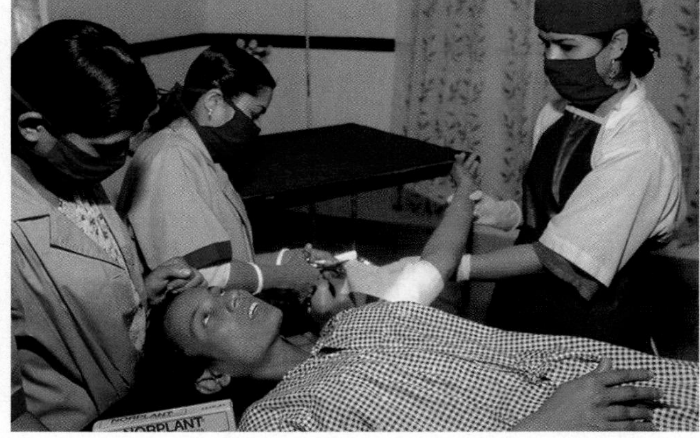

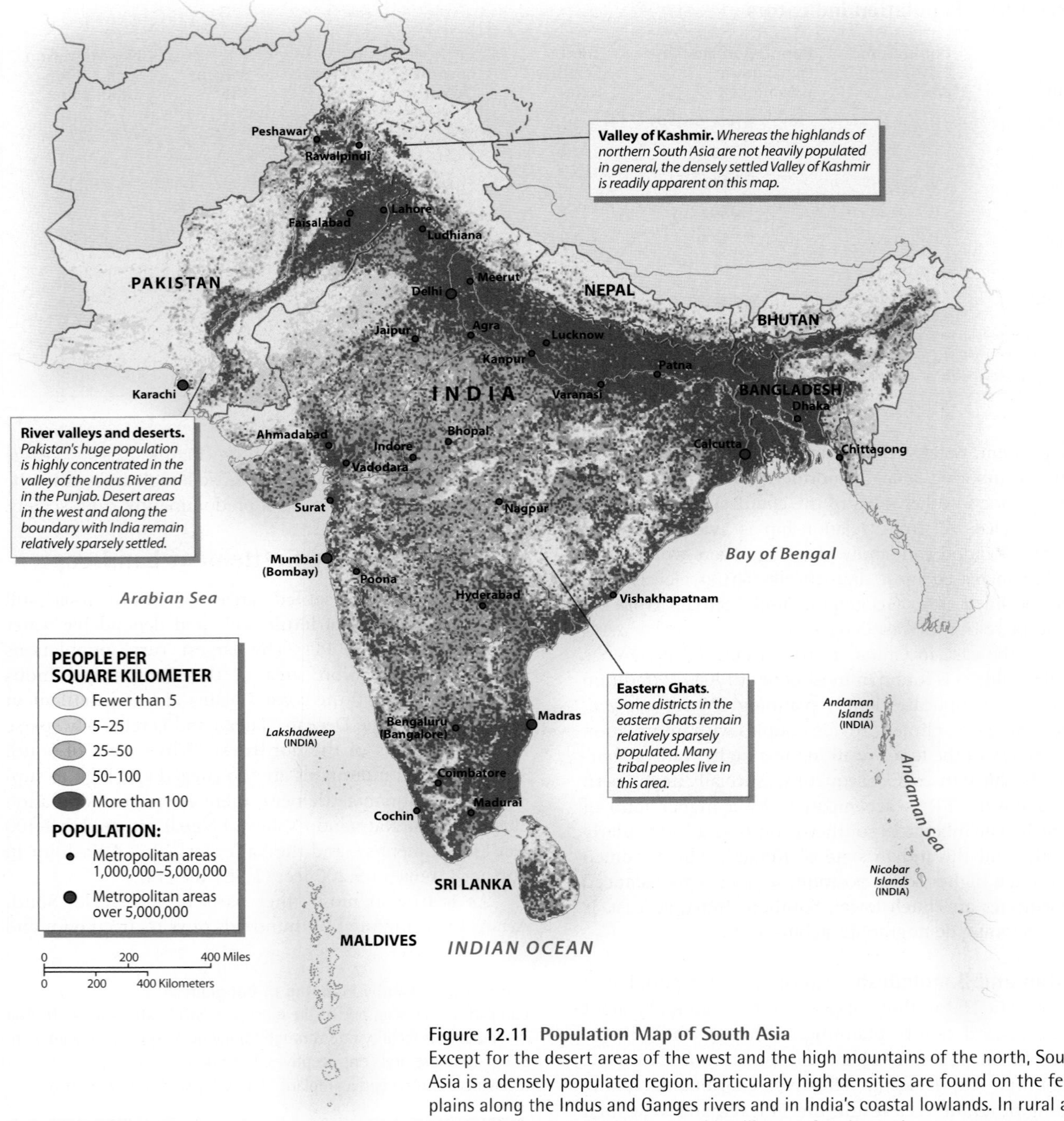

Figure 12.11 Population Map of South Asia
Except for the desert areas of the west and the high mountains of the north, South Asia is a densely populated region. Particularly high densities are found on the fertile plains along the Indus and Ganges rivers and in India's coastal lowlands. In rural areas, the population is typically clustered in villages, often located near water sources, such as streams, wells, canals, or small tanks that store water between monsoon rains.

densely populated areas to places that are either less densely populated or wealthier. This process is ubiquitous in South Asia today, but several areas stand out as zones of intensive out-migration: Bangladesh, the northern Indian states of Bihar, Punjab, and Rajasthan, and the northern portion of India's Andhra Pradesh (Figure 12.12). Migrants from Bangladesh are settling in large numbers in rural portions of adjacent Indian states, exacerbating ethnic and religious tensions. In Nepal, migrants have been moving over the past 50 years from crowded mountain valleys to formerly malaria-infested lowlands along the Indian border, which has also generated local ethnic problems. Sometimes migrants are forced out by war; sizable streams of people from northern Sri Lanka and from Kashmir have recently sought security away from their battle-scarred homelands.

South Asia is one of the least urbanized regions in the world, with less than 35 percent of its huge population living in settlements classified as cities. The majority of its people live in compact rural villages and small towns. Still, one of the important themes in South Asia's settlement and population geography is the rapid migration from villages to large cities, often resulting as much from desperate conditions in the countryside as from the attraction of employment in the city. A primary cause of movement to cities is changes in agriculture: the increased mechanization with resulting unemployment; the higher costs of farming modern crops; the expansion of large commercial farms at the expense of subsistence farming; and the environmental deterioration of over-used lands. The agricultural situation in many parts of India has become so desperate that the country is experiencing an epidemic of farmer suicides. In 2006, 1,447 farmers killed themselves in the state of Maharashtra alone.

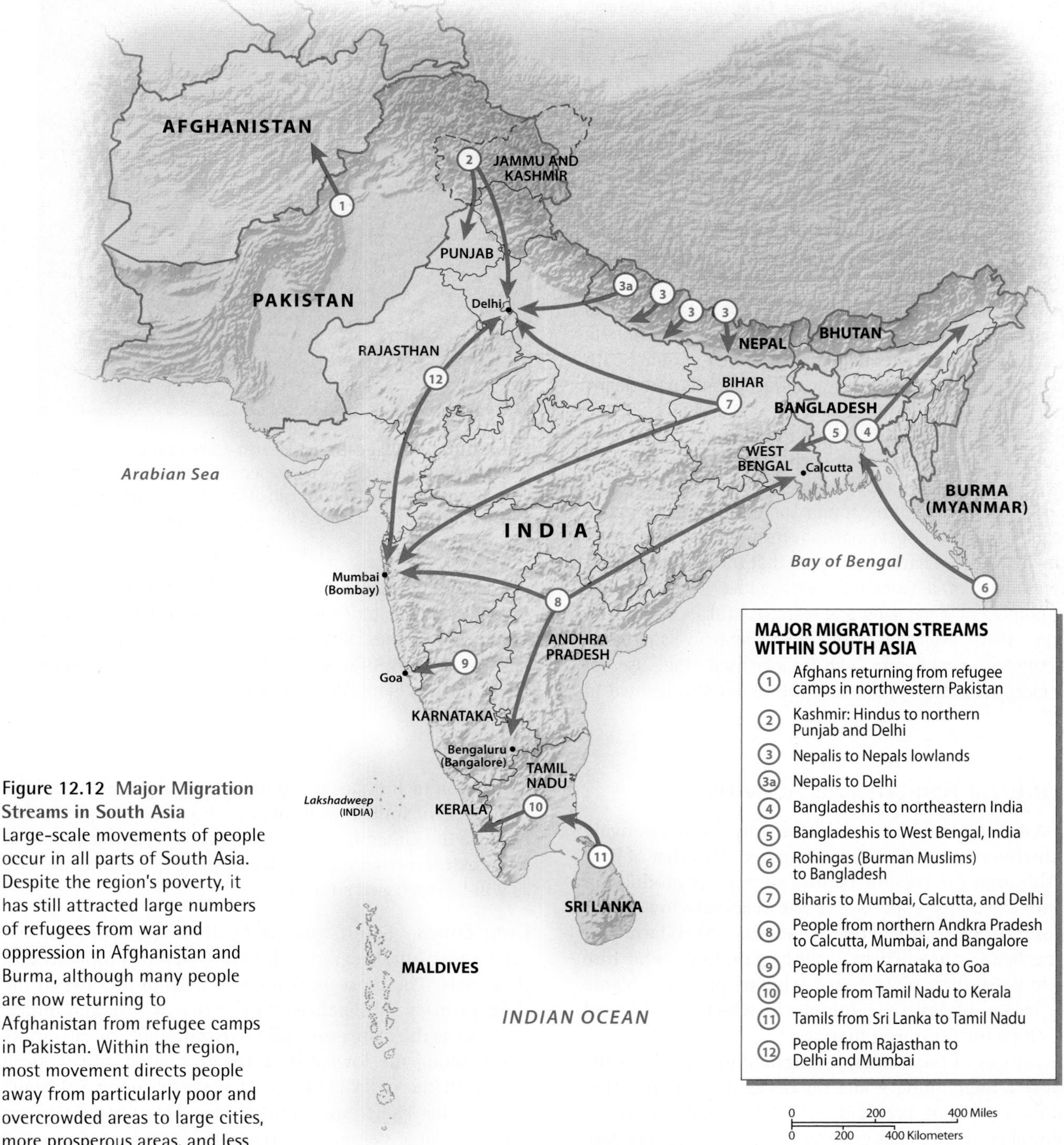

Figure 12.12 Major Migration Streams in South Asia Large-scale movements of people occur in all parts of South Asia. Despite the region's poverty, it has still attracted large numbers of refugees from war and oppression in Afghanistan and Burma, although many people are now returning to Afghanistan from refugee camps in Pakistan. Within the region, most movement directs people away from particularly poor and overcrowded areas to large cities, more prosperous areas, and less densely populated districts.

GEOGRAPHIC TOOLS

Survey Work on Contraceptive Use in West Bengal

Geographers, like other social scientists, often make extensive use of surveys in order to acquire raw data. Standardized surveys, in which sizable numbers of local people are asked identical and relatively easily answered questions, allow scholars to build quantitative databases. Statistical tests can then be conducted to determine correlations and to find relationships. Geographers often use such surveys to identify spatial patterns.

Geographers also sometimes follow anthropologists in using informal, open-ended interviews with select people who live in the area under investigation. Such a technique can give a deeper understanding of local attitudes and cultural perceptions and practices. They do not allow, however, for the kinds of statistical testing obtainable with standardized survey methods. As a result, geographers sometimes use both methods in order to obtain both quantitative and qualitative information.

One geographer who has effectively used both methods is Elizabeth Chacko, of the George Washington University, in her study of women's use of contraception in four villages in West Bengal, India. Determining why women in rural South Asia chose to use, or not to use, contraception is particularly important considering the demographic strains faced by the region. But it is a difficult subject to investigate, as women are often reluctant to discuss such issues, particularly with outsiders. Dr. Chacko therefore decided to use both the quantitative method of the formal survey and the qualitative method of in-depth, personal conversations and interviews.

In the end, Dr. Chacko discovered that the most important determinants of a woman's contraceptive decisions were age, number of living sons, and religious affiliation. She also found that availability of permanent, village-based government health-care clinics increased the use of contraceptives. Observations and interviews also pointed to the need to provide low-cost, culturally acceptable methods for achieving greater contraceptive usage in the community (Figure 12.2.1). These findings should be of great help to local health-care workers seeking both to empower rural women and to address India's demographic problems.

Figure 12.2.1 Village-Based Health-Care Clinics
A rural health-care worker poses behind her medical kit in rural West Bengal. Such local outreach efforts have resulted in a reduced birthrate in much of this area. *(Elizabeth Chacko)*

Agricultural Regions and Activities

South Asian agriculture has historically been relatively unproductive, especially when compared with that of East Asia. Although the reasons behind such poor production are complex, many experts cite the relatively low social status of most cultivators and the fact that much farmland has long been controlled by wealthy landlords who hire others to work their plots. Some scholars place the blame on the legacy of British colonialism, which emphasized export crops for European markets.

Regardless of the causes, low agricultural yields in the context of a huge, hungry, and rapidly growing population constitute a pressing problem. Since the 1970s, however, agricultural production has grown faster than the population, primarily because of the **Green Revolution**, agricultural cultivation techniques based on hybrid crop strains and the heavy use of industrial fertilizers and chemical pesticides. Because the Green Revolution also carries significant social and environmental costs, it has been highly controversial, as will be discussed shortly.

Crop Zones South Asia can be divided into several distinct agricultural regions, all with different problems and potentials. The most fundamental division is between the three primary subsistence crops of rice, wheat, and millet.

Rice is the main crop and foodstuff in the lower Ganges Valley, along the lowlands of India's eastern and western coasts, in the delta lands of Bangladesh, along Pakistan's lower Indus Valley, and in Sri Lanka. This distribution reflects the large volume of irrigation water needed to grow rice (Figure 12.13). The amount of rice grown in South Asia is impressive: India ranks behind only China in world rice production, and Bangladesh is the fourth largest producer.

Wheat is the principal crop of the northern Indus Valley and in the western half of India's Ganges Valley. South Asia's "breadbasket" is the northwestern Indian state of Punjab and adjacent areas in Pakistan. Here the Green Revolution has been particularly successful in increasing grain yields. In the less-fertile areas of central India, millet and sorghum are the main crops, along with root crops such as manioc. In general, wheat and rice are the preferred staples throughout South Asia, and it is generally poorer people who consume "rough" grains such as the various millets.

Many other crops are widely cultivated in South Asia, some commercially, others for local subsistence. Oil seeds, such as sesame and peanuts, for example, are grown in semiarid districts, while the humid southwestern state of Kerala (along with Sri Lanka) is noted for its coconut groves, spice gardens, and tea plantations. In both Pakistan and west central India, cotton is widely grown, while Bangladesh has long supplied most of the world's jute, a tough fiber used in the manufacture of rope.

Livestock Many if not most South Asians receive inadequate protein, and meat consumption is extremely low. This is partly a reflection of poverty, because meat is expensive to produce. In India, religion is equally important, as most Hindus are vegetarians. Despite this prohibition against eating meat, animal husbandry is vitally important throughout South Asia. India has the world's largest cattle population, partly because cattle are sacred in Hinduism but also because milk is one of South Asia's main sources of protein. While the cattle of India have traditionally yielded little milk, a so-called white revolution has increased dairy efficiency. As a result, India's milk production increased from roughly 30 million tons in 1980 to 87 million tons in 2004.

Figure 12.14 Green Revolution Farming
Because of "miracle" wheat strains that have increased yields in the Punjab area, this region has become the breadbasket of South Asia. India more than doubled its wheat production in the last 25 years and has moved from chronic food shortages to self-sufficiency. Increased production, however, has led to both social and environmental problems. *(Earl Kowall/Corbis)*

The Green Revolution The main reason South Asian agriculture has kept up with population growth is the Green Revolution, which originated during the 1960s in agricultural research stations established by international development agencies. One of the major problems researchers faced was the fact that higher yields could not be attained by simply fertilizing local seed strains, because the plants would simply grow taller and then fall to the ground before the grain could mature. The solution was to cross-breed new "dwarf" crop strains that would respond to heavy chemical fertilization by producing extra grain rather than longer stems.

By the 1970s, it was clear that these efforts had succeeded in reaching their initial goals. The more prosperous farmers of the Punjab quickly adopted the new "miracle wheat" varieties, solidifying the Punjab's position as the region's breadbasket (Figure 12.14). Green Revolution rice strains also were adopted in the more humid areas. As a result, South Asia was transformed from a region of chronic food deficiency to one of self-sufficiency. India more than doubled its annual grain production between 1970 and the mid-1990s, from 80 to 191 million tons.

Figure 12.13 Rice Cultivation
A large amount of irrigation water is needed to grow rice, as is apparent from this photo from Sri Lanka. Rice also is the main crop in the lower Ganges Valley and Delta, along the lower Indus River of Pakistan, and on India's coastal plains. *(Mahaux Photography/Getty Images Inc.—Image Bank)*

While the Green Revolution was clearly an agricultural success, many argue that it has been an ecological and social disaster. Serious environmental problems result from the chemical dependency of the new crop strains. Not only do they typically need large quantities of industrial fertilizer, which is both expensive and polluting, but they also require frequent pesticide applications because they often lack natural resistance to local plant diseases and insects.

Social problems have followed the Green Revolution. In many areas, particularly where wheat is cultivated, only the more prosperous farmers are able to afford the new seed strains, irrigation equipment, farm machinery, fertilizers, and pesticides necessary to support this new high-technology agriculture. As a result, poorer farmers often go deeply into debt and are then forced off the land when they fail to repay their loans.

Future Food Supply The Green Revolution has fed South Asia's expanding population over the past several decades, but whether it will be able to continue doing so in the near future remains unclear. Many of the crop improvements have seemingly exhausted their potential, and much of India's agricultural economy is currently in a state of crisis as farmers are unable to make ends meet.

Optimists believe, however, that South Asia's food production could be substantially increased. Improvements in highways and railroads, for example, would result in a decrease of wastage as well as higher profit margins for struggling farmers. Green Revolution techniques, moreover, might be profitably applied to secondary grain crops, and genetic engineering might provide another breakthrough, ushering in a second wave of increased crop yields. Most environmentalists, however, see serious dangers in this new technology.

Another option is expanded water delivery, because many fields remain unirrigated in South Asia's semiarid areas and even in the humid zone dry-season fallow is usually the norm. Irrigation, however, brings its own problems. In much of Pakistan and northwestern India, where irrigation has been practiced for generations, soil **salinization** (see Figure 12.3), or the buildup of salt in agricultural fields, is already a major constraint. Additionally, groundwater levels are falling in the Punjab, India's breadbasket, because double-cropping has pushed water use beyond the sustainable yield of the underlying aquifers.

Urban South Asia

Although South Asia is one of the least urbanized regions of the world (see Table 12.1), this does not mean that cities are a minor part of the settlement landscape; South Asia actually has some of the largest urban areas in the world. India alone lists more than 35 cities with populations greater than a million, most of which are growing rapidly. Mumbai's (Bombay) population is now roughly 13 million, with up to 25 million living in the city's metropolitan area, making it—by some measures—the most heavily populated urban agglomeration in the world, and South Asia's premier city (see "Cityscapes: Life on the Streets of Mumbai [Bombay]").

Because of this rapid growth, most South Asian cities have staggering problems with homelessness, poverty, congestion, water shortages, air pollution, and sewage disposal. Kolkata's (Calcutta) homeless are legendary, with perhaps half a million sleeping on the streets each night. In that city and others, sprawling squatter settlements, or bustees, mushroom in and around urban areas, providing temporary shelter for many urban migrants (Figure 12.15). A brief survey of the region's major cities gives clues to the problems and prospects of the region's urban areas.

Figure 12.15 Mumbai (Bombay) Hutments
Hundreds of thousands of people in Mumbai (Bombay) live in crude hutments, with no sanitary facilities, built on formerly busy sidewalks. Hutment construction is forbidden in many areas, but wherever it is allowed, sidewalks quickly disappear. *(Rob Crandall/www.robcrandall.com)*

Delhi Delhi, the sprawling capital of India, has almost 14 million people in its greater urban area. It consists of two contrasting landscapes expressing its past: Delhi (or old Delhi), a former Muslim capital, is a congested town of tight neighborhoods; New Delhi, in contrast, is a city of wide boulevards, monuments, parks, and expansive residential areas. It was born as a planned city when the British decided to move the colonial capital from Kolkata in 1911. Located here are the embassies, luxury hotels, government office buildings, and airline offices necessary for a vibrant political capital. South of the government area are more expensive residential areas, some of which are focused on expansive parks and ornamental gardens. Rapid growth, along with the government's inability to control auto and industrial emissions, has given Delhi some of the worst air pollution in the world. As a result, the government has attempted to shut down some of the city's small-scale industries, but local residents have banded together to protect their livelihoods.

Kolkata (Calcutta) To many, Kolkata is emblematic of the problems faced by rapidly growing cities in developing countries. Not only is homelessness widespread, but this city of some 14 million (in its metropolitan area) falls far short of supplying its residents with water, power, or sewage treatment. Electrical power is so inadequate that every hotel, restaurant, shop, and small business has to have some sort of standby power system. During the wet season, many streets are routinely flooded. No other Indian city has problems of this magnitude.

With continued rapid growth as migrants drain in from the countryside, a mixed Hindu-Muslim population that generates ethnic rivalry, a troubled economic base, and an overloaded infrastructure, Kolkata faces a problematic future. Yet it remains a culturally vibrant city noted for its fine educational institutions, theaters, and publishing firms.

Dhaka As the capital and major city of Bangladesh, Dhaka (also spelled *Dacca*) has experienced rapid growth due to migration from the surrounding countryside. In 1971, when the country gained independence from Pakistan, Dhaka had about 1 million inhabitants; today its metropolitan area numbers close to 11 million (Figure 12.16). Dhaka's economic vitality has increased since independence because it combines the administrative functions of government with the largest industrial concentration in Bangladesh. Cheap and abundant labor in Dhaka has made the city a global center for clothing, shoe, and sports equipment manufacturing. North American shoppers are readily reminded of this new role when looking through clothing tags at any department store.

Figure 12.16 Dhaka Street Scene
This busy capital of Bangladesh was a former colonial river port but is now a global center for the manufacturing of clothing, shoes, and sports equipment. As seen in this photo, a common mode of transportation is the human-powered pedicab. *(Dirk R. Frans/ The Hutchison Library)*

Karachi Karachi, a rapidly growing port city of approximately 13 million people, is Pakistan's largest urban area and its commercial core (Figure 12.17). It also served as the country's capital until 1963, when the new city of Islamabad was created in the northeast. Karachi, however, has suffered little from the exodus of government functions; it is the most cosmopolitan city in Pakistan, with its checkerboard street pattern lined with high-rise buildings, hotels, banks, and travel agencies. In many ways, its landscape conveys the sense that Karachi is a model metropolis for a developing country.

Karachi, however, suffers from serious political and ethnic tensions that have turned parts of the city into armed camps if not battlegrounds. During the worst of the violence in 1995, more than 200 people were killed on the city's streets each month, and army bunkers were lodged at major urban crossroads. Ethnic conflicts have generally pitted Sindis, the region's indigenous inhabitants, and the Muhajirs, Muslim refugees from India who settled in and around the city after independence in 1947. More recently,

Figure 12.17 Karachi Landscape
Karachi, Pakistan's largest city and main port, is noted for both its economic power and its ethnic violence. *(Peter/Georgina Bowater/ Creative Eye/MIRA.com)*

Bharatiya Janata Party (BJP), leading to widespread agitation against the country's Muslim minority in the mid-1990s. In several high-profile instances, Hindu mobs demolished Muslim mosques that were allegedly built on the sites of ancient

Musharraf, of betraying Islam in supporting the United States. Large amounts of money from Saudi Arabia, moreover, have gone into the development of fundamentalist religious schools that contribute to the growth of extremism.

Figure 12.18 Destruction of the Ayodhya Mosque
A group of Hindu nationalists are seen listening to speeches urging them to demolish the mosque at Ayodhya, allegedly built on the site of a former Hindu temple. The mosque was subsequently destroyed by Hindu fundamentalists. *(Sunil Malhotra/Reuters/ Getty Images Inc.—Hulton Archive Photos)*

Religious and political extremists in South Asia argue that the current struggles reflect deeply rooted historical divisions. Most scholars disagree, regarding religious violence as a more recent development. To weigh these contrasting claims, it is necessary to examine the cultural history of South Asian civilization.

Origins of South Asian Civilizations

Many scholars think that the roots of South Asian culture extend back to the Indus Valley civilization, which flourished more than 4,000 years ago in what is now Pakistan. This remarkable urban-oriented society vanished almost entirely in the second millennium BCE, after which the record grows dim. By 800 BCE, however, a new urban focus had emerged in the middle Ganges Valley (Figure 12.19). The social, religious, and intellectual customs associated with this civilization eventually spread throughout the lowlands of South Asia.

Hindu Civilization The religious complex of this early South Asian civilization was Hinduism, a tremendously complicated faith that incorporates diverse forms of worship and that lacks any standard system of beliefs (Figure 12.20). Certain deities are recognized, however, by all believers, as is the notion that these various gods are all manifestations of a single divine entity. All Hindus, moreover, share a common set of epic stories, usually written in the sacred language of Sanskrit. Hinduism is noted for its mystical tendencies, which have long inspired many men (and few women) to seek an ascetic lifestyle, renouncing property and sometimes all regular human relations. One of its hallmarks is a belief in the transmigration of souls from being to being through reincarnation, wherein the nature of one's acts in the physical world influences the courses of these future lives.

Scholars once confidently argued that Hinduism originated from the fusion of two distinct religious traditions: the mystical beliefs of the subcontinent's indigenous inhabitants (including the people of the ill-fated Indus Valley civilization) and the sky-god religion of the Indo-European

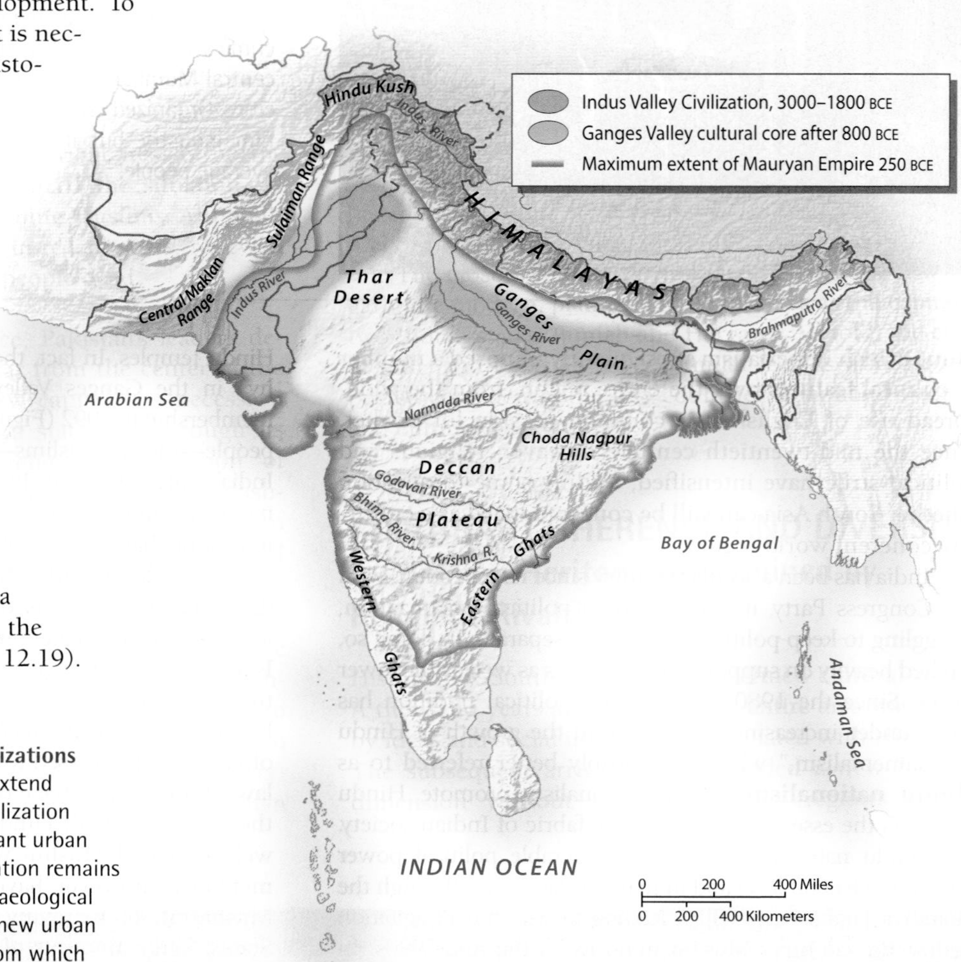

Figure 12.19 Early South Asian Civilizations
The roots of South Asian culture may extend back 5,000 years to an Indus Valley civilization based on irrigated agriculture and vibrant urban centers. What happened to that civilization remains a topic of conjecture, because the archaeological record grows dim by 1800 BCE. Later, a new urban focus emerged in the Ganges Valley, from which social, religious, and intellectual influences spread throughout lowland South Asia.

invaders who swept down into the region sometime in the second millennium BCE from Central Asia. Such a scenario also proved convenient for explaining India's **caste system**, the strict division of society into different hierarchically ranked hereditary groups. The elite invaders, according to this theory, wished to remain separate from the people they had defeated, resulting in an elaborate system of social division. Recent research, however, indicates that the caste system, like Hinduism itself, emerged through more gradual social and cultural processes.

Buddhism While a caste system of some sort seems to have existed in the early Ganges Valley civilization, it was soon challenged from within by Buddhism. Siddhartha Gautama, the Buddha, was born in 563 BCE in an elite caste. He rejected the life of wealth and power that was laid out before him, however, and sought instead to attain enlightenment, or mystical union with the cosmos. He preached that the path to such "nirvana" was open to all, regardless of social position. His followers eventually established Buddhism as a new religion. Buddhism spread through most of South Asia, becoming something of an official faith under the Mauryan Empire, which ruled much of the subcontinent in the third century BCE. Later centuries saw Buddhism expand through most of East, Southeast, and Central Asia.

But for all of its successes abroad, Buddhism never replaced Hinduism in India. It remained focused on monasteries rather than spreading through the wider society. Many Hindu priests, moreover, struggled against the new faith. One of their techniques was to embrace many of Buddhism's philosophical ideas and enfold them within the intellectual system of Hinduism. By 500 CE, Buddhism was on the retreat throughout South Asia, and within another 500 years it had virtually disappeared from the region. The only major exceptions were the island of Sri Lanka and the high Himalayas, both of which remain mostly Buddhist to this day.

Arrival of Islam The next major challenge to Hindu society—Islam—came from the outside. Arab armies conquered the lower Indus Valley around 700 CE but advanced no farther. Then around the year 1000, Turkish-speaking Muslims began to invade from Central Asia. At first they merely raided, but eventually they began to settle and rule on a permanent basis. By the 1300s, most of South Asia lay under Muslim power, although Hindu kingdoms persisted in southern India and in the arid lands of Rajasthan. Later, during the sixteenth and seventeenth centuries, the **Mughal Empire** (also spelled *Mogul*), the most powerful of the Muslim states, dominated much of the region from its power center in the upper Indus–Ganges Basin.

At first Muslims formed a small ruling elite, but over time increasing numbers of Hindus converted to the new religion, particularly those from lower castes who sought freedom from their rigid social order. Conversions were most pronounced in the northwest and northeast, and eventually the areas now known as Pakistan and Bangladesh became predominantly Muslim.

At first glance, Islam and Hinduism are strikingly divergent faiths. Islam is resolutely monotheistic, austere in its ceremonies, and spiritually egalitarian (all believers stand in the same relationship to God). Hinduism, by contrast, is polytheistic (at least on the surface), lavish in its rituals, and caste structured. Because of these profound differences, Hindu and Muslim communities in South Asia are sometimes viewed as utterly distinct, living in the same region, but not sharing the same culture or civilization. Increasingly, such a view is expressed in South Asia itself. Many residents of modern Pakistan stress the Islamic nature of their country and its complete separation from India.

Yet by overemphasizing the separation of Hindu and Muslim communities, one risks missing much of what is historically distinctive about South Asia. Until the twentieth century, Hindus and Muslims usually coexisted on amicable terms; the two faiths stood side by side for hundreds of years, during which time they came to influence each other in many ways. Moreover, aspects of caste organization have persisted among South Asian Muslims, and even among Indian Christians as well.

The Caste System

Although caste is one of the historically unifying features of South Asia, the system is not uniformly distributed across the subcontinent. It has never been significant in India's tribal areas; in modern Pakistan and Bangladesh its role is fading; and in the Buddhist society of Sri Lanka its influence has long been somewhat marginal. Even in India, caste is now deemphasized, especially among the more educated segments of society. But all said, caste continues to structure day-to-day social existence for hundreds of millions of Indians.

Caste is actually a rather clumsy term for denoting the complex social order of the Hindu world. The word itself, of Portuguese origin, combines two distinct local concepts: *varna* and *jati*. *Varna* refers to the ancient fourfold social hierarchy of the Hindu world, whereas *jati* refers to

Figure 12.20 Hindu Temple
Hindu temples dot India's landscape, taking many different forms, and are devoted to a wide range of deities. This is the Shantadurga temple in Ponda, Goa. *(Rob Crandall/www.robcrandall.com)*

the hundreds of local endogamous ("marrying within") groups that exist at each varna level (different jati groups are thus usually called *subcastes*). Jati, like varna, are hierarchically arranged, although the exact order of precedence is not so clear-cut.

It often has been argued that the essence of the caste system is the notion of social pollution. The lower one's position in the hierarchy, the more potentially polluting one's body supposedly is. Members of higher castes were traditionally not supposed to eat or drink with, or even use the same utensils as, members of lower castes.

The Main Caste Groups Three varna groups constitute the traditional elite of Hindu society. At the apex sit the Brahmins, members of the traditional priestly caste. Brahmins perform the high rituals of Hinduism, and they form the traditional intellectual elite of India. Almost all Brahmin groups value education highly, and today they are disproportionately represented among India's professional classes. Below the Brahmins in the traditional hierarchy are the Kshatriyas, members of the warrior or princely caste. In premodern India, this group actually had far more power and wealth than did the Brahmins; it was they who ruled the old Hindu kingdoms. Next stand the Vaishyas, members of the traditional merchant caste. In earlier centuries, a near monopolization of long-distance trade and moneylending in northern India gave many Vaishyas ample opportunities to accumulate wealth. The precepts of vegetarianism and nonviolence are particularly strong among certain merchant subcastes of western India. One prominent representative of this tradition was Mohandas Gandhi, the founder of modern India and one of the twentieth century's greatest religious and political leaders.

The majority of India's population fits into the fourth varna category, that of the Sudras. The Sudra caste is composed of an especially large array of subcastes (jati), most of which originally reflected occupational groupings. Most Sudra subcastes were traditionally associated with peasant farming, but others were based on craft occupations, including those of barbers, smiths, and potters.

While the Brahmins, Kshatriyas, Vaishyas, and Sudras form the basic fourfold scheme of caste society, another sizable group stands outside the varna system altogether. These are the so-called *untouchables*, or **dalits**, as they are now preferably called. Dalits were not traditionally allowed to enter Hindu temples. Such low-status positions were derived historically from "unclean" occupations, such as those of leather workers (who dispose of dead animals), scavengers, latrine cleaners, and swine herders. In the southern Indian state of Kerala until the nineteenth century, some groups were actually considered "unseeable"; these unfortunate persons had to hide in bushes and avoid walking on the main roads to protect members of higher castes from "visual pollution." Not surprisingly, many Indian dalits have converted to Islam, Christianity, and Buddhism in an attempt to escape from the caste system. Even so, they continue to suffer from discrimination.

The Changing Caste System The caste system is clearly in a state of flux in India today. Its original occupational structure has long been undermined by the necessities of a modern economy, and various social reforms have chipped away at the discrimination that it embodies. The dalit community itself has produced several notable national leaders who have waged partially successful political struggles. Owing to such efforts, the very concept of "untouchability" is now technically illegal in India, and in most Indian states a set number of university slots and governmental jobs are reserved for members of the lower castes and other groups that have suffered discrimination. Such "reservations," as they are called, can be quite controversial, as many people think that they unfairly penalize students and employees of higher caste background. As a result, in May 2006 some 100,000 people took to the streets of New Delhi to protest a proposal by India's government to increase the number of reserved positions in several of the country's most prestigious universities.

Contemporary Geographies of Religion

South Asia, as we have seen, has a predominantly Hindu heritage overlain by a substantial Muslim imprint. Such a picture fails, however, to capture the enormous diversity of modern religious expression in contemporary South Asia. The following discussion looks specifically at the geographical patterns of the region's main faiths (Figure 12.21).

Hinduism Less than 1 percent of the people of Pakistan are Hindu, and in Bangladesh and Sri Lanka Hinduism is a distinctly minority religion. Almost everywhere in India, however—and in Nepal, as well—Hinduism is very much the faith of the majority. In east-central India, more than 95 percent of the population is Hindu. Hinduism is itself a geographically complicated religion, with different aspects of faith varying between different parts of India. Even within a given Indian state, forms of worship often differ from region to region and from caste to caste.

Islam Islam may be considered a "minority" religion for the region as a whole, but such a designation obscures the tremendous importance of this religion in South Asia. With some 400 million members, the South Asian Muslim community is the largest in the world. Bangladesh and especially Pakistan are overwhelmingly Muslim. India's Islamic community, although constituting only some 15 percent of the country's population, is still roughly 150 million strong—a figure substantially larger than the total population of any country in the Muslim heartland of Southwest Asia and North Africa (Figure 12.22).

Muslims live in almost every part of India. However, as can be seen on Figure 12.21, they are concentrated in four main areas: in most of India's cities; in Kashmir, in the far north, particularly in the densely populated Vale of Kashmir (more than 80 percent of the population here follows Islam); in the upper and central Ganges plain, where Muslims constitute 15 to 20 percent of the population; and in the southwestern state of Kerala, which is approximately 25 percent Muslim. In northern India especially, Muslims tend to be poorer and less educated than their Hindu neighbors, leading some observers to call them the country's "new dalits."

Interestingly, the state of Kerala has a large Muslim minority even though it was one of the few parts of India that never experienced prolonged Muslim rule. Islam in Kerala is historically connected not to Central Asia, but rather to trade across the Arabian Sea. Kerala's Malabar Coast long supplied spices and other luxury products to Southwest Asia, enticing many Arabian traders to settle in

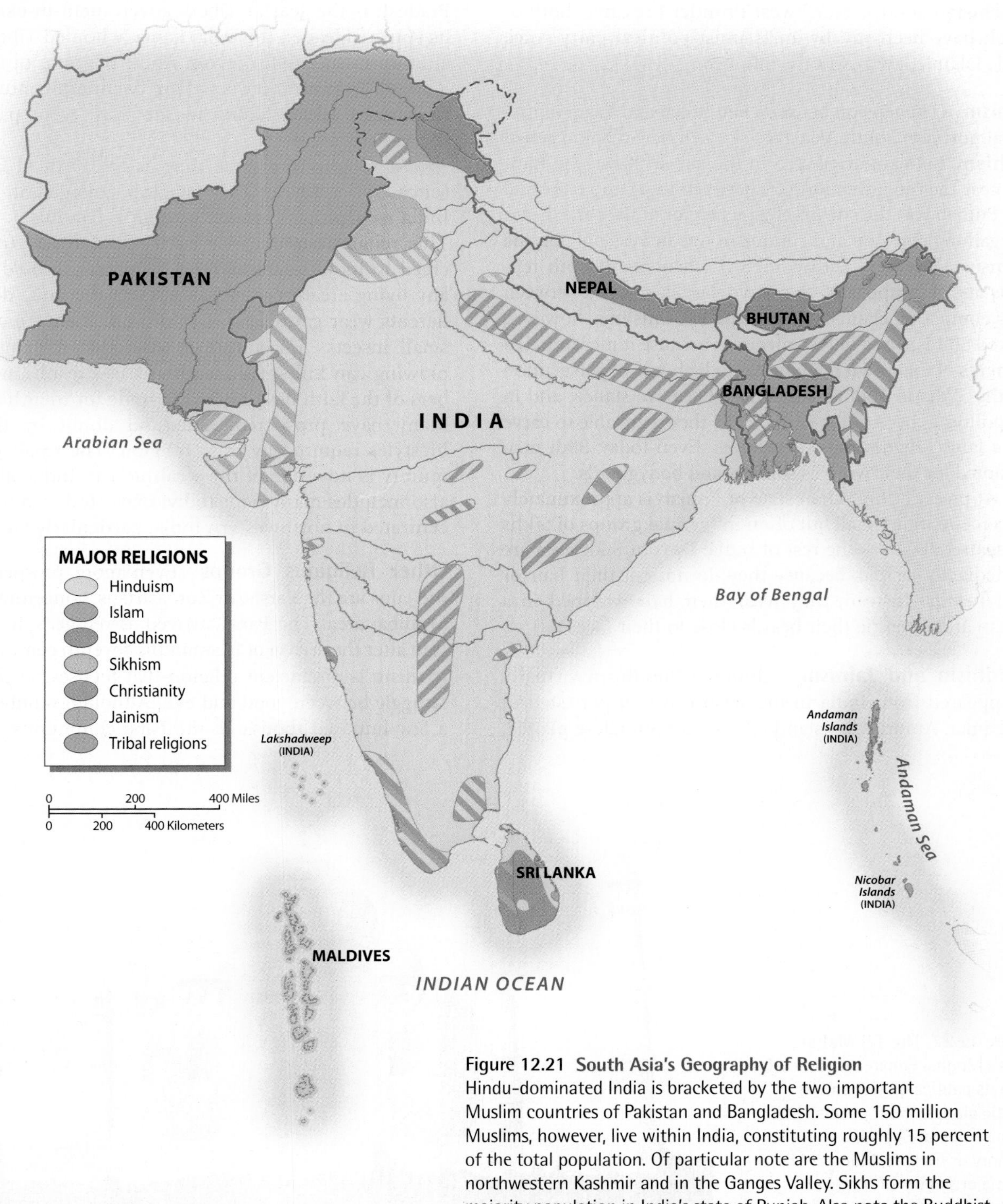

Figure 12.21 South Asia's Geography of Religion Hindu-dominated India is bracketed by the two important Muslim countries of Pakistan and Bangladesh. Some 150 million Muslims, however, live within India, constituting roughly 15 percent of the total population. Of particular note are the Muslims in northwestern Kashmir and in the Ganges Valley. Sikhs form the majority population in India's state of Punjab. Also note the Buddhist populations in Sri Lanka, Bhutan, and northern Nepal, the areas of tribal religion in the east, and the centers of Christianity in the southwest.

this part of India. Gradually many of Kerala's native inhabitants converted to the new religion as well. Owing to similar process, Sri Lanka is approximately 9 percent Muslim, and the Maldives is almost entirely Muslim.

As an overwhelmingly Muslim country, Pakistan officially calls itself an Islamic Republic. As such, Islamic law is technically supposed to override the country's secular laws, although in actuality the situation remains ambiguous. In Pakistan's two most conservative provinces, Balochistan and the Northwest Frontier Province, both of which have been run by an Islamist political party since 2001, Islamic law is strictly enforced.

Sikhism The tension between Hinduism and Islam in medieval northern South Asia gave rise to a new religion called **Sikhism**. Sikhism originated in the late 1400s in the Punjab, near the modern boundary between India and Pakistan. The Punjab was the site of religious fervor at the time; Islam was gaining converts and Hinduism was increasingly on the defensive. The new faith combined elements of both religions and thus appealed to many who felt trapped between their competing claims. Many orthodox Muslims, however, viewed Sikhism as a heresy precisely because it incorporated elements of their own religion. Periodic bouts of persecution led the Sikhs to adopt a militantly defensive stance, and in the political chaos of the early 1800s they were able to carve out a large kingdom for themselves. Even today, Sikh men are noted for their work as soldiers and bodyguards.

At present, the Indian state of Punjab is approximately 60 percent Sikh. Small but often influential groups of Sikhs are scattered across the rest of India. Devout Sikh men are immediately visible, because they do not cut their hair or their beards. Instead, they wear their hair wrapped in a turban and often tie their beards close to their faces.

Buddhism and Jainism Although Buddhism virtually disappeared from India in medieval times, it persisted in Sri Lanka. Among the island's dominant Sinhalese people, Theravada Buddhism developed into a virtual national religion, fostering close connections between Sri Lanka and mainland Southeast Asia. In the high valleys of the Himalayas, Buddhism also survived as the majority religion (Figure 12.23). Tibetan Buddhism, with its esoteric beliefs and huge monasteries, has been better preserved in Bhutan and in the Ladakh region of northeastern Kashmir than in Chinese-controlled Tibet itself. The small town of Dharmsala in the northern Indian state of Himachal Pradesh is the seat of Tibet's government-in-exile and of its spiritual leader, the Dalai Lama, who fled Tibet in 1959 after an unsuccessful revolt. Another group of Buddhists is found in central India, where twentieth-century efforts to convert Hindu dalits to the faith were particularly successful.

At roughly the same time as the birth of Buddhism (circa 500 BCE), another religion emerged in northern India as a protest against orthodox Hinduism: **Jainism**. This religion similarly stressed nonviolence, taking this creed to its ultimate extreme. Jains are forbidden to kill any living creatures, and as a result the most devout adherents wear gauze masks to prevent them from inhaling small insects. Agriculture is forbidden to Jains because plowing can kill small creatures. As a result, most members of the faith have looked to trade for their livelihoods. Many have prospered, aided no doubt by the frugal lifestyles required by their religion. The small Jain community is now one of the wealthiest in India, although it also includes many poor individuals. Today Jains are concentrated in northwestern India, particularly Gujarat.

Other Religious Groups Even more prosperous than the Jains are the Parsis, or Zoroastrians, concentrated in the Mumbai area. The Parsis arrived as refugees, fleeing from Iran after the arrival of Islam in the seventh century. Zoroastrianism is an ancient religion that focuses on the cosmic struggle between good and evil. Although numbering only a few hundred thousand, the Parsi community has had a

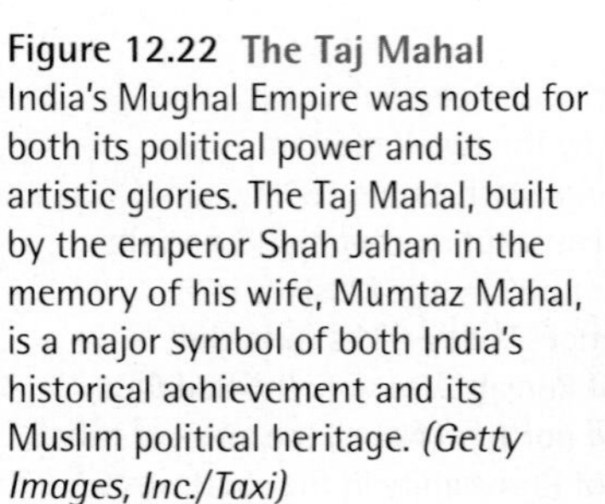

Figure 12.22 The Taj Mahal India's Mughal Empire was noted for both its political power and its artistic glories. The Taj Mahal, built by the emperor Shah Jahan in the memory of his wife, Mumtaz Mahal, is a major symbol of both India's historical achievement and its Muslim political heritage. *(Getty Images, Inc./Taxi)*

Figure 12.23 Buddhist Monastic Landscape
Buddhism is the dominant faith in most of the Himalayan districts of the far north and in central and southern Sri Lanka. Monasteries are particularly important in the north, where Tibetan Buddhism predominates. *(Linde Waidhofer/Getty Images, Inc.–Liaison)*

major impact on the Indian economy. Several of the country's largest industrial firms are controlled by Parsi families. Intermarriage and low fertility, however, now threaten the survival of the community.

Indian Christians are more numerous than either Parsis or Jains. Their religion arrived some 1,800 years ago; early contact between the Malabar Coast and Southwest Asia brought Christian as well as Muslim traders. A Jewish population also established itself but later declined; today it numbers only a few hundred. Kerala's Christians, by contrast, are counted in the millions, constituting some 20 percent of the state's population. Several Christian sects are represented, but the largest are historically affiliated with the Syrian Church of Southwest Asia. Another stronghold of Christianity is the small Indian state of Goa, a former Portuguese colony, where Roman Catholics and Hindus each make up roughly half of the population.

During the colonial period, British missionaries went to great efforts to convert South Asians to Christianity. They had very little success, however, in Hindu, Muslim, and Buddhist communities, where people were largely content with their own faiths and were wary of foreign missionaries associated with the ruling power. The remote tribal districts of British India, on the other hand, proved to be more receptive to missionary activity. In the uplands of India's extreme northeast, entire communities abandoned their traditional animist faith in favor of Protestant Christianity; 75 percent of the people of the state of Nagaland are now members of the Baptist Church. Christian missionaries are still active in many parts of India, but during the 1990s they began to experience severe pressure in many areas from Hindu nationalists. Pakistan's Christian community, some 3.8 million strong, has also been under pressure, in this case from Muslim radicals.

Geographies of Language

South Asia's linguistic diversity matches its religious diversity. In fact, one of the world's most important linguistic boundaries runs directly across India (Figure 12.24). North of the line, languages belong to the Indo-European group, the world's largest linguistic family. The languages of southern India, on the other hand, belong to the **Dravidian** family, a linguistic group that is unique to South Asia. Along the mountainous northern rim of the region, a third linguistic family, Tibeto-Burman, prevails, but this area is marginal to the South Asian cultural sphere. Scattered tribal groups in eastern India speak Austro-Asiatic languages related to those of mainland Southeast Asia. South Asia can thus be divided into two major linguistic zones, the Indo-European north and the Dravidian south. But within these broad divisions there are many different languages, each associated with a distinct culture. In most of South Asia, several languages are spoken within the same region or even city, and multilingualism is common everywhere.

How or when Indo-European languages came to South Asia is uncertain, but scholars have traditionally argued that they arrived with pastoral peoples from Central Asia who invaded the subcontinent in the second millennium BCE, largely replacing the indigenous Dravidian peoples. According to this hypothesis, offshoots of the same original cattle-herding people also swept across both Iran and Europe, bringing their language to all three places. Supposedly, the ancestral Indo-European tongue introduced to India was similar to Sanskrit. This rather simplistic scenario, however, is now regarded with some suspicion, and many scholars argue for a more gradual infiltration of Indo-European speakers from the northwest.

Any modern Indo-European language of India, such as Hindi or Bengali, is more closely related to English than it is to any Dravidian language of southern India, such as Tamil. That said, it should also be noted that South Asian languages on both sides of this linguistic divide do share a number of superficial features. Dravidian languages, for example, have borrowed many words from Sanskrit, particularly those associated with religion and scholarship.

The Indo-European North As can be seen in Figure 12.24, South Asia's Indo-European languages are themselves divided into two subfamilies: Iranian and Indo-Aryan. As might be expected, Iranian languages, such as Baluchi and Pashtun, are found in western Pakistan, near the border with Iran and Afghanistan. Languages of the much larger Indo-Aryan groups are all closely related to each other, yet are still quite distinctive, often written in different scripts. Each of the major languages of India is associated with one or more Indian states, because the country deliberately remapped its political subdivisions along linguistic lines in the 1950s. As a result, one finds Gujarati in Gujarat, Marathi in Maharashtra, Oriya in

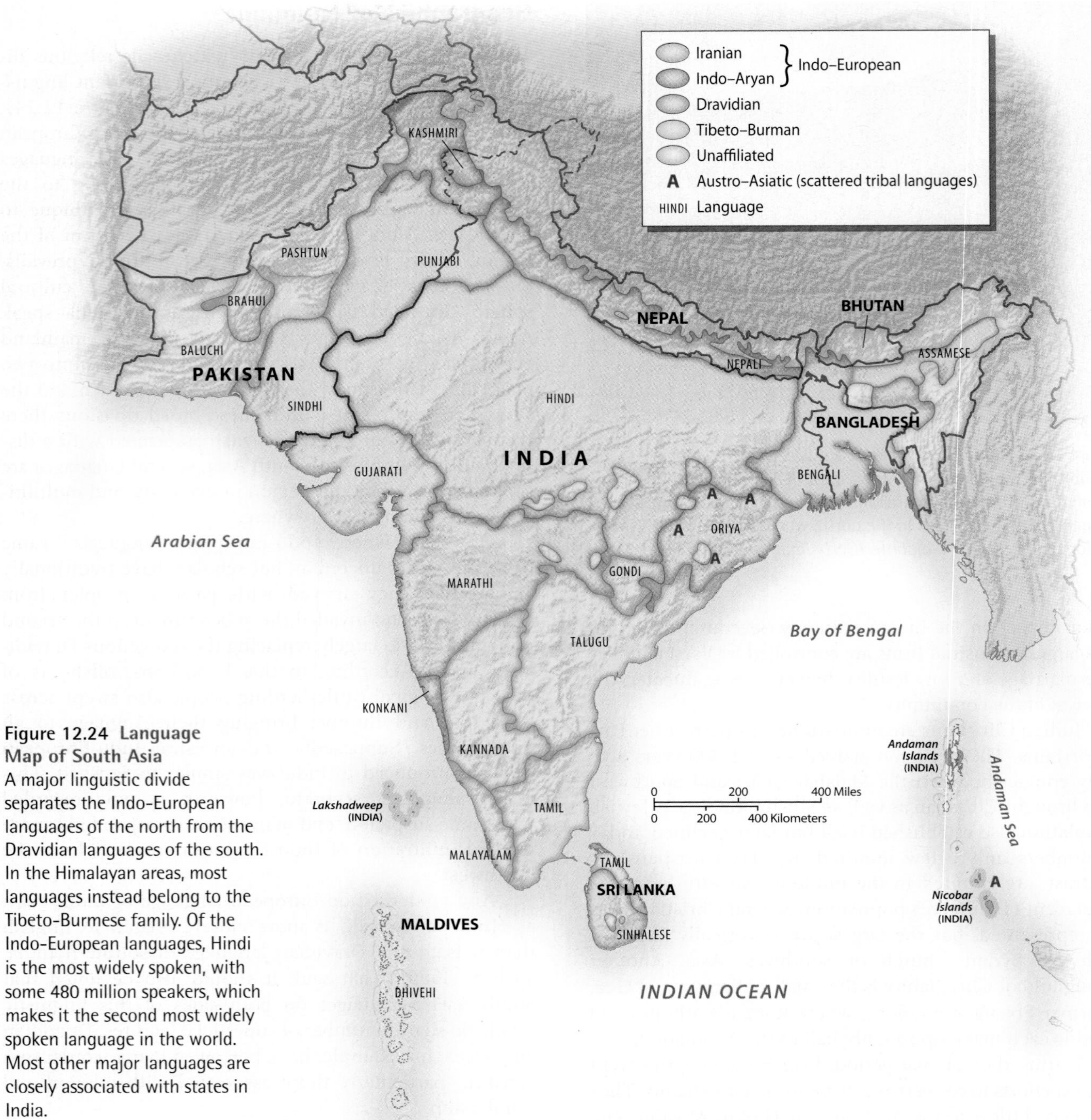

Figure 12.24 Language Map of South Asia A major linguistic divide separates the Indo-European languages of the north from the Dravidian languages of the south. In the Himalayan areas, most languages instead belong to the Tibeto-Burmese family. Of the Indo-European languages, Hindi is the most widely spoken, with some 480 million speakers, which makes it the second most widely spoken language in the world. Most other major languages are closely associated with states in India.

Orissa, and so on. Two of these languages, Punjabi and Bengali, span India's international boundaries to extend into Pakistan and Bangladesh, respectively, as these borders were established on religious rather than linguistic lines.

The most widely spoken language of South Asia is Hindi—not to be confused with the Hindu religion. With almost 500 million speakers, Hindi is the second most widely spoken language in the world. It occupies a prominent role in contemporary India, both because so many people speak it and because it is the main language of the Ganges Valley, the historical and demographic core of India. Unlike other Indian languages, Hindi is the dominant tongue of a number of Indian states, including Uttar Pradesh, Madhya Pradesh, and Haryana. In addition, the main forms of speech found in Rajasthan and Bihar are often considered to be dialects of Hindi. Many northerners would like Hindi to become the country's common language, but many other Indians, particularly those living in the south, strongly resist the idea. Still, most Indian students learn some Hindi, often as their second or third language.

Bengali is the second most widely spoken language in South Asia. It is the national language of Bangladesh and the main language of the Indian state of West Bengal. Spoken by more than 200 million persons, Bengali is the world's eighth or ninth most widely spoken language. Its significance extends beyond its official status in Bangladesh and its total numerical strength. Equally important is its extensive literature. West Bengal, particularly its capital of Kolkata (Calcutta), has long been one of South Asia's main literary and intellectual centers. Kolkata may be noted for its appalling poverty, but it also has one of the highest levels of cultural production in the world, as measured by the output of drama, poetry, novels, and film (Figure 12.25).

The Punjabi-speaking zone in the west was similarly split at the time of independence, in this case between Pakistan and the Indian state of Punjab. While an estimated 100 million persons speak Punjabi, it does not have the significance of Bengali. Although Punjabi is the main vehicle of Sikh religious writings, it lacks an extensive literary tradition. More importantly, Punjabi did not become the national language of Pakistan, even though it is the day-to-day language of some two-thirds of the country's inhabitants. Instead, that position was given to Urdu.

Urdu, like Hindi, originated on the plains of northern India. The difference between the two was largely one of religion: Hindi was the language of the Hindu majority, Urdu that of the Muslim minority, including the former ruling class. Owing to this distinction, Hindi and Urdu were from the start written differently—the former in the Devanagari script (derived from Sanskrit) and the latter in a modified version of the Arabic script. Although Urdu contains many words borrowed from Persian, its basic grammar and vocabulary are almost identical to those of Hindi.

With independence in 1947, millions of Urdu-speaking Muslims from the Ganges Valley fled to the new state of Pakistan. Since Urdu had a higher status than Pakistan's indigenous tongues, it was quickly established as the new country's official language. Karachi, Pakistan's largest city, is now largely Urdu-speaking, but elsewhere other languages, such as Punjabi and Sindhi, remain primary. Most Pakistanis, however, do speak Urdu as a second language.

Languages of the South Four thousand years ago, Dravidian languages were probably spoken across most of South Asia, even in the north. As Figure 12.24 indicates, a Dravidian tongue called *Brahui* is still found in the uplands of western Pakistan. The four main Dravidian languages, however, are confined to southern India and northern Sri Lanka. As in the north, each language is closely associated with an Indian state: Kannada in Karnataka, Malayalam in Kerala, Telugu in Andhra Pradesh, and Tamil in Tamil Nadu. Tamil is usually considered the most important member of the family because it has the longest history and the largest literature. Tamil poetry dates back to the first century CE, making it one of the world's oldest written languages.

Although Tamil is spoken in northern Sri Lanka, the country's majority population, the Sinhalese, speak an Indo-European language. Apparently the Sinhalese migrated from northern South Asia several thousand years ago. Although this movement is lost to history, the migrants evidently settled on the island's fertile and moist southwestern coastal and central highland areas, which formed the core of a succession of Sinhalese kingdoms. These same people also migrated to the Maldives, where the national language, Divehi, is essentially a Sinhalese dialect. The drier north and east of Sri Lanka, on the other hand, were settled many hundreds of years ago by Tamils. In the 1800s, British landowners imported Tamil peasants from the mainland to work on their tea plantations in the central highlands, giving rise to a second population of Tamil-speakers in Sri Lanka.

Figure 12.25 Kolkata (Calcutta) Bookstore
Although Kolkata (Calcutta) is noted in the West mostly for its abject poverty, the city is also known in India for its vibrant cultural and intellectual life, illustrated by its large number of bookstores, theaters, and publishing firms. *(Earl & Nazima Kowall/Corbis)*

Linguistic Dilemmas The multilingual countries Sri Lanka, Pakistan, and India are all troubled by linguistic conflicts. Such problems are most complex in India, simply because India is so large and has so many different languages. India's linguistic environment is changing in complicated ways, pushed along by modern economic and political forces.

Indian nationalists have long dreamed of a national language that could help forge the different communities of the country into a more unified nation. But this **linguistic nationalism**, or the linking of a specific language with nationalistic goals, meets the stiff resistance of provincial loyalty, which itself is intertwined with local languages. The obvious choice for a national language would be Hindi, and Hindi was indeed declared as such in 1947. Raising Hindi to this position, however, alienated speakers of important northern languages such as Bengali and Marathi, and even more so the speakers of the Dravidian tongues. As a result of this cultural tension, in the 1950 Indian constitution Hindi was demoted to sharing the position of "official language" of India with 14 other languages. Over time, additional languages were added to this list, such that India now has 23 separate official languages.

Regardless of opposition, the role of Hindi is expanding in India, especially in the Indo-European-speaking north. Here local languages are closely related to Hindi, which can therefore be learned without too much difficulty. Hindi is spreading through education, but even more significantly through popular media, especially television and motion pictures. Films and television programs are made in several

Indo-Aryan languages, but Hindi remains the primary vehicle. The most popular movies coming out of Mumbai's "Bollywood" film industry, however, tend to be delivered in a neutral dialect that is as close to Urdu as to Hindi—and with a good deal of English thrown in as well.

The Role of English Even if Hindi is spreading, it still cannot be considered anything like a common Indian language, even in the north. In the Dravidian south, more importantly, its role remains strictly secondary. National-level political, journalistic, and academic communication thus cannot be conducted in Hindi, or in any other indigenous language. Only English, an "associate official language" of contemporary India, serves this function.

Before independence, many educated South Asians learned English for its political and economic benefits under British colonialism. It therefore emerged as the de facto common tongue, albeit one largely limited to the upper and middle classes. Today a few extreme nationalists wish to deemphasize English, but most Indians, and particularly those of the south, advocate English as a neutral national language because all parts of the country have an equal stake in it. Furthermore, English confers substantial international benefits.

English is thus the main integrating language of India, and it remains widely used elsewhere in South Asia. Roughly a third of the people of India are able to carry out a conversation in English. English-medium schools abound in all parts of the region, and many children of the elite learn this global language well before they begin their schooling.

The English spoken in South Asia has forms and vocabulary elements that sometimes make cross-cultural communication difficult. As a result, information technology companies in India, particularly those that run international call centers, sometimes ask their employees to watch reruns of popular television shows from the United States in order to gain fluency in American pronunciation and slang.

South Asians in a Global Cultural Context

The widespread use of English in South Asia has not only facilitated the spread of global culture into the region, but it has also helped South Asians' cultural production to reach a global audience. The global spread of South Asian literature, however, is nothing new. As early as the turn of the twentieth century, Rabindranath Tagore gained international acclaim for his poetry and fiction, earning the Nobel Prize for Literature in 1913. In the 1980s and 1990s, such Indian novelists as Salman Rushdie and Vikram Seth became major literary figures in Europe and North America.

The spread of South Asian culture abroad has been accompanied by the spread of South Asians themselves. Migration from South Asia during the time of the British Empire led to the establishment of large communities in such far-flung places as eastern Africa, Fiji, and the southern Caribbean (Figure 12.26). Subsequent migration targeted the developed world; there are now several million people of South Asian descent living in Britain (mostly Pakistani), and a similar number in North America (mostly Indian). Many

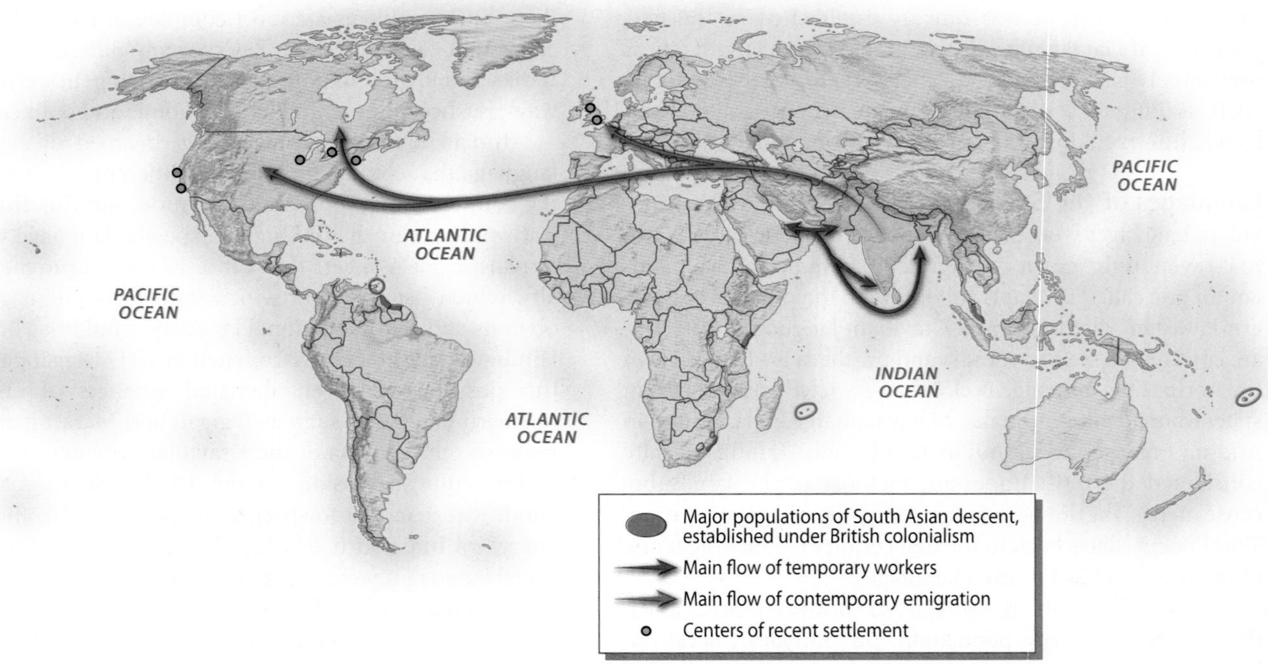

Figure 12.26 The South Asian Global Diaspora
During the British imperial period, large numbers of South Asian workers settled in other colonies. Today, roughly 50 percent of the population of such places as Fiji and Mauritius are of South Asian descent. More recently, large numbers have settled, and are still settling, in Europe (particularly Britain) and North America. Large numbers of temporary workers, both laborers and professionals, are employed in the wealthy oil-producing countries of the Persian Gulf.

Figure 12.27 **A Controversial Kiss**
Actor Richard Gere embraces and then kisses Indian actress Shilpa Shetty during an AIDS awareness program in 2007. Conservative Indian citizens were shocked by this act, which resulted in large-scale protests. *(Tanushree Punwani/Reuters/Landov)*

contemporary migrants to the United States are doctors, software engineers, and other professionals, making Indian Americans one of the country's wealthiest ethnic groups.

In South Asia itself, the globalization of culture has brought tensions as severe as those felt anywhere in the world. Traditional Hindu and Muslim religious customs frown on any overt display of sexuality—a staple feature of global popular culture. While romance is a recurrent theme in the often melodramatic Bollywood films of Mumbai, even kissing is considered risqué. In April 2007, actor Richard Gere sparked massive public protests when he kissed the Indian actress Shilpa Shetty at an AIDS awareness rally in New Delhi (Figure 12.27). Still, the pressures of internationalization are hard to resist. In the tourism-oriented Indian state of Goa, such cultural tensions are on full display. There, German and British sun-worshipers often wear nothing but thong bikini bottoms, whereas Indian women tourists go into the ocean fully clad. Young Indian men, for their part, often simply walk the beach and gawk, good-naturedly, at the outlandish foreigners.

Contemporary popular culture in South Asia thus reveals both global linkages as well as divisions. The same tensions can be seen, and in much stronger form, in the region's geopolitical framework.

GEOPOLITICAL FRAMEWORK: A Deeply Divided Region

Before the coming of British rule, South Asia had never been politically united. While a few empires at various times covered most of the subcontinent, none spanned its entire extent. Whatever unity the region had was cultural, not political. The British, however, brought the region into a single political system by the middle of the nineteenth century. A hundred years later, independence in 1947 witnessed the traumatic separation of Pakistan from India; in 1971, Pakistan itself was divided with the independence of Bangladesh, formerly East Pakistan. Serious internal tensions, moreover, began to rise in several South Asian countries in the late 1900s (Figure 12.28). Marxist revolutionary movements continue to be active in eastern India, and in Nepal they control large areas of the countryside. The most important geopolitical issue with global implications is the continuing tension between Pakistan and India, both of which are nuclear powers.

South Asia Before and After Independence in 1947

During the 1500s, when Europeans began to arrive on the coasts of South Asia, most of the northern subcontinent came under the power of the Mughal Empire, a mighty Muslim state ruled by people of Central Asian descent (Figure 12.29). Southern India remained under the control of a Hindu kingdom called Vijayanagara. European merchants, keen to obtain spices, textiles, and other Indian products, established a series of coastal trading posts. The Mughals and other South Asian rulers were little concerned with the growing European naval power, as their own focus was the control of land. The Portuguese carved out an enclave in Goa on the west coast, while the Dutch gained control over much of Sri Lanka in the 1600s, but neither was a threat to the Mughals.

The Mughal Empire grew stronger in the 1600s, whereas Hindu power declined until it was limited to the peninsula's extreme south. In the early 1700s, however, the Mughal Empire weakened rapidly. A number of contending states, some ruled by Muslims, others by Hindus, and a few by Sikhs, emerged in former Mughal territories. As a result, the eighteenth century in South Asia was a time of political and military turmoil, often bordering on chaos.

The British Conquest These unsettled conditions provided an opening for European imperialism. The British and French, having largely displaced the Dutch and Portuguese, competed for trading posts. Because Indian cotton textiles were the best in the world prior to the Industrial Revolution, British and French merchants needed to obtain huge quantities for their global trading networks. With Britain's overwhelming victory over France in the Seven Years' War (1756–63), France was reduced to a few marginal coastal possessions in southern India. Britain, or more specifically the **British East India Company**, the private firm that acted as an arm of the British government, now monopolized trade in the area and began to stake out a South Asian empire of its own.

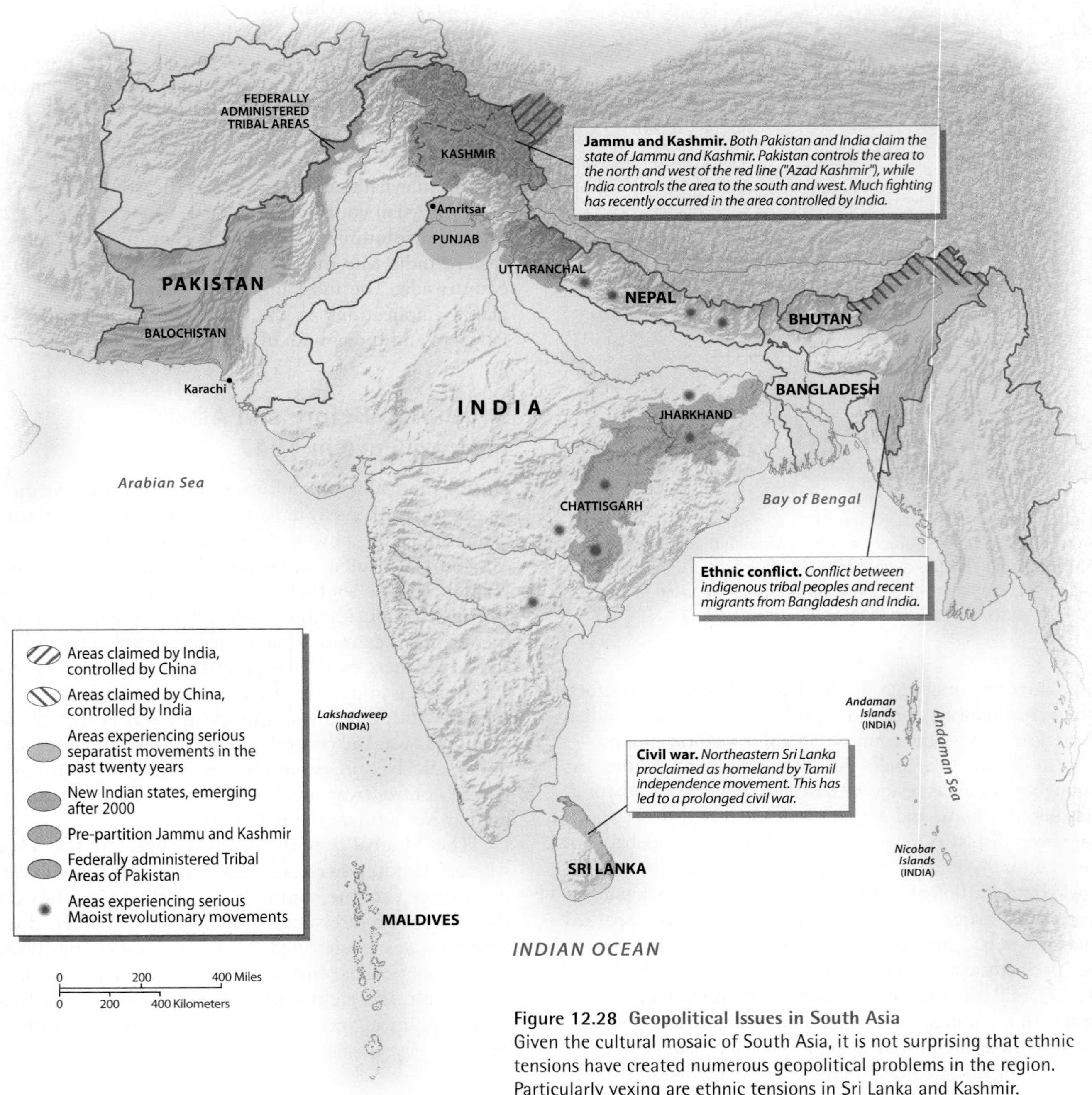

Figure 12.28 Geopolitical Issues in South Asia
Given the cultural mosaic of South Asia, it is not surprising that ethnic tensions have created numerous geopolitical problems in the region. Particularly vexing are ethnic tensions in Sri Lanka and Kashmir.

The company's usual method was to make strategic alliances with Indian states in order to defeat the latter's enemies, most of whose territories it would then grab for itself. As time passed, its army, largely composed of South Asian mercenaries, grew ever more powerful. Several Indian states put up strong resistance, but none could ultimately resist the immense resources of the East India Company. Valuable local allies, as well as a few former enemies, were allowed to remain in power, provided that they no longer threatened British interests. The territories of these indigenous states, however, were gradually whittled back, and British advisors increasingly dictated their policies.

From Company Control to British Colony The continuing reduction in size of the Indian states, coupled with the growing arrogance of British officials, led to a rebellion in 1857 across much of South Asia. When this uprising (called the *Sepoy Mutiny* by the British) was finally crushed, a new political order was implemented. South Asia was now to be ruled directly by the British government, with the monarch of England serving as its head of state.

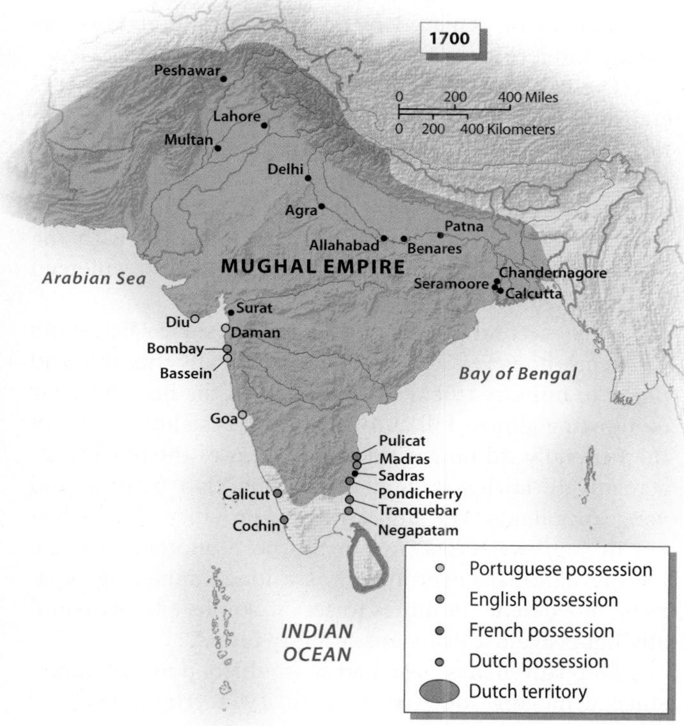

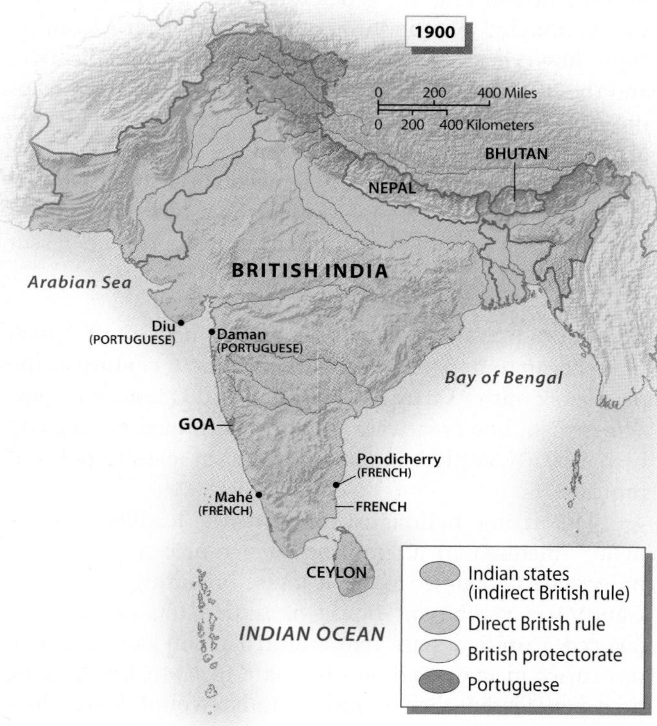

Figure 12.29 Geopolitical Change
At the onset of European colonialism before 1700, much of South Asia was dominated by the powerful Mughal Empire. Under Britain, the wealthiest parts of the region were ruled directly, but other lands remained under the partial authority of indigenous rulers. Independence for the region came after 1947, when the British abandoned their extensive colonial territory. Bangladesh, formerly East Pakistan, gained its independence in 1971 after a short struggle against centralized Pakistani rule from the west.

Until 1947, the British government maintained direct control over South Asia's most productive and densely populated areas, including virtually the entire Indus–Ganges Valley and most of the coastal plains. The British also ruled Sri Lanka directly, having supplanted the Dutch in the 1700s. The major areas of indirect rule, where Hindu, Muslim, and Sikh rulers retained their princely states under British advisors, were in Rajasthan, the uplands of central India, southern Kerala, and along the frontiers. The British administered this vast empire through three coastal cities that were largely their own creation: Bombay (Mumbai), Madras (Chennai), and, above all, Calcutta (Kolkata). In 1911, they began building a new capital in New Delhi, near the strategic divide between the Indus and Ganges drainage systems.

While the political geography of British India stabilized after 1856, the empire's frontiers remained unsettled. British officials worried about threats to their immensely profitable colony, particularly from the Russians advancing across Central Asia. In response, they attempted to expand as far to the north as possible. In some cases this merely entailed making alliances with local rulers. In such a manner Nepal and Bhutan retained their independence. In the extreme northeast, a number of small states and tribal territories, most of which had never been part of the

South Asian cultural sphere, were more directly brought into the British Empire, hence into India. A similar policy was conducted on the vulnerable northwestern frontier. Here, however, local resistance was much more effective, and the British-Indian army suffered defeat at the hands of the Afghans. Afghanistan thus retained its independence, forming an effective buffer between the British and Russian empires. The British also allowed the tribal Pashtun-speaking areas of what is now northwestern Pakistan to retain almost complete autonomy, thus forming a secondary buffer.

Independence and Partition The framework of British India began to unravel in the early twentieth century as the people of South Asia increasingly demanded independence. The British, however, were equally determined to stay, and by the 1920s South Asia was embroiled in massive political protests.

The rising nationalist movement's leaders faced a major dilemma in attempting to organize a potentially independent regime. Many leaders, including Mohandas Gandhi—a prominent figure of Indian independence—favored a unified state that would encompass all British territories in mainland South Asia. Most Muslim leaders, however, feared that a unified India would leave their people in a vulnerable position. They therefore argued for the division of British India into two new countries: a Hindu-majority India and a Muslim-majority Pakistan. In several parts of northern South Asia, however, Muslims and Hindus were settled in roughly equal proportions. A more significant obstacle was the fact that the areas of clear Muslim majority were located on opposite sides of the subcontinent, in present-day Pakistan and Bangladesh.

No longer able to maintain their world empire after World War II, the British withdrew from South Asia in 1947. As this occurred, the region was indeed divided into two countries: India and Pakistan. Partition itself was a horrific event, resulting in the death of some 200,000 to 1 million people. Roughly 7 million Hindus and Sikhs fled Pakistan, to be replaced by roughly 7 million Muslims fleeing India (Figure 12.30).

The Pakistan that emerged from partition was for several decades a clumsy two-part country, its western section in the Indus Valley, its eastern portion in the Ganges Delta. The Bengalis, occupying the poorer eastern section, complained that they were treated as second-class citizens. In 1971, they launched a rebellion, and, with the help of India, quickly prevailed. Bangladesh then emerged as a new country.

This second partition did not solve Pakistan's problems, however, as it remained politically unstable and prone to military rule. Pakistan retained the British policy of allowing almost full autonomy to the Pashtun tribes of the Federally Administered Tribal Areas of the northwest, a relatively lawless region marked by clan fighting and vengeance feuds. The large and poor province of Balochistan in southwestern Pakistan has been another problem for the national government, as a long-simmering separatist movement continues to set off bombs and occasionally fights against Pakistan's military forces.

Bangladesh has also had a troubled political career since achieving independence in 1971. High levels of corruption, growing Islamic radicalism, and street-level fighting between members its two major political parties have undermined its democratic institutions. As tensions mounted, the Bangladeshi army seized power in January 2007, declared a state of emergency, and detained up to 100,000 individuals. The country's top generals declared that democracy would be restored once order was reestablished and high-level corruption rooted out, but many observers remain skeptical.

Geopolitical Structure of India After independence in 1947, the leaders of India, committed to democracy, faced a

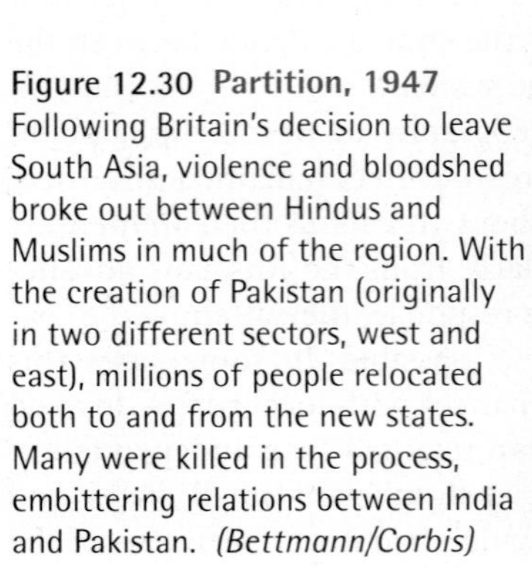

Figure 12.30 Partition, 1947 Following Britain's decision to leave South Asia, violence and bloodshed broke out between Hindus and Muslims in much of the region. With the creation of Pakistan (originally in two different sectors, west and east), millions of people relocated both to and from the new states. Many were killed in the process, embittering relations between India and Pakistan. *(Bettmann/Corbis)*

major challenge in organizing such a large and culturally diverse country. They decided to chart a middle ground between centralization and local autonomy. India itself was thus organized as a **federal state**, with a significant amount of power being vested in its individual states. The national government, however, retained full control over foreign affairs and a large degree of economic authority.

Following independence, India's constituent states were reorganized to match the country's linguistic geography. The idea was that each major language group should have its own state (with the massive Hindi-speaking population having several), hence a degree of political and cultural autonomy. Yet only the largest groups received their own territories, which has led to recurring demands from smaller groups that have felt politically excluded. Over time, several new states have been added to the map. Goa, after the Portuguese were forced out in 1961, became a separate state in 1987 over the objection of its large northern neighbor of Maharashtra. In 2000, three new states were added: Jharkand, Uttaranchal, and Chhattisgarh.

Ethnic Conflicts in South Asia

The movement for new states in India has been largely rooted in ethnic tensions. Unfortunately, violent ethnic conflicts persist in many parts of South Asia. Of these conflicts, the most complex—and perilous—is that in Kashmir.

Kashmir Relations between India and Pakistan were hostile from the start, and the situation in Kashmir has kept the conflict burning (Figure 12.31). During the British period, Kashmir was a large state with a primarily Muslim core joined to a Hindu district in the south (Jammu) and a Tibetan Buddhist district in the far northeast (Ladakh). Kashmir was then ruled by a Hindu **maharaja**, or a king subject to British advisors. During partition, Kashmir came under severe pressure from both India and Pakistan. Troops from Pakistan gained control of western and much of northern Kashmir, at which point the maharaja opted for union with India. India thus retained the core areas of the state, but neither country would accept the other's control over any portion of Kashmir. As a result, India and Pakistan have fought several inconclusive wars over the issue.

Although the Indo-Pakistani boundary has remained fixed, the struggle in Kashmir intensified, flaming out into an open insurgency in 1989. Many Muslim Kashmiris hope to join their homeland to Pakistan; others would rather see it emerge as an independent country. Moderate Muslim leaders prefer to work within the Indian democratic system for autonomy within India, but in doing so they risk assassination by extremists. Hindu militants and other Indian nationalists are similarly adamant that Kashmir remain part of India, and the Indian government agrees. India also has accused Pakistan of supporting training camps for Kashmiri militants and of helping them sneak across the border. As a result, India began the construction of a fortified fence along the so-called line of control that divides Kashmir. The result of all of this activity has been a low-level, but periodically brutal, war that has claimed over 40,000 lives and displaced one out of every six inhabitants of Kashmir.

The situation in Kashmir began to improve in 2004 when India and Pakistan initiated a serious round of negotiations. India now accepts that Pakistan must play a role in any Kashmir peace settlement, whereas Pakistan's

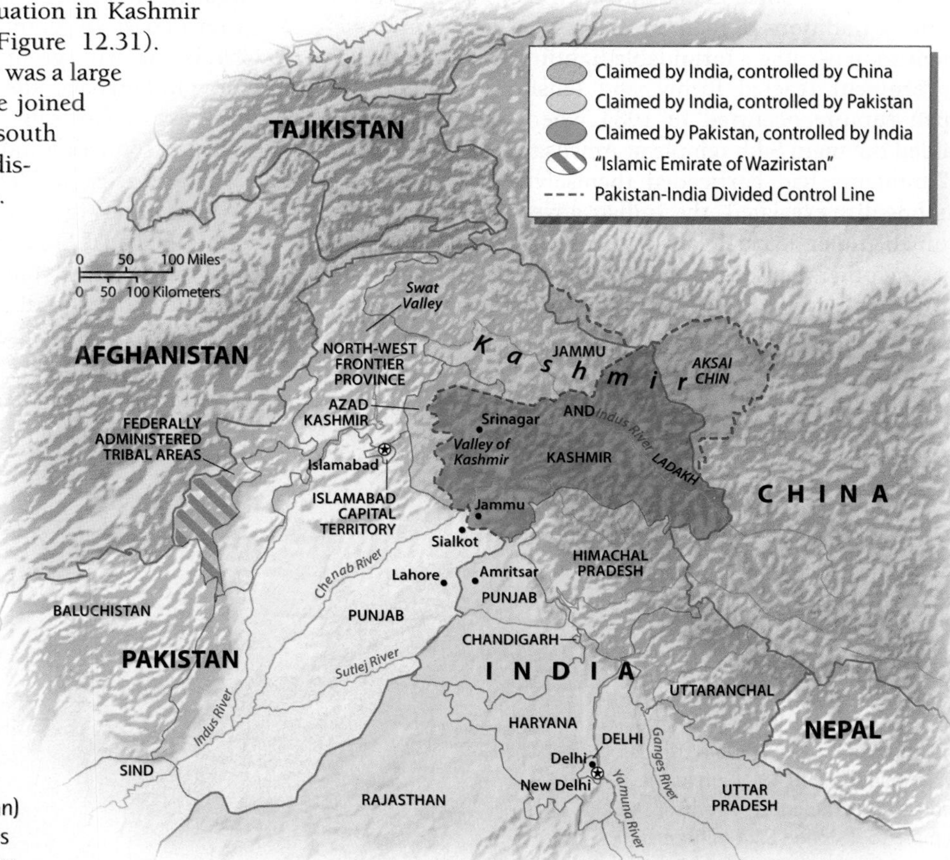

Figure 12.31 Conflict in Kashmir
Unrest in Kashmir inflames the continually hostile relationships between the two nuclear powers of India and Pakistan. Today many Kashmiris wish to join Pakistan, while many others argue for an independent state. Note also Pakistan's Federally Administered Tribal Areas (including the Islamic Emirate of Waziristan) and its North-West Frontier Province, areas that have experienced much recent fighting.

government—struggling with many problems in its own territory—has reduced its support of Muslim militants seeking to undermine Indian power. As a result, the daily death toll from political violence dropped from 10 in 2002 to 2 in early 2007. But as genuine peace still seems far off, India remains determined to keep a half-million-strong army in Kashmir.

The Vale of Kashmir, once one of South Asia's premier tourist destinations with its lush gardens and orchards nestled among some of the world's most spectacular mountains, is now a battle-scarred war zone. The war also has unleashed an internal diaspora, as hundreds of thousands of Kashmiris have fled their homeland to find security elsewhere in India.

The Punjab Religious conflict has also caused political tensions in India's Punjab. The original Punjab, an area of intermixed Hindu, Muslim, and Sikh communities, was divided between India and Pakistan in 1947. During partition, virtually all Hindus and Sikhs fled from Pakistan's allotted portion, just as Muslims were forced out of the zone awarded to India.

Partition did not solve the problems of India's portion of the Punjab. Most Sikhs resented the fact that the Indian government refused to recognize Sikhism as a separate religion, instead classifying it as a sect of Hinduism. In the 1970s and 1980s, as the area grew increasingly prosperous due to the Green Revolution, Sikh leaders began to strive for autonomy and Sikh radicals began to press for outright independence. Since many Sikh men had maintained the military traditions of their ancestors, this secession movement soon became a formidable fighting force. The Indian government reacted firmly, and tensions mounted. An open rupture occurred in 1984 when the Indian army raided the main Sikh temple at Amritsar, in which a group of militants had barricaded themselves (Figure 12.32). As hostility escalated, the Indian government placed the Punjab under martial law arresting many militants and driving others out of the country. Such a policy eventually proved successful in quelling political violence. Renewed conflict in the area, however, remains a possibility (see "People on the Move: Punjabis Leaving Agriculture and Leaving India").

The Northeastern Fringe A more complicated ethnic conflict emerged in the late 1900s in the uplands of India's extreme northeast, particularly in the states of Arunachal Pradesh, Nagaland, Manipur, and portions of Assam. One underlying problem stems from demographic change and cultural collision. Much of this area is still relatively lightly populated, and as a result has attracted many migrants from Bangladesh and adjacent provinces of India. Many local inhabitants consider this movement a threat to both their lands and their cultural integrity. In response to such perceived threats, fighters from a group called the United Liberation Front of Asom killed 70 poor migrants from northern India in January 2007, resulting in widespread panic among the area's Hindi-speaking minority.

In Manipur and Nagaland, several insurgent groups continue to demand independence from India. After 2000, the Indian government began to invest more money in this troubled region, hoping to reduce popular support for separatist movements. India is also eager to expand trade with Burma, and has been working with the Burmese government to secure the border zone. As a result of these and other initiatives, a number of local rebel movements have signed cease-fires with the Indian government. Several insurgent groups have even agreed to cooperate with the Indian army against those groups that continue to fight for independence.

Sri Lanka Interethnic violence in Sri Lanka is especially severe. Here the conflict has roots in both religious and linguistic differences. Northern Sri Lanka and parts of its eastern coast are dominated by Hindu Tamils, while the island's majority group is Buddhist in religion and Sinhalese in language. Relations between the two communities have historically been fairly good, but tensions mounted soon after independence.

The basic problem is that Sinhalese nationalists favor a unitary government, some of them going so far as to argue that Sri Lanka ought to be defined as a Buddhist state. Most Tamils, on the other hand, support political and cultural autonomy, and they have accused the government of discriminating against them. Overall, levels of education are higher among the Tamils, but the government has favored the Sinhalese majority. In 1983,

Figure 12.32 Sikh Temple at Amritsar
Separatism in Indian's Punjab region emerged because of hostility between the Sikh majority and the Indian government. These tensions were magnified in 1984 when the Indian army raided the Amritsar temple to dislodge Sikh militants. *(Daniel O'Leary/Panos Pictures)*

GEOGRAPHY IN THE MAKING Pakistan's Political Turmoil

By November 2007, the position of Pakistan's president and army chief, Pervez Musharraf, looked increasingly grim. Insurgencies continued to erupt in Baluchistan and in the tribal districts, and a radical Muslim cleric had just taken control over Swat, a vacation area near Islamabad. Earlier in the year, Musharraf had attempted to undermine Pakistan's Supreme Court, resulting in massive protests by suit-wearing lawyers. Musharraf's standing with the country's Islamic religious establishment, which had declined due to his alliance with the United States, further plummeted in July when he ordered a military attack on the Red Mosque in Islamabad, in which a number of radical Muslim leaders had barricaded themselves. On July 6, 2007, Musharraf survived yet another assassination attempt when an unknown group fired an anti-aircraft gun at his plane. In August, several public opinion polls showed that roughly two-thirds of Pakistan's people wanted Musharraf to relinquish power.

Musharraf's position seemed to improve in October 2007, when he won an indirect election reconfirming his presidency. His success, however, stemmed largely from the fact that most of the opposition boycotted the vote. Pakistan's Supreme Court was about to rule on the legality of the election when Musharraf struck. On November 3, 2007, he suspended the constitution, locked up Supreme Chief Justice Iftikhar Chaudhry, ordered the arrest of key political opponents, and essentially imposed martial law on the country.

Musharraf's power grab put the United States in a difficult position. The U.S. had long supported Pakistan's president while pressing him to follow a more democratic path. According to rumors in Pakistan, Musharraf had been about to declare a state of emergency in August 2007, but had been talked out of it by U.S. Secretary of State Condoleezza Rice. The U.S. subsequently encouraged Musharraf to reach an accommodation with an exiled former Prime Minister of Pakistan, Benazir Bhutto. The apparent plan was for Musharraf to retain the presidency but step down as military head while Bhutto would seek the position of prime minister. Bhutto returned to Pakistan on October 18; on the same day, a suicide-bomb attack on her motorcade, presumably set off by Islamic militants, killed 136 people but left her unharmed. After declaring the state of emergency, Musharraf placed Bhutto under house arrest, undermining the possibility of their future collaboration.

On November 25, 2007, Pakistani politics grew more complex when another former prime minister and Musharraf rival, Nawaz Sharif, returned from exile in Saudi Arabia. The fact that Sharif was allowed to return to Pakistan may indicate a loss of support for Musharraf within his own political party. Upon his return, Sharif called for a boycott of the parliamentary elections scheduled for January 8, 2008 unless Musharaff were to meet a series of demands. On December 19, 2007, Sharif and Bhutto—previously bitter rivals—announced that they would work together to try to restore democracy to Pakistan.

Benazir Bhutto's tentative collaboration with Nawaz Sharif came to a quick end on December 27, 2007 when she was assassinated at a campaign rally in Rawalpindi. The exact circumstances of her death remain unclear, but it appears that both gunshots and explosives were used. Pakistan's government blamed the attack on Al Qaeda, and an Al Qaeda commander claimed responsibility. Bhutto's supporters, however, were angered that the government had not provided adequate security. Many of them rioted after her death, destroying hundreds of cars as well as dozens of banks, gas stations, train cars, and election offices. As a result of Bhutto's assassination and its aftermath, Pakistan seems less politically stable than ever.

virtually uninhabited territory of Aksai Chin in northern Kashmir. Growing trade, however, has resulted in a gradual easing of hostilities between India and China, and in 1998 the two countries initiated a series of border talks. A declaration that 2006 would be an official "year of friendship," however, was undermined when China announced that it still considered India's northeastern state of Arunachal Pradesh to be Chinese territory. Most observers think that an eventual settlement will entail India recognizing Chinese sovereignty over Aksai Chin and China giving up its claims in the northeast. But as long as China maintains close military ties with Pakistan, its relationship with India will probably remain tense.

Pakistan After 9/11 Pakistan's geopolitical situation became more complex in the aftermath of the attacks in the United States on September 11, 2001. Until that time, Pakistan had strongly supported Afghanistan's Taliban regime (see Chapter 10). After the attack on the World Trade Center and Pentagon (orchestrated from Afghanistan by Osama bin Laden), the United States gave Pakistan a stark choice: either it would assist the United States in its fight against the Taliban and receive in return debt reductions and other forms of aid, or it would lose favor. Pakistan's president Pervez Musharraf quickly agreed to help, and Pakistan offered both military bases and valuable intelligence to the U.S. military.

Pakistan's decision to help the United States came with large risks. Osama bin Laden had gained popularity among Pakistan's Islamic fundamentalists. Both he and the Taliban, moreover, enjoyed significant support among the Pashtun population in Pakistan's North-West Frontier Province (NWFP) and especially in the virtually ungoverned Federally Administered Tribal Areas (see Figure 12.31). As the Taliban regrouped in northwestern Pakistan after being chased out of Afghanistan, the United States began to pressure Pakistan to act. Some 80,000 Pakistani troops marched into the tribal district of Waziristan, but after several years of bloody fighting, Pakistan withdrew its forces in 2006. In the resulting peace accord, Pakistan restored full autonomy to the region, which now calls itself the Islamic Emirate of Waziristan. The United States was not pleased with this development, as it claims that Waziristan is offering refuge to Taliban fighters seeking to overthrow the U.S.-backed government of Afghanistan.

The post-9/11 geopolitical situation also forced Pakistan to deal with its international nuclear connections. Under severe pressure from his own government, Pakistan's top nuclear scientist, Abdul Qadeer Khan, confessed that he had illegally transferred nuclear technology to Iran, North Korea, and Libya. Many foreign observers viewed Khan as a scapegoat, contending that Pakistan's government was itself deeply involved in these dealings. Many of Pakistan's people, on the other hand, saw the whole episode as a shameful sacrifice of one of their national heroes in order to placate what they saw as a fundamentally hostile U.S. government. Such feelings intensified in 2006 when President George W. Bush announced that the United States would offer India civil nuclear energy cooperation, even though India has never signed the Nuclear Non-Proliferation Treaty. As a result of these and other issues, the position of Pakistan's leader Pervez Musharraf grew increasingly insecure in 2007 (see "Geography in the Making: Pakistan's Political Turmoil").

India's Changing Geopolitical Situation The tension between India and Pakistan, and the complex relationships that this entails with China and the United States, overshadows all other international geopolitical issues in South Asia. Elsewhere in the region, the power of India is often overwhelming, if resented. Bangladesh, which owes its very existence to Indian support, long enjoyed relatively cordial relations with India. Their relations began to deteriorate in the late 1990s, however, owing to India's concerns about illegal Bangladeshi immigration and to its fears that Bangladesh was providing refuge to separatist fighters from India's northeastern periphery. As a result, India has recently completed a $1 billion security fence along the border between the two countries. In other parts of South Asia, India has used its power even more forcefully. For example, in 1975, it annexed the formerly semi-independent country of Sikkim in the Himalayas.

As the relationship between the United States and Pakistan has grown increasingly tense in recent years, the geopolitical connection between India and the United States strengthened. In 2006, the United States and India agreed to step up military cooperation, hold more bilateral exercises, and devise more effective ways to share information. One year later, the U.S. government announced plans for an unprecedented $1.1 billion sale of military aircraft to India. Many observers think that India is gradually forging a new military alliance with not only the United States but also with Australia and Japan. Such a development is highly controversial within India, in part because many Indians think that their government should focus more on issues of economic development.

ECONOMIC AND SOCIAL DEVELOPMENT: Burdened by Poverty

South Asia is a land of developmental paradoxes. It is, along with Sub-Saharan Africa, the poorest world region, yet it is also the site of some immense fortunes. It has a sizable and growing middle class, yet large areas remain virtually cut off from the processes of development. Many of South Asia's scientific and technological accomplishments are world-class, but it also has some of the world's highest illiteracy rates. While South Asia's high-tech businesses are closely integrated with centers of the global information economy, the South Asian economy as a whole was long one of the world's most self-contained and inward-looking.

South Asian Poverty

By most measures, South Asia is the second poorest world region, with only Sub-Saharan Africa having a higher percentage of impoverished people. Roughly 300 million Indian citizens live below their country's official poverty line, which is set at a very meager level, and up to 40 million are thought to be looking for work (Figure 12.35). Roughly 20 percent of India's citizens are seriously undernourished, as are 25 percent of the people of Pakistan. Bangladesh is poorer still, with some 30 percent of its inhabitants classified as undernourished. By measures such as infant mortality and average longevity, Nepal is in even worse condition. In urban slums throughout South Asia, rapidly growing populations have little chance of finding housing or basic social services. Estimates indicate that up to half a million South Asian children work as virtual slaves in carpet-weaving workshops and other small-scale factories.

Despite such deep and widespread poverty, South Asia should not be regarded as a zone of uniform misery. India especially has a large and growing middle class, as well as a small but wealthy upper class. Roughly 250 million Indians are able to purchase such modern consumer items as televisions, motor scooters, and washing machines. India's economy grew from the 1950s to the 1990s at a moderate but accelerating pace, and by 2007 it was

Figure 12.35 Poverty in India
India's rampant poverty results in a significant amount of child labor. In this photo, a 10-year-old boy is moving a large burden of plastic waste by bicycle. *(Deshakalyan Chowdhury/AFP/Getty Images)*

expanding at a booming annual rate of more than 9 percent. But if several Indian states have shown marked economic vitality (Figure 12.36), others have seen only stagnation or even deterioration.

Geographies of Economic Development

After independence, the governments of South Asia attempted to create new economic systems that would benefit their own people rather than foreign countries or corporations. As in most other parts of the world, planners stressed heavy industry and economic autonomy. While some major gains were realized, the overall pace of development remained slow. Since the 1990s, governments in the region, and especially that of India, have gradually opened their economies to the global economic system. In the process, core areas of economic development have emerged, surrounded by large peripheral areas that have lagged behind, creating landscapes of striking economic disparity (Table 12.2).

The Himalayan Countries Both Nepal and Bhutan are disadvantaged by their rugged terrain and remote locations, as well as by the fact that they remain relatively isolated from modern technology and infrastructure. But such measurements are somewhat misleading, especially for Bhutan, because they fail to take into account the fact that many areas in the Himalayas are still subsistence-oriented.

Bhutan has purposely remained largely disconnected from the modern world economy, its small population living in a relatively pristine natural environment. Bhutan is so isolationist that it has only recently allowed tourists to enter—provided that they agree to spend substantial amounts of money while in the country. Its government has made the unusual move of downplaying conventional measures of economic development, attempting, for example, to substitute "gross national happiness" for "gross national product."

Bhutan is not, however, completely cut off from the rest of the world. It exports substantial quantities of hydroelectric power to India, helping its economy grow by 12 percent in 2006. Such growth has brought immigrants; approximately 100,000 Indian laborers are now working on roads and other infrastructural projects in Bhutan. Local resentment against these and other newcomers from India and especially Nepal has resulted in serious ethnic tensions, forcing up to 60,000 people to flee from Bhutan to refugee camps in Nepal.

Nepal is more heavily populated and suffers much more severe environmental degradation than Bhutan. It also is more closely integrated with the Indian, and ultimately the world, economy, although three-quarters of its people still depend on small-scale agriculture for their livelihoods. Nepal's economy relies heavily on international tourism. Tourism has brought some prosperity to a few favored locations, but often at the cost of heightened ecological damage. Tourism in Nepal began to suffer, moreover, after the country entered a period of political turmoil in 2002, undermining the country's already fragile economy.

Bangladesh The economic figures for Bangladesh are not as low as those of the Himalayan countries, but they are more indicative of widespread hardship because most people there require cash to meet their basic needs. Partly because of the country's massive population, poverty is extreme and widespread.

Environmental degradation has contributed to Bangladesh's impoverishment, as did the partition of 1947. Most of prepartition Bengal's businesses were located in the western area, which went to India. The division of Bengal tore apart an integrated economic region, much to the detriment of the poorer and mainly rural eastern section. Bangladesh also has suffered because of its agricultural emphasis on jute, a plant that yields tough fibers useful for making ropes and burlap bags. Bangladesh failed to discover any major alternative export crops as synthetic materials undercut the global jute market.

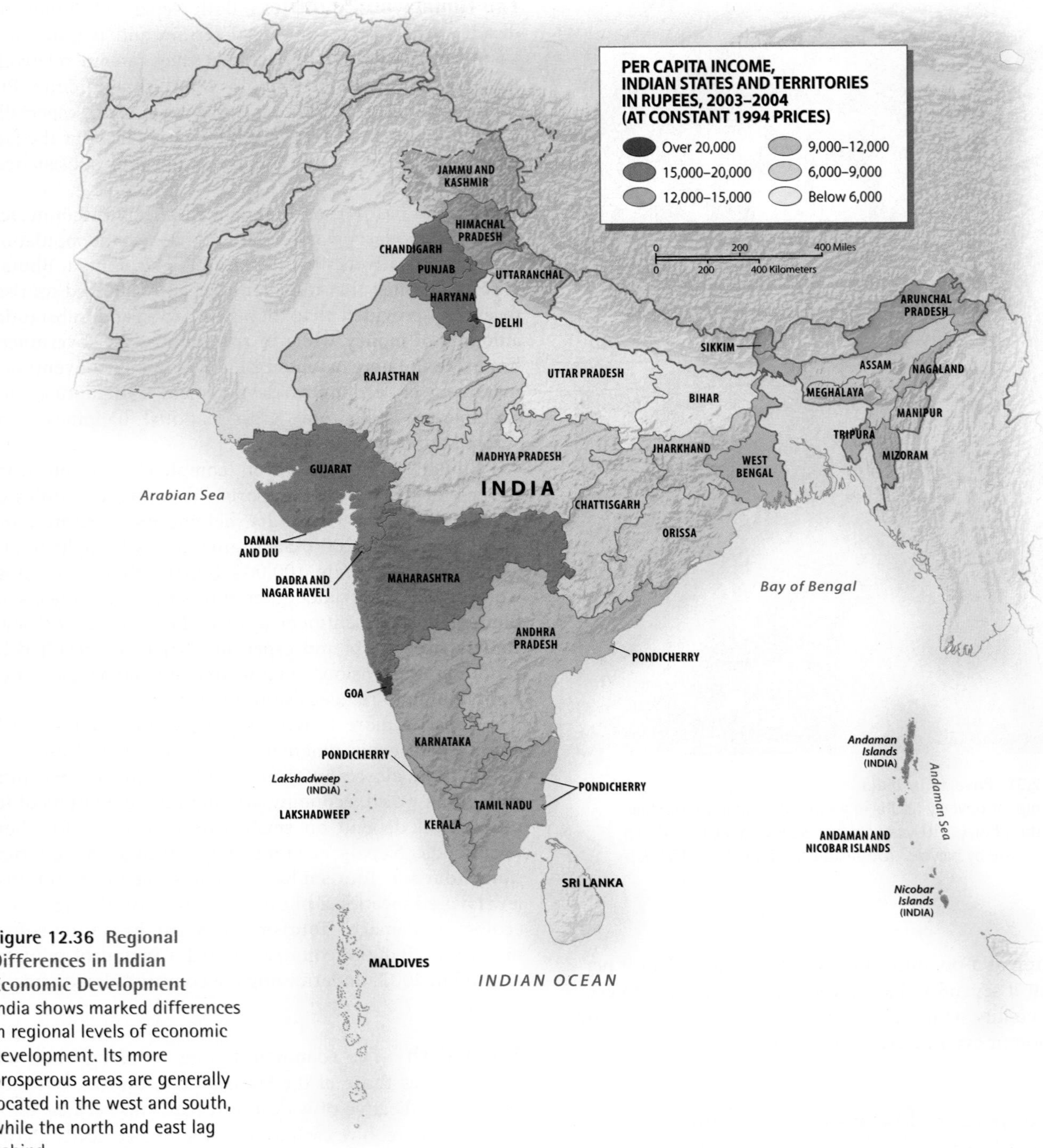

Figure 12.36 Regional Differences in Indian Economic Development India shows marked differences in regional levels of economic development. Its more prosperous areas are generally located in the west and south, while the north and east lag behind.

Not all of the economic news coming from Bangladesh is negative. The country is internationally competitive in textile and clothing manufacture, in part because its wage rate is so abysmally low. Low-interest credit provided by the internationally noted Grameen Bank has given hope to many poor women in Bangladesh, allowing the emergence of a number of vibrant small-scale enterprises (Figure 12.37). By 2000, with the country's birthrate steadily falling, the Bangladeshi economy was finally beginning to grow substantially faster than its population. By 2006, the economy was expanding at a healthy rate of 6.8 percent, but growing political instability in 2007 led many observers to question whether such a rate could be maintained.

Pakistan Pakistan also suffered the effects of partition in 1947. But unlike Bangladesh, Pakistan at least inherited a reasonably well-developed urban infrastructure. It has a productive agricultural sector, as it shares the fertile Punjab with India. Pakistan also boasts a large textile industry, based in part on its huge cotton crop. Several Pakistani cities, moreover, have developed important niches in the global economy (see "Global to Local: The

TABLE 12.2 Development Indicators

Country	GNI[a] per Capita, PPP[b] (2005)	GDP[c] Average Annual % Growth (2000–05)	Life Expectancy (2007)	Percent of Population Living on Less Than $2 a Day (2006)	Under Age 5 Mortality Rate 1990	Under Age 5 Mortality Rate 2005	Gender Equity[d]
Bangladesh	2090	5.4	62	83	149	73	101
Bhutan			64				
India	3460	7.0	64	80	123	74	87
Maldives			70				
Nepal	1530	2.8	62	69	145	74	88
Pakistan	2350	4.8	62	74	130	99	75
Sri Lanka	4520	4.2	74	42	32	14	102

[a]*Gross national income.*
[b]*Purchasing power parity.*
[c]*Gross domestic product.*
[d]*Ratio of female-to-male enrollments in primary and secondary school percentage. Numbers below 100 have more males in primary/secondary school; numbers above 100 have more females in primary/secondary schools.*
Source: World Bank, World Development Indicators, *2007; percentage of population living on $2 a day data from Population Reference Bureau,* World Data Sheet, *2006.*

Sialkot Industrial Complex"). Pakistan's economic growth accelerated after 2001, due in part to the concessions given to it by United States and international economic institutions for its cooperation in the fight against Afghanistan's Taliban regime. By 2006, Pakistan's economy was expanding at a brisk rate of 6.5 percent annually, its best showing in years.

Many experts argue, however, that Pakistan's economy is less dynamic than that of India, with a lower potential for growth. Part of the problem is that Pakistan is burdened by extremely high levels of military spending, yet at the same time it has experienced growing internal strife. Additionally, a small but powerful landlord class that pays virtually no taxes to the central government controls many of its best agricultural lands. Unlike India, moreover, Pakistan has not been able to develop a successful information-technology industry.

Like India, Pakistan has inadequate energy supplies, which has led it to look to both Central Asia and Southwest Asia for fossil fuels. One consequence of this policy has been the construction of a huge new deep-water port at Gwadar near the border with Iran, adjacent to some of the world's most important oil-tanker routes. Built with massive Chinese engineering and financial assistance, the port of Gwadar was turned over to be managed by the Port of Singapore when completed in 2007, showing the increasingly globalized nature of Pakistan's economic planning.

Figure 12.37 Grameen Bank This innovative institution lends money to rural women so they can buy land, purchase homes, or start cottage industries. In this photo, taken in Bangladesh, women proudly repay their loans to a bank official as testimony to their success. *(John Van Hasselt/Corbis/Sygma)*

GLOBAL TO LOCAL The Sialkot Industrial Complex

Pakistan does not have a highly globalized economy. Its exports are dominated by bulk cotton goods, mostly yarn, cloth, sheets, and towels rather than clothing, with rice, leather goods, and low-value chemicals rounding out the list. But one Pakistani city, Sialkot, stands out from all the rest, having carved a unique niche in the global economic system though its production of surgical steel implements and sporting goods (Figure 12.6.1). As a result, Sialkot is noted as Pakistan's most entrepreneurial city, enjoying a per capita income roughly double the national average.

During the Mughal period, Sialkot's weapon-makers were famous for their fine steel swords, but changes in warfare put them out of business by the late 1800s. When an American mission hospital was opened in the city, however, local artisans saw an opportunity to supply it with scalpels and other specialized implements. Before long, Sialkot was exporting such goods throughout British India, and then to the rest of the world.

The rise of the city's sporting goods industry is more difficult to explain. Legend says that it began in the late 1800s when an English tennis player broke his racket and asked a local artisan to repair it. He was evidently so impressed with the product that he commissioned additional replacements, and by 1895 a thriving new industry had emerged.

Today Sialkot's sporting goods industry is most famous for its hand-stitched soccer balls, exporting roughly 60 million each year. The city's highly competitive businesses, however, produce a huge array of other athletic goods, including equipment for horseback riding, motorcycle racing, boxing, and martial arts. The global nature of the industry is fully apparent in the production line of Saeed Classical Works (http://sportzbiz. pk/), which advertises such products as "Brazilian JiuJitsu Kimonos" sold under the "Bararian" brand name.

Despite its general success, Sialkot's sporting-goods business has faced a number of serious problems. In the 1990s, international criticism mounted over the use of child labor, particularly for hand-sewing soccer balls, an activity that rarely paid more than $1.50 a day. Since then, increased monitoring of factories has reduced labor violations, but abuses remain widespread. Another threat comes from China, which produces machine-assembled balls that dominate the lower end of the market. In response, Sialkot's producers have been striving to widen their product lines and to move into higher-end merchandise. The government of Pakistan offered help by opening a technical university in 2007. As befits the globalized nature of Sialkot's industrial complex, this new University of Engineering Science and Technology has been established in collaboration with Sweden's Royal Institute of Technology.

Figure 12.6.1 Sialkot Sporting Goods Firm Young employees work at a soccer-ball factory in Sialkot, Pakistan. Sialkot is noted for its exports of sporting goods and medical implements, but it has also been the site of much exploitative child labor. *(Ed Kashi/Corbis)*

Sri Lanka and the Maldives As can be seen in Table 12.2, Sri Lanka's economy is the second most highly developed in South Asia by conventional criteria. Its exports are concentrated in textiles and agricultural products such as rubber and tea. By global standards, however, it is still a very poor country. Its progress, moreover, has been undercut by its seemingly perennial civil war. Were it not for this conflict, Sri Lanka would benefit more from the prime location of the port of Colombo, from its high levels of education, and from its great tourism potential. In early 2007, however, renewed political violence resulted in a sharp decline in tourist receipts, further threatening an economy already suffering from high inflation and large budget deficits.

The Maldives is the most prosperous South Asian country based on per capita GNI, but its total economy, like its population, is very small. Most of its revenues are gained from fishing and international tourism. Eighty-eight of the Maldives' otherwise uninhabited islands have been turned over entirely to resorts, which are visited by some 600,000 tourists a year. The benefits from the tourist economy, however, flow mainly to the country's small elite population, resulting in large-scale public discontent and political repression.

India's Less-Developed Areas India's economy, like its population, dwarfs those of the other South Asian countries. While India's per capita GNI is comparable to that of Pakistan, its total economy is more than five times larger. As the region's largest country, India also exhibits far more internal variation in economic development than its neighbors. The most basic economic division is that between India's more prosperous west and its poorer districts in the east. A less clear-cut rift separates India's more dynamic south from its relatively stagnant north.

The so-called tribal states of India's northeastern fringe are generally near the bottom rungs of the economic ladder, as measured by per capita GNI, but the prevalence of subsistence economies makes such statistics misleading. More extreme deprivation is found in the lower Ganges Valley, where cash economies generally prevail. Bihar is India's poorest state by virtually all economic indicators, as well as its most politically corrupt.

Neighboring Uttar Pradesh, India's most populous state, also is extremely poor. Like Bihar, it is densely populated and has experienced little industrial development. Although both Bihar and Uttar Pradesh have fertile soils, their agricultural systems have not profited as much from the Green Revolution as have those of Punjab. In both states the caste system is deeply entrenched, tensions between Hindus and Muslims are bitter, and opportunities for most peasants are extremely limited. Ironically, South Asia's wealth was historically concentrated in the fertile lowlands of the Ganges Valley, yet today the area ranks among the poorest parts of an impoverished world region.

Other relatively poor states in eastern India include Orissa and West Bengal. The worst slums in India—and perhaps the world—are located in West Bengal's Kolkata (Calcutta). But Kolkata also supports a substantial and well-educated middle class, and it is the site of a large industrial complex. For most of the period of Indian independence, West Bengal has been governed by a leftist party that has fostered, without much success, heavy, state-led industry. In a dramatic turn-around in the 1990s, West Bengal's Marxist leaders began to advocate internationalization, encouraging large multinational firms to build new factories in "special economic zones" subjected to low levels of taxation. Such programs have generated substantial opposition. On May 2007, for example, 14 farmers were killed and over 100 were injured while protesting against the loss of farmland to a new chemical factory.

Although western India is in general much more prosperous than eastern India, the large western state of Rajasthan still ranks among the country's poorest areas. Rajasthan suffers from an arid climate; nowhere else in the world are deserts and semideserts so densely populated. It also is noted for its social conservatism. During the British period, almost all of this large state remained outside the sphere of direct imperial power. Here, in the courts of maharajas, the military and political traditions of Hindu India persisted up until recent times. Rajasthan's rulers not only maintained elaborate courts and fortifications, but also supported many traditional Indian arts. Because of this political and cultural legacy, Rajasthan is one of India's most important destinations for international tourists.

India's Centers of Economic Growth North of Rajasthan lie the Indian states of Punjab and Haryana, showcases of the Green Revolution. Their economies rest largely on agriculture, but recent investments in food processing and other industries have been substantial. Despite its relative prosperity, this region has seen rural unrest in recent years (see "People on the Move: Punjabis Leaving Agriculture and Leaving India"). On Haryana's eastern border lies the capital district of New Delhi, where India's political power and much of its wealth are concentrated.

India's west-central states of Gujarat and Maharashtra are noted for their industrial and financial clout, as well as for their agricultural productivity. Gujarat was one of the first parts of South Asia to experience substantial industrialization, and its textile mills are still among the most productive in the region (Figure 12.38). Gujaratis have long been famed as merchants and overseas traders, and they are disproportionately represented in the **Indian diaspora**, the migration of large numbers of Indians to foreign countries. As a result, cash remittances from these emigrants help to bolster the state's economy.

The state of Maharashtra is usually viewed as India's economic pacesetter. The huge city of Mumbai (Bombay) has long been the financial center, media capital, and

Figure 12.38 Gujarat Factory
A factory worker manages the flow of cotton as it blows out of the ginning equipment in a mill near Ahmedabad in the Indian state of Gujarat. Gujarat has long been a center of India's important textile industry. *(Aurora & Quanta Productions)*

manufacturing powerhouse of India. According to official figures, the Mumbai metropolitan area accounts for 10 percent of India's factory employment, 33 percent of its income tax collections, and 40 percent of its foreign trade. Major industrial zones are located in several other cities of Maharashtra, especially Pune and Nagpur. In recent years, Maharashtra's economy has grown more quickly than those of most other Indian states, reinforcing its primacy.

The center of India's fast-growing high-technology sector lies farther to the south, especially in Karnataka's capital of Bangalore, which in 2007 changed its name to Bengaluru. The Indian government selected the upland Bengaluru area, which is noted for its pleasant climate, for its fledgling aviation industry in the 1950s. Other technologically sophisticated ventures soon followed. In the 1980s and 1990s, a quickly growing computer software industry emerged, earning Bengaluru the label of "Silicon Plateau." In the 1980s, growth was spurred by the investments of U.S. and other foreign corporations eager to hire relatively inexpensive Indian technical talent. Since the 1990s, these multinational companies have been joined by an expanding group of locally owned firms. Unfortunately, Bengaluru's rapid growth has stretched the city's infrastructure to the breaking point. Roads are commonly jammed, electricity supplies are inadequate, and by 2007 most parts of the city could only count on three hours of water a day.

Partly because of Bengaluru's problems, other cities in southern India have recently emerged as high-tech centers. Hyderabad in Andhra Pradesh, often called "Cyberabad," is well known for its information technology and pharmaceutical firms. Chennai (Madras) in Tamil Nadu, recently voted as having the highest quality of life among India's major cities, is noted for its software production as well as its financial services and automobile industry.

India has proved especially competitive in software because software development does not require a sophisticated infrastructure; computer code can be exported via wireless telecommunication systems without the use of modern roads or port facilities. What is necessary, of course, is technical talent, and this India has in great abundance (Figure 12.39). Many Indian social groups are highly committed to education, and India has been a major scientific power for decades. With the growth of the software industry, India's brain power has finally begun to translate into economic gains. Whether such developments can spread benefits beyond the rather small high-tech enclaves they presently occupy remains to be seen. Most of India's rural areas are not prospering, and some evidence indicates that malnutrition has actually increased in recent years. What is certain, however, is that information technology has tightly linked certain parts of India to the global economy.

Globalization and India's Economic Future

As Figure 12.40 indicates, South Asia is not one of the world's most globalized regions by conventional economic criteria. The volume of foreign trade is not huge; foreign direct investment is still modest; and (with the exception of the Maldives) international tourists are few. But globalization in South Asia is advancing rapidly.

To understand the low globalization indicators for the region, it is necessary to look at its recent economic history. India's postindependence economic policy, like those of other South Asian countries, was based on widespread private ownership combined with high-tariff barriers and governmental control of planning, resource allocation, and certain heavy industrial sectors. This mixed socialist–capitalist system initially brought a fairly rapid development of heavy industry and allowed India to become virtually self-sufficient.

Figure 12.39 Indian Institutes of Technology
India's seven government-run institutes of technology provide world-class training for the country's top students in science and engineering, helping India develop globally oriented information technology industries. *(Balan Madhavan/Hornbil Images Pvt. Ltd.)*

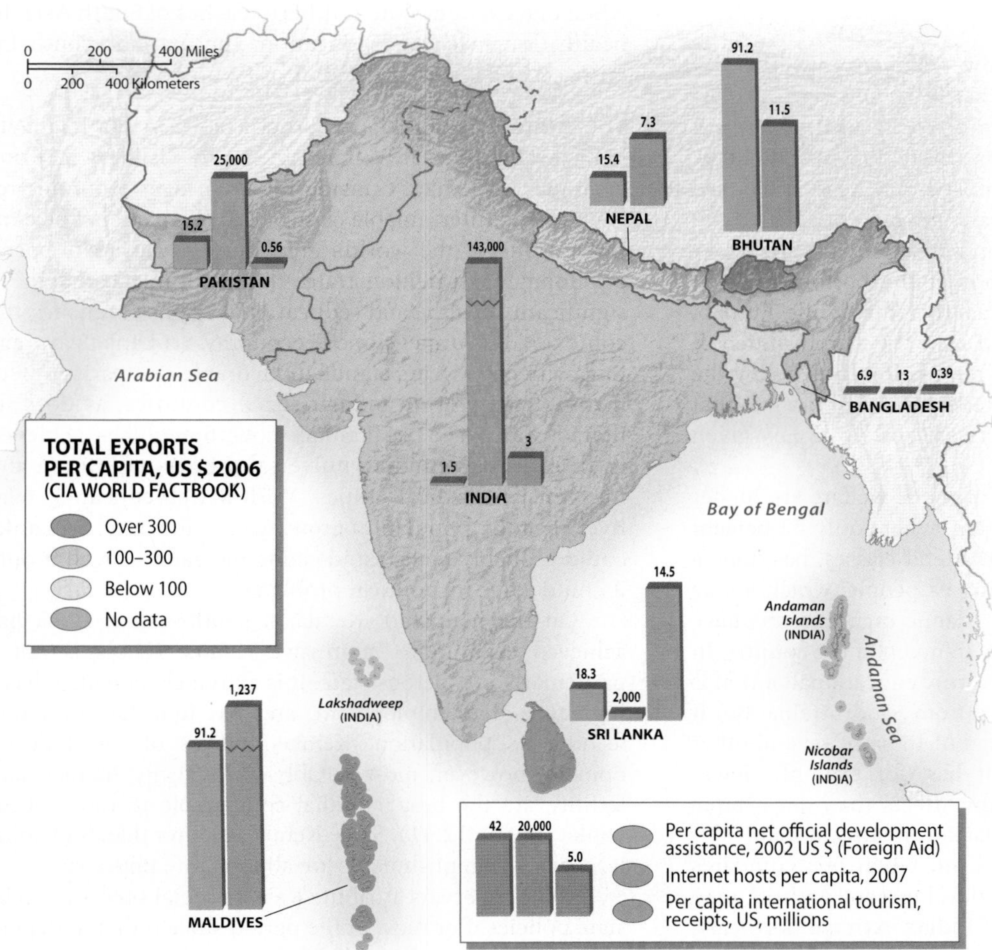

Figure 12.40 South Asia's Global Linkages
Despite South Asia's growing global connections, the region as a whole is still relatively self-contained, especially in regard to finance. The large countries of South Asia (especially India) receive very little foreign aid, whereas the small countries (Bhutan and the Maldives) receive substantial amounts. In regard to international tourism, the Maldives is simply off the chart. Internet use remains low, especially in Bangladesh and Nepal.

By the 1980s, however, problems with India's economic model were becoming apparent, and frustration was mounting among the business and political elite. While growth was persistent, it remained in most years only a percentage point or two above the rate of population expansion. The percentage of Indians living below the poverty line, moreover, remained virtually constant. At the same time, countries such as China and Thailand were experiencing far faster development after opening their economies to global forces. Many Indian businesspeople were irritated by the governmental regulations that undermined their ability to expand. In the 1980s, foreign indebtedness began to mushroom, putting further pressure on the economy.

In response to these difficulties, the Indian government began to liberalize its economy in 1991. Many regulations were modified and some were eliminated, and the economy was gradually opened to imports and multinational businesses. Other South Asian countries have followed a somewhat similar path. Pakistan, for example, began to privatize many of its state-owned industries in 1994, and in 2000 began to turn over the banking sector to private enterprise.

This gradual internationalization and deregulation of the Indian economy has generated substantial opposition. Foreign competitors are now seriously challenging domestic firms. Cheap manufactured goods from China are seen as an especially serious threat. India has a strong heritage of economic nationalism stemming from the colonial exploitation it long suffered. Agricultural liberalization, a central agenda of the World Trade Organization, became an especially contentious issue by the early 2000s because most of India's huge farming sector is not globally competitive. In 2006, mounting opposition led India's government to stall further plans for the privatization of state-controlled economic activities.

Although economic globalization is rapidly advancing in India, the country has not seen the levels of massive direct foreign investment that have recently transformed the Chinese economy. India also devotes a far smaller percentage of its national budget to infrastructural improvements than does China. Because of such investment shortages, India, like the rest of South Asia, does not have an adequate transportation system to meet its economic needs, and its supply of electricity is woefully short. As a result, many observers doubt that it will be able to match the growth rates that China has achieved, fearing that a rising inflation rate will force India to scale back its expectations.

Social Development

South Asia's social indices show relatively low levels of health and education, which is hardly surprising considering the region's poverty. Levels of social well-being, not surprisingly, vary greatly across the region. As might be expected, people in the more prosperous areas of western India are healthier, live longer, and are better educated, on average, than people in the poorer areas, such as the lower Ganges Valley. Bihar thus stands at the bottom of most social as well as economic measurements, while Punjab, Gujarat, and Maharashtra stand near the top. In much of Bihar, where almost half of the population lives below the poverty line, the situation is dire; considering the fact that 38 percent of the state's teachers are absent on any given day, progress will not be easy.

Several key measurements of social welfare are higher in India than in Pakistan. Pakistan, where only 52 percent of primary school-aged children attend classes, has done a particularly poor job of educating its people, which is one reason extreme fundamentalist Islamic organizations have been able to recruit so effectively in much of the country. In areas near the Afghan border, the only educational option is often the Islamic schools funded from Saudi Arabia. But it also is true that Pakistan is ahead of India in several other areas of social development. It has, for example, fewer beggars and people living on the streets than does India, thanks partly to its widespread Islamic charity system.

Several discrepancies stand out when one compares South Asia's map of economic development with its map of social well-being. Portions of India's extreme northeast, for example, have relatively high literacy rates despite their poverty, owing to the educational efforts of Christian missionaries. Kolkata (Calcutta) and its immediate environs also stand out as a relatively well-educated area, despite the general distress of the lower Ganges Basin. The most pronounced discrepancies, however, strike the eye when one examines the southern reaches of South Asia. In health, longevity, and education, the extreme south far outpaces the rest of the region.

The Educated South Southern South Asia's relatively high levels of social welfare are clearly visible when one examines Sri Lanka. Considering its meager economic resources and interminable civil war, Sri Lanka must be considered one of the world's great success stories of social development. It demonstrates that a country can achieve significant health and educational gains even in the context of an "undeveloped" economy. Sri Lanka's average longevity of 74 years stands in favorable comparison with many of the world's industrialized countries, as does its literacy rate. The Sri Lankan government has achieved these results by funding universal primary education and inexpensive medical clinics. With a well-educated, relatively healthy population growing at a slow rate, Sri Lanka could probably make rapid economic gains as well if only it could solve its political problems.

On the mainland, Kerala in southwestern India has achieved even more impressive results. Kerala is not a particularly prosperous state. It is extremely crowded, has a high rate of unemployment, and has long had difficulty feeding its population. Kerala's indices of social development, however, most notably in longevity, literacy, and fertility, are the best in India, comparable to those of Sri Lanka (Figure 12.41). Since Kerala is poorer than Sri Lanka, its social accomplishments are all the more impressive.

Some observers attribute Kerala's social successes to its state policies. For most of the period since Indian independence, Kerala has been led by a socialist party that has stressed mass education and community health care. Although no doubt an important factor, this does not seem to offer a complete explanation. West Bengal, for example, also has a socialist political heritage, but it has not been nearly as successful in its social programs. Kerala's

Figure 12.41 Education in Kerala India's southwestern state of Kerala, which has virtually eliminated illiteracy, is South Asia's most highly educated region. It also has the lowest fertility rate in South Asia. Because of this, many argue that women's education and empowerment is the best and most enduring form of contraception. *(Rob Crandall/ www.robcrandall.com)*

neighboring state of Tamil Nadu, on the other hand, also has made significant social progress despite having a different political environment. Some researchers suggest that one of the key variables for explaining the success of the far south is the relatively high social position of its women.

The Status of Women It is often said that South Asian women are accorded a very low social position in both the Hindu and Muslim traditions. In higher-class families of both religions throughout the Indus–Ganges Basin, women traditionally were secluded to a large degree, their social relations with men outside the family being severely restricted.

Some scholars have argued that the Hindu tradition is more limiting to women than the Muslim tradition. Hindu women are expressly forbidden to engage in certain economic activities (such as plowing), and in some areas they are excluded from inheriting land. Throughout northern India, women traditionally leave their own families shortly after puberty to join those of their husbands. As outsiders, often in distant villages, young brides have few opportunities, and it is not uncommon for them to be bullied by their parents-in-law. Widows in the higher castes, moreover, are not supposed to remarry, and instead are encouraged to go into permanent mourning.

Several social indices show that women in the Indus–Ganges Basin suffer discrimination. In Pakistan, Bangladesh, and such Indian states as Rajasthan, Bihar, and Uttar Pradesh, female levels of literacy are far lower than those of males. An even more telling statistic is that of gender ratios, the relative proportion of males and females in the population. All things being equal, there should be slightly more women than men in any population because women generally have a longer life expectancy. Northern South Asia, however, contains many more men than women.

Economics play a major role in India's biased sex ratios. In rural households, boys are usually viewed as a blessing because they typically remain with their families and work for their well-being. In the poorest groups, elderly people (especially widows) subsist largely on what their sons can provide. Girls, on the other hand, marry out of their families at an early age and must be provided with a dowry. They are thus seen as a net economic liability. Considering the economic desperation of the South Asian poor, it is perhaps not surprising to see such shortages of girls and young women.

Evidence suggests that the social position of women is improving, especially in the more prosperous parts of western India where employment opportunities outside the family context are emerging. But even in many of the region's middle-class households, women still suffer major disabilities. Dowry demands seem to be increasing in many areas, and there have been a number of well-publicized murders of young brides whose families failed to provide enough goods. In some parts of South Asia, gender ratios are growing even more male-biased now that technology allows gender-selective abortion. Laws against this practice have generally been ignored. In 2006, however, an Indian doctor was for the first time sentenced to prison for revealing the sex of a female fetus and then agreeing to perform an abortion.

While the social bias against women across northern South Asia is striking, it is much less evident in southern India and Sri Lanka. In Kerala especially, women have relatively high status, regardless of whether they are Hindus, Muslims, or Christians. Here the overall gender ratio shows the normal pattern, with a slight predominance of females over males. Female literacy is very high in Kerala, which is one reason the state's overall illiteracy rate is so low. Kerala's fertility rate is the lowest in India, which is another sign of women's social power. Not surprisingly, the high social position of women in southwestern India has deep historical roots. Among the Nairs—Kerala's traditional military and land-holding caste—all inheritance up to the 1920s had to pass through the female line. By 2000, however, census data showed that even in Kerala the number of girls being born relative to the number of boys was declining, showing some evidence of sex-selective abortion.

Summary

- South Asia, a large and complex area of more than a billion people, has in many ways been overshadowed by neighboring world regions: by the uneven globalization of Southeast Asia, by the size and political weight of East Asia, and by the geopolitical tensions of Southwest Asia. Much of that is changing, however, as South Asia now figures prominently in discussions of world problems and issues.

- Environmental degradation and instability pose particular problems for South Asia. The region's monsoon climate causes both floods and droughts to be more problematic here than in most other world regions. Rising sea level associated with global climate change directly threatens the low-lying Maldives, and changes in rainfall may play havoc with the monsoon-dependent agricultural systems of India, Pakistan, and Bangladesh.

- Continuing population growth in this already densely populated region demands attention. Although fertility rates have declined in recent years, Pakistan, northern India, and Bangladesh cannot easily meet the demands imposed by their expanding populations.

- South Asia's diverse cultural heritage, shaped by peoples speaking several dozen languages and following several major religions, makes for a particularly rich social environment. Unfortunately, cultural differences have often translated into political conflicts. Ethnically or religiously based separatist movements have severely challenged the governments of Pakistan, India, and Sri Lanka. In India, moreover, religious strife between Hindus and Muslims persists, whereas in Pakistan and Bangladesh Islamic radicals clash with the state.

- Geopolitical tensions within South Area are particularly severe, again demanding global attention. The long-standing feud between Pakistan and India escalated dangerously in the late 1990s, leading many observers to conclude that this was the most likely part of the world to experience a nuclear war. Although tensions between the two countries have lessened, the underlying sources of conflict—particularly the struggle in Kashmir—remain unresolved.
- Although South Asia remains one of the poorest parts of the world, much of the region has seen rapid economic expansion in recent years. Many argue that India in particular is well positioned to take advantage of economic globalization. Large segments of its huge labor force are well educated and speak excellent English, the major language of global commerce. But will these global connections help the vast numbers of India's poor or merely the small number of its economic elite? Advocates of free markets and globalization tend to see a bright future, while skeptics more often see growing problems.

Key Terms

British East India Company *(page 555)*
caste system *(page 547)*
cyclone *(page 530)*
dalit *(page 548)*
Dravidian language *(page 551)*
federal state *(page 559)*
forward capital *(page 544)*
Green Revolution *(page 540)*
Hindu nationalism *(page 545)*
Indian diaspora *(page 569)*
Jainism *(page 550)*
linguistic nationalism *(page 553)*
maharaja *(page 559)*
monsoon *(page 533)*
Mughal Empire (also spelled *Mogul*) *(page 547)*
orographic rainfall *(page 534)*
salinization *(page 542)*
Sikhism *(page 550)*
subcontinent *(page 532)*
Tamil Tigers *(page 561)*

Questions for Review

1. What are the four subregions of South Asia? Describe their similarities and differences.
2. What causes the South Asian monsoon? How does it affect different parts of South Asia?
3. What is orographic rainfall? Where is it important for South Asian agriculture?
4. How and why does the birthrate differ geographically within South Asia?
5. What are some of the infrastructural problems faced by South Asian cities? Give some specific examples.
6. Describe the geography of Islam within India. That is, where are the significant Muslim minorities located in India?
7. What are the major Indo-European languages in South Asia? Where are they located? Where are the non-Indo-European languages located?
8. Describe three different regions of geopolitical and ethnic tension within South Asia.
9. Where are the centers or core areas of economic development within the different South Asian countries?
10. What kinds of relationships are seen between women's literacy and different aspects of economic and social development?

Thinking Geographically

1. As a geographer, suggest different strategies for solving (or at least lessening) the serious flooding problems in Bangladesh. Consider the fact that this crowded country must maximize most of its area for agricultural production.
2. What are the advantages and disadvantages of expanding irrigated agriculture in India and Pakistan? How can the disadvantages be reduced to acceptable levels?
3. What are the pros and cons of the Green Revolution as a means of increasing South Asia's food supplies? What is the outlook for the next decade?
4. Discuss the conflict between wildlife protection and rural villages in India by evaluating the tension between preserving habitat and the needs of the rural poor.

5. What are the drawbacks and benefits of using English as a national language in India? Might it help or hinder unity? Would this increase or decrease India's links to the contemporary world?
6. Choose one of the areas of geopolitical tension (Kashmir, Punjab, Sri Lanka, etc.) and, after becoming better acquainted with the complex issues that underlie this conflict, evaluate the different proposals currently offered for solving (or at least ameliorating) the problem.
7. As a geographer, you work for an international arms reduction agency that is working to reduce tensions that could lead to nuclear war between Pakistan and India. What would you suggest?
8. From a geographical point of view, what is the best course of action for near-term future economic development in Pakistan, India, and Bangladesh?
9. Acquaint yourself with the environmental, social, and economic implications of Nepal's open-door policy toward trekking and other forms of tourism. Do the benefits seem to outweigh the costs?
10. Is the state of Kerala a good model for economic and social development in other parts of India? Why?

Regional Novels and Films

Novels

Anita Desai, *Fasting, Feasting* (1999, Houghton Mifflin)

E. M. Forster, *A Passage to India* (1924, Harcourt, Brace and Company)

Bharti Kirchner, *Darjeeling: A Novel* (2002, St. Martin's Press)

Arundhati Roy, *The God of Small Things* (1997, Random House)

Salman Rushdie, *Midnight's Children* (1981, Knopf)

Paul Scott, *The Raj Quartet* (1980, Morrow)

Vikram Seth, *A Suitable Boy* (1993, HarperCollins)

Bapsi Sidhwa, *Cracking India: A Novel* (1991, Milkweed)

Manil Suri, *The Death of Vishnu* (2001, Norton)

Thirty Umrigar, *Bombay Time* (2001, Picador)

Films

August Sun (Sri Lanka, 2003)

Bandit Queen (1994, India)

City of Joy (1992, UK)

Earth (1998, India)

East Is East (1999, UK)

Matir Moina (2002, Bangladesh)

Monsoon Wedding (2001, India)

Saalam Bombay (1988, India)

Saira (2007, India)

A Thousand Dreams Such As These (2003, India)

Bibliography

Bose, Sugata, and Jalal, Ayesha. 2002. *Modern South Asia: History, Culture, and Political Economy*. Delhi: Oxford University Press.

Eaton, Richard M. 1993. *The Rise of Islam and the Bengal Frontier, 1204–1760*. Berkeley: University of California Press.

Fox, Richard G., ed. 1977. *Realm and Region in Traditional India*. Durham, NC: Duke University Program in Comparative Studies on Southern Asia.

Hussain, Zahid. 2007. *Frontline Pakistan: The Struggle with Militant Islam*. New York: Columbia University Press.

Luce, Edward. 2007. *In Spite of the Gods: The Strange Rise of Modern India*. New York: Doubleday.

Schwartzberg, Joseph E. 1992. *A Historical Atlas of South Asia*. Oxford, UK: Oxford University Press.

Spate, O. H. K., and Learmonth, A. T. A. 1967. *India and Pakistan: A General and Regional Geography*. London: Methuen.

Steven, Stanley F. 1993. *Claiming the High Ground: Sherpas, Subsistence, and Environmental Change in the Highest Himalaya*. Berkeley: University of California Press.

Wolpert, Stanley. 1991. *India*. Berkeley: University of California Press.

Zurick, David, and Karan, P. P. 1999. *Himalaya: Life on the Edge of the World*. Baltimore: Johns Hopkins University Press.

Additional bibliographic resources are at the *Diversity Amid Globalization* Website: http://www. prenhall.com/rowntree/.

13 Southeast Asia

Large expanses of forest are being cleared in many parts of Southeast Asia in order to plant African oil palms. Oil palms yield large quantities of edible oil, much of which is now used for the production of bio-diesel fuel. Although bio-diesel is often considered to be an environmentally responsible form of energy, the creation of oil palm plantations of Southeast Asia releases large amounts of carbon dioxide and is thus a contributor to global climate change. *(Claire Leow/Bloomberg News/Landov)*

ENVIRONMENTAL GEOGRAPHY
Southeast Asia's rain forests are vital centers of biological diversity, but they are rapidly disappearing due to commercial logging and agricultural expansion.

POPULATION AND SETTLEMENT
Southeast Asia has a particularly uneven pattern of population distribution, with some areas experiencing serious crowding while others are noted for their very sparse settlement.

CULTURAL COHERENCE AND DIVERSITY
Culturally, Southeast Asia is characterized by much more diversity than coherence, hosting significant areas of Muslim, Buddhist, and Christian religions, as well as innumerable languages.

GEOPOLITICAL FRAMEWORK
Southeast Asia is one of the most geopolitically unified regions of the world, with all but one of its countries belonging to the Association of Southeast Asian Nations (ASEAN).

ECONOMIC AND SOCIAL DEVELOPMENT
Southeast Asia as a whole contains some of the world's most globalized and advanced economies and some of the most isolated and impoverished; it has also experienced marked periods of boom and bust in recent decades.

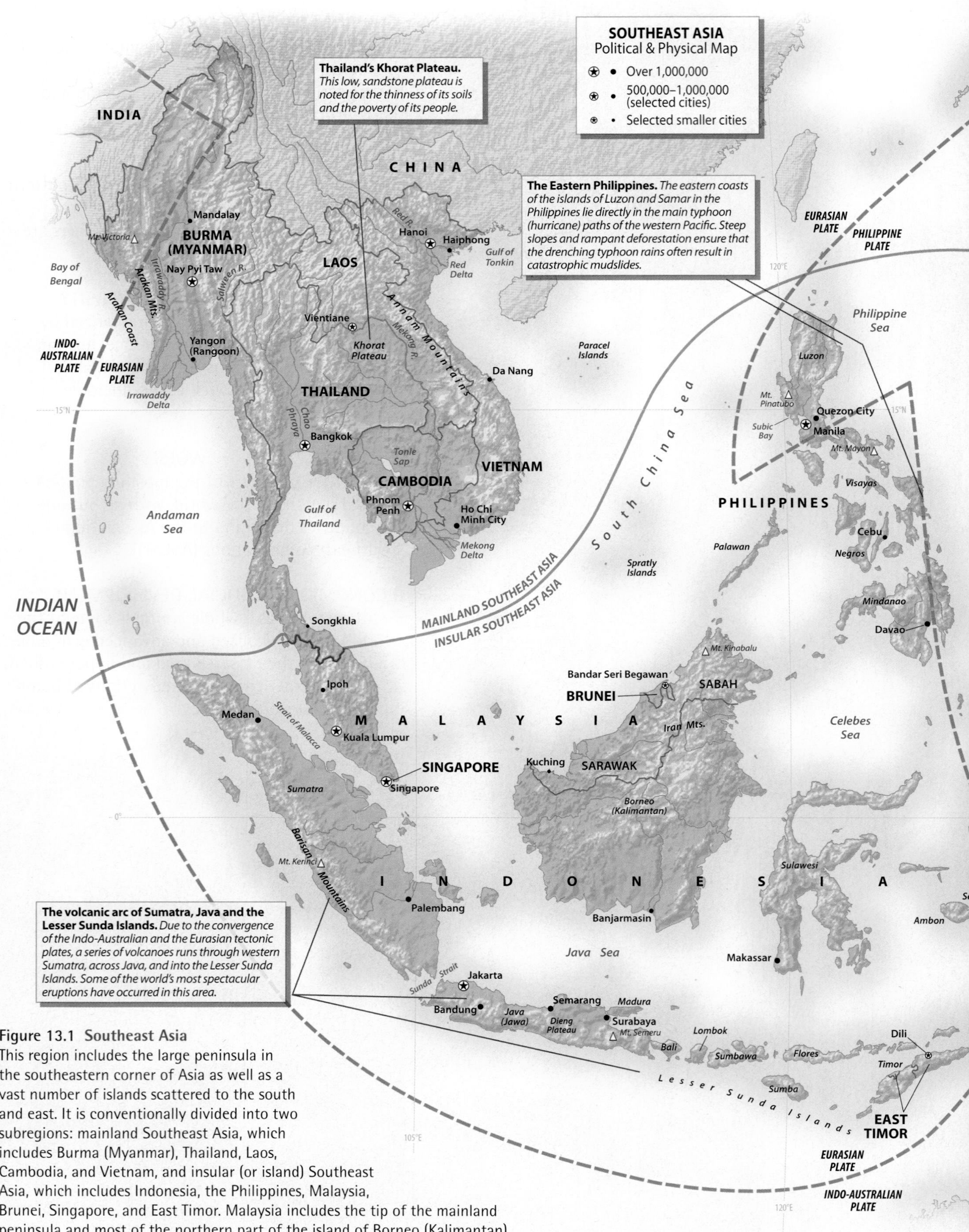

Figure 13.1 Southeast Asia
This region includes the large peninsula in the southeastern corner of Asia as well as a vast number of islands scattered to the south and east. It is conventionally divided into two subregions: mainland Southeast Asia, which includes Burma (Myanmar), Thailand, Laos, Cambodia, and Vietnam, and insular (or island) Southeast Asia, which includes Indonesia, the Philippines, Malaysia, Brunei, Singapore, and East Timor. Malaysia includes the tip of the mainland peninsula and most of the northern part of the island of Borneo (Kalimantan).

Southeast Asia occupies an important place in contemporary discussions of globalization (Figure 13.1). This highly diverse region includes some of the world's most globally networked countries, such as Singapore, as well as some of the countries most resistant to global economic and cultural forces, most notably Burma (Myanmar). Debates about the benefits and drawbacks of economic globalization have also often focused on Southeast Asia. Although human well-being has seen major gains in many parts of the region, the sweatshops associated with low-cost global production are criticized for their low wages and harsh labor practices. The globalization debate in regard to Southeast Asia grew especially heated in the 1990s, as the region's most important countries experienced a roller-coaster ride of economic boom and bust, descending almost overnight from the giddy heights of fast-growing "tiger economies" to severe recessions. More stable economic conditions have emerged in recent years, but Southeast Asia as a whole is still highly vulnerable to fluctuations in the global economy.

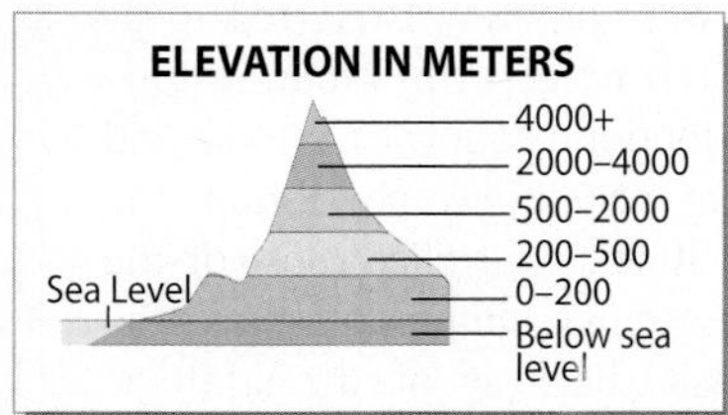

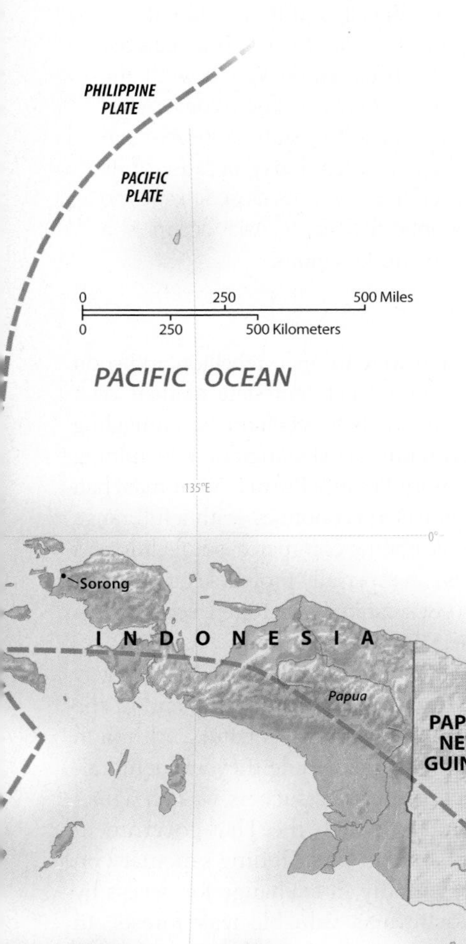

Southeast Asia's involvement with the larger world is not new. Chinese and especially Indian influences date back many centuries. Later, commercial ties with the Middle East opened the doors to Islam, and today Indonesia ranks as the most populous Muslim country in the world. More recently came the heavy-handed imprint of Western colonialism, as Britain, France, the Netherlands, and the United States administered large Southeast Asian colonies. During this period, national territories were rearranged, populations relocated, and new cities built to serve trade and military needs.

Southeast Asia's resources and its strategic location made it a major battlefield during World War II. Yet long after the end of this global conflict in 1945, warfare of a different sort continued in the region. Resistance to imperialism mounted, and after the colonial regimes were replaced by newly independent countries, Southeast Asia became a battleground for world powers and their different philosophical systems. In Vietnam, Laos, and Cambodia, communist forces, supported by China and the Soviet Union, struggled for control of local territory and people. Resisting this were the United States and several of its allies, concerned that communism would rapidly spread across the whole of Southeast Asia.

Ironically, while communist forces did prevail in Vietnam, Laos, and Cambodia, these countries subsequently opened their economies to global capitalism. With the end of the Cold War, competing political beliefs have taken a back seat to other problems in Southeast Asia. Accompanying the economic turmoil of the past decade has been an increase in ethnic and social tensions, especially in Indonesia. Geopolitically, however, the **Association of Southeast Asian Nations (ASEAN)**, which includes 10 of the region's 11 countries, has brought a new level of regional cooperation to the area (see "Setting the Boundaries").

ENVIRONMENTAL GEOGRAPHY:
A Once-Forested Region

The mountainous area along the border between northern Thailand and Burma (Myanmar) is a rugged place. Slopes are steep and thickly wooded in most places; rainfall is often torrential; leeches abound; and several strains of medicine-resistant malaria are prevalent. These uplands are home to one of the most distinctive Southeast Asian ethnic groups: the Karen, a people numbering roughly 7 million. Unfortunately, the story of the Karen and their homeland in eastern Burma (Myanmar) is not a happy one. The Burmese army has overrun much of their territory, forcing many of the Karen into squalid refugee camps in Thailand, where they still suffer periodic attacks from

Setting the Boundaries

Southeast Asia consists of 11 countries that vary widely in spatial extent, population, cultural attributes, and levels of economic and social development. Geographically, the region is commonly divided into an Asian mainland zone and the islands, or the insular realm. The mainland includes Burma (called *Myanmar* by the current government), Thailand, Cambodia, Laos, and Vietnam. Of these countries, Burma (Myanmar) is the largest in area, about the same size as Texas. Thailand is next in size, covering an area somewhat larger than California. Vietnam has the largest population of the mainland states, with 85 million people.

Insular Southeast Asia includes the sizable countries of Indonesia, the Philippines, and Malaysia, as well as the small countries of Singapore, Brunei, and East Timor. Although classified as part of the insular realm because of its cultural and historical background, Malaysia actually splits the difference between mainland and islands. Part of its national territory is on the mainland's Malay Peninsula and part is on the large island of Borneo (also known as Kalimantan), some 300 miles (480 kilometers) distant. Borneo also includes the Muslim sultanate of Brunei, a small but oil-rich island country of 400,000 people covering an area slightly larger than Rhode Island. Singapore is essentially a city-state, occupying a small island just to the south of the Malay Peninsula.

The island nation of Indonesia stretches 3,000 miles (4,800 kilometers, or about the same distance as from New York to San Francisco) from the large island of Sumatra in the west to New Guinea in the east, and contains more than 13,000 separate islands. Not only does it dwarf all other Southeast Asian states in size, but it is by far the largest in population. With more than 231 million people, Indonesia is the world's fourth most populous country. Lying north of the equator is the Philippines, a country of 91 million people spread over thousands of islands, both large and small.

Until the second half of the twentieth century, much of this world region was sometimes referred to as *Indochina*, a term that accurately reflects the strong historical influences of the large neighboring countries of India and China. Western colonial powers, including France, Britain, the Netherlands, and the United States, controlled most of the region until World War II, when Japan temporarily expanded its empire into Southeast Asia. Because of the strategic importance of this region, many heated battles were fought on its territory during World War II, and it was during this period that the geographic term *Southeast Asia* replaced *Indochina* and similar terms. After World War II, with the gradual and often reluctant withdrawal of the colonial powers' hold on territory, newly independent states appeared and the modern geopolitical map emerged. Today, because of its continued strategic value, coupled with its close linkages to the dynamic world economy, Southeast Asia occupies a prominent place in the list of world regions.

Burma (Myanmar). Furthermore, the Karen's loss of land has been accompanied by the logging of the magnificent teak forests of upland Burma (Myanmar). In the saga of the Karen, we can see the tragic meshing of cultural, political, economic, and environmental forces.

The Tragedy of the Karen

The Karen, like other tribal peoples of the upland areas of Burma (Myanmar), were never fully incorporated into the Burmese kingdom, which long ruled the lowlands of the country. With the imposition of British colonial rule in the 1800s, however, the Karen territory was joined to the Burmese lowlands. British and American missionaries educated many Karen and converted over 30 percent of them to Protestant Christianity. A number of Karen Christians obtained positions in Burma's colonial government. The Burmans of the lowlands, a strongly Buddhist people, resented this deeply, for they had long viewed the Karen as culturally inferior (according to conventional but confusing terminology, the term *Burmese* refers to all of the inhabitants of Burma [Myanmar], whereas *Burmans* refers only to the country's dominant, Burmese-speaking ethnic group). After independence, the Karen lost their favored position and soon grew to resent what they saw as Burman cultural and economic domination.

By the 1970s, the Karen were in open rebellion and soon managed to establish a semi-independent state of their own (Figure 13.2). They supported their rebellion by smuggling goods between Thailand and Burma (Myanmar) and by mining the gemstones of their territory. Because Burma (Myanmar) had one of the world's most protected economies, with a full range of governmental controls inhibiting commerce, smuggling was (and to some extent still is) an especially profitable occupation. Some estimates in the 1980s ranked the Karen economy as almost as large as the official Burmese economy.

The Burmese army, however, began to make headway against the rebels in the early 1990s, and by the end of the decade had overrun most of the Karen territory. Although some Karen continue to fight, as a whole they are demoralized and divided. Crucial to Burma's success was an agreement made with Thailand in which the Thai government agreed to prevent Karen soldiers from finding sanctuary on its side of the border, reportedly in exchange for access by Thai timber interests to Burma's valuable teak forests. In early 2007, a renewed round of fighting led large numbers of Karen to seek sanctuary across the border, but in several instances Thai border authorities denied entry to the refugees. All told, some 200,000 Karen people were driven from their homes by the Burmese military between 1994 and 2004.

Thailand's interests in Burma's forests stems from its own environmental problems. Thailand was once a major exporter

Figure 13.2 Karen Rebels
The Karen people have been in rebellion against Burma (Myanmar) since the 1970s. For several years they maintained a separate insurgent state, with its own capital city and regular army. The Burmese military advanced in the 1990s, however, forcing the Karen back to a guerrilla-style war. *(Dean Chapman/Panos Pictures)*

of teak and other tropical hardwoods, but by the 1990s the country had been largely deforested. The Thai government responded by banning commercial logging within the country. The only way Thai logging firms could stay in business was to move into the extensive forests of Burma (Myanmar), Laos, and Cambodia, countries that had previously been relatively isolated from the world economy, and therefore little touched by commercial logging operations. The price of admission in the case of Burma (Myanmar) was for Thailand to crack down on Karen guerrillas on its side of the border.

The Deforestation of Southeast Asia

While the story of the Karen and Burma's teak forests is unique, deforestation and related environmental problems are major issues throughout most of Southeast Asia (Figure 13.3). Globalization has had a particularly profound effect on the Southeast Asian environment. Export-oriented logging companies have reached deep into the region's forests, cutting trees, damaging watersheds, and severely reducing biodiversity.

Although colonial powers cut Southeast Asian forests for tropical hardwoods and naval supplies, and indigenous peoples have long cleared small areas of forest for agricultural use, rampant deforestation has come only in the past several decades with large-scale international commercial logging (Figure 13.4). This activity is largely driven by Asia's seemingly insatiable appetite for wood products such as plywood and paper pulp.

Most experts agree that Japan first globalized world forestry in the 1960s, but other Asian countries, such as Taiwan, South Korea, Malaysia, and Indonesia, have followed suit with their own wood-products firms. Although the headquarters, boardrooms, and profits of these companies may be global, the damaging effects of commercial logging are both local and global. Landscapes are denuded, watersheds destroyed, wildlife habitat devastated. Additionally, the costs are high for rural peoples who rely on forest resources for their traditional way of life.

Although countries such as Indonesia look to their forestlands for increasing food supplies through expanded agriculture and for relieving population pressure in their more densely settled areas, farming and population growth usually are not the main cause of deforestation. Most forests are cut so that the wood products can be exported to other parts of the world. Subsequently, some of the logged-off lands are replanted (often with fast-growing, "weedy" tree species), while others are opened for agricultural settlement.

Local Patterns of Deforestation Malaysia has long been a leading exporter of tropical hardwoods. In recent decades, 60 percent of these log exports went to Japan. The cost to the environment, however, has been high. Peninsular Malaysia was largely denuded by 1985, when a cutting ban was imposed. Since then, forest cutting has accelerated in the states of Sarawak and Sabah on the island of Borneo, where the granting of logging concessions to Malaysian and foreign firms has caused considerable problems with local tribal people by disrupting their traditional resource base.

Indonesia, the largest country in Southeast Asia, has fully two-thirds of the region's forest area, including about 10 percent of the world's true tropical rain forests. Its forest coverage, however, is deeply threatened; between 1990 and 2005, Indonesia lost an estimated 69 million acres (28 million hectares) of forest. Most of Sumatra's primary forests are gone, and those of Borneo (Kalimantan) will not last long at present rates of cutting. Indonesia's last forestry frontier is on the island of New Guinea. But Indonesia, which is still covered in 119 million acres (48 million hectares) of primary forests, is fortunate in comparison to the Philippines, which has lost most of its original forest coverage (see "Geographic Tools: Statistical Comparisons and the Assessment of Philippine Deforestation").

Thailand cut more than 50 percent of its forests between 1960 and 1980. This loss was followed by a series of logging bans, making forest cutting illegal by 1995. Damage to the landscape, however, was severe; flooding increased in lowland areas, and erosion on hillslopes led to such problems as the accumulation of silt in irrigation works and hydroelectric facilities. Increasingly, these cutover lands are being reforested with fast-growing Australian eucalyptus trees, a nonnative species that cannot support local wildlife. As a result of the Thai logging ban, the forestry frontier in mainland Southeast Asia has moved into the rugged uplands of Laos, Cambodia, and Burma (Myanmar).

Throughout the coastal areas of Southeast Asia, a more specific problem is the destruction of the mangrove forests that thrive in shallow and silty marine areas. Often mangrove forests are burned for charcoal, but may also be converted to fish and shrimp ponds as well as rice fields and oil palm plantations. As mangrove forests serve as nurseries for many fish species, their destruction threatens to undercut a number of important Southeast Asian fisheries.

Protected Areas Despite rampant deforestation, Indonesia, like most other Southeast Asian countries, has created a number of large national parks and other protected areas. Kutai National Park in the province of East Kalimantan, for example, covers more than 741,000 acres (300,000 hectares). Southeast Asian rain forests are among the most biologically diverse areas on the planet, containing a large number of species that are found

Figure 13.3 Environmental Issues in Southeast Asia
Southeast Asia was once one of the most heavily forested regions of the world. Most of the tropical forests of Thailand, the Philippines, peninsular Malaysia, Sumatra, and Java, however, have been destroyed by a combination of commercial logging and agricultural settlement. The forests of Borneo (Kalimantan), Burma (Myanmar), Laos, and Vietnam, moreover, are now being rapidly cleared. Water and urban air pollution, as well as soil erosion, are also widespread in Southeast Asia.

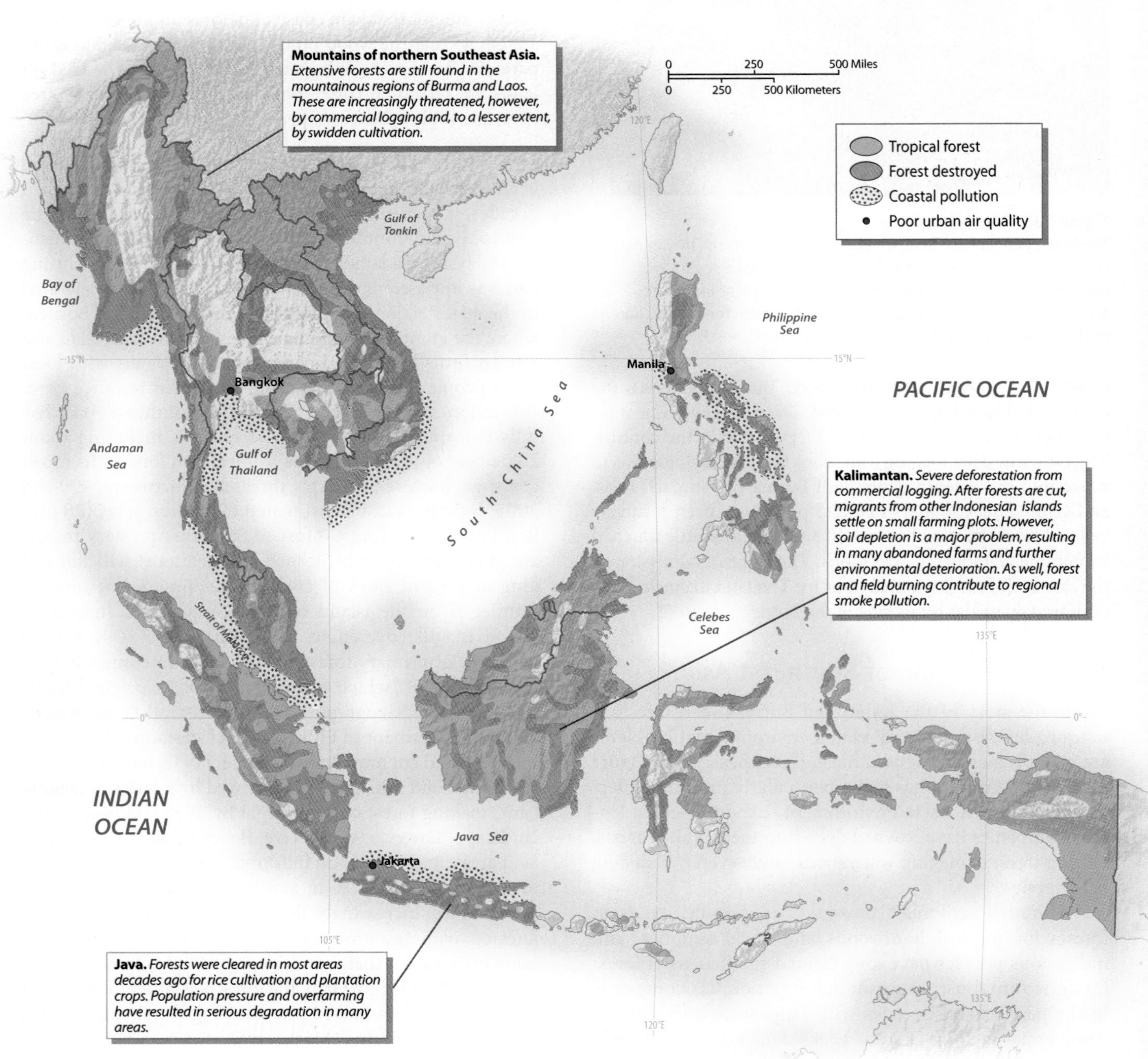

Figure 13.4 Commercial Logging
Southeast Asia has long been the world's most important supplier of tropical hardwoods. The logging process has destroyed most of the tropical forests of the Philippines and Thailand, as well as on the Indonesian islands of Java and Sumatra. *(Barry Calhoun/Redux)*

nowhere else. Conservation officials hope that protected areas will allow animals such as the orangutan, which survives now only in a small portion of northern Sumatra and a somewhat larger area of Borneo (Kalimantan), a chance to survive in the wild. Others are less optimistic, noting that Southeast Asian national parks often exist only on paper, receiving no real protection from loggers or immigrant farmers. Much of Kutai National Park, for example, has been logged and burned.

Fires, Smoke, and Air Pollution

Logging operations typically leave large quantities of wood (small trees, broken logs, roots, branches, and so on) on the ground. Exposed to the sun, this remaining "slash" becomes highly inflammable. Indeed, it is often burned on purpose in order to clear the ground for agriculture or tree replanting. Commercial forest cutting is responsible for most burning, even though the large logging firms commonly blame small farmers. Wildfires also are common in other Southeast Asian habitats. The most thoroughly deforested areas are often covered by rough grasses; these grasslands often are purposely burned every year so that cattle can graze on the tender shoots that emerge after a fire passes through. Grassland fires, in turn, prevent forest regeneration, further undercutting the region's biodiversity.

In the late 1990s, wildfires associated with both logging practices and a severe drought raged so intensively across much of insular Southeast Asia that the region suffered from two consecutive years of disastrous air pollution. During that period, a commercial airliner crashed because of poor visibility; countless road accidents resulted; two ferries collided in smoke-laden conditions; and hundreds of thousands of people were admitted to hospitals with life-threatening respiratory problems (Figure 13.5). The smoke crisis of the late 1990s led several Southeast Asian countries to devote more attention to air quality. Deforestation continues, however, and the fire threat remains, with a new smoke crisis occurring in the dry season of 2006.

Efforts to protect Southeast Asia's air quality, moreover, are also hampered by continuing industrial development along with a major increase in vehicular traffic. Southeast Asia's large metropolitan areas have extremely unhealthy levels of pollution, both of air and water. Several cities, including Bangkok and Manila, have built rail-based public

Figure 13.5 Urban Air Pollution
Air pollution has reached a crisis stage in the rapidly industrializing cities of Southeast Asia, particularly in Bangkok, Manila, and Jakarta. People sometimes resort to using face masks to filter out soot and other forms of particulate matter. Forest fires, which often follow logging, greatly exacerbate the problem. *(V. Miladinovic/Corbis/Sygma)*

GEOGRAPHIC TOOLS

Statistical Comparisons and the Assessment of Philippine Deforestation

The Philippines is one of the most thoroughly deforested countries of Southeast Asia. Forest coverage, which was roughly 90 percent at the time of the Spanish conquest in the sixteenth century, is less than 25 percent today (Figure 13.2.1). The remaining forests, most experts agree, are rapidly retreating and may disappear entirely.

Discussions of Philippine deforestation often are supported by precise statistics showing the steady decline of forest area. Whether such statistics are accurate, however, is a different matter. Geographer David Kummer has conducted research on national-level deforestation data in the Phillipines by examining how the data were derived, scrutinizing the methodologies used, and comparing the results. Kummer also traveled widely in the Philippines to confer with local experts and to inspect ongoing forestry practices. Through such work he demonstrates that we actually know surprisingly little about Philippine forest coverage. The numbers that are so confidently used by scholars, government agencies, and environmental activists generally turn out to be little better than educated guesswork.

Figure 13.2.1 Luzon, Phillipines
Part of northeastern Luzon in the Philippines, which appear dark green on this satellite image, are still thickly forested, but most of the rest of the island has experienced extensive deforestation. *(Image Works)*

There are several reasons for the poor quality of information on Philippine deforestation. Although we often assume that satellite images give accurate representations of land-use coverage and change, Kummer shows that this simply is not the case. Satellite images are expensive and often hard to obtain, but more importantly they can be difficult to interpret. Cloud coverage, image overlap, shadows in steep areas, and unclear classification categories all create uncertainty. As a result, ground surveys by foresters often result in more accurate data. In the Philippines, much of the survey work done on forests over the past half century has been lost or destroyed. Compounding the problem is the fact that the Philippine government seems to have misstated the extent of deforestation between 1950 and 1987 in order to encourage unregulated logging by politically well-connected firms.

A 2004 government press release indicates that "forest cover" has actually increased since the 1980s. However, the data on which this statement is based may not be reliable, and satellite images remain unverified by on-the-ground investigations. More troubling is the fact that this new survey employs land-cover categories different from those in previous work, making it impossible to evaluate the changes in the Philippines' primary forests—some of the most biologically diverse ecosystems in the world.

Taken as a whole, Kummer's available evidence indicates that primary tropical forests continue to decline as a result of both legal and illegal logging. Poor-quality secondary forests, however, seem to be increasing due to the abandonment of marginal farmland. Biologically impoverished "artificial forests" also are increasing as a result of both governmental reforestation efforts and the planting of fuelwood groves by small-scale farmers. Reforestation with a handful of nonnative, fast-growing tree species has now become a potential threat to biodiversity.

Kummer's work shows that, in poorer parts of the world and especially in countries with high levels of political corruption, official statistics may be questionable. The methods used for this type of statistical evaluation have potential application to other world regions, and to other environmental practices with global implications.

transportation systems in order to reduce traffic and vehicular emissions. Bangkok in particular has substantially reduced its levels of ozone and sulfur dioxide, but particulate matter remains at unhealthy levels. For urban Southeast Asia's air pollution problems to be fully addressed, many serious environmental reforms are needed.

Patterns of Physical Geography

To understand why forestry issues are so important in Southeast Asia, it is necessary to look in more detail at physical geography. The southern portion of the region—insular Southeast Asia—is one of the world's three main

Figure 13.6 **Delta Landscape**
Southeast Asia has some of the world's largest deltas. Deltaic environments are used for intensive irrigated rice cultivation, allowing very high rural population densities. Delta wetlands are also used for aquaculture (fish farming) and other forms of intensive food production. Most of mainland Southeast Asia's large cities are located in delta areas, resulting in periodic flooding and other environmental problems. *(John Elk III/Stock Boston)*

zones of tropical rain forest. The northern part of the region—mainland Southeast Asia—is located in the tropical wet-and-dry zone that is noted for particularly valuable timber species, such as teak. Distinguishing between the mainland and the islands is thus one of the keys for understanding the geography of Southeast Asia.

Mainland Environments Mainland Southeast Asia is an area of rugged uplands interspersed with broad lowlands and deltas associated with large rivers (Figure 13.6). The region's northern boundary lies in a cluster of mountains connected to the highlands of eastern Tibet and south-central China. In the far north of Burma (Myanmar), peaks reach 18,000 feet (5,500 meters). From this point, a series of distinct mountain ranges radiates out, extending through western Burma (Myanmar), along the Burma–Thailand border, and through Laos into southern Vietnam.

Several large rivers flow southward out of Tibet and its adjacent highlands into mainland Southeast Asia. The longest is the Mekong, which flows through Laos and Thailand, then across Cambodia before entering the South China Sea through an extensive delta in southern Vietnam. Second longest is the Irrawaddy, which flows through Burma's central plain before reaching the Bay of Bengal. This river also has a large delta. Two smaller rivers are equally significant: the Red River, which forms a sizable and heavily settled delta in northern Vietnam, and the Chao Phraya, which has created the fertile alluvial plain of central Thailand.

The centermost area of mainland Southeast Asia is Thailand's Khorat Plateau, which is neither a rugged upland nor a fertile river valley. This low, sandstone plateau averages about 500 feet (175 meters) in height and is noted for its thin, poor soils. Water shortages and periodic droughts make this extensive area difficult for settlement.

Monsoon Climates Almost all of mainland Southeast Asia is affected by the seasonally shifting winds known as the monsoon. The climate is characterized by a distinct hot and rainy season from May to October (Figure 13.7). This is followed by dry but still generally warm conditions from November to April. Only the central highlands of Vietnam and a few coastal areas receive significant rainfall during this autumn and winter period. In the far north, the winter months bring mild and sometimes rather cool weather.

Two tropical climate regions are dominant in mainland Southeast Asia. While both are affected by the monsoon (see the discussion in Chapter 12), they differ in the total amount of precipitation received during the year. Along the coasts and in the highlands, the tropical monsoon (Am) climate dominates. Rainfall totals for this climate usually register more than 100 inches (254 centimeters) each year.

The greater portion of mainland Southeast Asia falls into the tropical savanna (Aw) climate type. Here annual rainfall totals are about half those of the tropical monsoon region. In most cases, this can be explained by interior locations removed from the oceanic source of moisture, along with the lack of orographic lifting in these areas of more subdued topography. A good portion of Thailand, for example, receives only about 50 inches (127 centimeters) of rain during the year. Much of Burma's central Irrawaddy Valley is almost semiarid, with rainfall totals below 30 inches (76 centimeters). Forests in these areas are especially vulnerable, being easily converted by fire and agriculture into rough landscapes of brush and grass.

Insular Environments The signal feature of insular Southeast Asia is its island environment. Indeed, this is a region of countless islands of all sizes and shapes. Borneo (Kalimantan) and Sumatra are the third and sixth largest islands in the world, respectively, while many thousands of others are little more than specks of land rising at low tide from a shallow sea.

Indonesia alone is said to contain more than 13,000 islands, dominated by the four great land masses of Sumatra, Borneo (Kalimantan), Java, and the oddly shaped Sulawesi. This island nation also includes the western half of New Guinea and the Lesser Sunda Islands, which extend to the east of Java. A prominent mountain spine runs through these islands as a result of tectonic forces. In the western Indonesian islands, volcanic peaks of 10,000 feet (3,500 meters) are common.

The Philippines, another insular country, includes more than 7,000 islands. The two largest and most important are Luzon (about the size of Ohio) in the north and Mindanao (the size of South Carolina) in the south. Sandwiched between them are the Visayan Islands, which number roughly a dozen. Again because of tectonic forces that generate active volcanoes, the topography of the Philippines includes mountainous landscapes that reach elevations of 10,000 feet (3,500 meters).

Closely related to this impressive collection of islands is the world's largest expanse of shallow ocean. These waters cover the **Sunda Shelf**, which is an extension of the continental shelf extending from the mainland through the

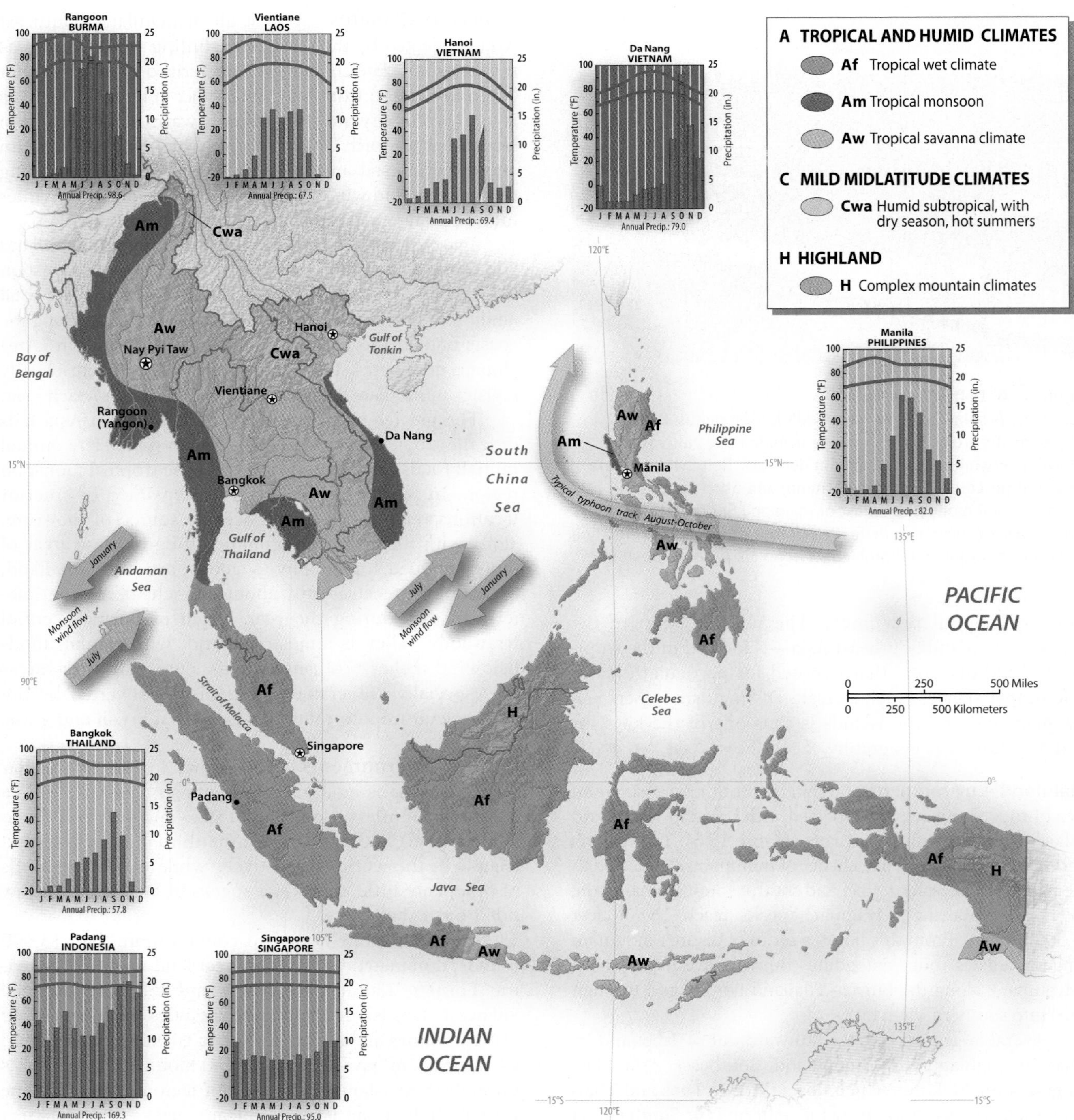

Figure 13.7 Climate Map of Southeast Asia
Most of insular Southeast Asia is characterized by the constantly hot and humid climates of the equatorial zone. Mainland Southeast Asia, on the other hand, has the seasonally wet and dry climates of the tropical monsoon and tropical savanna types. Only in the far north are subtropical climates, with relatively cool winters, encountered. The northern half of the region is strongly influenced by the seasonally shifting monsoon winds. Northeastern Southeast Asia—and especially the Philippines—often experiences typhoons from August to October.

Java Sea between Java and Borneo (Kalimantan). Here waters are generally less than 200 feet (70 meters) deep. Some insular Southeast Asian peoples have adopted lifestyles that rely on the rich marine life of this region, essentially living on their boats and setting foot on land only as necessary.

Insular Southeast Asia is less geologically stable than the mainland, and thus it is a region of more diverse landforms. Four of Earth's tectonic plates converge here: the Pacific, the Philippine, the Indo-Australian, and the Eurasian. As a result of this tectonic structure, earthquakes occur frequently near

Figure 13.8 Tsunami Damage
The December 26, 2004, tsunami killed some 230,000 people in South and Southeast Asia, hitting northern Sumatra especially hard. In this photograph, an Acehnese man chants the Quran over the grave of his sister in the devastated city of Banda Aceh. *(Kimimasa Mayama/Reuters/Corbis/Reuters America LLC)*

the plate boundaries. Large, often explosive volcanoes are another consequent feature of the insular Southeast Asian landscape. A string of active volcanoes extends the length of eastern Sumatra (in Indonesia) across Java and into the Lesser Sunda Islands, the small islands east of Java. Volcanic eruptions and earthquakes occasionally result in **tsunamis** (or "tidal waves"), which can devastate coastal regions over a vast area. A massive earthquake in northern Sumatra on December 26, 2004, for example, resulted in roughly 100,000 deaths in Indonesia alone (Figure 13.8). Other geological hazards also confront the region. In late 2006, 13,000 Indonesians had to be evacuated away from a new mud volcano near the city of Surabaya, which had already covered over a thousand acres (404 hectares) of farmland with hot and noxious mud (Figure 13.9).

Equatorial Island Climates The climates of insular Southeast Asia are somewhat more complex than those of the mainland because of three factors: a more complicated monsoon effect, the stronger influence of Pacific typhoons in the Philippines, and the equatorial location of the Indonesian islands.

Most of insular Southeast Asia, unlike the mainland, receives rain during the northern hemisphere's winter because the winter monsoon winds cross large areas of warm equatorial ocean, where they absorb moisture. As a result, these winds can bring heavy rains as the saturated air masses are lifted over island uplands. On Sumatra and Java, for example, where north winds blow between November and March, these air masses produce heavy rains on the northern side of these east–west-running islands. But during the May–September period, the heaviest rains are found on the southern flanks of these same islands because of the south winds associated with the Asian summer monsoon.

A second factor in the climate pattern of insular Southeast Asia are the tropical hurricanes, or **typhoons**, as they are called in the western Pacific, that bring heavy rainfall to the northeastern reaches of insular Asia during the period of August to October. These strong storms bring devastating winds and torrential rain. They develop east of the Philippines and then move westward into the South China Sea, where they often intensify. Each year a number of typhoons affect parts of the Philippines with heavy damage and loss of life through flooding and landslides. A single massive mudslide on the island of Leyte in 2006 resulted in more than 1,000 deaths. Strong tropical storms also commonly strike the eastern coast of Vietnam.

The third climate factor, the equatorial influence, results from the area's low latitude. More than mainland Southeast Asia, the islands experience very little seasonality. Temperatures remain high throughout the year, with very little variation. Also associated with the equatorial influence is the fact that rainfall is both higher and more evenly distributed during the year than on the mainland. As a result of this year-round precipitation, most of island Southeast Asia is placed into the tropical rainforest (Af)

Figure 13.9 Mud Volcano in Indonesia
Villages and farmland in east Java have been devastated by a mud volcano that continues to pour out hot, toxic sludge. The volcano itself was set off when a gas exploration drilling project tapped into a vent of extremely hot mud. *(Redux Pictures)*

climate category (Figure 13.7). The southeastern islands of Indonesia, however, do experience a distinct dry season from May to October, whereas most of the western Philippines are typically dry from November to April.

Global Warming and Southeast Asia

The fact that many, if not most, of Southeast Asia's people live in coastal and delta environments makes the region highly vulnerable to the rise in sea level associated with global warming. Periodic flooding is already a major problem in many of the regions' low-lying cities, particularly Bangkok, and will likely intensify in the near future. Southeast Asian farmland is also concentrated in delta environments, and thus could suffer from saltwater intrusion and heightened storm surges. It is also feared that higher temperatures themselves could reduce rice yields through much of the region.

Changes in precipitation across Southeast Asia brought about by global warming remain highly uncertain. Many experts foresee an intensification of the monsoon pattern, which could bring increased rainfall to much of the mainland. While enhanced precipitation would likely result in more destructive floods, it could bring some agricultural benefits to dry areas such as Burma's central Irrawaddy Valley. Complicating this scenario, however, is the prediction that global climate change could intensify the El Niño effect (discussed in Chapter 4), which would result in more extreme droughts, especially in the equatorial belt of Indonesia. Some observers think that the prolonged drought that Indonesia experienced in the late 1990s was a harbinger of future conditions.

All Southeast Asian countries except Laos, East Timor, and Brunei have ratified the 1997 Kyoto Accord. But since all are officially classified as developing countries—even wealthy Singapore—none are obligated to reduce their emissions of greenhouse gases. Still, Southeast Asia's overall emissions from conventional sources remain low by global standards. Indonesia, Malaysia, and Vietnam, however, are planning to increase their reliance on coal-fired power plants for electricity generation, which—combined with overall economic growth—would greatly increase the region's greenhouse gas emissions. In Thailand, local and international opposition has already led the government to reconsider its own coal-based expansion plans. One possible alternative is hydropower, especially in mountainous Laos, which is ambitiously planning to build a large number of new dams. But dam building results in its own ecological problems, and is opposed by most environmental groups.

When greenhouse gas emissions associated with deforestation are factored in, Southeast Asia's role in global climate change is revealed to be much larger than it appears at first glance. By some estimates, Indonesia is the world's third largest contributor to the problem, following only China and the United States. Vast quantities of carbon dioxide and methane are released into the atmosphere when slash is burned after cutting, a routine practice in much of the region.

A far greater problem, however, is the release of carbon from the burning and oxidation of peat, the partially decayed organic matter that accumulates in perennially saturated soils. The wetlands of coastal Indonesia hold 60 percent of the world's tropical peat, containing roughly 50 billion tons of carbon. During drought periods, peat soils sometimes burn, releasing the stored carbon. More worrisome is the fact that both Indonesia and Malaysia are actively draining coastal wetlands to make room for agricultural expansion, a process that results in the gradual oxidation of the peat. Already some 40 percent of Southeast Asia's peatlands have been lost. Ironically, many of these swamps are being destroyed in order to produce biodiesel from oil palms, a supposedly eco-friendly form of energy. While biodiesel does have many potential environmental benefits, its production in this region is problematic for its effects on greenhouse gas emissions and overall contribution to global warming.

POPULATION AND SETTLEMENT:
Densely Settled Lowlands amid Sparsely Settled Uplands

The scale of Southeast Asia's population is quite different from those of its giant neighbors, China and India. With a little over 500 million people today, Southeast Asia is relatively sparsely settled. In earlier periods, it was much less densely populated than either East Asia or South Asia. One of the reasons for low density is the extensive tracts of rugged mountains, which generally remain thinly inhabited. In contrast, however, relatively dense populations are found in the region's deltas, coastal areas, and zones of fertile volcanic soil (Figure 13.10).

Many of the favored lowlands of Southeast Asia have experienced striking population growth over the past half century. Demographic growth and family planning have thus become increasingly important concerns through much of the region. Several countries, especially Singapore and Thailand, have seen rapid reductions in birthrates in recent years, whereas fertility rates remain high in Cambodia and Laos. Indonesia and the Philippines, on the other hand, have experienced substantial migration to their outer islands in part because of population pressure in their core areas.

Settlement and Agriculture

Part of the reason for the historical lack of population in much of Southeast Asia is infertile soil, which is unable to support intensive agriculture and high population densities. The island rain forests, though lush and biologically rich, grow on poor soils. Plant nutrients are locked up in the vegetation itself, rather than being stored in the soil where they would easily benefit agriculture. Furthermore, the incessant rain of the equatorial zone tends to wash nutrients out of the soil. Agriculture must be carefully adapted to this limited soil fertility by constant field rotation or the application of heavy amounts of fertilizer.

There are, however, some notable exceptions to this generalization about soil fertility and settlement density in equatorial Southeast Asia. Unusually rich soils connected to

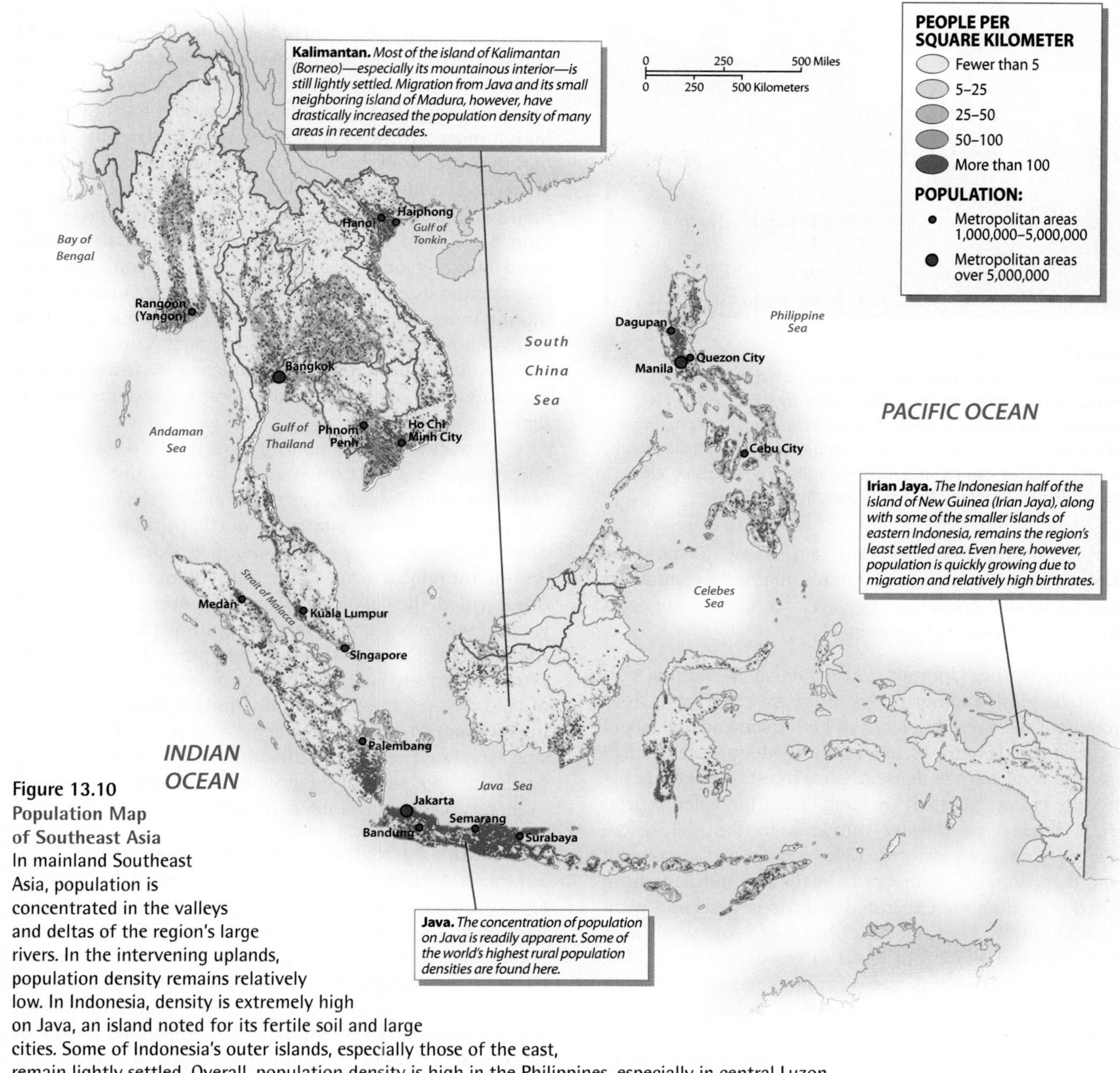

Figure 13.10 Population Map of Southeast Asia In mainland Southeast Asia, population is concentrated in the valleys and deltas of the region's large rivers. In the intervening uplands, population density remains relatively low. In Indonesia, density is extremely high on Java, an island noted for its fertile soil and large cities. Some of Indonesia's outer islands, especially those of the east, remain lightly settled. Overall, population density is high in the Philippines, especially in central Luzon.

volcanic activity are scattered through much of the region but are particularly prevalent on the island of Java. Java, with more than 50 volcanoes, is blessed with rich soils that support a large array of tropical crops and a very high population density. It has 124 million people—roughly half the total population of Indonesia—in an area smaller than the state of Iowa. Dense populations also are found in pockets of fertile alluvial soils along the coasts of insular Southeast Asia, where people have traditionally supplemented land-based farming with fishing and other commercial activities.

The demographic patterns in mainland Southeast Asia are less complicated than those of the island realm. In all the mainland countries, population is concentrated in the agriculturally intensive valleys and deltas of the large rivers (see Figure 13.10). The population core of Thailand, for example, is formed by the valley and delta of the Chao Phraya River, just as Burma's is focused on the Irrawaddy. In Vietnam, there are two distinct foci: the Red River Delta in the far north and the Mekong Delta in the far south. In contrast to these densely settled areas, the middle reaches of the Mekong River provide only limited lowland areas in Laos, which is one of the reasons that country has a much smaller population than its neighbors. In Cambodia, the largest population historically has clustered around Tonle Sap, a large lake with a very unusual seasonal flow reversal. During the rainy summer months the lake receives water from the Mekong drainage, but during the drier winter months it contributes to the river's flow.

Figure 13.11 Swidden Agriculture
In the uplands of Southeast Asia, swidden (or slash-and-burn) agriculture is widely practiced. When done by tribal peoples with low population densities, swidden is not environmentally harmful. When practiced by large numbers of immigrants from the lowlands, however, swidden can result in deforestation and extensive soil erosion. *(Paula Bronstein/Getty Images, Inc.–Liaison)*

Agricultural practices and settlement forms vary widely across the complex environments of Southeast Asia. Generally speaking, however, three farming and settlement patterns are apparent.

Swidden in the Uplands Also known as shifting cultivation or "slash-and-burn" agriculture, swidden is practiced throughout the rugged uplands of both mainland and island Southeast Asia (Figure 13.11). In the **swidden** system, small plots of several acres of dense tropical forest or brush are periodically cut or "slashed" by hand. Then the fallen vegetation is burned to transfer nutrients to the soil before subsistence crops are planted. Yields remain high for several years, then drop off dramatically as the soil nutrients from burned vegetation are exhausted and insect pests and plant diseases multiply. These plots are abandoned completely after a few years and allowed to revert to woody vegetation. The cycle of cutting, burning, and planting is moved to another small plot not far away—thus the term *shifting cultivation*. Villages generally control a large amount of territory so they can rotate their fields on a regular basis. After a period of 10 to 75 years, the farmers return to the original plot, which once again has nutrients in the dense vegetation.

Swidden is sustainable when population densities remain relatively low and when upland people control enough territory. Today, however, the swidden system is threatened for two reasons. First, it cannot easily support the increasing population that has resulted from relatively high human fertility and, in some cases, migration. With a higher population density, the rotation period must be shortened, undercutting soil resources. Second, the upland swidden system often is a casualty of commercial logging because of the devastation of forest environments. Road building often exacerbates the situation. Vietnam, for example, is constructing a highway system through its mountainous spine, designed to aid economic development and more fully integrate its national economy. Partly as a result of this activity, lowlanders are streaming into the mountains, disrupting both local ecosystems and indigenous societies.

When swidden can no longer support the local population, upland people often adapt by switching to a cash crop that will allow them to participate in the commercial economy. In the mountains of northern Southeast Asia, one of the main cash crops is opium, grown by local farmers for the global drug trade. This mountainous area is often called the "**Golden Triangle.**" Burma (Myanmar) is the world's second largest opium producer, after Afghanistan. Drug eradication programs, however, have proved relatively successful in recent years, with the area of land devoted to opium production supposedly decreasing by some 85 percent between 1998 and 2006. Evidence suggests, however, that methamphetamine manufacturing has been replacing opium growing and heroin production in many areas of the Golden Triangle.

Plantation Agriculture With European colonization, Southeast Asia became a focus for commercial plantation agriculture, growing high-value specialty crops ranging from rice to rubber. Even in the nineteenth century, Southeast Asia was closely linked to a globalized economy through the plantation system. Forests were cleared and swamps drained to make room for these plantations; labor was supplied, often unwillingly, by indigenous people or by contract laborers brought in from India or China.

Plantations are still an important part of Southeast Asia's geography and economy. Most of the world's natural

rubber, for example, is produced in Malaysia, Indonesia, and Thailand, while sugarcane has long been a plantation crop of the Philippines and parts of Indonesia. More recently, pineapple plantations have appeared in both the Philippines and Thailand, which are now the world's leading exporters. Indonesia is the region's leading producer of tea (Figure 13.12). Malaysia has traditionally dominated the production of palm oil, but by 2007 it was being surpassed by Indonesia, which plans to add 17 million acres (7 million hectares) of oil palm plantations by 2011. Coconut oil and **copra** (dried coconut meat) are widely produced in the Philippines, Indonesia, and elsewhere. In the early 2000s, Vietnam rapidly emerged as the world's second-largest coffee producer, its large harvests resulting in plummeting global coffee prices.

Rice in the Lowlands The lowland basins and deltas of mainland Southeast Asia are largely devoted to intensive rice cultivation. Throughout almost all of Southeast Asia, rice is the preferred staple food. Traditionally, rice was mainly cultivated on a subsistence basis by rural farmers. But as the number of wage laborers in Southeast Asia has grown as a result of economic development, so has the demand for commercial rice cultivation. Thailand is currently the world's largest rice exporter by a wide margin, with Vietnam occupying the number two position. Across most of Southeast Asia, the use of agricultural chemicals and high-yield crop varieties, along with improved water control and the expansion of irrigation to facilitate dry-season cropping, have allowed production to keep pace with population growth. Such techniques have, however, also resulted in environmental damage.

In those areas without irrigation, yields remain relatively low. Rice growing on Thailand's Khorat Plateau, for example, depends largely on the uncertain rainfall, without the benefits of the more sophisticated water control methods. In some lowland districts lacking irrigation, dry-field crops, especially sweet potatoes and manioc, form the staple foods of people too poor to buy market rice on a regular basis. In some of the poorest parts of Southeast Asia, population growth has combined with economic stagnation to force larger numbers of people into such meager diets. Elsewhere, economic growth and declining birthrates have significantly reduced the burden of poverty.

Recent Demographic Change

Because Southeast Asia is not facing the same kind of population pressure as East or South Asia, a wide range of government population policies is found. In countries with rapid demographic expansion, moreover, internal relocation away from densely populated areas to outlying districts is a common response.

Population Contrasts The Philippines, the second most populous country in Southeast Asia, has a worrisomely high growth rate (Table 13.1). Complicated internal politics in the Philippines tend to impede effective family planning. When a popular democratic government replaced a dictatorship in the 1980s, the Philippine Roman Catholic Church, which played an active role in the peaceful revolution, pressured the new government to cut funding for family planning programs. As a result, many clinics and centers that had dispensed family planning information were closed. Although high birthrates are not always associated with Catholicism, the Church's outspoken stand on birth control seems to inhibit the dispersal of family planning information.

Laos, a country of Buddhist religious tradition, has a particularly high total fertility rate (4.8). Here the high birthrate is best explained by the country's low level of economic and social development. Thailand, which shares cultural traditions with Laos yet is considerably more developed, demonstrates the other end of the spectrum. Here the TFR has dropped dramatically

Figure 13.12 Tea Harvesting in Indonesia
Plantation crops, such as tea, are major sources of exports for several Southeast Asian countries. Coconut, rubber, oil palms, and coffee are other major cash crops. Many of these crops require large amounts of labor, particularly at harvest time. *(Dermot Tatlow/Panos Pictures)*

TABLE 13.1 Population Indicators

Country	Population (Millions, 2007)	Population Density (per Square Kilometer)	Total Fertility Rate	Percent Urban	Percent <15[a]	Percent <65[b]	Net Migration (per 1,000, 2000–05)
Brunei	0.4	65	2.3	72	30	3	2.0
Burma (Myanmar)	49.8	74	2.3	29	27	6	0.3
Cambodia	14.4	79	3.4	15	37	3	−0.1
East Timor	1.0	70	7.0	22	45	3	19.2
Indonesia	231.6	122	2.4	42	28	6	−0.9
Laos	5.9	25	4.8	21	44	4	−0.2
Malaysia	27.2	82	2.9	62	33	4	1.2
Philippines	88.7	296	3.4	48	35	4	−2.3
Singapore	4.6	6,785	1.3	100	19	8	9.6
Thailand	65.7	128	1.7	33	23	7	−0.2
Vietnam	85.1	257	2.1	27	29	7	−0.5

[a]*Percentage of population younger than 15.*
[b]*Percentage of population older than 65.*
Sources: Population Reference Bureau, World Population Data Sheet, *2007; net migration rate data from* UN International Migration, *2006.*

within the last 30 years from 5.4 (in 1970) to 1.7, a figure that will soon bring population stability. But while economic growth may explain some of this decrease, it is also important to note that the Thai government has promoted family planning for both population and health reasons, including the high incidence of AIDS in the country.

The city-state of Singapore stands out on the demographic charts with its particularly low fertility rate. Unless the deficit is offset by immigration or a dramatic turnabout in the birthrate, Singapore's population will soon begin to decline. The government is concerned about this situation and is actively promoting marriage and child-bearing, particularly among the most highly educated segment of its population. As a result, Singaporeans can receive more than $10,000 in direct government subsidies for having a third or fourth child (Figure 13.13).

Figure 13.13 Pro-Reproduction Advertisement in Singapore Due to its extremely low birthrate, the population of Singapore will probably begin to decline soon. Shown here are perfumes created by local students and launched in a government-backed "romance" campaign. *(Ed Wray/AP World Wide Photos)*

Indonesia, with the region's largest population at 231 million, also has seen a dramatic decline in fertility in recent decades, although its fertility rate remains above the replacement level. If the present trend continues, however, Indonesia will reach population stability well before most other large developing countries. As with Thailand, this drop in fertility seems to have resulted from a strong government family planning effort, coupled with improvements in education.

Cambodia, like Laos, has a persistently high fertility rate. A partial explanation is the fact that Cambodia has one of the lowest life expectancy rates and one of the highest infant mortality rates in Southeast Asia. As in other world regions, high birthrates often are associated with high mortality. Cambodia's high mortality rates are linked to its recent history of civil strife and internal violence, factors that also have inhibited economic and social development. Although a semblance of stability has returned to Cambodia, the country has not yet seen any significant decline in its fertility rate.

Growth and Migration Until recently, Indonesia had an official policy of **transmigration**, with the government assisting in the relocation of people from one region to another within its national territory (Figure 13.14). Primarily because of migration from the densely populated islands of Java and Madura, the population of the outer islands of Indonesia has grown rapidly since the 1970s. The province of East Kalimantan, for example, experienced an astronomical growth rate of 30 percent per year

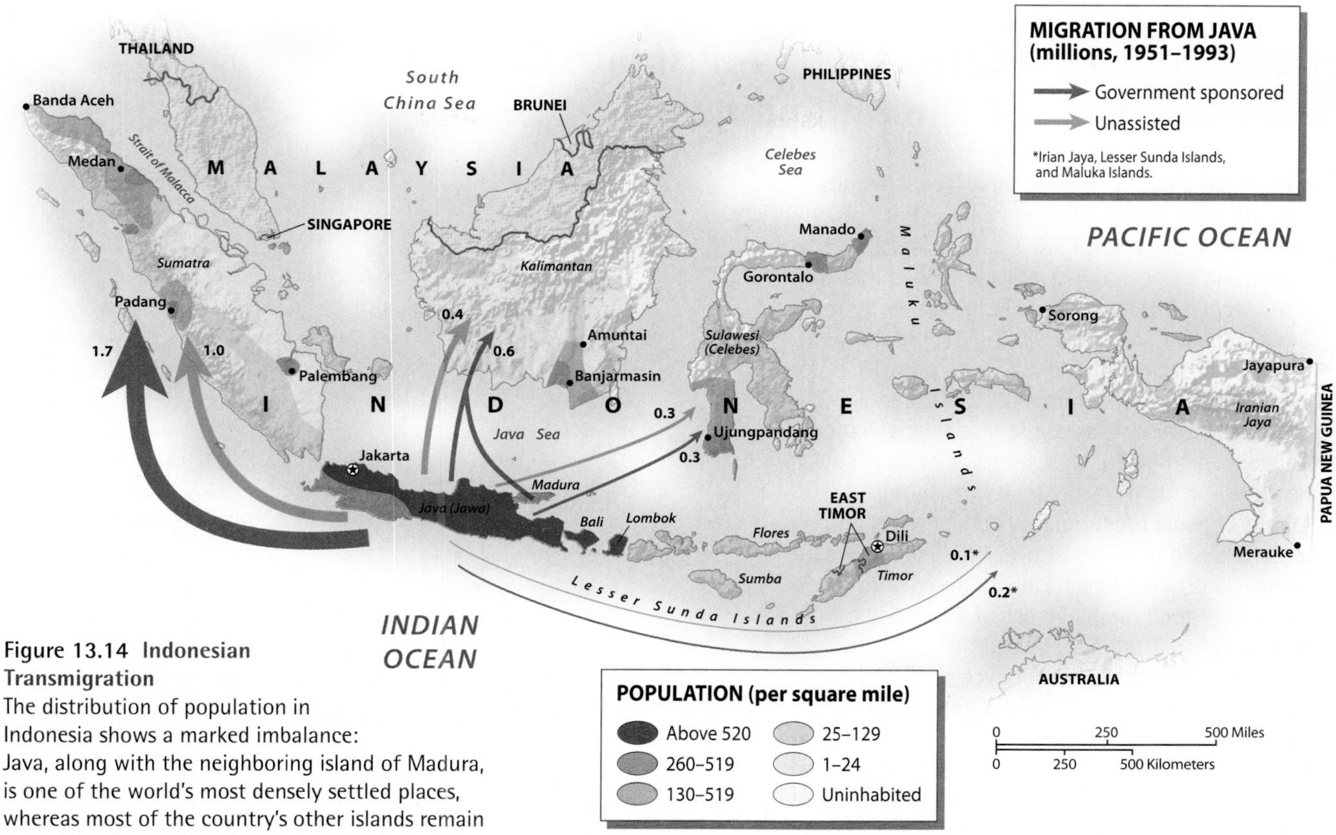

MIGRATION FROM JAVA (millions, 1951–1993)
Government sponsored
Unassisted
*Irian Jaya, Lesser Sunda Islands, and Maluka Islands.

POPULATION (per square mile)
Above 520
260–519
130–519
25–129
1–24
Uninhabited

THAILAND
Banda Aceh
Medan
Strait of Malacca
MALAYSIA
SINGAPORE
South China Sea
BRUNEI
PHILIPPINES
Celebes Sea
Manado
Gorontalo
Kalimantan
Sumatra
Padang
Palembang
Amuntai
Banjarmasin
Sulawesi (Celebes)
Ujungpandang
Jakarta
Java Sea
Madura
Java (Jawa)
Bali
Lombok
Flores
EAST TIMOR
Dili
Sumba
Timor
Lesser Sunda Islands
Maluku Islands
INDONESIA
Sorong
Jayapura
Iranian Jaya
Merauke
PAPUA NEW GUINEA
PACIFIC OCEAN
INDIAN OCEAN
AUSTRALIA
1.7
1.0
0.4
0.6
0.3
0.3
0.1*
0.2*
0 250 500 Miles
0 250 500 Kilometers

Figure 13.14 Indonesian Transmigration
The distribution of population in Indonesia shows a marked imbalance: Java, along with the neighboring island of Madura, is one of the world's most densely settled places, whereas most of the country's other islands remain rather lightly populated. As a result, the Indonesian government has encouraged the resettlement of Javanese and Madurese people to the outer islands, often paying the costs of relocation. This transmigration scheme has resulted in a somewhat more balanced population distribution pattern, but has also caused substantial environmental degradation and intensified several ethnic conflicts.

Figure 13.15 Migrant Settlement in Indonesia
Migration from densely settled to sparsely settled areas of Southeast Asia has resulted in the creation of thousands of new communities. Many of these communities have minimal infrastructure and services, and some of them struggle to survive. *(Charly Flyn/Panos Pictures)*

during the last two decades of the 1900s. As a result of this shift in population, many parts of Indonesia outside Java now have moderately high population densities, although many of the more remote districts remain lightly inhabited.

High social and environmental costs often accompany these relocation schemes. Javanese peasants, accustomed to working the highly fertile soils of their home island, often fail in their attempts to grow rice in the former rain forest of Borneo (Kalimantan). Field abandonment is high after repeated crop failures. In some areas, farmers have little choice but to adopt a semi-swidden form of cultivation, moving to new sites once the old ones have been exhausted, a process associated with further deforestation and soil degradation. The term **shifted cultivators** is sometimes used for these rural migrants who are moved from one area to another through relocation schemes (Figure 13.15). Partly because of these problems, the Indonesian government restructured and cut back on its transmigration program in 2000. As of 2006, it was relocating only about 15,000 people a year.

The Philippines also have used internal migration as a means of alleviating population pressure on the main islands. Beginning in the late nineteenth century, Philippine society has responded to increasing population pressure in the core areas of central Luzon by sending colonists to frontier zones. In the early twentieth century, this settlement frontier still lay in Luzon; by the postwar years it had

CITYSCAPES

Metro Manila, Primate City of the Philippines

With more than 10 million inhabitants, Manila is one of the world's largest cities. But, strictly speaking, Manila itself is a much smaller city, supporting fewer than 2 million inhabitants. The term *Manila* often refers to the urban entity more properly called *Metro Manila* (Figure 13.3.1). Metro Manila, consisting of Manila itself plus 16 other cities and towns, was formally constituted in 1975 as an administrative area, the Philippines' National Capital Region (NCR). Most forms of political power, however, remain decentralized among Metro Manila's constituent units, complicating policy initiatives and planning efforts. When a new mayor of Manila decided to crack down on prostitution in the 1990s, for example, most of the city's "go-go bars" simply relocated to Pasay City.

Metro Manila is economically as well as politically decentralized. Manila itself includes the old central business district, now somewhat decayed, as well as sizable slums, a congested "Chinatown" (Binondo) famous for its discount shopping, and several older, wealthy neighborhoods. It also contains the historic zone of Intramuros, the original fortified city of the Spanish colonial period (Figure 13.3.2). Although Manila was almost completely destroyed during World War II, many of Intramuros's buildings have been rebuilt, lending the area charm and historical significance.

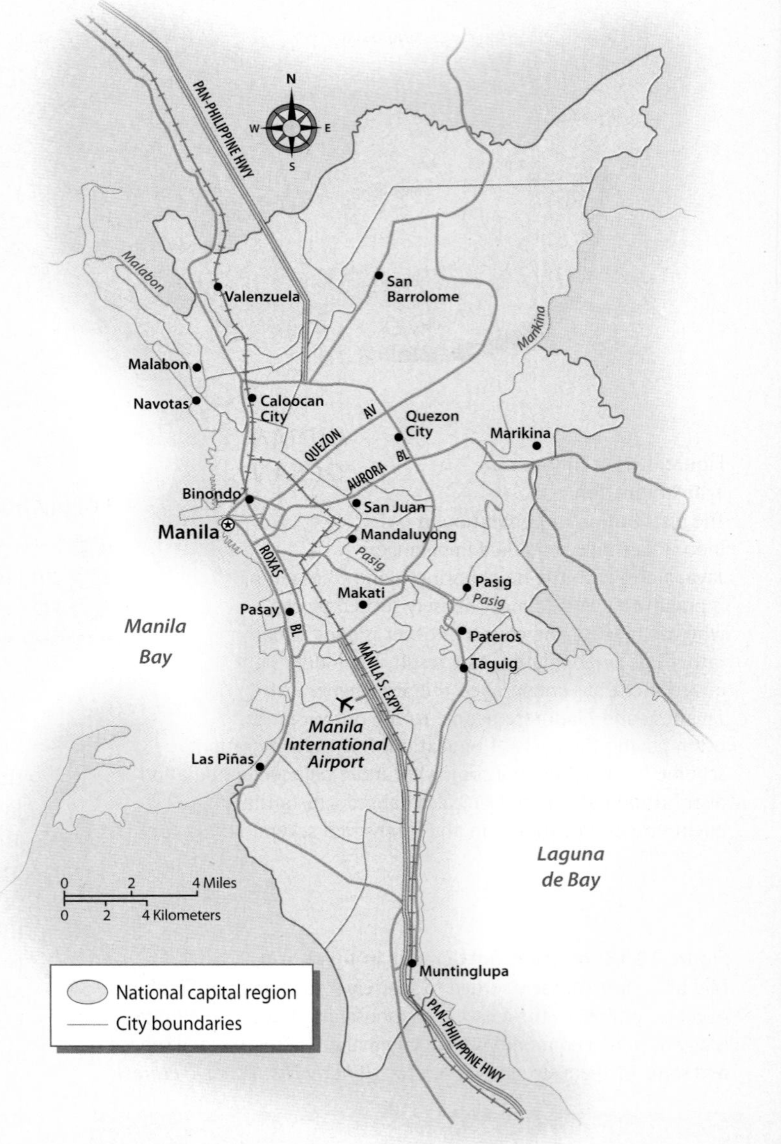

Figure 13.3.1 Metro Manila
The Philippines' National Capital Region (Metro Manila) consists of the city of Manila and a number of neighboring cities and suburbs, which together form a single metropolitan area. Such municipalities as Makati and Quezon City are of significance equal to Manila proper.

moved south to the island of Mindanao and subsequently into Mindoro and Palawan as well. Today, however, this option is less appealing, both because migration has reduced the population disparities between islands and because of political instability in the country's fringe areas. The emphasis is now on international migration, resulting in a Filipino overseas population that numbers some 8 million.

Population issues and policies thus vary greatly across Southeast Asia. Virtually every part of the region, however, has seen a rapid expansion of its urban population in recent decades.

Urban Settlement

Despite rapid urban growth and relatively high levels of economic development, Southeast Asia is still not heavily urbanized. Even Thailand retains a rural flavor, which is somewhat unusual for a country that has experienced so much recent industrialization. Within a few decades, however, Southeast Asia will become, like most of the rest of the world, a predominantly urban society.

Several Southeast Asian countries have **primate cities**, single, large urban settlements that overshadow all others.

Metro Manila's—and the Philippines'—commercial capital now is located in Makati, southeast of the old urban core. From a rural fringe zone in the 1940s, Makati has grown into a mature city of high-rise office buildings and hotels, shopping plazas, and gated communities that house the country's elite. More recently, Metro Manila has seen the rise of other areas that serve the wealthy, most notably Ortigas Center in neighboring Mandaluyong City. Particularly successful here is the SM Megamall, which opened in 1991. This gargantuan six-story enclosed mall contains 12 cinemas and an ice-skating rink, in addition to hundreds of stores and restaurants. SM Megamall attracts tens of thousands of visitors daily, some of whom come simply to experience the modern ambiance and enjoy the air conditioning.

Official government buildings of the Philippines are concentrated in Quezon City in northeast Metro Manila. Quezon City replaced Manila as the planned national capital between 1948 and 1976. This sprawling city of widely separated office buildings is also an educational center, supporting the country's two most highly regarded universities, the publicly run University of the Philippines (UP) and the Jesuit-run Ateneo de Manila University.

One of Metro Manila's main problems is an extreme shortage of decent housing. The rapid growth of the city has led to skyrocketing land values and rents, forcing roughly half of the city's residents to live in squatters' shacks. Some squatter settlements emerge almost overnight on vacant lots; others sprawl on for miles and have existed for decades. The inadequate infrastructure found in these communities is also characteristic of the rest of the city. Only an estimated 11 percent of Metro Manila's residences, for example, are connected to fully functioning sewer systems.

Figure 13.3.2 Intramuros Intramuros, the "city between the walls," was the original, fortified capital of the Spanish Philippines. It now is an important historic district. *(Jose Fuste Raga/AGE Fotostock America, Inc.)*

Thailand's urban system, for example, is dominated by Bangkok, just as Manila far surpasses all other cities in the Philippines (see "Cityscapes: Metro Manila, Primate City of the Philippines"). Both have recently grown into megacities with more than 10 million residents. More than half of all city-dwellers in Thailand live in the Bangkok metropolitan area (Figure 13.16), which is 25 times larger than the country's second city. In both Manila and Bangkok, as with other large cities in the region, explosive growth has led to housing problems, congestion, and pollution. In Manila, estimates indicate that more than half of the city's population lives in squatter settlements, usually without basic water and electricity service. Both cities also suffer from a lack of parks and other public spaces, which is one reason why massive shopping malls have become so popular. Bangkok's Paragon Mall has recently emerged as a major urban focus, complete with a conference center and a concert hall.

Thailand, the Philippines, and Indonesia are all making efforts to encourage growth of secondary cities by decentralizing economic functions. The goal is to stabilize the population of the primate cities. In the case of the Philippines, the city of Cebu has emerged in recent years as a

Figure 13.16 Bangkok
Bangkok saw the development of an impressive skyline during its boom years from the late 1970s through the late 1990s. Unfortunately, infrastructure did not keep pace with population and commercial growth, resulting in one of the most congested and polluted urban landscapes in the world. *(Saeed Khan/AFP/Getty Images)*

more dynamic economic center than Manila, leading to hopes that a more balanced urban system may be emerging.

Urban primacy is considerably less pronounced in other Southeast Asian countries. Vietnam, for example, has two main cities, Ho Chi Minh City (formerly Saigon) in the south, with 6 million people, and the capital city of Hanoi in the north, with slightly more than 3 million residents. Jakarta, a vast city, is the largest urban area in Indonesia, but the country has a host of other large and growing cities, including Bandung and Surabaya. Rangoon (Yangon), until recently the capital city of Burma (Myanmar), has doubled its population in the last two decades to more than 5 million residents. In Cambodia, the capital of Phnom Penh is still a relatively small city of around 2 million. Vientiane, the capital city of Laos, has a population of only about 200,000.

Kuala Lumpur, the largest city in Malaysia with some 2 million inhabitants, has received heavy investments from both the national government and the global business community in recent decades. This has produced a modern city of grandiose ambitions that is free of most infrastructural problems plaguing other Southeast Asian cities. The Petronas Towers, owned by the country's national oil company, were the world's tallest buildings at almost 1,500 feet (450 meters) when completed in 1996.

The independent republic of Singapore is essentially a city-state of 4.5 million people on an island of 240 square miles (600 square kilometers) (Figure 13.17). While space is at a premium, Singapore has successfully developed high-tech industries that have brought it great prosperity. Unlike other Southeast Asian cities, Singapore has no squatter settlements or slums. Only in the fast-disappearing Chinatown and historic colonial district does one find older buildings. Otherwise, Singapore is an extremely clean but rather sterile city of modern high-rise skyscrapers and space-intensive industry.

Singapore is unique in Southeast Asia not only because it is a city-state, but also because of the Chinese cultural background of most of its people. But as we shall see in the following section, strong Chinese influences also are found in most of the other large cities of the region.

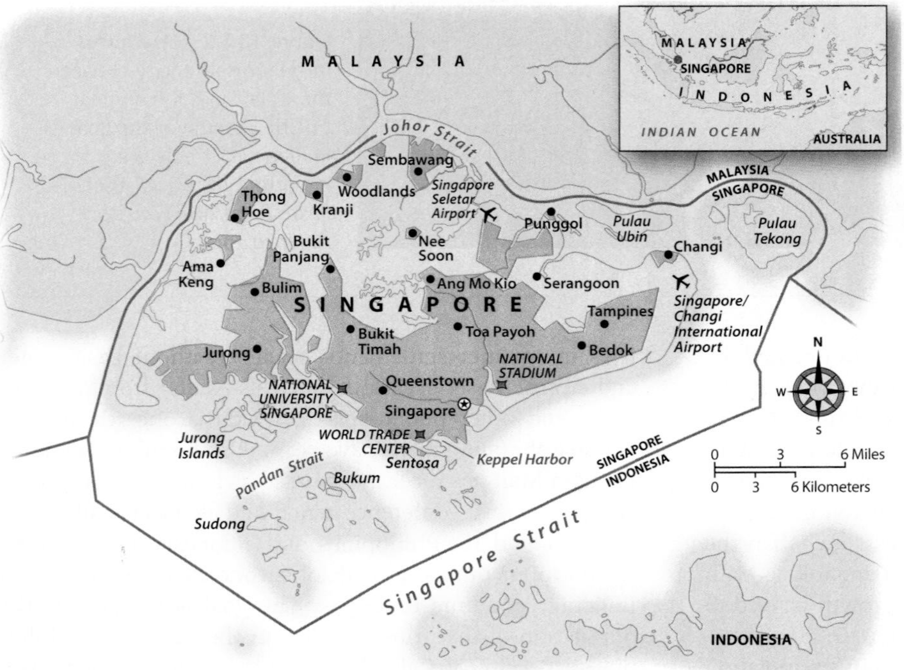

Figure 13.17 Singapore
Singapore remains the economic and technological hub of Southeast Asia. It is a very small and extremely crowded country, forming one of the world's few city-states. Despite its high population density, Singapore has devoted almost half of its area to open spaces. The country consists of one main island and some 60 smaller islands.

CULTURAL COHERENCE AND DIVERSITY: A Meeting Ground of World Cultures

Unlike many other world regions, Southeast Asia lacks the historical dominance of a single civilization. Instead, the region has been a meeting ground for cultural diffusion from South Asia, China, the Middle East, Europe, and even North America. Abundant natural resources, along with the region's strategic location on oceanic highways connecting major continents, have long made Southeast Asia attractive to outsiders. As a result, the modern cultural geography of this diverse region is in part a product of borrowing and combining external influences.

The Introduction and Spread of Major Cultural Traditions

In Southeast Asia, contemporary cultural diversity is embedded in the historical influences connected to the major religions of Hinduism, Buddhism, Islam, and Christianity (Figure 13.18).

South Asian Influences The first major external influence arrived from South Asia some 2,000 years ago when migrants from what is now India helped establish Hindu kingdoms in coastal locations in Burma (Myanmar), Thailand, Cambodia, southern Vietnam, Malaysia, and western Indonesia. Although Hinduism faded away in most locations, this tradition is still found on the Indonesian islands of Bali and Lombok, and vestiges

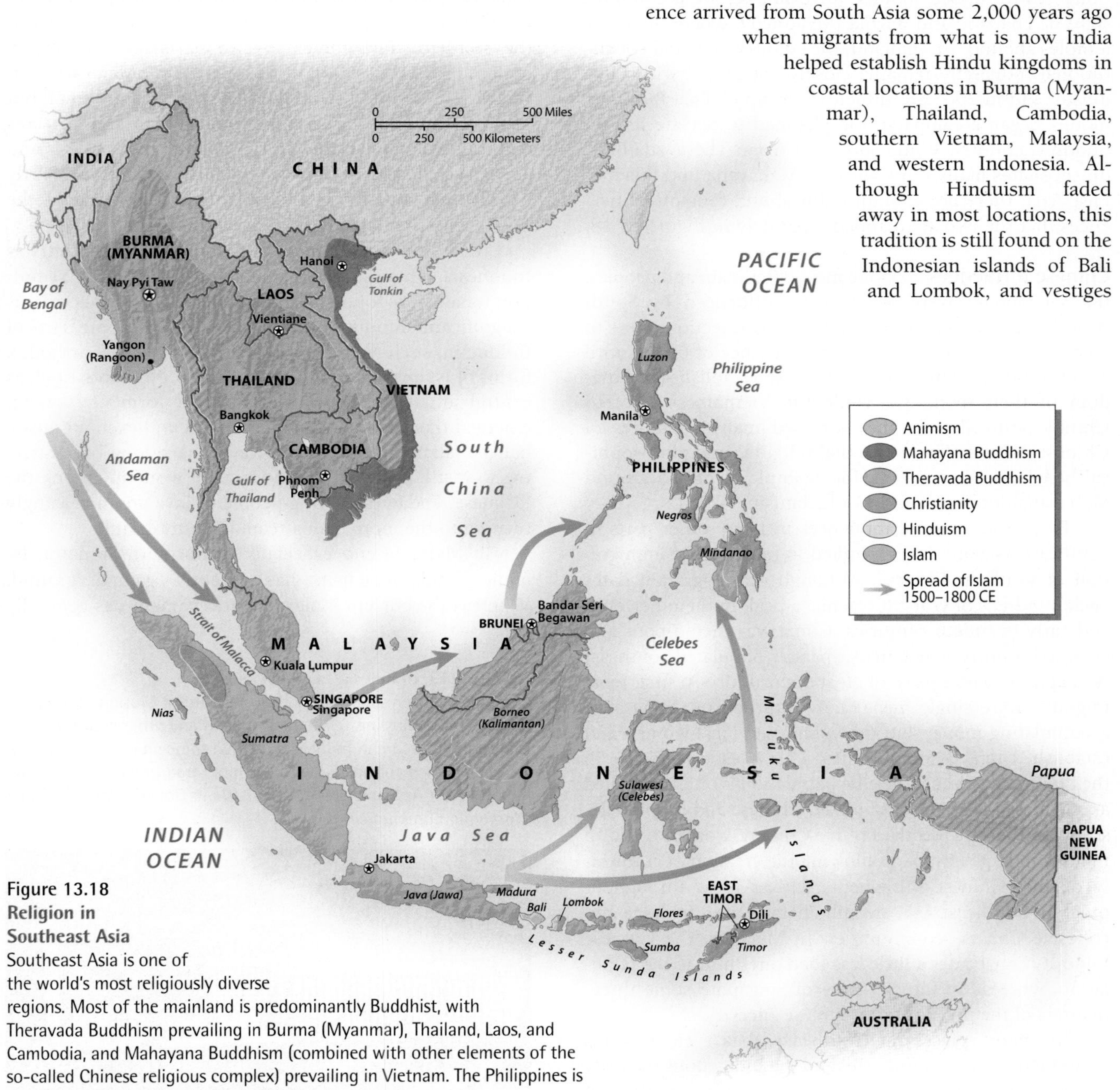

Figure 13.18 Religion in Southeast Asia
Southeast Asia is one of the world's most religiously diverse regions. Most of the mainland is predominantly Buddhist, with Theravada Buddhism prevailing in Burma (Myanmar), Thailand, Laos, and Cambodia, and Mahayana Buddhism (combined with other elements of the so-called Chinese religious complex) prevailing in Vietnam. The Philippines is primarily Christian (Roman Catholic), but the rest of insular Southeast Asia is primarily Muslim. Substantial Muslim minorities are found in the Philippines, Thailand, and Burma (Myanmar). Animist and Christian minorities can be found in remote areas throughout Southeast Asia, especially in Indonesia's portion of the large island of New Guinea.

remain in many other areas. The ancient Indian script formed the basis for many Southeast Asian writing systems, and in Muslim Java (in Indonesia) the Hindu epic called the **Ramayana** remains a central cultural feature to this day.

A second wave of South Asian religious influence reached mainland Southeast Asia in the thirteenth century in the form of Theravada Buddhism, which is closely associated with Sri Lanka. Virtually all of the people in lowland Burma (Myanmar), Thailand, Laos, and Cambodia converted to Buddhism at that time, and today this faith forms the foundation for many of their social institutions. Saffron-robed monks, for example, are a common sight in Thailand and Burma (Myanmar), where Buddhist temples abound. In Thailand, the country's revered constitutional monarchy remains closely connected with Buddhism, and in 2007 an influential group of Thai Buddhist monks began a campaign for Thailand to declare itself a Buddhist country. While Southeast Asian Theravada Buddhism shares many traits with the Mahayana Buddhism of East Asia, there are enough cultural and religious differences that the two are mapped separately in Figure 13.18.

Chinese Influences Unlike most other mainland peoples, the Vietnamese were not heavily influenced by South Asian civilization. Instead, their early connections were to East Asia. Vietnam was actually a province of China until about 1000 CE, when the Vietnamese established a kingdom of their own. But while the Vietnamese rejected China's political rule, they retained many attributes of Chinese culture. The traditional religious and philosophical beliefs of Vietnam are, for example, centered around Mahayana Buddhism and Confucianism.

East Asian cultural influences in many other parts of Southeast Asia are directly linked to more recent immigration of southern Chinese. Although this migration dates back hundreds of years, it reached a peak in the nineteenth and early twentieth centuries. China was then a poor and crowded country, which made sparsely populated Southeast Asia appear to be a place of great opportunity. At first, most migrants were single men. Many returned to China after accumulating money, but others married local women and established mixed communities. This is especially true in the Philippines, where the elite population is often described as "Chinese Mestizo," or of mixed Chinese and Filipino descent. In the nineteenth century Chinese women begin to migrate in large numbers, allowing the creation of ethnically distinct Chinese settlements. Urban areas throughout much of Southeast Asia are still characterized by large and cohesive Chinese communities. In Malaysia, the Chinese minority constitutes a little less than one-third of the population, whereas in the city-state of Singapore some three-quarters of the people are of Chinese ancestry.

In many places in Southeast Asia, relationships between the Chinese minority and the indigenous majority are strained. One significant source of tension is the fact that most overseas Chinese communities are relatively prosperous. Many Chinese emigrants prospered because they became small-scale merchants, a job that was often ignored by the local people. As a result, they often exert tremendous economic influence on local affairs—influence that is sometimes deeply resented by others. Anti-Chinese uprisings have occurred periodically throughout Southeast Asia. In 2000 and 2001, especially severe rioting broke out in several Indonesian cities.

The Arrival of Islam Muslim merchants from South and Southwest Asia arrived in Southeast Asia more than a thousand years ago, and by the 1200s their religion began to spread. From an initial focus in northern Sumatra, Islam diffused into the Malay Peninsula, through the main population centers in the Indonesian islands, and east to the southern Philippines. By 1650, Islam had largely replaced Hinduism and Buddhism in Malaysia and Indonesia. The only significant holdout was the small but fertile island of Bali, where thousands of Hindu musicians and artists fled from the courts of Java, giving the island an especially strong tradition of arts and crafts. Partly because of this artistic legacy, Bali is today one of the premier destinations of international tourism.

Today the world's most populous Muslim country is Indonesia, where some 88 percent of the nation's 231 million inhabitants follow Islam. This figure, however, masks a significant amount of internal diversity. In some parts of Indonesia, such as in northern Sumatra (Aceh), orthodox forms of Islam took root (Figure 13.19). In others, such as central and eastern Java, a more lax form of worship emerged that maintains certain Hindu and even animistic beliefs. Islamic reformers, however, have long been striving to instill more orthodox forms of worship among the Javanese. Today the young people of Java are increasingly turning to the more mainstream forms of Islam.

In Malaysia and especially in northern Sumatra, Islamic fundamentalism has recently gained ground. Whereas the current Malaysian government has generally

Figure 13.19 Indonesia's Largest Mosque
Indonesia is often said to be the world's largest Muslim nation, because more Muslims reside here than in any other country. Islamic architecture in Indonesia is often somewhat modernistic, especially when contrasted with the more traditional styles found in Southwest Asia and North Africa. *(Dana Downie/Dana Downie Marketing Communications)*

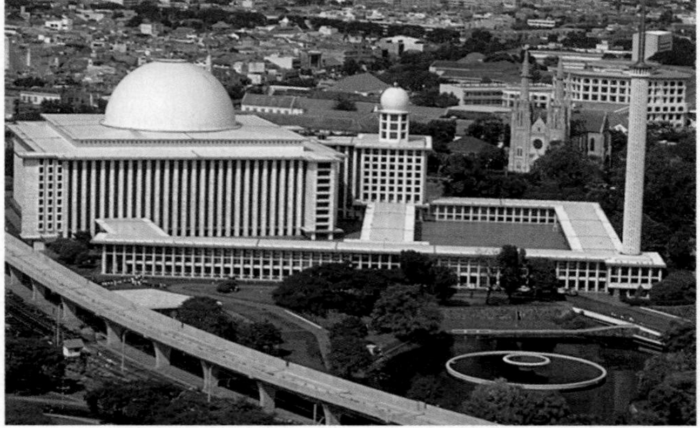

supported the revitalization of Islam, it is wary of the growing power of the fundamentalist movement. Some Malaysian Muslims, moreover, view the fundamentalists as advocating social practices derived from the Arabian Peninsula that are not necessarily religious in origin. Adding to Malaysia's religious tension is the fact that ethnic Malays, who narrowly form the country's majority population, are almost all Muslims, unlike the members of Malaysia's minority groups. Religious controversy reached a high point in 2007, when Malaysia's high court ruled that Lina Joy, a woman who had converted from Islam to Christianity, had no legal right to change her religion.

Islam was still spreading eastward through insular Southeast Asia when the Europeans arrived in the sixteenth century. When Spain claimed the Philippine Islands in the 1570s, it found the southwestern portion of the archipelago to be thoroughly Islamic. The Muslims resisted the Roman Catholicism introduced by the Spaniards and derived their cultural identity from their Islamic ties. To this day, most of the far southwest Philippines is thoroughly Muslim.

The Philippines as a whole is currently roughly 85 percent Roman Catholic, making it, along with East Timor, the only predominantly Christian country in all of Asia. Several Protestant sects have been spreading rapidly in the Philippines over the past several decades, however, creating a more complex religious environment.

Christianity and Indigenous Cultures Christian missions spread through other parts of Southeast Asia in the late nineteenth and early twentieth century when European colonial powers controlled the region. While French priests converted many people in Vietnam to Catholicism, they had little influence elsewhere. Beyond Vietnam, missions failed to make headway in areas of Hindu, Buddhist, or Islamic heritage. Missionaries were, however, far more successful in Southeast Asia's highland areas, where they found a wide array of tribal societies that had never accepted the major lowland religions. Instead, these people retained their indigenous belief systems, which generally focus worship on nature spirits and ancestors. Although some modern hill tribes retain such **animist** beliefs today, others were converted to Christianity. As a result, notable Christian concentrations are found in the Lake Batak area of north-central Sumatra, the mountainous borderlands between southern Burma (Myanmar) and Thailand, the northern peninsula of Sulawesi, and the highlands of southern Vietnam. Animism retains strongholds in the mountains of northern Southeast Asia, central Borneo, far eastern Indonesia, and the highlands of northern Luzon in the Philippines. In Indonesia, however, animist practices are technically illegal, as monotheism is one of the country's founding principles. Partly as a result, many tribal groups have been gradually converted to either Islam or Christianity.

Indonesia, more than anywhere else in the region, has experienced religious strife in recent years, especially between its Muslim majority and Christian minority. Relations between Muslims and Christians in Indonesia were generally quite good until the late 1990s. Indonesia, despite its strong Muslim majority, is a secular state that has always emphasized tolerance among country's officially recognized religions (Islam, Christianity, Buddhism, Hinduism, and Confucianism). With the Indonesian economic disaster of the late 1990s, however, relations quickly deteriorated, and religious fighting began to break out, especially in the Maluku Islands of eastern Indonesia and in Sulawesi. Between 1999 and 2003, an estimated 2,000 people were killed in Christian–Muslim clashes in the island of Sulawesi alone. Transmigration is also implicated in Indonesia's religious conflicts, as in many instances indigenous Christian and animist groups are now struggling against Muslim immigrants from Java and Madura.

Religion and Communism By 1975, communism had triumphed in Vietnam, Cambodia, and Laos. In all three countries, religious practices were then strongly discouraged. At present, Vietnam's officially communist government is struggling against a revival of faith among the country's Buddhist majority and its 8 million Christians. Buddhist monks are frequently detained, and the government reserves for itself the right to appoint all religious leaders. Also expanding is Vietnam's indigenous religion of Cao Dai, a *syncretic* (or mixed) faith that venerates not only the Buddha, Confucius, and Jesus, but also the French novelist Victor Hugo. Between 3 and 8 million Vietnamese currently belong to this religious community.

Geography of Language and Ethnicity

As with religion, language in Southeast Asia expresses the long history of external cultural influences and migration. The linguistic geography of the region is extremely complicated—far too complicated, in fact, to be adequately conveyed in a single map. The several hundred distinct languages of the region can, however, all be placed into five major linguistic families. These are *Austronesian*, which covers most of the islands from the Philippines to Indonesia along with the Malay Peninsula; *Tibeto-Burman*, which includes the languages of Burma (Myanmar); *Tai-Kadai*, centered on Thailand and Laos; *Mon-Khmer* encompassing most of the languages of Vietnam and Cambodia; and *Papuan*, a Pacific language family found also in eastern Indonesia (Figure 13.20; see Chapter 14 for further detail on Papuan).

The Austronesian Languages Austronesian is one of the world's most widespread language families, extending from Madagascar to Easter Island in the eastern Pacific. Linguists contend that this family originated prehistorically in Taiwan and adjacent areas of East Asia, then spread widely across the Indian and Pacific oceans by seafaring people who migrated from island to island.

Today almost all insular Southeast Asian languages are within the Austronesian family. This means that elements of both grammar and vocabulary are widely shared across the islands; thus, it is relatively easy for a person who speaks one of these languages to learn any other. But despite this linguistic commonality at the family level, more than 50 distinct Austronesian languages are spoken in Indonesia alone. And in far eastern Indonesia, a variety of languages falls into the completely separate family of Papuan, closely associated with New Guinea.

One language, however, overshadows all others in insular Southeast Asia: Malay. Malay is indigenous to the Malay Peninsula, eastern Sumatra, and coastal Borneo, (Kalimantan), yet was spread historically throughout the regions by merchants and seafarers. As a result, it became a common trade language, or **lingua franca**, understood and used by people of different languages throughout much of the insular realm. Dutch colonists in Indonesia eventually employed Malay as an administrative language, although they wrote it with the Roman alphabet, rather than in the Arabic-derived script used by native speakers. When Indonesia became an independent country in 1949, its leaders elected to use the lingua franca version of Malay as the basis for a new national language called *Bahasa Indonesia* (or more simply, *Indonesian*). Although Indonesian is slightly different from the Malaysian spoken in Malaysia, they essentially form a single, mutually intelligible language.

The goal of the new Indonesian government was to offer a common language that would overcome ethnic differences throughout the far-flung state. In general, this policy has been successful in that more than 80 percent of all Indonesians now understand the language. However, regionally based languages, such as Javanese, Balinese, and Sundanese, continue to be primary languages in most homes. More than 75 million people speak Javanese, which makes it one of the world's major tongues.

The eight major languages of the Philippines also belong in the Austronesian family. Despite more than 300 years of colonization by Spain, Spanish never became a unifying force for the islands. During

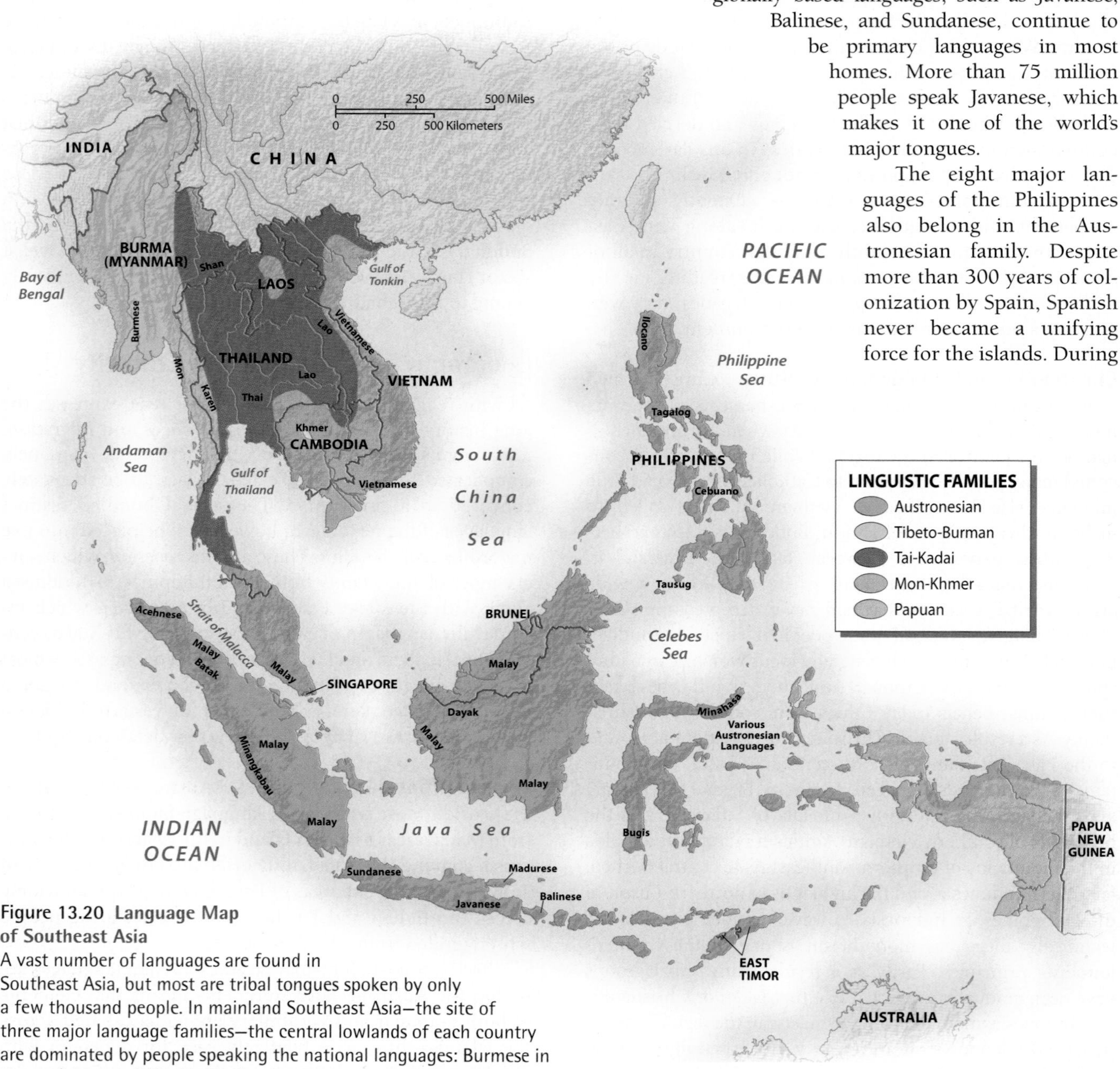

Figure 13.20 Language Map of Southeast Asia
A vast number of languages are found in Southeast Asia, but most are tribal tongues spoken by only a few thousand people. In mainland Southeast Asia—the site of three major language families—the central lowlands of each country are dominated by people speaking the national languages: Burmese in Burma (Myanmar), Thai in Thailand, Lao in Laos, and Vietnamese in Vietnam. Almost all languages in insular Southeast Asia belong to the Austronesian linguistic family. There were no dominant languages here before the creation of such national tongues as Filipino and Bahasa Indonesia in the mid-twentieth century.

the American period (1898–1946), English served as the language of administration and education. After independence following World War II, Philippine nationalists decided to create a national language that could replace English and help unify the new country. They selected Tagalog, the tongue of Manila and a language with a fairly well-developed literary tradition. The first task was to standardize and modernize Tagalog, which had many distinct dialects. After this was accomplished, it was renamed *Filipino* (alternatively, *Pilipino*). Today, mainly because of its use in education, television, and movies, Filipino is gradually becoming a unifying national language.

Tibeto-Burman Languages Each country of mainland Southeast Asia is closely identified with the national language spoken in its core territory. This does not mean, however, that all the inhabitants of these countries speak these official languages on a daily basis. In the mountains and other remote districts, nonnational languages are common. This linguistic diversity reinforces ethnic divisions despite national educational programs designed to foster unity.

A good example of ethnic diversity is Burma (Myanmar). Its national language is Burmese, which is closely related to Tibetan and perhaps, much more distantly, to Chinese. Today some 32 million people speak Burmese as their first language (Figure 13.21). Although the nationalistic military government of Burma (Myanmar) has attempted to force its version of unity on the population, a major schism developed with several non-Burman "hill tribes" that inhabit the rough uplands that flank the Burmese-speaking Irrawaddy Valley. Although most of these tribal groups speak languages in the Tibeto-Burman family, they are quite distinct from Burmese.

Tai-Kadai Languages The Tai-Kadai linguistic family probably originated in southern China and then spread into Southeast Asia starting around 1200 CE. Today, closely related languages within the Tai subfamily are found through most of Thailand and Laos, in the uplands of northern Vietnam, in Burma's Shan Plateau, and in parts of southern China. Most Tai languages are quite localized, spoken by small tribal groups. But two of them, Thai and Lao, are national languages of Thailand and Laos respectively.

Linguistic terminology in this part of the world is complicated. Historically, the main language of the Kingdom of Thailand, called *Siamese* (just as the kingdom was called *Siam*), was restricted to the lower Chao Phraya valley, which formed the national core. In the 1930s, however, the country changed its name to Thailand to emphasize the unity of all the peoples speaking closely related Tai languages within its territory. Siamese was similarly renamed *Thai*, and it has gradually become the unifying language for the country. There are still substantial variations in dialect, however, with those of the north often considered separate languages. Even more distinctive is Lao, the Tai language that became the national tongue of Laos. More Lao speakers reside in Thailand than in Laos, however, where they form the majority population of the relatively poor Khorat Plateau.

Figure 13.21 Burmese Script
The Burmese script, seen here on posters in Burma's primate city of Rangoon (Yangon), has characters with round shapes, supposedly because the palm leaves that were originally used as writing surfaces would have been split by straight lines. Burma's script, like those of Thailand, Laos, and Cambodia, can be traced back to a writing system from ancient India. *(Topham/The Image Works)*

Mon-Khmer Languages The Mon-Khmer language family probably covered virtually all of mainland Southeast Asia 1,500 years ago. It contains two major languages, Vietnamese (the national tongue of Vietnam) and Khmer (the national language of Cambodia), as well as a host of minor languages spoken by hill peoples and a few lowland groups. Because of the historical Chinese influence in Vietnam, the Vietnamese language was written with Chinese characters until the French imposed the Roman alphabet during their colonial reign, which remains in use. Khmer, like the other national languages of mainland Southeast Asia, is written in its own script, ultimately derived from India.

The most important aspect of linguistic geography in mainland Southeast Asia is probably the fact that in each country the national language is limited to the core area of densely populated lowlands, whereas the peripheral uplands are populated by tribal peoples speaking separate languages. In Laos, for example, up to 40 percent of the population is non-Lao, whereas in Vietnam, less than half of the national territory is occupied by Vietnamese speakers, even though they constitute a sizable majority of the country's population. This linguistic contrast between the lowlands and the uplands poses an obvious problem for national integration. Such problems are most extreme in Burma (Myanmar), where the upland peoples are numerous and well organized—and have strenuously resisted the domination of the lowland Burmans. The Burmese government has, in turn, devoted much of its energies to resisting the influences of global culture.

Figure 13.22 Monks in a Burmese Market
Buddhism remains important in mainland Southeast Asia, where many men become Buddhist monks. Religious traditionalism does not prevent engagement with modern global culture, as is evident in this photograph of monks in a Burmese market. *(Andres Hernandez/Getty Images, Inc.—Liaison)*

Southeast Asian Culture in Global Context

The imposition of European colonial rule ushered in a new era of globalization in Southeast Asia, bringing with it European languages, Christianity, and new governmental, economic, and educational systems. This period also deprived most Southeast Asians of their cultural autonomy. As a result, with decolonialization and political independence after World War II, many countries attempted to isolate themselves from the cultural and economic influences of the emerging global system. Burma (Myanmar), for example, retreated into its own form of Buddhist socialism, placing strict limits on foreign tourism, which the government viewed as a source of cultural contamination. Although the door has opened substantially since the 1980s, the government of Burma (Myanmar) remains wary of foreign practices and influences, an attitude that has become a major source of tension. Many of the country's student leaders and Buddhist monks, for example, resist governmental policies (Figure 13.22).

Other Southeast Asian countries have been more receptive to foreign cultural influences. This is particularly true in the case of the Philippines, where American colonialism may have predisposed the country to many of the more popular forms of Western culture. As a result, Filipino musicians and other entertainers are much in demand throughout East and Southeast Asia (Figure 13.23). Thailand, which was never subjected to colonial rule, is the most receptive mainland Southeast Asian country to global culture, with its open policy toward tourism, mass media, and economic interdependence.

But cultural globalization has also been challenged in some Southeast Asian countries. The Malaysian government, for example, is highly critical of many American films and television programs. Islamic revivalism, in both Malaysia and Indonesia, also presents a challenge to cultural globalization. In 2006, for example, an effort to start an Indonesian edition of *Playboy* magazine was quickly shot down by religious leaders. Although Singapore's leaders have also criticized Western culture, the city-state is now seeking to enhance tourism and bolster the economy by building Las Vegas–style casinos and resorts.

As the global language, English also causes ambivalence in many countries. On one hand, it is the language of questionable popular culture; yet, on the other hand, citizens need proficiency in English if they are to participate in global business and politics. In Malaysia, widespread proficiency in English was challenged in the 1980s as nationalists stressed the importance of their native tongue. This distressed the business community, which considers English vital to Malaysia's competitive position, as well as the influential Chinese communities, for which Malay is not a native language.

Figure 13.23 Filipino Entertainers
The people of the Philippines have adopted popular forms of Western culture more than most other Southeast Asians, in part because of their long experience with American colonialism. As a result, Filipino performers are often in demand in other Asian countries. This photograph shows a Filipina musician singing in Hong Kong. *(Andy Maluche Photography)*

PEOPLE ON THE MOVE Filipinos Take to the Seas

The Philippine economy is highly dependent on the money sent home by its citizens working abroad, with up to 10 percent of its GDP (gross domestic product) stemming from such remittances. Only the much larger countries of India and Mexico receive higher levels of remittance than the Philippines. More than 8 million Filipinos now live and work abroad, with the funds that they send back supporting up to 17 percent of all households in the country. Women from the Philippines traveling abroad primarily work as nurses or in domestic service, whereas men often find employment in construction or, increasingly, as crew members of large ocean-going ships.

Over the past several decades, seafaring has become a major Philippine specialization (Figure 13.4.1). Approximately 30 percent of the world's sea-going labor force, whether on container ships or luxury cruise liners, now comes from the Philippines. Almost 250,000 Filipinos gain their livelihoods at sea, earning an average of about $12,000 a year. Employers are required to send 80 percent of each worker's wages back to the Philippines in U.S. dollars, substantially boosting the Philippine economy. In 2003, $2 billion entered the country in such a manner.

The origins of the Philippine merchant marine force dates back to the Spanish period and beyond, but the maritime tradition was strengthened during the U.S. occupation. As early as 1901, the U.S. Navy began recruiting Filipinos through a special program that was not discontinued until 1992. In the early years of this program, Filipinos were restricted to low positions in the U.S. Navy, but they did gain valuable seafaring experience. The Philippine government has also long encouraged its men to take to the seas, maintaining the Philippine Merchant Marine Academy, Asia's oldest maritime institution. It has also set up special recruiting centers to help men begin their careers as ship-hands.

Much evidence suggests that young Filipino men are attracted to the sea more by the lack of other opportunities than by the actual rewards of a merchant marine career. On container ships, men are often forced to remain at sea for up to 10 months, during which time they work and socialize with the same 10 to 20 other men every day. Port liberties are rare and entertainment options on board ship are highly limited. Opportunities for advancement, moreover, are scarce, with fewer than 10 percent of seafarers from the Philippines rising to the position of senior officer. As a result, the large ships on which so many Filipinos work are sometimes jokingly referred to as "floating prisons." But as long as the Philippine economy as a whole continues to perform poorly, large numbers of Filipino men will sign up for jobs at sea.

Figure 13.4.1 Filipino Seaman
Hundreds of thousands of Filipinos work at sea, either as merchant marines on cargo ships or, as in this photo, as crew members of cruise ships. *(Jeff Greenberg/Image Works)*

In Singapore, the situation is more complex. Mandarin Chinese, English, Malay, and Tamil (from southeastern India) are all official languages of this country. Furthermore, the local languages of southern China are common in home environments because 75 percent of Singapore's population is of southern Chinese ancestry. The Singapore government now encourages the use of Mandarin Chinese and discourages the use of southern Chinese dialects. It also supports English as the language of global commerce, while discouraging *Singlish*, an English-based dialect containing many southern Chinese words and grammatical forms. Singaporean officials have been known to slap restrictive ratings on local films for "bad grammar" if they contain too much Singlish dialogue.

In the Philippines, nationalists have long decried the common use of English, even though widespread fluency has proved beneficial to the millions of Filipinos who have emigrated for better economic conditions or who work as crew members in the globalized shipping industry (see "People on the Move: Filipinos Take to the Seas"). The Philippine government is now gradually replacing English with Filipino in public schools and, as a result, some fear that competence in English is slowly declining. Others, however, counter that the spread of cable television, with its many U.S. shows, has enhanced the country's English-language skills.

While the diffusion of global cultural forms is beginning to merge cultures across Southeast Asia, religious revivalism and the spread of fundamentalism are spreading other cultures apart. This same combination of centripetal (merging together) and centrifugal (spreading apart) forces also is found in the geopolitical realm.

GEOPOLITICAL FRAMEWORK: War, Ethnic Strife, and Regional Cooperation

Southeast Asia is sometimes defined as a geopolitical entity of 10 different states that have joined together under the umbrella organization of the Association of Southeast Asian Nations, or ASEAN (Figure 13.24). Today ASEAN, more than anything else, gives Southeast Asia a geopolitical regional coherence. Within this framework, however, many states are still struggling with serious ethnic and regional tension. In 1999 Indonesia surprised much of the world by relinquishing its control of the eastern portion of the island of Timor, which emerged as a fully independent country in 2002. East Timor (or Timor-Leste, as the country is officially called) has applied for membership into ASEAN, but has not yet gained admission.

Before European Colonialism

The modern countries of mainland Southeast Asia all existed in one form or another as indigenous kingdoms before the onset of European colonialism. Cambodia emerged earliest, reaching its height in the twelfth century, when it controlled much of what is now Thailand and southern Vietnam. By the 1300s, independent kingdoms had been established by the Burman, Siamese (Thai), Lao, and Vietnamese people who were centered on the major river valleys and deltas. These kingdoms were in a nearly constant state of war with one another, often fighting more for labor than for territory. Victors would typically take home thousands of prisoners to settle

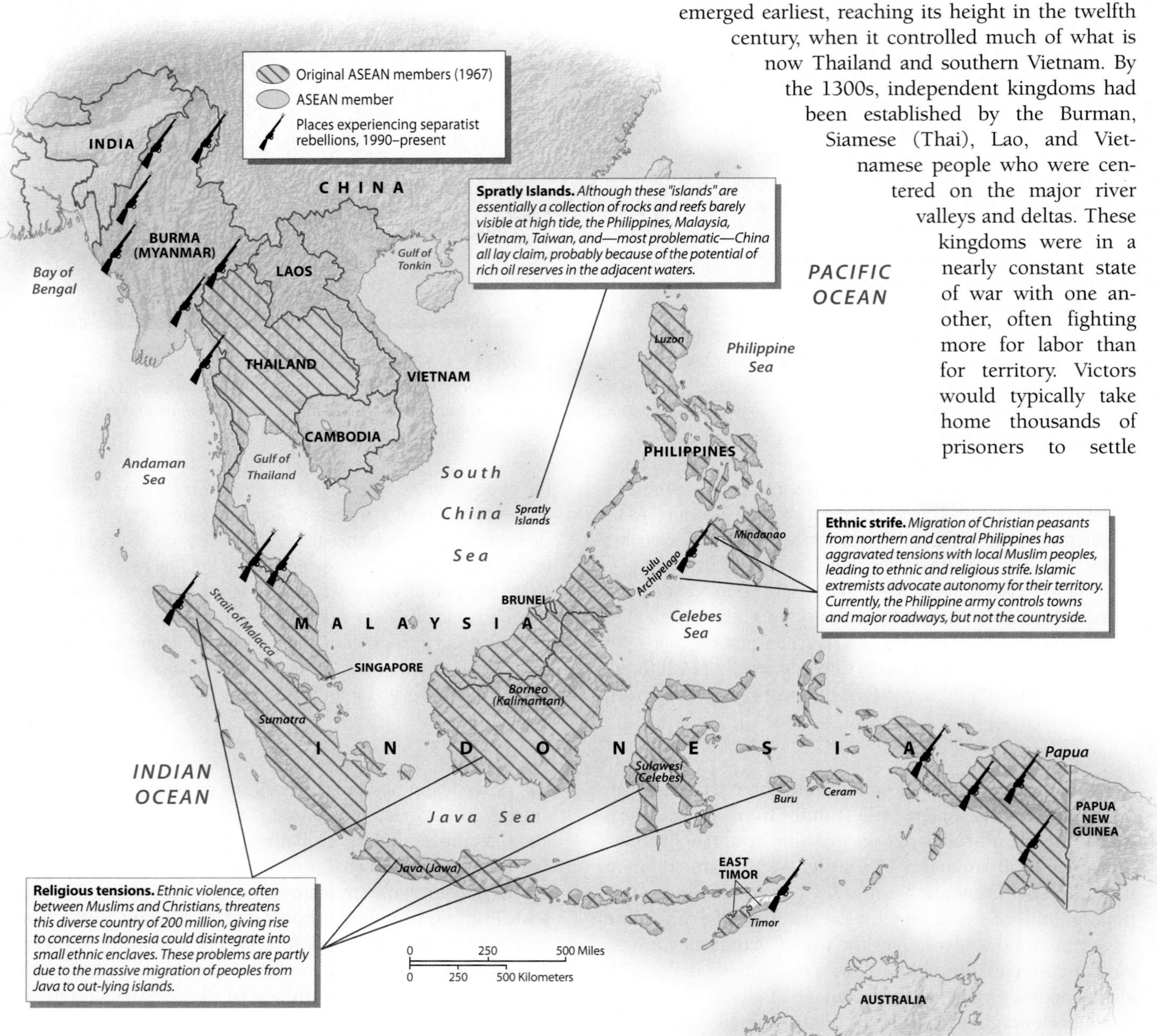

Figure 13.24 Geopolitical Issues in Southeast Asia
The countries of Southeast Asia have managed to solve most of their border disputes and other sources of potential conflicts through ASEAN (the Association of Southeast Asian Nations). Internal disputes, however, mostly focused on issues of religious and ethnic diversity, continue to plague several of the region's states, particularly Indonesia and Burma (Myanmar). ASEAN also experiences tension with China over the Spratly Islands of the South China Sea.

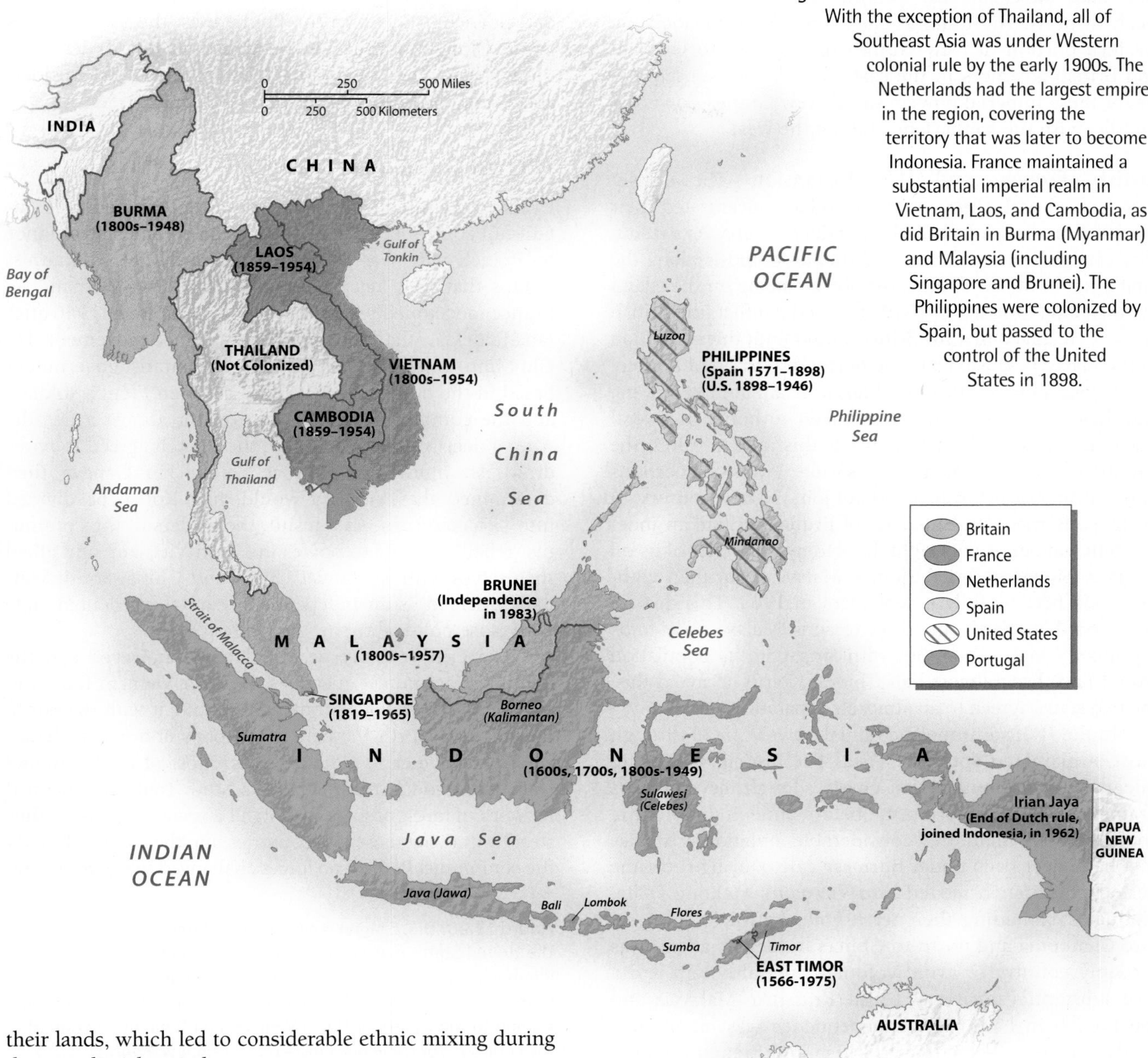

Figure 13.25 Colonial Southeast Asia With the exception of Thailand, all of Southeast Asia was under Western colonial rule by the early 1900s. The Netherlands had the largest empire in the region, covering the territory that was later to become Indonesia. France maintained a substantial imperial realm in Vietnam, Laos, and Cambodia, as did Britain in Burma (Myanmar) and Malaysia (including Singapore and Brunei). The Philippines were colonized by Spain, but passed to the control of the United States in 1898.

their lands, which led to considerable ethnic mixing during the precolonial period.

The situation in insular Southeast Asia was quite different from that of the mainland, with the premodern map bearing no resemblance to that of the modern nation-states. Many kingdoms existed on the Malay Peninsula and on the islands of Sumatra and Java, but few were territorially stable. In the Philippines, eastern Indonesia, and central Borneo, most societies were organized at the village level. The countries of Indonesia, the Philippines, and Malaysia thus owe their territorial configuration almost wholly to the European colonial powers (Figure 13.25).

The Colonial Era

The Portuguese were the first Europeans to arrive (around 1500), lured by the cloves and nutmeg of the Maluku Islands (formerly the Spice Islands) in what is now eastern Indonesia. In the late 1500s, the Spanish conquered most of the Philippines, which they used as a base for their silver and silk trade between China and the Americas. By the 1600s, the Dutch had started staking out Southeast Asian territory, followed by the British. With superior naval weaponry, the Europeans were quickly able to conquer key ports and strategic trading locales. Yet for the first 200 years of colonialism, except in the Philippines, the Europeans made no major geopolitical changes.

Dutch Power By the 1700s, the Netherlands had become the most powerful force in the region. As a result, a Dutch Empire in the "East Indies" (or Indonesia) began appearing on world maps. This empire continued to grow into the early twentieth century, when the Dutch defeated their last main adversary, the powerful Islamic sultanate of Aceh in northern Sumatra. Later, the Dutch invaded the western portion of New Guinea in response to German and British advances in the eastern half. In a subsequent treaty, these imperial powers sliced New Guinea down the middle, with the Netherlands taking the west.

British, French, and U.S. Expansion The British, preoccupied with their empire in India, concentrated their attention on the sea-lanes linking India to China. As a result, they established several fortified trading outposts along the vital Strait of Malacca, the most notable being on the island of Singapore, founded in 1819. To avoid conflict, the British and Dutch agreed that the British would limit their attention to the Malay Peninsula and the northern portion of Borneo. The British allowed Muslim sultans to retain limited powers, and their descendents still enjoy token authority in several Malaysian states. When Britain left this area in 1963, the country of Malaysia emerged in its wake. Two small portions of the former British sphere did not join the new country. In northern Borneo, the Sultanate of Brunei became an independent state in its own right, backed by the riches of its oil reserves. Singapore briefly joined Malaysia, but then withdrew and became fully independent in 1965. This divorce was carried out partly for ethnic reasons. Malaysia was to be a primarily Malay state, but with Singapore its population would have been almost half Chinese. Partly to avoid this, the two states agreed to an amicable separation.

In the 1800s European colonial power spread through most of mainland Southeast Asia. The British, seeking to safeguard their South Asian empire, fought several wars against the kingdom of Burma before annexing the entire area in 1885, including considerable upland territories that had never been under Burmese rule. During the same period, the French moved into Vietnam's Mekong Delta, gradually expanding their territorial control to the west into Cambodia and north to China's border. Thailand was the only country to avoid colonial rule, although it did lose substantial territories to the British in Malaysia and the French in Laos. Thai independence was maintained partly because it served both British and French interests to have a buffer state between their colonial empires.

The final colonial power to enter the area was the United States, which took the Philippines first from Spain and then, after a bitter war, from Filipino nationalists between 1898 and 1900. The U.S. army subsequently conquered the Muslim areas of the southwest, most of which had never been fully under Spanish authority.

Growing Nationalism Organized resistance to European rule began in the 1920s in mainland countries, but it took the Japanese occupation during World War II to shatter the myth of European invincibility. After Japan's surrender in 1945, agitation for independence was renewed throughout Southeast Asia. As Britain realized that it could no longer control its South Asian empire, it also withdrew from adjacent Burma, which achieved independence in 1948. Although the Netherlands attempted to reconquer Indonesia after WWII, it was forced to acknowledge Indonesian independence in 1949. In the Philippines, the United States granted long-promised independence on July 4, 1946, although it retained key military bases, as well as considerable economic influence, for several decades.

The Vietnam War and Its Aftermath

Unlike the United States, France was determined to maintain full control of its Southeast Asian colonies. Resistance to French rule was organized primarily by communist groups that were deeply rooted in northern Vietnam. As France reoccupied its Southeast Asian empire in 1946 after Japan's defeat, the leader of this resistance movement, Ho Chi Minh, emerged as president of a separatist government based in the north. Open warfare between French soldiers and the communist forces went on for almost a decade. After a decisive defeat in 1954, the French agreed to withdraw. An international peace council in Geneva then determined that Vietnam would, like Korea, be divided into two countries. As a result, the leaders of the communist rebellion came to power in North Vietnam, and allied themselves with the Soviet Union and China. South Vietnam emerged as an independent, capitalist-oriented state with close political ties to the United States.

The Geneva peace accord did not, however, end the fighting. Communist guerrillas in South Vietnam fought to overthrow the new government and unite it with the north. For its part, North Vietnam sent troops and war materials across the border to aid the rebels. Most of these supplies reached the south over the Ho Chi Minh Trail, an ill-defined network of forest passages through Laos and Cambodia, thus steadily drawing these two countries into the conflict. In Laos the communist Pathet Lao forces challenged the government,

Figure 13.26 U.S. Soldier and Vietcong Prisoners
The United States maintained a substantial military presence in Vietnam in the 1960s and early 1970s. Although U.S. forces claimed many victories, they were ultimately forced to withdraw, leading to the victory of North Vietnam and the reunification of the country. *(Getty Images Inc.—Hulton Archive Photos)*

while in Cambodia the **Khmer Rouge** guerrillas gained considerable power. In South Vietnam, the government gradually lost control of key areas, including much of the Mekong Delta, a region perilously close to the capital city of Saigon.

U.S. Intervention By 1962 the United States was sending large numbers of military advisors to South Vietnam. In Washington, DC, the **domino theory** became accepted foreign policy. According to this notion, if Vietnam fell to the communists, then so would Laos and Cambodia; once those countries were lost, Burma (Myanmar) and Thailand, and perhaps even Malaysia and Indonesia, would become members of the Soviet-dominated communist bloc. By 1965, thousands of U.S. troops had begun a ferocious land war against the communist guerrillas (Figure 13.26). But despite superiority in arms and troops—and total domination of the air—U.S. forces failed to gain control over much of the countryside. As casualties mounted and the antiwar movement back home strengthened, the United States held secret talks with North Vietnam in search of a negotiated settlement. U.S. troop withdrawals began in earnest by the early 1970s.

Communist Victory With the withdrawal of U.S. forces and financial support, the noncommunist governments of the former French colonial zone began to collapse. Saigon fell in 1975, and in the following year Vietnam was officially reunited under the government of the north. Reunification was a traumatic event in southern Vietnam. Hundreds of thousands of people fled from the new regime to other countries, especially the United States. The first wave of refugees consisted primarily of wealthy professionals and businesspeople, but later migrants included many relatively poor ethnic Chinese. Most of these refugees fled on small, rickety boats; large numbers suffered shipwreck or pirate attack.

Vietnam proved fortunate compared with Cambodia. There the Khmer Rouge installed one of the most brutal regimes the world has ever seen. City-dwellers were forced into the countryside to become peasants, and most wealthy and highly educated persons were summarily executed. The Khmer Rouge's goal was to create a wholly new agrarian society by returning to what they called "year zero." After several years of bloodshed that took an estimated 1.5 million lives, neighboring Vietnam invaded Cambodia and installed a far less brutal, but still repressive, regime. Fighting between different factions continued for more than a decade, but by the late 1980s the United Nations was able to broker a settlement to this civil war. Since that time, several unstable coalition governments have brought a tenuous peace to the shattered country. Although Cambodia is officially a constitutional monarchy with a democratic government, corruption is widespread and democratic institutions remain weak and unstable.

Vietnam stationed significant numbers of troops in Laos after 1975. Large numbers of Hmong and other tribal peoples—many of whom had fought on behalf of the United States—fled to Thailand and the United States. Elections were not held in Laos until 1989, when a people's assembly was formed to approve a new constitution. The Communist Party still maintains a monopoly over political power in Laos, although much of the economy has been opened to private firms. Many Laotian Hmong refugees who fled to the United States after the war, however, continue to seek political change at home. In 2007, the American Hmong community was shocked when U.S. federal agents arrested 11 of their leaders for plotting to overthrow the government of Laos (Figure 13.27).

Figure 13.27 Hmong General Vang Pao
General Vang Pao, a respected leader of Hmong-Americans and a suspect in the alleged plot to overthrow the government of Laos, is released on bail from Sacramento County jail on July 13, 2007. Because the Hmong community in Laos continues to suffer persecution, many Hmong-Americans want to see a new government in their original homeland. *(Carl Costas/MCT/Landov)*

Geopolitical Tensions in Contemporary Southeast Asia

A number of current conflicts in Southeast Asia are rooted in the region's colonial past. In several parts of the region, locally based ethnic groups are struggling against the centralized national governments that inherited territory from former colonial powers. Tension also results when tribal groups attempt to preserve their homeland from logging, mining, or interregional migration. Such conflicts are especially pronounced in the large, multiethnic country of Indonesia.

Conflicts in Indonesia When Indonesia gained independence in 1949, it encompassed all of the former Dutch possessions in the region except western New Guinea. The Netherlands retained this territory, arguing that its cultural background distinguished it from Indonesia. In 1962, Dutch authorities organized a referendum to see whether the people of western New Guinea wished to join Indonesia or form an independent country. The vote went for union, but many observers believe that the election was rigged by the Indonesian government.

Tensions in western New Guinea increased in the following decades as Javanese immigrants, along with mining and lumber firms, arrived in the area. Faced with the loss of their land and the degradation of their environment, a number of indigenous residents formed the separatist organization OPM (*Organisesi Papua Merdeka*) and launched a rebellion. Rebel leaders demanded independence, or at least autonomy, but they faced a far stronger force in the Indonesian army. This war is still a smoldering, sporadic, and occasionally bloody guerrilla affair. Indonesia is determined to maintain control of the region in part because it is home to one of the country's largest taxpayers, the highly polluting Grasberg copper and gold mine run by the New Orleans–based Freeport-McMoRan Corporation.

An even more brutal war erupted in 1975 on the island of Timor, in southeastern Indonesia. The eastern half of this poor island had been a Portuguese colony (the only survivor of Portugal's sixteenth-century foray into the region) and had therefore evolved into a largely Christian society. The East Timorese expected independence when the Portuguese finally withdrew. Indonesia, however, viewed the area as rightfully its own, largely by virtue of its geographical position, and immediately invaded. A ferocious war ensued, which the Indonesian army won in part by starving the people of East Timor into submission.

Figure 13.28 East Timorese Demonstration in Jakarta Indonesia invaded and claimed East Timor in 1975 when the Portuguese left their last colonial outpost in Southeast Asia. The East Timorese resisted the Indonesian takeover, and violence reached a climax in 1999 when the East Timorese voted for independence, after which pro-Indonesian militias attacked the Timorese leaders and large segments of Timorese society. *(V. Miladinovic/Corbis/Sygma)*

After experiencing an economic crisis in 1997, Indonesia's power in the region slipped. A new Indonesian government promised an election in 1999 to see whether the East Timorese wanted independence (Figure 13.28). At the same time, however, the Indonesian army began to organize loyal militias in an attempt to intimidate the people of East Timor into voting to remain within the country. When it was clear that the vote would be strongly for independence, the militias began rioting, looting, and slaughtering civilians. Under international pressure, Indonesia finally withdrew, United Nations forces arrived, and the East Timorese began to build a new country.

Considering the devastation caused by the militias, the reconstruction process in East Timor has not been easy. Even selecting an official language proved difficult; after some deliberation, both Tetum (one of 16 indigenous languages) and Portuguese were selected. In 2006, major rioting further damaged the country, requiring Australian intervention to reestablish peace. The disturbance broke out after 519 East Timorese soldiers were dismissed from the national army for complaining about being discriminated against because they were from the western—and more Indonesia-oriented—part of the island. A relatively successful election in 2007, however, indicated that East Timor has the potential to overcome its ethnic and political divisions.

Secession struggles have occurred elsewhere in Indonesia as well. In the late 1950s, central Sumatra and the northern portion of Sulawesi rebelled, but they were quickly defeated and eventually reconciled to Indonesian rule. Another small-scale war flared up in the late 1990s, this one in western and southern Borneo (Kalimantan). Here the indigenous Dayaks, a tribal people now partly converted to Christianity, began to clash with Muslim migrants from Madura, a densely inhabited island north of Java. The Indonesian military has restored order in Kalimantan's cities, but the countryside remains troubled. The southern Maluku Islands, especially Ambon and Seram, are another conflict zone between Muslims and Christians. Even the area's cities are now divided into Muslim and Christian sectors, and crossing to the wrong side can prove dangerous.

Over the past several decades, Indonesia's most serious regional conflict has been that of Aceh in northern Sumatra. Many Acehnese, in general the most orthodox Muslim people of Indonesia, have long demanded the creation of an independent Islamic state. While the Indonesian government has allowed Aceh a high degree of autonomy as a "special territory of Indonesia," it has been determined to do whatever is necessary to prevent actual independence. Ironically, the devastation caused by the December 2004 tsunami, which left over 500,000 Acehnese homeless, allowed a peace settlement to be finally reached, as the needs of the area were so great that the separatist fighters agreed to lay down their weapons. A 2006 election brought former rebel leaders into the heart of Aceh's government.

Indonesia has obviously had difficulties creating a unified nation over its vast, sprawling extent and across its numerous cultural and religious communities. As a creation of the colonial period, Indonesia has a weak historical

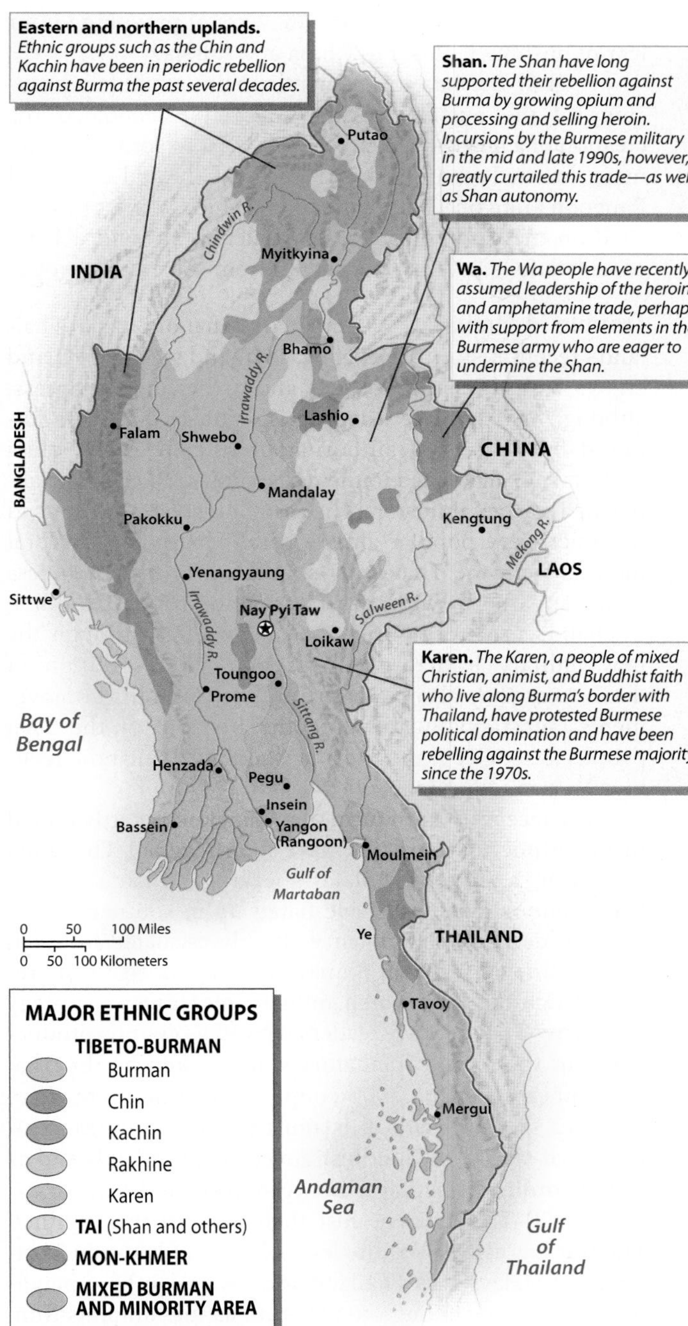

Figure 13.29 Ethnic Conflict in Burma (Myanmar) Although the central lowlands of Burma (Myanmar) are primarily populated by people of the dominant Burman ethnic group, the peripheral highlands and the southeastern lowlands are home to numerous non-Burman peoples. Most of these peoples have been in periodic rebellion against Burma (Myanmar) since the 1970s, owing to their perception that the national government was attempting to impose Burman cultural norms. Economic stagnation and political repression by the central government have intensified these conflicts.

foundation. Some critics have viewed it as something of a Javanese empire, but it is true that many non-Javanese individuals have risen to high governmental positions. While Indonesia now has the highest degree of political freedom in Southeast Asia, its military remains politically powerful and hostile to movements for regional autonomy.

Regional Tensions in the Philippines The Philippines also has suffered from a regional secession movement in its Islamic southwest. Although the government has made some headway in talks with the main rebel group by promising greater autonomy, the more extreme Muslim factions continue to demand independence. The migration of Christian peasants from northern and central Philippines into the Islamic southwest has exacerbated local tensions. The Philippine army generally maintains control over the area's main cities and roadways, but it has little power in the more remote villages. As radical Islamist groups reportedly have close ties with Osama bin Laden and his Al Qaeda network, the Philippines quickly became a key site in the U.S.-led struggle against global terrorism. Many Filipino nationalists, however, are concerned that their country is again falling under U.S. domination.

The Philippines also has faced a communist-oriented nationwide rebellion. In the mid-1980s, the NPA (New People's Army) controlled one-quarter of the country's villages scattered across all the major islands. Although the NPA's strength has declined since then, it remains a potent force in many parts of the country. Furthermore, the Philippine national government, although democratic, is far from stable, suffering from continual coup threats, corruption scandals, mass protests, and impeachment efforts.

The Quagmire of Burma Burma (Myanmar) has been one of the most war-ravaged countries of Southeast Asia. Burma's simultaneous wars of the 1970s, 1980s, and 1990s pitted the central government, dominated by the Burmese-speaking ethnic group (the Burmans), against the country's varied non-Burman societies. Fighting intensified gradually after independence in 1948, until almost half of the country's territory had become a combat zone (Figure 13.29). Burma's troubles, moreover, are by no means limited to the country's ethnic minorities. Since 1988, it has been ruled by a repressive military regime, one that is bitterly resented by many if not most Burmans. Democratic opposition to the government, however, has been vigorously suppressed, with its leader, Nobel Peace Prize winner Aung San Suu Kyi, remaining under house arrest (Figure 13.30).

The rebelling ethnic groups of Burma (Myanmar) have sought to maintain their cultural traditions, lands, and resources, and they generally see the national government as something of a Burman empire seeking to impose its language and, in some cases, religion—Theravada Buddhism—upon them. Most of the insurgent ethnic groups live in rugged and inaccessible terrain, and many of them are animists or Christians. But even some lowland groups have rebelled. The Muslim peoples of the Arakan coast (the Rohingyas) in far western Burma (Myanmar) have long

Figure 13.30 Aung San Suu Kyi
The noted Burmese democratic opposition leader Aung San Suu Kyi received the Nobel Peace Prize in 1991. As of 2007, she remained under house arrest. *(David Van der Veen/EPA/Corbis)*

faced discrimination. In the 1980s, many were forced to flee to neighboring Bangladesh, while more recently Burmese Muslims have been fleeing to Yunnan Province in China.

Several of Burma's ethnic insurgencies have been financed largely by opium-growing and heroin manufacture. The Shan, for example, a Tai-speaking people inhabiting a plateau area within the famous Golden Triangle of drug production, formed a breakaway state in the 1980s and 1990s based largely on the narcotics trade. Their efforts began to falter in the late 1990s after Burmese agreements with Thailand reduced the Shan heroin trade, while military operations cut off the supply of raw opium reaching the Shan factories. But even though the government apparently triumphed over the Shan, other ethnic groups have been able to move into the void. By 2001, the United Wa State Army (UWSA), having largely abandoned ethnic politics for the drug trade, emerged as the major supplier of illicit drugs. As of 2007, the UWSA controlled an estimated 50 heroin refineries and methamphetamine factories and maintained some 20,000 soldiers. Some reporters have alleged that Burma's military leaders tolerate the USWA both because it helps them in their struggles against other ethnic groups and because they get a share of the narcotics profits.

The fact that Burma (Myanmar) has a repressive and Burman-dominated government has resulted in serious international repercussions. Both the United States and the European Union maintain serious trade sanctions. Burma's military rulers do not seem very concerned with global opinion, however, and by joining ASEAN in 1997 they gained international credibility and regional support. Growing economic ties with India and especially China have also helped strengthen Burma's government. Russia also maintains ties with Burma, and in 2007 it agreed to help it build a small nuclear reactor. The other ASEAN countries, for their part, are losing patience with Burma's military government, but they are wary of provoking it for fear of pushing it closer to China and Russia.

Overall, Burma's government remains extremely suspicious of the outside world. In 2005 it mandated the creation of a new capital city at Nay Pyi Taw, located in a remote, forested area 200 miles north of the old capital of Rangoon (Yangon). The new capital, which was supposedly sited on the basis of astrological calculations and defense considerations, is largely closed off from the outside world.

Trouble in Thailand Compared to that of Burma, Thailand's recent history appears peaceful and stable. Thailand enjoys basic human freedoms and a thriving free press, although it does have a legacy of military takeovers followed by periods of authoritarian rule. In 2006, mass protests in Bangkok led to the resignation of the corruption-plagued prime minister, Thaksin Shinawatra, who had remained very popular among rural Thai voters. Several months later, the Thai army seized power in a bloodless coup, apparently with the blessing of King Bhumibol, Thailand's revered constitutional monarch. Although the new military leaders promised a quick restoration of democracy, they also banned a number of political movements, substantially reducing the freedom of the Thai people (See "Geography in the Making: Thailand's Troubled Return to Democracy").

The recent political turmoil in Bangkok is partly rooted in the ethnic tensions that have long plagued Thailand's far south, a primarily Malay-speaking, Muslim region. While minor rebellions have flared up in southern Thailand for decades, the violence sharply escalated in 2004. Prime Minister Thaksin Shinawatra responded to the renewed fighting with harsh military measures, a strategy that many Thai military leaders viewed as counterproductive. But the more conciliatory stance taken by the Thai government after the 2006 coup has been no more effective, as a shadowy radical Islamist group continues to attack Thai-speaking, Buddhist government officials and as well as ordinary civilians. Between 2004 and June 2007, over 2,100 Thai citizens lost their lives in this conflict. The war has also raised the level of tension between Thailand and Malaysia, as Thailand accuses its southern neighbor of not doing enough to prevent Islamic militants from slipping across the border.

International Dimensions of Southeast Asian Geopolitics

As southern Thailand shows, geopolitical conflicts in Southeast Asia can be complex affairs, involving several different countries as well as nonnational organizations. In earlier years, some of the most serious tensions emerged when two countries claimed the same territory. More recently, radical Islamist groups have posed the biggest challenge.

Territorial Conflicts In previous decades, several Southeast Asian countries quarreled over their common boundaries. The Philippines, for example, still maintains a

Figure 13.31 **The Spratly Islands**
The Spratly Islands are small and barely above water at high tide, but they are geopolitically important. Oil may exist in large quantities in the vicinity, heightening the competition over the islands. Southeast Asian countries have been especially concerned about China's military activities in the Spratlys. *(Nouvelle Chine/Getty Images, Inc.–Liaison)*

"dormant" claim to the Malaysian state of Sabah in northeastern Borneo (Kalimantan) based on the fact that in the 1800s the Islamic Sultanate of the Sulu Archipelago, which was centered in the southern Philippines, controlled much of its territory. The Philippines has suspended this claim as part of a general movement of Southeast Asian countries to reduce regional tensions. With the rise of the Association of Southeast Asian Nations (ASEAN), national leaders have concluded that friendly relations with one's neighbors are more important than the possible gain of additional territory.

More difficult has been the dispute over the Spratly Islands in the South China Sea, a group of insignificant rocks, reefs, and tiny islands that may well lie over substantial undersea oil reserves (Figure 13.31). The Philippines, Malaysia, Vietnam, and Brunei have all advanced territorial claims over the Spratly Islands, as have China and Taiwan. In 2002, however, all parties pledged to seek a peaceful solution.

ASEAN and Global Geopolitics The development and enlargement of ASEAN has helped reduce geopolitical tensions across Southeast Asia. Initially, ASEAN was an alliance of nonsocialist countries fearful of the communist regimes that had come to power in Vietnam, Cambodia, and Laos. Through the 1980s, the United States maintained naval and air bases in the Philippines, and U.S. military force helped bolster the anticommunist coalition. In the early 1990s, however, the United States, under mounting pressure from Filipino nationalists, withdrew from the Philippines. By that time, the struggle between communism and capitalism was no longer an international issue. In 1995, Vietnam itself gained membership in the organization, followed by Laos and Cambodia.

While ASEAN is on friendly terms with the United States, one purpose of the organization is to prevent the United States—or any other country—from gaining undue influence in the region. For that reason, ASEAN leaders are keen to include all Southeast Asian countries within the association, although East Timor (an ASEAN applicant) has yet to gain membership. ASEAN's ultimate international policy is to encourage conversation and negotiation over confrontation. For this reason, the ASEAN Regional Forum (ARF) was established in 1994 as an annual conference in which ASEAN leaders could meet with the leaders of both the East Asian and Western powers to attempt to ease tensions within the region. Seeking to further improve relations with East Asia, ASEAN leaders have also established an annual ASEAN+3 meeting where its foreign ministers confer with those of China, Japan, and South Korea.

One important issue faced by ASEAN is the rise of piracy in Southeast Asian seas. Southeast Asia is a key location for global shipping; roughly one-quarter of all of the world's maritime trade passes through the Strait of Malacca (between peninsular Malaysia and Sumatra) alone. The same strait experiences roughly a quarter of all global pirate attacks. Piracy recently has increased markedly throughout the region, due mainly to economic and political instability in Indonesia.

Global Terrorism and International Relations ASEAN has not defused all of the international political tensions of Southeast Asia. Most worrisome has been rise of radical Islamic fundamentalism in the Muslim parts of the region. One group, Jemaah Islamiya (JI), which calls for the creation of a single Islamic state across Indonesia, Malaysia, southern Thailand, and the southern Philippines, has cooperated extensively with Al Qaeda. In 2002, JI agents blew up a tourist-oriented nightclub in Bali, killing 202 people. The group followed this by bombing an American-run hotel in Jakarta in 2003, setting off explosions in Australia's Indonesian embassy in 2004, and attacking three tourist restaurants in Bali in 2005.

Most evidence shows that few Southeast Asian Muslims support radical groups such as JI. Since 2004, even mainstream Muslim political parties have not done particularly well in national elections in Indonesia and Malaysia. Some evidence also suggests that Indonesian authorities had by 2007 arrested most of JI's top leaders, thus crippling the organization. Radical Islamist groups in the southern Philippines and southern Thailand, on the other hand, remain militarily capable.

Although radical Islam has relatively few supporters in Southeast Asia, the people of the region have in general been wary of U.S. foreign policy initiatives designed to combat global terrorism. Resentment against the United States is especially pronounced in Indonesia and Malaysia, where most people see the war in Iraq as an assault on Islam. Perhaps as a result, Indonesia's courts have given lenient prison sentences to several JI operatives. The group's spiritual leader, Abu Bakar Basyir, for example, served only 26 months of a 30-month sentence for his alleged involvement in the 2002 Bali bombing.

ECONOMIC AND SOCIAL DEVELOPMENT: The Roller-Coaster Ride of Globalized Development

Until the downturn of the late 1990s, economic development in Southeast Asia was often held up to the world as a model for a new globalized capitalism. With investment capital flowing first from Japan and the United States, then from international investment portfolios, Thailand, Malaysia, and Indonesia experienced impressive economic booms. In the summer of 1997, however, regional economies suffered a profound crisis, with the currencies of both Thailand and Indonesia being devalued almost 50 percent (Figure 13.32). Subsequent years have seen economic ups and downs, but by 2007 most of the region had recovered and was experiencing relatively strong economic growth.

Recent efforts to enhance growth by fostering the integration of the region's economies have met with mixed success. In 2004, the core ASEAN countries (Brunei, Indonesia, Malaysia, the Philippines, Singapore, and Thailand) agreed to drop their tariffs against one another's products and pledged to create a Southeast Asian free trade area. The remaining ASEAN states have expressed interest in joining the pact in 2012, but some observers remain skeptical. ASEAN also agreed to pursue free trade arrangements with China in 2002, but progress toward implementation has been slow.

Figure 13.32 Bank Closings
The economic crisis of 1997–98 hit Southeast Asia's banking sector particularly hard. Many banks were closed because their liabilities were much greater than their assets. The economic crisis brought devastation to the poor and lower middle class. *(Marcus Rose/Panos Pictures)*

Uneven Economic Development

Southeast Asia today is a region of strikingly uneven economic and social development. While some countries, such as Indonesia, have experienced both boom and bust,

TABLE 13.2 Development Indicators

Country	GNI[a] per Capita, PPP[b]	GDP[c] Average Annual % Growth (2000–05)	Life Expectancy	Percent of Population Living on Less Than $2 a Day	Under Age 5 Mortality Rate 1990	Under Age 5 Mortality Rate 2005	Gender Equity[d]
Brunei			75				
Burma (Myanmar)		9.2	60		130	105	104
Cambodia	2,490	8.9	63	78	115	87	87
East Timor			58				
Indonesia	3,720	4.7	69	52	91	36	97
Laos	2,020	6.2	55	74	163	79	84
Malaysia	10,320	4.8	74	9	22	12	109
Philippines	5,300	4.7	69	48	62	33	106
Singapore	29,780	4.2	80		8	3	
Thailand	8,440	5.4	71		37	21	101
Vietnam	3,010	7.5	72		53	19	94

[a]*Gross national income.*
[b]*Purchasing power parity.*
[c]*Gross domestic product.*
[d]*Ratio of female-to-male enrollments in primary and secondary school percentage. Numbers below 100 have more males in primary/secondary school; numbers above 100 have more females in primary/secondary schools.*
Sources: World Bank, World Development Indicators, 2007; *percentage of population living on $2 a day data from* Population Reference Bureau, World Data Sheet, 2007; *gender equity data from* Millennium Development Goals.

GEOGRAPHY IN THE MAKING

Thailand's Troubled Return to Democracy

Politics in Thailand is noted for its instability. Since the abolition of absolute monarchy in 1932, Thailand has been under seventeen separate constitutions and charters, with periods of democracy alternating with military rule. Many observers thought that democratic governance was finally solidified with the 1997 Constitution, which guaranteed numerous human rights. The general election of 2001 was widely regarded as more open and free of corruption than any of its predecessors.

The victor in the 2001 election, however, proved to be a divisive figure. Prime Minister Thaksin Shinawatra, a wealthy mobile telephone entrepreneur, championed the poor, pushing for debt relief, rural development programs, and state-supported medical care. Such policies made Thaksin extremely popular in rural Thailand, especially in the poor northeastern region. Urban and middle-class Thai citizens more often complained about the Thaksin government's alleged corruption, hostility to freedom of the press, and human rights abuses. Many also objected to Thaksin's hard-line military approach to the rebellion in Thailand's Malay-speaking, Muslim dominated south.

Thaksin's landslide reelection in 2005 did not end Thailand's political unrest. In September 2006, a military junta— apparently conducted with the blessing of Thailand's revered king—overthrew the government, abrogated the constitution, and declared martial law. Military rule was partially suspended in January 2007, a new constitution was imposed, and new elections were promised. Thaksin's political party, Thai Rak Thai, was officially dissolved, while Thaksin himself remained in exile in Britain.

Despite the military junta's attempts to neutralize Thaksin, the former Prime Minister's movement continued to be an influential force in Thai politics. By mid-2007, Thaksin's supporters rallied around a new political organization, the People's Power Party (PPP), officially led by the fiery and sometimes vulgar former governor of Bangkok, Samak Sundaravej. Samak is noted for his strong support of Thailand's constitutional monarchy, a useful position considering the fact that Thaksin had often been accused of being disrespectful toward the king.

The People's Power Party emerged as the clear winner in Thailand's national election in December 2007, winning 232 out of 480 seats in Thailand's House of Representatives. Lacking an outright majority, the PPP had to seek allies to form a coalition government. It promised that once it did so, it would invite Thaksin to return home from exile. Any such move would anger Thailand's top military leaders, resulting perhaps in continued political unrest.

others, such as Burma (Myanmar), Laos, Cambodia, and East Timor, missed the expansion of the 1980s and 1990s and remain deeply impoverished. Oil-rich Brunei and technologically sophisticated Singapore, on the other hand, managed to largely avoid the economic crisis of the late 1990s and remain two of the world's more prosperous countries (Table 13.2). In the Philippines the situation is somewhat more complicated.

The Philippine Decline Fifty years ago, the Philippines was the most highly developed Southeast Asian country and was considered by many to have the brightest prospects in all of Asia. It boasted the best-educated populace in the region, and it seemed to be on the verge of sustained industrialization and economic development. Per capita GNI in 1960 was higher in the Philippines than in South Korea. By the late 1960s, however, Philippine development had been derailed. Through the 1980s and early 1990s, the country's economy failed to outpace its population growth, resulting in declining living standards for most. The Philippine people are still well educated and reasonably healthy by world standards, but even the country's educational and health systems declined during the dismal decades of the 1980s and 1990s.

Why did the Philippines fail so spectacularly despite its earlier promise? Although there are no simple answers, it is clear that dictator Ferdinand Marcos (who ruled from 1968 to 1986) squandered billions of dollars while failing to enact programs conducive to genuine development. The Marcos regime instituted a kind of **crony capitalism** in which the president's many friends were granted huge sectors of the economy, while those perceived to be enemies had their properties confiscated. After Marcos declared martial law in 1972 and suspended Philippine democracy, revolutionary activity intensified and the country began to fall into a downward spiral.

While it is tempting to blame the failure of the Philippines on the policies of the Marcos regime, such an explanation is only partially adequate. Indonesia and Thailand also have seen the development of crony capitalism, yet their economies have proved more competitive. And although the Marcos dictatorship was finally replaced by an elected democratic government in 1986, the Philippine economy continued to languish through the 1990s. Remittances by Filipinos working abroad have kept the economy afloat, but this exodus of labor represents in many respects a tragedy for the country as a whole (see "People on the Move: Filipinos Take to the Seas").

By the beginning of the new millennium, the Philippine economy showed some signs of revival. The government turned its attention to infrastructural problems, such as electricity provision, and foreign investments began to flow into the country. A vibrant local economy emerged in the former U.S. naval base of Subic Bay. Boasting both world-class facilities and a highly competent local government, Subic Bay has emerged as a major export-processing center. Cebu City, on the central Visayan island of Cebu, also has expanded quickly, giving rise to its local nickname of "Ceboom." By 2006, the Philippine economy was expanding at a relatively rapid annual rate of 5.4 percent.

Despite its recent recovery, the Philippine economy continues to be weakened by political and social problems. The Philippine political system, modeled on that of the United States, entails elaborate checks and balances between the different branches of government. Critics contend, however, that such "checks" are so effective that little is ever accomplished. The Philippine government habitually spends far more money than it takes in taxes, leading to a huge public debt that consumes a large portion of the national budget. Another obstacle is the fact that the Philippines has the least equitable distribution of wealth in the region. While many members of the elite are fantastically wealthy by any measure, roughly half of the country's people subsist on less than $2 a day. As a result, many Filipinos seek employment abroad (Figure 13.33).

The Regional Hub: Singapore If the Philippines has been the biggest economic disappointment in Southeast Asia over the past 60 years, Singapore has surely been the region's greatest developmental successes. Singapore has transformed itself from an **entrepôt** port city, a place where goods are imported, stored, and then transshipped, to one of the world's most prosperous and modern states. Singapore is now the communications and financial hub of Southeast Asia, as well as a thriving high-tech manufacturing center. The Singaporean government has played an active role in the development process but has also allowed market forces freedom to operate. Singapore has encouraged investment by multinational companies (especially those involved in technology) and has itself invested heavily in housing, education, and some social services (Figure 13.34). This system, however, is somewhat repressive and is far from fully democratic. Although elections are held, the government is able to manipulate the process, ensuring that the opposition never holds more than a few seats in parliament. While many Singaporeans object to such policies, others counter that they have brought fast growth as well as a clean, safe, and remarkably corruption-free society.

Figure 13.33 **Filipina Migrant Workers in Kuwait**
The long period of economic stagnation in the Philippines has resulted in an outflow of workers from the country. Women from the Philippines often work as domestic servants in the Persian Gulf region and in Singapore and Hong Kong. *(Penny Tweedie/Panos Pictures)*

Figure 13.34 **Housing in Singapore**
Despite its free-market approach to economics, the government of Singapore has invested heavily in public housing. Many Singaporeans live in buildings similar to the one depicted in this photograph. *(Dr. Pradeep Kumar/ProPhotoz)*

Although the Singaporean government has thus far been able to repress dissent, its authoritarian form of capitalism confronts a new technological challenge in the Internet. National leaders want the communication services that the Net provides, but they are worried about the free expression that it allows, fearing that it will lead to excessive individualism. It will be interesting to see how Singapore responds to this challenge, in part because the governments of China and Vietnam may be seeking to emulate the technocratic, authoritarian capitalism pioneered by Singapore.

Malaysia's Insecure Boom Although not nearly as prosperous as Singapore, Malaysia has experienced very rapid economic growth over the past several decades and has now broken into the ranks of the middle-income countries. Development was initially concentrated in agriculture and natural resource extraction, focused on tropical hardwoods, plantation products (mainly palm oil and rubber), and tin. More recently, manufacturing, especially in labor-intensive high-tech sectors, has become the main engine of growth. As Singapore prospers, moreover, many of its enterprises are spilling over into neighboring Malaysia. Increasingly, Malaysia's economy is multinational; many Western high-tech firms operate in the country, while several Malaysian companies are themselves establishing subsidiaries in foreign lands. As a result of its prosperity, Malaysia has attracted hundreds of thousands of illegal immigrants, mostly from Indonesia and the Philippines, generating some tension with those countries.

Malaysia's economy began to grow rapidly in the 1970s, but it was particularly hard hit by the Asian economic crisis of the late 1990s. The Malaysian government's response to the crisis was different from that of its neighbors. Spurning the advice of the International Monetary Fund, Malaysia instituted temporary currency controls, regulating the flow of international funds in and out of the country. Some observers believe that these policies helped Malaysia recover more rapidly than Thailand and Indonesia, but others are more inclined to credit Malaysia's tight fiscal policies, which have kept its external debt small and its foreign exchange reserves large. Highly dependent on the export of electronic goods, the Malaysian economy remains vulnerable to fluctuations in the North American, European, and Japanese markets.

The economic geography of modern Malaysia is not uniform. Most industrial development has occurred on the west side of peninsular Malaysia, with most of the rest of the country remaining largely dependent on agriculture and resource extraction. More important, however, are disparities based on ethnicity. The industrial wealth generated in Malaysia has been concentrated in the Chinese community. Ethnic Malays remain less prosperous than Chinese Malaysians, and those of South Asian descent tend to be poorer still. Many of the country's tribal peoples have suffered as development proceeds and their lands are taken away.

The disproportionate prosperity of the local Chinese community is a feature of most Southeast Asian countries. The problem is particularly acute in Malaysia, however, simply because its Chinese minority is so large (some 30 percent of the country's total population). The government's response has been one of aggressive "affirmative action," by which economic clout is transferred to the numerically dominant Malay, or **Bumiputra** ("sons of the soil"), community. This policy has been reasonably successful, although it has not yet reached its main goal of placing 30 percent of the nation's wealth in the hands of the Bumiputra community. Because the Malaysian economy as a whole has grown rapidly since the 1970s, the Chinese community has thrived even as its relative share of the country's wealth declined. Considerable resentment, however, is still felt by the Chinese, many of whom argue that both Malaysia's economic and educational systems are biased against them. One result has been a considerable "brain drain" of educated Chinese Malaysians emigrating to other countries, such as Canada and the United States.

Thailand's Ups and Downs Thailand, like Malaysia, climbed rapidly during the 1980s and 1990s into the ranks of the world's newly industrialized countries. Yet it also experienced a major downturn in the late 1990s that undermined much of this development through a devalued currency and the exodus of investment capital. The Asian crisis of 1997 actually began in the overheated Thai economy and real estate market. Recovery began in earnest after 2000, but the military coup of 2006 coupled with the intensifying insurgency in the far south resulted in another slowdown. Export growth, however, of both agricultural goods and machinery remains strong.

Japanese companies were leading players in the Thai boom of the 1980s and early 1990s. As Japan itself became too expensive for assembly and other manufacturing processes, Japanese firms began to relocate factories to such places as Thailand. They were particularly attracted by the country's low-wage, yet reasonably well-educated workforce. Thailand was also seen as politically stable, lacking the severe ethnic tensions found in many other parts of Asia. Although Thailand's Chinese population is large and economically powerful, relations between the Thai and the Chinese have generally been good.

Thailand's economic expansion has by no means benefited the entire country to an equal extent. Most industrial development has occurred in the historical core, especially in the city of Bangkok itself. Yet, even in Bangkok, the blessings of progress have been decidedly mixed. As the city begins to choke on its own growth, industry has begun to spread outward. The entire Chao Phraya lowland area shares to some extent in the general prosperity because of both its proximity to Bangkok and its rich agricultural resources. In northern Thailand, the Chiang Mai area has benefited from the fact that it attracts many international tourists.

Thailand's Lao-speaking northeast (the Khorat Plateau) is one of the country's poorest regions. Soils here are too thin to support intensive agriculture, yet the population is sizable. Because of the poverty of their homeland, northeasterners often are forced to seek employment in Bangkok. As Lao-speakers, they often experience ethnic discrimination. Men typically find work in the construction industry; northeastern women more often make their living as prostitutes (Figure 13.35).

Figure 13.35 Sex Tourism in Thailand
Thailand has one of the highest rates of prostitution in the world. While most prostitutes cater to a local clientele, those working in Bangkok's infamous Patpong district are usually hired by foreign men. Prostitution in Thailand is associated with high rates of HIV infection and with the brutal exploitation of women and girls. *(Agence France Presse/Getty Images)*

Indonesian Economic Development At the time of independence (1949), Indonesia was one of the poorest countries in the world. The Dutch had used their colony largely as an extraction zone for tropical crops and other resources and had invested little in infrastructure or education. The population of Java mushroomed in the nineteenth and early twentieth centuries, and serious land shortages burdened peasant communities.

The Indonesian economy finally began to expand in the 1970s. Oil exports fueled the early growth, as did the logging of tropical forests. But unlike most other oil exporters, Indonesia continued to grow even after oil prices plummeted in the 1980s. Oil production subsequently declined, and in 2004 Indonesia had to begin importing oil. But like Thailand and Malaysia, Indonesia proved attractive to multinational companies eager to export from a low-wage economy. Large Indonesian firms, some three-quarters of them owned by local Chinese families, also have capitalized on the country's low wages and abundant resources.

But despite rapid growth in the 1980s and 1990s, Indonesia remains a poor and economically troubled country. Its pace of economic expansion seldom matched that of Thailand, Singapore, and Malaysia, and it has remained much more dependent on the unsustainable exploitation of natural resources. The financial crisis of the late 1990s, moreover, hurt Indonesia more severely than any other country, and its banking sector is still considered fragile. Political instability continues to hamper economic recovery, as did a series of natural disasters in 2004, 2005, and 2006. When fiscal problems led Indonesia's government to raise fuel prices by 126 percent overnight in 2005, inflation intensified and interest rate hikes followed, further undermining the country's economy.

As in Thailand, development in Indonesia exhibits pronounced geographical disparities. Northwest Java, close to the capital city of Jakarta, has boomed, and much of the resource-rich and moderately populated island of Sumatra has long been relatively prosperous. Another resource-rich area, eastern Borneo (Kalimantan) also is relatively well off. In the overcrowded rural districts of central and eastern Java, however, many peasants have inadequate land, and thus remain on the margins of subsistence. Far eastern Indonesia has experienced little economic or social development, and throughout the remote areas of all the "outer islands," tribal peoples have suffered as their resources have been taken and their lands lost to outsiders.

Divergent Economic Paths: Vietnam, Laos, and Cambodia The three countries of former French Indochina—Vietnam, Cambodia, and Laos—experienced only modest economic expansion during Southeast Asia's boom years of the 1980s and 1990s. This area endured almost continual warfare between 1941 and 1975, and fighting persisted until the mid-1990s in Cambodia. Critics contend that the socialist economic system adopted by these countries prevented sustained economic growth. This debate is now moot, however, since a capitalist model of development has largely replaced socialism in all three countries.

Of these three countries, Vietnam has by far the strongest economy. The country's per capita GNI of $3,010, however, is still quite low by global standards. Postwar reunification in 1975 did not initially bring the anticipated growth, and economic stagnation ensued. Conditions grew worse in the early 1990s after the fall of the Soviet Union, Vietnam's main supporter and trading partner. Until the mid-1990s, Vietnam remained under embargo by the United States. Frustrated with their country's economic performance, Vietnam's leaders began to embrace the market while retaining the political forms of a communist state. They have, in other words, followed the Chinese model. Vietnam now welcomes multinational corporations, which are attracted by the extremely low wages received by its relatively well-educated workforce (Figure 13.36).

Such efforts seemingly began to pay off after 2000. By 2007 the Vietnamese economy was expanding at roughly 8 percent a year, one of the fastest rates of growth in the region. Foreign investment began to pour into the country, and exports—especially of textiles—began to surge. In January 2007, Vietnam joined the World Trade Organization, which will probably further boost exports while ensuring the continuation of market-oriented reforms.

Vietnam's recent economic rise, however, has not been untroubled. Many lowland peasants and upland tribal peoples have been excluded from the boom and are thus growing increasingly discontented. Tensions between the north, center of political authority, and the south, center of economic power, seem to be increasing. Trade disputes have arisen with the United States, which has accused Vietnamese exporters of dumping such products as farm-raised catfish and shrimp on American markets.

Laos and Cambodia face some of the most serious economic problems in the region. In Cambodia, the ravages of war followed by continuing political instability have undermined economic development, while Laos faces special difficulties owing to its rough terrain and relative isolation. Both countries also are hampered by a lack of infrastructure; outside the few cities, paved roads and reliable

Figure 13.36 **Capitalism in Vietnam**
Although Vietnam is a communist state, it has—like China—embraced many forms of capitalism. Private shops are now allowed and foreign investment is welcome. Large multinational corporations are increasingly investing in Vietnam, resulting in a recent economic boom. *(AP/Wide World Photos)*

electricity are rare. As a result, the economies of both Cambodia and Laos remain largely agricultural in orientation, while relying heavily on environmentally damaging logging and mining operations. In Laos, subsistence farming still accounts for more than three-quarters of total employment.

The Laotian government is pinning its economic hopes on dam and road building. The country is mountainous and has many large rivers, and could therefore generate large quantities of electricity, which is in high demand in neighboring Thailand. Up to 100 new dam projects have been proposed for the Mekong River system, most of them to be located in Laos. The Asian Development Bank also is supporting an ambitious road-building program that would pass through Laos in order to link China to Thailand. The Laotian government, however, remains repressive, discouraging foreign investment.

The Cambodian economy, for its part, is increasingly integrated with that of Thailand, a fact much resented in Cambodia. In 2003, rioting in Phnom Penh resulted in the destruction of the Thai embassy as well as 17 Thai-owned businesses. Since then, however, the ties between the two countries have only strengthened (see "Global to Local: Poipet and Southeast Asia's Other 'Sin Cities'").

Although still deeply impoverished, Cambodia has experienced a recent economic boom of surprising proportions. Foreign investment has led to the development of an expanding textile sector, tourism is thriving, mining is taking off, and in 2005 important oil and natural gas fields were discovered in Cambodian territorial waters. As a result of such developments, the Cambodian economy grew at the breathtaking rate of 13 percent in 2006. Such rapid growth was possible only because the country's economic starting point was at such a low level. A lack of both skills and basic infrastructure, as well as problems of political instability and corruption, could easily result in renewed economic difficulties in the near future. Unfortunately, the recent economic boom has led to an epidemic of "land-grabbing," in which politically connected elites take over the properties of impoverished peasants in order to develop them.

Despite their basic lack of development, Cambodia and Laos are not as miserable as one might expect from the official economic statistics. Both countries have relatively low population densities and abundant resources. Their low per capita GNI figures, more importantly, partially reflect the fact that many of their people remain in a subsistence economy. While the highland peoples of Laos make few contributions to GNI, most of them at least have adequate shelter and food.

Burma's Troubled Economy Burma (Myanmar) stands near the bottom of the scale of Southeast Asian economic development. For all of its many problems, however, Burma (Myanmar) is a land of great potential. It has abundant natural resources—including oil and other minerals, water, and timber—as well as a large expanse of fertile farmland. Its population density is moderate, and its people are reasonably well educated. The country, however, has seen little economic development, and as of 2006 its economy was growing at a slow rate of 2.6 percent.

Burma's economic woes can be traced in part to the continual warfare the country has experienced. Most observers also blame economic policy. Beginning in earnest in 1962, Burma (Myanmar) attempted to isolate its economic system from global forces in order to achieve self-sufficiency under a system of Buddhist socialism. While its intentions were admirable, the experiment was not successful; instead of creating a self-contained economy, Burma (Myanmar) found itself burdened by smuggling and black-market activities.

In the early 1990s, Burma began to open its economy to market forces and increase its involvement in foreign trade. By 2000, however, the reform measures had stalled out, and in 2003 a banking crisis resulted in another round of economic havoc. High rates of inflation and worrisome fiscal deficits undermine business confidence, but probably the most damaging economic impediment is Burma's exchange rate policy. In 2005, when the official exchange rate was 5.82 Burmese kyat to the U.S. dollar, one could usually get 1,075 kyat to the dollar in the street market. Although Burma's government is now hoping that enhanced trade with China and India will help solve its economic problems, its fiscal policies will probably continue to undermine genuine development.

Globalization and the Southeast Asian Economy

As the previous discussion shows, Southeast Asia has undergone rapid but highly uneven integration into the global economy (Figure 13.37). Singapore has thoroughly staked its future to the success of multinational capitalism, and several neighboring countries are following suit. According to one measurement, Malaysia and Singapore have, respectively, the fourth and fifth most trade-dependent economies in the world. Even Marxist Vietnam and once-isolationist Burma (Myanmar) have opened their doors to international commerce, although with far more success in the former case than the latter.

Much debate has arisen among scholars over the roots of Southeast Asia's economic gains, as well as its more recent economic problems. Those who credit primarily the diligence, discipline, and entrepreneurial skills of the Southeast Asian peoples are optimistic about future economic expansion. Some skeptics argue, however, that most of the region's growth has come from the application of large quantities of labor and capital unsupported by real advances in productivity. Regardless of how this debate turns out, it is clear that Southeast Asia's globalized economies are heavily dependent upon exports to the international market. In the 1990s, many observers thought that the booming economies of Southeast Asia were overly dependent on exports to the United States, but in recent years the rise of China has resulted in a more balanced trading regime. In 2005, for example, 15.4 percent of Thailand's exports went to the United States, while 13.9 percent went to China (including Hong Kong).

Globalized industrial production in Southeast Asia has become quite controversial. Consumers in the world's wealthy countries are increasingly aware that many of

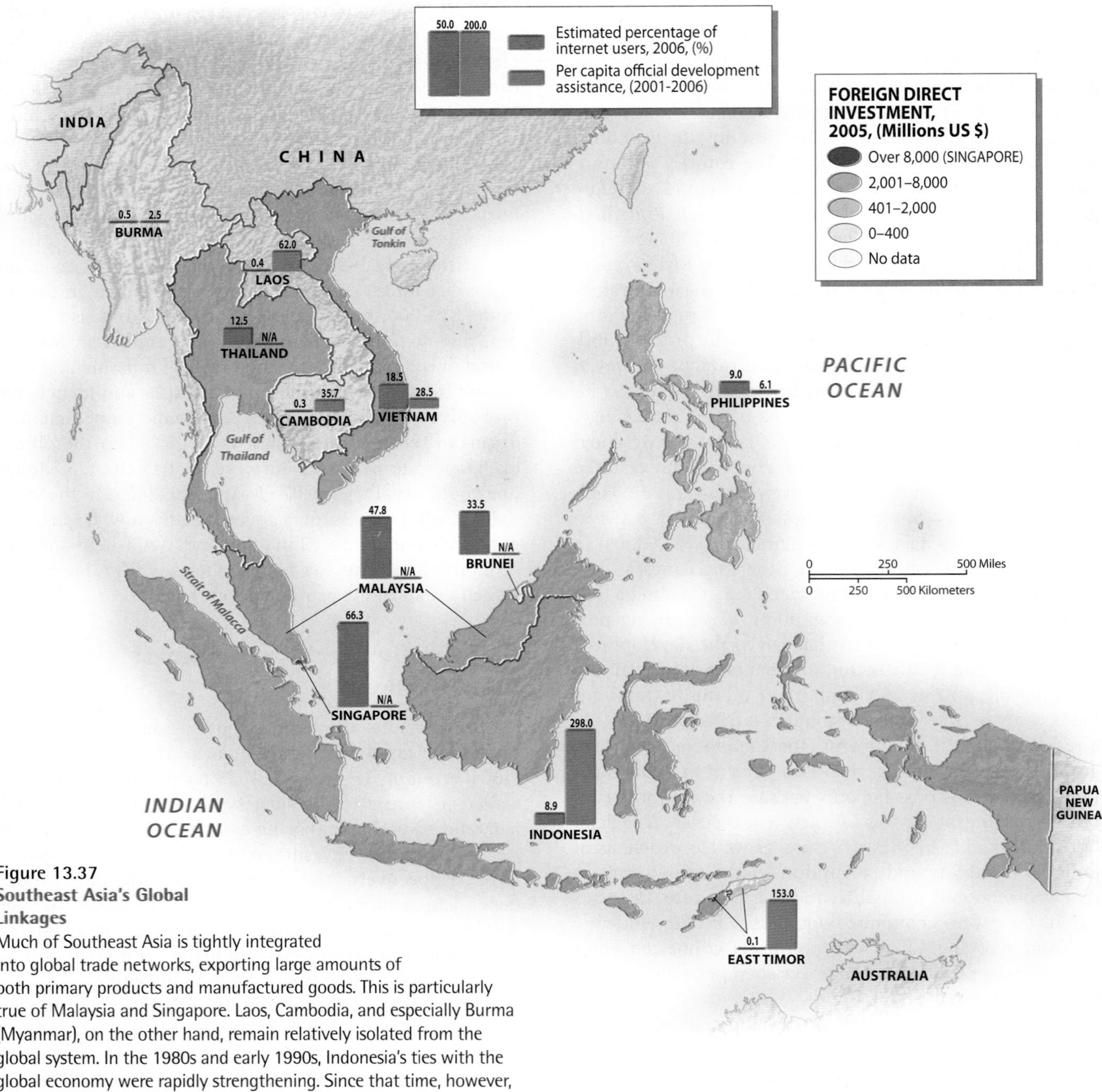

Figure 13.37
Southeast Asia's Global Linkages
Much of Southeast Asia is tightly integrated into global trade networks, exporting large amounts of both primary products and manufactured goods. This is particularly true of Malaysia and Singapore. Laos, Cambodia, and especially Burma (Myanmar), on the other hand, remain relatively isolated from the global system. In the 1980s and early 1990s, Indonesia's ties with the global economy were rapidly strengthening. Since that time, however, economic collapse, political instability, and ethnic violence have resulted in a tremendous flow of capital out of the country.

their basic purchases, such as sneakers and clothing, are produced under exploitative conditions (Figure 13.38). Movements have thus begun in Europe, the United States, and elsewhere to pressure both multinational corporations and Southeast Asian governments to improve the working conditions of laborers in the export industries. The multinational firms in question counter that such exploitative conditions occur not in their own factories, but rather in those of local companies that subcontract for them; activists, however, point out that multinationals have tremendous influence over their local subcontractors. Some Southeast Asian leaders object to the entire debate, accusing Westerners of wanting to prevent Southeast Asian development and protect their own home markets, under the pretext of concern over worker rights.

Issues of Social Development

As might be expected, several key indicators of social development in Southeast Asia correlate well with levels of economic development. Singapore thus ranks among the world leaders in regard to health and education, as does the small, yet oil-rich country of Brunei. East Timor, Laos, and Cambodia, not surprisingly, come out near the bottom for these measures. Note, however, that the people of Vietnam are healthier and better educated than might be expected on the basis of their country's overall economic performance. Burma (Myanmar), on the other hand, has

GLOBAL TO LOCAL

Poipet and Southeast Asia's Other "Sin Cities"

A number of Southeast Asian cities have prospered by providing socially objectionable services to an international clientele. Foremost among these is the resort city of Pattaya, located on the Gulf of Thailand 102 miles (165 kilometers) southeast of Bangkok. Although many of the approximately 5 million tourists who annually visit this small city (population 100,000) come primarily for the beaches and the shopping, a bigger draw is the hundreds of "go-go bars" that serve as fronts for prostitution. Although technically illegal, Thai prostitution is a big business, netting some $4.3 billion annually.

Pattaya is not the only small Southeast Asian city to find a niche in the global prostitution business. Before 1992, the economy of Angeles City in the Philippines was based on the nearby U.S. Air Force base of Clark Field. When the American military forces pulled out, Angeles City faced an uncertain economic future. Although the Philippine government created a special economic zone to allow businesses to take advantage of the base's extensive facilities, development has been slow. Angeles City's nightclub district, however, has experienced a successful expansion. Built initially to cater to American servicemen, the prostitution-oriented bars (numbering well over 100) now focus on an international clientele.

The exploitation of prostitutes in Thailand and the Philippines may be severe, but much worse conditions are encountered in the comparatively unregulated brothels of Laos, Cambodia, and Burma (Myanmar). In 2004, *New York Times* journalist Nicholas Kristof brought international attention to the horrific situation faced by child prostitutes in the Cambodian city of Poipet. Kristof went so far as to purchase (for $150 and $203, respectively) and then free two prostitute slave-girls. The sale of illegal drugs, especially methamphetamine, is also widespread in this tawdry Cambodian border town (Figure 13.6.1).

Figure 13.6.1 Poipet, Cambodia
The Holiday Palace casino is one of seven high-end gambling facilities recently constructed in the Cambodian border city of Poipet. *(Redux Pictures)*

The booming city of Poipet, however, owes its most recent growth to casino gambling. Propelling the rise of this city as the "Las Vegas of Southeast Asia" is the fact that casino gambling is extremely popular, although illegal, in neighboring Thailand, where Thais reportedly spend up to $11 billion annually on illegal gambling activities. In response, Cambodian authorities allowed the creation of a gambling enclave in Poipet directly on the Thai border, the first casino opening in 1999. Tourists from Thailand can enter the casino district without passing through Cambodian immigration, and the gambling halls prefer Thai rather than Cambodian currency. Although 90 percent of Poipet's tourists are from Thailand, increasing numbers are arriving from Malaysia, China, and Vietnam. A number of Poipet's casino managers, moreover, hail from North America, and many casino workers are from the Philippines. Overall, the venture is such an economic success that similar gambling reserves are being planned for several Laotian border towns.

Although Poipet has earned huge sums of money for investors and mangers, its broader social effects are disastrous. Poipet is a dangerous city with inadequate infrastructure that lures in desperately poor people, many of whom are forced into work as prostitutes or drug dealers. But considering the poverty and corruption found in Cambodia, it is not surprising that the rise of Poipet has met little national resistance.

seen a decline in some social indicators. Its per capita health budget is one of the lowest in the world, while its military budget, as a percentage of its overall economic output, is the fourth highest.

Figure 13.38 Indonesian Sweatshop
Much of Southeast Asia's recent economic growth has been derived from manufacturing jobs that pay low wages and demand rigid labor discipline. The products produced in factories such as the one shown in this photograph are mostly exported to wealthy countries. *(Reuters/Supri/Landov LLC)*

With the exception of Laos, Cambodia, Burma (Myanmar), and East Timor, Southeast Asia has achieved relatively high levels of social welfare. In Laos, however, life expectancy at birth hovers around a miserable 55 years (as compared to Thailand's 71 years), and female illiteracy rates remain above 40 percent. But even the poorest countries of the region have made some improvements in mortality under the age of 5, as is evident from the figures in Table 13.2. Note, however, that progress has been more pronounced in prosperous countries such as Malaysia and Singapore, whereas war-torn Cambodia has made relatively small gains. Thailand, on the other hand, has recently begun to provide a universal social safety net for its citizens, hoping to raise its level of social development to that found in the world's wealthier countries.

Most of the governments of Southeast Asia have placed a high priority on basic education. Literacy rates are relatively high in most countries of the region. Much less success, however, has been realized in university and technical education. As Southeast Asian economies continue to grow, this educational gap is beginning to have negative consequences, forcing many students to study abroad. High levels of basic education, along with general economic and social development, also have led to reduced birthrates through much of Southeast Asia. With population growing much more slowly now than before, economic gains are more easily translated into improved living standards.

Summary

- Some of the most serious problems created by globalization in Southeast Asia are environmental. Given the emphasis placed by global trade on wood products, it is perhaps understandable that Southeast Asia has sacrificed so many of its forests to support economic development. But in most of the region, forests are now seriously depleted. As the Chinese economy expands, market demand for Southeast Asian forest products is increasing rapidly.
- Deforestation in Southeast Asia is also linked to domestic population growth and changes in settlement patterns. As people move from densely populated, fertile lowland areas into remote uplands, both environmental damage and cultural conflicts often follow. Population movements in Southeast Asia also have a global dimension. This is particularly true in regard to the Philippines, which has sent some 8 million workers to more prosperous parts of the world.
- Southeast Asia is characterized today by tremendous cultural diversity. In recent years, conflicts over language and religion have seriously weakened several Southeast Asian countries, including Indonesia, East Timor, and Burma (Myanmar). However, the region has found a new sense of regional identity as expressed through ASEAN, the Association of Southeast Asian Nations.
- The relative success of ASEAN, however, has by no means solved all of Southeast Asia's political tensions. Many of its countries still argue about geographical, political, and economic issues, while global terrorism has generated major problems in Indonesia, the Philippines, and Thailand. Several of the region's countries, most notably Cambodia, Laos, and especially Burma (Myanmar), have also been held back by repressive and corrupt governments.
- Although ASEAN has played an economic as well as political role, its economic successes have been limited. Most of the region's trade is still directed outward toward the traditional centers of the global economy-North America, Europe, and East Asia. This orientation is not surprising, considering the export-focused policies of most Southeast Asian countries. A significant question for Southeast Asia's future is whether the region will develop an integrated regional economy. A more important issue is whether social and economic development will be able to lift the entire region out of poverty instead of benefiting just the more fortunate areas.

Key Terms

animism *(page 599)*
Association of Southeast Asian Nations (ASEAN) *(page 579)*
Bumiputra *(page 615)*
copra *(page 591)*
crony capitalism *(page 613)*
domino theory *(page 607)*
entrepôt *(page 614)*
Golden Triangle *(page 590)*
Khmer Rouge *(page 607)*
lingua franca *(page 600)*
primate cities *(page 594)*
Ramayana *(page 598)*
shifted cultivators *(page 593)*
Sunda Shelf *(page 585)*
swidden *(page 590)*
transmigration *(page 592)*
tsunamis *(page 587)*
typhoons *(page 587)*

Questions for Review

1. Why are river deltas so important in the settlement pattern of Southeast Asia?
2. Explain how and why the monsoon climates of Southeast Asia differ between mainland and islands.
3. What do we mean when we refer to the "globalization of world forestry"?
4. Explain why smoke and air pollution are so pronounced in Southeast Asia.

5. Compare and contrast the three major types of agriculture in the region—swidden, plantation, and rice cultivation.
6. How might "transmigration" solve and also aggravate a country's population problems?
7. What major religions are found in Southeast Asia? Describe the historical and contemporary patterns for each religion.
8. Why is there ambivalence about the English language in some Southeast Asian countries?
9. What are the ethnic tensions facing Indonesia? Locate the different problem areas on a map.
10. What are the goals of ASEAN, and how have those goals changed in the last several decades?

Thinking Geographically

1. Discuss the ramifications of state-sponsored migration in Indonesia from areas of high population density to areas of low population density in both the "sending" and the "receiving" areas.
2. Why should—or should not—citizens of the United States be concerned about deforestation in Southeast Asia? If they should be, what would be the proper ways they might show such concern?
3. What might the fate of animism be in the new millennium? Consider whether it might be doomed to extinction before the forces of modern economics and national integration, or whether it may persist as tribal peoples struggle to retain their cultural identities.
4. What should be the position of the English language in the educational systems of Southeast Asia? What should be the position of each country's national language? What about local languages?
5. How might ethnic tensions in countries such as Burma (Myanmar) and Indonesia be reduced? Does Indonesia have a reasonable claim to such areas as Irian Jaya?
6. What roles might ASEAN play in the coming years? Should it concentrate on economic or political issues?
7. How could Malaysia successfully integrate its economy into the global system and at the same time regulate the flow of global (or Western) culture? Evaluate whether Singapore can continue to experience economic growth while severely limiting basic freedoms.
8. Is the Southeast Asian economic path of integration into the global economy, which is marked by an openness to multinational corporations, going to prove wise in the long run, or do its potential hazards outweigh its benefits?

Regional Novels and Films

Novels

Anthony Burgess, *The Long Day Wanes: A Malaysian Trilogy* (1965, Norton)

Joseph Conrad, *Victory* (1936, Doubleday)

Graham Greene, *The Quiet American* (1956, Viking Press)

Ninfong Ho, *Rice without Rain* (1990, HarperCollins)

Mochtar Lubis, *A Road with No End* (1982, Graham Bush)

W. Somerset Maugham, *Borneo Stories* (1976, Heinemann)

Paul Theroux, *The Consul's Fire* (1977, Houghton Mifflin)

Films

The Bet Collector (2006, Philippines)

The Big Durian (2003, Malaysia)

Eliana, Eliana (2002, Indonesia)

Heaven and Earth (1993, U.S.)

Indochine (1992, France)

King of the Garbage Dump (2002, Vietnam)

Midnight, My Love (2005, Thailand)

Platoon (1987, U.S.)

The Scent of Green Papaya (1993, Vietnam)

Sunset at Chaopraya (1996, Thailand)

Bibliography

Acharya, Amitav. 2001. *Constructing a Security Community of Southeast Asia: ASEAN and the Problem of Regional Order*. New York: Routledge.

Broad, Robin, and Cavanagh, John. 1993. *Plundering Paradise: The Struggle for the Environment in the Philippines*. Berkeley: University of California Press.

Broek, Jan O. 1944. "Diversity and Unity in Southeast Asia." *Geographical Review 34*, 175–195.

Dixon, Chris. 1991. *South East Asia in the World Economy: A Regional Geography*. Cambridge, UK: Cambridge University Press.

Hefner, Robert W. 2001. *The Politics of Multiculturalism: Pluralism and Citizenship in Malaysia, Singapore, and Indonesia*. Honolulu: University of Hawaii Press.

Herring, George. 1995. *America's Longest War: The U.S. and Vietnam 1950–1975*. New York: McGraw-Hill.

Peleggi, Maurizio. 2007. *Thailand: The Worldly Kingdom*. London: Reaktion Books

Steinberg, David Joel. 2001. *Burma: The State of Myanmar*. Washington, DC: Georgetown University Press.

Taylor, Jean G. 2004. *Indonesia: Peoples and Histories*. New Haven, CT: Yale University Press.

Ulack, Richard, and Pauer, Gyula. 1989. *Atlas of Southeast Asia*. New York: Macmillan.

Additional bibliographic resources are at the *Diversity Amid Globalization* Website: http://www.prenhall.com/rowntree.

14 Australia and Oceania

Vast expanses of ocean, dotted with thousands of small islands, capture the sense of Oceania. This photo shows the village of Morobe, Hoskin Island, Papua New Guinea. *(Wolfgang Kaehler/Corbis)*

ENVIRONMENTAL GEOGRAPHY

Nonnative plants and animal species, including snakes, rabbits, and feral pigs, are damaging the region's unique biodiversity by preying upon local flora and fauna. Additionally, sea level rise from global warming threatens many low-lying islands.

POPULATION AND SETTLEMENT

While many small islands are crowded with people because of rapid population growth and few migration options, vast areas in both Australia and New Zealand remain virtually empty.

CULTURAL COHERENCE AND DIVERSITY

Until 1973, Australia protected its European ethnic roots with a White Australia Policy. But recent immigration from Asia and Africa is creating a new multicultural country.

GEOPOLITICAL FRAMEWORK

From Hawaii to Australia, native peoples are filing claims to regain ownership and use of their ancestral lands. More often than not, these land claims are fraught with controversy and tension.

ECONOMIC AND SOCIAL DEVELOPMENT

Increasing trade ties with China are bringing economic benefits to many countries in Oceania, particularly Australia. Yet those benefits also contain uncertainty should China's economy stumble.

This vast world region, dominated mostly by water, includes the island continent of Australia as well as **Oceania**, a sweeping collection of islands that reaches from New Guinea and New Zealand to the U.S. state of Hawaii in the mid-Pacific (Figure 14.1; also see "Setting the Boundaries"). While the area was settled by indigenous peoples thousands of years ago, more recent European, Asian, and North American colonization began the process of globalization that has now produced new and sometimes unsettled cultural and political geographies.

Ongoing political and ethnic unrest in Fiji illustrates how the heat of globalization has fired the cauldron of change (Figure 14.2). Currently, the country struggles to address new political and economic relationships between the two dominant ethnic groups, indigenous Fijians and the descendants of South Asian sugarcane workers (called Indo-Fijians) who were brought to the islands in the nineteenth century to solve labor shortages in the cane fields. Strong cultural and religious barriers exist between the two groups resulting in extremely low rates of intermarriage and interaction. Generally speaking, the Indo-Fijians dominate the country's commercial life and are better off economically than the Fijians, even though the indigenous

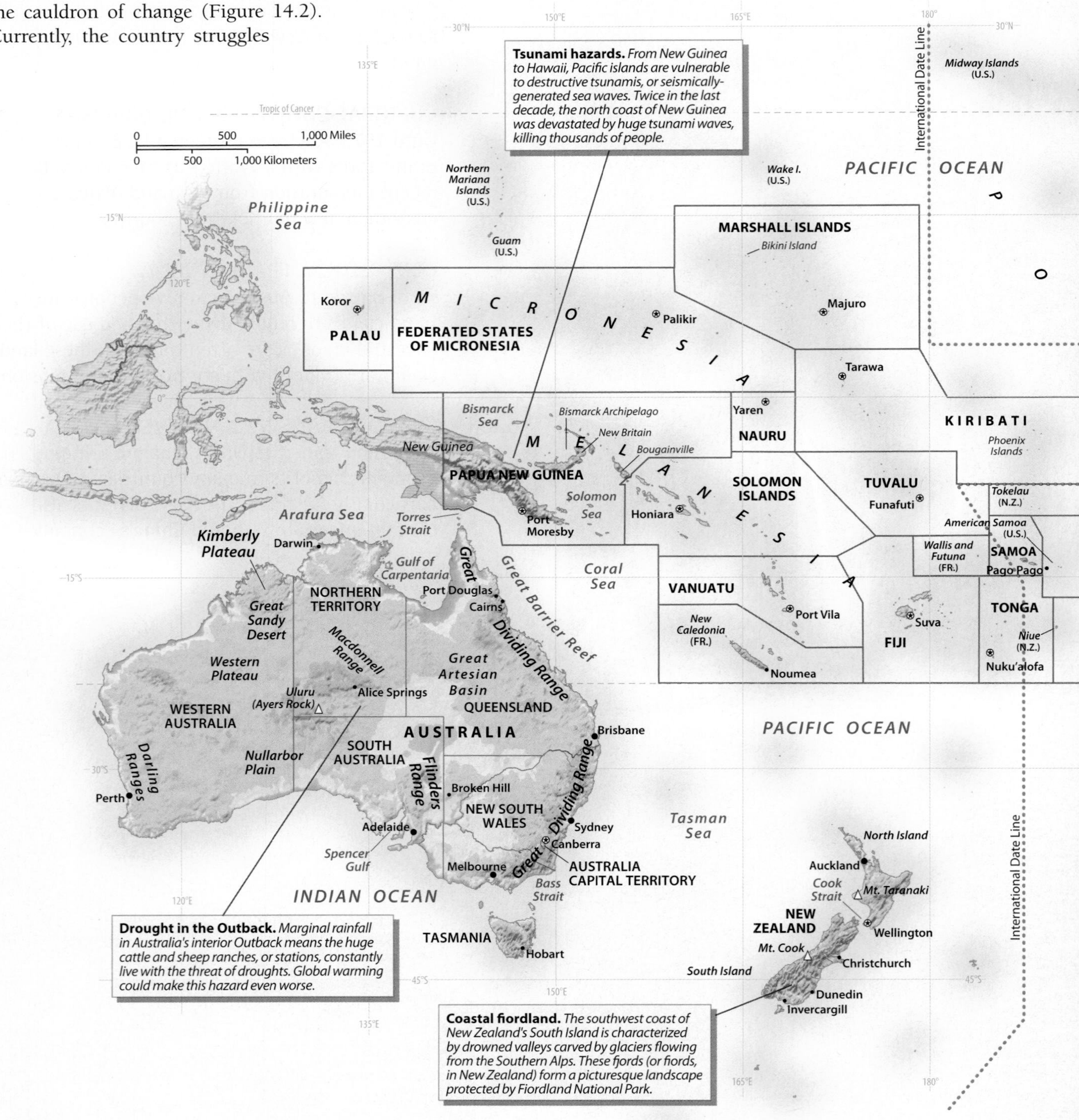

Setting the Boundaries

While the vast distances of the Pacific stretching from New Guinea to Hawaii help define the boundaries of this region, many of the national boundaries are born from political convenience. Others still remain rather ill defined. In contrast, Australia (or "southern land"), which is often thought of as a continent, forms a coherent political unit and subregion. To the east, New Zealand, a three-hour flight from Australia, and closely linked by shared historical ties to Britain, is usually considered part of Polynesia because of its native peoples, the Maori, the original inhabitants of the country. These cultural ties also act as a bridge between many of Oceania's island nations and New Zealand.

Hawaii, 4,400 miles (7,084 kilometers) northeast of New Zealand, shares the same Polynesian heritage and is thought of as the northeastern boundary of Oceania. The southeastern boundary of the region is usually delimited by the Polynesian islands of Tahiti, 3,000 miles (4,416 kilometers) to the southeast. Originally colonized by the French, Tahiti retains strong cultural and political ties to that European country.

Four thousand miles west of French Polynesia, well across the International Date Line, lies the island of New Guinea, the accepted yet sometimes confusing boundary between Oceania and Asia. Today, an arbitrary boundary line bisects the island, dividing the eastern half (Papua New Guinea, which is usually considered part of Oceania) from neighboring Irian Jaya (the western part), which, as part of Indonesia, is usually thought of as part of Southeast Asia.

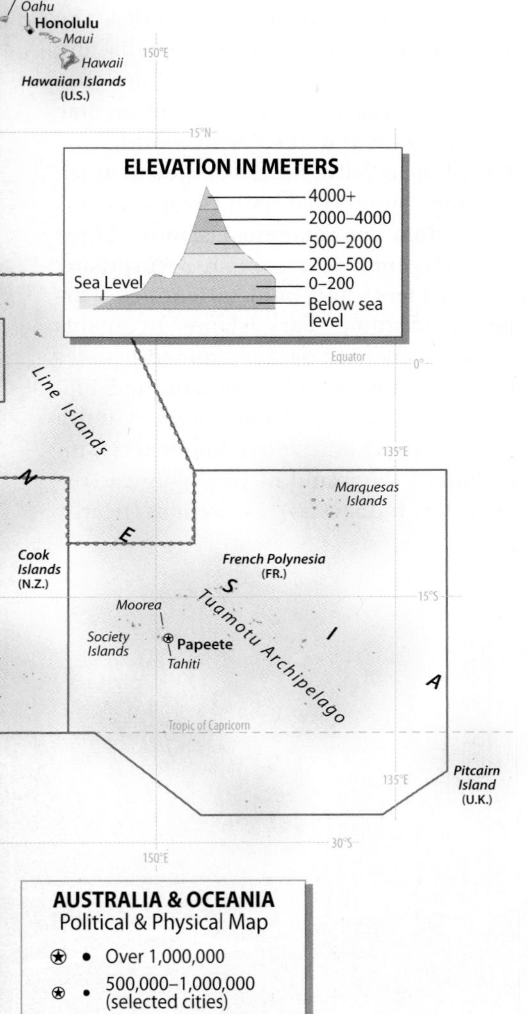

tribal communities own the land that produces much of the country's wealth. The fields are leased from Fijian tribes, and as these leases expire, the tribes tend to not renew them in hopes of improving their own economic position.

While tensions have long caused conflict between these two groups, globalization has accentuated this unrest as foreign investors push for economic restructuring and market development of the sugar industry. As a result, island governance has been unsettled by a series of coups and countercoups over the last two decades. In May 2000, armed indigenous Fijians took hostage the nation's first prime minister of Indian descent. Despite calls from the international community to restore the democratically elected Fijian government, leaders of the military coup retained power. Elections in 2001 and 2006 led to a Fijian-dominated government. Not surprisingly, large numbers of Indo-Fijians fled the islands, leaving the commercial and economic life of the country in shambles.

By all accounts these current tensions are more than a continuation of historic ethnic unrest between the two groups. Instead, observers attribute much of the problem to a reshuffling of status within the traditional Fijian community as its people become empowered by the spoils of economic globalization.

In many ways, the same kinds of tensions found in Fiji between indigenous people and "outsiders" (even if they arrived centuries ago) also are expressed in many other countries of Oceania including Australia and New Zealand. This unrest reminds us of the ways environmental, settlement, cultural, geopolitical, and economic activities are inseparably linked in the contemporary globalized world.

Australia and New Zealand share many geographical characteristics and dominate the region. Major population clusters in both countries are located in the middle

Figure 14.1 Australia and Oceania
More water than land, the Australia and Oceania region sprawls across the vast reaches of the western Pacific Ocean. Australia dominates the region, both in its physical size and in its economic and political clout. Along with New Zealand, Australia represents largely Europeanized settlement in the South Pacific. Elsewhere, however, the island subregions of Melanesia, Micronesia, and Polynesia contain large native populations that have mixed in varied ways with later European, Asian, and American arrivals.

latitudes. Australia's 21 million residents occupy a vast land area of 2.97 million square miles (7.69 million square kilometers), while New Zealand's combined North and South Islands (104,000 square miles, or 269,000 square kilometers) are home to 4.2 million people. Most residents of both countries live in urban settlements near the coasts (Figure 14.3). Australia's huge and dry interior, often termed the **Outback**, is as thinly settled as North Africa's Sahara Desert, and much of the New Zealand hinterland is a visually spectacular but sparsely occupied collection of volcanic peaks and rugged, glaciated mountain ranges. Taken together, the land areas of these two South Pacific nations almost equal that of the United States, but their populations total less than 10 percent of their distant North American neighbor. All three countries, however, share a pervasive European cultural heritage, the product of common global-scale processes that sent Europeans far from their homelands during the past three centuries. The highly Europeanized populations of both Australia and New Zealand also retain particularly close cultural ties with Britain as they enjoy relatively high levels of income and economic development within the Pacific world.

Figure 14.2 **Unrest in Fiji**
South Asians were brought to Fiji as workers in the sugarcane fields. Today, the Indo-Fijians generally control the sugar industry, which is a source of tension with indigenous Fijians (such as this cane cutter), many of whom support politicians who advocate Fijian control of the industry. *(Will Burgess/Reuters/Corbis/Bettmann)*

The blue waters of the tropical Pacific dominate much of the rest of the region, punctuated with isolated chains of sand-fringed and often mountainous islands. Three major subregions of Oceania each contain a surprising variety of human settlements and political entities. Farthest west, **Melanesia** (meaning "dark islands") contains the culturally complex, generally darker-skinned peoples of New Guinea, the Solomon Islands, Vanuatu, and Fiji. The largest of these countries, Papua New Guinea (179,000 square miles, or 463,000 square kilometers), includes the eastern half of the island of New Guinea (the western half is part of Indonesia), as well as nearby

Figure 14.3 **Sydney, Australia**
Most Australians live in cities, and the country's urban landscapes often resemble their North American counterparts. This view of Sydney features its world-famous harbor and displays the dramatic interplay of land and water. *(Rob Crandall/www.robcrandall. com)*

portions of the northern Solomon Islands. Its population of 6.3 million people is greater than that of New Zealand.

To the east, the small island groups, or **archipelagos**, of the central South Pacific are called **Polynesia** (meaning "many islands"). This linguistically unified subregion includes French-controlled Tahiti in the Society Islands, the Hawaiian Islands, and smaller political states such as Tonga, Tuvalu, and Samoa. New Zealand also is often considered a part of Polynesia because its native peoples, known collectively as the **Maori**, share many cultural and physical characteristics with the somewhat lighter-skinned peoples of the mid-Pacific region. Finally, the more culturally diverse region of **Micronesia** (meaning "small islands") is north of Melanesia and west of Polynesia and includes microstates such as Nauru and the Marshall Islands, as well as the U.S. territory of Guam.

Figure 14.4 The Australian Outback
Arid and generally treeless, the vast lands of the Australian Outback resemble some of the dry landscapes of the U.S. West. In this photo, wildflowers blossom along a dirt road near Tom Price, in the Pilbara region of Western Australia. *(Rob Crandall/www.robcrandall.com)*

ENVIRONMENTAL GEOGRAPHY: A Varied Natural and Human Habitat

The region's physical setting is dramatic testimony to the power of space: the geology and climate of the seemingly limitless Pacific Ocean define much of the physical geography of Oceania, and the expansive interior of Australia shapes the physical setting of that island continent.

Australian Environments

Curiously, Australia is one of the world's most urbanized societies, yet most people associate the country with its vast and arid Outback, a sparsely settled land of sweeping distances, scrubby vegetation, and exotic animals (Figure 14.4). Place-names such as the "Nullarbor [no trees] Plain," "Sandland," and the "Great Sandy Desert" suggest that Europeans exploring the region encountered a hostile and unfamiliar land, particularly as they probed its huge and arid interior. Indeed, distance and dryness have left a lasting mark on Australia's physical setting and shaped the ways in which people have subsequently settled its more-favored corners.

Landforms Travelers across Australia from west to east find three major landform regions that dominate the continent's physical geography. The Western Plateau occupies more than half of the continent and geologically represents an ancient shield land mass that once joined Antarctica. Most of the region is a vast, irregular plateau that averages only 1,000 to 1,800 feet (305 to 550 meters) in height. As one leaves the plateau, the Interior Lowland Basins stretch north to south for more than 1,000 miles from the swampy coastlands of the Gulf of Carpentaria to the Murray and Darling valleys, Australia's largest river system. Most of the region is a flat, featureless plain, punctuated occasionally by dry lake beds and by stream valleys where water is only a rare visitor. The Great Artesian Basin of western Queensland is included in this lowland region. Filled with sedimentary rocks, this vast basin, mostly below 500 feet (150 meters) in elevation, possesses a rich supply of underground water that has drained there from the higher country to the east.

Finally, more forested and mountainous country defines the horizon as one arrives in the Eastern Highlands along Australia's Pacific Ocean rim. The Great Dividing Range extends from the Cape York Peninsula in northern Queensland to southern Victoria. Nearby Tasmania is also dominated by mountains. These ancient eroded highlands include rugged tablelands but only a few peaks above 5,000 feet (1,525 meters). Nearby, off the eastern coast of Queensland, the Great Barrier Reef offers a final dramatic subsurface feature: Over the past 10,000 years, one of the world's most spectacular examples of coral reef-building has produced a living legacy now protected by the Great Barrier Reef Marine Park (Figure 14.5).

Climate and Vegetation Australia's varied climates strongly shape patterns of natural vegetation across the country (Figure 14.6). Generally, zones of somewhat higher precipitation encircle Australia's arid center. In the tropical low-latitude north, seasonal changes are dramatic and unpredictable. Localities such as Darwin experience drenching monsoonal rains in the summer (December to March), followed by bone-dry winters (June to September). Indeed, life across the region is shaped by this annual rhythm of "wet" and "dry." Much of the north is clothed in a mix of tropical woodlands and thorn forests interspersed with open grasslands. As one ventures south along the east coast of Queensland, precipitation remains substantial (60 to 100 inches, or 153 to 254 centimeters), but diminishes rapidly as one moves into the interior. Vegetation follows the pattern: coastal tropical rain forests give way to eucalyptus and acacia woodlands in the hill country, and eventually the trees thin out west of the mountains, replaced by the shortgrass and desert scrub vegetation of southwest Queensland's arid Outback. Indeed, some of the country's dry heartland is almost devoid of vegetation, and precipitation at interior locations such as the Northern Territory's Alice Springs averages less than 10 inches (25 centimeters) annually.

South of Brisbane, more midlatitude influences dominate eastern Australia's climate. Coastal New South Wales, southeastern Victoria, and Tasmania experience the country's most dependable year-round rainfall in a climatic regime that averages 40 to 60 inches (102 to 152 centimeters) per year. Forests cover most of the wetter southern coastlands and highlands, but tracts of more open heath dot the hills of southern Victoria, and drier conditions in the interior produce grasslands often suited to extensive grazing. Farther west, summers are hot and dry in much of South Australia and in the southwest corner of Western Australia, producing a distinctively Mediterranean climate. These zones of southern Mediterranean climate produce a mix of vegetation types. In moister hills near Perth, acacia, eucalyptus, and conifer forests cover the slopes, while portions of South Australia are dominated by **mallee** vegetation, a tough and scrubby eucalyptus woodland (Figure 14.7). As is the case throughout Australia, bushfires are a major hazard, particularly in the mallee vegetation zone (see "Geographic Tools: Fire Ecology and Management in Australia").

Figure 14.5 **The Great Barrier Reef**
Stretching along the eastern Queensland coast, the famed Great Barrier Reef is one of the world's most spectacular examples of coral reef-building. Threatened by varied forms of coastal pollution, much of the reef is now protected in a national marine park. *(Hilarie Kavanagh/Getty Images Inc.—Stone Allstock)*

New Zealand's Varied Landscape

Part of the Pacific Rim of Fire, New Zealand owes its geological origins to undersea mountain-building that produced two rugged and spectacular islands in the South Pacific more than 1,000 miles (1,600 kilometers) southeast of Australia. The active volcanic peaks and geothermal features of the North Island, in particular, reveal the country's fiery origins. Blessed with such indigenous names as Ruapehu, Ngauruhoe, and Tongariro, these volcanic peaks tower over nearby tablelands, reaching heights of more than 9,100 feet (2,775 meters). Even taller and more rugged mountains run down the western spine of the South Island. Mt. Cook (or Aoraki) is New Zealand's tallest peak, cresting at more than 12,000 feet (3,660 meters). Often mantled by high mountain glaciers and surrounded by steeply sloping valleys, the Southern Alps are known to the Maori as *Te Tapu Nui* (the Peaks of Intense Sacredness). They are one of the world's most spectacular mountain ranges, complete with narrow, fjordlike valleys that intricately indent much of the South Island's isolated western coast.

As in Australia, New Zealand's isolation offered opportunities for the development of unique plant and animal species. Eighty-five percent of the country's native trees and seed plants are found nowhere else on Earth. Bats are the region's only native mammals, while ancient tuatara reptiles, kiwi birds, and the flightless and now-extinct moas illustrate the country's special biological legacy.

North Island Environments Most of New Zealand's North Island is distinctly subtropical. The coastal lowlands near Auckland are mild and wet year-round. Still, local variations can be striking, as the area's volcanic peaks create their own microclimates. For example, 8,000-foot (2,440-meter) Mt. Taranaki, a volcano on the southwest side of the island, includes fern-enshrouded lowland forests, sagebrush-like subalpine slopes speckled with wildflowers, and the snow-clad peak itself, which is the haunt of high-country climbers and adventurous skiers (Figure 14.8).

South Island Environments Across the narrow Cook Strait, New Zealand's South Island offers an equally complex natural setting. Although the encircling waters of the Pacific moderate the climate, conditions become

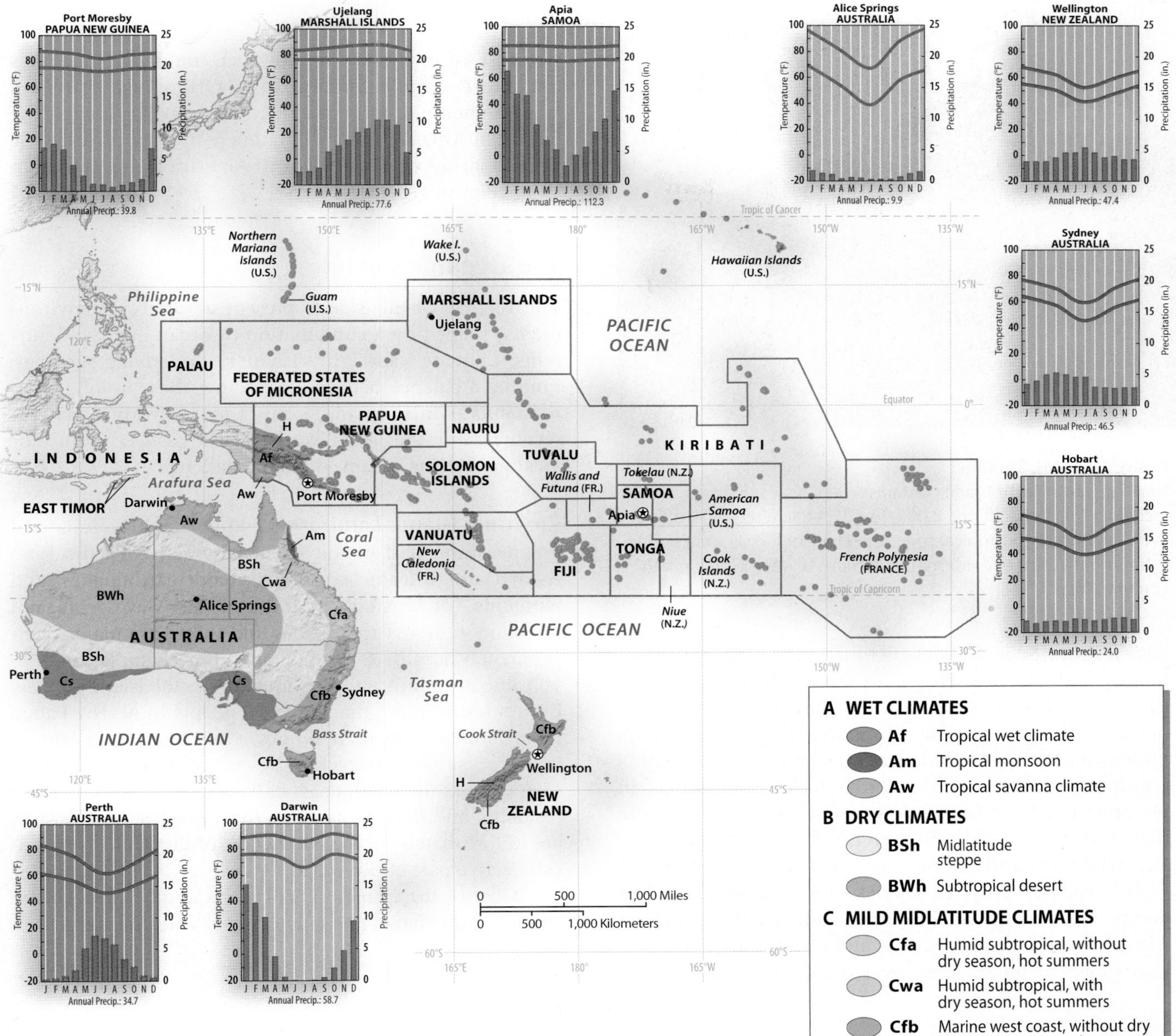

A WET CLIMATES

- **Af** Tropical wet climate
- **Am** Tropical monsoon
- **Aw** Tropical savanna climate

B DRY CLIMATES

- **BSh** Midlatitude steppe
- **BWh** Subtropical desert

C MILD MIDLATITUDE CLIMATES

- **Cfa** Humid subtropical, without dry season, hot summers
- **Cwa** Humid subtropical, with dry season, hot summers
- **Cfb** Marine west coast, without dry season, warm to cool summers
- **Cs** Mediterranean summer-dry

F HIGHLAND

- **H** Complex mountain climates

Figure 14.6 Climate Map of Australia and Oceania
Latitude and altitude shape the climatic patterns of the region. Equatorial portions of the Pacific bask in all-year warmth and humidity, while the Australian interior is predictably dry and under the dominance of subtropical high pressure. Cool and moisture-bearing storms of the southern Pacific Ocean provide midlatitude conditions across New Zealand and portions of Australia. More locally, mountain ranges dramatically raise precipitation totals in many highland zones.

distinctly cooler as one moves poleward south of the Cook Strait. Indeed, South Island's southern edge feels the seasonal breath of Antarctic chill, as it lies more than 46 degrees south of the equator (and more than 700 miles, or 1,120 kilometers, south of Auckland). It is no accident that the southern towns of Dunedin and Invercargill

Figure 14.7 Mallee Scrub Vegetation
The interior of South Australia and New South Wales features thousands of square miles of a tough, scrubby eucalyptus vegetation known as *mallee*. Farmers often remove this native cover in order to plant exotic grass and grain crops. *(Jaime Plaza Van Roon/Auscape International Pty. Ltd.)*

Figure 14.8 Mt. Taranaki
New Zealand's North Island contains several volcanic peaks, including Mt. Taranaki. The 8,000-foot (2,440-meter) peak offers everything from subtropical forests to challenging ski slopes and attracts both local and international tourists. *(Ken Graham/Ken Graham Agency)*

have a strong Scottish flavor, a function of both settlement history and climatic setting. To the north, milder conditions prevail on the Canterbury Plain in Christchurch, but nearby snow-covered mountains suggest the region's midlatitude character. Mountain ranges on New Zealand's South Island also display incredible local variations in precipitation: west-facing slopes are drenched with more than 100 inches (254 centimeters) of precipitation annually, while lowlands to the east (such as the Canterbury Plain) average only 25 inches (64 centimeters). The damp, west-facing or windward side of the Southern Alps, for instance, supports a dense, often impenetrable rain forest of misty, fern-covered canyons and forested slopes. East of the divide, however, much drier and less extreme conditions prevail. The Otago region, inland from Dunedin, sits partially in the rain shadow of the Southern Alps, and its rolling, open landscapes resemble the semiarid expanses of North America's Intermountain West (Figure 14.9).

Figure 14.9 Central Otago, South Island
On New Zealand's South Island, the Southern Alps capture rainfall on the west coast but leave areas to the east in a drier rain shadow. As a result, the Central Otago region has a semiarid landscape resembling portions of the U.S. West. *(John Lamb/Getty Images Inc.—Stone Allstock)*

The Oceanic Realm

The region's small islands present a variety of environments, shaped by a combination of ocean and wind currents, climate, and the Pacific's underlying geology. Finding a balance between protecting these unique environments while at the same time providing natural resources for the larger world and a cash economy for local people has became a major challenge.

Creating Island Landforms Much of Melanesia and Polynesia are part of the seismically active Pacific Rim. As a result, volcanic eruptions, major earthquakes, and **tsunamis**, or seismically induced sea waves, are not uncommon across the region, and they pose major environmental hazards for inhabitants. For example, in 1994 volcanic eruptions and earthquakes on the island of New Britain forced more than 100,000 people in nearby Papua New Guinea from their homes. Only four years later, a massive tsunami triggered by an offshore earthquake swept across the north coast of New Guinea, killing 3,000 residents and destroying numerous villages. Such events are unfortunately an inevitable part of life in this geologically active part of the world.

Most of the islands of Polynesia and Micronesia are truly oceanic, having originated from volcanic activity on

Figure 14.10 Bora Bora
The jewel of French Polynesia, Bora Bora displays many of the classic features of Pacific high islands. As the island's central volcanic core retreats, surrounding coral reefs produce a mix of wave-washed sandy shores and shallow lagoons. *(Paul Chesley/Getty Images Inc.—Stone Allstock)*

GEOGRAPHIC TOOLS

Fire Ecology and Management in Australia

Wildfires are common everywhere in Australia, from the "Top End" in the north to the island of Tasmania off the country's southeastern coast. In Queensland and the Northern Territory, tropical savanna fires scorch the countryside during the winter dry season, fueled by grass and shrub growth from the summer monsoon, whereas in southeast Australia, forest and brush fires singe the suburbs of major cities during the dry summer season. Each year, hundreds of thousands of acres burn across the country.

Fire, however, is not just a hazard that threatens lives and property. It is also a resource management tool for achieving environmental—and even social—goals. Since fire is a natural part of Australia's ecology, certain kinds of fire are necessary to enhance biodiversity and protect endangered plant and animal species. Fire also serves Aboriginal people in many different ways. For example, for some Aboriginal tribes, fire is part of a ritual to "clean up the countryside" by cleansing it of evil spirits. Fire also is important for manipulating vegetation to ensure that certain plant and animal species are available for these hunting and gathering peoples.

As a result, many different geographic tools are used to understand fire as both a resource and a hazard. Lesley Head, a geographer at the University of Wollongong, south of Sydney, is one of the leading researchers on Australia's fire ecology. To determine how often fires have occurred in prehistory, Professor Head and her students dig soil trenches several feet deep to find buried charcoal layers from past fires that can then be dated by laboratory analysis. For a more recent picture of burning, she and other fire ecologists commonly use remote sensing images to map and measure burned areas. Because satellites pass over Australia daily, they provide a continuing record of fires (Figure 14.2.1).

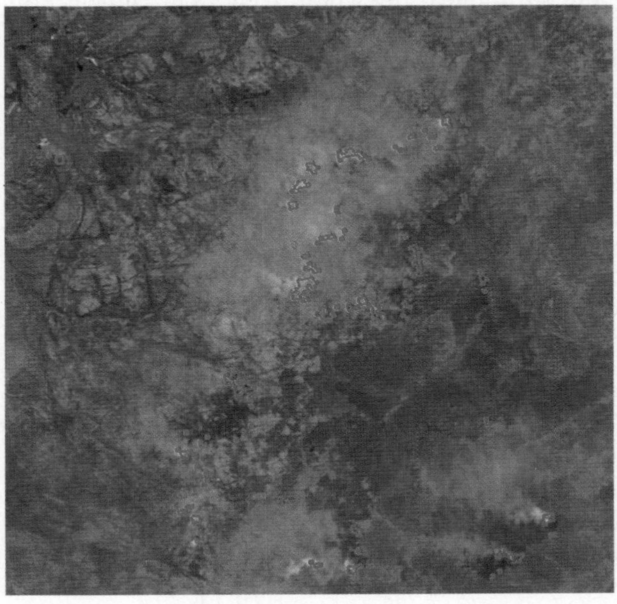

Figure 14.2.1 Australian Wildfires
This satellite photo taken in early October 2004 shows numerous fires burning in Australia's Northern Territory. Wildfires are common at the end of the dry season (October–November) before the monsoon rains begin in December. Using satellite images like this one, fire managers can decide which fires to let burn and which to suppress. *(NASA)*

These satellite pictures are used not only by researchers but also by fire managers who decide which fires to let burn and which to extinguish. By tracking fires on a daily basis, they decide which fires can provide resource benefits by being allowed to burn and which fires pose a hazard to Outback ranches and settlements and thus should be extinguished.

Even before fires start, various tools, from satellite images to ground measurements, are used to produce maps of potential fire risk. Satellite photos show the different vegetation types—grass, brush, trees—for an area that constitutes potential fuel for fires. In the field, fire managers collect information on the dryness of these different vegetation types. All of this information is then put into a geographic information system (GIS) that combines the different factors influencing fire behavior—vegetation types, fuel load and condition, and topography—with maps of settlement and structures that need to be protected. The resulting map provides an invaluable tool for everything from predicting where fires might occur to determining how to best manage them when they do start.

the ocean floor without any geological connection to larger land masses. The larger active and recently active volcanoes form **high islands**, which often rise to a considerable elevation and cover a substantial area. The island of Hawaii, the largest and youngest of the Pacific's high islands, is more than 80 miles (128 kilometers) across and rises to a height of more than 13,000 feet (3,980 meters). Indeed, the entire Hawaiian archipelago exemplifies a geological **hot spot** where moving oceanic crust passes over a supply of magma, thus creating a chain of volcanic uplifts. Many of the islands of French Polynesia, including Bora Bora, are smaller examples of high islands (Figure 14.10). In tropical latitudes, most high islands are ringed by coral reefs, which quickly establish themselves in the shallow waters near the shore.

High islands have a limited lifespan in geological terms. Gradually, the processes of volcanism that created them diminish; the islands subside, and erosion steadily wears them away. After a few hundred thousand years, only a few low peaks may rise out of a shallow lagoon surrounded by coral reefs. The nineteenth-century scientist Charles Darwin selected the island of Bora Bora to exemplify the geological evolution of these islands (Figure 14.11). Eventually, even remnant peaks erode away, leaving only a coral reef surrounding a shallow lagoon. Reefs tend to persist, however, because they are composed of living organisms that create new coral even as the (former) island base continues to subside. The top of the reef, therefore, tends to maintain its position at or just below sea level. **Low islands** are formed as large waves periodically break off and pulverize large pieces of coral, which are then deposited on adjacent sections of the reef to form narrow, wave-washed, and sandy islands.

The combination of narrow sandy islands, barrier coral reefs, and shallow central lagoons is also known as an **atoll**. The islands and reefs of the atoll characteristically form a circular or oval shape, although some are quite irregular. The world's largest atoll, Kwajalein in Micronesia's Marshall Islands, is 75 miles (120 kilometers) long and 15 miles (24 kilometers) wide. Polynesia and Micronesia are dotted with extensive atoll systems, and a number are found in Melanesia as well. Some extensive archipelagos, such as the Marshall Islands in Micronesia and the Tuamotus in Polynesia, are composed entirely of atolls.

Island Climates Many Pacific islands receive abundant precipitation, and high islands in particular are often noted for their heavy rainfall and dense tropical forests (see Figure 14.6). Much of the region is located in the rainy tropics or in a tropical wet–dry climate region where abundant summer rains and even tropical cyclones can bring heavy seasonal precipitation. Apia, Samoa, for example, has a dry fall and winter (April to August), but the town receives almost 18 inches (46 centimeters) of precipitation in the rainy summer month of January. In Melanesia's Papua New Guinea, Port Moresby is distinctly drier but still averages almost 50 inches (127 centimeters) of rain per year, most of it coming in the Southern Hemisphere summer (December to March). In the nearby highlands, however, conditions are often wetter, and local microclimates provide varied precipitation patterns.

Even frequent snows visit the taller 13,000-foot (3,980-meter) peaks. Some oceanic islands, however, experience significantly less precipitation. Low-lying atolls usually receive less precipitation than high islands and very often experience water shortages. This is partly because they have limited water-storage capacity, where a small "lens" of freshwater often "floats" above the salt water in the center of each sandy island. In dry periods, such limited stores are quickly depleted.

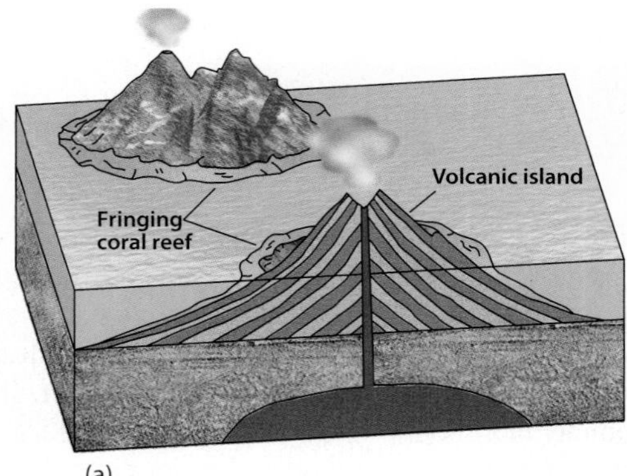

(a)

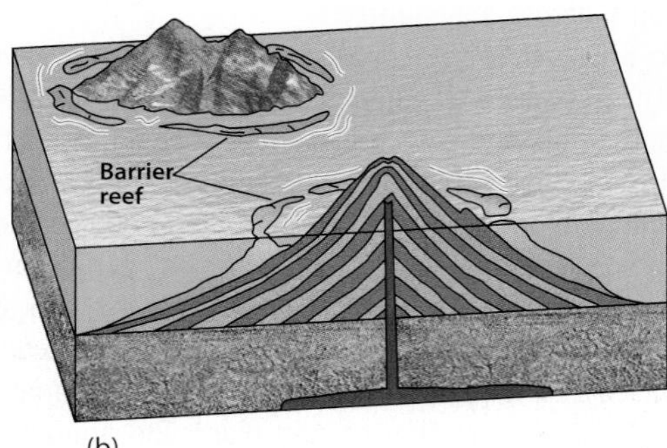

(b)

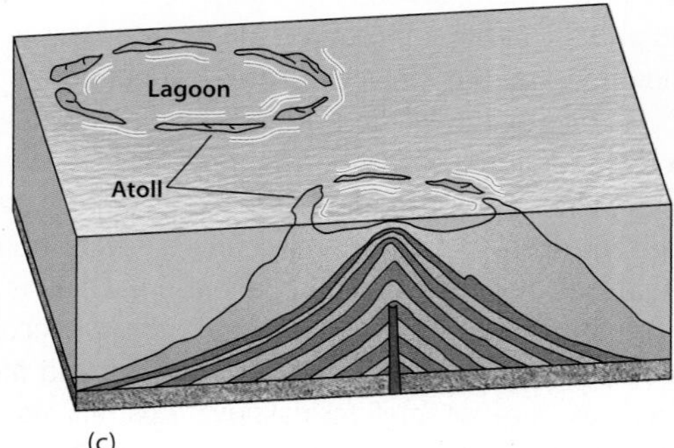

(c)

Figure 14.11 Evolution of an Atoll
Many Pacific islands begin as rugged volcanoes (a) with fringing coral reefs. However, as the extinct volcano subsides and erodes away, the coral reef expands, becoming a larger barrier reef (b). The term *barrier reef* comes from the hazards these features pose to navigation when approaching the island from sea. Finally, all that remains (c) is a coral atoll surrounding a shallow lagoon.

Environments at Risk

Even with relatively modest populations, many places in Australia and Oceania face significant human-induced environmental problems. Some challenges are simply the outcome of natural events that increasingly impact larger and more widely distributed human populations. For instance, Pacific Rim–related seismic hazards, periodic Australian droughts, and violent tropical cyclones now pose greater threats than they once did as new settlements have made increasing populations vulnerable to these problems. Other environmental issues, however, are even more directly related to human causes (Figure 14.12).

Global Resource Pressures Processes of globalization have exacted an environmental toll on Australia and Oceania. Specifically, the region's considerable base of natural resources has been opened to development, much of it by outside interests. While gaining some benefits from global investment, the region has also paid a considerable price for encouraging development, and the result is an increasingly threatened environment (see "Global to Local: Multinational Mining, Indigenous People, and Environmental Damage in Papua New Guinea").

Major mining operations have profoundly impacted Australia, Papua New Guinea, New Caledonia, and Nauru. Some of Australia's largest gold, silver, copper, and lead mines are located in sparsely settled portions of Queensland and New South Wales, but watersheds in these semiarid regions are highly susceptible to metals pollution. In Western Australia, huge open-pit iron mines dot the landscape, unearthing ore that is often bound for global markets, particularly Japan and China. To the north, Papua New Guinea's Bougainville copper mine has transformed the Solomon Islands, while even larger gold mining ventures have raised increasing environmental concerns on the island of New Guinea (Figure 14.13). Elsewhere, Micronesia's tiny Nauru has been virtually turned inside out as much of the island's jungle cover was removed to get at some of the world's richest phosphate deposits. Former Australian and New Zealand mine owners have already paid millions of dollars to settle environmental damage claims.

Deforestation is another major environmental threat across the region. Vast stretches of Australia's eucalyptus woodlands, for example, have been cleared to produce pastures. In addition, coastal rain forests in Queensland cover only a fraction of their original area, although a

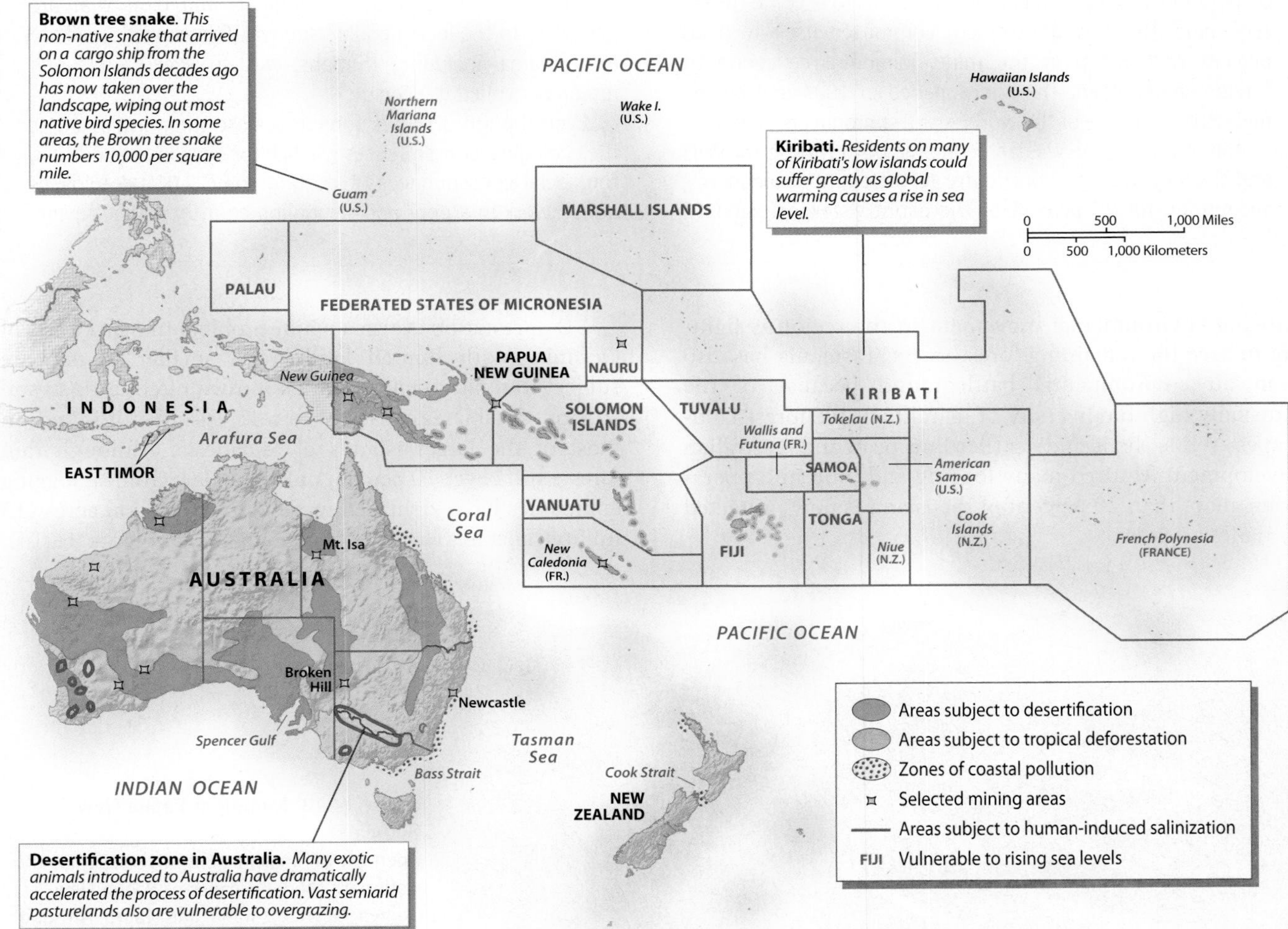

Figure 14.12 Environmental Issues in Australia and Oceania
Modern environmental problems belie the region's myth as an earthly paradise. Tropical deforestation, extensive mining, and a long record of nuclear testing by colonial powers have brought varied challenges to the region. Human settlements have also extensively modified the pattern of natural vegetation. Future environmental threats loom for low-lying Pacific islands as sea levels rise from global warming.

GLOBAL TO LOCAL

Multinational Mining, Indigenous People, and Environment Damage in Papua New Guinea

The story of the OK Tedi mine in Papua New Guinea (PNG) illustrates how local people and their environment can suffer when global corporations move in to exploit mineral resources. In addition, this case story shows how the governments of newly independent countries often are pressured to make financial decisions that may satisfy the global financial community, yet sacrifice the needs of their own people.

PNG became independent from Australia in 1975 and, following advice from the World Bank, drew up plans for further development of its mineral resources as a source of financial revenue from the international gold and copper trade. As part of that plan, the government authorized BHP (Broken Hill Proprietary), Australia's largest mining corporation, to prepare a plan for what was to become the OK Tedi mine. The new government committed itself to a 20 percent partnership in the mine, joining a collection of foreign shareholders. The mine opened in 1984 and by the mid-1990s was one of the world's largest producers of copper.

Initially, the mine was an economic powerhouse for PNG and the largest single contributor to the country's economy, accounting for 20 percent of the country's exports and 10 percent of the GDP. Within a short time, however, the ecological consequences of the OK Tedi mine began taking their toll on both the environment and PNG's indigenous people. The major problem was that each year the mine released some 40 million tons of waste rock and 30 million tons of tailings (fine sand) from its upland location into the Fly River drainage. As this waste material moved downstream, it did irreparable damage to ecosystems in the river itself as well as along its banks, reducing fish stocks by an estimated 70–90 percent in the last decade. Additionally, this sediment raised the Fly River's bed so that it flooded surrounding areas, devastating the taro and banana gardens local peoples used for subsistence, killing the surrounding forests, and driving away the animals and birds that are necessary to the local people's survival. The impact on all life in the area—including humans—was so severe that one researcher called it "ecocide."

Even though BHP has agreed to close the mine by 2010, the ecological consequences will last for decades. In addition, the social consequences of putting some 2,000 employees out of work in a poor and struggling country will be severe.

growing environmental movement in the region is fighting to save the remaining forest tracts. Tasmania has also been an environmental battleground because of the unusually high biodiversity of its midlatitude forest landscapes. While the island's earlier European and Australian development featured many logging and pulp mill operations, more than 20 percent of the island is now protected by national parks.

Deforestation is also a major problem in many parts of Oceania. With limited land areas, smaller islands are subject to rapid tree loss, which in turn often leads to soil erosion. Sizable expanses of rain forest are still found on most of the larger islands of Melanesia. Although rain forests still cover 70 percent of Papua New Guinea, logging companies have identified more than 37 million acres (15 million hectares) as suitable for removal (Figure 14.14).

Figure 14.13 Mining in Papua New Guinea
Open-pit mining for gold, silver, copper, and lead mark the landscapes of Papua New Guinea, New Caledonia, and Nauru. Although bringing some economic benefit to local peoples, these activities also cause immense environmental damage to the region. In New Guinea, for example, sediments from upland mines have severely damaged the Fly River ecosystem. *(Friedrich Stark/Das Fotoarchiv/Peter Arnold)*

Figure 14.14 Logging in Oceania
Foreign logging companies have made large investments in tropical Pacific settings such as Papua New Guinea and Samoa. While bringing new jobs, these ventures dramatically alter local environments as precious hardwood forests are harvested for export. *(David Austen/Woodfin Camp & Associates)*

Logging operations threaten some of the world's most biologically diverse environments, but landowners see the quick cash sales to loggers as attractive, even though the practice is a nonsustainable alternative to their traditional lifestyles. The global expansion of commercial logging also threatens nearby portions of the Solomon Islands, as well as many unique places in Polynesia. Increasingly, residents question unbridled development: Recently Solomon Islanders resisted a government resettlement plan that threw them off their land so that a Malaysian lumber company could clear the area of trees. Logging proceeded, despite continued protests.

Nonnative Plants and Animals

Introduced plant and animal species pose a major problem in many areas because they often threaten the viability of indigenous life-forms. In Australia, for example, nonnative rabbits successfully multiplied in an environment that lacked the diseases and predators that elsewhere kept their numbers in check. Before long, rabbit populations had reached plague proportions, and large sections of land were virtually stripped of vegetation. The animals were brought under control only through the purposeful introduction of the rabbit disease *myxomatosis*. Introduced sheep and cattle populations have also stressed the region's environment by accelerating soil erosion and contributing to desertification. In addition, feral livestock have spread. In various parts of Australia, one can find large numbers of once-domestic but now wild goats, horses, pigs, cattle, sheep, water buffalo, and even camels.

Many other islands have also witnessed the dramatic consequences of nonnative plant and animal invasion. For example, when rats, pigs, and other animals were introduced to islands that had no native land mammals, their vulnerable native bird and plant species were ravaged. The larger islands of the region, such as those of New Zealand, originally supported several species of large, flightless birds that filled some of the ecological niches held by mammals on the continents. The largest of these, the moas, were substantially larger than ostriches. In the first wave of human settlement in New Zealand some 1,500 years ago, moa numbers fell rapidly as they were hunted, their habitat burned, and their eggs consumed by nonnative rats. By 1800, the moas were completely exterminated.

A second wave of extinctions occurred in Polynesia and other portions of Oceania soon after the arrival of Europeans. As in Australia, introduced plants and animals often competed successfully with native species and threatened to eliminate very valuable species that occur nowhere else on the planet. In particular, the Hawaiian Islands suffered devastating extinctions following the arrival of European and American influences. The spread of nonnative species continues today, perhaps even at an accelerated pace. In Guam, for example, the brown tree snake, which arrived accidentally by cargo ship from the Solomon Islands in the 1950s, has taken over the landscape (Figure 14.15). In some forest areas, with more than 10,000 snakes per square mile, they have virtually wiped out all native bird species. The snakes also cause frequent power outages as they crawl along electrical wires. The brown tree snake threatens other islands as well because it can readily hide in cargo containers awaiting air or water shipment to island destinations. Because the snake is an agile climber, it has even hitched rides to other islands on airplane landing gear. Cargo handlers and custom agents throughout Oceania are on alert to keep this globalized pest from becoming established beyond Guam.

Figure 14.15 Island Pest
The brown tree snake, which arrived in Guam accidentally in the 1950s, has now taken over large parts of the island's forestlands and killed off most native bird species. Because these snakes, which reach 10 feet in length, climb along electrical wires, they frequently cause power outages throughout Guam. *(John Mitchell/Photo Researchers, Inc.)*

Global Warming in Oceania

Even though the Pacific world contributes relatively little atmospheric pollution, the harbingers of climate change are already widespread and problematic. In New Zealand, mountain glaciers are melting away while Australia suffers from frequent droughts and devastating wildfires. Warmer ocean waters have caused widespread bleaching of the Great Barrier Reef off Australia's coast, and rising sea levels are flooding several low-lying island nations, forcing residents to migrate to higher land (Figure 14.16). United Nations projections for the future are also highly disturbing—stronger tropical cyclones could devastate Pacific islands with widespread damage to land and life; island inhabitants will suffer from reduced coastal resources as ocean waters warm; and increased wildfire will threaten population centers in southeast Australia while agriculture will suffer from droughts and severe water shortages.

In response to these threats, the actions and policies taken by Oceania's countries vary considerably from full-on action to elusive disengagement. Until a recent change of government, Australia was the only industrial country besides the United States to not ratify the Kyoto Protocol. Perhaps the fact that Australia is the world's largest exporter of coal influenced that decision; not to be overlooked is that most of that country's emissions are produced by coal-fired power plants. However, while the former national government rejected Kyoto and downplayed the threat of global warming, as in the United States, Australia's states and cities compensated for this inaction with plans for local and regional emission control and carbon trading schemes.

In contrast, the national government of New Zealand has taken an aggressive stance toward global warming with specific plans for taxing carbon emissions. When New Zealand ratified the Kyoto agreement, it committed itself to a 5 percent reduction over its 1990 baseline but since then a booming economy has resulted in a 50 percent increase of emissions over the last several decades.

A major component of New Zealand's greenhouse gas pollution is methane emission from the country's large livestock population. These emissions, in fact, account for over half of the global warming pollution coming from that small country. As a result, New Zealand is discussing a much-publicized "flatulence tax" that would be levied on livestock owners.

The small, low-lying Pacific islands have emphasized adaptation to climate change problems rather than on curbing their miniscule carbon emissions. As noted earlier, several Pacific nations, most notably Tuvalu, Kiribati, and the Marshall Islands, are already experiencing disastrous flooding during high tides and storm surges, and, as a result, several thousand islanders have abandoned their homeland and migrated to New Zealand and Australia. Additionally, fish and reef resources have reportedly been degraded because of coral bleaching and warming waters.

Figure 14.16 Sea Level Rise in Tuvalu
These photos show the effect of recent sea level rise on the small island of Tuvalu. When the house was built it was rarely flooded at high tide, yet today it is often isolated by high waters. Scenes like this are behind the concern that this small island may soon be uninhabitable. As a result, many people are migrating to other islands, making them some of the first global warming refugees. *(Torsten Blackwood/AFP/Getty Images)*

To call attention to the plight of the Pacific Island peoples, Tuvalu threatened to sue the United States and Australia in the World Court, arguing because they were the only two industrial countries to not sign Kyoto, they should bear responsibility for island people forced from their homeland. To date, though, the World Court has not taken up this legal challenge, and, as mentioned, Australia finally ratified the Kyoto Protocol in December 2007.

POPULATION AND SETTLEMENT:
A Diverse Cultural Landscape

Modern population patterns across the region reflect the combined influences of indigenous and European settlement. In countries such as New Zealand, Australia, and the Hawaiian Islands, Anglo-European migration has structured the distribution and concentration of contemporary populations. In contrast, on smaller islands elsewhere in Oceania, population geographies are determined by the needs of native peoples. More recently, however, migration has taken place from outlying islands to Australia and New Zealand because of a combination of push forces, including unemployment, resource depletion, and the threat of flooding associated with global warming.

Contemporary Population Patterns

Despite the popular stereotypes of life in the Outback, modern Australia has one of the most highly urbanized populations in the world (Table 14.1). Indeed, 91 percent of the country's residents live within either the Sydney or Melbourne metropolitan areas. Australia's eastern and southern rimland is home to the overwhelming majority of its 21 million people (Figure 14.17). Most residents of Queensland, for example, live along the well-watered, amenity-rich coast, culminating in the state's capital city of Brisbane (1.8 million). Inland, population densities decline as rapidly as the rainfall: semiarid hills west of the Great Dividing Range still contain significant rural settlement, but the state's southwestern periphery remains sparsely peopled.

New South Wales is the country's most populous state, and its sprawling capital city of Sydney (4 million), focused around one of the world's most magnificent natural harbors, is the largest metropolitan area in the entire South Pacific. In the nearby state of Victoria, Melbourne's 3.8 million residents have long competed with Sydney for cultural and architectural supremacy over their slightly larger neighbor. Located between these two metropolitan giants, the much smaller federal capital of Canberra (325,000) represents a classic geopolitical compromise in the same spirit that created Washington, DC, midway between the populous southern and northern portions of the United States.

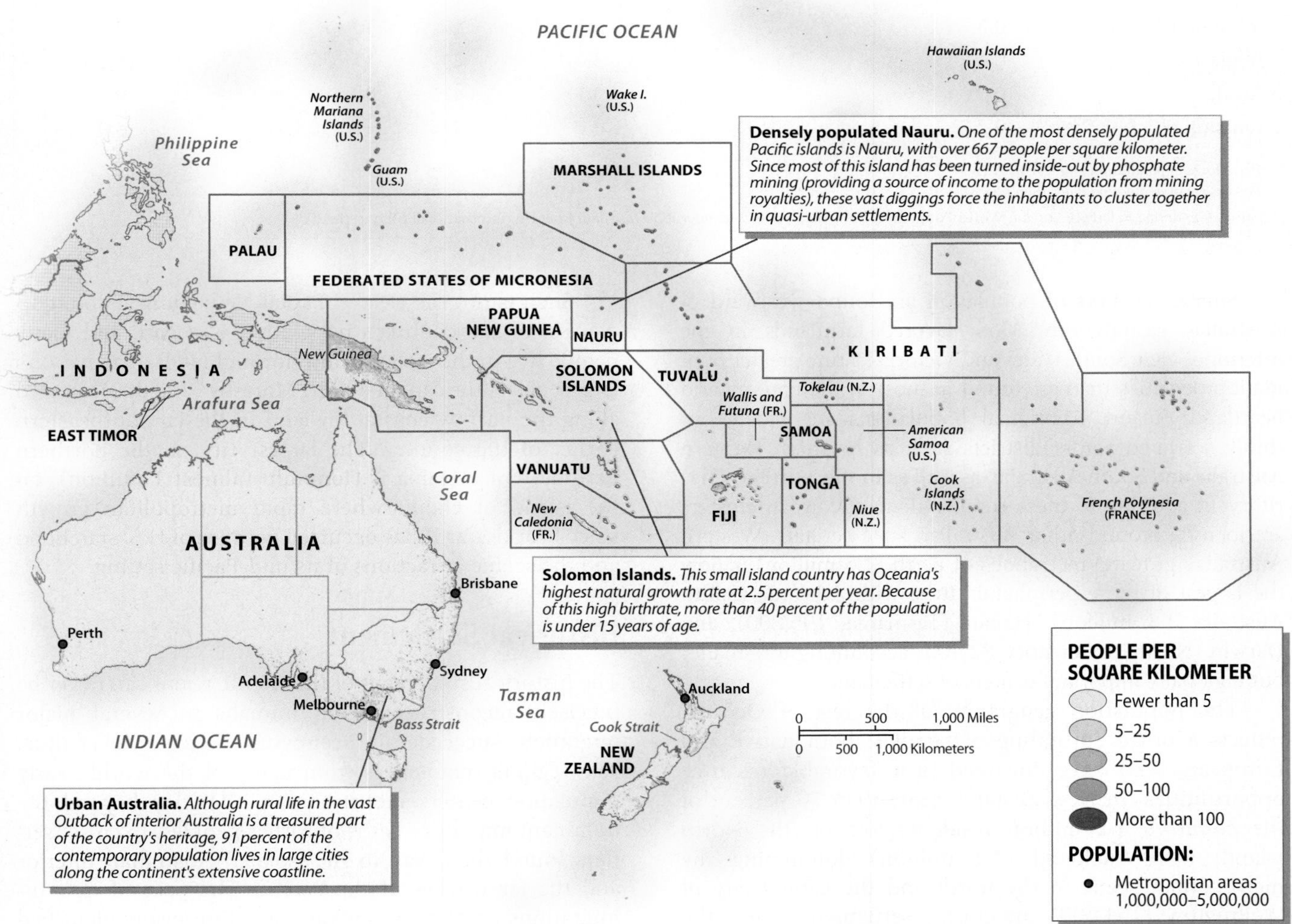

Figure 14.17 Population Map of Australia and Oceania
About 33 million people occupy this world region. While Papua New Guinea and many Pacific islands feature mainly rural settlements, most regional residents live in the large urban areas of Australia and New Zealand. Sydney and Melbourne account for almost half of Australia's population, and most New Zealand residents live on the North Island, home to the cities of Auckland and Wellington.

TABLE 14.1 Population Indicators

Country	Population (Millions, 2007)	Population Density (per Square Kilometer)	Total Fertility Rate	Percent Urban	Percent <15[a]	Percent >65[b]	Net Migration (per 1000, 2000–05)
Australia	21.0	3	1.8	91	20	13	5.1
Fed. States of Micronesia	0.1	154	4.1	22	39	3	−18.4
Fiji	0.9	47	2.5	51	31	4	−8.4
French Polynesia	0.3	65	2.2	53	28	5	2.1
Guam	0.2	315	2.8	93	30	6	0.0
Kiribati	0.1	132	4.2	47	39	3	−2.1
Marshall Islands	0.1	370	4.9	68	42	2	0.0
Nauru	0.01	667	3.4	100	39	2	8.7
New Caledonia	0.2	13	2.3	71	28	7	4.7
New Zealand	4.2	15	2.0	86	21	13	4.0
Palau	0.02	44	2.1	77	24	6	0.0
Papua New Guinea	6.3	14	4.1	13	41	2	0.0
Samoa	0.2	66	4.4	22	41	4	−15.5
Solomon Islands	0.5	17	4.5	17	40	3	0.0
Tonga	0.1	135	3.6	24	35	6	−14.1
Tuvalu	0.01	379	3.7	47	36	6	−9.7
Vanuatu	0.2	19	4.0	21	41	3	−6.0

[a]*Percentage of population younger than 15.*
[b]*Percentage of population older than 65.*
Source: Population Reference Bureau, World Population Data Sheet, 2007; *net migration rate data from* UN International Migration, 2006.

Smaller clusters of population are found westward of Australia's eastern rim. More favored farmlands in the interior of New South Wales and Victoria feature greater population densities than are found in most of the nation's arid heartland. Inland, Aboriginal populations are widely but thinly scattered across districts such as northern Western Australia and South Australia, as well as in the Northern Territory. In addition to these rural settlements, urban clusters are focused around other Australian state capitals. Western Australia's growing metropolis of Perth (1.5 million) is now the largest of these peripheral cities, with Adelaide (South Australia, 1.1 million), Hobart (Tasmania, 195,800), and Darwin (Northern Territory, 82,400) accounting for smaller but regionally important centers of settlement.

The population geography of the rest of Oceania reflects a broad sprinkling of peoples, both native and European, who have clustered near favorable resource opportunities. In New Zealand, more than 70 percent of the country's 4.2 million residents live on the North Island, with Auckland (1.1 million) dominating the metropolitan scene in the north and the capital city of Wellington (164,000) anchoring settlement along the Cook Strait in the south. Settlement on the South Island is strikingly clustered in the somewhat drier lowlands and coastal districts east of the mountains, with Christchurch (340,000) serving as the largest urban center. Elsewhere, rugged and mountainous terrain on both the North and South Islands feature much lower population densities.

Such is not the case in Papua New Guinea: about 13 percent of the country's population is urban, and many people live in the isolated, interior highlands. The nation's largest city is the capital of Port Moresby (200,000), located along the narrow coastal lowland in the far southeastern corner of the country. The largest city on the northern periphery of Oceania is Honolulu (almost 1 million), on the island of Oahu, where rapid metropolitan growth since World War II has occurred because of U.S. statehood and the scenic attractions of its mid-Pacific setting.

Historical Settlement

The historical settlement of the Pacific realm can never be precisely reconstructed, but humans in several major migrations succeeded in occupying the region over time. The region's remoteness from many of the world's early population centers meant that it often lay beyond the dominant migratory paths of earlier peoples. Even so, settlers found their way to the isolated Australian interior and the far reaches of the Pacific. The pace of new in-migrations accelerated greatly once Europeans identified the region and its resource potential.

Peopling the Pacific The large islands of New Guinea and Australia, given their proximity to the Asian land mass, were settled much earlier than the more distant islands of the Pacific, which could not be reached until the invention of sophisticated watercraft. Around 60,000 years ago, the ancestors of today's native Australian or **Aborigine** populations were making their way out of Southeast Asia and into Australia (Figure 14.18). The first Australians most likely arrived via some kind of watercraft. However, since such boats were probably not very seaworthy, the more distant islands remained inaccessible to humankind for tens of thousands of years. During glacial periods, however, sea levels were much lower than they are now, which would have allowed easier movement to Australia across relatively narrow spans of water. It is not known whether the original Australians came in one wave or in many, but the available evidence suggests that they soon occupied large portions of the continent, including Tasmania, which was then connected to the mainland by a land bridge.

Eastern Melanesia was settled much later than Australia and New Guinea. Distant islands could not be reached until better sailing craft were developed. By approximately 3,500 years ago, though, people gradually moved east to occupy New Caledonia, the Fiji Islands, and Samoa. From there, later movements took seafaring folk north into Micronesia, with areas such as the Marshall Islands occupied around 2,000 years ago.

Continuing movements from Asia further complicated the story of these migrating Melanesians. Some of the migrants mixed culturally and eventually reached western Polynesia, where they formed the nucleus of the Polynesian people. By 800 CE, they had reached such distant oceanic outposts as New Zealand, Hawaii, and Easter Island. Debate has centered on whether these Polynesian migrants purposefully set out to colonize new lands or whether they were blown off course during routine voyages, only to end up on new islands. Certainly, population pressures could quickly reach a crisis stage on relatively small islands, encouraging people to make spectacularly dangerous voyages. Equipped with sturdy outrigger sailing vessels and ample supplies of food, the Polynesians were quickly able to colonize most of the islands they discovered.

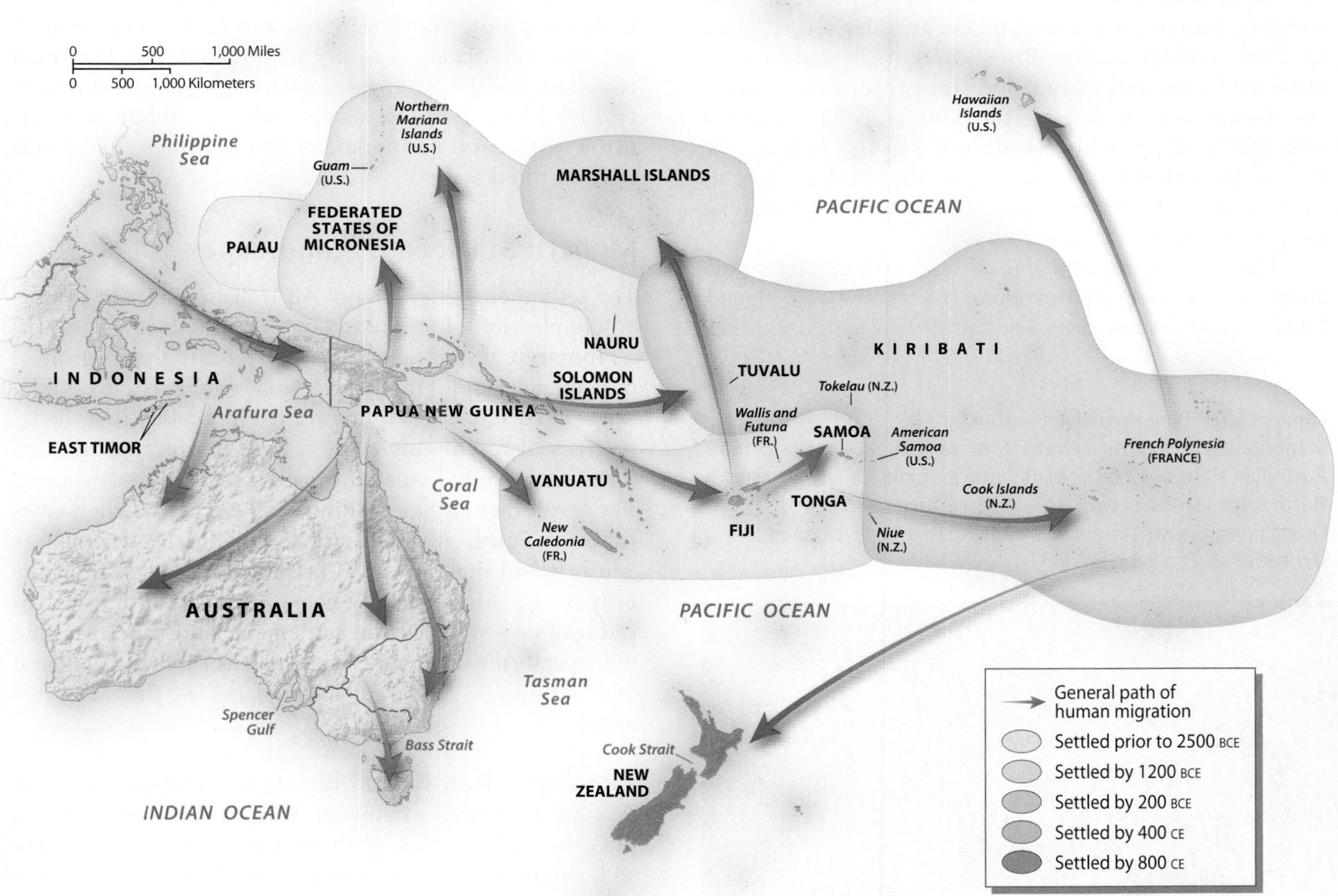

Figure 14.18 Peopling the Pacific
Ancestors of Australia's aboriginal population may have made their way into the island continent more than 60,000 years ago. Much more recent settlement of Pacific islands by Austronesian peoples from Southeast Asia shaped cultural patterns across the oceanic portions of the realm. Eastward migrations through the Solomon, Fiji, and Cook islands were followed by late movements to the north and south.

European Colonization About six centuries after the Maori settled New Zealand, navigator Abel Tasman spotted the islands on his global reconnaissance of 1642. Tasman's initial sighting marked the beginning of a new chapter in the human occupation of the South Pacific. Late in the following century, more lasting European contacts cemented ties to the region. British sea captain James Cook surveyed the shorelines of both New Zealand and Australia between 1768 and 1780. Cook and others believed that these distant lands might be worthy of European development. In addition, other expeditions were probing the Pacific, and most of Oceania's major island groups were assuming a place on European maps by the end of the eighteenth century.

Actual European colonization of the region began in Australia. The British needed a remote penal colony to which convicts could be exiled. The southeastern coast of Australia was selected as an appropriate site, and in 1788 the First Fleet arrived with 750 prisoners in Botany Bay near what is now modern Sydney. Other fleets and more convicts soon followed. Before long, however, free settlers outnumbered the convicts, who were themselves gradually gaining their freedom. The growing population of English-speaking people soon moved inland and also settled other favorable coastal locales. British and Irish settlers were attracted by the agricultural and stock-raising potential of the distant colony and by the lure of gold and other minerals (a major gold rush occurred in the 1850s). The British government also encouraged the emigration of its own citizens, often paying the transportation fare of those too poor to afford it themselves.

The new settlers came into conflict with the Aborigines almost immediately after arriving. No treaties were signed, however, and in most cases Aborigines were simply expelled from their lands. In some places, most notably Tasmania, they were hunted down and killed. By the mid-nineteenth century, Australia was primarily an English-speaking land as the native peoples were driven into submission.

Figure 14.19 **Downtown Melbourne**
Metropolitan Melbourne lies along the Yarra River. Capital of the Australian state of Victoria, Melbourne resembles many growing North American cities with its high-rise office buildings, entertainment districts, and downtown urban redevelopment. *(Fritz Prenzel/Peter Arnold, Inc.)*

British settlers were also attracted to the lush and fertile lands of New Zealand. European whalers and sealers arrived shortly before 1800, but more permanent agricultural settlement took shape after 1840 as the British formally declared sovereignty over the region. As new arrivals grew in numbers and the scope of planned settlement colonies on the North and South islands expanded, tensions increased with the native Maori population. Organized in small kingdoms or chiefdoms, the Maori were formidable fighters. In 1845 one indigenous group decided to resist further encroachments, leading to the more generalized Maori wars that engulfed New Zealand until 1870. The British eventually prevailed, however, and the Maori lost most of their land as well as control of their country.

The native Hawaiians also lost control of their lands to immigrants. Hawaii emerged as a united and powerful kingdom in the early 1800s, and for many years its native rulers limited U.S. and European claims to their islands. Increasing numbers of missionaries and settlers from the United States were allowed in, however, and by the late nineteenth century control of the Hawaiian economy had largely passed to foreign plantation owners. By 1898, U.S. forces were strong enough to overthrow the Hawaiian monarchy and to annex the islands to the United States.

Modern Settlement Landscapes

The settlement geography of Australia and Oceania offers a juxtaposition of local and colonial influences. The contemporary cultural landscape still reflects the imprint of indigenous peoples in many settings where native populations have remained numerically dominant. Elsewhere, patterns of recent colonization have produced a modern scene profoundly shaped by processes of Europeanization. The result includes everything from German-owned vineyards in South Australia to houses on New Zealand's South Island that appear to be plucked directly from the British Isles. More recently, economic and cultural globalization have reshaped the settlement landscape resulting in urban forms that make cities such as Perth or Auckland look strikingly similar to such places as San Diego or Seattle.

The Urban Transformation Both Australia and New Zealand are highly urbanized, Westernized societies, and thus the vast majority of their populations live in urban and suburban environments. As in Europe and North America, much of this urban transformation occurred during the twentieth century as the rural economy became less labor-intensive and as opportunities for urban manufacturing and service employment grew. The evolving urban landscapes took on many of the characteristics of their largely European populations but blended these

Figure 14.20 Sydney's Suburbs
Much like North American cities, the suburbs of major Australian cities sprawl into the surrounding countryside as people seek single-family dwellings and large backyards. *(Patrick Ward/Corbis)*

with a strong dose of North American influences, as well as with the unique settings native to each urban place. The result is an urban landscape in which many North Americans are quite comfortable, even though the varied local accents heard on the street and many features of the metropolitan scene are reminders of the strong and enduring attachments to British traditions.

All of the major cities of Australia and New Zealand are focused around vibrant and dynamic downtown areas that often resemble those of metropolitan North America. Coastal features, port settings, and waterfront districts shape the central city landscapes of every major urban area in the region and give each center a unique identity. Perth, Melbourne, Sydney, Brisbane, Wellington, and Auckland are anchored on the sea, a function of their important commercial connections to the global economy. Thus, central city skylines, complete with growing numbers of North American-style high-rises, are never far from their harbor settings (Figure 14.19). In addition, these urban places are often praised for their modest crime rates, lack of slums, relatively clean streets, spacious parks and open spaces, and efficient public transportation.

Suburbanization and peripheral commercial development are also changing these large cities of the Southern Hemisphere (Figure 14.20). For example, Sydney's metropolitan reach now extends west for more than 20 miles (32 kilometers) from the downtown area, and Melbourne's urban grasp is expanding inland toward the hills as well as farther east along Port Phillip Bay. Amenity-rich Perth (Western Australia) and Brisbane (Queensland) have seen some of the most rapid recent population gains, particularly in their expanding suburban fringe. Similarly, New Zealand's Auckland, given that city's water-bound limits between Manukau and Waitemata harbors, is spreading well beyond the old boundaries of the traditional city. Land uses and settlement landscapes in these dynamic urban peripheries parallel patterns across North America: low-density residential areas are linked by modern commercial highways and punctuated by "edge city" collections of office centers, industrial parks, and entertainment complexes.

The affluent Western-style urban settings in Australia and New Zealand offer a stunning contrast with the urban landscapes found in less-developed parts of the region. Walk the streets of Port Moresby in Papua New Guinea, and a very different urban landscape attests to the yawning gap between rich and poor within Oceania (Figure 14.21). Rapid growth in Port Moresby, the country's political capital and largest commercial center, has produced many of the classic challenges of urban underdevelopment: a shortage of adequate housing; an infrastructure that lags far behind the need for water and electricity; and rising street crime and alcoholism. Elsewhere, urban centers such as Suva (Fiji), Noumea (New Caledonia), and Apia (Samoa) also reflect the economic and cultural tensions generated as indigenous populations are exposed to Western influences (see "Cityscapes: Apia, Capital of Samoa").

Figure 14.21 Port Moresby, Papua New Guinea
Urban poverty and high crime haunt the city of Port Moresby, the capital of Papua New Guinea. The city's slums, many built out on the water, reflect stresses of recent urban growth as rural residents emigrate from nearby highlands. *(Chris Rainier/Corbis/Bettmann)*

Rapid growth is a common problem in these smaller urban Pacific settings because many native peoples from neighboring rural areas and nearby islands gravitate toward the job opportunities available in these areas. In the past 50 years, the spectacular global growth of tourism in places such as Fiji and Samoa has also transformed the urban scene—nineteenth-century village life has often been replaced by a landscape dominated by souvenir shops, honking taxicabs, and crowded seaside resorts.

The Rural Scene: Australia and New Zealand Rural landscapes across Australia and the Pacific region reflect a complex mosaic of cultural and economic influences. In some settings, Australian Aborigines or native Papua New Guinea Highlanders can still be found in their familiar homelands, their traditional lifeways and settlements barely changed from pre-European times. Yet such settlement landscapes are becoming increasingly rare. Global influences penetrate the scene as the cash economy, foreign tourism and investment, and popular culture work their way into the hinterlands. Relatively large areas of rural settlement still exist across many areas of the Australian and New Zealand interiors, as well as smaller examples sprinkled among the islands of the Pacific. Even here, however, land uses are often shaped not by the needs of local people but by the regional or global commercial agricultural economy or by the recreational values associated with such settings.

Much of rural Australia is too dry for farming or serves as only marginally valuable agricultural land. Although modest in size, the area in crops has doubled since 1960 as increased use of fertilizers, more widespread irrigation, and more aggressive rabbit eradication efforts have opened up new areas for development. Much of the remainder of the interior, however, features range-fed livestock, areas with no agricultural potential, and isolated areas where Aboriginal peoples still pursue their traditional forms of hunting and gathering.

Sheep and cattle dominate the livestock economy. Many rural landscapes in the interior of New South Wales, Western Australia, and Victoria, for example, are oriented around isolated sheep stations or ranch operations that move the flocks from one vast pasture to the next. Cattle can sometimes be found in these same areas, although many of the more extensive, range-fed cattle operations are concentrated farther north in Queensland. Cattle are even grazed on the tough tropical grasslands and woodlands of northern Australia, although environmental conditions in these settings are often marginal for any agricultural activity. More specialized dairy cattle operations, often on much smaller family-owned farms, are found in moister settings from the Queensland coast to Victoria.

Croplands also vary across the region (Figure 14.22). Sometimes mingling with the sheep country, a band of commercial wheat farming includes southern Queensland; the moister interiors of New South Wales, Victoria, and South Australia; and a swath of more-favored land east and north of Perth. Elsewhere, specialized sugarcane operations thrive along the narrow, warm, and humid coastal strip of Queensland. To the south and west, productive irrigated agriculture has developed in localities such as the Murray River Basin, allowing for the production of orchard crops and vegetables.

Viticulture, or grape cultivation (and other specialized horticulture), increasingly shapes the rural scene in places such as South Australia's Barossa Valley, New South Wales' Riverina district, and Western Australia's

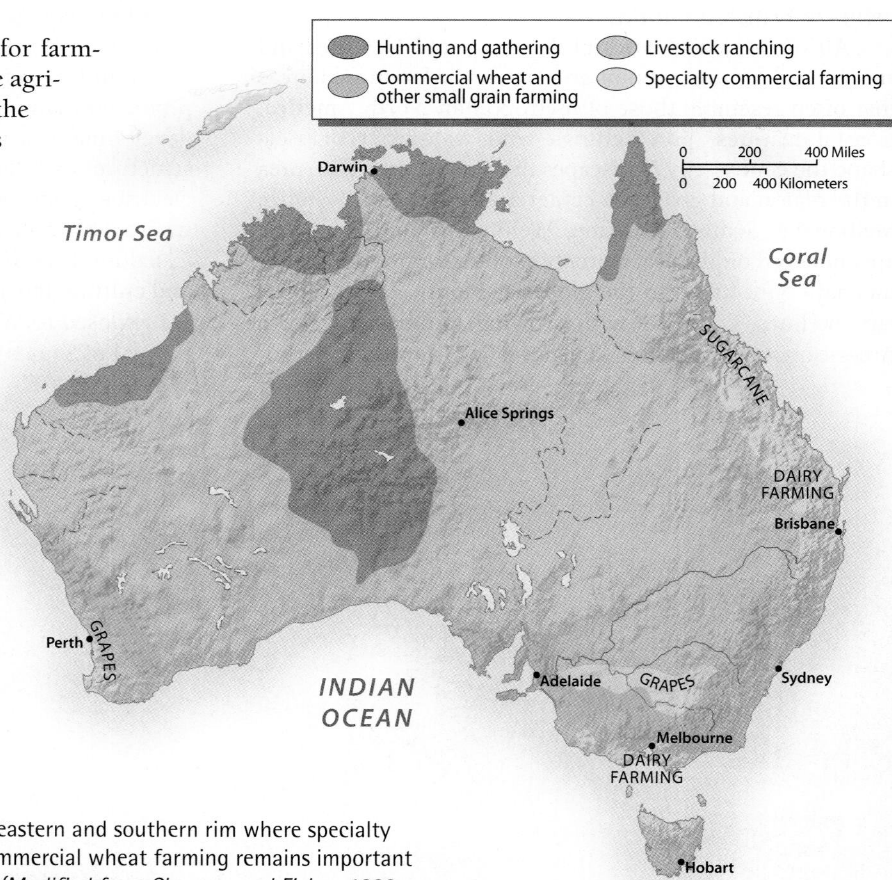

Figure 14.22 Australian Agriculture
Australia's best farmlands are found along its eastern and southern rim where specialty crops such as sugarcane and grapes thrive. Commercial wheat farming remains important in drier districts in the south and southwest. *(Modified from Clawson and Fisher, 1998,* World Regional Geography, *Upper Saddle River, NJ: Prentice Hall)*

Figure 14.23 Canterbury Plain
The varied agricultural landscape of South Island's Canterbury Plain offers a mix of grain fields, livestock, orchard crops, and vegetable gardens. The rugged Southern Alps are a dramatic backdrop to this productive region. *(Robert Frerck/Woodfin Camp & Associates)*

Swan Valley. Indeed, the area under grape cultivation grew by 50 percent in the 1990s as the popular Chardonnay, Cabernet Sauvignon, and Shiraz varieties propelled Australian wine exports beyond those of the United States.

Although much smaller in area, New Zealand's rural settlement landscape includes a variety of agricultural activities. Pastoral pursuits clearly dominate the New Zealand scene, with the vast majority of agricultural land devoted to livestock production, particularly sheep grazing and dairying. Commercial livestock outnumber people in New Zealand by a ratio of more than 20:1, and this is apparent everywhere on the rural scene. Sheep gnaw on largely imported European grasses from the isolated hinterlands of the South Island to the volcanic slopes and shorelines of the North Island. In addition, dairy operations dot the rural scene, mostly in the humid lowlands of the subtropical north, where they sometimes mingle with suburban landscapes in the vicinity of Auckland.

One of the largest zones of more specialized cropping spreads across the fertile Canterbury Plain near Christchurch (Figure 14.23). This spectacular South Island setting proved fertile ground for English settlement and continues to feature a varied landscape of pastures, grain fields, orchards, and vegetable gardens, all spread beneath the towering peaks of the Southern Alps.

Rural Oceania Elsewhere in Oceania, varied influences shape cultural landscapes. On better-watered high islands, inhabitants have more diverse agricultural opportunities than are found on the more barren low islands, where limited farming leads people to depend more on fishing. Several types of rural settlement can be identified across the island realm. In rural New Guinea, village-centered shifting cultivation dominates: farmers clear a patch of forest and then, after a few years, shift to another patch, thus practicing a form of land rotation. Subsistence foods such as sweet potatoes, taro (another starchy root crop), coconut palms, bananas, and other garden crops often are intercropped in the same field, and growing numbers of planters also include commercial crops such as coffee.

In other parts of Oceania, traditional agricultural patterns are similar: most rural settlements are organized around village-based societies surrounded by nearby fields in which varied crops are produced (Figure 14.24). Commercial plantation agriculture has also made its mark in many more accessible rural settings. In these localities, settlements consist of worker housing near crops that are typically controlled by absentee landowners. For example, copra (coconut), cocoa, and coffee operations have transformed many agricultural settings in places such as the Solomon Islands and Vanuatu. Sugarcane plantations have reshaped other island settings, particularly in Fiji and Hawaii.

Diverse Demographic Paths

Varied population-related issues face residents of the region today. In Australia and New Zealand, while populations grew rapidly (mostly from natural increases) in the twentieth century, today's low birthrates parallel the pattern in North America. Just as in the United States and Canada, however, significant population shifts within these countries continue to impose challenges.

Figure 14.24 Yam Harvest
These farmers on the Melanesian island of Vakuta (Papua New Guinea) are harvesting yams. Traditional tropical agriculture features a mix of crops, often grown in the same field. Where possible, fields are periodically rotated to maintain productivity. *(Peter Essick/Aurora & Quanta Productions)*

CITYSCAPES Apia, Capital of Samoa

The great cities of Australia and Oceania—Sydney, Auckland, and Honolulu—although exciting and attractive in their own ways, are essentially Western places, familiar to the North American eye with their array of skyscrapers, banks, and look-alike hotels connected by expressways to suburban shopping malls and neighborhoods of single-family homes.

But for a real Pacific Island experience, journey to the town of Apia, the capital of Samoa and home to 35,000 people. Here, in what is called the "heart of Polynesia"—about 2,500 (4,000 kilometers) miles southwest of Hawaii—the pulse beats more slowly. With globalization, however, the pace is faster now than it was just 20 years ago, and outside influences are changing the town's traditional landscape. On Beach Road, overlooking Apia harbor, the new eight-story government building and the neighboring seven-story Central Bank of Samoa seem out of place in a town dominated by colonial-era structures and thatch-roofed, open-sided Polynesian *fala*, or long houses (Figures 14.4.1 and 14.4.2).

The colonial past is apparent everywhere. Until moving into their new high-rise home in 1994, the Samoan government was housed in the wooden courthouse on Ififi Street built in the 1880s by German interests. Closer to the harbor, at the intersection of Beach Road and Vaea Street, is the World War I memorial, constructed by New Zealand and British colonials not only to honor their fallen compatriots but also to mark the end of the German period.

Perhaps the most enduring signs of American influence are the numerous churches built by Western missionaries. These churches, built in the late nineteenth century, attest to the strong role played by Christianity in Samoa's culture, both past and present. Visitors should be aware that all activity pauses about 6 P.M. for evening vespers. Cars stop on the street and drivers bow their heads; pedestrians cease walking; and conversations are halted while the population prays for 10 minutes. During the weekend, Apia falls silent as inhabitants respect the Sabbath.

West of the harbor, along the shoreline road leading to the airport and overlooking the harbor full of interisland ferries, transient yachts, and long-line tuna fishing boats, are several new industrial parks giving clue to Samoa's search for new economic activities to supplement its long reliance on copra (coconut) exports. Indicative of the country's emerging linkages to the globalized world is a Japanese auto parts factory employing 2,000 women who assemble electronic car parts. Farther down the road is the Vailima brewery, a local beer maker that now serves foreign markets, especially those in New Zealand and Australia.

Figure 14.4.1 Map of Apia
This map shows some of the major features of Apia, capital city of Samoa. The new auto parts factory and other industry is located along Mulinu'u Road, leading away from the town in the northwest.

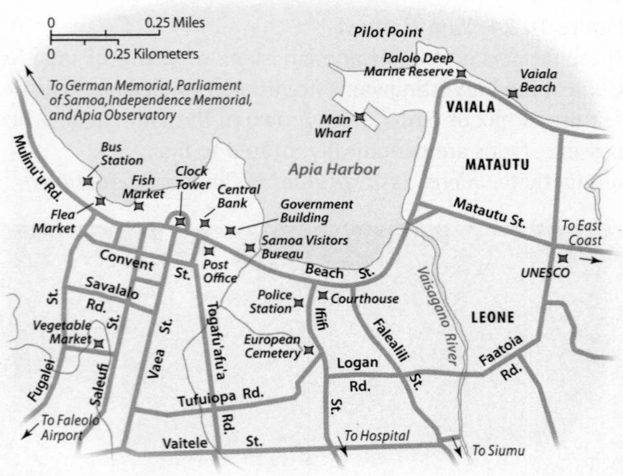

Figure 14.4.2 Apia Governmental Center
The Samoan parliament building gives a hint of the traditional architecture of that island nation with its thatched roof and longhouse appearance. *(Topham/The Image Works)*

For example, the exodus of farmers from Australia's wheat-growing and sheep-raising interior mirrors similar processes at work in the rural Midwest of the United States and in the Canadian prairies. Communities see many of their productive young people and professionals leave for the better employment opportunities of the city.

Older urban and industrial areas, particularly near Sydney and Melbourne, have also lost population. Conversely, other portions of Australia and New Zealand face rapid growth. As in North America, amenity-rich settings, retirement communities, and new upscale suburbs draw ever more people, thus pressuring infrastructure and public services. Some people in these two countries also complain about liberal national immigration policies allowing in too many new residents and workers, although such frustrations are often more rooted in cultural fears than they are in concerns about overall population numbers.

Different demographic challenges grip many of the less-developed island nations of Oceania. Population growth rates are above 2 percent per year in countries such as Vanuatu, the Marshall Islands, and the Solomon Islands. While the larger islands of Melanesia still have some room for settlement expansion, competitive pressures from commercial mining and logging operations limit the amount of new agricultural land that will probably be available in the future. On some of the smaller island groups in Micronesia and Polynesia, population growth is even more pressing. Nauru, for example, has about 10,000 inhibitants, but since they are crowded onto a small land area, the density is over 1500 people per square mile (7660 per square kilometer). Aggravating the situation is sea level rise, apparently from global warming, that is reducing the habitable area of Tuvalu. In atoll environments, a single island might be considered overpopulated if it supports only 100 people. Thus, even relatively small population centers in the Pacific Islands have little internal flexibility in coping with higher birthrates. Making matters worse, many young people are migrating to already crowded urban centers in these island nations.

CULTURAL COHERENCE AND DIVERSITY: A Global Crossroads

The Pacific world offers excellent examples of how culture is transformed as different groups migrate to a region, interact with one another, and then evolve over time. As European and other outsiders arrived in the region, colonization forced native peoples to adjust. More recently, worldwide processes of globalization have also redefined the region's cultural geography, provoking fears of homogenization from some quarters while, in response, native groups try to protect their heritage.

Multicultural Australia

Australia's cultural patterns illustrate many of these fundamental processes at work. Today, while still dominated by its colonial European roots, the country's multicultural character is becoming increasingly visible as native inhabitants assert their cultural identity and immigrant populations play larger roles in society, particularly within major metropolitan areas.

Aboriginal Imprints For thousands of years, Australia's Aborigines dominated the cultural geography of the continent. These indigenous peoples never practiced agriculture, opting instead for a hunting-gathering way of life that persisted up to the time of the European conquest. As the population consisted of foragers and hunters living in a relatively dry land, settlement densities remained low; tribal groups were often isolated from one another; and overall populations probably never numbered more than 300,000 inhabitants. To survive, Aborigines developed great adaptive skills, often subsisting in harsh environments that Europeans avoided. Their low densities meant linguistic fragmentation. Although precise counts vary, there were probably 250 languages spoken at the time of European contact, and almost 50 indigenous languages can still be found.

Radical changes accompanied the arrival of Europeans, and Aboriginal populations were decimated in the process. The geographical results of colonization were striking as Aboriginal settlements were relegated to the sparsely settled interior, particularly in northern and central Australia, where fewer Europeans competed for land. In most cases, the European attitude toward the Aboriginal population was even more prejudicial than it was toward native peoples of the Americas. As hunter-gatherers lacking centralized political organization, Aborigines were usually considered less than human and more often than not simply as another strange animal species. No treaties were signed and no concessions were made. Typically, Aborigines were simply run off any lands desired by Europeans.

Today Aboriginal cultures persevere in Australia, and a growing native peoples movement parallels similar activities in the Americas. Indigenous peoples represent approximately 2 percent (or 430,000) of Australia's population but their geographical distribution changed dramatically in the last century. Aborigines account for almost 30 percent of the Northern Territory's population (many of these in Arnhemland, near Darwin), and other sizable native reserves are located in northern Queensland and Western Australia. Most native peoples, however, live in the same large urban areas that dominate the country's overall population geography. Indeed, more than 70 percent of Aborigines live in cities, and very few of them still practice traditional hunting-gathering lifestyles. Processes of cultural assimilation are clearly at work: urban Aborigines are frequently employed in service occupations; Christianity has often replaced traditional animist religions; and only 13 percent of the native population still speaks an indigenous language.

The growing Aboriginal interest in preserving traditional cultural values suggests that forces of diversity are at work. Particularly in the Outback, a handful of Aboriginal languages retain their vitality and have growing numbers of speakers. In addition, cultural leaders are selectively preserving Aboriginal spiritualism, and these religious practices often link local populations to surrounding places and natural features that are considered sacred. In fact, a growing number of these sacred locations are at the center of land-use and resource controversies between Aboriginal populations and Australia's European majority. The future for Aboriginal cultures remains unclear: pressures for cultural assimilation will be intense as many native peoples relocate to more Western-oriented urban settlements and lifestyles. At the same time, more rapid rates of natural increase (almost twice the national average) and a growing cultural awareness of Aboriginal traditions will work to preserve elements of the country's indigenous cultures.

A Land of Immigrants Most Australians reflect the continent's more recent European-dominated migration history, but even these patterns have become more complex as a rising tide of Asian cultures becomes important. Overall, about 70 percent of Australia's population continues to reflect a British or Irish cultural heritage. These groups dominated many of the nineteenth- and early twentieth-century migrations into the country, and close cultural ties to the British Isles remain today. Sizable numbers of Italians, Greeks, and Germans added diversity to the European mix. Some of these migrants became involved in specialized agricultural activities, particularly in southeastern Australia, while others moved to the cities, where they added ethnic variety to the mostly Anglo mix. A need for laborers along the fertile Queensland coast also prompted European plantation owners to import inexpensive workers from the Solomons and New Hebrides. These Pacific Island laborers, known as **kanakas**, were socially segregated from their Anglo employers, yet further diversified the cultural mix of Queensland's "sugar coast." Historically, however, nonwhite migrations to the country were strictly limited by what is often termed a **White Australia Policy**, in which governmental guidelines promoted European and North American immigration at the expense of other groups. This remained national policy until 1973.

Recent migration trends have reversed this historical bias, and more diverse inflows of new workers and residents are now adding to the country's multicultural character. Beginning in the 1970s, the Migration Program offered visas to people on the basis of their educational background and potential for succeeding economically in Australian society instead of simply their country of origin. Today, for example, a growing number of families have come from places such as Africa, China, India, Malaysia, and the Philippines (see "People on the Move: Africans in Australia"). Smaller numbers have qualified as migrants through their New Zealand citizenship, while others have arrived as refugees from troubled parts of the world, such as Vietnam and the former Yugoslavia.

The result is a much more varied foreign-born population. Indeed, around 25 percent of Australia's people are now immigrants, reflecting the country's global popularity as a migration destination. In the early twenty-first century, almost 40 percent of the settlers arriving in the country were from Asia. Major cities offer particularly attractive possibilities: Sydney's Asian population exceeds 10 percent and is growing rapidly, while Perth's culture and economy are increasingly linked to its Asian neighbors.

Australian society, while enduringly Anglo, has been changed forever by its varied immigrant mix (Figure 14.25). For example, although the country is still largely Christian (26 percent Catholic and 20 percent Anglican), some of the nation's fastest-growing religions are Islam, Buddhism, and Hinduism. In similar fashion, while English is Australia's national language, more than 2.5 million residents now speak another language, including more than 100,000 speakers each of Italian, Greek, Cantonese, Arabic, and Vietnamese. Economic ties with Japan are also reshaping the country's cultural geography: many Japanese travel to Australia for business and pleasure, and a growing number of Australians are learning Japanese. However, a vocal minority of native-born whites resist the current cultural mosaic and advocate much more restrictive immigration policies on both racial and economic grounds.

Figure 14.25 Asian Immigrants in Australia
Unlike historical immigration, which was predominately from Britain and other European countries, today fully 40 percent of Australian immigrants are from Asia. As a result, neighborhoods in Sydney and other cities take on a distinct Asian flavor as they serve the needs of this new population. *(Time Wimborne/Reuters/Corbis)*

PEOPLE ON THE MOVE Africans in Australia

As Australia moves farther away from its historical White Australia immigration policy and becomes increasingly a multicultural society with a rainbow of immigrants, most notice goes to the country's new—and large—Asian population. Less publicized is an African connection that has created vibrant communities, in both city and countryside, with people from Ghana, Kenya, South Africa, and, most recently, Sudan and Somalia. Along with other immigrant groups, these new migrants are changing the cultural complexion of Australia.

True, the first wave of African migrants were white and English-speaking, refugees from Britain's African colonies, seeking a new homeland as the waves of African nationalism swept through the continent. In many ways, these migrants merely reinforced the dominant Australian culture of overseas British.

Following this, and coinciding with the end of the White Australia Policy in 1973, came middle-class black English-speaking Africans from former British colonies in Nigeria, Kenya, Ghana, Uganda, Botswana, and Zimbabwe. These immigrants, now well-integrated into Australian life in all occupations and professions, have also played a prominent role in cultural change by adding African music, literature, food, and speech to Australian culture. They've also built a strong social services infrastructure for the most recent wave of African migrants from the war-torn countries of Sudan and Somalia.

Currently, 70 percent of Australia's amnesty immigration quota is filled by Africans fleeing civil wars and civil unrest. Many of these refugees have been in refugee camps for a decade, awaiting a chance for relocation to an accepting country, be it Australia, North America, or Europe. While the earlier migrants from Africa spoke English as part of growing up in a British colony, that is not the case with these refugees. Instead, they are tribal people, many of them only recently removed from a nomadic life, and speak their own tribal language. Although many new migrants are enrolled in English classes provided by Somali and Sudanese refugee organizations, commonly they must draw upon a translator to help them navigate their way through Australian life as they visit doctors to solve health problems, shop for food in a cash economy far different from the bartering system they're used to, deal with endless bureaucratic matters related to their refugee status, and interact with teachers at children's schools.

Figure 14.5.1 Africans in Australia
Two Sudanese girls play at a SAIL (Sudanese-Australian Integrated Learning) party. This group provides community services for recent migrants from the Sudan. *(SAIL—Sudanese Australian Integrated Learning Program)*

Further insight to the issues and interests of Australia's newest inhabitants comes from browsing through Internet Websites designed for Africans in Australia. Here, one finds advice and resources for getting along in Australia—housing hints; employment possibilities; health matters, social events; even dating services. Although it's a bit of a stretch to think that a Somali migrant fresh from a refugee camp in Kenya might have the technological skills to draw upon these Internet resources (or even access a computer, for that matter), the very fact that these Websites exist speaks to one of the many ways that contemporary migrants adjust to a new homeland.

Global popular culture has also shaped Australian society, a relationship that is often a two-way street. The country's largely English-speaking population has embraced North American movies, music, and television. Many American television programs appear down under, although the Australian media are mandated to show at least 50 percent local programming from 6 A.M. to midnight in order to "keep the Australia in Australian television." Australian influences have also shaped global media. Australian ownership (News Corporation) of many British newspapers as well as the Fox television network in the United States is a reminder that global cultural influences move both in and out of the region.

Cultural Patterns in New Zealand

New Zealand's cultural geography broadly reflects the patterns seen in Australia, although the precise cultural mix differs slightly. Native Maori populations are more numerically important and culturally visible in New Zealand than their Aboriginal counterparts in Australia. While British colonization clearly mandated the dominance of Anglo cultural traditions by the late nineteenth century, Maori populations survived, although they lost most of their land in the process. After declining with initial European contacts and conflicts, native populations began rebounding in the twentieth century, and today the Maori account for about 8 percent of the country's approximately 4 million residents. Geographically, the Maori remain most numerous on the North Island, including a sizable concentration in metropolitan Auckland. While urban living is on the rise, many Maori, akin to their Aboriginal counterparts, are also committed to preserving their religion, traditional arts, and Polynesian lifeways (Figure 14.26). In addition, Maori is now an official language in the country, along with English.

Many New Zealanders still identify with their largely British heritage, but the country's twentieth-century cultural identity matured with an increasing sense of separateness from its British roots. Several processes have forged New Zealand's special cultural character. As Britain tightened its own links with the European continent after World War II, New Zealanders increasingly shaped a more independent identity. In many ways, popular culture ties the country ever more closely to Australia, the United States, and continental Europe, a function of increasingly global mass media. For example, the best known film ever made in New Zealand, *The Piano*, was produced by an Australian, financed with French capital, and reached its largest audiences in the United States.

Figure 14.26 Maori Artisans
New Zealand's native Maori population actively preserves its cultural traditions and has recently increased its political role in national affairs. These artisans are carving decorations for a traditional Maori canoe. *(Arno Gasteiger/Bilderberg/Aurora & Quanta Productions)*

Diversity also continues to shape the cultural setting. The nation's unique Polynesian roots impart a special regional character: In addition to its Maori population, more than 5 percent of New Zealand's people are Pacific Islanders, and Auckland has the largest Polynesian population of any city in the world. Adding further complexity to the cultural mix are growing numbers of Asians, who now make up another 5 percent of the country's residents and who are making their own dynamic contribution to the country's culture and economy, particularly in its larger urban centers. The end result is a national character truly forged at a Pacific crossroads, an accumulation of varied and vastly different cultural influences that have been amalgamated into a uniquely New Zealand identity that resists easy definition.

The Mosaic of Pacific Cultures

Native and exotic influences affect cultures across the islands of the South Pacific. In more isolated locales, traditional cultures maintain their integrity largely insulated from outside influences. In most cases, however, modern life in the islands revolves around an intricate cultural and economic interplay of local and Western influences. One thing is certain—the relative cultural insularity of the past is gone forever, and in its place is a Pacific realm rapidly adjusting to powerful forces of colonization, global capitalism, and popular culture.

Language Geography The modern language map reveals some significant cultural patterns that both unite and divide the region (see Figure 14.27). Most of the indigenous languages of Oceania belong to the Austronesian language family, which encompasses wide expanses of the Pacific, much of insular Southeast Asia, and Madagascar. Linguists hypothesize that the first great oceanic mariners spoke Austronesian languages, and thus disseminated them throughout this vast realm of islands and oceans. Within the broad Austronesian family, the Malayo-Polynesian subfamily includes most of the related languages of Micronesia and Polynesia, suggesting a common cultural and migratory history for these far-flung peoples.

Melanesia's language geography is more complex and still incompletely understood by experts. While coastal peoples often speak languages brought to the region by the seafaring Austronesians, more isolated highlander cultures, particularly on the island of New Guinea, speak varied Papuan languages. Indeed, the linguistic complexity of that island is so daunting—more than 1,000 languages have been identified—that many experts question whether they even make up a unified "Papuan family" of related languages. Some scholars estimate that half of New Guinea's languages are spoken by fewer than 500 persons, suggesting the cultural role played by the region's rugged topography in isolating cultural groups. These New Guinea highlands may hold some of the world's few remaining **uncontacted peoples**, cultural groups that have yet to be "discovered" by the Western world.

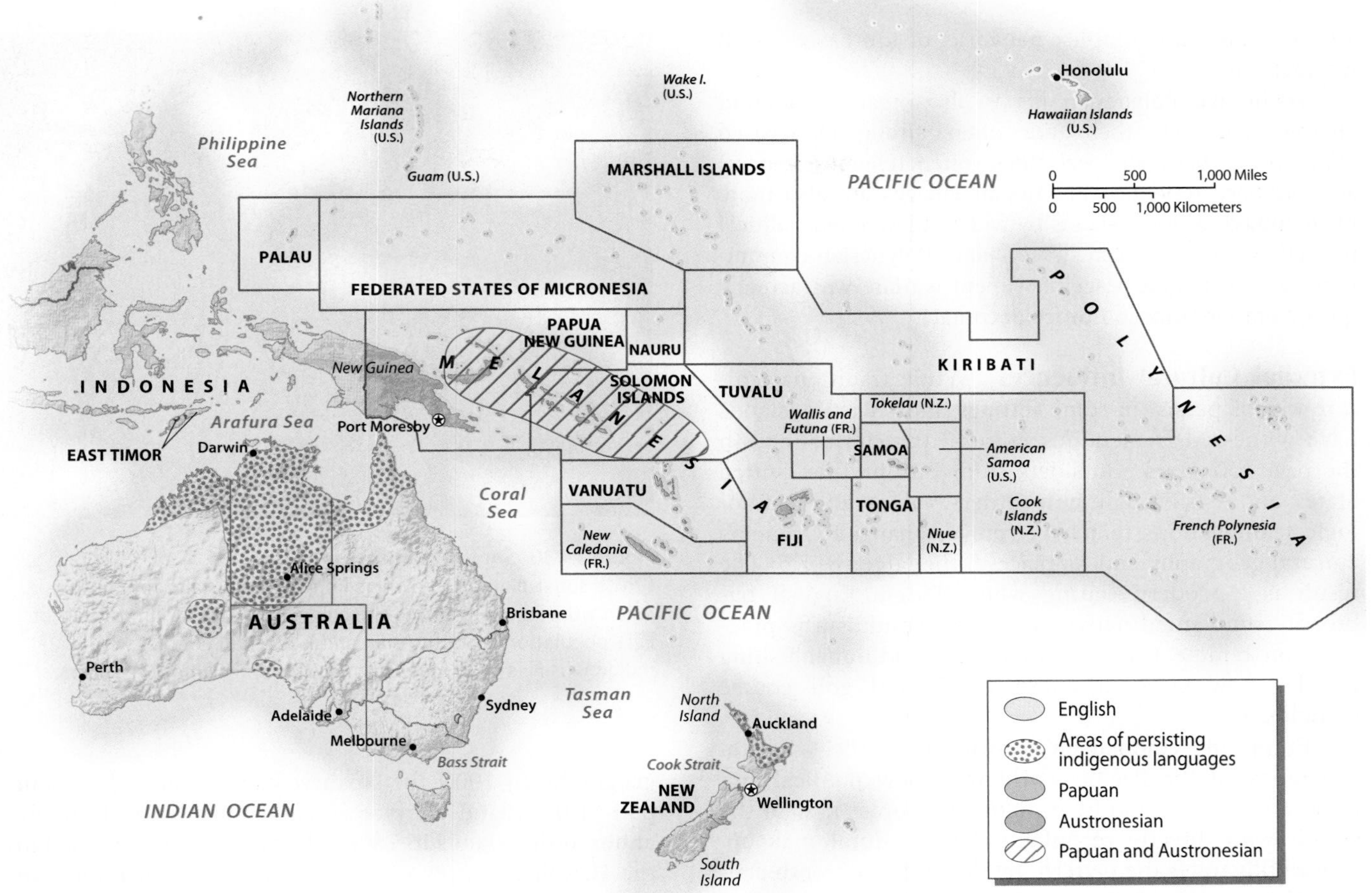

Northern Mariana Islands (U.S.)
Wake I. (U.S.)
Honolulu
Hawaiian Islands (U.S.)
Philippine Sea
Guam (U.S.)
MARSHALL ISLANDS
PACIFIC OCEAN
0 500 1,000 Miles
0 500 1,000 Kilometers
PALAU
FEDERATED STATES OF MICRONESIA
POLYNESIA
PAPUA NEW GUINEA
NAURU
KIRIBATI
New Guinea
MELANESIA
INDONESIA
SOLOMON ISLANDS
TUVALU
Tokelau (N.Z.)
Arafura Sea
Port Moresby
Wallis and Futuna (FR.)
EAST TIMOR
Darwin
SAMOA
American Samoa (U.S.)
Coral Sea
VANUATU
TONGA
Cook Islands (N.Z.)
New Caledonia (FR.)
FIJI
Niue (N.Z.)
French Polynesia (FR.)
Alice Springs
AUSTRALIA
Brisbane
PACIFIC OCEAN
Perth
Tasman Sea
North Island
Sydney
Adelaide
Auckland
Melbourne
Bass Strait
Cook Strait
NEW ZEALAND
Wellington
INDIAN OCEAN
South Island
English
Areas of persisting indigenous languages
Papuan
Austronesian
Papuan and Austronesian

Figure 14.27 Language Map of Australia and Oceania
While English is spoken by most residents, native peoples and their linguistic traditions remain an important cultural and political force in both Australia and New Zealand. Elsewhere, traditional Papuan and Austronesian languages dominate Oceania. The French colonial legacy also persists in select Pacific locations. Tremendous linguistic diversity has shaped the cultural geography of Melanesia, and more than 1,000 languages have been identified in Papua New Guinea.

Village Life Traditional patterns of social life are as complex and varied as the language map. In many cases, however, life revolves around predictable settings. For example, across much of Melanesia, including Papua New Guinea, most people live in small villages often occupied by a single clan or family group (Figure 14.28). Many of these traditional villages have fewer than 500 residents, although some larger communities may reach more than 1,000 people. Life often revolves around the gathering and growing of food, an annual round of rituals

Figure 14.28 Tonga Village
Although the economic and technological effects of globalization have arrived in Tonga, village life remains important in many Polynesian settings. Most village housing in these tropical environments reflects the use of locally available construction materials. *(Ted Streshinsky/Corbis/Bettmann)*

and festivals, and complex networks of kin-based social interactions.

Traditional Polynesian life is also organized around villages, although there are often strong class-based relationships between local elites (often religious leaders) and ordinary residents. Polynesian villages are also more likely linked to other islands by wider cultural and political ties. Despite the Western stereotype of Polynesian communities as idyllic and peaceful, violent warfare was actually quite common prior to European contact.

External Cultural Influences While traditional culture worlds persist in some settings, most Pacific islands have witnessed tremendous cultural transformations in the past 150 years. Outsiders from Europe, the United States, and Asia brought new settlers, values, and technological innovations that have forever changed Oceania's cultural geography and its place in the larger world. The result is a modern setting where Pidgin English has broadly supplanted native languages, Hinduism is practiced on remote Pacific islands, and traditional fishing peoples now work at resort hotels and golf course complexes.

European colonialism transformed the cultural geography of the Pacific world with new political and economic systems and by bringing in peoples new to the region who directly reconfigured its cultural makeup. Hawaii illustrates the pattern. By the mid-nineteenth century, Hawaii's King Kamehameha was already entertaining an assortment of whalers, Christian missionaries, traders, and navy officers from Europe and the United States. A small elite group of **haoles**, or light-skinned European and American foreigners, were successfully profiting from commercial sugarcane plantations and Pacific shipping contracts.

Labor shortages on the islands, however, prompted the importation of Chinese, Portuguese, and Japanese workers who further complicated the region's cultural geography. By 1900, the Japanese had become a dominant part of the island workforce. The United States formally annexed the islands in 1898. The cultural mix revealed in the Hawaiian census of 1910 suggests the magnitude of change: more than 55 percent of the population was Asian (mostly Japanese and Chinese), native peoples made up another 20 percent, and about 15 percent (mostly imported European workers) were white. By the end of the twentieth century, the Asian population was less dominant and more ethnically varied; the small number of remaining native Hawaiians had been joined by an increasingly diverse group of other Pacific Islanders; and about 40 percent of Hawaii's residents were white. In addition, ethnic mixing has produced a rich mosaic of Hawaiian creole cultures that offer a unique blend of North American, Asian, Pacific Island, and European influences (Figure 14.29).

Hawaii's story has been played out in many other Pacific Island settings. In the Mariana Islands, Guam was absorbed into America's Pacific empire as a spoils of the Spanish-American War in 1898. Thereafter, not only did native peoples feel the effects of Americanization (the island remains a self-governing U.S. territory today), but thousands of Filipinos were moved there to supplement its modest labor force. To the southeast, the British-controlled Fiji Islands offered similar opportunities for fundamentally redefining Oceania's cultural mix. The same sugar plantation economy that spurred changes in Hawaii prompted the British to import thousands of South Asian laborers to Fiji. The descendents of these Indians (most practicing Hinduism) now make up almost half the island country's population and often come into sharp conflict with the native Fijians (Figure 14.30).

Figure 14.30 South Asians in Fiji
British sugar plantation owners imported thousands of South Asian workers to Fiji during the colonial era. Today almost half of Fiji's population is South Asian and is often in conflict with indigenous Fijians. *(Frank Fournier/Woodfin Camp & Associates)*

Figure 14.29 Multicultural Hawaiians
Many residents of the Hawaiian Islands represent a blend of Pacific Island, Asian, and European influences. These young women express a mixture of Polynesian and Asian ancestors. *(Porterfield/Chickering/Photo Researchers, Inc.)*

In French-controlled portions of Oceania, small groups of traders and plantation owners filtered into the Society Islands (Tahiti), but a larger contingent of French colonial settlers (many originally a part of a penal colony) had a major impact on the cultural makeup of New Caledonia. Still a French colony, New Caledonia's population is more than one-third French, and its capital city of Noumea reveals a cultural setting curiously forged from French society and Melanesian traditions.

Given the frequency of contact between different island cultures, it is no surprise that people have generated shared forms of intercultural communication. For example, several forms of **Pidgin English** (also known simply as "Pijin") are found in the Solomons, Vanuatu, and New Guinea, where it is the major language used between ethnic groups. In Pijin, a largely English vocabulary is reworked and blended with Melanesian grammar. Conventional wisdom traces its origin to nineteenth-century Chinese sandalwood traders (*pijin* is the Chinese pronunciation of the word for "business"). Pijin is now becoming a "globalized" language of sorts in Oceania as trade and even political ties develop between different indigenous island groups.

A tidal wave of outside influences has engulfed the Pacific realm since World War II, producing further cultural changes as well as growing indigenous responses designed to preserve traditional values. Some groups, particularly in Melanesia, remain more isolated from the outside world, although even there growing demands for natural resources offer an avenue for increasing Western or Asian contacts. In many settings, however, the global growth of tourism has brought Oceania into the relatively easy reach of wealthier Europeans, North Americans, Asians, and Australians. The Hawaiian Islands, Fiji, French Polynesia, and American Samoa are being joined by an increasing number of other island tourist destinations. While offering tremendous economic benefits to certain localities, the onrush of tourists and their consumer-driven values have often come into sharp conflict with native cultures. Change is often seen in generational terms: grandparents may recall simpler times of more limited outside contact; parents might work long, hard hours at resort facilities or souvenir shops; and youngsters find themselves surrounded by fast-food restaurants and the glittery attractions of television.

Increasingly, however, selected native groups resist these pressures of cultural globalization. In Melanesia, a growing number of countries have joined the "Spearhead Group" of nations dedicated to preserving cultural traditions. For example, to keep native dialects from dying, many primary schools in Papua New Guinea now teach local languages to students in early grades before switching to Pidgin English thereafter. Even more dramatically, some native Hawaiians, paralleling indigenous movements elsewhere in the world, are advocating the return of one or more of the major islands to Polynesian control, thus preserving their cultural heritage.

GEOPOLITICAL FRAMEWORK: A Land of Fluid Boundaries

Pacific geopolitics reflect a complex interplay of local, colonial-era, and global-scale forces. Each political unit within the region reflects how these three types of forces have played out in particular places. The complexities become apparent in the story of Micronesia's Marshall Islands. This sprinkling of islands and atolls (covering 70 square miles, or 180 square kilometers, of land) historically held varied ethnic groups in essentially local political units before they were loosely incorporated into the Spanish and then German empires by the end of the nineteenth century. In 1914, the Japanese moved into the islands, and the area remained under their control until 1944, when U.S. troops occupied the region.

Following World War II, a United Nations trust territory (administered by the United States) was created across a wide swath of Micronesia, including the Marshall group. Demands for local autonomy grew during the 1960s and 1970s, resulting in a new constitution and independence for the Marshall Islanders by the early 1990s. Today, still benefiting from U.S. aid, government officials in the modest capital city on Majuro Atoll struggle to unite island populations, protect sprawling maritime sea claims, and adjudicate a generation of legal and medical problems that grew from U.S. nuclear bomb testing in the region (Figure 14.31). Such narratives are typical across the Pacific, suggesting a twenty-first-century political geography that is still very much in the making.

Shifting Geopolitical Patterns

Geopolitical space across Australia and Oceania has been reconfigured many times as different cultural groups and political powers have asserted themselves across the region. Present patterns of political organization are merely a snapshot in time and perhaps more prone to be redefined on future maps than in most areas of the world.

Indigenous Patterns Prior to European contact, the region's political geography was an intricate and ever-changing mosaic of indigenous territories. The indigenous political forms of Australia and Oceania varied widely. Aboriginal Australian society was organized around fluid bands of 30 to 60 people, most of whom were related by blood or marriage. Thus, the periodic movements of hunters and gatherers defined the shifting political space of the Aboriginal world. Melanesia was oriented around more sizable and sedentary communities, also based largely on kinship ties. In Micronesia, Polynesia, and some coastal sections of Melanesia, however, larger and more highly structured chiefdoms evolved. In Palau, for example, the 10 contemporary states of the main island of Babeldaob are based upon traditional village territories that were organized by chiefdoms. These villages were, in turn, given different rankings based upon the power of their chief. Archaeological and ethnographic

Figure 14.31 Marshall Islands
The political control of Micronesia has shifted numerous times during the last two centuries. While now part of the independent Marshall Islands, these palm-fringed beaches were once claimed by Spanish, German, Japanese, and U.S. interests. *(Douglas Peebles Photography)*

evidence suggests that a village's ranking affected its right to exploit natural resources, as well as its trade relations.

Centralized kingdoms in which power was concentrated in the hands of a single monarch existed on several of the larger volcanic islands of Polynesia. The most powerful of these kingdoms was based in Hawaii, which eventually encompassed all of the islands of the archipelago. While the days of ruling Hawaiian royalty have long since passed, these indigenous, frequently tribal affiliations still help native peoples define their political worlds, an orientation that sometimes runs counter to the boundaries of the region's modern states.

An Imposed Colonial Framework The European world turned the native political pattern on its head, replacing the nuanced, often fluid territorial boundaries of the indigenous world with more precise yet unstable colonial borders. Great Britain claimed and colonized both Australia and New Zealand. By the 1860s, the various colonial footholds in Australia, focused on widely separated coastal settlements, were formally delineated by the simple geometrical boundaries still apparent today. These separate administrative units functioned as distinctive colonial entities until they joined together in the formal Confederation of Australia in 1901 as an essentially independent commonwealth. New Zealand followed a similar path. Early on, isolated settlement clusters dotted the shoreline, marking the first tentative imprints of European political authority upon the Maori-controlled islands. After the Maori were conquered in the late nineteenth century, both islands were united in a formal and independent dominion under the British Crown in 1907.

The global reach of European colonialism also extended deep into the South Pacific during the nineteenth century. Interest in the area began to grow as early as 1842, when France claimed a "protectorate" over Tahiti. Most of Oceania was not formally colonized, however, until the end of the century, the same period that witnessed the scramble for African territory. By then, relatively few areas of the world were still open for annexation, and colonial powers were eager to acquire the parts that remained before their rivals could. Certain oceanic islands were also seen as particularly desirable because they could serve as coaling supply points for steamships and act as relay stations for submarine telegraph cables.

Germany played a major role in the division of Oceania, just as it did in the partition of Africa. German imperialists viewed New Guinea and the larger islands of the nearby Bismarck Archipelago as the main prize. Other colonial powers opposed the German initiatives, however. By 1885, New Guinea was divided into three territories: the Dutch got the western half (near their empire in Indonesia); the British received the southeastern quarter (near Australia); and the Germans acquired the northeastern quarter, along with the Bismarck Archipelago and a portion of the nearby Solomon Islands chain. Portions of Micronesia and Polynesian Samoa rounded out Germany's colonial holdings.

France, Great Britain, and the United States also had major territorial ambitions in Oceania. By the end of the nineteenth century, virtually all of eastern Polynesia lay under French control. Their domain included the Society Islands, the Marquesas Islands, and the Tuamotu Archipelago. Farther west, France colonized New Caledonia in Melanesia and began to move into the nearby New Hebrides Islands (now Vanuatu).

The British, however, having just acquired nearby territories to the north, were also interested in the archipelago. Rather than dividing it or fighting over it, the French and British simply agreed to an unprecedented "condominium" in the New Hebrides under which they shared power. Britain gained additional islands in the region, including the southern Solomon chain, the Fiji Islands, many of the smaller islands of western Polynesia, and a few outposts in eastern Polynesia (Pitcairn). Although most of these acquisitions became direct territories of the British Crown, Tonga (in western Polynesia) remained a protectorate with considerable local autonomy. Indeed,

Tonga was the only indigenous state of Oceania to retain its own political structure, and it remains a (constitutional) monarchy to this day.

For the United States, Hawaii was the main acquisition, but eastern Samoa also came under U.S. rule in 1900 as a result of an agreement signed with Britain and Germany. The United States also gained Guam, the largest island in Micronesia, as a result of its victory over Spain in the Spanish-American War.

Wars and treaty agreements among colonizing powers further changed Oceania's patterns of political geography. For example, Germany lost all of its foreign territories after World War I: those in Micronesia passed to Japan; those in Melanesia were ceded to Australia; and those in Polynesia went to New Zealand. Japan's gains proved crucial during World War II, as they provided the Japanese military with naval and air bases over a huge swath of the Pacific.

After World War II, however, Japan's Micronesian empire evaporated, passing to the United States under UN auspices as the Trust Territory of the Pacific. Meanwhile, following World War I, Australia inherited the British territory of southeastern New Guinea, to which it added the German territories in northeastern New Guinea and the Bismarck Archipelago. At this time the interior reaches of New Guinea were essentially unexplored, and the Australian government had little idea of what its large new colony actually contained. In the 1920s, however, gold was discovered, and by the 1930s Australian prospectors had penetrated the highlands. There they were shocked to discover broad highland plateaus inhabited by roughly a million people. The New Guinea Highlanders, for their part, were probably even more shocked at the intrusive habits and new technologies of the light-skinned interlopers who set about establishing Australian rule over the entire area.

Roads to Independence The modern political states of the region have arrived at independence along many different paths. But other political units remain colonial entities to this day. The newness and fluidity of the political boundaries are remarkable: the region's oldest independent states are Australia and New Zealand, and both were twentieth-century creations that are only now pondering the desirability of completing their formal political separation from the British Crown. Elsewhere, political ties between colony and mother country are even more intimate and enduring. Even many of the newly independent Pacific **microstates**, with their tiny overall land areas, retain special political and economic ties to countries such as the United States.

Independent Australia (1901) and New Zealand (1907) only gradually created their own political identities and still struggle with their ultimate political configuration. Although Australia became a commonwealth in 1901, it still acknowledges the British Crown as its sovereign (Figure 14.32). The Crown's role is strictly symbolic, and a national referendum in 1999 forced Australians to ponder whether they would like their country to drop this remaining tie to Britain and instead become a genuine republic with its own president replacing the British queen as head of state. However, a majority of 55 percent voted to retain Australia's ties to the Crown. Australia, like the United States, is a federal country, with each of its six states retaining substantial powers. The Northern Territory, however, remains directly under the authority of the central government, although it does enjoy limited self-rule. In New Zealand, formal legislative links with Great Britain were not severed until 1947, and in 1994 some New Zealand officials began discussing the same formal break with the British Crown being debated by the Australians.

Figure 14.32 Britain's Queen in Australia
By the voters' choice, Australia retains its links to the British Crown as their sovereign. Despite governmental action to make Australia a fully independent republic with its own president—instead of the queen—as head of state, in a national referendum in 1999, voters chose to retain traditional ties to the British Crown. Here, Queen Elizabeth II visits Aboriginal dancers at the Muda Aboriginal Language and Culture Centre in the Australian Outback. *(AP/Wide World Photos)*

Elsewhere in the Pacific, colonial ties were severed even more slowly, and the process has not yet been completed. In the 1970s, Britain and Australia began relinquishing their colonial empires in the Pacific. Fiji (Great Britain) gained independence in 1970, followed by Papua New Guinea (Australia) in 1975 and the Solomon Islands (Great Britain) in 1978. The small island nations of Kiribati and Tuvalu (Great Britain) also became independent in the late 1970s, even though some observers argued that they did not have adequate populations or territorial bases to be viable states.

The United States has recently turned over most of its Micronesian territories to local governments, while retaining considerable influence in the area. After gaining these islands from Japan in the 1940s, the U.S. government supplied large subsidies to islanders but also utilized a number of islands for military purposes. Bikini Atoll was obliterated by nuclear tests, and the large lagoon of Kwajalein Atoll was used as a giant missile target. A major naval base, moreover, was established in Palau, the westernmost archipelago of Oceania. By the early 1990s, both the Marshall Islands and the Federated States of Micronesia (comprising the Caroline Islands) had gained independence. Their ties to the United States, however, remain close. A number of other Pacific islands remain under loose U.S. sovereignty. Palau is an American "trust territory," although the Palauans have some local autonomy. The people of the Northern Marianas chose to become a "self-governing commonwealth in association with the United States," a rather ambiguous political position that allows them to become U.S. citizens. The residents of self-governing Guam and American Samoa also are U.S. citizens. Hawaii became a full-fledged U.S. state in 1959 and thus an integral part of the United States.

However, France and New Zealand seem less inclined to relinquish their oceanic possessions. New Zealand, for example, still controls substantial territories in Polynesia, including the Cook Islands, Tokelau, and the island of Niue. France has even more extensive holdings in the region. Its largest maritime possession is French Polynesia, encompassing a vast expanse of mid-Pacific territory. To the west, it retains the much smaller territory of Wallis and Futuna in Polynesia and the larger island of New Caledonia in Melanesia.

Persisting Geopolitical Tensions

Cultural diversity, colonial legacy, youthful states, and a rapidly changing political map contribute to continuing geopolitical tensions within the Pacific world (Figure 14.33). Indeed, some of these conflicts have consequences that extend far beyond the boundaries of the region. Others are more locally based but are still vivid

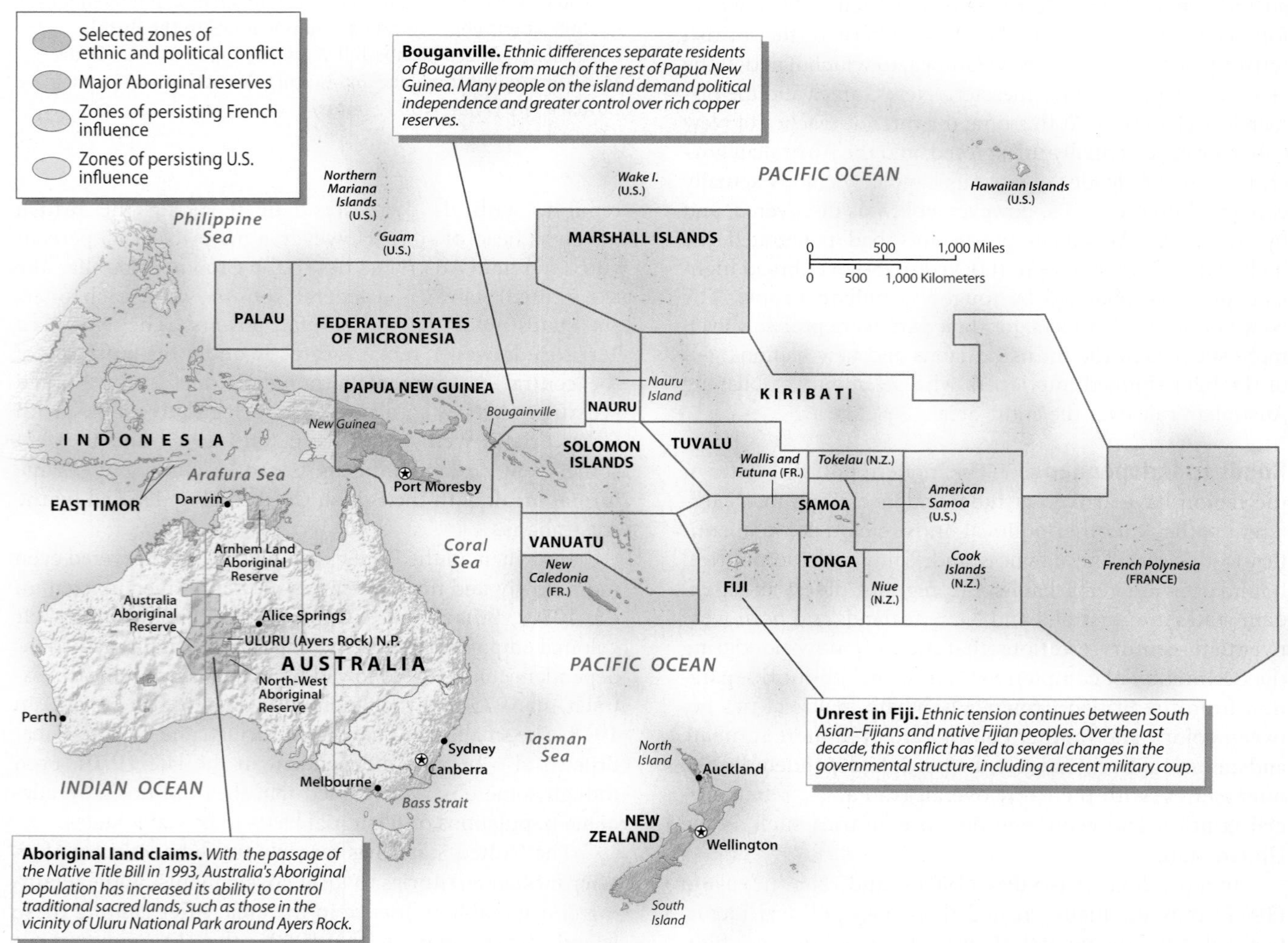

Figure 14.33 Geopolitical Issues in Australia and Oceania
Native land claim issues increasingly shape domestic politics in Australia and New Zealand. Elsewhere, ethnic conflicts have raised political tensions in settings such as Fiji and Papua New Guinea. Colonialism's impact endures as well: American and French interests remain particularly visible in the region, including a legacy of nuclear testing that continues to impact selected Pacific Island populations.

Figure 14.34 Aboriginals at Uluru National Park
Australian Aborigines gathered recently at Uluru National Park to celebrate their increased political control over the region. Since then, the Native Title Bill has promoted numerous land cessions and further legal settlements. *(Michael Jensen/Auscape International Pty. Ltd)*

reminders of the complexities involved as political space is radically redefined and refashioned across varied natural and cultural settings.

Native Rights in Australia and New Zealand Indigenous peoples in both Australia and New Zealand have used the political process to gain more control over land and resources in their two countries. Indeed, the strategies these native groups have used parallel efforts in the Americas and elsewhere. In Australia, Aboriginal groups are discovering newfound political power from both more effective lobbying efforts by native groups and a more sympathetic federal government. Since land treaties were generally not signed with Aborigines as whites conquered the continent, native peoples originally had no legal land rights whatsoever.

More recently, the Australian government established a number of Aboriginal reserves, particularly in the Northern Territory, and expanded Aboriginal control over sacred national parklands such as Uluru (Ayers Rock) (Figure 14.34). Further concessions to indigenous groups were made in 1993 as the government passed the **Native Title Bill**, which compensated Aborigines for already ceded lands, gave them the right to gain title to unclaimed lands they still occupied, and legally empowered them in dealings with mining companies in native-settled areas. Efforts to expand Aboriginal land rights have met vigorous opposition. Nevertheless, in 1996, an Australian court ruled that pastoral leases (the form of land tenure held by the cattle and sheep ranchers who control most of the Outback) do not take precedence over Aboriginal land rights (Figure 14.35). Grazing interests were infuriated, which led the government to weaken the court's ruling by declaring that Aboriginal claims might allow the visiting of sacred sites and some hunting and gathering, but not sustained political or economic control.

In New Zealand, Maori land claims have generated similar controversies in recent years. The Maori constitute a far larger proportion of the overall population, and the lands that they claim tend to be much more valuable, further complicating the issue. Recent protests include

Figure 14.35 Applications for Native Land Claims in Australia
This map shows the applications for native land claims in Australia filed by different Aboriginal groups as of 2004. Important to note is that these are applications only—not government-approved claims. Nevertheless, the widespread extent of the claims shows why the topic is so contentious and controversial. *(National Native Title Tribunal, Commonwealth of Australia)*

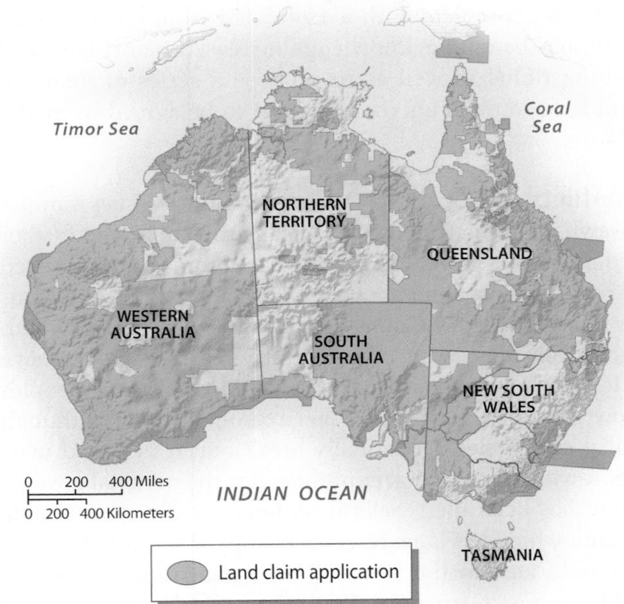

GEOGRAPHY IN THE MAKING

The Political and Economic Implications of Australia–China Trade

Until recently, the Australian economy was so dependent on U.S. trade that it was said only half-jokingly, "When America sneezes, Australia catches pneumonia." But this is no longer true since China and India have replaced America as Australia's major trading partners. By late 2007, with the U.S. economy staggering from huge trade deficits, an emaciated dollar, falling housing prices, and the mortgage meltdown, it's understandable that Australians have embraced these new economic ties with Asia.

However, there is considerable unease in political circles about these new allegiances because the political implications may be profound. Bluntly stated, China has not been shy about using its economic clout to seek political support for its claim on Taiwan. Until now, Australia has been one of the United State's staunchest allies in defending Taiwan's independence from China in what is referred to as the "two Chinas" policy. However, some members of the Australian government are not sure their country can continue trade with China while supporting an independent Taiwan. Their concerns are heightened as they hear Chinese officials emphasize that trade talks would be "delayed" with countries that support Taiwan. Whether China will force the issue with Australia remains to be seen.

Chinese diplomats have recently invited Australian politicians to help find a solution to the "Taiwan problem." These overtures have ominous overtones, for if Australia chooses to stand by the United States and support a separate and independent Taiwan, trade with China may suffer. On the other hand, China's rise as the customer of choice in Asia and Oceania will likely diminish the influence in those regions of other industrial nations, particularly the United States.

At the moment, though, Australian merchants and manufacturers cannot seem to work fast enough to satisfy the hunger of their biggest new customer. Both imports and exports between China and Australia have grown remarkably in the last decade. In 1990, for example, exports from China to Australia amounted to about $10 billion. By 2003, they had grown to 10 times that amount and have expanded substantially in the last five years. In the same time period, exports from Australia to China grew from $10 billion to $50 billion. Most of these exports are basic resources such as iron ore and natural gas from western and northern parts of the country. Not only are Australian firms shipping these resources to China, but now Chinese companies have been invited to invest in Australian mining firms so that production can be expanded.

periodic civic disobedience, growing Maori land claims over much of North and South islands, and a call to return the country's name to the indigenous **Aotearoa**, "Land of the Long White Cloud." The government response, complete with a 1995 visit from Queen Elizabeth, has been to acknowledge increased Maori land and fishing rights as well as to propose a series of financial and land settlements that have yet to be agreed to by the Maoris.

Conflicts in Oceania Other geopolitical issues simmer elsewhere in the Pacific, periodically threatening to further redefine the region's fluid territorial boundaries. Discussed earlier in this chapter are the cultural and political tensions in Fiji. Papua New Guinea (PNG) must also contend with ethnic tensions. Many of the country's cultural groups have a long history of mutual animosity. Most of these peoples now get along with each other reasonably well, although tribal skirmishes occasionally break out in highland market towns. A much bigger problem for the national government has been the rebellion on Bougainville. This sizable island, which has large reserves of copper and other minerals, is located in the Solomon archipelago but belongs to PNG because Germany colonized it in the late 1800s and it thus became politically attached to the eastern portion of New Guinea. Many of Bougainville's indigenous inhabitants believe that their resources are being exploited by foreign interests and by an unsympathetic national government, and they demand local control. Papua New Guinea has reacted with military force; about 5 percent of the island's entire population has already been killed in the conflict, and recent efforts have failed to create a more stable regional government.

The continued French colonial presence within the Pacific region has also created political uncertainties, both in relations with native peoples and between the French and other independent states within the region. Continued French rule in New Caledonia has provoked much local opposition. This large island has substantial mineral reserves (especially of nickel) and sizable numbers of French colonists. French settlement and exploitation of mineral resources angered many indigenous inhabitants. By the 1980s, a local independence movement was gaining strength, but in 1987 and 1998 the island's inhabitants (indigenous and French immigrants alike) voted to remain under French rule, at least until 2018. Still, an underground independence movement continues to operate.

Troubles also have reverberated in French Polynesia. Although the region receives substantial subsidies from the French and residents have voted to remain a colony, a large minority of the population opposes French control and demands independence. The independence movement was greatly strengthened in 1995 when France decided to resume nuclear testing in the Tuamotu Archipelago. Those activities angered not only the people of French Polynesia (there was antigovernment rioting in Tahiti), but also those of other Pacific countries and territories. Australian leaders called the tests "an act of stupidity"; New Zealand recalled its ambassador in protest; and other Pacific nations from Japan to Chile registered their disapproval. Future French nuclear policies in the Pacific will undoubtedly continue to impact geopolitical relations far beyond Polynesia.

ECONOMIC AND SOCIAL DEVELOPMENT: A Difficult Path to Paradise

As with all world regions, the Pacific realm contains a diversity of economic situations resulting in both wealth and poverty. Even within affluent Australia and New Zealand, for example, there are also pockets of pronounced poverty. Further, great economic disparities also exist between those countries with numerous trade ties and those small island nations lacking resources and external trade. While tourism offers some relief from abject poverty, the whims and fashions of foreign tourists can be fickle, resulting in a Pacific version of boom and bust economies. As result, the economic future of the Pacific realm remains uncertain because of its own small domestic markets, its peripheral position in the global economy, and a diminishing resource base.

Uncertain Avenues to Affluence

The per capita gross national incomes (PPP) of both Australia ($30,610) and New Zealand ($23,030) exemplify living standards well above the global norm. Indeed, these two nations are generally grouped among the world's developed countries in company with much of Europe, North America, and Japan. Their many economic assets include highly educated populations, a diverse base of natural resources, and a modern, highly integrated urban and industrial infrastructure. Even so, the relative economic affluence of these two South Pacific nations declined late in the twentieth century due to slower economic growth that was still heavily dependent on the extraction and export of raw materials. This trend has only recently been reversed with the blossoming of trade ties with China, a country that seems to have an insatiable appetite for raw materials such as lumber, coal, and iron ore.

The Australian Economy Much of Australia's economic affluence has been built upon the cheap extraction and export of abundant raw materials. Export-oriented agriculture, for example, has long been one of the key supports of Australia's economy. Australian agriculture is highly productive in terms of labor input, and it produces a wide variety of both temperate and tropical crops, as well as huge quantities of beef and wool for world markets. While farm exports are still important to the economy, the mining sector has grown much more rapidly in the last several decades. Today, Australia is one of the world's mining superpowers. Among Australia's many assets are coal; rich reserves of iron ore, particularly in Western Australia; and an assortment of other metals, such as bauxite (for aluminum), copper, gold, nickel, lead, and zinc. Indeed, the New South Wales–based Broken Hill Proprietary Company (BHP) is one of the world's largest mining corporations. Given the country's small population, much of that mineral base is consumed elsewhere, making Australia a huge source for raw materials for developed and developing Asian nations, particularly China and Japan.

Although the nation's mineral wealth served it well in the recent boom years, Australia's manufacturing sector continues to be a concern. Many of its industries are oriented around the simple processing of raw materials, but often these are not highly profitable operations. At the same time, small domestic markets discourage the blossoming of more specialized industries. Australia has lacked an entrepreneurial edge in the fast-growing high-technology and information-based industries that have powered recent economic expansion in North America, Japan, and Europe. But that may be changing: recent government policies have encouraged more domestic investment, higher savings rates, and more rapid economic growth. Growing numbers of Asian immigrants and economic links with potential Asian markets also bode well for the future. In addition, an expanding tourism industry is helping to diversify the economy. More than 7 percent of the nation's workforce is now devoted to serving the needs of more than 4 million visitors annually. Popular destinations include Melbourne and Sydney, as well as Queensland's resort-filled Gold Coast, the Great Barrier Reef, and the vast arid Outback. Along the Gold Coast, most luxury hotels are owned by Japanese firms and provide a bilingual resort experience for their Asian clientele (Figure 14.36).

Australia's affluence is distributed widely but unevenly around the country. As is typical in the developed world, much of the wealth is concentrated in the major cities, especially those of the southeast. The high-income suburbs of Melbourne and Sydney resemble those of Boston and Vancouver. Overall, however, no Australian state is significantly richer or poorer than any other, and the disparity of income between the most prosperous fifth of the population and the least prosperous fifth is relatively small in global terms. The Aborigine population, however, is mired in poverty. Whether living in cities or

Figure 14.36 Queensland's Gold Coast
Many of these luxury hotels in the Surfer's Paradise section of the Gold Coast are owned by Japanese firms specializing in accommodations for Asian tourists. *(Peter Hendrie/Getty Images Inc.—Image Bank)*

in the Outback, most Aborigines have low standards of living, and their incomes average only 65 percent of the national average. Although a small number of Aborigines still subsist off the land in the large reserves, most occupy a marginal place in the cash economy of the cities.

New Zealand's Economic Challenge New Zealand is also a wealthy country, but somewhat less prosperous than Australia. Before 1970, New Zealand relied heavily on exports to Great Britain, mostly agricultural products such as wool and butter. This relationship faltered, however, after Britain joined the European Union, which then adopted stringent agricultural protection policies. Unlike Australia, New Zealand lacked a rich base of mineral resources to export to global markets. By the 1980s, the country had slipped into a serious recession. Eventually, the New Zealand government enacted drastic reforms. The country had previously been noted for its lofty taxes, high levels of social welfare, and state ownership of large economic concerns. Suddenly "privatization" became the watchword, and most state industries were sold off to private parties. As a result, New Zealand has been transformed into one of the most market-oriented countries of the world. To diversify from its traditional export base, the nation has also promoted more aggressive development of its timber resources, fisheries, and tourist industry.

Although New Zealand's economy has boomed in the last two decades because of new privatization policies, its economic prospects remain uncertain. Critics also worry that growing income disparities will undermine New Zealand's social foundation. While suburbanites in the north benefit from some of the country's economic reforms, the country's hinterland populations, still often bound to the land, continue to struggle as global prices for their products decline. In addition, the Maori population suffers some of the same economic problems as Australia's Aborigines.

Oceania's Economic Diversity Varied economic activities shape the Pacific Island nations. One way of life is oriented around subsistence-based economies, such as shifting cultivation or fishing. In other places, the commercial extractive economy dominates, with large-scale plantations, mines, and timber activities often competing for land and labor with the traditional subsistence sector. Elsewhere, the tremendous growth in global tourism has transformed the economic geographies of many islands, forever changing the way that people make a living across much of the vast oceanic region. Many island nations also benefit from direct subsidies and economic assistance that flow from present and former colonial powers, money designed to promote development and stimulate employment.

Melanesia is, by most measures, the least-developed and poorest part of Oceania. Melanesian countries have benefited less from tourism or from subsidies from wealthy colonial and ex-colonial powers. Most Melanesians live in remote villages that remain somewhat isolated from the modern economy. The Solomon Islands, for example, with few industries other than fish canning and coconut processing, has a per capita GNI (PPP) of only $1,880 per year. Papua New Guinea's economy is somewhat more commercially oriented, supporting a per capita GNI (PPP) of $2,370. Outbound shipments of coconut products and coffee have increasingly been supplemented with forest products such as tropical hardwoods. Gold- and copper-mining ventures have also dramatically transformed the landscape, although political instability has intermittently suspended mineral production in settings such as Bougainville. Between sugar and tourism, Fiji remains the most prosperous Melanesian country with a per capita GNI (PPP) of almost $5,960. Among the smaller islands of Melanesia and Micronesia, mining economies dominate New Caledonia and Nauru. New Caledonia's nickel reserves, the world's second largest, are both a blessing and a curse: they sustain much of the island's export economy, yet they inevitably will dwindle in the future. Dramatic price fluctuations for the industrial economy also hamper economic planning for the French colony. Other activities include coffee growing, cattle grazing, and tourism. To the north, the tiny, phosphate-rich island of Nauru also depends on mining. The citizens of Nauru, for the most part, live directly off the royalties they receive from the mines. Much of the money gained in mining has been invested in a massive trust fund, and the Nauruan government has assured its citizens that they will be prosperous even after the phosphate deposits are exhausted. However, a good deal of the trust fund money found its way into the Asian real estate market in the 1990s, and these investments have proved rather risky. Should the trust fund money be lost, the Nauruans will end up with little more than an environmentally devastated island.

Elsewhere in Micronesia and Polynesia, economic conditions depend upon the viability of local subsistence economies or economic linkages to the world beyond. Many archipelagos export a few food products, but native populations survive mainly on fish, coconuts, bananas, and yams. Some island groups, however, enjoy substantial subsidies from either France or the United States. Change is also afoot in some of these islands: in 1999, Japan agreed to build a spaceport for its future shuttlecraft on Micronesia's Christmas Island (Kiribati). However, since Japan recently cancelled its plans for its own space shuttle, the future of this spaceport is questionable. Related, the Marshall Islands are the site of a planned industrial park financed by China.

Tourism has fundamentally transformed some island groups. In Hawaii, for example, more than one-third of the state's economy flows directly from tourist dollars. With almost 7 million visitors annually (including more than 1.5 million from Japan alone), Hawaii exemplifies all of the classic benefits and risks of the tourist economy. While job creation and economic growth have reshaped this island realm, congested highways, high prices, and the unpredictable spending habits of tourists have left the region vulnerable to future problems.

Elsewhere, French Polynesia has long been a favored destination of the international jet set (Figure 14.37). About a quarter of French Polynesia's GNI is derived from tourism, making it one of the most prosperous parts of the Pacific. More recently, Guam has emerged as a favorite destination of Japanese and Korean tourists, especially those on honeymoons. Indeed, on a smaller scale, tourism is on the rise across much of the island realm, and many economic planners see it as the avenue to future prosperity. Critics, however, warn that tourist jobs tend to be low-paying (outsiders are often imported for managerial positions); the local quality of life may in fact decline with the presence of tourists; and the natural environments of small islands can quickly be overwhelmed by an influx of demanding visitors.

The Global Economic Setting

Even as Australia and Oceania remain on the margins of global economic activity, their relationships with the world economy will increasingly shape the quality of life and the prospects for development within the region. Several intriguing questions remain. Will future trade patterns in the region shift away from Europe and North America in favor of closer links with Asia, China in particular? Will there be a move away from the traditional extractive economies that have shaped economic development in the region everywhere from giant Australia to tiny Nauru? Can growing economic linkages within the region make a difference in a part of the world where the total population is less than that of California? None of these questions have ready answers, but they suggest how residents of the region will need to ponder their entry into the twenty-first century's global economy.

Many international trade flows link the area to the far reaches of the Pacific and beyond. Australia and New Zealand dominate global trade patterns in the region. In the past 30 years, ties to Great Britain, the British Commonwealth, and Europe have weakened in comparison with growing trade links with Japan, East Asia, the Middle East, and the United States. Australia, for example, now imports more manufactured goods from Japan, China, and the United States than it does from Britain and Europe (Figure 14.38). Similarly, Australian exports, principally raw materials, increasingly flow toward destinations in Asia and North America, although Europe remains an important trading partner. Other global economic ties have come in the form of capital investment in the region. U.S. and Japanese banks and other financial institutions now dot the South Pacific landscape from Sydney to Suva. Both Australia and New Zealand also participate in the **Asia-Pacific Economic Cooperation Group (APEC)**, an

Figure 14.37 Tahitian Resort
Luxury resort settings in Tahiti (near Papeete) exemplify a growth industry in the South Pacific. While bringing important investment capital, such ventures reorder the region's social and economic structure as well as refashion its cultural landscape. *(Bob & Suzanne Clemenz)*

organization designed to foster economic development in Southeast Asia and the Pacific Basin.

Governments have also promoted economic integration within the region. In 1982, Australia and New Zealand signed the **Closer Economic Relationship (CER) Agreement**, which successfully slashed trade barriers between the two countries. New Zealand benefited from the opening of larger Australian markets to New Zealand exports, and Australian corporate and financial interests gained new access to New Zealand business opportunities. Since CER's signing, trade between the two countries has expanded almost 10 percent per year. Today, more than 20 percent of New Zealand's imports and exports come from Australia, and the pattern of regional free trade is likely to strengthen in the future. Smaller nations of Oceania, while often closely tied to countries such as Japan, the United States, and France, also benefit from their proximity to Australia and New Zealand. More than half of Fiji's imports come from those two nearby nations, and other countries, such as Papua New Guinea, Vanuatu, and the Solomon Islands, enjoy a similarly close trading relationship with their more-developed Pacific neighbors.

Enduring Social Challenges

Australians and New Zealanders enjoy high levels of social welfare but face some of the same challenges evident elsewhere in the developed world (Table 14.2). Lifespans average more than 80 years in both countries, and rates of child mortality have fallen substantially since 1960. Paralleling patterns in North America and Europe, cancer and heart disease are leading causes of death, and alcoholism is a persisting social problem, particularly in Australia. Unfortunately, Australia's rate of skin cancer is among the world's highest, the result of having a largely fair-skinned, outdoors-oriented population from northwest Europe in a sunny, low-latitude setting. Overall, Australia's Medicare program (initiated in 1984) and New Zealand's targeted system of social services provide high-quality health care to their populations. The position of women is also high in both countries, including participation in the workforce. Women have recently played key political roles in New Zealand, in particular.

Figure 14.38 Australia's Trade with China
Containers from Asia testify to the recent explosion of two-way trade between Australia and China. While raw materials, mainly iron ore, are exported to China, consumer goods flow from China into Australia, making that country a key beneficiary of China's recent economic growth. *(AFP Photo/Torsten Blackwood)*

Not surprisingly, the social conditions of the Aborigines and Maoris are much less favorable than those of the population overall. Schooling is irregular for many indigenous peoples, and levels of advanced postsecondary education for Aborigines (12 percent) and Maoris (14 percent) remain far below national averages (32 to 34 percent) in the two countries. Many other social measures reflect the pattern, as well. For example, fewer than one-third of Aboriginal households own their own homes, while more than 70 percent of white Australian households are homeowners. Furthermore, considerable discrimination against native peoples persists in both countries, a situation that has been aggravated and more publicized with the recent assertion of indigenous political rights and land claims. As with North American African American, Hispanic, and Native American populations, no simple social policies offer solutions to these enduring inequities.

Levels of social welfare in Oceania are higher than one might expect, based on the region's economic circumstances. Many of its countries and colonies have invested heavily in health and education services and have achieved considerable success. For example, the average life expectancy in the Solomon Islands, one of the world's poorer countries, is a respectable 62 years. By other social measures as well, the Solomon Islands and a number of other Oceania states have reached higher levels of human well-being than is the case in most Asian and African countries with similar levels of economic output. This is partly a result of successful policies, but it also reflects the relatively healthful natural environment of Oceania. Many of the debilitating tropical diseases so common in Africa simply do not exist within the region.

Papua New Guinea (PNG) is the major exception to these relatively high levels of social welfare. Here the average life expectancy is only 57 years, and a recent study suggests that 34 percent of its young people suffer from malnutrition, particularly protein deficiencies. Adult illiteracy (35 percent) is also more prevalent in PNG than elsewhere in the region. Unfortunately, PNG has found it difficult to provide even basic educational and health services to its people. The country possesses the largest expanse of land in Melanesia, and much of the population lives in relatively isolated villages in the rugged central highlands.

TABLE 14.2 Development Indicators

Country	GNI[a] per Capita, PPP[b] (2005)	GDP[c] Average Annual % Growth (2000–05)	Life Expectancy (2007)	Percent of Population Living on Less Than $2 a Day (2006)	Under Age 5 Mortality Rate 1990	Under Age 5 Mortality Rate 2005	Gender Equity[d]
Australia	30,610	3.2	81		10	6	102
Fed. States of Micronesia	2,300		67		23		
Fiji	5,960	4.1	68		20		
French Polynesia			74				
Guam			78				
Kiribati	1,210	1.8	62		65		
Marshall Islands	2,810	1.5	70		59		
Nauru			62				
New Caledonia			75				
New Zealand	23,030	3.7	80		11	6	113
Palau	7,120	2	71		27		
Papua New Guinea	2,370	1.6	57		103	93	87
Samoa	6,480	3.1	73		30		
Solomon Islands	1,880	5.5	62		56		
Tonga	8,040	4.3	71		25		
Tuvalu			64				
Vanuatu	3,170	2	67				

[a]*Gross national income.*
[b]*Purchasing power parity.*
[c]*Gross domestic product.*
[d]*Ratio of female to male enrollment in primary and secondary school percentage. Numbers below 100 have more males in primary/seconday school; numbers above 100 have more females in primary/secondary schools.*

Sources: World Bank, World Development Indicators, *2007; CIA World Factbook Website; percentage of population living on $2 a day data from Population Referece Bureau,* World Data Sheet, *2007; gender equity data from* Millennium Development Goals.

Summary

- The natural environment, which has been transformed over millennia by indigenous and colonial settlement, has witnessed accelerating change in the past 50 years as urbanization, tourism, extractive economic activities, exotic species, and climate change from global warming reconfigure the landscape and increase the vulnerability of island environments.
- Globalization has brought fresh interconnections to these island worlds: Japanese-financed golf courses pop up along tropical shores; Canadian and Chinese mining companies invest in the Australia uplands; and Korean newlyweds honeymoon beneath the coconut palms. How are native cultures being transformed in this process, and what new cultural hybrids will emerge as Pacific, European, North American, and Asian peoples mingle among the atolls and archipelagos of the South Pacific? These pivotal issues remain unresolved.
- The region's contemporary political geography reveals a still fluid character as countries struggle to disentangle themselves from colonial ties and assert their own political identities. Globalization complicates the process both culturally and politically: new people, many from Asia, are adding ethnic variety as well as providing flows of investment capital that cut across old colonial relationships.
- Australia remains dominant in the greater Pacific realm and should continue to play a pivotal role as a principal economic entry point into the region. Its large land area, resource base, and population are complemented by its increasing emergence as a regional finance center for the South Pacific, as well as nearby portions of Southeast and East Asia. Indeed, the Asian connection is playing a key part in Australia's future economic and even cultural identity. Asian migrants, many of them skilled and affluent, add a dynamic and creative component to Australia's largely European population. In addition, intermingled flows of money, raw materials, and manufactured goods also bind Australia to its Asian neighbors.
- If Australia sits at the edge of Asia, New Zealand might be viewed as sitting on the threshold of Polynesia. Today, its Maori population is growing quickly, joined by immigrants from other parts of Polynesia. Furthermore, New Zealand has taken an active role in the political affairs of the entire Pacific Basin, and it is in many respects Oceania's leading state.

Key Terms

Aborigine *(page 639)*
Aotearoa *(page 656)*
archipelagos *(page 627)*
Asia-Pacific Economic Cooperation Group (APEC) *(page 659)*
atoll *(page 632)*
Closer Economic Relationship (CER) Agreement *(page 660)*
haoles *(page 650)*
high islands *(page 631)*
hot spot *(page 631)*
kanakas *(page 646)*
low islands *(page 632)*
mallee *(page 628)*
Maori *(page 627)*
Melanesia *(page 626)*
Micronesia *(page 627)*
microstates *(page 653)*
Native Title Bill *(page 655)*
Oceania *(page 624)*
Outback *(page 626)*
Pidgin English (Pijin) *(page 651)*
Polynesia *(page 627)*
tsunamis *(page 630)*
uncontacted peoples (page 648)
viticulture *(page 642)*
White Australia Policy *(page 646)*

Questions for Review

1. Why does it make sense to group together Australia, New Zealand, and the other Pacific islands as a region? What complexities are involved in doing so?
2. What effects have nonnative plants and animals had on the environment of Australia and the Pacific islands?
3. What are the projected effects of global warming on Australia and Oceania? What measures are different countries taking to address these problems?
4. Describe how a Pacific high island is created and then transformed into a low island. How do these environments produce distinctive settings for human settlement?
5. What might be some notable similarities between Australian and North American cities?
6. Describe the types of farming you might observe in (a) South Australia, (b) the New Guinea Highlands, and (c) Fiji. Explain the differences, using both environmental and human variables in your discussion.
7. Discuss some of the key geopolitical issues faced by Pacific Islanders.
8. Why is there such a wide economic gap separating affluent Australia and New Zealand from the rest of Oceania?
9. How might it be argued that the social and environmental costs of tourism are even more important than the economic benefits?

Thinking Geographically

1. Should New Zealand welcome or be wary of Australian and Japanese investments in the country? Support your answer with specifics.
2. As a local economic development official in Papua New Guinea, argue both for and against a large new Malaysian-financed pulp mill being planned for your district.
3. Select a small Pacific island and briefly research its political history since 1800. What do your findings illustrate about more general processes of European colonization and political independence movements over the last 200 years?
4. Identify what you see as the three principal economic challenges facing Papua New Guinea. Discuss how the country's physical and human geography may affect its future prospects.
5. Does the continuing French colonial presence in the South Pacific help or hinder economic development within the region? Why?
6. As an Australian Aborigine in touch with your North American counterparts, what are some of the principal similarities and differences in the political and economic plight of your peoples?
7. It is 2042, and global warming has accelerated in the past 30 years. You are the ruler of a small Pacific island nation now being inundated by rising sea levels and altered climatic patterns. What is your strategy to survive, considering these new geographical circumstances?
8. As a maker of educational films, you have been hired to film a three-part series titled *The Essential South Pacific*. The episodes are to be set in (a) Australia, (b) New Zealand, and (c) Polynesia. Budgets are limited, and only a single location in each of those settings can be used to communicate the personality of the region. Identify three specific spots for filming, and explain your answers to the producer of the series.

Regional Novels and Films

Novels

Derek Hansen, *Sole Survivor* (1999, Simon and Schuster)

Keri Hulme, *The Bone People* (1985, Hodder and Stoughton)

James Michener, *Hawaii* (1959, Random House)

Silvia Watanabe, *Talking to the Dead* (1992, Doubleday)

Albert Wendt, *Leaves of the Banyan Tree* (1979, University of Hawaii)

Tim Winton, *Dirt Music* (2002, Scribner's)

Patrick White, *Riders in the Chariot* (1961, Viking Press)

B. Wongar, *Walg: A Novel of Australia* (1983, Dodd, Mead)

Films

The Adventures of Priscilla, Queen of the Desert (1994, Australia)

Crocodile Dundee (1986, U.S.)

Mauri (1988, New Zealand)

Once Were Warriors (1994, New Zealand)

Oscar and Lucinda (1997, U.S.)

The Piano (1993, Australia, filmed in New Zealand)

Picture Bride (1993, Hawaii, Japanese)

Rabbit-Proof Fence (2002, Australia)

Walkabout (1971, Australia)

Whalerider (2003, New Zealand)

Bibliography

Bambrick, Susan, ed. 1994. *The Cambridge Encyclopedia of Australia*. New York: Cambridge University Press.

Forster, Clive A. 1995. *Australian Cities: Continuity and Change*. Melbourne: Oxford University Press.

Head, Lesley. 2000. *Second Nature: The History and Implications of Australia as Aboriginal Landscape*. Syracuse, NY: Syracuse University Press.

Heathcote, R. L. 1994. *Australia*. New York: Wiley.

Kirch, Patrick V., and Hunt, Terry. 1997. *Historical Ecology in the Pacific Islands: Prehistoric Environmental and Landscape Change*. New Haven, CT: Yale University Press.

McKnight, Tom L. 1995. *Oceania: The Geography of Australia, New Zealand, and the Pacific Islands*. Upper Saddle River, NJ: Prentice Hall.

Pilger, John. 1991. *A Secret Country: The Hidden Australia*. New York: Knopf.

"Reborn: Melbourne." 1997. *The Economist*, August 23.

Rose, Deborah Bird. 1996. *Nourishing Terrains: Australian Aboriginal Views of Landscape and Wilderness*. Canberra: Australian Heritage Commission.

Stanley, David. 2000. *South Pacific Handbook*. 7th ed. Emeryville, CA: Avalon Travel Publishing.

Additional bibliographic references are at the *Diversity Amid Globalization* Website: http://www.prenhall.com/rowntree/.

Glossary

Aborigine An indigenous inhabitant of Australia.

acid rain Harmful form of precipitation high in sulfur and nitrogen oxides. Caused by industrial and auto emissions, acid rain damages aquatic and forest ecosystems in regions such as eastern North America and Europe.

African Union (AU) Founded in 1963, the organization grew to include all the states of the continent except South Africa, which finally was asked to join in 1994. In 2004 the body changed its name from the Organization of African Unity (OAU) to the African Union. It is mostly a political body that has tried to resolve regional conflicts.

agrarian reform A popular but controversial strategy to redistribute land to peasant farmers. Throughout the twentieth century, various states redistributed land from large estates or granted title from vast public lands in order to reallocate resources to the poor and stimulate development. Agrarian reform occurred in various forms, from awarding individual plots or communally held land to creating state-run collective farms.

agricultural density The number of farmers per unit of arable land. This figure indicates the number of people who directly depend upon agriculture, and it is an important indicator of population pressure in places where rural subsistence dominates.

alluvial fan A fan-shaped deposit of sediments dropped by a river or stream flowing out of a mountain range.

Altiplano The largest intermontane plateau in the Andes, which straddles Peru and Bolivia and ranges in elevation from 10,000 to 13,000 feet (3,000 to 4,000 meters).

altitudinal zonation The relationship between higher elevations, cooler temperatures, and changes in vegetation that result from the environmental lapse rate (averaging 3.5°F for every 1,000 feet [6.5°C for every 1,000 meters]). In Latin America, four general altitudinal zones exist: tierra caliente, tierra templada, tierra fria, and tierra helada.

animism A wide variety of tribal religions based on the worship of nature's spirits and human ancestors.

anthropogenic landscape A landscape heavily transformed by humans.

apartheid The policy of racial separateness that directed the separate residential and work spaces for white, blacks, coloureds, and Indians in South Africa for nearly 50 years. It was abolished when the African National Congress came to power in 1994.

archipelagos Island groups, often oriented in an elongated pattern.

areal differentiation The geographic description and explanation of spatial differences on Earth's surface; this includes physical as well as human patterns.

areal integration The geographic description and explanation of how places, landscapes, and regions are connected, interactive, and integrated with each other.

Asia-Pacific Economic Cooperation Group (APEC) An international group of Asian and Pacific Basin nations that fosters coordinated economic development within the region.

Association of Southeast Asian Nations (ASEAN) A supranational geopolitical group linking together the 10 different states of Southeast Asia.

asymmetrical warfare A term that aptly describes the differences between a superpower's military technology and strategy and the lower level technology and guerilla tactics used by Al Qaeda and the Taliban. Most superpower military strategists agree that the war on terrorism will be fought with this sort of battlefield asymmetry.

atoll Low, sandy islands made from coral, often oriented around a central lagoon.

autonomous areas Minor political subunits created in the former Soviet Union and designed to recognize the special status of minority groups within existing republics.

autonomous region In the context of China, provinces that have been granted a certain degree of political and cultural autonomy, or freedom from centralized authority, owing to the fact that they contain large numbers of non-Han Chinese people. Critics contend that they have little true autonomy.

Baikal-Amur Mainline (BAM) Railroad Key central Siberian railroad connection completed in the Soviet era (1984), which links the Yenisey and Amur rivers and parallels the Trans-Siberian Railroad.

Balfour Declaration Statement issued by Great Britain in 1917 pledging its support for establishing a home for the Jewish people in Palestine.

balkanization Geopolitical process of fragmentation of larger states into smaller ones through independence of smaller regions and ethnic groups. The term takes its name from the geopolitical fabric of the Balkan region.

barrios Urban Hispanic neighborhoods, often associated with low-income groups in North America.

Berlin Conference The 1884 conference that divided Africa into European colonial territories. The boundaries created in Berlin satisfied European ambition but ignored indigenous cultural affiliations. Many of Africa's civil conflicts can be traced to ill-conceived territorial divisions crafted in 1884.

biofuels Energy sources derived from plants or animals. Throughout the developing world, wood, charcoal, and dung are primary energy sources for cooking and heating.

biome Ecologically interactive flora and fauna adapted to a specific environment. Examples are deserts or tropical rain forests.

bioregion A spatial unit or region of local plants and animals adapted to a specific environment such as a tropical savanna.

Bodhisattva In the religion of Mahayana Buddhism, a spiritual being that helps others attain enlightenment.

Bolsheviks A faction within the Russian Communist movement led by Lenin that successfully took control of the country in 1917.

boreal forest Coniferous forest found in high-latitude or mountainous environments of the Northern Hemisphere.

brain drain Migration of the best-educated people from developing countries to developed nations where economic opportunities are greater.

British East India Company Private trade organization that acted as arm of colonial Britain—backed by the British army—in monopolizing trade in South Asia until 1857, when it was abolished and replaced by full governmental control.

bubble economy A highly inflated economy that cannot be sustained. Bubble economies usually result from rapid influx of international capital into a developing country.

buffer zone An array of nonaligned or friendly states that "buffer" a larger country from invasion. In Eurasia, keeping a buffer zone has been a long-term policy of Russia (and also of the former Soviet Union) to protect its western borders from European invasion.

Bumiputra The name given to native Malay (literally, "sons of the soil"), who are given preference for jobs and schooling by the Malaysian government.

Burakumin The indigenous outcast group of Japan, a people whose ancestors reputedly worked in leather-craft and other "polluting" industries.

capital leakage The gap between the gross receipts an industry (such as tourism) brings into a developing area and the amount of capital retained.

Caribbean Community and Common Market (CARICOM) A regional trade organization established in 1972 that includes former English colonies as its members.

caste system Complex division of South Asian society into different hierarchically ranked hereditary groups. Most explicit in Hindu society, but also found in other cultures to a lesser degree.

Central American Free Trade Association (CAFTA) A 2004 free trade agreement between the United States and five Central American countries—Guatemala, Nicaragua, El Salvador, Honduras, and Costa Rica—plus the Dominican Republic.

centralized economic planning An economic system in which the state sets production targets and controls the means of production.

central place theory A theory used to explain the distribution of cities, and the relationships between different cities, based on retail marketing.

centrifugal forces Those cultural and political forces, such as linguistic minorities, separatists, and fringe groups, that pull away from and weaken an existing nation-state.

centripetal forces Those cultural and political forces, such as a shared sense of history, a centralized economic structure, and the need for military security, that promote political unity in a nation-state.

chain migration A pattern of migration in which a sending area becomes linked to a particular destination, such as Dominicans with Queens, New York.

chernozem soils A Russian term for dark, fertile soil, often associated with grassland settings in southern Russia and Ukraine.

China proper The eastern half of the country of China where the Han Chinese form the dominant ethnic group. The vast majority of China's population is located in China proper.

circular migration Temporary labor migration in which an individual seeks short-term employment overseas, saves money, and then returns home.

clan A social unit that is typically smaller than a tribe or ethnic group but larger than a family, based on supposed descent from a common ancestor.

climate region A region of similar climatic conditions. An example would be the marine west-coast climate regions found on the west coasts of North America and Europe.

climograph Graph of average annual temperature and precipitation data by month and season.

Closer Economic Relationship (CER) Agreement An agreement signed in 1982 between Australia and New Zealand designed to eliminate all economic and trade barriers between the two countries.

Cold War The ideological struggle between the United States and the Soviet Union that was conducted between 1946 and 1991.

collective farms Group-farmed agricultural units organized around state-mandated production goals.

collectivization The agglomeration of small, privately owned agricultural parcels into larger, state-owned farms. This was a central component of communism in eastern Europe and the Soviet Union.

colonialism The formal, established (mainly historical) rule over local peoples by a larger imperialist government for the expansion of political and economic empire.

coloureds A racial category used throughout South Africa to define people of mixed European and African ancestry.

Columbian Exchange An exchange of people, diseases, plants, and animals between the Americas (New World) and Europe/Africa (Old World) initiated by the arrival of Christopher Columbus in 1492.

command economy Centrally planned and controlled economies, generally associated with socialist or communist countries, in which all goods and services, along with agricultural and industrial products, are strictly regulated. This was done during the Soviet era in both the Soviet Union and its eastern European satellites.

Commonwealth of Independent States (CIS) A loose political union of former Soviet republics (without the Baltic states) established in 1992 after the dissolution of the Soviet Union.

concentric zone model A simplified description of urban land use: a well-defined central business district (CBD) is surrounded by concentric zones of residential activity, with higher-income groups living on the urban periphery.

Confucianism The philosophical system developed by Confucius in the sixth century BCE.

connectivity The degree to which different locations are linked with one another through transportation and communication infrastructure.

continental climate Climate regions in continental interiors, removed from moderating oceanic influences, that are characterized by hot summers and cold winters. At least one month averages below freezing.

continentality Inland climates, removed from the ocean, with hot summers and cold winters such as those found in interior North America and Russia.

convection cells Large areas of slow-moving molten rock in Earth's interior that are responsible for moving tectonic plates.

convergent plate boundary Where two tectonic plates converge or are being forced together by convection cells. The Andes mountains, for example, are formed by the converging Nazca and South America plates.

copra Dried coconut meat.

core-periphery model According to this scheme, the United States, Canada, western Europe, and Japan constitute the global economic core, while other regions make up a less-developed economic periphery from which the core extracts resources.

Cossacks Highly mobile Slavic-speaking Christians of the southern Russian steppe who were pivotal in expanding Russian influence in sixteenth- and seventeenth-century Siberia.

Council for Mutual Economic Assistance (CMEA) The communist agency that coordinated economic planning and development between the Soviet Union and satellite countries in eastern Europe.

counterurbanization The movement of people out of metropolitan areas toward smaller towns and rural areas.

creolization The blending of African, European, and even some Amerindian cultural elements into the unique sociocultural systems found in the Caribbean.

crony capitalism A system in which close friends of a political leader are either legally or illegally given business advantages in return for their political support.

cultural assimilation The process in which immigrants are culturally absorbed into the larger host society.

cultural imperialism The active promotion of one cultural system over another, such as the implantation of a new language, school system, or bureaucracy. Historically, this has been primarily associated with European colonialism.

cultural landscape Primarily the visible and tangible expression of human settlement (house architecture, street patterns, field form, etc.), but also includes the intangible, value-laden aspects of a particular place and its association with a group of people.

cultural nationalism A process of protecting, either formally (with laws) or informally (with social values), the primacy of a certain cultural system against influences (real or imagined) from another culture.

cultural syncretism or hybridization The blending of two or more cultures, which produces a synergistic third culture that exhibits traits from all cultural parents.

culture Learned and shared behavior by a group of people empowering them with a distinct "way of life"; it includes both material (technology, tools, etc.) and immaterial (speech, religion, values, etc.) components.

culture hearth An area of historical cultural innovation.

cyclone A large storm, marked by well-defined air circulation around a low-pressure center. Tropical cyclones are typically called hurricanes in the Atlantic Ocean and typhoons in the western Pacific.

Cyrillic alphabet Based on the Greek alphabet and used by Slavic languages heavily influenced by the Eastern Orthodox Church. Attributed to the missionary work of St. Cyril in the ninth century.

dacha A Russian country cottage used especially in the summer.

dalit The currently preferred term used to denote the members of India's most discriminated-against ("lowest") caste groups, those people previously deemed "untouchables."

decolonialization The process of a former colony's gaining (or regaining) independence over its territory and establishing (or reestablishing) an independent government.

demographic transition A four-stage model of opulation change derived from the historical decline of the natural rate of increase as a population becomes increasingly urbanized through industrialization and economic development.

denuclearization The process whereby nuclear weapons are removed from an area and dismantled or taken elsewhere.

dependency theory A popular theory to explain patterns of economic development in Latin America. Its central premise is that underdevelopment was created by the expansion of European capitalism into the region that served to develop "core" countries in Europe and to impoverish and make dependent peripheral areas such as Latin America.

desertification The spread of desert conditions into semiarid areas owing to improper management of the land.

diaspora The scattering of a particular group of people over a vast geographical area. Originally, the term referred to the migration of Jews out of their original homeland, but now it has been generalized to refer to any ethnic dispersion.

divergent plate boundary A place where two tectonic plates move away from each other in opposite directions, magma often flows from Earth's interior, reaching the surface as volcanoes, such as in Iceland. In other places divergence creates large trenches or rift valleys.

dollarization An economic strategy in which a country adopts the U.S. dollar as its official currency. A country can be partially dollarized, using U.S. dollars alongside its national currency, or fully dollarized, when the U.S. dollar becomes the only medium of exchange and a country gives up its own national currency. Panama fully dollarized in 1904; more recently, Ecuador became fully dollarized in 2000.

domestication The purposeful selection and breeding of wild plants and animals for cultural purposes.

domino theory A U.S. geopolitical policy of the 1970s that stemmed from the assumption that if Vietnam fell to the communists, the rest of Southeast Asia would soon follow.

Dravidian language One of the earliest (from perhaps 4,000 years ago) language families and, unlike Hindi, not Indo-European. Once spoken throughout South Asia, Dravidian languages are now found only in southern India and part of Sri Lanka.

Eastern Orthodox Christianity A loose confederation of self-governing churches in eastern Europe and Russia that are historically linked to Byzantine traditions and to the primacy of the patriarch of Constantinople (Istanbul).

economic convergence The notion that globalization will result in the world's poorer countries gradually catching up with more advanced economies.

El Niño An abnormally large warm current that appears off the coast of Ecuador and Peru in December. During an El Niño year, torrential rains can bring devastating floods along the Pacific coast and drought conditions in the interior continents of the Americas.

entrepôt A city and port that specializes in transshipment of goods.

environmental lapse rate The decline in temperature as one ascends higher in the atmosphere. On average, the temperature declines 3.5°F for every 1,000 feet ascended, or 6.5°C for every 1,000 meters.

ethnicity A shared cultural identity held by a group of people with a common background or history, often as a minority group within a larger society.

ethnic religion A religion closely identified with a specific ethnic or tribal group, often to the point of assuming the role of the major defining characteristic of that group. Normally, ethnic religions do not actively seek new converts.

ethnographic boundaries State and national boundaries that are drawn to follow distinct differences in cultural traits, such as religion, language, or ethnic identity.

European Union (EU) The current association of 27 European countries that are joined together in an agenda of economic, political, and cultural integration.

exclave A portion of a country's territory that lies outside its contiguous land area.

exotic river A river that issues from a humid area and flows into a dry area otherwise lacking streams.

Fair Trade An international certification movement to identify primary commodities exported from the developing world in which farmers earn a better price for their product. Commodities such as coffee, tea, and forest products are certified "Fair Trade" when small-scale producers earn more for their product and production systems are viewed as environmentally and socially sustainable.

federal state Political system in which a significant amount of power is given to individual states. In India, these states were created upon independence in 1947 and were drawn primarily along linguistic lines, so that today state power is often associated with specific ethnic groups within the nation.

Fertile Crescent An ecologically diverse zone of lands in Southwest Asia that extends from Lebanon eastward to Iraq and that is often associated with early forms of agricultural domestication.

fjords Flooded, glacially carved valleys; in Europe, found primarily along Norway's western coast.

forward capital A capital city deliberately positioned near the international border of a contested territory, signifying the state's interest—and presence—in this zone of conflict.

fossil water Water supplies that were stored underground during wetter climatic periods.

free trade zone (FTZ) A duty-free and tax-exempt industrial park created to attract foreign corporations and create industrial jobs.

genocide The deliberate and systematic killing of a racial, political, or cultural group.

gentrification A process of urban revitalization in which higher-income residents displace lower-income residents in central-city neighborhoods.

geomancy The traditional Chinese and Korean practice of designing buildings in accordance with the principles of cosmic harmony and discord that supposedly course through the local topography.

geometric boundaries Boundaries of convenience drawn along lines of latitude or longitude without consideration for cultural or ethnic differences in an area.

ghettos Urban ethnic neighborhoods often associated with low-income groups in North America.

glasnost A policy of greater political openness initiated during the 1980s by Soviet president Mikhail Gorbachev.

global warming The increase in global temperatures as a result of the magnification of the natural greenhouse effect due to human-produced pollutants. The resulting climate change is causing profound—and often damaging—changes to Earth's environments.

globalization The increasing interconnectedness of people and places throughout the world through converging processes of economic, political, and cultural change.

Golden Triangle An area of northern Thailand, Burma, and Laos that is known as a major source region for heroin and is plugged into the global drug trade.

Gondwanaland The ancient megacontinent that included Africa, South America, Antarctica, Australia, Madagascar, and Saudi Arabia. Some 250 million years ago, it began to split apart due to plate tectonics.

grassification The conversion of tropical forest into pasture for cattle ranching. Typically, this process involves introducing species of grasses and cattle, mostly from Africa.

Greater Antilles The four large Caribbean islands of Cuba, Jamaica, Hispaniola, and Puerto Rico.

Great Escarpment A landform that rims southern Africa from Angola to South Africa. It forms where the narrow coastal plains meet the elevated plateaus in an abrupt break in elevation.

greenhouse effect The natural process of lower atmosphere heating that results from the trapping of incoming and reradiated solar energy by water moisture, clouds, and other atmospheric gases.

Green Revolution Term applied to the development of agricultural techniques used in developing countries that usually combine new, genetically altered seeds that provide higher yields than native seeds when combined with high inputs of chemical fertilizer, irrigation, and pesticides.

gross domestic product (GDP)—gross national product (GNP) GDP is the total value of goods and services produced within a given country (or other geographical unit) in a single year. GNP is a somewhat broader measure that includes the inflow of money from other countries in the form of the repatriation of profits and other returns on investments, as well as the outflow to other countries for the same purposes.

gross national income (GNI) The value of all final goods and services produced within a country's borders (gross domestic product, or GDP) plus the net income from abroad (formerly referred to as gross national product, or GNP).

gross national income (GNI) per capita The figure that results from dividing a country's GNI by the total population.

Group of Eight (G-8) A collection of powerful countries that confers regularly on key global economic and political issues. It includes the United States, Canada, Japan, Great Britain, Germany, France, Italy, and Russia.

guest workers Workers from Europe's agricultural periphery—primarily Greece, Turkey, southern Italy, and the former Yugoslavia—solicited to work in Germany, France, Sweden, and Switzerland during chronic labor shortages in Europe's boom years (1950s to 1970s).

Gulag Archipelago A collection of Soviet-era labor camps for political prisoners, made famous by writer Aleksandr Solzhenitsyn.

Hajj An Islamic religious pilgrimage to Makkah. One of the five essential pillars of the Muslim creed to be undertaken once in life, if an individual is physically and financially able to do it.

haoles Light-skinned Europeans or U.S. citizens in the Hawaiian Islands.

hierarchical diffusion The spread of an idea or cultural trait through adoption by leaders and other elite at the top of the social structure or hierarchy. In the case of religion, it was common for all members of a clan or tribe to convert if the leader adopted a new religion.

high islands Larger, more elevated islands, often focused around recent volcanic activity.

Hindu nationalism A contemporary "fundamental" religious and political movement that promotes Hindu values as the essential—and exclusive—fabric of Indian society. As a political movement, it appears to have less tolerance of India's large Muslim minority than other political movements.

homelands Nominally independent ethnic territories created for blacks under the grand apartheid scheme. Homelands were on marginal land, overcrowded, and poorly serviced. In the post-apartheid era, they were eliminated.

Horn of Africa The northeastern corner of Sub-Saharan Africa that includes the states of Somalia, Ethiopia, Eritrea, and Djibouti. Drought, famine, and ethnic warfare in the 1980s and 1990s resulted in political turmoil in this area.

hot spot A supply of magma that produces a chain of mid-ocean volcanoes atop a zone of moving oceanic crust.

hurricanes Storm systems with an abnormally low-pressure center sustaining winds of 75 mph or higher. Each year during hurricane season (July–October), a half dozen to a dozen hurricanes form in the warm waters of the Atlantic and Caribbean, bringing destructive winds and heavy rain.

hydropolitics The interplay of water resource issues and politics.

ideographic writing A writing system in which each symbol represents not a sound but rather a concept.

indentured labor Foreign workers (usually South Asians) contracted to labor on Caribbean agricultural estates for a set period of time, often several years. Usually the contract stipulated paying off the travel debt incurred by the laborers. Similar indentured labor arrangements have existed in most world regions.

Indian diaspora The historical and contemporary propensity of Indians to migrate to other countries in search of better opportunities. This has led to large Indian populations in South Africa, the Caribbean, and the Pacific islands, along with western Europe and North America.

informal sector A much-debated concept that presupposes a dual economic system consisting of formal and informal sectors. The informal sector includes self-employed, low-wage jobs that are usually unregulated and untaxed. Street vending, shoe shining, artisan manufacturing, and even self-built housing are considered part of the informal sector. Some scholars include illegal activities such as drug smuggling and prostitution in the informal economy.

insolation Incoming solar energy that enters the atmosphere adjacent to Earth.

internally displaced persons Groups and individuals who flee an area due to conflict or famine but still remain in their country of origin. These populations often live in refugee-like conditions but are harder to assist because they technically do not qualify as refugees.

Iron Curtain A term coined by British leader Winston Churchill during the Cold War that defined the western border of Soviet power in Europe. The notorious Berlin Wall was a concrete manifestation of the Iron Curtain.

irredentism A state or national policy of reclaiming lost lands or those inhabited by people of the same ethnicity in another nation-state.

Islamic fundamentalism A movement within both the Shiite and Sunni Muslim traditions to return to a more conservative, religious-based society and state. Often associated with a rejection of Western culture and with a political aim to merge civic and religious authority.

isolated proximity A concept that explores the contradictory position of the Caribbean states, which are physically close to North America and economically dependent upon that region. At the same time, Caribbean isolation fosters strong loyalties to locality and limited economic opportunity.

Jainism A religious group in South Asia that emerged as a protest against orthodox Hinduism about the sixth century BCE. Its ethical core is the doctrine of noninjury to all living creatures. Today, Jains are noted for their nonviolence, which prohibits them from taking the life of any animal.

kanakas Melanesian workers imported to Australia, historically often concentrated along Queensland's "sugar coast."

Khmer Rouge Literally, "Red (or communist) Cambodians." The left-wing insurgent group led by French-educated Marxists rebelled against the royal Cambodian government in the early 1960s and again in a peasants' revolt in 1967.

kibbutzes Collective farms in Israel.

kleptocracy A state where corruption is so institutionalized that politicians and bureaucrats siphon off a large percentage of a country's wealth for personal gain.

laissez-faire An economic system in which the state has minimal involvement and in which market forces largely guide economic activity.

latifundia A large estate or landholding.

Lesser Antilles The arc of small Caribbean islands from St. Maarten to Trinidad.

Levant The eastern Mediterranean region.

lingua franca An agreed-upon common language to facilitate communication on specific topics such as international business, politics, sports, or entertainment.

linguistic nationalism The promotion of one language over others that is, in turn, linked to shared notions of nationalism. In India, some Hindu nationalists promote Hindi as the national language, yet this is resisted by many other groups in which that language is either not spoken or does not have the same central cultural role, as in the Ganges Valley. The lack of a national language in India remains problematic.

location factors The various influences that explain why an economic activity takes place where it does.

loess A fine, wind-deposited sediment that makes fertile soil but is very vulnerable to water erosion.

low islands Low, small, sandy islands formed from eroding coral reefs. Generally less fertile and populated than high islands.

Maghreb A region in northwestern Africa, including portions of Morocco, Algeria, and Tunisia.

maharaja Regional Hindu royalty, usually a king or prince, who ruled specific areas of South Asia before independence, but who was usually subject to overrule by British colonial advisers.

mallee A tough and scrubby eucalyptus woodland of limited economic value that is common across portions of interior Australia.

Mandarin A member of the high-level bureaucracy of Imperial China (before 1911). Mandarin Chinese is the official spoken language of the country and is the native tongue of the vast majority of people living in north, central, and southwestern China.

Maori Indigenous Polynesian people of New Zealand.

maquiladora Assembly plants on the Mexican border built by foreign capital. Most of their products are exported to the United States.

marine west-coast climate Moderate climate with cool summers and mild winters that is heavily influenced by maritime conditions. Such climates are usually found on the west coasts of continents between latitudes of 45 to 50 degrees.

maritime climate Climate moderated by proximity to oceans or large seas. It is usually cool, cloudy, and wet, and lacks the temperature extremes of continental climates.

maroons Runaway slaves who established communities rich in African traditions throughout the Caribbean and Brazil.

Marxism The philosophy developed by Karl Marx, the most important historical proponent of communism. Marxism, which has many variants, presumes the desirability and, indeed, the necessity of a socialist economic system run through a central planning agency.

medieval landscape Urban landscapes from 900 to 1500 CE characterized by narrow, winding streets, three- or four-story structures (usually in stone, but sometimes wooden), with little open space except for the market square. These landscapes are still found in the centers of many European cities.

medina The original urban core of a traditional Islamic city.

Mediterranean climate A unique climate, found in only five locations in the world, that is characterized by hot, dry summers with very little rainfall. These climates are located on the west side of continents, between 30 and 40 degrees latitude.

megacity Urban conglomerations of more than 10 million people.

megalopolis A large urban region formed as multiple cities grow and merge with one another. The term is often applied to the string of cities in eastern North America that includes Washington, DC; Baltimore; Philadelphia; New York City; and Boston.

Melanesia Pacific Ocean region that includes the culturally complex, generally darker-skinned peoples of New Guinea, the Solomon Islands, Vanuatu, New Caledonia, and Fiji.

Mercosur The Southern Common Market established in 1991 that calls for free trade among member states and common external tariffs for nonmember states. Argentina, Paraguay, Brazil, and Uruguay are members; Chile is an associate member.

mestizo A person of mixed European and Indian ancestry.

Micronesia Pacific Ocean region that includes the culturally diverse, generally small islands north of Melanesia. Includes the Mariana Islands, Marshall Islands, and Federated States of Micronesia.

microstates Usually independent states that are small in both area and population.

mikrorayons Large, state-constructed urban housing projects built during the Soviet period in the 1970s and 1980s.

Millennium Development Goals Part of a group of programs implemented since 2000 to foster development in the world's poorest countries. The Millennium Development Goals are part of a global United Nations effort to reduce extreme poverty by 2015.

minifundia A small landholding farmed by peasants or tenants who produce food for subsistence and the market.

monocrop production Agriculture based upon a single crop.

monotheism A religious belief in a single God.

Monroe Doctrine A proclamation issued by U.S. president James Monroe in 1823 that the United States would not tolerate European military action in the Western Hemisphere. Focused on the Caribbean as a strategic area, the doctrine was repeatedly invoked to justify U.S. political and military intervention in the region.

monsoon The seasonal pattern of changes in winds, heat, and moisture in South Asia and other regions of the world that is a product of larger meteorological forces of land and water heating, the resultant pressure gradients, and jet-stream dynamics. The monsoon produces distinct wet and dry seasons.

moraines Hilly topographic features that mark the path of Pleistocene glaciers. They are composed of material eroded and carried by glaciers and ice sheets.

Mughal Empire (also spelled *Mogul*) The preeminent Islamic period of rule that covered most of South Asia during the early sixteenth to late seventeenth centuries and attempted to unify both Muslims and Hindus into a large South Asian state. The capital of this empire was Lahore, in what is now Pakistan. The last vestiges of the Mughal dynasty were dissolved by the British following the uprisings of 1857.

nation-state A relatively homogeneous cultural group (a nation) with its own political territory (the state).

Native Title Bill Australian legislation signed in 1993 that provides Aborigines with enhanced legal rights over land and resources within the country.

neocolonialism Economic and political strategies by which powerful states indirectly (and sometimes directly) extend their influence over other, weaker states.

neoliberal policies Economic policies widely adopted in the 1990s that stress privatization, export production, and few restrictions on imports.

net migration rate A statistic that assesses whether more people are entering or leaving a country by measuring the amount of immigration (in-migration) and emigration (out-migration). A positive figure means the population is growing because of in-migration, whereas a negative number means a population is shrinking because of out-migration.

North American Free Trade Agreement (NAFTA) An agreement made in 1994 between Canada, the United States, and Mexico that established a 15-year plan for reducing all barriers to trade among the three countries.

Oceania A major world subregion that usually includes New Zealand and the major islands of Melanesia, Micronesia, and Polynesia.

offshore banking Islands or microstates that offer financial services that are typically confidential and tax-exempt. As part of a global financial system, offshore banks have developed a unique niche, offering their services to individual and corporate clients for set fees. The Bahamas and Cayman Islands are leaders in this sector.

Organization of American States (OAS) Founded in 1948 and headquartered in Washington, DC, the organization advocates hemispheric cooperation and dialog. Most states in the Americas belong except Cuba.

Organization of Petroleum Exporting Countries (OPEC) An international organization of 12 oil-producing nations (formed in 1960) that attempts to influence global prices and supplies of oil. Algeria, Gabon, Indonesia, Iran, Iraq, Kuwait, Libya, Nigeria, Qatar, Saudi Arabia, UAE, and Venezuela are members.

orographic rainfall Enhanced precipitation over uplands that results from lifting (and cooling) of air masses as they are forced over mountains.

Ottoman Empire A large, Turkish-based empire (named for Osman, one of its founders) that dominated large portions of southeastern Europe, North Africa, and Southwest Asia between the sixteenth and nineteenth centuries.

Outback Australia's large, generally dry, and thinly settled interior.

outsourcing A business practice that transfers portions of a company's production and service activities to lower-cost settings, often located overseas.

overurbanization A process in which the rapid growth of a city, most often because of in-migration, exceeds the city's ability to provide jobs, housing, water, sewers, and transportation.

Palestinian Authority (PA) A quasi-governmental body that represents Palestinian interests in the West Bank and Gaza.

Pan-African Movement Founded in 1900 by U.S. intellectuals W. E. B. Du Bois and Marcus Garvey, this movement's slogan was "Africa for Africans," and its influence extended across the Atlantic.

pastoralists Nomadic and sedentary peoples who rely upon livestock (especially cattle, camels, sheep, and goats) for their sustenance and livelihood.

pastoral nomadism A traditional subsistence agricultural system in which practitioners depend on the seasonal movements of livestock within marginal natural environments.

perestroika A program of partially implemented, planned economic reforms (or restructuring) undertaken during the Gorbachev years in the Soviet Union designed to make the Soviet economy more efficient and responsive to consumer needs.

permafrost A cold-climate condition in which the ground remains permanently frozen.

physiological density A population statistic that relates the number of people in a country to the amount of arable land.

Pidgin English (Pijin) A version of English that also incorporates elements of other local languages, often utilized to foster trade and basic communication between different culture groups.

plantation America A cultural region that extends from midway up the coast of Brazil, through the Guianas and the Caribbean, and into the southeastern United States. In this coastal zone, European-owned plantations, worked by African laborers, produced agricultural products for export.

plate tectonics The theory that explains the gradual movement of large geological platforms (or plates) along Earth's surface.

podzol soils A Russian term for an acidic soil of limited fertility, typically found in northern forest environments.

pollution exporting The process of exporting industrial pollution and other waste material to other countries. Pollution exporting can be direct, as when waste is simply shipped abroad for disposal, or indirect, as when highly polluting factories are constructed abroad.

Polynesia Pacific Ocean region, broadly unified by language and cultural traditions, that includes the Hawaiian Islands, Marquesas Islands, Society Islands, Tuamotu Archipelago, Cook Islands, American Samoa, Samoa, Tonga, and Kiribati.

population pyramid A graph representing the structure of a population, including the percentage of young and old. The percentages of all different age groups are plotted along a vertical axis that divides the population into male and female. In a fast-growing population with a large percentage of young people and a small percentage of elderly, the graph will have a wide base and a narrow tip, giving it a pyramidal shape.

postindustrial economy An economy in which the tertiary and quaternary sectors dominate employment and expansion.

prairie An extensive area of grassland in North America. In the more humid eastern portions, grasses are usually longer than in the drier western areas, which are in the rain shadow of the Rocky Mountain range.

primate city The largest urban settlement in a country that dominates all other urban places, economically and politically. Often—yet not always—the primate city is also the country's capital.

privatization The process of moving formerly state-owned firms into the contemporary capitalist private sector.

protectorate During the period of global Western imperialism, a state or other political entity that remained autonomous but sacrificed its foreign affairs to an imperial power in exchange for "protection" from other imperial powers.

purchasing power parity (PPP) A method of reducing the influence of inflated currency rates by adjusting a local currency to a composite baseline of one U.S. dollar based upon its ability to purchase a standardized "market basket" of goods.

qanat system A traditional system of gravity-fed irrigation that uses gently sloping tunnels to capture groundwater and direct it to needed fields.

Quran (also spelled *Koran*) A book of divine revelations received by the prophet Muhammad that serves as a holy text in the religion of Islam.

Ramayana One of the two main epic poems of the Hindu religion, the Ramayana is also commonly performed in the shadow puppet theaters of the predominately Muslim island of Java.

rate of natural increase (RNI) The standard statistic used to express natural population growth per year for a country, region, or the world based upon the difference between birth and death rates. RNI does not consider population change from migration. Though most often a positive figure (such as 1.7 percent), RNI can also be expressed as a negative number for no-growth countries.

refugee A person who flees his or her country because of a well-founded fear of persecution based on race, ethnicity, religion, ideology, or political affiliation.

region A spatial unit or area that shares distinguishing traits, be they physical (such as climate or vegetation), cultural (a language area), or political (a nation-state or autonomous area).

regulatory lakes A term applied to a series of lakes in the middle Yangtze Valley of China. Regulatory lakes take excess water from the river during flood periods, and supply water to the river during dry periods.

remittances Money sent by immigrants to their country of origin to support family members left behind. For many countries, remittances are a principal source of foreign exchange.

Renaissance–Baroque landscape Urban landscapes generally constructed during the period from 1500 to 1800 that are characterized by wide, ceremonial boulevards, large monumental structures (palaces, public squares, churches), and ostentatious housing for the urban elite. A common landscape feature in European cities.

rift valley A surface landscape feature formed where two tectonic plates are diverging or moving apart. Usually, this forms a depression or large valley.

rimland The mainland coastal zone of the Caribbean, beginning with Belize and extending along the coast of Central America to northern South America.

rural-to-urban migration The flow of internal migrants from rural areas to cities that began in the 1950s and intensified in the 1960s and 1970s.

Russification A policy of the Soviet Union designed to spread Russian settlers and influences to non-Russian areas of the country.

Rust Belt Regions of heavy industry that experience marked economic decline after their factories cease to be competitive.

Sahel The semidesert region at the southern fringe of the Sahara, and the countries that fall within this region, which extends from Senegal to Sudan. Droughts in the 1970s and early 1980s caused widespread famine and dislocation of population.

salinization The accumulation of salts in the upper layers of soil, often causing a reduction in crop yields, resulting from irrigation with water of high natural salt content and/or irrigation of soils that contain a high level of mineral salts.

samurai The warrior class of traditional Japan. After 1600, the military role of the samurai declined as they assumed administrative positions, but their military ethos remained alive until the class was abolished in 1868.

Schengen Agreement The 1985 agreement between some—but not all—European Union member countries to reduce border formalities in order to facilitate free movement of citizens between member countries of this new "Schengenland." For example, today there are no border controls between France and Germany, or between France and Italy.

sectoral transformation The evolution of a labor force from being highly dependent on the primary sector to being oriented around more employment in the secondary, tertiary, and quaternary sectors.

secularization The widespread movement in western Europe away from regular participation and engagement with traditional organized religions such as Protestantism or Catholicism.

sediment load The amount of sand, silt, and clay carried by a river.

shield landscape Barren, mostly flat lands of southern Scandinavia that were heavily eroded by Pleistocene ice sheets. In many places, this landscape is characterized by large expanses of bedrock with little or no soil that resulted from glacial erosion.

shields Large upland areas of very old exposed rocks that range in elevation from 600 to 5,000 feet (200 to 1,500 meters). The three major shields in South America are the Guiana, Brazilian, and Patagonian.

shifted cultivators Migrants, with or without agricultural experience, who are transplanted by government relocation schemes.

shifting cultivation An agricultural system in which plots of land are farmed and then abandoned for a number of years until fertility is restored, at which point they are again brought under cultivation. See also *swidden agriculture.*

Shiites Muslims who practice one of the two main branches of Islam; especially dominant in Iran and nearby southern Iraq.

Shogun, Shogunate The true ruler of Japan before 1868, as opposed to the emperor, whose power was merely symbolic.

Sikhism An Indian religion combining Islamic and Hindu elements, founded in the Punjab region in the late fifteenth century. A long tradition of militarism continues today, and a large proportion of Sikh men are in the Indian armed forces.

Slavic peoples A group of peoples in eastern Europe and Russia who speak Slavic languages, a distinctive branch of the Indo-European language family.

social and regional differentiation "Social differentiation" refers to a process by which certain classes of people grow richer when others grow poorer; "regional differentiation" refers to a process by which certain places grow more prosperous while others become less prosperous.

socialist realism An artistic style once popular in the Soviet Union that was associated with realistic depictions of workers in their patriotic struggles against capitalism.

Spanglish A hybrid combination of English and Spanish spoken by Hispanic Americans.

Special Economic Zones (SEZs) Relatively small districts in China that have been fully opened to global capitalism.

spheres of influence In countries not formally colonized in the nineteenth and early twentieth centuries (particularly China and Iran), limited areas called "spheres of influence" were gained by particular European countries for trade purposes and more generally for economic exploitation and political manipulation.

squatter settlements Makeshift housing on land not legally owned or rented by urban migrants, usually in unoccupied open spaces within or on the outskirts of a rapidly growing city.

steppe Semiarid grasslands found in many parts of the world. Grasses are usually shorter and less dense than in prairies.

structural adjustment programs Controversial yet widely implemented programs used to reduce government spending, encourage the private sector, and refinance foreign debt. Typically, these IMF and World Bank policies trigger drastic cutbacks in government-supported services and food subsidies, which disproportionately affect the poor.

subcontinent A large segment of land separated from the main land mass on which it sits by lofty mountains or other geographical barriers. South Asia, separated from the rest of Eurasia by the Himalayas, is often called the "Indian subcontinent."

subduction zones Areas where two tectonic plates are converging or colliding. In these areas, one plate usually sinks below another. They are characterized by earthquakes, volcanoes, and deep oceanic trenches.

subnational organizations Groups that form along ethnic, ideological, or territorial lines that can induce serious internal divisions within a state.

subsistence agriculture Farming that produces only enough crops or animal products to support a farm family's needs. Usually, little is sold at local or regional markets.

Suez Canal Pivotal waterway connecting the Red Sea and the Mediterranean opened by the British in 1869.

Sunda Shelf An extension of the continental shelf from the Southeast Asia mainland to the outlying islands. Because of the shelf, the overlying sea is generally shallow (less than 200 feet [61 meters] deep).

Sunnis Muslims who practice the dominant branch of Islam.

superconurbation A massive urban agglomeration that results from the coalescing of two or more formerly separate metropolitan areas.

supranational organizations Governing bodies that include several states, such as trade organizations, and often involve a loss of some state powers to achieve the organization's goals.

sweatshops Small manufacturing units in developing countries where menial tasks are performed by largely unskilled and poorly paid workers.

swidden agriculture Also called "slash-and-burn agriculture." A form of cultivation in which forested or brushy plots are cleared of vegetation, burned, and then planted to crops, only to be abandoned a few years later as soil fertility declines.

syncretic religions The blending of different belief systems. In Latin America, many animist practices were folded into Christian worship.

taiga The vast coniferous forest of Russia that stretches from the Urals to the Pacific Ocean. The main forest species are fir, spruce, and larch.

Taliban A harsh, Islamic fundamentalist political group that ruled most of Afghanistan in the late 1990s. In 2001, the Taliban lost power, yet remnant groups continue to fight against U.S.-led forces in southern and eastern Afghanistan.

Tamil Tigers The common name of the rebel forces in Sri Lanka (officially known as the *Liberation Tigers of Tamil Eelam, or LTTE*) that have been fighting the Sri Lankan army since 1983.

tectonic plates The basic building blocks of Earth's crust; large blocks of solid rock that move very slowly over the underlying semimolten material.

theocratic state A political state led by religious authorities.

tonal language Language in which the same set of phonemes (or basic sounds) may have very different meanings depending on the pitch in which they are uttered.

total fertility rate (TFR) The average number of children who will be borne by women of a hypothetical, yet statistically valid, population, such as that of a specific cultural group or within a particular country. Demographers consider TFR a more reliable indicator of population change than the crude birthrate.

township Racially segregated neighborhoods created for nonwhite groups under apartheid in South Africa. They are usually found on the outskirts of cities and classified as black, coloured, or South Asian.

transhumance A form of pastoralism in which animals are taken to high-altitude pastures during the summer months and returned to low-altitude pastures during the winter.

transmigration The planned, government-sponsored relocation of people from one area to another within a state territory.

transnational corporation Firm or corporation that, although it may be chartered and have headquarters in one specific country, does international business through an array of global subsidiaries.

transnational migration Complex social and economic linkages that form between home and host countries through international migration. Unlike earlier generations of migrants, twenty-first century immigrants can maintain more enduring and complex ties to their home countries as a result of technological advances.

Trans-Siberian Railroad Key southern Siberian railroad connection completed during the Russian empire (1904) that links European Russia with the Russian Far East terminus of Vladivostok.

Treaty of Tordesillas A treaty signed in 1494 between Spain and Portugal that drew a north–south line some 300 leagues west of the Azores and Cape Verde islands. Spain received the land to the west of the line and Portugal the land to the east.

tribalism Allegiance to a particular tribe or ethnic group rather than to the nation-state. Tribalism is often blamed for internal conflict within Sub-Saharan states.

tribe A group of families or clans with a common kinship, language, and definable territory but not an organized state.

tsars A Russian term (also spelled *czar*) for "Caesar," or ruler; the authoritarian rulers of the Russian empire before its collapse in the 1917 revolution.

tsetse fly A fly that is a vector for a parasite that causes sleeping sickness (typanosomiasis), a disease that especially affects humans and livestock. Livestock is rarely found in those areas of Sub-Saharan Africa where the tsetse fly is common.

tsunamis Very large sea waves induced by earthquakes.

tundra Arctic region with a short growing season in which vegetation is limited to low shrubs, grasses, and flowering herbs.

typhoons Large tropical storms, similar to hurricanes, that form in the western Pacific Ocean in tropical latitudes and cause widespread damage to the Philippines and coastal Southeast and East Asia.

uncontacted peoples Cultures that have yet to be contacted and influenced by the Western world.

unitary state A political system in which power is centralized at the national level.

universalizing religion A religion, usually with an active missionary program, that appeals to a large group of people regardless of local culture and conditions. Christianity and Islam both have strong universalizing components. This contrasts with ethnic religions.

urban decentralization The process in which cities spread out over a larger geographical area.

urban form The physical arrangement or landscape of the city, made up of building architecture and style, street patterns, open spaces, housing types, and so on.

urbanized population That percentage of a country's population living in settlements characterized as cities. Usually, high rates of urbanization are associated with higher levels of industrialization and economic development, because these activities are usually found in and around cities. Conversely, lower urbanized populations (less than 50 percent) are characteristic of developing countries.

urban primacy A state in which a disproportionately large city, such as London, New York, or Bangkok, dominates the urban system and is the center of economic, political, and cultural life.

urban realms model A simplified description of urban land use, especially descriptive of the modern North American city. It features a number of dispersed, peripheral centers of dynamic commercial and industrial activity linked by sophisticated urban transportation networks.

urban structure The distribution and pattern of land use, such as commercial, residential, or manufacturing, within the city. Often, commonalities give rise to models of urban structure characteristic of the cities of a certain region or of a shared history, such as cities shaped by European colonialism.

viticulture Grape cultivation.

water stress An environmental planning tool used to predict areas that have—or will have—serious water problems based upon the per capital demand and supply of freshwater.

White Australia Policy Before 1975, a set of stringent Australian limitations on nonwhite immigration to the country. Largely replaced by a more flexible policy today.

World Trade Organization (WTO) Formed as an outgrowth of the General Agreement on Tariffs and Trade (GATT) in 1995, the WTO is a large collection of member states dedicated to reducing global barriers to trade.

Index

Page numbers in *italics* indicate figures.